nd Libr

HANDBOOK OF STATISTICS
VOLUME 12

Handbook of Statistics

VOLUME 12

General Editor

C. R. Rao

NORTH-HOLLAND
AMSTERDAM · LONDON · NEW YORK · TOKYO

Environmental Statistics

Edited by

G. P. Patil

Center for Statistical Ecology and Environmental Statistics
Department of Statistics, The Pennsylvania State University
University Park, PA, USA

C. R. Rao

Center for Multivariate Analysis
Department of Statistics, The Pennsylvania State University
University Park, PA, USA

1994

NORTH-HOLLAND
AMSTERDAM · LONDON · NEW YORK · TOKYO

ELSEVIER SCIENCE B.V.
Sara Burgerhartstraat 25
P.O. Box 211, 1000 AE Amsterdam, The Netherlands

Library of Congress Cataloging-in-Publication Data

Environmental statistics / edited by G. P. Patil, C. R. Rao.
p. cm. – (Handbook of statistics; v. 12)
Includes bibliographical references and index.
ISBN 0-444-89803-4
1. Environmental sciences – Statistical methods. I. Patil, Ganapati P. II. Rao, C. Radhakrishna (Calyampudi Radhakrishna), 1920– . III. Series.
GE45.S73E57 1994
363.7′0072 – dc20 93-33430
CIP

ISBN: 0-444-89803-4

This book is printed on acid-free paper.

Printed in The Netherlands.

Preface

Almost everyone interested wonders what environmental statistics is about and what it covers. The question becomes particularly important when it comes to the formulation of relevant research and training, whether in academics, agencies, or industries. A purpose of this volume is to help provide a beginning perception of environmental statistics with some examples that are of concern and interest today.

Environmental statistics is in a take-off stage both for reasons of societal challenge and statistical opportunity. It is becoming clear that environmental statistics is demanding more and more of non-traditional and innovative statistical approaches. This is partly because, unlike in many hard sciences, environmental studies involve speace, time, and locally ingenious environmental sampling and monitoring. Also, environmental statistics must satisfy environmental policy research in addition to disciplinary and interdisciplinary environmental research. Several chapters in this volume may provide reasonable pointers in these directions. The chapters specially prepared by several outstanding professionals involved in statistics and the environment discuss the current state of the art in different areas of environmental statistics and provide new perspectives and problems for future research, training, policy, and regulation involving statistics and the environment.

The chapters of this volume are classified under nine major headings:

I. Environmental Statistics as an Emerging Subject Area
II. Environmental Sampling and Design
III. Environmental Monitoring
IV. Statistical Ecology
V. Environmental Biometrics
VI. Systems Modeling and Analysis
VII. Spatial Statistics
VIII. Statistical Issues and Methods
IX. Statistics in Environmental Regulation

Finally, we wish to thank all the authors for the valuable contributions they have made, Barbara Freed for keeping it all together in a timely frame and Elsevier Science Publishers B.V. (North Holland) for their patience and cooperation in bringing out this volume.

G. P. Patil and C. R. Rao

Table of Contents

PART II. ENVIRONMENTAL SAMPLING AND DESIGN

PART V. ENVIRONMENTAL BIOMETRICS

PART VI. SYSTEMS MODELING AND ANALYSIS

PART VIII. STATISTICAL ISSUES AND METHODS

PART IX. STATISTICS IN ENVIRONMENTAL REGULATION

Contributors

M. Akritas, *Department of Statistics, Pennsylvania State University, University Park, PA 16802, USA* (Ch. 7)

J. J. M. Bedaux, *Department of Theoretical Biology, Vrije Universiteit Amsterdam, De Boelelaan 1087, 1081 HV Amsterdam, The Netherlands* (Ch. 16)

M. T. Boswell, *Department of Statistics, Penn State University, 416 Classroom Bldg., University Park, PA 16802, USA* (Ch. 24)

H. E. Burkhart, *Department of Forestry, VPI & SU, Blacksburg, VA 24061-0324, USA* (Ch. 11)

P. J. Catalano, *Dana Farber Cancer Institute, Division of Biostatistics, 44 Binney Street, Boston, MA 02115, USA* (Ch. 14)

C. R. Cothern, *US EPA/CES, 401 M Street, SW, Washington, DC 20460, USA* (Ch. 25)

C. B. Davis, *Environmetrics & Statistics Limited, 1853 Wellington Court, Henderson, NV 89014, USA* (Ch. 26)

C. A. Donnelly, *Department of Mathematics and Statistics, University of Edinburgh, James Clerk Maxwell Building, The King's Buildings, Mayfield Road, Edinburgh, EH9 3J2, UK* (Ch. 19)

R. O. Gilbert, *Pacific Northwest Lab, P.O. Box 999, Mail Stop K7-34, Richland, WA 99352, USA* (Ch. 27)

S. Gore, *Department of Statistics, Pennsylvania State University, University Park, PA 16802, USA* (Ch. 4)

J. Gove, *Research Forester, N.E. Forest Exp. Station, Cor. Concord & Mast Rds., P.O. Box 640, Durham, NH 03824, USA* (Ch. 12)

T. Gregoire, *Department of Forestry, VPI & SU, Blacksburg, VA 24061-0324, USA* (Ch. 11)

P. Guttorp, *Department of Statistics, GN-22, University of Washington, Seattle, WA 98195, USA* (Ch. 20)

V. Hasselblad, *Ctr for Hlth Policy Rsch & Education, Duke University, Erwin Square-Suite 230, 2200 W. Main Street, Durham, NC 27706, USA* (Ch. 21)

J. S. Hunter, *503 Lake Drive, Princeton, NJ 08540, USA* (Ch. 1)

S. A. L. M. Kooijman, *Biol. Laboratory, Vrije Universiteit Amsterdam, De Boelelaan 1087, 1081 HV Amsterdam, The Netherlands* (Ch. 16)

H. Lacayo, *U.S. EPA, PM 223, 401 M Street, Washington, DC 20460, USA* (Ch. 28)

N. M. Laird, *Department of Biostatistics, Harvard School of Public Health, 677 Huntington Avenue, Boston, MA 02115, USA* (Ch. 19)

W. S. Liggett, *National Institute of Standards and Technology, Admin A337, Gaithersburg, MD 20899, USA* (Ch. 3)

G. Lovison, *Dipartimento di Scienze, Statistiche, Universita Di Padova, via S. Francesco, 33, Padova 35121, Italy* (Ch. 4)

B. F. J. Manly, *Department of Mathematics and Statistics, University of Otago, P.O. Box 56, Dunedin, New Zealand* (Ch. 10)

J. H. Matis, *Department of Statistics, Texas A & M University, College Station, TX 77843-3143, USA* (Ch. 17)

W. L. Myers, *124 Land & Water Research Institute, Penn State University, University Park, PA 16802, USA* (Ch. 18)

P. M. North, *ASRU Ltd., Institute of Mathematics and Statistics, Cornwallis Building, The University, Canterbury, Kent CT2 7NF, UK* (Ch. 13)

J. S. O'Connor, *USEPA, Marine & Wetlands, Protection Branch, 26 Federal Plaza, New York, NY 10278, USA* (Ch. 24)

S. Overton, *SUNY-Coll of Env Sci & Forestry, 211 Marshall, Syracuse, NY 13210, USA* (Ch. 9)

G. P. Patil, *Department of Statistics, Penn State University, 421 Classroom Building, University Park, PA 16802, USA* (Chs. 2, 4, 5, 7, 12, 24)

W. W. Piegorsch, *Department of Statistics, University of South Carolina, Columbia, SC 29208, USA* (Ch. 15).

L. Ryan, *Division of Biostatistics, Dana-Farber Cancer Institute, 44 Binney Street, Boston, MA 02115, USA* (Ch. 14)

T. Ruscitti, *Department of Statistics, Pennsylvania State University, University Park, PA 16802, USA* (Ch. 7)

P. D. Sampson, *Department of Statistics, University of Washington, Seattle, WA 98185, USA* (Ch. 20)

G. A. F. Seber, *Department of Mathematics, University of Auckland, Private Bag, Auckland, New Zealand* (Ch. 6)

A. K. Sinha, *Department of Statistics, 421-B Classroom Bldg., Penn State University, University Park, PA 16802, USA* (Ch. 5)

E. P. Smith, *Department of Statistics, VPI & SU, Blacksburg, VA 24061, USA* (Ch. 8)

A. R. Solow, *Woods Hole Ocean. Institute, Woods Hole, MA 02543, USA* (Ch. 22)

S. V. Stehman, *College of Environmental Science and Forestry, State Univ. of New York, Syracuse, NY 13210, USA* (Ch. 9)

*B. Swindel, *Intensive Management Practices, School of Forest Resources, SE Forest Exp. Sta., University of Florida, Gainesville, FL 32611, USA* (Ch. 12)

C. Taillie, *Department of Statistics, Penn State University, 421C Classroom Bldg., University Park, PA 16802, USA* (Chs. 5, 12)

S. K. Thompson, *Department of Statistics, Penn State University, University Park, PA 16802, USA* (Ch. 6)

J. H. Ware, *Dean for Academic Affairs, Harvard School of Public Health, 677 Huntington Avenue, Boston, MA 02115, USA* (Ch. 19)

* Deceased.

T. E. Wehrly, *Department of Statistics, Texas A&M University, College Station, TX 77843-3143, USA* (Ch. 17)

J. H. Young, *Professor Emeritus, Entomology Department, Oklahoma State University, Stillwater, OK 74078, USA* (Ch. 23)

L. J. Young, *Biometry Department, University of Nebraska, 103 Miller Hall, E Campus, Lincoln, NE 68583-0712, USA* (Ch. 23)

G. P. Patil and C. R. Rao, eds., *Handbook of Statistics, Vol. 12*

Environmetrics: An Emerging Science*

J. Stuart Hunter

Abstract: Environmetrics finds its origins in the search for the understanding of the natural phenomena that surrounds mankind. In antiquity these studies led to the creation of the earliest instruments of measurement and to the beginning arts of mathematical description. Today's environmental studies combine the modern tools of physics and chemistry with mathematical modeling of great sophistication. But beyond measurement and mathematics, environmetrics has become a unique '*n*-science', a meeting ground for the ecologist, the natural and social scientist, the engineer and statistician, and ultimately the political scientist.

1. Origins

The word 'environmetrics' was first employed, we now believe, by Professor Philip Cox in 1971 at Iowa State University in the title of a proposal submitted to the NSF. Cox used the word again later in describing a session on Statistics in the Environmental Research at the 1972 meeting of the Biometric Society, ENAR, held at Iowa State University. The word was reborn quite independently and anew in discussions within the Committee on National Statistics of the National Academy of Sciences in 1976–77, and in testimony before the US House of Representatives Subcommittee on the Environment and the Atmosphere [1]. Over the ensuing years 'environmetrics' has not only become a common-place word, but the title of a technical journal, an international society, and even of an Australian consulting service. From initial creation to international recognition in a span of twenty years tells us something of the energy and vigor that environmetrics represents.

Environmetrics finds its roots in the global and diverse measurements of natural phenomena human beings have collected and stored since numbers were invented, all in an attempt to understand the natural world surrounding us. In the United States these data collection activities were initially scattered across different gov-

*Talk presented at the 8-th annual EPA Conference on Statistics, Philadelphia, PA (1992).

ernment organizations: for monitoring river flows, recording the weather, measuring soil erosion and estimating wildlife inventories. Much of this was changed by the National Environmental Policy Act (NEPA) of 1969, signed into law by President Nixon on 1 January 1970. The Council on Environmental Quality (CEQ) came into being on March 5-th of that same year followed in April by the nation's first 'earth day'. Both the Environmental Protection Agency (EPA) and National Oceanic and Atmospheric Administration (NOAA) became official entities on 2 December 1971. Of course, in 1971 Cox was already on board with the word 'environmetrics'.

2. Environmetrics as a science

Our question is whether environmetrics can be called a science. 'Science' is a team for both a type of knowledge and a form of activity. Since the ancients identified four natural environmental elements: air, water, earth and fire, let's consider these elements in the context of today's environmetrics. In air environmetricians study global weather, plume dispersions from power plants, automobile emissions in urban areas and oxygen levels in office buildings closed to outside influences. In water we find the environmetrician concerned with the maintenance of reservoirs, the ecology of lakes, wetlands, and marine life. They also study the influences of point and non-point sources of pollution in rivers, groundwater infusions and the build-up and persistence of life threatening contaminants in our drinking water. For the earth, the environmetrician finds himself engaged in range land studies, soil erosion investigations, forest management, and pest control. Air, water, earth and finally fire. For fire we can simply enumerate the many environmetric studies of the influences of power plant emissions, volcanoes and the monitoring of mankind's newest fire, the nuclear. There is no lack of either knowledge or activity in the science we call environmetrics.

A science can often be recognized by the language it uses and certainly environmetrics has generated its own language. Not only does it give special meaning to such word couplings as acid-rain, hazardous-waste and ozone-hole, it also creates new meanings for single words like 'bubble'. It influences the political arena with terms like 'impact statement', 'superfund' and the 'green party'. And it has its own very special three, four (and more!) letter acronyms. We can start with NEPA, EPA and NOAA and easily go from there to NPS, EMAP, NAAQS, PSI and on and on. Nor are there too many other philosophical arenas in which a novel word like 'kriging' can come into common use. No, the language of environmetrics, or shall I say its jargon, is sufficiently unique to qualify environmetrics as a science.

A science develops a repository of laws. Environmetrics entrains both the laws of nature and of man. The natural laws of environmetrics range from the simplest, for the case of particulates from smoke stacks – What goes up must come down – to much more sophisticated laws describing the photochemical dynamics of ozone concentration as a function of volatile organic compounds and nitrogen oxide leading to measures of smog. Of course, the man-made laws that derive from environmetrics are familiar to all. The recent passage of the 1990 Clean Air Act

Amendments and the associated National Ambient Air Quality Standards find their essential origins in evidence gathered and interpreted by environmetricians. Yes, environmetrics entrains the laws of both man and nature.

And certainly environmetrics as an emerging science has made special demands upon mathematics and statistics. In the beginning early days of the 1900s we find Hazen [2] employing the cumulative distribution function as a means for forecasting the frequency of the floods. The 1970 paper on stochastic modeling of river-flow time series by Lawrance and Kottegoda [3] appearing in the *J. Roy. Statist. Soc. Ser. A* listed over 150 mathematical-statistical references. One can even find the origins of fractals, that fascinating new mathematics, in a 1969 contributions of Mandelbrot in *Water Resources Res.* [4]. And today, in 1992, the latest issue of the new journal *Environmetrics* contains a paper by McLeod, Hipel and Bodo [5] with some thirty references expositing the latest developments in trend analyses for surface water quality time series.

Nor are the air and terrestrial environmetricians neglectful of the applications of mathematics and statistics. Who is to say when data-based weather and precipitation forecasting first began? Match that against today's efforts to model the dynamic and multivariate characteristics of global weather that strain to capacity of our largest and fastest computers. The use of kriging for mapping geological contours and the determination of three-dimensional distributions of minerals is now identified as a separate field called geostatics. Such studies employ vast amounts of data, both recorded and simulated. Similar mathematical-statistical studies involve the identification of pathways of pollutants in groundwater.

3. Mathematics and statistics

It is important now to discriminate between mathematics and statistics. Speaking broadly, it is safe to say that mathematicians are prone to start with a set of 'givens' and then, following carefully prescribed deductive steps, to go to wherever their formal logic proceeds. Their final product, mathematics, need not be related to anything in nature. Mathematicians are not limited except by the fertility of their imagination. Mathematics is the ultimate formal language, the purest of the philosophies. The astounding miracle is the adaptability of the machinery and language of mathematics as instruments to help us understand the world around us.

However, the philosophy we call statistics begins with data. Unfortunately, this has led many naive layman to think that statisticians are solely number librarians, equivalent to individuals who carefully label and store their autumn leaves. But the layman misses an essential point: data are not information. Information rests within data much like a rare earth within a matrix of rock. Of course, the ability to collect, organize, store and retrieve data is a commonplace for statisticians, as well as for laymen who call themselves statisticians, and frankly for many 'statistical' computer software programs. The distinguishing characteristic of the true statistician lies in his/her search for description and understanding, for a cognitive concentration of information derived from the welter of numbers. Statisticians search for pattern, for regularity, and for simplicity. Their objective is to uncover the scintilla of information resting within the data, the ore within the aggregate, and then to shape these essences into things they call 'statistics', or 'models' that communicate to the mind of the beholder. Therein begins the iterative process of learning, the game of catchball between the data and the analyst.

4. Information exposition

Exposition of information within data is the hall-mark of the true statistician. Given the remarkable ability of the mind to grasp and hold visual signals, graphical expositions of information are best, numeric and algebraic descriptions can always follow. The EPA in recent years has started to provide excellent graphical representations of its data. One illustration is the recent *National Air Quality and Emissions Trends Report* (1990) with its arrays of box and whisker plots arranged to display trends, four and six-variate pie wedges illustrating the levels of pollutants superimposed upon regional maps along with wind roses.

Good as these displays are, the graphical exposition of multivariate data still has a long way to go. We need better scratchings and marks to visually convey quantitative information, better 'metroglyphs' to quote a word originally used by Anderson [6]. Someone once raised the question as to why the Chinese civilization, with its long head start on the Western world, did not go further in the sciences. After all, the early Chinese had the rudiments of chemistry and physics well in hand, good mathematics, astronomers, and many fine engineers. One explanation of their failure to advance the sciences may lie in their use of glyphs, their method for recording information. It became a matter of pride amongst Chinese savants to shape their glyphs so that only the cognoscente could read. The beauty of the glyph became as important, or even more important, than its message. And so the transmission of information, most particularly of a scientific nature, ascended to an asymptote. Later, the western world developed a more manipulative glyph system, the alphabet and the numerals, and thus made greater strides. But what of today's world? Is there a danger that we too, now immersed in masses of data, may lose our direction and thus slow our scientific progress? We do seem often overwhelmed by a welter of numbers, a condition that gets steadily worse with better instrumentation coupled to larger computer memories. The floods of multivariate data that are a natural condition in environmetrics present a unique challenge.

Statistics both graphical and algebraic are constructed to speak to the mind. The mind in turn, particularly that of the subject matter expert, replies with explanations, with models, generalizations and structures that give succinct expression to the statistical evidence, and lead to predictions of future behavior. But as with the chicken and the egg, one cannot be sure whether the gathered statistics come before or after the models. They simply both occur as in a conversation. And therein lies an essential difference between statistics and mathematics. The logical path followed by a mathematician once the 'givens' are in hand will come to an end, a 'proven' result. The path can again be traced by the next mathematician, if not exactly, then different only in its furbishing of the details. However, the logical procedures of the statistician are not deductive, but inductive, and lead not down a single path but upwards through many paths, alternative models and hypotheses; to new unanswered questions, and always to the need for more data. Nor does one expect the next statistician or environmetrician working the same data to travel the same inductive paths. Statisticians and environmetricians are coupled to data and both are driven by an open-ended learning process called the scientific method.

5. Models

This chapter should contain, I suspect, as least one equation. Consider than a model for a single datum, a single observation y. This model quickly becomes the statistician's *two* model problem postulating that every individual observation y has two components, i.e., $y = \eta + \varepsilon$, a deterministic component η, and a stochastic component ε. At first, η is taken to be a constant and ε an independent random variable with zero expectation. It is easy now to move quickly to higher levels of complexity, the deterministic model for η may be expressed as a polynomial function of several factors, as an expression non-linear in its parameters, or even dynamic, while the stochastic model may become Gaussian, or non-Gaussian, be time-dependent or even non-stationary. More complex still are models multivariate in either or both η and ε. The new word 'chemometrics' is often employed to identify these multivariate situations.

Usually the deterministic models postulated by the environmetrician are empirical, i.e., derived from the data. Frequently, several models may be in competition for selection as *the* 'adequate' representation. Not uncommon is a sharing of the information content of the data by estimating multiple parameters simultaneously descriptive of both deterministic and stochastic aspects that go beyond the mean and the variance. In mind here are deterministic-stochastic type models characterized by the Box–Jenkins ARIMA system. The ultimate objective is to create a model as simple in structure as possible while simultaneously making the model useful both as a describer and as a forecaster. One art of good empiricism is to keep the number of parameters to a minimum, remembering that the purpose of a model is not to predict the history, but rather to forecast the future. The acronym KISS is familiar to many. Zellner (7) asks that we translate KISS to mean Keep It Sophisticatedly Simple. Sage advice.

But mechanistic models alone are not appropriate to most environmetric data. Needed are models for elucidating statistical variability. How does one best describe the heterogeneous appearance of a crop, a forest or populated areas perceived from a satellite many miles above the earth? How does one succinctly describe how spatially correlated characteristics change in time, what of their autocorrelations? Less global but no less urgent are models descriptive of the number, distribution and changing patterns of wildlife, models to describe the timing, growth and decay of insect infestations, and models to monitor the influence of pollutants on the size and health of migrating populations. The development of dynamic graphical models for these stochastic universes is a unique contribution of the science of environmetrics.

6. Measurement

Of course every applied science requires data, and data in turn begins with measurement. In the days of the pharaohs environmental water measurements were made using 'nilometers', deep wells placed along the shores of the Nile and dug especially to record the height of the river's water. We can only assume that the measurements were translated into estimates of crop yields and later, undoubtedly, into tax levies. Given these Nile flood water data, were they ever analyzed across time? We do know that the Egyptian astronomer Hipparchus on studying ancient records of the risings and settings of stars, was able to detect the earth's precession, a rotational wobble of the earth's axis which takes many thousands of years to complete. Hipparchus thus must have studied time series provided by his predecessors. But did the local Nile river hydrologist ever plot his data and study his time series? The mystery deepens when we realize that the first plotted time series presently extant is from the 10-th century AD, more than a thousand years later! One can only wonder at the linkages between data, models and hypotheses that vanished in the great fire that consumed the library of ancient Alexandria.

Measuring the environment is an awesome challenge, there are so many things to measure, and at so many different times and places. Data on the environment are often recorded over different time spacings, with different instruments, and are often missing. We are all aware of today's concerns over the quality of our manufacturing processes, and of the individual products produced. A similar concern pervades environmental data. Measurements define the product, and the system providing the measurements the process. Defining the measurements, entrains consideration of the appropriate metric, the use of surrogate measures, precision, calibration, specificity, interferences and complexity. Defining the measurement process involves traceability, sampling size and frequency, compatibility between different observers and laboratories, components of variance, auditing, data collection and processing. Both the structure of a measurement, and the liveliness of the measurement process are essential to the well being of the science of environmetrics. Unhappily, I fear these elements are often taken for granted and not seriously considered by many environmentalists. To be casual about so vital a matter as measurement is live dangerously. Poor data pollutes the ultimate environment, the mind.

7. The science environmetrics

Is environmetrics a science? Does it find applicability and search for new knowledge in air, water earth and fire. Is it developing a unique language? Does it influence the laws of man and nature? Does it make special demands of mathematics and statistics? Does it attempt to model both the deterministic and stochastic worlds? Does it require it own specialized measurements products and unique measurement processes? Is it an integral part of the scientific method? The answer to all these questions is an obvious 'yes'!

Certainly, environmetrics qualifies as a science, and happily one in which the statistician plays a recognized role. One could attempt a taxonomy of environmentrics. The body of environmetrics requires sinews drawn from physics and mathematics, strands extending through chemistry and the engineering sciences, essential elements from ecology and the natural sciences, components derived from the social and political sciences and the ability to handle measures of uncertainty borrowed from statistical science. Yet this taxonomy would not capture the full body of knowledge we now identify as environmetrics. Environmetrics is suffused with many sister sciences, it is as a minimum an '*n*-science', to paraphrase Bill Kruskal. And yet it is unique. Certainly, whenever a philosophy demonstrates uniqueness in its language and laws, in its applications of other sciences, in its demands for special data and associated applications of statistics and in its use of the scientific method, it qualifies as a science. Environmetrics is not an emerging science, the title to this chapter is incorrect. Environmetrics has arrived.

One final point. Given the special role that statistics plays within environmetrics, I am hopeful that the American Statistical Association will find therein a vigorous interactive role.

References

[1] Hunter, J. S. (1977). Hearings, Subcommittee on the Environment and the Atmosphere, U.S. House of Representatives, 95th Congress, **44**, 242–251.

[2] Hazen, A. (1914). Storage to be provided in improving reservoirs for municipal supply. *Trans. Amer. Soc. Civil. Eng.* **77**, 1539–1640.

[3] Lawrance, A. J. and N. T. Kottegoda (1977). Stochastic modelling of riverflow time series (with discussion). *J. Roy. Statist. Soc. Ser. A* **140**, 1–47.

[4] Mandelbrot, B. B. and J. R. Wallis (1969). Some long run properties of geophysical records. *Water Resources Res.* **5**, 228–267.

[5] McLeod, A. I., K. W. Hipel and B. A. Bodo (1992). Trend analysis methodology for water quality time series. *Environmetrics* **2**, 169–200.

[6] Anderson, E. (1960). A semigraphical method for the analysis of complex problems. *Technometrics* **2**, 387–392.

[7] Zellner, A. (1992). Statistics, science and public policy. *J. Amer. Statist. Assoc.* **87**, 1–6.

G. P. Patil and C. R. Rao, eds., *Handbook of Statistics, Vol. 12*

2

A National Center for Statistical Ecology and Environmental Statistics: A Center Without Walls

G. P. Patil

Abstract: It is now some twenty-five years that I have been involved with statistics, ecology, and the environment. Five years ago, in collaboration with several outstanding colleagues and friends devoted to the cause of statistics, ecology, and the environment, I had the pleasure to put together a concept of a national center without walls for statistical ecology and environmental statistics. The synergistic effort of conceptualization turned out to be extremely rewarding and fruitful for almost everyone involved. This chapter is adapted from the synergistic effort and from an invited address that I gave at the statistical ecology symposium of the fifth international congress of ecology held in Yokohama, Japan during August 1990. Several participants found the ideas stimulating and interesting. This chapter is prepared in hopes that it will help formulate timely approaches for research, training, and outreach in statistical ecology and environmental statistics, and also help continue a fruitful dialogue among those interested and concerned.

1. Introduction

The physical environment – air, water, fields, and forests and the ecological systems they support – is a major determinant of the quality of human life. Indeed, in the extreme, we are entirely dependent on our environment for life. At the same time, how we attempt to manage the environment at the margin – what we allow to be put into and taken out of the environment, when, and in what quantities – contributes to our ability to produce, distribute and to enjoy the services and material goods that also help to define our quality of life.

In the circumstances, it is vital for our society to understand the opportunities and constraints it faces in deciding how to relate to the natural world. But this understanding has in the past been and continues to be hampered by the difficulty of devising and conducting controlled experiments. Instead we face what has been called 'a cycle of no information, new information and non-information'. We encounter in one setting, in connection with a particular purpose, data gathered in another setting and for other purposes. We confront the results of advocacy in the design of data gathering efforts and choice of rules of analysis. And we face non-linear, dynamic systems with largely linear, static tools. More sophisticated statistical theory, translated into more widely applicable computational methodologies can help at every level. But the new tools must also relate to old problems, problems that have exercised other disciplines for decades. Whence comes the concept of a center without walls dedicated to contributing to better definition of problems, better identification of relevant data, better methodologies for drawing implications from available data, and better tools for predicting the results of particular human interventions in the environment.

2. Rationale for the center

A nationwide group of distinguished scholars and managers concerned about effective integration of issues and answers involving environment, ecology, and evidence have joined together to propose a truly national center without walls for statistical ecology and environmental statistics to meet a longtime critical need for the nation for accelerated economic well being and international competitiveness impacted by the unaffordable adversial processes and proceedings attempting to deal with non-standard evidence involving uncertainty.

2.1. Broadly speaking

The pressing problems in environmental and natural resource management, and more broadly human ecology and health, can be addressed on theoretical grounds, but empirical observations and experimentation are required to test the theories and, importantly, to answer specific real world questions. Experience has shown that in the general fields of ecology and environmental science, conventional research designs often are inadequate and the data involved usually have non-standard features due to spatial and temporal variations and to social, economic, and political constraints. We cannot change natural selection and evolution, the behavior of organisms, and the complexities of population–community–ecosystem dynamics. Nor do we have perfect measuring devices for the many variables involved. Statistics is the key to dealing with the probabilistic nature of natural processes and their responses to man-made perturbations and stress.

Statistical theory and techniques already have contributed greatly to advances in the natural sciences and, together with technological advances in computing and other fields such as remote sensing, to environmental and natural resource management. But serious problems remain. Definition and resolution of these problems require a multidisciplinary approach involving collaboration of basic biologists, ecologists, environmental scientists, social scientists and economists, and mathematical and applied statisticians. This collaboration needs to be given a coherent intellectual focus and organized appropriately to produce the synergistic results of which it is capable. The center mode, and the proposed national center for statistical ecology and environmental statistics, is intended to accomplish this.

Further, in a center mode the educational process can be stimulated and guided by the advances in research. Students and trainees can serve at once as participants and as a sounding board for the spectrum of research thrusts in progress. In keeping with the multidisciplinarity and dispersion of the research activities, the center mode will allow a corresponding diversity of faculty, locales, and facilities for education and training.

Also, the focus of the research and education on environmental and natural resource issues and problems will stimulate – require, if you will – interaction and cooperation between individuals and groups in the participating academic institutions and others in public agencies and private industry. Both expertise and funding will be shared in these endeavors. The center mode provides a uniquely logical and effective means of facilitating and supporting such collaboration.

2.2. Intellectual theme

Testing environmental issues against available and achievable evidence consisting of field data, published information, and expert opinion is often frustrating, contentious and inconclusive, becauase field populations are significantly affected by factors not incorporated into the models and the analyses. Such problems are exacerbated for ecological and environmental studies at the community level where the difficulties in setting up such studies often leads to subtle and unrecognized pseudo-replication. Choice of statistical method and assessment of statistical and substantive significance for non-experimental data have become bewildering topics in ecology and environmental sciences.

While there is no substitute for good, clever and improved data, many worthwhile studies out of necessity lack a clearly defined sampling frame for the populations under study. Observational mechanisms become procedures of unknown-and-unequal probability sampling without a sampling frame because of selection bias implicit in differential observing and recording.

The statistical analysis and interpretation of these kinds of observational studies and encountered data raise a variety of conceptual and methodological problems before the questions involving representativeness of the sample and robustness of substantive inferences about the true populations may be satisfactorily answered, whether for discovery or management.

It is natural that our unifying intellectual theme is: Intergrative excellence in ecological and environmental science in this computer age of information for which statistics is a key technology to break the unaffordable cycle of 'no information, new information, and non-information' pertaining to the environment and the ecological resources and habitats. It is essential to develop novel and innovative intellectual, methodological, and computational non-standard statistical techniques for sound planning, analysis, interpretation, and improvement in matters that relate to scientific evidence, environment, ecology, and economic competitiveness.

The proposed major themes are: probing encountered data in ecological and environmental work; spatial and temporal statistics and data interpretation and improvement; issues and evidence in biodiversity and community dynamics; resource assessment and prediction for management and conservation; environmental risk analysis and decision making; ecometric and environmetric computing; and cutting edge linkages with related areas of fundamental research.

2.3. *Major goals*

General. To establish a national facility for the development of a unified conceptual foundation and improved methodologies for integrative research, training, and knowledge transfer related to basic ecological, environmental, and natural resource issues and problems. Emphasis would be on multidisciplinary approaches to problem-solving, with statistical ecology and environmental statistics as the core.

Research. To develop and implement a program of interactive and integrated research in statistical ecology and environmental statistics; to establish a program structure of sufficient breadth and depth to effectively address specific issues and problems; to stimulate the creation and exchange of innovative ideas and approaches to the research process.

Training. To develop and conduct a program of graduate and post-doctoral training in both theoretical and applied aspects of statistical ecology and environmental statistics, with special emphasis on the merging of statistical understanding and skills with ecological perceptions and knowledge.

Knowledge transfer. To develop and carry out a broad, flexible outreach program to provide new and timely information to environmental and natural resource managers and policy-makers, and other potential user groups in private and public institutions and agencies, and to promote constructive interaction among scientists and practitioners at different stages and levels of implementation.

Impartial peer review capability. To provide impartial peer review capability on research and management issues that relate statistics, ecology, and environmental studies to discovery, and to policy research and evaluation.

2.4. *Integrative workplans*

(i) The center will encourage and support mutually related individual and integrative research. Cross-disciplinary issues and investigations will be pursued. Emphasis will be on true integration and integrative output, and not on juxtaposition. Thus the center research will add significantly to what is being done through other existing modes of research support and will accomplish what is practically impossible to do under the other existing modes of research support.

(ii) A major research focus of the center will be tied to real data sets and issues having substantive importance. Research investigators and participants will have worked with the data sets and the issues and shared the contents. Working thematic workshops will provide people-to-people interactions and synergism. The workshops will bring together cross-disciplinary and cross-institutional experts who may or may not be regular center investigators. Workshop reports will provide live materials for the classroom and the outreach. To accomplish these goals, the center will have a focal data interpretation and analysis laboratory (DIAL) to maintain a working network arrangement with interested ecological research sites and centers of ecological and environmental research in the nation.

(iii) The center will support 'mobile' postdocs and predocs, visiting faculty and scientists, sabbaticals, working seminars, integrated and/or team-taught special topics courses. Predocs will be encouraged to pursue integrative multidisciplinary training at one or more institutions. Postdocs will work in two or more thematic areas and at two or more participating institutions. And so also the participating faculty and scientists. Interdisciplinary faculty positions will be encouraged.

(iv) The management and the productivity of the center will be accomplished through a commensurate management and administration plan. The center will have the benefit of timely evaluation and guidance from its distinguished advisory board.

2.5. *Why a center without walls?*

(i) To provide an adequately sized pool of ecological scientists and environmental engineers with appreciation and expertise of what it takes to deal with real life non-standard evidence in the light of ecological and environmental science theories and practices.

(ii) To meet the contemporary critical need of society in its ecological research and environmental protection for economic development and competition. Current process is an unaffordable, expensive, and socially intolerable way to resolve environmental and related public health issues.

(iii) No single institution has the resources to form a center: Critical mass of researchers at no institution available or achievable; underminimal support and underminimal effort at every place, whether in academia, government or industry; and therefore a national focal point, umbrella facility, and national resource necessary as a center without walls.

3. Description of the intellectual focus and research

Surveys for monitoring changes and trends in our environment and its resources involve some unusual conceptual and methodological issues pertaining to the observer, the observed, and the observational process. Problems that are not typical of current statistical theory and practice arise. Everyone concerned needs to find innovative ways and means of not contributing to, but breaking into, the unaffordable cycles of 'no information, new information, and non-information'.

The motivation for and the goal of the proposed center is precisely that – to foster an institutional framework in which innovative methodologists and substantive scientists would link quantitative methods with empirical knowledge and mutual insights toward the resolution of societal problems. Areas of proposed research have been identified and are described under the following seven thematic headings:

Theme 1: Probing encountered data in ecological and environmental work.
Theme 2: Spatial and temporal statistics, data interpretation and improvement.
Theme 3: Issues and evidence in biodiversity and community dynamics.
Theme 4: Resource assessment and prediction for management and conservation.
Theme 5: Environmental risk analysis and decision making.
Theme 6: Ecometric and environmetric computing.
Theme 7: Cutting edge linkages.

We have emphasized describing what the problem is. Recognition is the first stage; solution is the final, more difficult stage. Some of the descriptions and references do point to solutions and developing methodology, but the objective is to describe general areas where better methods are needed and where research effort could be focused.

In the end, we must recognize that there is no substitute for good data, and in no way can we be complacent about it. Statistical thinking is an aid to the interpretation of data. It may help clarify seeming confusion; but it can also confuse seeming clarity. What is important is that the statistical approach contribute to the overall insight and perspective of the substantive issue and to its resolution in light of the evidence on hand, be it empirical, literature-assembled, or expert opinion.

3.1. Theme 1: Probing encountered data in ecological and environmental work

3.1.1. Probing encountered data using weighted distributions

Traditional statistics theory and practice have been occupied largely with statistics involving randomization and replication. But in ecological and environmental work, observations most often fall in the non-experimental, non-replicated, and non-random categories. Additionally, the problems of model specification and data interpretation acquire special importance and great concern. In several situations of diverse nature in statistical ecology and environmental statistics, the theory of weighted distributions provides a perceptive and unifying approach for the problems of model specification and data interpretation. For the growing literature, see Patil, Rao and Zelen (1987b).

Weighted distributions take into account the observer–observed interface, i.e., the method of ascertainment, by adjusting the probabilities of actual occurrence of events to arrive at a specification of the probabilities of those events as observed and recorded. Failure to make such adjustments can lead to wrong conclusions.

The problem of specification is a difficult one, requiring a detailed knowledge of the procedure employed in acquiring data. The situation is even more complicated with field observations and non-experimental data, where nature produces events which are then observed and recorded by investigators. An appropriate specification can be arrived at on the basis of past experience, information on the stochastic nature of events, knowledge of how events are ascertained and recorded, and an exploratory analysis of current and related data.

Inaccuracies in specification can lead to wrong inference. It is therefore important to study the data under different possible specifications to see how variant the conclusions could be and what the deficiencies in data are in terms of non-response, measurement error, contamination, misclassification, and extraneous variation.

3.1.2. Meta analysis – Combining data for stronger conclusions

Meta-analysis refers to the statistical integration of the results, and perhaps of the data, from separate studies. This quantitative approach to the synthesis of research efforts is a response to the growing need for broader and more rigorous summarization and integration of research literatures. See Hedges and Olkin (1985) for an overview and DuMouchel and Harris (1983) for application of empirical Bayes techniques.

Acid precipitation research. There are approximately 85 research projects in progress under the forest response program of the national acid precipitation assessment program. These projects are directed to a variety of scientific questions, none of which will be answered by any single project.

In their review of the environmental protection agency's acid precipitation research, Landau, Patil and Rawlings (1985) state: 'In an effort to facilitate the ultimate objective of providing dose-response data for the regional assessments, it is recommended that immediate attention be given to problems of combining the research results over time and space'. The most immediate candidate for such meta-analysis in the forest response program is the set of seedling exposure experiments. The joint interpretation of these experiments will not be a simple task, requiring active collaboration between statisticians and substantive scientists, and is an issue that can utilize some focal effort under the auspices of the center.

Recruitment of marine fish. Recruitment refers to the abundance of fish of a given stock that mature annually to an exploitable size. The recruitment process is governed by many factors, both environmental and biological, but there is currently little understanding of what those factors are, how they operate quantitatively, and how they interact. Recruitment series for a given stock, as well as corresponding environmental data, are generally quite short.

In an effort to overcome the data limitations, and seeking to identify recruitment patterns common to various stocks, Patil et al. (1986) have applied nonparametric density estimation to a data set obtained by combining recruitment values for 18 marine fish stocks from different oceans and from different species. Along these lines, potential activities of the proposed center include: (1) assessing the robustness of the conclusions to the choice of scaling parameters, (2) developing alternative stategies for combining data, and (3) extending the techniques to a multivariate setting that would permit the inclusion of environmental and other variables.

3.1.3. Heterogeneity, mixtures and empirical Bayes

A key difference between the so-called soft and hard sciences is in the heterogeneity of the objects or materials that are investigated. A primary concern of statistical ecology should be that of developing and evaluating methods for characterizing and dealing with this heterogeneity.

There are many situations where heterogeneity can be rationalized by a mixture approach to modeling. In this approach, the population of interest consists of a collection of homogeneous subpopulations P_i, whose proportions in the target population are denoted by π_i, with i in some index set. If we sample at random in a situation where the groups are mixed together, then a reasonable description for any measurement X is the mixture model $\sum \pi_i f(x;\theta_i)$, where f is the density for X in the i-th group and θ_i is a parameter that summarizes the characteristics of the i-th group. With appropriate measurements, the vectors π and θ can be estimated. Moreover, if we wish to make some inference about the whole population, say the total biomass, these estimates enable us to measure and account for the heterogeneity of the sampled population.

For a state-of-the-art review of mixture methodology, see Titterington et al. (1985). For application to ecological data, see Roeder et al. (1989) where mixture models are applied to answer questions concerning the paternity of seeds given partial genotypic information. Here are some areas of proposed research in which the gaps in current knowledge are a hindrance to the satisfactory application of mixture models.

Parametric modeling of mixing distributions. More flexible methods are needed for the parametric modeling and estimation of unknown mixing distributions. Current approaches typically employ either conjugate distributions (for reasons of convenience) or normal distributions (for reasons of familiarity). Lindsay (1986) is a step in this direction, but lacks adequate range.

Sampling distributions for non-parametric estimates. One important new method which has been developed for mixture problems is the non-parametric maximum likelihood estimator. It is useful both for non-parametric and semi-parametric modeling and as a diagnostic benchmark for comparing parametric modeling techniques. It has a feasible analytic and computational theory (Lindsay, 1983a, b). What the method currently lacks is an adequate sampling theory, and in particular a large sample asymptotic theory. This would be important in a practical sense for the development of confidence sets and for the assessment of standard errors.

Empirical Bayes. Empirical Bayes techniques (in which the mixing distribution plays the role of the Bayesian prior) offer considerable potential for improved estimation, and are particularly valuable in settings where information from a variety of sources is to be combined or where inferences are to be made for a repetitive series of observations. To date, most work has been restricted to normal theory models; it would be useful to develop these methods more fully in other important distributional contexts.

Improved algorithmic methods. Mixture modeling is computationally intensive. Newtonian methods can be unstable; the EM algorithm can be very slow. Since there is no feasible sampling distribution theory available, current approaches employ Monte Carlo simulation and bootstrapping. Improved algorithms are a prerequisite to handling larger data sets.

3.1.4. Environmental sampling and design

Model based sampling. Recent work by Fienberg and Tanur (1987) compares the evolutionary history of the design of experiments with that of sampling theory.

Strong parallels are threaded through these two fields, but the quite dissimilar terminology for similar concepts. Considerable benefits could be achieved by applying research results from the design of experiments to sampling problems, and vice versa.

Although the design of experiments literature considers the information content of estimators resulting from various designs, it usually relies on very simple cost functions. Conversely, the sampling literature has derived practical sampling schemes for quite complex cost structures. The sampling designs obtained by optimizing these cost structures have parallels in experimental design. When the observations or experimental responses are the result of several stages of nesting, the techniques used in sampling to minimize cost and maximize information can be applied to ecological experiments to achieve an efficient allocation of resources for each level of nesting.

Similarly, the application of optimal design theory (Mitchell, 1974; Galil and Kiefer, 1980), which depends fundamentally on the assumed model, has parallels and applications in model based sampling.

Messy data analysis. The term 'messy data' describes the non-standard reality of research data encountered in a broad spectrum of ecological and environmental studies. Thus, survey data is typically unbalanced. Field experiments, even when carefully designed, often result in poorly measured units due to the unpredictable nature of the field environment. Even when collected under ideal conditions, data is often based on multiple sampling and multiple experimental units, frequently with repeated measurements and pseudo-replication. These multiple issues arising in a single set of data can confound both the experienced researcher and the best of standard methodology. A recent book by Milliken and Johnson (1984) discusses many of the non-standard aspects of messy data that plague researchers when analyzing their data.

Repeated measures, a structured instance of pseudo-replication, arise in many situations, and would be a topic of investigation for the proposed center. The goal would be to develop model assumptions and corresponding procedures leading to more effective tests and improved estimation. For example, when the repeated measure is over time, e.g., to measure growth, then responses widely separated in time are less correlated than those close together. This calls for a family of covariance matrices that allow the variances to increase over time, and the correlations to decrease with increasing temporal spacing between responses. Restrictions such as these determine subregions within covariance parameter space. Estimating the covariance matrix within the appropriate subregion should provide improved statistical treatment of repeated measures data. Extension of such techniques to more general instances of pseudo-replication would also be pursued.

Biological sampling and survey designs. Utilization of sample selection probabilities is the key feature of scientific or probability sampling, providing the rationale for inference from the sample to the population. In conventional socioeconomic surveys, the selection probabilities are designed and controlled by the investigator. While such conventional survey designs may be applicable in selecting the physical sites for biological sampling, the biological measurements made at these sites commonly entail further sampling stages to enumerate the target organisms present.

The investigator's control over sample-inclusion probabilities is generally limited during these later stages. Designing, measuring, or modeling and estimating sample selection probabilities for the target organisms is a primary requirement of biological sampling methods.

Capture–recapture methods provide one approach for estimating the otherwise unknown sample-inclusion probabilities for elusive animals. Behavior modification may attend capturing and recapturing, however, and time lapses between samples with attendant changes in the target population, introduce an additional dimension to the analysis. Modeling, estimating, and testing the effects of such factors in capture–recapture experiments would be included within the research activities of the proposed center. Technological and practical feasibility of devices like radio telemetry and circular plots would likewise be evaluated.

Composite sampling for environmental monitoring. A composite sample is formed by taking a number of individual samples and physically mixing them. Composite sampling is the basis of some important sampling procedures for human health and environmental measurement (see Elder, 1977; Rhode, 1979; Garner et al., 1988; Boswell and Patil, 1987). The purpose may be to estimate the mean of a stochastic process in space or time, or to estimate the fraction of a population that possesses a given trait while maintaining the privacy of the individual. The following critical issues need to be resolved: situations amenable and not amenable to composite sampling, and statistical and substantive concerns involving sampling-plan procedures for (1) waste disposal and dump sites, (2) estuaries and rivers affected by non-point sources of pollution, (3) human populations with a rare trait, etc.

3.2. *Theme 2: Spatial and temporal statistics, data interpretation and improvement*

Space and time are fundamental to ecosystem structure and environmental processes. Consequently, temporal and spatial dependencies are inherent in virtually all ecological and environmental data sets. Spatial analysis and time series modeling are recognized research areas in statistics, with the resulting methodologies being applied in several contexts such as business decision making, governmental economic policy, and computer vision. There has been relatively limited penetration of statistical theory relating to these subjects, however, into ecological/environmental research and decision making. A major goal for the proposed center will be to appropriately extend the statistical concepts underlying spatial statistics and time series analysis to fit the context of ecological research and environmental management.

The lack of adequate interdisciplinary communication and cross-fertilization is a serious problem. This is illustrated by the contrast between the summary of the theoretical work in spatial and time series sampling methods as reviewed by Cambanis (1985) and the work on the design of hydrological data networks as reviewed by Moss and Tasker (1979). In both instances the general underlying question is: what is the most efficient sampling procedure for a process that is variable in both space and time. Of the 130 papers cited in the Moss and Tasker review, only one was from a standard statistical journal. Conversely, the Cambanis review did not cite any of the applied work in hydrology or environmental science.

The interdisciplinary thrust of a well conceived national program is ideally suited to generate the interchange between disciplines that history shows is not necessarily self-generating, especially in the present era of academic specialization. The proposed 'center without walls' structure will maximize such interchanges by formally linking multiple academic institutions with counter-part action agencies and industries.

A second major reason for a lack of statistical impact in the ecological/environmental arena is that the development of statistical methodology typically proceeds from a restrictive set of assumptions, such as normality, intended to make the analysis mathematically tractable. The relevance of a specific method is dependent on the underlying assumptions and the questions posed. While time series methods are well known and have been applied to ecological problems, often the time series observed are too short in length or too irregularly spaced to employ conventional approaches. Time series methods are also not well adapted to many ecological and environmental problems which are spatial in nature. This is an example of where classical statistical methods diverge from the needs of ecology and environmental science. Assumptions such as multivariate normality and random sampling are rarely valid for environmental or ecological data. In the absence of theoretical or empirical validation of the robustness of a methodology relative to its under-lying assumptions, uncertainty as to the applicability of the technique often deters potential application. Without the benefit of interdisciplinary communication statisticians are often not in a position to determine the appropriate assumptions and generalizations by themselves.

Failure to achieve correspondence between statistical methodology and application context leads to investigative inefficiency. In an attempt to ensure that standard statistical inference techniques are applicable, ecologists and environmental scientists often deliberately use randomization techniques to incorporate spatial and temporal components of variance into the error term of an analysis. The result of such inflated error terms is to compound the number of observations needed, increase decision uncertainty, and weaken inference.

This proposal will not review all the applications of spatial statistics and time series analysis to ecological and environmental work or for that matter all recent advances in the general theory. It describes an initial range of problems that have a direct impact on future ecological and environmental research, and their close relationship with the center's major goals.

3.2.1. Unresolved problems in the interpretation and analysis of spatial data

There are a number of characteristics of environmental and ecological data which are important for monitoring and assessment. These include intercorrelation of values for an ecological or environmental measurement as a function of position, correlation between different measurements, and the geostatistical support of the samples. Other attributes of concern are 'hard' numerical data versus 'soft' qualitative information or constraint intervals, the real extent of the region for which an inference is to be made, and the sample location pattern. Finally, there may be a temporal as well as a spatial dependence. While these are not the only important characteristics, they provide reasons why more standard statistical methods are inadequate for the appropriate data analysis.

Linear univariate geostatistics (Warrick, Myers and Neilsen, 1986) incorporates the spatial correlation as well as the sampling pattern information and support of the samples to provide estimates of spatial averages. In contrast multivariate techniques such as principal components, factor analysis and lesser known techniques like correspondence analysis or linear dependency analysis incorporate intervariable correlation but not sample support nor spatial dependence. Neither of these is efficient for estimating probabilities or probability distributions. A general perspective of multivariate spatial statistics ought to incorporate all of these data characteristics.

The following research areas and issues may be initially pursued:

Whichever form of geostatistical estimation, i.e., 'kriging', is utilized, the covariance or cross-covariance function (variogram, cross-variogram, etc.) must be estimated from the data. As yet, no fully reliable estimator has been identified, nor are the properties of the commonly used estimators well established. Moreover, measures of the adequacy of modeling the covariance are needed and the relationship of the covariance function determination to the design of efficient sample location networks needs to be investigated.

In as much as all forms of the kriging estimator are essentially linear, they are more or less sensitive to outliers and may be sensitive to inadequate determination of the variogram. This suggests a line of research pertaining to the robustness of these estimators as well as development of new estimators. Note that in the case of linear estimators, robustness with respect to the data and continuity of the weights are not synonymous.

As indicated above, many data sets arising out of ecological or environmental problems have temporal as well as spatial characteristics. As yet most methods focus on only one of these dependencies. A number of possible approaches for modeling and analysis of joint spatial-temporal dependencies have been suggested, but each has deficiencies and the matter needs further consideration.

All of the spatial and temporal statistical methods require an assumption of stationarity in some form. The methods currently in use for determining the presence of drift or obtaining a functional form for the drift are largely subjective and non-rigorous.

It is common practice in many instances to transform the data, e.g., logarithmic or probability transform. If the resulting estimates must be retransformed then an adjustment is required. Unless strong distributional assumptions are verified or invoked, the adjustments may be indeterminate. In the case of a probability transform, the data becomes multivariate and there may be order-relation inconsistencies in the resulting estimates. The transformation may also complicate the identification or determination of the presence of drift, i.e., of non-stationarity. These problems have been largely ignored, avoided or resolved by ad hoc solutions; more rigorous results and their practical exemplification are needed.

3.2.2. Issues of environmental design in space and time

In this section we give two examples of design problems that fit well within the theoretical context of spatial sampling and can be best approached by a careful merging of classical sampling theory and multivariate spatial statistics.

In the past few years, EPA has carried out many large scale surveys to determine, for example, concentration of radon in housing units, number of lakes affected by acid deposition, or the number of housing units in a local area affected by ground water or surface soil contamination. In practice the sampling design was determined using classical methods such as simple stratification with random sampling within strata.

Such situations can be abstracted by considering a finite population whose units are distributed in k-dimensional space and are described, by a binary variable, as either affected or unaffected. Possible goals are to estimate the number of affected units, or to predict which of the unsampled units are affected. This model fits exactly into the simple indicator kriging model. The sampling design problem then becomes: given the location of the units and a fixed sample size n find an efficient sampling plan. Under this setup one could compare the relative efficiencies of the classical sampling and estimation methods with those obtained using indicator kriging. The above problem differs markedly from the classical sampling designs for spatial processes as described in Cambanis (1985), in that selection of the points in space is confined to points where sampling units occur.

Of course real situations are more complex than the one just described. A common additional feature is the availability of covariates, such as soil type, elevation, and underlying rock type, that describe the characteristics of a point in space. These are the variables that would be used to stratify in the classical stratified sampling scheme. Incorporation of such covariates into the indicator kriging model is an example of a problem to be investigated.

Consider next a different kind of time series experimental design problem. The major cost of an environmental program is not always determined by the number of samples taken or even the frequency of sampling, but perhaps instead by the total time required for a program. For example, a researcher may be allotted a fixed amount of time to use an expensive research platform such as an oceanographic vessel. Large environmental assessment programs such as the EPA's acid deposition program are constrained by fixed time points at which decisions must be made. Also, improvements in instrumentation and measurement technology have made nearly continuous measurements relatively inexpensive in some cases. However, the emphasis in the statistical design of environmental programs has remained on the classical statistical problems of determining sample size and frequency of sampling.

Continuous time regression models can be used to evaluate research plans whose primary constraint is time. The statistical results are well known (Parzen, 1961); however, there has been little work in applying these results to the design problem. The basic approach will be to use mathematical and numerical analysis to evaluate the relative efficiencies of continuous designs with fixed time constraints. Like classical experimental designs, these continuous designs would be simplifications of real sampling and experimental problems. They should, however, provide meaningful guidelines for the practical design of an environmental program.

3.2.3. Conditional patterns of spatial distribution

Computer-based geographic information systems (GIS) have become a standard technology in the environmental sciences for recording in digital form the location of multiple environmental variables on a common coordinate base. The typical data base includes variables measured on both categorical and interval scales. Standard techniques for analyzing co-occurrence among the several variates comprising such a data base, however, are very simplistic and rely heavily on visual judgements made by an analyst upon examination of a video display superimposing the variates through a shading scheme entirely analogous to a simple man overlay. Extensive use is also made of heuristic clustering algorithms that yield results which are more suggestive than quantitative.

A major advance in the level of analytical quantification of co-occurrence with such systems could be achieved through the elaboration and algorithmic application of statistical approaches for analysis of spatial point patterns (Diggle, 1983; Ripley, 1977, 1979), with the distribution of subsets of dependent variables being conditional upon the distribution of subsets of independent variables. Such methodologies would provide powerful capabilities for generating and testing hypotheses concerning causalities and interactions of environmentally induced stresses for risk assessment. They could also be applied to the problem of spatial allocation of data collection activities for environmental monitoring.

3.3. Theme 3: Issues and evidence in biodiversity and community dynamics

3.3.1. Species diversity

Approximately 1.7 million species of plants, animals, and microorganisms have been formally identified; estimates of the number that actually exist range beyond 30 million (Wilson, 1985). Most biologists feel that species are currently being extinguished very rapidly and that, particularly in tropical forests, we are on the threshold of a mass extinction rivalling the Cretaceous and Premian extinctions of the geological past (RFF, 1983; Simberloff, 1983). This conclusion has been criticized, however, (e.g., Simon, 1986), on the grounds that it is inferred, rather than observed. It is true that extinctions are primarily inferred, but an inference of extinction can be strong if a species is well studied. The real problem is that most species thought to be going extinct are not even formally identified as species, so the inference cannot be nearly as strong. How can this problem be remedied?

A key necessity is greatly increased funding for biogeographic and systematic research; there are simply too few people locating organisms in nature and determining their specific status. Wilson (1985) feels that this research would require about 25 000 professional lifetimes without sophisticated computer and statistical aids, but that the latter would aid the project greatly.

However, even existing data can be used much more thoroughly than they have been, and it is important to establish a firmer statistical framework for treatment of new data gathered by the new army of systematists. The problem of estimating the probability of finding a new species, given the species-abundance curve of a sample from a community, has not been well studied statistically. Hill (1979) takes a Bayesian approach and reviews previous literature. One particular lacuna is that clumping of conspecific individuals in space is routine, yet existing stastistical approaches do not account for this contagion.

The data that may be most useful in estimating overall total diversity, and total diversity within particular taxa, are more likely to be geographic range data than species-abundance curves. For any taxon, it would be possible to gather from literature sources and specialists the estimated geographic ranges of all known species. Simply mapping these would allow a vast improvement in depicting geographic patterns of diversity and identifying regions of unusually high species diversity. Further, it should be possible to extrapolate a species-range curve to estimate the number of undiscovered species with ranges of a given size. This seems to be a classic encountered data problem; the probability of a large-ranged species having been detected is greater than that of a small-ranged species.

The species that are going extinct are, on average those with the smallest geographic ranges, yet these are the very species that are least likely to have been formally identified (Simberloff, 1986). Thus, estimation of numbers of unidentified species with different range sizes should aid greatly in estimation of extinction rates. Of course, data on habitat destruction in particular habitats and regions, in concert with maps of known and estimated diversity, should refine such estimation.

Yet another approach to estimating diversity derives from the study of taxonomic diversity – the hierarchical arrangement of species into genera, genera into families, etc. Geographical patterns of such statistics as species-to-genus ratios have been examined in both ecological settings (Simberloff, 1970) and evolutionary ones but not with an eye towards estimation of diversity. In short, given a sample of genera of which x_1 have one species, x_2 have two species, etc., what can we predict about the full collection of genera? Again this is an encounter problem, in that it is likely that genera with more species have a higher probability of being sampled (through this hypothesis needs testing, including with the proposed geographic range data set described above).

3.3.2. Community and ecosystem diversity

At the community level the problems are even greater. It is widely recognized that communities are entities worth saving in their own right and that, indeed, most species cannot be maintained outside a community. However, exactly what constitutes a community and how one knows whether two collections of species represent the same or different communities are unanswered questions. For want of an accepted method, organization such as the nature conservancy that are curretly attempting to preserve representatives of as many communities as possible use a rather coarse-grained, intuitive approach and recognize rather few communities in any geographic region. However, the ecological literature is rife with examples of subtle structure and dynamic differences between apparently similar communities.

From an ecological standpoint what is primarily required is empirical research on the forces that bind species together into communities and the extent to which a community's composition can be changed without the community's losing its integrity and identity. From a statistical standpoint this is a matter of classification and ordination. What is needed is consensus on the indices of similarity that are appropriate of classifying communities, a statistical examination of the extent to which communities in any oridination space are discrete, and an approach that will allow estimation of how many different communities there are, where they are located geographically, and how similar they are in an ordination space. Ordination methods describe species distributions and their interrelationships (covariance). Often these interrelations are non-linear, and yet most standard ecological methods used to study them are linear. Some even assume that the species distributions are Gaussian. There is a need to evaluate the sensitivity of the typically used methods to such assumptions as linearity and normality, and to investigate new methods that are less restrictive.

A variety of static measures have been developed for describing ecosystem and community structure. Ecosystem linkage or food web diagrams have been used as a basis for calculating trophic specialization, complexity, element cycling, and for finding keystone species. At present little can be said about the statistical properties of these measures or indices. For example, how does error in estimating state variables affect the accuracy of the indices?

Questions asked at the ecosystem level often involve a dynamic component. For impact analysis and conservation one wants to know to what extent disturbances propagate through the system. To this end, perturbation analysis of ecosystem models, loop analysis, and stability (e.g., uniqueness and existence of community equilibria, existence of discontinuous stability such as the cusp catastrophe) have been developed. For such methods serious questions arise about how to draw inferences from results of the analysis and how to compare predictions with the real world (Loehle, 1987).

3.3.3. Genetic diversity

In this age of genetic engineering, genes still have to come from existing organisms and the genetic structure of species is not thoroughly known for any higher organism. Such matters as the genetic differences between non-specific populations, and even the genetic differences between individuals within populations, require empirical research. For many years it has been recognized that most populations have significant polymorphism (ca. 10%–30% of loci), but only recently has it emerged (e.g., Lewontin, 1986) that polymorphisms are of several types, including those with many rare alleles. Rare and 'private' alleles have been used statistically to aid in estimation of gene flow (e.g., Slatkin, 1985), but there has been to attempt to use the distribution of allele frequencies, plus data on mutation rates and selection, to estimate total genetic diversity. Once again, the more common an allele is, the more likely its detection, so known alleles are a biased sample of existing alleles.

Can we explain the genetic diversity of individual locus-specific variations in natural populations based on the concept of stochastic ordering of diversity profiles (Patil and Taillie, 1977, 1979, 1982), instead of employing the summary measures like gene diversity (or heterozygosity)? In particular, we want to examine the patterns of diversity profiles of genetic variations of various protein-enzyme loci, categorizing them by their molecular structure and size, and functions served by them. The same can, of course, be done by stratification of the organisms by their habitat and biogeography.

While the population genetic works are mainly concerned with natural interpretations of the statistics used in terms of the biological quantities being measured, there has been considerable research to model the population genetic systems applying stochastic theory of diffusion equations, from which suitable sampling theory can be derived (e.g., Ewens, 1972, 1979; Ewens and Li, 1980). Such theories and various ramifications of these have been profitably used in interpreting data on patterns of genetic variations in natural populations (Fuerst et al., 1977; Chakraborty et al., 1980). Mechanistic explanations for the production and maintenance of genetic variability in natural populations had been a major objective of these studies.

Although these efforts were largely successful in predicting the pattern of genetic variation in terms of observed statistics of genetic variation, the power of discrimination of different hypotheses has been questioned (Watterson, 1977; Gillespie, 1978; Milkman, 1982). One of the difficulties in testing evolutionary models with genetic data is that the number of genetic variants at a locus is usually very small (often, less than five) and the mutation rate (the ultimate factor for producing genetic variants) is quite small (of the order 10^{-5} per generation, or less). Furthermore, statistical tests are generally made with summary measures like average heterozygosity, average number of alleles per locus, etc., which can only be reliably estimated when multiple loci are involved in such computations (Li and Nei, 1975; Nei, 1978). The unavailability of sampling distributions for such statistics are prevents the development of efficient testing procedures.

3.3.4. Formal similarities between community and genetic diversity

The literature on community diversity in ecological systems involve questions dealing with: (1) measures of species diversity in a habitat; (2) abundance of individual species within a habitat; (3) inter-habitat comparisons of species diversity; (4) treatment of diversity functions with replicated communities; (5) characterization of diversity profiles by species that are specialists versus generalists; and (6) mechanisms of production and maintenance of community composition (see e.g., Fisher, 1943; Watterson, 1974; May, 1975; Engen, 1978; Patil and Taillie, 1977, 1979, 1982). If we translate the terminologies of ecological diversity to those in population genetics (for example: a genetic locus to a habitat/community; an allele to a species; and allele frequency to the number of individuals within a species), Table 1 shows that these issues are respectively equivalent to some of the commonly

Table 1
Equivalence of some genetic and community diversity issues

Population genetic issues	Community diversity issues
(1) Measures of genetic variability within a locus	(1) Measures of species diversity within a community
(2) Number of copies of different alleles at locus observed in a sample	(2) Species-abundance in a particular habitat
(3) Measures of genetic variation across loci	(3) Inter-habitat comparison of species diversity profiles
(4) Treatment of genetic distance between populations	(4) Treatment of diversity functions with replicated communities
(5) Categorization of genetic loci by their function and molecular properties	(5) Categorization of species by specialists vs. generalists
(6) Adaptive vs. neutral evolution	(6) Mechanism of production of community composition

studied theoretical population genetic questions examined by Wright, Fisher, Haldane, Kimura, Nei, and Ewens (for review, see Wright, 1969; Crow and Kimura, 1970; Nei, 1975, 1987; Ewens, 1979).

The large number of variant species in an ecological community, and the availability of replicated communities in ecological systems allows a possible application of some of the population genetic models to ecological data, and seek for mechanistic explanations of the relationship of species-abundance with their equitability. Furthermore, the observations of inter- and intra-species competition can be used to define alternative models with adaptive significance of species diversity.

3.3.5. Patterns of genetic variation and environmental issues

The study of genetic variation in wild populations is currently limited by small sample sizes, spatial variation, and sampling problems which make statistical treatment non-standard. There are two basic areas in which genetic variation affects economic competitiveness directly: conservation, and introductions and invasions of species or genetically altered organisms. Conservation has a cost because detection of endangered species often results in establishment of preserves, delays in industrial projects and expensive monitoring of the endangered species. Introductions and invasions have a cost because both research and practice are being held up due to uncertainty about effects of introduced organisms. Damaging invasions have in the past had staggering costs (e.g., Bromus invasion of western ranges). This research unit will develop statistical methods to address such concerns.

A key element in a successful conservation program is the maintenance of breeding populations with significant genetic variability. There are various aspects to this problem, each with special statistical problems not currently treatable by standard methods. As the size of a population declines, a critical number is reached at which risk of extinction goes up sharply. This is the critical population size below which individuals may have trouble finding mates, genetic variation may be too low, and chance events may lead to extinction. A number of unresolved statistical issues are involved in estimating these quantities. Habitat fragmentation contributes to difficulty in finding mates and to chance extinctions, but statistical characterization of fragmentation does not rest on a secure footing. Fractal models and the spatial statistics of Theme 2 provide an avenue for studying this problem. Adequate statistical characterization of effective population genetic variance is a controversial issue to which the center could contribute. The role of chance events in extinctions using random walks or other stochastic models could also be addressed. Finally, it should be noted that all studies of endangered species are faced with estimating populations of rare and scattered individuals. This too is a non-standard problem.

The problem of introductions of genetically altered organisms is one that is generating a great deal of concern and public outcry. Currently, little sound basis exists for statistically estimating rates of spread and predicting impacts, for assessing risks, or for designing monitoring programs. Work at the center will address these issues. Historical data on consequences of past introductions provides a starting basis for constructing rational risk models and estimating likely rates of spread of introduced organisms. Proper statistical design of monitoring programs can ensure that escape of altered organisms from test plots or industrial facilities is detected without exessive cost. Criteria need to be established for safety or detectable impact.

3.3.6. Diversity issues in research and management

Among ecologists, 'diversity' invokes a plethora of fundamental and controversial issues. The issues are fundamental in that explaining the effects of environment on the distribution and abundance of species is the essence of many ecologists' work. The controversies arise from the intrinsic scientific importance of diversity theories, as well as from the broad economic and social ramifications of considering biodiversity in land use decisions. There is a sense of urgency about these issues. Global decline of biodiversity is widely perceived as reaching crisis proportions.

At the heart of the scientific and social controversies regarding diversity are problems of quantification, interpretation, and analysis. Existing inference methods are often inappropriate or insufficient for critical applications. For instance, estimating how many unknown species are present in a given region is a problem for which present statistical techniques are either unsatisfactory or have poorly understood properties. Another example concerns parameter estimation and hypothesis testing for species frequency distributions. Forty years of ecological practice with these models has been based on ad hoc fitting methods with no specification of the sampling frame.

An important reason why the inference methods need improvement is that sampling for diversity is non-standard. The more abundant a species is, the larger its range, or the more attracted it is to a light trap, the more likely it will be included in a sample. The smaller the extent or abundance of a species, the more likely it will be missing. The non-standard sampling prevalent in diversity studies implies that off-the-shelf statistical techniques are frequently inappropriate.

Another impediment to diversity research is a persisting communication gap between ecology and statistics. While such gaps occur between statistics and other fields, the situation is particularly acute in ecological diversity research. For instance, quantifying the concept of diversity has received an intense amount of statistical attention for the past decade (e.g., Patil and Taillie, 1982). Nevertheless, the issue is still debated and various diversity indexes are still invented by ecologists poorly equipped to consider problems of sampling, estimation, and testing, with little reference to the statistics literature. On the other hand, the statistical advances in diversity conceptualization received virtually no practical evaluation with real data sets in the statistics literature.

The research unit will seek to resolve quantitative problems that hinder progress in understanding community diversity. The problems will be undertaken based on the statistical theory of sampling, estimation, and testing, as guided by ecological insight. Examples of specific research questions to be considered are:

- Which ecological parameters are most suitable for characterizing community composition and its dynamics? What statistical methods are needed for monitoring these parameters?
- What are the appropriate uses and limitations of indexes of spatial diversity?
- How can species-abundance and species-range curves contribute to sample designs for rare species?
- What are the possible mechanistic explanations of Preston's canonical hypothesis (Preston, 1962)? More specifically, what sort of evolutionary mechanisms can explain the two intriguing observations on community diversity: (a) in an ecological system, the commonest species has its abundance near the mode of the individual curve (in logarithmic scale); and (b) the species richness is inversely related with species equitability.
- How can habitat diversity be characterized and operationalized for ecosystems management?

3.4. Theme 4: Resource assessment and prediction for management and conservation

Almost all of the aquatic living resources, and much of the land based living resources are in the public domain. Public agencies are responsible for the conservation and management of what is largely private sector harvesting. In general, the harvest of the resources is managed to provide sustained yields calculated on the basis of the rate of renewability. The Federal Government also includes considerations of social and economic factors such as multiple use, non-harvested dependent species, and profitability.

The government regulatory agencies are responsible for providing the scientific knowledge upon which management policy and regulations are based. The prime research activities are monitoring, assessment, modeling, and prediction of resource populations and harvest.

Resource assessments have traditionally been largely product-oriented, with emphasis on determining potential harvest in the near term. Survey designs have normally targeted individual resources, with uncertainty measures geared more toward large-area averages than to location-specific estimation. Current ecological, environmental, and economic concerns are causing such surveys to become increasingly inadequate. These concerns generate need for location-specific information on the dynamics of biological populations constituting the resource base, with uncertainty levels low enough to detect weak signals associated with incipient stresses due to environmental effects such as acid deposition. Monitoring of resource dynamics further requires joint quanitification of multiple species and environmental determinants constituting ecosystems. Standard resource survey designs are inadequate in one or more respects for meeting these special needs of modern resource monitoring. Research on non-standard approaches to these problems, with the goal of developing new and more comprehensive standards of resource assessment methodology is a major objective of this research theme of the proposed center.

Furthermore, modeling and prediction of future resource levels and future harvest pressures represent a weak link in the decision-making process. Much quantitative research within the National Marine Fisheries Service, U.S. Forest Service, and other agencies is devoted to construction of predictive mathematical models of living resources. The variability of existing data and lack of basic knowledge about ecological factors affecting resource growth rates has impeded the formulation of reliable models. Inroads into better prediction could potentially occur with greater integration of contemporary statistical theory and ecological science toward production of innovative and more useful modeling approaches. The accomplishment of such innovation is another major objective of this research theme.

Successful pursuit of these goals will require an institutional structure that fosters interchange between investigators and agencies. It must further permit the formation and effective operation of interdisciplinary research terms that cross boundaries of academic institutions. The focus on resource dynamics implies development of capabilities for model-based monitoring. Variances in time, space, species, and sites are the raw material for modeling. Progress in addressing important resource assessment and prediction issues has been fragmented and constrained by the fact that individual investigators working on selected species in specific environments encounter only a subset of the available quantitative techniques. An institutional structure that facilitates sharing of data bases and methods would greatly enhance progress. In fact, a cooperative and comparative approach to modeling is made necessary by the need for joint assessment of species comprising an ecosystem. Incorporation of interaction effects requires at the very least that subsystem models have compatible structures for input and output parameters. The proposed format of a ‘center without walls’ should be an idea mechanism for engendering the interdisciplinary mode of operation essential to these circumstances.

Achievement of the goals set forth in this research proposal have extremely important consequences for competitiveness of the United States economy and maintaining quality of life of both national and global scales. Accurate methodologies for resource assessment and prediction will support formulation of decision support systems that permit full economic development of national resources on a sustainable basis without adversely affecting the ecosystems that constitute the basis of our biological and social well-being.

3.4.1. Resource monitoring and assessment

Monitoring is largely a process of periodic sampling of the resource and harvest to estimate the standing stocks, the spatial and temporal distribution of species, and trends or changes. Assessment is the analysis of the sample along with estimates of rates of growth, mortality, and recruitment (the inflow of young members to the harvestable segment), to provide information on the current state of the resources relative to effects of past harvesting.

The characteristics of the resources and their environment create difficulties in making observations and, consequently, analysis of them. The traditional statistics based on controlled experiments do not apply. For living resources, it is true to say that we mostly have non-random, dependent, and non-replicated observations from populations the characteristics of which are difficult or impossible to define as a sample framework.

Renewable natural resources, whether managed or unmanaged, are in constant change. Because of their nature and value, continuous monitoring of natural resources is required. See Barnard et al. (1985) and references cited therein. Monitoring serves several purposes. Specifically, it:

(1) provides information, quantitative or qualitative, on the resource;

(2) generates data on changes in resource growth and productivity under a full range of conditions and management activities;

(3) discerns changes in the resource resulting from environmental shifts, perturbations and stresses;

(4) develops data to test or validate models and provides data for model development (e.g., ecological process models, population models);

(5) when properly designed, provides data for testing reliability of estimates of model parameters and/or model predictions.

Natural resource populations are highly variable in space and change in composition and structure over time. Consequently, monitoring and assessment require specialized techniques. The variables of interest often change as ecological, social, economic and environmental conditions are altered. New scientific findings and technology impact the need for and type of data collected. Continual innovation in assessment techniques and technology is required.

Although the general principles of assessment and monitoring apply across all natural resource populations, specific techniques and applications are unique for each resource. The aquatic and land resources have different attributes which generate different problems.

Aquatic (marine) resources. The marine living resources are diverse and dynamic. The reported global harvest lists hundreds of species of plants and animals. The resource occupies the entire troposphere, from depths of five hundred meters to the surface, and from shoreline to the mid-oceans. They are aggregated in time and space, the aggregations are governed by the interaction of their biologically driven behavior with the environment.

The ocean waters are dynamic and variable over a wide range of time and space scales. The recruitment to the harvestable stock can vary from one year to the next by factors up to thirty times. The fish can seldom be seen from the surface vessels from which sampling gear of diverse and usually unknown species-specific efficiency is deployed.

The marine ecosystems which are sampled are on the order of tens of thousands of square meters of surface area. Logistics constrains the fraction of water sampled to orders less than 10. In many cases, only one observation per year can be obtained, as, for example, recruitment which is an annual event.

Accurate and precise scientific knowledge is required by the resource agencies managers and fisheries to conserve, manage and utilize their public resources for best benefits. The agencies and their scientists have increasingly defined areas of scientific uncertainty because of the factors outlined above which leads to inaccurate and highly imprecise estimates. This leads to erroneous inferences and bad management. Many of the newly recognized statistical issues described in this proposal, and which form the relevance of the intellectual theme have stemmed from the difficulties in drawing valid inferences from the available data by the conservation and management agencies.

Forest resources. In the case of forests, individual trees and associated vegetation are fixed in place and occupy a finite space. However, the particular species or species groups, age or size classes, or condition classes to be monitored is not fixed, and the bounds or boundaries of the resources of interest change due to natural processes and man-caused impacts. Usually, estimates are required for multispecies complexes and sometimes of more than one life form (e.g., trees, and shrubs). Many 'standard' techniques, therefore, do not apply. Moreover, the wide range of management (including no management) objectives have correspondingly different requirement of scale, frequency and precision of measurements. Thus, sampling and monitoring techniques and procedures appropriate or acceptable for one situation may not be suitable for another.

The basic statistical issues raised here by forestry and fisheries are common, indeed, and apply also to other harvested and non-harvested living resources, e.g., birds and mammals. Effective and efficient assessment requires statistical skills plus knowledge of ecology and resource management. Statistical theory must be applied realistically to the resource situation. Input from other theme groups is crucial to the development of improved assessment and monitoring methods. The center mode best meets the need for integration and synthesis of statistical and ecological sciences for developing improved methods for assessing and monitoring natural resource populations.

The goal of this research unit is to provide a valid basis for drawing improved inferences and advice to support resource management. The benefit would be increased availability of fish and other living resources for consumption, improved conservation, and increased opportunity for commercial and recreational harvesting on a sustained basis.

3.4.2. Resource modeling and prediction

Most renewable resources are growing populations of organisms. Biomass harvested from fish, wildlife, forest, and other populations can be replaced indefinitely through reproduction and growth, provided the environment continues to sustain the organisms and the harvest rate is not too high. Large sectors of the economy depend on exploitation of renewable resources, and substantial protions of federal and state governments are devoted to managing the resources. Harvesting, regulatory, and investment decisions of social and economic consequence are based on predictions of future resource levels made with mathematical models of resource growth. See Dennis et al. (1985) and references cited therein.

An extensive literature has developed on the mathematical economics of decisions with resource growth models (Davis and Johnson, 1987). But many ecologists are convinced that the current state of resource growth modeling is not adequate to support a superstructure of major decisions. Species have complex adaptation to their environments, and unpredictable environmental changes have effects on population levels. Species interactions with each other are poorly understood. The result is a persistent, large, unexplained variability in observed population levels.

Many types of mathematical growth models, from differential equation systems, to detailed regression models, are presently used in ecology and resource management. Summaries of methods used for forest resources can be found in the books by Avery and Burkhart (1983) and Clutter et al. (1983). Often these models are deterministic, stemming from a strong, sixty year tradition of deterministic theories in population ecology. Moreover, little attention has been paid to the interface between models and data. Fitting models via least squares estimation is often done, but use and evaluation of more generalized error structures than independence or normality is seldom attempted. A stochastic approach to modeling would pay serious attention to the structure and nature of the variability, incorporating stochastic variability directly into the models and utilizing alternative estimation methods and model evaluation criteria.

This research unit will undertake projects linking statistical thinking with mathematical modeling of renewable resources. The unit will seek to combine concepts from statistics, population ecology, and resource management toward better understanding and prediction of renewable resources. One goal will be to help resource models achieve a more credible role in decision making. Another goal will be to catalyze better quantitative theories in population ecology. Finally, a goal will be to advance resource modeling in general by having quantitative fisheries, forestry, and wildlife scientists share ideas and approaches. Here are some examples of specific project topics submitted to the theme during preparation of this proposal.

- Stochastic models such as birth–death processes and stochastic differential equations have been proposed in the mathematical ecology literature. The research would develop and evaluate parameter estimation and hypothesis testing methods for stochastic population models, concentrating on how maximum likelihood, quasi-likelihood, and conditional least squares methods perform for specific models in the mathematical ecology literature, and on how the models themselves perform for given data sets.
- Many statistical issues are unresolved in forest stand dynamics modeling. Forest growth and yield models typically consist of large sets of regression equations for predicting growth increments under various sets of environmental conditions. Issues concern how to incorporate stochastic variation in time and space, and

how to build better statistical models of the risk of mortality, fire, or pest infestation. The research will address these questions by incorporating the latest techniques in generalized linear models and model diagnostics into forest growth models using large existing data sets.

- Modeling approaches among ecologists differ. Some investigators concentrate on single-species models, while others emphasize multispecies interactions. The work will contrast single species versus multispecies modeling approaches for existing data on various population systems and will result in a critical evaluation of the different approaches under different objectives.
- Major problems involved with the extreme variability of marine fish stocks. The research will concentrate on approaches to incorporating stochastic effects in fish life history components such as recruitment, spawning, and mortality, as well as in other forces such as predation and fishing. Research efforts will be targeted at producing improved recommendations for fishery management.
- The theory and application of stochastic compartment models has advanced in non-ecological fields such as pharmacology. There is tremendous potential for adapting these approaches to model organism life histories. Two specific systems are of current interest to research unit personnel and have important economic implications. First, there is concern that shrimp fishing may be a major source of mortality to sea turtles. Compartmental modeling of turtle life history, coupled with modeling efforts directed at the shrimp fishery, may help estimate the magnitude of this problem. Second, rangeland grasshoppers in the U.S. northern and central plains consume an estimated 25% of the production. Compartment models of the progression of grasshoppers through various development stages will greatly aid range managers in designing effective control strategies.

3.5. Theme 5: Environmental risk analysis and decision making

Over the course of the last decade, the word 'risk' has, subtly but persistently, been given a narrower meaning in environmental contexts than its primary dictionary meaning – the chance of injury, damage or loss. The change has left 'risk' meaning 'health damage', often with a strong implication of a toxic chemical causal agent. In other words, the notion of chance has been pushed aside and the nature of the 'injury, damage, or loss' has been taken to involve human health damage from a special kind of pollution.

This language change was probably an accident, in that no one set out to achieve it. It is likely, however, that much of the impetus for its development came from William Ruckelshaus, and in particular from his famous National Academy speech of June 1983, in which he discussed the distinction between risk assessment and risk management. There and in a subsequent speech at Princeton University (Ruckelshaus, 1984), his examples involved such problems as identifying carcinogenic chemicals, banning EDB, and regulating a copper smelter that emits arsenic in its stack gases. We need not be cynics to see how, in the worlds of politics and scientific grantsmanship, the narrower meaning could be emphasized for many reasons.

For the proposed center, however, we feel it is important to adopt a broad definition of risk, one that leaves the all important element of chance (probabilistic processes) in a central place and one that leaves room for the injury, damage, or loss, to be ecological as well as human. To say that our perceptions or risk and decision making are to be broad, however, does not imply that the center can deal with every aspect of either. Rather, it will be our intention to concentrate our efforts on particular topics with strong linkage to the relevant special strengths of the total center efforts and with endpoints achievable within the center's lifetime.

3.5.1. Environmental risk assessment – Its nature and needs

Environmental risk assessment encompasses methods that involve several levels of ecological organization. They are the organism level, community levels, and ecosystem level. The concepts of various disciplines are used in addressing these levels to determine environmental damage and the probability of damage. Among them are fishery management, ecotoxicology, statistics, ecology and computer science.

Current methods for estimating risks and defining safe levels of exposure do not take full advantage of the data and information from these disciplines. Rather, reliance is placed on over-simplified quantitative models, primarily for bioassasys. The current primary goal is to avoid underestimating risk and not necessarily to obtain realistic and best estimates. Advances in statistics, biology, and ecotoxicology render these methods as unnecessarily simplistic. The present models do not incorporate sufficient understanding of the ecological and biological processes involved in producing toxicity or of the protective mechanisms that are part of the natural environment. The complexity of the problem demands that an effort be made to integrate procedures, data and mechanisms with statistical models to obtain optimal estimates of risk. Improved models would reflect the growing data base evolving in this area.

Several examples can illustrate this process. The more efficient sampling procedures to determine fish-abundance in open and closed environments can lead to best estimates of survival in natural and impacted environments. This in turn can lead to a more precise indication of environmental effects.

Density dependent mechanisms are being studied to estimate survival to the first year of life. Statistical methods are needed to describe the density dependent mechanisms and to relate this process to survival.

Microcosm and macrocosm research are producing data on interactions in a controlled environment simulating an ecosystem. These approaches extend the bioassay and could lead to more realistic appraisals of damage. Statistical methods are necessary to address these problems.

Many applications of risk assessments have relied on only fragments of the full spectrum of available information and have omitted information on mechanisms, interactions, and ecological restoration. An ecological statistical center would substantially improve this state of affairs. The center will be useful in identifying the growing body of useful information and in developing methods to incorporate the data into best estimates of risk.

3.5.2. *Initial research areas and issues*

Environmental risk analysis. *Estimation of ecotoxicities of chemicals acting singly and jointly from quantitative structure-activity relationships (QSARs).* Hazardous chemicals can greatly affect natural ecosystems. This applies both to chemicals such as PCBs, DDT, dioxins, and dieldrin that are already wide-spread in the environment and to numerous new compounds that are introduced every year by chemical and pharmaceutical companies. While it is important to assess the ecotoxicity of these compounds, it is often impractical or prohibitively expensive to test all of them for their acute and chronic toxicity to, for example, aquatic life. It is, therefore, desirable to develop methodologies to estimate the toxicities of these chemicals from their general functional groups. Such correlations are called 'quantitative structure activity relationship' (QSARs).

For example, from the water-octanol partition coefficient (Kow) one can estimate the LC50 (Broderius, 1987; Veith et al. 1985). Regarding the joint action of chemicals, it was suggested by Broderius (1987) and shown quantitatively by Christensen (1987) that QSARs also can be used in the prediction of the joint response to chemicals. Furthermore, the work of Suter et al. (1986) and others indicated that the chronic effect concentrations (e.g., maximum acceptable toxic concentration (MATC)) are correlated with the acute effect concentration (LC50). But the reliability of these correlations is not well established. Also, the impact of different QSARs and biological species on the biological response needs further study.

On this issue, the following problems will be considered as a prototype for future studies:

- Investigate predictive correlations between water-octanol partition coefficients (Kow) and water solubilities on one hand, and acute toxicities (LC50) on the other. Chemical parameters will be classified according to QSAR principles.
- Investigate predictive correlations between acute effect concentrations (LC50) and chronic effect concentrations (MATC) for a variety of chemicals and aquatic biological species. Degree of correlation will be characterized in terms of mechanistic and statistical factors.
- Develop methodology to estimate the joint action of selected chemicals on aquatic species.

Extrapolation in ecotoxicology and ecological risk assessment. As discussed above, statistical models, primarily linear regression models, are commonly used in ecotoxicology to predict physicochemical and toxicological properties of chemicals from their molecular structure. In ecological risk assessment, similar models are used (1) to predict effects to toxicant exposures on untested species from toxicity tests performed on standard laboratory species and (2) to predict chronic effects of toxicants from data on acute effects. Standard statistical techniques such as ordinary least squares regression are unsuitable for these analyses because of violations of such critical assumptions as the independence of observations and lack of measurement errors in independent variables. The large-sample variance calculations currently employed can be improved upon considerably by application of computer-intensive methods, such as the bootstrap (e.g., Babu and Singh, 1983, 1984; Linder et al., 1986; Linder, 1987).

Extension of these methodologies to multivariate analysis would enable inclusion of test characteristics and test results for other species. A general theory of multivariate errors-in-variables regression has been developed (Fuller, 1987) but computational algorithms and software programs are not readily available.

Goodness-of-fit of models to environmental data. The need for validating models used in risk assessment appears in every list of research recommendations. Relatively little discussion is seen of the difficulty of performing validation studies. Regardless of how carefully the parameters are measured, and how scrupulously experiments are performed, the model rarely matches the data. There is always residual uncertainty about the model (both structural features and parameter values), and the measurements errors associated with the experimental observations are usually large. No single goodness-of-fit measure will universally apply to all models, or to the same model in all situations. Two major shortcomings of existing goodness-of-fit measures are (1) they fail to consider uncertainty in the observed data and (2) they do not deal effectively with multivariate, time-dependent model outputs. New statistical theory and corresponding application to specific model-data comparisons are required.

Propagation of uncertainties in environmental risk analysis. In most studies on environmental risk analysis, the endpoint is presented as a point estimate without uncertainty analysis. For a given structure in the study, we can identify all the nodal points and the variables that can be observed. One can then identify the variability of these quantities and propagate it through the structure to determine the uncertainty at the end point.

More formally, in the context of propagation of uncertainties in a system let $x_1, x_2, \ldots, x_n$ be the input variables characterizing the components of a system, and the output variable $y = f(x_1, \ldots x_n)$ may be characterized by a computer algorithm as opposed to a simple mathematical function. The problem of the propagation of uncertainties is the problem of finding the uncertainty in the output variable for given uncertainties in the input variable.

The research on this topic will be motivated by specific issues of interest to the ecologist. For a given structure, the first step will be screen the variables by using methods such as adjoining techniques. Then the next step will be to find approximating functions of the output variables using the techniques of response surface methodology. Then one determines the distributional properties of this approximating function; and determine the associated uncertainties. These approaches have been successfully applied in the probabilistic risk assessment associated with fault trees and decision trees as applied to nuclear safety. This methodology is applicable to ecological issues that can be studied using a risk analysis basis.

Decision making. The probabilistic risk analysis comprises one input into the decision making process of a regulatory agency. The overall decision problem typically involves (1) evaluating and combining the results of different studies (perhaps based on different scientific models) and subjective expert opinions;

(2) setting standards based on a compromise between the interests of different constituencies; and (3) devising a system to monitor and enforce compliance with these standards. The following paragraphs describe research to develop basic principles and methods for improving each of these components of the decision making process.

Combining personal probability assessments. Decision makers, in environmental and other areas, often solicit the opinions of experts who express divergent opinions, often based on models drawn from different scientific bases. Several methods have been proposed to deal with these conflicting opinions: (i) attempts to induce the experts to arrive at a consensus (the Delphi method), (ii) rules for weighting and combining probabilities (the linear and logarithmic opinion pools), and (iii) reduction of the combination problem to the assessment of a joint likelihood function in which expert opinions are treated as pieces of data.

Our work will focus on the selection and use of experts, noting that there are incentive compatibility problems involved in ensuring adequate information collection. We will use the recent developments in multiple-agent, principal-agent problems and the investigation of tournaments as an incentive device.

Expert systems. One possible way of combining insights from a multi-faceted program of research is the creation of an expert system for environmental risk analysis. This is an approach to inference that attempts to identify and extract known solutions summarized in a computerized 'knowledge base'. This framework will be developed for two different systems:

(i) An expert system for modeling toxicity: predicting the biological activity of untested or partially tested chemical compounds.

(ii) An expert system for environmental risk analysis: a training tool and preliminary screening protocol for environmental impact statements.

Environmental conflict resolution and mediation. Environmental problems affect different constituencies in different ways. For example, acid rain affects the environment of the eastern United States and Canada adversely, which stringent emission controls on sulphur dioxide would affect the utilities and industries of the midwestern states adversely. A similar conflict arises in the disposal of hazardous waste. The conflicts are especially complex since environmental problems are multi-attributed; apart from ecological receptors like forestry, crops, and lakes they also involve employment, consumer costs, industrial and state revenues, international competitiveness, and so on. Previous work (Anandalingam, 1987) has begun using multi-attribute utility theory to examine these problems. (See also the chapter on *environmental conflict resolution* in Raiffa, 1982.)

The center's research would address the multi-attributed conflict problem using the insights provided by the recent explosion in research in non-cooperative game theoretic models of bargaining. For a survey, see Chatterjee (1986). In a recent paper, Chaterjee and Samuelson (1987) explicitly modeled the bargaining process and, by analyzing the cost of conflict, have provided a way of determining the efficiency of different conflict resolution procedures. Recent work (see Myerson, 1987, for a non-technical survey) has also considered analytical approaches to mediation.

None of these models, however, has explicitly considered multiple attributed problems; nor is there any theory on the effect of third party intervenors. Our research proposes to remedy these deficiencies and to apply the resulting improvements to the environmental problem described earlier.

Monitoring and enforcement. By any standard, enforcement monitoring in environmental policy needs improvement (Russell, 1988). Improving the monitoring performance requires a multi-faceted approach including work on measurement technology, on methods of inference from available measurements, and on combining incentives and measurements capabilities to best advantage.

Some of the technical bases for improvement will be part of the program for research at the proposed center. Three areas seem especially promising: (1) The improvement of technology for monitoring discharges directly, as they occur (Hinkley, 1977; Patel, 1978; Pitt, 1981; Maugh, 1981, USEPA, 1979). (2) The improvement of indirect monitoring techniques – techniques that allow either post discharge identification of substances with sources; or that allow inferences about discharge rates from sets of ambient quality (Gordon, 1980; Courtney et al., 1981). (3) The development of linked systems of incentives and measurement techniques that make monitoring an easier problem for agencies (Russell, 1987).

Crystal cube for environmental quality. In view of the gravity of the gap between limited knowledge of 'environmental quality' and the information needed for better environmental decisions, decisionmakers must be excused if they wished for crystal balls. Though we will never possess crystal balls, nonetheless the requisite information can be gained and with increasing certainty. Major gains will almost certainly require significant charges in perspectives and technique. One such proposed change is a 'crystal cube'; one which promises much less than a crystal ball, but has the compensating advantage of practicality (Boswell and Patil, 1988). Conceptually the 'crystal cube' has on each face a truly useful representation (useful for decisionmaking) of an important parameter related to environmental quality. A crystal cube characterizes a suite of measurable or predictable environmental attributes; these indicators of environmental quality (relative to 'normal' environmental quality) offer promise as pragmatic aids to equitable environmental policy and decisionmaking (U.S. CEQ, 1984; O'Connor and Dewling, 1986).

The center's research aims along these lines are designed to develop and test a strategy for improving particular kinds of environmental policy and decisionmaking, by governments and by private enterprise. The aim is focused on maintaining or rehabilitating regional environmental quality and encouraging more economic efficiency in capital investment on regional space scales. Initially, this broad aim entails three objectives:

- Develop a limited number of quantitative measures and indicators sufficient to characterize 'environmental quality' for some regional policy purposes.
- Demonstrate, to the satisfaction of both science and relevant governments, the adequacy of such measures and indicators for environmental policymaking with regard to economically and environmentally contentious classes of enterprise.
- Using the same measures and indicators, develop quantitative criteria of 'unreasonable or unacceptable degradation' that are useful enough to become

authoritatively incorporated in existing laws and regulations. Such criteria or 'legislative endpoints' seems necessary to ensure 'environmental quality' in any particular sense (O'Connor and Dewling, 1986).

The work focuses on two geographic regions, although exceptionally informative data sets from other regions may be used to help assess the utility of potential measures (or indicators) or environmental quality within the study regions. The study regions are (1) the Great Lakes and their coastal zones, and (2) the industrialized coastal zones of Connecticut, New York and New Jersey, including marine areas to the edge of the continental shelf.

3.6. Theme 6: Ecometric and environmetric computing

To begin with, we note that there are questionable statistical routines, but no routine statistical questions. Care and caution are essential for creative and collaborative effort involving ecometric and environmetric computing. We present examples of three research areas where careful work in the center mode will be fruitful and productive. Basically, the issues and approaches for ecometric and environmetric computing throughout the center program are expected to be the same. The proposed data interpretation and analysis laboratory will provide the necessary cutting edge as a research and support unit of the center.

3.6.1. Long term ecological research: Interpretation and improvement

Many ecological phenomena, such as succession or the sporadic occurrence of stochastic events, can be understood only on a long term basis. These studies are important in many areas of ecological research and encompass both the analyses of historical records and long term field observations over decades or even centuries as has been done, for example, in agricultural systems of the Rothamsted Station, UK.

More attention should be paid to spatial and temporal scale heterogeneity in the design of research and the interpretation of results. Experimental manipulation in ecology should be encouraged, not to replace careful observations of patterns in nature, but to allow an understanding of mechanisms responsible for these patterns.

3.6.2. Ecosystem structure and function – A meta-analysis of spatio-temporal data

The impetus for improved mathematical approaches to the study of the structure and function of ecosystems has its foundations both in basic and in applied research. Theory without data is sterile, which data without theory is uninterpretable.

What is clear is that the solutions to the environmental problems we face are going to require the integration of experimental and theoretical approaches at a variety of levels of specialization. No single model, and no single level of description, will suffice. We will require multiple models, directed at different levels of detail, and ranging from the generic to the site- and situation-specific. That is, we will need to be concerned with the development both of special and of general theory.... Such understanding can emerge only from theoretical synthesis based on a comprehensive program of microcosm research and experimental manipulation coupled with retrospective studies.

In community theory, general models of population interactions have guided experimental investigations. The interplay between theory and experiment led to recognition of the importance of spatial heterogeneity, and the development of mathematical approaches to the dynamics of spatially distributed populations. More recently has come the recognition that patterns in the distribution both of the biotic and abiotic components of ecosystems are inter-related, and can be detected on almost every temporal and spatial scale. Major advances in the next few years will involve the development of methods for detecting and analyzing spatial and temporal pattern in the distribution of populations of materials, understanding how such pattern arises, and elaborating upon its consequences.

3.6.3. Long term adversary data

The existence of a formal statistical method for the analysis of sequential data is inadequate for non-standard sequential 'adversary data'. In a typical environmental project where sequential or 'long-term' data is collected to promote or abolish a project, the data becomes non-standard because of changing criteria about the data. Both the advocate and adversary alter one or more factors in the data aquisition procedures, thus changing the time series length available to support or reject a project. The standard statistical tests are with the assumption that there is a fixed sample test. A Bayesian approach uses the a priori conditions, but these differ for the advocate and adversary. As new criteria develop, new data accumulates, past data gets 'destroyed' or discounted with the claim that it is not good or valid, or the collection is not compatible with recent data. It is possible as this process of adversary and advocate continue to change criteria, the distribution of past data and accumulation of new data reaches an asymptote to the amount of 'real' data available for analysis.

This is an example of one type of environmental data that is non-standard through adversary relationship. Understanding how adversary factors affect data, how it makes it non-standard, and how to analyze such data becomes an important function of the center. A combining of activities in Themes 1 and 5 permit adequate integration of research knowledge and methods.

3.6.4. Computer age and ecometric computing

(i) The computer age has brought public policy and theoretical statistics together by bending into the circle what otherwise looked like a line with theoretical statistics and policy making at the distant ends.

(ii) Computers have opened the door to multiple levels of resolution, giving the potential to analyze and model complex systems at several levels of resolution at a time.

(iii) It is around the computer that diverse ecological specialists discover each other and feel the underlying unity in their intellectual and analytical work. The fish, the fowl, the forest and the fur find that the issues and approaches are so very similar. Each one begins to feel the individual pursuit as a part of one ecological science. The data related statistical computing initiates and strengthens this unifying bond. And so also the common disconcerting revelation that there can be an abundance of questionable statistical routines, and yet frustratingly, no routine statistical questions.

(iv) John Skellam comes to mind, saying: '... mathematical ecology is moving out of its classical phase carrying with it untold promise for the future, but, as H. A. L. Fisher, the historian, remarks, progress is not a law of nature. Without enlightenment and eternal vigilance on the part of both ecologists and mathematicians, there always lurks the danger that mathematical ecology might enter a dark age of barren formalism, fostered by an excessive faith in the magic of mathematics, blind acceptance of methodological dogma and worship of the new electronic gods. It is up to all of us to ensure that this does not happen'.

These challenges, opportunities, and concerns have led us to formulate an active data interpretation and analysis laboratory as an integral part of the center on the following lines.

3.6.5. Data interpretation and analysis laboratory (DIAL)

DIAL will provide the cutting edge of ecometric and environmetric computing. It is a research and support unit of the center with four goals: (1) research on computational statistics for environmental and ecological problems; (2) training of environmental and ecological scientists in new methods of computational statisitics; (3) education of undergraduate, graduate and post-doc students in the use of ecometric and environmetric computing; (4) fostering interaction between interested research groups in the country.

These goals will be accomplished by the use of electronic communication, by establishing courses, tutorials and workshops, and by the interaction of DIAL personnel with research and outreach groups across the nation.

3.7. Theme 7: Cutting edge linkages

A still untapped feature of statistical ecology and environmental statistics is that some of its important concepts, issues, approaches, and techniques arise naturally elsewhere in some manner. For example, the issue of the primary energy mineral exploration and discovery, the issue of public health and human ecology, and the issue of health resource and biomedical statistics. It is only appropriate that the necessary bridges be built and the cutting edge linkages exploited.

3.7.1. Primary energy mineral exploration and discover

The importance of credible projections of amounts of undiscovered oil, gas and uranium in formulating a coherent national energy policy became clear with OPEC's rise to market power in 1974. Gored by OPEC, the U.S. petroleum industry recognized this need as well.

These events provided a spur to researchers interested in the statistics of resource exploitation. Their investigations followed two distinct paths, one directed towards refinement of models of amounts of hydrocarbons discovered in very large geographic areas (the contiguous 48 states, for example) as a function of time or of exploratory effort (wells drilled, total footage drilled) and the other directed towards incorporation of the 'physics' of discovery into models for geologically homogeneous deposits in individual petroleum basins. The latter approach has more potential for improving our ability to provide useful information for policy evaluation and decision making.

Data describing the evolution of deposit discovery in petroleum basins is quintessentially encountered data: no sampling frame is available, a particular form of selection bias is present, and observations are not representative of the population of deposits of petroleum deposed by nature. Larger deposits tend to be found early in the exploration cycle and once discovered are not re-discovered. Visualized as a sampling process, discovery is akin to sampling proportional to some measure of deposit magnitude and without replacement from a finite population of deposits. Another complication is that, in contrast to survey sampling where a sample frame allows a priori calculation of the inclusion probabilities, no such frame exists for mineral resource discovery.

During the last decade and a half a small group of researchers has developed a variety of methods for estimation of parameters of models that incorporate these features of mineral exploration and discovery. While work to date has proven fruitful (for example, the geological survey of Canada uses a successive sampling model of discovery coupled with a super-population process as an aid in preparing Canadian petroleum resource projections) research to date has just scratched the surface. Much additional work should be devoted to clarifying porperties of estimators proposed in this literature. The impact of externalities that disturb model structure, such as sequestering of acreage from exploration, needs particular attention.

3.7.2. *Public health and human ecology*

During the past decade the link between human health and the environment has become an important social, economic and political issue through greater public awareness. From the toxic waste issues, to disease spread through biological vectors, to a long list of health/environment/ecology problems, the pressing need is for improved statistical analysis, sampling, and model development. These problems all have non-standard data and require non-standard analysis.

Joining these two disciplines requires a fundamental challenge to our perception of them. To begin, we must alter the view that disease results from an organism or contaminant attacking the body in isolation. Rather, it must be perceived within the broader social and ecological context. For example, an understanding of Schistosomiasis requires a study of human demography, the genetics of resistance in snails to infection, the geography of snail populations in relation to human activity, the demography of the parasite population including density-dependent effects, and the effects of changing ecological conditions on the growth of snails, their competitors and predators.

Each of these areas pose unusual sampling problems, produce data that often are not normally distributed, or in which adequate models have not been developed. In addition, new problems arise, such as determining the joint statistical distributions.

The joint distribution of the physiological state and the demographic conditions of an organism is one example. In this case, a positive correlation of a population density with the resource it consumes combined with the knowledge of the physiological state of the organism will determine whether during peak periods the population is vulnerable to environmental stress, or the resilience of the population at low numbers.

Some problems arise out of epidemiological studies. The childhood mortality statistics are frequently summarized as national or state averages, with little accounting of spatial and social components. The standard multivariate analysis does not provide a sampling frame to account for fine mesh distribution, nor insufficient sample sizes, missing data and other problems. Meta-analysis and the encountered data studies described under Theme 1 have direct application in this research area.

3.7.3. Health resource and biomedical statistics

At the outset we observe that the medical patient and the marine fish, say, have one thing in common. And that is the freedom of choice when it comes to the designs or desires of the research investigator in their regard. Clearly the resulting data are more in the nature of encountered data.

We further observe that both in biomedical and ecological/environmental studies the investigator gets to record information on an individual through his observational capability of a suitable lifestage of the individual. The resulting data base is non-random (and non-standard) depending on the nature of the implicit selection bias.

We also observe that large scale biomedical and environmental/ecological studies are conducted at various centers/sites and the data or the summarized data and conclusions are combined in a meta-analysis for purposes of stronger conclusions.

Thus there is much common ground between the health resource and biomedical statistics area on the one hand, and the living resource and environmental statistics area on the other. Cross fertilization has thus a sure place with the cutting edge linkage.

4. Educational features

The proposed national center provides for an integrated community of scientists engaged in a tripartite program of research, education, and knowledge transfer/outreach. The scientists involved represent a spectrum of multidisciplinary expertise and experience with a common commitment to the resolution of critical issues and problems of environmental quality and natural resource management. The education program will be organized broadly to develop an understanding among its students (graduate students and post-doctoral scholars) of the needs and benefits of multidisciplinary research, and provide opportunities for participation in research projects having broader dimensions than the traditional disciplinary objectives and standards of graduate study on university campuses. While advanced training in specialized subjects will be provided to improve technical skills and knowledge, the center mode further provides the basis for (a) continual interaction with other students, faculty, and outside experts with a broad range of related interests and experience, and (b) coordinated involvement in inter- and multidisciplinary research planning and education.

In a more specific context, the objectives of the educational component of the proposed center are to (a) organize and conduct a program of graduate and post-doctoral training in both theoretical and applied aspects of statistical ecology and environmental statistics, with special emphasis on merging statistical understanding and skills with ecological insights and knowledge, (b) provide training and experience in innovative non-standard approaches to research design, data acquisition, data analysis and interpretation, risk assessment and prognosis, and extrapolation to real-world problems in a multidisciplinary mode, and (c) develop unified criteria for evaluation and improvement of the educational process at the center. Current graduate programs at Penn State will furnish much of the course work needed by the students in residence on this campus. In the field of statistics, for example, M.S. and Ph.D. programs exist in statistics and biometrics. The ecology program crosses disciplines and colleges, with participating faculty members representing departments of statistics, biology, forest resources, wildlife, geography, entomology, and others. A quantitative ecology option provides training particularly relevant to the mission of the proposed center. Other universities committed to the center concept and program have similar basic capabilities for training students who will be located on their campuses. However, new courses, seminars, and other learning facilities will be needed as a program develops.

The participation and involvement of inquisitive, high calibre students is essential to the succes of the center. Efforts will be made to attract such students by informing them of interesting, valuable research opportunities and providing them with responsible and challenging roles in the conduct of the research.

The requested annual budget includes provision for financial support of the integrative graduate education program. These funds will be expanded in approximately equal amounts for graduate students and post-doctorates. Full support, however, for a critical mass in both categories will require additional funding. The additional funding for graduate students will come from cost-sharing with participating universities where the students will be located. Additional support for assistantships or fellowships will be solicited from outside agencies or organizations. Participating universities may also contribute to the post-doctoral training, through support of selected faculty and professional researchers, but funding for this level of training also will be expected to be provided by public agencies and industry desiring advanced training for their scientific personnel.

Graduate education is an essential element of the proposed national center. A new cohort of creative, motivated young scientists with highly developed skills in statistical ecology and environmental statistics is needed to surmount present attitudinal and institutional barriers to the resolution of critical environmental and natural resource problems. In concert with its research efforts, the center program of graduate education is intended to be dynamic and flexible so as to be at the forefront of this effort nationally and internationally.

5. Knowledge transfer, outreach and participation

Advances in scientific research and technology have a time dimension, which is more critical today, perhaps, then ever before. Scientific progress and economic competitiveness related to it are directly dependent on the timeliness and effectiveness of communication of new discoveries and technological improvements. Moreover, the complexity of issues affecting science and technology, involving sociopolitical and economic questions as well as purely scientific elements, requires a correspondingly broader base of communication. This is particularly true in any multidisciplinary field such as statistical ecology and environmental statistics focused on problems of environmental quality and natural resources, both of which have roots in human ecology. It is important, therefore, that the relevance and utility of new findings and developments from the research of the proposed national center be transmitted outward in a variety of forms to a wide spectrum of scientists and potential users. Of equal importance, positive feedback from the user groups, including environmental and natural resource managers and policymakers will be encouraged so that real-world insights and priorities can be integrated into the research and educational activities of the center. Achievement of its basic goals will be determined in large part by the efficiency and effectiveness of this two-way communication.

More specifically, the objectives of the knowledge transfer and outreach component of the center program are to (a) facilitate the exchange of accrued knowledge and experience among researchers engaged in related work within and outside the center community; (b) furnish new and timely information on innovative, non-standard approaches to research design, data acquisition, data analysis and interpretation, risk assessment and prognosis, and decisionmaking techniques to environmental and natural resource managers and policymakers, and to other potential users in public agencies and private organizations; (c) promote constructive interaction between scientists and practitioners at different stages and levels of implementation, and (d) provide a central forum for periodic reviews of critical issues and significant advances in relevant scientific and technological areas for high-level policymakers and legislators involved in the statutory aspects of environmental and natural resource management. A variety of mechanisms will be utilized for this knowledge transfer and outreach function, including sponsored work visits and personnel exchanges, short courses, seminars, symposia, special-topic workshops, conferences, and publications. In all instances, information on complementary developments in statistical ecology and environmental statistics by individuals and organizations in other countries will be included in these activities as needs and opportunities warrant.

Center sponsored technical workshops will bring together researchers from thematic research groups and invited researchers from outside the center on a regular basis so as to provide updates and discussions among participants on technical matters of importance, both to the scientific community and to administrators and policy makers.

From time to time workshops and training sessions will be conducted for science teachers, administrators, lawyers, and staff persons for judicial, legislative, or executive decision makers, and other responsible analysts who may benefit from discussions and study of the scientific issues surrounding controversial problems of major importance. Such workshops will be staffed by center collaborators and may take place either at center headquarters or at other locations convenient for all participants.

The scope and quality of its knowledge transfer and outreach activities will be a major determinant in measuring the achievements of this national center.

6. Data interpretation and analysis laboratory

DIAL will provide the cutting edge of ecometric and environmetric computing. It is a research and support unit of the center with four goals: (1) research on computational statistics for environmental and ecological problems; (2) training of environmental and ecological scientists in new methods of computational statistics; (3) educational of undergraduate, graduate and post-doc students in the use of ecometric and environmetric computing; (4) fostering interaction between interested research groups in the country. These goals will be accomplished by the use of electronic communication, by establishing courses, tutorials and workshops, and by the interaction of DIAL personnel with research and outreach groups across the nation.

DIAL is an important part of the center that will provide a focal point for the exchange of research ideas and results among persons of varied disciplinary backgrounds. DIAL personnel will not only conduct methodological statistical and computational research on the leading edge of developing ecological theories and methods, but will enhance the activities of center research through a variety of activities. DIAL will have a core of technical personnel with background in statistics, ecology and computer science, devoted to the improvement of methods for ecological and environmental research.

The importance of DIAL to the proper functioning of the center should not be underestimated. Existing at a single location, its members will be in regular contact, and by the very nature of their positions, DIAL personnel will be forced to cross disciplinary boundaries on a daily basis. Center participants through their contact with DIAL will be exposed to a naturally existing multidisciplinary environment, which in microcosm can serve as an example for the larger community of center researchers. DIAL's activities will be coordinated with and will serve the outreach, the educational and the transfer of knowledge functions of the center. It is clear that DIAL's role is not simply that of consulting.

DIAL will promote the reciprocal exchange and development of non-standard quantitative methodologies among: (1) quantitative ecologists and ecologists,

(2) ecological statisticians and statisticians, and (3) ecologists and statisticians. It is important that these exchanges take place in a manner sufficiently timely to alter the course of ongoing research projects. A main goal of DIAL is to facilitate development, dissemination and adoption of new statistical and computational procedures for all levels of ecological and environmental research.

Exchanges will be fostered in several ways. A major activity will center around the tutorial/workshop concept by which researchers (whether working under center auspices or not) will come to DIAL to present new or non-standard methods to groups of interested individuals drawn from industry, government and universities. Workshops will treat both theoretical and applied topics. A second activity will be to promote research collaboration among center participants: (1) by facilitating the rapid exchange of non-proprietary software, (2) by encouraging participants to share and mutually critique preliminary data analyses, and (3) by providing standard data sets for software validation, etc. Thus DIAL will enhance the exchange of expertise among ecologists, statisticians and computer scientists. DIAL itself might provide some limited assistance in data analysis or software evaluation.

To promote the timely exchange of information, DIAL will use electronic mail (e-mail) networks, such as BITNET, and establish a computer bulletin board which can be accessed over telephone lines. Information will be posted on the bulletin board as well as sent by e-mail to center participants. Requests for center publications received by e-mail or posted on the bulletin board will be answered by e-mail whenever possible. All aspects of DIAL will use e-mail and the computer bulletin board to expedite the exchange of information.

It is proposed that researchers who have developed or are well-along in the development of methods to solve non-standard problems be granted short-term fellowships at DIAL. Such researchers might naturally be expected to present a short tutorial/workshop. In other instances, a lecturer might be invited specifically to conduct a workshop. Presenters of workshops will be encouraged to prepare notes that DIAL will distribute. DIAL will publish a newsletter at irregular intervals.

Thematic group leaders will maintain direct ties with DIAL and participate in DIAL workshops. They will meet with DIAL personnel and other group leaders to discuss ongoing research and potential new directions. Thematic leaders will also be encouraged to organize workshops at DIAL based on their group's research. Here the goal will be to disseminate new methodologies in a timely manner and to become apprised of new methodological and computational needs arising from discovering of center researchers.

DIAL will also serve as a center in which research on computational aspects of methodologies would naturally lead to a capability to evaluate and recommend specific software for the needs of various center participants. While development of extensive software is not envisaged as part of DIAL's role, DIAL could serve in an expert advisory capacity including the possibility of overseeing contractural arrangements between thematic groups and software developers. DIAL will serve as a clearing house for the exchange of information and privately developed software. All of the above software functions are particularly important since computational approaches to ecological and environmental phenomenon are gaining popularity both as predictive and inferential tools. This development is natural owing to the complexity of ecological and environmental processes. Such processes are often intractable by standard modeling and statistical techniques.

The DIAL facilities and personnel will provide an excellent resource for the training of students in the application of statistical analysis to real life ecological problems. Students with a wide range of backgrounds will be accommodated. Students with the help of DIAL personnel will perform exploratory data analyses, use graphical methods, and will evaluate statistical models. The emphasis will be on advanced instruction, and students will be assigned research level case studies that will form the basis of group studies and tutorials.

7. Unifying visions, lenses, and perspectives for the rationale for the center

In support of the proposed center without walls, several unifying visions, lenses, and perspectives from a wide variety of concerned people have been received.

The following pages give a sampling of the underlying unity in the apparent diversity.

7.1. If Sir Ronald Fisher had gone to sea

The National Center for Statistical Ecology is an idea whose time has come from the standpoint of research on marine ocean fisheries from an ecosystem standpoint. Sampling ocean populations and making inferences from those samples as to their health and to their impact of people on that health is an extremely difficult task. Such sampling has often been likened to flying over the countryside on a cloudy day and throwing down a bucket to sample the statistics of the population while at the same time having to make management decisions requiring the accuracy of a study the size of a U.S. Census. I often wondered how much further we would be along if R. A. Fisher had gone to sea instead of to a farm or Snedecor had been at an oceanography university instead of Iowa State. While there are many statisticians who are called on to consult with researchers in the ecological area, this generally results in applying statistical developments from other fields to ecological problems. This has generally not been satisfactory and the results have been only an approximation of the desired models. What is needed is a think tank of significant critical mass that will address ecological statistical problems across the board developing new paradigms for the interpretation of ecological data from sampling.

I look forward from a standpoint of an applied fisheries research center moving towards greater involvement in ecosystem research to being able to bring problems to this center for their solution as well as being able to take output from a variety of research projects carried on at the center and applying the results to interpreting the data collected from the marine ecosystem.

7.2. *Mathematical statistical tower to open up for the environment*

Many areas of academic endeavor can and are carried out in relative solitude, in the classical tradition of the ivory tower. Mathematics and probability theory to a large extent, statistics to a lesser extent has progressed in this mode. But even mathematicians have felt a need to communicate, to meet and exchange ideas. Statistics has been much more dependent on the interchange with the areas of application but often this has been in the nature of consulting or simply as a data analyst. Often when the statistician has worked in a very applied area it has not been in direct collaboration with workers in the field. With the advent of the computer much of this has changed and there will be more changes. There are more opportunities and a greater need for statisticians to be involved as a full partner in collaborative research, there are also many instances of new statistical tools being developed by non-statisticians, the results often appearing in journals not commonly thought of as in the main stream of statistics. The real 'raison d'être' for a center for statistics is what it can cause or provoke to occur, developing new tools that are relevant to the specific problems, promolgating these new results to faculty, students and researchers as well as bringing together workers who would otherwise be unaware of their potential common interests.

The proposed title of the center is itself a provocative one because it reflects areas of application that have only recently begun to attract wider attention in the statistical community. The nature of the data and the kinds of questions demand answers are different from that for which the more classical methods were developed. The data is often spatially located and does not represent random sampling, it is frequently multivariate and not uncommonly is time dependent. In one sense the sample sizes may be small and in others the sample size may be very large, for example the number of sites used to monitor air polution in often small but the data is collected very frequently. Compositing in the form of spatial or time averages may complicate the mode of analysis. Classical methods frequently focus on the removal or partitioning of variability and rely heavily on asymptotic results such as the central limit theorem or strong distributional assumptions, techniques developed in the context of spatial statistics or in multivariate analysis treat the variability differently.

The strength of the center will be the breadth (between disciplines as well as between institutions) it can provide in order to provoke and organize interchange and communication. The activity in some areas is highly dispersed and on one institution could assemble the collective strength represented by the anticipated participants in this center. The potential for organizing short visits to other institutions, rotating post docs and graduates, writing new course material, supporting the development of new courses far transcends what an individual institution or even single department could do. The mechanisms, the funding and the orchestration of such interchanges must be an integral part of the center proposal and subsequent style of operation. In many instances the statistician is the link between workers in different fields working on related problems, this is especially true in the case of environmental monitoring and assessment.

Anyone familiar with university campuses knows that it is frequently very hard to provoke inter-disciplinary research because of departmental restrictions and demands as well as physical separation. It only happens when individuals are in an environment that is conducive and when they have sufficient exposure to understand the other person's perspective. This requires time, opportunity, knowledge and support. Ordinary organizational restrictions such as departmental demands, physical separation and lack of information as to others interested in the same problems kills the spark before it can grow. The center can provide the opportunities and the motivation to overcome these obstacles. It can create a broad community overlapping institutions of individuals who want to participate in inter-disciplinary research and who are kept aware of such opportunities. Moreover, this aura can be propagated to the graduate programs and research programs of the home institutions. It should not be overlooked that the center also proposes to incorporate collaboration with individuals working in various federal or other public agencies in a mode quite different from the traditional one of contracted research.

7.3. Forensic statistical ecology and environmental statisitcs

Private industry has been increasingly involved in litigation in defense of practices that impact the environment. Often industry is challenged by citizen coalitions or state or federal government agencies. Data collected by either side is often messy and the statistical methodology often confuses the issues even further. Each side will often hire a 'statistician-expert' with the jury forced to make decisions that even professionals in the field cannot make intelligently.

A national center for statistical ecology can serve in several roles, ultimately leading to an improvement in the way such cases are decided. The center will improve existing methodology for handling environmental data. This can include better guidelines for data collection and analysis. The latter can only take place if both biological and statistical experts are brought together to exchange ideas. Often models are created by statisticians with little understanding of the biological and physical processes involved. These models are often uncriticized because the biologist/environmentalist does not understand the mathematics and what it is predicting about the environment. Hence it is easy to get two statisticians, both considered competant by their peers to disagree in a courtroom. A center for statistical ecology would begin to help with this problem.

It should be noted that with the increased use of genetically engineered organisms, microbial contaminants may turn out to be a greater problem than chemical pollutants. Not only are existing models in need of improvement, but new models that deal with these risks will need to be developed.

Often federal and state agencies have similar goals that require data collection. For example, both state and federal agencies may need to estimate stocks of fish for determining management policies. By using different methodologies, different conclusions may be reached, resulting in different regulations. The center with an integrated data lab can investigate modes of resolving conflicts and perhaps define ways of combining different analyses.

Many of the issues dealing with the courtroom and regulation are issues of today. In order to understand what is coming down the road we need to ask more basic science questions as well. As the world becomes more crowded and more polluted, we need to understand the biological and physical processes that regulate the world's seas. Models that define major processes and those that integrate several types of information are certainly needed. Statistical methodology that deals with spatial data, community structure and other aspects of an ecosystem needs to be expanded upon. The center can both serve as an avenue for generating solutions as well as ideas for new research.

7.4. *Meta-analysis for acid precipitation research*

There are currently approximately 85 research projects being carried out under the auspices of the forest response program of the national acid precipitation assessment program. To these must be added projects already completed and others yet to be initiated. While these projects are directed to a variety of scientific questions, several are related to each of the individual questions. An individual project thus will not, in itself, be providing the answer to the specific question and a means must be found for integrating the information from all the relevant studies. This must also include any available results of studies performed outside the forest response program. As indicated earlier, in review of the environmental protection agency's acid precipitation research, Landau, Patil and Rawlings (1985) stated: 'In an effort to facilitate the ultimate objective of providing does-response data for the regional assessments, it is recommended that immediate attention be given to problems of combining the research results over time and space.'

The most immediate candidate for such meta-analysis is the set of seedling exposure experiments. The joint interpretation of these experiments will not be a simple task, and is an issue currently being addressed by the synthesis and integration team of the forest research program.

There is, however, an additional aspect that does not arise in the case of parallel experimentation with agricultural crops. The extrapolation of seedling results to mature trees may be precarious. Thus, to what extent the ideas of DuMouchel and Harris (1983) can be translated to this whole area is, at present, an open question.

7.5. *Ecological risk assessment needs a national center*

Ecological risk assessment (ERA) is a relatively more complicated concept than quantitative risk assessment. Whereas quantitative risk assessment is concerned only with the potential health problem of the individual, ERA is involved with the problems of the individual, population and ecosystem. Multivariate designs and corresponding analytic tools should be developed to improve the estimate of toxicity. A center with a dedicated staff can focus on needs of this research by combining resources, which are ordinarily not available.

At the ecosystem level, the complexity is enormous. Damage may be defined in many ways. Survival, change in biomass, nutrient loss, etc., have been used in large ecomodels. Recently, Monte Carlo approaches have been developed to estimate risk. The risk concept is the probability of damage. There are many statistical issues to be addressed. Prime among them is the degree of bias of the estimator, which is, in turn, related to the dimensionality of the system and the generation of outliers.

A center would be useful for relating the estimates at differing levels and for concentrating on fundamental statistical questions that arise at the various ecosystem levels. In addition, the center would create an atmosphere of exploration to integrate different ideas into a comprehensive outlook on risk assessment.

7.6. *Routine statistical techniques unsatisfactory, think tank essential*

The ecological research at our university would profit from a collaboration with statisticians from other universities. Routine statistical questions can be answered by our own consulting statisticians. However, much or the research that is presently being done in the general area of ecology has to deal with statistical questions that are presently fuzzy or difficult to deal with through routine statistical techniques. Yet our consulting statisticians do not have the time to work out new techniques that would be applicable to the type of questions that our ecologists are asking.

We foresee that collaboration could come about in several ways. One way is through postdoctoral fellowships, in which one or more statistically-oriented biologists or mathematicians would work with our ecologists and help develop the techniques that are appropriate to the sorts of questions that we are asking here. Another way is through establishing one or more visiting professorships, in which established statisticians or mathematical biologists could come to our University and work with our ecologists. A third way is through offering one or more graduate assistantships, in which promising students with a strong background in statistical mathematics would be recruited to work with our ecologists on biological problems. We have tried this latter approach in a limited way with excellent results.

7.7. *Unreasonable to expect reviewers to detect errors when they make the same errors themselves*

In many fields, replication of experiments is a key source of quality control. In much of ecology, however, exact replication is almost impossible because of the long duration of experiments, geographic isolation of study sites, and the uncontrolled nature of many studies. This places increased importance on the quality of what reaches print, and hence on the peer review process. Unfortunately, because of the diverse nature of subjects that end up combined in a study, it becomes less likely that any particular reviewer will be truly qualified to review the entire material.

Statistics is probably the most common problem for which more qualified reviewers could provide some assistance. Ecological data is notoriously complex, non-linear, and multivariate. Most ecologists are somewhat trained in statistics but few know all the subtleties of every method they are called upon to use. Common pitfalls include failure to present statistical results adequately, misuse of degrees of freedom, pseudo-replication, over-dimensioned analyses (too many independent variables relative to data points), autocorrelated samples, failure to consider statistical power, improper null models, violation of normality, and general abuse of multivariate methods. The studies cited document pervasive, continuing failure to apply statistics correctly. Under these circumstances, it seems unreasonable to expect reviewers to detect errors when they make the same errors themselves.

7.8. When is pseudoreplication not a sin? A need for the Mayo clinic of ecostatistics

The community of scientists could not fail to benefit from easier access to better methods: (i) with mechanisms for prompt dissemination of data analytical/statistical information, (ii) with a forum for debate on important current issues, such as when is pseudo-replication not a sin, (iii) with a center where novel or particularly knotty problems could be troubleshot (sort of a Mayo Clinic of ecostatistics, ... My experience is not strong in the development of statistical procedures, but rather, like the majority of ecologists/environmental scientists, as a frequent and frequently frustrated user of those methods.)

Are currently available statistical methods adequate for long-term studies?

Within the range of approaches available for investigating long-term phenomena, it is found that those which may be suitable for large spatial and long temporal scales often have the least potential for causal inference, while the textbook-style procedures of great inferential power may not yield insights applicable to many scales of interest. Furthermore, methods do exist for the design and analysis of studies across the continuum of potential study types, yet the graduate curriculum has been inadequate in preparing ecologists in these methods, and consequently researchers are often unaware of potentially useful approaches.

On one extreme are classical experiments, which have all of the following four elements: a manipulation, with a control, including replicated observations with randomization. In contrast are observational/comparative studies, which include no manipulation and may or may not have some replication or even randomization. Between these extremes is a wide variety of quasi-experimental approaches, which embody some manipulations, but which lack one or more of the necessary additional features of classical experiments. The nature of the inferences provided by the approaches varies across the spectrum of designs, from power to impute causality within the confines of the study (classical experiments), to causality with potentially unquantifiable significance (quasi-experiments), to solely correlative predictions (observational/comparative studies).

The difficulty for design, implementation, and analysis within each type of approach lies in matching the scale of the experiment or observation to the rates of the central processes considered. In determining the scale at which a research question is to be pursued, it is useful to consider whether the same information might be obtained at different scales, either by scaling down from larger/longer systems or by treating analogous systems of shorter scales.

7.9. *Hybrids of statistics and biology are the key*

A coordinated effort in this area of research would be beneficial.... The center would not be loaded with statisticians who do not have training in ecology, environmental science, or biology. Hybrids of statistics and biology are the key to having a successful program.

7.10. *I see a lot of good coming out of such a center*

I do see a lot of good coming out of such a center, mainly in increasing cooperation and reducing isolation further than can be done in the regular way, by journals, annual meetings, etc.

The first thing that occurs to me is its possibilities as a data resource, an archive. As you know, ecological experiments and observations are difficult and expensive to carry out. At the same time there is tremendous variation, both within an experiment or observational study and between similar experiments or studies carried out at different times or places. Thus generalization in ecology requires more data than generalization in more other sciences, but each individual datum costs a lot more to get. The present 'go-it-along' approach leads to something less than science: a vast collection of papers showing 'statistically significant' result from various manipulations carried out in various ways in small areas over limited periods of time. Whether these results permit generalization to other places or times is usually unknown, because experiments are rarely repeated and an integrated analysis of several data sets is rarely attempted. Even the meaning of 'significance' is unclear because of publication bias and other effects.

A well-documented archive, accessible by computer, would be a great encouragement for integrated meta-analyses, leading to much broader and perhaps deeper generalizations in ecology.

A second function for this center would be to make people more aware of each other's work. Ecology spans a range of disciplines. As a statistician working on modeling and assessment of environmental change, I can except to find ideas relevant to my work not only in statistics, mathematics and ecological journals, but also in management, econometrics, chemistry, toxicology, public health, geography, geology, computer science, ... There is no way to keep up with all this. In addition, there is a great deal of activity that does not appear in the journals: contracted risk and impact assessment reports, technical reports prepared for government departments, etc. When I find out new things about this non-journal work, I am usually led to new things in the journals too. Workshops, newsletters and electronic bulletin boards can all contribute to this. So too could cross-indexed, computer-accessible lists of reports and of people relevant to various areas and problems.

7.11. *Building bridges between theory and application*

Many of the problems are not simply problems for the statistician, but perhaps the statistician is in a good position to mediate a broad base of participation in resolving some of the key issues that limit our effectiveness in dealing with numerous ecological and environmental problems that require not only good-faith interactions among professionals of varied background, interests and concerns, but increased ability to perceive problems from a broad, integrative prespective rather than from a narrow disciplinary outlook.

The center for statistical ecology and environmental statistics should identify and develop processes that will best help to improve statistics of the environment in application.

An exceptionally important need at this time is to develop statistical guidelines and tools that are suitable for the realities of the way research is conducted and the way data are used in making environmental decisions.

7.12. *Linking ecologists in a new and effective way*

The idea for such a national center holds considerable promise, especially if it can be tied effectively into a network arrangement, such as with the LTERS (long term ecological research sites) and the developing association of ecosystem research centers. It is this possibility of linking ecologists in new and hopefully more effective ways that appeals to me particularly.

7.13. *Forwardlooking perspective and urgency at present*

As the world population grows, with the concomitant domestic and industrial development, the living resources of the world are being reduced, and species eliminated, due to harvest (e.g., marine fisheries), and degradation and elimination of their environment. Because the cost of the effects on the ecosystems are not included in the cost of development, attempting to redress or prevent them requires proof of effects before the perceived 'additional' costs can be justified.

The studies of ecological effects, therefore, must increasingly have a solid basis for statistical inference so that they can withstand attempts to discredit them; in a less perjurative manner, so that they are unbiased and convincing. The basis of decision must also include analysis of risk.

The rationale for the center then is to bring together ecologists and statisticians for the purpose of developing statistical methodology which meets the needs of drawing valid inferences from observations of the natural world which is effected by anthropogenic activities.

Statistics is an abstraction of physical phenomenon; ecology is an abstraction of the reality of natural living phenomenon. Both abstractions are severely limited; the interaction of them even more so.

The field of ecology, and its cottage industries of environmental impact, effects of harvest, effects of pests, etc., involves all of the major disciplines – biology, chemistry, physics, mathematics. An ecologist is not necessarily a good integrator of all the disciplines. There must be some means of close integration of specialists. This is particularly true of statistics, the relevance of which is required before observations are taken.

The proposed center will provide a means of achieving the integration of statistics and ecology which has hitherto been lacking. This will provide the twofold necessity of joint development of applicable statistical methods and theory. The center will enable statisticians to focus on the needs of the natural sciences, i.e., it will justify and elevate this to a necessary degree of recognition and will enable ecologists to help the statisticians formulate appropriate methodology.

The intellectual theme is to increase the reality of current abstractions of natural phenomena through development of useful and applicable statistics. The goal is to provide a basis for describing and measuring the impact of man's activities on the world ecosystem. This will be a continuing activity and the long term may be conceived as the viability of human life. The focus may change as the needs of humanity change; however, the goal will likely remain for some relatively long time. The need may change, for example, from evaluating effects to evaluating correction or necessary accommodation.

The development will have to adapt itself to the constraints imposed by a shrinking natural resource, and the rate at which the latter is being changed and consumed. It is, therefore, of particular importance that a valid technology is available for industry to utilize in meeting these constraints, and that the public in general is more assured that an adequate technological base exists or is being developed to protect their interests and dependency on a livable biosphere.

This will not occur within the necessary time, that is, before events overtake the ability to control them unless there is a national center to encourage and assure the interaction of the artifact with the natural world.

7.14. Unifying themes and requirements

Today one fact of human life is more certain than ever: that the stakes involved in finding solutions to man-made environmental problems is no less than the future and fate of our planet.

Three fundamental objectives have been defined for the conservation of the biosphere, namely: (1) Maintaining essential ecological processes and life-support systems. (2) Preserving genetic diversity. (3) Utilizing species and ecosystems sustainably.

However, in order to achieve these objectives, information is necessary. This information is obtained by observation, by survery and by experiments, within the traditional framework of the scientific method. Ecological processes are essentially complex, involving the interaction of many organisms with each other and with their physical environment. Interpretation of the information collected about these processes is often complicated by the non-linear nature of the relationships, and by the presence of feedback between causes and effects. Modeling of these relationships often aid in such interpretation.

Scientific research depends upon the scarce resources of trained manpower and, increasingly, on equipment with a high capital and maintenance cost. It is essential, therefore, that all research is conducted efficiently and without unnecessary waste of these resources. Good design of experiments and surveys is perhaps the most important benefit to come from the study of mathematical statistics, although this benefit is seldom emphasized in statistical texts, or in the papers contributed to statistical journals. Nevertheless, the application of the principles of survey and experiment design to environmental and ecological research would greatly increase the efficiency of that research and would make more effective use of the limited resources that are available. In this sense, 'design' is itself a unifying theme, and one which is more important than any other because, without good survey and experiment design, analysis and synthesis of ecological data is worthless.

Once data have been collected, it is important to ensure that they are analyzed efficiently and effectively. It is seldom realized, however, just how different are the methods used for the analysis of ecological data by experienced analysts from those which are given in statistical texts, or which are taught in university courses. Increasingly, traditional methods of analysis and interpretation are being augmented or replaced by new methods which depend on computers and the algorithms and graphics made possible by such machines. Teaching new generations of ecologists and environmentalists to use successful analytical strategies is therefore an essential task for the immediate future.

Finally, however strongly we feel, as scientists, about the importance of data and their correct interpretation for the solution of environmental problems, we also have to recognize that the majority of these problems arise from conflict between the policies and practices of many participants. Most of these conflicts are so complex, and depend on so many variables, that it is almost impossible to make rational decisions without the aid of modeling strategies. Increasingly, it is necessary to embed these modeling strategies in expert or knowledge-based systems which have the characteristics of being able to provide advice to decisionmakers as well as, or better than, human experts, and can justify their own conclusions. Much of our communication of statistical and ecological expertise must, in the future, be concentrated on the development of knowledge-based systems, as the only systems likely to be capable of transferring complex information to decisionmakers in a usable form.

7.15. *Proposed center as a national resource*

The pressing questions in natural resource management, environmental science, and more broadly human health and ecology require basic research and knowledge using empirical observations. In many cases, the data are non-standard, constrained by time, money, spatial, social, the adversarial nature of the problems, and/or political; and, sometimes the data are non-standard simply because of the nature of the problem – we cannot change the behavior of organisms, prevent evolution or selection, or have the perfect measuring device.

The application of statistics to ecological environmental science has greatly altered our capabilities and the results we have obtained so far are encouraging. Yet, we are looking to advance at a faster rate the use of statistics and models in non-standard problems. An integration of statisticians modelers, ecologists, and environmental scientists will foster new, innovative research. Questions asked in a center mode are not the same as for a single discipline, or even multidiscipline with a statistician attached to an environmental project, or an ecologist suggesting interesting 'unsolvable' problems for a statistician to examine.

A national center without walls is a collaboration among scientists, given actual non-standard data, with not only the capability to use their own individual knowledge, but cognizant that a focal point exists for gaining efficient access to the human resources of others. Such a center becomes a 'national resource', a place for creative research, and a place for government agencies, private enterprise and researchers to obtain and generate information.

7.16. *Proposed center as an effective meeting point*

By now it is a well-accepted fact that we derive our health, livelihood, and economic status from the environment in which we live. It is also clear that 'environmental data' has been at least as confusing as it has been enlightening. We have all shared in the frustration of watching the ecological aspects of an environmental impact statement being dismissed as at best inconclusive, if not outright irrelevant.

We recognize three main problems: (1) poorly acquired data, from statistical and ecological viewpoints. (2) Data that is insufficiently analyzed. (3) Statistically significant results with no apparent ecological meaning. The first problem has been adequately treated in many statistical textbooks yet it persists. What is reasonable to dismiss from the academic, scientific literature can not be pushed aside in legal or policy arenas.

The second and third problems are a historical consequence, With statistics developing in the main independently from ecology and environmental science. We envision a National Center for Statistical Ecology and Environmental Statistics an appropriate place to change this.

7.17. *A transcendant ecosystems science in modern society*

The transcendent task of ecosystems science in modern society is to develop a theoretical framework of understanding for the dynamics of physical and biotic processes expressed in a hierarchy of quantitative models that permit operational evaluation and projection of human environmental influences on all scales from the field or building site to the biosphere. All available sources of information must be exploited to collect, assemble, and integrate data bases required to drive such models so that specified levels of confidence can be attached to their outputs. Progress toward this goal is of utmost importance to sustained growth of economies and preserving the integrity of social structures.

In order to attain this goal it will be necessary to achieve synergistic integration of a broad spectrum of technologies ranging from advanced mathematics, computer systems, and artificial intelligence to ecology, engineering, and biotechnology. Individual scientists working on disciplinary projects can address particular aspects of such environmental understanding and model formulation, but progress is painfully show and limited in scope when each scientist seeks to attain a functional knowledge of several technologies that may be relevant to a problem in ecosystem science. It is only in extensive interdisciplinary intellectual exchange along with truly multidisciplinary projects conceived and pursued on a broader scope that there lies potential for quantum progress. The proposed national center for statistical ecology and environmental statistics, existing as a 'center without walls' is being conceived to foster this kind of interdisciplinary cooperation for both basic and applied research.

Shared insights and expertise with cooperative strategies of project planning are cornerstones of the center concept. Equally important, however, is research linkage through communal data bases coupled with an array of format conversion capabilities to accommodate differences among input structures for modeling and analysis software. In parallel with the way that temporal overlap is used to extend tree-ring record sequences over time and space, so also can temporal and spatial correlations be used to harmonize and combine data sets from diverse sensors and series. Multivariate spatial and temporal analysis coupled with geographic information systems (GIS) and remote sensing technology provide the means for creation of composite data bases that serve as testbeds for ecological models which address the all-important factor interactions that have proven so elusive in narrowly focused research.

7.18. Clean environment, rocks of ignorance and investment in knowledge

The rhetorical ship 'clean environment' has never gotten clear of the rocks of ignorance. The result is that the course the nation embarked on in the early 1970s has been and continues to be unnecessarily costly in the short-run and discouraging to environment-saving technology progress in the long-run. Put bluntly, we are almost certainly paying a very great deal more than we should for the environmental results we are getting. A few strategic investments in reducing ignorance in key areas could provide the foundation for policy improvements that could make a significant contribution to our nation's competitive position in world markets without risking environmental quality – indeed, while probably enhancing it.

What are the key environmental problems that require investment in knowledge? one version is: (1) Understanding current observations; (2) Predicting the future effects of current actions (and explaining current effects of past actions); (3) Monitoring the results of those actions.

How do these shoals of ignorance harm policy? First, a lack of perspective on current observations leads inexorably to a risk-averting strategy of assigning worst-case interpretations and taking extremely conservative actions. Second, inability to connect causes and effects means we may be going after the wrong pollutants or the wrong sources to achieve our ultimate goals. Third, our poor monitoring capabilities leave us prey to poor results even when our regulations are aimed at the right targets and have a reasonable structure. Frustration with these inadequacies is one important stimulus to such unfortunate public decisions as the adoption of meaningless zero-pollution goals and accompanying, but not-so-meaningless, requirements for rachetting down technology-based discharge standards.

The proposed national center for statistical ecology and environmental statistics can be seen as having three broad goals: (1) The establishment of temporal and geographical perspective within which society can view such observations as the Antarctic ozone 'hole'; the concentrations of CO at specific sites; global temperature averages; precipitation patterns; and ocean surface temperatures. (2) The construction of ecological (in the broadest sense) models aimed at practical prediction of the effects of planned or unplanned human interventions. While this goal requires a marriage of statistical and ecological knowledge and techniques it requires also a step toward what might be called 'ecological engineering' in which the niceties of complete understanding may be submerged in an act of self-discipline when a black box can be devised that will inform policy. The classic example may well be the Streeter–Phelps models of BOD-DO concentration changes in aerobic water bodies. (3) The enlargement and improvement of the set of monitoring tools available to responsible public agencies.

7.19. Problems serious in economic terms, but having clear paths to resolution within the life of the center

We focus some effort on selected problems that are serious in economic terms, and for which clear paths to resolution seem probable within the life of the center. More specifically, these problems (1) limit economic well being and competitiveness and (2) realistic solutions can be outlined and scheduled. The solutions necessarily entail improved or novel methods of data analysis.

An example problem: Non-productive expenditures on enterprises that are halted on environmental grounds.

Evident limitations upon economic well being and competitiveness are the investments wasted on ultimately non-productive enterprises; wasted because the developer or competing interests and governments halt the enterprises on grounds of unacceptable environmental impact. These enterprises are never-to-function chemical plants, nuclear power plants, waste management facilities, mining and logging facilities, roads, housing and office developments, etc. While the annual costs of all such 'false starts' are probably not available, they are probably large enough to justify serious reduction efforts. Even the few publicized costs are important for some sectors (e.g., hazardous waste disposal, nuclear power, and most classes of development in urban and suburban areas). Assuming only minor changes in the conduct of development planning, the cost of such false starts seems likely to continue its evident upward trend.

The principal cause of these costs is misjudgement by the governmental or private developer as to what sort of facility will be found environmentally acceptable. A realistic path to reducing these costs seems evident: clear guidance to developers as to what classes of development activity are unacceptable. Trends toward clearer guidance are already evident in densely populated regions such as Orange County, California, and regions with particularly strong public concerns about environmental quality such as the Northwest and the Great Lakes region. A realistic strategy to achieve such guidance in one or two evidently receptive regions would illustrate the economic benefits of reducing developmental false starts nationally.

The strategy entails four stages: (1) develop quantitative criteria for 'unreasonable degradation' within two specific regions; (2) quantify regional environmental quality and 'unreasonable degradation' as formalized by governments; (3) specify acceptable and unacceptable classes of development within particular areas; (4) provide formal governmental guidance regarding the probable acceptability of particular classes of development in particular areas (recognizing that the acceptability of some classes will remain uncertain at this regional planning stage).

7.20. New research and knowledge transfer

An important function of the center should be to assure that new ideas and information are shared among the working groups and that the knowledge and experience gained are effectively conveyed to a wide spectrum of potential user groups in academic, private, and public institutions and agencies. Positive feedback from environmental and natural resource managers and policymakers should be encouraged so that real-world insights and priorities can be integrated into the research and education activities of the center. Individuals and institutions in other countries will be included in the outreach effort, and both formal and informal lines of communication should be utilized to provide a continuing flow of information in both directions.

Achievement of the goals of the proposed center will be measured in large part by the quality and scope of the information on its research and educational accomplishments transmitted outward. It is vital, therefore, that this function be organized and financially supported at a scale adequate to meet the need.

References

Theme 1: Probing encountered data

Boswell, M. T. and G. P. Patil (1987). A perspective of composite sampling. Invited Paper for *Experiments in Factor Screening*: Special Issue of *Commn. Statist. Theory Methods* **16**, 3069–3093.

Buckland, S. T. (1987). On the variable circular plot method of estimating animal abundance. *Biometrics* **43**, 363–384.

Burnham, K. P., D. R. Anderson, G. C. White, C. Brownie and K. H. Pollock (1987). *Design and Analysis Methods for Fish Survival Experiments Based on Release-Recapture*, Monograph 5. Amer. Fisheries Soc., Bethesda, MD.

Cochran, W. G. (1983). *Planning and Analysis of Observational Studies*. Wiley, New York.

Dennis, B. and G. P. Patil (1984). The gamma distribution and weighted multimodal gamma distributions as models of population abundance. *Math. Biosci.* **68**, 187–212.

DuMouchel, W. H. and J. E. Harris (1983). Bayes methods for combining the results of cancer studies in humans and other species. *J. Amer. Statist. Assoc.* **78**, 293–308.

Eberhardt, L. L. (1978). Appraising variability in population studies. *J. Wildlife Management* **42**, 207–238.

Elder, R. S. (1977). Properties of composite sampling procedures. Ph.D. Dissertation. Virginia Polytechnic Institute and State University, Blacksburg, VA.

Fienberg, S. E. and J. M. Tanur (1987). Experimental and sampling structures: Parallels diverging and meeting. *Internat. Statist. Rev.* **55**(1), 75–96.

Fuller, W. A. (1987). *Measurement Error Models*. Wiley, New York.

Galil, Z. and J. Kiefer (1980). Time- and space-saving computer methods related to Mitchell's DETMAX for finding D-optimum experimental designs. *Technometrics* **22**, 301–313.

Garner, F. C., M. A. Stapanian and L. R. Williams (1988). Composite sampling for environmental monitoring. In: L. H. Keith, ed., *Principles of Environmental Sampling*. Amer. Chem. Soc., Washington, DC, 363–374.

Hedges, L. V. and I. Olkin (1985). *Statistical Methods for Meta-Analysis*. Academic Press, New York.

Hennemuth, R. C. and G. P. Patil (1983). Implementing statistical ecology initiatives to cope with global resource impacts. In: J. Bell and T. Atterbury, eds., *Proc. Conf. on Renewable Resource Inventories for Monitoring Changes and Trends*. Oregon State Univ. Corvallis, OR, 374–378.

Hennemuth, R. C., G. P. Patil and C. Taillie (1985). Can we design our encounters? Invited paper CM1985/D9, presented to the International Council for the Exploration of the Sea, London, UK.

Hennemuth, R. C., G. P. Patil and D. Simberloff (1986). Advanced research conference on frontiers of statistical ecology. *Intecol Newslett.* **16**(1), 4–5.

Hunter, J. S. (1977). Incorporating uncertainty into environmental guidelines. In: *Environmental Monitoring*, The NAS report to EPA from the Study Group on Environmental Monitoring. National Acad. Sci., Washington, DC, 139–153.

Hurlbert, S. H. (1984). Pseudoreplication and the design of ecological field experiments. *Ecological Monographs* **54**, 187–211.

Iyengar, S. and J. B. Greenhouse (1988). Selection models and the file-drawer problem (with discussion). *Statist. Sci.* **3**(1), 109–135.

Kremers, W. K. (1988). Estimation of survival rates from a mark-recapture study with tag loss. *Biometrics* **44**(1), 117–130.

Laird, N. M., N. Lange and D. Stram (1987). Maximum likelihood computations with repeated measures: Application of the EM algorithm. *J. Amer. Statist. Assoc.* **82**, 97–105.

Laird, N., G. P. Patil and C. Taillie (1988). Comment on S. Iyengar and J. B. Greenhouse, Selection models and file drawer problem. *Statist. Sci.* **3**(1), 126–128.

Landau, E., G. P. Patil and J. O. Rawlings (1985). ASA Reviews of EPA-funded acid precipitation research. *Amer. Statist.* **39**, 243–273.

Lindsay, B. G. (1983a). The geometry of mixture likelihoods: A general theory. *Ann. Statist.* **11**, 86–94.

Lindsay, B. G. (1983b). The geometry of mixture likelihoods: Part II, The exponential family. *Ann. Statist.* **11**, 783–792.

Lindsay, B. G. (1986). Exponential family mixture models (with least squares estimators). *Ann. Statist.* **14**, 124–137.

Louis, T. A. (1983). Estimating an ensemble of parameters using Bayes and empirical Bayes methods. *J. Amer. Statist. Assoc.* **78**, 47–59.

Louis, T. A. (1984). Estimating a population of parameter values using Bayes and empirical Bayes methods. *J. Amer. Statist. Assoc.* **79**, 393–398.

Madow, W. G. and I. Olkin, eds. (1983). *Incomplete Data in Sample Surveys*. Academic Press, New York.

Milliken, G. A. and D. E. Johnson (1984). *Analysis of Messy Data*. Lifetime Learning Publs., Belmont, CA.

Mitchell, T. J. (1974). An algorithm for the construction of "D-optimal" experimental designs. *Technometrics* **16**, 321–330.

Otis, D. L., K. P. Burnham, G. C. White and D. R. Anderson (1978). Statistical inference from capture data on closed animal populations. *Wildlife Monographs* **62**.

Patil, G. P. (1984). Studies in statistical ecology involving weighted distributions. In: J. K. Ghosh and J. Roy, eds., *Statistical Applications and New Directions: Proc. ISI Golden Jubilee International Conference*, December 1981. Statistical Publ. Soc., Calcutta, India, 478–503.

Patil, G. P. (1985). Fishery and forestry management: Preface. *Amer. Statist.* **39**, 361–362.

Patil, G. P., G. J. Babu, M. T. Boswell, K. Chatterjee, E. Linder and C. Taillie (1986). Statistical issues in combining ecological and environmental studies with examples in marine fisheries research and management. In: *Proc. ASA/EPA Conf. on Statistical Issues in Combining Environmental Studies*. EPA, Washington, DC.

Patil, G. P., G. J. Babu, R. C. Hennemuth, W. L. Myers, M. B. Rajarshi and C. Taillie (1988). Data-based sampling and model-based estimation for environmental resources. In: P. R. Krishnaiah and C. R. Rao, eds., *Handbook of Statistics*. Elsevier Science Publishers, Amsterdam, 489–513.

Patil, G. P., C. R. Rao and M. Zelen (1987a). Weighted distributions. In: S. Kotz and N. L. Johnson, eds., *Encyclopedia of Statistical Sciences*. Wiley, New York.

Patil, G. P., C. R. Rao and M. Zelen (1987b). A computerized bibliography of weighted distributions and related weighted methods for statistical analysis and interpretation of encountered data, observational studies, representativeness issues, and resulting inferences. Center for Statistical Ecology and Environmental Statistics, Penn State University.

Rao, C. R. (1985). Weighted distributions arising out of methods of ascertainment: What population does a sample represent? In: A. C. Atkinson and S. E. Fienberg, eds., *A Celebration of Statistics*. Springer, New York, Chapter 24.

Rhode, C. A. (1979). Batch, bulk and composite sampling. In: R. M. Cormack et al., eds., *Sampling Biological Populations*. ICPH, Fairland, MD, 365–377.

Roeder, K., B. Devlin and B. G. Lindsay (1989). Application of maximum likelihood methods to population genetic data for the estimation of individual fertilities. *Biometrics* **45**, 363–379.

Seber, G. A. F. (1986). A review of estimating animal abundance. *Biometrics* **42**, 267–292.

Stewart-Oaten, A., W. M. Murdoch and K. R. Parker (1986). Environmental impact assessment: Pseudoreplication in time? *Ecology* **67**(4), 929–940.

Titterington, M. D., A. F. M. Smith and U. E. Makov (1985). *Statistical Analysis of Finite Mixture Distributions*. John Wiley, Chichester, UK.

Tukey, J. W. (1986). Sunset salvo, 1–20. Personal communication.

Wainer, H., ed. (1986). *Drawing Inference from Self-Selected Samples*. Springer, New York.

Theme 2: Spatial and temporal statistics

Box, G. E. P. and G. C. Tiao (1975). Intervention analysis with applications to economic and environmental problems. *J. Amer. Statist. Assoc.* **70**, 70–79.

Cambanis, S. (1985). Sampling designs for time series. In: E. J. Hannan, P. R. Krishnaiah and M. M. Rao, eds., *Handbook of Statistics*, Vol. 5. Elsevier Science Publishers, Amsterdam, 337–362.

Cliff, A. D. and J. K. Ord, (1981). *Spatial Processes, Models, and Applications*. Methuen, New York.

Cressie, N. (1984). Towards resistant geostatistics. In: G. Verly et al., eds. *Geostatistics for Natural Resource Characterization*. Reidel, Dordrecht.

Cressie, N. (1986). Kriging nonstationary data. *J. Amer. Statist. Assoc.* **81**, 625–634.

Cochran, W. (1946). Relative accuracy of systematic and stratified random samples for a certain class of populations. *Ann. Math. Statist.* **17**, 164–177.

Diggle, P. (1983). *The Spatial Analysis of Point Patterns*. Academic Press, New York.

Hawkins, D. and N. Cressie (1984). Robust Kriging – A proposal. *Math. Geology* **16**, 3–18.

Journel, A. G. (1983). Nonparametric estimation of spatial statistics. *Math. Geology* **15**, 5.

Journel, A. G. (1986). Constrained interpolation and soft Kriging. In: R. V. Ramani, ed., *Proc. 19th APCOM*. Soc. Mining Eng., Littleton, CO, 15–30.

Kitanidis, P. (1983). Statistical estimation of polynomial generalized covariance functions and hydrologic applications. *Water Resources Res.* **19**, 909–931.

Lettenmaier, D. P., K. W. Hipel and A. I. McLeod (1978). Assessment of environmental impacts, Part II: Data collection. *Environ. Management* 537–554.

Mardia, K. V. and R. J. Marshall (1984). Maximum likelihood estimation of models for residual covariance in spatial regression. *Biometrika* **71**, 135–146.

Marshall, R. J. and K. V. Mardia (1985). Some minimum norm quadratic estimators of the components of spatial covariance. *Math. Geology* **17**(5), 517–525. Corrections in **18**(2), 267 (1986).

Moss, M. E. and C. D. Tasker, (1979). Progress in the design of hydrological-data networks. Rev. Geophys. Space Phys. **17**, 1298–1306.

Myers, D. E. (1982). Co-kriging: The matrix form. *Math. Geology* **14**, 628–644.

Myers, D. E. (1985). Some aspects of robustness. *Sci. Terre* **24**, 63–79.

Myers, D. E. (1986). The robustness and continuity of Kriging. Joint Statistics Meetings, Chicago, August 21.
Myers, D. E. (1988). Multivariate geostatistical analysis for environmental monitoring. *Sci. Terre* **27**, 411–427.
Parzen, E. (1961). Regression analysis of a continuous time series. In: *Proc. 4th Berkeley Sympos. on Mathematical Statistics and Probability*, Vol. 1. University of California Press, Berkeley, CA, 469–489.
Patil, G. P., W. L. Liggett, F. Ramsey and W. K. Smith (1985). Aquatic research review report. *Amer. Statist.* **39**(4), 254–258.
Quenouille, M. (1949). Problems in plane sampling. *Ann. Math. Statist.* **20**, 355–375.
Ripley, B. D. (1977). Modelling spatial patterns. *J. Roy. Statist. Soc. Ser. B* **39**, 172–212.
Ripley, B. D. (1979). Tests of randomness for spatial point patterns. *J. Roy. Statist. Soc. Ser. B* **41**, 368–374.
Smith, W. (1984). The design of efficient environmental surveys over time. In: S. M. Gertz and M. D. London, eds., *Statistics in Environmental Sciences*. Amer. Soc. Testing Mater., Philadelphia, PA, 90–97.
Smith, W. (1978). Environmental survey design: A time series approach. *Estuarine Coastal Marine Sci.* **6**, 217–224.
Warrick, A. W. and D. E. Myers (1987). Optimization of sampling locations for variogram calculations. *Water Resource Res.* **23**, 496–500.
Warrick, A. W., D. E. Myers and D. Neilsen (1986). Geostatistical methods applied to soil science. In: *Methods of Soil Analysis*, Part 1, Monograph No. 9. 2nd ed., Soil Sci. Soc. Amer., Madison, WI, 53–82.
Zirschky, J., G. P. Keary, R. O. Gilbert and E. J. Middlebrooks (1985). Spatial estimation of hazardous waste site data. *J. Environ. Eng.* **111**, 6.

Theme 3: Biodiversity and community dynamics

Anderson, S. (1974). Patterns of faunal evolution. *Ann. Rev. Biol.* **49**, 1–15.
Chakraborty, R., P. A. Fuerst and M. Nei (1980). Statistical studies on protein polymorphism in natural populations. II. Gene differentiation between populations. *Genetics* **88**, 367–390. III. Distribution of allele frequencies and the number of alleles per locus. *Genetics* **94**, 1039–1063.
Crow, J. F. and M. Kimura (1970). *An Introduction to Population Genetics Theory*. Harper, New York.
Digby, P. G. N. and R. A. Kempton (1987). *Multivariate Analysis of Ecological Communities*. Chapman and Hall, New York.
Engen, S. (1978). *Stochastic Abundance Models*. Chapman and Hall, London.
Ewens, W. J. (1972). The sampling theory of selectively neutral alleles. *Theoret. Population Biol.* **3**, 87–112.
Ewens, W. J. (1979). *Mathematical Population Genetics*. Springer, Berlin.
Ewens, W. J. and W. H. Li (1980). Frequency spectra of neutral and deleterious alleles in a finite population. *J. Math. Biol.* **10**, 155–166.
Fisher, R. A. (1943). The relation between the number of species and the number of individuals in a random sample from an animal population. *J. Anim. Ecol.* **12**, 54–57.
Fuerst, P. A., R. Chakraborty and M. Nei, (1977). Statistical studies on protein polymorphism in natural populations. I. Distribution of single locus heterozygosity. *Genetics* **86**, 455–483.
Gillespie, J. H. (1978). A general model to account for enzyme variation in natural populations. VI. The SAS-CFF model. *Theoret. Population Biol.* **14**, 1–45.
Hill, B. M. (1979). Posterior moments of the number of species in a finite population and the posterior probability of finding a new species. *J. Amer. Statist. Assoc.* **74**, 668–673.
Jenkins, R. (1976). Maintenance of natural diversity: Approach and recommendations. *Trans. N. Amer. Wildlife Natural Resources Conf.* **41**, 441–451.
Lewontin, R. C. (1986). Population genetics. In: P. J. Greenwood, P. H. Harvey and M. Slatkin, eds., *Evolution: Essays in Honor of John Maynard Smith*. Cambridge Univ. Press, Cambridge, 3–18.
Li, W. H. and M. Nei (1975). Drift variances of heterozygosity and genetic distance in transient states. *Genetics Res.* **25**, 229–248.
Loehle, C. (1987). Errors of construction, evaluation and inference: A classification of sources of error in ecological models. *Ecol. Model.* **36**, 297–314.

May, R. M. (1975). Patterns of species abundance and diversity. In: M. L. Cody and J. M. Diamond, eds., *Ecology and Evolution of Communities.* Harvard University Press, Cambridge, MA, 81–120.

Milkman, R. (1982). Toward a unified theory of selection. In: R. Milkman, ed., *Perspectives on Evolution.* Sinauer Associates, Sunderland, MA, 105–118.

Nei, M. (1975). *Molecular Population Genetics and Evolution.* Elsevier, New York.

Nei, M. (1978). Estimation of average heterozygosity and genetic distance from a small number of individuals. *Genetics* **89**, 583–590.

Nei, M. (1987). *Molecular Evolutionary Genetics.* Columbia Univ. Press, New York.

Patil, G. P. and C. Taillie (1977). Diversity as a concept and its implications for random communities. *Bull. Internat. Statist. Inst.* **42**, 497–515.

Patil, G. P. and C. Taillie (1979). Species abundance models, ecological diversity, and the canonical hypothesis. *Bull. Internat. Statist. Inst.* **44**, 1–23.

Patil, G. P. and C. Taillie (1982). Diversity as a concept and its measurement. *J. Amer. Statist. Assoc.* **77**, 548–567.

Preston, F. W. (1962). The cononical distribution of commonness and rarity. *Ecology* **43**, 186–215, 410–561.

RFF (1983). *Resources Future* **74**, 4–8.

Simberloff, D. (1970). Taxonomic diversity of island biota. *Evolution* **24**, 23–47.

Simberloff, D. (1986). Are we on the verge of a mass extinction in tropical rain forests? In: D. K. Elliott, ed., *Dynamics of Extinction.* Wiley, New York, 165–188.

Simon, J. (1983). Species preservation. *Resources* **74**, 5–6.

Slatkin, M. (1985). Gene flow in natural populations. *Ann. Rev. Ecol. System.* **16**, 393–430.

Watterson, G. A. (1974). Models for the logarithmic species abundance distributions. *Theoret. Population. Biol.* **6**, 217–250.

Watterson, G. A. (1977). Heterosis of neutrality? *Genetics* **85**, 789–814.

Wilson, E. O. (1985). The biological diversity crisis: A challenge to science. *Sci. Technology* **2**(1), 20–29.

Wright, S. (1969). *Evolution of the Genetics of Populations*, Vol. 2. Univ. of Chicago Press, Chicago.

Theme 4: Resource assessment and prediction

Avery, T. E. and H. E. Burkhart (1983). *Forest Measurements.* 3rd ed., McGraw-Hill, New York.

Barnad, J., W. Myers, J. Pearce, F. Ramsey, M. Sissenwine and W. Smith (1985). Surveys for monitoring changes and trends in renewable resources: Forest and marine fisheries. *Amer. Statist.* **39**(4), 363–373.

Bell, J. F. and T. Atterbury, eds. (1983). Renewable Resource inventories for monitoring Changes and Trends. Oregon State University, Corvallis, OR.

Clutter, J. L., J. C. Fortson, L. V. Piennaar, G. H. Brister and R. L. Bailey (1983). *Timber Management: A Quantitative Approach.* Wiley, New York.

Darr, D. R., and Hennemuth, R. C. (1985). Forecasting for resources in forestry and marine fisheries in the year 2000. *Amer. Statist.* **39**, 384–392.

Davis, L. S. and K. N. Johnson (1987). *Forest Management.* 3rd ed., McGraw-Hill, New York.

Dennis, B., B. E. Brown, A. R. Stage, H. E. Burkhart and S. Clark (1985). Problems of modeling growth and yield of renewable resources. *Amer. Statist.* **39**(4), 374–383.

Hennemuth, R. C. (1985). Implementing and maintaining management philosophies in marine resource inventories. In: J. F. Bell and T. Atterbury, eds., *Renewable Resource Inventories for Monitoring Changes and Trends.* Oregon State University, Corvallis, OR.

Holling, C. S., ed. (1978). *Adaptive Environmental Assessment and Management.* Wiley, New York.

Mangel, M. (1985). *Decision and Control in Uncertain Resource Systems.* Academic Press, New York.

Marine Technology Society (1986). *Monitoring Strategies Symposium Proceedings, Oceans 86.* IEEE and MTS, Washington, DC.

Matis, J. H., B. C. Patten and G. C. White, eds. (1979). *Compartmental Analysis of Ecosystem Models.* International Cooperative Publ. House, Fairland, MD.

McCullagh, P. and J. A. Nelder (1983). *Generalized Linear Models.* Chapman and Hall, New York.

Okubo, A. (1980). *Diffusion and Ecological Problems: Mathematical Models.* Springer, New York.

Patil, G. P. (1985). Fishery and forestry management: Preface. *Amer. Statist.* **39**(4), 361–362.

Patil, G. P., G. J. Babu, M. T. Boswell, K. Chatterjee, E. Linder and C. Taillie (1986). Statistical issues in combining ecological and environmental studies with examples in marine fisheries research and management. In: *Proc. ASA/EPA Conf. on Statistical Issues in Combining Environmental Studies.* EPA, Washington, DC.

Patil, G. P., G. J. Babu, R. C. Hennemuth, W. L. Myers, M. B. Rajarshi and C. Taillie (1987). Data-based sampling and model-based estimation for environmental resources. In: P. R. Krishnaiah and C. R. Rao, eds., *Sampling in the Series on Developments in Statistics.* North Holland, Amsterdam.

Rothschild, B. (1986). *Dynamics of Marine Fish Populations.* Harvard Univ. Press, Cambridge.

Ryman, N. and F. Utter, eds. (1987). *Population Genetics and Fishery Management.* Univ. of Washington Press, Seattle, WA.

Walters, C. (1986). *Adaptive Management of Renewable Resources.* MacMillan, New York.

Theme 5: Environmental risk analysis and decision making

Anandalingam, G. (1987). A multi-agent multi-attribute approach for conflict resolution in acid rain impact mitigation. *IEEE Trans. Systems Man Cybernet.* **19**(5), 1142–1153.

Babu, G. J. and K. Singh (1983). Nonparametric inference on means using bootsrap. *Ann. Statist.* **11**, 999–1003.

Babu, G. J. and K. Singh (1984). On one term Edgeworth correction by Efron's bootstrap. *Sankhyā, Ser. A* **46**, 219–232.

Boswell, M. T. and G. P. Patil (1988). Selection of endpoints for a crystal cube, and development of indices for coastal and estuarine environmental degradation for use in decisionmaking. Technical Report No. 87-0501, Center for Statistical Ecology and Environmental Statistics, Department of Statistics, Pennsylvania State University, University Park, PA.

Broderius, S. J. (1987). Joint aquatic toxicity of chemical mixtures and structure-activity relationships. In: *ASA/EPA Conf. on Interpretation of Environmental Data.* I. Current Assessment of Combined Toxicant Effects, May 5–6, 1986, EPA-230-03-87-027. USEPA, Washington, DC, 45–62.

Chatterjee, K. (1986). The theory of bargaining. In: L. Samuelson, ed., *Microeconomic Theory.* Kluwer-Nijhoff, Boston, 159–194.

Chatterjee, K. and L. Samuelson (1987). Bargaining under incomplete information: Infinite horizon models with alternating offers. *Rev. Econ. Stud.* **54**, 175–192.

Christensen, E. R. (1987). Development of models for combined toxicant effects. In: *ASA/EPA Conf. on Interpretation on Environmental Data.* I. Current Assessment of Combined Toxicant Effects, May 5–6, 1986, EPA-230-03-87-027. USEPA, Washington, DC, 66–74.

Christensen, E. R. and C. Y. Chen (1985). A general noninteractive multiple toxicity model including probit, logit, and Weibull transformations. *Biometrics* **41**, 711–725.

Christensen, E. R. and C. Y. Chen (1987). Modeling of combined toxic effects of chemicals. In: *Hazard Assessment of Chemicals: Current Developments*, Vol. 6. Hemisphere, Washington, DC.

Courtney, F. E., C. W. Frand and J. M. Powell (1981). Integration of modelling, monitoring, and laboratory observations to determine reasons for air quality violations. *Environ. Monitor. Assessment* **1**, 107–118.

DeGroot, M. H. (1974). Reaching a concensus. *J. Amer. Statist. Assoc.* **69**, 118–121.

DeGroot, M. H. and S. E. Fienberg (1986). Comparing probability forecasters: Basic binary concepts and multivariate extensions. In: P. K. Goel and A. Zellner, eds., Bayesian Inference and Desicion Techniques. North-Holland, Amsterdam, 247–264.

French, S. (1983). Modelling experts. In: J. M. Bernardo, ed., *Proc. 2nd Bayesian Statistics Conference*, Valencia.

Fuller, W. A. (1987). *Measurement Error Models.* Wiley, New York.

Genest. C. and J. V. Zidak (1986). Combining probability distributions: A critique and an annotated bibliography. *Statist. Sci.* **1**, 114–148.

Gleser, L. J. (1983). Estimation in a multivariate errors-in-variables regression model: Large sample results. *Ann. Statist.* **9**, 24–44.

Gordon, G. E. (1980). Receptor models. *Environ. Sci. Technol.* **14**, 792–800.
Hinkley, E. D. (1977). Air monitoring with turnable lasers. *Environ. Sci. Technol.* **11**, 564–567.
Linder, E. (1987). Inference in the linear errors-in-variable model using the bootstrap with applications in environmental risk analysis. Unpublished Ph.D. Thesis, Pennsylvania State University.
Linder, E., G. P. Patil, G. W. Sutter and C. Taillie (1986). Effects of toxic pollutants on aquatic resources using statistical models and techniques to extrapolate acute and chronic effects benchmarks. In: *Oceans 86 Proceedings*, Vol. 3, Monitoring Strategies Symposium. Washington, DC, 960–963.
Maugh, T. H., II (1981). New ways to measure SO remotely. *Science* **212**(4491), 152–153.
Muerson, R. (1987). Analysis of incentives for dispute resolution. Dispute Resolution Research Center Discussion Paper, Northwestern University.
O'Connor, J. S. and R. T. Dewling (1986). Indices of marine degradation: Their utility. *Environ. Management* **3**, 335–343.
Patel, C. K. N. (1978). Laser detection of pollution. *Science* **202**(4364), 157–173.
Patil, G. P. and C. Taillie (1987). Discussion of paper by Steven J. Broderius. In: *ASA/EPA Conf. on Interpretation of Environmental Data.* I. Current Assessment of Combined Toxicant Effects, May 5–6, 1986, US EPA-230-03-87-027. USEPA, Washington, DC, 63–65.
Pitt, W. W., Jr. (1981). Continuous monitoring, automated analysis, and sampling procedures. *J. Water Pollut. Control Feder.* **53**, 675–678.
Raiffa, H. (1982). *Art and Science of Negotiation.* Harvard Univ. Press, Boston.
Ruckelshaus, W. D. (1984). Risk in a free society. *Risk Anal.* **4**, 157–162.
Russell, C. S. (1987). Game theory lessons for structuring monitoring and enforcement systems. Prepared for the Association of Environmental and Resource Economists' Workshop on Environmental Monitoring and Enforcement: Theory and Practice. August, 1987.
Russell, C. S. (1988). Monitoring and enforcement. In: Portney, ed., *Environmental Regulation in the U. S.: Public Policies and Their Consequences.* Resources for the Future, Washington, DC.
Russell, C. S., G. P. Patil, E. Landau and K. P. Cantor, eds. (1985). Proc 8th Sympos. on Statistics, Law, and the Environment, Special Issue of *Amer. Statist.* **39**, 373–450.
Suter, G. W., II, S. E. Rosen and E. Linder (1986). Analysis of extrapolation error. In: L. W. Barnthouse and G. W. Suter II, eds., *User's Manual for Ecological Risk Analysis.* ORNL-6251. Oak Ridge National Laboratory, Oak Ridge, TN, 49–81.
U. S. CEQ (Council on Environmental Quality) (1984). Environmental Quality – 15th Annual Report. U. S. CEQ, Washington, 719 pp.
USEPA (1979). *Handbook: Continuous Air Pollution Source Monitoring Systems.* National Technical Information Service, Springfield, VA, EPA 625/6-79 005.
Veith, G. D.., D. DeFoe and M. Knuth (1985). Structure-activity relationships for screening organic chemicals for potential ecotoxicity effects. *Drug Metab. Rev.* **15**, 1295–1303.
Whyte, A. V. and I. Burton, eds. (1980). *Environmental Risk Assessment.* Wiley, New York.

Theme 7: Cutting edge linkages

Adelman, M. A., J. C. Houghton, G. M. Kaufman and M. B. Zimmerman (1983). *Energy Resources in an Uncertain Future: Coal, Gas, Oil and Uranium Supply Forecasting.* Ballinger, Cambridge, MA.
Andreatta, G. and G. M. Kaufman (1983). Estimation of finite population properties when sampling is without replacement and proportional to magnitude. *J. Amer. Statist. Assoc.*, MIT-EL82-027 working Paper.
Armitage, P. (1985). Biometry and medical statistics. *Biometrics* **41**, 823–833.
Arps, J. K. and T. G. Roberts (1958). Economics of drilling for cretaceous oil and gas on the east flank of the Denver–Julesberg basin. AAPG Bull. **42**, 2549–2566.
Barouch, E. and G. M. Kaufman (1977). Estimation of undiscovered oil and gas. In: *Proc. Sympos. in Applied Mathematics*, vol. 21. Amer. Math. Soc., Providence, RI, 77–91.
Barouch, E., S. Chow, G. M. Kaufman and T. Wright (1985). Properties of successive sample moment estimators. *Stud. Appl. Math.* **73**, 239–260.
Bloomfield, P. et al. (1979). Volume and area of oil fields and their impact on the order of discovery.

Resource estimation and validation project, Department of Statistics and Geology, Princeton University.
Chapman, D. G. (1951). Some Properties of the hypergeometric distributions with application to zoological sample censuses. *Univ. California Publ. Statist.* **1**, 1313–1316.
Cochran, W. G. (1953). *Sampling Techniques.* Wiley, New York.
Cox, D. R. (1969). *Some sampling problems in technology.* In: N. L. Johnson and H. Smith, Jr., eds., *Developments in Survey Sampling.* Wiley–Interscience, New York, 506–527.
Drew, L. J., J. H. Scheunemeyer and D. H. Root (1980). Petroleum-resource appraisal and discovery rate forecasting in partially explored regions, Part A: An application to the Denver basin. U. S. Geological Survey Professional Paper 1138-A,B,C, A1–A11.
Drew, L. B. J., J. H. Schuenemeyer and W. J. Bawiec (1982). Estimation of the Future Rates of Oil and Gas Discoveries in the Western Gulf of Mexico. U. S. Geological Survey Professional Paper 1252, 57.
Federer, W. T. (1984). Cutting edges in Biometry. *Biometrics* **40**, 827–839.
Gordon, L. (1981). Successive sampling in large finite populations. *Ann. Statist.* **11**, 702–706.
Gordon, L. (1983). Estimation for large successive samples with unknown inclusion probabilities. Manuscript.
Hajek, J. (1981). *Sampling from a Finite Population.* Marcel Dekker, New York.
Hardy, G. H., J. E. Littlewood and G. Polya (1967). *Inequalities.* 2nd ed., University Press, Cambridge, UK.
Holst, L. (1973). Some limit theorems with applications in sampling theory. *Ann. Statist.* **1**, 644–658.
Horwitz, D. G. and D. J. Thompson (1952). A generalization of sampling without replacement from a finite universe. *J. Amer. Statist. Assoc.* **47**, 663–685.
Kaufman, G. M., Y. Balcer and D. Kruyt (1975). A Probabilistic model of oil and gas discovery. In: J. Haun, ed., *Estimating the Volume of Undiscovered Oil and Gas Resources*, Amer. Assoc. Tulsa, OK, 113–142.
Kaufman, G. M. and J. Wang (1980). Model mis-specification and the Princeton study of volume and area of oil fields and their impact on the order of discovery. MIT Energy Laboratory Working Paper No. MIT-EL 80-003WP.
Kesten, H. (1987). Percolation theory and first passage percolation. *Ann. Probab.* **15**, 1231–1271.
Lagakos, S. W., B. J. Wesson and M. Zelen (1986). An analysis of contaminated well water and health effects in Woburn, Massachusetts (with discussion). *J. Amer. Statist. Assoc.* **81**, 583–596.
Laird, N. M. and T. A. Louis (1987). Empirical Bayes confidence intervals based on bootstrap samples (with discussion). *J. Amer. Statist. Assoc.* **82**, 739–750.
Lee, P. J. and P. C. C. Wang (1983). Probabilistic formulation of a method for the evaluation of petroleum resources. *Math. Geol.* **17**, 95–113.
Lee, P. J. and P. C. C. Wang (1985). Prediction of oil and gas pool sizes when discovery record is available. *Math. Geol.* **17**, 95–113.
Lee, P. J. and P. C. C. Wang (1986). Evaluation of petroleum resources from pool size distributions. In: D. D. Rice, ed., *Oil and Gas Assessment-Methods and Applications*, Studies in Geology, Vol. 21. Amer. Assoc. of Petroleum Geologists, Tulsa, OK, 33–42.
Levins, R. (1984). The significance of the variability of agricultural parameters. *Econom. Appl.* **37**, 377–392.
Levins, R. (1986). Perspectives in integrated pest management: From an industrial to an ecological model of pest management. In: M. Kogan, ed., *Ecological Theory and Integrated Pest Management Practice.* Wiley, New York, 1–18.
Louis, T. A. (1983). Estimating an ensemble of parameters using Bayes and empirical Bayes methods. *J. Amer. Statist. Assoc.* **78**, 47–59.
Louis, T. A. (1984). Estimating a population of parameter values using Bayes and empirical Bayes methods. *J. Amer. Statist. Assoc.* **79**, 393–398.
Meisner, J. and F. Demirmen (1981). The creaming method: A Bayesian procedure to forecast future oil and gas discoveries in mature exploration provinces. *J. Roy. Statist. Soc. Ser. A* **144**, 1–31.
Murphy, M. N. (1957). Ordered and unordered estimators in sampling without replacement. *Sankhyā* **18**, 378–390.

Nair, V. J. and C. C. Wang (1987). Maximum likelihood estimation under a successive sampling discovery model. Manuscript.

Olver, F. W. J. (1974). *Asymptotics and Special Functions.* Academic Press, New York.

Patil, G. P. and J. K. Ord (1976). On size-biased sampling and related form invariant weighted distributions. *Sankhyā Ser. B* **38**, 46–61.

Patil, G. P. and C. R. Rao (1977). The weighted distributions: A survey of their applications In: P. R. Krishnaiah, ed., *Application of Statistics.* North-Holland, Amsterdam, 383–405.

Patil, G. P. and C. R. Rao (1978). Weighted distributions and size-biased sampling with applications to wildlife populations and human families. *Biometrics* **34**, 179–189.

Puccia, C. J. and R. Levins (1985). *Qualitative Modeling of Complex Systems: An Introduction to Loop Analysis and Time Averaging.* Harvard Univ. Press, Cambridge.

Root, D. H. and J. H. Schunemeyer (1980). Petroleum-resource appraisal and discovery rate forecasting in partially explored regions: Forecasting in partially explored regions. Mathematical Foundations. U. S. Geological Survey Professional Paper 1138-B, B1-B11.

Rosen, B. (1972). Asymptotic theory for successive sampling with varying probabilities without replacement, I and II. *Ann. Math. Statist* **43**, 373–397, 748–776.

Schaeffer, R. L. (1972). Size-biased sampling. *Technometrics* **14**, 635–644.

Smith, J. L. and G. L. Ward (1981). Maximum likelihood estimates of the size distribution of North Sea oil fields. *Math. Geol.* **13**.

Uppuluri, V. R. R. and S. A. Patil (1979a). Limiting behavior of sampling proportional to random size. Contributed paper, 42nd Session, International Statistical Institute, Manila, Philippines.

Uppuluri, V. R. R. and S. A. Patil (1979b). Sampling proportional to random size. In: J. Xavier, R. Anula, R. Bellman, Y. L. Luke and A. K. Rigler, eds., *Proc. 2nd Internat. Conf. on Mathematical Modeling.* University of Missouri, Rolla, MO.

Vardi, Y. (1982). Nonparametric estimation in the presence of length bias. *Ann. Statist.* **10**, 616–620.

Wang, C. C. and P. J. Lee (1983a). Probalistic formulation of a method for the evaluation of petroleum resources. *Math. Geol.* **15**(1), 163–181.

Wang, C. C. and P. J. Lee (1983b). Conditional analysis for petroleum resource evaluations. *Math. Geol.* **15**(2), 349–361.

Wang, C. C. and V. J. Nair (1987). Statistical analysis of oil and gas discovery data. Manuscript.

Ware, J. H., F. Mosteller and J. A. Ingelfinker (1986). *P*-values. In: J. C. Bailor and F. Mosteller, eds., *Medical Uses of Statistics.* New England J. Medicine Books, Waltham, MA, 149–169.

Zelen, M. (1983). Biostatistical science as a discipline: A look into the future. *Biometrics* **39**, 827–837.

G. P. Patil and C. R. Rao, eds., *Handbook of Statistics, Vol. 12*

3

Replicate Measurements for Data Quality and Environmental Modeling

Walter Liggett

Abstract: Replicate measurements can be used to improve sampling and measurement procedures and thereby data quality and to select an environmental model that agrees reasonably with the data. The implications for study design of these uses of replicates and corresponding statistical analysis methods are presented. Design implications are important because successful use of replicates requires agreement between the physical procedures specified for replication and the assumptions of the statistical analysis. The statistical analysis methods presented for quality improvement and model selection are especially appropriate for environmental studies. The method for quality improvement is intended for observations each consisting of several measurements and for measurement techniques with error variance that depends on measured level. As a consequence, the quality improvement method discussed is a multivariate method generalized to include estimation of a power transformation for each constituent measurement. The method for model selection is intended for fitting methods with limited sensitivity to occasional large discrepancies between model and data and for observations with error probability distribution that has unknown form. As a consequence, the model selection method discussed employs a bootstrap based on replicates that is applicable to models fit by robust estimation.

1. Introduction

Repetition of chemical and biological measurements to catch blunders, which is often used in measurement quality control, seems needed in the sample collection and handling stage of environmental measurement because of practical limits on the use of reference materials. Replication to obtain an error assessment independent of model assumptions and thus a more dependable test of agreement between model and data, which is often part of designed experiments, seems needed in environmental studies because environmental models usually involve some questionable assumptions. These two uses of replicates are brought together in this chapter because the essential first step in both cases is specification of the replicates in the study design.

Replicate measurements have a physical procedures aspect and a statistical models aspect that must be in reasonable agreement for the replicates to be informative. For an environmental scientist, a replicate measurement is the result of a physical procedure whereby something is measured again. The scientist does not expect the second measurement to equal the first because both measurements are influenced by one or more unpredictable sources of variation. For a statistician, a replicate measurement is one that differs from its predecessor only in being subject to an independent realization of the measurement error. The correspondence between sources of variation and random components of error is part of the basis for the design and interpretation of environmental studies. In each study, this correspondence must be carefully considered because the match is usually not perfect. On one hand, scientists usually do not understand the sources of variation fully and may not be able to provide a complete list. On the other hand, statisticians often have no firm basis for treating error components as random. Consideration of this correspondence at the design stage generally leads to environmental measurements that are easier to interpret although the measurements themselves will often reveal something unanticipated about the correspondence.

The purpose of replicates is characterization of sources of variation, but replication procedures for all sources of variation may not be feasible. For example, characterization of the sampling and measurement error can be obtained through a replication procedure that specifies collection of a second sample at the same place and time as the first sample. However, sources of variation associated with the weather, although they are often treated as random, cannot generally be characterized through replication. In the design of an environmental study, consideration of what sources of variation can be characterized through replication is very important. One goal in study design is arrangement of the measurements so that the comparisons of interest are influenced only by sources of variation that can be characterized through replication. This, of course, is not always possible. Replication to characterize some but not all important sources of variation can have value. For example, replication to characterize just the sampling and measurement error can assure that some unusual feature apparent in the data is not merely due to a problem with field procedures and is therefore worthy of further investigation. In each study, the potential value of replication must be compared to the cost, which can be large when replication of field procedures is involved.

In this chapter, the use of replicate measurements is separated into the use for measurement quality control, which is detailed in Section 2, and the use for environmental modeling, which is detailed in Section 3. In both sections, statistical methods specially configured to meet the requirements of environmental studies are presented. Replicate measurements can be used in the construction of a variance control chart, called an s^2 chart by Ryan (1989, pp. 89–94). In Section 2, the s^2 chart is generalized to accommodate several measurements on each sample and the possibility that the error variance for each constituent measurement increases with measured level. The choice of a multivariate quality control procedure and the compensation for nonconstant variance are motivated by the needs of environmental studies. Replicate measurements can sometimes be used in environmental studies as replicates are used for lack-of-fit tests in regression (see Draper and Smith, 1981, pp. 33–42). In Section 3, lack-of-fit tests are generalized to accommodate nonnormal error and the consequent use of robust methods. The generalization is based on estimation of the probability distribution of the error from the replicates. Experience has shown that robust methods are often needed in the interpretation of environmental measurements.

2. Measurement quality control

2.1. Sample collection and handling

Consider environmental studies in which quantitative and perhaps multivariate measurements of air, water, soil, or biota are made at specified locations and times. As discussed by Gilbert (1987), the choice of where and when to measure is an important part of environmental study design. The topic of this section is the procedure whereby an environmental measurement is made at a chosen location and time. Specification of such a procedure in the study design is not enough. In part, because such procedures usually consist of many steps performed by different people over a considerable period of time, untested procedures often entail surprisingly large sources of error. For this reason, empirical investigation of procedures for making environmental measurements has for many years been recognized as essential. Optimization of such procedures through experimental adjustment, perhaps as part of a pilot study, is an attractive possibility. Statistical process control (SPC) for detection of problems that might be isolated and corrected is considered in this section.

The focus of this section is on the early stages of the measurement procedure since the last stages, the part of the procedure that is performed in the analytical or biological laboratory, has been discussed extensively. The statistical aspects of laboratory quality control are discussed by Mandel (1991). Practical investigation of the early stages of a measurement procedure cannot be done without familiarity with the potential problems in chemical and biological terms. Pio and Hall (1991) and Keith (1988) discuss the physical sampling of air, water, and soil for the purpose of making various measurements. Smith and James (1981) and Williams (1976) provide more details on the sampling and subsampling of bulk materials. Liggett, Inn and Hutchinson (1984) explore some seemingly inescapable problems in the measurement of trace quantities in bulk materials. From these references, one could compile a very long list of what can go wrong in sample collection and handling.

The following three examples should help make this section less abstract. The first example involves subsampling of bulk materials. Naively one might think that subsampling, the extraction of a quantity small enough for laboratory analysis from the sample obtained in the field, would not be a problem if sufficient mixing were specified. Problems arise because the ratio of the needed subsample to the total sample is often very large. One possible surprise is resegregation of samples consisting of nonuniform powders. The second example is contamination. In the measurement of trace quantities, precautions against contamination are obviously needed. The surprises are unexpected sources of contamination such as the materials used for sample processing. The third example is spoilage. Natural environmental processes, which often can be thought of as spoilage, must be stopped if some time will elapse before the sample is processed. Considerable experience exists with refrigeration and other methods of sample preservation, but no method exists that is foolproof and always practical.

Although statistical process control of laboratory measurement problems cannot be ignored, control of the measurement problems illustrated above that can occur enroute to the laboratory seem more challenging. One fundamental reason is that field procedures are inherently more diverse than laboratory procedures and are more likely to be created for the environmental study at hand. Another fundamental reason is that reference materials (materials with known values for the analyte in question to which the collection and handling procedure can be applied) are generally not available. Without reference materials, the only general approach seems to be replication. Of course, special approaches can be invented for specific problems. One example is the use of blanks (analyte-free samples) to detect contamination, an approach which is often followed in the sampling of water. However, blanks do not reveal sample spoilage or adsorption of the substance of interest on the walls of the sample container; yet, blanks can cost as much to measure as real samples. This example suggests that a combination of special approaches designed to detect suspected measurement problems may not be a satisfactory substitute for replication.

In statistical terms, measurement control based on replicates is more complicated than the measurement control usually recommended for the laboratory. A major reason for this is that replicate measurements occur at whatever levels are found where and when the samples are taken. Since the range in levels may be several orders of magnitude, dependence of the variance of the measurement error on the level must be considered. Moreover, the spacing of levels provided by the replicates cannot be chosen to make estimation of this dependence as efficient as possible. In particular, because the distribution of environmental measurements is often skewed to the right, one might find many replicates at the low end of the measurement range and only a few at the upper end. The fact that environmental measurements are often multivariate adds another complication. However, when problems in sample collection and handling effect more than one constituent measurement, this complication might be exploited to obtain additional sensitivity in problem detection. Other complications in the analysis of replicates will be mentioned later.

2.2. Methods for statistical process control

Of the various statistical process control (SPC) methods applicable to environmental measurements, four are presented in this section. First, SPC based on repeated measurement of the same reference material is considered. The method presented, which is based on familiar control charts, is the usual method for laboratory quality control. Next, the start-up phase of SPC, the phase when there are no established control limits to judge the observations, is discussed. This phase seems unusually important in the application of SPC to environmental studies. Third, the generalization to multivariate replicates is introduced. Finally, the use of data transformations to cope with the dependence of the standard deviation of the measurement error on the level is added.

When identical portions of a reference material are available, a standard type of SPC can be implemented (Ryan, 1989). Let the measurements on the reference material be made in groups of size r. Let the measurements be denoted by y_{ij}, where $i = 1, 2, \ldots$ indexes the groups and $j = 1, \ldots, r$ indexes measurements within the group. Standard practice specifies the creation of two control charts, one for the mean of the groups and the other for the spread of the groups. As the indication of spread, we choose the variance because this choice is consistent with the generalizations that follow. The first of these charts, called the $\bar{X}$ chart, contains a plot of

$$\bar{y}_i = \frac{1}{r} \sum_{j=1}^{r} y_{ij}$$

versus i. The second of these charts, called the s^2 chart, contains a plot of

$$s_i^2 = \frac{1}{r-1} \sum_{j=1}^{r} (y_{ij} - \bar{y}_i)^2$$

versus i. The patterns observed on control charts can be used in various ways to decide whether the measurement procedure should be examined for a special cause of variation that might be remedied.

Standard practice displays upper and lower control limits on control charts. A value of $\bar{y}_i$ or s_i^2 outside the control limits signals that the measurement procedure should be investigated. Control limits are set to reduce the probability of a futile investigation to an acceptably small level. In the simplest case, the mean and variance for measurements on the reference material, μ and σ^2, are known from previous experience and these can be used along with the assumption of normally distributed measurement error to compute control limits. The availability of a reference material and knowledge of what measurement error properties to anticipate often occur together. One way to limit the number of false alarms is to set the control limits so that each $\bar{y}_i$ or s_i^2 has only a small probability, say 0.002, of exceeding the control limits when the measurement system is performing as anticipated. In this case, the control limits for the $\bar{X}$ chart are $\mu - z_{.999}\sigma/\sqrt{r}$ and $\mu + z_{.999}\sigma/\sqrt{r}$ and the limits for the s^2 chart are $\sigma^2\chi^2_{.001}/(r-1)$ and $\sigma^2\chi^2_{.999}/(r-1)$, where $z_{.999}$ is the 99.9 percentile of the normal distribution and $\chi^2_{.001}$ and $\chi^2_{.999}$ are the 0.1 and 99.9 percentiles of the χ^2 distribution with $r-1$ degrees of freedom. With these control limits, the rate of unnecessary investigations will be less than 4 per

1000 groups of measurements. Note that the specification of physical procedures for reference material measurement requires care. One danger is that the measurements within groups will not be statistically independent, and consequently, numerous values of s_i^2 below the lower control limit will be observed.

The control limit approach to SPC decisions does not take advantage of trends in $\bar{y}_i$ or s_i^2 that may be visible on the control charts themselves. In other words, the control limit approach ignores the information in the time order of the observations. This omission can be remedied in the reference material case through CUSUM techniques (Ryan, 1989).

In the start-up phase of SPC, the control limits must be obtained from the same measurements that are being checked for evidence of measurement procedure problems. This circumstance complicates the statistical analysis. Yet, in environmental studies, the start-up phase is important, perhaps more important than in production where SPC is often supplemented at start-up by more careful testing of the first items manufactured. In environmental studies, the possibilities of more careful testing are limited. Although part of the sampling and measurement process can be tested with reference materials, the first real measurements from an environmental study usually cannot be subjected to more careful testing because there is little with which these measurements can be compared. In the case of multivariate observations, the start-up phase is even more important because it is prolonged by the need for enough data to obtain stable estimates of error properties.

The detection of out-of-control measurements in the start-up phase in SPC is closely related to the detection of outliers (Barnett and Lewis, 1984). In both cases, the measurement error that is inherent in the measurement procedure must be assessed so that evidence of additional, special causes of variation can be detected. This assessment is difficult because the measurements subject only to the inherent causes of variation cannot be distinguished unambiguously from those subject to special causes of variation. The aspects of outlier detection relevant to this section are a limited part of the far-ranging discussion provided by Barnett and Lewis (1984). First, when replicates are obtained from repetition of measurements made to satisfy the main objectives of the study, the mean of each group of replicates provides little help with SPC since little is available with which the mean can be compared. Thus, the s^2 control chart, to which the discussion in Barnett and Lewis (1984) of gamma distributed populations corresponds, is most relevant. Second, outlier detection methods are formulated in terms of the most discrepant observation in the entire data set. This is in contrast to the control limits discussed above which are determined from probabilities for a single group of r measurements. Outlier detection methods provide a critical point for the most discrepant observations and thus constrain the overall probability of a false indication of a measurement problem. For the start-up phase, this practice seems reasonable even though it is not the usual practice in SPC. Third, in SPC, the time order of the observation is considered an important source of information, whereas, outlier detection is usually considered in the context of random samples. Clearly, the time order in SPC observations should not be discarded. Fourth, Barnett and Lewis (1984) devote considerable space to the question of what model of the observations and what action is to be adopted if the null hypothesis in the test for outliers is rejected. In SPC, the action to be taken if outliers are detected is clear – a search for measurement procedure improvements. Fifth, because SPC does not require an estimate of the number of outliers, the problem of one outlier masking other outliers is simpler in SPC.

Consider SPC for the start-up phase based on replicates of a univariate measurement. Using the same notation as above, we denote the measurements by y_{ij}, where $i = 1, \ldots, m$ indexes the replicate group and $j = 1, \ldots, r_i$ indexes the measurement within the group. As above, the variance s_i^2 is estimated for each group. The quantity $(r_i - 1)s_i^2$ is gamma distributed with scale parameter $2\sigma^2$ and shape parameter $(r_i - 1)/2$. If the number of replicates is the same for every group, $r_i = r$, then the discordancy test for a single upper outlier in a gamma sample (Barnett and Lewis, 1984, p. 149) is appropriate. The test statistic is given by

$$s_{(m)}^2 / \sum_{i=1}^{m} s_i^2,$$

where $s_{(m)}^2$ is the largest of the variances. Barnett and Lewis (1984) provide a table of critical values for various values of m and shape parameter $(r - 1)/2$. Note that the performance of this test will be degraded if more than one group reflects a measurement problem and therefore has a variance larger than the rest.

For detection of error variance trends in the start-up phase, the following CUSUM-type method seems appropriate. Let the order in which the groups of replicates are observed be indexed by i, and let v_q denote the sequence of normalized partial sums

$$v_q = \sum_{i=1}^{q} (r_i - 1)s_i^2 \Big/ \sum_{i=1}^{m} (r_i - 1)s_i^2.$$

If there is an increasing trend in the variance then the sequence of partial sums will bow upwards. If there is a decreasing trend, then this sequence will bow downwards. Trend detection can be based on the statistic

$$c = \max_q \left| v_q - \sum_{i=1}^{q} (r_i - 1)/(N - m) \right|,$$

where

$$N = \sum_{i=1}^{m} r_i.$$

This statistic is related to the Kolmogorov–Smirnov test.

For $r_i = 3$, Durbin (1969), in the context of testing for serial correlation in time series, provides a table of critical values that are immediately applicable. The table of critical values requires that $(r_i - 1)s_i^2$ have a gamma distribution with shape parameter 1, which is the case if $r_i = 3$. If $r_i = 2$, then a similar statistic can be obtained by adding successive variances before forming the partial sums. Equivalently, in the case of m even, the maximization can be performed just over the even values of q. To proceed similarly in the case of m odd, the first group can be discarded without serious loss of information, it would seem. Thus, with available distributional results, we can cover the cases of both duplicates and triplicates, the most likely cases in SPC based on replicates.

Another important approach to the analysis of gamma distributed variates is the gamma probability plot (Gnanadesikan, 1977). On the basis of a probability plot, the possibility of multiple outliers that might result in masking can usually be discerned. However, as commonly used, gamma probability plots provide no guaranteed bound on the probability of a false alarm. Tests for multiple outliers among gamma-distributed variates are given by Barnett and Lewis (1984).

When as is often the case in environmental studies, sample collection, handling, and even some laboratory processing steps are common to the measurement of many constituents, multivariate SPC can be more sensitive to sample collection and handling problems (Ryan, 1989; Alt and Smith, 1988). In fact, configuration of the final measurement step to obtain additional results solely for SPC might be worthwhile. We denote the multivariate measurements by the p-dimensional column vector $\boldsymbol{y}_{ij}$, where as above i indexes the m groups and j indexes the r_i replicates within the group. The measurement on the k-th constituent, which is the k-th element of $\boldsymbol{y}_{ij}$, is denoted by y_{ijk}, where $k = 1, \ldots p$. SPC for the multivariate case is based on estimates of the covariance matrix given by

$$\boldsymbol{S}_i = \frac{1}{r_i - 1} \sum_{j=1}^{r_i} (\boldsymbol{y}_{ij} - \bar{\boldsymbol{y}}_i)(\boldsymbol{y}_{ij} - \bar{\boldsymbol{y}}_i)',$$

where

$$\bar{\boldsymbol{y}}_i = \frac{1}{r_i} \sum_{j=1}^{r_i} \boldsymbol{y}_{ij}.$$

The likelihood ratio criterion for testing whether there are no outliers against the alternative that the measurements in the i-th group are contaminated in unknown ways is given by the ratio of determinants

$$U_i = |\boldsymbol{A}^{(i)}| / |\boldsymbol{A}|, \tag{2.1}$$

where

$$\boldsymbol{A} = \sum_{i=1}^{m} (r_i - 1)\, \boldsymbol{S}_i, \qquad \boldsymbol{A}^{(i)} = \boldsymbol{A} - (r_i - 1)\boldsymbol{S}_i. \tag{2.2}$$

This criterion indicates the presence of a group containing outliers when its value is low. The test criterion for all m groups is the minium of U_i over i. This criterion is closely related to the criterion given by Wilks (1967) and discussed by Barnett and Lewis (1984). Barnett and Lewis (1984) show that the criterion given by Wilks is monotonically related to the Hotelling T^2 criterion usually considered in multivariate quality control (Alt and Smith, 1988). When $r_i = 2$, a similar relation exists in the case considered here.

For computation of appropriate critical points for min U_i, some results are available (Anderson, 1984). When $r_i = 2$, the ratio U_i follows the beta distribution Beta($(N - m - p)/2$, $p/2$). When $r_i = 3$, the ratio U_i is distributed as the product of two independent beta-distributed random variables, one with distribution Beta($(N - m - p)/2$, $p/2$) and the other with distribution Beta($(N - m - p - 1)/2$, $p/2$). To obtain critical points for min U_i, the Bonferroni inequality can be used. Consider the case in which all the r_i are the same, and let $\Pr(U_i < u)$ denote the probability that $U_i < u$, which can be computed from the beta distribution. The desired probability satisfies

$$\Pr(\min U_i < u) \leqslant m \Pr(U_i < u).$$

Thus, approximate critical points can be obtained. This is the approach followed by Wilks (1967).

Consider now the generalization of the test for trend discussed above. The multivariate generalization of this test is based on the partial sums over the groups in time order normalized by $\boldsymbol{A}^{-1}$

$$V_q = \mathrm{trace}\left[\sum_{i=1}^{q} (r_i - 1)\, \boldsymbol{S}_i \boldsymbol{A}^{-1}\right]. \tag{2.3}$$

We have chosen the trace but the determinant, the largest characteristic value, and perhaps other functions of the characteristic values are also possible. One argument for choice of the trace instead of the determinant is that the determinant may be too sensitive to the smallest characteristic values. The statistic V_q has the same form as the Bartlett–Nanda–Pillai trace test. As above, a test for trend can be based on the statistic

$$\max_q \left| V_q - p \sum_{i=1}^{q} (r_i - 1)/(N - m) \right|. \tag{2.4}$$

For a single value of i, tables of critical points for this statistic can be found in Anderson (1984). Perhaps the best approach to the computation of critical points for the maximum deviation shown in (2.4) is a Monte Carlo experiment. Such an experiment is not difficult since the statistic does not depend on the true covariance matrix. Thus, the experiment requires nothing more than the generation of independent normal variates with mean 0 and variance 1. From a set of Np such variates, values of $\boldsymbol{S}_i$, the value of $\boldsymbol{A}$, and finally the statistic (2.4) are computed. Independent repetitions of this provide the desired critical point.

Were the foregoing approach to SPC implemented in most environmental studies, replicates of high-level measurements would frequently, unnecessarily signal the need for corrective action. The reason is that for many measurement procedures, the variance of the measurement error increases with level. Consider the case in which the measurements are of nonnegative quantities and the measurements themselves are all nonnegative. In this case, one approach to this problem is use of the power transformation (Carroll and Ruppert, 1988). Before performing the SPC analysis, each measurement y_{ijk} is replaced by $h(y_{ijk}, \lambda_k, \delta_k)$, where

$$h(y_{ijk}, \lambda_k, \delta_k) = \begin{cases} [(y_{ijk} + \delta_k)^{\lambda_k} - 1]/\lambda_k & \text{if } \lambda_k \neq 0, \\ \log(y_{ijk} + \delta_k) & \text{if } \lambda_k = 0. \end{cases}$$

Note that a shift δ_k has been included. If the error in y_{ijk} is small compared to $\mu_{ik} + \delta_k$, where μ_{ik} is the actual level of which y_{ijk} is a measurement, then, as discussed by Carroll and Ruppert (1988),

$$\mathrm{Var}(h(y_{ijk}, \lambda_k, \delta_k)) \simeq (\mu_{ik} + \delta_k)^{2(\lambda_k - 1)} \mathrm{Var}(y_{ijk} - \mu_{ik}).$$

This shows that the transformation compensates for the change in error variance through a scale factor that decreases with level if $\lambda_k < 1$. The shift δ_k is important because in many cases the slope of the variance versus level curve decreases as the level approaches the limits on measurement imposed by the noise in the instrument. If $\delta_k > 0$, then the shift provides the appropriate behavior. Note that we allow the transformation parameters to depend on k so that they may differ from one element of the multivariate measurement to another.

Before discussing power transformations in SPC, we must consider estimation of the transformation parameters. The estimation method we choose is maximum likelihood based on the assumption that after transformation, the differences among replicates are normally distributed with the same variance regardless of the measurement level. Let $\boldsymbol{h}(\boldsymbol{y}_{ij})$ denote the transformation of $\boldsymbol{y}_{ij}$ so that $\boldsymbol{h}(\boldsymbol{y}_{ij})$ has components $h(y_{ijk}, \lambda_k, \delta_k)$. We assume that

$$\boldsymbol{h}(\boldsymbol{y}_{ij}) = \boldsymbol{h}(\boldsymbol{\mu}_i) + \boldsymbol{\varepsilon}_{ij},$$

where $\boldsymbol{\varepsilon}_{ij}$ is normal with mean 0 and covariance matrix $\boldsymbol{\Sigma}$. As discussed by Atkinson (1985), a difficulty arises in the estimation of the shift by maximum likelihood because the likelihood function increases without limit as the shift decreases toward the lowest measurement. In the case of nonnegative measurements, we can remedy this problem in part by requiring that the shift be nonnegative or if necessary, positive. This is reasonable in light of the reason for inclusion of the shift. The requirement that the shift be nonnegative leaves the possibility that the likelihood function will have two maxima the relative size of which depends on how small a shift is allowed. In practice, assessment of the contribution of the instrument noise to the error in low level measurements should provide a basis for setting a minimum level for the shift.

The likelihood function used in the estimation of the transformation parameters is obtained as in Carroll and Ruppert (1988) from the probability density function for

y_{ij}, which is

$$(2\pi)^{-p/2}|\boldsymbol{\Sigma}|^{-1/2}\exp\{\tfrac{1}{2}[\boldsymbol{h}(\boldsymbol{y}_{ij})-\boldsymbol{h}(\boldsymbol{\mu}_i)]'\boldsymbol{\Sigma}^{-1}[\boldsymbol{h}(\boldsymbol{y}_{ij})-\boldsymbol{h}(\boldsymbol{\mu}_i)]\}$$
$$\times\prod_{k=1}^{p}[y_{ijk}+\delta_k]^{\lambda_k-1}.$$

Let

$$\boldsymbol{S}_i=\frac{1}{r_i-1}\sum_{j=1}^{r_i}(\boldsymbol{h}(\boldsymbol{y}_{ij})-\bar{\boldsymbol{h}}_i)(\boldsymbol{h}(\boldsymbol{y}_{ij})-\bar{\boldsymbol{h}}_i)',$$

where

$$\bar{\boldsymbol{h}}_i=\frac{1}{r_i}\sum_{j=1}^{r_i}\boldsymbol{h}(\boldsymbol{y}_{ij});$$

and let $\boldsymbol{A}$ and $\boldsymbol{A}^{(i)}$ be defined as in (2.2). After substitution of the maximum likelihood estimates for $\boldsymbol{h}(\boldsymbol{\mu}_i)$ and $\boldsymbol{\Sigma}$, the log likelihood function up to an additive constant becomes

$$\begin{aligned}L_{\max}&=-\frac{1}{2}N\log|\boldsymbol{A}|+\sum_{i=1}^{m}\sum_{j=1}^{r_i}\sum_{k=1}^{p}(\lambda_k-1)\log(y_{ijk}+\delta_k)\\&=\sum_{k=1}^{p}\left\{-\frac{1}{2}N\log(a_{kk})+\sum_{i=1}^{m}\sum_{j=1}^{r_i}(\lambda_k-1)\log(y_{ijk}+\delta_k)\right\}\\&\quad-\frac{1}{2}N\log|\hat{\boldsymbol{\rho}}|,\end{aligned}\tag{2.5}$$

where $\hat{\boldsymbol{\rho}}$ is the correlation matrix estimated from $\boldsymbol{A}$. Note that only the term on the third line distinguishes this log likelihood from the sum of the univariate log likelihoods.

Maximization of this log likelihood can be performed with a general purpose function minimization routine such as the one published by Dennis and Schnabel (1983). In order to restrict the shift δ_k to positive values, we parameterize it as the square of a real number. In function minimization, the computation of first derivatives and even second derivatives is useful. In this case, only computation of the derivatives of $|\boldsymbol{A}|$ is not straight-forward. These derivatives can be obtained from the following formulas. Note that $\boldsymbol{A}$ has the form

$$\boldsymbol{A}=\sum_{i=1}^{m}\sum_{j=1}^{r_i}\boldsymbol{x}_{ij}\boldsymbol{x}_{ij}',$$

where

$$x_{ijk}=h(y_{ijk},\lambda_k,\delta_k)-\frac{1}{r_i}\sum_{j=1}^{r_i}h(y_{ijk},\lambda_k,\delta_k).$$

The transformation parameters λ_k and δ_k appear only in the k-th element of $\boldsymbol{x}$ and thus, in $\boldsymbol{A}$, only in row k and column k. Let

$$\boldsymbol{x}_{ij}^{(\alpha_k)} = \left(x_{ij1}, \ldots, x_{ijk-1}, \frac{\partial x_{ijk}}{\partial \alpha_k}, x_{ijk+1}, \ldots, x_{ijp} \right)',$$

where α_k is either λ_k or δ_k. Note that $\boldsymbol{x}_{ij}^{(\alpha_k)}$ differs from $\boldsymbol{x}_{ij}$ only in the k-th element. We obtain

$$\frac{\partial |\boldsymbol{A}|}{\partial \alpha_k} = 2 \left| \sum_{i=1}^{m} \sum_{j=1}^{r_i} \boldsymbol{x}_{ij}^{(\alpha_k)} \boldsymbol{x}_{ij}' \right|.$$

Similarly, we let

$$\boldsymbol{x}_{ij}^{(\alpha_k, \beta_k)} = \left(x_{ij1}, \ldots, x_{ijk-1}, \frac{\partial^2 x_{ijk}}{\partial \alpha_k \partial \beta_k}, x_{ijk+1}, \ldots, x_{ijp} \right)'$$

and for $k \neq h$, we let

$$\boldsymbol{x}_{ij}^{(\alpha_k, \beta_h)} = \left(x_{ij1}, \ldots, \frac{\partial x_{ijk}}{\partial \alpha_k}, \ldots, \frac{\partial x_{ijh}}{\partial \beta_h}, \ldots, x_{ijp} \right)',$$

where α_k is either λ_k or δ_k and β_h is either λ_h or δ_h. In either case, we obtain

$$\frac{\partial^2 |\boldsymbol{A}|}{\partial \alpha_k \partial \beta_h} = 2 \left| \sum_{i=1}^{m} \sum_{j=1}^{r_i} [\boldsymbol{x}_{ij}^{(\alpha_k, \beta_h)}][\boldsymbol{x}_{ij}]' \right| + 2 \left| \sum_{i=1}^{m} \sum_{j=1}^{r_i} [\boldsymbol{x}_{ij}^{(\alpha_k)}][\boldsymbol{x}_{ij}^{(\beta_h)}]' \right|.$$

In application of the above estimation procedure, one can easily come upon a case in which the estimates turn out to be a large negative value for λ_k and a large positive value for δ_k. Over the limited range of levels for which measurements are present in the set of replicates, such estimates may fit reasonably well even though extrapolation beyond the range may be unreasonable. For this reason, the SPC method suggested here may always have to be considered to be in the start-up phase. Adoption of the estimated transformation parameters for the ongoing phase might lead to false alarms whenever new replicate measurements are made at levels beyond those represented in the start-up phase. If the basis for an ongoing phase were required, then perhaps some restrictions on the estimates of the transformation parameters such as the restriction that λ lie between 0 and 1 would result in estimates that provide better accuracy when the transformation is extrapolated.

Note that even though we have adopted separate transformations for each component, we have presented a multivariate estimation procedure. Insight into the difference between the multivariate estimate and separate univariate estimates can be obtained from (2.5). This equation shows that in the maximization of $L_{\max}$, the maxima for the univariate likelihoods are modified so that $|\hat{\boldsymbol{\rho}}|$ is made smaller. In a sense, minimization of $|\hat{\boldsymbol{\rho}}|$ corresponds to maximization of the correlations among the transformed error components. Thus, when the error components are correlated as might be the case when the error arises in sample collection and handling procedures, better estimates of the transformation parameters may be obtained from the multivariate method. This can be confirmed through a Monte Carlo experiment.

We are now ready to include power transformations in SPC. Consider first the generalization of (2.1). In analogy with the derivation of (2.1), we form a likelihood ratio statistic from the likelihood functions used in estimation of the transformation parameters. Note that the ratio involves two separate estimates of the transformation parameters, one for the numerator and the other for the denominator. This seems necessary since if the group of replicates in question is an outlier, then we would not want it judged solely in terms of the transformation parameters estimated with the outlier included. On the other hand, removing a group of measurements made at a level near the extremes of the range could make the estimated transformation parameters very inaccurate for the excluded group and thus inappropriate for judging the group. The statistic we consider is

$$U_i = \frac{\min_{\lambda_1,\delta_1,\ldots,\lambda_p,\delta_p} |A^{(i)}| \left\{ \prod_{q=1}^{m} \prod_{j=1}^{r_q} \prod_{k=1}^{p} [y_{qjk} + \delta_k]^{\lambda_k - 1} \right\}^{-2/N}}{\min_{\lambda_1,\delta_1,\ldots,\lambda_p,\delta_p} |A| \left\{ \prod_{q=1}^{m} \prod_{j=1}^{r_q} \prod_{k=1}^{p} [y_{qjk} + \delta_k]^{\lambda_k - 1} \right\}^{-2/N}}. \tag{2.6}$$

Computation of this statistic requires the minimization algorithm discussed above.

Questions remain as to how best to obtain a critical point for the minimum over i of this statistic. Use of the critical point for the statistic in (2.1), which does not involve transformation, would seem to generally give a higher probability of false alarm than intended. Some indication of how much higher could be investigated by a Monte Carlo experiment. The proper critical point depends on the transformation parameters, the actual levels measured and the covariance matrix of the properly transformed measurements. These could be estimated from the data, and the critical point determined by a Monte Carlo experiment. This approach requires further investigation.

In the generalization of the test for a trend in the covariance structure of the replicates in (2.3) and (2.4), the use of only the transformation based on all the data seems reasonable. In the multivariate trend test discussed above, we merely substitute measurements transformed according to the maximum likelihood estimates of the transformation parameters. As above, the best approach to the determination of critical points has yet to be decided.

Diagnostic plots for the transformation parameters of each error component are important (Carroll and Ruppert, 1988; Atkinson, 1985). Ruppert and Aldershof (1989) suggest consideration of the squared residuals plotted versus the log of the fitted values. In the case considered here, a corresponding plot would be s_{ikk}, the k-th diagonal element of $\boldsymbol{S}_i$ versus $\log(\hat{\mu}_{ik})$, where $\hat{\mu}_{ik}$ satisfies

$$h(\hat{\mu}_{ik}, \lambda_k, \delta_k) = \frac{1}{r_i} \sum_{j=1}^{r_i} h(y_{ijk}, \lambda_k, \delta_k).$$

Note that if the transformation does indeed give independent, identically distributed normal measurement errors, then s_{ikk} is independent of $\hat{\mu}_{ik\cdot}$. This independence property should be helpful in interpreting such plots and can even be used to create tests and confidence intervals corresponding to such plots.

2.3. Example from Eastern Lake Survey

The Eastern Lake Survey, the first major field component of the National Acid Precipitation Assessment Program (Linthurst et al., 1986), collected duplicate samples and introduced blanks and audit samples as part of the quality assurance program. To illustrate the SPC method presented above, we apply it to the nitrate and sulfate measurements made by one of the analytical laboratories. Early in the Eastern Lake Survey, a problem with a laboratory procedure resulted in nitrate contamination. We consider the duplicates collected after that problem was solved. The measurements we consider are given in Table 1. We see that the nitrate measurements appear to range over three orders of magnitude and the sulfate measurements less.

The control chart based on equation (2.6) is shown in Figure 1. Recall that for the statistic chosen, outliers are indicated by values that are too small. A control limit for this chart is obtained from Wilks (1967, Table 2). By matching the beta distribution

Table 1
Duplicate concentrations (mg/L) from the Eastern Lake Survey

Index	Nitrate		Sulfate	
	Routine	Duplicate	Routine	Duplicate
1	0.026	0.018	5.560	5.630
2	0.857	0.844	5.053	5.004
3	0.017	0.015	2.692	2.689
4	0.007	0.007	4.390	4.415
5	0.092	0.093	6.250	6.247
6	0.021	0.021	4.763	4.772
7	0.015	0.018	9.229	9.192
8	0.001	0.001	2.916	2.964
9	0.010	0.009	3.351	3.441
10	0.012	0.013	7.456	7.450
11	1.145	1.175	6.272	6.130
12	1.057	1.043	10.601	10.433
13	0.036	0.036	4.880	4.826
14	0.004	0.004	6.738	6.722
15	0.070	0.072	7.284	7.355
16	1.010	1.022	6.923	7.087
17	0.059	0.018	5.158	5.236
18	0.137	0.153	11.013	10.921
19	0.035	0.043	7.066	7.048
20	0.122	0.117	5.824	5.771
21	0.036	0.040	14.710	14.594

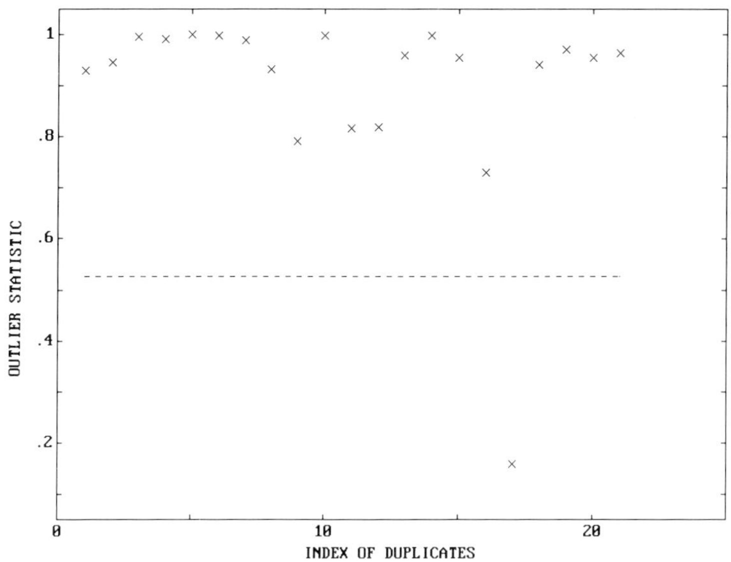

Fig. 1. Nitrate–sulfate control chart.

underlying this table with the beta distribution given for U_i (equation (2.1)), we select a table entry even though this is not exactly correct even for the statistic without the transformation. We see that one value is below the control limit and thus should lead to an investigation. The problem with this value seems to be in the chemical characteristics of the lake sampled. This lake has very high turbidity which indicates unusual chemical characteristics. Various chemical processes associated with the character of this lake could be the reason why the duplicate nitrate values shown in Table 1 are so discrepant. Although this lake is unusual, one can still, of course, ask whether the sample collection and handling procedure could be improved so that it is adequate for the sampling of lakes with such unusual characteristics.

From the point of view of illustrating what can happen with the SPC method presented here, these data give some other interesting results. The transformation parameters estimated from the entire data set are (0.64, 0.0) for nitrate and (0.01, 1.5) for sulfate. When the outlier is excluded, the estimates become (0.40, 0) and (0.34, 0). Were the transformation extrapolated beyond the measurement levels for which observations are available, these two sets of estimates would be quite different. This suggests that the available measurements are inadequate for obtaining estimates stable enough for the ongoing phase of SPC. Moreover, we note that in computing U_{16}, the transformation parameters (λ_2, δ_2) for the numerator turned out to be -40, the most negative value the program used allowed, and 20.4. Despite these estimates, the statistic for outlier detection did not signal an outlier. The reason is that in this case, the likelihood function for the transformation parameters changes little with a large change in the transformation parameters. This suggests that the above SPC method is not necessarily limited by a need for enough measurements to obtain accurate estimates of the transformation parameters.

Diagnostic plots for the transformation parameters obtained when the outlier is excluded are shown in Figures 2 and 3. As discussed above, these are s_{ikk} versus $\log(\hat{\mu}_{ik})$. These figures look different because of the outlier in Figure 2, which might

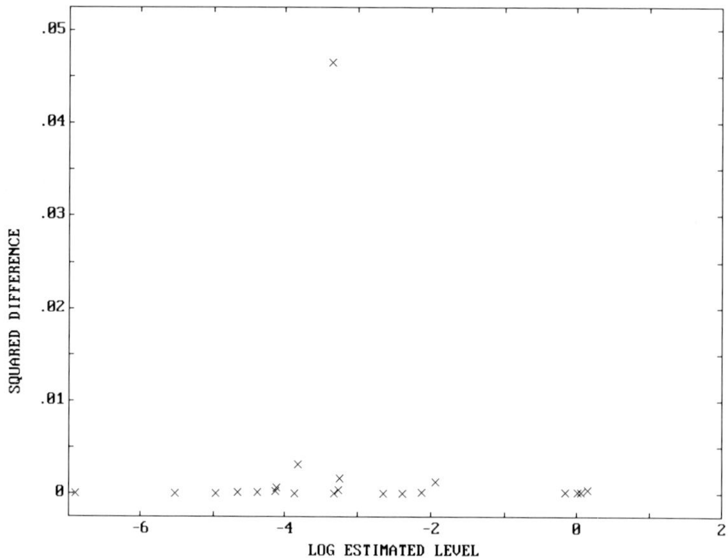

Fig. 2. Nitrate transformation diagnostic plot.

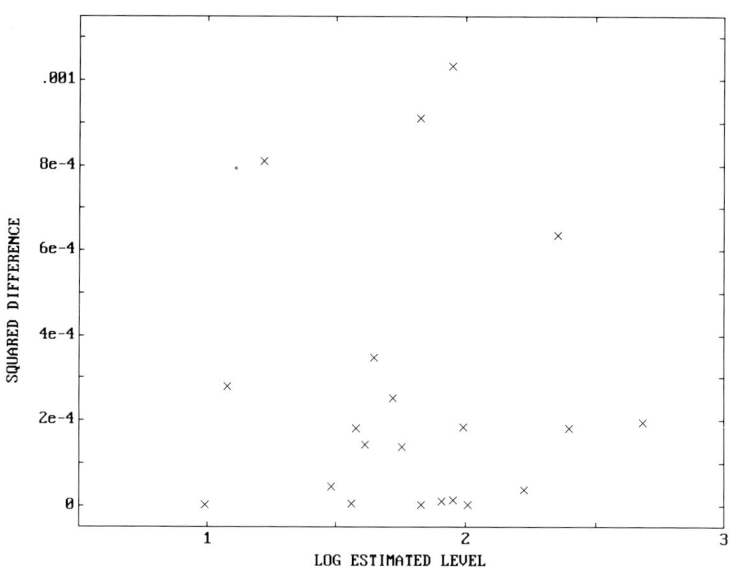

Fig. 3. Sulfate transformation diagnostic plot.

be removed to obtain a better look at the other points. Neither figure shows a problem with the estimate of the transformation parameters.

2.4. *Error components*

The foregoing development ignores the fact that sometimes it is necessary to consider the various sources of variation, that is, components of the measurement error, separately. Potentially, every step in the sample collection, handling, and measurement procedure has a separate error component. Moreover, some sources of variation may make the same contribution to several measurements. For example, laboratory procedures may process samples in batches with each batch run under the same instrument calibration curve. The Eastern Lake Survey measurements were run in batches with each duplicate pair considered above run in the same batch. Thus, a batch-to-batch error cannot be assessed from the duplicates (Liggett, 1986). This complication effects how study conclusions are drawn, and how measurement problems are diagnosed.

One aspect of measurement problem diagnosis is the need to make a measurement procedure work properly one step at a time. Smith and James (1981) discuss this. They consider a procedure that involves subsampling and instrumental analysis. Clearly, one should have the instrument performing properly before starting experiments on the subsampling procedure.

A further complication of multiple error components in environmental measurements arises when more than one component has variance that increases with level (Liggett, 1988). Specification of replicates in this case requires considerable care.

Finally, we note that the managers of some environmental studies employ more than one analytical laboratory. Although there may be advantages to this, the use of many analytical laboratories makes the SPC task more difficult since laboratories not only have different biases but also have different error variances and different relations between error variance and measured level. The SPC method presented here could not be implemented if the replicate measurements were from many laboratories. Separate SPC analyses would have to be done for each laboratory. One problem is that separate SPC analyses would reduce the sensitivity to sample collection and handling problems common to all laboratories.

3. Lack-of-fit tests

3.1. *Sources of variation*

In addition to measurement quality control, replicates can be used to test hypothesized environmental models. Consider the case of univariate measurements believed to satisfy $y_{ij} = f(\boldsymbol{x}_i, \boldsymbol{\beta}) + \varepsilon_{ij}$ where $\boldsymbol{x}_i$ is a known vector that characterizes the circumstances under which measurement (i, j) is made and $\boldsymbol{\beta}$ is a vector of unknown parameters. As above, replicate measurements are indexed by j. In this section, we present statistical methods for detecting lack of fit of the model $f(\boldsymbol{x}_i, \boldsymbol{\beta})$. Such lack of fit becomes apparent in the residuals $y_{ij} - f(\boldsymbol{x}_i, \hat{\boldsymbol{\beta}})$, where $\hat{\boldsymbol{\beta}}$ is an appropriate estimate of $\boldsymbol{\beta}$. The methods for testing lack of fit considered here are based on comparison of $y_{ij} - f(\boldsymbol{x}_i, \hat{\boldsymbol{\beta}})$ with characteristics of the residuals that would be expected were $f(\boldsymbol{x}_i, \boldsymbol{\beta})$ a valid model and were the probability properties of ε_{ij} like those estimated from the replicates. Since the lack-of-fit tests to be considered require adequate characterization of ε_{ij} on the basis of the replicates, we begin by discussing how to meet this requirement.

Sometimes the required replicates can be obtained by moving the point where the replication begins to the earliest point practical in the measurement process. For example, in bulk sampling, duplicate subsamples extracted after mixing might be replaced by repetition of the entire sample collection process including the mixing. In ground water sampling, division of a single sample pumped from the well might be replaced by a second visit to the sampling site a few days later at which time the entire process of preparing the well for sampling would be repeated. In these examples, more complete replication adds a potentially important component to the error characterization.

Sometimes the quantities that are to be the focus of the statistical analysis can be chosen so that the required replicates can be obtained. For example, consider the concentrations of heavy metals in the particulate fraction of samples from the Chesapeake Bay as reported by Kingston et al. (1983) and Kingston and Greenberg (1984). As observed in this study, the concentrations, each of which is the elemental mass of some metal in the particles in a unit volume of Bay water, are influenced by many factors and vary over more than two orders of magnitude. Moreover, although the concentrations may appear random, they are influenced by factors such as currents, tides, wind, and runoff from storms and thus cannot be properly replicated. Interestingly, there are some metals that occur with nearly constant relative abundance, the natural crustal abundance for the geographical area being studied. These are metals that are generally not used in refined form by industry, do not have variable natural abundance over the geographical region, and do not shift from the dissolved to the particulate phase as the result of chemical processes in the water. These metals can be used for normalization in the investigation of how other metals that are partly the result of anthropogenic sources vary within the Bay. Thus, concentration is replaced by the ratio of concentrations in the statistical analysis.

The metals that are hypothesized to occur with nearly constant abundance are scandium, thorium, and perhaps iron. Iron is, of course, refined but also occurs in such high concentrations naturally that it occurs at nearly constant relative concentration. Since iron is also subject to shifts from dissolved to particulate in the deeper parts of the Bay, we consider only samples collected near the surface. Comparison of iron, scandium, and thorium provide a basis for the claim that such normalization is appropriate. Let y_{ijk} denote the measurement of the k-th element ($k = 1, 2, 3$ for iron, scandium, and thorium, respectively) in the j-th replicate of the i-th group. Consider the model $\log(y_{ijk}) = \beta_k + \gamma_{ij} + \varepsilon_{ijk}$, where the term γ_{ij} represents the sample to sample variation in the amount of inorganic particulate, that is, particulate with heavy metal content. If we think of the last two terms of this model as the error term, we see that the entire error cannot be properly replicated. However, the contribution of the error to differences among the $\log(y_{ijk})$, $k = 1, 2, 3$,

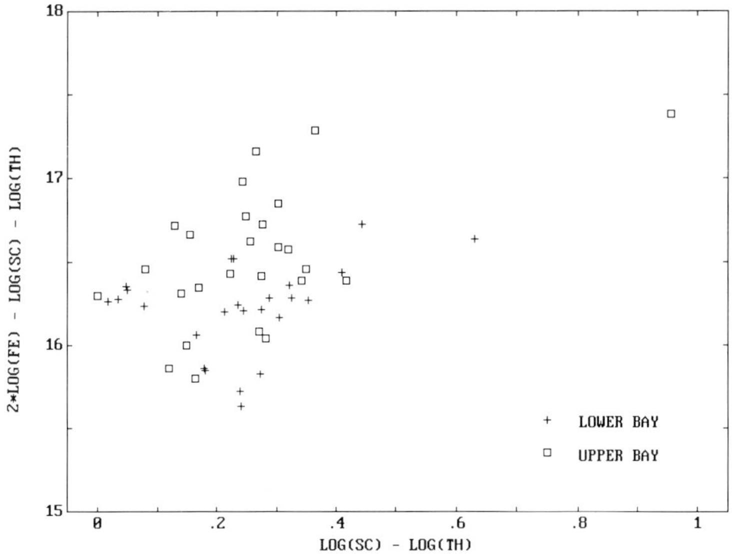

Fig. 4. Iron (Fe), scandium (Sc), and thorium (Th) in particles suspended in the Chesapeake Bay.

does not involve the γ_{ij} and thus might be properly replicated. Consider the hypothesis that the relative concentrations of iron, scandium, and thorium are the same throughout the Chesapeake Bay against the alternative that the upper half of the Bay differs from the lower half. We plot $2\log(y_{ij1}) - \log(y_{ij2}) - \log(y_{ij3})$ versus $\log(y_{ij2}) - \log(y_{ij3})$ using two symbols to distinguish the two parts of the Bay. The result is shown in Figure 4. We see that there is little to indicate that the relative concentrations in the two parts are different although one might argue that there is some indication of a difference in iron. Moreover, the spread of the measurements is nearly what would be expected from the measurement error (Kingston et al., 1983). Thus, observed concentrations when taken in combinations that cancel the contribution of the γ_{ij} seem to be replicates. That this holds for such a large geographic region suggests that studies of smaller regions can be based on properly selected ratios of concentrations.

The implications of this normalization for statistical inference is simply shown through the comparison of two locations, $i = 1, 2$, on the basis of replicate samples. Consider concentration measurements for a fourth metal y_{ij4} made along with measurements of scandium y_{ij2}. A one-way analysis of variance applied to $\log(y_{ij4})$ would be meaningless because the many causes for differences between the locations cannot be distinguished. In part, this results from inability to specify replicates that characterize the uninteresting causes. On the other hand, the Chesapeake measurements suggest that one way analysis of variance could be applied to $\log(y_{ij4}) - \log(y_{ij2})$ to test the hypothesis that the particulate fraction at both locations is from the same natural source. This example is a simple application of a class of environmental models called source apportionment models (Currie, 1992; Hopke, 1985).

More complicated studies based on these models might also be designed so that replicates can be used to test the model for lack of fit.

3.2. Replicates density estimate

Consider univariate measurements that satisfy

$$y_{ij} = \mu_i + \varepsilon_{ij}, \quad i = 1, \ldots, m, \ j = 1, \ldots, r_i,$$

where the error ε_{ij} is independently and identically distributed with mean zero. In this subsection, we present an algorithm for estimation of the distribution of the ε_{ij}. The derivation of this algorithm is given in Liggett (1989). The algorithm requires that $r_i \geqslant 3$ for several values of i but requires no assumptions about the μ_i. The μ_i might be $f(\boldsymbol{x}_i, \boldsymbol{\beta})$ as above or might themselves be random variates drawn from a distribution with unknown parameters. The density estimate for the ε_{ij} can be used for inference on the model assumed for the μ_i.

For reasons given in Liggett (1989), we select as the form for the density estimate a form based on Hermite functions. Hermite functions, denoted $\phi_q(x)$, are products of Hermite polynomials and $\exp(-x^2/2)$ and are scaled so that

$$\int_{-\infty}^{\infty} \phi_p(x)\phi_q(x)\,\mathrm{d}x = \begin{cases} 1 & \text{if } p = q, \\ 0 & \text{if } p \neq q. \end{cases}$$

Most important for computations based on the estimated density is the recursive definition given by

$$\phi_0(x) = \pi^{-1/4} \exp(-x^2/2), \qquad \phi_1(x) = 2^{1/2}\pi^{-1/4}\, x \exp(-x^2/2),$$
$$\phi_q(x) = \left(\tfrac{2}{q}\right)^{1/2} x\phi_{q-1}(x) - \left(\tfrac{q-1}{q}\right)^{1/2} \phi_{q-2}(x).$$

So that a density estimate with a moderate number of terms will suffice, we scale the observations based on an estimate of spread. This estimate is based on absolute (first) differences between replicates $|y_{ij} - y_{ik}|$, where $j < k$. The estimate is based on the median of these absolute differences of which there are $\sum_i r_i(r_i - 1)/2$. The estimate of spread is given by

$$s_r = (0.6745)^{-1}\,(2)^{-1/2} \operatorname{median} |y_{ij} - y_{ik}| \quad (1 \leqslant j < k \leqslant r_i \ \ i = 1, \ldots, m).$$

The use of the median makes s_r insensitive to large outliers.

The form of the density selected is

$$p(x) = s_r^{-1} \left| \sum_{q=0}^{Q/2} z_q \phi_q \{(x + \alpha)/(\sqrt{2}s_r)\} \right|^2, \tag{3.1}$$

where Q is even, z_0 is real, and z_q $(q = 1, \ldots, Q/2)$ is complex. As discussed below, the values of z_q are initially determined by nonlinear least squares and then scaled so that the density integrates to 1. The value of α is determined so that the density has mean 0.

The form for the density in (3.1) never gives negative values, which is an advantage over the usual Hermite function expansion (Schwartz, 1967). The form in (3.1) can be rewritten in the form of a Hermite function expansion, however. We obtain

$$p(x) = s_r^{-1} \sum_{q=0}^{Q} a_q \phi_q\{(x+\alpha)/s_r\}, \tag{3.2}$$

where

$$a_q = \sum_{j=\max(0,q-Q/2)}^{Q/2} \sum_{k=\max(0,q-j)}^{Q/2} (z_j^{(r)} z_k^{(r)} + z_j^{(i)} z_k^{(i)}) B_d(j,k,q) \tag{3.3}$$

and $z_j^{(r)}$ and $z_j^{(i)}$ are the real and imaginary parts of z_j. The coefficients B_d, and also the coefficients B_c which appear below, arise from Hermite function expansions of products of Hermite functions (Liggett, 1989). From this context, we see that we can define $B_d(j,k,q)=0$ if $j+k+q$ is odd or if $j+k<q$. Moreover, since $B_d(j,k,q)=B_d(k,j,q)$, we can specify $B_d(j,k,q)$ just for $j \leqslant k$. We have for $j \leqslant k$

$$B_d(j,k,q) = 2^{q/2-j-k}\left(\frac{q!}{j!k!}\right)^{1/2} \frac{(j+k-q)!}{\{(j+k-q)/2\}!}\pi^{-1/4}$$
$$\times \sum_{n=\max(0,(k-j-q)/2)}^{\min([(k-j)/2],(k+j-q)/2)} (-1)^n \binom{j}{(j+k-q-2n)/2}\binom{k-j}{2n},$$

where in the limits on sum, $[x]$ denotes the greatest integer less than or equal to x.

From any set of values for the z_q and in particular from those that come from the nonlinear least squares fitting described below, we can obtain a proper density (one that integrates to 1) that has zero mean. The values of $\hat{a}_q$ for the proper density are given by

$$\hat{a}_q = a_q \Big/ \left\{ (2\pi)^{1/2} \sum_{n=0}^{Q/2} (-1)^n a_{2n} \phi_{2n}(0) \right\}.$$

Scaling z_q by the square root of the above factor gives the corresponding $\hat{z}_q$. The value for α is given by

$$\alpha = (2\pi)^{1/2} s_r \sum_{n=0}^{Q/2-1} (-1)^n \hat{a}_{2n+1} \{(n+\tfrac{1}{2})^{1/2}\phi_{2n}(0) - (n+1)^{1/2}\phi_{2n+2}(0)\}.$$

We denote the proper zero-mean density by $\hat{p}(x)$.

One way to make use of the estimated density $\hat{p}(x)$ is to generate random numbers from $\hat{p}(x)$ and use these in Monte Carlo experiments, that is, use these as the basis for a bootstrap method. To generate random numbers, we generate uniform (0, 1) variates U^* and solve

$$U^* = \int_{-\infty}^{E^*} \hat{p}(x)\, \mathrm{d}x$$

for E^*. This requires a root finding algorithm and a way to integrate $\hat{p}(x)$. The integration is conveniently performed through the recursion

$$\int_b^a \phi_q(x)\, \mathrm{d}x = (2/q)^{1/2}\{\phi_{q-1}(a) - \phi_{q-1}(b)\} + \{(q-1)/q\}^{1/2} \int_a^b \phi_{q-2}(x)\, \mathrm{d}x.$$

When many independent realizations of E^* are needed, the iterations in the root finding can be reduced by ordering the realizations of U^*, starting each root finding iteration from the result of the previous iteration, and finally putting the realizations of E^* in the original order of the realizations of U^*.

The nonlinear least squares approach to the estimation of the z_q adopted here is formulated in terms of the coefficients of Hermite function expansions of the densities of the first and second differences between replicates, the densities of $\varepsilon_{ij} - \varepsilon_{ij'}$, where $j \neq j'$, and $2\varepsilon_{ih} - \varepsilon_{ij} - \varepsilon_{ij'}$, where $h \neq j \neq j' \neq h$. The estimates of these coefficients are given by

$$\hat{d}_{2q} = (N-m)^{-1} \sum_{i=1, r_i>1}^{m} \frac{2}{r_i} \sum_{j=2}^{r_i} \sum_{j'=1}^{j-1} \phi_{2q}\{(y_{ij} - y_{ij'})/(\sqrt{2}s_r)\} \tag{3.4}$$

and

$$\hat{c}_q = (N-m)^{-1} \sum_{\substack{i=1,\\ r_i>2}}^{m} \frac{2}{r_i(r_i-2)} \sum_{h=1}^{r_i} \sum_{\substack{j=2,\\ j\neq h}}^{r_i} \sum_{\substack{j'=1,\\ j'\neq h}}^{j-1} \phi_q\{(2y_{ih} - y_{ij} - y_{ij'})/(\sqrt{6}s_r)\}, \tag{3.5}$$

where $N = \sum_i r_i$. These coefficients are given in terms of the parameters that describe the density estimate by

$$d_{2q} = (2\pi)^{1/2} \sum_{j=\max(0,2q-Q)}^{Q} \sum_{k=\max(0,2q-j)}^{Q} (-1)^{(j-k)/2-q} a_j a_k B_d(j,k,2q) \tag{3.6}$$

and

$$c_q = 2\pi \sum_{j=\max(0,q-2Q)}^{Q} \sum_{k=\max(0,q-j-Q)}^{Q} \sum_{p=\max(0,q-j-k)}^{Q} (-1)^{(p-j-k-q)/2}$$

$$\times\, a_j a_k a_p B_c(j,k,p,q), \tag{3.7}$$

where the coefficient $B_c(j,k,p,q)$ is 0 if $j+k+p+q$ is odd or if $j+k+p<q$, satisfies $B_c(j,k,p,q) = B_c(k,j,p,q)$, and for $j \leqslant k$ is given by

$$B_c(j,k,p,q) = 2^{q/2-j-k}3^{-q/2}\left(\frac{q!}{j!k!\,p!}\right)^{1/2}\pi^{-1/2}$$

$$\times \sum_{h=0}^{\min(j,L)} \sum_{i=0}^{\min([(k-j)/2],L-h)} \sum_{n=\max(0,[(p-q)/2])}^{\min([p/2],L-h-i)} (-1)^{L-h}$$

$$\times 2^{L-h-i-2n} 3^{h+i-L}$$

$$\times \binom{j}{h}\binom{k-j}{2i}\binom{p}{2n}\binom{j+k+1-2h-2i}{2L-2h-2i-2n}$$

$$\times \frac{\{2(h+i)\}!}{(h+i)!} \frac{\{2(L-h-i)\}!}{(L-h-i)!}$$

$$\times \left\{\frac{j+k+1-2(L-n)}{j+k+1-2(h+i)} - \binom{p-2n}{2n+1}\left(\frac{L-h-i-n}{j+k+1-2(h+i)}\right)\right\},$$

where $L=(j+k+p-q)/2$. The defferences $\hat{d}_{2q}-d_{2q}, q=0,\ldots,Q$ and $\hat{c}_q-c_q$, $q=0,\ldots,3Q$ could be used as they stand for nonlinear least squares fitting with an algorithm that does not require derivatives. However, we have chosen to compute derivatives and use a weight matrix.

Equations (3.4) and (3.5) show that the $\hat{d}_{2q}$ and $\hat{c}_q$ are correlated and that the use of a properly chosen weight matrix will give better estimates than unweighted least squares. The optimum weight matrix is the inverse of the covariance matrix for $\hat{d}_{2q}$ and $\hat{c}_q$, but this depends on the unknown density that is to be estimated. The approach of using a preliminary estimate of the density to estimate the needed covariance matrix is possible although perhaps computationally burdensome. Instead, we base our weight matrix on computation of the covariance matrix for the case in which ε_{ij} is normally distributed with mean 0 and standard deviation s_r. Equations (3.4) and (3.5) show that $\hat{d}_{2q}$ and $\hat{c}_q$ are linear combinations of $\phi_q(x)$ and therefore that the covariance matrix for $\hat{d}_{2q}$ and $\hat{c}_q$ can easily be obtained once the covariance between $\phi_p(x_1)$ and $\phi_q(x_2)$ is obtained. Note that under the assumption under which we are computing the weight matrix, x_1 and x_2 are normal with mean 0 and variance 1. The correlation $\rho = \mathrm{E}(x_1x_2)$ depends on which of the first or second differences x_1 and x_2 represent. When $p+q$ is odd, the covariance between $\phi_p(x_1)$ and $\phi_q(x_2)$ is 0. When $p+q$ is even and $p \leqslant q$, the covariance is given by

$$\mathrm{E}\{\phi_q(x_1)\} = \mathrm{E}\{\phi_q(x_2)\} = \begin{cases} \pi^{-1/4}2^{-1/2} & \text{if } q=0, \\ 0 & \text{if } q \neq 0, \end{cases}$$

$$\mathrm{E}\{\phi_p(x_1)\phi_q(x_2)\} = 2^{-p-q}(p!q!)^{-1/2}\pi^{-1/2}\{4-\rho^2\}^{-1/2}$$

$$\times \sum_{i=0}^{p} \sum_{j=0}^{(q-p)/2} (-1)^{p-i}\binom{p}{i}\binom{q-p}{2j}$$

$$\times \frac{(2i+2j)!}{(i+j)!} \frac{(p+q-2i-2j)!}{\{(p+q)/2-i-j\}!}$$

$$\times \left\{\frac{\rho}{2+\rho}\right\}^{i+j} \left\{\frac{-\rho}{2-\rho}\right\}^{\{(p+q)/2\}-i-j}.$$

The computation of the covariance matrix of $\hat{d}_{2q}$ and $\hat{c}_q$ from these equations requires accounting for the combinations of first and second differences that are possible for those values of the r_i that appear in the set of measurements. To the $4Q+2$ by $4Q+2$ covariance matrix thus obtained, we add a small positive quantity to each diagonal element since the assumptions underlying the covariance matrix do not hold exactly. We denote the resultant matrix by $\boldsymbol{V}$.

We compute the estimates of coefficients of the error density with the algorithm of Dennis and Schnabel (1983). The $Q+1$ parameters to be estimated are $z_0^{(r)}, \ldots, z_{Q/2}^{(r)}, z_1^{(i)}, \ldots, z_{Q/2}^{(i)}$. Let the vector $\boldsymbol{e}$ with components e_j denote the $4Q+2$ differences $\hat{d}_{2q} - d_{2q}$ $(q = 0, \ldots, Q)$, $\hat{c}_q - c_q$ $(q = 0, \ldots, 3Q)$. The computational objective is to find the values of $z_0^{(r)}, \ldots, z_{Q/2}^{(r)}, z_1^{(i)}, \ldots, z_{Q/2}^{(i)}$ that minimize

$$f = \frac{1}{2} \sum_{i=1}^{4Q+2} \sum_{j=1}^{4Q+2} e_i V^{ij} e_j,$$

where V^{ij} is the i, j-th element of the inverse of $\boldsymbol{V}$. The evaluation of f is carried out using (3.3) and then (3.6) and (3.7). We use the version of the minimization algorithm based on analytically computed gradients and Hessians. The element of the gradient corresponding to $z_k^{(r)}$ is given by

$$\sum_{i=1}^{4Q+2} \sum_{j=1}^{4Q+2} \frac{\partial e_i}{\partial z_k^{(r)}} V^{ij} e_j$$

and the element corresponding to $z_k^{(i)}$ is similar. When the derivatives in the elements of the gradient are expressed by means of the chain rule

$$\frac{\partial e_i}{\partial z_k^{(r)}} = \sum_{q=0}^{Q} \frac{\partial e_i}{\partial a_q} \frac{\partial a_q}{\partial z_k^{(r)}},$$

the way to use (3.3), (3.6), and (3.7) in the evaluation of the gradient becomes clear. The element of the Hessian corresponding to $z_h^{(r)}$ and $z_k^{(r)}$ is

$$\sum_{i=1}^{4Q+2} \sum_{j=1}^{4Q+2} \left\{ \frac{\partial e_i}{\partial z_h^{(r)}} V^{ij} \frac{\partial e_i}{\partial z_k^{(r)}} + \frac{\partial^2 e_i}{\partial z_h^{(r)} \partial z_k^{(r)}} V^{ij} e_j \right\}.$$

Again, the chain rule can be used to evaluate the derivatives

$$\frac{\partial^2 e_i}{\partial z_h^{(r)} \partial z_k^{(r)}} = \sum_{p=0}^{Q} \sum_{q=0}^{Q} \left\{ \frac{\partial^2 e_i}{\partial a_p \partial a_q} \frac{\partial a_p}{\partial z_h^{(r)}} \frac{\partial a_q}{\partial z_k^{(r)}} \right\} + \sum_{q=0}^{Q} \left\{ \frac{\partial e_i}{\partial a_q} \frac{\partial^2 a_q}{\partial z_h^{(r)} \partial z_k^{(r)}} \right\}.$$

Finally, we note that as a starting point for the iteration, we let $z_0 = (\hat{d}_0 \pi^{1/4} / \sqrt{2})^{1/4}$, let $z_1^{(i)}$ have a small positive value, and let all the other values be 0.

One other important aspect of computing the replicates density estimate is the storage of $B_c(j, k, p, q)$, which is needed for $0 \leqslant j, k, p \leqslant Q$ and $0 \leqslant q \leqslant 3Q$. If only the values of $B_c(j,k,p,q)$ for which $j+k+p+q$ is even and for which $j+k+p \geqslant q$ are stored, then $B_c(j,k,p,q)$ can be stored in the location $(j,k,p,[q/2])$ if $q \leqslant 3Q/2+1$ and in $(Q-j,Q-k,Q-p,[(3Q+2-q)/2])$ if $q > 3Q/2+1$. The values of $B_c(j,k,p,q)$ not stored are all zero. This approach reduces the memory needed for $B_c(j,k,p,q)$ by a factor of 4. Similarly, if only the values of $B_d(j,k,q)$ for which $j+k+q$ is even and for which $j+k \geqslant q$ are stored, then $B_d(j,k,q)$ can be stored in $(j,k,[q/2])$ if $q \leqslant Q+1$ and in $(Q-j,Q-k,[(2Q+2-q)/2])$ if $q > Q+1$.

3.3. Performance

A perspective on the performance of this density estimation method is provided by two Monte Carlo experiments. In both experiments, we let $Q=14$. In the first experiment, we let the error be distributed as χ^2 with three degrees of freedom and generated triplicate measurements on 24 items, $r_i=3$, $m=24$. In the second, we let the error be normally distributed with variance 1 and generated triplicate measurements on 48 items. In each experiment, we performed 100 independent trials.

To assess the estimates graphically, we use quantile–quantile plots. These plots show for each probability between 0.005 and 0.995, the quantile for the estimate versus the quantile for the true density. (Evaluation of the cumulative distribution function at the quantile gives the corresponding probability.) Since we cannot show all 100 estimates, we select plots by ranking the estimates from best to worst according to $\int|p-\hat{p}|\,dx$. Figure 5 shows quantile–quantile plots for the best 10 estimates, the middle 10, and the worst 10 for the case of χ^2 error. Note that we have centered the χ^2 at its mean. This figure shows that although the true density is zero for $x<-3$, the estimate is not configured to imitate this behavior. Thus, near

a

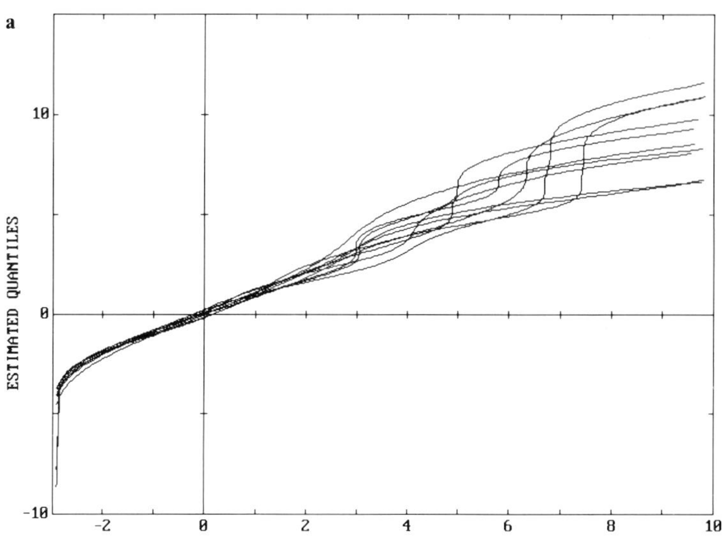

(continued on next page)

b

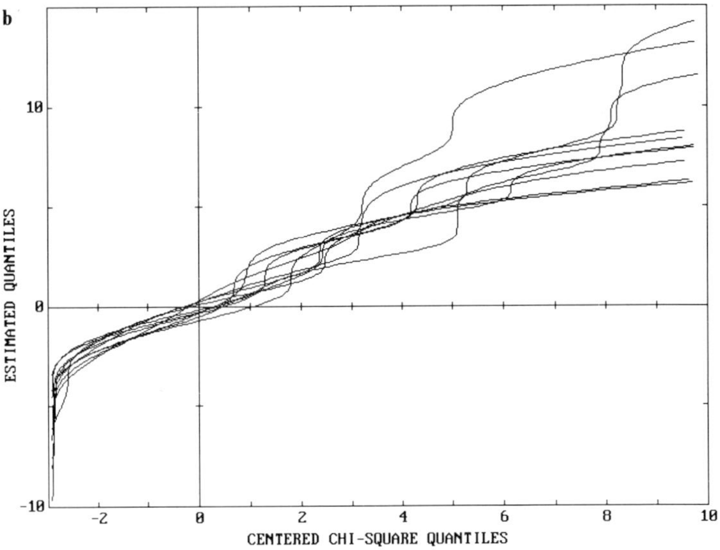

c

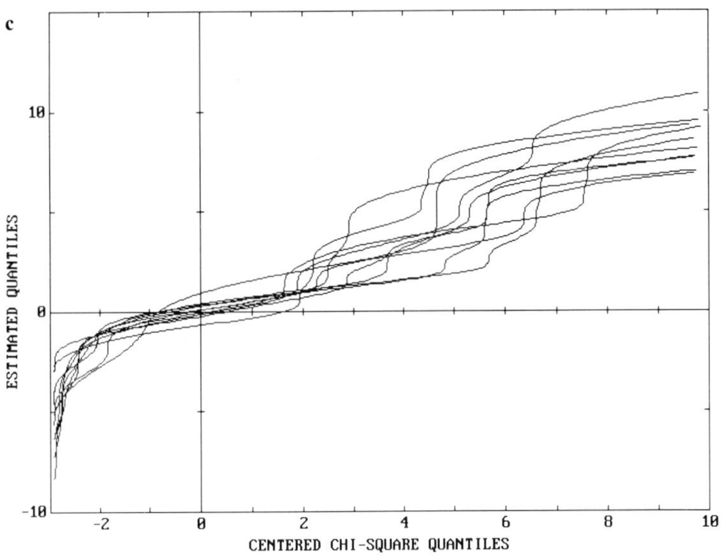

Fig. 5. Estimates of a chi-square density – Out of 100, (a) best 10, (b) middle 10, (c) worst 10.

$x = -3$, the plot proceeds downward. The best 10 and middle 10 estimates have slopes near 1 in the middle of the distribution and thus estimate the middle of the distribution reasonably well. The worst 10 have slopes less than 1 in the middle and compensating jumps in the upper tail. Thus, the worst have too little probability in the center. One consequence of this in lack-of-fit tests is demonstrated in Section 3.4.

We conclude that this density estimation method has value for estimation of the middle of the density but should not be used for estimating the behavior of the density in the tails. Development of a density estimation method based on replicates that performs better in the tails seems to be an open problem.

In contrast to Figure 5, Figure 6 shows quantile–quantile plots for the case of normal error and 48 triplicates. Here we see much better results. We conclude that

a

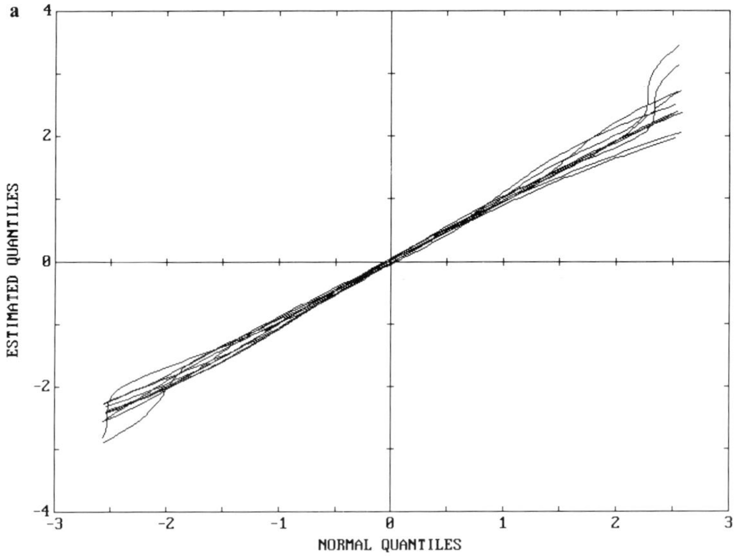

b

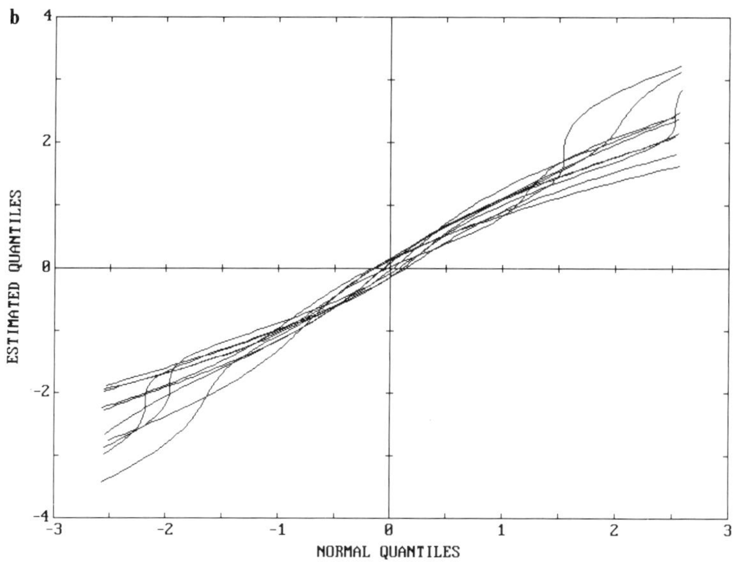

(continued on next page)

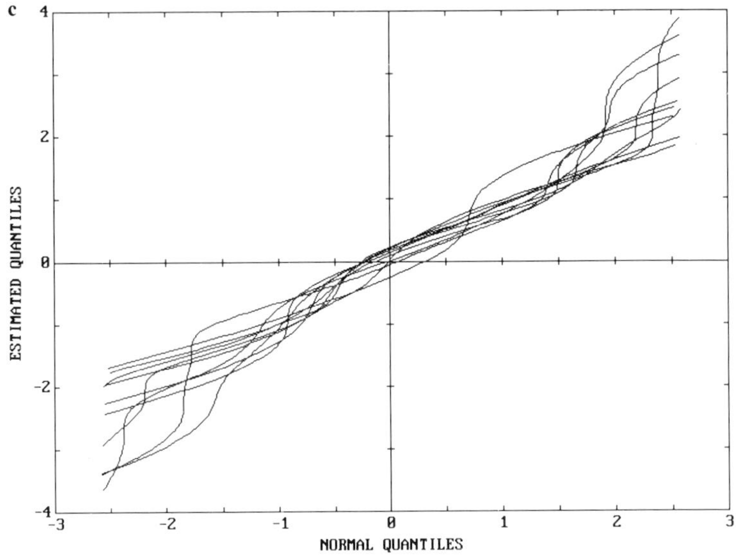

Fig. 6. Estimates of a normal density – Out of 100, (a) best 10, (b) middle 10, (c) worst 10.

for error that is closer to normal, a case where the usual normal theory approach is also useful, our density estimation method performs better.

When one thinks of density estimation, one often thinks of the estimation of tail behavior or the answer to the question of whether the error is really normal. The foregoing density estimation method is not suitable for either of these purposes. However, the method is useful for judging the results of model fitting by robust methods. Robust methods are insensitive to the tail behavior, and thus when these methods are involved, a density estimate that is not right in the tails should still be useful for assessing the quality of the fit.

3.4. Robust methods for two-way tables

Lack-of-fit tests for models fit by an algorithm closely related to median polish (Tukey, 1977) is an appropriate illustration of the use of the replicates density estimate because environmental designs that give two-way tables are frequently selected and estimators such as median polish have often been found necessary in environmental data analysis (Wu and Zidek, 1989). Monitoring design is often based on the selection of a number of monitoring sites and the selection of times for simultaneous sampling of these sites. The resulting data then fit into a two-way table in which one dimension indexes site and the other time of sampling. In the choice of estimator for an environmental study, high breakdown point is often essential. The breakdown point of an estimator is an indication of how many points in the data set can be changed by an arbitrarily large amount without an arbitrarily large change in the estimate (Rousseeuw and Leroy, 1987). The median is an example of a high breakdown point estimate. The median may change when data values are changed, but an arbitrarily large change in the median does not occur unless half the data values are changed.

Simultaneous measurement at several locations performed at a sequence of times provides the basis for the detection of local environmental changes. Local changes are those that occur at some but not all of the monitoring locations. Such changes are of interest because their causes are also local and therefore might be understood. Local changes must be distinguished from differences between monitoring locations that are constant over the study period. Such differences have complicated origins that often cannot be untangled. Local changes must also be distinguished from variations with time that are the same at all monitoring locations. Such variations are usually caused by weather changes and other seasonal phenomena.

The study design we consider calls for replicate observations to be made at each of K locations and H sampling times. We compare the measurements made in the two periods that consist of the first half of the sampling times $(1, \ldots, H/2)$ and the second half $(H/2 + 1, \ldots, H)$, respectively. In particular, we would like to know whether the change between the two periods is greater at some locations than at other locations. Among-location differences in the period-to-period change indicate local changes.

The change detection is based on fitting a central value for each location to the measurements in each period. These central values are fit by means of an algorithm closely related to median polish that gives the least (sum of) absolute deviations (LAD) (Siegel, 1983). In many environmental studies, median polish has been used instead of the least-squares algorithm because least-squares gives residuals that are difficult to interpret when some measurements are outliers. The advantage of median polish and the LAD algorithm is that if at some location and time, all the measurements are unusual, the central values will not be greatly affected.

In addition to data collection blunders, other causes for unusual measurements might lead to use of the LAD algorithm. One such cause is an environmental event that we want to ignore in the comparison of one period with another. For example, at some location and time, the measurements might be made on a stream that was flooded due to a recent storm. Note that in this case, although the measurements are valid, the model to be used in the comparison is not because of the storm runoff. This example illustrates that the LAD algorithm can help when the available environmental models are not sufficiently sophisticated.

The method used is based on fitting for each period the model

$$y_{hkj} = \beta_{Th} + \beta_{Sk} + \varepsilon_{hkj},$$

where y_{hkj} is the j-th replicate at time h and location k. Our objective is to estimate the β_{Sk}, which we do by fitting $\beta_{Th} + \beta_{Sk}$ to the measurements. For each time period, we compute the LAD fit that has residuals with the property that the low medians of the rows equal 0. The low median of a set of n values is the middle value when n is odd and the lower of the two middle values when n is even. Let the estimates of β_{Sk} for the two periods be $\hat{\beta}_{Sk}^{(1)}$ and $\hat{\beta}_{Sk}^{(2)}$. Let the difference in the value for location k between the first and second half of the data be given by $\Delta_k = \hat{\beta}_{Sk}^{(1)} - \hat{\beta}_{Sk}^{(2)}$. The statistic we use for comparison is

$$\sum_{k=1}^{K} \left\{ \Delta_k - \frac{1}{K} \sum_{k=1}^{K} \Delta_k \right\}^2 .$$

Note that if the change in β_{Sk} from period to period is the same for all k, then this statistic depends only on the probability density of the error. To test the hypothesis that all the Δ_k are the same, we use the estimated density to obtain the critical point with which to compare this statistic.

The following Monte Carlo experiment indicates how the foregoing method might perform. We consider 3 locations each sampled 4 times in the first half and 4 times in the second half. At each point in space and time, the measurement is replicated to give $r_i = 3$. The measurements used in this study are the chi-square distributed values used above in the Monte Carlo experiment on replicates density estimation. For each of the 100 density estimates, we obtained 200 sets of bootstrap resamples, took advantage of symmetry to obtain 1200 somewhat dependent values for the statistic, and ranked these to determine the critical point. Also, we took advantage of the symmetry in other ways to make the Monte Carlo experiment more efficient. In the results, when we intended to obtain a 10 percent probability of Type I error, we obtained 16 percent (with a Monte Carlo standard deviation of 2 percent). When we intended to obtain a 5 percent Type I error, we obtained 10 percent (with a standard deviation of 2 percent). The probability of Type I error realized is larger than intended but perhaps close enough to that intended for the method to be useful. A comparison of these Monte Carlo results with the density estimation results in Section 3.3 shows that the poorly fitting density estimates are at least partly to blame for the discrepancy between the intended and the realized Type I error. The failure of the density estimates shown in Figure 5c to portray the middle of the χ^2 density seems to be the cause.

4. Conclusions

Environmental problems put a new focus on familiar statistical methods; they cause a rethinking of these methods and motivate searches for new methods. This chapter illustrates one aspect of this, the consequences of lack of trust in environmental models. Because of this lack of trust, comparison of a measurement with a model prediction tells us almost nothing about the validity of the measurement. Moreover, model-free testing of assumed environmental models for lack of fit is necessary. For these reasons, we look for ways to assess measurement error without use of a model and we are led to replicates.

Environmental studies in which there is a serious intention to assess the validity of models through replicates must be very carefully planned. This planning is expensive as is the processing of replicates. The risk is that some totally unforeseen circumstance will arise and make the plans inapplicable. This risk is higher in environmental studies than in some other fields. It is nevertheless true that there is experimental information about the environment that would be very valuable could it be obtained. In particular, it is often the case that we know that the environment is changing but we do not know why. Thus, people seriously interested in understanding environmental mechanisms might do well to take the risk.

References

Alt, F. B. and N. D. Smith (1988). Multivariate process control. In: P. R. Krishnaiah and C. R. Rao, eds. *Handbook of Statistics*, Vol. 7. North-Holland, Amsterdam, 333–351.

Anderson, T. W. (1984). *An Introduction to Multivariate Statistical Analysis.* 2nd ed., Wiley, New York.

Atkinson, A. C. (1985). *Plots, Transformations, and Regression.* Oxford Univ. Press, Oxford.

Barnett, V. and T. Lewis (1984). *Outliers in Statistical Data.* 2nd ed., Wiley, New York.

Carroll, R. J. and D. Ruppert (1988). *Transformation and Weighting in Regression.* Chapman and Hall, New York.

Currie, L. A. (1992). Source apportionment of atmospheric particles. In: J. Buffle and H. P. van Leeuwen, eds., *Environmental Particles*, Vol. 1. Lewis, Boca Raton, FL, 3–73.

Dennis, J. E. and R. B. Schnabel (1983). *Numerical Methods for Unconstrained Optimization and Nonlinear Equations.* Prentice-Hall, Englewood Cliffs, NJ.

Draper, N. R. and H. Smith (1981). *Applied Regression Analysis.* 2nd ed., Wiley, New York.

Durbin, J. (1969). Tests for serial correlation in regression analysis based on the periodogram of least-squares residuals. *Biometrika* **56**, 1–15.

Gilbert, R. O. (1987). *Statistical Methods for Environmental Pollution Monitoring.* Van Nostrand Reinhold, New York.

Gnanadesikan, R. (1977). *Methods for Statistical Data Analysis of Multivariate Observations.* Wiley, New York.

Hopke, P. K. (1985). *Receptor Modeling in Environmental Chemistry.* Wiley, New York.

Keith, L. H. ed., (1988). *Principles of Environmental Sampling.* Amer. Chem. Soc., Washington, DC.

Kingston, H. M., R. R. Greenberg, E. S. Beary, B. R. Hardas, J. R. Moody, T. C. Rains and W. S. Liggett, (1983). *The Characterization of the Chesapeake Bay: A Systematic Analysis of Toxic Trace Elements*, NBSIR 83–2698. National Institute of Standards and Technology, Gaithersburg, MD.

Kingston, H. M. and R. R. Greenberg (1984). An elemental ratioing technique for assessing concentration data from a complex water system. *Environ. Internat.* **10**, 153–161.

Liggett, W. S. (1986). Designs for assessment of measurement uncertainty: Experience in the Eastern Lake Survey. In: *Proc. 13th Internat. Biometric Conf.: Invited Papers.* Biometric Soc., Washington, DC.

Liggett, W. S. (1988). Assessment of measurement uncertainty: Designs for two heteroscedastic error components. In: L. H. Keith, ed., *Principles of Environmental Sampling.* Amer. Chem. Soc.,Washington, DC, 191–208.

Liggett, W. S. (1989). Estimation of an asymmetrical density from several small samples. *Biometrika* **76**, 13–21.

Liggett, W. S., K. G. W. Inn and J. M. R. Hutchinson (1984). Statistical assessment of subsampling procedures. *Environ. Internat.* **10**, 143–151.

Linthurst, R. A., D. H. Landers, J. M. Eilers, D. F. Brakke, W. S. Overton, E. P. Meier and R. E. Crowe, (1986). *Characteristics of Lakes in the Eastern United States*, Vol. 1, Population Descriptions and Physico-Chemical Relationships, EPA/600/4-86/007a. U.S. Environmental Protection Agency, Washington, DC.

Mandel, J. (1991). *Evaluation and Control of Measurements.* Marcel Dekker, New York.

Pio, C. A. and A. Hall, (1991). Sampling methods. In: C. N. Hewitt, ed., *Instrumental Analysis of Pollutants.* Elsevier Science Publishers, London, 1–41.

Rousseeuw, P. J. and A. M. Leroy (1987). *Robust Regression and Outlier Detection.* Wiley, New York.

Ruppert, D. and B. Aldershof (1989). Transformations to symmetry and homoscedasticity. *J. Amer. Statist. Assoc.* **84**, 437–446.

Ryan, T. P. (1989). *Statistical Methods for Quality Improvement.* Wiley, New York.

Schwartz, S. C. (1967). Estimation of probability density by an orthogonal series. *Ann. Math. Statist.* **38**, 1261–1265.

Siegel, A. F. (1983). Low median and least absolute residual analysis of two-way tables. *J. Amer. Statist. Assoc.* **78**, 371–374.

Smith R. and G. V. James (1981). *The Sampling of Bulk Materials.* Royal Society of Chemistry, London.

Tukey, J. W. (1977). *Exploratory Data Analysis.* Addison-Wesley, Reading, MA.

Wilks, S. S. (1967). Multivariate statistical outliers. In: T. W. Anderson, ed., *S. S. Wilks: Collected Papers, Contributions to Mathematical Statistics.* Wiley, New York. Reprinted from *Sankhyā-Ser. A* **25** (1963) 407–426.

Williams, J. C. (1976). The segregation of particulate materials. A review. *Powder Technol.* **15**, 245–251.

Wu, S. and J. V. Zidek (1989). Selected methods for analyzing trends in environmental data series. Technical Report 127, SIMS Publications, New Canaan, CT.

G. P. Patil and C. R. Rao, eds., *Handbook of Statistics, Vol. 12*

4

Design and Analysis of Composite Sampling Procedures: A Review*

G. Lovison, S. D. Gore and G. P. Patil

Abstract: Composite sampling is emerging as a cost efficient sampling method in environmental studies. Treating the process of sample selection (i.e., identification of sampling units to be included in the sample) as distinct from that of sample quantification (i.e., making measurements on the selected sample units), it is possible to achieve two conflicting goals simultaneously. On one hand, one can select a large number of sample units to ensure the desired sample support, so that the results of the analysis will be sufficiently reliable. On the other hand, by compositing several individual samples into fewer composites, one reduces the number of analytical measurements so that a study can be carried out without affecting the available financial resources. In this paper, we review some prominent composite sample techniques to design and analyze composite sample data, with special emphasis on environmental applications.

1. Introduction

Composite sampling involves taking several individual sampling units and mixing them before measuring, counting or otherwise analyzing the composite sample. Mixing is usually physical, by a blender, homogenizer or other blending device, but other types of mixing are possible. Biological compositing is achieved in infection transmission studies, where host plants or animals are inoculated with insects carrying the infection and the observed variable consists of whether the host becomes diseased. In group testing, the analog of composite sampling in industrial applications, the status of components of a system is evaluated by observing the status of the system as a whole, and mixing is therefore mechanical.

* Prepared with partial support from the Statistical Analysis and Computing Branch, Environmental Statistics and Information Division, Office of Policy, Planning, and Evaluation, with Brenda Odom as EPA Project Officer. The contents have not been subjected to Agency review and therefore do not necessarily reflect the views of the Agency and no official endorsement should be inferred.

In general, several subsamples can be taken from each composite sample and several measurements can be made on each subsample. Although this possibility is not always exploited, owing to cost or time constraints, it can be shown to improve the statistical properties of the composite sample data as compared to the measurement made on a single composite sample.

Composite sampling is used for drawing inference on the parameters of a (possibly multivariate) random variable in the population of interest or for classifying the individual sampling units as having or not a (possibly multivariate) trait.

The advantages of analyzing composite samples are as follows:

- *cost savings*: when the cost of measurement is substantially higher than the cost of sampling and compositing, composite sampling reduces total costs for classifying all the individual sampling units or for estimating parameters with a given precision;
- *precision*: for fixed costs, composite sampling often permits estimation with greater precision;
- *timeliness*: when measurement is time consuming, composite sampling can give a fast estimate or classification by reducing the number of laboratory tests;
- *confidentiality*: composite sampling permits estimation of prevalence of a trait in the population while preserving privacy of the individuals (for an application concerning AIDS, see Gastwirth and Hammick, 1989).

There are also limitations to the use of composite sampling:

- *cost*: when the cost of sampling and compositing is higher than the cost of measurement, composite sampling may not be cost effective;
- *technical constraints*: composite sampling may be inappropriate in the presence of hardly mixable materials or of dilution effects.

A recent review on composite sampling can be found in Boswell, Burnham and Patil (1988). The annotated bibliography of composite sampling (Boswell, Gore, Lovison and Patil, 1992) may also be useful for a review of composite sampling methods and applications. After setting up a unifying terminology, this entry revises some of the traditional results and reviews some recent developments in both classification (in Section 2) and estimation (in Section 3) with composite data.

1.1. Terminology

Terms used in papers on both theory and applications of composite sampling have usually been specific to particular areas of application. Owing to this specificity, terminology is very diversified and sometimes confusing, with the same term used for different concepts or different terms used for the same concept. Therefore, it is useful to standardize the terminology to make it appropriate for a fairly general presentation like the one contained in this entry.

The framework for the terminology used in this paper comes from considering composite sampling as comprising five phases: identification of target population; statistical sampling; compositing; subsampling and measurement. Not all these phases need be present in a particular application, but it helps to think of them as potential steps in the conceptualization of compositing procedures.

The main distinction at the population identification level is between a *finite population* and a *bulk population*. According to Duncan (1962), Bicking (1967) and Rohde (1979), a *finite population* is a population which consists of identifiable units, which exist before sampling takes place. On the other hand, a *bulk population* is one in which the units to be selected in the sample are themselves created by the sampling procedure. As pointed out by Rohde (1979) every concept of finite sampling theory becomes difficult to apply when the population of units to which results must be generalized is not well defined before sampling takes place.

For the statistical sampling phase, we shall adopt the terminology of finite population sampling theory up to the final selection of sampling units before compositing. Deviating from this standard terminology, we shall call the unit which is taken from the final sampling unit to enter a composite sample the *compositing object*. It is clear that such units are actually the final sampling units, being the last selection in the sampling hierarchy, but there are at least two reasons to give them a distinct name. First, the name 'compositing object' suggests that, unlike what happens in a 'traditional' sampling situation, the ultimate product of thc sampling scheme is not going to be measured but to be composited. Second, the features of the compositing object (in particular its size and shape) and the procedure through which it is selected are usually dictated by operational considerations regarding the compositing phase rather than by statistical considerations regarding their sampling properties.

At the compositing, sub-sampling and measurement stages, it is useful to introduce the following terminology. We shall call '*composited object*' the portion of a composite sample which is actually measured and '*contributing object*' the portion of a compositing object which ultimately contributes to a composited object. The number of compositing objects which are used to form a composite is called *composite sample size*, and the physical size (volume, amount, number, etc.) of the contributing object is called *contributing weight* or, shortly, *weight*. Since such weights are usually divided by the total size of the composited object, i.e., they sum up to one over the contributing objects, they are also termed *contributing proportions*, or simply *proportions*.

In the simplest case, the 'compositing objects' and the 'contributing objects' coincide and are just original samples of the material (or population or lot) under investigation and the 'composited objects' are just the composite samples. In these cases, we shall also use the more traditional terms *individual sampling unit* and *composite sample* respectively. But, in more general situations, this simple equivalence does not hold and a more precise terminology becomes important.

As for the former, we may have various sampling stages, so that the term 'individual sampling unit' might be ambiguous. On the other hand, the individual sampling unit is often subdivided into portions, out of technical reasons associated with the mixing procedure and/or to permit retesting. Such portions, which are actually the compositing objects in this case, are variously named by terms like 'increments', 'aliquots', 'specimens'. These terms are strongly associated with specific areas of application (industrial, biomedical, etc.) of composite sampling, and therefore are not completely appropriate for a general presentation.

As for the latter, in many cases the composite samples are not entirely measured. On subsampling stage is very common, and some authors (see, for example, Duncan, 1962 and Elder, 1977) have addressed the usefulness of two subsampling stages. Using the term 'composited object', we concentrate on the fact that this is the statistical unit on which the final measurement is made, regardless of the possibly complicated procedure which has produced it.

2. Classification

Group testing for identification of members possessing a certain trait was initiated by Dorfman (1943) to identify U.S. servicemen infected with syphilis. Subsequent applications of group testing include screening for pollutants, testing for leaking containers, identifying faulty components in a flow test, identifying active users in a communications system, recognizing the pattern of a binary code, and screening experimental factors affecting yield. If the purpose of an investigation is to identify every individual element with a rare attribute, then the number of tests can be greatly reduced by first testing groups of several items and then testing constituent individual members only when a group tests positive (Feller, 1968; Garner, Stapanian and Williams, 1988).

Traditionally, classification of individual items is desired when the trait is either present or absent, both in a group and in an individual item. However, the trait may also be defined by exceedance of some critical level by a measurement. This case must be treated separately since compositing causes averaging of continuous measurements. Ignoring dilution below the detection limit, presence of the trait in any individual item signifies its presence in the group; but a positively testing group may not possess its sample possessing the trait. In either case, considerable savings can be realized by group testing, provided the trait is relatively rare.

In group testing, also called composite sampling, where individual members in a group are physically pooled and mixed, individual samples are collected and prepared for testing. Composite samples are formed and tested. If a composite sample tests positive, then further testing (called retesting) is necessary to classify every constituent individual sample. Retesting implies making additional measurements and may involve forming more composite samples. If the cost of forming a composite sample is negligible compared to the cost of a laboratory test, then the problem can be studied in terms of the relative cost, which is defined as the number of measurements per individual sample classified. Exhaustive testing of all individual samples results in making one measurement per sample. In order for composite sample techniques to be cost effective, their relative cost must be less than unity. Considerable cost savings can be realized if the trait is rare. For example, classifying items from a binomial distribution with $p < 0.01$ will result in an average of one measurement per five items classified by the Dorfman retesting procedure with the optimal composite sample size.

For presence/absence measurements, if any one item possesses the trait, then the group also possesses the trait. On the other hand, if none of the members possesses the trait, then the group also does not possess the trait. For continuous measurements made without error, if any one sample value exceeds the criterion level c, then the measurement on a composite sample of k individual samples will exceed c/k. Of course, it is possible that none of the individual sample measurements exceed c and the compoite sample measurement still exceeds c/k. If the composite sample measurement exceeds c/k, then retesting would be necessary to identify individual samples with measurements exceeding c. On the other hand, if the composite measurement does not exceed c/k, then none of the k individual sample values can exceed c and hence no retesting is necessary.

Since the two types of measurements require separate considerations, Section 2.1 discusses the presence/absence measurement case, while Section 2.3 discusses the continuous measurement case. Since the optimal procedure for classification depends on the unknown prevalence rate, it may be difficult, often impossible, to implement an optimal retesting procedure. Section 2.2 discusses a Bayesian method to overcome this problem. Under the assumption of no measurement error (that is, no misclassification), all the procedures described here give correct classifications. However, the relative cost is affected by model assumptions such as independence of individual sample values. If an assumption such as spatial autocorrelation among individual sample values is imposed, then the relative cost may change, but still a correct classification is obtained.

Derivations of the formulas presented in the remainder of this section can be found in Boswell, Gore and Patil (1990).

2.1. Presence/absence measurements

The cost of selecting, taking and preparing samples prior to sending them to the laboratory for testing is constant since the number of individual samples is assumed to be predetermined. The relative cost of classification is defined as the expected number of measurements required to classify all samples divided by the number of such samples. When the cost of measurement is higher than the costs of sampling and forming composite samples, the relative cost can be reduced by the use of composite sample techniques. Processing of one composite sample results in a correct classification of all its constituent individual samples. The number of individual samples used to form a composite sample is defined as the composite sample size. We assume an infinite number of individual samples to be classified, and hence it is always possible to form a composite sample of any desired size. The relative cost computed under this assumption is then called the asymptotic relative cost.

2.1.1. The Dorfman procedure

All individual samples are assumed to be independent with a common probability p of possessing the trait. The Dorfman retesting procedure utilizes a two-stage sampling scheme. The first stage comprises making measurements on all groups, or

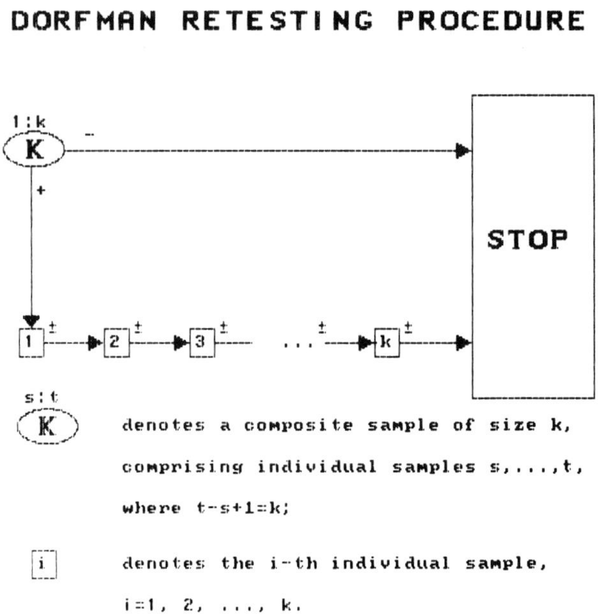

Fig. 1. Description of the Dorfman retesting procedure.

on composite samples. The second stage involves no retesting for negative testing composites and exhaustive testing of all constituent individual samples of positive testing composites. Figure 1 shows the two stages of the Dorfman procedure. For a composite sample of size k, the number of measurements T_k is either 1 or $k + 1$. The expected number of measurements required to classify all the k individual samples is then given by

$$E(T_k) = k + 1 - k(1 - p)^k.$$

The relative cost, defined by

$$RC = E(T_k)/k$$

is then given by

$$RC = 1 + 1/k - (1 - p)^k.$$

Samuels (1978) showed that for all $p \leqslant 1 - (\frac{1}{3})^{1/3} \times 0.307$, the optimal composite sample size is either $1 + [p^{-1/2}]$ or $2 + [p^{-1/2}]$, depending on which of these two minimizes the relative cost. Here $[x]$ represents the integer part of x. Table 1 shows the optimal composite size k as a function of p.

Table 1
Optimal composite sample size k for different prevalences

p	k_{opt}	RC*	p	k_{opt}	RC*
(0.3066, 1]	1	(1.00)	(0.0039, 0.0044]	16	(0.12, 0.13]
(0.1239, 0.3066]	3	(0.66, 1.00)	(0.0035, 0.0039]	17	0.12
(0.0656, 0.1239]	4	(0.49, 0.66)	(0.0031, 0.0035]	18	0.11
(0.0411, 0.0656]	5	(0.39, 0.49)	(0.0028, 0.0031]	19	0.10
(0.0283, 0.0411]	6	(0.32, 0.39)	(0.0025, 0.0028]	20	0.10
(0.0207, 0.0283]	7	(0.28, 0.32)	(0.0023, 0.0025]	21	0.09
(0.0158, 0.0207]	8	(0.24, 0.28)	(0.0021, 0.0023]	22	0.09
(0.0124, 0.0158]	9	(0.22, 0.24)	(0.0019, 0.0021]	23	0.09
(0.0101, 0.0124]	10	(0.20, 0.22)	(0.0017, 0.0019]	24	0.08
(0.0083, 0.0101]	11	(0.18, 0.20)	(0.0016, 0.0017]	25	0.08
(0.0070, 0.0083]	12	(0.16, 0.18)	(0.0015, 0.0016]	26	0.08
(0.0059, 0.0070]	13	(0.15, 0.16)	(0.0014, 0.0015]	27	0.07
(0.0051, 0.0059]	14	(0.14, 0.15)	(0.0013, 0.0014]	28	0.07
(0.0044, 0.0051]	15	(0.13, 0.14)	(0.0012, 0.0013]	29	0.07

* RC: Relative cost.

2.1.2. The curtailed Dorfman procedure

The second stage of the Dorfman procedure can be modified if sequential retesting is possible. Suppose a composite sample of size k tested positive and therefore exhaustive testing is undertaken. If the first $k-1$ individual samples test negative, then without making a measurement we can infer that the k-th individual sample must test positive. Figure 2 shows how the curtailed Dorfman procedure differs from the classical Dorfman procedure. The number T_k of measurements in this case takes values 1, k or $k+1$. The expected number of tests is then given by

$$E(T_k) = k + 1 - k(1-p)^k - p(1-p)^{k-1}.$$

The (asymptotic) relative cost is therefore given by

$$RC = 1 - (1-p)^k + [1 - p(1-p)^{k-1}]/k.$$

2.1.3. The Sterrett procedure

Assuming a binomial model, Sterrett (1957) proposed a modification to the Dorfman procedure. When a composite sample tests negative, then no further testing is required and the Sterrett procedure is the same as the Dorfman procedure. If a composite sample tests positive, and therefore exhaustive testing is undertaken, then the Sterrett procedure is the same as the Dorfman procedure until a positive testing individual sample is encountered. At this point, the remaining unclassified individual samples are composited to obtain a single composite measurement, since no information is available as to whether these contain any positive testing individual sample. This procedure is repeated (as many times as is required) to classify all the individual samples. Figure 3 shows how the Sterrett procedure is implemented, and

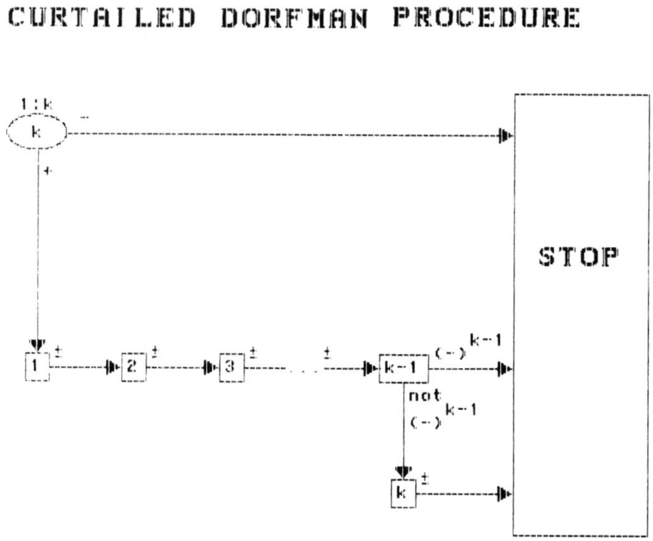

s:t (k) denotes a composite sample of size k, comprising individual samples s, s+1, ..., t, where t − s + 1 = k;

[i] denotes the i-th individual sample, i=1, 2, ..., k;

$(-)^{k-1}$ denotes the event that all of the first k−1 individual samples test negative;

$\mathrm{not}(-)^{k-1}$ denotes the event that at least one of the first k−1 individual samples tests positive.

Fig. 2. Description of the Dorfman curtailed retesting procedure.

Figure 4 shows the relative costs of classification for the Dorfman and the Sterrett procedures as functions of composite sample sizes for $q = 1 - p = 0.9$, 0.95, 0.99 and 0.999.

Let T_k be the number of measurements required to correctly classify all the k individual samples in a single composite. A recurrence formula can be obtained for $\mathrm{E}(T_k)$ by conditioning on the number of retests J to the first positive testing individual samples. Thus, $J = 0$ if the composite sample tests negative; $J = 1$ if the composite sample tests positive and the first individual sample tests positive at the

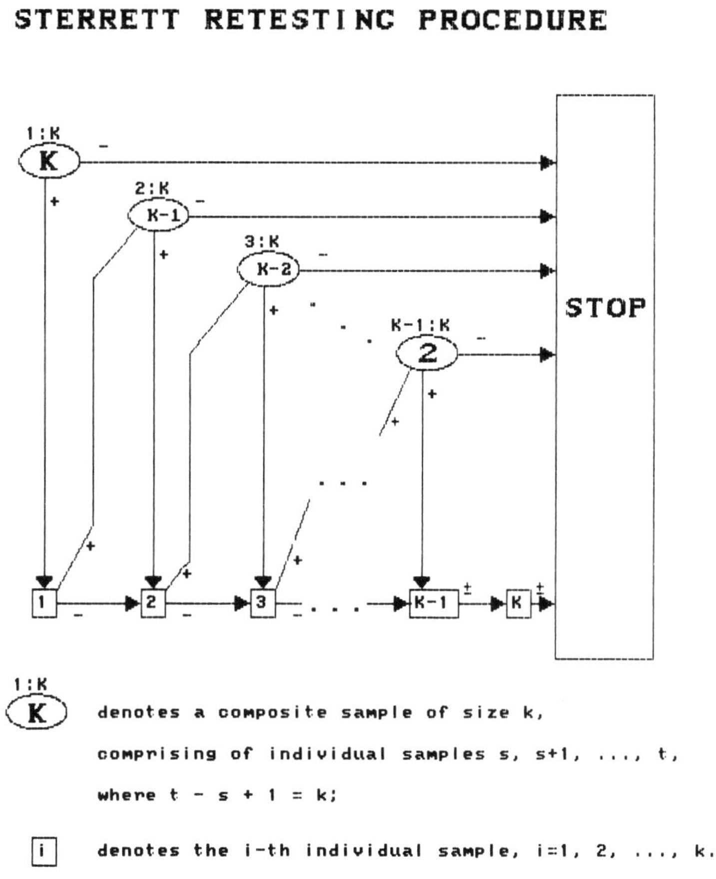

Fig. 3. Description of the Sterrett retesting procedure.

second stage of exhaustive testing; and so on. Now,

$$\begin{aligned}
E(T_k) &= E[(T_k \mid J)] \\
&= \sum_{j=0}^{k} E[T_k \mid J=j] \cdot P[J=j] \\
&= E[T_k \mid J=0] \cdot (1-p)^k + \sum_{j=1}^{k} E[T_k \mid J=j] p(1-p)^{j-1} \\
&= q^k + \sum_{j=1}^{k} [j+1+E(T_{k-j})] p(1-p)^{j-1}, \quad k=2,3,\ldots,
\end{aligned}$$

since, when a composite sample tests positive, given that the first $j-1$ individual samples tested negative and that the j-th individual sample tested positive, the

a

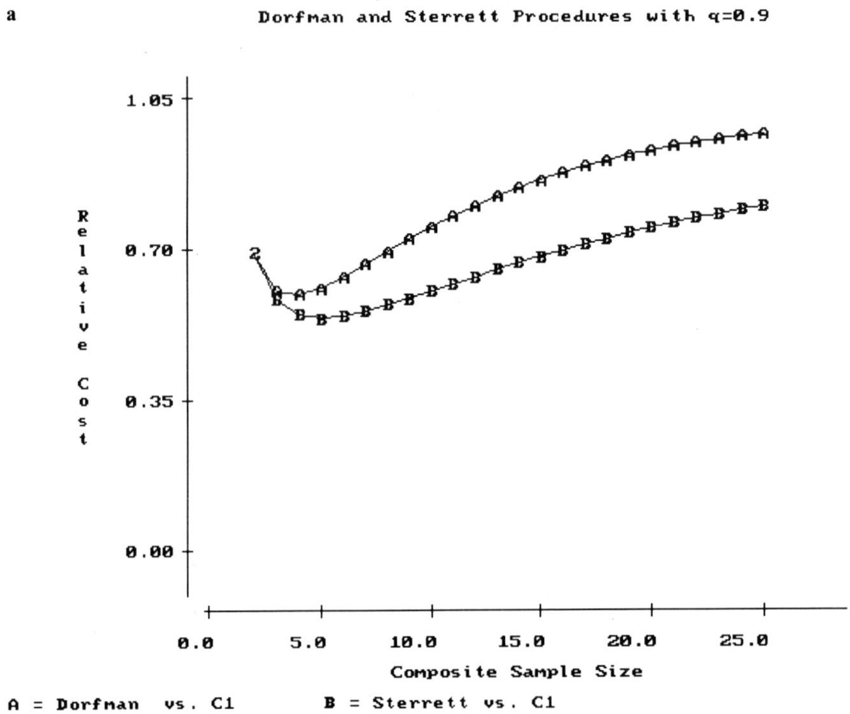
Dorfman and Sterrett Procedures with q=0.9
Relative Cost
1.05
0.70
0.35
0.00
0.0
5.0
10.0
15.0
20.0
25.0
Composite Sample Size
A = Dorfman vs. C1
B = Sterrett vs. C1

b

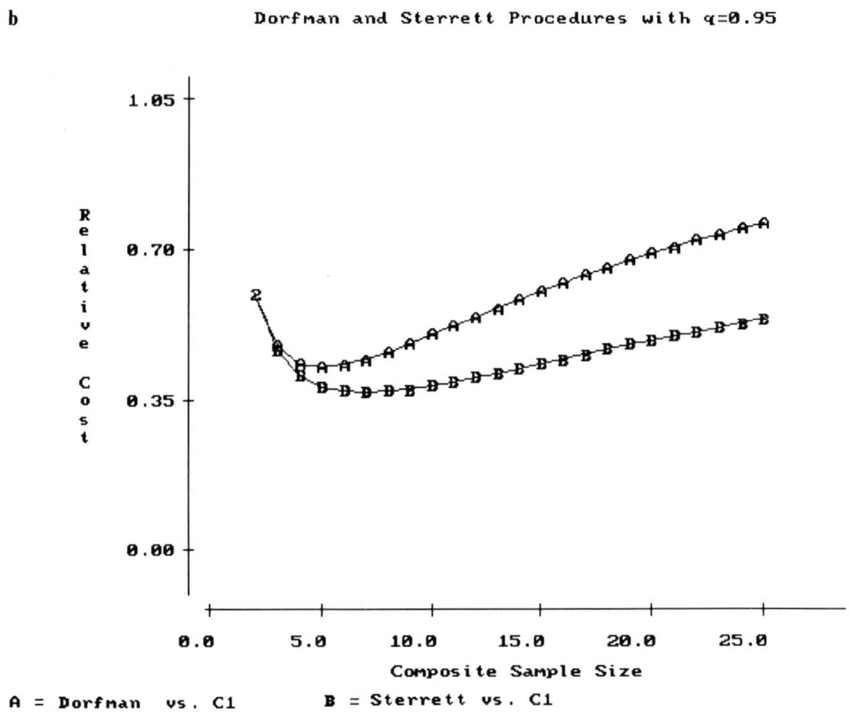
Dorfman and Sterrett Procedures with q=0.95
Relative Cost
1.05
0.70
0.35
0.00
0.0
5.0
10.0
15.0
20.0
25.0
Composite Sample Size
A = Dorfman vs. C1
B = Sterrett vs. C1

c

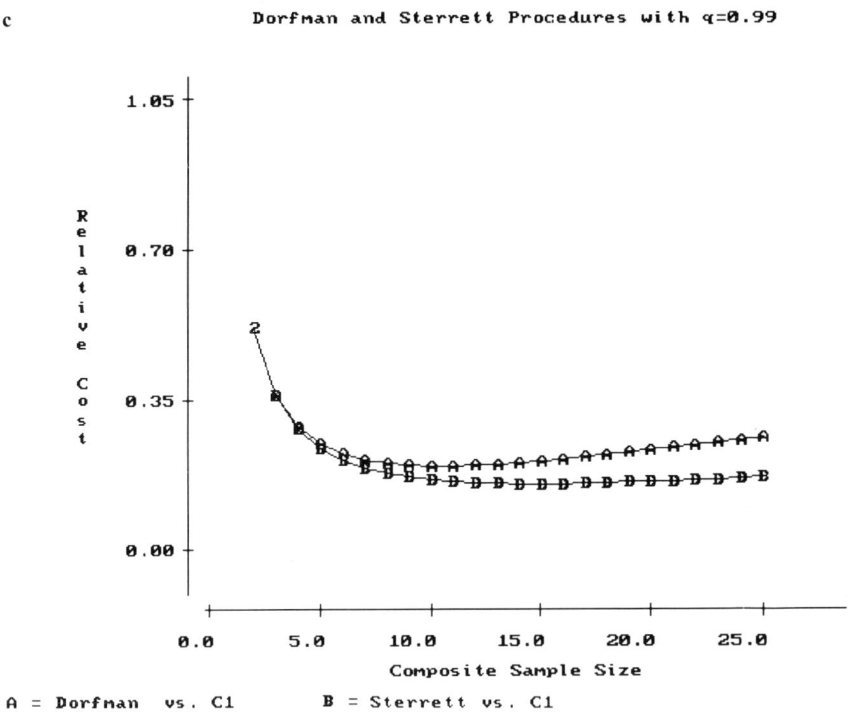

d

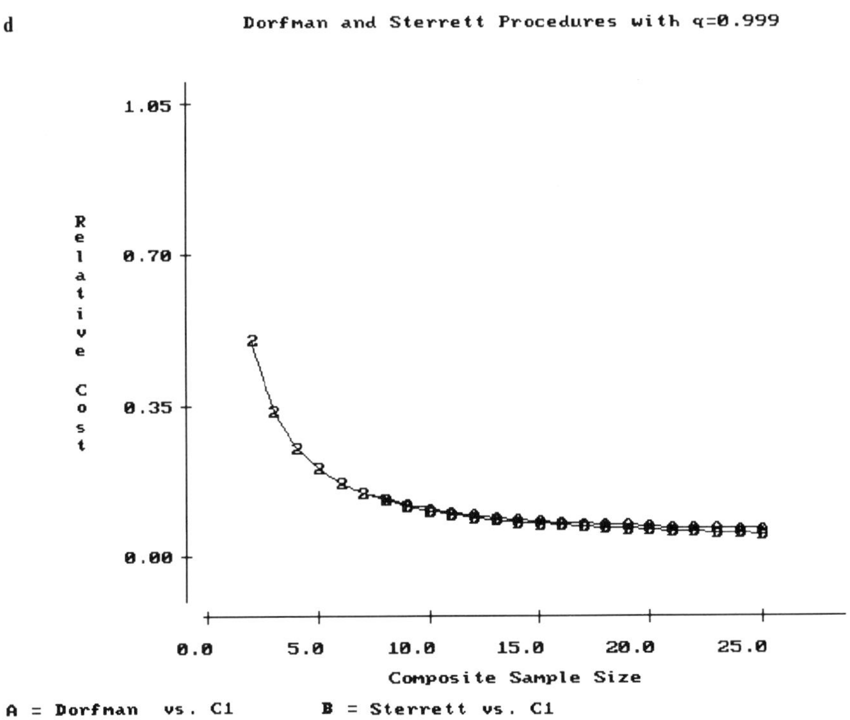

Fig. 4. Relative costs of Dorfman and Sterrett retesting procedures. A: Dorfman procedure; B: Sterrett procedure. (a) $q = 0.9$, (b) $q = 0.95$, (c) $q = 0.99$, (d) $q = 0.999$.

remaining $k-j$ individual samples are independent Bernoulli random variables with parameter p. Therefore,

$$\mathrm{E}(T_k) = q^k + \sum_{j=1}^{k} (j+1)p(1-p)^{j-1} + \sum_{j=1}^{k} \mathrm{E}[T_{k-j}]p(1-p)^{j-1}$$

$$= q^k + \sum_{j=0}^{k-1} (j+2)p(1-p)^j + \sum_{j=0}^{k-1} \mathrm{E}[T_j]p(1-p)^{k-j-1}.$$

It is then easy to see that

$$q\mathrm{E}[T_{k-1}] = q^k + \sum_{j=1}^{k-1} (j+1)p(1-p)^j + \sum_{j=0}^{k-2} \mathrm{E}[T_j]p(1-p)^{k-j-1},$$

$$k = 3, 4, \ldots .$$

Then we have

$$\mathrm{E}[T_k] - q\mathrm{E}[T_{k-1}] = 2 - q - q^k, \quad q = 1-p.$$

Recursive application of this formula gives

$$\mathrm{E}[T_k] = 2k - (k-3)q - q^2 - \frac{1-q^{k+1}}{1-q}, \quad k = 2, 3, \ldots .$$

The (asymptotic) relative cost is

$$\mathrm{RC} = 2 - q + \frac{1}{k}\left[3q - q^2 - \frac{1-q^{k+1}}{1-q}\right].$$

2.1.4. The curtailed Sterrett procedure

The modification suggested to the Dorfman procedure can also be made to the Sterrett procedure. Figure 5 shows the change in the Sterrett procedure due to curtailing. Let J be as defined in Section 2.1.3 above. Then

$$\mathrm{E}(T_k) = \sum_{j=0}^{k} \mathrm{E}[T_k | J = j]p[J = j]$$

$$= q^k + kpq^{k-1} + \sum^{k-2} pq^j + \sum_{j=1}^{k-1} \mathrm{E}[T_j]pq^{k-j-1}.$$

Then, as in Section 2.1.3,

$$q\mathrm{E}[T_{k-1}] = q^k + (k-1)pq^{k-1} + \sum_{j=0}^{k-3} (j+2)pq^{j+1} + \sum_{j=0}^{k-2} \mathrm{E}[T_j]pq^{k-j-1},$$

$$k = 3, 4, \ldots .$$

CURTAILED STERRETT PROCEDURE

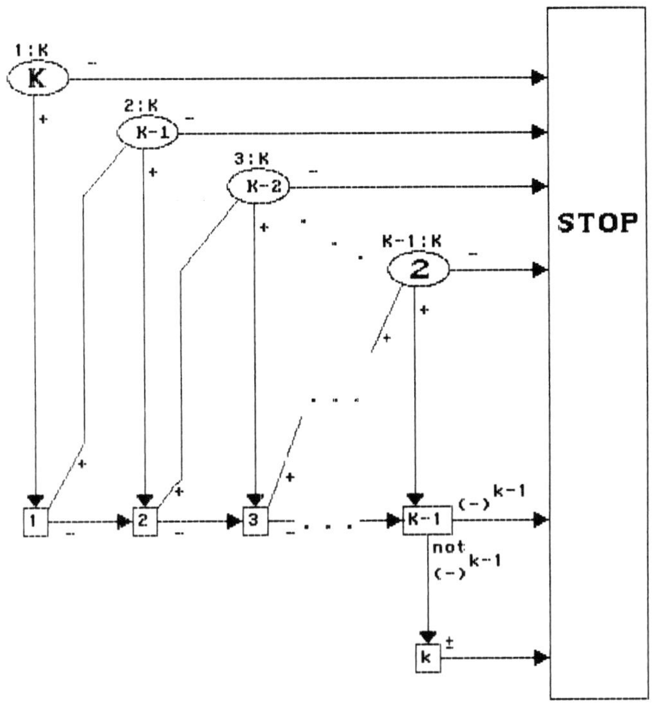

1:K (K) denotes a composite sample of size k, comprising of individual samples s, s+1, ..., t, where t − s + 1 = k;

[i] denotes the i-th individual sample, i=1, 2, ..., k.

$(-)^{k-1}$ denotes the event that all of the first k-1 individual samples tested negative;

not$(-)^{k-1}$ denotes the event that at least one of the first k-1 individual samples tested positive.

Fig. 5. The curtailed Sterrett retesting procedure.

We therefore obtain

$$\mathrm{E}[T_k] - q\mathrm{E}[T_{k-1}] = qp^{k-1} + 2p + q(1 - q^{k-2}), \quad k = 3, 4, \ldots .$$

Solving as in Section 2.1.3 gives

$$\mathrm{E}[T_k] = 2k - (k-2)q - \frac{1 - q^{k+1}}{1 - q}, \quad k = 2, 3, \ldots .$$

The relative cost is then given by

$$\mathrm{RC} = 2 - q + \left[2q - \frac{1 - q^{k+1}}{1 - q} \right] / k.$$

It is interesting to note that the number of tests may be any number from 1 through $2k - 1$, except 2.

2.1.5. The Gill and Gottlieb procedure

Assuming a binomial model, Gill and Gottlieb (1974) proposed that positive testing groups be divided into two as nearly equal subgroups as possible, and positive testing subgroups be recursively subjected to division and testing. Figure 6 indicates how the Gill and Gottlieb procedure works, while Figures 7 and 8 illustrate the Gill and Gottlieb procedure for the cases $k = 4$ and $k = 8$ respectively. The graphs in Figure 9 show the relative costs of classification for the Dorfman and the Gill and Gottlieb procedures as functions of composite sample sizes for prevalences $q = 0.9$, 0.95, 0.99, and 0.999.

Let T_k be the number of tests required to classify the k individual samples in a composite. When retesting is required, the composite of k individual samples is divided into two groups of sizes k_1 and $k_2 = k - k_1$, where $k_1 = k_2 = k/2$ if k is even and $k_1 = k_2 - 1 = (k-1)/2$ if k is odd. One can write $k_1 = [k/2]$ and $k_2 = [(k+1)/2]$ where $[x]$ denotes the integer part of x.

As earlier, conditioning on the random variable J, the expression for $\mathrm{E}[T_k]$ is derived below. $J = 0$ if no retesting is required; $J = 1$ if the composite of size k_1 tests positive; and $J = 2$ otherwise. Let $Y_k = 0$ or 1 according as the composite sample of size k tests negative or positive, respectively. Using conditional expectations, then,

$$\begin{aligned} \mathrm{E}[T_k] = q^k &+ \{1 + \mathrm{E}[T_{k_1} \mid Y_{k_1} = 1] + \mathrm{E}[T_{k_2}]\}(1 - q^{k_1}) \\ &+ \{2 + \mathrm{E}[T_{k_2} \mid Y_{k_2} = 1]\} q^{k_1}(1 - q^{k_2}). \end{aligned}$$

Now, for $i = 1, 2$,

$$\mathrm{E}[T_{k_i}] = \mathrm{E}[T_{k_i} \mid Y_{k_i} = 0] q^{k_i} + \mathrm{E}[T_{k_i} \mid Y_{k_i} = 1](1 - q^{k_i});$$

and so

$$\mathrm{E}[T_{k_i} \mid Y_{k_i} = 1](1 - q^{k_i}) = \mathrm{E}[T_{k_i}] - q^{k_i}.$$

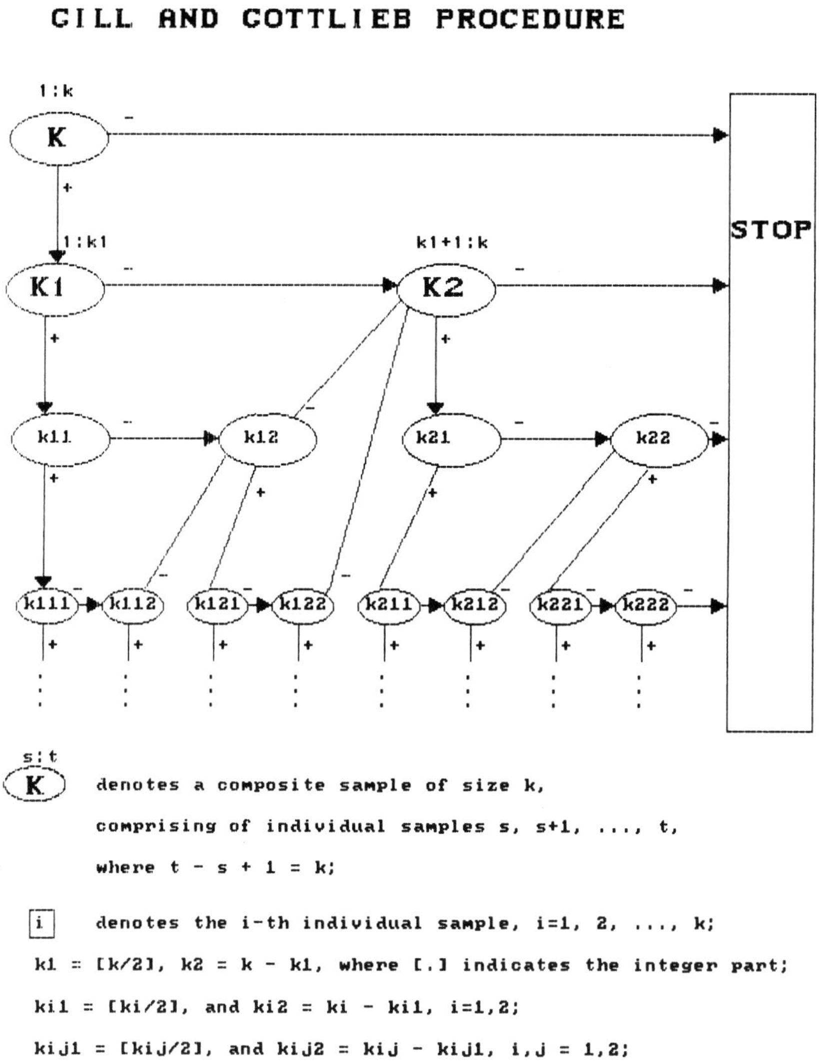

Fig. 6. The Gill and Gottlieb retesting procedure.

Therefore,

$$E[T_k] = E[T_{k_1}] + E[T_{k_2}] + 1 - 2q^k.$$

2.1.6. The curtailed Gill and Gottlieb procedure

As in the case of the Dorfman and Sterrett procedures, the Gill and Gottlieb procedure can also be modified to eliminate redundant tests, assuming that measurements are made without error. Suppose that the composite sample of size k

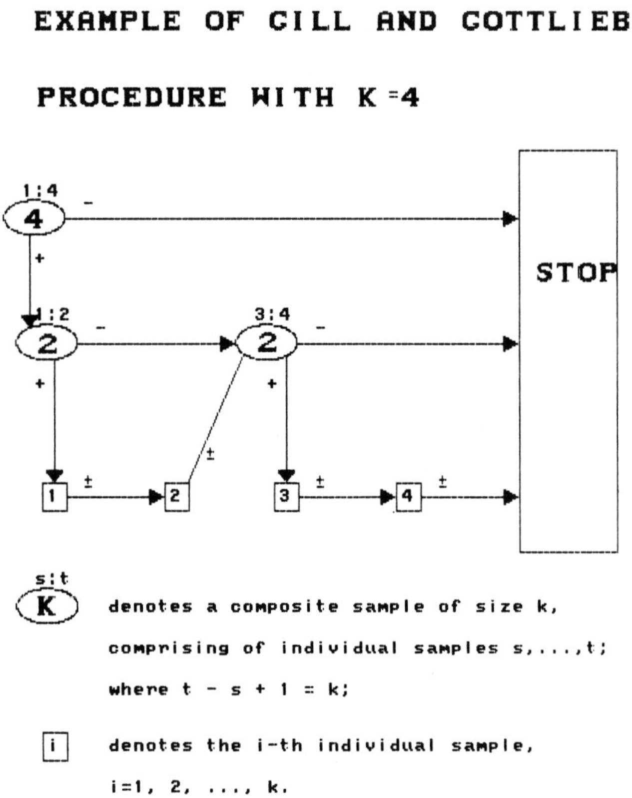

Fig. 7. Example of the Gill and Gottlieb retesting procedure with $k = 4$.

tested positive, and therefore is split into a composite sample of size $k_1 = [k/2]$ and the other composite sample of size $k_2 = [(k+1)/2]$. If the composite sample of size k_1 tests negative, then it can be inferred that the composite sample of size k_2 must test positive, and testing that composite sample may be avoided. The following recursion formula can be obtained in the same way as one for the uncurtailed procedure:

$$\mathrm{E}[T_k] = \mathrm{E}[T_{k_1}] + \mathrm{E}[T_{k_2}] + 1 - q^{k_1} - q^k.$$

2.1.7. The Hwang procedure

Hwang (1984) assumed a binomial model and modified the Gill and Gottlieb procedure so that the feedback from completed tests can be used to determine successive test groups. As in the Gill and Gottlieb procedure, positive testing groups of size k are divided into two subgroups as nearly equal as possible, and one subgroup of size k_1, say, is tested. If the result of the test is negative, then all the k individual samples are classified as not possessing the trait and the complimentary subgroup of size $k_2 = k - k_1$, is automatically judged to test positive and is therefore

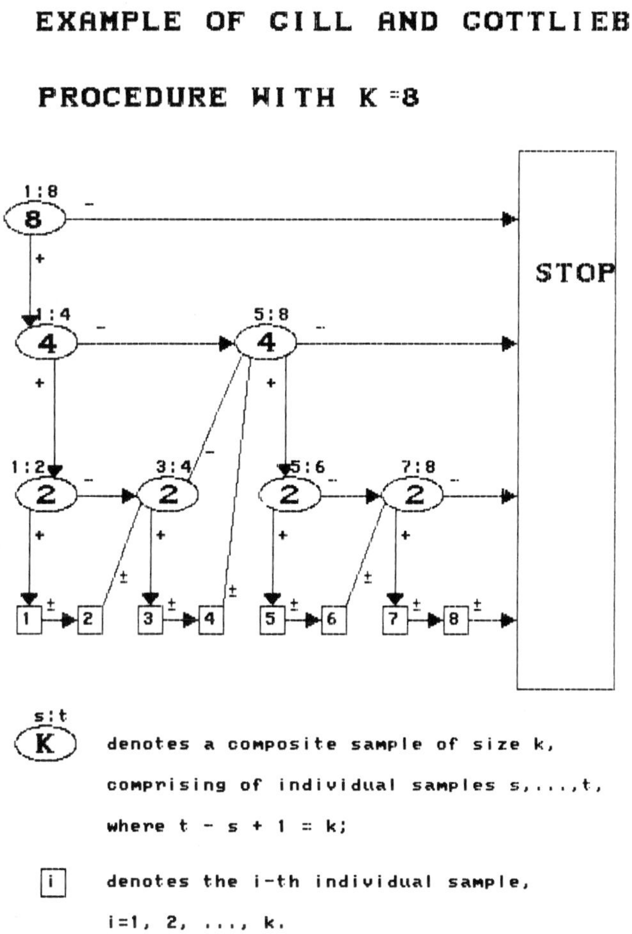

Fig. 8. Example of the Gill and Gottlieb retesting procedure with $k = 8$.

subdivided for further retesting. Conversely, if the first subgroup tests positive, then the second subgroup is not classified and, in fact, is returned to the pool of untested samples. The positive testing subgroup is divided and the same procedure is repeated. In this case, it should be noted that the first positive testing subgroup is divided for testing, as before. At each retesting stage, in this way, there are two groups, one of which is either judged to test positive and used to continue the subdivisions, or is returned to the pool of untested samples. The result of every successive test is either to classify all the items of the group as testing negative and proceed to subdivide the other subsample, or, retain the subsample for subdivision and return the other subsample to the pool of untested sample. Since the number of individual samples that get classified is random, it is not enough to merely calculate the expected number of tests for a single composite sample.

a

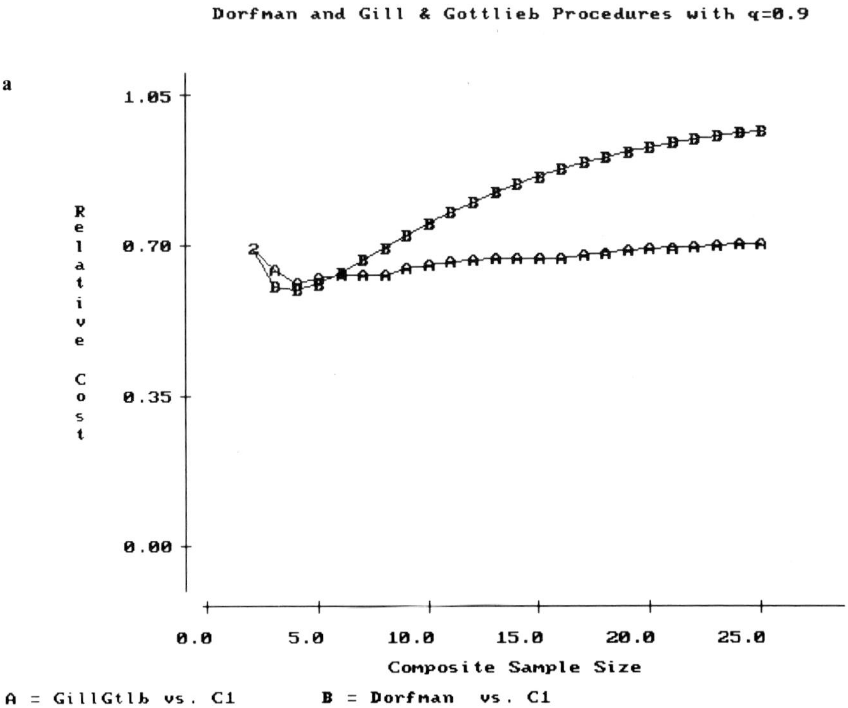
Dorfman and Gill & Gottlieb Procedures with q=0.9
Relative Cost
1.05
0.70
0.35
0.00
0.0
5.0
10.0
15.0
20.0
25.0
Composite Sample Size
A = GillGtlb vs. C1
B = Dorfman vs. C1

b

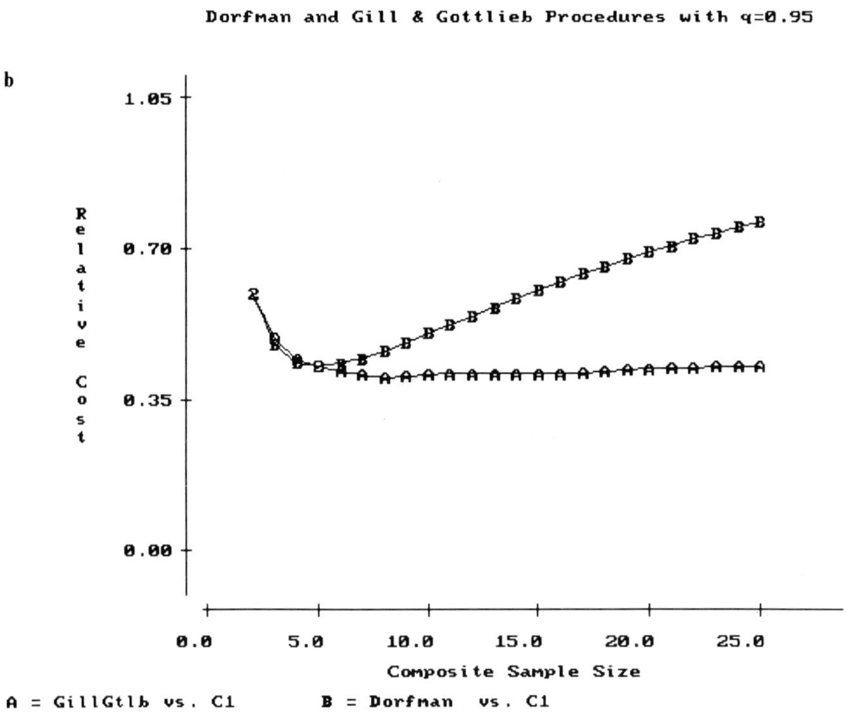
Dorfman and Gill & Gottlieb Procedures with q=0.95
Relative Cost
1.05
0.70
0.35
0.00
0.0
5.0
10.0
15.0
20.0
25.0
Composite Sample Size
A = GillGtlb vs. C1
B = Dorfman vs. C1

c

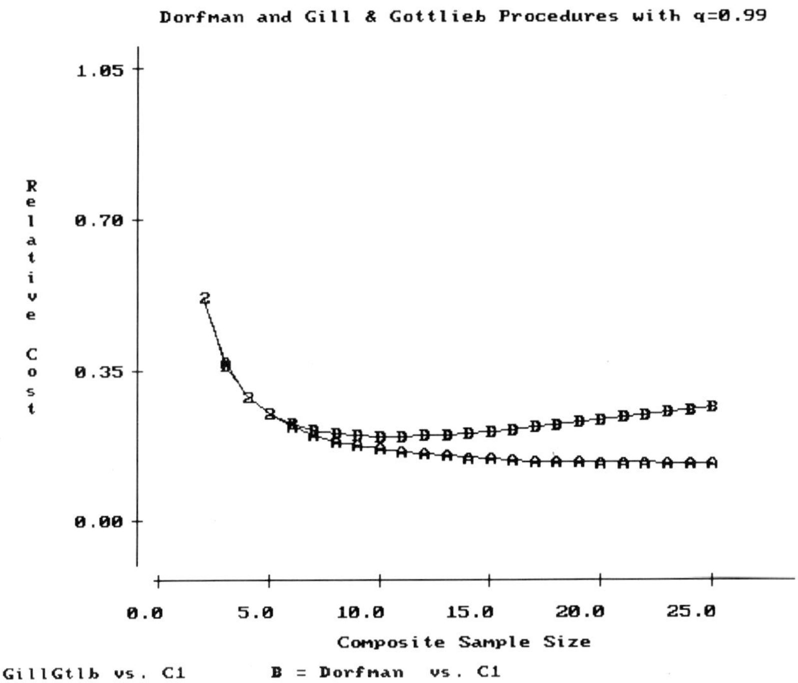

d

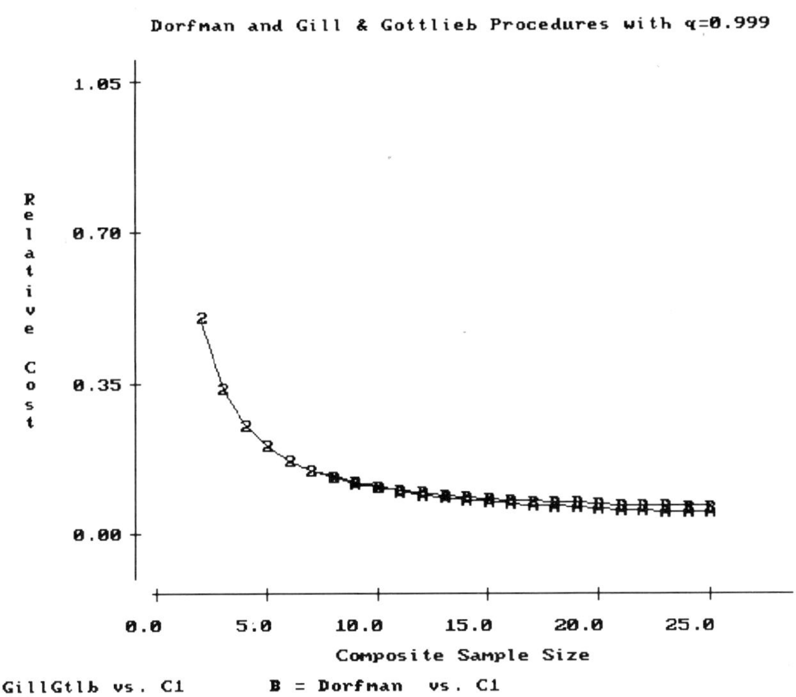

Fig. 9. Relative costs of Dorfman and Gill and Gottlieb procedures. A: Gill and Gottlieb procedure; B: Dorfman procedure. (a) $q = 0.9$, (b) $q = 0.95$, (c) $q = 0.99$, (d) $q = 0.999$.

When a composite sample is processed completely, including any retesting, then either all the k constituent individual samples are classified as testing negative, or exactly one individual sample is classified as testing positive. All individual samples tested, individually or in a composite sample, before the positive testing samples are classified as testing negative, and all individual samples not tested before identifying the positive testing sample remain unclassified. For a composite sample of size $k = 2^{m-1}$, either 1 or m tests are required for this classification. For a composite sample of size greater than 2^{m-2} and less than 2^{m-1}, either $1, m-1$ or m tests are required, depending on the location J of the first positive testing sample. (Note $J = 0$ if the composite sample tests negative). Let T_r be the number of tests to classify a total of r individual samples in a composite sample of a fixed size k. Note that T_r depends on k. The following cases illustrate a method to derive $\mathrm{E}(T_r)$ for some small values of k.

Case 1: *Composite samples of size* 2 *are used.* There are either 1 or 2 tests for each composite sample, and either one or both the samples are classified. In this case, one finds that

$$\mathrm{E}[T_m] - \mathrm{E}[T_{m-1}] = (2 - 2q - q^2 + q^3) + q^2\{\mathrm{E}[T_{m-2}] - \mathrm{E}[T_{m-3}]\},$$
$$m = 3, 4, \ldots .$$

The difference on the left is the average number of tests to classify one individual sample when there are m individual samples to be classified. For large m, this is essentially a constant, giving us the asymptotic relative cost

$$\mathrm{RC} = (2 - q^2)/(1 + q).$$

Case 2: *Composite samples of size* 3 *are used.* There are 1, 2 or 3 tests for each composite sample and 1, 2 or 3 individual samples are classified. In this case, one has

$$\mathrm{E}[T_m] - \mathrm{E}[T_{m-1}] = (2 - q - q^2 - 2q^3 + 2q^4) + q^3\{\mathrm{E}[T_{m-3}] - \mathrm{E}[T_{m-4}]\},$$
$$m = 4, 5, \ldots .$$

Allowing $m \to \infty$, the asymptotic relative cost is obtained as

$$\mathrm{RC} = (2 + q - 2q^3)/(1 + q + q^2).$$

Case 3: *Composite samples of size* 4 *are used.* There are either 1 or 3 tests. Again, as earlier, we have

$$\mathrm{E}[T_m] - \mathrm{E}[T_{m-1}] = (3 - 3q - 2q^4 + 2q^5) + q^4\{\mathrm{E}[T_{m-4}] - \mathrm{E}[T_{m-5}]\},$$
$$m = 5, 6, \ldots .$$

The asymptotic relative cost is then given by

$$\mathrm{RC} = (3 - 2q^4)/(1 + q + q^2 + q^3).$$

Table 2

m	RC
7	$(3+q-3q^7)/(1+q+q^2+\cdots+q^6)$
8	$(4-3q^8)/(1+q+q^2+\cdots+q^7)$
9	$(4+q^7-4q^9)/(1+q+q^2+\cdots+q^8)$
10	$(4+q^3-q^5+q^8-4q^{10})/(1+\cdots+q^9)$

Case 4: *Composite samples of size* 5 *are used.* There are 1, 3 or 4 tests and 1, 2, 3, 4 or 5 samples are classified. In this case, we obtain

$$\mathrm{E}[T_m]-\mathrm{E}[T_{m-1}]=(5-7q+2q^2+q^3-q^4 3q^5+3q^6)$$
$$+q^5\{\mathrm{E}[T_{m-5}]-\mathrm{E}[T_{m-6}]\},\quad m=7,8,\ldots.$$

Table 2 summarizes the asymptotic relative cost when composite samples of size $m=7,8,9$, or 10 are used.

The optimal composite sample size depends on both m and q. Table 3 gives the optimal composite sample size as a function of q for values of $m=2$ through 16. Figure 10 shows the asymptotic relative cost ($m=\infty$) for composite sample sizes of 2 through 12.

2.2. *A Bayesian approach to the classification problem*

Consider the Dorfman retesting procedure for classifying individual samples. Table 1 gives the optimal composite sample size k for any prevalence p, where $0.0012 \leqslant p \leqslant 1$. What is important to note, however, is the fact that the prevalence p is usually unknown and therefore the optimal composite sample size cannot be determined accurately. One possible method of choosing k is to predict the value of p using the available information, and then determine k from Table 1. The predicted value of p will generally be different from its true value and can therefore lead to two possible errors: a strategic error and a tactical error. A strategic error is committed if either we employ composite sampling when its relative cost exceeds unity, or we do not employ composite sampling when its relative cost would be below unity. For example, consider a situation with the true prevalence of $p=0.5$. In this case, composite sampling with any composite sample size $k \geqslant 3$ incurs a relative cost exceeding unity, the relative cost of exhaustive testing. The relatlve cost of composite sampling for the composite sample size k varying from 3 through 50 is shown in Figure 11a. Next, consider another situation where the true prevalence is $p=0.05$. Suppose a strategic error of not employing composite sampling is committed, thereby incurring a relative cost of unity. The Dorfman retesting procedure with any composite sample size k between 3 and 50 would have incurred a relative cost below unity, as shown in Figure 11b. A tactical error, on the other hand, is an error of choosing a nonoptimal composite sample size, and thus not attaining the smallest

Table 3
Range of q for the optimal values of k, $m = 1:16$

k	$m=1$	$m=2$	$m=3$	$m=4$	$m=5$	$m=6$	$m=7$	$m=8$
1	0.0 to 1.0	0.0 to 0.62	0.0 to 0.62	0.0 to 0.62	0.0 to 0.62	0.0 to 0.62	0.0 to 0.62	0.0 to 0.62
2	–	0.62 to 1.0	0.62 to 0.71	0.62 to 0.76	0.62 to 0.73	0.62 to 0.75	0.62 to 0.75	0.62 to 0.75
3	–	–	0.71 to 1.0	–	0.73 to 0.81	0.75 to 0.83	0.75 to 0.79	0.75 to 0.81
4	–	–	–	0.76 to 1.0	–	0.83 to 0.84	0.79 to 0.86	0.81 to 0.88
5	–	–	–	–	0.81 to 1.0	–	–	–
6	–	–	–	–	–	0.84 to 1.0	–	–
7	–	–	–	–	–	–	0.86 to 1.0	–
8	–	–	–	–	–	–	–	0.88 to 1.0

k	$m=9$	$m=10$	$m=11$	$m=12$	$m=13$	$m=14$	$m=15$	$m=16$
1	0.0 to 0.62	0.0 to 0.62	0.0 to 0.62	0.0 to 0.62	0.0 to 0.62	0.0 to 0.62	0.0 to 0.62	0.0 to 0.62
2	0.62 to 0.75	0.62 to 0.78	0.62 to 0.75	0.62 to 0.75	0.62 to 0.75	0.62 to 0.75	0.62 to 0.75	0.62 to 0.76
3	0.75 to 0.82	0.75 to 0.81	0.75 to 0.81	0.75 to 0.81	0.75 to 0.82	0.75 to 0.82	0.75 to 0.82	0.76 to 0.81
4	–	0.81 to 0.83	0.81 to 0.87	0.81 to 0.86	0.82 to 0.84	0.82 to 0.84	0.82 to 0.85	0.81 to 0.87
5	0.82 to 0.89	0.83 to 0.90	–	0.86 to 0.87	0.84 to 0.88	0.84 to 0.88	0.85 to 0.89	–
6	–	–	0.87 to 0.88	–	–	–	–	–
7	–	–	0.88 to 0.91	0.87 to 0.92	0.88 to 0.92	0.88 to 0.93	–	0.87 to 0.89
8	–	–	–	–	–	–	–	0.89 to 0.90
9	0.89 to 1.0	–	–	–	–	–	0.89 to 0.93	–
10	–	0.90 to 1.0	–	–	–	–	–	–
11	–	–	0.91 to 1.0	–	–	–	–	0.90 to 0.93
12	–	–	–	0.92 to 1.0	–	–	–	–
13	–	–	–	–	0.92 to 1.0	–	–	–
14	–	–	–	–	–	0.93 to 1.0	0.93 to 1.0	–
15	–	–	–	–	–	–	–	–
16	–	–	–	–	–	–	–	0.93 to 1.0

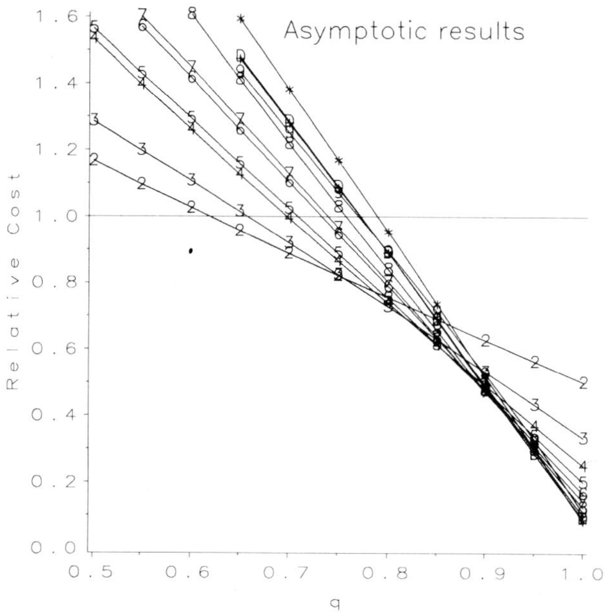

Fig. 10. Shows the relative cost (expected number of tests per individual sample classified) for composite sample sizes $k = 2$ through 12. These are functions of q^{k-1} (the probability an individual sample will be classified as negative. The smaller the relative cost the better. For values of q^{k-1} less than 0.62 it is cheaper to use $k = 1$ (no composite sampling). For values of q^{k-1} 0.62 through 0.78 composite samples of size $k = 2$ are the most efficient.

achievable relative cost. For instance, consider a situation with the prevalence of $p = 0.01$. The optimal composite sample size in this case is 11 with a relative cost of 0.1956. An inaccurate prediction of p would result in a nonoptimal composite sample size, thus incurring a relative cost above 0.1956, though still below unity. Figure 11c depicts this situation, while Figure 11d puts these three cases together on a single graph.

The consequences of an inaccurate prediction of the prevalence p become more serious if the total number of individual samples to be classified is very large. However, in such a case, it may be reasonable to make appropriate changes in the predicted value of the prevelance p from time to time, as the results of classifications become progressively available. As observed by West (1986), this allowance makes the problem suitable for a sequential Bayes approach. What we propose is the following method: An initial prediction, p_0 say, of the prevalence is based purely on the prior information, which may include results of previous studies, expert knowledge, and any other information that may be available. The initial composite sample size, k_0 say, is the optimal composite sample size corresponding to the prevalence p_0. After a composite sample of size k_0 is classified, suppose there are X_0 individual samples that are classified as testing positive. Using this number X_0, we update the p = 0.01

a

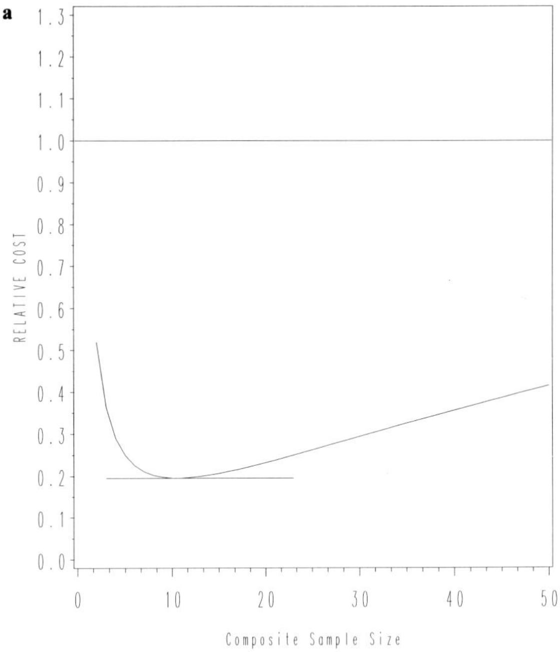
1.3
1.2
1.1
1.0
0.9
0.8
0.7
0.6
0.5
0.4
0.3
0.2
0.1
0.0
RELATIVE COST
0
10
20
30
40
50
Composite Sample Size

p = 0.05

b

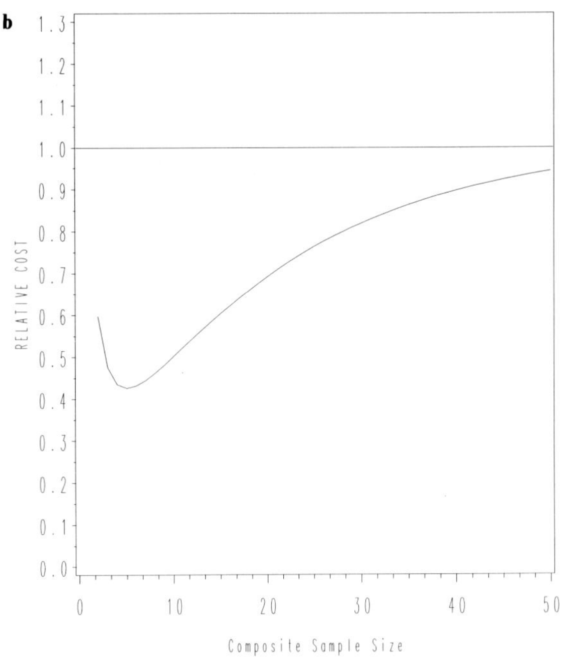
1.3
1.2
1.1
1.0
0.9
0.8
0.7
0.6
0.5
0.4
0.3
0.2
0.1
0.0
RELATIVE COST
0
10
20
30
40
50
Composite Sample Size

c

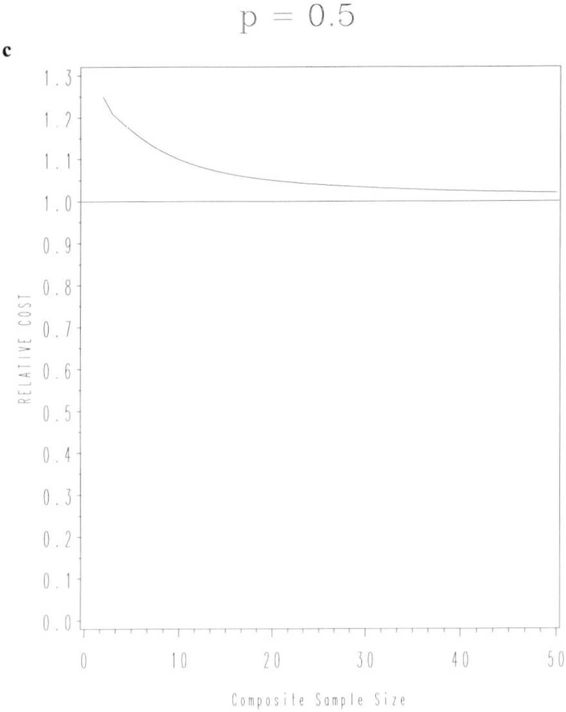

d

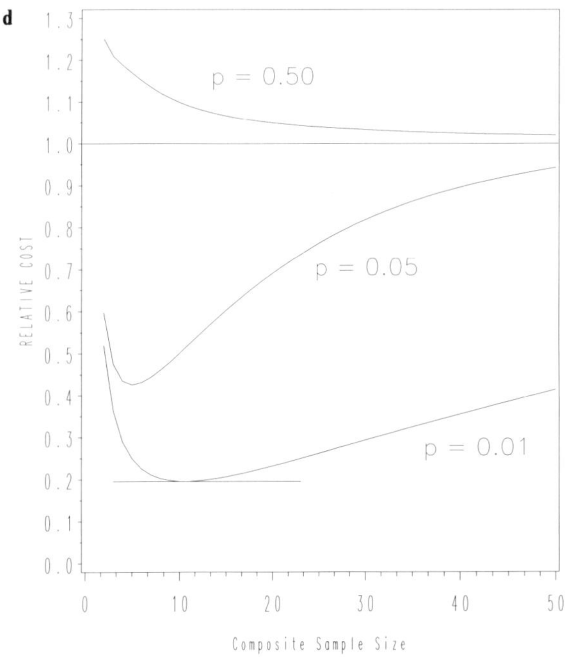

Fig. 11. (a) Effects of a tactical error. The small horizontal line represents the relative cost if the optimal composite sample size is used. The curve represents the actual relative cost of composite sampling as a function of the composite sample size. (b) and (c) Effects of strategic errors. The horizontal line at relative cost = 1 represents the relative cost of exhaustive testing. The curve represents the actual relative cost of composite sampling as a function of the composite sample size. (d) Shows (a), (b) and (c) on a single graph to facilitate comparison. Note: In (d), $p = 0.05$ near the curve in the center $p = 0.01$ near the curve at the bottom is required.

predicted prevalence from p_0 to p_1 which, in turn, updates the composite sample size to k_1, say. After a composite sample of size k_1 is classified, resulting in X_1 positive testing individual samples, we further update the prevalence from p_1 to p_2, and so on. It, however, remains to determine how the predicted prevalence should be updated using the prior prediction and the observed proportion of positive testing individual samples.

2.2.1. A Moment updating procedure

The problem described in the foregoing discussion is to obtain an updating formula for the predicted value of p. After a sampling stage is carried out, a new predicted value of p is found, and a new sampling stage is undertaken. Now, X_i is the number of samples classified as polluted and n_i is the total number of samples classified in the i-th sampling stage. Here it is not necessary to assume any particular composite sampling technique as being used to classify the individual samples; any method may have been used. We obtain the first two posterior moments of the predicted prevalence rate p_i, using its prior moments and the data. Then it is possible to obtain $\mathrm{E}[p_{i+1} \mid X_i]$ in terms of $\mathrm{E}[p_i]$, $\sigma^2(p_i)$ and X_i. We define the dispersion ratio by

$$\tau_i = \frac{k_i^2 \sigma^2(p_i)}{\operatorname{var}(X_i)}.$$

It is easy to see that

$$\begin{aligned}
\operatorname{Var}(X_i) &= \mathrm{E}[\operatorname{var}(X_i \mid p_i)] + \operatorname{var}[\mathrm{E}(X_i \mid p_i)] \\
&= \mathrm{E}[n_i \cdot p_i(1-p_i)] + \operatorname{var}[n_i \cdot p_i] \\
&= n_i \mathrm{E}[p_i(1-p_i)] + n_i^2 \sigma^2(p_i).
\end{aligned}$$

Thus

$$\tau_i = \frac{n_i^2 \sigma^2(p_i)}{\operatorname{var}(X_i)} = \frac{n_i^2 \sigma^2(p_i)}{n_i \mathrm{E}[p_i(1-p_i)] + n_i^2 \sigma^2(p_i)}$$

is the proportion of variation in X_i explained by variation in p_i (due to its random nature).

Our interest is in obtaining the posterior mean of p_i given the data X_i. It is easy to obtain

$$\mathrm{E}[p_i \mid X_i] = (1-\tau_i)\mathrm{E}[p_i] + \tau_i X_i / n_i.$$

In order to proceed, we also need the posterior dispersion ratio, and this is where the knowledge of only the first two prior moments of p_i is found not to be sufficient. The third moment of p_i is required to obtain the posterior dispersion ratio, τ_{i+1} say. Thus, at the second stage, computation of τ_2 will require the third moment of the prior distribution of p_1. Then the third stage will require the fourth moment of the prior distribution of p_1. In this way, every stage of updating will require a moment of one order higher than that required at the preceding stage. This requirement may not be practicable since it involves specifying the prior information in terms of moments of p_i without assuming any family of distributions.

2.2.2. Bayesian updating of p

The Bayesian approach is convenient, as noted in West (1986), and routine periodic monitoring is an appropriate situation for a sequential Bayesian approach. If we assume the full prior distribution rather than first few moments of p, then observed data allow us to update our prior distribution and calculate a new predicted value for p. The only situation where the first two moments of the prior are sufficient for obtaining the first two moments of the posterior is when a conjugate prior distribution is used. This implies a beta prior distribution and results in a beta posterior distribution. The first two moments of a beta distribution characterize the entire distribution, and hence the assumption of a conjugate prior eliminates the difference between the two approaches.

Let the initial beta prior on p have parameters a_1 and b_1. This distribution has the mean given by

$$p_1 = \frac{a_1}{a_1 + b_1}.$$

For the Dorfman retesting scheme, the composite sample size used for the first stage k_1 is the value which minimizes the relative cost for $p = p_1$. That is, composite samples of size k_1 are formed and tested. Suppose X_1 individual samples out of n_1 samples are classified as testing positive in the first sampling stage. The posterior distribution of p given X_1 is beta with parameters $a_2 = a_1 + X_1$ and $b_2 = b_1 + n_1 - X_1$. This distribution is used as the prior distribution of p for determining the composite sample size for the second sampling stage. For Dorfman retesting scheme, the optimal composite sample size, k_2, corresponds to the predicted prevalence,

$$p_2 = \frac{a_2}{a_2 + b_2} = \frac{a_1 + X_1}{a_1 + b_1 + n_1}.$$

Composite samples of size k_2 are then formed and tested, yielding X_2 out of n_2 individual samples that test positive for pollution. In general, after r sampling stages when the corresponding individual samples have been classified, the posterior distribution of p becomes the beta distribution with parameters

$$a_{r+1} = a_1 + \sum_{i=1}^{r} X_i$$

and

$$b_{r+1} = b_1 + \sum_{i=1}^{r} n_i - \sum_{i=1}^{r} X_i.$$

This probability distribution is then used as the prior distribution of p for the $(r+1)$-th sampling stage.

The relative cost at the i-th sampling stage depends on the true prevalence p and the composite sample size k_i, which in turn depends on X_{i-1}, n_{i-1} and k_{i-1}. Since closed form formulas are not available for the relative costs at successive stages, computer calculations are used to calculate the binomial probabilities in all possible cases and to evaluate the relative cost, for the first 10 updates. The number of cases increases geometrically with the number of sampling stages. The updating is done after each composite sample has been processed. That is, n_i is taken to be k_i.

Since composite sample techniques are cost efficient only for relatively small prevalences, it was decided to use J-shaped beta prior distributions, and to restrict the prevalence to be smaller than 0.5. If the mean p is restricted to be smaller than 0.5, then the J-shaped beta distribution with mean p and the largest possible variance has $b = 1$ and $a = p/(1-p)$, that is $\alpha = 1.0$ below. If the mean is $p = 0.5$, then the distribution is the uniform distribution. The J-shaped distribution with mean $p < 0.5$ and the smallest possible variance has $a = 1$ and $b = (1-p)/p$, that is $\alpha = 0.0$ below. Then J-shaped curves with intermediate variances are obtained by taking the following linear combinations:

$$a = \alpha \cdot \left(\frac{p}{1-p}\right) + (1+\alpha) \cdot 1,$$

$$b = \alpha \cdot 1 + (1-\alpha)\left(\frac{1-p}{p}\right).$$

The variances of these distributions with $\alpha = 0.0, 0.5, 0.6, 0.7, 0.8, 0.9$, and 1.0 are shown in Figure 12. Large variances cause large changes when updating. When the initial predicted prevalence is far from the true prevalence, large changes are desirable. However, when the initial predicted prevalence is close to the true prevalence, large changes cause the procedure to deviate from the optimal procedure, and it may take several sampling stages for it to approach the optimal procedure.

Although the emphasis here is on the relative costs, the average composite sample sizes are also calculated. This paper assumes that a composite sample of any size can be formed. Small predicted prevalences result in large composite sample sizes. When $\alpha = 1.0$, the composite sample size shoots up dramatically for small predicted prevalences and small true prevalences, and takes several sampling stages to come back down. The use of larger predicted prevalence and/or the use of smaller values of α may avoid this problem.

The relative costs corresponding to the case $\alpha = 1.0$ are given in Figure 13a–p. Each graph corresponds to a different value of the true prevalence. The individual curves correspond to different initial prevalences. For example, Figure 13c shows the results of the computations starting with predicted prevalences of 0.001, 0.01, 0.05, 0.1, 0.3, and 0.5 when the true prevalence is 0.01. The curve delineated by '+' has an estimated prevalence of 0.01, which is also the true prevalence. The initial

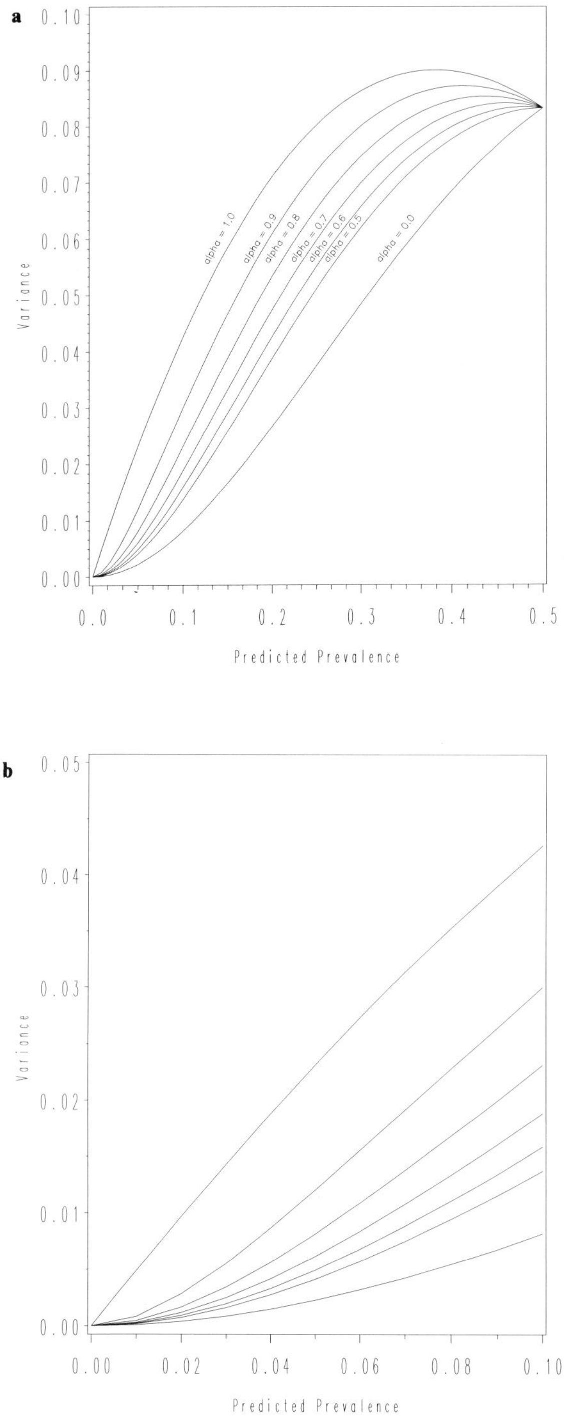

Fig. 12. Variances for J-shaped beta distributions; $a = \alpha \cdot p/(1-p) + (1-\alpha) \cdot 1$, $b = \alpha \cdot 1 + (1-\alpha)(1-p)/p$; $\alpha = 1.0$ gives maximum variance; $\alpha = 0.0$ gives minimum variance.

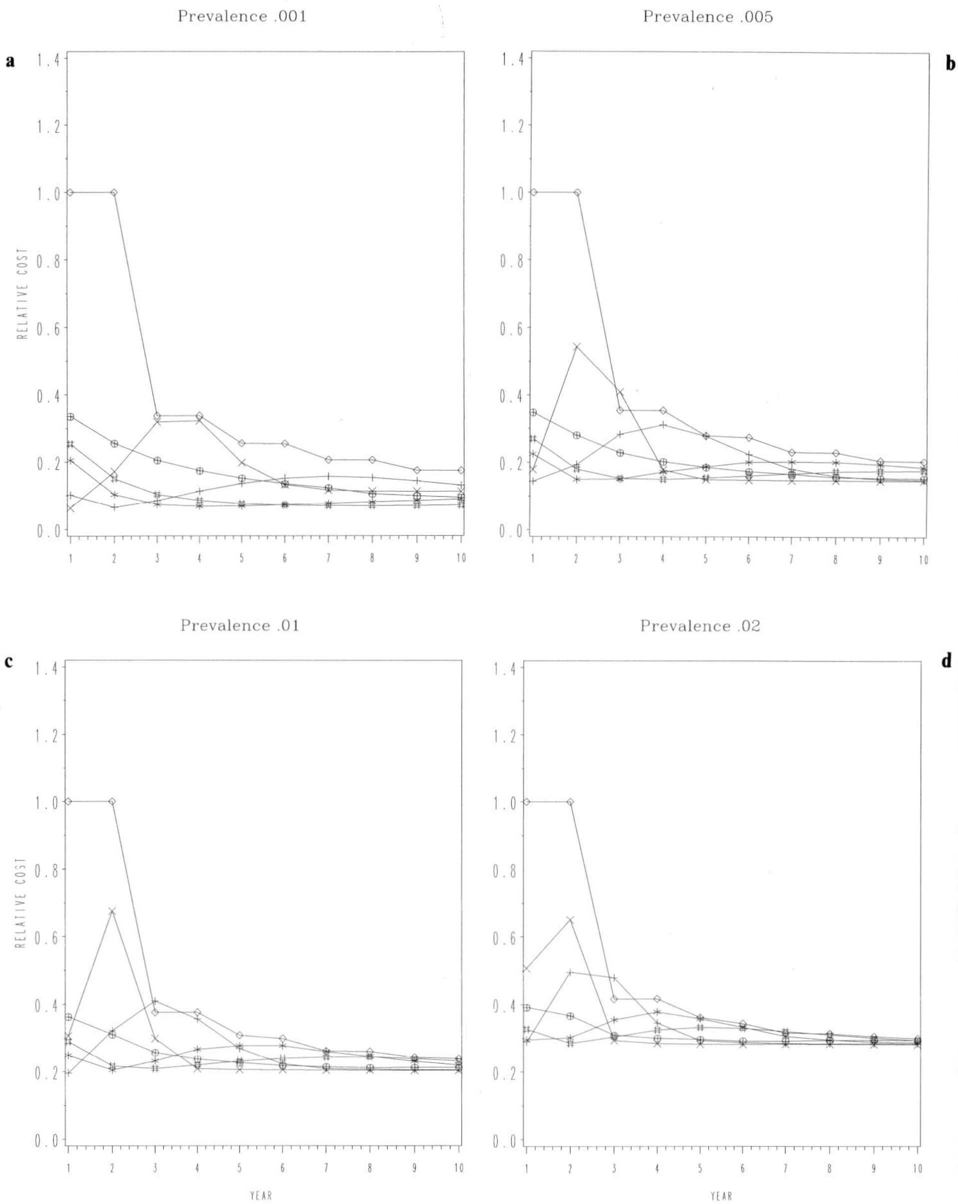

a
Prevalence .001
b
Prevalence .005
c
Prevalence .01
d
Prevalence .02
RELATIVE COST
YEAR
0.0
0.2
0.4
0.6
0.8
1.0
1.2
1.4
1
2
3
4
5
6
7
8
9
10

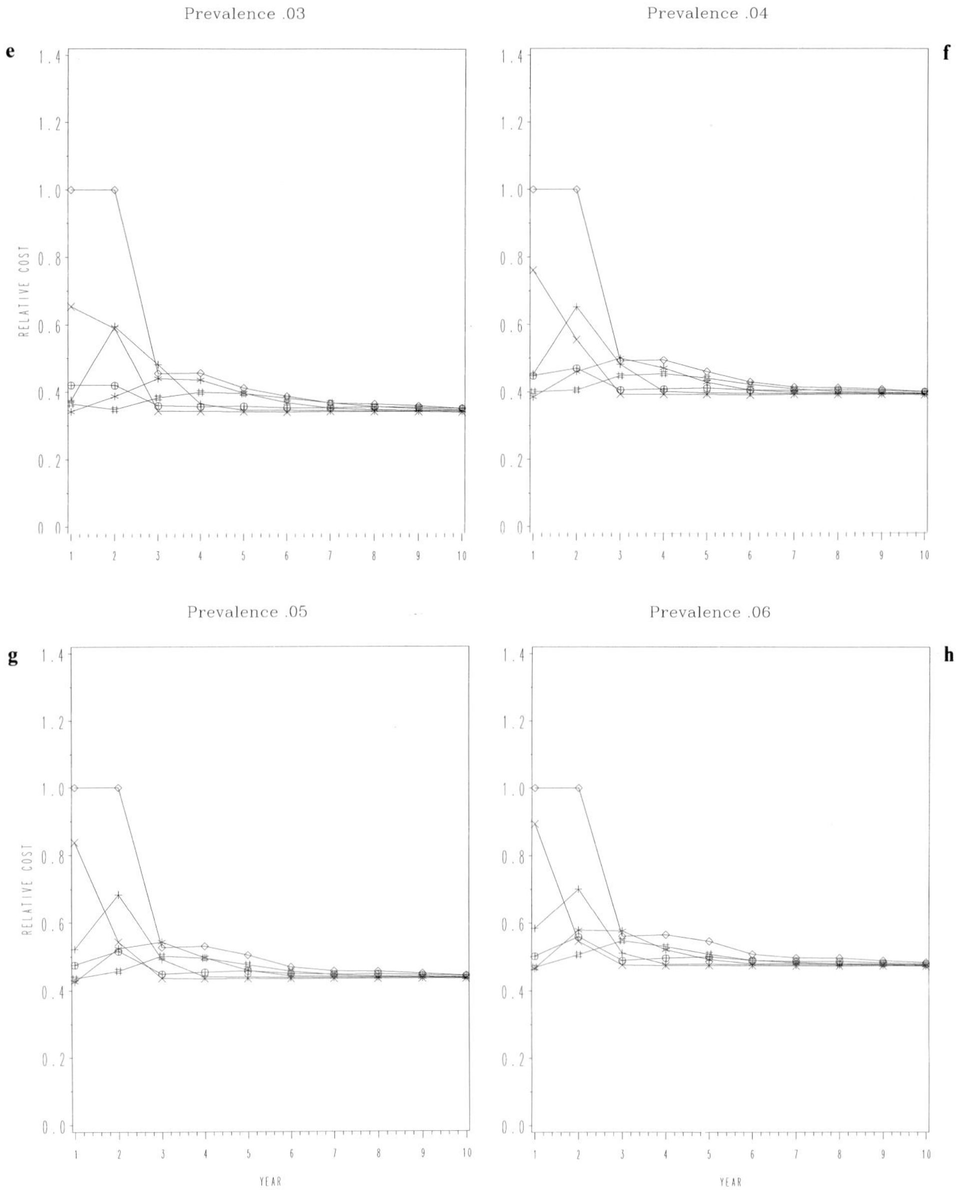

(continued on next page)

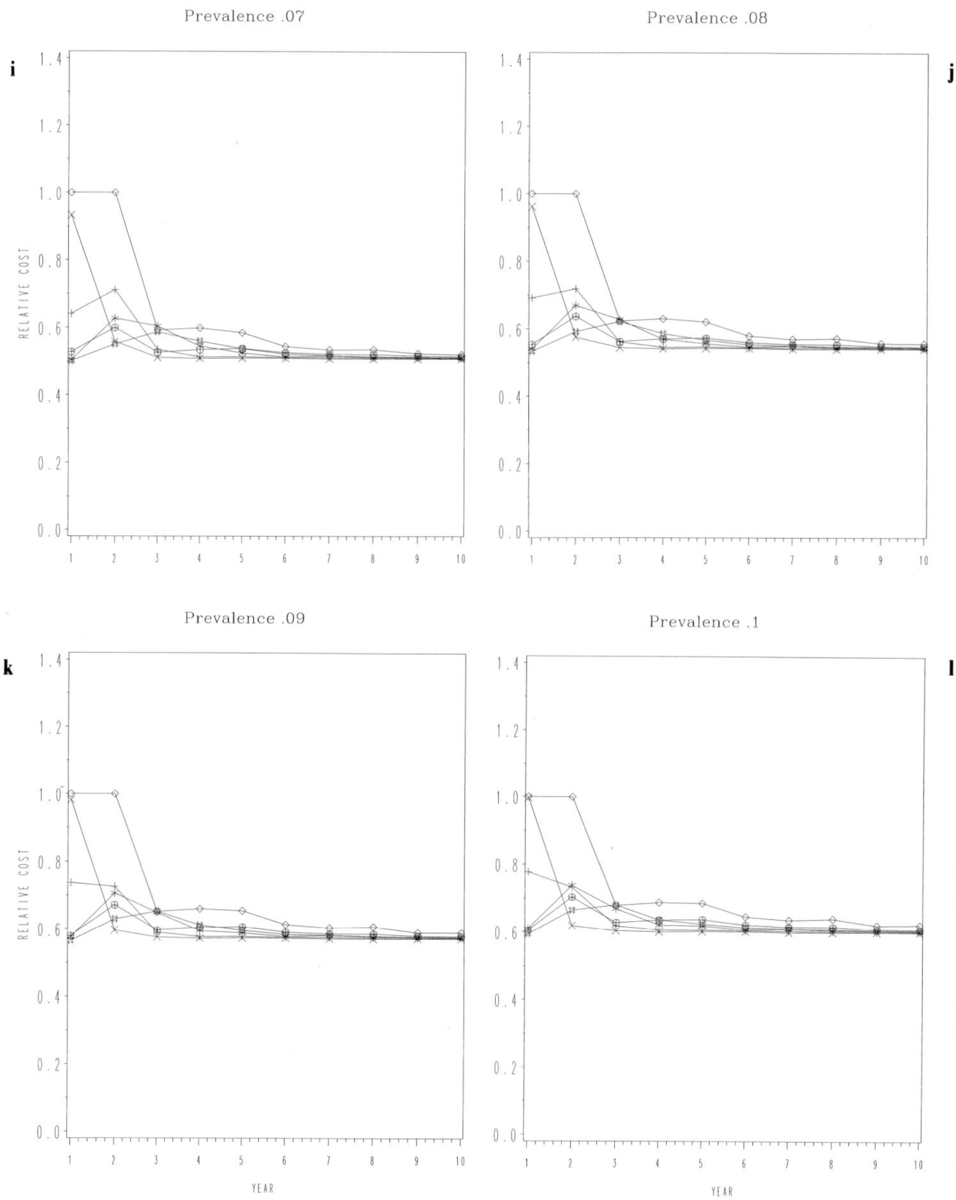

Prevalence .07
Prevalence .08
Prevalence .09
Prevalence .1
i
j
k
l
RELATIVE COST
RELATIVE COST
YEAR
YEAR
1.4
1.2
1.0
0.8
0.6
0.4
0.2
0.0
1
2
3
4
5
6
7
8
9
10

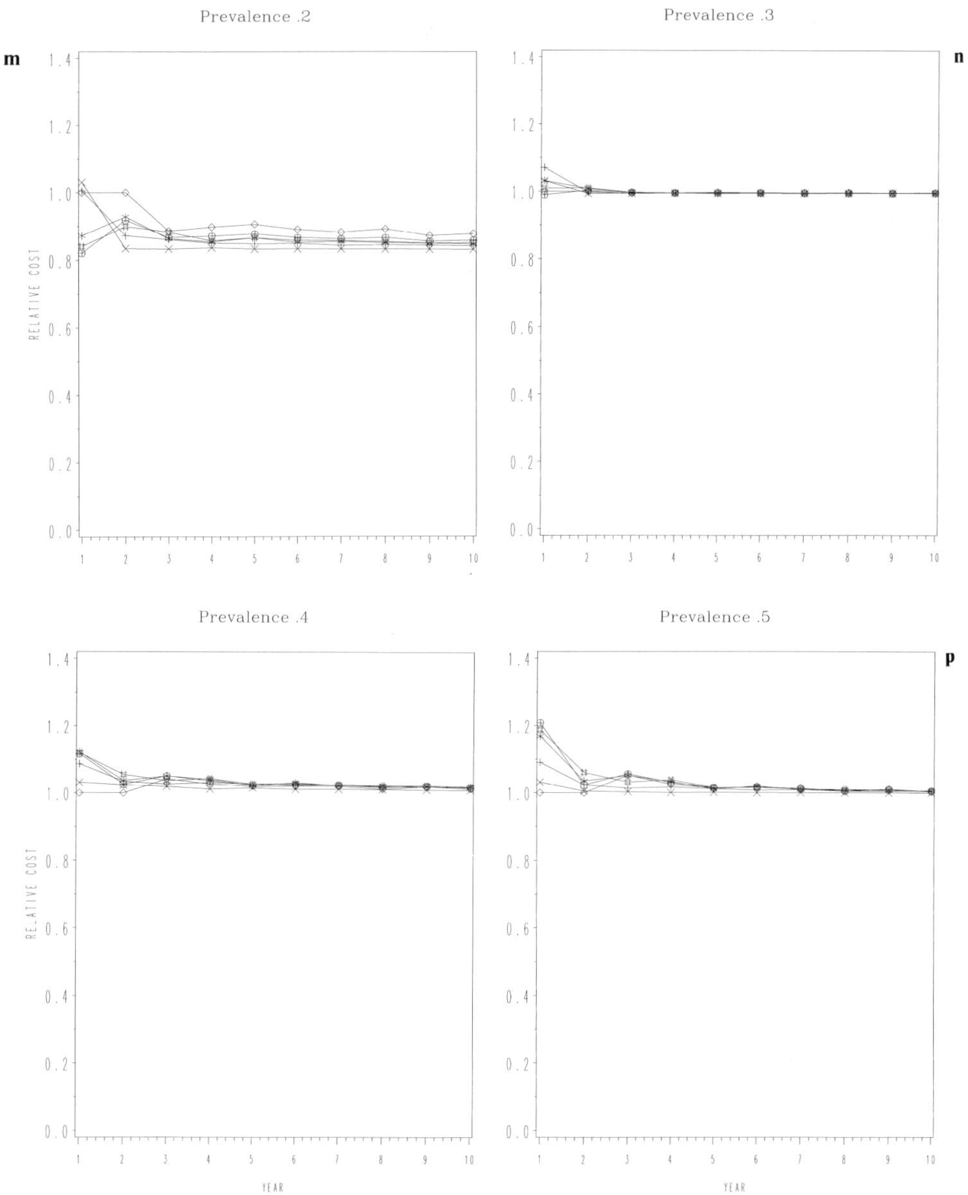

Fig. 13. Relative cost for estimated prevalences 0.001, 0,01, 0.05, 0.1, 0.3 and 0.5 for a given true prevalence, using the model with $\alpha = 1.0$.

composite sample size is optimal, and therefore the relative cost is minimal. Over the next two sampling episodes the process moves away from the optimal and then starts to approach the optimal procedure again.

Similarly, the relative costs corresponding to the case $\alpha = 0.0$ are given in Figure 14a–p. For small predicted prevalences the model for $\alpha = 0.0$ avoids the peak of large relative cost for the first few sampling episodes, and exhibits good convergence towards the optimal procedure. For larger true prevalences the convergence to the optimal procedure is slower than in the case of $\alpha = 1.0$. These observations suggest that it might be best not to use initial predicted values less than 0.01 and to use the model with $\alpha = 1.0$.

2.2.3. A Bayesian approach to minimize the expected relative cost

Recall the relative cost for the Dorfman retesting procedure, given by $\mathrm{RC} = 1 + 1/k - q^k$, is a function of the composite sample size k and the prevalence $p = 1 - q$ of the trait under study. In decision theoretic notation, since the composite sample size is a decision of the statistician and the prevalence of the trait is beyond the statistician's control, the relative cost is a loss function, which could be written as

$$L(k, q) = 1 + 1/k - q^k.$$

The approach in the foregoing discussion imposed a conjugate prior on the prevalence p, giving the expected prevalence $\hat{p}$, say, and then the composite sample size, $\hat{k}$ say, that minimizes $L(k, \hat{q})$ was determined. That is, $\hat{k}$ satisfies

$$L(\hat{k}, \hat{q}) = \min_k L(k, \hat{q}).$$

In this section, we consider a decision theoretical approach, minimizing the expected loss rather than the loss function at the expected prevalence. Thus, we use the prior distribution with a density f, say, to compute the risk function

$$R(k, f) = \mathrm{E}[L(k, q)] = \int_0^1 L(k, q) f(p)\, \mathrm{d}p,$$

and then determine the composite sample size k that minimizes $R(k, f)$. The conjugate prior distribution of the prevalence p is a beta distribution, which has two parameters, a and b. Thus, the risk function can be expressed as

$$\begin{aligned} R(k, \theta) &= \int_0^1 L(k, q) \frac{1}{B(a, b)} p^{a-1} q^{b-1}\, \mathrm{d}p \\ &= 1 + \frac{1}{k} - \frac{B(a, b+k)}{B(a, b)}, \end{aligned}$$

where $\theta = (a, b)$, and $B(a, b) = \Gamma(a)\Gamma(b)/\Gamma(a + b)$. The composite sample size k that

a

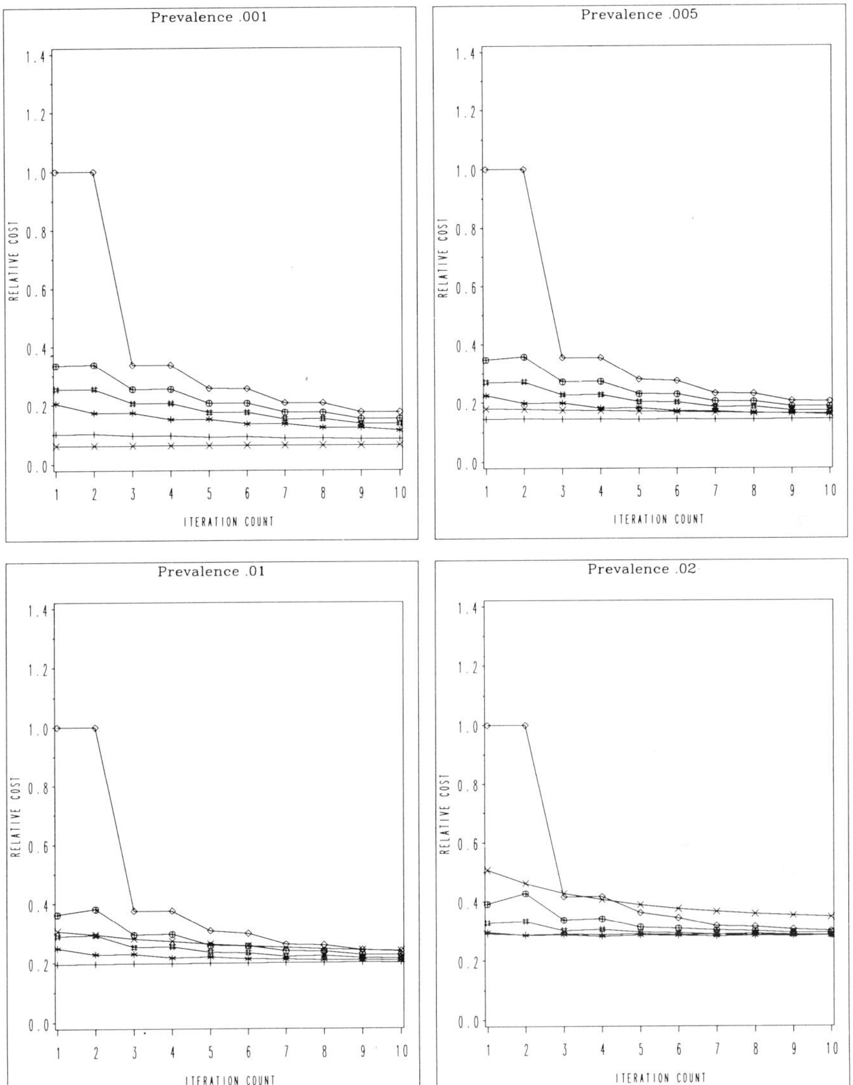

(continued on next page)

b

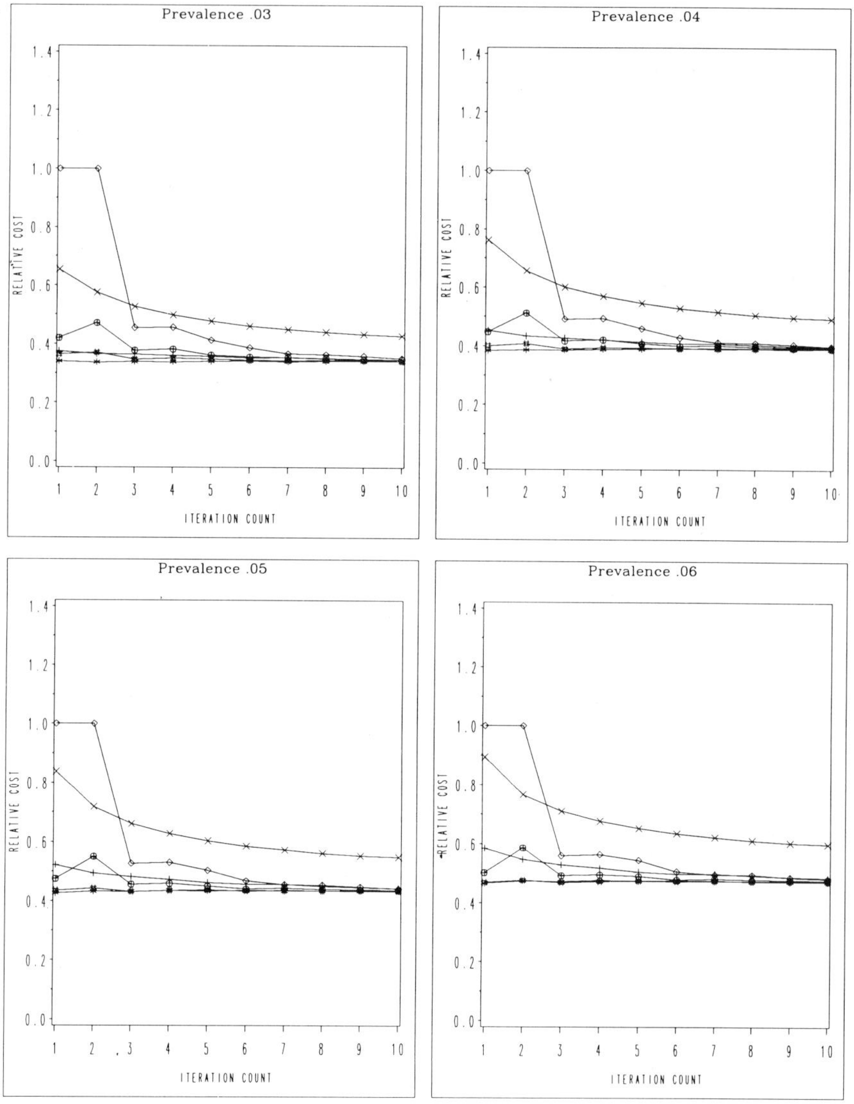
Prevalence .03
Prevalence .04
Prevalence .05
Prevalence .06
RELATIVE COST
ITERATION COUNT
1.4
1.2
1.0
0.8
0.6
0.4
0.2
0.0
1
2
3
4
5
6
7
8
9
10

c

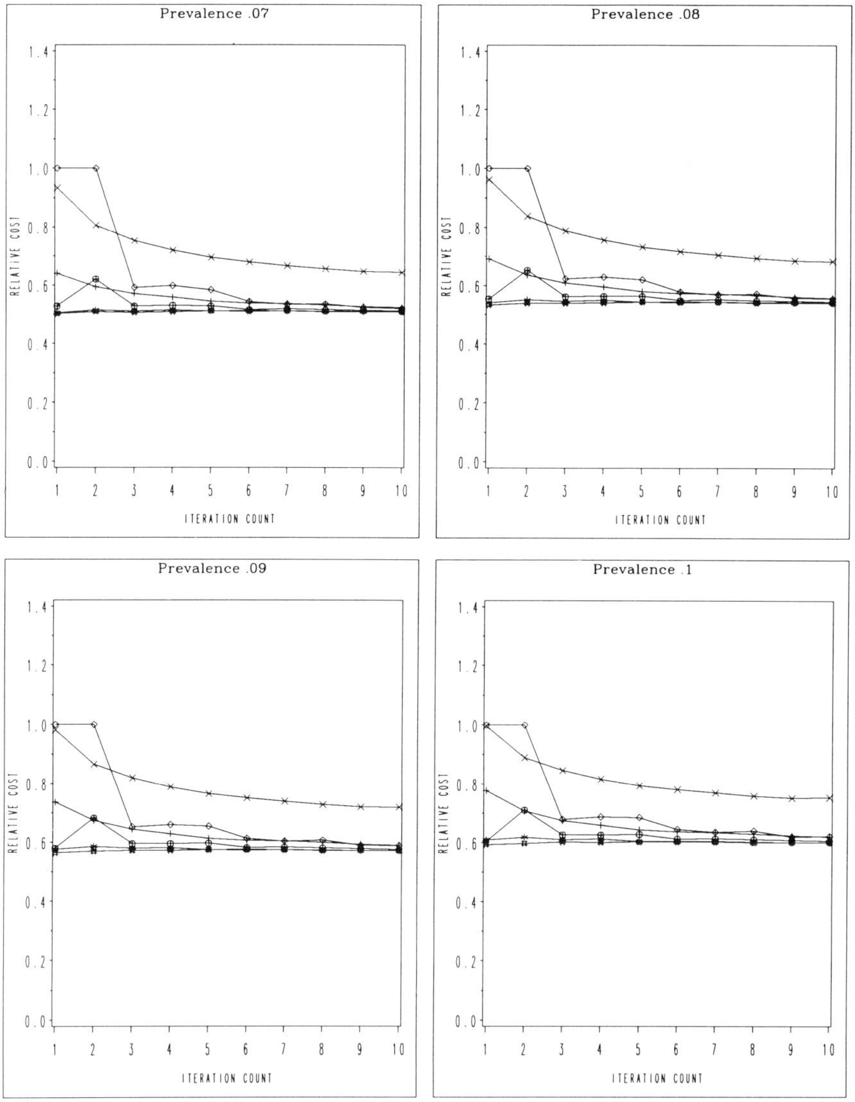

(continued on next page)

d

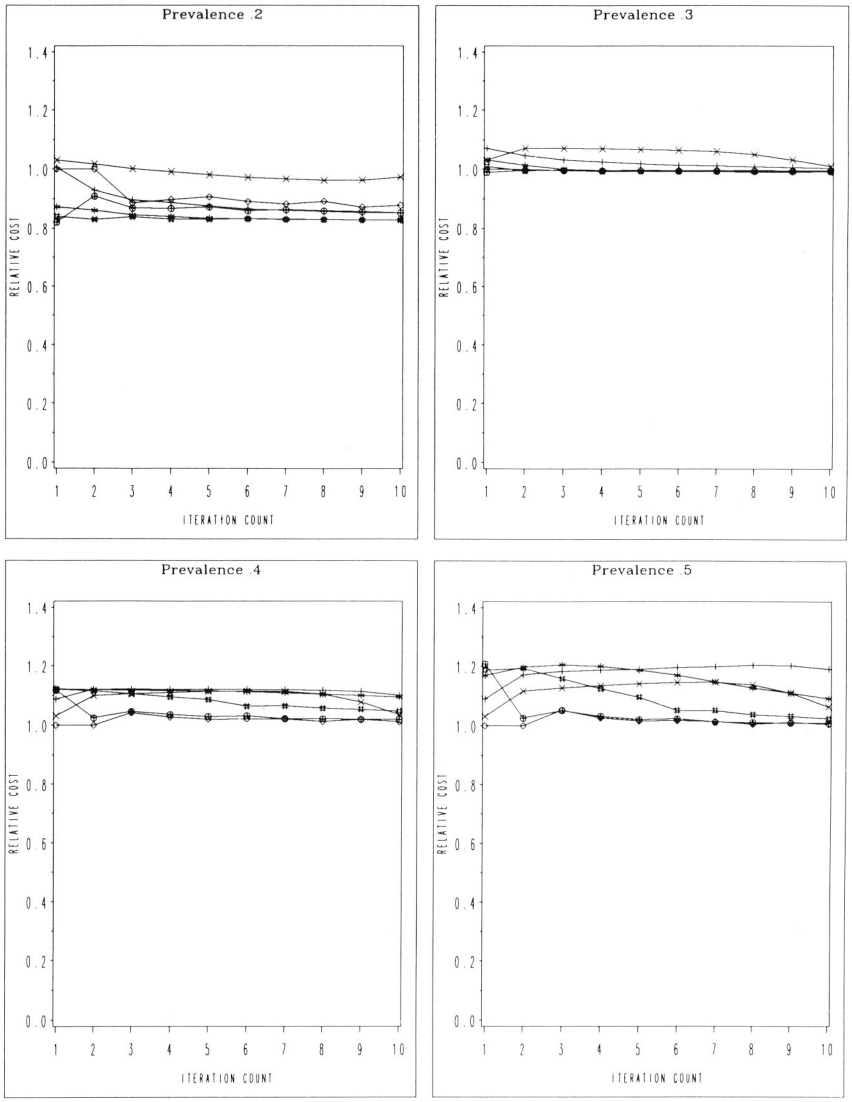

Fig. 14. Relative cost for estimated prevalences 0.001, 0,01, 0.05, 0.1, 0.3 and 0.5 for a given true prevalence, using the model with $\alpha = 0.0$.

minimizes $R(k, \theta)$ will be the optimal composite sample size since it minimizes the expected relative cost under the beta prior distribution.

The composite sample size k determined by this method is a step function of a and b, which are the parameters of the beta prior distribution. Note that the composite sample size k determined by the method described in Section 4 was a function of the mean of the prior distribution, which is given by $a/(a+b)$. Thus, the method of Section 4 would return the same composite sample size for all pairs (a, b) which have $a/(a+b)$ constant. To get an idea as to how the values of k determined by minimizing $R(k, \theta)$ change with a and b, some numerical results were obtained. These are shown in Figure 15. Note that the contour lines in these figures are not straight lines.

The sequential updating of the prior information can be described as follows. Let the initial beta prior on p have the parameters a_1 and b_1. For the Dorfman retesting scheme, the composite sample size for the first stage, k_1 say, is the value of k which minimizes $R(k, \theta_1)$, where $\theta_1 = (a_1, b_1)$. Then composite samples of size k_1 are formed and tested. Suppose X_1 individual samples out of n_1 are classified as polluted in the first stage. The posterior distribution of p given X_1 is then beta with parameters $a_2 = a_1 + X_1$ and $b_2 = b_1 + n_1 - X_1$. This distribution is used as the prior distribution at the second stage. For Dorfman retesting scheme, the composite sample size k_2 corresponds to the risk function $R(k, \theta_2)$, where $\theta_2 = (a_2, b_2)$. Composite samples of size k_2 are then formed and tested, yielding X_2 individual samples that test positive for pollution. In general, after r sampling stages, as in Section 4, the posterior distribution of p is the beta distribution with parameters

$$a_{r+1} = a_1 + \sum_{i=1}^{r} X_i$$

and

$$b_{r+1} = b_1 + \sum_{i=1}^{r} n_i - \sum_{i=1}^{r} X_i.$$

This distribution is then used as the prior distribution of p for the $(r+1)$-th stage.

2.3. *Retesting strategies for continuous measurements made without error*

The classical retesting strategies were developed for presence/absence measurements. They need to be modified when the measurements are continuous, indicating the level or concentration of the characteristic of interest in a sample, either individual or composite. In case of a continuous measurements, a criterion level, sometimes called an action level c is specified, and the problem is to identify all individual samples that exceed the criterion level. Let $X_1, X_2, \ldots$ indicate individual sample values and let Y be a composite sample value. If the composite sample size is k, then for continuous measurements made without error,

$$kY = X_1 + \cdots + X_k.$$

If $kY < c$, then $X_i < c$ for all $i = 1, \ldots, k$, and the corresponding k individual samples

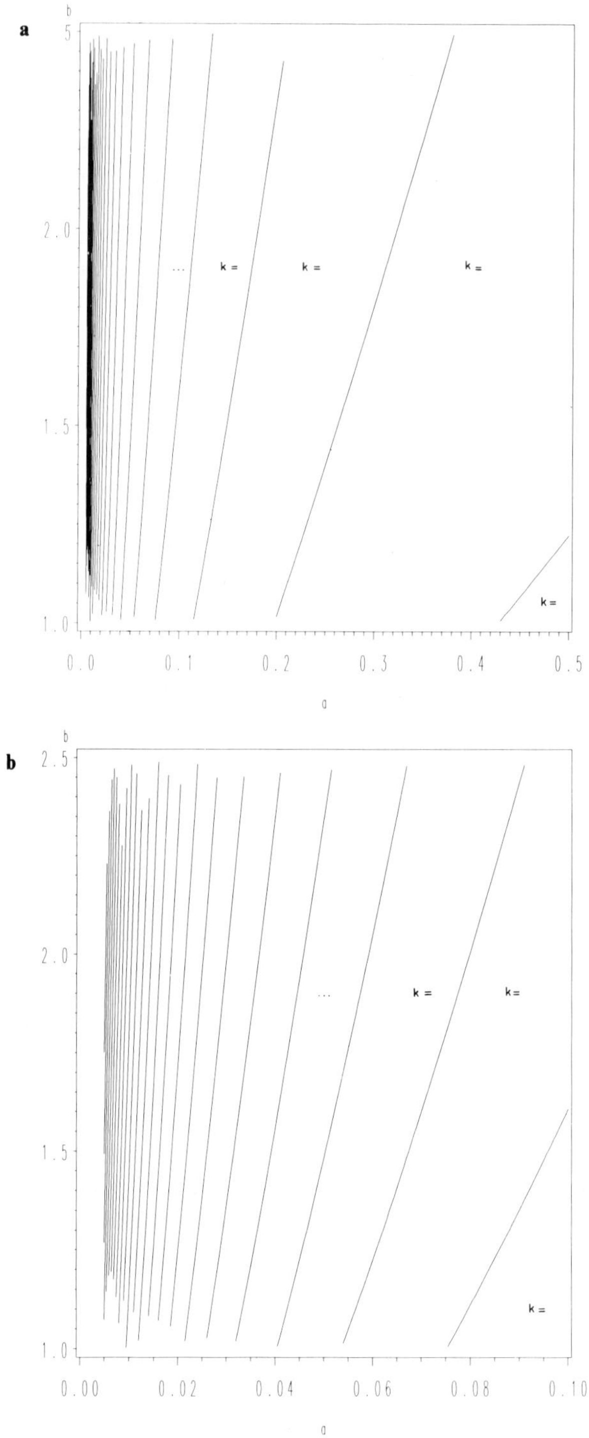

Fig. 15. Regions of optimal composite sample size k as a function of a and b.

can be classified as not exceeding the criterion level. On the other hand, if $kY \geqslant c$, then there may be at least one sample value that equals or exceeds the action level. In this case some type of retesting is necessary. In the case of presence/absence measurements, a retesting procedure under the assumption that the individual sample values are independent and identically distributed results in a relative cost that depends on the prevalence of samples possessing the trait and the composite sample size used. In the case of continuous measurements, assuming the individual sample values are independent and identically distributed, the relative cost depends on the statistical distribution of the trait and the composite sample size. For example, the probability that exactly one measurement is required to classify k individual samples is

$$q_k = \mathrm{P}[X_1 + \cdots + X_k < c].$$

Since different retesting schemes result in the use of composite samples of varying sizes, the distribution of the sums of varying numbers of X_i is all that is needed. Moreover, these convolution distribution functions need be known only at the action level c. That is, in the case of continuous measurements the relative cost of any retesting procedure may depend on some or all of $q_1, q_2, \ldots, q_k$ as well as on k, where

$$q_i = \mathrm{P}[X_1 + \cdots + X_i < c], \quad i = 1, 2, \ldots, k.$$

Table 4 lists the values of $q_i, i = 2, \ldots, 24$, when the individual sample values are assumed to follow the standard normal distribution.

Unlike the case of presence/absence measurements, the case of continuous measurements has not attracted much attention of researchers, and hence there is hardly any mention of this case in the literature on composite sampling. Patil, Gore and Taillie (1994) have reported some interesting results on the classical retesting procedures suitably modified for continuous measurements.

3. Estimation

Important applications of composite sampling involve drawing inference on population parameters. Traditionally, most of these applications have been characterized by the following features:

(a) the parameter of interest is the average (and possibly the variance) of a continuous variable, or the probability of success (or the prevalence) of a binary variable;

(b) such parameters are estimated using multiple homogeneous composite samples which are formed from individual sampling units of a fixed physical size. The individual sampling units are randomly drawn from an unstructured population. The composite samples are perfectly mixed, (possibly) subsampled and then measured with or without error.

We will refer to this set of features as the ‘basic setting’ throughout this section.

Table 4
Table of q_i, $i = 2, \ldots, 24$

k	$q = 0.9$	$q = 0.95$	$q = 0.975$	$q = 0.99$	$q = 0.995$	$q = 0.999$
2	0.817583	0.877603	0.917112	0.950013	0.965726	0.985560
3	0.770321	0.828857	0.871096	0.910383	0.931513	0.962800
4	0.739166	0.794583	0.836453	0.877621	0.901112	0.938841
5	0.716721	0.769013	0.809627	0.850917	0.875328	0.916513
6	0.699579	0.749052	0.788189	0.828875	0.853503	0.896451
7	0.685942	0.732929	0.770592	0.810374	0.834866	0.878596
8	0.674760	0.719563	0.755830	0.794601	0.818772	0.862707
9	0.665377	0.708252	0.743226	0.780963	0.804722	0.848513
10	0.657358	0.698520	0.732304	0.769030	0.792335	0.835769
11	0.650401	0.690032	0.722724	0.758480	0.781315	0.824265
12	0.644291	0.682545	0.714233	0.749068	0.771434	0.813823
13	0.638869	0.675877	0.706640	0.740605	0.762512	0.804299
14	0.634016	0.669889	0.699799	0.732944	0.754406	0.795570
15	0.629638	0.664472	0.693593	0.725967	0.746999	0.787534
16	0.625663	0.659542	0.687930	0.719578	0.740199	0.780108
17	0.622032	0.655030	0.682735	0.713699	0.733926	0.773220
18	0.618698	0.650879	0.677948	0.708266	0.728117	0.766808
19	0.615624	0.647045	0.673517	0.703226	0.722718	0.760822
20	0.612777	0.643489	0.669402	0.698533	0.717683	0.755217
21	0.610130	0.640179	0.665565	0.694151	0.712973	0.749954
22	0.607661	0.637087	0.661977	0.690045	0.708555	0.745001
23	0.605351	0.634192	0.658613	0.686189	0.704400	0.740329
24	0.603183	0.631472	0.655450	0.682558	0.700483	0.735912
25	0.601144	0.628911	0.652468	0.679131	0.696781	0.731728

3.1. Estimating the probability of occurrence of a binary r.v.

Let X be a binary random variable:

$$X = \begin{cases} 1 & \text{with } \operatorname{pr}\{X = 1\} = \pi_x, \\ 0 & \text{with } \operatorname{pr}\{X = 0\} = 1 - \pi_x, \end{cases}$$

representing presence/absence of a trait. Typical environmental examples could be presence of contamination in water wells or need for remedial action of locations at a hazardous waste site.

Let x_i, $i = 1, \ldots, m$, be the presence/absence values characterizing a random sample of size m from the population under study. Assume these m individual sampling units are randomly divided into n sets of size k each and denote by S_j the set containing the indexes of the individual sampling units included in the i-th set. These k units are the compositing objects that are mixed to form the j-th composite. Here, the assumptions made in the basic setting imply that there is no overdilution effect, i.e., that the compositing objects do not interact when mixed. If the composite is measured without (or with negligible) error, then:

$$y_j = 1 - \prod_{i \in S_j} (1 - x_i), \quad j = 1, \ldots, n,$$

is the presence/absence observation on the j-th composite sample.

The statistical problem is estimation of π_x using the y_j rather than the x_i. This problem arises in many applications, especially in biology and epidemiology:

- estimation of the infection transmission rate of vectors on plants (Watson, 1936; Gibbs and Gower, 1960; Chiang and Reeves, 1962; Thompson, 1962; Fine, 1978) or animals (Mahoney and Mirre, 1971);
- estimation of prevalence of a disease in a population (Worlund and Taylor, 1983; Gastwirth and Hammick, 1989).

Statistical aspects of both design and analysis of composite data to estimate a probability have been discussed by several researchers: Gibbs and Gower (1960), Chiang and Reeves (1962), Thompson (1962), Kerr (1971), Griffiths (1972), Sobel and Elashoff (1975), Loyer (1983), Swallow (1985, 1987), Burrows (1987), Gastwirth and Hammick (1989), Garner et al. (1988), Chen and Swallow (1990).

3.1.1. The ML estimator

Traditionally, estimation of π_x, has been undertaken by the principle of maximum likelihood. Let $\pi_y = 1 - (1 - \pi_x)^k$ be the probability that a composite sample tests positive and $r = \sum_{j=1}^{n} y_j$ the number of composite samples testing positive out of n. If the individual sampling units do not interact when composited, the likelihood, as a function of π_y, is

$$L(\pi_y; y_1, y_2, \ldots, y_n) = \binom{n}{r} \pi_y^r (1 - \pi_y)^{n-r}.$$

The ML estimator of π_y is $\hat{\pi}_y = r/n$ and, by the invariance property, the MLE of π_x is

$$\hat{\pi}_x = 1 - (1 - r/n)^{1/k}. \tag{3.1}$$

Clearly, while $\hat{\pi}_y$ is an unbiased estimator of π_y, $\hat{\pi}_x$ is a biased estimator of π_x.

Taylor expansion of $\hat{\pi}_x$ as a function of $\hat{\pi}_y$ around π_y gives approximate expressions for the bias and the mean squared error (MSE):

$$\begin{aligned} \text{bias}(\hat{\pi}_x) = \pi_y(1 - \pi_x)\left(\frac{k-1}{2k^2}\right) \\ \times \left[\frac{1}{n(1-\pi_y)} + \frac{(2k-1)[(k-3)\pi_y + 4k]}{n^2(1-\pi_y)^2 3k} \right. \\ \left. + \, \mathrm{o}((n(1-\pi_y))^{-3}) \right] \end{aligned} \tag{3.2}$$

and

$$\mathrm{MSE}(\hat{\pi}_x) = \frac{\pi_y(1-\pi_x)^2}{k^2} \times \left[\frac{1}{n(1-\pi_y)} + \frac{7(\frac{k-1}{2k})\pi_y + 2(\frac{k-1}{2k})(1-\pi_y)}{n^2(1-\pi_y)^2} + o((n(1-\pi_y))^{-3})\right]. \tag{3.3}$$

When $k = 1$, the composited ML estimator is equivalent to the uncomposited ML estimator. When $k > 1$, inspection of (3.2) shows that the composited estimator is positively biased but consistent, as $n \to \infty$, with asymptotic variance:

$$\mathrm{var}(\hat{\pi}_x) = \frac{1-(1-\pi_x)^k}{nk^2(1-\pi_x)^{k-2}}. \tag{3.4}$$

Various extensions of the composited ML estimator have been developed: for the case of unequal composite sample sizes (Chiang and Reeves, 1962; Walter, Hildreth and Beaty, 1980; Le, 1981); for the case when each individual sampling unit can be retested by providing homogeneous compositing objects to more than one composite (Sobel and Elashoff, 1975; Chen and Swallow, 1990); for situations where the composite samples that are classified as positive by a screening test are given a confirmatory test (Gastwirth and Hammick, 1989). This latter reference and Garner et al. (1988) also allow for measurement errors which result in false positive and false negative tests. Chen and Swallow (1990) investigate the robustness of the ML estimator to such misclassification errors, and the use of unequal composite sample sizes to test the validity of the binomial model. Finally, Bhattacharyya (1979) has considered estimation of π_x by composite sampling when the population of interest is finite.

3.1.2. Alternative estimators

Burrows (1987), adapting the approach found in Haldane (1955) and Anscombe (1956) for the estimation of nonlinear functions of binomial probabilities, proposed an adjusted estimator of π_x:

$$\tilde{\pi}_x = 1 - \left[1 - \left(\frac{2kn}{2kn+k-1}\right)\frac{r}{n}\right]^{1/k}$$

with

$$\mathrm{bias}(\tilde{\pi}_x) = \pi_y(1-\pi_x)\left(\frac{k-1}{2k^2}\right) \times \left[\frac{(2k-1)[(k-3)\pi_y+4k]}{n^2(1-\pi_y)^2 3k} + o((n(1-\pi_y))^{-3})\right], \tag{3.5}$$

$$\mathrm{MSE}(\tilde{\pi}_x) = \frac{\pi_y(1-\pi_x)^2}{k^2} \times \left[\frac{1}{n(1-\pi_y)} + \frac{2(\frac{k-1}{2k})^2 \pi_y}{n^2(1-\pi_y)^2} + \mathrm{o}((n(1-\pi_y))^{-3}) \right] \tag{3.6}$$

By removing the term dependent on π_y in the bias, a dramatic improvement is achieved: in Burrows' paper a large number of simulations for different values of n and π_x show that $\tilde{\pi}_x$ has a bias which ranges from 1% to 5% of that of $\hat{\pi}_x$. This reduction in bias is not obtained sacrificing precision, since the same numerical work shows that $\mathrm{var}(\tilde{\pi}_x)$ is uniformly less than $\mathrm{var}(\hat{\pi}_x)$. As a result of this superiority in both bias and variance terms, $\mathrm{MSE}(\tilde{\pi}_x)$ is also uniformly less than $\mathrm{MSE}(\hat{\pi}_x)$.

A simple, moment-type estimator has been proposed by Gastwirth and Hammick (1989) in the context of the particular double-state testing procedure mentioned above and allowing for measurement errors. Under the simpler conditions considered in this entry, this estimator reduces to

$$\bar{\pi}_x = \sum_{j=1}^{n} y_j/nk \tag{3.7}$$

with

$$\mathrm{bias}(\bar{\pi}_x) = \frac{[1 - k\pi_x - (1-\pi_x)^k]}{k}, \tag{3.8}$$

$$\mathrm{MSE}(\bar{\pi}_x) = \frac{1}{k^2 n} \cdot [(n-1)(1-\pi_x)^{2k} + (-2kn\pi_x + 2n + 1)(1-\pi_x)^k - k^2 n\pi_x(\tfrac{2}{k} - \pi_x) + n]. \tag{3.9}$$

As suggested by the authors, the bias and MSE performances of this simple estimator are satisfactory only for very low levels of prevalence ($\pi_x \leqslant 0.02$) and small composite sample sizes ($k \leqslant 10$).

3.1.3. Design considerations

Many authors have addressed design considerations concerning estimation of a probability with composited data: Gibbs and Gower (1960), Chiang and Reeves (1962), Thompson (1962), Kerr (1971), Griffiths (1972), Loyer (1983), Swallow (1985, 1987), Burrows (1987). The two main design questions in this context are as follows:

(a) How does compositing compare with traditional, uncomposited, sampling?

(b) How many composites should be formed and what composite size should be used?

As for the first question, comparisons should be made in terms of precision of estimates for fixed costs. If measurement costs are much greater than sampling and compositing costs, as is often the case, this means to keep n fixed.

Let $\hat{\pi}_x^* = \sum_{i=1}^{n} x_i/n$ denote the uncomposited ML estimator of π_x obtained by the traditional binomial approach.

If n is large, the composited ML estimator $\hat{\pi}_x$ is consistent, there is no bias concern and relative efficiency can be evaluated comparing the asymptotic variance (3.4) of $\hat{\pi}_x$ and the exact variance $\pi_x(1-\pi_x)/n$ of $\hat{\pi}_x^*$. Compositing is more efficient than traditional individual testing whenever $\pi_x < \pi_x(\infty, k)$, where $\pi_x(\infty, k)$ is the unique root between 0 and 1 of the equation

$$\frac{1-(1-\pi_x)^k}{k^2(1-\pi_x)^{k-2}} = \pi_x(1-\pi_x).$$

The last column of Table 5, reproduced from Loyer (1983), gives such solutions for selected values of k. They range from $\pi_x(\infty, k) = \frac{2}{3}$ for $k = 2$ to 0.063 for $k = 100$. Hence, for a wide range of values of π_x, depending on k, $\hat{\pi}_x$ is asymptotically more efficient than $\hat{\pi}_x^*$.

When n is small, comparisons should be based on MSEs rather than variances. Loyer (1983) and Swallow (1985) investigate the relative efficiency of $\hat{\pi}_x$ and $\hat{\pi}_x^*$ for various values of n. The results are not qualitatively different from the asymptotic case, as can be seen from Table 5, which shows the values of π_x below which compositing is more efficient than individual testing. Swallow concludes that 'although group testing is often discussed as being appropriate when π_x is very small and indeed its benefits are greatest then, it is useful for larger π_x than is often supposed'.

The answer to the second design question must take into account statistical as well as technical and cost considerations. As inspection of (3.3), (3.4), (3.6), (3.9) shows, whatever estimator is used, increasing n decreases the variance or MSE of the estimator of π_x, hence giving a more reliable estimate. But usually this is impractical, since the number of composites that can be formed and tested is limited by time and/or cost constraints. As for the composite sample size k, although time and/or cost constraints are usually less stringent, there may be other technical consider-

Table 5
$\hat{\Pi}_x(n, k)$, the value of Π_x below which $\hat{\Pi}_x$, calculated on n composites of size k, is more efficient than $\hat{\Pi}_x^*$, calculated on n individual sampling units. (Reproduced from Loyer, 1983)

k	n					
	5	10	20	50	100	∞
2	0.330	0.461	0.555	0.625	0.656	0.667
3	0.208	0.321	0.416	0.508	0.547	0.579
4	0.148	0.243	0.330	0.422	0.469	0.514
5	0.112	0.194	0.271	0.359	0.414	0.464
10	0.047	0.092	0.138	0.197	0.240	0.321
15	0.028	0.058	0.091	0.135	0.166	0.251
20	0.020	0.042	0.067	0.101	0.127	0.208
25	0.015	0.032	0.053	0.080	0.102	0.179
50	0.006	0.015	0.025	0.039	0.051	0.108
100	0.002	0.006	0.012	0.109	0.024	0.063

ations which suggest putting an upper bound on it, since after some critical threshold in the number k of compositing objects mixed in a composite, two undesirable effects can take place. First, there may be overdilution effects, by which the measurement process is not so sensitive to classify as positive a composite formed by a single positive contributing object and a large number of negative objects. Secondly, especially in biological compositing, there may be interactions between the compositing microorganisms, either in the direction of increasing or decreasing their probability of transmitting a disease when caged together in a large number.

On the grounds of these considerations, the most typical design situation is one in which n is fixed and the composite sample size k has to be chosen, usually below some specified upper bound, in order to maximize the precision of the chosen estimator. If n is large, this implies minimizing the asymptotic variance (3.4) with respect to k. This was the preferred approach in early papers on this topic (Gibbs and Gower, 1960; Chiang and Reeves, 1962; Thompson, 1962; Kerr 1971; Griffiths, 1972), since under the reasonable assumption that k be large in relation with π_x it gives the following approximate, but easily computable, analytical solution for the optimal k,

$$k^+ = \frac{1.5936}{\pi_x}.$$

With the availability of powerful computing facilities, optimization of k has become feasible for any values of n and for alternative estimators. For small to moderate n, this implies to minimize the MSE of the selected estimator. Optimal composite sample sizes for this situation are given by Loyer (1983) and Swallow (1985) for $\hat{\pi}_x$ and by Burrows (1987) for $\tilde{\pi}_x$. The main results are summarized in Table 6; the four entries for each combination of n and π_x are the optimal composite sample size k^+, bias$(\hat{\pi}_x(n, k^+))$ and MSE$(\hat{\pi}_x(n, k^+))$ of the ML composited estimator calculated with n composites of optimal size k^+ and var $(\hat{\pi}_x^*(n))$ of the uncomposited ML estimator calculated with n individual sampling units.

Finally, Swallow (1985) investigates the case when the estimator of interest is $\hat{\pi}_x$, $n \times k = m$, the number of individual sampling units, is fixed and the problem is to choose the optimal combination of n and k. Clearly, if there are no cost constraints, the best choice is always $k = 1$ and $n = m$. When, as is usually the case, sampling, compositing and measurement costs are markedly different, some other combinations can be found, which yield MSE(π_x)s just above the minimum and permit large cost savings.

Notice that all these proposals for the design of k under different conditions depend on π_x which is the unknown parameter we are estimating. There seems to be no completely satisfactory theoretical solution to this problem in the literature as yet, but useful empirical procedures have been suggested by various authors (see for example Griffiths, 1972; Swallow, 1985). They are based on the reasonable assumption that the researchers usually have some prior knowledge on π_x, either in the form of an upper bound or of a prior distribution, and exploit this information to suggest a conservative choice for k.

Table 6
Optimal composite sample size k^+, bias $[\hat{\Pi}_x(n,k^+)]$. MSE$[\hat{\Pi}_x(n,k^+)]$ and var$[\hat{\Pi}_x(n)]$ for various combinations of n and $\hat{\Pi}_x$. (Abridged from Swallow, 1985)

π_x	n						
	10	15	20	30	40	50	100
	35	50	50	50	50	50	50
	0.0007	0.0005	0.0003	0.0002	0.0002	0.0001	0.0001
0.01	0.000046	0.000021	0.000014	0.000009	0.000007	0.000005	0.000003
	0.000990	0.000660	0.000495	0.000330	0.000248	0.000198	0.000099
	19	30	35	45	50	50	50
	0.0013	0.0010	0.0008	0.0006	0.0004	0.0004	0.0002
0.02	0.000162	0.000078	0.000048	0.000027	0.000019	0.000015	0.000007
	0.001960	0.001307	0.000980	0.000653	0.000490	0.000392	0.000196
	14	20	25	30	35	40	45
	0.0020	0.0014	0.0012	0.0008	0.0007	0.0006	0.0003
0.03	0.000337	0.000164	0.000104	0.000059	0.000041	0.000031	0.000015
	0.002910	0.001940	0.001455	0.000970	0.000728	0.000582	0.000291
	11	16	19	25	30	30	35
	0.0026	0.0019	0.0015	0.0012	0.0010	0.0008	0.0004
0.04	0.000565	0.000281	0.000180	0.000102	0.000072	0.000055	0.000026
	0.003840	0.002560	0.001920	0.001280	0.000960	0.000768	0.000384
	9	13	16	20	23	25	30
	0.0032	0.0024	0.0020	0.0015	0.0012	0.0010	0.0006
0.05	0.000842	0.000424	0.000274	0.000157	0.000110	0.000084	0.000040
	0.004750	0.003167	0.002375	0.001583	0.001188	0.000950	0.000475
	5	7	8	10	12	12	14
	0.0057	0.0045	0.0036	0.0027	0.0023	0.0018	0.0010
0.10	0.002807	0.001482	0.000987	0.000579	0.000409	0.000317	0.000149
	0.009000	0.006000	0.004500	0.003000	0.002250	0.001800	0.000900
	3	4	4	5	6	6	7
	0.0099	0.0083	0.0059	0.0047	0.0042	0.0033	0.0019
0.20	0.008356	0.004764	0.003284	0.001975	0.001416	0.001100	0.000523
	0.016000	0.010667	0.008000	0.005333	0.004000	0.003200	0.001600
	2	3	3	3	4	4	4
	0.0104	0.0118	0.083	0.0053	0.0056	0.0044	0.0021
0.30	0.014516	0.008861	0.006000	0.003769	0.002716	0.002114	0.001009
	0.021000	0.014000	0.010500	0.007000	0.005250	0.004200	0.002100
	2	2	2	2	3	3	3
	0.0169	0.0100	0.0072	0.0047	0.0065	0.0051	0.0025
0.40	0.020264	0.011981	0.008632	0.005589	0.004032	0.003143	0.001506
	0.024000	0.016000	0.012000	0.000800	0.006000	0.004800	0.002400
	1	2	2	2	2	2	2
	0.0000	0.0157	0.0108	0.0067	0.0049	0.0039	0.0019
0.50	0.025000	0.015722	0.001759	0.006722	0.004932	0.003901	0.001911
	0.025000	0.016667	0.012500	0.008333	0.006250	0.005000	0.002500

3.2. *Estimating the parameters of a continuous r.v.*

Let X now be a continuous r.v., with a distribution characterized by the first two moments: $X \sim (\mu_x, \sigma_x^2)$, and let again $x_i, i = 1, \ldots, m$ be the m observations on a random sample from the population under study. Then, if measurement is made without error:

$$y_j = \sum_{i=1}^{k} w_{ji} x_i$$

is the observation on the j-th composite sample formed mixing k out of the m individual sampling units. The traditional problem is estimation of μ_x, while estimation of σ_x^2 is either of direct interest or necessary for estimating the precision of the estimate of μ_x. Classical results related to this problem are briefly recalled in the next section.

In recent years, sample compositing has been increasingly used in a wide range of applications involving estimation. This has called for the development of new statistical methods to handle new features which have added complexity to the basic setting described above, in particular:

(1) estimation of effects of explanatory variables, particularly of differential effects of specific subpopulations of interest;

(2) use of complex sampling schemes before compositing, particularly in case of segmented populations;

(3) randomness of weights.

Recent developments related to these features are reviewed in Sections 3.2.2., 3.2.3 and 3.2.4. More details can be found in Lovison, Gore and Patil (1991). Lovison (1993a) proposes a unified linear model which handles both the basic setting and these complications.

3.2.1. *Estimating the mean and variance of a continuous r.v.*

When there is no measurement error, it is a well-known result (see, for example, Boswell and Patil, 1987, 1990) that the measurement on one composite of size k with uniform weights ($w_{ji} = 1/k$) is equivalent to the sample mean of the measurements of k individual sampling units. With just one composite the possibility of estimating the variance, and hence the precision of the estimate of μ_x, is lost. However, by processing $c > 1$ composite samples, the variance may be estimated, still with considerable savings.

To complete an overview of the use of composite sampling for estimation of the mean and variance of a continuous r.v. in the basic setting, it is worth considering the effect of subsampling from the composite and of the presence of nonnegligible measurement error.

Suppose a random sample of $m = ck$ individual sampling units is randomly assigned to c composites of size k. From each composite s subsamples are drawn and finally measured with error.

Let x_{jh} be the X-value taken by the h-th individual sampling unit selected for the j-th composite, $h = 1, \dots, k; j = 1, \dots, c$. Let x_{ijh} and v_{ijh} denote the X-value and the physical size (volume, mass, amount, etc.) characterizing the ijh-th contributing object, i.e., the portion of the h-th individual sampling entering the i-th subsample drawn from the j-th composite. Then $w_{ijh} = v_{ijh}/\sum_{i=1}^{s} v_{ijh}$ is the contributing weight. In the basic setting, the assumptions of homogeneous composites and fixed weights are made, which imply

$$x_{ijh} = x_{i'jh} = x_{jh} \quad \text{and} \quad w_{ijh} = w_{i'jh} \qquad \forall i \neq i'.$$

Finally, let $y_{ij} = \sum_{h=1}^{k} w_{ijh} x_{ijh} + \eta_{ij}$ be the measurement, made with error, on the ij-th composited object, i.e., the i-th subsample from the j-th composite.

Under the following assumptions on the distribution of x_{jh} and η_{ij},

$$x_{jh} = \mu_x + \varepsilon_{jh} \quad \text{with } \varepsilon_{jh} \text{ i.i.d.} \sim (0, \sigma_\varepsilon^2), \ \forall j, h,$$

$$\eta_{ij} \text{ i.i.d.} \sim (0, \sigma_\eta^2) \quad \text{with } \mathrm{E}(\varepsilon_{jh}, \eta_{ij}) = 0, \ \forall i, j, h$$

and with the side-condition of uniform weights, i.e., $w_{jih} = 1/k$, $\forall i, j, h$, we get

$$\begin{aligned}
\mathrm{E}(y_{ji}) &= \mu_x, \\
\mathrm{var}(y_{ji}) &= \sigma_\varepsilon^2/k + \sigma_\eta^2, \\
\mathrm{cov}(y_{ji}, y_{ji'}) &= \sigma_\varepsilon^2/k, \quad i \neq i' \\
\mathrm{cov}(y_{ji}, y_{j'i'}) &= 0, \quad j \neq j', \ \forall i, i'.
\end{aligned}$$

Under these conditions, the composited mean $\bar{y} = (1/cs)\sum_{i=1}^{s}\sum_{j=1}^{c} y_{ij}$ has expected value and variance:

$$\begin{aligned}
\mathrm{E}(\bar{y}) &= \mu_x, \\
\mathrm{var}(\bar{y}) &= \sigma_\varepsilon^2/ck + \sigma_\eta^2/cs.
\end{aligned}$$

Hence:

- $\bar{y}$ is an unbiased estimator of μ_x;
- σ_ε^2 and σ_η^2 are separately estimable;
- if s increases only the variance component due to η decreases;
- if k increases only the variance component due to ε decreases;
- if c increases the total variance of the composited mean decreases.

Consequently, for fixed total cost, it is convenient to allocate more resources on increasing the number of composites rather than the composite sample size or the number of subsamples extracted from each composite.

Finally it is useful to compare the composited estimator with the traditional, uncomposited mean that would have been obtained using measurements on n

individual sampling units. Denote by $x_l^* = x_l + \eta_l$ the measurement, affected by error, on a randomly drawn individual sampling unit, $l = 1,\ldots,n$, and by $\bar{x}^*$ the sample mean. Clearly,

$$\mathrm{E}(\bar{x}^*) = \mu_x,$$
$$\mathrm{var}(\bar{x}^*) = (\sigma_\varepsilon^2 + \sigma_\eta^2)/n.$$

Comparison of $\bar{y}$ and $\bar{x}^*$ depends on the relative magnitude of the error components involved. Choosing

$$c = n(\lambda + (1 - \lambda)/k), \quad \text{where } \lambda = \sigma_\eta^2/(\sigma_\varepsilon^2 + \sigma_\eta^2)$$

the composited mean has the same estimation variance as the uncomposited mean, with cost savings.

Alternatively, if $n = cs$, $k > s$ and the cost of sampling, compositing and subsampling is lower than the cost of measurement, the composited mean is more precise then the uncomposited, at the same cost.

3.2.2. Estimation of fixed effect parameters

Compositing can still be applied when the goal of analysis is to estimate not only the general mean μ_x in the population, but also the effects of explanatory variables $F_1, F_2,\ldots, F_p$ on X. It is assumed that the values of these explanatory variables are known for the original compositing objects. Although use of composite sampling for estimating fixed effect parameters is not yet very frequent, interesting examples can be found in the literature. Mekonnen et al. (1976) and Sharma et al. (1983) used composite samples to estimate the effects of climatological and genetic factors on the characteristics of milk. Mack and Robinson (1985) and Orban, Lordo and Schwemberger (1990) discuss estimation and hypothesis testing of the effects of four variables (census region, age, sex, race) on the log-concentration of chemicals in the adipose tissue of the U.S. population.

The major problem in this context is that of retaining estimability of the fixed effect parameters with composite data. As mentioned by Rohde (1976), mixing in the same composite compositing object characterized by different sets of such parameters can destroy the possibility of estimating them. A simple illustrative example can show this point but also the fact that estimability of linear models parameters from the composited measurements, as well as desirable statistical properties of the estimators, can be achieved by carefully designing the compositing procedure.

3.2.2.1. Estimating the effect of a binary factor. For illustration purposes, we shall introduce the basic problems encountered in the estimation of fixed-effects parameters in linear models with composited data in the context of the simplest possible example, that of a binary factor.

Suppose that four individual sampling units, two characterized by level 1 and two by level 2 of the binary factor F, are available and that two composite samples are formed mixing two of these individual sampling units in each composite. The composite samples are then completely tested, without subsampling, at the laboratory to determine the concentration of X. The measurements are taken with an error which is so negligible, compared with natural variability of X, that they may be considered to be free of errors. This situation may be modeled as follows. Let

$$x_{hi} = \mu_x + \beta_h + \varepsilon_{hi}, \quad h = 1, 2, \ i = 1, 2$$

with β_1, β_2 unknown fixed parameters, which satisfy the constraint

$$\sum_{h=1}^{2} \beta_h = 0$$

and

$$\varepsilon_{hi} \text{ i.i.d.} \sim (0, \sigma_\varepsilon^2),$$

be the model expressing the effect of the two levels of the binary factor F on the variable X of interest in the individual sampling units.

Let

$$y_j = \sum_{h=1}^{2} \sum_{i=1}^{2} u_{jhi} x_{hi}, \quad j = 1, 2,$$

where

$$u_{jhi} = \begin{cases} 0 < w_{jhi} \leqslant 1 & \text{if the } hi\text{-th individual sampling unit contributes to} \\ & \text{the } j\text{-th composite with weight } w_{jhi}, \\ 0 & \text{otherwise,} \end{cases}$$

be the compositing model, which describes the way the four individual sampling units are composited into two composites of size two.

In matrix form:

$$\boldsymbol{x} = \boldsymbol{F}\boldsymbol{\beta} + \boldsymbol{\varepsilon}, \quad \text{with } \mathrm{E}(\boldsymbol{\varepsilon}) = \boldsymbol{0}, \ \mathrm{V}(\boldsymbol{\varepsilon}) = \sigma_\varepsilon^2 \boldsymbol{I}_m, \tag{3.10}$$

$$y = \boldsymbol{U}\boldsymbol{x}, \tag{3.11}$$

where

$$\boldsymbol{x} = \begin{bmatrix} x_{11} \\ x_{12} \\ x_{21} \\ x_{22} \end{bmatrix}, \quad \boldsymbol{F} = \begin{bmatrix} 1 & 1 \\ 1 & 1 \\ 1 & -1 \\ 1 & -1 \end{bmatrix}, \quad \boldsymbol{\beta} = \begin{bmatrix} \mu_x \\ \beta_1 \end{bmatrix},$$

$$\boldsymbol{y} = \begin{bmatrix} y_1 \\ y_2 \end{bmatrix}, \quad \boldsymbol{U} = \begin{bmatrix} u_{111} & u_{112} & u_{121} & u_{122} \\ u_{211} & u_{212} & u_{221} & u_{222} \end{bmatrix}.$$

Notice that all the design and model features concerning the individual sampling units are summarized in $\boldsymbol{F}$, which $\boldsymbol{U}$ summarizes all the features of the compositing design (assignment of individual samples to the composites and weights).

To see how the estimation of $\boldsymbol{\beta}$ can be handled in this context, observe that (3.10) and (3.11) may be rewritten as a unique model:

$$\boldsymbol{y} = \boldsymbol{U}\boldsymbol{F}\boldsymbol{\beta} + \boldsymbol{U}\boldsymbol{\varepsilon} = \boldsymbol{D}\boldsymbol{\beta} + \boldsymbol{\zeta}.$$

Since $\mathrm{V}(\boldsymbol{y}) = \boldsymbol{U}\mathrm{V}(\boldsymbol{\varepsilon})\boldsymbol{U}' = \sigma_\varepsilon^2\,\boldsymbol{U}\boldsymbol{U}'$ depends on the structure of $\boldsymbol{U}$ and may therefore display heteroscedasticity even if $\mathrm{V}(\boldsymbol{\varepsilon})$ does not, we must use a generalized least squares approach to estimate $\boldsymbol{\beta}$. The normal equations are

$$(\boldsymbol{D}'[\boldsymbol{U}\boldsymbol{U}']^{-1}\boldsymbol{D})\hat{\boldsymbol{\beta}} = \boldsymbol{D}'[\boldsymbol{U}\boldsymbol{U}']^{-1}\boldsymbol{y}.$$

The basic problem in the estimation of fixed linear models with composited data can now be addressed more clearly: are $\boldsymbol{D}$ and $\boldsymbol{\zeta}$ such that $\boldsymbol{\beta}$ is still identifiable? and: which is the best choice of $\boldsymbol{D}$ among all the $\boldsymbol{D}$ matrices which ensure identifiability of $\boldsymbol{\beta}$?

To answer the first question, consider the design matrix $\boldsymbol{D}$ of the composited observations

$$\boldsymbol{D} = \begin{bmatrix} 1 & u_{11.} - u_{12.} \\ 1 & u_{21.} - u_{22.} \end{bmatrix} = \begin{bmatrix} 1 & \delta_1 \\ 1 & \delta_2 \end{bmatrix}, \tag{3.12}$$

where $u_{jh.} = \sum_{i=1}^{2} u_{jhi}$ is the total proportion with which contributing objects characterized by level h of F contribute to the j-th composite.

Clearly, inspection of (3.12) suggests that $\boldsymbol{D}$ will be of full rank, and hence $\boldsymbol{\beta}$ will be identifiable, as long as $\delta_1 \neq \delta_2$, i.e., the contributing objects characterized by different levels of the factor F enter the two composites in different proportions.

Let us now restrict attention to vectors of weights which preserve identifiability of $\boldsymbol{\beta}$. The choice of the weights can then be made in terms of D-optimality, i.e., the weights can be chosen in such a way to minimize the generalized variance of the estimator $\hat{\boldsymbol{\beta}}$. Denote the squared norms of the weight vectors by v_1 and v_2 respectively, i.e., let $v_j = \sum_{h=1}^{2}\sum_{i=1}^{2} u_{jhi}^2$. The variance/covariance matrix of $\hat{\boldsymbol{\beta}}$ is

$$\mathrm{V}(\hat{\boldsymbol{\beta}}) = \sigma_\xi^2[\boldsymbol{D}'(\boldsymbol{U}\boldsymbol{U}')^{-1}\boldsymbol{D}]^{-1}$$

$$= \sigma_\varepsilon^2 \begin{bmatrix} \dfrac{v_1\delta_2^2 + v_2\delta_1^2}{(\delta_1 - \delta_2)^2} & \dfrac{v_1\delta_2 + v_2\delta_1}{(\delta_1 - \delta_2)^2} \\[2ex] \dfrac{v_1\delta_2 + v_2\delta_1}{(\delta_1 - \delta_2)} & \dfrac{v_1 + v_2}{(\delta_1 - \delta_2)} \end{bmatrix}$$

with determinant

$$|\mathrm{V}(\hat{\boldsymbol{\beta}})| \propto \frac{(v_1 v_2)}{(\delta_1 - \delta_2)^2}. \tag{3.13}$$

Inspection of (3.13) shows that

- the D-optimal weights must be sought among those which maximize $(\delta_1 - \delta_2)^2$; this quantity is maximized when the compositing procedure is *segregated*, i.e., when all the compositing objects forming each composite are characterized by the same level of the binary factor F;
- among all the segregated procedures, the D-optimal choice of weights is attained when the numerator $v_1 v_2$ is minimized; since $\sum_{h=1}^{2} \sum_{i=1}^{2} u_{jhi} = 1$, this quantity is minimized when the nonzero weights are *uniform*, i.e., when $w_{jhi} = 1/k$, $+ j, h, i$.

Summarizing, as long as possible, the weights should be chosen to ensure segregation and to be uniform. If segregation cannot be achieved, it can be shown (Lovison, 1993b) that a sub-optimal strategy is to choose the contributing weights of compositing objects characterized by different levels of the explanatory variables to be 'as different as possible' in different composites.

3.2.3. *Complex sampling schemes before compositing*

In many applications of composite sampling, the population of interest is structured and/or practical reasons make simple random sampling inadvisable. In all these cases, a complex sampling scheme, i.e., a sampling design with two or more sampling stages, with stratification and/or selection of clusters (or segments) at some of such stages, must be employed for drawing the final individual sampling units to be composited. The features of such complex sampling scheme must be taken into account in the analysis of the final measurement on the composite samples, since they usually inflate the variance of each final measurement and sometimes introduce correlation between final measurements.

One powerful and flexible approach for taking the complex sampling scheme into consideration is to account for the effects of the various complications of the sampling design through random-effect parameters in the model for x_i, the value characterizing the i-th individual sampling unit. We illustrate this approach in the simple case when the only complexity in the sampling scheme used to draw the individual sampling units is some form of stratification; its use in this simple case dates back to Brown and Fisher (1972). More complex sampling designs may be taken into consideration by extending this technique: there is a vast literature on this topic, especially from researchers in the area of analytical use of sample survey data from finite populations. A good review is in Skinner, Holt and Smith (1989).

3.2.3.1. Estimating the mean in segmented populations. A simple and very common instance of structured population is provided by situations in which the population of interest is naturally divided into *segments*, i.e., into clusters of spatially, temporally or physically aggregated objects. Examples of such populations include bales of wool (Cameron, 1951; Brown and Fisher, 1972), sites in a water body (Rohde, 1979), fertilizer in bags (Duncan, 1962).

Let us first suppose that the population is composed of a very large number, $B \to \infty$, of segments. cb segments are randomly selected and grouped into c subsets.

From each segment in each subset, a constant number a of compositing objects is drawn randomly. The $k = ab$ compositing objects are finally composited; therefore, there are c composite samples of size $k = ab$, and each is measured without error.

This layout may be modeled by a random-effects linear model expressing the X-value in each individual sample as an additive function of the general mean μ_x (the parameter of interest), an effect due to the random selection of segments and an error component:

$$x_{jhi} = \mu_x + \gamma_{jh} + \varepsilon_{jhi} \tag{3.14}$$

with

$$\gamma_{jh} \text{ i.i.d. } (0, \sigma_\gamma^2), \quad j = 1, \ldots, c, \; h = 1, \ldots, b,$$

$$\varepsilon_{jhi} \text{ i.i.d. } (0, \sigma_\varepsilon^2), \quad j = 1, \ldots, c, \; h = 1, \ldots, b, \; i = 1, \ldots, a,$$

$$\mathrm{E}(\gamma_{jh}, \varepsilon_{jhi}) = 0, \quad \forall j, h, i.$$

Hence

$$\mathrm{var}(x_{jhi}) = \sigma_\gamma^2 + \sigma_\varepsilon^2,$$

$$\mathrm{cov}(x_{jhi}, x_{jhi'}) = \sigma_\gamma^2, \quad i \neq i',$$

$$\mathrm{cov}(x_{jhi}, x_{jh'i'}) = 0, \quad h \neq h', \; \forall i, i',$$

$$\mathrm{cov}(x_{jhi}, x_{j'h'i'}) = 0, \quad j \neq j', \; \forall h, h', \; \forall i, i'.$$

Denote the final measurement on the j-th composite sample by

$$y_j = \sum_{h=1}^{b} \sum_{i=1}^{a} w_{jhi} x_{jhi}, \quad j = 1, \ldots, c. \tag{3.15}$$

If $w_{jhi} = 1/k, \forall j, h, i$, then

$$\mathrm{E}(y_j) = \mu_x, \quad \forall j,$$

$$\mathrm{var}(y_j) = \frac{\sigma_\gamma^2}{b} + \frac{\sigma_\varepsilon^2}{k}, \quad \forall j,$$

$$\mathrm{cov}(y_j, y_{j'}) = 0, \quad j \neq j'$$

and the composited mean $\bar{y} = \sum_{j=1}^{c} y_j / c$ has expected value

$$\mathrm{E}(\bar{y}) = \mu_x \tag{3.16}$$

and variance

$$\mathrm{var}(\bar{y}) = \frac{\sigma_\gamma^2}{cb} + \frac{\sigma_\varepsilon^2}{ck}, \tag{3.17}$$

i.e., the composited mean is an unbiased estimator of μ_x, but the variance of $\bar{y}$ is inflated, when compared to that of a composited mean in unsegmented populations, by a quantity σ_γ^2/cb. Notice that it is still possible to obtain an unbiased estimator of var($\bar{y}$), since

$$\mathrm{E}\left[\frac{\sum_{j=1}^{c}(y_j-\bar{y})^2}{c(c-1)}\right]=\frac{\sigma_\gamma^2}{cb}+\frac{\sigma_\varepsilon^2}{ck}=\mathrm{var}(\bar{y})$$

but σ_γ^2 and σ_ε^2 are not separately estimable.

3.2.3.2. Estimating variance components in segmented populations. If the variability of X in the parent population is of direct interest, then careful design of the compositing procedure makes it possible to get an estimate of σ_ε^2.

To show this, let us now suppose that gb segments are randomly selected and randomly grouped into g groups of b segments each. From each segment in each group a compositing objects are randomly drawn and composited; the procedure is repeated until d composites are formed within each group. So, there are $c = gd$ composites of size $k = ab$ each.

The X-value taken by the i-th compositing object from the h-th segment of the l-th group which enters the j-th composite, may be modeled as follows,

$$x_{ljhi}=\mu_x+\gamma_{lh}+\varepsilon_{lhi} \tag{3.18}$$

with

$$\begin{aligned}
&\gamma_{lh}\ \text{i.i.d.}\ (0,\sigma_\gamma^2),\quad l=1,\ldots,g,\ \ h=1,\ldots,b,\\
&\varepsilon_{lhi}\ \text{i.i.d.}\ (0,\sigma_\varepsilon^2),\quad l=1,\ldots,g,\ h=1,\ldots,b,\ i=1,\ldots,a,\\
&\mathrm{E}(\gamma_{lh}),\ \varepsilon_{lhi}=0,\ \ \forall\, l,\ h,\ i.
\end{aligned}$$

Therefore

$$\begin{aligned}
\mathrm{var}(x_{ljhi})&=\sigma_\gamma^2+\sigma_\varepsilon^2,\\
\mathrm{cov}(x_{ljhi},x_{ljhi'})&=\sigma_\gamma^2,\quad i\neq i',\\
\mathrm{cov}(x_{ljhi},x_{ljh'i'})&=0,\quad h\neq h',\ \forall i,i',\\
\mathrm{cov}(x_{ljhi},x_{lj'hi'})&=\sigma_\gamma^2,\quad j\neq j',\ \forall\, i,\ i',\\
\mathrm{cov}(x_{ljhi},x_{lj'h'i'})&=0,\quad h\neq h',\\
\mathrm{cov}(x_{ljhi},x_{l'j'h'i'})&=0,\quad l\neq l'.
\end{aligned}$$

Let

$$y_{lj}=\sum_{h=1}^{b}\sum_{i=1}^{a}w_{ljhi}\,x_{ljhi} \tag{3.19}$$

be the final measurement, without error, in the j-th composite in the l-th group. Then

$$\mathrm{var}(y_{lj}) = \frac{\sigma_\gamma^2}{b} + \frac{\sigma_\varepsilon^2}{k}, \quad \forall l, j,$$

$$\mathrm{cov}(y_{lj}, y_{lj'}) = \frac{\sigma_\gamma^2}{b}, \quad j \neq j',$$

$$\mathrm{cov}(x_{ij}, y_{l'j'}) = 0, \quad l \neq l', \ \forall j, j'.$$

The composited general mean is now

$$\bar{y} = \sum_{l=1}^{g} \sum_{j=1}^{d} y_{lj}/gd$$

and it is also possible to define the group means

$$\bar{y}_l = \sum_{j=1}^{d} y_{lj}/d, \quad l = 1, \ldots, g.$$

The total composited sum of squares may be decomposed into

$$\sum_{l=1}^{g} \sum_{j=1}^{d} (y_{lj} - \bar{y})^2 = \sum_{l=1}^{g} \sum_{j=1}^{d} (y_{lj} - \bar{y}_l)^2 + \sum_{l=1}^{g} d(\bar{y}_l - \bar{y})^2$$

and it is possible to show that

$$\mathrm{E}\left[\sum_{l=1}^{g} \sum_{j=1}^{d} (y_{lj} - \bar{y}_l)^2 \right] = \sigma_\varepsilon^2 \left(\frac{d(g-1}{k} \right), \tag{3.20}$$

i.e., $(k/d(g-1)) \sum_{l=1}^{g} \sum_{j=1}^{d} (y_{lj} - \bar{y}_l)^2$ is an unbiased estimator of σ_ε^2.

3.2.4. Randomness of weights

There are two different situations which may lead to random weights:

(a) When the weights are generated by a truly random mechanism not under the researcher's control. An example of genuinely random mechanism generating the weights is the compositing of water samples taken with volumes proportional to flow (Schaeffer, Kerster and Janardan, 1980; Schaeffer et al., 1983).

(b) When the weights are fixed by the sample, but the compositing apparatus fails to mix them thoroughly; then, if the composite is subsampled, the proportion in each subsample contributed by each compositing object varies from subsample to subsample according to some probability distribution. The less satisfactorily the compositing device blends the compositing objects, the larger the variance of such a distribution will be. Imperfect physical mixing leading to random weights is more likely to happen when subsampling from discrete, countable materials (for example,

(a) Neglecting the randomness of the weights may be misleading in assessing the variance of the composited estimator $\bar{y}$ of μ_x, which is never smaller than the corresponding variance in the 'fixed weights' case, namely σ_ε^2/ck.

(b) Writing the upper bound on the variance as

$$\operatorname{var}(\bar{y}) \leqslant \frac{\sigma_\varepsilon^2}{cs}\left[1 + \frac{s-1}{k}\right],$$

it can be seen that, in the presence of random weights, the composited estimator cannot be worse, in variance terms, than the comparable 'uncomposited' estimator $\bar{x} = \sum_{i=1}^{cs} xi/cs$ as long as $s = 1$, i.e., one subsample is selected from each composite. If $s > 1$, the composited estimator may have a larger variance than the uncomposited one, and some knowledge of σ_w^2 is necessary in order to assess the effect of compositing.

(c) The effect on σ_w^2 of increasing k within any finite interval is contradictory. On one hand, since in general a composite is harder to mix the larger it is, σ_w^2 may increase with an increasing k; on the other hand, the larger the composite, the smaller each proportion and the smaller σ_w^2 will be (actually, as $k \to \infty$, $\sigma_w^2 \to 0$). Therfore, in the random weights case, it is difficult to predict how $\operatorname{var}(\bar{y})$ varies with the composite sample size k.

Notice that both estimation of $\operatorname{var}(\bar{y})$ and evaluation of the effects on it of changes in the compositing design require some knowledge of σ_w^2. Furthermore, σ_w^2 could be of direct interest, especially at the early stage of a new study, as it may be thought of as a measure of mixability of the material under investigation. In some cases, researchers in the substantive area of application do have prior knowledge on σ_w^2, at least in terms of a reasonable range for it. But often this knowledge is vague, and estimation of σ_w^2 from the data becomes desirable. This topic seems to have received insufficient attention in the literature. The most comprehensive account appears to be in Elder (1977). This author also presents a procedure for estimation of σ_w^2 using a tracer method, which is theoretically interesting but of little practical use due to its technical sophistication and cost.

References

Anscombe, F. J. (1956). On estimating binomial response relations. *Biometrika* **43**, 461–464.

Bhattacharyya, G. K., M. G. Karandinos and G. R. DeFoliart (1979). Point estimates and confidence intervals for infection rates using pooled organisms in epidemiologic studies. *Amer. J. Epidemiol.* **109**(2), 124–131.

Bicking, C. A. (1967). The sampling of bulk materials. *Mater. Res. Standards* **7**(3), 95–116.

Boswell, M. T., K. P. Burnham and G. P. Patil (1988). Role and use of composite sampling and capture-recapture sampling in ecological studies. In: P. R. Krishnaiah and C. R. Rao, eds., *Handbook of Statistics*, Vol. 6. North-Holland, Amsterdam, 469–488.

Boswell, M. T., S. Gore, G. Lovison and G. P. Patil (1992). Annotated bibliography of composite sampling. Technical Report 92-0802, Center for Statistical Ecology and Environmental Statistics, Dept. of Statistics, Penn State University, University Park, PA.

Boswell, M. T., S. D. Gore and G. P. Patil (1990). Efficiency of various composite sample retesting schemes to classify samples with presence/absence measurements. Technical Report 90-0901, Center for Statistical Ecology and Environmental Statistics, Dept. of Statistics, Penn State University, University Park, PA.

Boswell, M. T. and G. P. Patil (1987). A perspective of composite sampling. *Comm. Statist. Theory Methods* **16**, 3069–3093.

Boswell, M. T. and G. P. Patil (1990) Composite sampling using spatial autocorrelation for Palmerton hazardous waste site: A preliminary report. In: H. Lacayo, R. J. Nadeau, G. P. Patil and L. Zaragoza, eds., *Proc. Workshop on Superfund Hazardous Waste: Statistical Issues in Characterizing a Site: Protocols, Tools, and Research Needs*, Penn State University, University Park, PA, 20–42.

Brown, G. H. and N. I. Fisher (1972). Subsampling a mixture of sampled material. *Technometrics* **14**(3), 663–668.

Burrows, P. M. (1987). Improved estimation of pathogen transmission rates by group testing. *Phytopathology* **77**(2), 363–365.

Cameron, J. M. (1951). The use of components of variance in preparing schedules for sampling of baled wool. *Biometrics* **7**, 83–96.

Chen, L. C. and W. H. Swallow (1990). Using group testing to estimate a proportion, and to test the binomial model. Biometrics **46**, 1035–1046.

Chiang, C. L. and W. C. Reeves (1962). Statistical estimation of virus infection rates in mosquito vector populations. *Amer. J. Hygiene* **75**, 377–391.

Dorfman, R. (1943). The detection of defective members of large populations. *Ann. Math. Statist.* **14**, 436–440.

Ducan, A. J. (1962). Bulk sampling: Problems and lines of attack. *Technometrics* **4**(2), 319–344.

Elder, R. S. (1977). Properties of composite sampling procedure. Ph.D. Dissertation, Virginia Polytechnique Institute and State University, Blacksburg, VA.

Elder, R. S., W. O. Thompson and R. H. Myers (1980). Properties of composite sampling procedures. *Technometrics* **22**(2), 179–186.

Fabius, J. (1973). Two characterizations of the Dirichlet distribution. *Ann. Statist.* **1**, 583–587.

Feller, W. (1968). *An Introduction to Probability Theory and Its Applications*. 3rd Ed., Wiley, New York.

Fine, P. E. M. and E. S. Sylvester (1978). Calculation of vertical transmission rates of infection, illustrated with data on an aphid-borne virus. *Amer. Naturalist* **112**, 781–786.

Garner, F. C., M. A. Stapanian and L. R. Williams (1988). Composite sampling for environmental monitoring. In: *Principles of Environmental Sampling*. Amer. Chem. Soc., Washington, DC, 363–374.

Gastwirth, J. L. and P. A. Hammick (1989). Estimation of the prevalence of a rare disease, preserving the anonymity of the subjects by groups testing: Application to estimating the prevalence of AIDS antibodies in blood donors. *J. Statist. Plann. Inference* **22**, 15–27.

Gibbs, A. J. and J. C. Gower (1960). The use of a multiple-transfer method in plant virus transmission studies – Some statistical points arising in the analysis of results. *Ann. Appl. Biol.* **48**, 75–83.

Gill, A. and D. Gottlieb (1974). The identification of a set by successive intersections. In: *Information and Control*. Ellis Horwood, Chichester, 20–35.

Griffiths, D. A. (1972). A further note on the probability of disease transmission. *Biometrics* **28**, 1133–1139.

Haldane, J. B. S. (1955). The estimation and significance of the logarithm of a ratio of frequencies. *Ann. Human Genetics* **20**, 309–311.

Hwang, F. K. (1984). Robust group testing. *J. Qual. Tech.* **16**, 189–195.

Kerr, J. D. (1971). The probability of disease transmission. *Biometrics* **27**, 219–222.

Le, C. T. (1981). A new estimator for infection rates using pools of variable size. *Amer. J. Epidemiol.* **114**, 132–135.

Lovison, G. (1993a). A unified linear model for estimation with composite sample data. In: G. P. Patil and C. R. Rao, eds., *Multivariate Environmental Statistics*. Elsevier Science Publishers, Amsterdam, to appear.

Lovison, G. (1993b). Optimal design of fixed weights composite samples, with environmental applications. In: *Proc. Internat. Conf. on Quantitative Methods for Applied Sciences*, June 11–13, 1992, Siena, Italy, to appear.

Lovison, G., S. D. Gore and G. P. Patil (1991). Composite sampling for estimating population parameters: Some extentions and a unifying model. Technical Report 91-0201, Center for Statistical Ecology and Environmental Statistics, Dept. of Statistics, Penn State University, University Park, PA.

Loyer, M. W. (1983). Bad probability, good statistics, and group testing for binomial estimation. *Amer. Statist.* **37**, 57–59.

Mack, G. A. and P. E. Robinson (1985). Use of composited samples to increase the precision and probability of detection of toxic chemicals. In: J. J. Breen and P. E. Robinson, eds., *Environmental Applications of Chemometrics.* Amer. Chem. Soc., Washington, DC, 174–183.

Mahoney, D. F. and G. B. Mirre (1971). Bovine babesiasis: Estimation of infection rates in the tick vector Boophilus microplus (Canestrini). *Ann. Trop. Med. Parasitology* **65**(3), 309–317.

Mekonnen, G., C. J. Wilcox, K. C. Bachman, W. W. Thatcher and F. G. Martin (1976). Genetic parameters for a.m.–p.m. composite monthly milk samples. *J. Dairy Sci.* **59**(1), 15–16.

Orban, J. E., R. Lordo and J. Schwemberger (1990). Statistical methods for analyzing composite sample data applied to EPA's human monitoring program. Manuscript.

Patil, G. P., S. Gore and C. Taillie (1994). Composite sampling: A novel approach to accomplish observational economy in environmental studies. Technical Report 92-1001, Center for Statistical Ecology and Environmental Statistics, Dept. of Statistics, Penn State University, University Park, PA, in preparation.

Rohde, C. A. (1976). Composite sampling. *Biometrics* **32**, 273–282.

Rohde, C. A. (1979). Batch, bulk and composite sampling. In: R. M. Cormack, G. P. Patil and D. S. Robson, eds., *Sampling Biological Populations*, International Co-operative Publishing House, Fairland, MD, 365–377.

Samuels, S. M. (1978). The exact solution to the two-stage group-testing problem. *Technometrics* **20**, 497–500.

Schaeffer, D. J., H. W. Kerster, D. R. Bauer, K. Rees and S. McCormick (1983). Composite samples overestimate waste loads. *Pollut. Control Fed J.* **55**(11), 1387–1392.

Schaeffer, D. J., H. W. Kerster and K. G. Janardan (1980). Grab versus composite sampling: A primer for the manager and engineer. *Environ. Management* **4**(2), 157–163.

Sharma, A. K., L. A. Rodriguez, G. Mekonnen, C. J. Wilcox, K. C. Bachman and R. Collier (1983). Climatological and genetic effects on milk composition and yield. *Dairy Sci.* **66**, 119–126.

Skinner, C. J., D. Holt and T. M. F. Smith (1989). *Analysis of Complex Surveys.* Wiley, New York.

Sobel, K. M. and R. M. Elashoff (1975). Group testing with a new goal, estimation. *Biometrika* **62**, 181–193.

Sterrett, A. (1957). On the detection of defective members of large populations. *Ann. Math. Statist.* **28**, 1033–1036.

Swallow, W. H. (1985). Group testing for estimating infection rates and probabilities of disease transmission. *Phytopathology* **75**, 882–889.

Swallow, W. H. (1987). Relative mean squared error and cost considerations in choosing group size for group testing to estimate infection rates and probabilities of disease transmission. *Phytopathology* **77**, 1376–1381.

Thompson, K. H. (1962). Estimation of the proportion of vectors in a natural population of insects. *Biometrics* **18**, 568–578.

Walter, S. D., S. W. Hildreth and B. J. Beaty (1980). Estimation of infection rates in populations of organisms using pools of variable size. *Amer. J. Epidemiol.* **112**(1), 124–128.

Watson, M. A. (1936). Factors affecting the amount of infection obtained by aphis transmission of the virus. *Philos. Trans. Roy. Soc. London Ser. B* **226**, 457–489.

West, M. (1986). Bayesian model monitoring. *J. Roy. Statist. Soc. B* **48**, 70–78.

Worlund, D. D. and G. Taylor (1983). Estimation of disease incidence in fish populations. *Canad. J. Fish. Aquat. Sci.* **40**, 2194–2197.

G. P. Patil and C. R. Rao, eds., *Handbook of Statistics, Vol. 12*

5

Ranked Set Sampling*

G. P. Patil, A. K. Sinha and C. Taillie

Abstract: McIntyre (1952) proposed a method of sampling to estimate mean pasture yields with greater efficiency than simple random sampling. This method of sampling has come to be known as ranked set sampling (RSS) because it involves a preliminary ranking of randomly selected units from the population, after which only a certain few of these sampled units are actually quantified. The method was used first by Halls and Dell (1966) to estimate the weight of browse and herbage in a pine hardwood forest. Although McIntyre (1952) had mentioned an upper bound on the relative precision (RP) of the RSS estimator of a population mean relative to that of the simple random sample (SRS) estimator with the same sample size, Takahasi and Wakimoto (1968) gave a rigorous mathematical derivation of this bound. Since then, a number of papers discussing the RSS method have appeared. These papers may broadly be classified into three groups: (i) theory, (ii) methods, and (iii) applications. We have, in this chapter, made an attempt to review these various aspects in a single unified notation. This study also illustrates the performance of RSS compared to that of SRS in determining the level of PCB contamination at a hazardous waste site. In this context, we also demonstrate the use of RSS methods for improving the formation of composite samples.

1. Introduction

In environmental and ecological sampling one may encounter a situation where the exact measurement (or quantification) of a selected unit is either difficult or expensive in terms of time, money or labor, but where the ranking of a small set of selected units according to the characteristic of interest can be done with reasonable

* The present chapter has been prepared with partial support from the Statistical Analysis and Computing Branch, Environmental Statistics and Information Division, Office of Policy, Planning, and Evaluation, with Brenda Odom as EPA Project Officer. The contents have not been subjected to Agency review and therefore do not necessarily reflect the views of the Agency and no official endorsement should be inferred.

success on the basis of visual inspection or other rough method not requiring actual measurement. For example, if interest lies in estimating the mean height of trees, then measuring the height of the sampled trees could pose a problem, but it would be relatively easy to rank small sets of trees on the basis of visual inspection of their heights. In situations where visual inspection is not directly available, ranking can sometimes be done on the basis of a covariate that is more accessible than, but correlated with, the character of interest. Thus, if we are interested in the volumes of trees, we may use the ranking by diameter to approximate the ranking by volume. Likewise, at hazardous waste sites, the ranking of sampling units (locations) according to their contamination levels can be indicated by such visual cues as defoliation and soil discoloration. Other potential ordinal indicators in environmental work include the use of special chemically-responsive papers, electromagnetic readings, and remotely sensed data. The method of ranked set sampling (RSS) was developed by McIntyre (1952) to improve upon simple random sampling for situations such as these where some preliminary ranking of sampled units is feasible.

In fact, McIntyre's motivation was for the estimation of mean pasture and forage yields. In agricultural sampling, it is usually easy to assess relative crop yields by visual inspection and this can lead (deliberately or not) to purposive sampling with a corresponding risk of estimation bias. McIntyre's goal was to maintain the unbiasedness of simple random sampling while effectively incorporating into the estimates the outside information provided by visual inspection. The basic idea is that the collection of sampled units is randomly partitioned into small groups. A group size (called the *set-size*) of two, three, or four units is typical. The members of each given group are then ranked relative to one another and, on the basis of the ranking, exactly one member of each group is chosen for quantification. Even if there are ranking errors, the method is shown to be at least as efficient as simple random sampling with the same number of quantifications. Relative efficiencies of several hundred percent are possible depending upon factors such as ranking accuracy, set-size, and the underlying population distribution. Since only a fraction of the sampled units are quantified, the method presumes that the physical acquisition of units is cheap as compared with their quantification.

McIntyre's proposal appears to have lain dormant for over a decade, until Halls and Dell (1966) conducted a field trial evaluating the method's applicability to estimating forage yields in a pine hardwood forest. They found empirically that RSS was more efficient than simple random sampling (SRS), and that the subjective rankings of two separate observers gave nearly the same efficiency as perfect rankings. Halls and Dell also coined the name *ranked set sampling* that is in current use. Evans (1967) used the technique for obtaining the average number of longleaf pine (*Pinus palustris mill.*) seedlings per milacre plot and reported on the empirical effectiveness of the method. But the much needed mathematical foundation for RSS was provided by Takahasi and Wakimoto (1968) and, independently, by Dell (1969) who also discussed some applications of RSS.

It is apparent that errors may creep in while ordering the randomly selected units of a set because of dependence upon the ranker's judgement. This aspect of RSS was first considered theoretically by Dell and Clutter (1972) who have shown that, regardless of ranking errors, the RSS estimator of a population mean is unbiased and is at least as efficient as the SRS estimator with the same number of quantifications. The relative precision (RP) of the two methods is equal to unity only if the ranking is no better than random. Dell and Clutter emphasize that the RP depends upon the characteristics of the population as well as the magnitude of errors in ranking.

David and Levine (1972) consider the case where the ranking is done on the basis of a numerical covariate instead of judgement. Under certain assumptions, including normality, they obtain a formula expressing RP in terms of the squared coefficient of correlation between the covariate and the variate of interest. Stokes (1976, 1977) has explored this model further.

Ridout and Cobby (1987) point out that, in addition to ranking errors, another source of error can arise in the practical implementation of ranked set sampling. While the protocol calls for a random grouping into subsets, it often happens for reasons of convenience or ease of ranking that the members of each subset are spatially proximate or otherwise clustered. The effects of such intraclass correlation on RSS have been studied by Takahasi (1969).

While discussing the measurement of grassland vegetation, Mannetje (1978) observes that the most suitable method of sampling depends on the type of vegetation, the total area to be sampled, the topography, and the available facilities; but he suggests that RSS can improve sampling efficiency in most circumstances. In his last paper on vegetation sampling, McIntyre (1978) emphasizes the unbiasedness of RSS while pointing out that it can give 'worthwhile gain for little extra effort'. Further, he notes that the method could be employed most effectively to improve the estimates of strata means in the presence of gradients within the plots.

Martin et al. (1980) have reported on the effectiveness of the RSS procedure for estimating shrub phytomass in forest stands. They infer that the method is more efficient than SRS without replacement in both the actual and the simulated sampling trails. Discussing sward measurement, Jewiss (1981) has demonstrated the suitability of ranked set sampling for situations in which there are substantial changes in the botanical composition of the herbage within a plot. Jewiss also mentions that Curll (1980) has used the method for the measurement of tiller number on mixed swards of perennial ryegrass and white clover. The applicability of RSS to the estimation of herbage mass and clover content in grazed swards has also been examined by Cobby et al. (1985).

Stokes (1980) has proposed an estimator of the population variance based on a ranked set sample. She shows that the estimator is asymptotically unbiased even in the presence of errors in ranking. Stokes (1986) has also given a useful review of some theoretical aspects of RSS.

In this chapter we have made an attempt to study mainly (i) theory, (ii) methods, and (iii) applications of the method in a single unified notation. Apart from mentioning the real trials based on it we have tried to include several other situations where it could be used advantageously as pointed out by several authors. Also, an attempt has been made to illustrate the superiority of the protocol over SRS to determine the level of PCB at a hazardous waste site, and its usefulness in forming better composite samples. In fact, this study may be of particular interest in environmental studies related to characterization, evaluation, and remediation of a hazardous waste site because the RSS method can be considerably more cost effective than SRS.

2. The method of ranked set sampling

The procedure of selection of a ranked set sample involves the drawing from an infinite population of m random samples with m units in each sample. Equivalently, m^2 units may be drawn and randomly partitioned into m equal samples. In either case, m is a design parameter to be chosen by the investigator. Next, the units of each sample are ranked by a visual inspection or by any other method not requiring the exact measurement of the variable of interest. The unit with the smallest rank is quantified from the first sample, the unit having the second smallest rank is quantified from the second sample, and this process of quantification continues until the unit with the largest rank is quantified from the m-th sample. This procedure involves the quantification of m units out of the m^2 units originally selected. In practice m is generally chosen to be small in order to facilitate the judgemental ranking, which may be easier when there are fewer units to compare. To provide enough quantifications for inference, the entire process (or *cycle*) can be replicated r times, thus yielding $n = mr$ quantified units out of m^2r selected units. These n quantified units constitute the ranked set sample. The foregoing procedure of sampling is also referred to as MTW sampling in recognition of McIntyre's (1952) original formulation and the subsequent mathematical contributions of Takahasi and Wakimoto (1968).

In order to illustrate the method, let us take a set-size of $m = 3$ with $r = 4$ replications or cycles. Then the sampling scheme is shown in the following diagram.

Cycle	Rank		
	1	2	3
1	⊙	·	·
	·	⊙	·
	·	·	⊙
2	⊙	·	·
	·	⊙	·
	·	·	⊙
3	⊙	·	·
	·	⊙	·
	·	·	⊙
4	⊙	·	·
	·	⊙	·
	·	·	⊙

Here each row denotes a judgement ordered sample for each cycle, and the circled units indicate the units which are to be quantified. In all, 36 units have been drawn randomly in 4 cycles but only 12 units (circled in the diagram) are quantified to get the required ranked set sample.

Let $X_{11}, X_{12}, \ldots, X_{1m}, X_{21}, X_{22}, \ldots, X_{2m}, \ldots, X_{m1}, X_{m2}, \ldots, X_{mm}$ be independent random variables all having the same cumulative distribution function $F(x)$. Also let $X_{i(1)}, X_{i(2)}, \ldots, X_{i(m)}$ denote the corresponding order statistics of $X_{i1}, X_{i2}, \ldots, X_{im}$ $(i = 1, 2, \ldots, m)$. Then $X_{1(1)}, X_{2(2)}, \ldots, X_{i(i)}, \ldots, X_{m(m)}$ is the ranked set sample (considering one cycle only), since $X_{i(i)}$ is the i-th order statistic in the i-th sample. For the moment, we suppose that there are no ranking errors.

The values X_{ij} for the randomly drawn units can be arranged as in the following diagram:

Set				
1	X_{11}	X_{12}	$\cdots$	X_{1m}
2	X_{21}	X_{22}	$\cdots$	X_{2m}
$\vdots$				
m	X_{m1}	X_{m2}	$\cdots$	X_{mm}

After ranking the units appear as shown below:

Set				
1	$X_{1(1)}$	$X_{1(2)}$	$\cdots$	$X_{1(m)}$
2	$X_{2(1)}$	$X_{2(2)}$	$\cdots$	$X_{2(m)}$
$\vdots$				
m	$X_{m(1)}$	$X_{m(2)}$	$\cdots$	$X_{m(m)}$

The quantified units are presented as given below:

Set				
1	$X_{1(1)}$	*	$\cdots$	*
2	*	$X_{2(2)}$	$\cdots$	*
$\vdots$				
m	*	*	$\cdots$	$X_{m(m)}$

Still considering only one cycle, the mean of the ranked set sample is denoted by $\bar{X}_{(m)}$ where

$$\bar{X}_{(m)} = \frac{1}{m} \sum_{i=1}^{m} X_{i(i)}.$$

For convenience, we also write $X_{i(i)}$ as $X_{(i:m)}$. This notation exphasizes that $X_{(i:m)}$ is an observation on the $i:m$-th order statistic from the population; parentheses are used to surround $i:m$ in the subscript since the different $X_{(i:m)}$ are (obviously) independent as contrasted with the usual order statistics which are positively correlated. Now,

$$\bar{X}_{(m)} = \frac{1}{m} \sum_{i=1}^{m} X_{(i:m)}, \tag{1}$$

so that

$$\mathrm{E}[\bar{X}_{(m)}] = \frac{1}{m} \sum_{i=1}^{m} \mathrm{E}[X_{(i:m)}] = \mu .$$

This shows that $\bar{X}_{(m)}$ is an unbiased estimator of the population mean μ.

Let us compare this estimator with the sample mean based on m iid quantifications. The latter can be written as

$$\bar{X} = \frac{1}{m} \sum_{i=1}^{m} X_{i:m},$$

where $X_{i:m}$ are the order statistics of the m quantifications. Since, as pointed out, the $X_{i:m}$ are positively correlated, it follows that $\bar{X}_{(m)}$ is more efficient than $\bar{X}$ for estimating μ. In essence, the RSS quantifications $X_{(i:m)}$ are more regularly spaced with less clustering than is a simple random sample of size m. See Figure 1.

When the whole procedure of drawing random samples is repeated r times, the i-th order statistic from the i-th sample in the j-th cycle will be denoted by $X_{(i:m)j}$. The

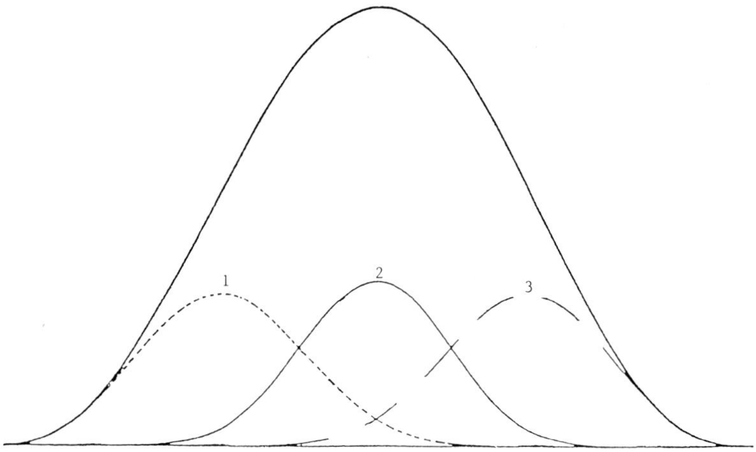

Fig. 1. Parent distribution is portrayed by the bigger figure and distribution of samples of each rank is shown by the smaller figures.

unbiased estimator of the population mean μ is given by

$$\bar{X}_{(m)r} = \frac{\sum_{i=1}^{m}\sum_{j=1}^{r} X_{(i:m)j}}{mr}. \tag{2}$$

We write $\mu_{(i:m)}$ and $\sigma^2_{(i:m)}$ for the mean and the variance of $X_{(i:m)}$; in the absence of ranking errors, these are also the first two moments of the order statistic $X_{i:m}$. The variance of $\bar{X}_{(m)r}$ is given by

$$\mathrm{var}(\bar{X}_{(m)r}) = \frac{1}{mr}\sum_{i=1}^{m}\frac{\sigma^2_{(i:m)}}{m}. \tag{3}$$

An equivalent expression is

$$\mathrm{var}(\bar{X}_{(m)r}) = \frac{1}{mr}\left\{\sigma^2 - \frac{1}{m}\sum_{i=1}^{m}(\mu_{(i:m)} - \mu)^2\right\}, \tag{4}$$

where σ^2 denotes the population variance.

2.1. *Within-set dependence*

The RSS protocol requires the units within each sample (set) be selected independently prior to ranking. This requirement can lead to severe implementation

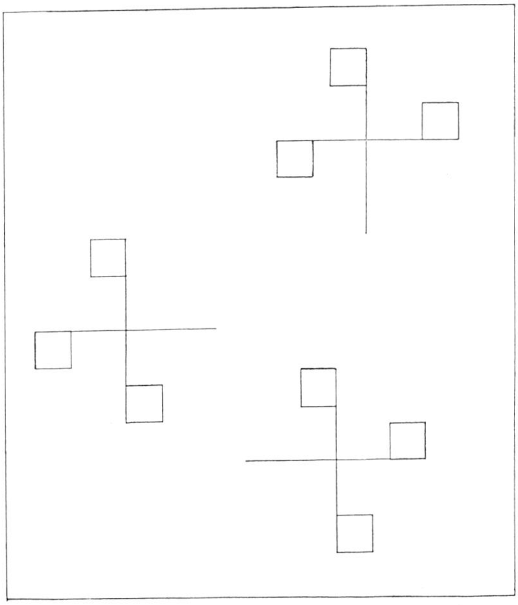

Fig. 2. Square crosses with quadrats.

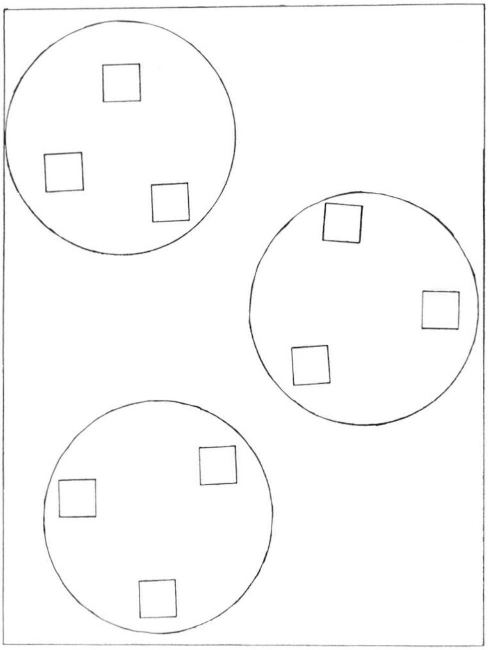

Fig. 3. Quadrats within circles.

difficulties in the case of judgemental ranking because randomly selected units may be so widely separated in space that visual comparisons among them are difficult. Often, then, RSS is implemented in a two-stage fashion in which some locations are first randomly selected within target area and then the units to be ranked are selected at these locations. In a typical scheme, a 'square cross' or a 'circular frame' is thrown down at each location (Figures 2 and 3). Quadrats of fixed size are either put at the marked positions on the square cross or placed randomly within the circle. Halls and Dell (1966) used such a method and found that it did not substantially reduce the benefits of RSS for their investigation. But Cobby et al. (1985) noticed in their study that this approach caused higher variation between sets than within sets and ultimately diminished the superiority of RSS over SRS. In order to deal with the situation encountered by Cobby et al. (1985), McIntyre (1952) suggested that the target area be partitioned into portions of equal size with no well defined gradients within each portion and that an equal number of samples be selected within each portion. In any case, McIntyre points out that the 'local' RSS protocol will be efficient whenever local variation accounts for a substantial portion of total variation.

2.2. Ranking by concomitant variables

Sometimes it is difficult or impossible to order the selected units by visual inspection. Then it may be possible to accomplish the ranking by means of a covariate that is easily measured and is correlated with the variable of interest X. The covariate is called a concomitant variable and will be denoted by Y. To carry out the ranking, m bivariate samples each having m units are drawn from the population. The X associated with the smallest ordered Y is quantified from the first sample, the X associated with the second smallest Y is quantified from the second sample and so on. Lastly, the X associated with the largest Y is quantified from the m-th sample. The whole cycle is repeated r times to obtain mr quantifications of the variable X.

3. Precision of the RSS estimator of a population mean relative to the SRS estimator

We compare the variance of the mean of a ranked set sample with that of a mean based on a random sample of mr observations and not with a random sample based on m^2r observations. This is because the cost of acquiring and ranking samples is not taken into account but only the cost of quantification is considered. Using equation (4), the relative precision (RP) of the two estimators is given by

$$\begin{aligned} \mathrm{RP} &= \frac{\text{variance of mean with random sampling}}{\text{variance of mean with ranked set sampling}} \\ &= \frac{\operatorname{var}(\bar{X})}{\operatorname{var}(\bar{X}_{(m)r})} \\ &= \frac{\sigma^2/mr}{\frac{1}{mr}\{\sigma^2 - \frac{1}{m}\sum_{i=1}^{m}(\mu_{(i:m)} - \mu)^2\}} \\ &= \frac{1}{1 - \frac{1}{m}\sum_{i=1}^{m}(\tau_{(i)}/\sigma)^2}, \end{aligned} \tag{5}$$

where $\tau_{(i)} = \mu_{(i:m)} - \mu$. Equivalent, but useful, measures of relative efficiency are the relative cost (RC) and the relative savings (RS) as given by

$$\mathrm{RC} = \frac{1}{\mathrm{RP}}, \qquad \mathrm{RS} = 1 - \mathrm{RC}.$$

In this context, the relative savings assumes the particularly simple form,

$$\mathrm{RS} = \frac{1}{m}\sum_{i=1}^{m}\left(\frac{\tau_{(i)}}{\sigma}\right)^2. \tag{6}$$

Since this expression is positive, ranked set sampling is always more cost efficient than simple random sampling with the same number of quantifications.

3.1. *Effect of the population distribution*

Based on some numerical computations with specific distributions, McIntyre (1952) had conjectured that for 'typical unimodal distributions' RP would not be much less than $(m+1)/2$ under the assumption of perfect ranking. This conjecture was answered in the affirmative by Takahasi and Wakimoto (1968), who showed that RP is bounded below by 1 and above by $(m+1)/2$ for all continuous distributions with finite variances. They also proved that the upper bound is realized only in case of a rectangular distribution.

McIntyre also concluded that RP appeared to decrease with increasing skewness of the population. Dell (1969) studied this issue for a wide range of distributions. According to formula (5), the RP depends upon the expected order statistics for the population. Here, Dell drew upon the work of Harter (1964) for the gamma $\frac{1}{2}(\frac{1}{2})4$ and Weibull $\frac{1}{2}(\frac{1}{2})4(1)8$, Teichroew (1956) for the normal, Gupta (1960) for the gamma 1(1)5, Hastings et al. (1947) for the rectangular, McIntyre (personal communication to Dell) for the beta with parameter (4, 4), (7, 4) and (13, 4), Govindarajulu (1962) for the chi-square with one degree of freedom, Sarhan (1954) for the double exponential and the triangular, Sarhan and Greenberg (1962) for the exponential, and Leiblein and Salzer (1957) for the extreme value type 1. Dell's tabulation also gives the values of μ, σ^2, $\sqrt{\beta_1}$ and β_2 for these distributions. Our Table 1 is based on Dell (1969) and

Table 1

μ, σ^2, $\sqrt{\beta_1}$, β_2 and relative precision (RP) for some distributions. The relative precision is given for set-sizes of $m = 2, 3, 4, 5$

Distribution	μ	σ^2	$\sqrt{\beta_1}$	β_2	2	3	4	5
Rectangular	0.500	0.083	0.000	1.80	1.500	2.000	2.500	3.000
Exponential	1.000	1.000	2.000	9.00	1.333	1.636	1.920	2.190
Gamma (0.5)	0.500	0.500	2.828	15.00	1.254	1.483	1.696	1.898
Gamma (1.5)	1.500	1.500	1.633	7.00	1.370	1.710	2.030	2.336
Gamma (2.0)	2.000	2.000	1.414	6.00	1.391	1.753	2.096	2.424
Gamma (3.0)	3.000	3.000	1.155	5.00	1.414	1.801	2.169	2.524
Gamma (4.0)	4.000	4.000	1.000	4.50	1.427	1.827	2.210	2.575
Gamma (5.0)	5.000	5.000	0.894	4.20	1.434	1.843	2.236	2.615
Normal	0.000	1.000	0.000	3.00	1.467	1.914	2.347	2.770
Beta (4, 4)	0.500	0.028	0.000	2.45	1.484	1.958	2.425	2.890
Beta (7, 4)	0.636	0.019	−0.302	2.70	1.475	1.936	2.389	2.834
Beta (13, 4)	0.765	0.010	−0.557	3.14	1.460	1.903	2.333	2.765
Weibull (0.5)	2.000	20.000	6.619	87.72	1.127	1.236	1.334	1.425
Weibull (1.5)	0.903	0.376	1.072	4.39	1.422	1.822	2.205	2.578
Weibull (2.0)	0.886	0.215	0.631	3.24	1.458	1.897	2.325	2.744
Weibull (3.0)	0.893	0.105	0.168	2.73	1.476	1.936	2.387	2.830
Weibull (4.0)	0.906	0.065	−0.087	2.75	1.474	1.932	2.380	2.820
Weibull (5.0)	0.918	0.044	−0.254	2.88	1.469	1.921	2.361	2.793
Weibull (6.0)	0.928	0.032	−0.373	3.04	1.464	1.909	2.341	2.764
Weibull (7.0)	0.935	0.025	−0.463	3.19	1.459	1.898	2.324	2.740
Weibull (8.0)	0.942	0.019	−0.534	3.33	1.456	1.890	2.309	2.718
Chi-square 1 DF	0.789	0.363	0.995	3.87	1.430	1.841	2.239	–
Double Exponential	0.000	2.000	0.000	6.00	1.381	1.730	2.038	2.327
Triangular	0.500	0.042	0.000	2.40	1.485	1.961	2.430	2.896
Extreme value, Type I	0.577	1.645	1.300	5.40	1.413	1.793	2.153	2.497

Source: Dell (1969) and Dell and Clutter (1972).

Table 2
Relative precision of the RSS estimator compared to the SRS estimator when the parent population follows a lognormal distribution

m	Sigma (logarithmic standard deviation)									
	0.1	0.2	0.3	0.4	0.5	0.6	0.7	0.8	0.9	1.0
2	1.46	1.45	1.42	1.40	1.37	1.33	1.29	1.26	1.22	1.19
3	1.90	1.87	1.83	1.77	1.70	1.62	1.55	1.47	1.40	1.34
4	2.33	2.28	2.21	2.11	2.00	1.89	1.78	1.67	1.56	1.47
5	2.75	2.68	2.57	2.44	2.29	2.14	1.99	1.84	1.71	1.59
6	3.15	3.07	2.93	2.76	2.57	2.37	2.18	2.00	1.84	1.70
7	3.56	3.45	3.28	3.07	2.83	2.60	2.40	2.16	1.96	1.80
8	3.95	3.86	3.61	3.36	3.09	2.81	2.55	2.30	2.08	1.89
9	4.34	4.19	3.95	3.66	3.34	3.02	2.72	2.44	2.20	1.98
10	4.73	4.54	4.28	3.94	3.58	3.22	2.88	2.57	2.30	2.06

Dell and Clutter (1972), who have pointed out the RP has a better numerical relationship with kurtosis β_2 than with skewness $\sqrt{\beta_1}$.

In view of the importance of the lognormal distribution in applied work (Crow and Shimizu, 1988), we have examined the efficiency of RSS relative to SRS for lognormal populations. Table 2 gives the values of RP for a range of values of the shape parameter σ^2. Here, it may be noted that the squared coefficient of variation for the lognormal is

$$CV^2 = \exp(\sigma^2) - 1,$$

so that, for example, $\sigma = 1$ corresponds to a coefficient of variation of 131 percent. In general, Table 2 shows that the performance of ranked set sampling for lognormal distributions decreases rather rapidly as the skewness increases (i.e., as σ increases). Also, the lognormal converges to normality as σ goes to zero, so the appropriate entries in Table 1 and Table 2 match up.

3.2. *Effect of the set-size*

An important design issue in applying ranked set sampling is the choice of the set-size m. Takahasi and Wakimoto (1968) have shown that $\text{var}(\bar{X}_{(m)r})$ is a decreasing function of m when the number of quantifications $n = mr$ is fixed. Since RP does not depend upon r, this implies that the relative efficiency of RSS is an increasing function of the set-size for a given population. Mathematically, then, the RSS estimator may be improved by increasing m, and decreasing r, for a fixed sample size. It further means that the best RSS estimator could be obtained with a single cycle, i.e., with $r = 1$. However, this is usually not practical because of the difficulty in ranking the sets when they contain many units. Choice of m will thus involve some tradeoff between the efficiency of estimation and the convenience and accuracy of ranking. Table 1 and Table 2 provide some sense of how efficiency grows with increasing m for various populations.

4. Practical difficulties in implementing RSS: Solutions and consequences

The performance of RSS depends, apart from other factors, on how perfectly the judgemental ordering of the randomly selected units is accomplished. The ordering, in turn, helps in achieving stratification without quantification and utilizes the experience and the expertise of the investigators. But the judgmental ordering may not always coincide with the true ordering. This causes ranking error which adversely affects the performance of RSS. Besides this, in a field investigation where the randomly selected units are found located usually far apart, it becomes difficult to do the ordering. As a solution to this problem, a 'square cross' or a 'circular frame' is used to secure a sample consisting of closely located units. Although this helps in judgmental ranking, the sample may not be considered as a random sample drawn from the entire target area. In fact, sometimes this leads to larger variation between sets than within sets, and ultimately the RP of RSS relative to SRS gets diminished. Still another factor on which the magnitude of RP depends is the statistical character of the population under consideration.

Thus we have identified the following three factors which affect the performance of RSS:

(a) Error in ranking and use of a concomitant variable.
(b) Nonrandom selection of units within sets.
(c) Characteristics of the population under consideration.

Let us discuss these factors separately.

4.1. *Error in ranking*

Accurate ranking is the most common difficulty encountered in attempting to carry out the RSS protocol. Errors in ranking cause units to be assigned ranks that are different from their true ranks according to the variable of interest. This, in turn, leads to a measurement difference between the unit that is quantified and the unit that ought to have been quantified. Dell and Clutter (1972) were the first to study this effect formally. In this situation of inaccurate ranking, the quantified unit from the i-th sample in the j-th cycle does not represent the i-th order statistic in that sample but rather the i-th 'judgement order statistic'. If we denote the 'judgement order statistic' and the 'true order statistic' by $X_{[i:m]j}$ and $X_{(i:m)j}$, respectively, then Dell and Clutter use the model

$$X_{[i:m]j} = X_{(i:m)j} + \varepsilon_{ij}, \tag{7}$$

where $X_{(i:m)j}$ and ε_{ij} are independent and where $\varepsilon_{ij} \sim \mathrm{N}(0, \sigma_\varepsilon^2)$, in order to study the impact of ranking errors. We may note that (7) is essentially the same as the ranking model introduced by the psychologist Thurstone (see Diaconis, 1988, p. 174). Dell and Clutter show that the estimator

$$\bar{X}_{[m]r} = \frac{\sum_{j=1}^{r} \sum_{i=1}^{m} X_{[i:m]j}}{mr} \tag{8}$$

is unbiased despite the presence of ranking error and that

$$\mathrm{RP} = \frac{\mathrm{var}(\bar{X})}{\mathrm{var}\, \bar{X}_{[m]r}} \geqslant 1,$$

where $\bar{X}$ denotes the SRS estimator of the population mean. The equality sign holds only when the ranking is no better than random selection.

In order to investigate the numerical effect of ranking error on RP, Dell and Clutter conducted 5000 computer simulation trials with the rectangular, exponential, and normal as the population distributions and with 2, 3, and 5 as the set-sizes. The random errors in (7) were taken to be normally distributed with mean zero and with variances σ_ε^2 of 0.05, 0.15, 0.30 and 0.50. The resulting RP values are given in Table 2 of their paper and are reproduced in Table 3. The column for $\sigma_\varepsilon^2 = 0$ corresponds to the case of perfect ranking and may be used for benchmarking.

Dell and Clutter point out that the model (7) may not be valid for describing all cases of practical interest. Additivity of the error may often be questioned. In addition, judgmental comparisons can be influenced by the set-size as well as by the magnitudes of the units themselves. Keeping these facts in view, David and Levine

Table 3
RP values based on 5000 computer simulation trials

m	σ_ε^2				
	0	0.05	0.15	0.30	0.50
	Rectangular ($\mu = 0.5$ and $\sigma^2 = 0.0833$)				
2	1.50	1.28	1.13	1.08	1.05
3	2.00	1.48	1.22	1.12	1.08
5	3.00	1.72	1.32	1.16	1.10
	Exponential ($\mu = 1.0$ and $\sigma^2 = 1.0$)				
2	1.33	1.33	1.24	1.23	1.19
3	1.64	1.56	1.50	1.42	1.34
5	2.19	1.99	1.86	1.69	1.56
	Normal ($\mu = 0$ and $\sigma^2 = 1.0$)				
2	1.47	1.48	1.37	1.34	1.28
3	1.91	1.86	1.74	1.59	1.48
5	2.77	2.64	2.29	1.98	1.76

Source: Dell and Clutter (1972).

Table 4
RP of ranked set sampling from the normal distribution with inaccurate ranking

m	σ_ε^2									
	0.00	0.05	0.10	0.15	0.20	0.25	0.30	0.40	0.50	1.00
2	1.47	1.44	1.41	1.38	1.36	1.34	1.32	1.29	1.27	1.19
3	1.91	1.83	1.77	1.71	1.66	1.62	1.58	1.52	1.47	1.31
4	2.35	2.21	2.09	2.00	1.92	1.85	1.79	1.69	1.62	1.40
5	2.77	2.55	2.39	2.25	2.14	2.05	1.97	1.84	1.74	1.47
6	3.19	2.89	2.66	2.48	2.34	2.22	2.12	1.96	1.84	1.52
7	3.59	3.20	2.91	2.69	2.51	2.37	2.25	2.06	1.93	1.56
8	4.00	3.50	3.14	2.87	2.67	2.50	2.36	2.15	2.00	1.60
9	4.40	3.79	3.36	3.05	2.81	2.62	2.47	2.23	2.06	1.63
10	4.79	4.06	3.56	3.21	2.94	2.73	2.56	2.30	2.12	1.65

Source: David and Levine (1972).

(1972) have suggested a theoretical approach for obtaining the values of $\mathrm{E}[X_{[i:m]}]$ and then of RP in the presence of judgement error. For the case of an underlying normal distribution, they obtain the expression

$$\mathrm{E}[X_{[i:m]}] = \frac{\mu_{(i:m)}}{(1+\sigma_\varepsilon^2)^{1/2}} \tag{9}$$

where $\mu_{(i:m)}$ denotes the expected value of the i-th order statistic in a random sample of size m from a unit normal distribution. The values of RP were computed using the values of $\mu_{(i:m)}$ given by Teichroew (1956). The values of Table A1 of David and Levine are reproduced in Table 4. On comparing the values of RP with those given in Table 3, we find good agreement between them.

4.2. Use of a concomitant variable

Extending the work of Dell and Clutter (1972) and David and Levine (1972) on ranking errors, Stokes (1977) proposed the use of a concomitant variable Y for the purpose. In this model, the ranking is done according to the concomitant Y which, to be effective, should be easily observed and enjoy a high rank-correlation with the variable of interest X. Following Stokes, we suppose that (i) the regression of X on Y is linear and (ii) the standardized variates $(X-\mu_X)/\sigma_X$ and $(Y-\mu_Y)/\sigma_Y$ have the same cumulative distribution function. Notice that (ii) is automatic when (X, Y) is bivariate normal.

Let $X_{[1:m]1}, \ldots, X_{[m:m]r}$ denote the ranked set sample of size mr drawn with the help of the concomitant variable Y. Since the linear regression relationship is maintained when we rank according to the concomitant, the assumption (i) implies that

$$\mathrm{E}\{X_{[i:m]} | Y_{(i:m)}\} = \mu_X + \frac{\rho\sigma_X}{\sigma_Y}(Y_{(i:m)} - \mu_Y), \tag{10}$$

where the $Y_{(i:m)}$ are the correctly ranked order statistics on Y. Equation (10) implies the fundamental relationship

$$\frac{\mathrm{E}X_{[i:m]}-\mu_X}{\sigma_X}=\rho\frac{\mathrm{E}Y_{(i:m)}-\mu_Y}{\sigma_Y}. \tag{11}$$

The estimator of the population mean μ_X based upon the $X_{[i:m]}$ is

$$\bar{X}_{[m]r}=\frac{1}{mr}\sum_{i=1}^{m}\sum_{j=1}^{r}X_{[i:m]j}$$

and, as in equation (4), its variance is given by

$$\mathrm{var}(\bar{X}_{[m]r})=\frac{1}{mr}\left[\sigma_X^2-\frac{1}{m}\sum_{i=1}^{m}\{\mathrm{E}X_{[i:m]}-\mu_X\}^2\right]. \tag{12}$$

The relative precision of $\bar{X}_{[m]r}$ is

$$\begin{aligned}\mathrm{RP}_X^{\mathrm{c}}&=\frac{1}{1-\frac{1}{m}\sum_{i=1}^{m}\left\{\frac{\mathrm{E}X_{[i:m]}-\mu_X}{\sigma_X}\right\}^2}\\&=\frac{1}{1-\frac{\rho^2}{m}\sum_{i=1}^{m}\left\{\frac{\mathrm{E}Y_{(i:m)}-\mu_Y}{\sigma_Y}\right\}^2}.\end{aligned} \tag{13}$$

Here the superscript c indicates that the ranking is on the basis of a concomitant variable. The formula (13) takes on a particularly elegant form when expressed in terms of the relative savings:

$$\mathrm{RS}_X^{\mathrm{c}}=\rho_{X,Y}^2\times\mathrm{RS}_Y. \tag{14}$$

Despite its elegance, formula (14) has only limited practical usefulness since the righthand side is in terms of RS_Y and does not permit a direct comparison of $\mathrm{RS}_X^{\mathrm{c}}$ with RS_X. This is where the assumption (ii) enters since, if X and Y have the same standardized distributions, RP_X and RP_Y are equal to one another (by equation (5)) and we have

$$\mathrm{RS}_X^{\mathrm{c}}=\rho_{X,Y}^2\times\mathrm{RS}_X. \tag{15}$$

Stokes (1977) has calculated $\mathrm{RP}_X^{\mathrm{c}}$ for the case where (X, Y) is bivariate normal. The results are summarized in Table 5. Notice that the RP values in each column steadily increase with m. This has an important consequence for the sample design: When the ranking is done by concomitant variable, then the set-size should be chosen as large as is practical since the adverse effect of ranking errors never fully

Table 5
$\mathrm{RP}_X^c = \mathrm{var}(\bar{X})/\mathrm{var}(\bar{X}_{[m]r})$

m	ρ			
	± 0.25	± 0.50	± 0.75	± 0.90
2	1.02	1.09	1.22	1.35
4	1.04	1.17	1.48	1.87
6	1.05	1.21	1.63	2.25
8	1.05	1.23	1.73	2.55
10	1.05	1.25	1.80	2.79

Source: Stokes (1977).

offsets the benefits of increasing the set-size. This feature does not hold for all ranking methods; but it does hold for *coherent* ranking methods. Coherence means that whenever a set is ranked then the ranking induced on each subset is the same as if the subset had been directly ranked by the method. Equivalently, coherence means that a set can be consistently ranked by all pairwise comparisons of its members. It is obvious that ranking by a concomitant variable is coherent.

It is also useful to compare the performance of RSS under concomitant ranking with that under perfect ranking. In Stokes' notation, the relative precision is

$$\begin{aligned}\mathrm{RP}_1 &= \frac{\mathrm{var}(\bar{X}_{(m)r})}{\mathrm{var}(\bar{X}_{[m]r})}\\ &= \frac{\mathrm{RP}_X^c}{\mathrm{RP}_X}\\ &= \frac{\sigma_X^2 - \frac{1}{m}\sum_{i=1}^m\{\mathrm{E}X_{(i:m)} - \mu_X\}^2}{\sigma_X^2 - \frac{1}{m}\sum_{i=1}^m\{\mathrm{E}X_{[i:m]} - \mu_X\}^2}. \end{aligned} \tag{16}$$

Some values of RP_1 are given in Table 6, again in the case of the bivariate normal distribution. The magnitudes of the values in Table 6 would seem to indicate that

Table 6
$\mathrm{RP}_1 = \mathrm{var}(\bar{X}_{(m)r})/\mathrm{var}(\bar{X}_{[m]r})$

m	ρ			
	± 0.25	± 0.50	± 0.75	± 0.90
2	0.70	0.74	0.83	0.92
4	0.44	0.50	0.63	0.80
6	0.33	0.38	0.51	0.71
8	0.26	0.31	0.43	0.64
10	0.22	0.26	0.38	0.58

Source: Stokes (1977).

ranking error can significantly reduce the benefits of ranked set sampling below what is theoretically attainable under perfect ranking. Although RSS can still be usefully applied, one has to guard against relying on variance formulae that have been derived on the assumption of perfect ranking. The table indicates that this would seriously underestimate the variance even when ρ is as large as 0.9. Further, the underestimation becomes more pronounced as the set-size m increases.

Ridout and Cobby (1987) have unified the work of David and Levine (1972) and Stokes (1977). In order to describe their result we require a more systematic notation. Let us denote by RP$(X:Y)$ the RP when ranking is based on the concomitant variable Y, and by RP(X) the RP under perfect ranking of the variable of interest X. If the regression of X of Y is linear then (equation (13))

$$\mathrm{RP}(X:Y)=\frac{1}{1-\frac{\rho^2}{m\sigma_Y^2}\sum_{i=1}^{m}[\mathrm{E}(Y_{(i:m)})-\mu_Y]^2}.$$

Since

$$\mathrm{RP}(Y)=\frac{1}{1-\frac{\rho^2}{m\sigma_Y^2}\sum_{i=1}^{m}[\mathrm{E}(Y_{(i:m)})-\mu_Y]^2}.$$

we obtain

$$\mathrm{RP}(X:Y)=\frac{\mathrm{RP}(Y)}{\rho^2+(1-\rho^2)\mathrm{RP}(Y)}. \tag{17}$$

If the standardized variates $(X-\mu_Y)/\sigma_Y$ and $(Y-\mu_Y)/\sigma_Y$ are identically distributed, then RP$(X)=$ RP(Y) and so we get

$$\mathrm{RP}(X:Y)=\frac{\mathrm{RP}(X)}{\rho^2+(1-\rho^2)\mathrm{RP}(X)}.$$

Let us consider the model of Dell and Clutter (1972) as $Y_i=X_i+\varepsilon_i$ $(i=1,\ldots,m)$ where $\varepsilon_i\sim \mathrm{N}(0,\sigma_\varepsilon^2)$. If X follows the normal distribution, then

$$\rho=\frac{\sigma_X}{(\sigma_X^2+\sigma_\varepsilon^2)^{1/2}}$$

and

$$\mathrm{RP}(X:Y)=\frac{(\sigma_X^2+\sigma_\varepsilon^2)\mathrm{RP}(X)}{\sigma_X^2+\sigma_\varepsilon^2\,\mathrm{RP}(X)}. \tag{18}$$

In particular, if $\sigma_X^2=1$, then

$$\mathrm{RP}(X:Y)=\frac{(1+\sigma_\varepsilon^2)\mathrm{RP}(X)}{1+\sigma_\varepsilon^2\,\mathrm{RP}(X)}.$$

Table 7
RP($X:Y$) with the corresponding values of RP obtained by Dell and Clutter (1972) shown in parentheses

Set-size (m)	σ_ε^2 (error variance)				
	0	0.05	0.15	0.30	0.50
	Rectangular ($\mu = 0.5$ and $\sigma^2 = 0.0833$)				
2	1.50	1.26	1.14	1.08	1.05
		(1.28)	(1.13)	(1.08)	(1.05)
3	2.00	1.45	1.22	1.12	1.08
		(1.48)	(1.22)	(1.12)	(1.08)
5	3.00	1.71	1.31	1.17	1.11
		(1.72)	(1.32)	(1.16)	(1.10)
	Exponential ($\mu = 1.0$ and $\sigma^2 = 1.0$)				
2	1.33	1.31	1.28	1.24	1.20
		(1.33)	(1.24)	(1.23)	(1.19)
3	1.64	1.59	1.51	1.43	1.35
		(1.56)	(1.50)	(1.42)	(1.34)
5	2.19	2.07	1.90	1.72	1.57
		(1.99)	(1.86)	(1.69)	(1.56)

Source: Ridout and Cobby (1987).

One could get the values of RP given by David and Levine in Table A1 for values of σ_ε^2 and m. Ridout and Cobby (1987) have further stated that the values of RP given in Table 2 of Dell and Clutter (1972) for the rectangular and exponential distributions could also be computed by RP($X:Y$) though the assumptions adopted in its derivation do not hold. In fact, RP($X:Y$) closely approximates the values obtained by Dell and Clutter (1972) in their Table 2 using the computer simulation. Table 1 of Ridout and Cobby (1987) gives the values of RP($X:Y$) under the model of ranking errors for the rectangular and the exponential distributions with the corresponding values obtained by Dell and Clutter (1972). We have reproduced it in our Table 7.

4.3. *Non-random selection of samples within sets*

Cobby et al. (1985) have noticed that the need to rank the units of a set may prevent the use of random sampling from the full target area in forming the set. They have also compared the effect of this source of error with that of ranking error in reducing the effectiveness of RSS. Ridout and Cobby (1987) adopt the following model of cluster sampling

$$X_{ik} = U_i + V_{ik}, \quad i = 1, \ldots, m \text{ and } k = 1, \ldots, m,$$

where X_{ik} denotes the value on k-th unit in the i-th set and

$$E(U_i)=0, \quad E(V_{ik})=E(X_{ik})=\mu_X,$$

$$\text{var}(U_i)=\sigma_u^2, \qquad \text{var}(V_{ik})=\sigma_v^2.$$

Further, all U and V are independent so that

$$\text{var}(X_{ik})=\sigma_u^2+\sigma_v^2.$$

Since U is constant within each set, the ranking is determined by the ranking of the V and

$$X_{(i:m)}=U_i+V_{(i:m)}.$$

With a single cycle, the RSS estimator of μ is obtained as

$$\begin{aligned}\bar{X}_{(m)}&=\frac{1}{m}\sum_{i=1}^{m}X_{(i:m)}\\&=\frac{1}{m}\sum_{i=1}^{m}\{U_i+V_{(i:m)}\}\\&=\bar{U}+\bar{V}_{(m)}.\end{aligned}$$

Thus

$$\text{var}(\bar{X}_{(m)})=\frac{\sigma_u^2}{m}+\text{var}(\bar{V}_{(m)}).$$

The relative precision of the RSS relative to the SRS is given by

$$\text{RP}(X)=\frac{\text{var}(\bar{X})}{\text{var}(\bar{X}_{(m)})}=\frac{\sigma_u^2+\sigma_v^2}{\sigma_u^2+m\,\text{var}(\bar{V}_{(m)})}.$$

But

$$\text{RP}(V)=\frac{\sigma_v^2/m}{\text{var}(\bar{V}_{(m)})},$$

so we can eliminate $\text{var}(\bar{V}_{(m)})$ to get

$$\text{RP}(X)=\frac{(1+\sigma_u^2/\sigma_v^2)\,\text{RP}(V)}{1+(\sigma_u^2/\sigma_v^2)\,\text{RP}(V)}. \tag{19}$$

Ridout and Cobby (1987) have computed some values of RP when U and V are both normally distributed. We have reproduced their Table 2 as Table 8 of this chapter.

Table 8
RP of RSS from a normal population in the presence of within-set clustering

Set-size (m)	σ_u^2/σ_v^2						
	0.00	0.10	0.25	0.50	1.00	2.00	5.00
2	1.47	1.41	1.34	1.27	1.19	1.12	1.06
3	1.91	1.77	1.62	1.47	1.31	1.19	1.09
4	2.35	2.09	1.85	1.62	1.40	1.24	1.11
5	2.77	2.39	2.05	1.74	1.47	1.27	1.12

Source: Ridout and Cobby (1987).

It is apparent from Table 8 that RSS is superior to SRS even in the presence of clustering. Although this result was obtained only for the normal population by Ridout and Cobby (1987), it holds regardless of the nature of the population because $\mathrm{RP}(V) \geqslant 1$.

In order to study jointly the impact of ranking error and clustering on RP, Ridout and Cobby (1987) considered the model given as

$$Y_{ik} = X_{ik} + \varepsilon_{ik},$$

$$Y_{ik} = U_i + V_{ik} + \varepsilon_{ik} = U_i + W_{ik},$$

where $X_{ik} = U_i + V_{ik}$ and $W_{ik} = V_{ik} + \varepsilon_{ik}$. Now, equation (19) remains true when the ranking is according to the concomitant variable Y. Also, ranking according to Y is the same as ranking according to W since $Y_{ik} = U_i + W_{ik}$ and the U are the same within each set to be ranked. Thus, (19) gives

$$\mathrm{RP}(X:Y) = \frac{(1 + \sigma_u^2/\sigma_v^2)\,\mathrm{RP}(V:W)}{1 + (\sigma_u^2/\sigma_v^2)\,\mathrm{RP}(V:W)}. \tag{20}$$

Since V and W are related by the model of Dell and Clutter, equation (18) gives, assuming joint normality,

$$\mathrm{RP}(V:W) = \frac{(1 + \sigma_\varepsilon^2/\sigma_v^2)\,\mathrm{RP}(W)}{1 + (\sigma_\varepsilon^2/\sigma_v^2)\,\mathrm{RP}(W)}.$$

On putting this expression for $\mathrm{RP}(V:W)$ into (20), we obtain

$$\mathrm{RP}(X:Y) = \frac{(1 + \sigma_\varepsilon^2/\sigma_v^2)(1 + \sigma_u^2/\sigma_v^2)\,\mathrm{RP}(W)}{1 + (\sigma_u^2/\sigma_v^2 + \sigma_\varepsilon^2/\sigma_v^2 + (\sigma_u^2/\sigma_v^2)(\sigma_\varepsilon^2/\sigma_v^2))\,\mathrm{RP}(W)}. \tag{21}$$

Since (21) has been obtained under the assumption of normality, $\mathrm{RP}(W)$ is simply the relative precision value for the normal distribution. Equation (21) shows how the RP depends jointly upon the cluster sampling effect, σ_u^2/σ_v^2, and the ranking error effect, $\sigma_\varepsilon^2/\sigma_v^2$. Interestingly, (21) is a symmetric function of these two variance ratios. In any case, $\mathrm{RP}(X:Y) \geqslant 1$ since $\mathrm{RP}(W) \geqslant 1$. This means that RSS is superior to SRS for the estimation of a population mean even in the presence of errors in ranking as well as cluster sampling. Of course, this result has been obtained for a very specific model.

Takahasi (1969) has also studied the effect of within-set clustering on the performance of ranked set sampling. He considers a model with set-to-set independence but with an arbitrary exchangeable joint distribution for the units within each set. Under this model, he demonstrates that the RSS estimator is an unbiased estimator of the population mean, and that the RSS estimator has a smaller variance than the corresponding SRS estimator with the same sample size irrespective of the within-set dependence structure. Takahasi also compares the performance of this 'clustered' RSS procedure with the RSS procedure having independently selected units. He is able to show that the performance of the RSS estimator is degraded when the within-set joint distribution exhibits a certain positive dependence somewhat similar to positive quadrant dependence. Performance of RSS is enhanced in the opposite case of negative dependence.

4.4. Characteristics of the population under consideration

We have seen that skewness of the population can adversely impact upon the performance of RSS. To deal with asymmetric distributions McIntyre (1952) suggested an unequal allocation of quantification effort to the different ranks. In particular, ranks corresponding to the long tail would be quantified more frequently than the other ranks. McIntyre also pointed out that, for the optimal allocation, the frequency of quantification of a rank should be proportional to the standard deviation of the corresponding order statistic (Neyman allocation). Numerical calculation by McIntyre showed that considerable benefit would obtain for the exponential distribution, but that little was to be gained for beta distributions with integral shape parameters. Halls and Dell (1966) applied the method in the field and found substantial reduction in the variance of the estimator. But contrary to this finding, Martin et al. (1980) obtained little or no gain from an unequal allocation in their investigation.

In order to describe the method, let $r_1, r_2, \ldots, r_m$ denote the numbers of sets whose quantified unit has the respective rank $1, 2, \ldots, m$. As usual, one and only one member from each set is quantified. We have to require that each r_i be nonzero. The total number of quantifications is

$$n = r_1 + r_2 + \cdots + r_m.$$

The optimal value of r_i is proportional to $\sigma_{(i:m)}$, the standard deviation of $X_{(i:m)}$. Let T_i be the sum of the quantifications for the units having the i-th rank. Then the RSS estimator $\bar{X}_{(m)u}$ of the population mean is given by

$$\bar{X}_{(m)u} = \frac{1}{m} \sum_{i=1}^{m} \frac{T_i}{r_i}$$

and is easily seen to be unbiased for μ. Further,

$$\operatorname{var}(\bar{X}_{(m)u}) = \frac{1}{m^2} \sum_{i=1}^{m} \frac{\sigma^2_{(i:m)}}{r_i}$$

and

$$\mathrm{RP} = \frac{\operatorname{var}(\bar{X})}{\operatorname{var}(\bar{X}_{(m)u})}.$$

There is no analogue of equation (4) expressing the variance in terms of $\mu_{(i:m)}$, but it may be noted that $\sigma^2_{(i:m)}$ is estimable from the quantifications provided $r_i \geqslant 2$ for all i. If $r_1 = r_2 = \cdots = r_m = r$ then the RSS design said to be *balanced*, otherwise *unbalanced*. Inappropriate allocations are possible, so the unbalanced RSS estimator can have a relative precision less than unity. Takahasi and Wakimoto (1968) have shown that

$$0 \leqslant \mathrm{RP} \leqslant m, \tag{22}$$

where the upper bound is approached for a certain sequence of highly skew power function distributions. The upper bound in (22) may be compared with the corresponding bound of $(m+1)/2$ that applies in the case of a balanced design. In effect, with an appropriate choice of replication factors r_i, it is possible to use ranked set sampling to actually *exploit* skewness to obtain highly precise estimates of the population mean.

5. Estimation of a population variance

5.1. The RSS estimator

Stokes (1980) has proposed an estimator of the population variance σ^2 based on a ranked set sample. It is asymptotically unbiased irrespective of the presence of errors in ranking and is more efficient than the conventional estimator based on a random sample of the same size. Let $X_{[i:m]j}$ denote the quantified value of a unit of the j-th cycle having the i-th rank. Here, $i = 1, \ldots, m$ and $j = 1, \ldots, r$. Let X_{ij} denote a sample of mr iid random variables selected from the same population with $\operatorname{var}(X) = \sigma^2$. The usual estimator of σ^2 based on the simple random sample is given by

$$s^2 = \frac{1}{(mr-1)} \sum_{i=1}^{m} \sum_{j=1}^{r} (X_{ij} - \bar{X})^2,$$

where

$$\bar{X} = \frac{\sum_{i=1}^{m} \sum_{j=1}^{r} X_{ij}}{mr}.$$

Stokes (1980) has suggested the following estimator based on the ranked set sample

$$\hat{\sigma}^2 = \frac{1}{(mr-1)} \sum_{j=1}^{r} \sum_{i=1}^{m} (X_{[i:m]j} - \bar{X}_{[m]r})^2,$$

where

$$\bar{X}_{[m]r} = \frac{\sum_{j=1}^{r} \sum_{i=1}^{m} X_{[i:m]j}}{mr}.$$

Now

$$\mathrm{E}(\hat{\sigma}^2) = \mathrm{E}\left[\sum_{j=1}^{r} \sum_{i=1}^{m} X_{[i:m]j}^2 - mr\bar{X}_{[m]r}^2\right] / (mr-1)$$

which simplifies to

$$\mathrm{E}(\hat{\sigma}^2) = \sigma^2 + \frac{\sum_{i=1}^{m} \tau_{[i]}^2}{m(mr-1)},$$

where $\tau_{[i]} = \mu_{[i:m]} - \mu$. This shows that $\hat{\sigma}^2$ is a positively biased estimator but is asymptotically unbiased as r increases.

5.2. *RP of the RSS estimator of the population variance relative to its SRS estimator*

Stokes (1980) shows that

$$\begin{aligned}\mathrm{MSE}(\hat{\sigma}^2) = {} & \frac{1}{mr}\mu_4 + \frac{(mr-3)(mr+1)}{(mr-1)^2 m^2 r}\sum_i \tau_{[i]}^4 + \frac{4}{m^2 r(mr-1)} \\ & \times \sum_i \tau_{[i]} \mathrm{E}(X_{[i:m]} - \mu)^3 + \frac{6-2(mr)^2}{m^2 r(mr-1)^2}\sum_i \tau^2_{[i]} \mathrm{E}(X_{[i:m]} - \mu)^2 \\ & - \frac{(mr)^2 - 2mr + 3}{(mr-1)^2 m^2 r}\sum_i \sigma_{[i:m]}^4 + \frac{2}{m^2(mr-1)^2}\left(\sum_i \sigma_{[i:m]}^2\right)^2 \\ & + \frac{1}{(mr-1)^2 m^2}\left(\sum \tau_{[i]}^2\right)^2, \end{aligned} \tag{23}$$

$$\mathrm{var}(s^2) = \frac{\mu_4}{mr} - \frac{(mr-3)\sigma^4}{mr(mr-1)}, \tag{24}$$

$$mr \times \mathrm{MES}(\hat{\sigma}^2) = \mu_4 - \left(\frac{1}{m}\right)\sum_i (\sigma_{[i:m]}^2 + \tau_{[i]}^2)^2 + \mathrm{O}\{(mr)^{-2}\}, \tag{25}$$

$$mr \times \mathrm{var}(s^2) = \mu_4 - \sigma^4 + \mathrm{O}\{(mr)^{-2}\}. \tag{26}$$

Therefore the limit of the RP, i.e., $\mathrm{var}(s^2)/\mathrm{MSE}(\hat{\sigma}^2)$ cannot be less than one. See

Stokes (1980). The RP for finite r can be computed from (23) and (24), while the asymptotic expression as $r \to \infty$ is obtained from (25) and (26) as follows

$$\lim_{r \to \infty} \mathrm{RP} = \frac{\mu_4 - \sigma^4}{\mu_4 - \frac{1}{m}\sum_{i=1}^{m} (\sigma^2_{[i:m]} + \tau^2_{[i]})^2}.$$

Table 1 of Stokes (1980) is given as our Table 9 and gives the values of RP for $r = 1$ and $r \to \infty$. Stokes (1980) also discusses the estimation of σ^2 when ranking is subject to error. It is evident from the table that any benefits of RSS in estimating σ^2 are small. So it is not worthwhile to draw a ranked set sample for the primary purpose of estimating the variance. Instead this method of sampling is most useful when both

Table 9
$\mathrm{RP} = \mathrm{var}(s^2)/\mathrm{MES}(\hat{\sigma}^2)$ for $r = 1$ and $r \to \infty$

Distribution	m	RP	
		$r = 1$	$r \to \infty$
(i) U-shaped	2	0.85	1.00
$f(x) = \begin{cases} 3x^2/2, & -1 \leqslant x \leqslant 1 \\ 0, & \text{otherwise} \end{cases}$	3	1.22	1.10
	4	1.28	1.21
(ii) Uniform	2	0.72	1.00
	3	0.92	1.11
	4	1.09	1.25
	5	1.20	1.40
(iii) Normal (Bose and Gupta, 1959)	2	0.68	1.00
	3	0.81	1.08
	4	0.93	1.18
	5	1.03	1.27
(iv) Gamma (Gupta, 1960)	2	0.71	1.02
$f(x) = \begin{cases} x^4 \exp(-x)/\Gamma_{(5)}, & x \geqslant 0 \\ 0, & x < 0 \end{cases}$	3	0.81	1.08
	4	0.91	1.16
	5	1.00	1.23
	6	1.09	1.35
(v) Exponential	2	0.78	1.03
	3	0.84	1.08
	4	0.91	1.12
	5	0.97	1.17
	6	1.02	1.22
(vi) Lognormal (Gupta et al., 1974)	2	0.93	1.00
	3	0.95	1.01
	4	0.96	1.01
	5	0.97	1.02
	6	0.98	1.02
	7	0.99	1.03
	8	1.00	1.03
	9	1.01	1.04

Source: Stokes (1980).

the population mean and variance are to be estimated. McIntyre (1952) had also discovered numerically for several distributions that the gain due to RSS for estimating μ_2, μ_3 and μ_4 is small.

6. Reported and suggested applications of RSS

6.1. Reported applications

(i) Although McIntyre had suggested the RSS method of sampling in 1952, the first reported field trial was conducted by Halls and Dell (1966) for estimating the weights of browse and of herbage in a pine hardwood forest of east Texas. The authors found the RSS method to be considerably more efficient than SRS. In carrying out the experiment, sets of three closely grouped quadrats were formed on a 300-acre tract of the Stephen F. Austin Experimental Forest. The locations for the sets were first decided. Then, at each selected location, metal frames of 3.1 feet square were placed at three randomly selected points within a circle having a 13 foot radius. The quadrats were ranked as highest, intermediate, or lowest according to the perceived weight of browse and (separately) of herbage. Then, after clipping and drying, the quantity of browse and of herbage was determined for each quadrat.

In order to simulate the SRS estimator for the mean weight of browse, one quadrat was randomly selected from each set without taking into account its rank. This yielded the mean browse weight as 14.9 grams per quadrat and the variance of the mean as 4.55. In order to obtain the RSS estimate, the rank to be quantified was chosen at random for each set. The mean yield (13.2 grams) was computed on the basis of 43 highest ranks, 46 intermediate ranks, and 37 lowest ranks. The variance of the mean was 2.18 which is almost half of the corresponding value based on SRS.

Halls and Dell also examined McIntyre's suggestion that an unbalanced design might improve the efficiency of estimation. Since the standard deviations for the order statistics were 17.7 for the high-yield, 13 for the intermediate, and 7 for the low rank, i.e., in a ratio of 4:2:1, they again selected 72 quadrats in the high, 40 in the intermediate, and 14 in the low group. The mean was 12.9 grams and its variance was found to be 1.91. This result supported McIntyre's contention. A similar pattern was found for the herbage estimates.

The experiment had, in fact, employed two observers with different professional backgrounds to independently rank the quadrats. But this had practically no effect on the results which were nearly the same as what would have been obtained for perfect ranking.

(ii) With the aim of investigating the effectiveness of RSS for improving the sampling precision of seedling counts, Evans (1967) conducted a study on an area in central Louisiana seeded to longleaf pine (*Pinus palustris mill.*). The target area was divided into 24 blocks. The blocks so obtained were again partitioned into 25 one-milacre plots. For computing the population mean and variance, all 600 plots were used. For the RSS protocol, three plots were randomly selected from each one of the 24 blocks (sets) for a visual ranking. The seedlings were counted in the highest ranking plot for eight sets. Similarly, seedlings in the medium ranking plots were counted for another eight sets. Seedlings in the lowest ranking plots were counted for the remaining eight sets. Thus each rank was quantified eight times. The whole procedure was repeated three times to conduct three separate field trials.

In addition, 80 RSS trials were conducted by a random selection procedure. For this, three plots in each of the 24 blocks were randomly selected and then the rank (high, medium or low) to be quantified was chosen at random for each block. For the first trial, it turned out that eight high ranking, nine medium ranking, and seven low ranking plots were selected for quantification. The whole procedure was repeated 80 times. For SRS one plot from each block was randomly selected for quantification. Again, the whole procedure was carried out 80 times. The means based on RSS and on SRS were not significantly different, but the computed variances of the means were very different for the two schemes of sampling. In fact, the variance among the 80 means for SRS was four times that for RSS. The author has noted that he found practically no difference between the ranks based on visual inspection and those based on detailed quantification.

Evans makes two suggestions for further studies of this type: (i) the size of the plot (sampling unit) should be increased and (ii) sample sizes should be proportional to the standard deviations in each group when sampling is from skewed populations, as suggested by McIntyre (1952).

(iii) The performance of RSS for estimating shrub phytomass in forest stands was evaluated by Martin et al. (1980). The investigation was carried out on the southeast slope of Potts Mountain in Craig County, Virginia, and studied four major vegetation types: mixed hardwood, mixed oak, mixed oak and pine, and mixed pine. For each of the four vegetation types an area was located which was further divided into 16 plots of equal size.

In order to carry out RSS, four sets of four plots were randomly selected from the 16 plots in each vegetation type. The plots in each set were ranked by a visual inspection. The lowest ranking plot was quantified from the first set, the second lowest from the second set, and the process was continued till the highest ranking plot was quantified from the fourth set. For SRS, four plots out of the 16 plots in each vegetation type were randomly selected without replacement. Each such selected plot was quantified for shrub phytomass. Then, the quantity of shrub phytomass was also determined for all the plots to determine the mean phytomass and variance of the mean on the basis of all 64 plots. In order to compare the sampling methods with a large sample size, a simulation trial of RSS and SRS was conducted for each vegetation type. For unequal allocation, the sample sizes were taken to be proportional to the variances. The authors found that RSS was more efficient than SRS for both the actual and the simulated trials. But they did not find any appreciable gain by resorting to unequal allocation of samples in different ranks.

(iv) In order to study the effectiveness of RSS compared to SRS for estimation of herbage mass in pure grass swards and of herbage mass and clover content in mixed grass-clover swards Cobby et al. (1985) conducted four experiments at Hurley (UK) during 1983. The authors have found three main factors which tend to reduce the advantages of RSS over SRS. These factors are:

(a) Imperfect ranking within sets.

(b) Greater variation between sets than within sets.

(c) Asymmetric distribution of the quantified values.

Out of the four experiments, the first two experiments were conducted by randomly selecting 15 locations and then randomly selecting three quadrats at each location. For the last two experiments, 45 quadrats were drawn at random from the entire target area. On the basis of analysis of data, the authors found that all the three factors contributed towards reducing the advantages of RSS. The main adverse factor was found to be the presence of within-set clustering due to the closeness of quadrats in each set. Nonetheless, the authors recommend the RSS procedure when sampling grass and grass-clover swards provided the quadrats are placed as far apart as practicable.

(v) Yanagawa and Chen (1980) mention that the RSS technique is regularly employed at the Pastoral Research Laboratory, CSIRO at Armidale, NSW, Australia. A plate with four holes is randomly thrown on a field, the pasture in the four holes is ranked by eye, and a hole is selected for quantification of pasture.

(vi) Yanagawa and Chen (1980) also mention that the method has been used to estimate rice crops in Okinawa, Japan. They attribute this information to H. Mizuno at the 'Mathematical Method in Sampling' symposium held at Chiba University (Japan), 1974.

(vii) In order to carry out a vegetation study which involves a determination of the number of stems, the size of the shrubs may be considered as a concomitant variable for ranking the number of stems (the main variable of interest). Muttlak and McDonald (1990a, b) have used the method in their study and have combined RSS with size-biased sampling to obtain improved estimators for the mean number of stems.

(viii) Attempts have also been made to incorporate the ranked set sampling protocol into the area of nonparametric studies. Stokes and Sager (1988) have obtained an RSS estimator of the population distribution function $F(\cdot)$. They have shown that an empirical distribution function based on RSS in an unbiased estimator of $F(\cdot)$ and the estimator possesses smaller variance than the one based on SRS. Stokes and Sager have also shown how to use the Kolmogorov–Smirnov statistic to obtain a confidence band for $F(\cdot)$ based on RSS.

Using the RSS empirical distribution function, Bohn and Wolfe (1992) have developed an RSS version of the Mann–Whitney statistic and have obtained some of its distributional properties as well as critical values for selected sample sizes. They have also compared the asymptotic relative efficiency of the RSS Mann–Whitney test with its SRS counterpart and come to the conclusion that the RSS approach would be preferable to the usual MWW (Mann–Whitney–Wilcoxon) procedure for the same numbers of measured observations.

In discussing possible extensions of this approach they have developed the natural Hodges–Lehmann point estimator for the shift parameter and the confidence intervals which are based on the RSS Mann–Whitney statistic.

6.2. *Suggested applications*

In addition to the reported applications of RSS there are some more areas where this technique could be used advantageously. These areas are given below.

(i) Evans (1967) has pointed out that the method would prove time-saving in the determination of cell wall thickness of different species of wood. In the same area of application, Dell (1969) has mentioned that the RSS procedure should be efficient for estimating averages for various properties of cells in a cross section of wood chips.

(ii) The RSS method of sampling could be useful in determining the average length of various kinds of bacterial cells. Also, it may be used to determine the average number of bacterial cells per unit volume. This is possible because it is convenient to order the tubes containing the cell suspension on the basis of concentration with the help of an optical instrument without knowing the exact number of the bacterial cells. Takahasi and Wakimoto (1968) have suggested these applications.

(iii) The technique may also be used to determine the average height of trees because it is easy to rank the heights of several nearby trees by a visual inspection. This application has been mentioned by Takahasi and Wakimoto (1968).

(iv) Stokes and Sager (1988) have suggested that the method of sampling could also be used to investigate a difficult-to-measure characteristic in human populations. They have, for example, referred to the consumer expenditure survey. The results of this survey are used for the construction of the consumer price index. But this survey requires detailed record keeping by the participating household as well the services of professional interviewers. In this situation a pre-measurement ranking could be performed on the basis of a cheaper screening interview and the technique of RSS may prove to be a timely innovation.

7. Illustration

7.1. *Determination of level of PCB*

In order to illustrate the performance of RSS compared to that of SRS in estimating the level of concentration of PCB we have used the measurements of the contaminant collected at Armagh site along the gas pipeline of the Texas Eastern Gas Pipeline Company. The data sets were provided by the Pennsylvania State Department of Environmental Resources to the Penn State Center for Statistical Ecology and Environmental Statistics. We have used grids A and C with the data of Phase I and II combined. The data are highly skewed. See Figures 4 and 5. With this fact in view, we have examined the unbalanced as well as the balanced allocation of samples. We have examined the following schemes in our investigation:

(a) Balanced allocation of samples using all possible combinations of each set size.

(b) Balanced allocation of samples for a particular sample.

(c) Unbalanced allocation of samples.

We have, in this illustration, considered only set-sizes 2, 3, and 4. Table 10

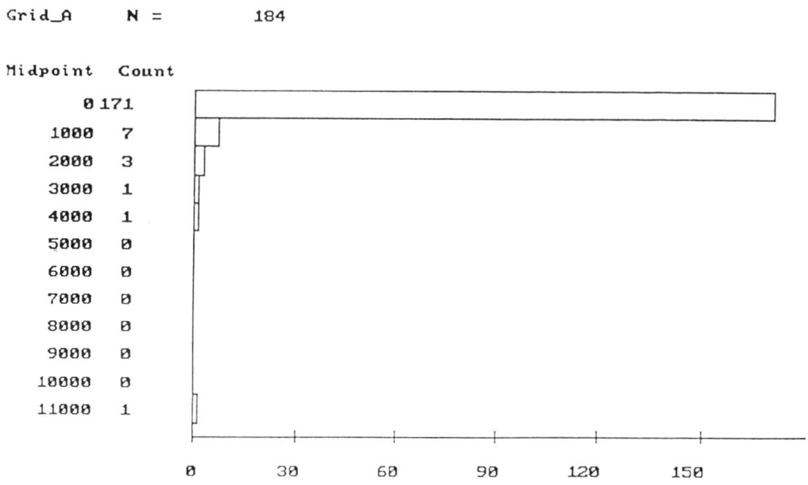

Fig. 4. Histogram of PCB values in grid *A*.

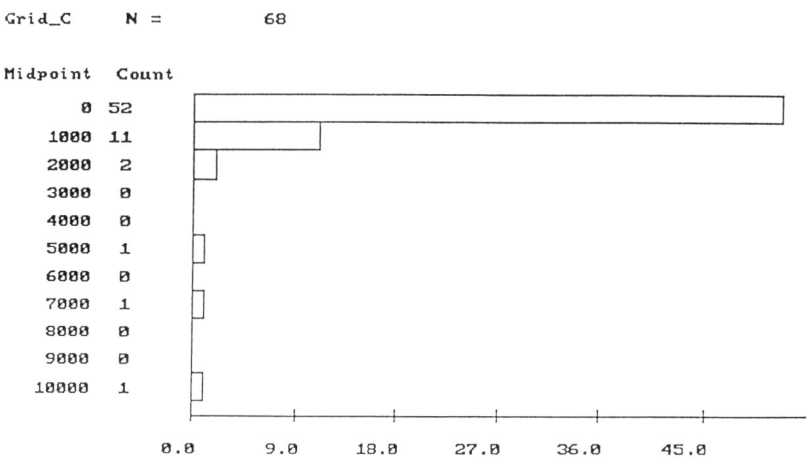

Fig. 5. Histogram of PCB values in grid *C*.

provides the number of observations, mean, standard deviation, coefficient of variation, coefficients of skewness and kurtosis of PCB values in grids *A* and *C*. Relative savings due to RSS protocol have been computed taking into consideration all possible combinations of each set size for both the grids under balanced allocation scheme. The results are given in Table 11. It is evident that the relative savings (RS) increase with the set-size but that the magnitude of RS is higher for grid *C* than grid *A* because both the coefficients of skewness and kurtosis are smaller for grid *C*. For a particular RSS sample, the quantities $\bar{X}_{(m)r}$, RP and RS have been computed for both grids. Table 12 shows these values. It is obvious that the results

Stokes, S. L. (1980). Estimation of variance using judgment ordered ranked set samples. *Biometrics* **36**, 35–42.

Stokes, S. L. (1986). Ranked set sampling. In: S. Kotz et al., eds., *Encyclopedia of Statistical Sciences.* Wiley, New York, 585–588.

Stokes, S. L. and T. W. Sager (1988). Characterization of a ranked-set sample with application to estimating distribution functions. *J. Amer. Statist. Assoc.* **83**, 374–381.

Takahasi, K. (1969). On the estimation of population mean based on ordered samples from an equicorrelated multivariate distribution. *Ann. Inst. Statist. Math.* **21**, 249–255.

Takahasi, K. and K. Wakimoto (1968). On unbiased estimates of the population mean based on the sample stratified means of ordering. *Ann. Inst. Statist. Math.* **20**, 1–31.

Teichroew, D. (1956). Tables of expected values of order statistics and products of order statistics from samples of size twenty and less from the normal distribution. *Ann. Math. Statist.* **27**, 410–426.

Yanagawa, T. and S. H. Chen (1980). The MG procedure in ranked set sampling. *J. Statist. Plann. Inference* **4**, 33–44.

G. P. Patil and C. R. Rao, eds., *Handbook of Statistics*, *Vol. 12*

6

Environmental Adaptive Sampling

G. A. F. Seber and S. K. Thompson

Abstract: Adaptive sampling designs are designs in which the procedure for selecting sites or units at which to observe the variable of interest – such as the level of a pollutant or the abundance of a species – depends on values observed during the survey. For example, additional observations may be made in the vicinity of sites at which high levels of the pollutant or high abundance of the species are encountered. In adaptive cluster sampling whenever the variable of interest satisfies a specified condition units in the neighborhood of that unit are added to the sample. With adaptive allocation designs sample sizes in strata or primary units depend on observed values. When detectability is imperfect in adaptive sampling, estimation methods taking the imperfect detectability into account are available. For rare, clustered or spatially uneven populations adaptive designs can produce substantial gains in precision compared to conventional designs of equivalent sample size. Additional advantages of the adaptive sampling include increasing the yield of the sample – so that, for example, more individuals of a rare species are observed – and helping to find local maxima – the 'hot spots' of environmental pollution surveys.

1. Introduction

In environmental pollution studies the following situation is commonly encountered. In most places sampled the pollution is light or negligible, but a few scattered pockets of high pollution are encountered. Two questions of interest are, first, what is the average level of pollution for the whole area, and, second, where are the pockets or hot spots likely to be? A traditional approach would be to select a random or systematic sample of sites or units and measure the pollutant at each of the selected sites or units. The average of these measurements provides an unbiased estimate of the population average, and the individual observations might then provide a contour map so that pollution peaks could be located. However, in environmental sampling situations there are problems with this procedure. In the extreme case where the pollution is negligible over most of the area, a majority of the measurements will record a zero level of pollution, and there is a good chance that most of the pockets will be missed. This means that although the sample average will give an unbiased estimate, it will have a large variance and a very wide associated confidence interval. In addition, the contour map will have little accuracy in the areas of high concentration, which are of particular interest.

Adaptive sampling could provide a better approach to the problem. In adaptive sampling the direction taken by the sampling procedure at any stage is determined by what information has come in so far from the sampling. For the pollution study we could begin, initially, with a smaller random sample. Provided we do not strike all zeros, we can then go back to the sites corresponding to, say, the three highest readings, and sample all plots or sites adjacent to these. If the reading in any of these additional plots exceeds the fourth largest reading in the initial sample then we add its neighboring plots as well. Each of the initial plots then gives rise to a cluster of plots with each cluster tending to grow towards the locations of highest pollution. Not surprisingly, this method of sampling will produce biased results if the traditional estimates like the sample average are used. In this article we will summarize the basic models and theory for adaptive sampling that have been developed so far. The estimators described are unbiased and have the potential for giving much smaller variances then the classical sample survey estimates. The main disadvantage of the adaptive method is that the sample size is random and it is more difficult to determine the cost of such a survey in advance.

Adaptive sampling is clearly useful in any situation where there is a sparse but highly aggregated population. Examples of such populations occur in fisheries (e.g., shrimps cluster in large but scattered schools), mineral investigation (in which the concentrations of ore are very unevenly distributed), and the epidemiology of rare diseases (e.g., in surveying for AIDS we would investigate all the friends and acquaintances of a person who is found to be HIV positive), as well as in environmental pollution studies. Animal and plant populations are important in environmental studies as their density estimates can provide useful indications of environmental changes. Of particular interest are rare and endangered species. Using adaptive sampling, the finding of an animal would lead to a search in the neighborhood, thus increasing the chance of finding more animals. An added advantage of adaptive sampling is that more animals are found so that more measurements – such as size, weight, and fecundity – can be made, thus increasing the overall usefulness of the survey. In surveys of marine mammals, rare bird species, and other animal species, visibility problems arise. Although an animal may be present, it may not be visible. Adjustments for handling imperfect detectability in adaptive sampling are described in Section 4.

The key elements of an adaptive sampling scheme are (i) choosing an initial sample of units, and (ii) choosing a rule for determining which neighboring units are to be added. The rule for the pollution example above is complicated as it is based on order statistics. A much simpler rule, particularly applicable to living populations, might add adjacent plots if there is at least one object present and continue doing this until the cluster is bordered by empty plots. In Figure 1a we have a sparse but

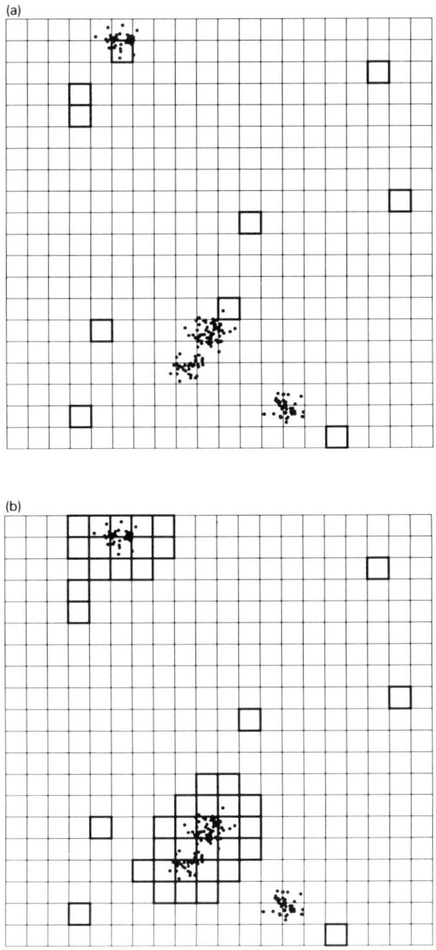

Fig. 1. (a) An initial random sample of 10 units. (b) The final adaptive sample. (From Thompson (1990). With permission from the American Statistical Association.)

clustered population. A random sample of 10 plots is shown there and only two of the sample plots are not empty. For each of these two plots, we sample the four plots with adjacent boundaries (i.e., forming a cross with the original plot). If any of these neighboring plots are nonempty we do the same again. Using this procedure we end up with the 45 sampled plots shown in Figure 1b. Clearly a variety of rules (e.g., three or more animals, two or more species, etc.) can be used for determining the formation of the clusters.

If something is known about the pollution levels or population density it may be possible to stratify the population, with the initial allocation of sampling effort among strata based on the prior knowledge. Adaptive methods can then be used in each stratum (Section 3.3). Even if population levels are unknown, it may still be convenient to divide the population up into strata and then allocate the sample adaptively within each stratum. These designs are discussed in Section 5.

2. Some theory

Athough the literature on sample survey theory is very extensive, there is relatively little on adaptive sampling. Basu (1969, p. 452) suggested that the potentially most efficient designs would be adaptive (informative) in the sense that the selection probabilities were conditional on the observed values. Zacks (1969) developed an optimal design of this type, but Solomon and Zacks (1970) felt it was too complicated to use and advocated the development of much simpler sequential or two-phase designs. Cassel et al. (1977, p. 13) commented briefly on informative designs but concluded that they were 'of rather limited interest from a practical point of view'. Sequential methods for finite populations are reviewed in Sinha (1991). For further references and a brief review see Thompson (1992).

The emphasis of this paper is expository rather than mathematical and the reader will be referred to various papers for the theoretical details. Such details will also be given in a monograph currently being written by the authors. However, some notation and theory is needed at this stage to explain some of the underlying ideas.

Suppose we have a population of N units (plots) labeled $1, 2, \ldots, N$. Associated with the i-th unit is the value y_i of a variable of interest; e.g., the number of animals on the i-th plot. The entire sequence of y-values, $\theta = (y_1, y_2, \ldots, y_n)$ is regarded as a (vector) parameter of the population, and interest focuses on estimating some function of it such as the population total $\tau = \sum_{i=1}^{N} y_i$ or the population mean $\mu = \tau/N$. Suppose we take a random sample from the population leading to the observed sequence $(y_{i_1}, y_{i_2}, \ldots, y_{i_n})$ of y-values. If simple random sampling without replacement (SRS) is used, then all the labels are distinct and represent an ordered subset $s = (i_1, i_2, \ldots, i_n)$ of the N labels. The information from the sample then consists of the ordered pairs $D = ((y_{i_1}, i_1), (y_{i_2}, i_2), \ldots, (y_{i_n}, i_n))$ giving the y-value for each sampled unit and the unit's label. For example, in the case of animal populations, we record not only the number of animals on the plot but also where the plot is and the order in which the plots were chosen. When sampling is with replacement (SRSR), then some of the labels might be repeats.

As the above notation using double subscripts can be awkward we shall use alternative notations such as $(y_i \colon i \in s)$ or $(y_i \colon i = 1, 2, \ldots, n)$. The second notation is taken to mean any n units in the sample, and not the first n units in the population. The sample average then takes the form

$$\bar{y} = \frac{1}{n} \sum_{i \in s} y_i = \frac{1}{n} \sum_{i=1}^{n} y_i .$$

The sampling design in the procedure by which the sample s is selected and is defined by its associated probability function $P_\theta(s)$. In SRS, all $C_n = N(N-1) \cdots (N-n+1)$ possible orderings of n distinct units have the same probability of being the sample selected; i.e., $P_\theta(s)$ is constant. For many conventional designs $P_\theta(s)$ does not depend on the y-values and therefore does not depend on θ. Furthermore, for SRS $P_\theta(s)$ is also nonzero for all values of θ, $\theta \in \mathbb{R}_N$. Such a design is said to be noninformative (after Cassel et al., 1977). However, if $P_\theta(s)$ does depend on θ, then the design is said to be informative. Sequential sampling is like this. For example, suppose we use SRS and continue sampling until a y-value exceeds some prechosen level c. Then n is random and the probability of getting a particular sample will depend on the y-values in the sample and therefore on the value of θ. This example warrants a closer investigation. Suppose that only the first unit in the population has a y-value exceeding c; i.e., $y_1 > c$. Then it is readily seen that, with distinct labels in s, $P_\theta(s) = 1/N$ for $s = (1)$ and $P_\theta(s) = 1/[(N(N-1)\cdots(N-k)]$ for $s = (i_1, i_2, \ldots, i_k, 1)$, $i_j \neq 1$, for $j = 1, 2, \ldots, k$ $(k \leqslant N-1)$, and $P_\theta(s) = 0$ otherwise. If θ was such that $y_2 > c$ instead of y_1, the $P_\theta(s) = 0$ for $s = (1)$ as θ is not compatible with the sample consisting of just the first unit. Hence for each s there is a set Θ_s of values of θ which are compatible with s. On this set $P_\theta(s)$ is constant and positive for a given s, but zero otherwise so that $P_\theta(s)$ is not directly dependent on θ. A similar observation was made by Basu (1969 p. 447).

The situation is similar in certain adaptive designs. For example, suppose that a random sample s_0 is selected and this is adaptively augmented in a prescribed order to determine the final sample s in a way that depends on the y-values in the sample. Although the final sample size is random and the augmented part of the sample depends on θ, the probability of getting the final sample is the same as that for the initial sample; i.e., $P_\theta(s) = p_\theta(s_0)$. In the use of SRS, $P_\theta(s_0) = 1/C_{n_1}$, where n_1 is the initial sample size. However, as in the above sequential procedure, not every θ is compatible with s so that, once again $P_\theta(s)$ is a nonzero constant for $\theta \in \Theta_s$ and zero otherwise. Given θ, we see that for SRS $P_\theta(s)$ is the same for every s compatible with θ. In addition to $P_\theta(s)$ being sometimes zero there is one other important difference between SRS and its adaptive augmentation. In SRS the probability that unit i is included in the sample is $\pi_i = n/N$, which does not depend on θ. In adaptive sampling π_i will depend on the clustering criterion and therefore on θ, and π_i will no longer be constant.

We noted that with SRSR some of the units may be repeated in the sample. The same thing can happen with adaptive sampling even if the initial sample is SRS. For example, if two units from SRS turn up in the same cluster then the cluster is included twice in the final sample. It is therefore convenient to define $s_u = \{i_1, i_2, \ldots, i_v\}$ as the unordered set of v distinct labels. Attaching the associated y-values we have the distinct unordered pairs $D_u = \{(y_i, i): i \in s_u\}$. Here D_u ignores information about order and multiplicity of labels. Then under fairly general conditions which include the sequential and adaptive schemes described above, D_u is a minimal sufficient statistic for the parameter θ (Basu, 1969; Cassel et al., 1977, pp. 36–38). Therefore, by the Rao–Blackwell theorem, any unbiased estimator of a parameter that is not a function of D_u can be improved upon by taking its conditional expectation given D_u. Unfortunately with finite population sampling, such an improved estimator is generally not a unique minimum variance unbiased estimator (UMVUE) as D_u is not complete (see Cassel et al., 1977 for details). This means that in adaptive sampling, more than one distinct estimator can be obtained by the Rao–Blackwell method. Such competing unbiased estimators need to be compared for efficiency.

In some sampling situations the labels are unavailable or ignored. If this is the case with SRS, then D_u without the labels, namely $\{y_i: i \in s_u\}$ with the y_i not necessarily distinct, is complete and a UMVUE exists (e.g., Royall, 1968; Lehmann, 1983 p. 209, with the ordered y-values being equivalent to $\{y_i: i \in s_u\}$). Although omitting the labels removes some of the complications, the use of labels allows for the application of adaptive strategies which give estimates with smaller variances than those obtained without the use of labels.

Two estimators from classical sample survey theory are particularly useful as they provide ways of developing adaptive counterparts. Suppose the probability π_i that unit i is included in the sample is greater than zero for each unit in the population. Then for any design with or without replacement, if π_i is known for each unit in the sample, then an unbiased estimator of μ is the Horvitz–Thompson (HT) estimator

$$\hat{\mu}_{\mathrm{HT}} = \frac{1}{N} \sum_{i=1}^{v} \frac{y_i}{\pi_i}, \tag{1}$$

where $y_1, y_2, \ldots, y_v$ represent the v distinct units in the sample. If sampling is with replacement and p_i is the probability of selecting the i-th unit at each draw, then we can use the Hansen–Hurwitz (HH) estimator

$$\hat{\mu}_{\mathrm{HH}} = \frac{1}{nN} \sum_{i=1}^{v} \frac{y_i}{p_i}. \tag{2}$$

The unbiasedness of the above two estimators follows from the nature of sampling design used. Thus if t is such an estimator, then

$$\mathrm{E}[t] = \sum_{s} t_s P_\theta(s) = \mu,$$

where t_s is value of t arising from sample s and the summation is over all possible samples. An estimator with this property is said to be design unbiased as the unbiasedness is a property of the design.

In some situations random sampling or other probability design may not be possible. However, it may be possible to formulate an underlying model for the distributions of the y-values. For example suppose there are y_i animals on the i-th plot (unit) and τ animals altogether. It is assumed that all N plots have the same area. If the animals move randomly and independently over the population area, then, conditional on τ, the y_i have a multinomial distribution with probabilities $1/N$ for each unit. It follows that $\sum_{i=1}^{n} y_i \sim \mathrm{Binomial}(\tau, n/N)$ and

$$\mathrm{E}[\bar{y}] = \tau/N = \mu, \tag{3}$$

where μ is the mean number of animals per unit. Here $\bar{y}$ is unbiased because of the underlying model and not because of any particular sampling scheme. Such an estimate is said to be model unbiased as its unbiasedness depends on the assumptions of the model proposed for the population. Although design unbiased estimators are the ones generally used in adaptive sampling, there are occasions when a model unbiased estimator is appropriate (see end of Section 5.2).

3. Adaptive cluster sampling

3.1. *Initial random sample (with or without replacement)*

We now look at an adaptive procedure suggested by Figure 1 and assume that either SRS or SRSR is used to obtain the initial sample of size n_1. For each unit i in the population we define a neighborhood of units A_i consisting of a collection of units including unit i which does not depend on the value of θ. A neighborhood can consist of a single pattern of adjacent units (forming a cross) as described for Figure 1b or a more complicated pattern including a larger contiguous grid of units or even a noncontiguous set such as a systematic pattern around the original unit. The neighborhood relationship is assumed to be symmetric in the sense that if unit j is in the neighborhood of i then i is in the neighborhood of j. If y_i satisfies some criterion C (typically $y_i > c$), then the rest of A_i is added to the sample. Furthermore, if any other units in A_i satisfy C then their neighborhoods are also added to the sample. This process continues until eventually a cluster of units is obtained with a boundary of units, called edge units, which do not satisfy C. The sample then consists of all the units and clusters obtained in this manner.

A cluster without its edge units is called a network. Any unit, including an edge unit, which does not satisfy C is said to be a network of one unit as its selection does not lead to the inclusion of any other units. The networks have two properties. First, given C, the networks form a partition of the N population units. Second, if a unit selected by the initial sample is anywhere in the network then the whole network is selected. This means that if two initial units come from the same network then the network is selected twice.

For any unit i in the initial sample, let B_i be its network with m_i elements and mean

$$w_i = \frac{1}{m_i} \sum_{r \in B_i} y_r. \tag{4}$$

Note that some of the B_i may be the same. By changing the meanings of π_i and p_i we can modify the HH and HT estimators to obtain the following unbiased estimators of μ, namely

$$\hat{\mu} = \frac{1}{n_1} \sum_{i=1}^{n_1} w_i \tag{5}$$

and

$$\hat{\mu} = N^{-1} \sum_{j=1}^{v} y_j I_j / \pi'_j. \tag{6}$$

Hence v is the number of distinct units in the whole sample (i.e., the effective sample size) and I_j is an indicator variable taking the value zero if the j-th unit in the sample does not satisfy C and was not selected in the initial sample, and takes the value one otherwise (with probability π'_j). We note that edge units are not used in these estimators as $I_j = 0$ for an edge unit in a cluster. Hence π'_j is not the same as π_j, the probability that unit j is included in the sample. Rather, π'_j is the probability that unit j is *utilized* in the estimator. It is also the probability that the initial sample intersects unit j. With adaptive cluster sampling, the actual inclusion probability π_i is not known for every unit in the sample, so the ordinary Horvitz–Thompson estimator cannot be used.

An alternative expression for $\tilde{\mu}$ which is useful for finding variances is

$$\tilde{\mu} = \sum_{k=1}^{K} J_k\, y_{k.}/\alpha_k, \tag{7}$$

where K is the number of distinct networks in the population, J_k is an indicator variable taking the value 1 with (intersection) probability α_k if the initial sample contains (intersects) one or more units from the k-th network and zero otherwise, and $y_{k.}$ is the sum of the y-values for the k-th network. If there are x_k units in the k-th network then α_k, the intersection probability of the k-th network, is

$$\left[1 - \binom{N - x_k}{n_1} \Big/ \binom{N}{n_1}\right]$$

when sampling is without replacement and $1 - (1 - x_k/N)^{n_1}$ when sampling is with replacement.

A natural question to ask is: why not use the cluster data rather than the network data in $\hat{\mu}$ and $\tilde{\mu}$, thus utilizing information from edge units. The answer is that if an edge unit is not included in the initial sample, its probability of inclusion is unknown as we do not know for which of clusters it forms an edge unit. By ignoring information from these units we are able to adapt the HH and HT methods.

We now have three unbiased estimators of μ, namely $\hat{\mu}$, $\tilde{\mu}$, and $\bar{y}_1$, the mean of the original sample. All three can be used when sampling is either with or without replacement. However, none of the three is a function of the sufficient statistic D_u, the unordered set of distinct units (including the edge units). All three depend on the order of selection ($\bar{y}_1$ depends on the order as we have to know which members of the final sample are in the initial sample); $\hat{\mu}$ also depends on repeat selections; and, when the initial sample is with replacement, $\bar{y}_1$ also depends on repeat selections. We can therefore improve on all three estimators using the Rao–Blackwell method. Let t be any of the three estimators and consider $t_{\mathrm{RB}} = \mathrm{E}[t \mid D_u]$. When the initial sample is selected without replacement define $\eta = \binom{v}{n_1}$, the number of combinations of n_1 distinct units from the v in the sample. Let these combinations be indexed in an arbitrary way by the label r, and let t_r denote the value of the estimator obtained when the initial sample consists of combination r. Let $\widehat{\text{var}}_r[t]$ denote the value of the unbiased estimate of $\text{var}[t]$ computed with combination r.

The key to evaluating $E[t \mid D_u]$ is to find all the t_r that come from initial samples which are compatible with D_u, that is which give rise to D_u. To this end define the indicator variable I_r to be one if the r-th combination of n_1 units from the sample is compatible with D_u and zero otherwise. Then the number of compatible units is $\xi = \sum_{i=1}^{\eta} I_r$ and

$$t_{\text{RB}} = \frac{1}{\xi} \sum_{r=1}^{\eta} t_r I_r. \tag{8}$$

Also $\text{var}[t_{\text{RB}}] = \text{var}[t] - E[\text{var}(t \mid D_u)]$ and

$$\widehat{\text{var}}[t_{\text{RB}}] = \frac{1}{\xi} \sum_{r=1}^{\eta} \{\widehat{\text{var}}_r[t] - (t_r - t_{\text{RB}})^2\} I_r. \tag{9}$$

It should be noted that although an edge unit is not utilized in t it can be utilized in t_{RB} as there may be a t_r for which an edge unit becomes included in one of the possible initial samples giving rise to D_u.

If the initial sample is selected with replacement, then the above formulas hold with $\eta = v^{n_1}$, the number of sequences of n_1 units, distinguishing order and allowing repeats, from the v distinct units in the sample and with label r identifying a sequence in the collection. Further algebraic details and variance formulas are given by Thompson (1990) with notational changes α_k^* and π_k instead of π_k and α_k, and by Thompson (1992). He shows that the Rao–Blackwell method gives the same estimate when applied to $\bar{y}_1$ and $\hat{\mu}$, but a different estimate when applied to $\tilde{\mu}$. Since there is no UMVUE of μ we are not surprised to find neither of the two estimators obtained by conditioning on D_u is uniformly better than the other. In comparing adaptive estimators with $\bar{y}_1$ it is appropriate to use $E[v]$ (the expected effective sample size of the adaptive procedure) instead of n_1 in calculating $\text{var}[\bar{y}_1]$. Thompson gives two examples which demonstrate that adaptive sampling can provide much more efficient estimators than simple random sampling, and that $\hat{\mu}$ and $\tilde{\mu}$ can lead to estimators with very different efficiencies with respect to $\bar{y}_1$.

The above theory readily extends to the multivariate situation where y_i is replaced by a vector of characteristics $\boldsymbol{y}_i$ with a corresponding population total $\boldsymbol{\tau}$. The criterion C for adding a neighborhood is now determined by some function of $\boldsymbol{y}$; e.g., its first element, the sum of its elements, and so on. The criterion creates a partition of networks which is used to estimate the elements of $\boldsymbol{\tau}$ separately. For further details see Thompson (1993b).

In the remainder of this section we give several alternative schemes for obtaining the initial sample in adaptive cluster sampling.

3.2. Using primary and secondary units

Suppose we have MN (secondary) units in the population and these are divided into N primary units, each consisting of M secondary units. Let y_{ij} be the y-value for the j-th secondary unit in the i-th primary unit. The initial sample will now consist of a selection of n primary units and we then add further (secondary) units using the method described above. This means that if a sample primary unit intersects a network then all the units in the corresponding cluster are added to the sample. However, only those units in the network are utilized in the estimates. For example in Figure 2 the primary unit is a strip and there is an initial random selection of five strips. Neighborhoods are defined as for Figure 1 so that if a secondary unit intersected by the primary unit contains animals then adjacent units are added to the sample. The final sample is outlined in Figure 2.

The units in the primary unit need not be contiguous. For example we could use a spatial systematic sample with several randomly selected starting points as in Figure 3.

We assume the primary units are selected by random sampling without replacement. Once again we have three estimators similar in form to $\bar{y}_1$, $\hat{\mu}$ and $\tilde{\mu}$ of equation (7). Algebraic details and variance formulas are given by Thompson (1991a) who again uses an example to demonstrate that substantial gains in precision can be obtained using adaptive samples, particularly with initial systematic samples.

Although the Rao–Blackwell method still applies with the designs organized into primary and secondary units, it is unlikely to be useful. In the previous section it was generally possible to find several initial samples which lead to the same D_u, i.e., $\eta \neq 1$. With strip sampling, however, it is unlikely that more than one initial selection of

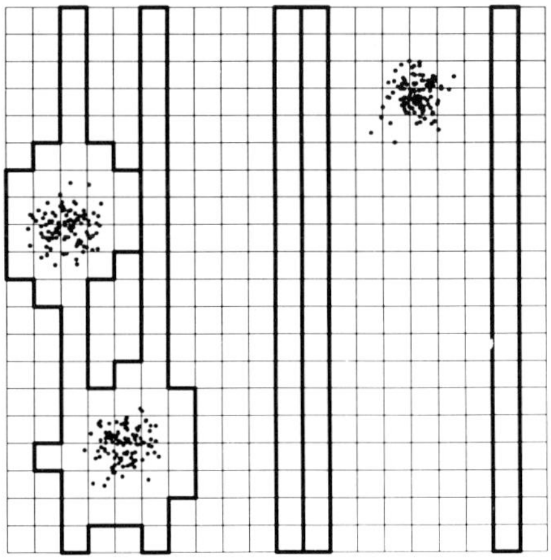

Fig. 2. Adaptive cluster sample with initial random selection of five strip plots. The final sample is outlined. (From Thompson (1991a). With permission from the Biometric Society.)

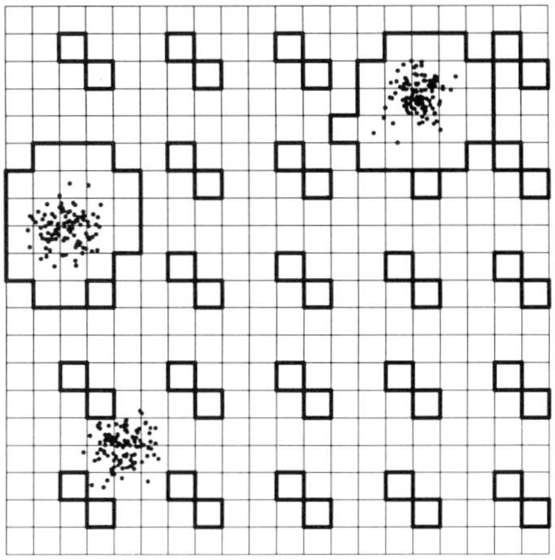

Fig. 3. Adaptive cluster sample with initial random selection of two systematic samples. The final sample obtained is outlined. (From Thompson (1991a). With permission from the Biometric Society.)

strips will lead to the same D_u so that $\eta = 1$ and $E[t \mid D_u] = t$ for t one of the three estimators.

Adaptive strip sampling is particularly useful in sampling large areas. For example in aerial or ship surveys the strip observed in each selected transect forms a primary unit. If objects or animals are sighted, then additional searching is carried out on either side of the transect. Adaptive cluster sampling with an initial systematic sample (with a random starting point) appears to be a very efficient design for rare, clustered populations.

3.3. Stratification

In some populations there may be some prior information about where aggregations are likely to be so that stratification can be used. The method of Section 3.1 can be used with an initial sample in each stratum. If growing clusters are truncated at the stratum boundaries, then stratum estimates are independent and can be combined in the usual fashion to provide an overall (weighted) estimate of the population mean. However, truncating clusters leads to a loss of efficiency and a more efficient scheme would be to allow clusters to overlap stratum boundaries as in Figure 4. Thompson (1991b) introduces a number of unbiased estimators $\bar{y}_{\mathrm{st}1}$, $\hat{\mu}$, $\hat{\mu}'$, $\hat{\mu}''$ and $\tilde{\mu}$ of the population mean for both schemes. Here $\bar{y}_{\mathrm{st}1}$ is the usual stratified mean of the initial (stratified) sample; $\hat{\mu}$ is a modified HH estimator when clusters are allowed to overlap strata; $\hat{\mu}'$ is related to a certain stratified multiplicity-type estimator of network sampling (Sirken, 1972; Levy, 1977), $\hat{\mu}''$ is a counterpart ignoring overlap between strata, and $\tilde{\mu}$ is a modified HT estimator with overlapping allowed. In comparing the variances of these estimators, $\mathrm{var}[\bar{y}_{\mathrm{st}1}]$ is replaced by $\mathrm{var}[\bar{y}^*_{\mathrm{st}}]$, the latter being based on a total sample size of $E[\nu]$, the expected sample size under the adaptive design. For his example Thompson showed that $\hat{\mu}$, $\hat{\mu}'$ and $\hat{\mu}''$

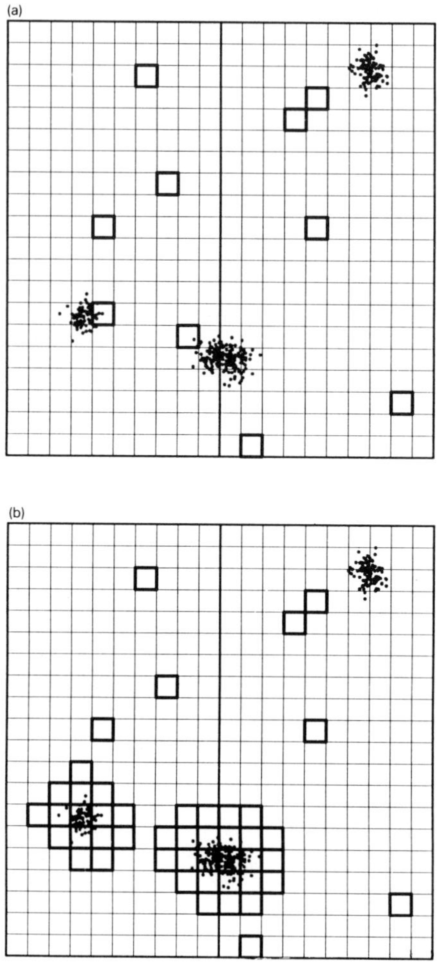

Fig. 4. (a) An initial stratified random sample of five units in each of two strata. (b) The final adaptive sample with the overlapping of strata allowed. (From Thompson (1991b). With permission from the Biometrika Trustees.)

are fairly similar, which is to be expected if aggregations tended to be away from stratum boundaries. He also demonstrated that substantial variance reductions can be achieved relative to simple random sampling when adaptive sampling is used; 1/50 in one case using $\tilde{\mu}$.

As in Section 3.1, the Rao–Blackwell method can be applied to all the estimators to reduce their variances. However, much more computation is involved when there is stratification. For example, within each stratum there are a number of initial samples which are compatible with D_u. Any such combination in the stratum can be combined with any of the combinations in the other strata to give an overall initial stratified sample compatible with D_u. Clearly a good algorithm is needed for computing the Rao–Blackwell versions.

3.4. *Using order statistics*

We return now to the pollution example at the start of this chapter. The adaptive approach suggested there was based on the ordered y-values. For example, suppose pollution readings taken on an initial sample of n_1 sites yielded the ordered readings $y_{(1)} < y_{(2)} < \cdots < y_{(n_1)}$. We then decide to sample the neighborhoods of those sites whose y-values exceed the r-th value $y_{(r)}$, namely, $y_{(r+1)}, \ldots, y_{(n_1)}$. When $r = n_1 - 1$ we sample the neighborhood of just the largest. If any of the added sites give readings higher than $y_{(r)}$, then their neighborhoods are added as well. This scheme differs radically from the previous scheme in which the criterion for determining cluster membership (e.g., number of animals on a unit) is fixed and determined before the experiment is begun. There the clusters are determined in advance. However, using order statistics, the clusters are determined by the data so that the usual adaptive estimators are no longer appropriate. Such an approach is applicable to the situation where one does not known prior to the survey whether 10 parts per million of the pollutant, 100 000 microroentgens of radiation, one animal of the species, or another fixed criterion is appropriate. If the wrong criterion is used it may turn out that every unit in the study region or none of them satisfies the criterion. A criterion based on the highest values encountered during the survey is then very appealing. In what follows we assume that the initial sample is without replacement.

The edge units of the cluster are those units whose y-values do not exceed $y_{(r)}$, and the networks are the clusters without their edge units. Any unit whose y-value does not exceed $y_{(r)}$ is a network of size one. Let A_{is} denote the network, determined by the overall sample s, containing unit i. Let m_{is} be the number of units in A_{is} and define the network mean

$$w_{is} = \frac{1}{m_{is}} \sum_{j \in A_{is}} y_j.$$

Then if i is a unit in the initial sample, an unbiased estimator of μ is given by

$$\hat{\mu} = \frac{1}{n_1} \sum_{i=1}^{n_1} w_{is}. \tag{10}$$

This has the alternative representation

$$\hat{\mu} = \frac{1}{n_1} \sum_{i \in s} \frac{f_{is} y_i}{m_i},$$

where f_{is} denotes the number of units of A_{is} included in the initial sample. Note that $f_{is} = 0$ if unit i is not in the initial sample. As in previous sections, $\hat{\mu}$ is not a function of the minimal sufficient statistic D_u as $\hat{\mu}$ depends on the order in which the initial sample is selected. The Rao–Blackwell method can therefore be used to obtain an improved estimate $\hat{\mu}_{RB} = E[\hat{\mu} \mid D_u]$. Further details and variance formulas are given by Thompson (1993a).

By analogy with previous sections, $\hat{\mu}$ can be regarded as an adapted HH estimator. However, in contrast to the previous work, the probability that the network including unit i is intersected by the sample depends on the other units in the sample. Therefore the usual adapted HT estimator is not unbiased when the adaptive cluster sampling is based on order statistics.

4. Adaptive sampling with imperfect detectability

Adaptive sampling is particularly useful for sampling rare or elusive populations. In such applications it is not uncommon for individuals to be missed so that the probability of detecting an individual, given that it is present, is less than one. Examples are aerial or ocean surveys, a feasibility survey for mining diamonds, and surveys of artifacts in archeological sites. The methods described in this section are completely general and can be used for any type of sampling scheme, conventional or adaptive.

4.1. Constant detectability

Suppose that the probability of detection g is the same for every object or animal. Then a usual estimator of the population total τ would in fact estimate $g\tau$, so that we would divide our estimator by g. For conventional designs, the mean and variance of such an estimator could be conveniently derived by conditioning on the sample s of units actually selected (as in Steinhorst and Samuel, 1989; Thompson, 1992), and taking expectations and variances, first with respect to visibility given the sample of units and second with respect to the sampling design. However, the situation is much more complicated with adaptive designs. Here the detectability not only alters what is seen but it also changes the selection probabilities for different possible samples. For example, suppose we use the design of Section 3.2 is which the primary unit is a strip of plots (secondary units). We assume that neighborhoods are added to those plots in which one or more animals are detected. If one of the initial strips intersects a network of plots but none of the animals in that part of the initial sample are detected, then no additional units will be added and the whole cluster will be missed. If with the same initial sample some of the animals are detected, then additional units of the associated cluster will be added. Thus the probability of including an animal depends not only on its detection probability but also on the detection probabilities of the animals in nearby plots. Adaptive samples therefore require a different approach, in fact the reverse procedure to that for conventional designs. Instead of conditioning on the sample selected, we now condition on a realization of the population set of potential detections as this determines the range of possible samples.

Let τ_i be the number of objects in unit i and let y_i be the number of objects detected by the observer in unit i. Then using the y_i from the sample we wish to estimate $\tau = \sum_{i=1}^{N} \tau_i$, the total number of objects in the study region. Although y_i is, in practice, observed only for units in the sample, it is mathematically convenient to define the vector $\boldsymbol{y} = (y_1, y_2, \ldots, y_N)$ of potential detections, thus recognizing the hypothetical possibility of observing every unit in the population. Let $Y = \sum_{i=1}^{N} y_i$ be the (hypothetical) population total of the y-values. We note that $y_i \sim \text{Binomial}(\tau_i, g)$ and the y_i are mutually independent as the detections of different objects are assumed to be independent.

Now suppose that we have an adaptive estimator $\hat{\tau}_0$ (such as $\hat{\tau} = N\hat{\mu}$ or $\tilde{\tau} = N\tilde{\mu}$ of Section 3) which would be design unbiased for τ if the detectability was perfect. However, as the detectability is not perfect, we condition on $\boldsymbol{y}$ so that the role of τ is replaced by Y. Hence

$$\begin{aligned} \mathrm{E}[\hat{\tau}_0] &= \mathrm{E}\{\mathrm{E}[\hat{\tau}_0 \mid Y]\} \\ &= \mathrm{E}[Y] \\ &= g\tau, \end{aligned}$$

and $\hat{\tau}_0/g$ is an unbiased estimator of τ. A similar method can be used to find $\mathrm{var}[\hat{\tau}_0]$ and an unbiased estimator $\widehat{\mathrm{var}}[\hat{\tau}_0]$. Further details are given by Thompson and Seber [1993].

When g is estimated by a consistent estimator $\hat{g}$, obtained independently through such methods as capture–recapture, double sampling, or regression techniques (cf. Seber 1982, 1986, 1992), then $\hat{\tau}_0/\hat{g}$ is approximately unbiased for τ and an additional term is added to the variance formula (Thompson and Seber, 1993).

4.2. *Variable detectability*

Instead of g being assumed constant we now assume that g_{ij} is the detection probability for the j-th object in the i-th unit and that y_{ij} is the value of the associated variable of interest. Let M_i be the number of objects in unit i and redefine $\boldsymbol{y}$ to be $\boldsymbol{y} = (y_{11}, y_{12}, \ldots, y_{1M_1}, \ldots, y_{N1}, y_{N2}, \ldots, y_{NM_N})$, the population vector of y-values for every object in the population. For estimating numbers of objects, as in Section 4.1, $y_{ij} = 1$ for each object. Then $y_i = \sum_j y_{ij} z_{ij}$, where z_{ij} is an indicator variable taking the value of one with probability g_{ij} if object ij is detected, and zero otherwise. Our vector of potential realizations for the whole population is now $\boldsymbol{z} = (z_{11}, z_{12}, \ldots, z_{NM_N})$, the vector of detection indicators, instead of $\boldsymbol{y}$. Let $\boldsymbol{y}_s$ denote the y_{ij} values in the sample.

Now suppose that we have an estimator $\hat{\tau}_0(\boldsymbol{y}_s)$, adaptive or conventional, which would be design unbiased for $\tau = \sum\sum y_{ij}$ if the detectability was perfect, i.e.,

$$\mathrm{E}[\hat{\tau}_0(\boldsymbol{y}_s)] = \sum_{i=1}^{N} \sum_{j=1}^{M_i} y_{ij} = \tau. \tag{11}$$

For the case of imperfect detectability we associate with every object in the population a new variable of interest $w_{ij} = y_{ij}x_{ij}/g_{ij}$ which conditional on z is a fixed value. Let $\boldsymbol{w}_s$ be the corresponding sample vector based on the components of $\boldsymbol{y}_s$. Then, from (11),

$$\mathrm{E}[\hat{\tau}_0(\boldsymbol{w}_s) \mid z] = \sum_{i=1}^{N} \sum_{j=1}^{M_i} w_{ij} = \sum_{i=1}^{N} \sum_{j=1}^{M_i} \frac{y_{ij}}{g_{ij}} z_{ij}$$

and

$$\mathrm{E}[\hat{\tau}_0(\boldsymbol{w}_s)] = \sum_{i=1}^{N} \sum_{j=1}^{M_i} \frac{y_{ij}}{g_{ij}} \mathrm{E}[z_{ij}] \tag{12}$$

$$= \sum_{i=1}^{N} \sum_{j=1}^{M_i} y_{ij} = \tau. \tag{13}$$

We can therefore obtain an unbiased estimator very simply as follows. Given an unbiased estimator for the perfect detectability situation we replace each y_{ij} in the estimator by w_{ij}. Variance formulas and examples of $\hat{\tau}_0$ are given by Thompson and Seber (1993). The g_{ij} can be estimated in various ways (e.g., Steinhorst and Samuel, 1989).

5. Adaptive allocation

5.1. Sample sizes based on initial observations in each stratum

In Section 3 the adaptive procedures used were essentially determined by the initial sample. Clusters were then adaptively formed around each sampled unit without any further random sampling being involved. A different approach is to divide up the population area into strata and use a 'two-phase' procedure. In the first phase, a random selection of units is made in each stratum. In the second phase one returns to those strata with, say, the highest y-values and samples more units. We call this type of procedure adaptive allocation in contrast to the adaptive cluster sampling of Section 3. In what follows we assume that simple random sampling without replacement is used for both phases. This problem was considered by Kremers (1987) and we use his approach.

Suppose there are N_h units in stratum h $(h = 1, 2, \ldots, H)$, and $N = \sum_h N_h$ units altogether. In phase j $(j = 1, 2)$, a simple random sample of size n_{hj} is taken without replacement from stratum h. Let $n_j = \sum_h n_{hj}$. Then the n_{h1} and n_1 are fixed, and n_2 is random. However, the n_{h2} will depend on the allocation or stopping rule C. Let $\bar{y}_{h1}$ and s_{h1}^2 be the sample mean and variance for the first phase sample from stratum h. One criterion might be: if $\bar{y}_{h1} > c$ then stratum h undergoes second phase sampling using, for example, proportional allocation, i.e., $n_{h2}/n_2 = N_h/N_s$, where N_s is the sum of the N_h which are candidates for second phase sampling. Our aim is to estimate each μ_h, the mean of the y-values for each stratum. Such estimates can then be combined to give an estimate of μ, the overall mean of the N units.

Since unit labels within strata are not used for estimation we find that Y_h, the set of y-values corresponding to the unordered distinct units for the total sample (phase 1 and 2) in stratum h, is not only sufficient for μ_h, but, using the methods of Kremers (1986), it can be shown to be complete as well. Hence using $\bar{y}_{h1}$ and s_{h1}^2 we can apply the Rao–Blackwell method to construct minimum variance unbiased estimators $\hat{\mu}_{\mathrm{RB}} = \mathrm{E}[\bar{y}_h | Y_h]$ and $s_{\mathrm{RB}}^2 = \mathrm{E}[s_h^2 | Y_h]$ to estimate μ_h and the stratum variance σ_h^2 respectively.

To simplify the notation we drop the subscript h and work with just one stratum. Let $n = n_1 + n_2$ be the total sample size for the stratum. Then, for every one of the $\eta = n!$ permutations of Y, we can compute the mean $\bar{y}_r$ of the first n_1 units $(r = 1, 2, \ldots, \eta)$. Let I be an indicator variable taking the value of one when C is satisfied and zero otherwise. Since all η permutations are equally likely it follows that all $\xi = \sum_{r=1}^{\eta} I_r$ permutations satisfying C are also equally likely so that

$$\hat{\mu}_{\mathrm{RB}} = \frac{1}{\xi} \sum_{r=1}^{\eta} \bar{y}_r I_r.$$

This equation is the same as equation (8) so that an unbiased estimator of $\mathrm{var}[\hat{\mu}_{\mathrm{RB}}]$ follows from (9). The same method can be used to find s_{RB}^2. Once these estimators are available from each stratum they can be combined in the usual fashion (see equations (14) and (15) below). If we use $\bar{y}$ calculated from both phases, then Kremers (1986) proves that $\bar{y}$ is negatively biased.

A variation on the above scheme was proposed earlier by Francis (1984) for estimating fish density. After the initial stratified sample is taken (using an approximate Neyman optimal allocation based on prior data) an algorithm is used to add units one at a time from different strata. The basis of the algorithm is to estimate stratum variances with the initial stratified sample and to allocate the rest of the total sample size among the strata so as to minimize the estimated final variance. Introducing the subscript h again, the usual unbiased stratified estimate of μ is

$$\bar{y}_{\mathrm{st}1} = \sum_h \frac{N_h}{N} \bar{y}_{h1}, \tag{14}$$

where $\bar{y}_{h1}$ is the sample mean of the initial sample in stratum h. An unbiased variance estimator based only on the initial sample is

$$v_{\mathrm{st}1} = \sum_h \frac{N_h^2}{N^2} s_{h1}^2 \left(\frac{1}{n_{h1}} - \frac{1}{N_h} \right). \tag{15}$$

We note that the estimators $\bar{y}_{\mathrm{st}}$ and v_{st} based on (14) and (15) but with the total sample will be (negatively) biased. However, Francis (1984) used these estimators with n_2 much less than n_1 on the grounds that the biases are then small, which appears to be the case. Choices of different values of n_1 and n_2 were compared in a simulation study.

A slightly different allocation scheme was proposed by Jolly and Hampton (1990) who used two phase sampling for estimating fish density from acoustic surveys. Once again Neyman-type allocation based on previous data was used to allocate sample sizes to the strata. A certain proportion (about 75%) of the total sample of size n was used for the first outward-going (eastward) phase. Based on the results of this phase 1 survey the Neyman allocation was reworked to determine the return (westward) phase 2 allocation. As with other adaptive schemes, the traditional stratified estimate of fish density is negatively biased for the above allocation plan, a fact noted by Francis (1991) who compared the above two schemes. In a reply to Francis, Jolly and Hampton (1991) mentioned that they had a method for estimating the bias.

5.2. Sample sizes based on observations from previous strata

One problem with the methods of the previous section is that they require two 'passes' over the population area. The first pass provides the phase 1 sample and this determines the phase 2 sampling effort in the second pass. Another approach which requires only a single pass proceeds sequentially. Here the strata are sampled one at a time in a sequential fashion and the level of sampling in a particular stratum depends on what happened in the previous stratum. An example of this is as follows.

In a design used in trawl surveys of shrimp the net is towed for a measured distance along the ocean flow and shrimp density is estimated from the weight of shrimp caught and the area swept by the net, taking into account the width of the net and the distance towed. However, because shrimp tend to aggregate in large schools, there is a tendency for many consecutive tows to have zero or near zero catches of shrimp. The catches then increase fairly steadily as an aggregation is approached. Two adaptive procedures which make use of this knowledge are now described.

In the first procedure, m_1 random half-mile tows are made in the first stratum. If an average of 50 pounds or less per mile are caught, then m_1 tows are carried out in the second stratum. However, if more than 50 pounds per mile are caught, then m_2 tows ($m_2 > m_1$) are carried out in the second stratum. If less than 50 pounds per mile are caught in the second stratum then m_1 tows are taken in the third stratum; otherwise m_2 are taken, and so on. For the second procedure we set $m_1 = 1$ and $m_2 = 2$ but with the m_2 tows merged into a single one-mile tow.

We will assume that simple random sampling without replacement is carried out in each stratum and use the notation of the previous section. If the procedure was not adpative then we would use the stratified estimator (14). Now if c_h denotes 'the set of all $\bar{y}_k$ for $k < h$' then c_h determines n_h. Hence by the properties of simple random sampling with a fixed sample size we have

$$\mathrm{E}[\bar{y}_h | c_h] = \mathrm{E}[\bar{y}_h | n_h] = \mu_h,$$

the population mean of stratum h, and

$$\mathrm{E}[\bar{y}_h] = \mu_h.$$

Therefore

$$\begin{aligned}\mathrm{E}[\bar{y}_{\mathrm{st}}] &= \sum_h \frac{N_h}{N}\mathrm{E}[\bar{y}_h] \\ &= \sum_h \frac{N_h \tau_h}{N} \\ &= \mu,\end{aligned}$$

and $\bar{y}_{\mathrm{st}}$ is unbiased in spite of the adaptive procedure. Furthermore, it can be shown that the variance estimate v_{st} (equation (15) using the whole sample) is unbiased also. We conclude, therefore, that even with this particular type of adaptive allocation, the traditional estimators are still appropriate.

The theory for the above procedure is discussed in detail by Thompson, Ramsey and Seber (1992) but with an emphasis on estimating population density rather than population numbers. In this case $N_h/N = A_h/A$, where A_h is the area of stratum h and A is the total population area.

In the second adaptive scheme discussed above where tows are either half-mile or one-mile we cannot use the above variance estimate as we only have one observation per stratum. To handle this Thompson et al. (1992) developed a model-unbiased estimation procedure.

Acknowledgement

This work was partially supported by National Science Foundation grant DMS-9016708.

References

Basu, D. (1969). Role of the sufficiency and likelihood principles in sample survey theory. *Sankhyā Ser. A* **31**, 441–454.

Cassel, C. M., C. E. Särndal and J. H. Wretman (1977). *Foundations of Inference in Survey Sampling*. Wiley, New York.

Francis, R. I. C. C. (1984). An adaptive strategy for stratified random trawl surveys. *New Zealand J. Marine Fresh-water Res.* **18**, 59–71.

Francis, R. I. C. C. (1991). Statistical properties of two-phase surveys: Comment. *Canad. J. Fish. Aquat. Sci.* **48**, 1128.

Jolly, G. M. and I. Hampton (1990). A stratified random transect design for acoustic surveys of fish stocks. *Canad. J. Fish. Aquat. Sci.* **47**, 1282–1291.

Jolly, G. M. and I. Hampton (1991). Reply to Comment by R. I. C. C. Francis, *Canad. J. Fish. Aquat. Sci.* **48**, 1128–1129.

Kremers, W. K. (1986). Completeness and unbiased estimation for sum-quota sampling. *J. Amer. Statist. Assoc.* **81**, 1070–1073.

Kremers, W. K. (1987). Adaptive sampling to account for unknown variability among strata. Preprint No. 128. Institut für Mathematik, Universität Augsburg, Germany.

Lehmann, E. L. (1983). *Theory of Point Estimation.* Wiley, New York.

Levy, P. S. (1977). Optimum allocation in stratified random network sampling for estimating the prevalence of attributes in rare populations. *J. Amer. Statist. Assoc.* **72**, 758–763.

Royall, R. (1968). An old approach to finite population sampling theorem. *J. Amer. Statist. Assoc.* **63**, 1269–1279.

Seber, G. A. F. (1982). *The Estimation of Animal Abundance and Related Parameters.* 2nd ed., Griffin, London.

Seber, G. A. F. (1986). A review of estimating animal abundance. *Biometrics* **42**, 267–292.

Seber, G. A. F. (1992). A review of estimating animal abundance II. *Internat. Statist. Rev.* **60**, 129–166.

Sinha, B. K. (1991). Sequential methods for finite populations. In: B. K. Ghosh and P. K. Sen, eds., *Handbook of Sequential Analysis.* Marcel Dekker, New York, 313–329.

Sirken, M. G. (1972). Stratified sample surveys with multiplicity. *J. Amer. Statist. Assoc.* **67**, 224–227.

Solomon, H. and S. Zacks (1970). Optimal design of sampling from finite populations: A critical review and indication of new research areas. *J. Amer. Statist. Assoc.* **65**, 653–677.

Steinhorst, R. K. and M. D. Samuel (1989). Sightability adjustment methods for aerial surveys of wildlife populations. *Biometrics* **45**, 415–425.

Thompson, S. K. (1990). Adaptive cluster sampling. *J. Amer. Statist. Assoc.* **85**, 1050–1059.

Thompson, S. K. (1991a). Adaptive cluster sampling: Designs with primary and secondary units. *Biometrics* **47**, 1103–1115.

Thompson, S. K. (1991b). Stratified adaptive cluster sampling. *Biometrika* **78**, 389–397.

Thompson, S. K. (1992). *Sampling.* Wiley, New York. In press.

Thompson, S. K. (1993a). Adaptive cluster sampling based on order statistics. Submitted for publication.

Thompson, S. K. (1993b). Multivariate aspects of adaptive cluster sampling. In: G. P. Patil and C. R. Rao, eds., *Multivariate Environmental Statistics.* North Holland/Elsevier Science Publishers, New York. In press.

Thompson, S. K., F. L. Ramsey and G. A. F. Seber (1992). An adaptive procedure for sampling animal populations. *Biometrics* **48**, 1195–1199.

Thompson, S. K. and G. A. F. Seber (1993). Detectability in adaptive and conventional sampling. *Biometrics.* In press.

Zacks, S. (1969). Bayes sequential designs of fixed size samples from finite populations. *J. Amer. Statist. Assoc.* **64**, 1342–1349.

G. P. Patil and C. R. Rao, eds., *Handbook of Statistics, Vol. 12*

7

Statistical Analysis of Censored Environmental Data

Michael G. Akritas, Thomas F. Ruscitti and G. P. Patil

Abstract: Censored data commonly occurs in environmental studies when pollutant levels fall below the detection (or reporting) limits of instrumentation. Estimation of population parameters, inference, and other analyses of censored data sets are problematic. Various methods for parameter estimation are surveyed, including simple substitution of detection limits; maximum likelihood estimators; and probability plotting. Estimation of location difference in the 2-sample case is presented in the framework of extensions of the nonparametric Hodges–Lehmann estimator. Various regression methods for censored data are discussed, including maximum likelihood; Buckley–James; least absolute deviations; and Theil–Sen regression. Selected examples, proposed new methods, and an extended bibliography are included throughout.

1. Introduction

The topic of censored and truncated data has a rich literature. Cohen (1991) traces the first analysis of truncated samples to Sir Francis Galton in 1897; credits R. A. Fisher with applying the method of maximum likelihood to truncated samples in 1931; and cites Stevens (1937) as first deriving maximum likelihood equations for type I singly and doubly censored samples. Censored and truncated data analyses find applications in survival analysis and, especially within the last 25 years, many aspects of environmental science. Chemists and statisticians have been steadily adapting the language and methods of classical censored data analysis to fit the unique problems of environmental measurements. This chapter will review current censored data analysis methods from an environmental viewpoint.

Censoring of environmental data most typically arises when the level of a trace chemical in a sample is less than the 'limit of quantitation' (LOQ) or 'limit of detection' (LOD) of analytical instruments. The physical meanings of LOQ and LOD differ with the analytical technology being applied (e.g., gas chromatography, mass spectroscopy, on-line field instruments). Such observations are usually reported as 'less than detectable' (LTD), signifying that a measurement was made but that its low level precluded reporting a quantitative value. Type I, left censoring is most often encountered; that is, each detection limit is fixed and a random (but known) number of censored observations occur below each limit. The data in a given sample may be subject to a single detection limit (singly censored) or multiple detection limits (multiply censored). Right censoring is also a conceivable problem in environmental monitoring, but remediation decisions are unlikely to be ambiguous in the presence of 'greater than detectable' observations.

The literature on censored environmental data analysis is largely driven by the issues raised by water quality analysis. Water quality problems are pervasive and negative health impacts are associated with ever lower concentrations of certain chemicals. Costly remediation decisions, regulatory action, and litigation hinge on data sets with censored observations. Also, differences between laboratory technology and procedures, along with improving instrument technology which lowers detection limits, all but guarantee that any long-term monitoring program faces the problem of multiple detection limits. Inevitably, more importance is placed on extracting information from 'less than detectable' observations. A number of articles surveying methods for analyzing censored water quality data have already been published; many are cited here to provide a unified bibliography of the subject. One that is particularly well suited as an introduction, and is detailed enough to serve as a handbook, is by Atwood et al. (1991). In addition to outlining commonly applied methods, this chapter highlights useful procedures that have received limited attention in other reviews, especially for the 2-sample and regression settings.

One question remains before proceeding: If every bit of information is critical, why censor at all? The literature on environmental censoring is beginning to include more discussion of whether and how to use measurement values which fall below the reporting limit. These approaches are designed to extract more information from such uncertain results than would be available by treating them as simply 'less than detectable'. Without downplaying the importance of 'uncensored' methods, this chapter does not address this issue; until quantitative scientists and statisticians agree on the importance of seeing *all* of the data, censoring will plague environmental data sets. Accordingly, this chapter is a survey of current methods for treating the inevitable censored data sets. For more information and bibliographies regarding LOQ, LOD, and the inclusion of uncertain measurements, see Porter et al. (1988) and Lambert et al. (1991).

2. 1-sample case: Parameter estimation

Methods in this section deal with the estimation of population parameters (μ, σ^2, and quantiles) from censored data sets. The methods are first presented for the case of singly censored data; adaptations for multiply censored data are noted if they are available. Unless otherwise noted, μ and σ^2 are the parameters of interest in the methods under consideration.

Helsel (1990) classifies existing methods into three general areas:

- *Simple substitution*: Here, censored observations are replaced with fabricated values, typically zero, the detection limit (DL), or one-half the detection limit (DL/2). Alternatively, values in the censored region (zero to DL) may be assumed to fit some distribution, e.g., uniform, normal, or lognormal, so that censored values may be substituted with random values. The substituted values are combined with the uncensored data to produce parameter estimates.
- *Parametric*. Most parametric methods are variations of maximum likelihood estimation (MLE) methods developed by Cohen (1959, 1991) for censored and truncated data. There are also techniques which use parametric methods to estimate μ and σ^2 from the uncensored data, then 'fill in' the expected values of the censored observations on the assumption that the entire data set is normal (or transformed to normality). Then, overall estimates of μ and σ^2 are calculated by the usual methods on the 'completed' data set. The initial parameter estimates are obtained from the uncensored data using either MLE methods or probability plots. These latter techniques are referred to in the literature as *fill-in* or *pseudo-completion* methods.
- *Robust parametric*: A variation of the fill-in method is the use of probability plot regression to fit a line to the uncensored data, then extrapolating the line back into the censored region to produce expected values for the censored data. The expected censored values are combined with the uncensored data so that parameter estimates $\hat{\mu} = \bar{x}$ and $\hat{\sigma}^2 = s^2$ may be calculated. Note that this plotting approach is different than the parametric version, where the intercept and slope of the line are interpreted as $\hat{\mu}$ and $\hat{\sigma}^2$.

A more detailed description of these methods is given below.

2.1. *Simple substitution methods*

The simplicity of substitution methods has led to wide use, despite well-known bias problems. Here, each censored value is replaced by zero, DL, or DL/2; the substitution choice and resulting bias is entirely arbitrary. Substituting zero generally imposes a downward bias on estimates of μ; the reverse is true for substitution with DL. One enhancement to simple substitution involves assuming that censored observations are uniformly distributed between zero and the detection limit; substituted values then follow a uniform distribution centered on DL/2. This method estimates μ identically to substituting all DL/2, but produces a better estimate of σ^2 (Gilliom and Helsel, 1986). Note, however, that the true performance of any simple substitution method depends entirely on the actual (usually unknown) underlying distribution. Occasionally, the bias of a simple substitution method is used purposively, such as when a regulatory agency stipulates the substitution of DL for censored values in order to produce a high, conservative estimate (in the sense of forcing action to protect public health) of toxin levels (Gleit, 1985).

Better methods exist, and the availability of powerful, inexpensive computers undermines the 'simplicity' argument for simple substitution. Helsel (1990) sums it up: 'estimates from simple substitution are not defensible'. Since simple substitution methods for multiply censored data share the same origins and faults as those in the singly censored case, no further discussion of them will be included here.

2.2. *Parametric methods*

Maximum likelihood estimation for censored environmental data is largely based on methods proposed by Cohen (1959, 1991) for normal samples. Obviously, the sample must either be from an underlying normal distribution or be transformed to normality. Gilliom and Helsel (1986) and Helsel and Gilliom (1986) applied Cohen's MLE methods to the logarithm of lognormally distributed data, and Shumway et al. (1989) proposed more general power transformations to normality. When the assumed (or transformed) distribution is correct and sample sizes are large, the MLE estimates can be quite good. They have the added advantage of providing straightforward asymptotic variance–covariance properties of the estimators by inverting the Fisher information matrix (Cohen, 1991; Schneider, 1986).

EXAMPLE 1. Cohen's MLE method (Cohen, 1991, pp. 8–30) is not difficult to apply. One criticism has been that tabulated values are required for a quantity involved in the estimation procedure (it is called an 'auxiliary estimation function' by Cohen); however, a computational solution to the auxiliary function is easily developed. Consider a sample $\vec{X}$ of $n = 50$ from a 2-parameter lognormal distribution, such that $\log(\vec{X})$ is distributed N(0, 1). The example data are summarized in Table 1.

Censoring at the 20-th sample percentile excludes 10 data points; the censoring limit at this level is 0.4 in the original units. Once the data are log-transformed to normality, applying the MLE procedure yields the estimators $\hat{\mu} = -0.0712$ and $\hat{\sigma}^2 = 1.0215$. Note that the usual transformation bias corrections would be necessary

Table 1
Summary of example data

X				
0.1007	0.1113	0.1167	0.1848	0.2531
0.2621	0.2747	0.2915	0.3529	0.3629
0.4014	0.4136	0.4154	0.4182	0.4443
0.4696	0.5282	0.5386	0.5648	0.6152
0.6779	0.7485	0.7508	0.8170	1.1029
1.1210	1.1938	1.3856	1.4381	1.4526
1.5144	1.5220	1.5470	1.5472	1.6761
1.7095	2.0705	2.2106	2.2118	2.2966
2.4249	2.4444	2.4615	2.7230	2.7461
3.3692	4.4418	4.8487	4.8613	6.9258

	$\bar{x}$	$s^2(x)$	median
X	1.467	2.064	1.112
$\log(X)$	−0.088	1.088	0.106

to express $\hat{\mu}$ and $\hat{\sigma}^2$ in the original lognormal units. The variance of $\hat{\mu}$ and $\hat{\sigma}$ may also be estimated by Cohen's method. In this example, $V(\hat{\mu}) = 0.0217$ and $V(\hat{\sigma}^2) = 0.0141$.

Note, however, that MLE estimators are quite sensitive to incorrect distributional assumptions. As environmental datasets are often small and difficult to classify distributionally, MLE estimates quickly run into trouble. Also, even when the distribution is correctly identified, estimators are also biased for small sample sizes ($N < 20$ is a guideline suggested by Cohen (1991)). Schneider (1988) and Cohen (1988) have proposed methods for correcting biased MLE estimates.

Another set of parametric methods involves estimating μ and σ^2 from the uncensored data, then computing expected values for the censored observations under normality with the newly estimated parameters. The expected values are combined with the uncensored observations, and μ and σ^2 are recalculated. Gleit (1985) proposed an MLE based technique which iteratively adjusts the parameter estimates, stopping when consecutive estimates differ by less than a tolerance level. The probability plotting method estimates μ and σ^2 from the slope and intercept of a line fit to a probability plot of the uncensored values (Travis and Land, 1990).

Helsel and Cohn (1988) evaluated the performance of a multiple threshold MLE method developed by Cohen (1976). A bias-adjusted MLE estimator (Cohn, 1988) is also proposed for use in the multiple threshold case.

2.3. *Robust parametric methods*

These methods are variations on the fill-in technique, using probability plot regression. Here, a normal (or lognormal) probability plot is prepared, ranking the uncensored observations from $n_c + 1$ to N, where n_c is the number of censored observations and N is the total sample size. A line is fit through the uncensored data using simple linear regression, then extrapolated back into the censored region. On the assumption that the censored observations follow the zero to DL portion of a normal (or lognormal) distribution, the censored observations are placed on the extrapolated line at plotting positions corresponding to ranks 1 to n_c. The expected values of the censored observations may then be obtained from the abscissa, combined with the uncensored values, and used to estimate μ and σ^2 in the usual way. This method is attributed to Hashimoto and Trussell (1983). When the data are assumed normal, the method is referred to as *normal probability regression* (NR); for lognormal, it is *lognormal probability regression* (LR) (Gilliom and Helsel, 1986). Note that the extrapolation may produce inadmissible negative values, depending on the slope of the fit through the uncensored data.

There is some debate about the plotting positions used in the probability regression methods. Helsel and Cohn (1988) give the general plotting position formula as

$$PP_i = \frac{i - \alpha}{n + 1 - 2\alpha}. \tag{1}$$

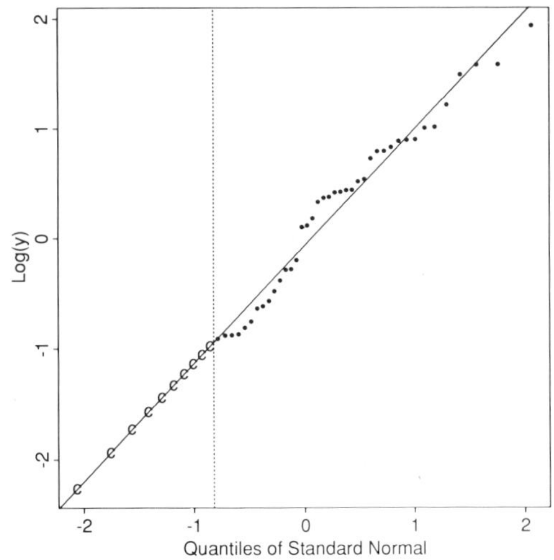

Fig. 1. Lognormal probability regression method. Dots represent uncensored data points, while 'c' indicates predicted values of censored data points. The censored region lies to the left of the vertical dotted line.

Common values for α include zero (Weibull method), 0.375 (Blom), and 0.5 (Hazen). Helsel and Cohn (1988) conclude from simulation experiments that, for the LR method, errors due to choice of α are small compared to all other sources.

EXAMPLE 2. Consider again the log-transformed data of Example 1. First, the 40 uncensored data points are assigned ranks from 11 to 50, since there are 10 censored values. Next, these data are used to create a normal probability plot (the Weibull plotting position is used here). Regressing the uncensored data points on the resulting quantiles of the standard normal distribution gives a slope of 1.064 with an intercept of -0.0689. The fitted line may be extrapolated back into the censored

Table 2
Summary of results from Example 2: LR method

$y_{(i)}$	$\hat{y}_{(i)}$
<0.4	0.1040
<0.4	0.1434
<0.4	0.1765
<0.4	0.2068
<0.4	0.2357
<0.4	0.2639
<0.4	0.2917
<0.4	0.3194
<0.4	0.3472
<0.4	0.3754

$\hat{\mu} = 1.4703$, $\hat{\sigma}^2 = 2.0558$

region and used to predict the censored values at the plotting positions corresponding to ranks 1 to n_c (see Figure 1). The expected censored values, re-transformed to the original units, are listed in Table 2 along with the estimates $\hat{\mu}$ and $\hat{\sigma}^2$ calculated from the sample mean $\bar{x}$ and variance s^2. Note thqat re-transforming the expected censored values back to the lognormal scale before calculating the parameter estimates avoids any transformation bias corrections.

El-Shaarawi (1989) proposed a modification to the LR method that includes information about the detection limit itself. Here, the DL is added to the set of uncensored values which are used to produce the regression line.

A method to modify the lognormal probability regression method for the multiple-threshold case (MLR) was proposed by Hirsch and Stedinger (1987) and Helsel and Cohn (1988). This method uses probabilities of non-exceedence of each of the detection limits to produce plotting positions. These probabilities are calculated using a Kaplan–Meier approach, which divides the probability mass of a censored observation (or group of observations) among all other observations of lower rank. In the usual right-censored survival analysis case, the Kaplan–Meier estimator of the survival function $S(t)$ may be defined as

$$\hat{S}(t) = \prod_{y(i) \leq t} \left(\frac{n-i}{n-i+1} \right)^{\delta_{(i)}}, \tag{2}$$

where $y_{(i)}$ are the ordered survival times and $\delta(i) = 1$ if $y_{(i)}$ is uncensored and 0 if censored (Miller, 1981). The estimator is easily adapted to the multiple detection limit, left-censored case,

$$\hat{P}(\mathrm{DL}_j) = \prod_{y(i) > \mathrm{DL}_j} \left(\frac{i-1}{i} \right)^{\delta_{(i)}}. \tag{3}$$

Here, $\hat{P}(\mathrm{DL}_j)$ is the probability of non-exceedence of the j-th detection limit; $y(i)$ is the i-th order statistic of the data set; and $\delta_{(i)}$ is the censoring indicator variable defined above. The resulting estimates may be used to calculate plotting positions by the method of Hirsch and Stedinger (1987) and Helsel and Cohn (1988), where plotting positions are determined by the following rules:

(1) Censored values recorded as 'less than DL_j' are plotted at probabilities equally spaced between zero and $\hat{P}(\mathrm{DL}_j)$.

(2) Uncensored values greater than the highest detection limit are plotted at probabilities equally spaced between $\hat{P}(\max\{\mathrm{DL}_j\})$ and 1.0.

(3) Uncensored values lying between detection limits ($\mathrm{DL}_i < y_j < \mathrm{DL}_k$) are plotted at probabilities equally spaced between $\hat{P}(\mathrm{DL}_i)$ and $\hat{P}(\mathrm{DL}_k)$.

EXAMPLE 3. The dataset of Example 1 may be modified so that three intermediate points are recorded as 'less than 1.0', giving a total of 13 censored data points with $\mathrm{DL}_1 = 0.4$ and $\mathrm{DL}_2 = 1.0$.

The Kaplan–Meier non-exceedence probabilities for the two detection limits are $\hat{P}(\mathrm{DL}_1) = 0.229$ and $\hat{P}(\mathrm{DL}_2) = 0.48$. The new data set is listed in Table 3, along with the censoring indicator $\delta_{(i)}$, plotting probabilities $p_{(i)}$, and the corresponding normal

Table 3
Dataset for Example 3: MLR method with two detection limits

$y_{(i)}$	$\delta_{(i)}$	$p_{(i)}$	$q_{(i)}$	$y_{(i)}$	$\delta_{(i)}$	$p_{(i)}$	$q_{(i)}$
<0.4	0	0.0208	−2.0379	1.121	1	0.5185	0.0464
<0.4	0	0.0416	−1.7329	1.1938	1	0.5378	0.0948
<0.4	0	0.0623	−1.5354	1.3856	1	0.557	0.1435
<0.4	0	0.0831	−1.3844	1.4381	1	0.5763	0.1924
<0.4	0	0.1039	−1.2597	1.4526	1	0.5956	0.2419
<0.4	0	0.1247	−1.1519	1.5144	1	0.6148	0.2919
<0.4	0	0.1455	−1.0561	1.522	1	0.6341	0.3427
<0.4	0	0.1662	−0.9692	1.547	1	0.6533	0.3943
<0.4	0	0.187	−0.889	1.5472	1	0.6726	0.4471
<0.4	0	0.2078	−0.8141	1.6761	1	0.6919	0.5011
0.4014	1	0.2495	−0.676	1.7095	1	0.7111	0.5566
0.4136	1	0.2705	−0.6114	2.0705	1	0.7304	0.6139
0.4154	1	0.2914	−0.5492	2.2106	1	0.7496	0.6733
0.4182	1	0.3124	−0.4891	2.2118	1	0.7689	0.7352
0.4443	1	0.3333	−0.4307	2.2966	1	0.7881	0.8
0.4696	1	0.3543	−0.3738	2.4249	1	0.8074	0.8684
0.5282	1	0.3752	−0.318	2.4444	1	0.8267	0.9411
0.5386	1	0.3962	−0.2632	2.4615	1	0.8459	1.0191
0.5648	1	0.4171	−0.2092	2.723	1	0.8652	1.1039
0.6152	1	0.4381	−0.1558	2.7461	1	0.8844	1.1975
0.6779	1	0.459	−0.1028	3.3692	1	0.9037	1.3029
<1.0	0	0.12	−1.175	4.4418	1	0.923	1.4253
<1.0	0	0.24	−0.7063	4.8487	1	0.9422	1.5737
<1.0	0	0.36	−0.3585	4.8613	1	0.9615	1.7681
1.1029	1	0.4993	−0.0019	6.9258	1	0.9807	2.0693

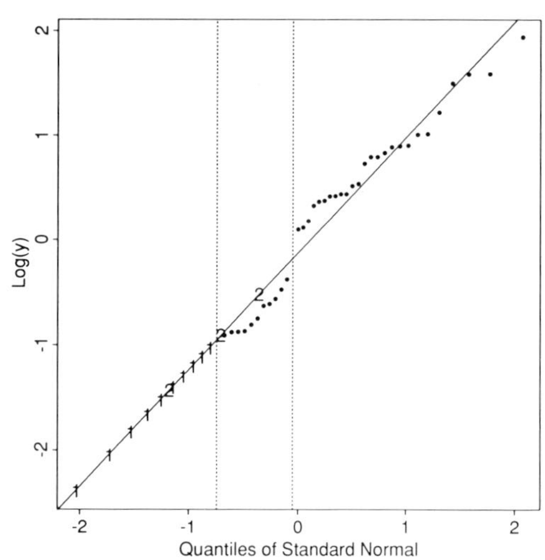

Fig. 2. Lognormal probability regression method with two detection limits. Dots represent uncensored data points, while '1' and '2' indicate predicted values of censored data points corresponding to DL_1 and DL_2 (indicated by vertical dotted lines).

Table 4
Summary of results from Example 3: MLR method

$y_{(i)}$	$\hat{y}_{(i)}$
< 0.4	0.0922
< 0.4	0.1292
< 0.4	0.1607
< 0.4	0.1900
< 0.4	0.2181
< 0.4	0.2457
< 0.4	0.2732
< 0.4	0.3009
< 0.4	0.3288
< 0.4	0.3572
< 1.0	0.2396
< 1.0	0.4025
< 1.0	0.5915

$\hat{\mu} = 1.4453$, $\hat{\sigma}^2 = 2.1036$

quantiles $q_{(i)}$ used to produce the MLR plot in Figure 2. The predicted values for the censored data points are listed in Table 4 with the resulting estimates $\hat{\mu}$ and $\hat{\sigma}^2$.

More in line with the traditional meaning of 'robust parametric', James (1986) introduced *M*-estimators for censored data. Lee (1991) generalized the work of James by studying a wider class of *M*-estimators.

2.4. Variance of estimators

As noted previously, MLE methods provide expressions for the variance of the estimated parameters μ and σ^2 through the inversion of the Fisher information matrix (Cohen, 1991; Schneider, 1986).

There are currently no analytical expressions available for the variance of estimators produced with the probability regression methods. Helsel and Gilliom (1986) proposed simulation based methods for constructing confidence intervals around LR parameter estimates. These methods used the root mean squared error (rmse) and bias results from a simulation study involving skewed distributions designed to emulate typical water-quality data sets used to evaluate trace constituents.

2.5. Comparison of methods

Simulation studies have been used to compare the various single detection limit methods, examining the bias and mean squared error of each under various circumstances. Each study involves different sample sizes, distributions, and censoring levels, and should be examined for their specific applicability; however, some general conclusions may be drawn. Gleit (1985) found the iterative, MLE-based fill-in method (see Section 2.2) to be superior to the traditional MLE methods for estimating μ and σ for small, normal data sets ($N \leqslant 15$). El-Shaarawi (1989) compared MLE, LR, a modified LR (see Section 2.3), and Gleit's MLE based fill-in method; normal, lognormal, and normal mixture distributions were used with $N = 10$, 25 and 50. Here, the recommendation was to use either traditional MLE with bias correction or the MLE based fill-in method for the larger sample sizes, and the modified LR method for small, lognormal samples. The MLE methods were favored, at least in part, for the availability of estimator variances. Gilliom and Helsel (1986) undertook one of the more comprehensive comparison studies for singly censored data, using 16 parent distributions chosen to represent the skew and variation patterns encountered in environmental data ($N = 10$, 25 and 50). Each method was ranked on the basis of root mean squared error (rmse) for each estimator. They found the LR method to be the most robust overall, in the sense that it had the lowest overall sum of rmse ranks across all tested distributions. When performance was examined for individual estimators, the LR method was the best for moment estimators, while the lognormal maximum likelihood method was recommended for quantile estimators. Helsel and Gilliom (1986) went on to verify their simulation results by applying censoring limits to complete water quality data sets. Newman et al. (1989) also used actual water quality data to evaluate eight methods. When the underlying distribution is known, a restricted MLE method (Schneider, 1986) with bias correction was recommended; again, the robust parametric methods performed best for ambiguous distributions.

Helsel and Cohn (1988) used the parent data sets of Gilliom and Helsel (1986) with $N = 25$ to examine the multiple detection limit case. Results were similar to those of single detection limit studies. The most robust method overall was the LR, modified for multiple detection limits (see Section 2.3); when the parent distribution is known, the bias-adjusted MLE method of Cohn (1988) performs well.

2.6. *Estimation of quantiles*

Environmental measurements are quite often from skewed distributions, and moment estimators may be biased even without censoring, especially for small samples. Hence, there is strong motivation for characterizing such distributions with quantiles rather than moments.

In case of singly censored data, sample quantiles can be used for those quantiles that are greater than the censoring level. Caution should be exercised, however, as simulation studies have shown that sample quantiles close to the censoring level will tend to overestimate the true population quantiles. For such cases, as well as for the estimation of quantiles below the censoring level, a parametric approach should be used, keeping in mind the dependence on correct distributional assumptions. Alternatively, the various 'fill-in' procedures (including the robust parametric) permit direct calculation of sample quantile estimates once the expected values of the censored observations are obtained. Unlike the estimation of the mean, simulation studies have shown that MLE methods perform better than the robust parametric methods (Gilliom and Helsel, 1986).

The estimation of quantiles from multiply censored data poses no additional difficulties. One can use sample quantiles based on the Kaplan–Meier estimator, or a new enumeration method proposed in Akritas and Morder (1993) for the estimation of quantiles that are not too close to the lowest censoring limit. For quantiles near or below the lowest censoring limit, the only available methods are parametric.

The p-th quantile based on the Kaplan–Meier estimator $\hat{F}$ is commonly defined as

$$\inf\{x: \hat{F}(x) \geqslant p\}, \tag{4}$$

but in practice one interpolates between the two consecutive uncensored data points where the Kaplan–Meier estimator is above and below p (Miller, 1981; Lawless, 1982). Note that any censored observations that may exist between the consecutive uncensored observations are not taken directly into account. The enumeration method of Akritas and Morder finds the p-th quantile by first counting all observations as if they were uncensored and then adjusts for the presence of censoring. In particular, let

$$T_p(z) = \frac{1}{n}\{\#\text{data} \leqslant z\} - \frac{1}{n}\{\#\text{data} > z\} - [(2p-1)(1-\hat{G}(z))] - \hat{G}(z), \tag{5}$$

where $\hat{G}$ is the Kaplan–Meier estimate of censoring distribution. When the censoring point is always known (as is the case with water quality data), $\hat{G}$ can be the usual empirical distribution function.

Interpolating between the two points were $T_p(z)$ crosses zero gives the enumeration estimate $\hat{Q}_p$ of the p-th quantile. The enumeration estimate is always between the same two consecutive uncensored data points that the usual Kaplan–Meier based estimate lies, and coincides with the usual estimate when there are no censored data points between the two consecutive uncensored points. An interesting feature of the enumeration method is that it allows an easy estimate of the asymptotic variance of the quantiles bypassing the need for density estimation. Indeed, an estimate of the asymptotic variance of $\hat{Q}_p$ is

$$\frac{4[1-\hat{G}(\hat{Q}_p)]^2\, \hat{V}(\hat{Q}_p)}{[(n^{1/2-\delta})\, T_p(n^{-1/2+\delta} + \hat{Q}_p)]^2}, \tag{6}$$

where $\hat{V}(t)$ is the usual estimate of the asymptotic variance of $\hat{F}(t)$ and δ is any number between zero and 0.5. Simulation studies are underway to determine the best value of δ for small samples.

3. Two- and k-samples

3.1. General comments

There is an extensive literature on nonparametric testing methods for the 2- and k-sample testing problem with censored data, deriving primarily from biostatistical applications. See Jones and Crowley (1989) for a recent, very general formulation. A recent discussion (including limited comparisons of various methods) for the environmental community may be found in Helsel (1990). In the same reference, Helsel also suggests tobit regression as an alternative to MLE methods for estimation and testing in the 2-sample case. A simulation study comparing the performance of various nonparametric tests with multiply censored data may be found in Millard and Deverel (1988). Our review will focus on estimation of the location difference in the 2-sample case, and briefly discuss multiple comparisons in the k-sample case.

3.2. *Estimation of location difference*

Most nonparametric, or rank-based, estimation methods with censored data are extensions of the Hodges–Lehmann estimator; see Louis (1981), Padgett and Wei (1982), Wei and Gail (1983), Meng et al. (1991) and Akritas et al. (1992a). With uncensored data, the Hodges–Lehmann estimator (Hodges and Lehmann, 1963) is the median of pairwise differences $x_{1i} - x_{2j}$, where x_{1i}, $i = 1, \ldots, n_1$, and x_{2j}, $j = 1, \ldots, n_2$ are the two samples. The Hodges–Lehmann estimator is obtained by 'inverting' the Wilcoxon rank sum test. By 'extensions' of the Hodges–Lehmann estimator we mean either a direct extension of the median of pairwise differences (Meng et al., 1991) or by an inversion of a suitable test statistic (all other methods). Though the direct method of Meng et al. is elegant, it will not be described here, as it is quite involved. The method of Akritas et al., using Theil–Sen regression, is easy to implement and has a simple estimate of the asymptotic variance. It is described in detail in Section 4.2.4 as a general regression method. To apply it in the 2-sample context, we use a covariate that takes values 0 or 1 for the first and second samples, respectively; an example is provided in Section 4.2.4.

Wei and Gail present a general theory for inverting 2-sample statistics with censored data: let (X_{1i}, δ_{1i}), $i = 1, \ldots, n_1$, (X_{2j}, δ_{2j}), $j = 1, \ldots n_2$ be the two samples (where δ is an indicator of censoring), and let $t_n = t_n(\vec{X}_1, \vec{X}_2, \vec{\delta}_1, \vec{\delta}_2)$ be a statistic for testing inequality of the two underlying distributions satisfying certain conditions. Then set $h_n(a) = t_n(\vec{X}_1, \vec{X}_2/a, \vec{\delta}_1, \vec{\delta}_2)$ and define the Hodges–Lehmann estimate of the ratio of the two scale parameters (the difference of location parameters is estimated analogously) as

$$\theta_n(t_n) = [\theta_n^{(1)}(t_n) + \theta_n^{(2)}(t_n)]/2\,, \tag{7}$$

where $\theta_n^{(1)}(t_n) = \sup\{a\colon h_n(a) > \mu_n\}$ and $\theta_n^{(2)}(t_n) = \inf\{a\colon h_n(a) < \mu_n\}$. For many of the common statistics, including Gehan's statistic and the log-rank statistic, $\mu_n = 0$. It is important to note that for these statistics, the function $h_n(a)$ is monotone, so the estimator $\theta_n(t_n)$ is as simple to obtain as the estimator of Akritas et al. mentioned above. However, estimation of the asymptotic variance is different. See Wei and Gail (1983) for details and for simulation results.

3.3. *Multiple comparisons*

The multiple comparisons procedure discussed here is based on the k-sample statistic proposed by Peto and Peto (1972). This statistic was derived under the assumption of equal censoring distributions, but Akritas (1992) showed it is robust to small deviations from that assumption. In the latter paper, it was also shown that:

(1) The statistic of Peto and Peto is equivalent to the one-way ANOVA F-statistic evaluated on logistic scores.

(2) Any multiple comparisons procedure (such as the one based on Scheffé-type simultaneous confidence intervals), when performed on the logistic scores, yields valid multiple comparisons procedures.

Other score functions may also be used. The simulation studies in Akritas (1992) indicate that even with unequal sample sizes and unequal censoring distributions the 2-sample t-test on logistic scores performs well. It is particularly powerful against location alternatives.

4. Regression with censored data

The problem of fitting straight lines with censored data has received considerable attention over the past fifteen years. In this section we will briefly review some of the proposed methods. One of the procedures we describe in detail, Theil–Sen regression, seems particularly well suited for data arising from ecological monitoring studies in the sense that it easily extends to doubly censored data. By doubly censored data, in the regression setting, we mean that both the explanatory and response variables are subject to random censoring. For regression with only the response variable censored, we will assume that the observed data are of the form (Y_i, δ_i, x_i), $i = 1, \ldots, n$, where $Y_i = \min\{T_i, C_i\}$, T_i is the variable of interest, C_i is the censoring variable, and δ_i is a variable indicating whether Y_i is censored or not. While this pertains to right censoring, note that the left censoring case, which is common in environmental monitoring studies, can be converted to right censoring by multiplying the variables by -1. It is assumed that

$$T_i = \alpha + \beta x_i + \varepsilon_i, \tag{8}$$

where the ε_i are i.i.d. error terms. The procedures we review will be broadly classified into parametric and nonparametric.

4.1. *Parametric methods*

Perhaps the most common, and mathematically straightforward, regression method is maximum likelihood (ML) regression where the researcher specifies the form of the error distribution. It is described in detail in Lawless (1982). The popularity of ML regression stems partly from its availability in such leading software packages as

SAS; see, for instance, Donald (1992). It is a very flexible method for both simple and multiple regression that allows simultaneous inference on subsets of the parameter vector. Care must be exercised, however, as the results may depend on the assumed error distribution. This contrasts with the uncensored case where inferences based on the assumption of a normal error distribution remain asymptotically valid under a wide range of symmetric or asymmetric distributions. (See Section 5.3 for a bibliography of methods for checking the validity of assumptions on the error distribution.) Cohn et al. (1992) discuss an adjusted MLE approach, applied to the problem of estimating pollutant transport in rivers.

Another parametric regression method, popular among econometricians, is tobit regression. This applies to type I or singly censored data and is based on the assumption of normal error distribution. See Amemiya (1985) for a detailed description.

4.2. *Nonparametric methods*

4.2.1. *Buckley–James regression*

Perhaps the first successful regression method that does not require specification of the error distribution is the method proposed by Buckley and James (1979). Buckley and James define the pseudo random variables Y_i^* as

$$Y_i^* = Y_i \delta_i + \mathrm{E}[T_i | T_i > Y_i](1 - \delta_i). \tag{9}$$

They note that

$$\mathrm{E}[Y_i^* | x_i] = \alpha + \beta x_i, \tag{10}$$

which means that Y_i^* follows the same regression model as T_i. Using residuals obtained from preliminary estimators of α and β (e.g., from ordinary least squares on the pairs (Y_i, x_i)), they estimate the conditional expectation $\mathrm{E}[T_i | T_i > Y_i]$ from the Kaplan–Meier estimator of these residuals. The estimated conditional expectation produces an approximation, $\hat{Y}_i^*$, of Y_i^*, which is used to update the preliminary estimates of α and β by ordinary least squares on the pairs $(\hat{Y}_i^*, x_i)$, $i = 1, \ldots, n$. The procedure is then iterated until convergence. See, also, Miller and Halpern (1982). The simulation studies conducted by McLeish (1983), Moon (1989) and Schneider (1986) all indicate that the Buckley–James estimator is unbiased. However, when the variance of the error term is not bounded, the Buckley–James estimator becomes biased. For example Moon (1989) showed that when the Cauchy distribution is used for the error term, the Buckley–James estimator deviates seriously from the true coefficient. The consistency of the estimators was proven by James and Smith (1984). Recent work by Ritov (1990) and Lai and Ying (1991a) have also established the asymptotic normality. Worth mentioning is the fact that the Buckley–James method is a nonparametric version of the iterative least squares method devised independently by Schmee and Hahn (1979); see also Aitken (1981). Finally, related is the method of 'synthetic data' regression proposed by Leurgans (1987) and studied further by Zhou (1992).

4.2.2. Schmitt's weighted least squares regression

This method was proposed by the astronomer Schmitt (1985) to deal with the problem of fitting a straight line to doubly censored data. This is essentially a moment estimator that imitates the uncensored data least squares estimator (LSE). Indeed, in the absence of censoring, the LSE of the slope parameter is defined as the ratio of the sample covariance of the response and explanatory variables over the sample variance of the explanatory variable. Schmitt's idea was to obtain an estimate of the bivariate distribution of the response and explanatory variables and use this to obtain a censored data version of the sample covariance and sample variance. The ratio of these is Schmitt's weighted least squares estimator (the name 'weighted least squares estimator' derives from the fact that the Schmitt estimator of the slope can be obtained by the weighted least squares principle, with weights determined from the mass the bivariate distribution assigns to points). Improved versions of this method can be obtained by using more recent estimates of the bivariate distribution under censoring. Such versions of the Schmitt estimator were included in a Monte Carlo comparison in Akritas et al. (1992a). It is noted there that the consistency of the Schmitt estimator requires certain restrictions on the censoring mechanism. It is believed, however, that a suitably modified Schmitt estimator can be a useful data analysis tool, valid under a general censoring mechanism and for regression models with heteroskedastic errors. it should also be noted that the extension of this to multiple regression models requires only the bivariate distribution of the response variable with each explanatory variable.

4.2.3. Least absolute deviations regression

With uncensored data, the method of least absolute deviations estimates the parameters α and β by minimizing

$$\sum |T_i - a - bx_i| \tag{11}$$

with respect to a and b. Powell (1984) proposed an interesting variation on the least absolute deviations estimation for the case of type I or singly censored data. For simplicity of presentation, assume that T_i is observed only if $T_i \geqslant 0$; otherwise, only the information that $T_i < 0$ is available (zero may be replaced by any other constant; note also that the description here involves left censoring). In this setting, Powell estimated the parameters by minimizing

$$\sum |T_i^+ - (a + bx_i)^+|, \tag{12}$$

where, for any number c, $(c)^+ = \max\{c, 0\}$, and gave conditions for consistency and asymptotic normality. Using modern techniques from the theory of empirical distributions, Pollard (1990) rederived the asymptotic normality of Powell's estimator under weaker assumptions. Powell's estimator does not extend to general censoring mechanisms since use is made of the knowledge of the censoring limit. (See Ying et al., 1992.) In environmental data, however, the censoring limit is always known (even in multiply censored data) and thus the extension should be possible.

4.2.4. Theil–Sen regression

With uncensored data, the Theil–Sen estimator (Theil, 1950; Sen, 1968) is easy to compute, has a bounded influence function, and is consistent and asymptotically normal under a wide range of error distributions. It is intuitively defined as the median of the pairwise slopes. It is also obtained by inverting Kendall's tau statistic; i.e., as the value b that makes Kendall's tau statistic between the residuals $T_i - bx_i$ and x_i (approximately) equal to zero. This method was extended to the case of randomly censored response variable by Ireson and Rao (1985), and to the case of doubly censored data by Akritas et al. (1992a). Ireson and Rao provided a permutation variance that requires the (rather unrealistic) assumption that the censoring variable follows a regression model on X with the same slope as the regression model of T on X. In addition to extending the method to doubly censored data, Akritas et al. (1992a) provided an easy estimate of the asymptotic variance that relaxes the unrealistic assumption of Ireson and Rao. To describe this estimator for doubly censored data, we first extend the delta-notation to this case. In particular we let δ_i^x and δ_i^y denote the censoring indicators for the covariate and response variable respectively. Set

$$\begin{aligned} T_n(b) &= \sum_{i<j} [\delta_i^x \delta_j^x (I\{x_i < x_j\} - I\{x_j < x_i\}) \\ &\qquad \times (\delta_i^y I\{r_i(b) < r_j(b)\} - \delta_j^y I\{r_j(b) < r_i(b)\})] \\ &= \sum_{i<j} h_{i,j}(b), \end{aligned} \tag{13}$$

where $r_i(b) = Y_i - bx_i$, and $h_{i,j}$ is defined implicitly from the above equation. This is a natural concordance statistic between the residuals and the covariate in the presence

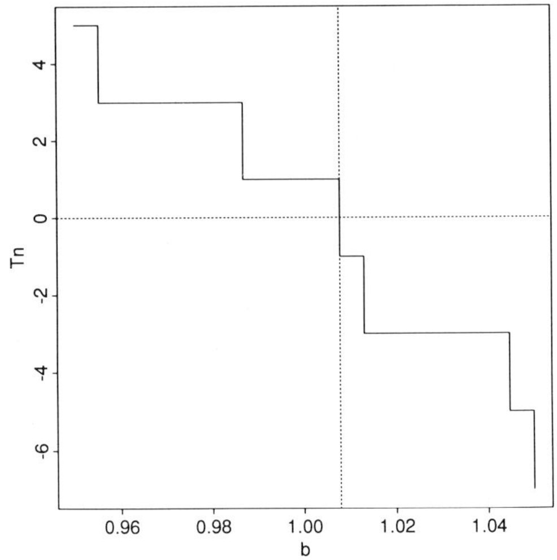

Fig. 3. Form of the function $T_n(b)$ for Theil–Sen regression.

of doubly censored data. $T_n(b)$ is monotone non-increasing in b, so it is rather straightforward to find the point at which it crosses zero.

EXAMPLE 4. The form of the function $T_n(b)$ is not immediately obvious; in fact, it is a step function. Figure 3 shows an example of $T_n(b)$ for the simple case of $Y = X + \varepsilon$, $\varepsilon \sim N(0, 1)$, and $n = 10$. The root of $T_n(b)$ is straightforward in this example, and is found to be 1.008. However, the horizontal part of a step in T_n may coincide with zero; in this case, the root is taken to be the center of the step (see the following definition of the Theil–Sen estimator).

The Theil–Sen estimator, in this context, is defined as

$$\hat{\beta} = (\hat{b}_1 + \hat{b}_2)/2, \tag{14}$$

where $\hat{b}_1 = \sup\{b\colon T_n(b) > 0\}$ and $\hat{b}_2 = \inf\{b\colon T_n(b) < 0\}$. It is asymptotically normal with mean β and variance $4n^{-1}\zeta/\theta$. A consistent estimate of ζ is

$$\hat{\zeta} = (n-1)^{-1} \sum_i \left[(n-1)^{-1} \sum_{j \neq i} h_{i,j}(\hat{\beta}) \right]^2. \tag{15}$$

For each δ in $(0, 0.5)$, there is a consistent estimator of θ given by

$$\hat{\theta}(\delta) = n^{0.5-\delta} \binom{n}{2}^{-1} T_n(n^{-0.5+\delta} + \hat{\beta}). \tag{16}$$

Simulation studies in Akritas et al. (1992a) have shown that for moderate sample sizes ($n = 50$ or 80) the choice of $\delta = 0.01$ is preferable in the sense that the achieved confidence levels are closer to the nominal level. The extension of this method to multiple regression (more than one explanatory variable) is possible. This work is currently in progress at Penn State.

EXAMPLE 5. Consider estimating the location difference between two censored lognormal datasets. The first dataset was presented in Table 1: a 2-parameter lognormal sample of $n = 50$ such that $\log(\vec{X}) \sim N(0, 1)$. The second dataset is another 2-parameter lognormal sample of $n = 50$, such that $\log(\vec{X}) \sim N(3, 1)$ (see Table 5). Censoring both datasets at the 20% level gives a censoring limit of 0.40 for the first dataset and 8.95 for the second.

Applying the Theil–Sen scheme to the 2-sample case requires coding the covariate to be 0 for all data points in the first sample, and 1 for the second sample. The data may be kept in the original units for the analysis. The slope estimator $\hat{\beta}$ then provides an estimate of the location difference between the two samples, corresponding to the median of all pairwise differences. In this example, $\hat{\beta} = 22.77$ and $\hat{\sigma} = 2.79$, using $\delta = 0.01$.

See Akritas et al. (1992a) for examples of Theil–Sen regression applied to singly and doubly censored regression data.

the setting of spatial point processes (where the locations of events are the variables of interest).

Acknowledgement

The authors would like to thank Michael LaValley for a number of helpful discussions, Cory Atwood and N. K. Nagaraj for commenting on an early version of the manuscript.

References

Aitken, M. (1981). A note on the regression analysis of censored data. *Technometrics* **23**, 161–163.

Akritas, M. (1988). Pearson-type goodness-of-fit tests: The univariate case. *J. Amer. Statist. Assoc.* **83**, 222–230.

Akritas, M. (1992). Rank transform statistics with censored data. *Statist. Probab. Lett.* **13**, 209–221.

Akritas, M. and V. Morder (1993). Work in progress.

Akritas, M., S. Murpy, M. LaValley and E. Feigelson (1992a). The Theil–Sen estimator with doubly censored data and applications to astronomy. *J. Amer. Statist. Assoc.*, to appear.

Akritas, M., A. Torbeyns and E. Feigelson (1992b). Pearson-type goodness-of-fit tests for regression. Submitted for publication.

Amemiya, T. (1985). *Advanced Econometrics*. Harvard Univ. Press, Cambridge, MA.

Atwood, C., L. Blackwood, G. Harris and C. Loehr (1991). Recommended methods for statistical analysis of data containing less-than-detectable measurements. Tech. Report prepared by Idaho National Engineering Lab. for U.S. Dept. of Energy.

Buckley, J. and I. James (1979). Linear regression with censored data. *Biometrika* **66**(3), 429–436.

Cohen, A. C. (1959). Simplified estimators for the normal distribution when samples are singly censored or truncated. *Technometrics* **1**(3), 217–237.

Cohen, A. C. (1991). *Truncated and Censored Samples, Theory and Applications*. Marcel Dekker, New York.

Cohen, A. (1976). Progressively censored sampling in the three parameter log-normal distribution. *Technometrics* **18**(1), 99–103.

Cohn, T. A. (1988). Adjusted maximum likelihood of the moments of lognormal populations from Type-I censored samples. Open-file report 88-350, U.S. Geological Survey, Reston, VA.

Cohn, T., E. Gilroy and W. Baier (1992). Estimating fluvial transport of trace constituents using a regression model with data subject to censoring. In: *Proc. Section on Statistics and the Environment*. American Statistical Association, Alexandria, VA.

Cressie, N. (1991). *Statistics for Spatial Data*. Wiley, New York.

Davis, C. (1993). Environmental Regulatory Statistics. In: G. P. Patil and C. R. Rao, eds., *Environmental Statistics*, Handbook of Statistics, Vol. 12. Elsevier, Amsterdam.

Donald, M. (1992). Regression analysis of censored heavy metal concentrations in effluent from four primary sewage treatments: A report of a pilot study at Malabar STP. Presented at the International Environmental Biometrics Conference, Sydney, Australia.

El-Shaarawi, A. (1989). Inferences about the mean from censored water quality data, *Water Resources Res.* **25**(4), 685–690.

El-Shaarawi, A. (1991). Fitting regression models to censored and autocorrelated water quality data. Invited presentation at *3rd Internat. Conf. on Environmetrics*, Madison, WI.

El-Shaarawi, A. (1992). Environmental monitoring, assessment, and prediction of change. Invited presentation at *Internat. Environmental Biometrics Conf.* Sydney, Australia. (To appear in *Environmetrics*, 1993).

Gilbert, R. (1993). An overview of statistical issues related to environmental cleanup. In: G. P. Patil and C. R. Rao, eds., *Environmental Statistics*, Handbook of Statistics, Vol. 12. Elsevier, Amsterdam.

Gilliom, R. and D. Helsel (1986). Estimation of distributional parameters for censored trace level water quality data; 1. Estimation techniques. *Water Resources Res.* **22**(2), 135–146.

Gleit, A. (1985). Estimation for small normal data sets with detection limits. *Environ. Sci. Technol.* **19**(12), 1201–1206.

Hashimoto, L. and R. R. Trussell (1983). Evaluating water quality data near the detection limit. In: *Proc. Amer. Water Works Assoc. Advanced Technology Conf.* Las Vegas, Nev. American Water Works Assoc.

Helsel, D. (1990). Less than obvious: Statistical treatment of data below the detection limit. *Environ. Sci. Technol.* **24**(12), 1767–1774.

Helsel, D. and T. Cohn (1988). Estimation of descriptive statistics for multiply censored water quality data. *Water Resources Res.* **24**(12), 1997–2004.

Helsel, D. and R. Gilliom (1986). Estimation of distributional parameters for censored trace level water quality data; 2. Verification and applications. *Water Resources Res.* **22**(2), 147–155.

Hirsch, R. and J. Stedinger (1987). Plotting positions for historical floods and their precision. *Water Resources Res.* **23**(4), 715–727.

Hodges, J. and E. Lehmann (1963). Estimates of location based on rank tests. *Ann. Math. Statist.* **34**, 598–611.

Hollander, M. and E. Pena (1992). A chi-squared goodness-of-fit test for randomly censored data. *J. Amer. Statist. Assoc.* **87**, 458–463.

Ireson, M. and P. Rao (1985). Interval estimation of slope with right censored data. *Biometrika* **72**, 601–608.

James, I. (1986). On estimating equations with censored data. *Biometrika* **73**, 35–42.

James, I. and P. Smith (1984). Consistency results for linear regression with censored data. *Ann. Statist.* **12**, 590–600.

Jones, M. and Crowley, J. (1989). A general class of nonparametric tests for survival analysis. *Biometrics* **45**, 157–170.

Lai, T. and Z. Ying (1991a). Large sample theory of a modified Buckley–James estimator for regression analysis with censored data. *Ann. Statist.* **19**, 1370–1402.

Lai, T. and Z. Ying (1991b). Rank regression methods for left truncated and right censored data. *Ann. Statist.* **19**, 531–554.

Lambert, D., B. Peterson and I. Terpenning (1991). Nondetects, detection limits, and the probability of detection. *J. Amer. Statist. Assoc.* **86**(414), 266–277.

Lawless, J. (1982). *Statistical Models and Methods for Lifetime Data.* Wiley, New York.

Lebreton, J., K. Burnham, J. Clobert and D. Anderson (1991). Modeling survival and testing biological hypothesis using marked animals: A unified approach with case studies. *Ecological Monographs* **62**, 67–118.

Lee, M. (1991). Robust M-estimation under type I.5 censoring and its application to accelerated life testing. Ph.D. Thesis, Penn State University.

Leurgans, S. (1987). Linear models, random censoring and synthetic data. *Biometrica* **74**, 301–309.

Lin, D. and C. Geyer (1992). Computational methods for semiparametric linear regression with censored data. *J. Comput. Graph. Statist.* **1**, 77–90.

Louis, T. (1981). Non-parametric analysis of an accelerated failure time model. *Biometrics* **68**, 381–390.

Massam, H. (1991). Numerical computation of the maximum likelihood estimates in an AR(p) model. Invited presentation at *3rd Internat. Conf. on Environmetrics*, Madison, WI.

McLeish, D. (1983). A note on nonparametric censored regression. *J. Statist. Comput. Simulation* **18**, 1–6.

Meng, X., Y. Bassiakos and S. Lo (1991). Large-sample properties for a general estimator of the treatment effect in the two-sample problem with right censoring. *Ann. Statist.* **19**, 1786–1812.

Millard, S. and S. Deverel (1988). Nonparametric statistical methods for comparing two sites based on data with multiple nondetect limits. *Water Resources Res.* **24**(12), 2087–2098.

Miller, R. (1981). *Survival Analysis.* Wiley, New York.

Miller, R. and J. Halpern (1982). Regression with censored data. *Biometrika* **69**, 521–531.

Moon, C. (1989). A Monte Carlo comparison of semiparametric Tobit estimators. *J. Appl. Econometrics* **4**, 361–382.

Nagaraj, N. and S. Brunenmeister (1993). A new approach for accommodation of below detection limit data in trend analysis of water quality. In: R. Cothern and N. Ross, eds., *Environmental Statistics, Assessment and Forecasting*. Lewis, Ann Arbor, MI.

Newman, M., P. Dixon, B. Looney and J. Pinder (1989). Estimating mean and variance for environmental samples with below detection limit observations. *Water Resources Bull.* **25**(4), 905–916.

Padgett, W. and L. Wei (1982). Estimation of the ratio of scale parameters in the two sample problem with arbitrary right censorship. *Biometrika* **69**, 252–256.

Peto, R. and J. Peto (1972). Asymptotically efficient rank invariant test procedures (with discussion). *J. Roy. Statist. Soc. Ser. A.* **135**, 185–206.

Pollard, D. (1990). *Empirical Processes: Theory and Applications.* Vol. 2 of NSF-CBMS *Regional Conference Series in Probability and Statistics.* Institute of Mathematical Statistics, Hayward, CA.

Pollock, K. (1991). Modeling capture, recapture, and removal statistics for estimation of demographic parameters for fish and wildlife populations: Past, present, and future. *J. Amer. Statist. Assoc.* **86**, 225–238.

Pollock, K., J. Nicholas, C. Brownie and J. Hines (1990). Statistical inference for capture–recapture experiments. *Wildlife Monographs* **107**, 1–97.

Porter, S., R. Ward and H. Bell (1988). The detection limit. *Environ. Sci. Technol.* **22**(8), 856–861.

Powell, J. (1984). Least absolute diviations estimation for the censored regression model. *J. Econometrics* **25**, 303–325.

Ritov, Y. (1990). Estimation in a linear regression model with censored data. *Ann. Statist.* **18**, 303–328.

Schmee, J. and G. J. Hahn (1979). A simple method for regression analysis with censored data. *Technometrics* **21**, 417–432.

Schmitt, J. (1985). Statistical analysis of astronomical data containing upper bounds: General methods and examples drawn from X-ray astronomy. *Astrophys. J.* **293**, 178–191.

Schneider, H. (1986). *Truncated and Censored Samples from Normal Populations.* Marcel Dekker, New York.

Sen, P. (1968). Estimates of the regression coefficient based on Kendall's tau. *J. Amer. Statist. Assoc.* **63**, 1379–1389.

Shumway, R., A. Azari and P. Johnson (1989). Estimating mean concentrations under transformation for environmental data with detection limits. *Technometrics* **31** (3), 347–355.

Stevens, W. (1937). The truncated normal distribution (Appendix to paper by C. I. Bliss: The calculation of the time mortality curve). *Ann. Appl. Biol.* **24**, 815–852.

Theil, H. (1950). A rank-invariant method of linear and polynomial regression analysis. I, II and III. *Nederl. Akad. Wetensch. Proc.* **53**, 386–392, 521–525 and 1397–1412.

Torbeyns, A. (1992). Pearson-type goodness-of-fit tests for regression. Ph.D. Thesis, Penn State University.

Travis, C. and M. L. Land (1990). Estimating the mean of data sets with nondetectable values. *Environ. Sci. Technol.* **24**, 961–962.

Tsiatis, A. (1990). Estimating regression parameters using linear rank tests for censored data. *Ann. Statist.* **18**, 354–372.

Wei, L. and M. Gail (1983). Nonparametric estimation for a scale change with censored observations. *J. Amer. Statist. Assoc.* **78**, 382–388.

Wei, L., Z. Ying and D. Lin (1990). Linear regression analysis of censored survival data based on rank tests. *Biometrika* **19**, 417–442.

Ying, Z., S. Jung and L. Wei (1992). Survival analysis with median regression models. Preprint.

Zhou, M. (1992). Asymptotic normality of the 'synthetic data' regression estimator for censored data. *Ann. Statist.* **20**, 1002–1021.

Zimmerman, D. (1991). Censored distance-based intensity estimation of spatial point processes. *Biometrika* **78**(2), 287–294.

G. P. Patil and C. R. Rao, eds., *Handbook of Statistics, Vol. 12*

8

Biological Monitoring: Statistical Issues and Models

Eric P. Smith

Abstract: Biological monitoring refers to the collection of data on biological organisms to establish that explicit quality criteria are being met. Biological monitoring is used to provide early warning of deleterious effects, detect episodic events, detect trends or change, evaluate information redundancy and assess effects of biological and physical stressors on biological communities. Statistical methodology and models are important in monitoring of biological systems. Some of the statistical methodologies used to aid biological monitoring programs are discussed. Models that are used to evaluate change in biological systems are described. Further needs and directions in research in biomonitoring are addressed.

1. Introduction

The term *biological monitoring* has been widely used to include almost any type of data involving biological organisms or measurements on biological organisms gathered to assess environmental impact of discharges and the biological health of the environment. In a strict sense, the term biological monitoring should be limited to a continuing collection of data to establish whether explicitly stated quality control conditions are being met (Moriarty, 1983). Researchers distinguish between biological monitoring, biological surveys and biological surveillance. A survey typically involves a single time point while surveillance implies repeated surveys. The focus of this paper is more on the stricter definition of biological monitoring, although the distinctions with regard to statistical methodology are not great.

Statistical methodology for biological monitoring depends on the purpose of biological monitoring, the type of data collected and the temporal and spatial scales that are involved in the monitoring. Statistics must play a more important role in biological monitoring because of the need to make explicit statements of confidence in the biological monitoring results. With appropriate statistical evaluation of the data, professional judgement on whether to initiate immediate action or wait for more confirming data will be more objective and reliable.

The purposes of biological monitoring and related statistical methodology are discussed. As many of the statistical tests and methods are general, specific tests will not be discussed here, rather the focus is to review methodologies that are applicable. However, some methodology that is more specifically designed to address biological monitoring problems will be described in greater detail. Also some trends in approaches to monitoring and analysis are mentioned. Finally, some future needs are discussed.

2. Purposes of biological monitoring

Biological monitoring plans have several possible purposes:

(1) To provide an early warning of a violation of quality control systems in time to avoid deleterious effects to ecosystems.

(2) To detect episodic events such as accidental spills, failure of predictive models, failure of early warning systems, or illegal disposal of wastes at night, etc.

(3) To detect trends or cycles.

(4) To determine information redundancy.

(5) To evaluate environmental effects associated with the introduction of genetically engineered organisms into natural systems.

These purposes and all statistical considerations entering into each type of application are discussed next.

2.1. Monitoring to provide early warning

The purpose of early warning systems is to alert decision makers to the impending entry of deleterious materials into natural ecosystems or impending harm to the biota before it occurs. These studies are characterized by having a known source of pollutant and the health of biological organisms are used to signal if a pollutant is present. A compilation of the literature on the older early warning systems may be found in Cairns and Van der Schalie (1980). A more recent overview of early warning systems is found in Gruber et al. (1991). A simple approach is to place a fish or other organism in a cage in the receiving water system and to monitor the organism in much the same manner that an industrial process is monitored for quality. The response in the fish might be the respiration rate. An example of this approach and the analysis of data is given in Thompson et al. (1982). Another approach is based on the movement of fish in a tank. The response variables are the location of the fish in the tank and the number of movements. As pollution stress increases fish may become hyperactive or may be inactive. Hall et al. (1975) describe a statistical approach to the analysis of data from such an experiment.

Guidance for the statistical analysis of early warning systems can be found in the statistical quality control literature. One approach is to view the system as stable, then some intervention causes a change in the system. The purpose of monitoring is to detect the change as quickly as possible and this type of monitoring is sometimes referred to as monitoring for complicance. Specifically, there is some pre-set standard that the end point is not to exceed. A variety of methods may be used in this context, including methods from quality control (Schaeffer et al., 1980; Loftis et al., 1987; Bailar, 1987). The system is conceptualized as a process that is in 'control', and a standard value is used to judge deviations from control. Then graphical and testing methods can be used to assess the system. One difficulty in using this approach with biological data is that standards for ecological systems are difficult to set, may be multivariate and may depend on the nature of the ecological insult.

Statistical methods can also aid in designing the monitoring study (Prince, 1987). The most commonly used approach is based on a fixed sampling interval and some measure of the stability of the system. For example, in a system using a fish species, the respiration of the fish might be an indication of overall quality. Standard quality control methods are used to assess compliance. A variation on this approach is to use a variable sampling interval (Arnold, 1970). Some recent work (Reynolds et al., 1988) indicates that the variable sampling approach can be improved by using two sets of intervals and two sampling rates. For example, when the fish respiration or other end point is close to the norm, sampling should be done infrequently. When the fish respiration is farther away from the norm, but is not great enough to signal trouble, sampling should be done more frequently. A flexible system, in which changes to what is sampled as well as when it is sampled can be made, may be the most cost-effective approach for detection of change. Several levels of testing at differing frequencies can be used in a comprehensive monitoring system. These techniques illustrate the importance of feedback control as a characteristic of effective biological monitoring systems. Analysis of previous data determines what is done next in an ongoing process. The information gathered is used to make decisions and not just archived.

Biological early warning systems are difficult to design in situations where the polluter or pollutant are unknown. Setting a standard (for example, drinking water standards have maximum contaminant levels) with chemical data is often easy but for ecological data are more difficult. Biological indices of environmental health will fluctuate naturally in time and space: over season, over succession, over substrate, over ecoregion. While it is possible to allow a standard to fluctuate, the variance in the system may be quite large, and the test for a violation may be weak. Thus far, simple measures that are accepted as related to overall system health, such as BOD or DO, have been used most frequently. These measures may be useful for signaling gross effects (such as fish kills) but may not be as useful as indicators of subtle deteriorations. It may be difficult to design a system of end points that are high in ecological complexity yet also capable of providing early warnings. Appropriate choice of end points is essential but is subject to a great deal of controversy (see Suter, 1989; Plafkin et al., 1989). In general, as the hierarchical complexity of the system increases and the physical scale of the end point increases, the ability to detect the change of interest will decrease. The time scale for fish community changes is considerably larger than that for diatoms. While diatoms may be useful for early warning systems, they may not be a good when interest is in long-term changes or changes in diet of fish or energy flows.

Three issues must be resolved in the selection of biological early warning systems. The first consideration is the question of detection. A good system needs to be cost-effective and detect the change rapidly so that corrective action can be taken. Although field data are usually the most informative and relevant index of environmental effects, experience suggests that ecosystem studies tend to have higher costs and variability than toxicity tests. However, Karr (1991) indicates that the index of biotic integrity (IBI) can be cost competitive with toxicity testing. End points that are sensitive, inexpensive to measure, and have lower variance can be identified through continued evaluation and standardization. The second issue is the question of type of system. Researchers are divided over single species versus multispecies systems. Single species systems are more common (early versions of this were the canary in the mine and the king's wine taster), but recent research has focused on more complex systems such as microcosms and mesocosms because of increased realism. The argument against single species tests (e.g., Cairns et al., 1970), is that the species chosen might well be either more sensitive or more tolerant to a particular chemical than the organisms in the receiving system and hence only a limited set of species is protected. Cairns (1986) provides some additional concerns.

The third issue is whether the decision is correct, which is the problem of validity. Single species tests signal quickly and efficiently but have high error rates when used as the sole indicators of ecosystem health. It is difficult to find a single measure of ecosystem health. Measures that combine several end points may be more accurate, but these indices tend to be subjective and no combination of variables enjoys wide support among investigators (Suter, 1989; Plafkin et al., 1989). Very little work has been done on the validation of these systems in terms of error rates.

2.2. Monitoring to detect episodic events

Episodic events such as an oil spill, accidental or intentional dumping occur with some frequency. For assessing the effects of these events it is important to have baseline data. The Valdez oil spill in Alaska provides a good example of the value of baseline biomonitoring data. Here, there was a major ecosystem insult for which there was no explicit and detailed data on the structure or function of the Prince William Sound ecosystem. The existence of a pre-accident data base would give a better indication of the effects and possible aid in managing the recovery of the system. Emphasis without such data usually results in a clean-up effort rather than a restoration effort.

2.3. Monitoring to detect trends or cycles

The original role of biological monitoring was to detect effects from single point source discharges. More recent focus is on larger areas such as drainage basins, regions and even at a global level. Methods for assessing trend and change in a basin are similar to those for smaller areas but methods of assessing change in regional and global monitoring studies are still being developed. Besides the usual statistical question of sample size for these studies, concern must also focus on the scale (or size of sampling unit) needed to address the change and the correlation between units (Jeffers, 1989). Scales that are too small may be too erratic, while a larger scale may not exhibit enough variation to detect the change (see for example, Carlile et al., 1989). Legendre et al. (1989) indicate that a two-year program to collect baseline data is needed to address the question of scale for the important processes in a bay. Larger scales require longer times to establish baselines. The problem of correlation is also important. Studies in space and time typically do not yield independent observations.

2.4. Monitoring to determine information redundancy

An important issue in monitoring is what to measure and what to use as the 'endpoint' or summary of biological information. As biological and chemical monitoring is costly, attention is required to the level of biological organization (population, community or ecosystem) and what to use to summarize the biological information. However, in the initial stages of the development of a biological monitoring network, it is important to determine which types of data are most informative under each set of circumstances. Several benefits can be derived from this exercise.

(1) The most informative and important end points can be selected from among an array of attributes. These attributes will have a high correspondence in response with other crucial attributes.

(2) Attributes that are unique and have low redundancy with other attributes can be determined and their significance to the continuation of data gathering can be evaluated.

(3) Consistency of trends from multiple lines of evidence can be determined in order to increase the confidence in the implications of the changes being measured.

The fact that certain types of information are redundant in a particular ecosystem does not mean that they will be so in every ecosystem. As a consequence, for the widespread use of biological monitoring, the types of information that are redundant must be determined and then this correspondence must be validated in a variety of ecosystems. Out of this analysis should emerge the attributes of ecosystems that are worth measuring on a global basis because of their high correspondence, or even because they are unique representations of particular ecosystem response. Estimation of redundancy must be done carefully because the information that appears redundant for short-term monitoring may not be so for long-term monitoring to establish trends, cycles, and the like. Nevertheless, information redundancy analysis provides the surest way of reducing the volume of monitoring data generated while at the same time ensuring that the quality remains high.

Finally, uncertainty is reduced when there is high information correspondence from a diversity of methodologies and attributes. Therefore, even when redundancy is established, it may be well to continue gathering multiple lines of corresponding evidence, especially during the early stages of a monitoring program. An interesting statistical problem in this area is the problem of combining information from different sources. One approach is meta-analysis, in which a variety of tests (usually independent) are combined to provide a single test statistic for assessing an effect. This approach has been useful in combining medical studies where sample sizes are not adequate but has not received much attention in ecological studies (but see Neuhold, 1987 and Suter et al., 1983). A recent paper by Mann (1990) provides a good introduction and some examples of meta-analysis (see also Chapter 20 in this volume).

2.5. Monitoring to assess effects of genetically engineered organisms

All the forms of monitoring previously discussed are appropriate when genetically engineered organisms (GEMs) are introduced, either deliberately or accidentally, into a natural system. Any ecological disequilibrium these organisms might cause could presumably be determined in the ways previously described. Additionally, however, the persistence of these organisms should also be monitored, especially if they are microorganisms. Methodology for determining the presence of GEMs in extremely low numbers are available (e.g., Scanferlato et al., 1990). Determination of die-off curves in the natural systems should be compared to the estimate of die-off measured under laboratory conditions. These same techniques can be used to monitor the introduction of a naturally occurring species exotic to a particular ecosystem. For example, monitoring microorganisms in bioremediation of an oil spill can determine whether the microbial community returns to normal or whether the introduced organisms have displaced organisms naturally occurring in that particular ecosystem. The transport of GEMs or exotic organisms from the ecosystem where they were accidentally or deliberately introduced into ecologically different ecosystems adjacent to the area of introduction could also be monitored.

A statistical problem that occurs in the monitoring of GEMs is the issue of rareness. For the community to return to normal, is it required to have all the genetically altered organisms below a certain density, or completely eliminated from the system? Claiming that no altered organisms are left is difficult using sampled data. The problem is one of sampling rare organisms, which is a rather difficult problem (see for example, Kalton and Anderson, 1986). An additional complication occurs as genetic material may have been transferred so that the genome persists even though the monitored organism does not.

3. Statistical models and the detection of change

A variety of statistical techniques can guide the design of monitoring and analysis of data from biological monitoring programs. Some specific considerations have been discussed above; however, many methods are applicable to biological monitoring programs for more than one purpose. While many of the methods used in the analysis of biological data are common to applied statistics, some methods are more commonly used because of the special nature of biological data. Several of the models in use in biomonitoring are described below.

3.1. Models of change

Episodic events are commonly associated with a type of trend called the step change or trend. The step change model occurs when data are collected before and after the start of the impact (or in some cases restoration) process. Thus, the data prior to the change are used to set the standard for the post change data. The step change is often thought of in terms of the hypothesis that the mean (or median or variance) of the measurements in the before period is the same as the mean (or variance) in the after period. Many statistical methods may be used to test for a step change (e.g., Green, 1979; Gilbert, 1987; Hirsch, 1988).

Models for assessing change in biological communities are usually based on the ANOVA approach. While early researchers used simple models that were not appropriate, recent methods place more importance on control information and the design of the study. Eberhardt and colleagues (Eberhardt, 1976a, b; Gore et al., 1977; Thomas et al., 1978; Skalski and McKenzie, 1982) suggest a variety of models for analyzing impact assessment data. For example, Thomas et al. (1978) used an ANOVA model to analyze the effect of the thermal plume of a nuclear plant on the ecosystem. Factors included plant status (preoperational, operational), year within status (three years prior, five years post), sites (upstream control, exposed, downstream control), station within sites and substrate type. This model is quite complex as some factors are nested and the design is unbalanced.

An additional model discussed by Thomas et al. (1978) involves pairing of sites. Sampling sites are selected in unimpacted sites to serve as controls. These sites are matched to the habitat and ecological characteristics of the sampling stations in the impact site. Data are collected in the different periods associated with impact (this may include pre-stress, pre-construction, pre-operational as well as the period of potential impact). The ANOVA model is similar to the previous model except that the dependent variable is difference between the control and impact site pair (possibly following transformation).

A variation on this model is the BACI (before–after control impact) test (Stewart-Oaten et al., 1986). Two or more sites are needed to use the BACI method, and measurements must be taken on each site, before and after the plant has started. Measurements are typically taken over time. The model implied by the BACI design is

$$Y_{ijk} = \mu + \alpha_i + \tau_{j(i)} + \beta_k + \varepsilon_{ijk}, \tag{3.1}$$

where m is the overall mean, α_i is the effect of plant status (i = before or after), $\tau_{j(i)}$ is the effect of time of sampling ($j = 1, 2, \ldots, n$ for $i = 1$ and $j = 1, 2, \ldots, m$ for $i = 2$), β_k is the effect of site (k = control or impact) and ε_{ijk} is the error term. Rather than using an ANOVA approach for the analysis, Stewart-Oaten et al. (1986) suggest taking differences between the control and impact sites for each sampling time. The test that the event has no impact on the system is obtained by testing the hypothesis that the mean of the before differences is the same as the mean of the after differences. If a t-test is used, the test is equivalent to the F-test for interaction in the ANOVA model. Nonparametric methods can also be used to carry out the test (see Stewart-Oaten et al., 1986). One problem with this model and all models for impact assessment is the potential limitation in interpretation resulting from a lack of the linkage between the design and analysis. Problems due to poor design, such as confounding and lack of adequate controls lead to difficulties in interpreting results of models.

Eberhardt and colleagues (Eberhardt, 1976a,b; Thomas et al., 1978 and Eberhardt and Thomas, 1991) provide some of the best guidance on the design of impact assessment studies. These papers describe a number of approaches to assess impact and a number of the problems that might be encountered in designing and implementing a biological impact assessment. For example, one of the difficult problems is the choice of adequate sample sizes to detect biologically important changes (Millard and Lettenmaier, 1984). Another problem is the lack of adequate controls; for example, the upstream site may be primarily sand, while the downstream is primarily silt. Under these conditions, some statistical methods may lead to statistically significant results that are biologically irrelevant (see also Hurlbert, 1984; Green, 1984, 1987). Also, statistical models may necessarily be relatively complex to account for all the sources of variation. This statistical complexity may lead to problems in the analysis and interpretation of the data. Some other difficulties are described in Smith et al. (1993) and Finney (1978).

An approach to assessing change that does not seem successful in biological monitoring studies is intervention analysis. (However, some successful examples are given in Musters et al., 1988 and Thompson et al., 1982.) The lack of use of this model results from the requirements of the model. First is the length of the series, which needs to be long enough to characterize the time-series and estimate parameters. For hydrologic variables, such as flow, this may not be a problem as they may be inexpensive to measure and long series can be obtained. However, other variables, such as the abundance of a species, are more expensive to measure and cannot be measured as frequently. Second, it is preferable to have evenly spaced measurements. While unevenly spaced measurements can be analyzed, the approach is not as simple. Biological data are again problematic to work with because sampling is often (and should be) done in an irregular manner. For example, why sample aquatic macrophytes in the winter if they are not in the streams? Wetlands should be sampled more frequently during important hydrological events as water levels are important in determining ecological events.

Another possible drawback to the use of the intervention model is the lack of power. As pointed out by Woodfield (1987) and Lettenmaier et al. (1978), the power of the intervention test for a step change is typically controlled by the number of data points prior to the intervention. This number is fixed. Although the number of points after the intervention can be increased, the increase has very little effect on the power (once the number of points after is larger than that in the before period). Thus, unless the number of pre-intervention points is large, the power may be small. If, however, the effect of the intervention is a shift followed by a decay, then the emphasis should be placed on data collection in the post-intervention period. Thus, different types of interventions lead to different approaches to biomonitoring. This problem may be serious, especially when funds are limited and the expected change is unknown.

Some other discussions on the design of time-series studies are given in Smith (1984) and the references therein.

A recent variation of the intervention approach (Carpenter et al., 1989) uses a randomization approach to test for changes in manipulated lakes. The approach is similar to a BACI model in which one lake is used as a reference lake to be compared with a manipulated lake. Differences are formed, the effect of the manipulation is summarized using the mean difference before minus the mean difference after manipulation and the differences permuted. The permutation distribution is then used to assess if the manipulation affects the lake. This approach has the same limitations of the BACI model (Smith et al., 1993; Stewart-Oaten et al., 1992).

Another application of the permutation distribution to assess ecological change due to pollutants is based on distance measures (Van Belle and Fisher, 1977; Smith and Conquest, 1981). Data on species abundance were collected for a number of stations in Lake Michigan at two times and interest focused on if there was an association between a city on the lake and the abundance of the species. Van Belle and Fisher suggested using the statistic

$$X = \sum_{j=1}^{L} d_j f(N_j), \tag{3.2}$$

where d_j is the distance from station j to the suspected source of the pollution, $f(N_j)$ is the distance between the vectors of ranks of the species frequency lists for the two times, and L is the number of stations. The statistic measures the relationship between change in the abundance of species and the distance to the source of pollution. The statistic is evaluated using a permutation test or an approximation. Some further tests of this type are given in Smith et al. (1990).

3.2. Models for trend assessment

Another common approach to assessing change is to test for a trend. Trend assessment has been frequently reported in the environmental literature (Hirsch and Slack, 1984; Hirsch et al., 1982). The methods frequently assume that the trend is linear, although nonlinear trends can also be assessed. Other related models for analysis of change, such as analysis of covariance, can also be useful (Mathur et al., 1980). Trends can also be incorporated into intervention analysis as described above. A commonly used approach by water quality scientists for assessing trend which is resistant to many of the problems with environmental measurements is based on the Mann–Kendall trend statistic (Gilbert, 1987). If observations $x_1, x_2, \ldots, x_n$ are collected at n time points, then the Mann–Kendall statistics is given by

$$k = \sum_{i=1}^{n-1} \sum_{j=i+1}^{n} \operatorname{sign}(x_j - x_i), \tag{3.3}$$

where the sign function assigns a $+1$, -1 or zero depending on whether the difference is positive, negative or zero respectively. Tables for the statistic are given in the book by Gilbert (1987) for use with smaller sample sizes. With larger samples sizes, inference is based on the approximately normal statistic

$$z = \frac{k}{\operatorname{var}(k)^{1/2}}. \tag{3.4}$$

Schmitt et al. (1990) used this method to test for trends in organochlorine chemicals in freshwater fish in the United States.

Extensions of the method are available to deal with problems and questions due to multiple stations or times. For example, if there are multiple measurements taken at p seasons over the year, then correlation between adjacent seasons may be of concern. One approach is to view the seasons as variables and compute a Mann–Kendall statistic for each season (denote these by k_i, $i = 1, 2, \ldots, n$). The seasons can be tested individually for trend or the information combined to obtain an overall test. Three statistics have been used to test for trend. The first test is often referred to as the 'covariance inversion' test

$$\chi^2_{\text{ci}} = \boldsymbol{k}'S^{-1}\boldsymbol{k}, \tag{3.5}$$

where $\boldsymbol{k}' = (k_1, k_2, \ldots, k_p)$ is the vector of Mann–Kendall statistics and S the estimated covariance matrix. The statistic is asymptotically chi-square with p degrees of freedom. The accuracy of the chi-square approximation has been criticized (Loftis et al., 1991) and two alternatives were suggested. Hirsch and Slack (1984) noted that since the trend should be similar in each season, the statistics could be summed. Thus they recommend using

$$\chi^2_{\text{cs}} = (\mathbf{1}'\boldsymbol{k})'(\mathbf{1}'S\mathbf{1})^{-1}(\mathbf{1}'\boldsymbol{k}), \tag{3.6}$$

where $\mathbf{1}$ is a vector of ones of length p. This statistic is approximately chi-square with one degree of freedom. This test is sometimes referred to as the 'covariance sum' test. The test will have low power when the trends are not of the same sign. Because of this possibility and to extend the statistic to the more general multivariate case, Lettenmaier (1988) suggests using

$$\chi^2_{\text{cc}} = \boldsymbol{k}'\boldsymbol{k}. \tag{3.7}$$

To approximate significance values for this statistic, Lettenmaier suggests computing the moments and using a noncentral chi-square or three parameter gamma distribution. As this approximation involves computing eigenvalues of S, the method is referred to as the 'covariance eigenvalue' method.

A further extension of the methodology is based on decomposing the chi-square statistics into homogeneity components. For example if the data are collected for four seasons (i.e., $p = 4$) then create a matrix

$$C = \begin{bmatrix} 1 & 1 & 1 & 1 \\ 1 & -1 & 0 & 0 \\ 0 & 1 & -1 & 0 \\ 0 & 0 & 1 & -1 \end{bmatrix}.$$

The matrix C can be divided into $C = (C_1 \quad C_2)$ where C_1 is a $(1 \times p)$ matrix of ones and

$$C_2 = \begin{bmatrix} 1 & -1 & 0 \\ 0 & 1 & 0 \\ 0 & 0 & -1 \end{bmatrix}.$$

Then

$$\chi^2_{ci} = (C\boldsymbol{k})'(CSC')^{-1}(C\boldsymbol{k}) = \chi^2_{c1} + \chi^2_{c2}, \tag{3.8}$$

where

$$\chi^2_{cj} = (C_j\boldsymbol{k})'(C_jSC_j')^{-1}(C_j\,\boldsymbol{k})$$

is asymptotically chi-square distribution with 1 degree of freedom for C_1 and $p-1$ degrees of freedom for C_2. These types of decompositions are discussed in Van Belle and Hughes (1984).

A different type of trend is encountered in problems of restoration and the biomonitoring of these systems. Systems that are created or restored often undergo successional change. In some cases, the successional change may be toward a climax state. If the change toward climax is rapid, the climax state can be compared with stable, control systems using standard statistical approaches. In many cases, the system does not reach a climax and the evaluation process is more difficult. Loehle and Smith (1990) describe some methods that might be useful for evaluation of succession in this case.

The approaches discussed above imply a movement toward a goal. In many systems, the interest is more on obtaining a system that can persist through time. Natural wetlands are often in a state of flux. The state of the system depends strongly on external factors, primarily hydrologic. For such wetlands, stability is not the same as that for terrestrial systems but is a 'pulsed stability' (Odum, 1971). What is more important than reaching a climax state in a successional process is the ability for the system to respond to fluctuating environmental conditions and persist. Thus, the goal of interest is not to duplicate another system but to make the created wetland a persistent system (Carter, 1986; Niering, 1989).

There are two problems with applying statistical tests to the restoration problem. First is the philosophical problem that the tests that are used were designed for experimental situations and do not apply to data collected on recovery. This problem is referred to in the ecological literature as the pseudoreplication problem (Hurlbert, 1984). In typical experimental design, a researcher has control over the way the treatments are allocated to the experimental units and the way that the experimental units are selected. In restoration problems, the researcher does not have this control. The treatment is the restoration process. There is typically one control site and one restoration or treatment site. One of Hurlbert's criticisms of this approach is that if more and more samples are taken in each site, then eventually differences will emerge because no two sites are identical. Use of multiple control sites is not feasible because of expense, and it is foolish to spend money on the controls when the recovering site is of much more interest. Replication when costs are high may cause reduction in frequency of sampling and a weakening of effort. Thus, the control site may be statistically different from the recovery site, but the difference may not be scientifically important. The limitations of the statistical method in this approach must be recognized.

Another problem with the testing approach is that the methods do not test for recovery. Testing for recovery requires a specific definition of recovery. As recovery is often dynamic and recovered systems are not fixed systems, it may be difficult to specify recovery in terms of the value of a single parameter or to define recovery simply. When this is the case, expert judgement is of great value. In much the same manner that a medical specialist reviews tests made on a subject and combines the results with his experience, so would a restoration expert view the results of the statistical analysis and interpret results based on the tests and his interpretation of the system.

3.3. Indices of change and quality

An important aspect of biomonitoring is the measurement of changes that may occur and assessing the biological quality of the system. Data that is typically collected in a biomonitoring study is either in the form of presence and absence of species or the abundance of species (although rank abundance is sometimes used). Summary measures of quality are then functions of the data. Commonly used measures fall into two general categories depending on whether the data is weighted or not. The unweighted summary measures include the number of species, diversity measures and similarity measures. There are a large number of these measures (see Fausch et al., 1990, for example); the most commonly used are the number of species in the sample (S) and the Shannon diversity measure

$$H = \sum_{i=1}^{S} p_i \log(p_i), \tag{3.9}$$

where p_i is the proportion of species i in the sample. The statistical properties of these types of measures are well studied under a variety of conditions. For example, some properties of the estimate of the number of species assuming a random distribution for individuals and random sampling are discussed in Heck et al. (1975) and under quadrat sampling in Smith and Van Belle (1984) and in Smith et al. (1985). Similar results are available for other diversity measures (Zahl, 1977; Smith and Grassle, 1977 and many others). An interesting approach to assessment of diversity is Rao's (1986) ANOVA-like analysis of diversity in which diversity is decomposed in components. This has yet to be applied in environmental studies.

The other type of measure of quality may be viewed as a weighted measure. These include measures based on indicator organisms, combinations of measures and indices of integrity. For example, instead of using the Shannon–Weaver measure, a weighted measure

$$H = \sum_{i=1}^{s} w_i p_i \log(w_i p_i) \tag{3.10}$$

would be used. Here w_i is a weight based on tolerance to pollutants. Measures based on indicator species are also examples. The trend in biomonitoring is to use these weighted measures. Some recent popular measures are those based on the concept of biological integrity (Fausch et al., 1990). In this approach, streams or rivers that are unpolluted are assessed. Then categories of species are developed based on ecological knowledge. For example, for an index involving fish, there may be a categories for trout abundance, the number of individuals and the proportion of individuals with evidence of disease. Sites are scored based on ecological expectation for each category (on a five point scale) and the scores summed (over the categories) to form a measure. The final score is then used to classify the site in terms of biological integrity. Indices are typically based on fish or benthic invertebrates (Karr, 1991; Plafkin et al., 1989). The role of sampling variability and the statistical properties of these measures has not been thoroughly investigated.

3.4. Using organisms to predict pollution

Although the above measures can be used in a monitoring program to measure the biological health of a site, the link to a pollutant is not a direct one. There is a good deal of research that has focused on using biological data to predict the level of a pollutant. This is the calibration problem in statistics but is different from the usual application because the species data is in the form of counts or is binary. The primary application of the methodology has been to predict the pH of lakes over time based on the fossil history of organisms (Charles, 1984). The model is based on the idea of reciprocal averaging and has led to the development of canonical correspondence analysis (Ter Braak, 1986). Canonical correspondence analysis is similar to canonical correlation analysis but requires adjustment due to the nature of species abundance data. Specifically, there are usually a large number of species (variables) and the data are in the form of counts. Additionally the response of species to environmental variables is usually assumed to be nonlinear (Jongman et al., 1987). Ecological evidence indicates that a species will be intolerant to low and high levels of an environmental variable and have an optimum survival for a level in a middle area. For example, animals cannot survive in low or high pH. Survival depends on the level of pH and there is a level that produces optimal survival (unimodal assumption). The optimum level is assumed to be different for different species. The usual model is a Gaussian model for each species.

Canonical correspondence analysis relates the species data to the environmental vairables. Let Y be the $m \times n$ matrix of data on m species collected at n sites and Z be the $(p+1) \times n$ matrix of p environmental variables collected at the sites, augmented by a column of ones. Define $M = \text{diag}(y_{k+})$ and $N = \text{diag}(y_{+j})$. Canonical correspondence analysis maximizes the equation

$$\delta = \boldsymbol{x}'Y'M^{-1}\boldsymbol{x}/\boldsymbol{x}'N\boldsymbol{x}, \tag{3.11}$$

where $\boldsymbol{x}$ is the $(n \times 1)$ vector of scores for each site. The score for site i depends on the environmental variables and is given by $x_i = \boldsymbol{c}'\boldsymbol{z}_i$ where $\boldsymbol{c}$ is a vector of regression coefficients. Further, $\boldsymbol{x}$ is constrained so that $\mathbf{1}'N\boldsymbol{x} = 0$. The above model can be written in the form of canonical correlation analysis and results in an eigenanalysis. The result is an equation based on the eigenvector that is used to predict a site score based on the biological data and hence a prediction of pollution. Additional details on the methodology and extensions are given in Jongman et al. (1987) and Ter Braak (1988). The methodology provides a powerful method for relating the species in a system to the environmental conditions. A program for this analysis is given in Ter Braak (1987).

3.5. Encountered data and sampling bias

Sampling of biological organisms requires a sampling device. Researchers are aware of biases associated with sampling devices but little attention is given to adjusting for the sampling bias. Recent work by Patil and Rao (1978) provide a statistical foundation for adjusting for bias. Suppose X is the abundance of a species with distribution $f(x;\theta)$ where θ is the natural parameter. Then assume there is a weight or recording function $w(x,\beta)$ that is associated with the sampling device. The observation that is recorded is not from the distribution f, but can be viewed as having come from f after alteration by the weight function. Patil and Rao (1978) show the observations can be modeled using a random variable X^w with distribution function

$$f^w(x;\theta,\beta) = \frac{w(x;\beta)f(x;\theta)}{\omega}, \tag{3.12}$$

where ω is a normalizing factor. A common model would be size biased sampling, in which $w(x,\beta) = x$ and $\omega = \text{E}(x)$. This model has many applications in biomonitoring but has not been applied. It may be a useful model for example, for comparing data in which different sampling gear is used or for comparing data from different stream orders.

4. Some statistical issues and directions in biomonitoring

Environmental data are often complex and techniques may not be available to address the problems inherent in the data. Some statistical problems related to biological monitoring are described below. These problems are viewed not only as areas of concern but also as areas for research.

Goals of monitoring must be clearly defined because any monitoring program has limitations as to the questions that can be answered. If the researchers have clear ideas as to why the monitoring program is needed, then a very powerful design of the monitoring program is possible. Powerful as used here means a minimum cost design is available that will answer the questions of interest with high reliability. Often the goals of the research change or the data collected are inadequate to answer the necessary questions. There may be little or no data on specific chemicals, important species, or important environmental processes for a region or time period during which important events are known to have occurred. Changes in the monitoring program will often compromise the effort. The best intentions often run afoul, and programs that are set up with adequate funding may in the future be a victim of a funding cut. The result is that the sampling plan changes, and, hence, the questions that can be answered change.

A related problem is one of compatibility of data. When data are collected over time, there are possible effects due to changing the method of collection or changing the researchers. For example, switching from collecting fish in nets to collecting fish using electrofishing can create considerable problems in assessing change. Changes in the detection levels can have an effect on the inference for chemicals. Improved accuracy of chemical measurements means that the variance of the measures is decreasing (which conflicts with the usual assumption of homogeneity of variance).

Multiple interventions are common in the real world, yet more difficult to assess. The simplest view for assessing change is that the change is caused by a single source, but many environmental effects are due to multiple sources. When the magnitude of the sources is small, it is not difficult to assign a single source as the cause, and these interventions sometimes can be dealt with using special programs. For example, if interest is in the effect of a manufacturing plant on the biota and a sewage treatment facility is also added in the vicinity of the plant, a design to separate the potential effects of the two potential impacts is needed.

Over the last 30 years, a number of environmental regulations have been issued. Environmental data on large scales should reflect the effects of the programs put into effect to meet the regulatory challenges. Current statistical methods do not accommodate these types of interventions. The biological aspects of multiple stresses is only recently receiving attention (Cairns and Niederlehner, 1989). It thus becomes difficult to assess the effects of policies and regulations. There is also some feeling amongst ecologists that statistics is developed to address simple agricultural experiments and methods to assess questions in monitoring studies are not available (Strayer et al., 1986). This problem is more apparent in large scale studies (either long term studies or whole lake experiments) with multiple objectives and measurements.

5. Summary

The term *biological monitoring* has been widely used in this discussion to include almost any type of data gathered to assess environmental impact of discharges. In a strict sense, biological monitoring is limited to a continuing collection of data to establish whether explicitly stated quality control conditions are being met. If these conditions are not being met, there will be an immediate decision to take corrective action. Purposes of biological monitoring include providing early warnings of hazards, detecting spills, detecting environmental trends or cycles, determining the best and least redundant information for monitoring, and evaluating the environmental effects associated with the introduction of genetically engineered organisms into natural systems. The variety of purpose means that there is no best design for biological monitoring, but, if the researchers have clearly defined goals for the monitoring program, powerful designs are possible. A variety of methods are available for analyzing data from monitoring programs. The choice of the analysis depends on the objectives and design of the program. Several new approaches to analysis of biomonitoring data are described and some statistical challenges discussed.

References

Arnold, J. C. (1970). A Markovian sampling strategy applied to quality monitoring of streams. *Biometrics* **26**, 739–747.

Bailar, J. C., III (1987). *ASA/EPA Conference on Compliance Sampling*, EPA-230-03-047. U.S. Environmental Protection Agency, Washington, DC.

Cairns, J., Jr. (1986). The myth of the most sensitive species. *BioScience* **36**(10), 670–672.

Cairns, J., Jr. and B. R. Niederlehner (1989). Adaptation and resistance of ecosystems to stress: A major knowledge gap in understanding anthropogenic perturbations. *Speculations Sci. Technol.* **12**, 23–40.

Cairns, J., Jr. and W. H. Van der Schalie (1980). Biological monitoring. Part I: Early warning systems. *Water Res.* **14**, 1179–1196.

Cairns, J., Jr., K. L. Dickson, R. E. Sparks and W. T. Waller (1970). A preliminary report on rapid biological information systems for water pollution control. *J. Water Pollution Control Feder.* **42**, 685–703.

Carlile, D. W., J. R. Skalski, J. E. Baker, J. M. Thomas and V. I. Cullinan (1989). Determination of ecological scale. *Landscape Ecol.* **2**, 203–213.

Carpenter, S. R., T. M. Frost, D. Heisey and T. K. Kratz (1989). Randomized intervention analysis and the interpretation of whole ecosystem experiments. *Ecology* **70**, 1142–1151.

Carter, V. (1986). An overview of the hydrologic concerns related to wetlands in the United States. *Canad. J. Botany* **64**, 364–374.

Charles, D. F. (1984). Recent pH history of Big Moose Lake (Adirondack Mountains, New York, U.S.A.) inferred from sediment diatom assemblages. *Ver. Theor. Angew. Limnol. Verh.* **22**, 559–566.

Eberhardt, L. L. (1976a). Some quantitative issues in ecological impact evaluation. In: R. L. Sharma, J. D. Buffington and J. T. McFadden, eds., *Proc. Workshop on Biological Significance of Environmental Impacts*, NRC-CONF-002. U.S. Nuclear Regulatory Commission, Washington, DC, 307–315.

Eberhardt, L. L. (1976b). Quantitative ecology and impact assessment. *J. Environ. Management* **4**, 27–70.

Eberhardt, L. L. and J. M. Thomas (1991). Designing environmental field studies. *Ecol. Monographs* **61**, 185–199.

Fausch, K. D., J. Lyons, J. R. Karr and P. L. Angermeier (1990). Fish communities as indicators of environmental degradation. *Amer. Fish. Soc. Sympos.* **8**, 123–144.

Finney, D. J. (1978). Testing the effect of intervention in sequential ecological data. *Biometrics* **34**, 706–708.

Gilbert, R. O. (1987). *Statistical Methods for Environmental Pollution Monitoring*. Van Nostrand Reinhold, New York.

Gore, K. L., J. M. Thomas, L. D. Kannberg and D. G. Watson (1977). Evaluation of millstone nuclear power plant environmental impact prediction, based on monitoring programs. BNWL-2152, NRC-1. Battelle, Pacific Northwest Laboratories, Richland, WA. 120 pp.

Green, R. H. (1979). *Sampling Design and Statistical Methods for Environmental Biologists*. Wiley, New York.

Green, R. H. (1984). Statistical and nonstatistical considerations for environmental monitoring studies. *Environ. Monitor. Assess.* **4**, 293–301.

Green, R. H. (1987). Statistical and mathematical aspects: Distinction between natural and induced variation. In: V. B. Bouk, G. C. Butler, A. C. Upton, D. V. Parker and S. C. Asher, eds., *Methods for Assessing the Effects of Mixtures of Chemicals*, SCOPE. Wiley, New York, 335–354.

Gruber, D., J. M. Diamond and M. J. Parson (1991). Automated biomonitoring. *Environ. Auditor* **2**, 229–238.

Hall, J. W., J. C. Arnold, W. T. Waller and J. Cairns Jr. (1975). A procedure for the detection of pollution by fish movements. *Biometrics* **31**, 11–18.

Heck, K. L., G. Van Belle and D. Simberloff (1975). Explicit calculation of rarefaction diversity measurement and the determination of sufficient sample size. *Ecology* **56**, 1459–1461.

Hirsch, R. M. (1988). Statistical methods and sampling design for estimating step trends in surface-water quality. *Water Resources Bull.* **24**, 493–503.

Hirsch, R. M. and J. R. Slack (1984). A nonparametric trend test for seasonal data with serial dependence. *Water Resources Res.* **20**, 727–732.

Hirsch, R. M., J. R. Slack and R. A. Smith (1982). Techniques for trend analysis for monthly water quality data. *Water Resources Res.* **18**, 107–121.

Hurlbert, S. H. (1984). Pseudoreplication and the design of ecological field experiments. *Ecol. Monographs* **54**, 187–211.

Jeffers, J. N. R. (1989). Statistical and mathematical approaches to issues of scale in ecology. In: T. Rosswall, R. G. Woodmansee and P. G. Risser, eds., *Scales and Global Change*, (SCOPE 35). Wiley, New York, 47–56.

Jongman, R. H. G., C. J. F. Ter Braak and O. F. R. Van Tongeren (1987). *Data Analysis in Community and Landscape Ecology*. Centre for Agricultural Publishing and Documentation (Pudoc), Wageningen.

Kalton, G. and D. W. Anderson (1986). Sampling rare populations. *J. Roy. Statist. Soc. Ser. A* **149**, 65–82.

Karr, J. R. (1991). Biological integrity: A long-neglected aspect of water resource management. *Ecol. Appl.* **1**, 66–84.

Legendre, P., M. Trousslier, V. Jarry and M. Fortin (1989). Design for simultaneous sampling of ecological variables: From concepts to numerical solutions. *Oikos* **55**, 30–42.

Lettenmaier, D. O. (1988). Multivariate nonparametric tests for trend in water quality. *Water Resources. Bull.* **24**, 505–512.

Lettenmaier, D. P., K. W. Hipel and A. I. McLeod (1978). Assessment of environmental impacts, part 2: Data collection. *Environ. Management* **6**, 537–554.

Loehle, C. and E. P. Smith (1990). An assessment methodology for successional systems II. Statistical tests and specific examples. *Environ. Management* **14**, 259–268.

Loftis, J. C., C. H. Taylor and P. L. Chapman (1991). Multivariate tests for trend in water quality. *Water Resources Res.* **27**, 1419–1429.

Loftis, J. C., P. S. Porter and G. Settembre (1987). Statistical analysis of industrial wastewater monitoring data. *J. Water Pollution Control Feder.* **59**, 145–151.

Mann, C. (1990). Meta-analysis in the breech. *Science* **249**, 476–480.

Mathur, D., T. W. Robbins and E. J. Purdy Jr. (1980). Assessment of thermal discharges on zooplankton in Conowingo Pond, Pennsylvania. *Canad. J. Fish. Aqua. Sci.* **37**, 937–944.

Millard, S. and D. P. Lettenmaier (1984) Optimal design of biological sampling programs using analysis of variance. *Estuarine Coastal Shelf Sci.* **22**, 637–656.

Moriarty, F. (1983). *Ecotoxicology: The Study of Pollutants in Ecosystems*. Academic Press, New York.

Musters, C. J. M., H. C. Van Latesteijn and W. J. Ter Keurs (1988). The effect of the dumping of waste-acid from the titanium-dioxide production on the chlorophyll-a concentration of North Sea water: A time series analysis. *Environ. Monitor. Assess.* **10**, 181–203.

Neuhold, J. M. (1987). The relationship of life history attributes to toxicant tolerances of fishes. *Environ. Toxicol. Chem.* **6**, 709–716.

Niering, W. A. (1989). Vegetation dynamics in relation to wetland creation. In: J. A. Kusler, ed., *Wetland Creation and Restoration: The Status of the Science*, U.S. Environmental Protection Agency, Washington, DC, 67–74.

Odum, E. P. (1971). *Fundamentals of Ecology*, Saunders, Philadelphia, PA.

Patil, G. P. and C. R. Rao (1978). Weighted distributions and size biased sampling with applications to wildlife populations and human families. *Biometrics* **34**, 179–189.

Plafkin, J. L., M. T. Barbour, K. D. Porter, S. K. Gross and R. M. Hughes (1989). *Rapid Bioassessment Protocols for Use in Streams and Rivers: Benthic Macroinvertebrates and Fish*, EPA 444/4-89-001. National Technical Information Service, Springfield, VA.

Price, B. (1987). Quality control issues in testing compliance with a regulatory standard: Controlling statistical decision error rates. In: J. C. Bailar III, ed., *ASA/EPA Conf. on Compliance Sampling*, EPA-230-03-047. U.S. Environmental Protection Agency, Washington, DC.

Rao, C. R. (1986). ANODIV: Generalizations of ANOVA through entropy and cross entropy functions. Technical report 86-04. Center for Multivariate Statistics, Pennsylvania State University, University Park, PA.

Reynolds, M. R., Jr., R. W. Amin, J. C. Arnold and J. A. Nachlas (1988). Xbar Charts with variable sampling intervals. *Technometrics* **30**, 181–192.

Scanferlato, V. S., G. H. Lacy and J. Cairns Jr. (1990). Persistence of genetically-engineered *Erwinia carotovora* in perturbed and unperturbed aquatic microcosms and effect on recovery of indigenous bacteria. *Microb. Ecol.* **20**, 11–20.

Schaeffer, D. J., K. G. Janardan, H. W. Kerster and M. S. Shekar (1980). Graphical effluent quality control for compliance monitoring: What is a violation? *Enviorn. Management* **4**, 241–245.

Schmitt, C. J., J. L. Zajicek and P. H. Peterman (1990). National contaminant biomonitoring program: Residues of organochlorine chemicals in U.S. freshwater fish, 1976–1984. *Arch. Environ. Contam. Toxicol.* **19**, 748–781.

Skalski, J. R. and D. H. McKenzie (1982). A design for aquatic monitoring programs. *J. Environ. Management* **14**, 237–251.

Smith, E. P., D. Orvos, and J. Cairns Jr. (1993). Impact assesment using the before-after control-impact (BACI) model: Concerns and comments. *Canad. J. Fish. Aqua. Sci.* **50**, 627–637.

Smith, E. P., K. Pontasch and J. Cairns Jr. (1990). Community similarity and the analysis of multispecies environmental data. *Water Res.* **24**, 507–514.

Smith, E. P. and G. Van Belle (1984). Nonparametric estimation of species richness. *Biometrics* **40**, 119–129.

Smith, E. P., P. M. Stewart and J. Cairns Jr. (1985). Similarities between rarefaction methods. *Hydrobiologia* **120**, 167–170.

Smith, E. P. and L. L. Conquest (1981). On the van Belle–Fisher statistic for water pollution monitoring via ecological change. *J. Water Pollution Control Feder.* **61**, 1648–1650.

Smith, W. D. 1984. Design of efficient environmental surveys over time. In: S. M. Getz and M. D. London, eds., *Statistics in the Environmental Sciences*, Amer. Soc. for Testing and Materials, Philadelphia, PA, 90–97.

Smith, W. D. and J. F. Grassle (1977). Sampling properties of a family of diversity measures. *Biometrics* **33**, 283–292.

Stewart-Oaten, A., W. R. Murdoch and K. R. Parker (1986). Environmental impact assessment: 'Pseudoreplication' in time? *Ecology* **67**, 929–940.

Stewart-Oaten, A., J. R. Bence and C. W. Osenberg (1992). Assessing effects of unreplicated perturbations: No simple solutions. *Ecology* **73**, 1396–1404.

Strayer, D., J. S. Glitzenstein, C. G. Jones, J. Kolasa, G. E. Likens, M. J. McDonnell, G. G. Parker and S. T. A. Pickett (1986). Long term ecological studies: An illustrated account of their design, operation and importance to ecology. Occasional Publication Number 2, Institute of Ecosystem Studies, Millbrook, NY.

Suter, G. (1989). Ecological endpoints. In: W. Warren-Hicks, B. Parkhurst and S. Baker Jr., eds., *Ecological Assessment of Hazardous Waste Sites: A Field and Laboratory Reference*, EPA 600/3-89-013. National Technical Information Service, Springfield, VA, 2.1–2.26.

Suter, G. W., II, D. S. Vaughn and R. H. Gardner (1983). Risk assessment by analysis of extrapolation error: A demonstration for effects of pollutants on fish. *Environ. Toxicol. Chem.* **2**, 369–378.

Ter Braak, C. J. F. (1986). Canonical correspondence analysis: A new eigenvector technique for multivariate direct gardient analysis. *Ecology* **67**, 1167–1179.

Ter Braak, C. J. F. (1987). CANOCO–A FORTRAN program for canonical community ordination by [partial][detrended] [canonical] correspondence analysis, principal components analysis and redundance analysis (version 2.1). ITI–TNO, Wageningen.

Ter Braak, C. J. F. (1988). Partial canonical correspondence analysis. In: H. H. Bock, ed., *Classification and Related Methods of Data Analysis*. North-Holland, Amsterdam, 551–558.

Thomas, J. M., J. A. Mahaffey, K. L. Gore and D. G. Watson (1978). Statistical methods used to assess biological impact at nuclear power plants. *J. Environ. Management* **7**, 269–290.

Thompson, K. W., M. L. Deaton, R. V. Foutz, J. Cairns Jr. and A. C. Hendricks (1982). Application of time-series intervention analysis to fish ventilatory response data. *Canad. J. Fish. Aqua. Sci.* **39**, 518–521.

Van Belle, G. and J. P. Hughes (1984). Nonparametric tests for trend in water quality. *Water Resources Res.* **20**, 127–236.

Van Belle, G. and L. Fisher (1977). Monitoring the environment for ecological change. *J. Water Pollution Control Feder.* **49**, 1671–1679.

Woodfield, T. J. (1987). Statistical power considerations in time series intervention analysis. In: *ASA Proc. Statist. Comput. Section*. American Statistical Association, Alexandria, VA, 102–106.

Zahl. S. (1977). Jackknifing an index of diversity. *Ecology* **58**, 907–913.

G. P. Patil and C. R. Rao, eds., *Handbook of Statistics, Vol. 12*

9

Environmental Sampling and Monitoring

Stephen V. Stehman and W. Scott Overton

Abstract: Methods of probability sampling provide a rigorous protocol by which scientifically reliable information on environmental issues may be obtained. We review fundamentals of probability sampling from the perspective of monitoring environmental resources. We first describe basic concepts and definitions of sampling, such as the sampling universe, population, and design-based inference. As the natural, general estimator for probability sampling designs, the Horvitz–Thompson estimator provides the basis of a coherent strategy of consistent estimation. When auxiliary information is available, difference, ratio, and regression estimators are viable alternatives to the Horvitz–Thompson estimator, providing enhanced precision in many circumstances. Rather than repeat descriptions of basic sampling designs available in many texts, we discuss practical implications of a few standard designs useful in environmental sampling. Topics in spatial sampling and monitoring over time are discussed, as are the role and implications of analytic uses of surveys.

1. Introduction

Sampling theory and methods, which have traditionally played a strong role in the health sciences, forestry, agriculture, opinion research, economics, and business and labor statistics, have become increasingly important in environmental studies. Providing statistically sound environmental data is critical to decision making and scientific progress. We outline some of the fundamental principles and methods necessary for a basic understanding of sampling, and attempt to highlight some of the special needs of environmental sampling. Many of the definitions, concepts, and formulas needed for scientific sampling are described in standard sampling texts such as Cochran (1977), Jessen (1978), Kish (1965), Murthy (1967), Raj (1968), Särndal, Swensson and Wretman (1992) and Stuart (1962). Yates' (1981) approach to the theory and methods of sampling is of special interest in environmental sampling, while Gilbert (1987) describes some specific applications in the environmental area.

The statistical handbook edited by Krishnaiah and Rao (1988) serves as an introduction to the details of many of the topics discussed in this chapter.

1.1. Examples of environmental surveys

Before presenting basic concepts and definitions, we briefly describe several surveys that demonstrate objectives and methods of environmental sampling. The three examples, all surveys sponsored by the U.S. Environmental Protection Agency (EPA), will frequently be used as illustrations throughout this chapter. The National Lake Survey (NLS) and the National Stream Survey (NSS) were components of the National Acid Precipitation Assessment Program, while the Environmental Monitoring and Assessment Program (EMAP) is a current EPA initiative.

EXAMPLE 1.1. The primary objectives of the National Lake Survey were to: (1) determine the numbers of lakes (by various classes and location) that are acidic in potentially sensitive regions of the U.S.; (2) determine the percentage (by number and area) and location of lakes that have low acid neutralizing capacity in potentially sensitive regions of the U.S.; and (3) determine the chemical characteristics of lakes in potentially sensitive regions of the U.S. (Overton et al., 1986). This survey was designed to provide a snapshot of the condition of lakes at a fixed point in time. Typical of these kinds of sampling problems, an important consideration of the survey design was how to represent the condition of a lake selected in the sample. That is, a lake was the sampling unit of primary interest, but a lake possesses spatial and temporal variability, so that sampling within a particular lake represented another sampling problem within the main survey.

EXAMPLE 1.2. An objective of the National Stream Survey was 'To determine the percentage, extent (number, length, and drainage area), location, and chemical characteristics of streams in the United States that are presently acidic, or that have low ANC (acid neutralizing capacity) and thus might become acidic in the future' (Kaufmann et al., 1988). The target universe of stream reaches consisted of those reaches having drainage areas less than 60 square miles at their lower ends, but which were large enough to be represented as blue lines on 1 : 250 000-scale U.S. Geological Survey topographic maps. Similar to the NLS, the NSS was a multipurpose survey of conditions at a fixed point in time, and also required consideration of the spatial pattern of response variables. The question of how to characterize a sampled stream, given spatial and temporal variability, was also considered in this survey.

EXAMPLE 1.3. An objective of EMAP is to estimate current status and trends of indicators of the 'condition of the nation's ecological resources on a regional basis with known confidence' (Messer, Linthurst and Overton, 1991). The ecological resources monitored by EMAP include forests, lakes, wetlands, streams, arid lands, the Great Lakes, estuaries, and agroecosystems. Many different response variables will be measured; therefore, the sampling design must be able to accommodate special sampling problems for the different resources. As was the case with the previous two examples, EMAP is a multipurpose survey, and spatial patterns of variables are of interest. For most resources, a complete list of all units of that resource is not available, so a spatial sampling plan must be used. Because trends in resources are to be evaluated, the EMAP design must accommodate sampling and analysis over time.

These examples illustrate several important features of environmental sampling. One of these features is that the problems of interest are usually concerned with real populations, because a population of lakes, trees, or streams is a tangible, real-world entity, as is a well-identified spatial region such as an estuary or grassland. This strong identity of a real population in a sampling problem is in contrast to conventional statistical inference, in which 'the population is not well defined and the scientist starts with a sample' (Smith, 1990). The difference between conventional statistical inference and sampling inference is manifested in the different objectives addressed by the two approaches. Sampling problems have traditionally focused more on description. In environmental surveys, description of the population, of spatial pattern, of trends over time, or of associations, such as by regression or tables of means, are often of interest. Typically, hypothesis testing plays no more than a supporting role. Priority of objectives varies for different studies, and the choice of a sampling design and estimation strategy will depend on which objectives have highest priority. No sampling strategy will be good for all purposes, so the sampling practitioner must be aware of the trade-offs among objectives that choice of strategy entails. Although we focus on the descriptive element of surveys, we must keep in mind that survey data may be subjected to many analyses.

Samples may be selected according to a probability sampling design (Section 2) or a nonprobability sampling design, such as a judgment sample (Section 4.5). The three examples given contain elements of both types of sampling. Although we recognize some special applications of nonprobability samples, we advocate probability sampling as the basis for rigorous scientific inference for environmental sampling and monitoring. Probability sampling allows statistical inferences to be made for the real population of interest, and the population to which the inferences apply is clearly identifiable from the sampling protocol (Overton, 1992).

In addition to the three examples given, many other examples are, of course, available. The book, *Sampling Biological Populations*, edited by Cormack, Patil and Robson (1979) contains several applications of sampling to environmental problems. Gilbert (1987) provides examples and references of sampling for soil and groundwater contamination and for water quality monitoring, Frayer et al. (1983) describe methods used in the National Wetlands Inventory, and Baker et al. (1990) present results of a survey of Adirondack lakes. Webster and Oliver (1990) describe many examples and issues pertinent to soil sampling.

A rigorous monitoring design requires not only a rigorous sampling protocol, but also well-defined protocols for observation, sample taking, sample handling, measurement, data management, and analysis. A rigorous program further requires an additional protocol of verification, to ensure that the specified protocols were followed. Special data are often required to provide estimates of measurement and handling error and to provide calibration to standards of measurements. None of these topics will be covered. Several other topics often associated with environmental sampling also will not be discussed in this paper, including sampling designs for estimation of the size of animal populations, line transect sampling methods, and sampling designs for determining habitat utilization or home range size. Our discussion of continuous and extensive populations is relevant to issues of sampling a selected population unit, but we do not attempt a definitive treatment of those issues.

1.2. The sampling universe and populations

The fundamental structures of sampling are the universe and the populations defined on a universe. The sampling universe, denoted $\mathcal{U}$, is the collection of objects or spatial extent of interest in a sampling problem. It is useful to recognize several types of universes in environmental sampling. The universe in classical finite population sampling is a finite point set of discrete objects, such as lakes, trees, or animals, and is denoted by $\{u_1, u_2, \ldots, u_N\}$. In contrast to a finite universe, a continuous universe is a spatial volume or area of interest, such as the space occupied by a single large lake or forest. A continuous spatial universe may be partitioned into a finite number of areas, thus converting the continuous universe into a finite set of these area segments. This conversion allows the use of the finite sampling model, but it is also possible to sample continuous universes directly by several designs.

Strictly speaking, the population is the response or attribute defined on the universe, but frequently, the terms population and universe are used interchangeably in the sampling literature. We will denote the population values by y_u, where u may refer to an element of a finite point set or to a position in space for a continuous universe. Several types of populations are defined. A finite population is a population described on a finite point set, the classical population of traditional survey sampling. A continuous population is defined on a continuous universe, and the measured value is expected to vary continuously over the universe; that is, discontinuities are not expected. For example, if the universe is a single large lake, dissolved oxygen, acid neutralizing capacity, pH, and temperature would also be continuous populations defined on that universe. Extensive populations are also defined on continuous universes, but the values of the population are not continuous. For example, in EMAP, wetlands are treated as an extensive population for a variable such as woody plant biomass because the spatial pattern of biomass over the continuous universe will display discontinuities. Volume of trees in a forest stand also represents an extensive population. Recognizing the different kinds of universes and populations is important because a different sampling strategy may be required for each. Sampling considerations may apply to several scales of resolution. This concept was illustrated earlier, in the NLS for example, in that a lake may be part of a population of lakes, yet a single lake may itself represent a universe of interest.

Given that a conceptual definition of the target universe and populations has been formulated, construction of a frame that unambiguously represents the universe is necessary to provide the basis for sample selection. The frame should also reference any auxiliary variables required by the intended analysis. The sampling units in the frame may be defined without identifying a response variable, which demonstrates another distinction between 'sampling' and conventional statistics: the sample is a subset of the universe, not just a file or list of data. Then observations are made on the sample to provide inferences about specific populations defined on the universe. The identity of the unit on which an observation was made will be a permanent part of the data record.

Probably the most common type of frame in traditional survey work is a list frame, which is simply a list of the units in the universe. Another type of frame commonly used in environmental studies is a map frame. A map frame may be a topographic map of the spatial region of interest, or it may be a map displaying the spatial locations of the objects in the universe. A map of the objects could be converted into a list frame, but the cost may be prohibitive, and a list frame retaining the spatial pattern of the objects in the universe may be difficult to construct; for this reason, an alternative frame, based on a grid, was chosen for the National Stream Survey (see Section 4.4). Some authors make a distinction between a target universe and a frame universe – the former being the universe of actual interest as specified by the study objectives, and the latter being the universe for which a frame is available. The identity of the sampled and target universes should be readily apparent from the description of the frame and objectives of the study, respectively. It may be prohibitively expensive to extract the target universe from the frame universe, even though the target universe is conceptually well defined. Sometimes the frame does not cover the intended target universe; in such cases, it may be desirable to redefine the target universe to comply with the achieved coverage.

Classical finite population sampling is primarily concerned with description of populations, particularly estimation of collective or aggregate properties, such as means, totals, proportions, numbers, and ratios. For environmental applications, estimation of distribution functions often provides an excellent description of the population. Distribution functions (Section 3.2) display the estimated portion of the population above and below any specified value of the response variable (Figure 1). Frequently, population means or totals give an incomplete picture of the biologically important features of the population, because two populations with the same mean may have very different distribution functions and very different import. Estimated distribution functions formed the backbone of population description in the NLS and NSS, and are expected to play a major role in EMAP. Several important variations were used in the NLS and NSS including distribution of the number of lakes, of the area of lakes, and of stream miles with a value of the response variable below a specified critical value.

Subpopulation estimation is another important feature of environmental surveys. A subpopulation is simply the population defined on a subset of the universe. Examples of subpopulations include lakes within certain size classes of surface area or volume, lakes located within certain elevation zones, streams of different reach orders, forests of different community types or ownership, and wetlands of certain

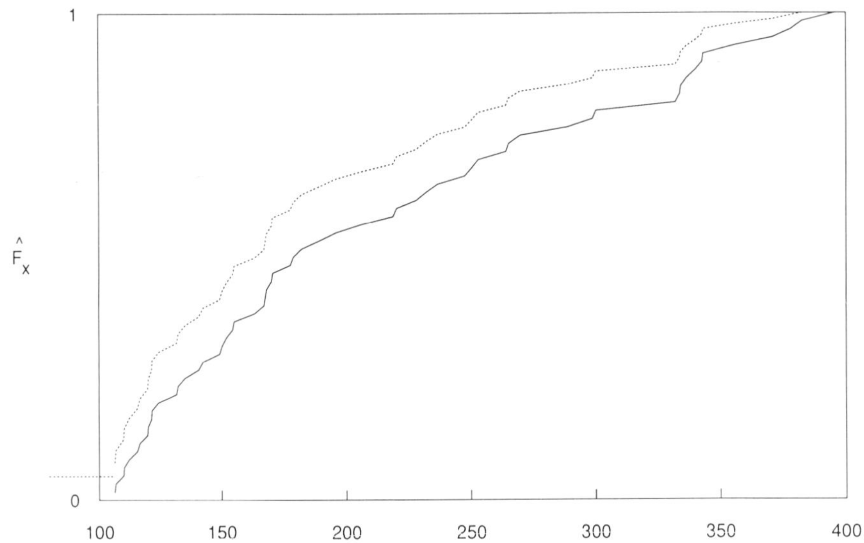

Fig. 1. Estimated distribution function of a variable. Dashed line is an upper 95% confidence bound, solid line is the point estimate of the distribution function.

hydrology types. Sometimes subpopulation classes become part of the sampling design as strata, but stratification is not necessary for subpopulation estimation. Estimates for a large number of subpopulations may be of interest, so the sampling plan must be able to accommodate these estimates. Although the term 'domain' is often used in the same manner as subpopulation, we find it helpful to reserve domain to describe a spatially defined subuniverse. Some subpopulations are defined on domains, but others may be intermixed over the universe.

1.3. Types of sampling units and methods of selection

Discrete objects such as lakes, trees, stands of trees, stream reaches, or watersheds may be sampled by a list frame, if such a frame for the target universe is available. Constructing a list frame from maps or computerized spatial databases is an expensive, exacting, and time-consuming process. Additionally, even if a list frame is available, spatially distributed populations should usually be sampled in a manner to ensure good spatial distribution of the sample, so some spatial device will be necessary. Consequently, sampling directly from a map or other spatial representation of the universe is useful even for discrete universes.

The basic device for sampling a spatially distributed population is to use a 'frame' that corresponds to the spatial extent of the universe, and to sample this space. One possibility is to use point sampling in this context. A key feature of such a design is that each discrete object must be identifiable by a unique spatial location, such as the upper node of a stream reach, the centroid of a forest stand, or the centroid of a lake.

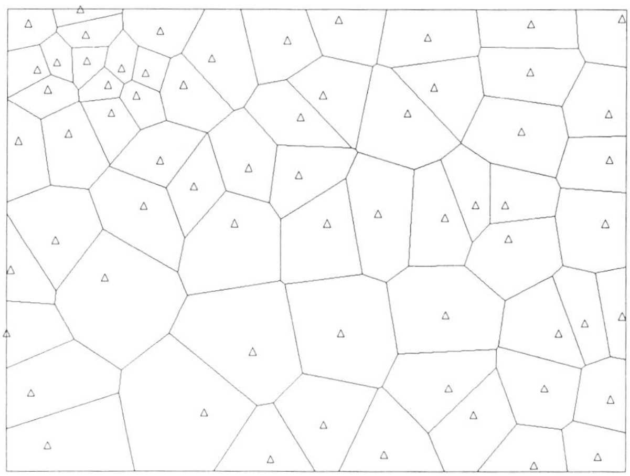

Fig. 2. Thiessen polygons for selecting the nearest object to a random point. The Thiessen polygon encloses the region surrounding an object (triangle symbol) in which any point is closer to that object than to any other object. A procedure selecting the closest object to a random point will select objects with probability proportional to the area of the object's Thiessen polygon.

The sample elements are selected using a point-to-point rule of association between the object's location and a spatial sampling point identified using a systematic grid, an unrestricted random sample, or some other spatial point sampling scheme. For example, a common procedure is to select the closest object to the sampling point. The probability of selecting a given object is then the probability of a point falling into the area of the object's Thiessen polygon, where the Thiessen polygon encloses the region closer to that object's spatial location than to any other object (Figure 2). The probability of selecting a given unit is then proportional to the area of the Thiessen polygon. Such a design results in a variable probability sample, in which different elements of the universe have different probabilities of being included in the sample (Section 4.4).

Discrete objects may also be sampled by an areal sampling unit, which is a circular, square, hexagonal, or other enclosed area centered on a randomly located point. Discrete objects whose spatial positions are within the areal unit would be selected into the sample. Extensive populations may also be sampled in this manner. The area in the spatial universe falling within the areal sampling unit constitutes the sample of extensive populations on that areal segment. For example, in EMAP, a 40 km^2 hexagon centered on the grid points of a triangular grid is an areal sampling unit for the first phase of the sample (Figure 3). In forestry, circular plots of 1/1000 of an acre ('milacre' plots) are often used. Estimation for areal samples is presented in Section 3.4.

Partitioning of the universe into areal segments is still another approach to sampling spatially identifiable finite populations or extensive populations (Murthy, 1967, Sections 6.8 and 6.10d). These areal segments may be fields or other natural

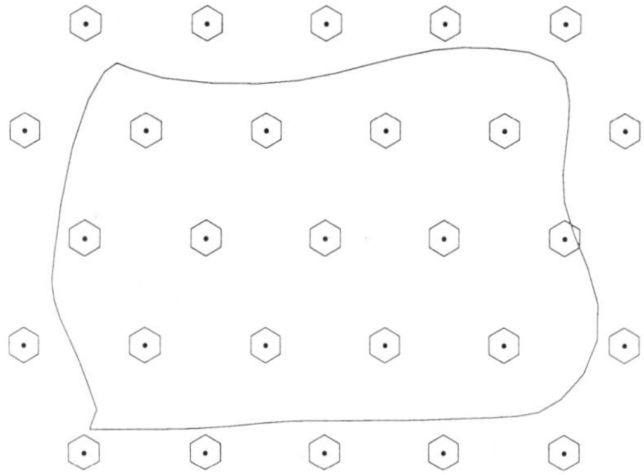

Fig. 3. Diagram of the EMAP phase I areal sample. The areal sampling unit is a 40 km^2 hexagon centered on the grid points of a triangular grid.

features of the landscape, or simply arbitrary figures such as rectangular areas superimposed over the region of interest. Areal segments may be selected by randomly located points or a systematic grid of points, or by constructing a list frame of areal segments, if the universe has been partitioned into such sets. Subsampling of the elements identified in areal segments will often be necessary.

Point or areal samples may also be used to sample continuous universes, with either extensive or continuous populations. For point sampling, the values of the population are measured at the sampling point identified, and the 'weight' associated with the observation in the estimator is a function of grid density. Several options for point sampling of continuous populations are described in Section 5. For areal samples, the observations used in estimation can be integrals of the population value over the areal unit, or they may be qualitative categories of the attributes of interest. In contrast to point samples, additional information is available from the areal sampling unit. For example, information on spatial patterns within the sampling unit may be recorded for description or for future use in monitoring change over time (Figure 4).

1.4. Notation

For easier reference, some of the notation frequently appearing in this chapter is listed below:

$\mathscr{A}$ = subpopulation
f = sampling fraction, n/N
n = sample size
n_a = sample size in subpopulation $\mathscr{A}$
N = number of objects in universe

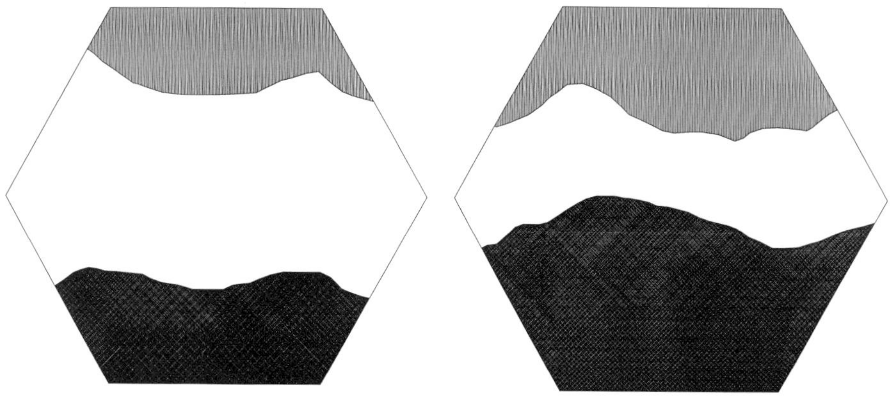

Fig. 4. Monitoring change in land use with an areal sampling unit. The two hexagons represent the same areal sampling unit at two points in time. The shaded and clear regions represent different land uses. The areal sampling unit contains information on the total area and change in area of each land use, and also spatial information on change in spatial pattern of land use.

N_a = number of objects in subpopulation $\mathscr{A}$
π_u = inclusion probability of element u
π_{uv} = pairwise inclusion probability of elements u and v
s = sample
$\mathscr{S}$ = sample space
$\sum_s$ = summation over elements in the sample ($u \in s$)
$\sum_{\mathscr{U}}$ = summation over elements in the universe ($u \in \mathscr{U}$)
$T_y(T_x)$ = population total of the variable $y(x)$
$\hat{T}_y$ = Horvitz–Thompson estimator of the y-population total
$\mathscr{U}$ = universe
V_y = finite population variance of y
x_u = value of an auxiliary variable on element u
y_u = value of a response variable on element u
$\bar{y}_s$ = sample mean of y

2. Fundamentals of probability sampling

A sampling design will be defined as a mathematical function $p(s)$ by which the probability of selecting any sample s is given. The set of all possible samples $\mathscr{S}$ is called the sample space. Note that $\sum_{s \in \mathscr{S}} p(s) = 1$. A probability sample may be defined as a sample in which $p(s)$ is known for all s, and each element in the universe has a positive probability of being included in the sample (Overton, 1992). Commonly used sampling designs such as simple random sampling, systematic sampling, cluster sampling, and double sampling are all probability sampling designs. Whereas a probability sample requires some random mechanism for selection of the sample, a nonprobability sample may be selected without any random mechanism. A simple example of a nonprobability sample is a sample chosen purposively by a scientist to be 'representative' of a class of interest.

The classical approach to inference in sampling is design-based inference. In this approach, the response variable y_u defined on each element u of the universe is treated as a fixed quantity measured without error. The relevant uncertainty in design-based inference is then due to the randomization mechanism of the design. That is, the variability of concern for a statistic is the variability of that statistic over all possible samples identified by the sample space.

Design-based properties of estimators in finite population sampling such as bias, variance, and mean-square error may be determined from the following formulas, in which t is a statistic, and t_s is the value of t for sample s:

$$\mathrm{E}(t) = \sum_{s \in \mathscr{S}} p(s) t_s, \tag{2.1}$$

$$\mathrm{V}(t) = \sum_{s \in \mathscr{S}} p(s)[t_s - \mathrm{E}(t)]^2. \tag{2.2}$$

Then if we designate t as an estimator of a parameter T, we can define

$$\mathrm{Bias}(t) = \sum_{s \in \mathscr{S}} p(s)[t_s - T], \tag{2.3}$$

$$\mathrm{MSE}(t) = \sum_{s \in \mathscr{S}} p(s)[t_s - T]^2. \tag{2.4}$$

Stuart (1962) provides an excellent introduction to these basic ideas of sampling.

Confidence intervals are often constructed using the normal distribution and the formula $t \pm z\sqrt{\hat{V}(t)}$, where z is a percentile from the standard normal distribution, and $\hat{V}(t)$ is an estimate of the variance of the statistic t. These confidence intervals require justification via a central limit theorem idea (Särndal et al., 1992, Section 2.11; Wolter, 1985, Appendix B). In some cases, exact confidence bounds are available, a simple example being the use of the hypergeometric distribution with simple random sampling of a binary attribute.

2.1. Inclusion probabilities

Characteristics of sampling designs and estimators may also be determined via representation of a probability sample by inclusion probabilities. Derivation of most finite population sampling results is simpler in this representation than it is in the general approach to probability sampling. Common sampling designs are simply special cases of probability sampling, and analysis using inclusion probabilities leads to the standard results for these designs. Our discussion of inclusion

If the total area A is known, the variance of the estimated areal distribution function is estimated by

$$\hat{V}(\hat{G}(A)) = \hat{V}(\hat{A}(c))/A^2, \tag{3.14}$$

and $\hat{V}(\hat{A}(c))$ is estimated by replacing y_u with z_u in equation (3.3) or (3.4). Confidence bounds are set by normal approximations, but these bounds are not expected to perform well at the ends of the distribution.

When N must be estimated by $\hat{N}$, $\hat{F}(c)$ is a ratio (equation (3.7)) and methods suitable to ratio estimation are required. A useful variance approximation (Särndal et al., 1992, Result 5.6.2) is given by

$$\hat{V}_{\mathrm{HT}}(\hat{F}(c)) = \frac{1}{\hat{N}^2}\left[\sum_{u \in s}\left(\frac{d_u - \hat{F}(c)}{\pi_u}\right)^2 (1 - \pi_u) + \sum_{u \in s}\sum_{\substack{v \in s \\ v \neq u}}\left(\frac{\pi_{uv} - \pi_u\pi_v}{\pi_{uv}}\right)\frac{[d_u - \hat{F}(c)]}{\pi_u}\frac{[d_v - \hat{F}(c)]}{\pi_v}\right]. \tag{3.15}$$

Again, the normal approximation is used to set confidence bounds.

For simple random sampling, confidence bound estimation becomes much simpler. Given the subpopulation size N_a the conditional sample in subpopulation $\mathscr{A}$ is also SRS, and the ratio estimator variance is unnecessary. Conditional on the observed n_a, the sample number of observations in $\mathscr{A}$, such that $y_u \leqslant c$, is a hypergeometric random variable with parameters N_a, n_a, and $N_a(c)$, and the exact hypergeometric confidence bounds for $N_a(c)$ can be computed (cf. Cochran 1977, p. 57). If N_a is unknown, the observed number of elements in the sample in subpopulation $\mathscr{A}$ such that $y_u \leqslant c$ may be considered binomial with parameters n_a and $F_a(c)$. Confidence bounds for $F_a(c)$ can be obtained from this binomial distribution rather than from the normal approximation to the ratio estimator.

3.3. Estimation using auxiliary information

Alternative estimators to $\hat{T}_y$ may be constructed using a model specifying the relationship between y_u and one or more auxiliary variables. Such an alternative estimator can use a prediction of y_u obtained either from information external to the sample or from regression estimators of the model parameters developed from the sample. Difference estimators are an example of the former type of estimator, while ratio and regression estimators are examples of the latter. These estimators may be used with any probability sampling design, but they usually require additional frame information, such as the population size N; the population total of the x's, T_x; or the values of the auxiliary variables on the entire frame. In environmental sampling, elevation, temperature, annual rainfall, soil classification, ecoregion, and geologic substrate are examples of readily available auxiliary variables of potential benefit. The utility of the estimators depends on how well the model represents the population relationship or how closely an externally generated predictor predicts the actual y_u. Inference is usually design based, with the models serving to 'assist' in the analysis by improving precision (Rao and Bellhouse, 1990; Särndal et al., 1992).

3.3.1. Difference estimator

The difference estimator may be used if a theoretical value, prediction, or even a guess of the value of y_u for each element of the universe can be constructed from information external to the sample y's. For example, suppose the parameter of interest is total volume of trees in a forest. Volume of a tree is difficult to measure, but height and diameter at breast height (dbh) are more easily obtained. Each tree in the forest is visited, and volume is calculated according to a specific formula from dbh and height. Denote the theoretical or predicted value as y_u^*, and let $e_u = y_u - y_u^*$. Then

$$T_e = \sum_{\mathscr{U}} e_u = T_y - T_{y^*}, \tag{3.16}$$

where $T_{y^*} = \sum_{\mathscr{U}} y_u^*$. T_e may be estimated from the sample data using the Horvitz–Thompson estimator, $\hat{T}_e = \sum_s e_u/\pi_u$. Because T_{y^*} is known, an unbiased estimator of T_y is given by the difference estimator,

$$\hat{T}_{y,\mathrm{D}} = T_{y^*} + \hat{T}_e, \tag{3.17}$$

with $\mathrm{V}(\hat{T}_{y,\mathrm{D}}) = \mathrm{V}(\hat{T}_e)$. For SRS,

$$\hat{T}_{y,\mathrm{D}} = T_{y^*} + \frac{N}{n}\sum_s e_u \tag{3.18}$$

and

$$\mathrm{V}(\hat{T}_{y,\mathrm{D}}) = N^2(1-f)\frac{V_e}{n}, \tag{3.19}$$

where V_e is the finite population variance of the differences e_u. For SRS, $\mathrm{V}(\hat{T}_y)$ is given by equation (3.19) with V_e replaced by V_y. Thus, if y^* is a good approximation to y, then $V_e < V_y$, and $\hat{T}_{y,\mathrm{D}}$ results in a substantial reduction in variance over the direct Horvitz–Thompson estimator of T_y.

For variable probability sampling, the difference estimator may be written as

$$\hat{T}_{y,\mathrm{D}} = \sum_{\mathscr{U}} y_u^* + \sum_s \frac{e_u}{\pi_u}. \tag{3.20}$$

This estimator is often called the 'generalized difference estimator'. The general variable probability sampling formulas for $\mathrm{V}(\hat{T}_{y,\mathrm{D}})$ and $\hat{\mathrm{V}}(\hat{T}_{y,\mathrm{D}})$ are obtained by replacing y_u by e_u in equation (3.1) or (3.2) and equation (3.3) or (3.4), respectively.

3.3.2. Ratio estimator

If the population total of the x's, T_x, is known, a common ratio estimator of T_y is

$$\hat{T}_{y,\mathrm{R}} = T_x \frac{\hat{T}_y}{\hat{T}_x}. \tag{3.21}$$

Several special cases of this ratio estimator are important. For equal probability sampling, the estimator simplifies to

$$\hat{T}_{y,\mathrm{R}} = T_x \frac{\bar{y}_s}{\bar{x}_s}, \tag{3.22}$$

the ratio of means estimator. If $x_u = 1$ for all $u \in \mathcal{U}$, the estimator becomes

$$\hat{T}_{y,\mathrm{R}} = N \frac{\hat{T}_y}{\hat{N}}, \tag{3.23}$$

and an estimator of the population mean is then $\hat{T}_y/\hat{N}$. The estimator given by equation (3.23), sometimes called Hajek's estimator, has been recommended for use with variable probability sampling when the inclusion probabilities are not highly correlated with the y, a case in which the usual Horvitz–Thompson estimator has poor precision. The ratio estimator may also be used for subpopulation and distribution function estimation (Sections 2.2 and 3.2). For example, when N_a is unknown, the mean of the subpopulation, $\bar{Y}_a$, is estimated by

$$\hat{\bar{Y}} = \hat{T}_{ya}/\hat{N}_a, \tag{3.24}$$

and the proportion in the subpopulation, P_a, is estimated by

$$\hat{P}_a = \hat{N}_a/\hat{N}. \tag{3.25}$$

Formula (3.24) is preferred to $\hat{T}_{ya}/N_a$ even if N_a is known (Särndal et al., 1992).

To obtain a computationally simple variance approximation for the ratio estimator (equation (3.21)), Särndal et al. (1992) suggest

$$\hat{V}_{\mathrm{HT}}(\hat{T}_{y,R}) = \frac{T_x^2}{\hat{T}_x^2} \sum_{u \in s} \sum_{v \in s} \left(\frac{\pi_{uv} - \pi_u \pi_v}{\pi_{uv}} \right) \frac{e_u}{\pi_u} \frac{e_v}{\pi_v}, \tag{3.26}$$

or, for fixed sample sizes,

$$\hat{V}_{\mathrm{YGS}}(\hat{T}_{y,R}) = \frac{1}{2} \frac{T_x^2}{\hat{T}_x^2} \sum_{u \in s} \sum_{\substack{v \in s \\ v \neq u}} \left(\frac{\pi_u \pi_v - \pi_{uv}}{\pi_{uv}} \right) \left(\frac{e_u}{\pi_u} - \frac{e_v}{\pi_v} \right)^2, \tag{3.27}$$

where $e_u = (y_u - (\hat{T}_y/\hat{T}_x)x_u)$. Royall and Cumberland (1981) evaluated several other estimators of variance for use with the ratio estimator and SRS.

3.3.3. Generalized regression estimator

The generalized regression estimator may be viewed as a form of a difference estimator based on model predictions. Instead of using a theoretical or externally developed predictor, y_u is predicted from the sample data via regression. Let $\hat{y}_u$ denote the prediction of y_u constructed from the regression estimates obtained from the sample data. The general form of the regression estimator may be written, analogously to equation (3.17), as

$$\hat{T}_{y,\mathrm{GRE}} = T_{\hat{y}} + \hat{T}_e \tag{3.28}$$

$$= \sum_{\mathscr{U}} \hat{y}_u + \sum_{s} (y_u - \hat{y}_u)/\pi_u, \tag{3.29}$$

where $e_u = y_u - \hat{y}_u$. Any number of auxiliary variables may be used to establish the prediction $\hat{y}_u$, although a single predictor variable is commonly used. Equation (3.29) illustrates that the generalized regression estimator consists of two components, the population total of the predicted $\hat{y}$'s derived from the model, and a sample-based estimate $\hat{T}_e$, applied to the residuals to 'correct' for model misspecification (Little 1983). Särndal (1980) investigated two schemes for estimating the regression coefficients required for $\hat{y}_u$. He concluded that weighting by either the π_u or the best linear unbiased weights from the specified model resulted in equal efficiency. We have extended this investigation empirically and found no strong reason not to use ordinary least squares in this estimator.

To estimate the variance of the generalized regression estimator, e_u may be substituted for y_u in equation (3.3) or (3.4). These variance estimators are exactly of the form given for the difference estimator. As usual, the Yates–Grundy–Sen form of the variance estimator requires a fixed sample size design. Särndal et al. (1989) present a detailed investigation of variance estimation using weighted residuals for generalized regression estimators.

3.3.4. Simple regression estimator

Estimators based on regression models may also be derived from a fully model-based, prediction perspective. The values of y are known for the sample, so estimation of T_y may be viewed as a problem of predicting the y values for the unobserved elements of $\mathscr{U}$. That is, if the population total is written as

$$T_y = \sum_{s} y_u + \sum_{\mathscr{U}-s} y_u, \tag{3.30}$$

then the prediction estimator, ignoring design features, is the simple regression estimator,

$$\hat{T}_{y,\mathrm{SRE}} = \sum_{s} y_u + \sum_{\mathscr{U}-s} \hat{y}_u \tag{3.31}$$

$$= \sum_{\mathscr{U}} \hat{y}_u + \sum_{s} (y_u - \hat{y}_u). \tag{3.32}$$

As with the generalized regression estimator, the prediction of $\hat{y}_u$ may be obtained by simple or multiple linear regression, nonlinear regression, or any other procedure deemed useful. Little (1983) describes several such prediction estimators. When ordinary least squares is used, $\sum_s(y_u - \hat{y}_u) \equiv 0$, so the estimator is simply

$$\hat{T}_{y,\mathrm{SRE}} = \sum_{\mathcal{U}} \hat{y}_u \tag{3.33}$$

(see also Särndal et al., 1992, Section 6.5). Comparison of equation (3.33) to equation (3.29) reinforces the model misspecification correction feature of the generalized regression estimator.

The model-assisted estimators presented in Section 3.3 often provide more precise inferences than the Horvitz–Thompson estimator, but this advantage is achieved at the cost of obtaining the additional frame information. The increased precision often justifies this effort. Sometimes, however, it is not possible to obtain all the population values of x, so that the general forms of the regression estimators are unavailable. In these cases, simplification of the models may reduce the demand on frame data. To illustrate, under the linear model

$$y = \alpha + \beta x + \varepsilon, \tag{3.34}$$

equation (3.33) can be written as

$$\hat{T}_{y,\mathrm{SRE}} = N\hat{\alpha} + \hat{\beta} T_x, \tag{3.35}$$

and equation (3.29) can be written as

$$\hat{T}_{y,\mathrm{GRE}} = N\hat{\alpha} + \hat{\beta} T_x + \sum_s (y_u - \hat{y}_u)/\pi_u. \tag{3.36}$$

These forms of the estimators require knowledge of N and T_x, but not of the x-values for the entire population.

The actual regression estimator selected for a given situation depends on the conjectured relationship between the response and auxiliary variables and on the circumstances of sampling. Because estimation of an intercept term produces little additional cost, regression estimators appear advantageous to ratio estimation. The correction for model misspecification present with $\hat{T}_{y,\mathrm{GRE}}$ conveys a potential 'robustness' advantage to this estimator over $\hat{T}_{y,\mathrm{SRE}}$. Model-based estimation of distribution functions (Chambers and Dunstan, 1986) may require more careful attention to model specification so that ordinary least squares is not necessarily uniformly adequate. Little (1983) and Cassel et al. (1977) provide detailed assessments of various model-assisted estimators.

3.4. *Estimation for areal samples*

In Section 1.3, areal sampling of a continuous universe was described. Suppose that an areal sampling unit is centered on each point of a randomly located grid of points so that the ratio of size of areal unit to the inverse of grid density is $1/k$. If the grid is randomly positioned in an appropriate manner, then the probability that any point in the spatial universe is included in the sample area is $1/k$. The weighting factor for all observations on the sampling units is k. Let the subscript r denote a particular resource class, for example high elevation lakes, and let a_{ru} be the area covered by class r in the areal sample surrounding grid point u, and η_{ru} be the integral of the value of y over the area a_{ru}. Then for a sample of n grid points,

$$\hat{A}_r = k \sum_s a_{ru} \tag{3.37}$$

is unbiased for A_r, the total area in class r, and

$$\hat{T}_{yr} = k \sum_s \eta_{ru} \tag{3.38}$$

is unbiased for T_{yr}, the total of y over the resource. An estimator of the variance of $\hat{A}_r$ is

$$\hat{V}(\hat{A}_r) = k^2 n \left(\frac{k-1}{k}\right) s_a^2, \tag{3.39}$$

where s_a^2 is the sample variance of the a_{ru} and $(k-1)/k$ is a finite population correction factor. Similarly, the variance of $\hat{T}_{yr}$ is estimated by

$$\hat{V}(\hat{T}_{yr}) = k^2 n \left(\frac{k-1}{k}\right) s_\eta^2, \tag{3.40}$$

where s_η^2 is the simple variance of the η_{ru}. These variance estimators are approximations because they do not account for the systematic nature of the grid design described. Unless reasonably strong spatial autocorrelation exists for the response variable, such approximations are justified. Section 4.1 provides further discussion of variance estimation for systematic sampling.

4. Designs useful in environmental sampling

A complete review of designs useful in environmental sampling is beyond the scope of this chapter. However, some issues of application of standard sampling designs to environmental problems merit discussion. In this section, we comment on four commonly used probability sampling designs, and discuss applications of non-probability sampling.

4.1. Systematic sampling

Systematic sampling is often used in environmental sampling because it is practical and convenient to implement in the field, and often provides better precision and more complete spatial coverage than unrestricted random sampling. Practitioners usually desire the advantages of systematic sampling, but sometimes hesitate to employ this design because of concern over the lack of a generally satisfactory variance estimator, and concern with the poor precision inherent when sampling systematically in phase with periodicity in a population. Although technically legitimate, these concerns are often overemphasized. The advantages of systematic sampling for spatial analyses and model-assisted inference should not be overlooked.

Systematic samples with a random start do permit design-unbiased estimation of the population parameter of interest, often with high precision. A potential problem is that estimation of precision of such estimators is approximate: the greater the gain in precision, the poorer the variance estimate. Standard recommendations for variance estimation with systematic sampling include the following: (1) use of a variance estimator appropriate for random sampling, implicitly assuming that the ordering of the population is such that the systematic sample behaves similarly to a simple random sample; or (2) use of an alternative estimator based on the mean-square successive difference (Von Neuman et al., 1941), such as the following formula used in one-dimensional, equal probability systematic sampling (cf. Cochran, 1977, p. 225):

$$\hat{V}_{\mathrm{sys}\,1} = \frac{1 - n/N}{2n(n-1)} \sum_{u=1}^{n-1} (y_{u+1} - y_u)^2. \tag{4.1}$$

Summarizing his study of variance estimators for equal probability systematic sampling, Wolter (1985) concluded that estimators based on successive differences were 'broadly useful for a variety of populations found in practice'. Unbiased estimation of variance is important in practice to the extent that a badly biased variance estimator will likely result in poor confidence interval properties, but unbiased estimation of variance, available with a simple random sample, is often attained at the price of lower precision.

Periodicity in a population may be either a help or a hindrance to efficiency of a systematic design. Selecting a sampling interval that is an odd multiple of the half period of the population periodicity is very favorable to systematic sampling (Cochran, 1977, p. 218), but if information is insufficient to avoid an unfavorable sampling interval, and the periodicity is dominant, systematic sampling would not be recommended (Matérn, 1986, p. 66). The prevalence of periodicity in natural populations is a matter of debate. Many periodic phenomena are known, for example, diurnal and seasonal fluctuations in lake chemistry and temperature, or characteristics of forest stands in an alternating ridge and valley topography; these phenomena should be evident to the investigator. Milne (1959) stated that he could not find 'reasonable grounds for expecting spatial periodicity anywhere on this earth except where man himself, either directly or indirectly, has imposed periodic conditions sufficiently accurate to override the natural environmental irregularity', and man-made periodicity would almost always be suspected. Matérn (1986) believed Milne's position extreme, but stated that he could find no clear case of periodicity reported in forest surveys. An often overlooked remedy to the concern with periodicity is the option of adding supplementary random points, even after the original systematic sample has been selected. Zinger (1980) describes how the supplementary observations may be used for variance estimation. Our position is that an informed investigator who is familiar with the ecological process or material being sampled will seldom make a serious error using systematic sampling.

Selection of a systematic design often involves a choice between two competing criteria of design efficiency and unbiased variance estimation. Often the choice is simply whether to risk a poor estimate of variance or an inferior estimate of the parameter of interest. Efficient estimation of the parameter of interest would usually be a higher priority. Systematic designs should always be given strong consideration in any circumstance in which spatial patterns are prevalent, because systematic design enhances description of those spatial patterns (Section 5). Even when these spatial patterns are not addressed by the objectives, their influence on the precision of estimated population distribution functions and other parameters may be important, particularly if model-assisted methodology is employed. Moreover, systematic sampling provides the sample balance that lends an advantage to model-assisted estimation.

4.2. Stratified sampling

Grouping relatively homogeneous elements of the population into strata is a common design tactic in environmental monitoring. For example, sampling units consisting of stream segments may be stratified by habitats, such as pools and riffles, or by size, as measured by drainage area. Forest stands may be stratified by forest type, ownership, or vitality class (De Vries, 1986). Lakes may be stratified by surface area or drainage type. Because strata partition the universe, they must be unambiguously represented on the frame. The examples of stratification variables given so far require acquisition of those classification variables for all elements in the universe, presenting a difficult task in frame development. In most environmental studies, variables such as elevation, ecoregion, and resource domain are particularly useful criteria for spatial strata because they can be readily identified from map materials. Stratification does not restrict choice of sampling design within strata, and different designs may be used for different strata, as long as each stratum is sampled independently of the other strata.

The strongest reason for stratification in environmental surveys is to obtain precise estimates for each stratum. Although strata are not necessary to obtain estimates for subpopulations of interest, they may prove desirable, especially if one subpopulation is quite rare. Strata may also be constructed to improve precision of estimates of the overall population parameter and to provide administrative convenience; even if stratification is highly efficient for estimates of the entire population, however, it may not be nearly as efficient when estimates are made for a small subpopulation (Durbin, 1958). Selection of a stratification variable is problematic, because in a multipurpose survey, strata selected to enhance precision of one or a few response variables will likely reduce precision for other response variables. Gains in precision because of stratification may not be large, such as when the response variable is binary and the proportions vary little among strata (Cochran, 1977, Section 5.11). Stevens and Olsen (1991) present an example in which stratification yields little or no improvement in precision because the information from which strata are constructed contains errors. Further, analytic uses of survey data (Section 6) are complicated by stratified sampling. Although some degree of stratification is probably useful in many environmental applications, excessive stratification will often lead to more problems than benefits, particularly in multiple objective surveys and when many subpopulations are of interest.

4.3. Two-phase sampling

Two applications of two-phase (or double) sampling to environmental sampling will be described. In the first application, two-phase sampling is used to assist in sample selection, while the second application is for improved estimation. Sometimes the frame does not contain enough information to identify the sampling units of interest adequately. For example, Barnard et al. (1985) describe two-phase sampling for stratification in regional and state forest resource inventories. In the first phase, aerial photos are used to construct strata defined by the photo-interpreted points. The second-phase sample of ground plots is then selected from these photo-interpreted points, and data collected on these ground plots. In an application of two-phase sampling proposed for EMAP, the first-phase sample is a large areal sample of 40 km^2 hexagons centered on a point of a systematic, triangular grid. Aerial photography or another remote sensing technique is then used to classify the area within the sampled hexagon according to resource classes such as type of wetland, forest, or agricultural use. The first-phase sample provides estimates of the number and area of discrete resources, such as lakes, or area of extensive resources, such as wetlands or forests. The classification of areas permits stratification of the first-phase sample, so that resource units identified at the first phase are then subsampled, and actual field visits are made to these second-phase sampling units. Data from the second-phase sample are used to generate more detailed population descriptions, including distribution functions of the variables measured during the field visits.

Two-phase sampling is also commonly combined with ratio or regression estimation, an application particularly suitable when destructive sampling must be used. For example, accurate measurement of tree volume often requires cutting down the tree. A rough estimate of volume can be obtained by using diameter at breast height, perhaps along with height or other measurements. Suppose on a first-phase sample of trees, $x = \text{dbh}$ is measured. Then a second-phase sample, often a subset of the first-phase sample, is selected, and an accurate measure of volume y is obtained by felling the tree. A regression equation is developed relating the difficult and expensive to measure variable y, to the easy to measure variable x. A similar application of two-phase sampling could be used to sample plant roots in a wetland survey. Digging up the plants would destroy the plots or quadrats being used, and these plots could not be revisited in a long-term monitoring program. Instead of digging up all sample plots, a subset of the first phase sample could be selected, and regression estimates could then be made within the usual two-phase sampling approach. Cochran (1977, Chapter 12) and Särndal et al. (1992, Chapter 9) present details on the analysis of two-phase samples.

4.4. Variable probability sampling

Stratified random sampling is a familiar design in which elements of the universe may have different inclusion probabilities. In most sampling texts, more general variable probability sampling designs are introduced in the context of sampling primary sampling units (clusters) according to a measure of size of the clusters (Cochran, 1977; Hansen et al., 1953; Raj, 1968). Large units such as cities, villages, or farms, are grouped into strata, and a small number of units, often two, are sampled from each stratum. Applications of variable probability sampling in forestry illustrate a common objective of these designs, which is to improve precision. Probability proportional to prediction sampling and variable plot (or Bitterlich) sampling (De Vries, 1986) are designed to improve precision of estimates, usually of total volume of a stand of trees, by sampling with probability proportional to a variable highly correlated with volume, such as dbh or a predicted volume for individual trees. These forestry examples are typically single objective surveys focusing on one variable, so the design is tailored to that variable; disadvantages that often accompany variable probability sampling in multiple objective surveys do not appear.

In environmental applications, variable probability sampling is more likely to arise because of practical considerations than for reasons of efficiency. Thus, variable probability sampling has a very different flavor in environmental surveys than in the survey sampling applications often referenced in standard texts and in the forestry examples above. The sampling design selected for the National Stream Survey (Kaufmann et al., 1988) illustrates this point. Constructing a list frame of all stream reaches in the target universe of the NSS was too expensive and time consuming to be practical, and such a list frame likely would have contained many imperfections. In place of a list frame, sampling units were selected using a square dot-grid, with density of one grid point per 64 square miles, imposed on 1 : 250 000-scale topographic maps of the target area. A target stream reach was selected into the sample if a grid point fell into the direct watershed of that reach. The direct watershed of a reach was that portion of the landscape in which a raindrop would flow directly into the stream reach, without first entering another reach. This protocol resulted in reaches being sampled with probability proportional to direct watershed area. Figure 5 illustrates a typical pattern of direct watersheds of target

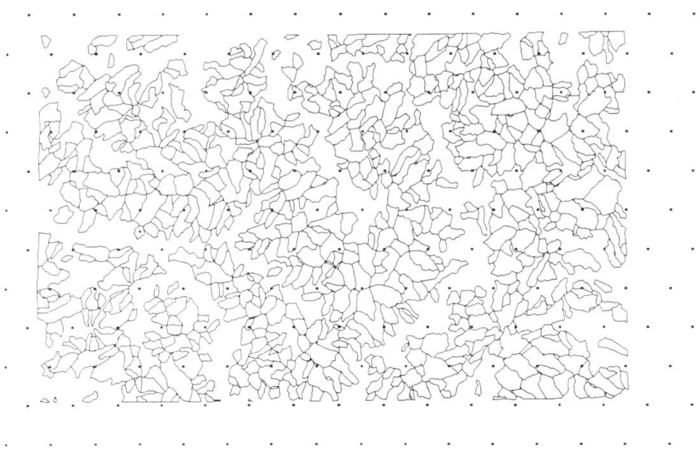

Fig. 5. Direct watersheds and dot grid for a single map quadrangle in the National Stream Survey. A stream reach within a direct watershed containing a grid point is sampled by the grid. Many grid points do not sample a unit. Grid density shown is much higher than that of the actual NSS.

reaches in a map quadrangle of part of the study area, with a square sampling grid imposed. Hankin (1984) describes another environmental application of variable probability sampling to selection of stream segments for estimating fish abundance.

Numerous procedures have been cataloged for selecting a variable probability sample from a list frame (Brewer and Hanif, 1983; Chaudhuri and Vos, 1988). Many of these procedures are of limited utility for environmental applications because they are designed for selecting a few primary sampling units in each stratum. Of these selection procedures, variable probability systematic (VPS) sampling (Brewer and Hanif, 1983, methods 1 and 2) is the easiest to implement for samples of size greater than two so that the sample size is fixed and π_u is proportional to x_u (Sunter, 1986; Wolter, 1985). Ordered VPS sampling differs from randomized VPS sampling in that the elements of the universe are randomly ordered prior to selection of the randomized version of the design, whereas the elements are sampled in their fixed position, or nonrandom order, for the ordered VPS design. An important special case of ordered VPS sampling orders the elements by an auxiliary variable prior to sampling, which ensures sample balance on that auxiliary variable.

Estimation for variable probability sampling is still based on the Horvitz–Thompson estimator (equation (2.13)) and the accompanying variance estimators (equation (3.3) or 3.4)). However, when an ordered VPS design is used, the usual problems with variance estimation for a systematic sample arise. As stated in Section 4.1, a common approach to variance estimation is to assume the universe was randomized prior to selection of the sample, and to use a variance estimator appropriate for randomized VPS sampling. Under this randomized design, or the randomized model for the ordered design, the pairwise inclusion probabilities may be calculated exactly (Hidiriglou and Gray, 1980) or approximated by the Hartley and Rao formula (1962). However, both of these methods require that the inclusion probabilities for the entire universe are known. In cases such as the National Stream Survey example, when only the inclusion probabilities for the sample elements are available, the following approximation suggested by Overton (1985) may be used:

$$\pi_{uv} \doteq \frac{2(n-1)\,\pi_u \pi_v}{2n - \pi_u - \pi_v}. \tag{4.2}$$

Stehman and Overton (1990) evaluated the performance of this approximation when used with the variance estimators given by equations (3.3) and (3.4). If the randomized VPS model is considered inappropriate as an approximation to the variance of an ordered VPS design, variance approximations suggested by Wolter (1985) and Stehman and Overton (1989) may be applied.

Although variable probability sampling may provide improved precision for some response variables, it is unlikely that more than a few of the many response variables of a multipurpose, environmental survey will be highly correlated with the inclusion probabilities. Thus variable probability sampling would be potentially inefficient for many variables. If the correlation between the inclusion probabilities and the response variable is low, the generalized regression estimator (Section 3.3.3) recovers virtually all of the precision lost by an inefficient variable probability design relative to equal probability sampling. Variable probability sampling has additional disadvantages in that variance estimation is more complex due to the computation of the pairwise inclusion probabilities, and analytic uses of survey data are more complicated (Section 6). Consequently, variable probability sampling is not recommended for general application in environmental sampling. For single variable surveys and for cases in which it is the most practical design, variable probability sampling retains an important role.

4.5. *Nonprobability sampling*

Practical considerations, such as when limited resources are available to characterize sample units, sometimes justify use of a nonprobability sample. For example, in the National Lake Survey, it was impractical to field visit each sampled lake more than once, and a probability sample selecting a random date and a random location in the lake would likely yield a poor representation of the condition of the sampled lake. Alternatively, selecting an index period and location for characterizing the lake may provide much better information about the condition of the lake. In the NLS, the center of the lake was selected as the index location, and the index period was just after lake turnover in the fall. Different resources will require a different index sample. For example, an index period selected for an estuary may be late summer, when the estuary is at its most biologically critical stage; the index location may be the deposition zone of the estuary, as nearly as this location can be determined. Although an index sample is still a judgment sample, the criteria for choosing the index sample are clearly identified and objectively applied. An index sample often will provide a better measure of the characteristics of a sampling unit than a small random sample, and it may be used in situations in which a probability sample is impractical. Conversely, in other situations, a simple random sample may allow meaningful assessment that cannot be obtained from an index sample, even if the number of samples per resource unit is very small.

Another example of nonprobability sampling is line transect sampling of animal populations. The probability of obtaining a particular animal in such a survey depends on the detectability of the animal, which may be a function of many variables, such as observer skill, weather conditions, and the animal's behavior. Because the true inclusion probabilities are unknown, special estimators have been developed, usually based on a model of the detectability function (Burnham et al., 1980).

Nonprobability samples do have a place in some other common statistical applications. For example, suppose a study is being designed to evaluate use of a biological indicator of ecological condition for application to lakes in New York State. Availability of both 'clean' and 'polluted' lakes is desirable for testing the indicator's response to widely varying environmental conditions. If a probability sample is used to select lakes for the study, variability among the lakes selected may be so high that any differences in the response of the indicators is obscured. Further, a probability sample may be impractical because it would allow little control over travel and other field expenses, and a costly effort may be needed to stratify the universe of lakes to ensure that a sufficient number of lakes in the 'clean' and 'polluted' classes was obtained. For this type of comparative study, particularly if only a small set of lakes will be used, selecting a judgment sample of conveniently located lakes representative of the conditions desired would be a more practical approach. If a large number of such lakes is available, a probability sample of these lakes would be reasonable.

In the experimental design setting, experimental units are usually purposively selected, so identifying a population for which this sample is representative is difficult (Mead, 1988, p. 214). Fienberg and Tanur (1987) discuss the role of probability sampling in experimental design, and Eberhardt and Thomas (1991) advocate integrating sampling of ecological systems into the framework of experimental field studies. Use of a probability sample allows clear identification of the population for which the sample is representative, but such a probability sample may be impractical or inefficient, and it is not wise to let probability sampling dominate the sampling protocol in this setting.

5. Sampling continuous spatial regions

For many environmental sampling problems, the universe is a continuous two-dimensional region. Suppose that a continuous response variable is defined over this surface. The objectives for sampling such a region may be population description, requiring estimation of means, totals, or distribution functions of populations defined on $\mathscr{U}$; alternatively, the objectives may require description of spatial pattern, such as by fitting explicit surface models (e.g., 'trend surface analysis' of geostatistics

(Koch and Link, 1971)) or prediction by kriging. Population description and spatial analyses represent different modes of inference, and the representation of variability is different in the two approaches. For the standard estimators of population description, design-based inference is still applicable, and the relevant variance is that generated by the randomization distribution of the design. For description of spatial pattern, the variability of concern is usually specified by a model, so that inferences are model dependent. De Gruijter and Ter Braak's (1990) discussion of design-based and model-based inference in the context of spatial sampling is highly recommended for clarification of these issues.

Several spatial sampling designs based on point samples may be considered. In unrestricted random sampling, the sample points are obtained by selecting n independent, random locations in the study domain. This approach is the continuous spatial analog of a simple random sample. Systematic sampling may be implemented by a rectangular grid or a triangular grid with a randomly positioned starting point. A third design of interest, called a tessellation-stratified design (Figure 6), is obtained by selecting one random point per stratum, the strata being the squares or hexagons formed by the tessellation of a rectangular or triangular grid, respectively. For these designs, the grid may be purposively placed, or it may be randomly placed. Olea (1984) lists a variety of other spatial sampling designs.

From the design-based perspective, comparisons of the precision of the estimated mean $\bar{y}$ for systematic, tessellation-stratified, and unrestricted random sampling have been made for various population models, including linear trend, periodic variation, and spatially correlated observations (Bellhouse, 1977, 1981a, b, 1985, 1988; Matérn, 1986; Stehman and Overton, 1991). In general, recommendations follow those found for systematic sampling in one dimension. Bellhouse (1981b), Koop (1990) and Matérn (1986) investigated precision of various grid sampling designs for estimation of area.

Variance estimation for both the systematic and tessellation-stratified designs is problematic. For the systematic design, a common approach to variance estimation is again to treat the systematic sample as an unrestricted random sample and to estimate variance by

$$\hat{V}(\bar{y}) = \sum_{s}(y_u - \bar{y})^2/(n-1). \tag{5.1}$$

This formula usually overestimates variance (Ripley, 1981). Yates (1981) and Matérn (1986) describe several alternative variance estimators for systematic sampling from a square grid. Stehman and Overton (1991) examined the bias of variance estimators derived from the mean-square-successive difference adapted to a triangular grid and found that these estimators worked better for the tessellation-stratified design than for the strict systematic design.

Some investigations of spatial sampling designs have focused on objectives other than precision of the estimated mean. Gilbert (1987, Section 8.7) recommended a square or triangular systematic grid if the objective is to estimate the spatial pattern and total amount of pollution in a region. Cressie (1991, Section 5.6.1) reviewed the

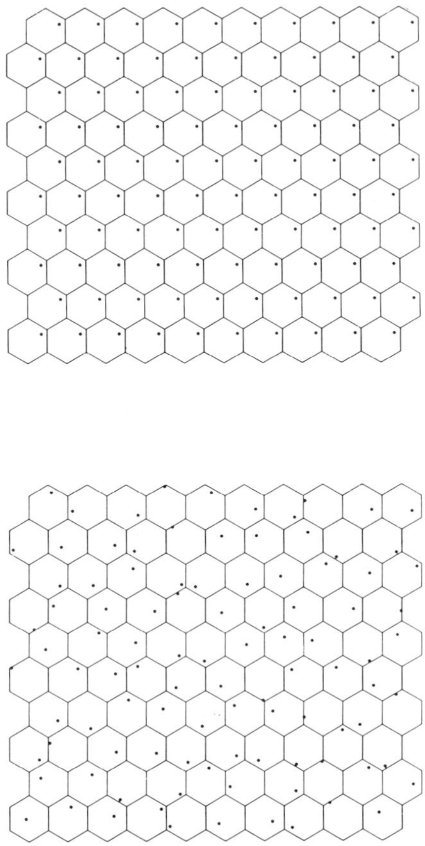

Fig. 6. Tessellation-stratified and systematic sampling designs for a triangular grid. The hexagonal strata represent the tessellation of the triangular grid. The systematic sample is shown in the upper figure. For the tessellation-stratified design (bottom figure), one random location is selected within each tessellation hexagon.

merits of different sampling designs according to the criterion of minimizing kriging variance, while Olea (1984) compared several sampling designs on the basis of precision for estimating spatial functions by universal kriging. An advantage of the tessellation-stratified design over the systematic design is that estimation of the variogram, a necessary step in many spatial analyses, is improved by having a greater variation in distances between neighboring sample points, including some distances which are small.

6. Analytic uses of survey data

Deming (1950, 1975) distinguished between enumerative studies, 'in which action will be taken on the material in the frame studied', and analytic studies, 'in which action will be taken on the process or cause-system that produced the frame studied'. Enumerative studies focus primarily on description of finite populations, whereas analytic studies focus primarily on comparisons and superpopulation inference. However, occasionally comparisons in the finite population mode are of interest, as are descriptions in the superpopulation mode. Enumerative and analytic studies cannot be distinguished on the basis simply of whether the objectives require description or comparison. Deming's distinction is closer to the general issue.

Analyses such as simple and multiple linear regression, analysis of variance, partial and multiple correlation, contingency table, and multivariate, are uses of survey data that are often labeled as analytic. If these standard analyses are conducted on survey data collected via a systematic, cluster, variable probability, or any probability sampling scheme other than simple random sampling, the analyses will often lead to inappropriate inferences (Kish, 1965; Rao and Bellhouse, 1990). Further, it may be difficult or even impossible to modify the analyses to yield appropriate treatment. Modifications of some common analyses for certain survey designs have been developed. Rao and Bellhouse (1990) and Särndal et al. (1992, Section 13.3) review some of these analytic uses of survey data. Skinner et al. (1989) present a comprehensive review of categorical data and regression analyses for complex survey data. We will limit our discussion of analytic surveys to a few specific points relevant to environmental sampling.

Finite population inferences frequently are of concern in environmental sampling because of the importance of resource management. For example, in the National Lake Survey, the objectives specified description of the current status of lakes at a particular point in time, and focused on numbers of lakes 'at risk'. Conversely, ecological questions may be directed toward a more general population question, such as whether the proportion of lakes at risk differs between two geographic regions. The question in such a comparison may be whether the *processes* generating lakes in the two regions differ, so that the real populations represent single realizations of different processes in each region. The null hypothesis is that the two samples came from the same superpopulation process. Comparison of the processes would be made in an infinite population mode, and no finite population correction factor is needed for the variance estimates (Deming, 1950). This point is obvious when it is considered that, if the two populations are completely enumerated (i.e., if $s \equiv \mathcal{U}$ in both populations), a statistical test is still required to evaluate the hypothesis.

Finite population comparison addresses the question, say, of which of two finite populations has the larger mean, whereas analytic comparison addresses the question of whether the two finite populations may be reasonably regarded as samples from the same superpopulation, or if one superpopulation has a larger mean than the other. If, given total finite population enumeration, a question of equality of the means remains, the population of concern is a superpopulation, and superpopulation inference should be applied. Conversely, if no uncertainty remains about the equality of the means, given complete population enumeration, then the comparison is a finite population comparison. Generally, management questions are answered in the finite mode, and scientific questions are addressed in the infinite mode. Statements about population numbers or totals are invariably finite population statements, while statements about proportions and means can be made in either mode.

Most sampling designs are focused on objectives of population description. However, stratified designs are well suited for surveys in which the objective is to compare populations of different strata. The common 'optimum allocation' for stratified sampling is not relevant, and an alternate allocation is optimal for the comparison objective (Cochran, 1977, Section 5A.13). If more sophisticated analyses are desired, then the class of lattice designs is appropriate (Jessen, 1978, Chapter 11). Typically, these designs lead to analysis of variance.

We can illustrate several key issues involved in the analysis of a stratified survey with an example which is motivated by analyses used in the National Lake Survey. The National Lake Survey design was based on a two-dimensional lattice design with approximately equal allocation in each of four strata, denoted $A1$, $B1$, $A2$ and $B2$, where A and B represent subregions, and 1 and 2 represent alkalinity classes. Suppose a population of 1400 lakes is partitioned into the strata with the population number in each stratum N_h shown in Table 1. An SRS of 50 lakes is obtained from each stratum, and description and comparison of distribution functions over strata is of interest. Distribution functions are compared by a contingency table analysis with the categories identified from the estimated distribution of the complete population; a feature of the analysis is that the estimated number of population units in each category is $N/5$, where N is a known parameter.

The Pearson χ^2 test may be used to test for differences among the distribution functions in the four strata. The SRS sample in each stratum is independently and identically distributed (iid) on the superpopulation appropriate for the stratum, resulting in four multinomial samples, each with $n = 50$. This simple χ^2-test is available only on the original strata, however, and any comparative question requiring combined strata encounters difficulties. For example, testing the combination of strata $A1$ and $B1$ against the combination of strata $A2$ and $B2$ would require a much more complicated analysis (see Särndal et al., 1992, Section 13.5).

Table 1
Contingency table for comparing distribution functions from a stratified random sample. The categories of the contingency table are constructed from the estimated population quintiles of the complete population. The subscript h is used to denote strata, while the subscript q denotes a category determined by the quintiles

Stratum	Quintile					n_h	N_h	$w_h = 1/\pi_h$
	1	2	3	4	5			
$A1$	0	0	10	20	20	50	200	4
$B1$	6	9	8	10	17	50	300	6
$A2$	13	12	9	10	6	50	400	8
$B2$	14	13	12	6	5	50	500	10
Sum	33	34	39	46	48	200	1400	
$\hat{N}_q$	280	280	280	280	280			

A simple design change eliminates the complexity inherent when combining strata. Let the sample be unstratified, with an SRS of size 200 out of the total population size of 1400 (Table 2). Now the observed sample in $A1$ is, conditional on $n = 32$ and given $N = 200$, an SRS. Similarly, the sample in the combination of $A1$ and $B1$ is conditionally an SRS, given $n = 72$ and $N = 500$. Any identifiable subset of the sample is also conditionally an SRS. If the necessary subpopulation N are unknown, the analytic model invokes conditional multinomials, directly from the iid sample from the full superpopulation. Comparison of the distribution functions of these disjoint sample subsets are readily made by contingency table analyses (Table 3), with inferences applying to the corresponding superpopulations. Stratified random sampling with proportional allocation could be analyzed by the same techniques for analysis and inference as the SRS (Skinner et al., 1989, p. 92).

Table 2
Contingency table for comparing distribution functions from a simple random sample. The strata of Table 1 are now identified as subpopulations, and the subscript h is used to denote the subpopulation

'Stratum'	Quintile					n_h	N_h	$\hat{N}_h$
	1	2	3	4	5			
$A1$	0	0	6	12	14	32	200	224
$B1$	4	8	9	7	12	40	300	280
$A2$	16	14	12	8	5	55	400	385
$B2$	20	18	13	13	9	73	500	511
Sum	40	40	40	40	40	200	1400	
$\hat{N}_q$	280	280	280	280	280			

Table 3
Contingency table for comparing distribution functions in combined subpopulations from a simple random sample. The subpopulations (identified as strata in Table 1) $A1$ and $B1$ are combined to form one subpopulation, and subpopulations $A2$ and $B2$ are combined to form another subpopulation. The sample data are as listed in Table 2. $\hat{N}_q$ is now the estimated number of units in the category identified by quintile q. Estimates of $\hat{N}_q$ are given for the two combined subpopulations

Stratum	Quintile					n_h	N_h	$\hat{N}_h$
	1	2	3	4	5			
$A1$ and $B1$	4	8	15	19	26	72	500	504
$A2$ and $B2$	36	32	25	21	14	128	900	896
Sum	40	40	40	40	40	200	1400	
$\hat{N}_q$, $A1 + B1$	28	56	105	133	182			
$\hat{N}_q$, $A2 + B2$	252	224	175	147	98			

Changing from the stratified to the unstratified design results in easier comparison of the distribution functions; the penalty is lowered precision of estimates for the smaller strata, and lowered precision of comparisons of two strata. When analyses of combined subpopulations are anticipated, stratification should be avoided if possible, unless sampling is done with proportional allocation. A similar position is taken with respect to variable probability sampling, either with or without strata, in that analytic comparisons cannot readily be made unless the design is equal probability, and preferably a simple random sample. That survey data are often available for public access places a premium on ease of appropriate analyses.

Another difficulty with standard analyses is posed when the selected units are clustered. Whether intracluster correlation is positive, as in attributes of grains of wheat sampled via entire heads, or negative, as in sex and age structure of quail sampled via family groups, clustering will often result in inappropriate identification of 'sample size' because of the lack of independence among individual observations. Many χ^2 tests conducted on contingency tables generated from cluster samples are suspect on this count. A simple, often effective way of dealing with this problem is to treat the cluster as the basic sampling unit, and to work either with cluster means or totals. This analysis may appear to lose observations, but that loss simply reflects the effect of the lack of independence. Correlation among sampling units often arises because environmental surveys have a spatial component. The effects of spatial pattern on standard analytic methods should also be considered, and Cressie (1991) and Upton and Fingleton (1985, 1989) provide detailed descriptions of procedures for analysis of spatial data.

Lastly, analysis of associations is a common analytic use of survey data. While associations among variables may be discovered via exploratory analysis of, say, correlations, following collection of the survey data, a more productive approach is to plan the survey with anticipated associations in mind. For example, if variables measured on wetlands are anticipated to be associated with conditions of bordering rivers or streams, the sampling unit may be constructed so that data are collected for wetland/river or wetland/stream locations in close proximity. Assessment of associations is still difficult because the effect of a variable on the response of the system may be widely separated in space and/or time from the original stimulation. Further, biologically unimportant or spurious associations are likely to be found simply by the sheer number of associations examined. Even if strong associations are discovered, assessment of causal relationships is impossible, because survey designs have no randomization mechanism assigning conditions or characteristics to the sampling units. Survey data are observational (Cochran, 1983, Chapter 1.1) regardless of the probability structure of the design.

7. Monitoring over time

Assessing change (difference between two time periods) and trend (consistent change over several time periods) in a population is frequently an objective of a monitoring program. Annual changes in condition of ecological resources are expected simply because of natural annual variation in ecological populations. A long-term monitoring project may be able to detect changes in conditions beyond those expected from normal annual variation. Estimation of trend generates issues not encountered in estimation of status, and as usual, the approach depends on the objectives of the survey. Smith (1990) stated that 'If the primary purpose is to produce descriptive statistics of the state of the system at each time period then the surveys can be considered as repetitions of a cross-section survey and each one can be analyzed independently'. In such cases, standard design-based inference may be applied. But if the survey objectives call for procedures attempting to explain changes in estimates over time or to predict future conditions, model-based methods would be required. These latter procedures would be classified as analytic uses of survey data, and as such would be subject to the difficulties introduced by a complex sampling design (Section 6).

In a monitoring program, the sampling design should reflect the dual concerns of description of trends and estimation of current condition. Estimation of change or trend leads to the desirability of having some repeat observations over time. Procedures such as rotating panel designs and sampling with partial replacement (Patterson, 1950) have been developed to provide for repeat observations for at least part of the sample. Duncan and Kalton (1987) and Binder and Hidiriglou (1988) provide a general review of surveys for sampling over time. The design proposed for EMAP (Overton et al., 1991) to estimate status and trends illustrates an environmental application of sampling over time. Two studies have compared the EMAP design to other alternatives. Lesser and Kalsbeek (1991) compared three types of panel survey designs for use in the agroecosystem component of EMAP, while Urquhart et al. (1991) evaluated rotating designs on the basis of a linear model formulation that included random effects for sites and years.

Analyses of change and trend from survey data will only be briefly discussed. Gilbert (1987) reviews a variety of statistical techniques for evaluating trend, all of which fit into the conventional statistical inference framework. Yates (1981, Sections 6.21 and 6.22) provides formulas for estimating the mean at each time period, and for estimating change in the mean response over two time periods. As stated previously, distribution functions are often more important than the mean as descriptors of the population.

A method for evaluating change in distribution functions was applied in the phase II National Lake Survey. The method assumes an equal probability sample and employs contingency table analyses of observations paired over time. A one degree of freedom chi-square test for shift in the distribution function is conducted by counting the number of observations plotted above and below a diagonal line through the origin for the data plotted as pairs, with the x-value the observation recorded at the first time period, and the y-value the observation recorded at the second time period (Figure 7). This test is essentially a sign test. Following the test for shift in the distribution function, a test for change in shape of the distribution function may be conducted using a 2×2 contingency table. The rows of the table account for whether the observation is above or below the diagonal line, and the columns account for whether the observations are above or below the median of a

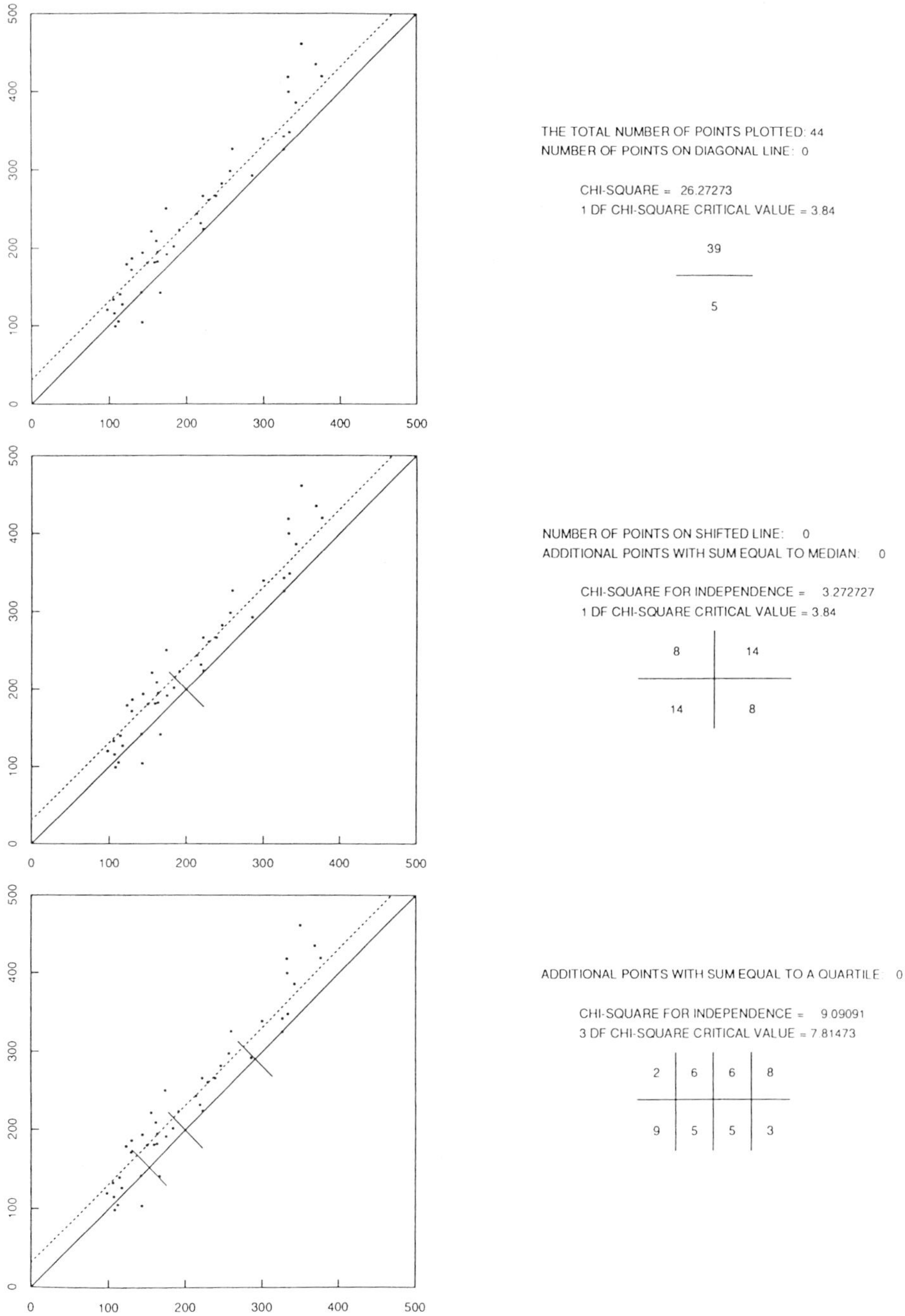

Fig. 7. Analysis of change in distribution functions over two time periods. The upper diagram displays the data for testing shift of the distribution function. The main diagonal is the solid 45 degree line through the origin, and the dashed line represents the shifted distribution of the pooled data. In the middle diagram, a solid line perpendicular to the diagonal divides the distribution of the sum of paired observations at the median. In the lower diagram the sums of the paired observations are divided at the quartiles. The plotted points represent the values of the variable for the two time periods on the same sampling unit.

variable which is the sum of the paired observations. A simple extension partitions the observations by the quartiles of the distribution of this sum. In practice, the diagonal line may be shifted to provide equal numbers above and below the main diagonal. A chi-square test is then calculated to test if a change in shape of the distribution function has occurred. The motivation in a monitoring program for such a test is that detection of changes via a simple screening test is an essential first step. Then, given evidence of real change, the process should be modeled and studied in sufficient depth to describe the nature and regularity of the change or trend observed.

8. Summary

For environmental surveys, our position is that a probability sample is always preferred for the basic survey design so that rigorous statistical inference can be made to the real population of concern. Nonprobability sampling may be used to characterize sampling units selected in the main survey and is particularly valuable when limited resources are available for measuring each sampling unit. Design-based inference should be applied, with the Horvitz–Thompson estimator providing the basis for a general, consistent strategy of estimation. In many instances, model-assisted estimation improves, sometimes substantially, the precision of description, and these estimators should routinely be considered in survey planning, and implemented when possible.

Because of their practical utility, advantage for description of spatial pattern, and advantage when model-assisted methods are used, systematic sampling designs should be given strong consideration in environmental sampling. Stratified sampling resulting in unequal inclusion probabilities, and variable probability sampling in general, are usually not recommended because of the complications introduced into analytic uses of survey data, and because precision of estimates for variables not highly correlated with the inclusion probabilities may be even worse than would have been obtained with equal probability sampling. However, variable probability sampling is sometimes unavoidable for practical reasons and is potentially valuable in single or limited objective surveys focused on description. Model-assisted inference can usually recover any precision lost by an inefficient variable probability design. Two-phase sampling should routinely be considered when destructive sampling is necessary and, in general, is a valuable design feature that can often be employed to advantage.

Simplicity of design is a key asset of any environmental sampling plan, particular for a multipurpose survey. In practice, the questions generated by the survey objectives must be carefully considered to determine the relative importance of analytic versus enumerative uses of the data, and to identify the populations that are of particular interest. Secondary analyses, carried out after the survey has been completed, are often a major use of survey data. If analytic uses dominate, equal probability designs lead to simpler analyses. Equal probability designs are also preferable if distribution functions are of primary interest. Effects of spatial autocorrelation and clustering on subsequent analyses should be considered when planning a sampling design.

Population description has long been a cornerstone of sampling theory and practice, so this area of environmental sampling is well established. Integrating spatial analyses and descriptions with probability sampling has recently begun (Cressie and Ver Hoef, 1992; De Gruijter and Ter Braak, 1990; Jager and Overton, 1992), but this area is not yet well developed. Similarly, design and analysis of probability samples for the purpose of describing temporal trends, and analytic uses of survey data from complex sampling designs, will likely see further advances in the years ahead.

Acknowledgements

We thank Cliff Cordy for reviewing portions of this manuscript, Perry Suk for technical editing, and George Weaver for graphics. This work has been supported by cooperative agreements CR815422 and CR816721 between the Environmental Protection Agency and Oregon State University.

References

Baker, J. P., S. A. Gherini, S. W. Christensen, C. T. Driscoll, J. Gallagher, R. K. Munson, R. M. Newton, K. H. Reckhow and C. L. Schofield (1990). *Adirondack Lakes Survey: An Interpretive Analysis of Fish Communities and Water Chemistry*, 1984–1987. Adirondack Lakes Survey Corporation, Ray Brook, NY.

Barnard, J., W. Myers, J. Pearce, F. Ramsey, M. Sissenwine and W. Smith (1985). Surveys for monitoring changes and trends in renewable resources: Forests and marine fisheries. *Amer. Statist.* **39**, 363–373.

Binder, D. A. and M. A. Hidiriglou (1988). Sampling in time. In: P. R. Krishnaiah and C. R. Rao, eds., *Handbook of Statistics*, Vol. 6: Sampling. Elsevier Science Publishers, Amsterdam, 187–211.

Bellhouse, D. R. (1977). Some optimal designs for sampling in two dimensions. *Biometrika* **64**, 605–611.

Bellhouse, D. R. (1981a). Spatial sampling in the presence of a trend. *J. Statist. Plann. Inference* **5**, 365–375.

Bellhouse, D. R. (1981b). Area estimation by point-counting techniques. *Biometrics* **37**, 303–312.

Bellhouse, D. R. (1985). Systematic sampling of periodic functions. *Canad. J. Statist.* **13**, 17–28.

Bellhouse, D. R. (1988). Systematic sampling. In: P. R. Krishnaiah and C. R. Rao, eds., *Handbook of Statistics*, Vol. 6: Sampling. Elsevier Science Publishers, Amsterdam, 125–145.

Brewer, K. R. W. and M. Hanif (1983). *Sampling with Unequal Probabilities*. Springer, New York.

Burnham, K. P., D. R. Anderson and J. L. Laake (1980). Estimation of density for line transect sampling of biological populations. *Wildlife Monographs* **72**.

Cassel, C.-M., C.-E. Särndal and J. H. Wretman (1977). *Foundations of Inference in Survey Sampling*. Wiley, New York.

Chambers, R. L. and R. Dunstan (1986). Estimating distribution functions from survey data. *Biometrika* **73**, 597–604.

Chaudhuri, A. and J. W. E. Vos (1988). *Unified Theory and Strategies of Survey Sampling*. North-Holland, Amsterdam.

Cochran, W. G. (1977). *Sampling Techniques*. 3rd Ed., Wiley, New York.

Cochran, W. G. (1983). *Planning and Analysis of Observational Studies*. Wiley, New York.

Cormack, R. M., G. P. Patil and D. S. Robson (1979). *Sampling Biological Populations*. International Co-operative Publishing House, Fairland, MD.

Cressie, N. A. C. (1991). *Statistics for Spatial Data.* Wiley, New York.

Cressie, N. A. C. and J. M. Ver Hoef (1992). Spatial statistical analysis of environmental and ecological data. In: *Proc. Conf. on Integrating Geographic Information Systems and Environmental Modelling,* Boulder, CO, September 1991.

Cumberland, W. G. and R. M. Royall (1981). Prediction models and unequal probability sampling. *J. Roy. Statist. Soc. Ser. B* **43**, 353–367.

De Gruijter, J. J. and C. J. F. Ter Braak (1990). Model-free estimation from spatial samples: A reappraisal of classical sampling theory. *Math. Geol.* **22**, 407–415.

Deming, W. E. (1950). *Some Theory of Sampling.* Wiley, New York.

Deming, W. E. (1975). On probability as a basis for action. *Amer. Statist.* **29**, 146–152.

De Vries, P. G. (1986). *Sampling Theory for Forest Inventory.* Springer, New York.

Duncan, G. J. and G. Kalton (1987). Issues of design and analysis of surveys across time. *Internat. Statist. Rev.* **55**, 97–117.

Durbin, J. (1958). Sampling theory for estimates based on fewer individuals than the number selected. *Bull. Internat. Statist. Inst.* **36**, 113–119.

Eberhardt, L. L. and J. M. Thomas (1991). Designing environmental field studies. *Ecol. Monographs* **61**, 53–73.

Fienberg, S. E. and J. M. Tanur (1987). Experimental and sampling structures: Parallels diverging and meeting. *Internat. Statist. Rev.* **55**, 75–96.

Fisher, R. A. (1956). *Statistical Methods and Scientific Inference.* Oliver and Boyd, London.

Frayer, W. E., T. J. Monahan, D. C. Bowden and F. A. Graybill (1983). Status and trends of wetlands and deepwater habitats in the conterminous United States, 1950 to 1970. Dept. of Forest and Wood Sciences, Colorado State University, Ft. Collins, 32 pp.

Gilbert, R. O. (1987). *Statistical Methods for Environmental Pollution Monitoring.* Van Nostrand Reinhold, New York.

Hankin, D. G. (1984). Multistage sampling designs in fisheries research: Applications in small streams. *Canad. J. Fish. Aqua. Sci.* **41**, 1575–1591.

Hansen, M. H., W. N. Hurwitz and W. G. Madow (1953). *Sample Survey Methods and Theory,* Vol. I: Methods and Applications. Wiley, New York.

Hansen, M. H., W. G. Madow and B. J. Tepping (1983). An evaluation of model-dependent and probability-sampling inferences in sample surveys (with discussion). *J. Amer. Statist. Assoc.* **78**, 776–807.

Hartley, H. O. and J. N. K. Rao (1962). Sampling with unequal probability and without replacement. *Ann. Math. Statist.* **33**, 350–374.

Hidiriglou, M. A. and G. B. Gray (1980). Construction of joint probability of selection for systematic p.p.s. sampling. *Appl. Statist.* **29**, 107–112.

Horvitz, D. G. and D. J. Thompson (1952). A generalization of sampling without replacement from a finite universe. *J. Amer. Statist. Assoc.* **47**, 663–685.

Jager, H. I. and W. S. Overton (1992). Explanatory models for ecological response surfaces. In: *Proc. Conf. on Integrating Geographic Information Systems and Environmental Modelling*, Boulder, CO, September 1991.

Jessen, R. J. (1978). *Statistical Survey Techniques.* Wiley, New York.

Kalton, G. (1983). Models in the practice of survey sampling. *Internat. Statist. Rev.* **51**, 175–188.

Kaufmann, P. R., A. T. Herlihy, J. W. Elwood, M. E. Mitch, W. S. Overton, M. J. Sale, J. J. Messer, K. A. Cougan, D. V. Peck, K. H. Reckhow, A. J. Kinney, S. J. Christie, D. D. Brown, C. A. Hagley and H. I. Jager (1988). *Chemical Characteristics of Streams in the Mid-Atlantic and Southeastern United States.* Vol. I: Population Descriptions and Physico-Chemical Relationships (EPA/600/3-88/021a). U. S. Environmental Protection Agency, Washington, DC.

Kish, L. (1965). *Survey Sampling.* Wiley, New York.

Koch, G. S. and R. F. Link (1971). *Statistical Analysis of Geological Data:* Vol. II. Wiley, New York.

Koop, J. C. (1990). Systematic sampling of two-dimensional surfaces and related problems. *Comm. Statist. Theory Methods* **19**, 1701–1750.

Krishnaiah, P. R. and C. R. Rao eds. (1988). *Handbook of Statistics*, Vol. 6: Sampling. North-Holland, New York.

Kuk, A. Y. C. (1988). Estimation of distribution functions and medians under sampling with unequal probabilities. *Biometrika* **75**, 97–103.
Lesser, V. M. and W. D. Kalsbeek (1991). A comparison of periodic survey designs employing multi-stage sampling. *Proc. Sec. Statist. Environ. Amer. Statist. Assoc.*, 186–190.
Little, R. J. A. (1983). Estimating a finite population mean from unequal probability samples. *J. Amer. Statist. Assoc.* **78**, 596–604.
Matérn, B. (1986). *Spatial Variation.* 2nd ed., Springer, New York.
Mead, R. (1988). *The Design of Experiments: Statistical Principles for Practical Application.* Cambridge Univ. Press, New York.
Messer, J. J., R. A. Linthurst and W. S. Overton (1991). An EPA program for monitoring ecological status and trends. *Environ. Monitor. Assess.* **20**, 67–78.
Milne, A. (1959). The centric systematic area-sample treated as a random sample. *Biometrics* **15**, 270–297.
Murthy, M. N. (1967). *Sampling Theory and Methods.* Statistical Publ. Soc., Calcutta.
Olea, R. A. (1984). Sampling design optimization for spatial functions. *Math. Geol.* **16**, 369–392.
Overton, W. S. (1985). A sampling plan for streams in the national stream survey. Technical Report 114, Dept. of Stat., Oregon St. Univ., Corvallis, OR.
Overton, W. S. (1992). Probability sampling and population inference in monitoring programs. In: *Proc. Conf. on Integrating Geographic Information Systems and Environmental Modelling*, Boulder, CO, September 1991.
Overton, W. S., P. Kanciruk, L. A. Hook, J. M. Eilers, D. H. Landers, D. F. Brakke, D. J. Blick, R. A. Linthurst, M. D. DeHaan and J. M. Omernik (1986). *Characteristics of Lakes in the Eastern United States.* Vol. II: Lakes Sampled and Descriptive Statistics for Physical and Chemical Variables (EPA/600/4-86/007b). U.S. Environmental Protection Agency, Washington, DC.
Overton, W. S., D. White and D. L. Stevens (1991). *Design Report for EMAP: Environmental Monitoring and Assessment Program.* (EPA/600/3-91/053). U.S. Environmental Protection Agency, Washington, DC.
Patterson, H. D. (1950). Sampling on successive occasions with partial replacement of units. *J. Roy. Statist. Soc. Ser. B* **12**, 241–255.
Raj, D. (1968). *Sampling Theory.* McGraw-Hill, New York.
Rao, J. N. K. and D. R. Bellhouse (1990). History and development of the theoretical foundations of survey based estimation and analysis. *Survey Methodol.* **16**, 3–29.
Rao, J. N. K., J. G. Kovar and H. G. Mantel (1990). On estimating distribution functions and quantiles from survey data using auxiliary information. *Biometrika* **77**, 365–375.
Ripley, B. D. (1981). *Spatial Statistics.* Wiley, New York.
Royall, R. M. (1970). On finite population sampling theory under certain linear regression models. *Biometrika* **57**, 377–387.
Royall, R. M. and W. G. Cumberland (1981). An empirical study of the ratio estimator and estimators of its variance. *J. Amer. Statist. Assoc.* **76**, 66–77.
Särndal, C. E. (1980). On π-inverse weighting versus best linear unbiased weighting in probability sampling. *Biometrika* **67**, 639–650.
Särndal, C. E., B. Swensson and J. H. Wretman (1989). The weighted residual technique for estimating the variance of the general regression estimator of the finite population total. *Biometrika* **76**, 527–537.
Särndal, C. E., B. Swensson and J. Wretman (1992). *Model Assisted Survey Sampling.* Springer, New York.
Sedransk, J. and P. J. Smith (1988). Inference for finite population quantiles. In: P. R. Krishnaiah and C. R. Rao, eds., *Handbook of Statistics*, Vol. 6: Sampling. North-Holland, Amsterdam, 267–289.
Sen, A. R. (1953). On the estimate of the variance in sampling with varying probabilities. *J. Indian Soc. Agric. Statist.* **7**, 119–127.
Skinner, C. J., D. Holt and T. M. F. Smith eds. (1989). *Analysis of Complex Surveys.* Wiley, New York.
Smith, T. M. F. (1990). Comment on 'History and development of the theoretical foundations of survey based estimation and analysis', by J. N. K. Rao and D. R. Bellhouse. *Survey Methodol.* **16**, 26–29.
Stehman, S. V. and W. S. Overton (1989). Variance estimation for fixed-configuration, systematic sampling. Technical Report 134, Dept. of Stat., Oregon St. Univ., Corvallis, OR.
Stehman, S. V. and W. S. Overton (1990). An empirical, general population assessment of the variance

and variance estimators of the Horvitz–Thompson estimator under variable probability sampling. *Proc. Sec. Surv. Res. Meth. Amer. Statist. Assoc.*, 213–218.

Stehman, S. V. and W. S. Overton (1991). Properties of designs for sampling spatial continuous resources. *Proc. Sec. Statist. Environ. Amer. Statist. Assoc.*, 182–187.

Stevens, D. L., Jr. and A. R. Olsen (1991). Statistical issues in environmental monitoring and assessment. *Proc. Sec. Statist. Environ. Amer. Statist. Assoc.*, 76–85.

Stuart, A. (1962). *The Basic Ideas of Scientific Sampling.* 1984 ed., Charles Griffin, London.

Sunter, A. (1986). Solutions to the problem of unequal probability sampling without replacement. *Internat. Statist. Rev.* **54**, 33–50.

Upton, G. J. G. and B. Fingleton (1985). *Spatial Data Analysis by Example.* Vol. I: Point Pattern and Quantitative Data. Wiley, New York.

Upton, G. J. G. and B. Fingleton (1989). *Spatial Data Analysis by Example.* Vol. II: Categorical and Directional Data. Wiley, New York.

Urquhart, N. S., W. S. Overton and D. S. Birkes (1991). Comparing sampling designs for monitoring ecological status and trends: Impact of temporal patterns. Technical Report 149, Dept. of Stat., Oregon St. Univ., Corvallis, OR.

Von Neuman, J., R. H. Kent, H. R. Bellinson and B. I. Hart (1941). The mean square successive difference. *Ann. Math. Statist.* **12**, 153–162.

Webster, R. and M. A. Oliver (1990). *Statistical Methods in Soil and Land Resource Survey.* Oxford Univ. Press, New York.

Wolter, K. M. (1985). *Introduction to Variance Estimation.* Springer, New York.

Yates, F. (1981). *Sampling Methods for Censuses and Surveys.* 4th ed., Charles Griffin, London.

Yates, F. and P. M. Grundy (1953). Selection without replacement from within strata with probability proportional to size. *J. Roy. Statist. Soc. Ser. B* **15**, 235–261.

Zinger, A. (1980). Variance estimation in partially systematic sampling. *J. Amer. Statist. Assoc.* **75**, 206–211.

G. P. Patil and C. R. Rao, eds., *Handbook of Statistics, Vol. 12*

10

Ecological Statistics

Bryan F. J. Manly

Abstract: This chapter provides a review of statistical ecology, which is interpreted to include all statistical methods that are of value in the study of the relationship between animals and plants and their environment, but more particularly those statistical methods that have been specially developed for use in this area. The review is framed in terms of three broad topics: single species problems, community ecology problems, and general questions concerning observational and experimental studies in ecology.

1. Introduction

This chapter provides a review of some statistical methods that have been developed primarily for applications in studies of animal and plant ecology. These methods are considered within the three broad categories of (a) those that are applicable to populations of a single species, including the estimation of population parameters, studies of distributions in time and space, and the use of resources by the species; (b) those that are applicable to communities, including the measurement of characteristics such as species diversity, the interaction between different species, and the extent to which the use of resources overlaps for different species; and (c) general matters related to the design of observational and experimental studies. The approach adopted for individual topics within these categories is to provide a brief review including a description of important recent developments, followed in some cases by an example to illustrate how methods are used in practice. Because of the wide range of methods covered by the description 'ecological statistics', it is inevitable that some important topics are omitted, or only given a cursory mention.

In order to clarify what should and what should not be included in the chapter, a distinction has been made between mathematical ecology (which is concerned with the use of mathematical models to elucidate the effects of ecological factors, without these models necessarily being related to data collected on real populations), and statistical ecology (which is concerned primarily with the analysis of data). The emphasis is firmly on the latter area, with mathematical models only being considered to the extent that they aid with data analysis.

2. Single species problems

The estimation of basic demographic parameters is crucial to studies of the ecology of a single animal population. To that end, a variety of techniques have been developed, including mark–recapture methods for estimating sizes, survival rates and entry rates for mobile populations, and other methods for estimation based on encounter sampling. Sometimes it is the spatial structure of a population that is of interest, including the question of whether the members of the population are in random locations in the study area, and the manner in which mobile organisms disperse. Time changes may also be important, including the assessment of whether the size of a population is controlled by density-dependent factors, or whether changes in an ecological time series can be attributed to the impact of some disturbance. Many studies are also concerned with the use of resources by a population. Typical questions here relate to the selection of food and habitat by animals, but the study of the locations of plants raises similar problems.

Obviously other topics related to the ecology of single populations can be and are studied using statistical methods. However, those that have been mentioned are sufficient to convey what is involved with single species problems and it is these that are discussed in the following six sections.

2.1. Mark–recapture and related methods

Mark–recapture sampling methods have been developed mainly for the estimation of the number of individuals in a mobile animal population. Early uses of the principles involved were by Laplace in 1786 to estimate the human population in France, by Petersen in 1896 to estimate the size of a fish population, and by Lincoln in 1930 to estimate the wildfowl population in North America (Cormack, 1968).

The basic idea is a simple argument of proportions. For example, suppose that it is desired to estimate N, the number of fish in a lake. This can be done by taking a sample of n_1 fish from the lake, tagging each of them, releasing them back into the lake, waiting long enough for the tagged fish to mix freely with the untagged fish, taking another sample of n_2 fish, and recording m, the number of tagged fish. Then, assuming that (a) the population is 'closed' (so that the size N does not change during the sampling period), (b) the second sample is a simple random sample from the population, (c) individuals do not lose their tags between sample times, and (d) all tags can be recognized in the second sample, it can be argued that the proportion of tagged fish in the lake should be approximately the same as the proportion in the second sample. Thus $n_1/N \approx m/n_2$, and an estimator of the population total (the 'Lincoln–Petersen index') is

$$\hat{N} = \frac{n_1 n_2}{m}. \tag{2.1}$$

Since it is possible to get no marked animals in the second sample it is apparent that equation (2.1) can produce an infinite estimate of N, and may therefore be unsatisfactory. For this reason, the modified estimator

$$\hat{N}^* = \frac{(n_1+1)(n_2+1)}{m+1} - 1, \tag{2.2}$$

is better, where this is unbiased when $n_1 + n_2 \geqslant N$ and approximately unbiased otherwise, and has the estimated standard error

$$\widehat{\mathrm{SE}}(\hat{N}^*) = \sqrt{\frac{(n_1+1)(n_2+1)(n_1-m)(n_2-m)}{(m+1)^2(m+2)}}, \tag{2.3}$$

(Seber, 1982, p. 60). A 'large sample' $100(1-\alpha)\%$ confidence interval is then $\hat{N}^* \pm z_{\alpha/2}\widehat{\mathrm{SE}}(\hat{N}^*)$, where $z_{\alpha/2}$ is the value that is exceeded with probability $\alpha/2$ for the standard normal distribution. For the small sample sizes often met with in practice this confidence interval may be rather poor and a good deal of recent research has been directed toward obtaining better limits (see below).

As an example, suppose that before the start of the hunting season, a wildlife organization wants to estimate the size of a deer population. A sample of 300 deer taken, and the deer are tagged and released. A second sample of 200 deer is taken two weeks later and found to contain 62 tagged deer. Then the nearly unbiased estimate is $\hat{N}^* = 301 \times 201/63 - 1 = 959.3$, with

$$\widehat{\mathrm{SE}}(\hat{N}^*) = \sqrt{\frac{301 \times 201 \times (300-62) \times (200-62)}{63^2 \times 64}} = 88.4.$$

This suggests that the population size is probably within the range $959 \pm 1.96 \times 88.4$, i.e., 959 ± 173.

From the starting point of the Lincoln–Petersen index a large body of statistical methods have been developed for the estimation of the size of a closed population from two or more samples, with the capture probabilities for individual animals allowed to vary with time or individual or capture history, as well as some combinations of these three factors (Seber, 1982, 1986; Pollock et al., 1990; Pollock, 1991).

A major advance was the development of the theory for a set of eight models for a study involving k successive samples being taken, with unmarked animals being marked in the first $k-1$ samples in such a way that their recaptures can be recorded. This theory was summarized by Otis et al. (1978), who also facilitated its use by developing a computer program CAPTURE to carry out the calculations. Briefly, the eight models are as follows:

M_o. *Equal catchability*: All animals have the same probability p of being captured on each sampling occasion;

M_h. *Heterogeneity*: The j-th animal has its own unique capture probability p_j, which remains constant for all sampling occasions;

M_b. *Trap response*: Every unmarked animal has the same capture probability p, which changes to c after the first capture;

M_{bh}. *Heterogeneity and trap response*: The j-th animal has its own unique capture probability p_j before it is captured, which changes to c_j after the first capture;

M_t. *Time variation*: All animals have the capture probability p_i for the i-th sample;

M_{th}. *Time variation and heterogeneity*: The j-th animal has its own unique capture probability, which varies from sample to sample;

M_{th}. *Time variation and trap response*: The probability of capture is p_i for uncaptured animals and c_i for captured animals in the i-th sample;

M_{tbh}. *Time variation, trap response and heterogeneity*: The j-th animal has its own unique capture probability, which varies with time, and changes after the first capture. The program CAPTURE includes a model selection algorithm for choosing between these models on the basis of goodness of fit test, but this facility needs to be used with caution with small sets of data (Pollock et al., 1990).

Many of the developments in mark–recapture methodology have been designed for 'open' animal populations, where new animals are entering through 'births' and immigration, and animals are leaving through deaths and emigration. Here studies usually involve sampling the population at several times $t_1, t_2, \ldots, t_k$, with animals being suitable marked when they are first captured so that they can be recognized when they are recaptured and a record obtained of the captures and recaptures of individual animals. A major milestone was the publication of the Jolly–Seber model (Jolly, 1965; Seber, 1965) which provides explicit parameter estimators with variances on the assumption that (a) every animal present in the population has the same probability p_i of being captured in the i-th sample; (b) every marked animal present in the population immediately after the i-th sample has the same probability ϕ_i of surviving until the $(i+1)$-th sample is taken; (c) marks are not lost or not recognized; and (d) all samples are effectively instantaneous, and releases are made immediately after sampling.

The Jolly–Seber estimators were derived using the method of maximum likelihood. However, they can also be justified intuitively in the manner that will now be described. First, consider the estimation of the number M_i of marked animals in the population just before time t_i. This consists of the number m_i of marked animals seen in the i-th sample, plus those that were in the population but not captured. To estimate the latter, it can be argued that the probability of an animal that is in the population at time t_i being recaptured in a later sample can be estimated by $z_i/(M_i - m_i)$, where z_i is the number of animals seen before and after the i-th sample but not in the i-th sample, and also by r_i/R_i, where R_i is the number of animals released at time t_i, of which r_i are recaptured later. Hence it is expected that $z_i/(M_i - m_i) \approx r_i/R_i$, so that $M_i \approx m_i + R_i z_i/r_i$. This then suggests the estimator

$$\hat{M}_i = m_i + \frac{R_i z_i}{r_i} \tag{2.4}$$

for M_i, which can be evaluated for $i = 2, 3, \ldots, k-1$ since for these values of i all the values on the right-hand side can be computed from the data.

Once the M_i is estimated, the population size N_i at time i can be estimated since the proportion of marked animals in the i-th sample should be approximately equal to the proportion in the population, i.e., $m_i/n_i \approx M_i/N_i$. Thus an estimator of N_i is

$$\hat{N}_i = \frac{n_i \hat{M}_i}{m_i}. \tag{2.5}$$

This can be evaluated for $i = 2, 3, \ldots, k-1$.

To estimate ϕ_i, the survival rate over the period t_i to t_{i+1}, it can be noted that there are $M_i + R_i - m_i$ marked animals in the population just after the release from the i-th sample, and that M_{i+1} of these are still alive at time t_{i+1}. Thus $\phi_i \approx M_{i+1}/(M_i + R_i - r_i)$, suggesting the survival estimator

$$\hat{\phi}_i \approx \frac{\hat{M}_{i+1}}{\hat{M}_i + R_i - r_i}, \tag{2.6}$$

which can be evaluated for $i = 1, 2, \ldots, k-2$, taking $\hat{M}_i = 0$ because there are no marked animals in the population just before the time of the first sample.

Finally, to estimate B_i, the number of new entries to the population between time t_i and time t_{i+1} that are still alive at time t_{i+1}, this can be set equal to the estimated difference between the population size at time t_{i+1} and the expected number of survivors from time t_i, i.e.,

$$\hat{B}_i = \hat{N}_{i+1} - \hat{\phi}_i(\hat{N}_i - n_i + R_i). \tag{2.7}$$

This can be evaluated for $i = 2, 3, \ldots, k-2$.

One important feature of the estimation equations (2.4) to (2.7) is that it is not necessary for the number of animals released from the i-th sample (R_i) to equal the sample size (n_i). This means that damaged animals do not need to be released, and extra animals can be added to the population if desired. Of course, if the number released does not equal the number captured then this will change population sizes from what they would otherwise have been.

In practice the Jolly–Seber estimators will be biased to some extent. In fact, infinite estimates are possible. However, approximately unbiased estimators can be obtained by replacing $\hat{M}_i$ and $\hat{N}_i$ by

$$\hat{M}_i^* = m_i + \frac{(R_i + 1)z_i}{r_i + 1}, \tag{2.8}$$

and

$$\hat{N}_i^* = \frac{(n_i + 1)\hat{M}_i}{m_i + 1} \tag{2.9}$$

respectively (Seber, 1982, p. 304). The other estimation equations can be left unchanged.

Variance equations were provided by Jolly (1965) and Seber (1965), where in using these it is necessary to distinguish between sampling errors (the difference between estimated parameters and the parameters occurring in the realized population) and stochastic errors (the difference between the realized parameters and their mean values from hypothetical repeated generations of the population). For the population size it seems to be generally accepted that only sampling errors are relevant so that the variance of $\hat{N}_i$ given N_i is needed, where this is approximately

$$\mathrm{Var}(\hat{N}_i \mid N_i) = N_i[N_i - \mathrm{E}(n_i)]\left[\{M_i - \mathrm{E}(m_i) + R_i\}\left\{\frac{1}{\mathrm{E}(r_i)} - \frac{1}{R_i}\right\}\Big/ M_i + \frac{(N_i - M_i)}{N_i \mathrm{E}(m_i)}\right], \quad (2.10)$$

with $\mathrm{E}(\cdot)$ indicating an expected value from repeatedly resampling the same population. On the other hand, it is usually considered that it is the survival probability that is being estimated rather than the proportion surviving in the population. Therefore, the variance of $\hat{\phi}_i$ is taken to include stochastic errors. It can then be approximated by

$$\begin{aligned}\mathrm{Var}(\hat{\phi}_i) = \phi_i^2 \Bigg[\Bigg[&\{\mathrm{E}(M_{i+1}) - \mathrm{E}(m_{i+1})\}\{\mathrm{E}(M_{i+1}) - \mathrm{E}(m_{i+1}) \\ &+ \mathrm{E}(R_i)\}/\mathrm{E}(M_{i+1})^2\Bigg]\left\{\frac{1}{\mathrm{E}(r_{i+1})} - \frac{1}{\mathrm{E}(R_{i+1})}\right\} \\ &+ \{\mathrm{E}(M_i) - \mathrm{E}(m_i)\}/\{\mathrm{E}(M_i) - \mathrm{E}(m_i) + \mathrm{E}(R_i)\} \\ &\times \left\{\frac{1}{\mathrm{E}(r_i)} - \frac{1}{\mathrm{E}(R_i)}\right\}\Bigg] \\ &+ \frac{\phi_i(1-\phi_i)}{\mathrm{E}(M_i) - \mathrm{E}(m_i) + \mathrm{E}(R_i)}. \end{aligned} \quad (2.11)$$

In practice $\mathrm{Var}(\hat{N}_i|N_i)$ and $\mathrm{Var}(\phi_i)$ can be estimated by substituting observed data values for the expected values in equations (2.10) and (2.11). Variance equations for $\hat{M}_i$, $\hat{B}_i$ can be found in the original papers of Jolly (1965) and Seber (1965), and in reviews such as those of Seber (1982) and Pollock et al. (1990). Covariance equations are also available from the same sources.

Accurate estimates can only be obtained from the Jolly–Seber model if the data are extensive since so many parameters are estimated. For this reason, some later work has been directed towards models with fewer parameters, such as models with a constant survival rate per unit time. Generalizations of the Jolly–Seber model allowing for trap response, the analysis of data from several cohorts of animals, and age-dependent survival rates are also available (Seber, 1986; Pollock et al., 1990; Pollock, 1991). Computer programs JOLLY and JOLLYAGE for carrying out the calculations for these different models are described by Pollock et al. (1990).

At the same time that methods have been developing for the design and analysis of mark–recapture experiments there has been a parallel development of models for estimating survival rates from experiments that involve recovering dead animals, principally in the context of bird banding (Brownie, et al., 1985; North, 1987), with much of this work being summarized in the proceedings of the EURING conferences (Morgan and North, 1985; North, 1987). The study of the movement of birds from band-recovery and recapture experiments has also received a good deal of attention in this area (see below).

An understanding of the different uses of the mark–recapture method is aided by considering the classification of mark–recapture models that is shown in Figure 1. Here the most basic distinction is between situations where marked animals are recaptured with probability one (if they are alive), and situations where recapture is not certain. The first of these two categories includes studies where individual animals are fitted with radio transmitters, and studies where eggs in nests are marked and then revisited later to see whether they have survived. The second of the main categories includes studies involving the recapture of live animals or the recovery of dead animals, with the former category subdivided according to whether the population is open or closed, whether two or more samples are taken, etc.

Recent work on mark–recapture methods for closed populations has included the development of modified estimators and sampling schemes (Castledine, 1981; McDonald et al., 1983; Gazey and Staley, 1986; Bonett et al., 1987; Chao, 1987, 1989; Chao et al., 1992; P. J. Smith, 1988; Yip, 1989, 1991a, b, 1992; Cormack, 1989; Minta and Mangel, 1989; Eberhardt, 1990; Wolter, 1990); the development of alternative confidence intervals for the true population size (Viveros and Sprott, 1986; Wong and Mcdonald, 1987; Jensen, 1989; McDonald and Palanacki, 1989; Garthwaite and Buckland, 1990; Cormack, 1992), including computer intensive ones based on bootstrapping and related methods (Buckland, 1984; Buckland and Garthwaite, 1991; Garthwaite and Buckland, 1992); the development and testing of improved methods for allowing for the movement of animals (Anderson et al., 1983; Parmenter et al., 1989; Whitehead, 1990); and the development of methods for allowing for capture probabilities to depend on other variables measured either at the individual or the population level (Pollock et al., 1984; Huggins, 1989, 1991; Alho, 1990).

For open populations recent work has included the development of a robust sampling design for long term studies that assumes that a population is approximately closed during short periods of intensive sampling, but is open between these intensive sampling periods (Pollock, 1982; Pollock et al., 1990); the use of this model to obtain separate estimates of immigration and reproduction parameters (Nichols and Pollock, 1990); the development of methods for comparing the survival rates of different groups of fish and other animals (Burnham et al., 1987); the development of methods for allowing for the movement of animals between geographical areas

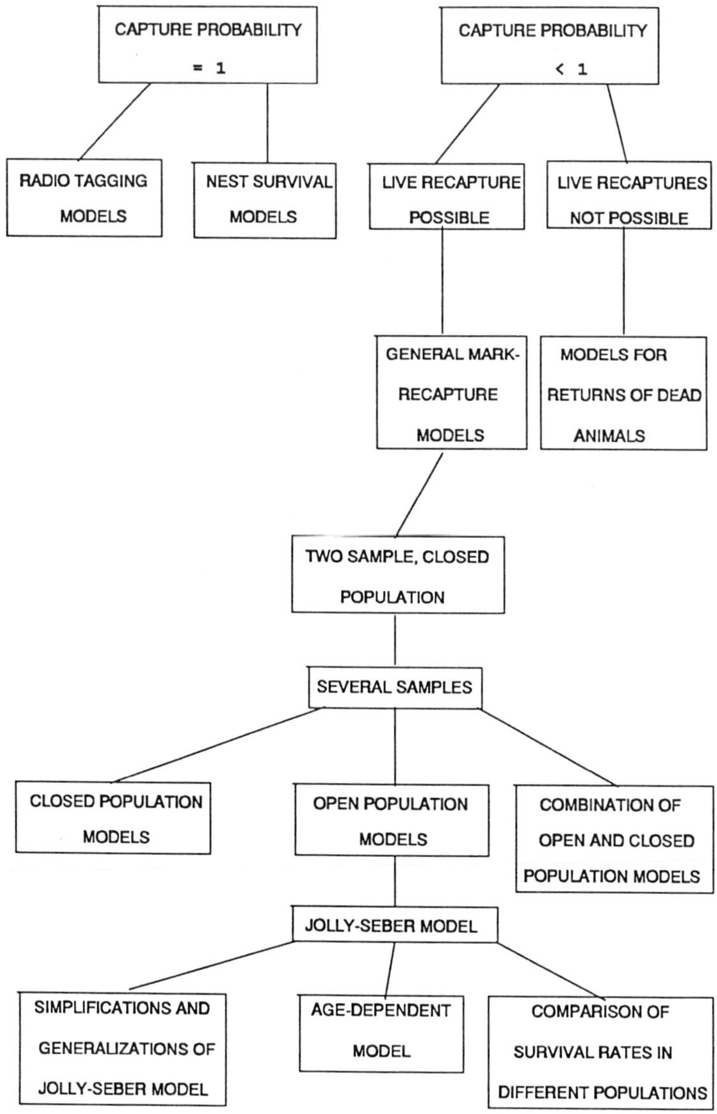

Fig. 1. Relationships between models for mark–recapture data. Based on two figures provided by Pollock (1991).

(Dorazio and Rago, 1991; Hestbeck et al., 1991) or between different development stages (Nichols et al., 1991); the allowance for movement using diffusion models (Vogt and Morton, 1991; Manly and Chatterjee, 1992); the extension of log-linear models from closed to open populations (Cormack, 1989); the development of estimators of breeding probabilities (Lebreton et al., 1990); and the development of computer intensive methods for calculating confidence intervals (Buckland and

Garthwaite, 1991). A recent review (Lebreton et al., 1992) summarizes many of these advances.

For data collected on the returns of dead animals, recent research has included work on methods for allowing survival and recovery probabilities to vary with strata that are determined after the release of animals, such as strata based on recovery locations (Schwarz et al., 1988), work on methods for allowing for the survival rate of young animals to vary with time (Morgan and Freeman, 1989; Freeman and Morgan, 1992), and studies of the properties of estimators when assumptions do not hold (Barker, 1992).

One of the major factors influencing the ease with which modern methods for analyzing mark–recapture data can be used is in the availability of suitable computer programs. Some computer programs have been mentioned above. Useful information about some others is provided by Nichols (1992) and Lebreton et al. (1992).

EXAMPLE 2.1 (*Estimating parameters for a moth population*). To illustrate the use of mark–recapture methods, consider Manly and Parr's (1968) data obtained from sampling an isolated population of the burnet moth *Zygaena filipendulae* that are shown in Table 1. This was an open population so that any analysis should allow for ingress and egress. The Jolly–Seber estimation equations will therefore be applied to the data. Because the study involved five samples taken on 19, 20, 21, 22 and 24 July, 1968, this permits the estimation of the population size on 20, 21 and 22 July, survival rates for the periods 19–20, 20–21 and 21–22 July, and the number of new entries to the population for the periods 20–21 and 21–22 July.

From these data the values m_i, R_i, z_i and r_i needed for equations (2.4) to (2.9) can be obtained, and hence estimates of population parameters with their estimated standard errors obtained as shown in Table 2. The 95% confidence intervals shown in this table are not simply ESTIMATE $\pm$ 1.96 (STANDARD ERROR), as is often used. Rather, they are obtained using transformations that have been derived to obtain improved limits. See Manly (1984) for more details.

Having obtained estimates, the next stage in a full analysis of the data would involve carrying out goodness of fit tests along the lines suggested by Pollock et al. (1990). However, this will not be pursued here.

2.2. *Estimating population size by other methods*

If animals cannot be marked and recaptured then there are some alternative approaches that can be used to estimate population numbers. One such is the removal method which relies on the fact that if the animals found in a sample are removed from a population then the number available for capture in future samples is reduced. Hence if a series of samples are taken then the number captured is likely to show a downward trend with time. Thus, suppose that whenever a sample is taken from a (closed) population then approximately a fraction p of the animals are captured. This means that if the initial population size is N then the number of animals captured in the i-th of a series of samples will be approximately given by the

Table 1
Manly and Parr's data from sampling a population of the burnet moth *Zygaena filipendulae* in Dale, Pembrokeshire, Wales, in July 1968. There were 25 distinct capture patterns obtained, as indicated by the last five columns of the table, where 0 indicates no capture and 1 indicates capture, and the second column shows how many moths displayed these patterns. Thus, one moth was seen in all five samples, one moth was seen in the first four samples only, nine moths were seen in the first three samples only, etc.

Capture pattern	Number of moths	Sample date				
		19	20	21	22	24
1	1	1	1	1	1	1
2	1	1	1	1	1	0
3	9	1	1	1	0	0
4	4	1	1	0	0	1
5	10	1	1	0	0	0
6	5	1	0	1	1	1
7	2	1	0	1	0	1
8	1	1	0	1	0	0
9	2	1	0	0	1	1
10	1	1	0	0	0	1
11	1	0	1	1	1	0
12	4	0	1	1	0	1
13	4	0	1	1	0	0
14	1	0	1	0	1	1
15	2	0	1	0	1	0
16	2	0	1	0	0	1
17	1	0	0	1	1	1
18	3	0	0	1	1	0
19	7	0	0	1	0	1
20	5	0	0	0	1	1
21	21	1	0	0	0	0
22	12	0	1	0	0	0
23	13	0	0	1	0	0
24	9	0	0	0	1	0
25	19	0	0	0	0	1

equation $Y_i \approx p(N - X_i)$, where X_i is the number found in earlier samples, and removed from the population. Hence, one way to estimate N involves using linear regression to fit the equation $Y_i = a + bx_i$ to observed values of X_i and Y_i for a series of samples, and then applying the equation $\hat{N} = -a/b$ on the assumption that $a \approx pN$ and $b \approx p$. This and similar approaches to population estimation have been suggested independently by a number of authors over the last 50 years or so (Otis et al., 1978), and maximum likelihood estimation was investigated by Moran (1951) and Zippin (1956, 1958). More recent research has included a procedure for estimating the proportions of a closed population in different regions where animals can move between regions so that an increase in one area is matched by decreases in one or more other regions (Heimbuch and Hoenig, 1989) and the development of methods for incorporating auxiliary information such as values for environmental variables or measures of signs of activity (Pollock et al., 1984; Routledge, 1989).

Table 2
Estimates obtained by the Jolly–Seber equations from the data in Table 1. Standard errors are calculated using equations (2.10) and (2.11) with observed values of $n_i = R_i$, z_i, r_i and m_i replacing their expected values

Sample statistics

Date	n_i	x_i	t_i	m_i
19	57	0	36	0
20	51	11	29	25
21	52	12	25	28
22	31	20	15	17
24	54	0	0	35

Estimates of population parameters

Date	N	SE	Approx. 95% confidence limits	ϕ	SE	Approx. 95% confidence limits	B	SE
19	–	–	–	0.77	0.10	0.60 to 0.97	–	–
20	88.1	12.0	73 to 121	0.75	0.11	0.55 to 0.99	–	–
21	95.9	14.0	73 to 130	0.75	0.15	0.50 to 1.09	29.9	11.3
22	101.3	22.1	80 to 169	–	–	–	29.9	16.1

Another alternative to mark–recapture methods for estimating the size of animal populations is the line transect method, where this involves an observer moving along a straight line through a study area and noting the distance from the line for all animals seen. A key consideration is then that all the animals that are close enough to the line to be detected will not necessarily be seen. Rather, it can be expected that all the animals that are on the transect line will be seen, but the probability of seeing animals that are not on the line will decrease as the distance from the line increases, as indicated in Figure 2.

The mathematical basis of line transect sampling is fairly straightforward. Thus, suppose that an observer follows a randomly located transect line in the study region and records objects that are seen within a strip of width $2w$ centered on the line. Then any object within the strip is equally likely to be at any distance from 0 to w from the transect line, so that the probability of detecting it is

$$\begin{aligned} p &= \int_0^w \frac{1}{w} g(x)\,\mathrm{d}x \\ &= a/w, \end{aligned} \tag{2.12}$$

where $g(x)$ is the probability of detection at a distance x from the transect line and a is the integral of $g(x)$ from 0 to w. Therefore, if there are N objects within the strip then a proportion pN of these are expected to be detected. If n objects are in fact

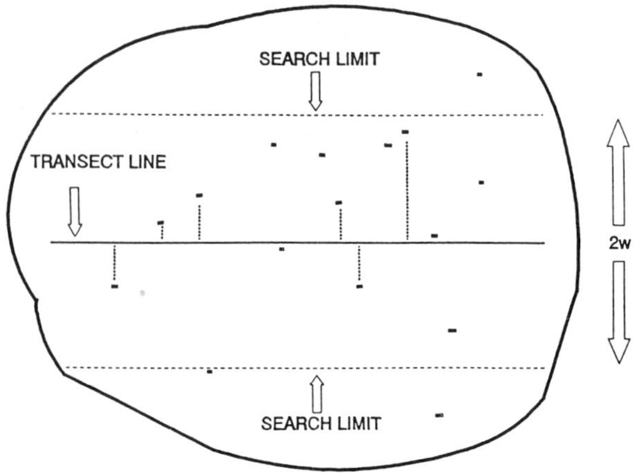

Fig. 2. Line transect sampling in a study area within which animals (or other objects) are distributed. Animals that are detected are indicated by the broken line from the point to the transect line. The probability of detection decreases as the distance from an animal to the line increases. The search distance w on either side of the transect line need not be finite. Based on Figure 2.1 of Burnham et al. (1980).

detected then there is the approximate relationship $n \approx (a/w)N$, or $N \approx nw/a$. Thus if the transect line is of length L then the density of object is $D = N/2wL \approx n/2La$, from which it follows that an estimate of the density is

$$\hat{D} = \frac{n}{2L\hat{a}}, \tag{2.13}$$

where $\hat{a}$ is an estimate of a. Since this equation does not involve w, the distance searched from the transect line can be whatever is convenient, and can in fact be regarded as being infinite.

A slightly different way to write equation (2.13) is in terms of the probability density function (pdf) $f(x)$ of the distribution of observed distances of objects from the transect line, i.e., the distribution of distances conditional on objects being detected. It can be shown (Burnham et al., 1980) that $f(x) = g(x)/a$ which, coupled with the assumption that all objects are detected on the transect line so that $g(0) = 1$, gives the relationship $f(0) = 1/a$. Therefore equation (2.13) becomes

$$\hat{D} = \frac{n\hat{f}(0)}{2L}, \tag{2.14}$$

where $\hat{f}(0)$ is an estimate of the height of the pdf of the distribution of object to line distances corresponding to a distance of zero. A Taylor series approximation gives the equation

$$\mathrm{Var}(\hat{D}) \approx D^2\left[\frac{\mathrm{Var}(n)}{\mathrm{E}(n)^2}+\frac{\mathrm{Var}\{\hat{f}(0)\}}{f(0)^2}\right] \tag{2.15}$$

for the variance of $\hat{D}$, where this can be estimated by substituting $\hat{D}$, n and $\hat{f}(0)$ for D, $\mathrm{E}(n)$ and $f(0)$, respectively. Values for $\mathrm{Var}(n)$ and $\mathrm{Var}\{\hat{f}(0)\}$ depend on the circumstances for particular examples and the method used to estimate $\hat{f}(0)$.

The major problem with using equation (2.14) is deciding on the form of the function $f(x)$. At present a Fourier series seems to be the most popular choice, with

$$f(x)=\frac{1}{w^*}+\sum_{k=1}^{m} a_k \cos\left(\frac{k\pi x}{w^*}\right), \tag{2.16}$$

where $w^* \leqslant w$ and m is large enough to give a good fit to the data. The coefficients a_k can be estimated as

$$\hat{a}_k=\frac{2}{nw^*}\sum_{i=1}^{n} \cos\left(\frac{k\pi x_i}{w^*}\right), \tag{2.17}$$

using a method of moments argument, and the estimator of $\hat{f}(0)$ for equation (2.14) is then

$$\hat{f}(0)=\frac{1}{w^*}+\sum_{k=1}^{m} \hat{a}_k, \tag{2.18}$$

with

$$\mathrm{Var}\{\hat{f}(0)\} \approx \sum_{j=1}^{m}\sum_{k=1}^{m} \mathrm{Cov}(\hat{a}_j, \hat{a}_k), \tag{2.19}$$

where

$$\mathrm{Var}(\hat{a}_k)=\mathrm{Cov}(\hat{a}_k, \hat{a}_k)=\frac{(a_{2k}+2/w^*)w^*-a_k^2}{\mathrm{E}(n)-1},$$

and

$$\mathrm{Cov}(\hat{a}_j, \hat{a}_k)=\frac{(a_{j+k}+a_{j-k})/w^*-a_j a_k}{\mathrm{E}(n)-1},$$

for $j > k$. For a more detailed discussion of this Fourier series method, alternative choices for $f(x)$, variance equations etc., see the monograph of Burnham et al. (1980) and the more recent review by Ramsey et al. (1988).

Although the line transect method has been described for use with animal populations it is apparent that it could be used equally well with plants, although this is not done in practice since alternative sampling schemes are more convenient. The method can also be used with variable circular plot sampling where the observer stops at regular intervals and records the distance from where they are standing to all objects that can be detected (Reynolds et al., 1980; Buckland, 1987).

Recent research has been on improved forms of the pdf $f(x)$ (Buckland, 1985; Johnson and Routledge, 1985; Zahl, 1989; Routledge and Fyfe, 1992); methods to adjust for the fact that if animals occur in groups then the probability of detecting a group depends on its size as well as its distance from the transect line (Drummer and McDonald, 1987; Drummer et al., 1990); simulation studies of line transect and variable circular plot sampling (Roeder et al., 1987); the use of bootstrapping to obtain confidence intervals and remove bias (Johnson and Routledge, 1985; Quang, 1990; Routledge and Fyfe, 1992); the verification that line transect sampling works on populations of known size (Bergstedt and Anderson, 1990; Otto and Pollock, 1990); the allowance for a 'blind strip' below the aircraft used for an aerial survey (Allredge and Gates, 1985; Quang and Lanctot, 1991; see also Zahl, 1989); and adjustments for the movement of animals in response to the observer (Engeman and Bromaghin, 1990; Turnock and Quinn, 1991).

Removal, line transect, and variable circular plot sampling are by no means the only alternatives to mark–recapture for estimating population sizes. There are many special purpose sampling methods that have been developed for particular populations such as Żeh et al.'s (1986) modification of the removal method to estimate the size of the western Arctic population of bowhead whales from sightings in two census camps, and Rivest et al.'s (1990) two-phase (airplane and helicopter) sampling plan to estimate the size of the moose population in southern Quebec. Also, in some cases it is possible to carry out a sample survey that simply involves counting the animals or plants found in sample units and multiplying the total count obtained by the population area divided by the sampled area.

A problem with the sample survey approach is the excessive clustering that occurs with many biological populations so that sample units tend to either have a zero count or a high count. This seems to be particularly extreme with marine surveys, with the result that estimates of abundance are highly variable. Two solutions to this problem have been proposed. First, model-based inference (which assumes some specific distribution for sample counts) can replace design-based inference (which merely assumes random sampling). Second, the sampling scheme adopted can be modified in an attempt to get larger samples in the most variable parts of a population.

At first sight, model-based inference seems to hold some promise in this area. It might, for example be assumed that sample counts follow the Δ-distribution (a log-normal distribution modified to include a finite probability of observing zero), in which case the minimum variance unbiased estimator of the population mean is not the sample mean as suggested by design-based theory (S. J. Smith, 1988). Unfortunately, Smith (1990) has shown that model-based estimators may not be robust to deviations from the assumed models so that their apparent advantages over design-based estimators may be an illusion. He suggests that models relating sample counts to measured covariates offer more promise.

Proposed modifications to sampling schemes include two-phase stratified sampling, where some of the sample units are allocated to strata using the data obtained

Table 3
Line transect data from a study conducted by J. L. Laake using an artificial population of wooden stakes (Burnham et al., 1980, Table 6, with values rounded to one decimal place). The values shown are distances from the transect line for the 68 stakes detected in walking $L = 1000$ m and searching $w = 20$ m on each side of the line. The starred value has arbitrarily been changed to 13.3 from the value of 31.31 given by Burnham et al. (1980) since 31.3 is clearly an error

2.0	5.7	8.3	10.7	10.1	3.3	3.1
0.5	3.0	4.9	3.9	4.4	8.5	6.6
10.4	4.0	1.5	6.1	1.3	6.1	4.4
3.6	0.1	18.6	6.4	13.7	0.4	5.0
0.9	11.8	0.4	3.8	6.3	9.3	3.2
1.0	14.2	0.4	15.2	3.6	0.5	7.7
3.4	2.4	0.2	3.5	9.0	1.2	18.2
2.9	1.6	11.6	3.1	7.7	1.7	4.1
8.2	13.3*	3.2	7.9	4.9	4.5	
6.5	6.5	7.1	18.2	9.1	3.1	

from the first part of the survey (Francis, 1984, 1991; Jolly and Hampton, 1990); and adaptive sampling, where extra sample units are added in the neighborhood of existing sample units with counts that are over a predetermined level (Thompson, 1991a, b).

EXAMPLE 2.2 (*Line transect sampling of wooden stakes*). To illustrate the use of line transect sampling and estimation, consider some data from a study by J. L. Laake of how the method works on known populations (Burnham et al., 1980). As part of this study, a population consisting of 150 wooden stakes was set up in a sagebrush meadow in a strip of width 40 m and length 1000 m. A student then walked a transect line along the center of the strip and found 68 stakes within 20 m of the path, with the path to stake distances that are shown in Table 3.

The Fourier series method can be used to fit the pdf $f(x)$ of distances, using equation (2.17) to estimate the a_k coefficients of equation (2.16). To do this, the maximum search distance w^* must be specified. Often data are truncated by making w^* smaller than the distance actually used in a study to avoid a few very large distances having an undue influence on the analysis. However, here the actual search distance will be used by taking $w^* = 20$ m. On this basis it is found that $\hat{a}_1 = 0.05138$ and $\hat{a}_2 = 0.01225$ for a two term Fourier fit. Hence the estimate of $f(x)$ is

$$\hat{f}(x) = 0.05 + 0.05138 \cos(\pi x/20) + 0.01225 \cos(2\pi x/20).$$

Figure 3 shows that this function fits the observed distribution quite well, and it will therefore be accepted. Tests for the goodness of fit with different numbers of Fourier series terms are available (Pollock et al., 1980) but will not be discussed here.

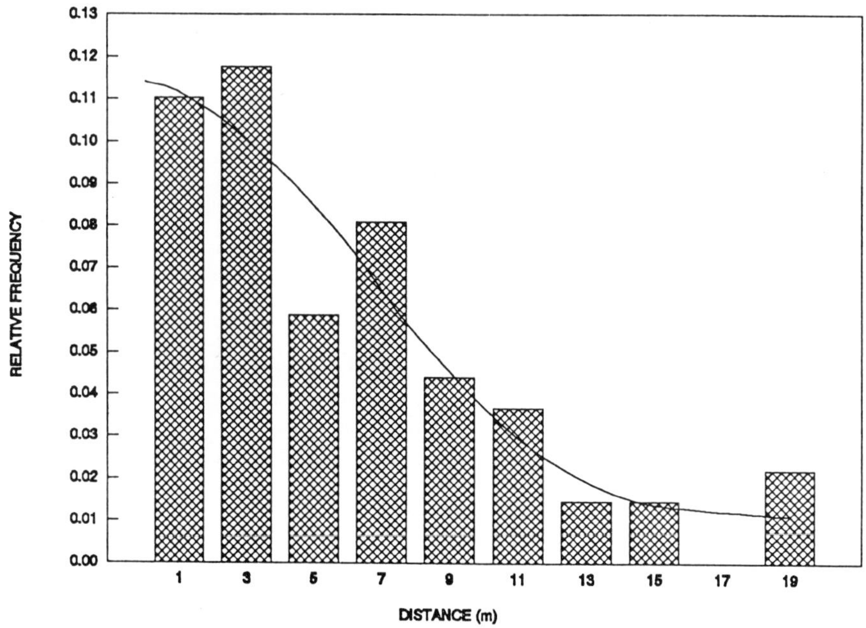

Fig. 3. Comparison between the Fourier series approximation with two estimated coefficients (the continuous curve) and the relative frequency histogram for Laake's data on an artificial population of wooden stakes.

From equation (2.18) $\hat{f}(0) = 0.11364$. Substitution into equation (2.14) then yields $\hat{D} = 0.00386$ stakes per m^2 (which compares with the true density of 0.00375 per m^2). Multiplying by the area of the strip searched gives an estimate of the total number of stakes $\hat{N} = 2 \times 20 \times 1000 \times 0.00386 = 154.6$, say 155 (which compares with the actual of 150).

The variance of $\hat{D}$ can be estimated using equation (2.15), where this requires estimates of $\text{Var}(n)$ and $\text{Var}\{\hat{f}(0)\}$ to be known. As far as $\text{Var}(n)$ is concerned, an obvious possibility is to regard n as a binomially distributed random variable for which there are estimated to be $\hat{N} = 155$ 'trials' (i.e., wooden stakes), with an estimated probability $\hat{p} = 68/155 = 0.439$ of a 'success' (i.e., detecting the stake). On this basis, $\text{Var}(n) \approx \hat{N}\hat{p}(1 - \hat{p}) = 38.16$. Alternatively, n might be regarded as a Poisson variable so that $\text{Var}(n) = \text{E}(n) \approx n = 68$. In fact, using either 38.16 or 68 for $\text{Var}(n)$ gives about the same variance for $\hat{D}$ in this example, so that $\text{Var}(n) = 68$ will be used, although this is probably somewhat too large. Turning to $\text{Var}\{\hat{f}(0)\}$, this can be approximated by substituting estimates in place of population values where required in equation (2.19) to obtain $\widehat{\text{Var}}\{\hat{f}(0)\} = 0.0001746$. Equation (2.15) then yields the estimated variance of $\hat{D}$ to be

$$\widehat{\text{Var}}(\hat{D}) = 0.00386^2 \left\{ \frac{68^2}{68} + \frac{0.0001746}{0.11364^2} \right\} = 0.000000421.$$

The standard error is then estimated by the square root of this quantity, $\widehat{\mathrm{SE}}(\hat{D}) = 0.000649$, and an approximate 95% confidence interval for the true density, assuming that $\hat{D}$ is normally distributed, is $\hat{D} \pm 1.96 \cdot \widehat{\mathrm{SE}}(\hat{D})$, i.e., 0.00386 ± 0.00127. This can be converted to an approximate 95% confidence interval for the population size by multiplying by the area searched (40 000 m^2) to obtain 154.4 ± 50.8.

The method that has been used to obtain variances and confidence intervals is approximate and relies on assumptions such as a normal distribution for $\hat{D}$ that may not be satisfied. For this reason it is best whenever possible to estimate $\mathrm{Var}(\hat{D})$ directly by running a reasonably large number of randomly placed transects of approximately the same length through the study region and obtaining an estimate of density from each line. The density estimates can then be treated as a random sample from a population for which the mean is to be estimated. If desired, stratification and other standard techniques of classical sampling theory can be used when placing transects (Pollock et al., 1980). See also the references on bootstrap methods for obtaining confidence intervals that are provided above.

2.3. Spatial pattern analyses

The spatial distribution of a population may be of interest for a number of reasons. If individuals appear to be located at random positions, independent of each other, then this may indicate a lack of interaction. The alternatives are clustering (which could be due to such factors as the reproductive process, habitat differences, the need for similar resources, or gregarious behavior) and regular spacing (which could be due to competition for resources, or aggressive behavior).

A classification of spatial pattern analyses is shown in Figure 4. Here the first important distinction is between studies with natural sampling units (e.g., the leaves on a tree or the trees in a plantation), and those where the sampling unit must be

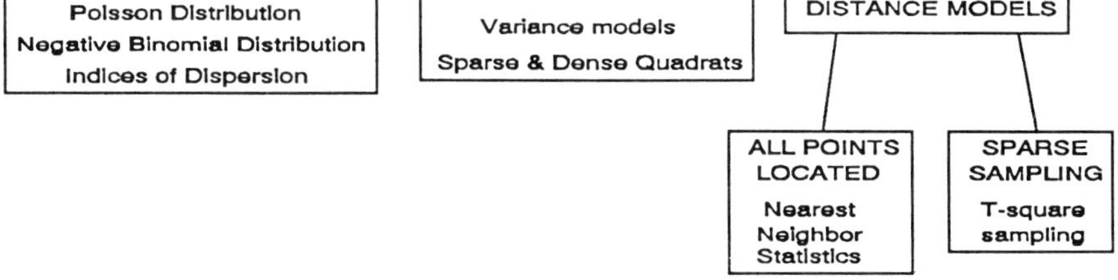

Fig. 4. Types of spatial pattern analysis after the choice of the sampling unit (following Ludwig and Reynolds, 1988, Figure 2.3, with modifications).

fixed by the researcher (e.g., quadrats of a certain size laid out in grassland). In the first case it is assumed that the sample units are a random sample from a population of such units and that spatial relationships between units do not have to be taken into account. In the second case there are two different types of analysis possible. One is based on quadrat sampling, while the other is based on measuring distances between random points and the nearest individual or between individuals and their nearest neighbors.

With natural sampling units and no spatial considerations statistical methods have mainly concerned the fitting of distributions for the number of individuals per unit and the measurement of various aspects of clustering. The Poisson distribution provides a natural null model in this context for the counts in the sample units (providing that all the units are the same size) because this is what can be expected if there is a large number of individuals in a population and each of them has a small probability of occurring in any one sample unit, independent of how many other individuals are present. However, the Poisson distribution is often not a good fit to observed counts, which display more variation than this distribution allows. For the Poisson distribution the variance σ^2 equals the mean μ. There has therefore been interest in distributions for which $\sigma^2 > \mu$. Several distributions with this property have been used to fit distributions, but the negative binomial with probabilities given by

$$P(x) = \frac{(k+x-1)!}{\{x!(k-1)!\}\{\mu/(\mu+k)\}^x\{+\mu/k\}^{-k}} \tag{2.20}$$

for $x = 0, 1, 2, \ldots$ is the one that is now most favored by ecologists. This has mean μ and variance $\sigma^2 = \mu + \mu^2/k$. This approaches the Poisson distribution for large k $(1/k \to 0)$, and can be readily extended to include the positive binomial distribution for negative integers k. Hence the distribution (2.20) can cover uniform $(\sigma^2 < \mu)$, random $(\sigma^2 = \mu)$, and clustered $(\sigma^2 > \mu)$ distributions (Kuno, 1991). A number of mechanisms gives rise to the negative binomial distribution. Consequently, the fact that this distribution fits a particular set of data does not 'explain' how the data arose (Pielou, 1969, p. 88).

Naturally enough, ecologists have felt the need to measure the extent of aggregation using single indices. This has led to a surprisingly large number of possibilities including the index of dispersion or variance-to-mean ratio $s^2/\bar{x}$ (where s^2 is the variance and $\bar{x}$ is the mean for a sample of n units), David and Moore's (1954) index of clumping $s^2/\bar{x} - 1$, Green's (1966) index $(s^2/\bar{x} - 1)/(n-1)$, Lloyd's (1967) indices of patchiness, $\bar{x} + s^2/\bar{x} - 1$, and clumping, $(\bar{x} + s^2/\bar{x} - 1)/\bar{x}$, Morisita's (1971) index $\{n/(n-1)\}\{\bar{x} - s^2/\bar{x} - 1\}$, the negative binomial parameter k, and Downing's (1991) index $\log(s^2/\bar{x}^{1.5})$.

For moderately large values of n, $(n-1)s^2/\bar{x}$ approximately follows a chi-squared distribution with $n-1$ degrees of freedom for samples from a Poisson distribution. Therefore a test for the Poisson distribution sometimes uses this statistic. A significantly small value indicates a regular distribution, a significantly large value indicates clustering, and a nonsignificant value provides no evidence against the hypothesis of randomness. This test seems to be more powerful than a standard goodness of fit test for detecting likely alternatives to randomness (Mead, 1974), although if there are many small quadrats an alternative test based on the number of empty quadrats may be better (Heltshe and Ritchey, 1984).

It is commonly found that when populations of a species are sampled at a number of times or places then the variance and mean seem to have a functional relationship, where this has been approximated by Taylor's (1961) power law $\sigma^2 = a\mu^b$ or Iwao and Kuno's (1968) quadratic function $\sigma^2 = A\mu + B\mu^2$. There is a continuing debate about which of these relationships is most satisfactory (Kuno, 1991; Routledge and Swartz, 1991). One advantage of the power law is that it forms the basis for the Adés family of distributions that can be used to model a series of distributions obtained by sampling a species a number of times (Perry and Taylor, 1985, 1988).

The analysis of the spatial distribution of a population becomes more flexible when there is no natural sampling unit and the spatial relationship of the sampling units must be taken into account. As shown in Figure 4, there are then two approaches that can be used, with one being based on quadrat counts and the other being based on distances. Of course, the methods already mentioned for natural sample units can be used equally well with units with arbitrary sizes, but the results obtained are likely to depend on the size that is used. This is true with Taylor's power law, although it is sometimes claimed that the parameter a is a sampling factor and the parameter b a population characteristic (Sawyer, 1989; Yamamura, 1990).

One way to use quadrats involves following Greig-Smith (1952) and dividing the study area into a grid of squares and counting the number of individuals in each square. Pairs of contiguous squares are then combined to form oblong blocks of two units, of which there are half as many as the original squares. Contiguous blocks of two units are then combined to form squares of size four, and so on. A nested analysis of variance is then carried out to see which block sizes, if any, display large mean squares. These are then considered to indicate clustering at the scale of those block sizes. An alternative design uses a line of contiguous quadrats, grouping them first in blocks of two, then in blocks of four, etc.

Greig-Smith's analysis of variance has been criticized on several grounds, including its inability to detect patterns for block sizes that are not a power of 2 (Pielou, 1969, p. 105). As a result, a number of alternative procedures have been proposed (Hill, 1973a; Goodall, 1974; Mead, 1974; Ripley, 1978; Ludwig and Reynolds, 1988, Chapter 4). The situation is complicated by the need to detect different types of regularity and clustering, which makes it difficult to define a suitable test statistic. At present it appears that Hill's (1973a) two-term local variance approach is most satisfactory, possibly with some recently proposed modifications (Galiano et al., 1987; Dale and MacIsaac, 1989).

Another way of considering the results obtained from taking a line or grid of quadrat samples is in terms of the pattern of presences and absences of quadrats with certain specific characteristics, such as those that are sparse or dense. Pielou (1969, p. 107) discusses this situation for a rectangular grid of quadrats in terms of whether the number of adjacent dense cells is consistent with the hypothesis that these cells are randomly placed. A computer programme ZORRO provided by Romesburg

(1989) can be used in this context for any configuration of quadrats, using a nearest neighbor statistic to test whether similar types of quadrat tend to be close together or separated in space. For quadrats set out along a line, standard tests for randomness in a time series such as the runs above and below the median test are applicable (Manly, 1991a, p. 171). See also the patch-gap analysis of Dale and MacIsaac (1989).

Distance methods for studying spatial patterns can be divided into two categories according to whether or not the positions are determined for all individuals in the study region, i.e., whether or not all individuals are mapped. If all positions are known then the main question of interest is whether these positions indicate a random, regular, or clustered distribution. On the other hand, if only some individuals are located, so that the study area is sparsely sampled, then the same question can be asked, but it may also be of interest to estimate the total number of individuals in the study area.

If a map is available then a simple but computer intensive method for assessing the nature of the spatial point pattern involves comparing the observed distribution of distances between individuals and their nearest neighbors with the distribution obtained by randomly placing the same number of individuals in the study area a large number of times. This can be done for second nearest neighbors, third nearest neighbors, etc., as well as first nearest neighbors, as demonstrated by Manly (1991a, p. 147) for the distribution of some Swedish pine samplings. Alternatives test statistics can be used, and an important feature of this approach is that irregular boundaries can be accommodated fairly easily. A FORTRAN program for rectangular study areas is provided by Manly (1991a, p. 160). For square study areas enough distribution theory is known so that randomization can be avoided if desired. A computer program for handling the calculations in this case is described by Koen (1991).

Sparse sampling methods involve measuring distances between random points and their nearest objects in the study region, distances between random objects and their nearest neighboring objects, or both of these types of distances. A method called T-square sampling has found favor in this situation since it uses both distances in such a way that they are independent. This involves (a) selecting a random point in the study region, (b) finding the distance X to the nearest individual, and then (c) finding the distance Y from that point to its nearest neighbor in a half circle in the direction away from the original random point (Figure 5). This is repeated n times to obtain random point to individual distances $x_1, x_2, \ldots, x_n$, and individual to individual distances $y_1, y_2, \ldots, y_n$.

If this sampling scheme is used on a population for which individuals occur independently at random positions at the rate of D per unit area then it can be shown the $2A_{xi}D = 2\pi D x_i^2$ and $2A_{yi}D = \pi D y_i^2$ have independent chi-squared distributions with two degrees of freedom, where A_{xi} is the circular area searched to find the individual closest to the i-th initial random point and A_{yi} is the half-circular area subsequently searched to find a second individual. It follows that the statistic

$$U = \frac{1}{n} \sum_{i=1}^{n} \frac{x_i^2}{x_i^2 + \frac{1}{2} y_i^2} = \frac{1}{n} \sum_{i=1}^{n} \frac{A_{xi}}{A_{xi} + A_{yi}} \tag{2.21}$$

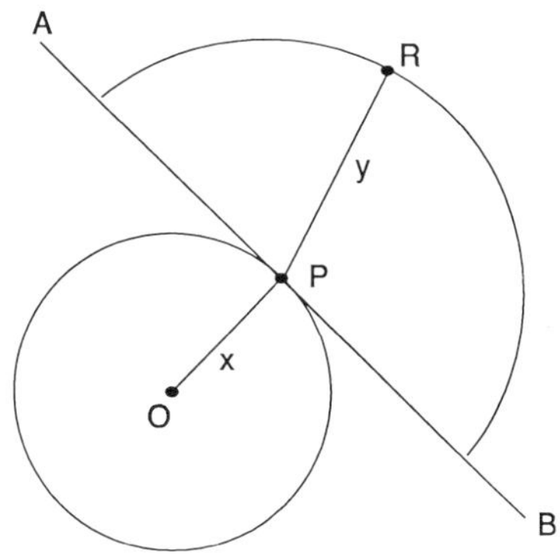

Fig. 5. The T-square sampling method. A random point O is chosen and the distance x to the nearest individual P is measured. A search is then made on the opposite side of the line AB for the nearest neighbor to P, and the distance Y to this nearest neighbor determined. The T is formed by the lines OP and AB.

approximately has a normal distribution with mean 0.5 and variance $1/12n$ for a randomly distributed population (Diggle and Cox, 1983). If an observed value of U is significant and high then this indicates that the area that must be searched to find an individual starting at a random point is greater than the area that has to be searched to find the closest neighbor to an individual, which indicates some clustering. If U is significant and low instead then this indicates some regularity in the distribution.

An estimate of the density of a population (number of individuals per unit area) can be obtained from the x_i distances by noting that the average area searched to find one individual is $\sum A_{xi}/n = \pi \sum x_i^2/n$. A density estimator is therefore the reciprocal

$$\hat{D}_x = \frac{n}{\pi \sum_{i=1}^{n} x_i^2}. \tag{2.22}$$

A similar argument shows that

$$\hat{D}_y = \frac{n}{\pi \sum_{i=1}^{n} y_i^2} \tag{2.23}$$

is a density estimator based on the y_i distances. Both estimators are valid if points are in random locations, but if this is not the case then neither will be valid. Various ad hoc methods have been suggested for overcoming this problem, such as using $\sqrt{\hat{D}_x \hat{D}_y}$ (Diggle and Cox, 1983), but this seems to be an area where more work is still required. Bootstrapping seems to offer some possibilities (Solow, 1989a).

Ludwig and Reynolds (1988, Chapter 5) discuss indices to measure the extent to which a spatial distribution is nonrandom. The statistic U of equation (2.21) can be used for this purpose but this requires T-square sampling. Generally indices have just been based only on distances from points to the nearest individual.

Spatial data analysis is ecology for single species encompasses a large range of methods, of which only a few have been discussed above. This section therefore concludes with the mention of some other methods and some recent developments with quadrat and distance sampling. Important methods not discussed here include the use of Mantel's (1967) matrix randomization test for relating geographical distances between colonies of animals to environmental and biological differences (Manly, 1991a, Chapter 7), the use of spatial autocorrelation methods as developed by Sokal and his colleagues (Sokal and Thomson, 1987; Sokal, 1990), and the use of methods imported from geostatistics (Cressie and Read, 1989). Recent work on quadrat sampling has included a suggestion for measuring the aggregation of animal populations on the basis of the amount of effort needed to achieve complete crowding with the amount of effort needed to achieve complete randomness (Perry and Hewitt, 1991), and a method for detecting aggregation using data on several variables (Solow and Smith, 1991). Recent work on distance-based analyses of spatial data has included new methods for density estimation with sparse sampling (Delincé, 1986), and new methods for quantifying clustering with mapped populations (Getis and Franklin, 1987; Shaw, 1990).

EXAMPLE 2.3 (*Site locations in prehistoric Thailand*). The example that will be used to illustrate spatial pattern analysis falls within the area of archaeology rather than ecology. However, the data are very similar to data that ecologists sometimes collect. Moreover, it can be justified as an example related to human ecology. It concerns part of an investigation into the factors governing the positions and sizes of prehistoric settlement sites in north-east Thailand that is described by Higham et al. (1982). Here the question addressed is whether sites appear to have been located at random positions within the Khorat Basin which is bounded by the Mekong River to the east and north, and the Petchabun and Dang Raek mountains to the west and south, respectively. The example only concerns moated sites within an area of about 225 by 225 km bounded by longitudes 102°30′, and 104°30′, and latitudes 14°30′ and 16°30′ that were apparently occupied between about 400 BC and about 1100 AD, at which time the sites were abandoned. Figure 6 shows the positions of 95 sites within the area of interest, with an indication of whether the site is 'small' (under 35 ha) or 'large' (over 35 ha). The size of the sites is important here since the largest sites may well have been social centers arounds which smaller sites tended to cluster.

A simple analysis of the distribution of the sites can be based on quadrat counts. For example, if the study region is divided into $n = 16$ equal size quadrats, then the counts of all sites within these squares are as follows:

2	20	2	0
0	4	12	2
4	22	11	4
3	4	5	0.

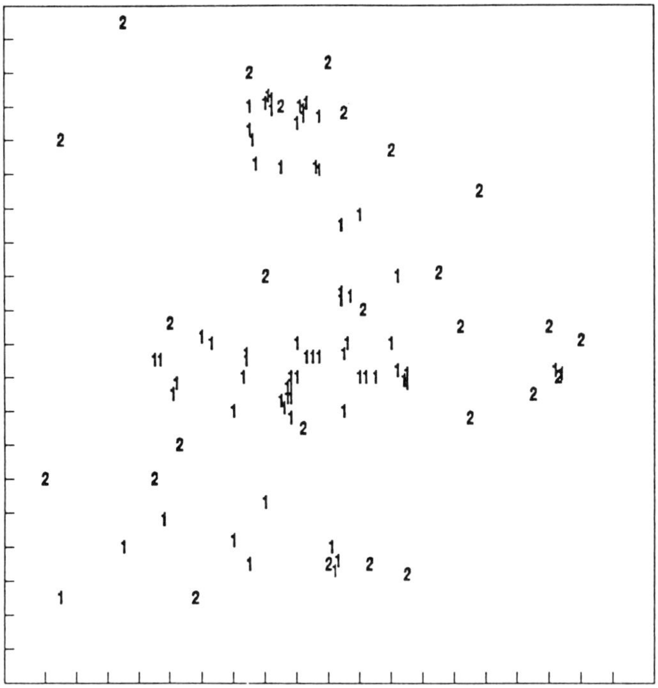

Fig. 6. Positions of sites within a 225 by 225 km region within the Khorat basin of Thailand, with sites of less than 35 ha indicated by '1' and larger sites by '2'.

The mean number of sites per quadrat is $\bar{x} = 5.94$, and the variance is $s^2 = 46.32$. The variance to mean ratio is therefore $s^2/\bar{x} = 7.80$. As mentioned in the last section, a test for randomness can be based on comparing $(n-1)s^2/\bar{x} = 116.97$ with the chi-squared distribution with $n - 1 = 15$ degrees of freedom. Clearly the result is highly significant, which indicates some clustering of sites.

If quadrat counts are only considered for the large sites, then a different situation is found. The quadrat counts are

2	3	2	0
0	2	4	2
2	2	1	2
0	2	2	0.

The mean is 1.63, the variance is 1.32, the variance to mean ratio is 0.81, and the test statistic for randomness is 12.15 with 15 degrees of freedom. Here there is no evidence of nonrandomness. Hence it seems from this analysis of quadrat counts that the large sites could well have been in random positions, but if all sites are considered then there is clustering. In fact Figure 6 indicates that the small sites are indeed clustered together and tend to be close to the large sites.

Another way to approach the question of the clustering of sites is through an analysis based on distances. Thus it can be noted that if the distance from each of the 95 sites to its nearest neighbor is determined, then the average of these distances is 7.7 km. The question can then be asked as to whether this average distance is reasonable if sites were in random positions. This can be answered by comparing the observed 7.7 km with the distribution obtained by placing sites at random positions within the study area using, for example, the FORTRAN computer program that is provided by Manly (1991a, Chapter 8). To this end, the process of randomly allocating 95 sites to random positions in the study region was repeated 5000 times on a computer. It was found that on no occasion was the mean nearest neighbor distance as small as the 7.7 km found for the actual sites. On this basis it is clear that a mean distance as small as 7.7 km was extremely unlikely to occur by chance, so that there is clear evidence that sites were not in random positions with respect to each other. Instead, it appears that there was some tendency for sites to be close to each other. In other words, there is evidence of clustering.

If the 26 large sites are considered on their own then it is found that the average distance from each sites to its nearest neighbor is 23.0 km. When 5000 random allocations of sites to the study region were made by computer it was found that on about 45% of occasions the mean nearest neighbor distance was less than 23.0 km. Hence the observed mean nearest neighbor distance is well within the range of values that are likely to occur from randomly placed sites and there is no reason to reject the hypothesis of random positioning for the real sites. The conclusion from the distance analysis is therefore the same as was obtained from the analysis of quadrat counts: large sites could well have been located in random positions but small sites are clustered to some extent.

Of course, common sense suggests that settlement sites are not generally randomly located since factors such as the type of soil and the distance to water must have been of some importance. Indeed, sites are always fairly close to water. For this reason, the analysis of Higham et al. (1982) was more concerned with determining the characteristics of a region that gave a high probability of a site being present than with the question of whether the positions of sites appeared to be random. Logistic regression was used to study this.

Another interesting question that can be explored with this set of data concerns whether the large and small sites appear to have been distributed independently. Using distances between sites, this could be investigated with Romesburg's (1989) computer program ZORRO, or using other tests that are described by Hamill and Wright (1986), Manly (1991a, Chapter 8), and Diggle and Chetwynd (1991) for assessing relationships between two point patterns. However, these possibilities will not be investigated here.

2.4. Dispersion and movement

Dispersion is a fundamental process in biology that is closely related to the spatial pattern that is observed for an organism. Early studies in this area took the random walk as a basic model where this implies that if an organism starts at the point

(x_0, y_0) at time 0, then the position (X_t, Y_t) at some time $t > 0$ will have a bivariate normal distribution with mean (x_0, y_0), $\mathrm{Var}(X_t) = \mathrm{Var}(Y_t) = \sigma^2 t$, and $\mathrm{Cov}(X_t, Y_t) = 0$ (Dobzhansky and Wright, 1943; Skellam, 1973; Kareiva, 1983). However, the comparison of field data with this model has made it clear that the random walk is often not altogether realistic for various reasons, including the need to allow for geographical constraints (Barraclough, 1978; Smith et al., 1990), and the fact that organisms often tend to move in one direction and are attracted or repelled from the starting point or each other (Manly, 1977; Taylor, 1981a, b; Kareiva and Shigesada, 1983; Thórarinsson, 1986; Bovett and Benhamou, 1988; Marsh and Jones, 1988; Othmer et al., 1988; Routledge, 1990). An important area that requires more work is the incorporating of these models into field methods for estimating population parameters (Manly and Chatterjee, 1992).

The concept of the home range within which an animal usually moves has led to a range of methods for estimating the bounds of this region or a bivariate distribution for the probability that the animal is in different positions (Worton, 1987, 1989; Jaremovic and Croft, 1987; Boulanger and White, 1990). Such estimates are useful, for example, when assessing the resources available to animals for comparison with the resources used (Section 2.6). The data analyzed in this application consist of repeated observations on the location of individual animals, and it is important when assessing the accuracy of home range parameters that these locations should be independent observations. In practice independent observations are not easy to obtain, and often it is a question of how far apart in time two observations have to be in order for this property to hold.

Schoener (1981) suggested that the statistic t^2/r^2 can be used as a measure of the independence between observations on the location of an animal, where t^2 is the mean squared distance between these successive observations and r^2 is the mean squared distance from the center of activity. Methods for determining whether values of this statistic indicate correlation between observations have been provided by Swihart and Slade (1987), but Solow (1989b) observed that Swihart and Slade's test is dependent on the home range being effectively circular and proposed testing t^2/r^2 against the distribution obtained by randomizing the order of observations. This is one of many examples of how statistical tests with minimal assumptions can be obtained at the expense of a considerable amount of computing.

2.5. *Time series*

One of the most controversial questions of ecology since the early part of this century concerns the extent to which natural populations are regulated by density-dependent processes. On the one hand, some (e.g., Hassell et al., 1989; Berryman, 1991) argue that a negative feedback process must be involved in the dynamics of most populations. On the other hand, others (e.g., Den Boer, 1991; Wolda, 1989, 1991) argue that this idea is not supported by the available data when analyzed in an appropriate manner.

A statistical problem at the heart of this argument is about how to determine whether the changes in a time series are a function of earlier values in the series. To be precise, suppose that $N_1, N_2, \ldots, N_G$ are the sizes for G generations of a population. Then one question of interest is whether the changes $N_{i+1} - N_i$ are negatively related to N_i, so that a large population size tends to be followed by a negative change and a small population size tends to be followed by positive change.

Since overlapping generations confuse the situation, most studies have concentrated on insect populations where the individuals emerge at the start of one generation, produce the eggs for the next generation, and then die. A complicating factor is that in practice population sizes can only be estimated for natural populations and sampling errors in the estimates may lead to apparent density-dependent changes.

A variety of tests have been suggested in this situation (Reddingius and Den Boer, 1989; Den Boer and Reddingius, 1989; Den Boer, 1990). However, it has been found that some of these have relatively low power, or require complicated corrections to allow for estimation errors in population sizes. As a result, the best approach at the present time appears to be the one proposed by Pollard et al. (1987), which involves the consideration of three potential models. For the first model it is assumed that $N_{i+1} = N_i \exp(e_i)$, where e_i is a random value from a normal distribution with mean zero. Taking logarithms then gives the linear equation

$$x_{i+1} = x_i + e_i, \tag{2.24}$$

where $x_i = \log_e(N_i)$. This is a random walk model for which the variance of x_i increases indefinitely with i, although the mean is constant. For the second model it is assumed that $N_{i+1} = \exp(r) N_i \exp(e_i)$, so that

$$x_{i+1} = r + x_i + e_i. \tag{2.25}$$

This is an unregulated model where the variance of x_i increases with i and a trend in the mean is also allowed. For the third model $N_{i+1} = \exp(r) N_i^{\beta} \exp(e_i)$, so that

$$x_{i+1} = r + \beta x_i + e_i. \tag{2.26}$$

This is a first-order autoregressive series which is stationary providing that $|\beta| < 1$.

To compare models (2.24) and (2.25), Pollard et al. (1987) suggested testing the statistic $(x_n - x_1)/\sqrt{\sum (x_{i+1} - x_i)^2}$ against the t-distribution with $n-1$ degrees of freedom, where the summation is from 2 to n. This has been criticized by Link and Hoover (1991) who show that it is better to compare a modified statistic with the t-distribution with $n-2$ degrees of freedom. However, the more interesting test in the context of the arguments for and against density-dependent regulation of a population is between models (2.25) and (2.26). For this, Pollard et al. propose comparing the observed correlation between x_i and $d_i = x_{i+1} - x_i$ with the distribution obtained for this statistic when new time series are constructed on a computer by starting from x_1 and then adding on the observed values of d_i in a random order. This test has the advantage of not being affected by a trend in population sizes. However, it is worrying that sampling errors in populations estimates will have the same effect as density-dependence. This is obvious when it is considered that a large over-estimate in x_i will almost certainly be followed by an under-estimate of $d_i = x_{i+1} - x_i$. Therefore if sampling errors are not taken into account it is not possible to know whether a significant negative correlation between d_i and x_i is due to density-dependent population changes rather than sampling errors.

Recently, Crowley (1992) has developed a new computer intensive test based on simulation that does take sampling errors into account and therefore overcomes this problem. This test (the random walk attraction test), and another new test devised by Crowley (the randomized attraction test) are designed to see whether population sizes tend to remain within a restricted range that they are attracted back to whenever they leave it.

In practice the arguments about density-dependence are complicated by the fact that even the best tests for the effect of this factor have low power with the lengths of time series that are usually available (Solow and Steele, 1990). It can therefore be argued that the lack of significant results is at least partly the results of lack of data (Gaston and Lawton, 1987; Vickery and Nudds, 1991). A further problem is that the first-order autoregressive model $x_{i+1} = e + \beta x_i + e_i$ that is usually used for density-dependence may not be correct but the tests being carried out do not have much power to detect other forms of density-dependence (Solow, 1990; Reddingius, 1990). Furthermore, it now seems possible that the models that have been considered are of the completely wrong type. It may be that population variation that appears to be random is in fact chaos produced by nearly deterministic processes. For example, Sugihara et al. (1990) have shown that time series of the incidence of measles cases on the scale of single cities are better accounted for in terms of low-dimensional chaos than in terms of autoregressive models. Clearly, much more research is required on the importance of density-dependence on real populations.

An area of statistical ecology that is closely related to the detection of density-dependence is key factor analysis. This can be applied to any population with discrete generations where the life cycle within a generation can be divided into a series of stages through which each individual must pass providing it survives long enough. The key factor in the life cycle is then the stage that contributes most to the variation in the number in the final stage. The data for a key factor analysis consist of counts or estimates of the numbers entering stages for a series of generations and a statistical analysis is required to elucidate the relationship between survival and reproduction rates and population sizes. A review of this area is provided by Manly (1990a, Chapter 7).

In order to produce the data for a key factor analysis it is often necessary to estimate the numbers entering stages in a generation from data on the numbers observed in each stage in field samples. To do this it is necessary to set up models for the dynamics of stage-structured populations for which the parameters can be estimated from the data. This area is reviewed by Manly (1990a, Chapters 4 and 5). A related area is the use of matrix models to describe populations with overlapping generations with individual members categorized either by their age or the development stage that they are in (Legendre and Legendre, 1983, Chapter 11; Manly, 1990a, Chapter 6).

Of course, time series methods in ecology are by no means restricted to the study of density-dependence. Other important applications are in the area of intervention analysis, where the question of interest concerns whether or not a particular change to the environment has had an adverse effect, and the study of apparent cycles in ecological series. Recent reviews of many of the issues in the area of intervention analysis are provided in a series of papers edited by Matson and Carpenter (1990), and a paper by Underwood (1991a). Research from the literature on behavior change is useful in this area (Manly, 1992a, Chapter 6), although this does not appear to be widely known to ecologists. Computer intensive methods for testing for periodicity and other types of nonrandomness in time series are discussed by Manly (1991a, Chapter 6). Finally, it can be noted that general introductions to time series methods are provided by Legendre and Legendre (1983, Chapter 10) and Diggle (1990).

Example 2.4 (*The winter moth in Wytham wood*). One of the most discussed sets of data relevant to the controversy over the importance of density-dependent factors in population dynamics is from the winter moth (*Operophtera brumata*) in Wytham wood near Oxford in the UK. This moth has one generation per year with the individuals in one year coming from eggs laid in the previous year. Logarithms to base 10 of adult population densities provided by Varley et al. (1973) are shown in part (a) of Figure 7, and differences between these values are shown in part (b) of the same figure. In this example the question of whether this series displays density-dependent changes will be discussed. A much fuller analysis taking into account different stages of the life cycle is provided by Manly (1990a, Chapter 7).

It might be thought that a regression of the values of $d_i = x_{i+1} - x_i$ against x_i will give a test for density-dependence in a case like this, where, as in the previous section, x_i is the logarithm of the density in generation i. There is indeed some suggestion of a relationship on this basis, as shown by Figure 8 which includes the plots of d_i values against x_i values, together with the fitted regression line $d_i = 0.45 - 0.53x_i$. However, this relationship is not significant if tested using Pollard et al.'s (1987) method, which involves comparing the correlation between x_i and d_i with the distribution obtained by adding the d_i to the x_i in a random order. The observed correlation is -0.496. This seems quite low, but when 5000 randomizations were carried out it was found that a value this low or lower was obtained about 25.3% of the time. From this point of view, the observed correlation is very typical of the values expected if there is no density-dependence. The apparent density-dependence is merely an artefact due to the value of x_i being involved in the calculation of d_i. This example indicates that it is not surprising to find disagreements in the interpretation of data related to density-dependence.

2.6. Resource selection

The final aspect of statistical ecology relating to single species populations concerns resource selection by animals. In studies with a relatively simple design this is investigated by defining a unit of a resource, and then comparing the distribution of

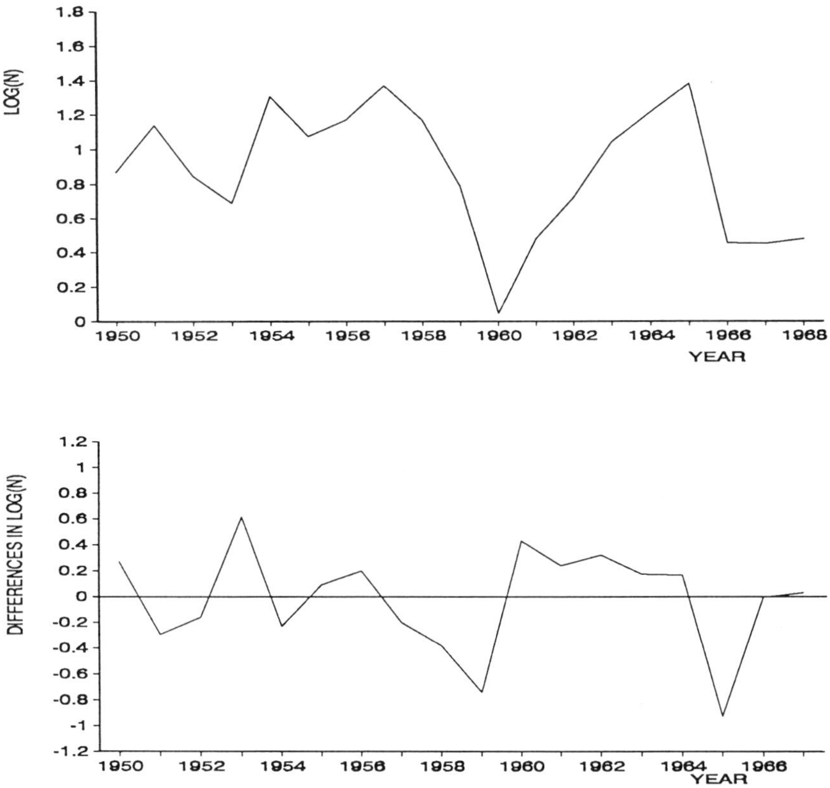

Fig. 7. The winter moth population in Wytham wood: (a) logarithms of densities (frequencies per square metre), (b) differences between logarithms of densities.

units observed in a sample of available or unused resources with the distribution observed in a sample of utilized resources. Thus, the trees in a forest might be characterized by their height, the amount of vegetation cover, the species of the tree, and so on. A random sample from the population of all trees in the forest could then be taken, and compared with a random sample of trees that are used for nesting by a particular species of birds. In this way it is possible to determine which characteristics of the trees seem to be most important for determining nest sites. Many field studies in the literature on wildlife research are of this nature. Apart from the scientific interest of knowing which food and habitat are selected by animals, there are obvious practical applications for the results of studies of resource selection related to areas such as the protection of the important habitat and food of endangered species, and the management of wildlife populations for human use.

One approach to the analysis of data on resource selection is through the estimation of resource selection functions. With this approach it is assumed that the resource units can be characterized by the values that they possess for variables

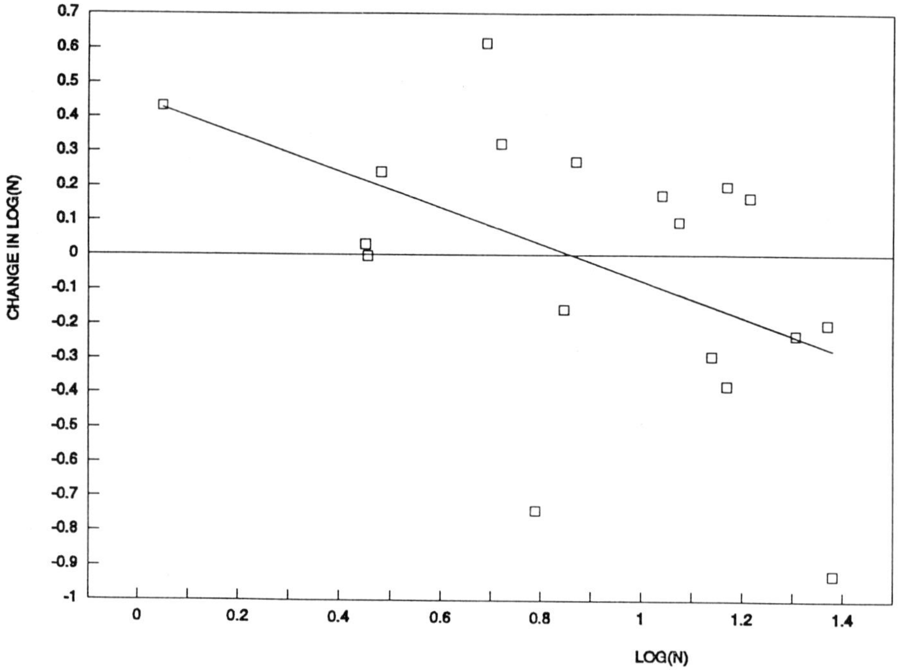

Fig. 8. Plot of changes $d_i = x_{i+1} - x_i$ in logarithms of densities against logarithms of densities x_i for the winter moth data.

$\boldsymbol{x} = (x_1, x_2, \ldots, x_p)$, and that the probability of a unit being used be time t is a function $w^*(\boldsymbol{x}; t)$ of these variables. Any other function $w(\boldsymbol{x}; t)$ that is proportional to $w^*(\boldsymbol{x}; t)$ is then defined to be a resource selection function. The analysis of resource selection data then becomes a question of estimating $w(\boldsymbol{x}; t)$ taking into account the way that data are collected. This approach is considered at length by Manly et al. (1993), who show that a wide range of statistical tools can be applied including logistic regression, log-linear modelling, proportional hazards and other models for survival data, and general maximum likelihood methods.

EXAMPLE 2.5 (*Nest site selection by fernbirds*). A typical example of a resource selection study is provided by a study of nest selection by fernbirds (*Bowdleria puncta*) in Otago, New Zealand, that is described by Harris (1986). Harris measured nine variables on 24 nest sites found during the 1982–83 and 1983–84 seasons and found comparative available sites by choosing 25 random points in the study area and locating a polystyrene model nest at the centre of the nearest clump of vegetation. Harris concluded that only three of the nine variables that he measured showed important differences between the two samples. The data for these variables (x_1, the canopy height; x_2, the distance from the outer edge of the nest to the nearest outer surface of the clump of vegetation in which the nest is situated; and x_3, the perimeter of the clump of vegetation) are shown in Table 4.

Table 4
Comparison of variables measured on fernbird nest sites and randomly located sites in the same area

	Nest sites			Available sites		
	Canopy height (m)	Distance to edge (cm)	Perimeter of clump (m)	Canopy height (m)	Distancce to edge (cm)	Perimeter of clump (m)
	1.20	14.0	8.90	0.47	13.5	3.17
	0.58	25.0	4.34	0.62	8.0	3.23
	0.74	14.0	2.30	0.75	19.0	2.44
	0.70	12.0	5.16	0.52	5.0	1.56
	1.36	14.5	2.92	0.73	8.0	2.28
	0.78	17.0	3.30	0.62	16.0	3.16
	0.45	15.0	3.17	0.60	17.0	2.78
	0.78	15.0	4.81	0.26	4.5	3.07
	0.63	16.0	2.40	0.46	15.0	3.84
	0.75	12.0	3.74	0.28	12.0	3.33
	0.55	12.0	4.86	0.53	11.0	2.80
	0.45	20.0	2.88	0.42	17.0	2.92
	1.56	16.0	4.90	0.47	20.0	4.40
	0.85	23.0	4.65	0.50	13.0	3.86
	0.58	12.0	4.02	0.54	16.0	3.48
	0.75	13.0	4.54	0.56	18.0	2.36
	0.55	18.0	3.22	0.32	7.0	3.08
	0.56	18.0	3.08	0.62	16.0	5.07
	0.57	19.5	4.43	0.39	15.0	2.02
	0.41	16.0	3.48	0.56	8.0	1.81
	0.65	18.0	4.50	0.27	9.0	2.05
	0.78	14.0	2.96	0.42	11.0	1.74
	0.64	18.0	5.25	0.70	13.0	2.85
	0.71	18.0	3.07	0.26	14.0	3.64
				0.34	9.5	2.40
Mean SD	0.73	16.3	4.04	0.49	12.6	2.93
	0.28	3.4	1.37	0.15	4.4	0.84

Here, standard t-tests indicate that there are differences between the means of the three variables for available and used nest sites, and F-tests indicate variance differences for the canopy height and the perimeter of clumps of vegetation. These differences are also confirmed by multivariate tests on the mean and covariance matrices.

From the data available there is no way of estimating the absolute probability of a potential nest site being used. However, it is possible to estimate a resource selection function by assuming that it takes a specific parametric form such as $w(\boldsymbol{x}) = \exp(\beta_0 + \beta_1 x_1 + \beta_2 x_2 + \beta_3 x_3)$. The selection time does not need to be considered since it is constant. This model can be estimated by log-linear modelling or logistic regression (Manly et al., 1992, Examples 7.1 and 8.1). Using either method of

Table 5
Estimated values of the resource selection function for the fernbird data, after scaling to have a mean of 1 for the sample of available sites. The sites are in the same order as in Table 4

Available sites

0.3	0.3	4.3	0.0	0.3	1.6	1.2	0.0	0.7	0.1	0.2	0.3	3.4	0.6	1.1	0.7
0.0	8.5	0.1	0.1	0.0	0.0	1.2	0.1	0.0							

Mean = 1.0, Standard deviation = 1.9, Minimum = 0.0, Maximum = 8.5

Nest sites

15030.4	21.5	1.2	7.4	294.7	7.6	0.3	19.0	0.9	3.1	1.8	0.8	11021.8
152.0	1.1	7.8	1.5	1.4	6.8	0.4	9.8	3.0	17.6	4.4		

Mean = 1109.0, Standard deviation = 3718.3, Minimum = 0.3, Maximum = 15030.4

estimation it is found that the fitted equation is $w(\boldsymbol{x}) = \exp(7.80x_1 + 0.21x_2 + 0.88x_3)$, where this provides the estimated values shown in Table 5 for the sampled nest sites after they have been scaled to have a mean of 1 for the available sites. It appears that some of the used sites had a very much higher probability of use than the sites in the available sample. Hence the nest birds appear to have used strong selection in their choice of sites.

3. Community ecology problems

Community ecology is concerned with the relationship between populations of different species that are sharing the same resources. Studies in this area are concerned with questions related to the distribution of species abundance and diversity, the similarities and differences between species in their use of resources, the grouping of different communities in terms of their species compositions, the extent to which relationships between different communities can be represented by graphs in two or more dimensions, and accounting for differences between communities using environmental variables. All of these questions are of a quantitative nature, and they can only be properly addressed by statistical methods. Some of the more important of these methods are reviewed in the following six sections.

3.1. Species abundance and diversity

If a sample is taken from a community then it may be of interest to consider the distribution of the abundances that are observed for the s species found, to see whether this distribution seems to follow some specific model. There are two reasons for carrying out such an analysis. First, if the same model fits in different communities (possibly with different parameters for each community) then the model is a summary of the data that may be useful for understanding differences between communities. Second, it becomes possible to test hypotheses concerning the manner in which species abundances are determined.

Many models have been proposed for species abundances, based either on ideas of resource apportioning or simply on particular statistical distributions. These include:

(a) the broken stick model, which says that the resource available to the species in a community can be considered to be analogous to a stick which is broken at $s - 1$ random points to produce the resources (and hence the relative numbers) for s species, and predicts that the i-th most frequent species has the expected abundance $A_i = (1/i + 1/(i+1) + \cdots + 1/s)/s$ (Pielou, 1975, p. 24);

(b) the geometric model, which predicts that the i-th most frequent species has the expected abundance $A_i = A_1(1-k)^{i-1}$ (Pielou, 1975, p. 21);

(c) the overlapping niche model that predicts that the i-th most frequent species has expected abundance $A_i = N[1 - s!\Gamma(i-0.5)/(i-1)!\Gamma(s+1.5)]$, where N is the total abundance (Pielou, 1975, p. 27);

(d) the logarithmic series model for which the frequency of species with r individuals is given by $f_r = \alpha x^r/r$ (Pielou, 1975, p. 43);

(e) the lognormal model for which the number of individuals in a species is a random value from the lognormal distribution (Pielou, 1975, p. 47);

(f) the Zipf model that predicts that the abundance of the i-th most frequent species will be $A_i = A_1 i^{-\gamma}$ (Wilson, 1991); and

(g) the Zipf–Mandelbrot model, which is the Zipf model modified to $A_i = A_1(i+\beta)^{-\gamma}$ (Wilson, 1991).

Originally, models (d) and (e) were derived as purely statistical descriptions of data without being based on assumptions about resource use. However, mechanisms to account for these distributions are now available (Kolasa and Strayer, 1988).

Recently Wilson (1991) reviewed existing models for species abundance and noted that there has been some confusion about how these models should be fitted to data, which is something that needs to be clarified before the value of the models can be ascertained for real communities. He proposed standard methods for fitting six models based on dominance-diversity curves, which are plots of the logarithm of species numbers against the species rank order in terms of abundance. The models were then compared for their goodness of fit on two sets of plant data with replicated sampling of quadrats. Significant differences in the goodness of fit were found for the six models, with the log-normal giving the best fit, followed by the geometric. However, the models were generally a relatively good or poor fit with different replicates depending on the smoothness of the observed dominance-diversity plot. Wilson concluded his paper by noting a number of questions that need to be addressed by future research including what effect changing the spatial scale used for sampling has on species abundance distributions, and what should be made of changes in the rank ordering of species between different replicates, since for the purpose of fitting data this ordering is treated as being error free.

The broken stick model mentioned above has a special place in statistical ecology since it is often thought of as the basic model for 'randomness'. An early test for the goodness of fit of this model was proposed by Barton and David (1956). However, a more recent test that has been proposed by Hopf and Brown (1986) seems to be more powerful. This latter test is based on the statistic

$$\mathrm{R} = \left\{ \sum_{i=1}^{s} f_i^2 - \frac{1}{s} \right\} \Big/ \left(1 - \frac{1}{s} \right), \tag{3.1}$$

Sometimes the abundance of species depends in a nonlinear way on an environmental variable and a suitable model for this abundance is thought to be of the Gaussian form $E(Y_{ij}) = c_j \exp\{-\frac{1}{2}(x_i - \mu_j)^2/t_j^2\}$, where Y_{ij} is the abundance of species j at site i, c_j is the maximum abundance for species j, x_i is the value of an environmental variable at site i, μ_j is the optimum of the environmental variable for species j, and t_j is the tolerance for species j (Gauch and Chase, 1974). If the Y values are counts then the model can be fitted to data by log-linear modelling and if the Y values are 0 and 1 indicating presence and absence then logistic regression can be used. Alternatively, the ordination technique of correspondence analysis (see Section 3.5) gives approximately the same solution (Ter Braak, 1985; Ter Braak and Looman, 1986; Stein, 1988).

Pielou (1984) described an interesting distribution-free test for the effect of an environmental gradient which she suggests might be carried out before direct gradient analysis. The example considered by Pielou concerns the situation where an insecticide was injected into a river and the question of interest was whether this affected the abundances of different invertebrate genera further downstream. It was assumed that if any effects were present then it would mean that if the samples were ordered in terms of their distance from the point of injection, then they would also be approximately in order in terms of some unknown linear combination of the measured abundances of genera. This linear combination was then searched for by generating large numbers of linear combinations with random coefficients for the different abundance measures ('random skewers') and finding the distribution of the rank correlation between the values produced by these combinations and the ordering of the samples. The distribution of the rank correlation found in this way was quite different from the distribution for random data, presumably because some 'random skewers' by chance were close to linear combinations displaying real effects. See Manly (1991a, p. 252) for a further discussion of this method.

Computer intensive methods have also been used to search for evidence that species are staggered or clustered along a time or space axis. For example, a major question has been whether the flowering times of plants are staggered, clustered or random within a growing season. This is another of the areas in ecology where it has proved difficult to get agreement on appropriate statistical methods, as shown in Example 3.2 below.

Methods for relating species abundances or presences and absences to environmental variables are an area of continuing research at the present time, that is closely linked with research on the ordination methods that are reviewed in Section 3.5. Some recent developments not already mentioned are Ter Braak's (1986) canonical correspondence analysis which relates species abundances to linear combinations of environmental variables, randomization testing for relationships between species distributions and environmental variables (Ter Braak, 1988; Rejmankova et al., 1991) and Mercier et al.'s (1992) simultaneous use of correspondence analysis on species abundance data and multiple correspondence analysis on environmental data. See Ter Braak and Prentice (1988) for a review of the different types of problem that arise within the area of gradient analysis.

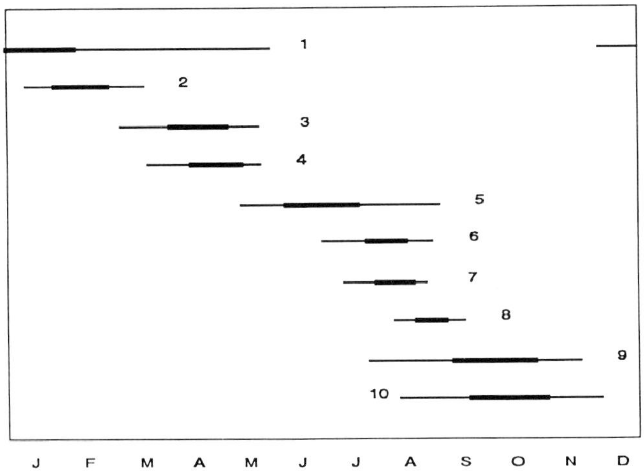

Fig. 10. Flowering times of plants pollinated by hummingbirds in a Costa Rican forest in 1970. Horizontal lines represent periods of peak blooming (thick) and good blooming (thinner). Species are identified by numbers. Modified from Figure 1 of Stiles (1977).

EXAMPLE 3.2 (*Flowering times in Costa Rica*). Figure 10 shows the first of four years data (1971–74) upon which Stiles (1977) based a claim that flowering times are staggered for plants pollinated by hummingbirds in a Costa Rican forest. The other years showed similar results, although the plant species flowered in a different order. On the face of it the claim for staggering seems rather reasonable. However, Poole and Rathcke (1979) challenged this by comparing the separation observed in flowering peaks with what is expected if these are randomly placed on the time axis.

Poolc and Rathke adopted two conventions for their analysis. First, they defined the peak flowering time as the center of the period defined as the 'peak bloom' by Stiles for each of the ten species being considered. Second, they defined the growing season as being from the median flowering time of the first species to the median flowering time of the last species in each year, which is about the middle of January to the end of October. Their approach to the analysis of Stiles' data involved calculating the differences between flowering times as a fraction of the length of the total growing season. This gave then nine differences $d_1, d_2, \ldots, d_9$ between the ten flowering times, plus the difference d_0 between the start of the season and the first flowering time, and the difference d_{10} from the last flowering time to the end of the season. They then asked how the observed variance of these differences compared with its expectation if flowering times are uniformly distributed over the flowering season.

In the general case of s species, the variance of the difference d_0 to d_{k+1} is given by the equation $\sigma_d^2 = s/(s+1)^2(s+2)$ if each flowering time is uniformly distributed, and Poole and Rathcke suggested that a test for whether an observed variance V_d differs significantly from σ_d^2 can be based on treating sV_d/σ_d^2 as a chi-squared variable with k degrees of freedom. In fact, they found that for each of the years 1971 to 1974 the observed variance was about twice as large as σ_d^2 for Stiles' data. They therefore argued that, if anything, flowering times are clustered rather than staggered.

This example indicates the difficulties in this area very well. It turns out that the ten species being considered are probably in two groups, corresponding to dry and wet season flowering, and this in itself can quite easily account for the observed clustering. Treating these groups separately produces evidence of staggering (Cole, 1981). Also, the Poole and Rathcke analysis does not use what is presumably useful information on the full flowering periods of different species. It is the use of this information that suggests the application of computer intensive methods involving, for example, the comparison of the observed amount of overlap in flowering times with the distribution obtained by randomly placing the flowering periods of different species on the time axis (Cole, 1981).

The debate on how to analyze data of this type is summarized by Pleasant (1990), who notes that there are still three problems to be resolved: (a) deciding on the endpoints of the axis being considered, (b) deciding whether all points on the axis are equally likely to occur, and (c) deciding whether the flowering curve for a species should be regarded as being fixed. He also points out the close relationship between testing for random placement of species on an axis and testing for constant body size ratios as discussed in Section 3.1.

3.3. Relationships between distribution of species

If data are available on the presence and absence or the abundance of each of several species at a number of sites then it is possible to examine the extent to which species spatial distributions are dependent or independent. Analyses can include calculating coefficients that measure the extent to which pairs of species have similar distributions, testing these coefficients for randomness, using multivariate methods such as cluster analysis or multidimensional scaling to summarize differences between species, and comparing observed data with data generated on the computer using null models for which species have independent distributions.

There are a large number of coefficients designed to measure the similarity between the distribution of two species when only presence and absence data are available (Legendre and Legendre, 1983, Chapter 6; Digby and Kempton, 1987; Ludwig and Reynolds, 1988, Chapter 11). In this case a survey of n sites provides count data of the form shown in Table 7 and some popular indices are the simple matching coefficient $S_1 = (a + d)/n$, the Ochiai index $S_2 = a/\sqrt{(a + b)\ (a + c)}$, the Dice index $S_3 = 2a/(2a + b + c)$, and the Jaccard index $S_4 = a/(a + b + c)$. There has been a good deal of debate about whether the number of joint absences d should count in similarity indices, as it does in S_1. The problem is that there is a danger of finding high similarity between two species since they are both absent from many sites, but the ecological meaning of this can be questioned. Hubalek (1982) defined five conditions that he considered an index should satisfy, of which one was that d should not be included. On this basis he found only six of 43 indices to be admissible, including those denoted here by S_2 to S_4; see also Janson and Vegelius (1981). More

Table 7
Presence and absence data obtained from two species at n sites, where a, b, c and d are counts

Species 1	Species 2		
	Present	Absent	Total
Present	a	b	$a+b$
Absent	c	d	$c+d$
Total	$a+c$	$b+d$	n

recently, Jackson et al. (1989) have suggested that many similarity indices effectively incorporate a component measuring the frequency of occurrence of species that is distorting the results obtained from applying them, although Ochiai's index is an exception. Another exception is the C_8 coefficient of Cole (1949) and Hurlbert (1969) that has been found to give good results by Wilson et al. (1990).

If the abundance of species is recorded for sites, rather than just presence or absence, then a correlation coefficient can be used to measure the similarity between two species. It may still be necessary to address the question of whether a high correlation based on many shared zero counts is allowed.

Naturally enough, the ecologist will often wish to know whether the association that is observed between species is consistent with the hypothesis that the species distributions are quite unrelated. But this is subject to some difficulties and has led to another area of considerable debate, particularly in the context of whether species occur independently of each other on islands. At first sight testing for independence appears to be straightforward since with a moderately large sample of sites the significance of the observed association between two species based on presences and absences (Table 7) can be tested using the 2×2 chi-squared test, and correlation coefficients for counts of abundance can be tested in the usual way. Unfortunately, this leads to a potential problem of multiple testing because of the possibly large number of pairs of species that may be involved. For example, with 10 species there are 45 pairs, some of which may be expected to be significant at the 5% level (say) due to chance alone. This suggests that what is needed is an overall test of the hypothesis that species distributions are unrelated.

One such test for presence and absence data was suggested by Schluter (1984), where this is based on the ratio of the variance of the total number of species at each site to the expected value of this variance based on the presences and absences of individual species. This ratio is compared to the critical values of a chi-squared distribution. Unfortunately, however, this test is not satisfactory since it ignores the fact that a high variance in the number of species at different sites may just be due to the fact that some sites are simply more attractive than others (McCulloch, 1985).

This is an area where Monte Carlo methods hold more promise than anything else. An early contribution of this nature was by Harper (1978) who proposed comparing an observed measure of the clustering of localities (based on a standard method of cluster analysis) with the distribution of this measure that is obtained by randomly reallocating the presences and absences of species to sites. This could be used equally well to test for clustering of species, but it was pointed out by Strauss (1982) that the test is unsatisfactory since it maintains the total number of occurrences of each species at the observed level but allows the number of species at each site to vary randomly. Thus it has the same problem as Schluter's (1984) test in not allowing for the fact that some sites are more attractive than others.

Strauss suggested that in fact both the total number of species at each site and the total number of occurrences should be considered to be fixed, and subsequently, several algorithms to carry out randomizations subject to this restriction have been developed (Manly, 1991a, p. 233; Jackson et al., 1992), but are usually not evaluated in terms of how they perform on data that are actually generated from a null model. In this respect, Wilson's (1987) study is a useful guide to what should be done.

Recent research (Roberts and Stone, 1990; Stone and Roberts, 1991) suggests that this is an area where a generalized Monte Carlo test along the lines suggested by Besag and Clifford (1989) is the best approach. Such a test involves starting with the observed configuration of presences and absences and changing this in a stepwise manner by swapping a 0 (absence) and 1 (presence) for one species with a 1 0 for another species in such a way as to maintain the toal number of presences at each site, with the swap being randomly chosen from potential swaps. Besag and Clifford suggest two algorithms for carrying out a valid tests in this situation. See Manly (1992b) for a further discussion of this procedure.

Monte Carlo tests have also been used for cases where the abundances of different species are recorded at a number of sites. The hypothesis of no relationship between species is then related to the species diversity as discussed in Section 3.1. It has been claimed that a community with no structure means one where the species abundances at different sites are determined by the broken stick model, in which case Hopf and Brown's (1986) statistic R of equation (3.1) can be used as the basis of a test. However, this ignores the fact that species distributions may be independent but nevertheless some species tend to be more abundant than others. In fact, it seems that the generalized Monte Carlo test for no species association based on presence and absence data can be extended to the situation where species abundances are recorded, by using exactly the same process of swapping the randomly chosen records for two species. As yet, this does not seem to have been tried.

One of the problems with studying species associations is the fact that competing species require similar resources. Therefore any tendency for the presence of one species to reduce the probability that the other species is present is likely to be more than offset by the fact that both species will have a high probability of occurrence in the same locations. As Diamond (1992) puts it, 'enemies are doomed to associate', which makes the assessment of species associations in terms of competition very difficult. Recently, Schoener and Adler (1991) have come up with a solution to this problem that involves using logistic regression to relate the probability of a species being present to ecological variables as well as the presence of other species. It has been found in this way that positive associations between species become negative in most cases after ecological covariates are taken into account.

Most studies of the relationship between species based on abundance data have considered single samples taken from different sites. However, another possibility is to observe time series of species abundances at a single site. A question of some interest is then whether the assemblage of species being considered shows stability over time. The tests devised by Crowley (1992) for assessing the stability of a single population, that have already been discussed in Section 2.5, have applications here. Crowley suggested adding up test statistics for each species and testing the total by comparison with the distribution of this total obtained either by simulated random walks (the random walk attraction test), by randomizing the order of the observed data (the randomized attraction test), or by generalizing Pollard et al. 's (1987) test. In this manner, Crowley and Johnson (1992) found evidence of stability for a group of 13 dragonfly taxa studied over 12 years and for a group of 25 fruit fly polulations observed for nine generations.

A rather different approach to studying stability was applied by Prins and Douglas-Hamilton (1990) who analyzed nine census counts of large herbivore species in Lake Manyara National Park, Tanzania, over the period 1959–84. They used a particular theoretical model for stability in multi-species ecosystems that leads to an index of stability $s\sqrt{mC}$ that is less than 1 for stability and greater than 1 for instability, where s is the standard deviation of species competition coefficients that are estimated by regression, m is the number of species, and C is the proportion of species that directly interact. Prins and Douglas-Hamilton claimed that since their observed stability index was less than 1, the species assemblage was stable. However, Rejmanek (1992) has noted several problems with this analysis that throw strong doubts about its validity either with the particular example or in general. Hallett (1991) used a similar analysis to the one used by Prins and Douglas-Hamilton, but based on data collected from the same species at different sites, rather than the same species at different times from one site.

The use of regression to estimate coefficients that are assumed to measure the degree of competition between species was suggested by Schoener (1974) as an alternative to the use of the niche overlap indices that are discussed in the next section. Subsequent uses of this method are reviewed by Carnes and Slade (1988) and Morris (1989). It appears that regression can be successful in this context, but only under fairly restricted conditions.

The spatial pattern of interacting species is obviously an important aspect of their relationship. Methods for studying this aspect are reviewed by Andersen (1992).

3.4. Niche overlap and niche breadth

The previous section was concerned with the analysis of data on species presences and absences, or species abundances, at a number of different sites or at a number of different times as a means of determining the relationships between species. The present section is also concerned with the relationship between species. But now the question concerns how individuals or species are related in terms of their use of resources. Thus the basic data consists of a record of the amounts of different types of resources used either by a number of different individual or a number of different species.

Suppose that it is species that are being considered, and that resources can be considered to be in a number of distinct categories. Let f_{ij} denote the amount of resource j used by species i ($i = 1, 2, \ldots, s; j = 1, 2, \ldots, R$), and let p_{ij} be the fraction of the total resources used by species i that are in category j. Then one of the simplest measures of the overlap between species j and species k in terms of the use that they make of resources (their niche overlap) is

$$S_{jk} = 1 - \tfrac{1}{2} \sum_{i=1}^{R} | p_{ij} - p_{ik} |. \tag{3.7}$$

This is often attributed to Schoener (1970), although Hurlbert (1978) suggests that it was first used by Renkonen (1938). At one extreme this index takes the value 1 for two species that use resources in exactly the same proportions ($p_{ij} = p_{ik'}$, complete overlap). At the other extreme it is zero for two species that only use resources from different categories (no overlap).

Many other indices have been proposed for use with resources in distinct categories, including Morisita's (1959) index

$$M_{jk} = \frac{2 \sum (p_{ij} p_{ik})}{\sum p_{ij}^2 + \sum p_{ik}^2}, \tag{3.8}$$

Pianka's (1973) index

$$P_{jk} = \frac{\sum (p_{ij} p_{ik})}{\sqrt{\sum p_{ij}^2 \sum p_{ik}^2}}, \tag{3.9}$$

where in both cases the summations are over the R resource categories, and rather more complicated indices based on information theory (Horn, 1966) and likelihood functions (Petraitis, 1979). Thirteen such indices are reviewed and discussed by Schatzmann et al. (1986). It seems that to a large extent the choice of the overlap measure used in any study is arbitrary. The situation is therefore comparable to that for similarity indices, as discussed in the previous section. Clearly, what is needed in both cases is a clearer understanding of what are the implications of using each index so that the ecologist has good reasons for using the one chosen, and so that the choice can be made before the data are collected.

An alternative to the situation where resources are considered to be in discrete categories occurs when the resources are measured on a continuous scale. For example, prey use is sometimes considered in terms of the size of the prey. The overlap between two species is then the extent to which they utilize prey of a similar size. One obvious way to measure this is by the common area shared by the usage distributions, as illustrated in Figure 11, which can be expressed by an explicit equation for some distributions. Thus, suppose that species i selects prey in such a way that the eaten individuals have a size that is normally distributed with mean μ_i and standard deviation σ_i. It can then be shown that the overlap between the normal

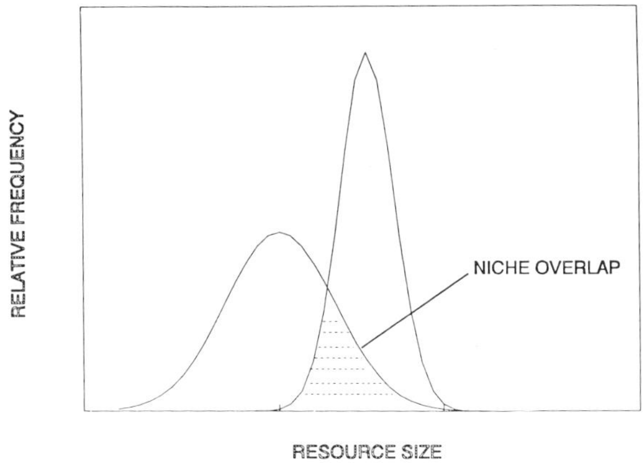

Fig. 11. Niche overlap between two species with different distributions for the size of a resource.

distributions for species i and species j is

$$N_{ij} = \exp\left[-\frac{(\mu_i - \mu_j)^2}{2(\sigma_i + \sigma_j)} \right] \tag{3.10}$$

(MacArthur and Levins, 1967). Of course, many prey size distributions are not normally distributed and, in particular, highly skewed distributions are common. This led Manly and Patterson (1984) to suggest the use of the Weibull distribution instead of the normal for usage distributions. Estimating the overlap between two distributions is then more complicated than it is with normal distributions, but still quite feasible.

The concept of measuring overlap as the area shared by two usage distributions generalizes easily to the situation where the used resources are described by several variables rather than one. With two multivariate normal distributions with mean vectors $\boldsymbol{\mu}_i$ and $\boldsymbol{\mu}_j$ and the same covariance matrix $\boldsymbol{\Sigma}$ this leads to the overlap measure

$$N^*_{ij} = \exp\{ -(\boldsymbol{\mu}_i - \boldsymbol{\mu}_j)' \boldsymbol{\Sigma}^{-1}(\boldsymbol{\Sigma}_i - \boldsymbol{\Sigma}_j)/4\} \tag{3.11}$$

(Harner and Whitmore, 1977) and several generalizations of this result are discussed by Lu et al. (1989).

It has been suggested that niche overlap measures should incorporate information about the relative amounts of different types of resource since if two species both make high use of a rare resource then this should mean more overlap than if they make the same use of a common resource (Colwell and Futuyama, 1971; Hurlbert, 1978). This leads to the idea of weighted overlap measures, where the most commonly used weight is the inverse of the availability (Smith, 1985a).

Early uses of overlap measures ignored the problem of assessing the likely level of sampling errors, which is complicated by the extremely nonnormal distributions that tend to be observed in practice. Recent work in this area has indicated that bias is a potential problem, and that jackknifing and bootstrapping are useful techniques (Mueller and Altenberg, 1985; Smith, 1985a, b; Lu et al., 1989). Within-species replication (observations from several individuals for each species being compared) is important here to avoid assumptions about the nature of the data that will quite likely be incorrect. For example, the assumption that the observed numbers of different resource units used by a species is an observation from a multinomial distribution seems reasonable at first sight, but may not be true because units are not selected independently from the population of available units.

Niche overlap indices are calculated for a variety of reasons. Sometimes the motivation is just the desire to quantify one aspect of the relationship between species or individuals, whereas at other times the indices are used as the basis for testing hypotheses about the structure of ecological communities. Example of the latter are a randomization test for whether species select prey in the same manner (Patterson, 1986), randomization and bootstrap tests for whether levels of overlap between different groups change with time (Linton et al., 1989; Manly, 1990b), and randomization tests for whether the observed levels of overlap in a community are consistent with a null model of no interaction between species (Sale, 1974; Lawlor, 1980a; Jaksic and Medel, 1990).

The use of overlap indices for studying interaction between species is not without its difficulties. Equating the niche overlap between two species with current levels of competition may be misleading since competition may tend to reduce overlap by leading the species to use alternative resources (Lawlor, 1980b; Thomson and Rusterholz, 1982; Petrais, 1989). However, as was noted in the last section, alternative procedures based on regression relationships between species numbers such as were used by Prins and Douglas-Hamilton (1990) to analyze census counts of large herbivore species in Lake Manyara National Park are also open to criticism. This area is clearly one where more study is required.

The concept of niche breadth is related to the concept of niche overlap since the number of related species that can coexist in a community can be thought of as depending on the niche breadth of the individual species, and the extent to which these breadths overlap. On this basis, a community of species with wide niche breadths and little overlap will not contain as many species as another community for which either the niche breadths are narrower or there is more niche overlaps. However, quantifying these concepts has not been simple. As has already been shown, many alternative measures of niche overlap have been proposed. The same is true for niche breadths.

Early measures of the niche breadth of a species were based on diversity measures such as Shannon's statistic given by equation (3.2) above, but calculated for proportions of different types of resources used rather than proportions of different individuals (Levins, 1968). This was later criticized since it does not take into account the availability of different types of resources, and alternative indices have been proposed that compare use with availability (Smith, 1982). For example, Feinsinger et al. (1981) proposed using equation (3.7) but with the resource proportions used by a species compared with the proportions available rather than with the proportions used by another species. This leads back to the measurement of resource selection that has already been discussed in Section 2.6. It seems possible that the idea that each species operates with a different resource selection function, as discussed by Manly et al. (1992) may therefore illuminate the measurement of niche breadth. However, that has not been investigated at the present time.

3.5. Ordination

When ecologists collect data on the abundance of species at different sites they will probably be interested in the relationships between the sites in terms of which are relatively similar and which are relatively different. This can be seen most easily if each site can be assigned a value for two or three index variables that reflect key aspects of the variation between sites, and the sites are plotted against these variables. The process of determining the index variables is then called ordination.

Ordination is related to direct gradient analysis, as discussed in Section 3.2. With direct gradient analysis, the sites are considered to be in a predetermined order along an environmental gradient and species abundances are related directly to the values of the environmental gradient. With ordination, which is sometimes called indirect gradient analysis, no initial assumptions are made about the relationships between the sites. However, once an ordination has been carried out the relationships between the sites can be examined in terms of environmental variables.

There is a large literature on the uses of ordination in ecology, particularly in terms of applications in botany. Extensive reviews are provided by Gauch (1982, Chapter 4), Digby and Kempton (1987, Chapters 3–4), and Ludwig and Reynolds (1988, Chapters 17–21). Here only some of the most commonly used ordination methods will be briefly described.

The simplest ordination method, which is really a type of direct gradient analysis, is weighted averaging, which dates back to the work of Ellenberg (1948) and Whittaker (1948). This involves the i-th species being assigned a value w_i according to its perceived position along some ecological gradient. Each site is then allocated a score which is equal to the weighted average of the species abundances at the site, with the i-th species being given the weight w_i. The sites are then ordered according to their scores. The major limitation of this method is clearly the somewhat arbitrary choice of weights.

Polar ordination was introduced by Bray and Curtis (1957). This begins with the calculation of measures of distances between sites. Two sites are then chosen to represent the two extremes along an ecological axis, and the remaining sites are ordered according to their relative distances to the two polar sites. Ordination on more than one axis is made possible by choosing further different pairs of polar sites. Like weighted averaging, this technique has the limitation of including an element of arbitrariness, although this is in the choice of polar sites rather than in the weights given to sites.

Principal components analysis seems to have been first applied to ordination by Goodall (1954). This standard method of multivariate analysis involves finding linear combinations of variables (in this case the abundances of different species) such that the first combination has the maximum variance, the second combination has maximum variance given that it is uncorrelated with the first linear combination for the sample of cases (in this case the different sites), the third combination has maximum variance given that it is uncorrelated with the first two combinations, and so on. The solution involves finding the eigenvalues and eigenvectors of the covariance matrix of the species abundance variables, usually after these variables have been standardized.

A problem with that has been found with this method of ordination is that responses to environmental gradients tend to show up as curves linking sites when they are plotted against the values of the first two or three principal components. This is a result of individual species favoring particular regions on gradients, so that typically a species shows its maximum abundance at some particular point on the gradient, with lower abundances on either side of this point (Gauch, 1982, p. 139). This 'arch problem' or 'horseshoe effect' can be overcome by an adjustment to the analysis that is called detrended principal components (Ludwig and Reynolds, 1988, p. 260).

Principal coordinates analysis is a generalization of principal components analysis that allows the distances between sites to be measured by something other than the Euclidean distances based on species abundances (Gower, 1966; Digby and Gower, 1986). According to Gauch (1982, p. 162) tests with simulated and field data show that this method performs slightly better than principal components analysis if distances between sites are measured as percentage differences in species composition.

At the present time the most popular methods for ordination (at least with plant ecologists) are correspondence analysis and detrended correspondence analysis because they seem to be very good at recovering the main trend in data if one exists. Correspondence analysis, which is also called reciprocal averaging, was developed by Hirschfeld (1935) and Fisher (1940). The term 'reciprocal averaging' comes about because it is possible to obtain ordination scores by an iterative process. First, arbitrary scores are assigned to species, say in the range 0–100. Next, a weighted average of these scores is calculated for each site, with weights equal to species abundances. A new score is then calculated for each species as a weighted average of the site scores. The species scores are then scaled to the range 0–100, and used for the start of a new iteration. The ordination values are found when there are no changes in site or species scores from one iteration to the next. Values for other ordination axes can be obtained in a similar manner (Hill, 1973b).

Formally, the calculations just described are an iterative method for extracting eigenvalues of matrices. As such, correspondence analysis is similar to principal component analysis, although correspondence analysis is generally considered to give better ordinations (Gauch, 1982, p. 148; Ludwig and Reynolds, 1988, p. 255).

The 'arch problem' is present with correspondence analysis. This has led to the development of a modification called detrended correspondence analysis which is described in an informal way by Gauch (1982, p. 153) and more formally by Hill and Gauch (1980). Basically, detrending consists of ensuring that the several ordination axes considered give scores for sites that are completely unrelated as well as being uncorrelated. More recently, a new method of detrending by polynomials has been suggested by Ter Braak and Prentice (1988), but Knox (1989) found this to give disappointing results.

Finally, nonmetric multidimensional scaling can be mentioned. This standard multivariate method was developed as a general means of producing an ordination from any matrix of distances between cases, which are sites in the present context. The description 'nonmetric' is used because the only requirement that is placed on the ordination is that the distances in the initial matrix are in the correct order of size in the sense that if the distance between cases i and j is greater than the distance between cases r and s then the cases i and j should be further apart when plotted against the ordination axes. Metric multidimensional scaling is also possible where this incorporates a regression type relationship between the plotted distances and the distances that the analysis begins with.

Nonmetric multidimensional scaling was developed by Sheppard (1962) and Kruskal (1964a, b) and seems to have been first used for the ordination of ecological communities by Anderson (1971). Gauch (1982, p. 161) concludes that this method has little advantage over the other methods that have been described above and, in particular, detrended correspondence analysis usually gives better results. A similar view is also expressed by Digby and Kempton (1987, p. 103). However, other authors are less negative, suggesting that the method is useful in cases where trends in the data are highly nonlinear (Legendre and Legendre, 1983, p. 293; Kenkel and Orloci, 1986; Ludwig and Reynolds, 1988, p. 273).

Whatever method of ordination is employed, it is important to gauge the extent to which the results are effected by sampling variation. This is an area where bootstrapping and other computer intensive methods offer the most promise. Applications of bootstrapping have been discussed for principal component analysis (Diaconis and Efron, 1983), correspondence analysis (Greenacre, 1984), multidimensional scaling (Weinberg et al., 1984), and detrended correspondence analysis (Oksanen, 1988; Knox and Peet, 1989). See also Grossman et al.'s (1991) study of the ability of tests on eigenvalues to distinguish between field collected species abundance data and computer generated data with no structure.

Modern methods of ordination require suitable computer programs. Many are available, including DECORANA for detrended correspondence analysis (Hill, 1979a), CANOCO for a range of ordination and direct gradient analyses (Ter Braak, 1988), programs included with the book by Ludwig and Reynolds (1988) for many of the analyses that they describe, PATN for a range of ordination and related analyses (Belbin, 1989), and MVSP for a range of ordination analyses (Kovach, 1990). Standard statistical packages also often include options that can be used for ordination.

EXAMPLE 3.3 (*Ordination of plots from vegetation data*). Table 8 shows the abundances of the 25 most abundant plant species on 17 plots from a grazed meadow in Steneryd Nature Reserve in Sweden as measured by Persson (1981) and used for an example by Digby and Kempton (1987). Each value in the table is the sum of cover values in a range from 0 to 5 for nine sample quadrats, so that a value of 45 corre-

Table 8
Abundance measures for 25 plant species on 17 plots from a grazed meadow in Steneryd Nature Reserve, Sweden

Species	Plot																
	1	2	3	4	5	6	7	8	9	10	11	12	13	14	15	16	17
1. *Festuca ovina*	38	43	43	30	10	11	20	0	0	5	4	1	1	0	0	0	0
2. *Anemone nemorosa*	0	0	0	4	10	7	21	14	13	19	20	19	6	10	12	14	21
3. *Stallaria holostea*	0	0	0	0	0	6	8	21	39	31	7	12	0	16	11	6	9
4. *Agrostis tenuis*	10	12	19	15	16	9	0	9	28	8	0	4	0	0	0	0	0
5. *Ranunculus ficaria*	0	0	0	0	0	0	0	0	0	0	13	0	0	21	20	21	37
6. *Mercurialis perennis*	0	0	0	0	0	0	0	0	0	0	1	0	0	0	11	45	45
7. *Poa pratenis*	1	0	5	6	2	8	10	15	12	15	4	5	6	7	0	0	0
8. *Rumex acetosa*	0	7	0	10	9	9	3	9	8	9	2	5	5	1	7	0	0
9. *Veronica chamaedrys*	0	0	1	4	6	9	9	9	11	11	6	5	4	1	7	0	0
10. *Dactylis glomerata*	0	0	0	0	0	8	0	14	2	14	3	9	8	7	7	2	1
11. *Fraxinus excelsior* (juv.)	0	0	0	0	0	8	0	0	6	5	4	7	9	8	8	7	6
12. *Saxifraga granulata*	0	5	3	9	12	9	0	1	7	4	5	1	1	1	3	0	0
13. *Deschampsia flexuosa*	0	0	0	0	0	0	30	0	14	3	8	0	3	3	0	0	0
14. *Luzula campestris*	4	10	10	9	7	6	9	0	0	2	1	0	2	0	1	0	0
15. *Plantago lanceolata*	2	9	7	15	13	8	0	0	0	0	0	0	0	0	0	0	0
16. *Festuca rubra*	0	0	0	0	15	6	0	18	1	9	0	0	2	0	0	0	0
17. *Hieracium pilosella*	12	7	16	8	1	6	0	0	0	0	0	0	0	0	0	0	0
18. *Geum urbanum*	0	0	0	0	0	7	0	2	2	1	0	7	9	2	3	8	7
19. *Lathyrun montanus*	0	0	0	0	0	7	9	2	12	6	3	8	0	0	0	0	0
20. *Campanula persicifolia*	0	0	0	0	2	6	3	0	6	5	3	9	3	2	7	0	0
21. *Viola riviniana*	0	0	0	0	0	4	1	4	2	9	6	8	4	1	6	0	0
22. *Hepatica nobilis*	0	0	0	0	0	8	0	4	0	6	2	10	6	0	2	7	0
23. *Achillea millefolium*	1	9	16	9	5	2	0	0	0	0	0	0	0	0	0	0	0
24. *Allium* sp.	0	0	0	0	2	7	0	1	0	3	1	6	8	2	0	7	4
25. *Trifolium repens*	0	0	6	14	19	2	0	0	0	0	0	0	0	0	0	0	0

sponds to the complete cover by the species being considered. Note that the species are in order from the most abundant (1) to the least abundant (25) and the plots are in the order given by Digby and Kempton (1987, Table 3.2), which corresponds to variation in certain environmental factors such as light and moisture.

The results of three ordinations are shown in Figure 12. In part (a) of the figure the plots are plotted against the values that they have for the first two principal components, as calculated using the MVSP program of Kovach (1990). Plots are labelled from 1 to 17 as in Table 8, and it can be seen that there is a tendency for the low numbered plots to appear in the middle right-hand part of the graph, the intermediate numbered plots to appear in the top left-hand part of the graph, and the highest numbered plots to appear in the bottom left-hand side. Hence there is something of an 'arch effect', although this is hidden by other variation to a large extent.

In part (b) of the figure the plots are plotted against the first two axes found by correspondence analysis. This was also carried out using the program MVSP, with the relationship between plots measured by the percentage similarity. The pattern shown is much clearer here, with an arch consisting of the low numbered plots on the right-hand side and high numbered plots on the left-hand side. In a way it is like an upside-down version of the principal components plot with much less 'random' variation.

Part (c) of the figure shows the result of plotting the samples against two axes obtained by multidimensional scaling using the program NMDS provided with the book by Ludwig and Reynolds (1988). The relationship between plots is not clear on the vertical axis here, but the low numbered plots are still on the right-hand side and the high numbered plots on the left-hand side.

The relationship between the three ordinations, and the time trends displayed by the ordination axes are shown in Figure 13, which shows all six ordination axes plotted against each other, and also against the plot number. Here it can be seen, for example, that the plot number is related to both principal component axes, both correspondence analysis axes, and the first nonmetric scaling axis. Also, although the three methods are not estimating quite the same things there are some clear relationships. Thus the first axis is measuring much the same effect for all three methods of ordination. Also, the second axes from principal components analysis and correspondence analysis are rather similar.

This example demostrates the difficulties associated with ordination. Three reasonable methods have been used on the data. They have all produced ordinations that show the tendency of the closeness of plots to be related to their plot number, but there are some clear inconsistencies. For example, in Figure 12, plots 6 and 7 appear very similar in plots (a) and (b), but rather different in plot (c).

3.6. Classification

Classification methods (also called cluster analyses) are used to produce groups of objects that are relatively similar. In community ecology the objects are usually either samples taken at different places that are grouped on the basis of their similarity in terms of species compositions, or species grouped on the basis of the similarity of their distributions in different samples. However, any objects can be grouped on the basis of their similarity in terms of a set of variables that are measured on each object in order to describe it.

Reviews of classification methods in ecology are provided by Gauch (1982, Chapter 5), Legendre and Legendre (1983, Chapter 7), Digby and Kempton (1987, Chapter 5) and Ludwig and Reynolds (1988, Chapters 13–16). Many different methods have been proposed and the treatment here will be restricted to a discussion of some of the most popular ones only.

One possibility involves basing a classification on one of the ordination methods that were discussed in the last section. The objects of interests are plotted against the ordination axes and then clusters of close objects are defined. This is relatively simple, but suffers from the disadvantage that the ordination axes do not use all of the information that is originally available. Hence the apparent distance between objects from the ordination may appear to be either greater than or less than the real distance.

An early approach to clustering that was developed by Czekanowski (1909) is called seriation in archaeology. This involves ordering the objects of interest so that

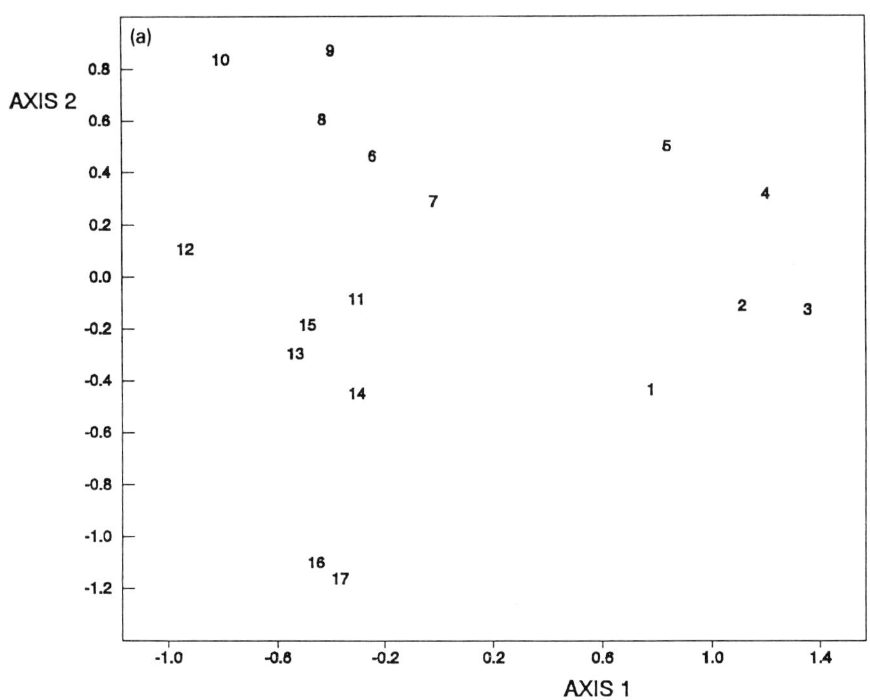

Fig. 12. Ordinations obtained for 17 sites from the Steneryd Nature Reserve: (a) sites plotted against principal component scores; (b) sites plotted against axes found by correspondence analysis; and (c) sites plotted against two axes found by nonmetric multidimensional scaling.

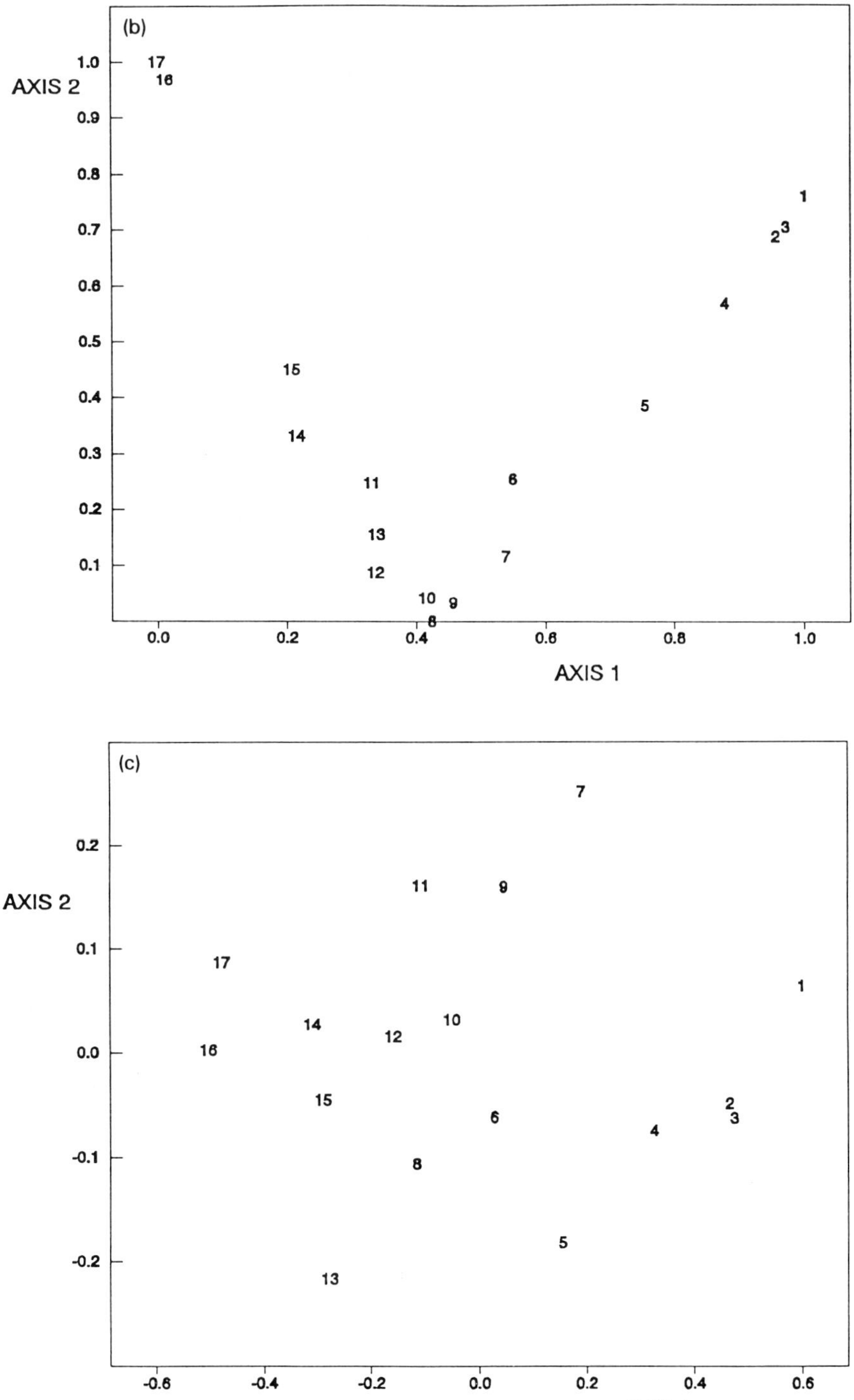
(b)
AXIS 2
1.0
0.9
0.8
0.7
0.6
0.5
0.4
0.3
0.2
0.1
0.0
0.2
0.4
0.6
0.8
1.0
AXIS 1
(c)
AXIS 2
0.2
0.1
0.0
-0.1
-0.2
-0.6
-0.4
-0.2
0.0
0.2
0.4
0.6
AXIS 1

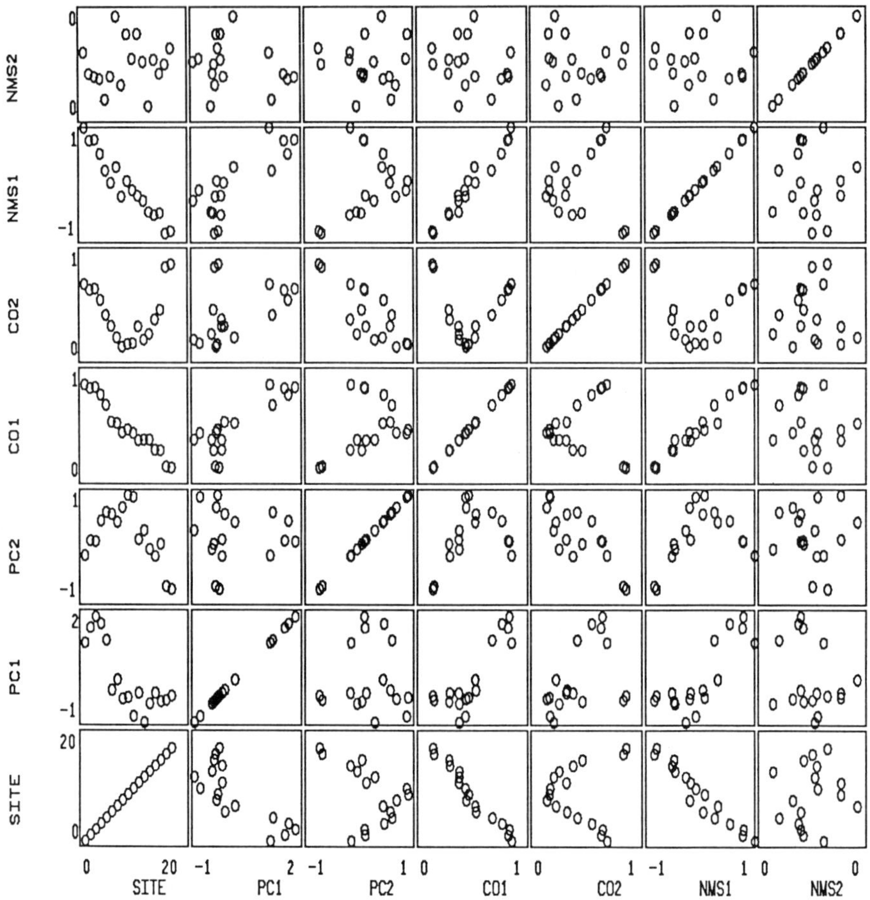

Fig. 13. Scores for ordination axes plotted against each other and against the site number for the three ordinations shown in Figure 12 (PC1 and PC2 are principal component axes 1 and 2; CO1 and CO2 are correspondence analysis axes 1 and 2, NMS1 and NMS2 are nonmetric dimensional scaling axes 1 and 2; SITE is the sample number).

the matrix of similarities between the objects has high values close to the diagonal as far as possible. Then the objects that are close in the ordering will tend to be similar and groups can be formed accordingly. A practical algorithm for ordering the objects was proposed by Beum and Brundage (1950), and a computer intensive alternative which maximizes the correlation between the distance between objects in the ordering and another measure of distance was used by Manly (1991b).

With a table of species abundances at a number of sites, seriation can be used first to order the species and then to order the sites. The reordered data table should then display blocks of similar sites and similar species that can be treated as groups. This then achieves the same result as the more subjective Braun-Blanquet (1932) table-work method that has been widely used by plant ecologists in Europe (Gauch, 1982, p. 182), as is implemented in the computer program TABORD (Van der Maarel et al., 1978). Of course, this can be thought of as a type of joint ordination of species and sites instead of clustering.

Association analysis is another early method used by plant ecologists to classify sites. This was developed by Goodall (1953) and Williams and Lambert (1959) and subsequently used for classifying species as well as sites (Williams and Lambert, 1961; Lambert and Williams, 1962). It can be described as a monothetic divisive method since it starts with all sites in a single cluster and then divides this cluster into smaller and smaller clusters on the basis of the presences and absences of single species. The procedure, which uses sums of chi-squared values for deciding on divisions, is described by Ludwig and Reynolds (1988, Chapter 15), who also provide a computer program.

More generally, classification methods can be described as hierarchic, in which case the results can be displayed as a dendrogram, or partitioning, which may not permit an easy graphical representation. Hierarchic methods may be either divisive (as in the case of association analysis), so that all objects are initially in a single cluster that is subsequently split into smaller clusters, or agglomerative, with objects starting in clusters of size one that are subsequently merged. Partitioning methods are nonhierarchic in the sense that the algorithms used allow objects to move freely between clusters in the process of determining the final allocation to a specified number of clusters.

Algorithms for hierarchic agglomerative clustering include the single linkage (or nearest neighbor) method which combines clusters providing that an object in one cluster is sufficiently close to at least one object in the other cluster, the furthest neighbor method, which combines two clusters if the largest of the distances from the objects in one cluster to the objects in the other cluster is small enough, and the group average method, which combines two clusters if the average distance from the objects in one cluster to the objects in the other cluster is small enough. Many variations of these types of algorithm have been proposed (Gauch, 1982, p. 197; Legendre and Legendre, 1983, p. 226; Digby and Kempton, 1987, p. 125; Ludwig and Reynolds, 1988, Chapter 16), including the Lance and Wailliams (1966, 1967) flexible clustering algorithm that covers most other methods as special cases.

Hierarchic divisive methods seem to be used less often than the agglomerative ones. Association analysis has already been mentioned. Other proposed methods are based on an initial ordination (Gauch, 1982, p. 200), or are based on considering distances between the clusters that are found when clusters are split in all possible ways, or by removing objects one by one (Legendre and Legendre, 1983, p. 245). Indicator species analysis, which is based on an initial ordination using correspondence analysis and can be implemented using the program TWINSPAN, seems to be used for much of the vegetation classification that is being currently carried out (Hill et al., 1975; Hill, 1979b).

Partitioning methods of classification are described by Hartigan (1975, Chapter 3), Gauch (1982, p. 193), Legendre and Legendre (1983, p. 249), and Seber (1984). A fundamental difficulty here is the fact that the number of possible partitions is huge even with modest sets of data. For example, there are about 1.9×10^8 ways to place 19 objects into three clusters (Seber, 1984, p. 379). Therefore, all the 'quick partition' algorithms attempt to find the best clusters on the basis of some chosen criterion without considering all the possibilities.

Most methods that have been proposed for classification insist that each object is placed in a single cluster. However, 'fuzzy' clustering is also possible, where this involves each object being associated with different clusters by amount that are measured by indicator variables between 0 and 1. 'Fuzzy' clustering of ecological data is discussed by Equihua (1990), who concludes that it is useful since ecological communities are better represented this way than by clusters of individuals in rigidly defined groups.

The assessment of the results of a classification analysis is not particularly straightforward. Clearly, significance tests for mean differences between clusters for either single variables or all the variables used for clustering are invalid since the analysis itself necessarily finds clusters that show differences even when there are no clusters. Therefore such tests provide no real evidence that clusters exist. Computer intensive methods are better, where these involve the comparison of the cluster analysis on a real set of data with the results obtained from randomly reordered data (Harper, 1978), or simulated sets of data (Aldenderfer and Blashfield, 1984, p. 66), or bootstrapped data (Nemec and Brinkhurst, 1988).

EXAMPLE 3.4 (*Classification of plant species*). Example 3.3 was concerned with the ordination of 17 plots from a grazed meadow in Steneryd Nature Reserve in Sweden.

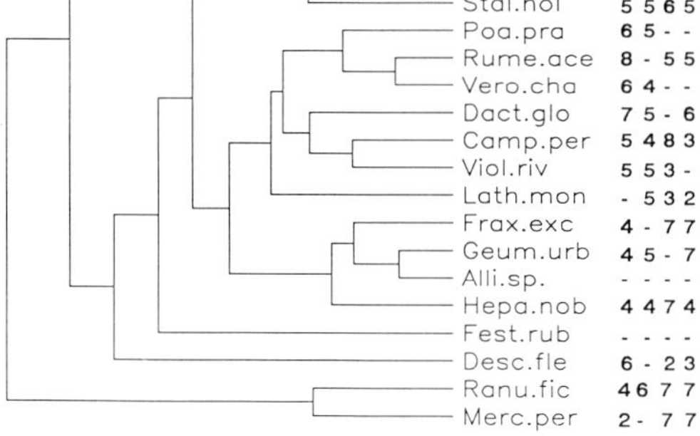

Fig. 14. Dendrogram produced by clustering the species shown in Table 8 using percentage differences to measure the distance between the species, and group average hierarchic agglomerative clustering. Species indicator values for the response to environmental factors (where known) are L = Light, M = Moisture, R = Reaction, and N = Nitrogen. Species names are abbreviated in an obvious way.

The example now being considered involves the use of cluster analysis to group the 25 plant species on the basis of their distributions in the plots. This is then in a sense the inverse of the classification of the plots in terms of their species compositions, which will not be considered here.

The data to be analyzed are provided in Table 8. From these data, a 25 by 25 matrix of percentage similarities was calculated in which the element in the i-th row and j-th column was $100(1 - 0.5\sum|p_{ir} - p_{jr}|)$, where p_{ir} is the percentage of the total frequency for species i that was found in the r-th plot, and the summation is over the 17 plots. The program MVSP (Kovach, 1990) was used to produce similarity matrix and also to produce the dendrogram shown in Figure 14, using the group average agglomerative method.

Obviously it is desirable that a classification of the type shown in Figure 14, which is based entirely on the distribution of species over plots, can be related to specific similarities between the species. In the present case this can be done by considering how the grouping of species relates to the values that the species have for four variables that are believed to indicate the response to light, moisture, soil reaction and nitrogen. Some values for these indicator variables for the species being considered have been provided by Digby and Kempton (1987), and these are reproduced in Figure 14. However, the relationship between the classification and these variables will not be pursued further here.

4. Observational and experimental studies in ecology

Many different study designs can be used in ecology. A basic distinction is between observational studies, that are essentially passive, and experimental studies, that involve the researcher manipulating some variables while at the same time trying to keep other variables constant. However, this is a fairly coarse classification, and Eberhardt and Thomas (1991) have suggestesd that eight types of study can be recognized. Three of these involve events being controlled by the observer: replicated experiments, unreplicated experiments, and sampling for modelling. The other five involve the study of events that are not controlled by the observer: intervention analysis, group comparisons, analytical sampling, descriptive sampling, and sampling for pattern. The characteristics of these different study designs will not be discussed in detail here. The point being made is the variety of designs that exist. Intervention analysis has already been covered in Section 2.5.

There are many books and papers available on the design of observational and experimental studies in general (Manly, 1992a). There is also a reasonably large literature on sampling designs for observational studies in ecology (e.g., Green, 1967; Southwood, 1978; Gilbert, 1987). However, the literature on experiments in ecology is quite limited. The papers that are available in this area include one by Hurlbert (1984) in which he emphasizes the need for true replication in experiments, as distinct from pseudoreplication obtained by resampling at points that are close in space or time. Later, Eberhardt and Thomas (1991) reinforced Hurlbert's comments on replication and noted that many experiments have had sample sizes that are simply too small to detect effects of a realistic size. At about the same time, Underwood (1991b) discussed the logical basis of ecological experiments in the context of a particular case study in the area of marine biology. The only books that are available on experiments in ecology seem to be the ones by Hairston (1989), Mooney et al. (1991) and Scheiner and Gurevitch (1993).

Eberhardt and Thomas' (1991) comments on inadequate sample sizes are an echo of essentially the same message given to ecologists earlier by Toft and Shea (1983). Similiar comments have also been made with respect to clinical trials, fisheries management, the study of natural selection, and no doubt in other areas as well (Manly, 1992a, p. 325; Fairweather, 1991). There is surely little doubt that in future more attention should be devoted to ascertaining the power of study designs before they are used, in order to ensure that sample sizes are adequate.

5. Concluding remarks

This review has concentrated on statistical methods that have been developed primarily for use in animal and plant ecology. These are interpreted here to form the body of methods that can be called 'statistical ecology', although it is interesting to note that to some extent this description has in the past been used to describe a narrower range of methods. This seems to be due to animal ecologists and plant ecologists tending to use different methods. Thus Legendre and Legendre (1983) emphasize the material covered in Section 3 in this review, in a book that is described in a preface by Rohlf as a reference to a variety of important topics in mathematical and statistical ecology. Similarly, the books of Ludwig and Reynolds (1988) and Digby and Kempton (1987) concentrate mainly on ordination and classification.

The situation is complicated, of course, by the fact that all of the standard methods of statistical design and analysis have applications in ecology so that a quantitative ecologist needs a good knowledge of these standard methods as well as the narrower range of methods specially developed for use in this area. But in reality there is little common agreement at present about what precisely statistical ecology encompasses. Thus when six researchers working in statistics and ecology were asked to define the field, the answers were surprisingly varied, ranging from the one upon which this review is based to suggestions that present statistical methods are inadequate for solving ecological problems (Dixon, 1992).

Statistical ecology can also be thought of as forming part of a wider field of 'environmental statistics'. This wider field, which can be thought of as being defined by the contents of this *Handbook of Statistics*, is concerned with all applications of statistics to environmental problems, including those that are not directly related to animal and plant populations.

Whatever definition is used for statistical ecology, it is obvious from the topics reviewed in this chapter that there is much scope for statisticians and quantitative ecologists to develop new and improved methods for collecting and analyzing ecological data.

Acknowledgements

I wish to thank Philip Crowley and Bastow Wilson for reading an earlier draft of this paper and making helpful suggestions for its improvement. In particular, Bastow helped to reduce my 'animal bias'.

References

Aldenderfer, M. S., and R. K. Blashfield (1984). *Cluster Analysis.* Sage, Beverley Hills, CA.

Alho, J. M. (1990). Logistic regression in capture–recapture models. *Biometrics* **46**, 623–635.

Allredge, J. R., and C. E. Gates (1985). Line transect estimators for left-truncated distributions. *Biometrics* **41**, 273–280.

Andersen, M. (1992). Spatial analysis of two-species interactions. *Oecologia* **91**, 134–140.

Anderson, A. J. B. (1971). Ordination methods in ecology. *J. Ecol.* **59**, 713–726.

Anderson, D. R., K. P. Burnham, G. C. White and D. L. Otis (1983). Density estimation of small mammal populations using a trapping web and distance sampling. *Ecology* **64**, 674–680.

Arrhenius, O. (1921). Species and area. *J. Ecol.* **9**, 95–99.

Barker, R. J. (1992). Effect of heterogenous survival on bird-banding model confidence interval coverage rates. *J. Wildlife Management* **56**, 111–116.

Barraclough, G. F. (1978). Sampling bias in dispersal studies based on finite area. *Bird Band.* **49**, 333–341.

Barton, D. E. and F. N. David (1956). Some notes on ordered random intervals. *J. Roy. Statist. Soc. Ser. B* **18**, 79–94.

Belbin, L. (1989). *PATN, Pattern Analysis Package.* Division of Wildlife and Ecology, CSIRO, Lyneham, ACT, Australia.

Bergstedt, R. A. and D. R. Anderson (1990). Evaluation of line transect sampling based on remotely sensed data from underwater video. *Trans. Amer. Fish. Soc.* **119**, 86–91.

Berryman, A. A. (1991). Stabilization or regulation: What it all means! *Oecologia* **86**, 140–143.

Besag, J. and P. Clifford (1989). Generalized Monte Carlo significance tests. *Biometrika* **76**, 633–642.

Beum, C. O. and E. G. Brundage (1950). A method for analyzing the sociomatrix. *Sociometry* **13**, 141–145.

Bonett, D. G., J. A. Woodward and P. M. Bentler (1987). Estimating the size of closed populations using inverse multiple-recapture sampling. *Biometrics* **43**, 985–991.

Boulanger, J. G. and G. C. White (1990). A comparison of home-range estimators using Monte Carlo simulation. *J. Wildlife Management* **54**, 310–315.

Bovet, P. and S. Benhamou (1988). Spatial analysis of animals' movements using a correlated random walk model. *J. Theor. Biol.* **131**, 419–433.

Braun-Blanquet, J. (1932). *Plant Sociology: The Study of Plant Communities.* Hafner Press, London.

Bray, J. R. and J. T. Curtis (1957). An ordination of the upland forest communities of southern Wesconsin. *Ecol. Monogr.* **27**, 325–349.

Brownie, C., D. R. Anderson, K. P. Burnham and D. S. Robson (1985). *Statistical Inference from Band Recovery Data – A Handbook.* Resource Publication 156, U.S. Fish and Wildlife Service, Washington, DC.

Buckland, S. T. (1984). Monte Carlo confidence intervals. *Biometrics* **40**, 811–817.

Buckland, S. T. (1985). Perpendicular distance models for line transect sampling. *Biometrics* **41**, 177–195.

Buckland, S. T. (1987). On the variable circular plot method of estimating animal density. *Biometrics* **43**, 363–384.

Buckland, S. T. and P. H. Garthwaite (1991). Quantifying precision of mark–recapture estimates using the bootstrap and related methods. *Biometrics* **47**, 255–268.

Burnham, K. P., D. R. Anderson and J. L. Laake (1980). Estimation of density from line transect sampling of biological populations. *Wildlife Monogr.* **72**.

Burnham, K. P., D. R. Anderson, G. C. White, C. Brownie and K. H. Pollock (1987). *Design and Analysis*

Methods for Fish Survival Experiments Based on Release–Recapture, Monograph 5. Amer. Fisheries Soc. Bethesda, MD.

Carnes, B. A. and N. A. Slade (1988). The use of regression for detecting competition with multicollinear data. *Ecology* **69**, 1266–1274.

Castledine, B. J. (1981). A Bayesian analysis of multiple-recapture sampling for a closed population. *Biometrika* **67**, 197–210.

Chao, A. (1987). Estimating the population size for capture–recapture data with unequal catchability. *Biometrics* **43**, 783–791.

Chao, A. (1989). Estimating population size for sparse data in capture–recapture experiments. *Biometrics* **45**, 427–438.

Chao, A., S. Lee and S. Jeng (1992). Estimating population size for capture–recapture data when capture probabilities vary by time and individual animal. *Biometrics* **48**, 201–216.

Cole, B. J. (1981). Overlap, regularity, and flowering phenologies. *Amer. Nat.* **117**, 993–997.

Cole, L. C. (1949). The measurement of interspecific association. *Ecology* **30**, 411–424.

Colwell, R. K. and D. J. Futuyama (1971). On the measurement of niche breadth and overlap. *Ecology* **52**, 567–576.

Cormack, R. M. (1968). The statistics of capture–recapture methods. *Ann. Rev. Oceanog. Mar. Biol.* **6**, 455–506.

Cormack, R. M. (1989). Log-linear models for capture–recapture. *Biometrics* **45**, 395–413.

Cormack, R. M. (1992). Interval estimation for mark–recapture studies of closed populations. *Biometrics* **48**, 567–576.

Cressie, N. and T. R. C. Read (1989). Spatial data analysis on regional counts. *Biom. J.* **31**, 699–719.

Crowley, P. H. (1992). Density dependence, boundedness, and attraction: Detecting stability in stochastic systems. *Oecologia* **90**, 246–254.

Crowley, P. H. and D. M. Johnson (1992). Variability and stability of a dragonfly assemblage. *Oecologia* **90**, 260–269.

Cushman, J. H., J. H. Lawton and B. F. J. Manly (1993). Latitudinal patterns in European ant assemblages: Variation in species, richness and size. *Oecologia* (in press).

Czekanowski, J. (1909). Zur differentialdiagnose der Neandertalgruppe. *Korresp. Deutsch. Ges. Anthropol. Urgesch.* **40**, 44–47.

Dale, M. R. T. and D. A. MacIsaac (1989). New methods for the analysis of spatial pattern in vegetation. *J. Ecol.* **77**, 78–91.

David, F. N. and P. G. Moore (1954). Notes on contagious distributions in plant populations. *Ann. Bot.* **18**, 47–53.

Delincé, J. (1986). Robust density estimation through distance measurements. *Ecology* **67**, 1576–1581.

Den Boer, P. J. (1990). On the stabilization of animal numbers. Problems of testing. 3. What do we conclude from significance test results? *Oecologia* **83**, 38–46.

Den Boer, P. J. (1991). Seeing the trees for the wood: Random walk or bounded fluctuations of population size? *Oecologia* **86**, 484–491.

Den Boer, P. J. and J. Reddingius (1989). On the stabilization of animal numbers. Problems of testing. 2. Confrontation with data from the field. *Oecologia* **79**, 143–149.

Diaconis, P. and B. Efron (1983). Computer intensive methods in statistics. *Sci. Amer.* **248**, 116–130.

Diamond, J. M. (1992). Enemies doomed to associate. *Nature* **355**, 501–502.

Digby, P. G. N. and J. C. Gower (1986). *Ordination and Classification.* Montreal University Press, Montreal.

Digby, P. G. N. and R. A. Kempton (1987). *Multivariate Analysis of Ecological Communities.* Chapman and Hall, London.

Diggle, P. J. (1990). *Time Series: A Biostatistical Introduction.* Oxford Univ. Press, Oxford.

Diggle, P. J. and A. G. Chetwynd (1991). Second-order analysis of spatial clustering for inhomogeneous populations. *Biometrics* **47**, 1155–1163.

Diggle, P. J. and T. F. Cox (1983). Some distance-based tests of independence for sparsely-sampled multivariate spatial point patterns. *Internat. Statist. Rev.* **51**, 11–23.

Dixon, P., ed. (1992). What is statistical ecology/ecological statistics? A discussion. *Ecol. Soc. Amer. Newslett.*, Statistical Ecology Section, Spring.

Dobzhansky, T. and S. Wright (1943). Genetics of natural populations. X. Dispersion rates in *Drosophila pseudoobscura*. *Genetics* **28**, 304–340.

Dorazio, R. M. and P. J. Rago (1991). Evaluation of a mark–recapture method for estimating mortality and migration rates of stratified populations. *Canad. J. Fish. Aqua. Sci.* **48**, 254–260.

Downing, J. A. (1991). Biological heterogeneity in aquatic ecosystems. In: S. T. J. Pickett, ed., *Ecological Heterogeneity*. Springer, New York, 160–180.

Drummer, T. D., A. R. Degange, L. L. Pank and L. L. McDonald (1990). Adjusting for group size influence in line transect sampling. *J. Wildlife Management* **54**, 511–514.

Drummer, T. D. and L. L. McDonald (1987). Size bias in line transect sampling. *Biometrics* **43**, 13–21.

Eberhardt, L. L. (1990). Using radio-telemetry for mark–recapture studies with edge effects. *J. Appl. Ecol.* **27**, 259–271.

Eberhardt, L. L. and J. M. Thomas (1991). Designing environmental field studies. *Ecol. Monog.* **61**, 53–73.

Ekbohm, G. and H. Rydin (1990). On estimating the species–area relationship: Commenting Loehle. *Oikos* **57**, 145–146.

Ellenberg, H. (1948). Unkrautgesellschaften als Mass für den Säuregrad, die Verdichtung und andere Eigenschaften des Ackerbodens. *Ber. Landtech. Kuratorium Tech. Bauwesen Landwirtschaft* **4**, 130–146.

Engeman, R. M. and J. F. Bromaghin (1990). An approach to estimating density from line transect data where the animals move in response to the observer. *J. Statist. Comput. Simulation* **34**, 145–154.

Equihua, M. (1990). Fuzzy clustering of ecological data. *J. Ecol.* **78**, 519–534.

Fairweather, P. G. (1991). Statistical power and design requirements for environmental modelling. *Austral. J. Mar. Freshwater Res.* **42**, 555–567.

Feinsinger, P., E. E. Spears and R. W. Poole (1981). A simple measure of niche breadth. *Ecology* **62**, 27–32.

Fisher, R. A. (1940). The precision of discriminant functions. *Ann. Eugen.* **10**, 422–429.

Francis, R. I. C. C. (1984). An adaptive strategy for stratified random trawl surveys. *New Zealand J. Mar. Freshwater Res.* **18**, 59–71.

Francis, R. I. C. C. (1991). Statistical properties of two-phase surveys: Comment. *Canad. J. Fish. Aqua. Sci.* **48**, 1128–1129.

Freeman, S. N. and B. J. T. Morgan (1992). A modelling strategy for recovery data from birds ringed as nestlings. *Biometrics* **48**, 217–235.

Galiano, E. F., I. Castro and A. Sterling (1987). A test for spatial pattern in vegetation using a Monte Carlo simulation. *J. Ecol.* **75**; 915–924.

Garthwaite, P. H. and S. T. Buckland (1990). Analysis of a multiple-recapture census by computing conditional probabilities. *Biometrics* **46**, 231–238.

Garthwaite, P. H. and S. T. Buckland (1992). Generating Monte Carlo confidence intervals by the Robbins–Monro process. *Appl. Statist.* **41**, 159–171.

Gaston, K. J. and J. H. Lawton (1987). A test of statistical techniques for detecting density dependence in sequential censuses of animal populations. *Oecologia* **74**, 404–410.

Gauch, H. G. (1982). *Multivariate Analysis in Community Ecology*. Cambridge Univ. Press, Cambridge.

Gauch, H. G. and G. B. Chase (1974). Fitting the Gaussian curve to ecological data. *Ecology* **55**, 1377–1381.

Gazey, W. J. and M. J. Staley (1986). Population estimation from mark–recapture experiments using a sequential Bayes algorithm. *Ecology* **67**, 941–951.

Getis, A. and J. Franklin (1987). Second-order neighborhood analysis of mapped point patterns. *Ecology* **68**, 473–477.

Ghent, A. W. (1991). Insights into diversity and niche breadth analyses from exact small-sample tests of the equal abundance hypothesis. *Amer. Midl. Natur.* **126**, 213–255.

Gilbert, R. O. (1987). *Statistical Methods for Environmental Pollution Monitoring*. Van Nostrand Reinhold, New York.

Gitay, H., S. H. Roxburgh and J. B. Wilson (1991). Species–area relations in a New Zealand tussock grassland, with implications for nature reserve design and for community structure. *J. Veg Sci.* **2**, 113–118.

Goodall, D. W. (1953). Objective methods for the classification of vegetation. I. The use of positive interspecific correlation. *Austral. J. Bot.* **1**, 39–63.
Goodall, D. W. (1954). Objective methods for the classification of vegetation. III. An essay on the use of factor analysis. *Austral. J. Bot.* **2**, 304–324.
Goodall, D. W. (1974). A new method for the analysis of spatial pattern by random pairing of quadrats. *Vegetatio* **29**, 135–146.
Gower, J. C. (1966). Some distance properties of latent root and vector methods used in multivariate analysis. *Biometrika* **53**, 325–338.
Green, R. H. (1966). Measurement of non-randomness in spatial distributions. *Res. Popul. Ecol.* **8**, 1–7.
Green, R. H. (1967). *Sampling Design and Statistical Methods for Environmental Biologists.* Wiley, New York.
Greenacre, M. J. (1984). *Theory and Applications of Correspondence Analysis.* Academic Press, London.
Gregorius, H. (1990). A diversity-independent measure of evenness. *Amer. Nat.* **136**, 701–711.
Greig-Smith, P. (1952). The use of random and contiguous quadrats in the study of the structure of plant communities. *Ann. Bot. New Ser.* **16**, 293–316.
Grossman, G. D., D. M. Nickerson and M. C. Freeman (1991). Principal component analysis of assemblage structure data: Utility of tests based on eigenvalues. *Ecology* **72**, 341–347.
Hairston, N. G. (1989). *Ecological Experiments: Purpose, Design and Execution.* Cambridge Univ. Press, Cambridge.
Hallett, J. G. (1991). The structure and stability of small mammal faunas. *Oecologia* **88**, 383–393.
Hamill, D. N. and S. Wright (1986). Testing the dispersion of juveniles relative to adults: A new analytical method. *Ecology* **67**, 952–957.
Harner, E. J. and R. C. Whitmore (1977). Multivariate measures of niche overlap using discriminant function analysis. *Theor. Popul. Biol.* **12**, 21–36.
Harper, C. W. (1978). Groupings by locality in community ecology and paleoecology: Tests of significance. *Lethia* **11**, 251–257.
Harris, W. F. (1986). The breeding ecology of the South Island fernbird in Otago Wetlands. Ph.D. Thesis, University of Otago, New Zealand.
Harrison, S., S. J. Ross and J. H. Lawton (1992). Beta diversity on geographical gradients in Britain. *J. Anim. Ecol.* **61**, 151–158.
Hartigan, J. A. (1975). *Clustering Algorithms.* Wiley, New York.
Hassell, M. P., J. Latto and R. M. May (1989). Seeing the wood for the trees: Detecting density dependence from existing life-table studies. *J. Anim. Ecol.* **58**, 883–892.
Hestbeck, J. B., J. D. Nichols and R. A. Malecki (1991). Estimates of movement and site fidelity using mark-resight data on wintering Canada geese. *Ecology* **72**, 523–533.
Heimbuch, D. G. and J. M. Hoenig (1989). Change-in-ratio estimators for habitat usage and relative population size. *Biometrics* **45**, 439–451.
Heltshe, J. F. and T. A. Ritchey (1984). Spatial pattern detection using quadrat samples. *Biometrics* **40**, 877–885.
Higham, C. F. W., A. Kijngam and B. F. J. Manly (1982). Site location and site hierarchy in prehistoric Thailand. *Proc. Prehist. Soc.* **48**, 1–27.
Hill, M. O. (1973a). The intensity of spatial pattern in plant communities. *J. Ecol.* **61**, 225–235.
Hill, M. O. (1973b) . Reciprocal averaging: An eigenvector method of ordination. *J. Ecol.* **61**, 237–249.
Hill, M. O. (1979a). DECORANA – A FORTRAN program for detrended correspondence analysis and reciprocal averaging. Section of Ecology and Systematics, Cornell University, Ithaca, NY.
Hill, M. O. (1979b). TWINSPAN – A FORTRAN program for arranging multivariate data in an ordered two-way table by classification of the individuals and attributes. Section of Ecology and Systematics, Cornell University, Ithaca, New York.
Hill, M. O., R. G. H. Bunce and M. W. Shaw (1975). Indicator species analysis, a divisive polythetic method of classification, and its application to a survey of native pinewoods in Scotland. *J. Ecol.* **63**, 597–613.
Hill, M. O. and H. G. Gauch (1980). Detrended correspondence analysis, an improved ordination technique. *Vegetatio* **42**, 47–58.

Hirschfeld, H. O. (1935). A connection between correlation and contingency. *Proc. Cambridge Philos. Soc.* **31**, 520–524.

Hopf, F. A. and J. H. Brown (1986). The bull's-eye method for testing randomness in ecological communities. *Ecology* **67**, 1139–1155.

Horn, H. (1966). Measurement of overlap in comparative ecological studies. *Amer. Nat.* **100**, 419–424.

Hubalek, Z. (1982). Coefficients of association and similarity based on binary (presence–absence) data. *Biol. Rev.* **57**, 669–689.

Huggins, R. M. (1989). On the statistical analysis of capture experiments. *Biometrika* **76**, 133–140.

Huggins, R. M. (1991). Some practical aspects of a conditional likelihood approach to capture experiments. *Biometrics* **47**, 725–732.

Hurlbert, S. H. (1969). A coefficient of interspecific association. *Ecology* **50**, 1–9.

Hurlbert, S. H. (1978). The measurement of niche overlap and its relatives. *Ecology* **59**, 67–77.

Hurlbert, S. H. (1984). Pseudoreplication and the design of ecological field experiments. *Ecol. Monog.* **54**, 187–211.

Iwao, S. and E. Kuno (1968). Use of the regression of mean crowding on mean density for estimating sample size and the transformation of data for the analysis of variance. *Res. Popul. Ecol.* **10**, 210–214.

Jackson, D. A., K. M. Somers and H. H. Harvey (1989). Similarity coefficients: Measures of co-occurrence and association or simply measures of occurrence? *Amer. Nat.* **133**, 436–453.

Jackson, D. A., K. M. Somers and H. H. Harvey (1992). Null models and fish communities: Evidence of nonrandom patterns. *Amer. Nat.* **139**, 930–951.

Jaksic, F. M. and R. G. Medel (1990). Objective recognition of guilds: Testing for statistically significant species clusters. *Oecologia* **82**, 87–92.

Janson, S. and J. Vegelius (1981). Measures of ecological association. *Oecologia* **49**, 371–376.

Jaremovic, R. V. and D. B. Croft (1987). Comparison of techniques to determine eastern grey kangaroo home range. *J. Wildlife Management* **51**, 921–930.

Jensen, A. L. (1989). Confidence intervals for nearly unbiased estimators in single-mark and single-recapture experiments. *Biometrics* **45**, 1233–1237.

Johnson, E. G. and R. D. Routledge (1985). The line transect method: A nonparametric estimator based on shape restrictions. *Biometrics* **41**, 669–679.

Jolly, G. M. (1965). Explicit estimates from capture–recapture data with both death and immigration – Stochastic model. *Biometrika* **52**, 225–247.

Jolly, G. M. and I. Hampton (1990). A stratified random transect design for acoustic surveys of fish stocks. *Canad. J. Fish. Aqua. Sci.* **47**, 1282–1291.

Kareiva, P. M. (1983). Local movement in herbivorous insects: Applying a passive diffusion model to mark–recapture field experiments. *Oecologia* **57**, 322–327.

Kareiva, P. M. and N. Shigesada (1983). Analyzing insect movement as a correlated random walk. *Oecologia* **56**, 234–238.

Kenkel, N. C. and L. Orloci (1986). Applying metric and nonmetric multidimensional scaling to ecological studies: Some new results. *Ecology* **67**, 919–928.

Knox, R. G. (1989). Effects of detrending and rescaling on correspondence analysis: Stability and accuracy. *Vegetatio* **83**, 129–136.

Knox, R. G. and R. K. Peet (1989). Bootstrapped ordination: A method for estimating sampling effects in indirect gradient analysis. *Vegetatio* **80**, 153–165.

Koen, C. (1991). A computer program package for the statistical analysis of spatial point processes in a square. *Biom. J.* **33**, 493–503.

Kolasa, J. and D. Strayer (1988). Patterns of the abundance in species: A comparison of two hierarchic models. *Oikos* **53**, 235–241.

Kovach, W. L. (1990). MVSP, A multivariate statistics package. Institute of Earth Studies, University College of Wales, Aberystwyth.

Kruskal, J. B. (1964a). Multidimensional scaling by optimizing goodness of fit to a nonmetric hypothesis. *Psychometrika* **29**, 1–27.

Kruskal, J. B. (1964b). Nonmetric multidimensional scaling: A numerical method. *Psychometrika* **29**, 115–129.

Kuno, E. (1991). Sampling and analysis of insect populations. *Ann. Rev. Entomol.* **36**, 285–304.

Kvålseth, T. O. (1991). Note on biological diversity, evenness, and homogeneity measures. *Oikos* **62**, 123–127.

Lambert, J. M. and W. T. Williams (1962). Multivariate methods in plant ecology. IV. Nodal analysis. *J. Ecol.* **50**, 775–802.

Lance, G. N. and W. T. Williams (1966). A generalized sorting strategy for computer classifications. *Nature* **212**, 218.

Lance, G. N. and W. T. Williams (1967). A general theory of classificatory sorting strategies. I. Hierarchical systems. *Comput. J.* **9**, 373–380.

Lawlor, L. R. (1980a). Structure and stability in natural and randomly constructed competitive communities. *Amer. Nat.* **116**, 394–408.

Lawlor, L. R. (1980b). Overlap, similarity, and competition coefficients. *Ecology* **61**, 245–251.

Lebreton, J., K. P. Burnham, J. Clobert and D. R. Anderson (1992). Modelling survival and testing biological hypotheses using marked animals: A unified approach with case studies. *Ecol. Monogr.* **62**, 67–118.

Lebreton, J., G. Hemery, J. Clobert and H. Coquillart (1990). The estimation of age-specific breeding probabilities from recaptures or resightings in vertebrate populations. I. Transversal Models. *Biometrics* **46**, 609–622.

Legendre, L. and P. Legendre (1983). *Numerical Ecology*. Elsevier, Amsterdam.

Levins, R. (1968). *Evolution in Changing Environments*. Princeton Univ. Press, Princeton, NJ.

Link, W. A. and B. A. Hoover (1991). The detection of density-dependence from a series of annual censuses – A comment. *Ecology* **72**, 1514–1515.

Linton, L. R., E. S. Edgington and R. W. Davis (1989). A view of niche overlap amenable to statistical analysis. *Canad. J. Zool.* **67**, 55–60.

Lloyd, M. (1967). Mean crowding. *J. Anim. Ecol.* **36**, 1–30.

Loehle, C. (1990). Proper statistical treatment of species–area data. *Oikos* **57**, 143–145.

Losos, J. B., S. Naeem and R. K. Colwell (1989). Hutchinsonian ratios and statistical power. *Evolution* **43**, 1820–1826.

Lu, R., E. P. Smith and I. J. Good (1989). Multivariate measures of similarity and niche overlap. *Theor. Popul. Biol.* **35**, 1–21.

Ludwig, J. A. and J. F. Reynolds (1988). *Statistical Ecology*. Wiley, New York.

MacArthur, R. H. and R. Levins (1967). The limiting similarity, convergence and divergence of coexisting species. *Amer. Nat.* **101**, 377–385.

Magurran, A. E. (1988). *Ecological Diversity and its Measurement*. Princeton Univ. Press, Princeton, NJ.

Manly, B. F. J. (1977). A model for dispersion experiments. *Oecologia* **31**, 119–130.

Manly, B. F. J. (1984). Obtaining confidence limits on parameters of the Jolly–Seber model for capture–recapture data. *Biometrics* **40**, 749–758.

Manly, B. F. J. (1990a). *Stage-Structured Populations: Sampling, Analysis and Simulation*. Chapman and Hall, London.

Manly, B. F. J. (1990b). On the statistical analysis of niche overlap data. *Canad. J. Zool.* **68**, 1420–1422.

Manly, B. F. J. (1991a). *Randomization and Monte Carlo Methods in Biology*. Chapman and Hall, London.

Manly, B. F. J. (1991b). The statistical analysis of shell species counts. In: C. F. W. Higham and R. Bannanurag, eds., *The Excavation of Khok Phanom Di*. 2. The Biological Remains. (Part 1). Society of Antiquaries, London, 321–331.

Manly, B. F. J. (1992a). *The Design and Analysis of Research Studies*. Cambridge Univ. Press, Cambridge.

Manly, B. F. J. (1992b). A review of computer intensive multivariate methods in ecology. In: G. P. Patil and N. P. Ross, eds., *Proc. 7th Internat. Conf. on Multivariate Analysis*. North-Holland/Elsevier Science Publishers, New York.

Manly, B. F. J. and C. Chatterjee (1992). A model for mark-recapture data allowing for animal movement. In: J. D. Lebreton and P. M. North, eds., *Marked Individuals in the Studies of Bird Populations*. Birkhäuser Verlag, Basel, 309–322.

Manly, B. F. J., L. L. McDonald and D. L. Thomas (1992). *Resource Selection by Animals: Statistical Design and Analysis for Field Studies*. Chapman and Hall, London.

Manly, B. F. J. and M. J. Parr (1968). A new method of estimating population size, survivorship, and birth rate from capture–recapture data. *Trans. Soc. Brit. Ent.* **18**, 81–89.

Manly, B. F. J. and G. B. Patterson (1984). The use of Weibull curves to measure niche overlap. *New Zealand J. Zool.* **11**, 337–342.

Mantel, N. (1967). The detection of disease clustering and a generalized regression approach. *Cancer Res.* **27**, 209–220.

Marsh, L. M. and R. E. Jones (1988). The form and consequences of random walk models. *J. Theor. Biol.* **133**. 113–131.

Matson, P. A. and S. R. Carpenter eds. (1990). Special feature: Statistical analysis of ecological response to large-scale perturbations. *Ecology* **71**, 2037–2068.

McCulloch, C. E. (1985). Variance tests for species association. *Ecology* **66**, 1676–1681.

McDonald, J. F. and D. Palanacki (1989). Interval estimation of the size of a small population from a mark–recapture experiment. *Biometrics* **45**, 1223–1231.

McDonald, J. F., M. A. Selby, C. S. Wong, L. D. Favro and P. K. Kuo (1983). Tests, point estimations, and confidence sets for a capture–recapture model. *J. Amer. Statist. Assoc.* **78**, 913–919.

McIntosh, R. P. (1967). An index of diversity and the relation of certain concepts of diversity. *Ecology* **48**, 392–404.

Mead, R. (1974). A test for spatial pattern at several scales using data from a grid of continuous quadrats. *Biometrics* **30**, 295–307.

Mercier, P., D. Chessel and S. Dolédec (1992). Complete correspondence analysis of an ecological profile data table: A central ordination method. *Acta Oecol.* **13**, 25–44.

Minta, S. and M. Mangel (1989). A simple population estimate based on simulation for capture–recapture and capture–resight data. *Ecology* **70**, 1738–1751.

Molinari, J. (1989). A calibrated index for the measurement of evenness. *Oikos* **56**, 319–326.

Mooney, H. A., E. Medina, D. W. Schindler, E. Schulze and B. H. Walker (1991). *SCOPE 45: Ecosystem Experiments.* Wiley, Chichester.

Moran, P. A. P. (1951). A mathematical theory of animal trapping. *Biometrika* **38**, 307–311.

Morgan, B. J. T. and S. N. Freeman (1989). A model with first-year variation for ring-recovery data. *Biometrics* **45**, 1087–1101.

Morgan, B. J. T. and P. M. North eds. (1985). *Statistics in Ornithology.* Lecture Notes in Statistics, Vol. 29. Springer, Berlin.

Morisita, M. (1959). Measuring of interspecific association and similarity between communities. *Mem. Fac. Sci. Kyushu Univ.* **E3**, 64–80.

Morisita, M. (1971). Composition of the I_d index. *Res. Popul. Ecol.* **13**, 1–27.

Morris, D. W. (1989). Habitat-dependent estimates of competitive interaction. *Oikos* **55**, 111–120.

Mueller, L. D. and L. Altenberg (1985). Statistical inference on measures of niche overlap. *Ecology* **66**, 1204–1210.

Nemec, A. F. L. and R. O. Brinkhurst (1988). Using the bootstrap to assess statistical significance in the cluster analysis of species abundance data. *Canad. J. Fish. Aqua. Sci.* **45**, 965–970.

Nichols, J. D. (1992). Capture recapture models. *Bioscience* **42**, 94–102.

Nichols, J. D. and K. H. Pollock (1990). Estimation of recruitment from immigration versus in situ reproduction using Pollock's robust design. *Ecology* **71**, 21–26.

Nichols, J. D., J. R. Sauer, K. H. Pollock and J. B. Hestbeck (1992). Estimating transition probabilities for stage-based population projection matrices using capture–recapture data. *Ecology* **73**, 306–312.

North, P. M., ed. (1987). *Ringing recoveries analytical methods. Acta Ornithol.* **23** (1).

North, P. M. (1987). Ring recovery analytical methods: A brief review. *Acta Ornithol.* **23**, 7–12.

Oksanen, J. (1988). A note on the occasional instability of detrending in correspondence analysis. *Vegetatio* **74**, 29–32.

Othmer, H. G., S. R. Dunbar and W. Alt (1988). Models of dispersal in biological systems. *J. Math. Biol.* **26**, 263–298.

Otis, D. L., K. P. Burnham, G. C. White and D. R. Anderson (1978). Statistical inference from capture data on closed populations. *Wildlife Monogr.* **62**.

Otto, M. C. and K. H. Pollock (1990). Size bias in line transect sampling: A field test. *Biometrics* **46**, 239–245.

Palmer, M. W. (1990). The estimation of species richness by extrapolation. *Ecology* **71**, 1195–1198.
Palmer, M. W. (1991). Estimating species richness: The second-order jackknife reconsidered. *Ecology* **72**, 1512–1513.
Paramenter, R. R., J. A. MacMahon and D. R. Anderson (1989). Animal density estimation using a trapping web design: Field validation experiments. *Ecology* **70**, 169–179.
Patterson, G. B. (1986). A statistical method of testing for dietary differences. *New Zealand J. Zool.* **13**, 113–115.
Payne, C. D., ed. (1987). *The GLIM System, Release 3.77*. Numerical Algorithms Group, Oxford.
Perry, J. N. and M. Hewitt (1991). A new index of aggregation for animal counts. *Biometrics* **47**, 1505–1518.
Perry, J. N. and L. R. Taylor (1985). Adès: New ecological families of species specific frequency distributions that describe repeated spatial samples with an intrinsic power-law variance-mean property. *J. Anim. Ecol.* **54**, 931–953.
Perry, J. N. and L. R. Taylor (1988). Families of distributions for repeated samples of animal counts. *Biometrics* **44**, 881–890.
Persson, S. (1981). Ecological indicator values as an aid in the interpretation of ordination diagrams. *J. Ecol.* **69**, 71–84.
Petrais, P. S. (1989). The representation of niche breadth and overlap on Tilman's consumer-resource graphs. *Oikos* **56**, 289–292.
Petraitis, P. S. (1979). Likelihood measures of niche breadth and overlap. *Ecology* **60**, 703–710.
Pianka, E. R. (1973). The structure of lizard communities. *Ann. Rev. Ecol. Systems* **4**, 53–74.
Pielou, E. C. (1969). *An Introduction to Mathematical Ecology*. Wiley, New York.
Pielou, E. C. (1975). *Ecological Diversity*. Wiley, New York.
Pielou, E. C. (1984). Probing multivariate data with random skewers: A preliminary to direct gradient analysis. *Oikos* **42**, 161–165.
Pleasant, J. M. (1990). Null model tests for competitive displacement: The fallacy of not focusing on the whole community. *Ecology* **71**, 1078–1084.
Pollard, E., K. H. Lakhani and P. Rothery (1987). The detection of density-dependence from a series of annual censuses. *Ecology* **68**, 2046–2055.
Pollock, K. H. (1982). A capture–recapture design robust to unequal probability of capture. *J. Wildlife Management* **46**, 752–757.
Pollock, K. H. (1991). Modelling capture, recapture, and removal statistics for estimation of demographic parameters for fish and wildlife populations: Past, present, and future. *J. Amer. Statist. Assoc.* **86**, 225–238.
Pollock, K. H., J. E. Hines and J. D. Nichols (1984). The use of auxiliary variables in capture–recapture and removal experiments. *Biometrics* **40**, 329–340.
Pollock, K. H., J. D. Nichols, C. Brownie and J. E. Hines (1990). Statistical inference for capture–recapture experiments. *Wildlife Monogr.* **107**.
Poole, R. W. and B. J. Rathcke (1979). Regularity, randomness, and aggregation in flowering phenologies. *Science* **203**, 470–471.
Prins, H. H. T. and I. Douglas-Hamilton (1990). Stability in a multi-species assemblage of large herbivores in East Africa. *Oecologia* **83**, 392–400.
Quang, P. X. (1990). Confidence intervals for densities in line transect sampling. *Biometrics* **46**, 459–472.
Quang, P. X. and R. B. Lanctot (1991). A line transect model for aerial surveys. *Biometrics* **47**, 1089–1102.
Ramsey, F. L., C. E. Gates, G. P. Patil and C. Taillie (1988). On transect sampling to assess wildlife populations and marine resources. In: P. R. Krishnaiah and C. R. Rao, eds., *Handbook of Statistics*, Vol. 6. North-Holland, Amsterdam, 515–532.
Reddingius, J. (1990). Models for testing: A secondary note. *Oecologia* **83**, 50–52.
Reddingius, J. and P. J. Den Boer (1989). On the stabilization of animal numbers. Problems of testing. 1. Power estimates and estimation errors. *Oecologia* **78**, 1–8.
Rejmanek, M. (1992). Stability in a multi-species assemblage of large herbivores in East Africa: An alternative interpretation. *Oecologia* **89**, 454–456.
Rejmankova, E., H. M. Savage, M. Rejmanek, Arredondo-Jimenez and D. R. Roberts (1991). Multivari-

ate analysis of relationships between habitats, environmental factors and occurrence of anopheline mosquito larvae *Anopheles albimanus* and *A. pseudopunctipennis* in southern Chipas, Mexico. *J. Appl. Ecol.* **28**, 827–841.

Renkonen, O. (1938). Statistisch-ökologische Untersuchungen über die terrestrische Käferwelt der finnischen Brunchmoore. *Ann. Zool. Soc. Zool. Bot. Fenn. Vanamo* **6**, 1–231.

Reynolds, R. T., J. M. Scott and R. A. Nussbaum (1980). A variable circular plot method for estimating bird numbers. *Condor* **82**, 309–313.

Ripley, B. D. (1978). Spectral analysis and the analysis of pattern in plant communities. *J. Anim. Ecol.* **66**, 965–981.

Rivest, L., H. Crépeau and M. Crête (1990). A two-phase sampling plan for the estimation of the size of a moose population. *Biometrics* **46**, 163–176.

Roberts, A. and L. Stone (1990). Island sharing by archipelago species. *Oecologia* **83**, 560–567.

Roeder, K., B. Dennis and E. O. Garton (1987). Estimating density from variable circular plot censuses. *J. Wildlife Management* **51**, 224–230.

Romesburg, H. C. (1989). ZORRO, a randomization test for spatial pattern. *Comput. Geosci.* **15**, 1011–1017.

Routledge, R. D. (1983). Evenness indices: Are any admissible? *Oikos* **40**, 149–151.

Routledge, R. D. (1984). Estimating ecological components of diversity. *Oikos* **42**, 23–29.

Routledge, R. D. (1989). The removal method for estimating natural populations: incorporating auxiliary information. *Biometrics* **45**, 111–121.

Routledge, R. D. (1990). Spatial patterns arising from plant dispersal as modelled by a correlated random walk. *J. Appl. Prob.* **27**, 1–13.

Routledge, R. D. and D. A. Fyfe (1992). Confidence limits for line transect estimates based on shape restrictions. *J. Wildlife Management* **56**, 402–407.

Routledge, R. D. and T. B. Swartz (1991). Taylor's power law re-examined. *Oikos* **60**, 107–112.

Sale, P. F. (1974). Overlap in resource use, and interspecific competition. *Oecologia* **17**, 245–256.

Sawyer, A. J. (1989). Inconstancy of Taylor's b: Simulated sampling with different quadrat sizes and spatial distributions. *Res. Popul. Ecol.* **31**, 11–24.

Schatzmann, E., R. Gerrard and A. D. Barbour (1986). Measures of niche overlap, I. *Inst. Math. Appl. J. Math. Appl. Med. Biol.* **3**, 99–113.

Scheiner, S. M. and J. Gurevitch, eds. (1993). *The Design and Analysis of Ecological Experiments.* Chapman and Hall, London.

Schluter, D. (1984). A variance test for detecting species associations, with some example applications. *Ecology* **65**, 998–1005.

Schoener, T. W. (1970). Non-synchronous spatial overlap of lizards in patchy habitats. *Ecology* **51**, 408–418.

Schoener, T. W. (1974). Competition and the form of habitat shift. *Theor. Popul. Biol.* **6**, 265–307.

Schoener, T. W. (1981). An empirically-based estimate of home range. *Theor. Popul. Biol.* **20**, 281–325.

Schoener, T. W. and G. H. Adler (1991). Greater resolution of distributional complementarities by controlling for habitat affinities: A study with Bahamian lizards and birds. *Amer. Nat.* **137**, 669–692.

Schwarz, C. J., K. P. Burnham and A. N. Anderson (1988). Post-release stratification in band-recovery models. *Biometrics* **44**, 765–785.

Seber, G. A. F. (1965). A note on the multiple-recapture census. *Biometrika* **52**, 249–259.

Seber, G. A. F. (1982). *The Estimation of Animal Abundance and Related Parameters.* 2nd ed., Griffin, London.

Seber, G. A. F. (1984). *Multivariate Observations.* Wiley, New York.

Seber, G. A. F. (1986). A review of estimating animal abundance. *Biometrics* **42**, 267–292.

Shaw, M. W. (1990). A test of spatial randomness on small scales, combining information from mapped locations within several quadrats. *Biometrics* **46**, 447–458.

Sheppard, R. N. (1962). The analysis of proximities: Multidimensional scaling with an unknown distance function. *Psychometrika* **27**, 125–139.

Simberloff, D. and W. Boecklen (1981). Santa Rosalia reconsidered: Size ratios and competition. *Evolution* **35**, 1206–1228.

Simpson, E. H. (1949). Measurement of diversity. *Nature* **163**, 688.
Sinclair, D. F., J. E. Mosimann and D. A. Meeter (1985). Tests for character displacement. *Biometrics* **41**, 835–846.
Skellam, J. G. (1973). The formulation and interpretation of mathematical models of diffusionary processes in population biology. In: M. S. Bartlett and R. W. Hiorns, eds., *The Mathematical Theory of the Dynamics of Biological Populations*. Academic Press, London, 63–85.
Smith, B. D., G. A. McFarlane and A. J. Cass (1990). Movement and mortality of tagged male and female Lingcod in the Strait of Georgia, British Columbia. *Trans. Amer. Fish. Soc.* **119**, 813–824.
Smith, E. P. (1982). Niche breadth, resource availability, and inference. *Ecology* **63**, 1675–1681.
Smith, E. P. (1985a). Statistical comparison of weighted overlap measures. *Trans. Amer. Fish. Soc.* **114**, 250–257.
Smith, E. P. (1985b). Estimating the reliability of diet overlap measures. *Environ. Biol. Fish.* **13**, 125–138.
Smith, P. J. (1988). Bayesian methods for multiple capture–recapture surveys. *Biometrics* **44**, 1177–1189.
Smith, S. J. (1988). Evaluating the efficiency of the Δ-distribution mean estimator. *Biometrics* **44**, 485–493.
Smith, S. J. (1990). Use of statistical models for the estimation of abundance from groundfish trawl survey data. *Canad. J. Fish. Aqua. Sci.* **47**, 894–903.
Sokal, R. S. (1990). Genetics and language in European populations. *Amer. Natur.* **135**, 157–175.
Sokal, R. S. and D. Thomson (1987). Applications of spatial autocorrelation in ecology. In: P. Legendre and L. Legendre, eds., *Developments in Numerical Ecology*, Springer, New York, 431–466.
Solow, A. R. (1989a). Bootstrapping sparsely sampled point patterns. *Ecology* **70**, 379–382.
Solow, A. R. (1989b). A randomization test for independence of animal locations. *Ecology* **70**, 1546–1549.
Solow, A. R. (1990). Testing for density dependence: A cautionary note. *Oecologia* **83**, 47–49.
Solow, A. R. and W. Smith (1991). Detecting cluster in a heterogenous community sampled by quadrats. *Biometrics* **47**, 311–317.
Solow, A. R. and J. H. Steele (1990). On sample size, statistical power, and the detection of density dependence. *J. Anim. Ecol.* **59**, 1073–1076.
Southwood, T. R. E. (1978). *Ecological Methods with Particular Reference to Insect Populations.* Chapman and Hall, London.
Stein, G. Z. (1988). Modelling counts in biological populations. *Math. Sci.* **13**, 56–65.
Stiles, F. G. (1977). Coadapted competitors: The flowering seasons of hummingbird-pollinated plants in a tropical forest. *Science* **198**, 285–301.
Stone, L. and A. Roberts (1991). The checkerboard score and species distributions. *Oecologia* **85**, 74–79.
Strauss, R. E. (1982). Statistical significance of species clusters in association analysis. *Ecology* **63**, 634–639.
Sugihara, G., B. Grenfell and R. M. May (1990). Distinguishing error from chaos in ecological time series. *Philos. Trans. Roy. Soc. London Ser. B* **330**, 235–251.
Swihart, R. K. and N. A. Slade (1987). A test for independence of movements as shown by live trapping. *Amer. Mild. Natur.* **117**, 204–207.
Taylor, L. R. (1961). Aggregation, variance and the mean. *Nature* **189**, 732–735.
Taylor, R. A. J. (1981a). The behavioral basis of redistribution. I. The Δ-model concept. *J. Anim. Ecol.* **50**, 573–586.
Taylor, R. A. J. (1981b). The behavioral basis of redistribution. II. Simulations of the Δ-model. *J. Anim. Ecol.* **50**, 587–604.
Ter Braak, C. J. F. (1985). Correspondence analysis of incidence and abundance data: Properties in terms of a unimodal response model. *Biometrics* **41**, 859–873.
Ter Braak, C. J. F. (1986). Canonical correspondence analysis: A new eigenvalue technique for multivariate direct gradient analysis. *Ecology* **67**, 1167–1179.
Ter Braak, C. J. F. (1988). CANOCO – A FORTRAN program for canonical community ordination by (partial) (detrended) (canonical) correspondence analysis, principal components analysis and redundancy analysis, Ver. 2.1. Rep. LWA-88-02, Agricultural Mathematics Group, Wageningen, The Netherlands.
Ter Braak, C. J. F. and C. W. N. Looman (1986). Weighted averaging, logistic regression and the Gaussian response model. *Vegetatio* **65**, 3–11.
Ter Braak, C. J. F. and I. C. Prentice (1988). A theory of gradient analysis. *Adv. Ecol. Res.* **18**, 271–317.
Thompson, S. K. (1991a). Stratified adaptive cluster sampling. *Biometrika* **78**, 389–397.

Thompson, S. K. (1991b). Adaptive cluster sampling: Designs with primary and secondary units. *Biometrics* **47**, 1103–1115.

Thomson, J. D. and K. A. Rusterholz (1982). Overlap summary indices and the detection of community structure. *Ecology* **63**, 274–277.

Thórarinsson, K. (1986). Population density and movement: A critique of Δ-models. *Oikos* **46**, 70–81.

Toft, C. A. and P. J. Shea (1983). Detecting community-wide patterns: Estimating power strengthens statistical inference. *Amer. Natur.* **122**, 618–625.

Tonkyn, D. W. and B. J. Cole (1986). The statistical analysis of body-size ratios. *Amer. Natur.* **128**, 66–81.

Turnock, B. J. and T. J. Quinn (1991). The effect of responsive movement on abundance estimation using line transect sampling. *Biometrics* **47**, 701–705

Underwood, A. J. (1991a). Beyond BACI: Experimental designs for detecting human environmental impacts on temporal variations in natural populations. *Austral. J. Mar. Freshwater Res.* **42**, 569–587.

Underwood, A. J. (1991b). The logic of ecological experiments: A case history from studies of the distribution of macro-algae on rocky intertidal shores. *J. Mar. Biol. Assoc.* **71**, 841–866.

Van der Maarel, E., J. G. M. Janssen and J. M. W. Louppen (1978). TABORD, a program for structuring phytosociological tables. *Vegetatio* **38**, 143–156.

Varley, G. C., G. R. Gradwell and M. P. Hassell (1973). *Insect Population Ecology*. Blackwell, Oxford.

Vickery, W. L. and T. D. Nudds (1991). Testing for density-dependent effects is sequential censuses. *Oecologia* **85**, 419–423.

Viveros, R. and D. A. Sprott (1986). Conditional inference and maximum likelihood in a capture–recapture model. *Comm. Statist. Theor. Methods* **15**, 1035–1046.

Vogt, W. G. and R. Morton (1991). Estimation of population size and survival of sheep blowflies, *Lucilia cuprina*, in the field from serial recoveries of marked flies affected by weather, dispersal and age-dependent trappability. *Res. Popul. Ecol.* **33**, 141–163.

Washington, H. G. (1984). Diversity, biotic, and similarity indices: A review with special relevance to aquatic ecosystems. *Water Res.* **18**, 653–694.

Weinberg, S. L., J. D. Carroll and H. S. Cohen (1984). Confidence regions for INDSCAL using the jackknife and bootstrap techniques. *Psychometrika* **49**, 475–491.

Whittaker, R. H. (1948). A vegetation analysis of the Great Smoky Mountains. Ph.D. Thesis, University of Illinois, Urbana, IL.

Whittaker, R. H. (1972). Evolution and measurement of species diversity. *Taxon* **21**, 213–251.

Whitehead, H. (1990). Mark–recapture estimates with emigration and reimmigration. *Biometrics* **46**, 473–479.

Williams, W. T. and J. M. Lambert (1959). Multivariate methods in plant ecology. I. Association analysis in plant communities. *J. Ecol.* **47**, 83–101.

Williams, W. T. and J. M. Lambert (1961). Multivariate methods in plant ecology. II. Inverse association analysis. *J. Ecol.* **49**, 717–729.

Williamson, M. (1988). Relationship of species number to area, distance and other variables. In: A. A. Myer and P. S. Giller, eds., *Analytical Biogeography*. Chapman and Hall, London, 91–115.

Wilson, J. B. (1987). Methods for detecting nonrandomness in species co-occurrences: a contribution. *Oecologia* **73**, 579–582.

Wilson, J. B. (1991). Methods for fitting dominance/diversity curves. *J. Veg. Sci.* **2**, 35–46.

Wilson, J. B., T. R. Partridge and M. T. Sykes (1990). The use of the Cole/Hurlbert C_8 association coefficient in inverse ecological classification. *J. Veg. Sci.* **1**, 367–374.

Wilson, M. V. and A. Shmida (1984). Measuring beta diversity with presence–absence data. *J. Ecol.* **72**, 1055–1064.

Wolda, H. (1989). The equilibrium concept and density dependence tests. What does it all mean? *Oecologia* **81**, 430–432.

Wolda, H. (1991). The usefulness of the equilibrium concept in population dynamics. *Oecologia* **86**, 144–145.

Wolter, K. M. (1990). Capture–recapture estimation in the presence of a known sex ratio. *Biometrics* **46**, 157–162.

Wong, C. S. and J. F. McDonald (1987). Confidence intervals for a population size based on a capture–recapture model. *Comm. Statist. Theor. Methods* **16**, 79–91.

Worton, B. J. (1987). A review of models of home range for animal movement. *Ecol. Model.* **38**, 277–298.

Worton, B. J. (1989). Kernel methods for estimating the utilization distribution in home-range studies. *Ecology* **70**, 164–168.

Yamamura, K. (1990). Sampling scale dependence of Taylor's power law. *Oikos* **59**, 121–125.

Yip, P. (1989). An inference procedure for a capture–recapture experiment with time-dependent capture probabilities. *Biometrics* **45**, 471–479.

Yip. P. (1991a). Estimating population size from a capture–recapture experiment with known removals. *Theor. Popul. Biol.* **40**, 1–13.

Yip, P. (1991b). A martingale estimating equation for a capture–recapture experiment in discrete time. *Biometrics* **47**, 1081–1088.

Yip, P. (1992). Goodness of fit tests with heterogeneity in a capture–recapture experiment. *Austral. J. Statist.* **34**, 265–271.

Zahl, S. (1989). Line transect sampling with unknown probability of detection along the transect. *Biometrics* **45**, 453–470.

Zeh, J. E., D. Ko, B. D. Krogman and R. Sonntag (1986). A multinomial model for estimating the size of a whale population from incomplete census data. *Biometrics* **42**, 1–14.

Zippin, C. (1956). An evaluation of the removal method of estimating animal populations. *Biometrics* **12**, 163–189.

Zippin, C. (1958). The removal method of population estimation. *J. Wildlife Management* **22**, 82–90.

G. P. Patil and C. R. Rao, eds., *Handbook of Statistics, Vol. 12*

11

Forest Biometrics

Harold E. Burkhart and Timothy G. Gregoire

Abstract: Reliable quantitative information is required to manage forest lands. This chapter gives an overview of statistical procedures used to quantify tree volume and form, site quality (height–age curves), site occupancy (stand density), and growth–yield relationships.

The quantification of tree volume and form is basic to assessing the volume and value of forest stands and to evaluating response to silvicultural treatments. Regression has been used to relate tree volume to easily measured variables, such as diameter at breast height and total tree height. The usual assumptions of normality and homogeneity of error variance frequently are not satisfied. As an alternative to predicting volume directly, taper functions, which predict diameter at specified heights, can be fitted and integrated for volume estimates. While estimates of volume from integrated taper functions have proven to be quite accurate in many instances, they are likely to bc biased. Because foresters employ regression for a variety of prediction purposes (volume, diameter at given heights, etc.), simultaneous estimation of tree volume and taper functions under a joint mean square error criterion has been proposed.

The site index concept (average height at a specified reference age) is used to quantify site quality. Basic assumptions of the site index method often are not met and numerous statistical problems arise when estimating the parameters in height–age curves. A natural selection bias oftentimes occurs which invalidates the usual assumption of age–site quality independence when applying the guide curve method; lack of independence between observations and the absence of an appropriate sampling protocol cause difficulty when fitting height–age curves to stem analysis data (multiple observations on each individual in a sample of trees).

Stand density has been quantified by a variety of measures – many of which are mathematically equivalent. Stand density is dynamic through time and must be considered as such when projecting stand dynamics, growth and yield.

A wide variety of growth and yield models – ranging from models which predict whole-stand volumes only to simulators that project the growth of each individual tree – have been developed. The most appropriate modeling approach depends on the detail needed when the models are applied for specified objective(s). Regardless of the modeling approach taken, a system of interrelated equations arise. Estimation of the parameters in these systems of equations presents a variety of challenging statistical problems.

1. Introduction

It has been said of biometry in general that the evolution of statistical practice has been driven by the needs of biologists to solve problems and to answer questions arising from day-to-day applications in the discipline in which they labor. This certainly has been true in forestry where reliable quantitative information is required to manage forest lands. This chapter gives an overview of statistical procedures used to quantify tree volume and form, site quality (height–age curves), site occupancy (stand density), and growth–yield relationships. All four areas of discussion are long-standing subjects of concern within forestry and are likely to remain so wherever forests are actively managed and regulated. Because of needs for increasingly precise information about forest resources, improved statistical analysis and understanding remain an area of active research and development.

This chapter deals mainly with modeling aspects of forest biometry, yet it is hardly a comprehensive exposition of modeling needs and practices within the discipline. Important topics such as growth models for individual trees, dendrochronology and tree-ring analysis, models for categorical data, and physiologically based process models of tree and stand functions constitute a few selected areas not discussed. Nor is the broad field of forest inventory broached, where, in the argot of forestry, the term inventory connotes a sample-based survey of the forest resource. The long tradition of sampling practice within the field has an extensive literature (see, for example, DeVries, 1986 and Schreuder, Gregoire and Wood, 1993), which is beyond the scope of the present chapter.

2. Prediction of standing tree volume

A substantial portion of the biomass of an individual tree collects as woody material in its main stem or bole. It is for this reason that a significant concern of those who manage, inventory, and use the forest resource for the production of products and services has focused on this aspect of tree size, and, by extension, to the collective stemwood material of the forest. This content has been expressed in terms of the amount of end-product; for example, board-foot measure is still widely used in North America as an expression of lumber yield. However, a more broadly useful measure is the volume (m^3) or weight (kg) of raw product. Accordingly the principal measure of size that will be discussed in this section is the volume or cubic content of the stem or bole of a tree. Many of the allometrical and statistical relations between volume and other tree dimensions apply, too, when bole weight is the primary variable of interest, and to a somewhat lesser extent when end-product volume or weight is of interest. In concert with established jargon of professional forestry, tree volume is often used synonymously with stem volume and bole volume, despite the fact that there are numerous other volumetric components of a tree in addition to its stem.

Direct determination of tree volume is a costly and laborious task which typically entails felling and destructive sampling. For routine forestry operations, an imputation or prediction of stem volume is sought, which is based on measurements of size that leave the tree standing.

Tabular summaries of individual tree volumes, indexed by some measure of stem diameter or girth and possibly by stem length, date back to the mid-eighteenth century Europe (cf. Meyer, 1953; Spurr, 1952). Similar tables were developed for indigenous tree species as forest management was initiated in North America. With the advent of modern statistical methods and as training in their use became more widespread, volume tables gradually gave way to simple regression equations for bole volume. Improvements in computing capabilities have resulted in a spectrum of more complicated model forms and fitting procedures. Irrespective of the complexity of the model, however, the chief purpose of a volume equation is the accurate prediction of stem volume based upon one or more measured dimensions of standing tree size or shape. Were it not for the requirement that the tree remain standing following measurement, volume could be assayed more directly: both the burden of measurement would be less and the subsequent prediction of volume would be more accurate if done after a tree is felled and bucked into logs. The mensurational and statistical considerations pertinent to the ascertainment of log volume, however, is outside the scope of this section which deals exclusively with volume prediction made on the basis of standing tree measurement(s) of size.

The bole volume of a tree is an expression of cumulative cambial and apical growth. In plant biology the modeling of cumulative growth has a long and extensive history, and models of cumulative growth almost invariably are driven by some measure of the age of the organism. In contrast and with a few exceptions, as for example Sadiq and Smith (1983), age is conspicuously absent from tree volume equations. The substantial effect of inter-tree competition on the size attained at a particular age confounds any attempt to correlate volume with age without also incorporating some measure of competition or density. In addition, the measurement of tree age is more difficult than the measurement of dimensional aspects of the tree that correlate more directly with stem volume.

In the evolution and development of tree volume equations, great emphasis has been placed upon unbiasedness of prediction, because even a small bias accumulates to sizeable over- or under-prediction of volume when the offending volume equation is applied repetitively to hundreds of thousands of trees. For that reason, imprecision in a volume equation resulting from random prediction error is more tolerable than systematic inaccuracy of prediction of even slight magnitude.

Tree volume will be denoted generally by the variable V. In a particular application it is usually necessary to adopt a more comprehensive notation in order to discern whether the volume is reckoned inside-bark or outside-bark, and to be

Table 1
Notation for stem volume from stump to an indicated point on the upper stem

Symbol	Description
$V_i(d_i)$	Inside-bark volume to an upper-stem inside-bark diameter d
$V_i(d_o)$	Inside-bark volume to an upper-stem outside-bark diameter d
$V_o(d_i)$	Outside-bark volume to an upper-stem inside-bark diameter d
$V_o(d_o)$	Outside-bark volume to an upper-stem outside-bark diameter d
$V_i(h)$	Inside-bark volume to height h on the upper stem
$V_o(h)$	Outside-bark volume to height h on the upper stem

explicit about the portion of the stem to which it applies. To this end, the symbology of Table 1 is introduced. For example, $V_i(d_o)$ denotes inside-bark volume (m^3) from the ground or stump level to the point on the upper portion of the stem where its outside-bark diameter has diminished to the value d_o (cm). When $d_o = d_i = 0$, $V_i(0)$ is the inside-bark volume of the entire tree, and $V_o(0)$ is the corresponding outside-bark volume, which will alternatively be denoted by V_{it} and V_{ot}, respectively.

Merchantable volume to an upper-stem diameter d_i when $d_i > 0$, is the conventional term that is used to distinguish this portion of tree volume from total tree volume. Merchantable volume can alternatively be specified with respect to an upper-stem outside-bark diameter $d_o > 0$ or to some above-ground height $h < H_t$, where H_t is the total height of the stem.

To even the most casual observer it is apparent that tree stems taper in cross-sectional area from ground level upwards, at least for most of the globe's temperate-zone species. Upon closer inspection of many tree boles it is found that the tapering is not strictly monotonic, as swelling may occur in the vicinity of major branches and near a whorl of branches that is a morphological characteristic of some coniferous tree species.

Many coniferous species and some broad-leaved deciduous species grown under forested conditions exhibit an excurrent branching morphology with a main stem which extends more or less vertically to tree tip. In contrast, other species have a characteristic deliquescent morphology such that the main stem effectively splits into a number of major branches which support the crown of the tree. In the latter case, total tree height exceeds the height of the main stem, whereas the two are the same in the former. An idealized excurrent stem form is displayed in Figure 1. As the annotations in this figure indicate, the geometric form that is approximated by the vertical profile or shape of the tree stem characteristically changes from ground level to tree tip. Generally speaking, the same characteristics are exhibited by deliquescent stems up to the point where the stem disperses into crown.

While it is clear that no simple geometric form can be assumed that will exactly model stem form, geometrical considerations nevertheless have strongly influenced the models used to predict stem volume. Two classes of models have evolved. In one class, stem volume is fitted as a function of stem diameter, of stem diameter and height jointly, or of stem diameter, height, and some expression of taper jointly. In

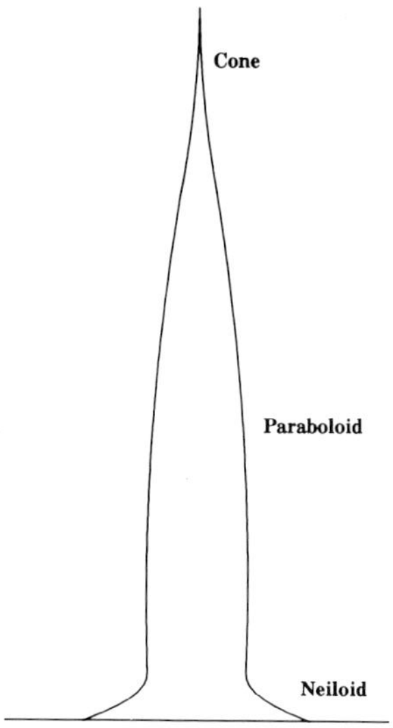

Fig. 1. Idealized shape of an excurrent tree stem.

the forest biometrics literature, equations in this class are those that traditionally have been called direct volume equations because volume is predicted from one or more input variables. In the other class, stem diameter is fitted as a function of height aboveground, providing a diameter taper equation. Volume is then predicted indirectly by integrating the square of the taper equation and multiplying by appropriate constants. Variations on this theme include replacing diameter with the horizontal cross-sectional area of the stem, or fitting a height-diameter taper equation.

3. Direct volume equations

In temperate zones tree diameter is customarily measured at breast height rather than at ground level or stump height. Breast height is a convenient measurement point and its use generally avoids inclusion of swelling and mechanical buttressing that, if included, would tend to increase variation and decrease correlation between a single stem diameter measurement and the usable volume in that stem. While the definition of breast height may vary from one continental region to another, it is within the range 1.3–1.4 m aboveground universally. The outside-bark diameter at breast height D may be measured directly by calipering the stem, or indirectly from a measurement of bole girth outside-bark. An important measure of stem size is basal area B which is the outside-bark cross-sectional area at breast height. Because B is impossible to measure directly without destructive sampling, it is usually deduced from D under an assumption of circularity of the tree bole. Empirical examinations of the validity of this assumption have been made by Gregoire et al. (1990) and Matérn

(1990). In general they found that for conifers the circularity assumptions results in a bias of 1–3 percent in estimated basal area.

The relation between $V = V_o(d_o)$ and D among 427 loblolly pine (*Pinus taeda* L.) trees is shown in Figure 2a, and that between $V = V_i(d_o)$ and D among 860 red oak (*Quercus rubra* L.) trees in Figure 2b. The upward concavity of the trends exhibited in these figures is characteristic of nearly all tree species. The curvilinearity is nearly eliminated when V is plotted against stem basal area B or, equivalently, against D^2 as in Figures 2c–2d. For a fixed value of D, the principal reason for dispersion in V owes to the differences in height among trees of the same diameter. If height is taken into account, as demonstrated in Figures 2e–2f, the conditional variation in volume is decreased. In these latter graphs, V is shown in relation to squared-diameter multiplied by stem height H. This combined variable, D^2H, has formed the basis for innumerable tree volume equations since its presentation by Spurr (1952).

A 'standard volume equation' is one in which D and some measure of tree height is used in order to predict V. For example the simple linear regression

$$V = \beta_0 + \beta_1 D^2 H + \varepsilon \tag{1}$$

is a specific model of a standard volume equation. Generalizations of (1) include

$$V = \beta_0 + \beta_1 D^2 + \beta_2 H + \beta_3 D^2 H + \varepsilon, \tag{2}$$

$$V = \beta_0 + \beta_1 D^{\beta_2} H^{\beta_3} + \varepsilon. \tag{3}$$

In (1) it is evident that β_1 can be regarded as an adjustment to the cylindrical volume D^2H, and that if D^2H and V are expressed in identical units, then $0 < \beta_1 < 1$. Because a tree bole is not a smooth geometrical shape for reasons exposited earlier, there is no intrinsic merit to D^2H as the base volume as opposed to the volume given by the product of D and H when both are raised to arbitrary exponents, as embodied in (3). Model (2) has a somewhat less direct geometrical interpretation; implicit in (2) is an assumption of additive 'effects' of D^2 (or B) and H and their interaction.

When a tree is shorter than breast height, $D = 0$ yet, $V_{ot} > V_{it} > 0$. One would expect, therefore, that $\beta_0 > 0$ when (1) is used to model total volume, and this reasoning is usually supported by statistical estimates of β_0 for this model. Conversely it is reasonable to expect $\beta_0 < 0$ when modeling merchantable volume by (1). In the latter case, the fitted equation may yield illogically negative estimates for small diameter, short trees.

Other forms of standard volume equations that have been used include

$$\ln V = \beta_0 + \beta_1 \ln D + \beta_2 \ln H + \varepsilon \tag{4}$$

introduced by Schumacher and Hall (1933), and

$$V = D^2 H(\beta_0 + \beta_1 H)^{-1} + \varepsilon \tag{5}$$

introduced by Honer (1965).

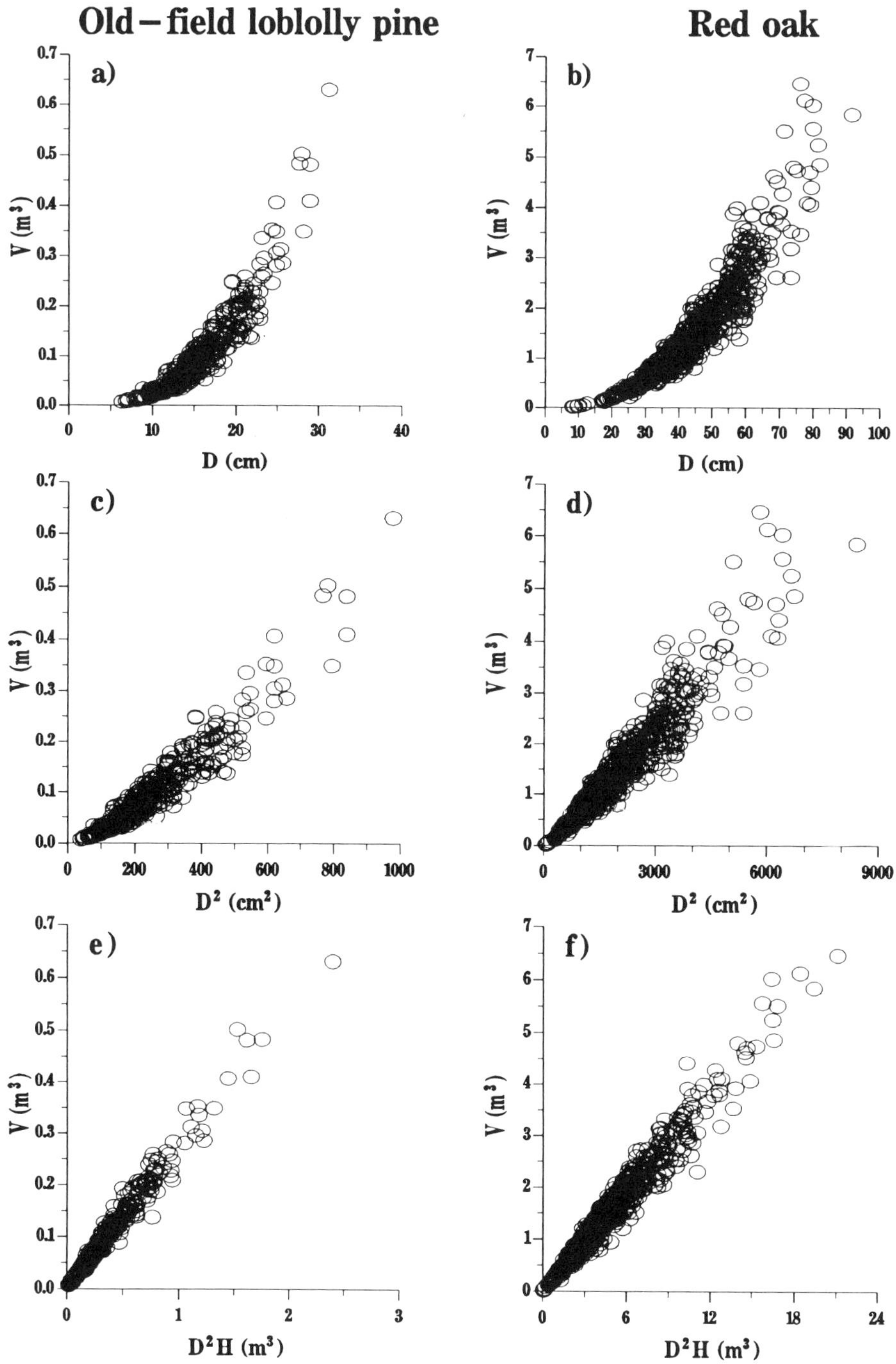

Fig. 2. Relationship of total tree volume to diameter and height. Frames (a), (c) and (e) show inside-bark volume for old-field loblolly pine; frames (b), (d) and (f) show outside-bark volume for red oak.

When (1)–(5) are used to model total volume, viz. $V = V_{ot}$ or V_{it}, the measure of height H is usually the total height of the tree. In equations to predict merchantable volume, i.e., $V = V_o(d)$ or $V_i(d)$ for $d = d_o > 0$ or $d = d_i > 0$, H may represent the height to the point on the bole where its diameter is d, viz. $H = H(d)$, or it may be total tree height, $H = H_t = H(0)$. Even though $V(d)$ usually has a slightly higher correlation with $H(d)$ than H_t, the latter may be easier to measure accurately on a standing tree. In other words a better fitting model may be achievable with $H = H(d)$ instead of $H = H_t$; but when the fitted equation is applied to measurements taken under field conditions, any gain achieved by using $H(d)$ to fit the model may be more than offset by measurement error introduced in the application of the equation to predict standing tree volume.

Alternatives to models (1)–(5) for merchantable volume are those that include the upper-stem diameter limit d in the model structure. Two models that have been used for loblolly pine trees are

$$V_o(d_o) = (\beta_o + \beta_1 D^2 H)(1 + \beta_2 d_o^{\beta_3}/D^{\beta_4}) + \varepsilon \tag{6}$$

from Burkhart (1977) and Amateis and Burkhart (1987), and

$$V_o(d_o) = \beta_0 D^{\beta_1} H^{\beta_2} - \beta_3 (d_o/D)^{\beta_4} D^2 (H - 1.37) + \varepsilon \tag{7}$$

from Lenhart et al. (1987). Evidently when $d_o = 0$, model (7) serves as a model for total volume of the form

$$\begin{aligned} V_o(0) &= \beta_o D^{\beta_1} H^{\beta_2} + \varepsilon \\ &= V_{ot} + \varepsilon \end{aligned} \tag{7a}$$

whereas model (6) reduces to the combined-variable model, (1). When $d_o > 0$, model (6) entails a multiplicative adjustment to V_{ot} whereas the latter involves an additive adjustment. Prior to the introduction of models such as these it had been customary to fit a new merchantable volume equation whenever utilization capabilities changed, thereby changing dimensional merchantability criteria. For example, advances in machinery and technology have made it economically feasible to process stemwood to an upper-stem diameter of $d = 10$ cm, whereas economic considerations in the past imposed diameter limits of $d = 20$ or 15 cm, say. Not only is the practice of refitting models an expensive process, the plethora of separately fitted merchantable volume equations allows for the possibility of inconsistent volume predictions among equations. Models such as (6) and (7) are preferable because both are applicable to any arbitrary upper limit d and because both ensure that $V_o(d_o) \geqslant V_o(d'_o)$ when $d_o \leqslant d'_o$, they are logically consistent (Avery and Burkhart, 1994).

Regardless of the model chosen to portray the relation between volume D and H, least squares has almost invariably been the statistical method used to estimate the model parameters. Because stem taper is not explicitly included in a standard volume equation, $E[V|D, H]$ implicitly represents the expected volume corresponding to a tree of average form. As has long been recognized (cf. Behre, 1927), unless the stem form or taper characteristics of the population to which a volume equation is applied matches that of the population which was sampled to fit the model, biased predictions will result. Furthermore, because of inherent morphological differences among species, it generally is necessary to fit separate standard volume equations for each species or closely related species group.

Sample trees are usually chosen such that their range in diameters and heights span the range in tree sizes to which the fitted equation will be applied. Typically the sample trees will be felled and cut into short lengths (≈ 1 m); the volume of each length will be accurately determined, and these volumes will be cumulated to get the volume of the entire stem. In short, the data to which a volume equation is fitted are carefully collected and contain negligible measurement error.

Because prediction rather than inference has been of primary importance, the behavior of the error term in the regression models fitted to volume data has been a secondary concern. Nonetheless, Meyer (1953) advocated weighted regression in order to account for the increased variation in volume with increasing tree size, which is a phenomenon exhibited by many tree species. The objective is to obtain more precise estimators of model parameters by appropriately accounting for error heteroscedasticity within a generalized least squares (GLS) framework. Gedney and Johnson (1959) assumed that $\mathrm{Var}(\varepsilon|D, H) \propto D^{\gamma}$, and tabulated appropriate weights for a wide range of equally spaced values of D. They observed that the weights were fairly constant across several softwood and hardwood species of northwestern U.S. Cunia (1964) suggested weighting inversely by the sample variances which are computed from contiguous, discrete groupings of sample trees by D or D^2H.

Gregoire and Dyer (1989) examined model (1) with $\mathrm{Var}(\varepsilon|D, H) = \sigma^2 X^{\gamma}$, where $X = D^2H$. The heterogeneity parameter γ was estimated by fitting the model

$$R = \sigma^2 + \gamma \ln X + v,$$

where $R = \ln e^2$, e is the residual from an ordinary least squares (OLS) fit of (1), and v is the model error. Subsequently model (1) was refitted by GLS with $w = X^{-\hat{\gamma}}$ as the weight. This method, due to Harvey (1976), is sensitive to small OLS residuals, as explained by Davidian and Carroll (1987).

As an alternative to two-step procedures to account for error heteroscedasticity, one can estimate parameters by maximum likelihood. Again using model (1) as an example if $\varepsilon \sim \mathrm{N}(0, \sigma^2 X^{\gamma})$ and $X = D^2H$, the log-likelihood is

$$\ln L = \phi - \tfrac{1}{2} n \ln \sigma^2 - \tfrac{1}{2} \sum_{k=1}^{n} \ln X_k^{\gamma} - \tfrac{1}{2} \sum_{k=1}^{n} \frac{(V_k - \beta_0 - \beta_1 X_k)^2}{\sigma^2 X_k^{\gamma}}, \tag{8}$$

where $\phi = -\tfrac{1}{2} n \ln 2\pi$. Its derivatives with respect to the four unknown parameters, when evaluated at zero, provide the maximum likelihood (ML) estimating equations (cf. Gregoire and Dyer, 1989). Magnus (1978) demonstrated that iterative solution of the estimating equations converges to the maximum likelihood solution. The variance–covariance matrix of the ML estimates can be estimated from the inverse of the information matrix evaluated at the ML solutions (Kmenta, 1986).

For conventional volume equation models it is reasonable to assume that volume observations are realizations of uncorrelated random variables unless a correlation is induced by the procedure used to select trees into the sample to fit the equation. This assumption is not very tenable when values of $V(d)$ are measured for a succession of different upper diameters on each tree sampled in order to fit model (6) or (7). In this situation it may be more reasonable to assume that observed values within a tree are more or less correlated depending on the distance separating them along the stem. Let $\boldsymbol{V}_k$ denote the T_k element vector of observed volumes on the k-th tree, $k = 1, \ldots, n$; and let

$$V_{kt}(d_{kt}) = f(D_k, H_k, d_{kt}) + \varepsilon_{kt}$$

denote its t-th element, $t = 1, \ldots, T_k$. The normal log-likelihood of $\boldsymbol{V}_k$ is

$$\ln L_k = \phi_k - \tfrac{1}{2} \ln |\Omega_k^{-1}| - \tfrac{1}{2} \boldsymbol{\varepsilon}_k' \Omega_k^{-1} \boldsymbol{\varepsilon}_k, \tag{9}$$

where

$$\phi_k = \tfrac{1}{2} T_k \ln 2\pi,$$

$$\boldsymbol{\varepsilon}_k = \boldsymbol{V}_k - \boldsymbol{f}_k,$$

$$\boldsymbol{f}_k = [f(D_k, H_k, d_{k1}), \ldots, f(D_k, H_k, d_{kT_k})]'$$

and Ω_k is the $T_k \times T_k$ covariance matrix with st-th element given by $\mathrm{E}[\varepsilon_{ks}\varepsilon_{kt} | D_k, H_k, d_{ks}, d_{kt}]$. To account for size-induced error heteroscedasticity among trees we re-express Ω_k as $\sigma_k^2 P_k$, where $\sigma_k^2 = \sigma^2 X_k^{\gamma}$, and P_k is the corresponding correlation matrix.

If the multiple measurements d_{kt}, $t = 1, \ldots, T_k$, are made at equally spaced intervals, Δ_{kt} say, along the bole, a first-order spatial-autocorrelation error structure might be reasonable. Under this structure $\varepsilon_{kt} = \rho_k \varepsilon_{k,t-1} + \omega_t$, where $\mathrm{E}[\omega_t] = \mathrm{E}[\omega_t \omega_s] = 0$ for $t \neq s$, $\mathrm{E}[\omega_t^2] = \sigma_\omega^2 = (1 - \rho_k^2)\sigma_k^2$, and $0 \leqslant \rho_k < 1$. Consequently Ω_k^{-1} is a fairly simple banded matrix (cf. Judge et al., 1985), and $|\Omega_k^{-1}| = (1 - \rho_k^2)[(1 - \rho_k^2)\sigma_k^2]^{-T_k}$ (Amemiya, 1985).

For the entire sample of n trees the log-likelihood is

$$\ln L = \phi - \tfrac{1}{2} \sum_{k=1}^{n} \ln |\Omega_k^{-1}| - \tfrac{1}{2} \sum_{k=1}^{n} \boldsymbol{\varepsilon}_k' \Omega_k^{-1} \boldsymbol{\varepsilon}_k, \tag{10}$$

where $\phi = -\tfrac{1}{2} T \ln 2\pi$, and $T = T_1 + \cdots + T_n$. A simplification to (10) occurs when $\rho_k = \rho_j$, $\forall\, k, j$; this specification of the model may be especially justified when $\Delta_{kt} = \Delta_{js}$, $\forall\, k, j, \forall\, t, s$. In situations where measurements occur at unevenly spaced intervals, i.e., $\Delta_{kt} \neq \Delta_{js}$ within a tree, a continuous autocorrelation process may be a more tenable specification than the discrete process described above. In a continuous process the autocorrelation coefficient is parametrized in terms of Δ. One form that this might take is, after Diggle (1988), $\rho(\Delta) = e^{-\alpha\Delta}$, where α is the estimable parameter.

To apply a standard volume equation in order to predict the volume of a tree, its height H must be measured. In comparison to the time it takes to measure D, much more time is required to measure H accurately under normal forest conditions. The difficulty arises from the obscured view of tree stems within the crowns when the forest canopy is closed. As an alternative to the measurement of H on every tree sampled in a forest inventory, height is sometimes measured on a subsample of trees only. Later in the office a regression relating H onto D is fitted to the subsample data. For those trees outside the subsample, volume is predicted from a standard volume equation using the height, $\hat{H}$ say, predicted from the H–D regression in place of the unmeasured value of H. While the use of $\hat{H}$ instead of H may be viewed as an 'errors-in-variables' problem, it differs from the situation that commonly is referenced by that term and that has been comprehensively treated by Fuller (1987) and Carroll (1989). The focus here is on the error injected into the prediction of tree volume by a volume equation that has been fitted previously with measured, essentially error-free values of the explanatory variables, but is applied subsequently with values of an explanatory variable that may contain prediction error.

The prediction bias introduced by this practice is negligible providing that (1) H is unbiasedly predicted by the fitted regression on D, and (2) the range of environmental conditions encountered in the broad forest inventory matches those in the subsample. An upshot of these provisions is that prediction bias may not be negligible if the H–D prediction equation was obtained in a different inventory from the one in which it is applied. While the bias may be insignificant, the effect on the variance of volume predictions may not be. Gertner (1986) examined the latter issue using a Taylor-series approximation to the component of variance due to this 'post-calibration error'. Gregoire and Williams (1992) derived exact mean square prediction error results for a particular choice of standard volume equation and H–D regressions. In a companion Monte Carlo study involving a population of 14 381 loblolly pine trees, they observed an increase of roughly 35–38% in root mean square prediction error which was attributable to the ancillary prediction of tree height and its subsequent use to predict tree volume.

The implied imbedding of the H–D regression into the volume equation was made explicit in a mixed-model formulation by Lappi (1991). Lappi viewed the practice as a way to localize a standard volume equation and he derived the best linear unbiased predictors, or BLUPs (see Goldberger, 1962; Robinson, 1991) of the correlated random parameters.

The prediction of H from D essentially makes the volume prediction solely dependent on D. There is, thus, an implicit trade-off of precision in volume prediction for savings in time and effort in the field. In contradistinction to this is the incorporation into the volume equation of some point measure of stem taper with the concomitant requirement of additional field work required to obtain that measure. Among trees with the same D and H, variation in volume is due mainly to variation in stem taper, hence the rationale for its inclusion into the prediction equation.

The most prevalent point measure of form is the ratio of an upper-stem diameter to D, $d(h)/D$ say, where h represents a specified height on the stem. Schmid-Haas et al. (1971) determined $6 \leqslant h \leqslant 9$ m to be the optimal height based on an empirical study of 14 000 spruce trees. The range in spruce tree heights was unreported by Schmid-Haas, and it is likely that different ranges in the location of $d(h)$ would prove optimal for species with substantially taller height at maturity. Even if $d(h)$ is not optimally located, however, the inclusion of an upper-stem diameter measurement considerably improves the prediction of volume (cf. Loetsch et al., 1973; McTague, 1992). Instead of an upper-stem diameter at a fixed height on the stem, one at a stipulated relative height, e.g., $h = 0.3 H_t$, or at the height where a fixed taper ratio is achieved, have both been suggested as alternatives.

In North America the most common point measure used for this purpose is the Girard form class F which is the ratio of inside-bark diameter at 5.3 m aboveground to D. Examples of a form-class volume equation are, after Spurr (1952),

$$V = \beta_0 + \beta_1 F + \beta_2 D^2 H + \beta_3 F D^2 H, \tag{11}$$

and from Burk et al. (1989),

$$V = \beta_0 D^{\beta_1} H^{\beta_3} (1 + \operatorname{sgn}(F - \bar{F}) \beta_4 |F - \bar{F}|^{\beta_5}), \tag{12}$$

where $\operatorname{sgn}(\cdot)$ indicates algebraic sign and $\bar{F}$ is algebraic average form class.

4. Volume predictions from taper equations

The ratio of $d(h)$ to D will depend jointly on h and H_t. If the $d(h)/D$ relationship over the range $0 < h \leqslant H_t$ can be approximated by a smooth function, e.g.,

$$d(h)/D \approx g(h, H_t) \tag{13}$$

then

$$V(h) \approx kD^2 \int_0^h \{g(\tau, H_t)\}^2 \, d\tau, \tag{14}$$

where $k = 2.5\pi \times 10^{-5}$ m^2/cm^2. This establishes the basic rationale for prediction of tree volume as the definite integral of a bole profile equation. While interpolating splines have been used in forestry to derive the profile function (Sloboda, 1977; Lahtinen and Laasasenaho, 1979), statistically fitted regression functions have been much more prevalent in the published literature.

The earliest use of a taper equation generally is credited to Höjer (1903; cited by Behre, 1927). It appears that the early applications of taper equations were restricted to the development of volume tables indexed by D and H (cf. Wright, 1923). Empirical taper curves were also used and volume was obtained planimetrically (Reineke, 1926). More sophisticated use followed advances in statistical methods, and increasingly more complex functions to describe taper were introduced as computing capabilities were developed to estimate the unknown parameters. Because of the flexibility they provide, taper functions as the basis for volume equations are increasingly replacing direct volume equations (Kilkki and Lappi, 1987).

Behre (1923) introduced a hyperbolic formula to express tree taper

$$\mathrm{E}[d(h)/D]=\frac{(h'/H_t)}{\beta_0+\beta_1(h'/H_t)}, \tag{15}$$

where $h' = H_t - h$, that has been the basis of considerable subsequent modeling of tree taper (Newnham, 1988; Monserud, 1981).

Polynomials in powers of h/H_t or h'/H_t have often been used to describe taper. Fries and Matérn (1966) performed principal component analyses of birch tree taper with powers up to the 64-th. The first principal component accounted chiefly for average taper, the second was an expression of basal (butt) swelling, and the third an expression of form.

Variations of a quadratic polynomial presented by Kozak et al. (1969)

$$\mathrm{E}[(d/D)^2]=\beta_0+\beta_1(h/H_t)+\beta_2(h/H_t)^2 \tag{16}$$

have been used widely. Oftentimes the polynomial coefficients are constrained in such a way to ensure that $d/D = 0$ when $h = H$. Since D is, by definition, the outside-bark diameter at breast height, h_b say, H_t is often replaced by $H_t - h_b$ to further ensure consistency. In addition to pure polynomials of various degrees, there are mixed functions such as

$$\mathrm{E}[d(h')/D]=\sum_{k=0}^{4}\beta_0(h/H)^k+\beta_5 Z+\beta_6 Z^2 \tag{17}$$

where $Z = \ln(h/H)$ (Laasasenaho, 1982).

Omerod's (1973) taper function

$$\mathrm{E}[d/D]=\left[\frac{h'}{H_t-h_b}\right]^{\beta} \tag{18}$$

and Forslund's (1982)

$$\mathrm{E}[d/D]=[1-(h/H)^{\beta_1}]^{\beta_2} \tag{19}$$

are attempts to express taper simply as a smooth geometric form. For example in (19), when $\beta_1 = 1$ and $\beta_2 = 1.5$, the implied form is intermediate between a cone and a paraboloid. This 'paracone' shape has been a useful compromise between precision and complexity.

Thomas and Parresol (1991) noted that the relation of d/D to h/H_t was directly analogous to the behavior of some trigonometric functions on the unit circle.

Accordingly they examined the model

$$\mathrm{E}\left[\left(\frac{d(h)}{D}\right)^2\right] = \beta_1\left(\frac{h'}{H_t}\right) + \beta_2 \sin\left(\frac{\beta_3 \pi h}{H_t}\right) + \beta_4 \cot\left(\frac{0.5\pi h}{H_t}\right). \tag{20}$$

The authors advocate (20) as a parsimonious alternative to high-order polynomials.

Building upon Gallant and Fuller (1973), Max and Burkhart (1976) fitted segmented polynomial models with estimated join points. The model that best fitted data from 490 plantation-grown and natural-stand loblolly pine trees was

$$\mathrm{E}\left[\left(\frac{d_i(h)}{D}\right)^2\right] = \beta_1\left(\frac{h}{H_t} - 1\right) + \beta_2\left(\frac{h^2}{H_t^2} - 1\right) + \beta_3\left(\alpha_1 - \frac{h}{H_t}\right)^2 I_1 + \beta_4\left(\alpha_2 - \frac{h}{H_t}\right)^2 I_2, \tag{21}$$

where $I_1 = 1$ if $\alpha_1 \geqslant h/H_t$, $I_1 = 0$ otherwise; and $I_2 = 1$ if $\alpha_2 \geqslant h/H_t$, $I_2 = 0$ otherwise. For this three-segment model fitted to loblolly pines, the estimated join points were located at $\alpha_1 \approx 0.11$ and $\alpha_2 \approx 0.74$. With reference to Figure 1 one can surmise that the lower segment captured the characteristic butt swell of the tree, whereas the third segment served to characterize the conic profile of the upper stem. The Max–Burkhart model for tree taper has been implemented for a wide range of species, as cited in Solomon et al. (1989).

Both Sloboda (1977) and Lappi (1986) transformed $d(h)$ and fitted taper functions with dimensions determined in polar coordinates based on a series of angles from the tree base. Lappi (1986) proposed a mixed linear model in which variation in stem dimensions was partitioned into variation between stands and within stands. The first few principal components of the between-stand and within-stand covariance matrices were derived to describe random variation in stem form more economically. When applied to a particular stand, BLUPs of the random stand effects are obtained in order to calibrate the taper equation to the particular stand. Korhonen (1991) investigated the use of this methodology in concert with subsampled height-diameter data from forest inventory. He concluded that these models can be calibrated efficiently with fewer sample trees than conventional volume equations.

The above models may be fitted by least squares, and inference concerning linear combinations of parameters follow directly from linear or nonlinear model theory as appropriate. The statistical evaluation of volume estimates computed from the integral of a fitted taper equation is more problematic. It seems certain that the resulting predictor of volume will be biased, and the asymptotic behavior of these predictors has not been studied. Empirical checks have generally shown the bias to be negligible in cross-validation studies. It seems likely that linearizing techniques can be developed to approximate the standard error of volume predictors, but this area, too, still needs to be explored.

5. Evaluation of site quality

In parallel to land capability estimation for agricultural crops, site quality in forestry is an assessment of the ability of currently or potentially forested land to grow trees (Roth, 1916; Carmean, 1975). In contrast to agricultural practice and owing to the much longer time for crop rotation in forestry, forest site quality cannot be assessed directly on the basis of observable sustained yield under a fixed planting or management regime. Moreover, for forest planning purposes it is desirable to have a measure of site quality that is not linked or conditioned on one particular management scheme. In both fields of endeavor, the influence of physiographic and edaphic factors on the capacity of a site to produce a crop of a particular plant species or variety limits the precision of site quality assessment to regions within which these factors are roughly homogeneous. While there are extremes of site conditions that are inimical to virtually all plant growth, between these extremes exist a broad spectrum of intermediate conditions at which site conditions may be quite favorable for one or more species, yet quite unfavorable to others. Hence there is the implicit assumption in any discussion of forest site quality that its evaluation pertains to a specific species or closely-knit species group. Efforts to relate site quality of one species to that of another have had limited success (Doolittle, 1958; Olson and Della-Bianca, 1959).

Ever since the convincing appeal made by Frothingham (1918), the average total height of dominant trees at a specified age has been the most widely accepted index of forest land capacity. The rationale for this measure is the empirical observation that height growth of forest-grown trees is relatively unaffected by stand density, and thus is a manifestation of the inherent productivity of a site. On trees that have been overtopped by neighboring trees, this indication of site quality will be confounded by the inhibiting effect on growth caused by reduced light interception. Hence the trees whose height is used to index site quality are restricted to those whose position in the canopy has been dominant or codominant. This restriction usually limits the applicability of site index, the name by which this measure of site quality has come to be known universally, to even-aged stands whose height development has not been disturbed by natural catastrophe or human intervention.

The idea of using site index as a measure of site quality neither originated with Frothingham nor is it free of recurrent criticism. But because of its prevalence and the attendant statistical aspects of its estimation and prediction, the ensuing discussion is limited to site index. Historical surveys of the development and comparative advantages of site index plus other measures of site quality have been written by Jones (1969), Carmean (1975), Hägglund (1981) and Grey (1989).

6. Site index

Let $H(A)$ denote the total height of dominant trees at stand age A. Let A_I denote the predetermined age for which $S = H(A_I)$ is intended to serve as an index of site quality. $A_I = 10, 25, 50$ or 100 years are common choices depending on the height growth of the particular species for which site index is to be predicted. Development of a site index equation customarily has entailed estimating the parameters of an assumed regression relating $H(A)$ to A, and then eliminating one of the parameters by re-expressing it in terms of S and A_I. This time-worn method in forestry has been termed elsewhere as an expected value parametrization by Ratkowsky (1980), stemming from a suggestion made by Ross (1970) for deriving stable estimates of parameters in nonlinear models. As an example of such a site index equation consider the simple linear model first proposed by Schumacher (1939)

$$H = e^{\alpha + \beta/A}\varepsilon \tag{22}$$

such that at A_I,

$$S = e^{\alpha + \beta/A_I}\varepsilon'$$

which leads to

$$S = He^{-\beta(1/A - 1/A_I)}\left(\frac{\varepsilon'}{\varepsilon}\right). \tag{23}$$

With observations of S, H, A on a sample of forested stands, (23) can be fitted directly. Because of the difficulty of assembling a sample for which $S = H(A_I)$ can be observed directly, the usual alternative has been to fit the guide curve (22) to sample observations of H, A only. Expected site index at age A_I can be predicted subsequently from

$$\hat{S} = H^* e^{-\hat{\beta}(1/A^* - 1/A_I)} \tag{24}$$

for some future observations H^*, A^*.

While (24), or the corresponding prediction equation for an alternative choice of the guide curve (22), is convenient for computational purposes, site index curves are often presented graphically. A sheaf of site index curves from (24) can be constructed by arrangement to $H^* = \hat{S}e^{\hat{\beta}(1/A^* - 1/A_I)}$, and solution of this form at predetermined values of $\hat{S}$.

The intended purpose of a height–age regression, such as (22), which serves as the basis for a site index curve, is to portray the trend of expected height growth. In many forested populations there exists a negative correlation between site index and stand age (young stands occur preponderantly on better sites than older stands) for reasons elucidated in Walters et al. (1989). The mechanism is analogous to that encountered in epidemiological and wildlife research which leads to intercepted data as exposited by Vardi (1988). When the sample data available to fit the regression model are cross-sectional, namely isolated H–A observations from a cross-section of stands, a form of self-selection may bias conventional parameter estimates. Beck and Trousdell (1973) report biases from -2 to -10 m at $A_I = 50$ years for *Pinus strobus* L.; Monserud (1985) reports negative biases randing from -4 to -8 m for

Pseudotsuga menziesii (Mirb.) Franco. An instrumental variable can be used to provide consistent parameter estimates in some cases (cf. Walters et al., 1989). A preferred practice, however, is to fit the H–A regression to longitudinal data wherein the trend of dominant height with age is observed and measured for each of a cross-sectional sample of forest stands. Furnival and Wilson (1971) provide a heuristic explanation for the elimination of bias when ΔH is modeled as a function of ΔA.

When longitudinal or repeated-measurement data are available not only can the actual H–A trend be modeled directly, but also a much broader family of models can be fitted. Reflective of this, a large number of different models and modeling tactics have been used within forestry, all with the aim toward enabling site index predictions. It is widely believed that the relative rate of height growth is greater on better sites than it is on lesser sites. This has led to the purposeful development of polymorphic site index curves in preference to anamorphic curves, where the latter exhibit a proportionality as explained by Avery and Burkhart (1994).

One approach is to assume that the coefficients of the H–A regression are the result of a stochastic process whose parameters depend, inter alia, upon site. King (1966) modeled the fitted coefficients simply as a linear function of site index, whereas Biging (1985) employed tactics elucidated by Swamy (1970) for feasible GLS estimation of the random coefficients.

Variants of the Von Bertalanffy (1957) model have appeared frequently in reports of site index research; see, for example, Newnham (1988) and references therein. Not only has this basic model form fitted height–age data exceedingly well for an astonishingly wide array of species and physiographic regions, but the biological interpretation that can be given to its parameters have contributed to its widespread acceptance. The extensive application of the Von Bertalanffy model notwithstanding, numerous other flexible curve forms have been fitted with success. For example, Bailey and Clutter (1974) proposed a polymorphic site index model based upon

$$H = \beta_1 \left(\frac{S}{\beta_1} \right)^{(A_1/A)^{\beta_2}} + \varepsilon.$$

Amateis and Burkhart (1985) fitted

$$H = S e^{(A_1/A)^{\beta_1} e^{\beta_2 (1/A - 1/A_1)}} + \varepsilon$$

to loblolly pine data from throughout the southeastern United States. García (1983) modified the differential form of the well-known monomolecular function by substituting a power transformation H^{γ} for H, and then adding a Wiener process to represent the stochastic perturbations in the environment (site) that fluctuate over time. Devan and Burkhart (1982) examined a segmented polynomial model. All of these models are base-age invariant, a property which precludes the otherwise undesirable characteristic of site index curve estimation that the arbitrary choice of index age A_1 leads to different parameter estimates.

The notion of calibrating a general H–A equation to local conditions is a recurrent theme. Lloyd et al. (1982) proposed a model that included site-specific parameters explicitly. Namely, the vector of parameters $\boldsymbol{\theta}$ of the H–A regression is partitioned as $\boldsymbol{\theta}=(\gamma,\boldsymbol{\beta})$, where γ is the subvector of site-specific parameters which identifies the individual growth curve from a family of curves that have $\boldsymbol{\beta}$ in common. It is possible to estimate $\boldsymbol{\theta}$ from historical growth data, but for prediction of site index for a stand outside the sample data, indeed for the prediction of height at any arbitrary age, further observations are necessary only to calibrate the site-specific indexing parameter γ.

Lappi and Bailey (1988) forwarded a model of height growth which included random tree and stand parameters and which accounts for covariance among multiple measurements of height garnered within the same forest stand or plot. In application the BLUPs of the random effects is used to combine information from sample tree measurements with the fixed site curve model.

In yet another direction. Walters et al. (1991) took a Kalman-filter approach to localizing a general H–A equation, wherein local H–A data are used to modify a regional site index curve by essentially weighting inversely proportional to variance.

7. Measures of stand density

The growth rate achieved by any given forest stand is largely determined by the productive capacity of the site and the degree to which the site is occupied. Site index (Section 6) is the most commonly applied method for quantifying site quality. A large number of methods for quantifying site occupancy or stand density have been proposed and several have been widely applied. The degree to which a forest stand of a given species composition utilizes the resources of a specific site depends on the numbers and sizes of the trees present and on the spatial distribution of the trees.

There are two general categories of density expressions: (1) stand density measures which provide an estimate of average crowding, and (2) point density measures which provide an estimate of the competitive situation for individual trees. We will confine our attention to stand-level measures of density which are used in predicting growth and yield of forest stands, the subject of the next section. Information on and comparisons of various measures of point density can be found in Alemdag (1978), Daniels et al. (1986), Gerrard (1969), Johnson (1973), Martin and Ek (1984), Noone and Bell (1980) and Tomé and Burkhart (1989).

Stand density measures are of two types: (1) those that are direct functions of measured stand characteristics, and (2) those that involve comparison of the existing stand to some norm or with some previously established limiting relationship. One obvious measure of density that can be estimated through complete enumeration or sampling in the stand of interest is the number of trees per unit area. While number of trees per unit area has been found to be reasonably well correlated with growth in planted stands (where stems typically are more uniformly distributed), it is a less useful measure in naturally reproduced stands. Numbers of stems alone do not include any information on tree size. A measure that incorporates tree size and that has been widely applied is the basal area per unit area, defined as the sum of the cross-sectional areas at breast height and expressed as ft^2/acre or m^2/ha. Numerous investigators (for example, Nelson and Brender, 1963) have found stand basal area to be highly correlated with growth and yield.

Measures of density based on the two component parts of basal area – number of trees per unit area and diameter of the tree of average basal area – have been called stand-density indices (Spurr, 1952). The most widely used variant of these indices is Reineke's (1933) stand-density index. Reineke used a simple linear regression between logarithm of number of trees per unit area and logarithm of the average diameter to define the limits of maximum stocking. He found that the same slope could be used for the limiting line but that the level varied by species. Subsequent to Reineke's work there has been much interest in the so-called '3/2 law of self thinning'. Like Reineke's stand-density index, the 3/2 relationship is based on the concept of a maximum size–density relationship. In the case of the 3/2 law of self-thinning, the logarithm of mean tree volume or weight is plotted against the logarithm of the number of trees per unit area. Although Reineke's stand-density index has been popular in forestry and the 3/2 self-thinning line has been widely used by plant ecologists, the two are mathematically equivalent (Bredenkamp and Burkhart, 1990).

Other measures of stand-density that have been applied in forestry include tree–area ratio, crown competition factor, and relative spacing (Clutter et al., 1983). As pointed out by Curtis (1970) and West (1983), most of the commonly used measures of stand density can be shown to be equivalent. All of these measures can be used to quantify stand density, although none have been found universally superior to number of trees and basal area per unit area as explanatory variables in models developed for the purpose of predicting stand growth. Hence, subsequent discussion will be limited to models which incorporate number of trees or basal area as a predictor. When making predictions of stand growth the site quality (site index) can be assumed to remain constant over a rotation. By contrast, stand density must be considered as a dynamic component of the stand development prediction system.

8. Overview of growth and yield prediction

Forecasting volume growth (the increase in volume over a given period of time) and yield (the accumulated volume at a given time) is implicit in the process of forest management. The primary uses of growth and yield information can be categorized as follows: (i) inventory updating, (ii) evaluation of silvicultural alternatives, (iii) management planning, and (iv) harvest scheduling. Although yield tables for even-aged stands have been in use in the United States since about 1900, we still lack adequate growth and yield information for many purposes and circumstances. Consequently, growth and yield estimation continues to be an active area of research.

Because growth and yield information is used for such a variety of purposes, no single data base or modeling approach serves all purposes equally well. The approaches taken in numerous modeling efforts have varied to serve the developer's objectives, the background and research interests of the modeler, and the analytical methods, data base and computational machinery available. The latter factor, namely the three-way interaction of analytical methods-data base-computational technology has had a profound influence on growth and yield modeling.

Stand dynamics (i.e., the growth, mortality, and associated changes in the stand) can be predicted through direct or indirect methods. Direct methods, such as the stand-table projection technique, utilize field observations of past growth and mortality trends in existing stands to infer future trends in the same stands (additional detail is contained in Avery and Burkhart, 1994 and Clutter et al., 1983).

There are many situations in which direct observation of past growth and mortality is not feasible in each stand for which prediction of future yield is desired, however. Moreover, diameter growth, mortality, and ingrowth relationships developed through stand-table projection techniques are not reliable for long periods of time, nor do they permit foresters to evaluate a range of management alternatives which differ from past management activities because inferences from past growth are limited to the conditions under which that growth occurred. Thus, when management actions are changing the characteristics of a forest, stand-table projection may not provide a realistic estimate of future growth.

Owing to the inferential difficulties when using past growth directly to deduce future growth of forests, that possibly will be subjected to changing conditions, foresters have increasingly relied on indirect methods of predicting stand dynamics. That is, growth, mortality, and related quantities of a stand are inferred from the study of other stands through the application of growth and yield models. The predictor variables in models appropriate for the yield of even-aged stands generally consist of age, site index, a measure of stand density (usually basal area or number of trees per unit area), and management treatments. Application of growth models is the preferred option for forecasting growth when management practices and changing conditions evolve the forest rapidly thus rendering measurement of past growth of limited value for estimating future growth directly.

A variety of approaches has been taken to growth and yield prediction. In American terminology, an 'empirical yield equation' supposedly applies to 'average' stocking. Because density is not explicitly included as a predictor variable, empirical yield functions apply only to the average density levels found on the sample plots used to estimate the parameters of the equation(s). However, most growth and yield equations include density as a dynamic part of the stand projection system. Such models are commonly termed 'variable-density' equations (or models).

Many investigators have used multiple regression techniques to predict growth and/or yield for the total stand or for some merchantable portion of the stand (such as Beck and Della-Bianca, 1972; Bennett, 1970; Burkhart and Sprinz, 1984; Farrar, 1979; Sullivan and Clutter, 1972). Stand level variables such as age, site index, basal area or number of trees per unit area are utilized in the whole stand approach to predict some specified aggregate stand volume.

A number of models have been developed which consider the stand in terms of the distribution of the number of trees per unit area by size-class. In most cases dbh classes have been used. A common modeling approach in this general category is one which fits a posited distribution function to observed distributions of diameter data (for example, Bennett and Clutter, 1968; Burkhart and Strub, 1974; Dell et al., 1979; Schreuder et al., 1979; Smalley and Bailey, 1974).

The various diameter distribution models differ chiefly in the function used to describe the diameter distribution. In this approach, the number of trees per unit area in each diameter class is estimated through the use of a probability density function (pdf) which provides the relative frequency of trees by diameters. Mean total tree heights are predicted for trees of given diameters growing under given stand conditions. Volume per diameter class is calculated by substituting the predicted mean tree heights and the diameter class midpoints into tree volume equations. Yield estimates are obtained by summing the diameter classes of interest. Although only overall stand values (such as age, site index, and number of trees per unit area) are needed as input, detailed stand distributional information is obtainable as output.

Regardless of the probability density function used, the procedure involves estimating the pdf parameters for each plot in the data set (usually by the method of moments or maximum likelihood) and then, in a subsequent stage, developing regression equations to relate these parameter estimates to stand characteristics such as age, site index and number of trees per unit area. Owing to their low precision, functions for relating the pdf parameters to stand characteristics typically have not been fully satisfactory. Consequently an alternative method, sometimes called a 'parameter recovery method', has been developed and applied. The parameter recovery method consists of forecasting overall stand attributes related to the moments of the distribution (such as average diameter, total basal area) and solving for the parameters of the diameter distribution model (such as the beta or Weibull) that will give rise to the overall stand attributes. This approach provides a direct mathematical link between the overall stand volume and the distribution of that volume. Additional information on parameter recovery methods can be found in the papers by Hyink and Moser (1983), Matney and Sullivan (1982), Cao et al. (1982) and Knoebel et al. (1986).

Approaches to predicting stand growth and yield which use individual trees as the basic unit are referred to as 'individual-tree models'. The components of tree growth in these models are commonly linked together through a computer program which simulates the growth of each tree and then aggregates these to provide estimates of stand growth and yield. Models based on individual-tree growth provide detailed information about stand dynamics and structure, including the distribution of stand volume by size classes.

Individual-tree models may be divided into two classes, distance independent and distance dependent, depending on whether or not individual tree locations are required tree attributes. Distance-independent models project tree growth either individually or by size classes, usually as a function of present size and stand level variables (e.g., age, site index, and basal area per unit area). It is not necessary to know individual tree locations when applying these models. Typically, distance-independent models consist of three basic components: (1) a diameter–growth component, (2) a height–growth component (or a height/diameter relationship to predict heights from dbh values), and (3) a mortality component. Examples of this approach include Belcher et al. (1982), Harrison et al. (1986) and Amateis et al. (1989).

Distance-dependent models vary in detail but are quite similar in overall concept and structure. Initial stand conditions are input or generated, and each tree is assigned a coordinate location. The growth of each tree is simulated as a function of its attributes, the site quality, and a measure of competition from neighbors. The competition index varies from model to model but in general is a function of the size of the subject tree and the size of and distance to competitors. Survival is controlled either stochastically or deterministically as a function of competition and/or individual-tree attributes. Yield estimates are obtained by summing the individual-tree volumes and multiplying by appropriate expansion factors. Distance-dependent models include those of Newnham and Smith (1964), Ek and Monserud (1974), Hegyi (1974) and Burkhart et al. (1987).

9. Stand-level growth and yield prediction

Although a wide variety of approaches to growth and yield modeling have been taken, it is not possible to provide a detailed discussion of all approaches here. Consequently we will confine ourselves to an indepth review of stand-level growth and yield prediction.

The basic model form proposed by Schumacher (1939) has been adopted and used extensively by growth and yield scientists. The variant of the Schumacher model applied by Clutter (1963),

$$\mathrm{E}(\ln Y) = b_0 + b_1 \ln S + b_2 A^{-1} + b_3 \ln B, \quad \text{where}$$

Y = volume per unit area,
A = stand age,
S = site index,
B = basal area per unit area,

is perhaps the most commonly applied stand-level yield equation.

Early work did not attempt to relate growth analyses to yield analyses, although the two quantities are biologically and mathematically related. Buckman (1962) and Clutter (1963) were the first researchers in the United States to recognize explicitly the mathematical relationships between growth and yield in their analyses. Clutter derived compatible growth and yield models for loblolly pine by ensuring that the algebraic form of the yield model could be obtained by mathematical integration of the growth model. Subsequently, Sullivan and Clutter (1972) extended Clutter's models by estimating yield and cumulative growth as a function of initial stand age, initial basal area, site index, and future age.

The models used by Sullivan and Clutter (1972) for yield at an initial and a subsequent time and for projected basal area can be written as

$$\mathrm{E}(\ln y_1) = \beta_0 + \beta_1 S + \beta_2 A_1^{-1} + \beta_3 \ln B_1, \tag{25}$$

$$\mathrm{E}(\ln y_2) = \beta_0 + \beta_1 S + \beta_2 A_2^{-1} + \beta_3 \ln B_2, \tag{26}$$

$$\mathrm{E}(\ln B_2) = (A_1/A_2) \ln B_1 + \alpha_1(1 - A_1/A_2) + \alpha_2(1 - A_1/A_2)S, \tag{27}$$

where

y_i = stand volume per unit area at age A_i,
S = site index,
B_i = basal area per unit area at age A_i,
A_i = stand age at time i (A_1 = present age, A_2 = projected age).

Replacing $\ln B_2$ in equation (26) by its expected value (equation (27)) and simplifying the resulting expression results in

$$\begin{aligned}\mathrm{E}(\ln y_2) = {} & \beta_0 + \beta_1 S + \beta_2 A_2^{-1} + \beta_3 (A_1/A_2) \ln B_1 \\ & + \beta_3 \alpha_1 (1 - A_1/A_2) + \beta_3 \alpha_2 (1 - A_1/A_2) S.\end{aligned} \tag{28}$$

By letting $\beta_4 = \beta_3 \alpha_1$ and $\beta_5 = \beta_3 \alpha_2$, equation (28) can be written as

$$\begin{aligned}\mathrm{E}(\ln y_2) = {} & \beta_0 + \beta_1 S + \beta_2 A_2^{-1} + \beta_3 (A_1/A_2) \ln B_1 \\ & + \beta_4 (1 - A_1/A_2) + \beta_5 (1 - A_1/A_2) S.\end{aligned} \tag{29}$$

Equation (29) is a model for projected volume as a function of site index, initial age, projected age and initial basal area. When $A_2 = A_1$, i.e., the projection period is zero years, equation (29) reduces to the yield model (25). Thus equation (29) is simultaneously a yield model for observations at the initial time and a projection or growth model for subsequent times.

Sullivan and Clutter (1972) showed that equation (29) is invariant for different combinations of projection lengths. That is, if yield is projected from A_1 to A_2 and then from A_2 to A_3, the result is identical with a projection from A_1 to A_3.

A second desirable property is that substituting projected basal area into the volume yield equation to calculate yield at A_2 should be equivalent to using initial basal area in the volume projection equation. To ensure this equivalency property, Sullivan and Clutter (1972) noted that the definitions $\beta_4 = \beta_3 \alpha_1$ and $\beta_5 = \beta_3 \alpha_2$ in the derivation of their simultaneous growth and yield model imply that $\alpha_1 = \beta_4/\beta_3$ and $\alpha_2 = \beta_5/\beta_3$. Given estimates for the parameters of (29), they suggested that estimates of α_1 and α_2 in (27) be computed as $\hat{\alpha}_1 = \hat{\beta}_4/\hat{\beta}_3$ and $\hat{\alpha}_2 = \hat{\beta}_5/\hat{\beta}_3$. While this procedure ensures the desired equivalency property, it does not provide the most statistically efficient estimates of α_1 and α_2 (indeed none of the basal area growth information in the data is used). Furthermore, the estimates of α_1 and α_2, and thus the basal area projection equation, are dependent on the volume units and merchantability limits chosen for the dependent variable in (29).

Burkhart and Sprinz (1984) selected the models specified by Sullivan and Clutter (1972) for fitting with data from thinned loblolly pine plantations because these models have the desired properties of analytic compatibility between growth and yield, invariance for projection length, and numeric equivalency between alternative applications of the equations. In applying these models, however, they sought to develop more efficient and stable (i.e., not dependent on volume units or merchantability standards) estimators for the basal area equation.

As an alternative to deriving coefficients for the basal area projection equation from parameter estimates in the volume projection equation, Burkhart and Sprinz (1984) fitted both functions simultaneously. When defining the criterion or loss function to be minimized in the simultaneous fitting, equal weights were assigned to volume and basal area projection. This led to the standardized loss function,

$$F = \frac{\Sigma(y_i - \hat{y}_i)^2}{\hat{\sigma}_y^2} + \frac{\Sigma(B_i - \hat{B}_i)^2}{\hat{\sigma}_B^2}, \tag{30}$$

where y_i and $\hat{y}_i$ are the observed and predicted volume values, B_i and $\hat{B}_i$ are the observed and predicted basal area figures, and $\hat{\sigma}_y^2$ and $\hat{\sigma}_B^2$ are estimates of the variance around the regression lines for volume and basal area, respectively. These variance estimates, $\hat{\sigma}_y^2$ and $\hat{\sigma}_B^2$, were computed as the mean square error from ordinary least squares regression fits of equations (29) and (27), respectively.

When simultaneously fitting models (29) and (27), the restrictions $\hat{\alpha}_1 = \hat{\beta}_4/\hat{\beta}_3$ and $\hat{\alpha}_2 = \hat{\beta}_5/\hat{\beta}_3$ were imposed. Beginning with coefficient estimates from the ordinary least squares fit of (29), the coefficients of models (29) and (27) were adjusted through an iterative process until F in (30) was minimized. By using the simultaneous fitting procedure, coefficients in the basal area projection equations were nearly identical regardless of the merchantability definitions employed in the volume equation. Further, with the imposed restriction, this process results in a system of equations that gave numerically consistent results. As expected, there was a slight increase in the sum of squared error for volume when fitting by the simultaneous approach. However, this small increase in SSE for volume was more than offset by the decrease in SSE for basal area in the simultaneous fit versus deriving basal area coefficients from the volume equation. Subsequently, Van Deusen (1988) has shown that the equations employed by Burkhart and Sprinz (1984) can be solved by applying the methods of seemingly unrelated regression.

Growth and yield models often employ a system or set of equations to describe stand development. Generally the individual equations in the system are fitted one at a time by the method of least squares, but this may not be entirely satisfactory. The same variable may be dependent in one equation and independent in another. Further, certain coefficients in one equation may be functionally related to coefficients in another equation. And the residuals from the several equations are likely correlated. Because of these considerations, there is increasing interest in treating growth and yield models as a system of equations and simultaneously estimating all the coefficients in the systems.

The use of these methods, developed largely in the field of econometrics, to simultaneously estimate the coefficients for a growth and yield model was first applied by Furnival and Wilson (1971). Murphy and Sternitzke (1979) and Murphy and Beltz (1981) applied three-stage least squares to develop an equation system to estimate current volume, projected basal area, and projected volume for loblolly and shortleaf pine stands, respectively. Borders and Bailey (1986) developed a system of compatible and inter-relating equations to predict growth and yield in volume, basal area and dominant height. When fitting their system of equations they reviewed and tested a number of techniques from econometrics and found three-stage least squares gave the best results. In another study, Gregoire (1987) examined the combined time-series cross-sectional nature of remeasurement data from permanent plots with the objective of improving the precision of yield models fitted with these data. Alternative error covariance structures were specified and the yield models were fitted using two-stage generalized least squares and, in one case, full maximum likelihood estimation.

When considering the adequacy of various procedures for estimating parameters in systems of equations, several points must be kept in mind (Burkhart, 1986). The primary purpose of a growth and yield system is usually prediction, and there has been little work done to illustrate how well the simultaneously estimated systems perform for this objective. Simultaneous estimation involves a compromise, with none of the components as well estimated as they could be individually. It should also be pointed out that many of the simultaneous equation fitting techniques from econometrics were developed for determining underlying functional relations, and they may not be fully appropriate for biological systems where the objective is prediction.

When attempting to apply extant methodology for simultaneous fitting of equations to growth and yield data, other problems arise. Many of the systems in which we are interested involve nonlinear components and nonlinear constraints. While work has been done on simultaneous fitting of nonlinear models, considering nonlinear models generally makes the problem much more complex. Sample size may also be a problem. Although asymptotic statistical properties have been established for many econometric procedures, the small sample properties are often unknown.

There have been many simultaneous estimation techniques developed, and it behooves biometricians to be aware of these powerful tools. It is tempting, however, to make one's problem conform to a well-established procedure, rather than carefully examine the problem at hand. We must be careful to avoid such temptation. It is important that biometricians characterize the biological systems of interest as realistically as possible, commensurate with the modeling objectives. Experience with developing growth and yield prediction tools indicates that model specification is far more important than the procedures used to estimate the parameters (given that all of the procedures considered basically involve minimization of a squared error loss function). Parameter estimates in models specified may or may not be readily obtained – it may be necessary to develop new methodology in some instances. Prior to embarking to such developments, however, it is prudent to become thoroughly versed on relevant past work in all areas of statistics. It is suggested that we make the fullest possible use of theory and methodology that has been developed, but that we not be limited to it.

10. Incorporation of stochastic variation

Past work in growth and yield modeling has concentrated on deterministic techniques (Dennis et al., 1985). Investigators have generally studied the mean behavior of the system and have come to accept large amounts of unexplained variability as part of the occupational hazard of quantifying forest biology. There has been limited incorporation of stochastic elements in stand simulators, typically through the mortality or the growth increment components. Mortality may be generated stochastically through Bernoulli trials. For growth increments, it is common to add a normally distributed random component to the predicted growth from a regression equation. This random component generally is specified to have mean zero and variance equal to the residual mean square from the fitted regression.

Such stochastic models, however, do not incorporate some major sources of variation in forest growth and yield. In particular, temporal variation and spatial variation may be present in ways not accounted for in typical models. Temporal variation can be present in the system because of differences between past and future climatic and other environmental variables. Spatial variation is generally not accounted for in growth and yield models but is frequently present in the system.

A recent development in modeling methodology worthy of mention is the use of stochastic differential equations. García (1983, 1984) has developed a comprehensive growth and yield system in which the components are represented as stochastic differential equations. The sophisticated mathematical and statistical techniques required in the stochastic differential equation approach have limited past applications, but the efforts of García, Sloboda (1984), and others should help to pave the way for future work.

References

Alemdag, I. S. (1978). Evaluation of some competition indexes for the prediction of diameter increment in planted white spruce. Canad. Forest Service Forest Management Inst. Inf. Rep. 108.

Amateis, R. L. and H. E. Burkhart (1985). Site index curves for loblolly pine plantations on cutover site-prepared lands. *South. J. Appl. Forest.* **9**, 166–169.

Amateis, R. L. and H. E. Burkhart (1987). Cubic-foot volume equations for loblolly pine trees in cutover, site-prepared plantations. *South. J. Appl. Forest.* **11**, 190–192.

Amateis, R. L., H. E. Burkhart and T. A. Walsh (1989). Diameter increment and survival equations for loblolly pine trees growing in thinned and unthinned plantations on cutover, site-prepared lands. *South. J. Appl. Forest.* **13**, 170–174.

Amemiya, T. (1985). *Advanced Econometrics.* Harvard Univ. Press, Cambridge, MA.

Avery, T. E. and H. E. Burkhart (1994). *Forest Measurements.* 4th ed., McGraw-Hill, New York.

Bailey, R. L. and J. L. Clutter (1974). Base-age invariant polymorphic site curves. *Forest Sci.* **20**, 155–159.

Beck, D. E. and L. Della-Bianca (1972). Growth and yield of thinned yellow-poplar. U.S.D.A. Forest Service, Research Paper SE-101.

Beck, D. E. and K. B. Trousdell (1973). Site index: Accuracy of prediction. U.S.D.A. Forest Service, Research Paper SE-108.

Behre, C. E. (1923). Preliminary notes on the study of tree form. *J. Forest.* **21**, 507–511.

Behre, C. E. (1927). Form-class taper curves and volume tables and their applications. *J. Agric. Res.* **35**, 673–744.

Belcher, D. W., M. R. Holdaway and G. J. Brand (1982). A description of STEMS – The stand and tree evaluation and modeling system. U.S.D.A. Forest Service, Gen. Tech. Report NC-79.

Bennett, F. A. (1970). Variable-density yield tables for managed stands of natural slash pine. U.S.D.A. Forest Service, Research Note SE-141.

Bennett, F. A. and J. L. Clutter (1968). Multiple-product yield estimates for unthinned slash pine plantations – pulpwood, sawtimber, gum. U.S.D.A. Forest Service, Research Paper SE–35.

Biging, G. S. (1985). Improved estimates of site index curves using a varying parameter model. *Forest Sci.* **31**, 248–259.

Borders, B. E. and R. L. Bailey (1986). A compatible system of growth and yield equations for slash pine fitted with restricted three-stage least squares. *Forest Sci.* **32**, 185–201.

Bredenkamp, B. V. and H. E. Burkhart (1990). An examination of spacing indices for *Eucalyptus grandis*. *Canad. J. Forest Res.* **20**, 1909–1916.

Buckman, R. E. (1962). Growth and yield of red pine in Minnesota. U.S.D.A. Forest Service, Tech. Bull. 1272.

Burk, T. E., R. P. Hans and E. H. Wharton (1989). Individual tree volume equations for the northeastern United States: Evaluation and new form quotient board foot equations. *North. J. Appl. Forest.* **6**, 27–31.

Burkhart, H. E. (1986). Fitting analytically related models to forestry data. In: *Proc. Invited Papers, 13th Internat. Biometric Conf.*, Seattle, WA.

Burkhart, H. E. (1977). Cubic-foot volume of loblolly pine to any merchantable top limit. *South. J. Appl. Forest.* **1**, 7–9.

Burkhart, H. E., K. D. Farrar, R. L. Amateis and R. F. Daniels (1987). Simulation of individual tree growth and stand development in loblolly pine plantations on cutover, site-prepared areas. School of Forestry and Wildlife Resources, Virginia Polytechnic Institute and State University, Publ. FWS-1–87.

Burkhart, H. E. and P. T. Sprinz (1984). Compatible cubic volume and basal area projection equations for thinned old-field loblolly pine plantations. *Forest Sci.* **30**, 86–93.

Burkhart, H. E. and M. R. Strub (1974). A model for simulation of planted loblolly pine stands. In: J. Fries, ed., *Growth Models for Tree and Stand Simulation.* Royal College of Forestry, Stockholm, Sweden, 128–135.

Cao, Q. V., H. E. Burkhart, R. C. Lemin (1982). Diameter distributions and yields of thinned loblolly pine planatations. School of Forestry and Wildlife Resources, Virginia Polytechnic Institute and State University, Publ. FWS-1–82.

Carmean, W. H. (1975). Forest site quality evaluation in the United States. *Adv. in Agron.* **27**, 209–269.

Carroll, R. J. (1989). Covariance analysis in generalized linear measurement error models. *Statist. Medicine* **8**, 1075–1093.

Clutter, J. L. (1963). Compatible growth and yield models for loblolly pine. *Forest Sci.* **9**, 354–371.

Clutter, J. L., J. C. Fortson, L. V. Pienaar, G. H. Brister and R. L. Bailey (1983). *Timber Management: A Quantitative Approach.* Wiley, New York.

Cunia, T. (1964). Weighted least squares method and construction of volume tables. *Forest Sci.* **10**, 180–191.

Curtis, R. O. (1970). Stand density measures: an interpretation. *Forest Sci.* **16**, 403–414.

Daniels, R. F., H. E. Burkhart and T. R. Clason (1986). A comparison of competition measures for predicting growth of loblolly pine trees. *Canad. J. Forest. Res.* **16**, 1230–1237.

Davidian, M. and R. J. Carroll (1987). Variance function estimation. *J. Amer. Statist. Assoc.* **82**, 1079–1091.

Dell, T. R., D. P. Feduccia, T. E. Campbell, W. F. Mann and B. H. Polmer (1979). Yields of unthinned slash pine plantations on cutover sites in the West Gulf Region. U.S.D.A. Forest Service, Res. Paper SO-147.

Dennis, B., B. E. Brown, A. R. Stage, H. E. Burkhart and S. Clark (1985). Problems of modeling growth and yield of renewable resources. *Amer. Statist.* **39**, 374–383.

Devan, J. S. and H. E. Burkhart (1982). Polymorphic site index equations for loblolly pine based on a segmented polynomial differential model. *Forest Sci.* **28**, 544–555.

DeVries, P. (1986). *Sampling Theory for Forest Inventory*. Springer, Berlin.

Diggle, P. J. (1988). An approach to the analysis of repeated measurements. *Biometrics* **44**, 959–971.

Doolittle, W. T. (1958). Forest soil-site relationships and species comparisons in the Southern Appalachians. *Soil Sci. Soc. Amer. Proc.* **22**, 455–458.

Ek, A. R. and R. A. Monserud (1974). FOREST: A computer model for simulating the growth and reproduction of mixed species forest stands. University of Wisconsin School of Natural Resources, Res. Report R2635.

Farrar, R. M. (1979). Growth and yield predictions for thinned stands of even-aged natural longleaf pine. U.S.D.A. Forest Service, Research Paper SO-156.

Forslund, R. R. (1982). A geometrical tree volume model based on the location of the centre of gravity of the bole. *Canad. J. Forest. Res.* **12**, 215–221.

Fries, J. and B. Matérn (1966). On the use of multivariate methods for the construction of tree taper curves. Advisory Group of Forest Statisticians of the IUFRO, Sect. 28, 2nd Conf. Proc., Stockholm, Sweden, 27 Sept.–1 Oct. 1965. *Rapporter Uppsatser Inst. Skolig Mat. Stat. Skogshögskolan* **9**, 85–117.

Frothingham, E. H. (1918). Height growth as a key to site. *J. Forest.* **16**, 754–760.

Fuller, W. A. (1987). *Measurement Error Models*. Wiley, New York.

Furnival, G. M. and R. W. Wilson (1971). Systems of equations for predicting forest growth and yield. In: G. P. Patil, E. C. Pielou and W. E. Waters, eds., *Statistical Ecology*, Vol. 3. Penn. Stat. Univ. Press, University Park, PA.

Gallant, A. R. and W. A. Fuller (1973). Fitting segmented polynomial regression models whose join points have to be estimated. *J. Amer. Statist. Assoc.* **82**, 144–147.

García, O. (1983). A stochastic differential equation model for the height growth of forest stands. *Biometrics* **39**, 1059–1072.

García, O. (1984). New class of growth models for even-aged stands: *Pinus radiata* in Golden Downs Forest. *New Zealand J. Forest. Sci.* **14**, 65–88.

Gedney, D. R. and F. A. Johnson (1959). Weighting factors for computing the relation between tree volume and D.B.H. in the Pacific Northwest. U.S.D.A. Forest Service, Research Note PNW 174.

Gerrard, D. J. (1989). Competition quotient: A new measure of the competition affecting individual forest trees. *Mich. Agric. Expt. Sta. Res. Bull.* **20**, 1–30.

Gertner, G. (1986). Postcalibration sensitivity procedure for regressor variable errors. *Canad. J. Forest. Res.* **16**, 1120–1123.

Goldberger, A. A. (1962). Best linear unbiased prediction in the generalized linear regression model. *J. Amer. Statist. Assoc.* **57**, 369–375.

Gregoire, T. G. (1987). Generalized error structure for forestry yield models fitted with permanent plot data. *Forest Sci.* **33**, 423–444.

Gregoire, T. G. and M. E. Dyer (1989). Model fitting under patterned heterogeneity of variance. *Forest Sci.* **35**, 105–125.

Gregoire, T. G., S. M. Zedaker and N. S. Nicholas (1990). Modeling relative error in stem basal area estimates. *Canad. J. Forest. Res.* **20**, 496–502.

Gregoire, T. G. and M. Williams (1992). Identifying and evaluating the components of non-measurement error in the application of standard volume equations. *Statistician* **41**, 509–518.

Grey, D. C. (1989). Site index – A review. *South. Afric. J. Forest.* **148**, 28–32.

Hägglund, B. (1981). Evaluation of forest site productivity. *Forest. Abstracts* **42**, 515–527.

Harrison, W. C., T. E. Burk and D. E. Beck (1986). Individual tree basal area increment and total height equations for Appalachian mixed hardwoods after thinning. *South. J. Appl. Forest.* **10**, 99–104.

Harvey, A. C. (1976). Estimating regression models with multiplicative heteroscedasticity. *Econometrica* **44**, 461–465.

Hegyi, F. (1974). A simulation model for managing jack-pine stands. In: E. J. Fries, ed., *Growth Models for Tree and Stand Simulation.* Royal College of Forestry, Stockholm, Sweden, 74–90.

Höjer, A. G. (1903). Tallens och granens tillväxt (Growth of Scots pine and Norway spruce). Stockholm.

Honer, T. G. (1965). A new total tree cubic foot volume function. *Forest. Chron.* **41**, 476–493.

Hyink, D. M. and J. W. Moser Jr. (1983). A generalized framework for projecting forest yield and stand structure using diameter distributions. *Forest Sci.* **29**, 85–95.

Johnson, E. W. (1973). Relationship between point density measurements and subsequent growth of southern pines. *Alabama Agric. Expt. Statist. Bull.* **447**, 1–109.

Jones, J. R. (1969). Review and comparison of site evaluation methods. U.S.D.A. Forest Service, Research Paper RM-51.

Judge, G. G., W. E. Griffiths, R. C. Hill, H. Lütkepohl and T.-C. Lee (1985). *The Theory and Practice of Economics.* 2nd ed., Wiley, New York.

Kilkki, P. and J. Lappi (1987). Estimation of taper curve using stand variables and sample tree measurements. *Scand. J. Forest Res.* **2**, 121–126.

King, J. E. (1966). Site index curves for Douglas-fir in the Northwest. Forestry Paper No. 8, Weyerhaeuser Company, Centralia, WA.

Kmenta, J. (1986). *Elements of Econometrics.* 2nd ed., MacMillan, New York.

Knoebel, B. R., H. E. Burkhart and D. E. Beck (1986). A growth and yield model for thinned stands of yellow-poplar. *Forest Sci. Monograph*, Vol. 27. Society of American Foresters, Washington, DC.

Korhonen, K. T. (1991). Using taper curve models based on mixed linear models in forest inventory. *Folia Forestalia* **774**, 1–27.

Kozak, A., D. D. Munro and J. H. G. Smith (1969). Taper functions and their application in forest inventory. *Forest. Chron.* **45**, 278–283.

Laasasenaho, J. (1982). Taper curve and volume functions for pine, spruce and birch. *Comm. Inst. Forest. Fennae* **108**, 1–74.

Lahtinen, A. and J. Laasasenaho (1979). On the construction of taper curves by using spline functions. *Comm. Inst. Forest. Fenniae* **98**, 1–63.

Lappi, J. (1986). Mixed linear models for analyzing and predicting stem form variation of Scots pine. *Comm. Inst. Forest. Fenniae* **134**, 1–69.

Lappi, J. and R. L. Bailey (1988). A height prediction model with random stand and tree parameters: An alternative to traditional site index methods. *Forest Sci.* **34**, 907–927.

Lappi, J. (1991). Calibration of height and volume equations with random parameters. *Forest Sci.* **37**, 781–801.

Lenhart, J. D., T. L. Hackett, C. J. Laman, T. J. Wiswell and J. A. Blackard (1987). Tree content and taper functions for loblolly and slash pine trees planted on non-old-fields in East Texas. *South. J. Appl. Forest.* **11**, 147–151.

Lloyd, F. T., H. D. Muse and W. L. Hafley (1982). A regression application for comparing growth potential of environments at different points in the growth cycle. *Biometrics* **38**, 479–484.

Loetsch, F., F. Zöhrer and K. E. Haller (1973). *Forest Inventory*, Vol. 2. BLV Verlagsgesellschaft, Munich.

Magnus, J. R. (1978). Maximum likelihood estimation of the GLS model with unknown parameters in the disturbance covariance matrix. *J. Econometrics* **7**, 281–312.

Martin, G. L. and A. R. Ek (1984). A comparison of competition measures and growth models for predicting plantation red pine diameter and height growth. *Forest Sci.* **30**, 731–743.

Matérn, B. (1990). On the shape of the cross-section of a tree stem: An empirical study of the geometry of mensurational methods. Swedish Univ. Agric. Sciences, Sect. Forest Biometry, S-901 83 Umeå, Sweden, Report 28.

Matney, T. G. and A. D. Sullivan (1982). Compatible stand and stock tables for thinned and unthinned loblolly pine stands. *Forest Sci.* **28**, 161–171.

Max, T. A. and H. E. Burkhart (1976). Segmented polynomial regression applied to taper equations. *Forest Sci.* **22**, 283–289.

McTague, J. P. (1992). Enahcned estimates of total volume with any single upper-stem measurement. *Forest. Ecol. Management* **48**, 55–67.

Meyer, H. A. (1953). *Forest Mensuration.* Penn. Valley, State College, PA.

Monserud, R. A. (1981). Estimating truncated tree volumes with the Behre hyperboloid and existing total volume equations. *Forest Sci.* **27**, 253–265.

Monserud, R. A. (1985). Comparison of Douglas-fir site index and height growth curves in the Northwest. *Canad. J. Forest. Res.* **15**, 673–679.

Murphy, P. A. and H. S. Sternitzke (1979). Growth and yield estimation for loblolly pine in the West Gulf. U.S.D.A. Forest Service, Research Paper SO-154.

Murphy, P. A. and R. C. Beltz (1981). Growth and yield of shortleaf pine in the West Gulf region. U.S.D.A. Forest Service, Research Paper SO-169.

Nelson, T. C. and E. V. Brender (1963). Comparison of stand density measures for loblolly pine cubic-foot growth prediction. *Forest Sci.* **9**, 8–14.

Newnham, R. M. and J. H. G. Smith (1964). Development and testing of stand models for Douglas-fir and lodgepole pine. *Forest. Chron.* **40**, 492–502.

Newnham, R. M. (1988). A variable-form taper equation. Petawawa National Forestry Institute Information Report PI-X-83, Chalk River, Ont.

Newnhman, R. M. (1988). A modification of the Ek–Payandeh nonlinear regression model for site index curves. *Canad. J. Forest. Res.* **18**, 115–120.

Noone, C. S. and J. F. Bell (1980). An evaluation of eight intertree competition indices. Oregon State Univ., Forest Research Laboratory, Research Note 66.

Olson, D. F., Jr. and L. Della-Bianca (1959). Site index comparisons for several tree species in the Virginia-Carolina Piedmont. U.S.D.A. Forest Service, Paper 104.

Omerod, D. P. (1973). A simple bole model. *Forest. Chron.* **49**, 136–138.

Ratkowsky, D. A. (1980). Statistical properties of alternative parametrizations of the von Bertalanffy growth curve. *Canad. J. Fish. Aquat. Sci.* **43**, 742–747.

Reineke, L. H. (1926). The determination of volume by planimeter. *J. Forest.* **24**, 183–189.

Reineke, L. H. (1933). Perfecting a stand-density index for even-aged forests. *J. Agric. Res.* **46**, 627–638.

Robinson, G. K. (1991). That BLUP is a good thing; the estimation of random effects. *Statist. Sci.* **6**, 15–51.

Ross, G. J. S. (1970). The efficient use of function minimisation in non-linear maximum-likelihood estimation. *Appl. Statist.* **19**, 205–221.

Roth, F. (1916). Concerning site. *Forest. Quart.* **16**, 749–753.

Sadiq, R. A. and V. G. Smith (1983). Estimation of individual tree volumes with age and diameter. *Canad. J. Forest. Res.* **13**, 32–39.

Schmid-Haas, P., P. Roiko-Jokela, P. Mingard and M. Zoberry (1971). The optimal determination of the volume of standing trees. *Mitteil. Forstl. Bundesv. Wien* **91**, 33–54.

Schreuder, H. T., W. L. Hafley and F. A. Bennett (1979). Yield prediction for unthinned natural slash pine stands. *Forest. Sci.* **25**, 25–30.

Schreuder, H. T., T. G. Gregoire and G. B. Wood (1993). *Sampling Methods for Multiresource Forest Inventory*. Wiley, New York.

Schumacher, F. X. (1939). A new growth curve and its application to timber yield studies. *J. Forest.* **37**, 819–820.

Schumacher, F. X. and F. S. Hall (1933). Logarithmic expression of timber-tree volume. *J. Agric. Res.* **47**, 719–734.

Sloboda, B. (1977). Spline und Bitterlich'sche Interpolation bei der Baumschaftform. *Mitteil. Forstl. Bundesv. Wien* **120**, 53–60.

Sloboda, B. (1984). Possibilities of mathematically predicting timber production in commercial forests. *Bull. Nagoya Univ. Forests* **7**, 261–279.

Smalley, G. W. and R. L. Bailey (1974). Yield tables and stand structure for loblolly pine plantations in Tennessee, alabama, and Georgia Highlands. U.S.D.A. Forest Service, Research Paper SO-96.

Solomon, D. S., T. D. Droessler and R. C. Lemin Jr. (1989). Segmented quadratic taper equations for spruce and fir in the Northeast. *North. J. Appl. Forest* **6**, 123–126.

Spurr, S. H. (1952). *Forest Inventory*. Ronald Press, New York.

Sullivan, A. D. and J. L. Clutter (1972). A simultaneous growth and yield model for loblolly pine. *Forest Sci.* **18**, 76–86.

Swamy, P. A. V. B. (1970). Efficient estimation in a random coefficient regression model. *Econometrica* **38**, 311–323.

Thomas, C. E. and B. R. Parresol (1991). Simple, flexible, trigonometric taper equations. *Canad. J. Forest. Res.* **21**, 1131–1137.

Tomé, M. and H. E. Burkhart (1989). Distance-dependent competition measures for predicting growth of individual trees. *Forest Sci.* **35**, 816–831.

Van Deusen, P. C. (1988). Simultaneous estimation with a squared error loss function. *Canad. J. Forest. Res.* **18**, 1093–1096.

Vardi, Y. (1988). Statistical models for intercepted data. *J. Amer. Statist. Assoc.* **83**, 193–197.

Von Bertalanffy, L. (1957). Quantitative laws in metabolism and growth. *Quart. Rev. Biol.* **32**, 217–231.

Walters, D. K., T. G. Gregoire and H. E. Burkhart (1989). Consistent estimation of site index curves fitted to temporary polt data. *Biometrics* **45**, 23–33.

Walters, D. K., H. E. Burkhart, M. R. Reynolds Jr. and T. G. Gregoire (1991). A Kalman filter approach to localizing height-age equations. *Forest. Sci.* **37**, 1526–1537.

West, P. W. (1983). Comparison of stand density measures in even-aged regrowth eucalypt forest of southern Tasmania. *Canad. J. Forest. Res.* **13**, 22–31.

Wright, W. G. (1923). Investigation to taper as a factor in measurement of standing timber. *J. Forest.* **21**, 569–581.

G. P. Patil and C. R. Rao, eds., *Handbook of Statistics, Vol. 12*

12

Ecological Diversity and Forest Management*

Jeffrey H. Gove, Ganapati P. Patil, Benee F. Swindel[†]
and Charles Taillie

Abstract: The diversity of an ecological community is defined in terms of the average species rarity of that community using both dichotomous- and rank-type rarity measures. Common diversity indices and profiles are developed using this definition and the concept of intrinsic diversity ordering is presented. The use of these diversity measures is illustrated in case studies involving the plant communities in two distinct forested ecosystems of the southeastern and northeastern United States. In these case studies, the objective is the comparative assessment of diversity changes over time and between treatments. Diversity profiles are further utilized in nonlinear mathematical programming models for uneven-aged stand management. The models present strategies for maximization and maintenance of structural diversity within uneven-aged stands. Finally, a principal components regression technique is presented which facilitates prediction of plant species diversity; the user need only be able to classify an individual into one of five life-forms – significantly reducing the taxonomic skills required for diversity assessment.

1. Introduction

Community diversity, diversity measurement, and diversity comparisons have been important in community ecology studies for more than four decades; with the passage of the National Forest Management Act (NFMA) of 1976 (Federal Register, 1982, 219.26, 219.27(g)) for the National Forest System of the US, diversity has

*This paper was prepared with partial support from the Statistical Analysis and Computing Branch, Environmental Statistics and Information Division, Office of Policy, Planning, and Evaluation, United States Environmental Protection Agency, Washington, DC under a Cooperative Agreement Number CR-815273, with Drs. Pepi Lacayo and Robert O' Brien as EPA Project Officers. The contents have not been subjected to Agency review and therefore do not necessarily reflect the views of the Agency and no official endorsement should be inferred.

become recognized as a critical area of research in forest management. Boyce and Cost (1978) were among the first to articulate the need to incorporate diversity considerations into the management of forests with multiple-use objectives; their work subsequently evolved into a detailed decision-making framework for multiple-benefit forestry management (Boyce, 1985). More recently, diversity research in forest management has focused on developing methodology for achieving specific mandates in the NFMA (Gove et al., 1992; Swindel and Grosenbaugh, 1988; Swindel et al., 1984, 1987), and on improving our understanding of the interplay among forest diversity, forest harvesting practices, and economic goals (Gove et al., 1991, Niese and Strong, 1992). Unfortunately, there is still a paucity of research on forest-related diversity and its significance in forest management. The 'state-of-the-art' studies reviewed in this chapter show that research and application in this subject area are still in their infancy.

Much has been made in recent years over the issue of biological diversity, or biodiversity, and its conservation. As McMinn (1991) notes, a distinction must be made between the terms 'biodiversity' and 'diversity'. *Biodiversity* is an omnibus concept that incorporates the variety of life in all its forms from organic molecules, to species, to communities, landscapes, and biomes (Hunter, 1990; McMinn, 1991; Office of Technology Assessment, 1987). *Diversity*, on the other hand, is a generic term referring to the condition of being different; a useful synonym for diversity is variety. It summarizes for qualitative characteristics (e.g., tree species) what the variance summarizes for quantitative measurements (e.g., tree diameter) (Pielou, 1975, p. 5).

In this chapter, diversity is looked upon as a community property and, accordingly, the trophic, taxonomic or structural levels that comprise the community must be specifically delimited. Most often, it is unrealistic to identify the components of a community as all the living organisms within a region. Instead, the limited class of organisms for which diversity assessment is desired needs to be clearly determined; in reference to this class, Pielou (1974, p. 289) has coined the term *taxocene*, which may be applied with equal validity to taxonomic groups, structural components, or even communities within communities (e.g., forest stands within a landscape mosaic).

This chapter is not concerned with biodiversity per se, except that the communities described in the studies presented here make up component parts of that all-inclusive concept. Biodiversity, as it relates to forest management, has been admirably covered by Hunter (1990). This chapter is concerned with elucidating several of the more recent studies dealing with the issues of *diversity* measurement and comparison for specifically defined *communities* in the context of forest management. Our intent is to facilitate 'technology transfer' and to encourage future research in this topic, with a view toward improved integration of diversity considerations into forest management decisions and policy analysis. To this end, the complete data sets for the communities discussed are reproduced in the appendices. Also, formulations and model specifications are often presented in fuller detail than in the original papers to allow complete reproducibility of the results.

In making diversity comparisons, it is not uncommon to find that diversity has increased according to one index, but decreased according to a second index. This merely reflects the fact that diversity is a complex multidimensional property of a community. To view diversity through the lens of a single index is to project that multidimensional complexity onto a one-dimensional ordinal scale with distorted perception and possibly misleading conclusions.

Section 2 confronts the multidimensional character of diversity by establishing an intrinsic, and index-free, diversity ordering. An implication of the preceding discussion is that a given pair of communities may not be comparable according to this intrinsic diversity ordering. In effect, diversity may appear to have decreased when viewed from one vantage point (i.e., index), and increased when viewed from a different perspective.

In view of the inadequacy of a single index, we quantify diversity by means of *diversity profiles*. A diversity profile is a curve depicting the simultaneous values of a large collection of diversity indices. Thus, the profile portrays the views of diversity from many different vantage points simultaneously and in a single picture.

Changes, or differences, in community diversity are studied by comparing profiles. If the two communities are intrinsically comparable, then one profile will lie uniformly above the other. Conversely, when the communities are not intrinsically comparable, their profiles may intersect. But even here, the profiles can reveal which portions of the community have undergone opposing diversity changes.

Section 3 discusses two case studies in which diversity profiles were used to determine temporal changes and site-to-site differences in plant diversity. The first study is concerned with a pair of slash pine watersheds of northern Florida. Both sites were clearcut, but were subjected to opposite extremes of silvicultural treatment after cutting. The profiles portray clearly the temporal trend in diversity at each site, and also reveal marked differences in the course of the diversity trends at the two sites.

The second study examines data from two watersheds in the Hubbard Brook experimental forest of New Hampshire. One site was block clearcut, while the other site was progressively stripcut. A notable feature of the study is that a jackknife procedure was used to adjust for a negative bias in estimated species diversity. This statistical treatment also allows one to assess the statistical significance of any diversity differences and diversity changes that are indicated by the profiles.

Section 4 takes up the issues of optimizing and of maintaining diversity in uneven-aged forest management. For concreteness, the discussion is in terms of diversity of the diameter-class distribution, but other forms of diversity could also be used. In uneven-aged forest management, there is the idealized notion of a 'target' stand: a forest, initially at the target stand, is allowed to grow and then is cut back to the target stand after a fixed number of years, known as the length of the cutting cycle (cc). The cycle then repeats itself with cuttings every cc years. The problem is to determine the target stand according to some optimality criterion. The growth equations also place limits on the permissible target stands, so it is a constrained optimization problem. Traditionally, the optimality criteria have been based upon economic considerations. Section 4 brings diversity considerations into this framework in two different ways. In a first scenario, the optimality criterion is diversity-based and the economic considerations are incorporated in the form of additional constraints. For the second scenario, the criterion is economic but the maintenance of diversity is imposed as an additional constraint.

The slash pine sites discussed in Section 3 involved large numbers of species (nearly 100) and required considerable taxonomic skill and effort to classify the organisms to the species level. Section 5 investigates whether it would be possible to eliminate much of this taxonomic effort while retaining most of the information needed for diversity assessment. The different species are collapsed into five life-form categories of which two are overstory ('trees' and 'shrubs') and three are understory. These categories are such that field-classification is quick and dependable. The analysis of Section 5 demonstrates that the *life-form* abundances are able to account for most of the variability in *species* diversity, suggesting that diversity-related management decisions might be based upon the more accessible life-form data.

2. Measuring diversity – An overview

This section presents the background material needed to understand the different diversity measures used in the applications throughout the rest of the chapter. The literature on diversity measurement is immense with numerous indices and other methods available for use in measuring diversity. For this reason, only the measures encountered in the rest of the chapter are reviewed – no attempt has been made at being encyclopedic. In addition, problems with misapplication of diversity measures evidently are commonplace in the literature. Other sources such as Grassle et al. (1979), Hurlbert (1971), Magurran (1988) and Pielou (1974, 1975, 1977) discuss many of these other measures and their correct use; so lengthy and accessible is the treatment of this subject by these authors that it need not be repeated here.

The semantics of a few terms used throughout this chapter need to be addressed straightway. The term *species* as used when referring to diversity is simply a convenient label for a set of distinct and countable categories comprising the community; therefore, 'species' could mean a set of vegetative height strata (e.g., 0–20′, 21–40′, 41′+), forest stands, or true biological species. The *community*, then, is used in an equally general sense to mean a recognizable collection of species in a space–time continuum.

2.1. *Average species rarity*

The diversity of a community consists of two components: the number of species, or species richness, and the evenness, or equitability of species as given by their relative abundance. Let s be the species richness for a community, and define the relative abundance vector for this community as $\boldsymbol{\pi} = (\pi_1, \ldots, \pi_s)$, such that $\sum_{i=1}^{s} \pi_i = 1$. The relative abundance vector may be computed from any one of a number of different measures such as number of individuals, percent cover, or biomass per unit area. The quantity used will depend on the definition of species (i.e., the composition of the s categories of interest) and there is, as of yet, no consensus on which measure to use in a given situation (Pielou, 1974; Magurran, 1988). A conceptual community may be written as $C(\mu)$, where μ may be some function of s and $\boldsymbol{\pi}$, or μ may have some space–time connotation. The *completely even community* is a special case where $\pi_i = 1/s = \pi_E$ for all i, so that s alone determines the relative abundance vector. The completely even community is denoted as $C_E(s)$.

Community diversity is defined here as the average species rarity within the community. If a numerical measure of rarity can be associated with each species in the community, then the community diversity can be determined as its expected value. Thus, if the rarity of the i-th species in community $C(\boldsymbol{\pi})$ is given as $R(i; \boldsymbol{\pi})$, then

$$\Delta = \Delta(\boldsymbol{\pi}) = \sum_{i=1}^{s} R(i; \boldsymbol{\pi})\pi_i \tag{2.1}$$

is the community diversity. The choice of how to assign a rarity value to each individual species will determine the diversity measure.

Patil and Taillie (1979a, 1982) discuss two types of rarity measures: *dichotomous-type* and *rank-type*. With the dichotomous-type rarity measures, the rarity of the i-th species depends only on its own relative abundance π_i. In this case, rarity is denoted $R(\pi_i)$, and the measure of community diversity is defined as $\Delta = \sum_{i=1}^{s} R(\pi_i)\pi_i$. For rank-type measures, however, species rarity depends only on its rank. Consider the *ranked* relative abundance vector, $\boldsymbol{\pi}^{\#}$, which is determined by ranking the components of $\boldsymbol{\pi}$ from greatest to least so that $\boldsymbol{\pi}^{\#} = (\pi_1^{\#}, \ldots, \pi_s^{\#})$, where $\pi_1^{\#} \geqslant \cdots \geqslant \pi_s^{\#}$. Then if the rarity of the i-th-ranked species is denoted as $R(i)$, community diversity is given as $\Delta = \sum_{i=1}^{s} R(i)\pi_i^{\#}$. Examples of both types of rarity measures and their associated diversity functions are given in the following sections.

2.2. *Diversity indices*

Several of the most frequently used diversity indices in forest management may be conveniently expressed under the umbrella of average species rarity through judicious choice of rarity functions. Species richness, species count, Shannon's, and Simpson's indices all may be derived from this theory as follows

$$\Delta_{SR} = \sum_{i=1}^{s} \left\{\frac{1}{\pi_i}\right\} \pi_i = s \qquad \text{species richness,} \tag{2.2}$$

$$\Delta_{SC} = \sum_{i=1}^{s} \left\{\frac{1}{\pi_i} - 1\right\} \pi_i = s - 1 \qquad \text{species count,} \tag{2.3}$$

$$\Delta_{Sh} = \sum_{i=1}^{s} \{-\log \pi_i\} \pi_i = -\sum_{i=1}^{s} \pi_i \log \pi_i \qquad \text{Shannon,} \tag{2.4}$$

$$\Delta_{Si} = \sum_{i=1}^{s} \{1 - \pi_i\} \pi_i = 1 - \sum_{i=1}^{s} \pi_i^2 \qquad \text{Simpson,} \tag{2.5}$$

where the term in brackets denotes the species rarity function used in each case.

Table 1
Three hypothetical forest stand communities composed of five or fewer species of trees

Species	Stand		
	1	2	3
	Relative abundance		
Pinus strobus	0.50	0.25	0.35
Quercus rubra	0.30	0.25	0.35
Tsuga canadensis	0.10	0.25	0.30
Acer rubrum	0.05	0.25	0.00
Betula papyrifera	0.05	0.00	0.00
Total	1.00	1.00	1.00
	Diversity index		
Δ_{SR}	5	4	3
Δ_{SC}	4	3	2
Δ_{Sh}	1.24	1.39	1.10
Δ_{Si}	0.65	0.75	0.67

Table 1 presents a hypothetical example of three forest stands composed of just five or fewer species of trees. The relative abundances of these tree species based on some quantitative measure of abundance are given, and the diversity indices (2.2) through (2.5) calculated from these relative abundances also are shown for each community. The example clearly shows the inconsistency of the different indices in their ranking of these three communities. For example, $\Delta_{\mathrm{SC}}(\text{stand 1}) > \Delta_{\mathrm{SC}}(\text{stand 2})$, but $\Delta_{\mathrm{Sh}}(\text{stand 1}) < \Delta_{\mathrm{Sh}}(\text{stand 2})$ and $\Delta_{\mathrm{Si}}(\text{stand 1}) < \Delta_{\mathrm{Si}}(\text{stand 2})$. This is an interesting comparison because it illustrates how one may be led to the conclusion that a community with fewer species (stand 2) can be more diverse than one with more species (stand 1) using either Shannon's or Simpson's index. Similar inconsistencies among the indices may be found by comparing stands 1 and 3. The only comparison that is consistently ordered with all indices is $\Delta(\text{stand 2}) > \Delta(\text{stand 3})$. This inconsistency of different diversity indices evidently is quite common when making comparisons between communities (Hurlbert, 1971) and arises from a lack of intrinsic diversity ordering between the communities being compared (Section 2.3).

2.3. *Diversity profiles*

Diversity profiles allow the graphical comparison of diversity between communities. One set of profiles that incorporates indices (2.3) through (2.5) as point estimates along the curve are the so-called Δ_β profiles of Patil and Taillie (1979a, 1982). Since the Δ_β profile incorporates indices developed from dichotomous-type rarity measures, it too may be developed in the same manner:

$$\Delta_\beta = \sum_{i=1}^{s} \left\{ \frac{(1 - \pi_i^\beta)}{\beta} \right\} \pi_i = \frac{1 - \sum_{i=1}^{s} \pi_i^{\beta+1}}{\beta}, \quad \beta \geqslant -1. \tag{2.6}$$

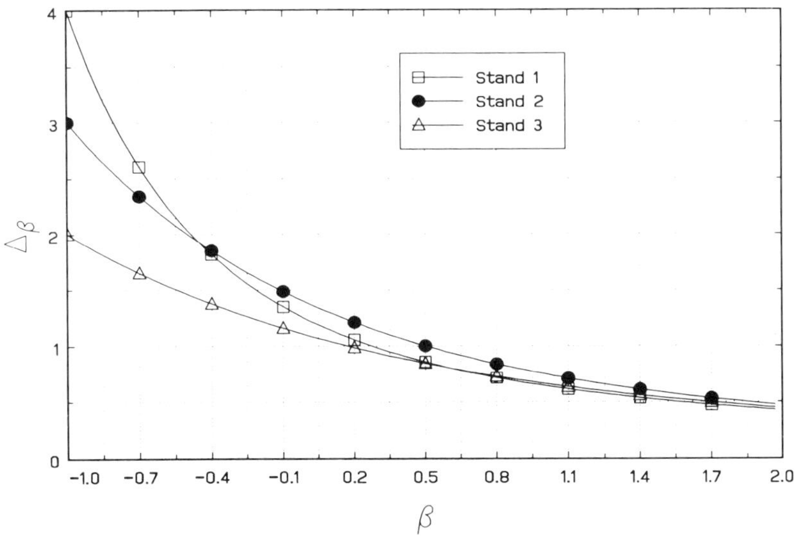

Fig. 1. Δ_β profiles for the three hypothetical forest stand communities in Table 1.

The restriction that $\beta \geqslant -1$ assures that Δ_β has certain desirable properties (Patil and Taillie, 1979a, 1982). The species count, Shannon and Simpson indices are related to Δ_β by $\Delta_{SC} = \Delta_{-1}$, $\Delta_{Sh} = \Delta_0$, and $\Delta_{Si} = \Delta_1$.

The Δ_β diversity profiles for the three stands in Table 1 are presented in Figure 1. Note that the profile for stand 1 crosses both profiles for stands 2 and 3. The profile for stand 1 crosses that of stand 2 at $\beta = -0.45$, which explains why both Δ_{Sh} and Δ_{Si} rank diversity of these two communities differently from Δ_{SC}. On the other hand, the profiles for stands 1 and 3 cross at $\beta = 0.62$ showing how Δ_{SC} and Δ_{Sh} rank these two communities differently from Δ_{Si}. In general, it also is possible for two Δ_β profiles to cross at $\beta > 1$ or for them to cross more than once; in either case, even calculating all three indices (Δ_{SC}, Δ_{Sh}, and Δ_{Si}) alone may not be enough to show the inconsistent ranking of communities at larger β. Calculating and plotting Δ_β profiles for $\beta > 1$ may not be helpful either because the profiles tend to converge quickly beyond this point and intersections do not resolve – an algorithm for numerically finding the intersections of any two Δ_β profiles is required in this case.

Perhaps the most useful way to compare diversity between communities is by the concept of *intrinsic diversity ordering*. This concept may be defined as follows (Patil and Taillie, 1982):

> Community C' is *intrinsically more diverse* than community C (written $C' \overset{\text{imd}}{>} C$) provided C leads to C' by a finite sequence of (1) introducing a species, (2) transferring abundance from more to less abundant species without reversing the rank-order of the species, and (3) relabeling species (i.e., permuting the components of the abundance vector).

Patil and Taillie (1982) note that this ordering is only partial and, like Solomon (1979), they stress that two given communities need not be intrinsically comparable.

A diversity profile approach has been developed by Patil and Taillie (1979a, b, 1982) using a rank-type rarity measure of $\pi^{\#}$ that incorporates the concepts of intrinsic diversity ordering defined above. Let

$$R(i) = \begin{cases} 1 & \text{if } i > j, \\ 0 & \text{if } i \leqslant j, \end{cases}$$

for $1 \leqslant j \leqslant s$. Then average species rarity is given as

$$T_j = \sum_{i=j+1}^{s} \pi_i^{\#}, \quad j = 1, \ldots, s-1, \tag{2.7}$$

where $T_s = 0$ and $T_0 = 1$. The quantity in (2.7) is termed the right tail-sum of the ranked relative abundance vector $\pi^{\#}$, and when a plot of the (j, T_j) pairs is constructed for each community, the resulting profiles are termed *intrinsic diversity profiles* (Patil and Taillie, 1979b). Any intrinsic orderings of the communities, if they exist, can be determined with the intrinsic diversity (T_j) profiles.

The right tail-sum profiles for the three stands in Table 1 are plotted in Figure 2. Notice that the profile for stand 1 crosses both those for stands 2 and 3, but that the profile for stand 2 is everywhere above that for stand 3. It follows that the only intrinsic diversity ordering for these stands is $C(\text{stand } 2) \overset{\text{imd}}{>} C(\text{stand } 3)$. This is consistent with the findings of the indices in Section 2.2 and the Δ_β profiles. The Δ_β profiles are isotonic to intrinsic diversity ordering in that, if an intrinsic diversity

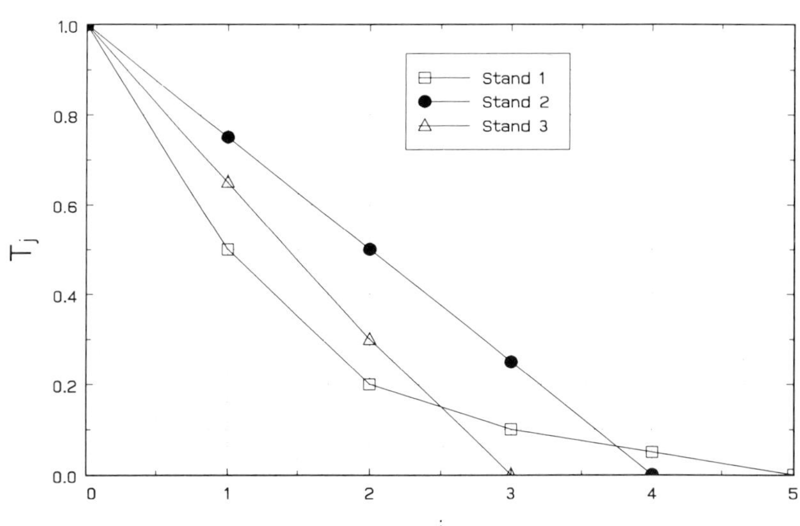

Fig. 2. Right tail-sum (T_j) profiles for the three hypothetical forest stand communities in Table 1.

ordering exists, they will preserve it. However, the Δ_β profiles may not cross even if the T_j profiles do; therefore, the Δ_β profiles do not necessarily reflect intrinsic diversity ordering. Since the diversity indices discussed have the same properties as the Δ_β profiles, it should be emphasized that, of the methods presented thus far, the T_j profiles are the most reliable measure of intrinsic diversity ordering between communities.

Swindel et al. (1987) presented another profile approach that is based on the intrinsic diversity (T_j) profiles. These profiles are termed *comparative diversity profiles* by Swindel et al. (1987) and use the *left* tail-sums of the ranked relative abundance vector $\boldsymbol{\pi}^\#$; viz.,

$$L_j = \sum_{i=1}^{j} \pi_i^\#, \quad j = 1, \ldots, s-1, \tag{2.8}$$

where $L_0 = 0$ and $L_s = 1$. Instead of plotting the (j, L_j) pairs as in the T_j profiles, pairs of left tail-sums are plotted for community comparisons. That is, given two communities C and C' with ranked relative abundance vectors $\boldsymbol{\pi}^\#$ and $\boldsymbol{\pi}^{\#'}$, the points

$$\begin{aligned} p_0 &= (L_0, L_0') = (0, 0), \\ p_1 &= (L_1, L_1') = (\pi_1^\#, \pi_1^{\#'}), \\ p_2 &= (L_2, L_2') = (L_1 + \pi_2^\#, L_1' + \pi_2^{\#'}), \\ &\vdots \\ p_k &= (L_k, L_k') = (L_{k-1} + \pi_k^\#, L_{k-1}' + \pi_k^{\#'}), \\ &\vdots \\ p_z &= (L_z, L_z') = (1, 1), \end{aligned} \tag{2.9}$$

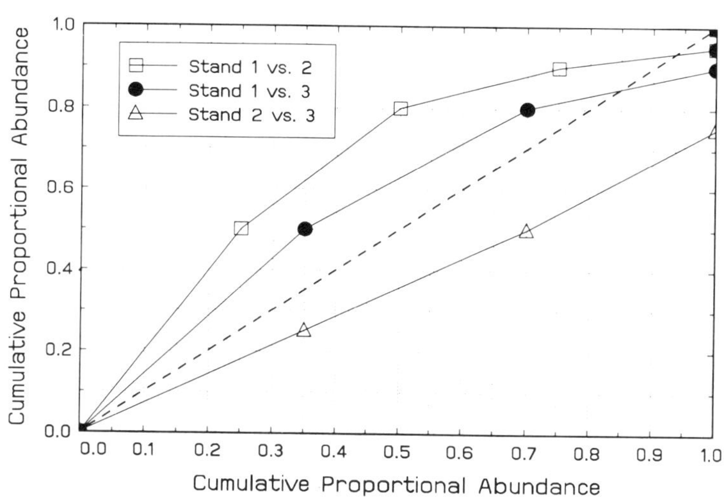

Fig. 3. Comparative (L_j) diversity profiles for the three hypothetical forest stand communities in Table 1.

diversity has not progressed as rapidly or as far on WS 2 as the decline on WS 1 described above.

3.1.7. *Direct comparisons between watersheds*

Figure 6 gives comparative diversity profiles for the second growth and the plantation at ages 1, 5 and 7 years on WS 1 (y-axis) against stands of similar ages on WS 2 (x-axis). These profiles show that the second growth on WS 1 was intrinsically more diverse than on WS 2. At plantation age 1 year, this difference was maintained and much increased as evidenced by the profile being a greater distance from the diagonal. By age 5 years, WS 1 was only more diverse in the tail; that ordering persisted at age 7 years, but the profile is approaching the diagonal, indicating converging diversity on the two watersheds. Widely disparate regeneration regimes, therefore, produced 7-year-old plantations of similar diversity in these sites.

3.1.8. *Discussion*

The results described in this section are index-free. Thus, the conclusion that the young (ages 1–5 years on WS 1 and ages 3–5 years on WS 2) slash pine plantations on the two watersheds described were more diverse than the antecedent second growth would be confirmed by any computed univariate diversity index (Patil and Taillie, 1982). In addition to the intrinsic orderings established by Figures 4–6, comparative diversity profiles display other qualitative and quantitative information about community composition, including contributions of relatively rare species, about successional trends, and about treatment effects on static and dynamic characteristics of plant communities. Techniques for gleaning such infor-

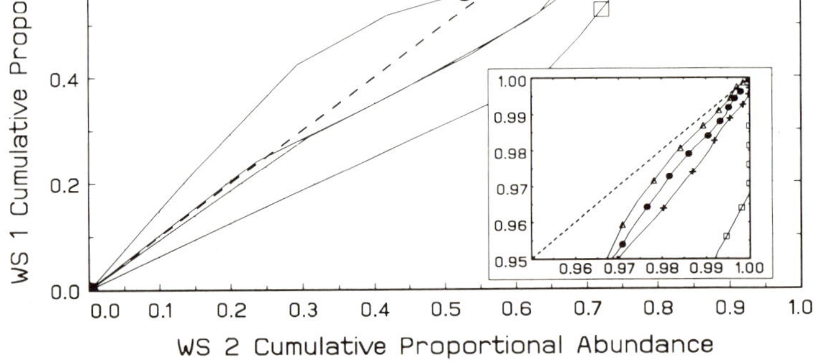

Fig. 6. Comparative diversity profiles for WS 1 (y-axis) versus WS 2 (x-axis) for the antecedent second-growth and plantation communities at 1, 5 and 7 years using percent crown cover data.

mation from comparative diversity profiles, including examination of large-scale graphs of only the tails of profiles have been described elsewhere (Swindel et al., 1987; Swindel and Grosenbaugh, 1988). Here we wish to mention just one other property of comparative diversity profiles. A profile that plots always on one side of the diagonal $y = x$ establishes an intrinsic ordering between two communities, as we have seen. Moreover, two profiles that do not cross each other establish an additional intrinsic diversity ordering. Thus, the plantation on WS 2 was intrinsically less diverse at age 7 years than at age 5 (Figure 5). Similarly, crossing profiles can establish orderings in the tail. The plantation on WS 1 was less diverse in the tail at age 7 years than at age 5 (Figure 4).

3.2. Northern hardwoods, Hubbard Brook, New Hampshire

Gove et al. (1992) presents a study of plant species diversity using the concept of intrinsic diversity ordering (Patil and Taillie, 1979a, 1982) to compare pre- and posttreatment plant communities on two watersheds within the northern hardwood ecosystem. The treatments on these watersheds were initiated in 1970 and the plant data used had already been collected to meet other objectives (Martin and Hornbeck, 1989). These initial objectives included determining the impacts of even-aged harvests on hydrologic and nutrient cycles, and the composition of tree species in the regenerating forest. The diversity comparisons on the watersheds presented by Gove et al. (1992) laid the foundation for long-term monitoring of diversity in response to temporally changing environmental and ecological conditions on the treated watersheds. This objective is consistent with the overall goal of the long-term ecological research at the Hubbard Brook experimental forest; that is, to develop a better understanding of the response of northern hardwood ecosystems to large-scale disturbance.

In this section, the details of the analytical methods used to make diversity comparisons in the Hubbard Brook study are presented. These comparisons, while similar to those of Section 3.1, differ in that a nonparametric jackknifing technique is used in an attempt to correct for possible biases that may occur when sample-based community abundance vectors are simply substituted in place of the unknown population vectors. The jackknifing procedure also allows simple hypotheses to be tested among communities so that inferences can be made about intrinisic diversity orderings.

3.2.1 Study site

Two watersheds located at Hubbard Brook experimential forest within the White Mountain National Forest in central New Hampshire were used for the study. Watershed 101 is a 12 ha block clearcut that ranges in elevation from 470 to 595 m. Watershed 4 is a 36 ha progressive strip cut that ranges in elevation from 440 to 730 m. Both watersheds have southerly aspects and slopes of 20 to 30%. Other details including pre-harvest forest conditions and soil characteristics are given elsewhere (Martin and Hornbeck, 1989).

In preparation for harvest, watershed 4 was surveyed into 49 strips, each 25 m wide, oriented from east to west roughly parallel to the watershed contour. Harvest began in the autumn of 1970 by cutting every third 25 m strip. After a 2-year period for establishment of regeneration, a second series of 25 m strips was cut on the southernmost side of the original cuts. The remaining strips were then cut after a second establishment period of 2 years, completing the harvest over a 4-year interval. After the first strip was cut, it received shade from the trees on its southerly border, providing favorable conditions for germination and early survival. When the second strip was cut, it received shade for initial establishment, but also allowed the first strip to receive direct sunlight for growth. The block clearcut on watershed 101 was carried out in a single operation immediately upon completion of the first strip cut in late autumn of 1970.

On both the strip cut and block clearcut, trees and snags whose diameter at breast height (dbh) was > 2.5 cm were felled and commercial portions removed. Scarification was encouraged during logging to provide exposed mineral soil for seedling establishment. Regeneration on the block clearcut and strip cut has been surveyed on permanent plots at 1- to 4-year intervals since harvest. On the strip cut, 57 25 × 25 m plots were selected. These plots were cross-classified by year of cut and elevation and were randomly allocated. The plot allocation produced 19 plots for each year of cut, and 19 plots at low elevation (440–550 m), 18 at mid-elevation (550–650 m) and 20 at high elevation (650–730 m). On the block clearcut, 48 25 × 25 m plots were chosen at random with no stratification since the harvest was completed in a single operation and the elevation range of the entire harvest area was about the same as for the mid-elevation plots on the strip cut. Surveys were conducted before cutting on each of the strips, but not on the block clearcut (Table 2).

On each watershed the entire 25 × 25 m plot was used to sample trees whose dbh was ⩾ 10 cm. Trees between 2 and 10 cm dbh were tallied on a 1 × 25 m strip plot running along either the east or west border (decision by coin toss) of each plot. Saplings measured on the 1 × 25 m nested strip plots before cutting were considered to be advanced regeneration. There were a total of 57 1 × 25 m strips on watershed 4 and 48 on watershed 101. Lesser vegetation was surveyed on four 1 m^2 subplots spaced evenly along each strip plot, for a total of 228 subplots on watershed 4 and 192 on watershed 101. In addition, stem counts were made by species for trees, shrubs, and herbs in two height classes, < 0.5 m and > 0.5 m tall. Data from the

Table 2
Pre- and posttreatment community measurement dates in the two watersheds

Treatment		Precut	1 year post	10 year post
WS 4	Strip 1	1970	1971	1980
	Strip 2	1972	1973	1982
	Strip 3	1974	1975	1984
WS 101	All	–[1]	1971	1980

[1] No data was available in this period.

subplots were averaged to give stems m^{-2} for each 1×25 m strip plot. Individual stems of all species of trees, shrubs, and herbs were counted on each subplot except for sedges, grasses, liverworts, rushes, clubmosses, mosses, oxalis, and spikenard, which were inventoried as percent cover and were not included in this analysis.

3.2.2. Analysis methods

Diversity indices such as Shannon's index and diversity profiles such as T_j and Δ_β were developed for use when the $\boldsymbol{\pi}$ have been calculated from complete enumeration of the population. Patil and Taillie (1979b) note that 'most diversity indices, when evaluated on a sample, consistently underestimate community diversity by an amount varying with sample size'. They presented a jackknifing method that may help to correct for this bias and, in addition, provide estimates of precision that can be used for hypothesis testing. In particular, the jackknifing method was used to test whether intersections that occur in the T_j profiles are real or simply a sampling effect. Patil and Taillie's (1979b) jackknife approach is used on all Δ_β profiles for bias correction and intersection testing. The procedure is summarized as follows:

(1) Calculate and plot the T_j profiles with all plots included for the community of interest and identify any intersections.

(2) Calculate and plot the Δ_β profiles with all plots included.

(3) Jackknife the Δ_β profiles for each community as follows:

(i) Calculate $\Delta_\beta^{(-i)}$, the Δ_β profile with the i-th plot data omitted.

(ii) Calculate a pseudo value as given in (3.1) at each β,

$$\psi_\beta^{(-i)} = n\Delta_\beta - (n-1)\,\Delta_\beta^{(-i)}. \tag{3.1}$$

(iii) Repeat steps (i) and (ii) n times, where n is the number of plots.

(iv) Calculate the average of the pseudo values and the estimated variance at each β,

$$\hat{\Delta}_\beta = \frac{1}{n}\sum_{i=1}^{n} \psi_\beta^{(-i)}, \tag{3.2}$$

$$\widehat{\mathrm{var}}(\hat{\Delta}_\beta) = \frac{1}{n(n-1)}\sum_{i=1}^{n} (\psi_\beta^{(-i)} - \hat{\Delta}_\beta)^2. \tag{3.3}$$

(4) Plot the jackknifed $\hat{\Delta}_\beta$ profiles for each community.

In implementing the above procedure, $m = 100$ evenly spaced β values in the range $-1 \leqslant \beta \leqslant 2$ were generated in (3) and (4) for each profile. The parameter β was allowed to range to two since some intersections were identified in the range $+1 \leqslant \beta \leqslant 2$. The actual intersections for the Δ_β profiles (nonjackknifed) were determined with a root-finding algorithm. The approximate intersections for the $\hat{\Delta}_\beta$ profiles were found by subtraction of pairs of profiles. The reason for using subtraction in the case of jackknifing is that the jackknifing procedure above does not provide an estimate $\hat{\boldsymbol{\pi}}$ to evaluate (2.6) analytically; it provides only m discrete points on the $\hat{\Delta}_\beta$ curve.

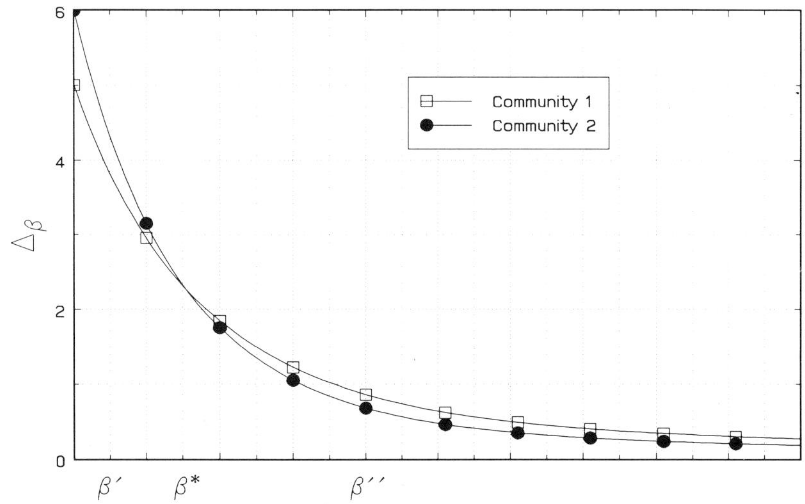

Fig. 7. Δ_β profiles illustrating an intersection and hypothetical points (β', β^*, β'') for the composite null hypothesis described in the text.

The estimation phase entailed applying the jackknifing algorithm, matching up the intersections on the T_j profiles with those on the $\hat{\Delta}_\beta$ profiles and subsequently testing the significance of these intersections. In so doing, it is helpful to keep in mind that small j in the T_j profiles corresponds to large β in the $\hat{\Delta}_\beta$ or Δ_β profiles. The inference procedure involves forming the composite null hypothesis given by Patil and Taillie (1979b). To illustrate, Figure 7 shows $\hat{\Delta}_\beta$ profiles for two communities (C_1 and C_2) with an intersection found at the point β^* . In this case, the procedure would be to choose β' and β'' so that $\beta' < \beta^* < \beta''$, and form the composite null hypothesis:

$$\mathcal{H}'_0: \hat{\Delta}_{\beta'}(C_1) \geqslant \hat{\Delta}_{\beta'}(C_2),$$
$$\mathcal{H}''_0: \hat{\Delta}_{\beta''}(C_1) \leqslant \hat{\Delta}_{\beta''}(C_2).$$

Simultaneous rejection of the composite null hypothesis at some specified level of significance would establish that $\Delta_{\beta'}(C_1) < \Delta_{\beta'}(C_2)$ and $\Delta_{\beta''}(C_1) > \Delta_{\beta''}(C_2)$; that is, the two community profiles $\hat{\Delta}_\beta(C_1)$ and $\hat{\Delta}_\beta(C_2)$ *do* intersect and, therefore, the two communities are *not* intrinsically comparable.

To test the composite null hypothesis calculate the two-sample t statistics,

$$t_\beta = \frac{\hat{\Delta}_\beta(C_1) - \hat{\Delta}_\beta(C_2)}{[\widehat{\text{var}}(\hat{\Delta}_\beta(C_1)) + \widehat{\text{var}}(\hat{\Delta}_\beta(C_2))]^{1/2}}, \quad \beta = -1, -0.97, -0.94, \ldots, 2. \tag{3.4}$$

If at least one t_β (viz. the maximum t_β) on each side of the intersection is significant when compared to the one-tailed Bonferonni-adjusted $t(1-\alpha/g;\, n-1)$ value (where $g=2$ and $\alpha=0.05$ here), then the implication is that the profiles intersect, and the communities C_1 and C_2 are *not* intrinsically comparable. If none of the t_β values calculated are significant, then it may be inferred that there is no difference in diversity between the two communities being compared.

This procedure must be viewed as only an approximate test because it involves 'difficult and unresolved questions of simultaneous inference' (Patil and Taillie, 1979b). However, even with this qualification, it remains a valuable procedure over simply treating the $\boldsymbol{\pi}$ as population values. In addition, it by no means diminishes the usefulness of the bias correction aspect of the jackknife.

3.2.3. Community diversity comparisons

Both temporal and spatial community comparisons were of interest. The temporal comparisons tracked individual communities within a treatment over time, while the spatial comparisons contrasted treatments at certain relative (i.e., since time of treatment) points in time. For example, comparisons within the first set of cut strips fall into the within-treatment category, while comparisons between strips, 1 year after cut, fall into the between-treatment category. Four within-treatment comparisons were made corresponding to the rows in Table 2. Between-treatment comparisons were made at 1 and 10 years postharvest. Right tail-sum, Δ_β, and jackknifed $\hat{\Delta}_\beta$ profiles were generated for each comparison. The following presents a detailed analysis of the temporal comparisons on strip 2, watershed 4; the results of the other temporal and spatial comparisons are summarized at the end of this section.

Appendix B presents the relative abundances ($\boldsymbol{\pi}$) for the precut, 1- and 10-year postcut plant communities in strip 2. These relative abundances may be ranked from greatest to least for each year to form the ($\boldsymbol{\pi}^\#$) vectors and the T_j profiles (Figure 8). Examination of the T_j profiles without the benefit of the inset graph may tend to suggest that the profiles do not cross because of the resolution at large j. However, the inset clearly shows that each pair of profiles intersects once. The points of intersection may easily be found by examining the inset graph or by taking the differences between the right tail-sums for each pair of communities and noting where the change in sign occurs (Table 3). Note that the T_j profiles all cross in the tail – where j is most sensitive to rare species.

The population Δ_β profiles (i.e., those constructed by treating $\boldsymbol{\pi}$ as the population values), and the jackknifed $\hat{\Delta}_\beta$ profiles for strip 2 are shown in Figure 9. Comparison of these two sets of profiles shows that they are very similar in shape, but that the jackknifed profiles have changed considerably for β near -1, illustrating the effect of the jackknife bias correction. The crossing points for the two sets of profiles, however, have not been greatly affected by jackknifing, as the following shows.

Δ_β profile	Precut vs. 1-year post	Precut vs. 10-year post	1-year vs. 10-year post
Population	−0.768	−0.852	–
Jackknife	−0.655	−0.775	–

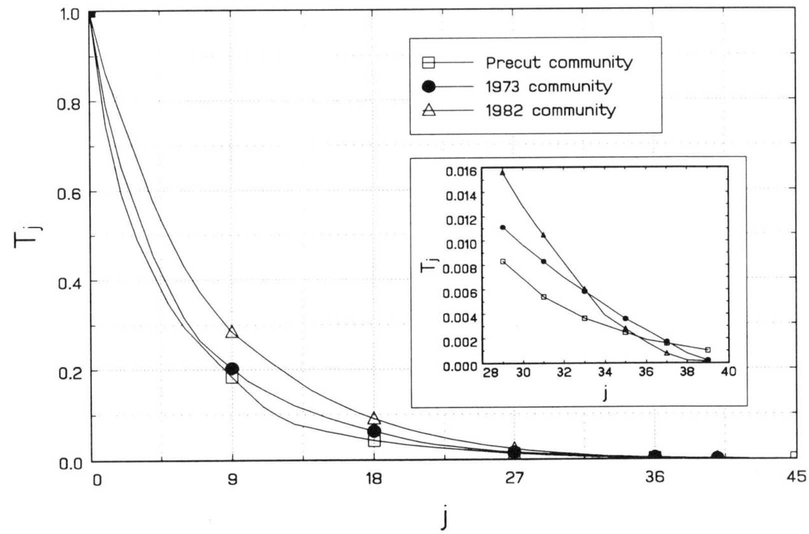

Fig. 8. Right tail-sum (T_j) diversity profiles for the antecedent, 1- and 10-year postharvest communities on strip 2, watershed 4 (inset shows details in the tail of the profiles where intersections occur).

Table 3
Right tail-sum differences for strip 2 on watershed 4 of the Hubbard Brook experimental forest[1]

j	$T_j(1972) - T_j(1973)$	$T_j(1972) - T_j(1982)$	$T_j(1973) - T_j(1982)$
0	0.00000000	0.00000000	0.00000000
1	−0.04369470	−0.11794593	−0.07425123
2	−0.05931226	−0.17395762	−0.11464536
3	−0.06430045	−0.18215802	−0.11785757
4	−0.03960770	−0.16102352	−0.12141582
5	−0.03770408	−0.15449268	−0.11678861
6	−0.02088950	−0.13618703	−0.11529754
⋮	⋮	⋮	⋮
32	−0.00244935	−0.00371484	−0.00126549
33	−0.00219867	−0.00237613	**−0.00017746**
34	−0.00165556	−0.00081152	**0.00084404**
35	−0.00111245	**−0.00028838**	0.00082408
36	−0.00075734	**0.00023476**	0.00099211
37	**−0.00012092**	0.00083978	0.00096069
38	**0.00052662**	0.00110135	0.00057473
39	0.00079818	0.00089755	0.00009937
40	0.00069375	0.00069375	0.00000000
41	0.00040132	0.00040132	0.00000000
42	0.00010890	0.00010890	0.00000000
43	0.00004445	0.00004445	0.00000000
44	0.00000222	0.00000222	0.00000000
45	0.00000000	0.00000000	0.00000000

[1] Bold entries indicate points of intersection for the respective T_j profiles.

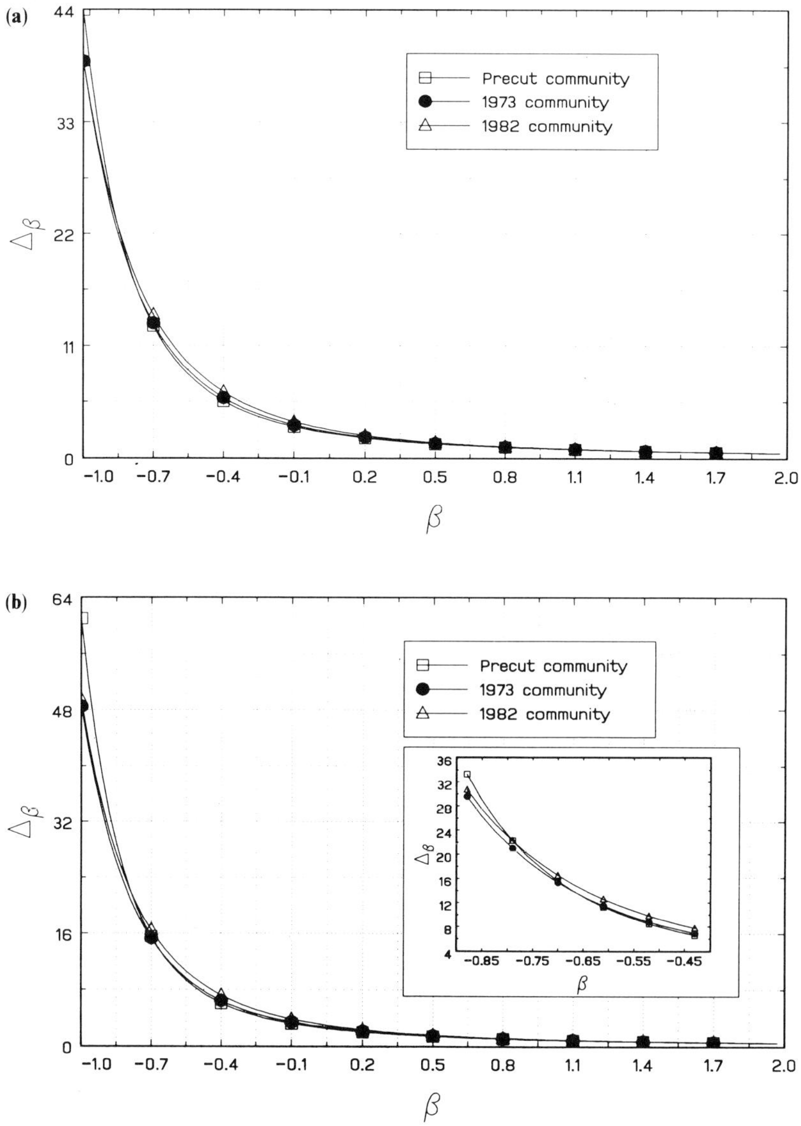

Fig. 9. Δ_β profiles for the antecedent, 1- and 10-year postharvest communities on strip 2, watershed 4: (a) population profiles; (b) jackknifed profiles (inset shows details at points of intersection).

Even though the two estimates of the species count index (Δ_{-1}) are quite different, jackknifing seems to preserve the approximate crossing points in β – where β is sensitive to rare species. Therefore, both the Δ_β and $\hat{\Delta}_\beta$ profiles are consistent with the T_j profiles in Figure 8 in that they all cross where their respective parameters are sensitive to rare species.

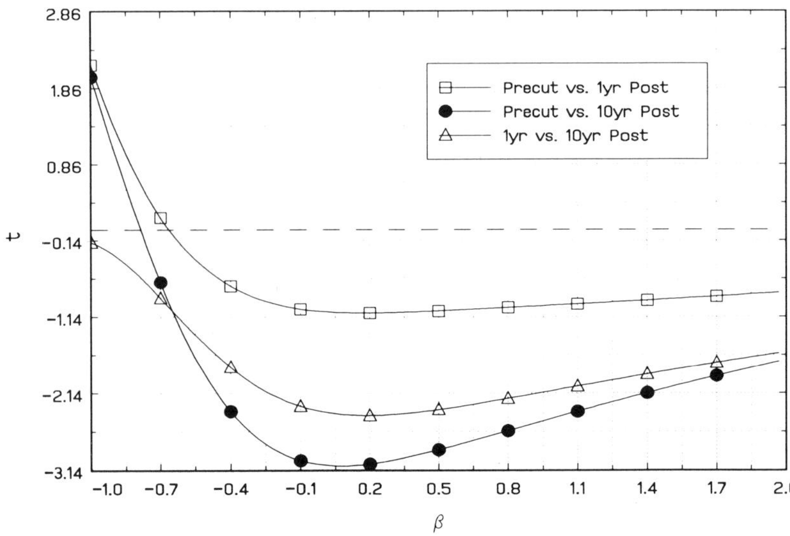

Fig. 10. Two-sample t_β statistics for the jackknifed community comparisons showing intersection points where the profiles cross zero; significant and nonsignificant differences in the profiles also are shown (see text for details).

In order to evaluate the significance of the intersections on the $\hat{\Delta}_\beta$ profiles, t_β values were calculated using (3.4) and are plotted in Figure 10. Comparing the precut community with the 10-year postharvest community, the precut profile is above the 1982 profile for $-1 \leqslant \beta < -0.775$ and the reverse is true for $\beta > -0.775$. The largest t_β statistics found to the left and right of the intersection were $t_{-1} = 2.00$ and $t_{0.08} = -3.09$, respectively. Comparison with the Bonferroni-adjusted $t(0.025, 18) = 2.101$ leads to a significant difference only for $\beta > -0.775$. This also may be seen on Figure 10. Therefore, given the current data, there is not enough evidence to reject the hypothesis that $C(1982) \overset{\text{imd}}{>} C(\text{precut})$.

The intersection between the precut and 1-year postharvest communities finds a significant difference only at the species count index ($t_{-1} = 2.16$). This would seem to suggest a second apparent ordering: $C(\text{precut}) \overset{\text{imd}}{>} C(1973)$. However, since only one extreme test statistic (t_{-1}) is significant out of all 100 calculated, it seems instructive to scrutinize the graphs before settling on this apparent ordering. The T_j, Δ_β, and $\hat{\Delta}_\beta$ profiles all show that $C(1973)$ plots above $C(\text{precut})$ over most of their lengths. This would seem to contradict the intrinsic diversity ordering above. In addition, there really seems to be very little difference between the two community profiles (see especially Figure 8). Therefore, a more reasonable interpretation would be that the apportionment of individuals within these two communities probably is quite similar for the major (most dominant) species. However, the significant difference in species richness between these two communities suggests that $C(\text{precut})$ is significantly more diverse in the tail than $C(1973)$. That is, the rarer species are more diverse in $C(\text{precut})$ than in $C(1973)$, but there seems to be no difference in diversity between the two communities overall; i.e., $C(\text{precut}) = C(1973)$.

The t_β curves (Figure 10) illustrate that the intersections for the $\hat{\Delta}_\beta$ profiles may be found by noting where these curves cross zero, indicating the point where the numerator in (3.4) becomes zero; viz., the point where the profiles are the same. Neither the Δ_β nor the $\hat{\Delta}_\beta$ profiles shows an intersection for $C(1973)$ and $C(1982)$. Yet the T_j profiles *do* record such an intersection. This anomaly points out a potential problem with this technique of intersection testing: the Δ_β and $\hat{\Delta}_\beta$ profiles may not cross even though the T_j profiles do, or the Δ_β or $\hat{\Delta}_\beta$ profiles may cross more than once (this problem was alluded to in Section 2.3). Patil and Taillie (1979b) recognized this problem and in order to draw inferences about the T_j intersections, used other profiles to provide test statistics in such cases. In this case, that is an option too; however, an alternative approach also is possible. It was noted above that $C(1982) \overset{\text{imd}}{>} C(\text{precut})$, and that $C(\text{precut}) = C(1973)$ through the testing procedure. In addition, Figure 10 shows that there are numerous significant t_β when comparing $C(1973)$ and $C(1982)$; that is, there seems to be some difference between the two community profiles. These observations would strongly support the final intrinsic ordering for strip 2: $C(1982) \overset{\text{imd}}{>} C(\text{precut}) = C(1973)$.

It is interesting to note that the species count and Shannon indices rank communities $C(\text{precut})$ and $C(1982)$ in the opposite sense. This may be seen from the $\hat{\Delta}_\beta$ profile since $\hat{\Delta}_{-1}(\text{precut}) > \hat{\Delta}_{-1}(1982)$, while $\hat{\Delta}_0(\text{precut}) < \hat{\Delta}_0(1982)$. Had only these two point estimates of Δ_β been calculated, without the benefit of a large range of β values (and hence diversity indices) or hypothesis testing through the jackknife, contradictory conclusions would have been reached concerning the ordering of these two communities. A subsequent check of Simpson's index at Δ_1 might have been enough to sway the investigator toward the same conclusions drawn here with jackknifing, however, since $\hat{\Delta}_1(\text{precut}) < \hat{\Delta}_1(1982)$; but it would have been based on subjective judgment. Similar results may be found for the $C(\text{precut})$ and $C(1973)$ profiles.

Applying this same procedure to the rest of watershed 4 and to watershed 101 yielded similar results. On strip 1, $C(1980) \overset{\text{imd}}{>} C(1970) \overset{\text{imd}}{>} C(1971)$; on strip 3 $C(1984) \overset{\text{imd}}{>} C(1974) = C(1975)$. Unfortunately, no inventory data were taken for the precut community before the block clearcut. Therefore, all that could be done was to study the trend in diversity following clearcutting on watershed 101; this trend showed $C(1980) \overset{\text{imd}}{>} C(1971)$. As in the strip cut, diversity seems lowest right after the cut, then increases.

Between-treatment (spatial) comparisons were made within watershed 4 (i.e., between strips) at 1 and 10 years postharvest. The results showed a change in the spatial diversity ranking over time. In the 1-year communities the intrinsic ordering $C(\text{strip 2}) \overset{\text{imd}}{>} C(\text{strip 3}) \overset{\text{imd}}{>} C(\text{strip 1})$. In the 10-year communities, however, there has been a change in ranking so that $C(\text{strip 2}) = C(\text{strip 3}) \overset{\text{imd}}{>} C(\text{strip 1})$.

Comparisons also were made between strip 1 and the block clearcut for both the 1- and 10-year postharvest communities. After intersection testing, it was found that there was no significant difference between the 1-year communities, but that after 10 years, $C(\text{strip 1}) \overset{\text{imd}}{>} C(\text{block clearcut})$. Since strips 2 and 3 were both intrinsically more diverse than strip 1 for the 1- and 10-year comparisons mentioned, it is easily argued that all three strips were intrinsically more diverse than the block clearcut at both points in time; indeed, graphical analysis (not shown) supported this argument.

3.2.4. Concluding remarks

The jackknifing procedure of Patil and Taillie (1979b) provides a useful, approximate method for testing intrinsic diversity ordering when relative abundance data are based on samples rather than complete enumeration. Other assumptions and uses of the jackknife have been discussed in detail in relation to general forestry usage by Gregoire (1984), and in relation to diversity indices by Zahl (1977) and Palmer (1991). Quenouille (1956) originally introduced the jackknife as a bias reduction technique. The effect of bias reduction on diversity indices can be seen using the easily comprehensible species count index ($s - 1$) as an example. The sample-based ($s - 1$) and jackknifed ($\hat{s} - 1$) estimates are presented in Table 4 along with the variances estimated through jackknifing. Comparison of $s - 1$ and $\hat{s} - 1$ supports the a priori expectation that the simple sample-based estimates will underestimate the jackknifed estimates of the species count index. These comparisons give strong evidence that the species count index is a negatively biased estimate of the true species richness (minus 1) when sampling variability is not accounted for in the communities on these watersheds. Unfortunately, the true population species richness remains unknown; therefore, it is not possible to judge how closely either of these estimates are to the true population values in this analysis.

The jackknife's usefulness in correcting for bias also is apparent in the between-treatment comparisons. For example, Table 4 shows that it would be difficult to compare the two watersheds on the basis of the uncorrected species counts alone. The main reason for this is the disparity in sample size between watersheds: $n = 19$ on the strip cut and $n = 48$ on the block clearcut. Though the differences look real,

Table 4
Sample-based ($s - 1$) and jackknifed ($\hat{s} - 1$) estimates of the species count index in the two watersheds[1]

Treatment		Precut	1 year post	10 year post
		$s - 1$		
WS 4	Strip 1	35	33	40
	Strip 2	44	39	39
	Strip 3	41	39	46
WS 101	All	–[2]	36	35
		$\hat{s} - 1$		
WS 4	Strip 1	43.5 (6.382)	38.7 (7.679)	49.5 (15.856)
	Strip 2	61.1 (17.950)	48.5 (15.856)	49.4 (15.756)
	Strip 3	50.5 (10.172)	48.5 (12.066)	59.3 (60.332)
WS 101	All	–[2]	45.8 (15.585)	40.9 (7.099)

[1] Jackknife estimates of the variance for the species count index are given in parentheses.
[2] No data were available in these periods.

especially for strips 2 and 3 when compared to block clearcutting, the differences in sample size leave doubt without the jackknife correction.

The profiles and comparisons presented in this section clearly show the dynamic nature of plant diversity in managed northern hardwood forests even over time-spans as short as 10 years. To what extent the different regeneration methods influenced the diversity orderings encountered here, and to what extent natural factors were responsible, cannot be tested with present techniques.

4. Diversity in forest management decisions

Structural diversity is a very important component of the forest community. The vertical diversity of foliage in spruce-fir forests, for example, was found by MacArthur and MacArthur (1961) to be a good predictor of bird species diversity. In uneven-aged forest management the shape of the diameter distribution, maximum tree diameter, and stocking level are of paramount importance to the forest manager. Two mathematical programming models are presented in this section that allow the forest manager to (1) optimally determine a diameter distribution which maximizes diameter class diversity within an uneven-aged forest stand; and (2) to ensure that the diameter class distribution after cutting is at least as diverse as that found before harvest in the antecedent community. Because the models present the optimal diameter class distribution at the *end* of a cutting cycle and the growth dynamics are known (Appendix C), cutting guides along with economic criteria also can be calculated easily from these results.

4.1. Maximizing diameter class diversity

The model presented here finds the diameter distribution which maximizes diameter-class diversity in an uneven-aged northern hardwood stand under certain constraints (Gove et al., 1991). It addresses only the horizontal aspect of structural diversity and does not explicitly consider the other components. The model is completely general and may be adapted to even-aged stand management as well. In addition, many other types of constraints not considered in this section can be added to the model. Finally, extensions to the model could incorporate wildlife concerns too; for example, tree diameter and height relationships could be used to incorporate some measure of foliage height diversity into the model which would allow the maximization of bird species diversity using relationships similar to those of MacArthur and MacArthur (1961).

4.1.1. Maximizing diversity: General

In Section 2.3 it was noted that for any given vector of relative abundances π, a diversity profile could be generated by allowing β to vary in (2.6). In this section $\Delta_\beta(\pi)$ is viewed in the opposite sense: β is held fixed and π is allowed to vary subject to the constraints that $\beta \geqslant -1$ and $\sum_{i=1}^{s} \pi_i = 1$. When this is done a *diversity surface* is generated at β.

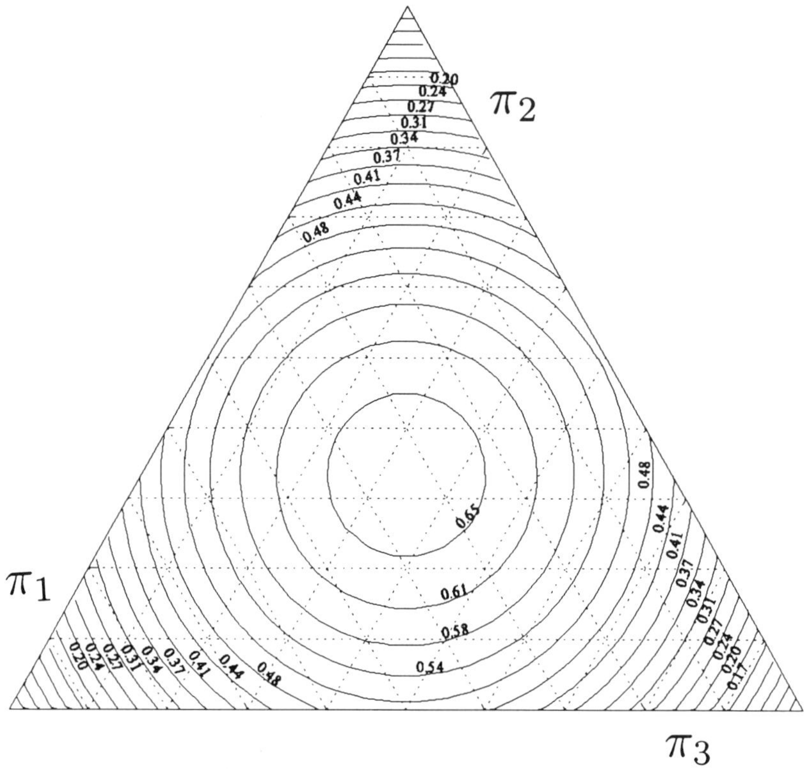

Fig. 11. Δ_β diversity surface at $\beta = 1$ for all 3-species communities.

Figure 11 presents a triangular chart of a three-species ($s = 3$) community with each of the three axes of the chart scaled such that $0 \leqslant \pi_i \leqslant 1, i = 1, \ldots, 3$. This type of chart is useful for envisioning the diversity surface since it automatically incorporates the constraint $\sum_{i=1}^{s} \pi_i = 1$. The contours plotted on the interior of the chart represent the level curves of the $\Delta_\beta(\boldsymbol{\pi})$ diversity surface when $\beta = 1$ (Simpson's index). Any corner point on the chart represents a single-species community, edges are two-species communities, and interior points are three-species communities. The chart clearly shows for $\beta = 1$ that the $\Delta_\beta(\boldsymbol{\pi})$ diversity surface reaches its maximum at the center – the completely even community.

A similar chart is shown in Figure 12 for the $\Delta_\beta(\boldsymbol{\pi})$ surface at $\beta = 2$. Note the slight difference in the shape of the diversity surface level curves when compared with Figure 11. In Figure 12, the curves are less circular and are beginning to become somewhat triangular in shape. Indeed, if the $\Delta_\beta(\boldsymbol{\pi})$ surface is plotted as $\beta \to \infty$, the level curves become more and more triangular. The maximum, again, is clearly seen to occur at the completely even community when $\beta = 2$.

That the completely even community maximizes diversity for a given number of species s is well known (Patil and Taillie, 1979a, 1982; Pielou, 1974; Solomon, 1979).

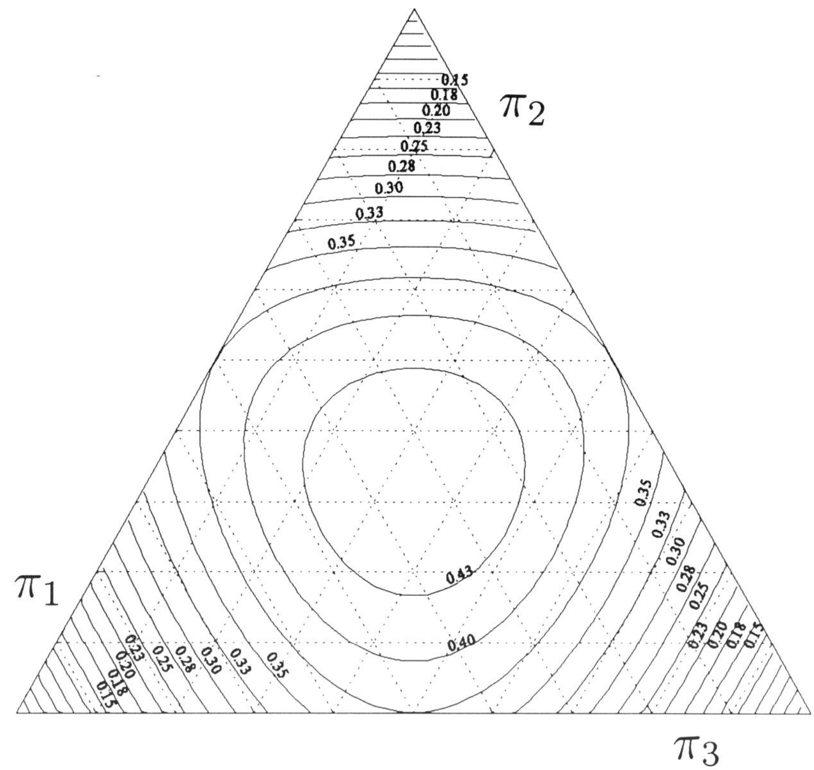

Fig. 12. Δ_β diversity surface at $\beta = 2$ for all 3-species communities.

In general, the problem may be formulated for $\Delta_\beta(\boldsymbol{\pi})$ as

$$\begin{aligned} &\max_{\{\pi\}} \Delta_\beta(\boldsymbol{\pi}) \\ &\text{subject to} \quad \sum_{i=1}^{s} \pi_i = \boldsymbol{\pi}'\mathbf{1} = 1. \end{aligned} \tag{4.1}$$

It is straightforward to show that the solution to (4.1) is $C_{\mathrm{E}}(s)$. This is an interesting finding because it allows the introduction of an alternative objective function into model (4.1); maximizing (4.1) should, therefore, be equivalent to the following problem

$$\begin{aligned} &\min_{\{\pi\}} \sum_{i=1}^{s} |\pi_{\mathrm{E}} - \pi_i| \\ &\text{subject to} \quad \sum_{i=1}^{s} \pi_i = \boldsymbol{\pi}'\mathbf{1} = 1. \end{aligned} \tag{4.2}$$

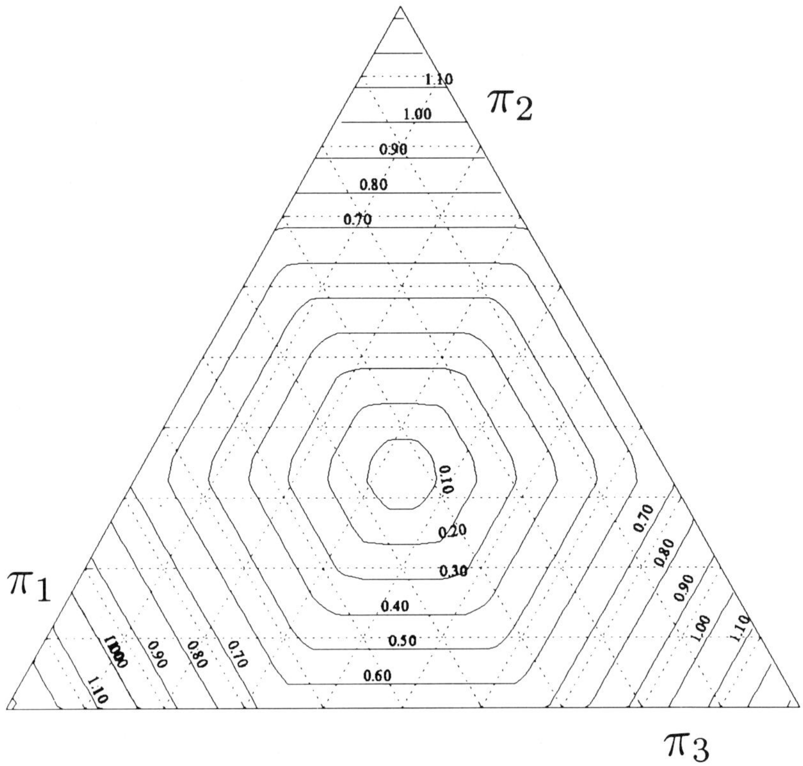

Fig. 13. Model (4.2) diversity surface for all 3-species communities. (Note that the surface is actually piecewise linear.)

The *unevenness* measure $\sum_{i=1}^{s} |\pi_E - \pi_i|$ in (4.2) is known as the Pietra index of income inequality in economics (Arnold, 1987). The diversity surface for this formulation is presented in Figure 13. Note the difference in the shape of the level curves in this surface when compared to Figures 11 and 12; the level curves for (4.2) are hexagonal. This surface also is minimized at the completely even community, implying that diversity is at its maximum.

4.1.2. Maximizing diameter class diversity: Model formulation

Consider the following question for an uneven-aged northern hardwood stand: Given certain stocking, economic, and biological growth constraints, what is the diameter distribution that maximizes diameter class diversity? Based on the results of the last section, the intuitive answer to this question is the diameter distribution that is most nearly even. However, the result will not be a completely even diameter distribution if the constraints used in the model impose any true restriction on the diversity surface.

Under the current scenario, notice that 'species' has now become synonymous with *diameter class*; therefore, s is now the number of diameter classes which is held constant. Relative abundances composing $\boldsymbol{\pi}$ are determined in terms of number of trees per acre.

A mathematical programming model structure first introduced by Adams and Ek (1974) is adopted here. Two general model formulations are presented and solved as nonlinear programs. The concepts discussed in the previous section may be extended to $s > 3$ in these two models to maximize diameter class diversity. The first formulation in equation set (4.3) maximizes the diameter-class diversity using the objective function from (4.2) and is termed model I.

$$\min_{\{\pi\}} \sum_{i=1}^{s} |\pi_{\mathrm{E}} - \pi_i|$$

$$\begin{aligned} \text{subject to} \quad & \sum_{i=1}^{s} \pi_i = \boldsymbol{\pi}'\mathbf{1} = 1, \\ & N_i(t) - N_i(t-1) \geqslant 0, \quad i = 1, \ldots, s+1, \\ & 0 < \pi_i \leqslant 1, \\ & \mathrm{BPA} = \mathrm{PSL}, \\ & \mathrm{LEV} = \mathrm{ESL}. \end{aligned} \tag{4.3}$$

s = The number of diameter classes ('species'); $s = 9$ in this study. The diameter class width used was 2 inches, with a minimum diameter class of 6 inches.

$N_i(t-1)$ = The number of trees in diameter class i in the optimal stand at the *beginning* of the 5-year growth period.

$N_1(t)$ = The number of trees in diameter class i in the optimal stand at the *end* of the 5-year growth period given the growth dynamics predicted by the growth model.

BPA = The total basal area per acre in the optimal stand (taken at the midpoint of the 2-inch diameter classes – see Appendix C).

LEV = $\mathrm{VG}/((1 + \mathrm{arr})^{\mathrm{cc}} - 1) - \mathrm{VGS}$ = The land expectation value for the optimal stand.

VGS = $\sum_{i=1}^{s} V_i N_i(t-1)$ = The value of initial growing stock.

VG = $\sum_{i=1}^{s+1} V_i N_i(t) - \mathrm{VGS}$ = The 5-year value growth.

cc = The cutting cycle (time between harvests) length in years (cc = 5 in this study).

V_i = The individual tree value in dollars for the i-th diameter class. The individual tree values used in the computation of LEV are Martin's (1982) fair site (site index 55) values (see Table 5).

arr = The alternative rate of return (3% in this study).

PSL = Some physical stocking level of basal area per acre.

ESL = Some economic stocking level in present value dollars per acre.

Model II maximizes $\Delta_\beta(\boldsymbol{\pi})$ with the added constraint that β must be fixed, the rest

Table 5
Maximum diversity diameter distributions for model I at different basal area stocking levels

Diameter class (inches)	Basal area per acre			Value per tree[1]
	60 ft^2	80 ft^2	120 ft^2	
	Trees per acre			
6	18.44	30.03	64.63	0.11
8	13.66	21.90	45.62	0.30
10	10.67	16.86	34.16	0.54
12	8.63	13.47	25.02	3.83
14	7.16	11.05	20.08	6.15
16	6.05	9.25	16.35	8.61
18	5.20	7.87	4.80	11.23
20	4.52	3.01	0.11	14.66
22	1.47	0.56	0.04	17.79
Total TPA	75.80	114.00	210.81	
LEV$/acre	0.00	0.00	0.00	
Unevenness[2]	0.467401	0.554299	0.718561	

[1] The tree values used are from Martin's (1982) fair site guides, values for 24" and 26" trees used by Martin were $ 21.19 and $ 24.97 respectively.
[2] Unevenness is defined as $\sum_{i=1}^{s} |\pi_E - \pi_i|$.

of the model is the same as model I. The complete formulation is given in equation set (4.4):

$$\max_{\{\pi\}} \Delta_\beta(\boldsymbol{\pi})$$

$$\begin{aligned} \text{subject to} \quad & \sum_{i=1}^{s} \pi_i = \boldsymbol{\pi}'\mathbf{1} = 1, \\ & N_i(t) - N_i(t-1) \geqslant 0, \quad i = 1, \ldots, s+1, \\ & 0 < \pi_i \leqslant 1, \\ & \text{BPA} = \text{PSL}, \\ & \text{LEV} = \text{ESL}, \\ & \beta = b, \end{aligned} \tag{4.4}$$

where β may be fixed at any value b such that $-1 \leqslant b < \infty$.

The growth dynamics for both formulations are modeled with a simple set of nonlinear whole-stand diameter-class growth equations given in Appendix C. Therefore, the diameter distribution recovered in the optimal stand is for the entire community composed of all species – no individual species distributions are available.

The physical stocking constraint on basal area and the economic constraint on LEV keep the solutions from both models feasible from a biological perspective. If both of these are set simply to be positive, model I will lead to a degenerate solution; model II may find a feasible solution with nonzero π, however, the stand basal area and trees per acre effectively will be zero. Therefore, the growth constraints alone do little to determine a biologically reasonable solution. In addition, if only one of these two constraints is used and it is restricted between lower and upper bounds (e.g., $BPA_L \leqslant BPA \leqslant BPA_U$), both models always seem to find a solution at the lower bound.

4.1.3. Model results

Models I and II were optimized using the generalized reduced gradient program GRG2 (Lasdon and Waren, 1986). Solutions were found at several economic and physical stocking combinations; all solutions presented satisfied the Kuhn–Tucker stationary conditions.

Table 5 presents solutions to model I with LEV constrained only to be positive, but with stand basal area set at several stocking levels. Note that the diameter distributions in Table 5 are not completely even; this is a consequence of the constraints on growth and basal area which are all binding in both model formulations. A plot of the $\Delta_\beta(\pi)$ profiles for these three communities is shown in Figure 14. The distribution at 60 ft^2 is the most diverse community according to the diversity

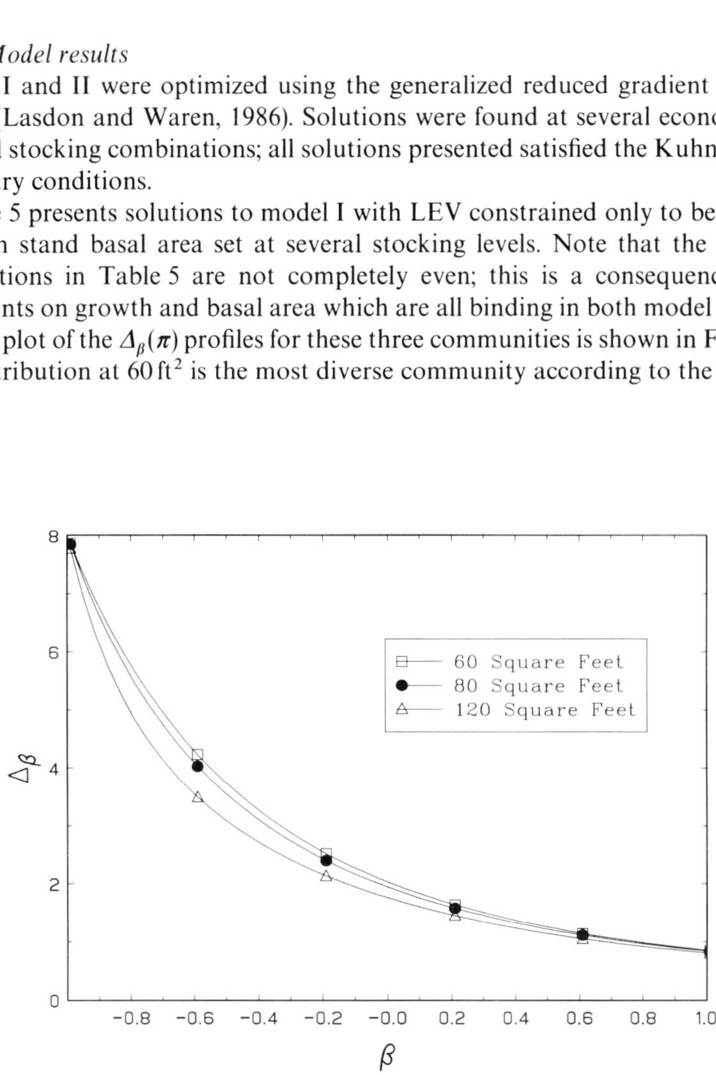

Fig. 14. Δ_β profiles of maximum diversity diameter distributions for model I at different basal area per acre stocking levels.

ordering of the $\Delta_\beta(\pi)$ profiles. In addition, the unevenness criterion correctly orders each community with respect to diversity in this example.

The most striking aspect of the distributions in Table 5 is that LEV is zero for all solutions. The reason for this is that LEV and evenness work against each other in these formulations. In order to even out a distribution (i.e., maximize diversity), as many trees as possible are put into the larger diameter classes. This happens in accordance with satisfying the growth and BPA constraints until LEV reaches its lower bound of zero. However, trees in the sawtimber size classes ($\geqslant 12$ inches) contribute substantially more to holding costs in the calculation of LEV; therefore, few trees are needed in these classes to drive LEV to zero. Thus, the positive constraint on LEV is a mechanism that works against evening out the distribution at optimality. Indeed, if LEV were left unconstrained, it would be driven negative in the optimal solution, resulting in a more diverse community than the 60 ft^2 solution in Table 5. Such a result may be reasonable if financial considerations are of no concern to the landowner.

Maximizing model II is not as straightfoward as minimizing model I. The reason for this was pointed out earlier: inasmuch as different diversity surfaces are generated at each β, $\Delta_\beta(\pi)$ should be optimized several times, each at a different level of β to allow for comparison of the resulting distributions. Table 6 presents the results of this process for four levels of β at 80 ft^2 of basal area per acre. All of the constraints (with the exception of β) have remained the same in each of these solutions; therefore, the solution space has not changed – the only difference contributing to the slightly different results is the shape of the diversity surface at each β. This phenomenon can be envisioned readily by imagining one or two simple linear constraints in Figures 11–13. Note that depending upon the arrangement of the

Table 6
Maximum diversity diameter distributions with 80 ft^2 of basal area per acre for model II at different β

Diameter class (inches)	β			
	0.01	1.0	10.0	100.0
	Trees per acre			
6	30.53	29.99	30.03	34.78
8	22.26	21.87	21.90	25.29
10	17.13	16.84	16.86	18.51
12	13.69	13.45	13.47	10.29
14	11.22	11.04	11.05	8.47
16	7.66	9.24	9.25	7.11
18	6.53	7.86	7.87	6.06
20	3.03	2.66	3.01	5.24
22	2.01	0.89	0.56	1.91
Total TPA	114.06	113.84	114.00	117.66
LEV $/acre	0.00	0.00	0.00	0.00
Unevenness	0.577137	0.554539	0.554299	0.669038

constraints in these figures, the optimal solution may be slightly different in each case. This same reasoning applies to the results in Table 6.

The results in Table 6 show a range of only four trees per acre difference between the resulting stands; therefore, from both a biological and practical perspective, there is no difference between the resulting distributions at different β. Theoretically, the value of the unevenness statistic might be used to judge which of these distributions is, in fact, the most diverse at the 80 ft^2 level. However, the unevenness criterion is only a one-dimensional statistic and has not been shown to order communities consistently for a given s as have the $\Delta_\beta(\pi)$ profiles. Indeed, a plot of the four $\Delta_\beta(\pi)$ profiles (not shown) reveals that they are not intrinsically comparable since the profiles cross. In this case, it seems reasonable to pick the community at $\beta = 10$ since it has the smallest unevenness and is the solution to model I (see Table 5). Given the practical considerations, nothing is compromised in this decision.

One further cautionary note is in order when optimizing model II. If β is set equal to -1, the diversity surface generated by $\Delta_\beta(\pi)$ is a constant at $s-1$ for *all* π, as noted earlier. In this case, the only factors restricting the solution are the constraints, and any point which satisfies the constraints within the feasible solution space may be chosen as a solution. Thus, model II should never be optimized at $\beta = -1$ as solution vectors having little relation to the results of model I may result. In addition, if β is left unconstrained in model II, the same result occurs since $\beta \rightarrow -1$ in this case.

The results of adding a LEV constraint different from zero to model I are shown in Table 7. The first distribution constrains LEV to be \$100 per acre while allowing BPA to go free. The other two distributions constrain both LEV and BPA.

Table 7
Effects of constraining LEV on maximum diversity diameter distributions for model I

Diameter class (inches)	Basal area per acre		
	50.7[1] ft^2	60 ft^2	80 ft^2
	Trees per acre		
6	26.78	25.56	42.38
8	19.79	18.84	30.72
10	15.42	14.64	23.53
12	11.34	11.80	18.71
14	9.40	9.75	15.28
16	4.98	8.22	4.64
18	2.25	5.06	3.62
20	0.07	0.15	1.93
22	0.03	0.00	0.00
Total TPA	90.06	94.02	140.81
LEV \$/acre	100.00	60.00	80.00
Unevenness	0.739646	0.618121	0.749250

[1] The BPA constraint was free in this distribution.

Comparing the 60 ft^2 and 80 ft^2 distributions in Tables 5 and 7 clearly shows that the effect of the LEV constraint is to add more trees to the smaller diameter classes while removing trees from the larger sawtimber classes. This causes a decrease in the holding costs, allowing LEV to increase over the distributions in Table 5. It also decreases the unevenness statistic and, therefore, the diversity, as expected. This illustrates the interplay between evenness and LEV alluded to earlier.

4.1.4. Concluding remarks

The solutions of models I and II suggest that both models will give approximately the same answers. However, because of the nature of the diversity surfaces generated in these models and the uncertainty of diversity ordering based on a single index like unevenness alone, it is recommended that both models be solved, as was done in the previous section. In addition, one of the profile methods discussed in Section 2.3 should *always* be used when comparing models for diversity ordering.

Each of the tables gives the absolute abundance vectors because this measure is necessary for management. It is therefore possible to incorrectly interpret the results of these tables by trying to judge evenness based merely on the diameter distributions alone while not taking into account the respective total trees per acre. In addition, 'species richness' was held constant for each distribution to facilitate comparison. If new distributions were generated with different numbers of diameter classes, this also would enter into the subsequent evaluation of diversity ordering (see Section 2).

Finally, $\boldsymbol{\pi}$ has been treated as a deterministic vector in both models for simplicity. Actually, $\boldsymbol{\pi}$ is a random vector with unknown sampling distribution due to the stochastic nature of the underlying growth equations. Gove and Fairweather (1992) discuss this in detail and present a method based on the nonparametric bootstrap and multivariate normal distribution to find confidence limits for the optimal diameter distribution.

4.2. Maintaining diameter class diversity

It may be of considerable interest to the forest manager to be able to ensure that an adopted harvesting plan will not diminish the diversity of the forest or even of individual stands within the forest over the cutting cycle. In this section as in the last, diversity is defined in terms of the diameter distribution for an uneven-aged northern hardwood stand. A model is reviewed (Gove et al., 1994) which ensures that the residual stand diameter distribution will be at least as diverse as some target condition, whether it be the antecedent stand, or some neighboring stand. The part of the model which ensures diversity maintenance is a constraint set that is completely general and may be applied to numerous other problems such as maintaining landscape diversity in larger forest areas.

4.2.1. Model formulation

The mathematical definition of intrinsic diversity ordering presented in Section 2.3 provides a vehicle for writing a set of constraints in a mathematical programming formulation to ensure that the solution vector at optimality yields a community that is at least as diverse as the antecedent community. Let $\boldsymbol{\pi}$ be the relative diameter-class distribution based on number of trees per acre so that $\boldsymbol{\pi}^{\#}$ is the ranked relative diameter-class distribution. The generic Adams and Ek (1974) model used in Section 4.1.2 yields a diameter distribution that maximizes some physical stocking, growth, or economic quality, subject to certain constraints of similar type. The basic formulation used is from Adams (1976), which yields investment-efficient stocking guides at optimality by maximizing the land expectation value and is given in equation set (4.5):

$$\begin{aligned} &\max_{\{N_j(t-1)\}} \ \text{LEV} \\ &\text{subject to} \quad N_j(t) - N_j(t-1) \geqslant 0, \quad j = 1, \ldots, s+1, \\ &\qquad\qquad\quad \text{BPA} = \text{PSL} \ \{\text{optional}\}, \\ &\qquad\qquad\quad N_j(t-1) \geqslant 0, \quad j = 1, \ldots, s, \end{aligned} \tag{4.5}$$

where all variables are defined as in Section 4.1.2, and the growth model used is from Appendix C.

Now consider the antecedent community with relative abundance vector $\boldsymbol{\varphi}$ and ranked relative abundance vector $\boldsymbol{\varphi}^{\#}$ based on its two-inch diameter distribution. Let $\boldsymbol{\pi}$ denote the relative abundance vector for the optimal stand where the individual components of $\boldsymbol{\pi}$ are given by $\pi_j = N_j(t-1)/\text{TPA}$, and $\text{TPA} = \sum_{j=1}^{s} N_j(t-1)$. It follows that the ranked relative abundance vector for the optimal stand is $\boldsymbol{\pi}^{\#}$. Given these definitions and the definitions of intrinsic diversity ordering from Section 2.3, the following set of $s-1$ constraints may be added to equation set (4.5):

$$d_j = T_j(\boldsymbol{\pi}) - T_j(\boldsymbol{\varphi}) \geqslant 0, \quad j = 1, \ldots, s-1. \tag{4.6}$$

Providing that a feasible solution to this new model exists, this set of constraints guarantees that the optimal diameter distribution for $\boldsymbol{\pi}$ will be intrinsically more diverse than the diameter distribution for $\boldsymbol{\varphi}$. In the special case where $\boldsymbol{\pi}^{\#} = \boldsymbol{\varphi}^{\#}$, there has been no change in diversity. This case will arise if the antecedent community already exactly follows the solution to the optimization model used *without* constraint set (4.6); that is, if the antecedent community already is being managed according to the investment-efficient guide produced from (4.5). Model (4.5) with constraint set (4.6) included has been termed the *diversity-constrained* model by Gove et al. (1994). The solution to the diversity-constrained model requires that the $\boldsymbol{\pi}$ be sorted at *each* step in the solution algorithm before the constraints are evaluated. The sorting requirement is consistent with the definition of intrinsic diversity ordering in Section 2.3. This ability to relabel species (diameter classes) is central to the ability to produce a feasible solution from the diversity-constrained model.

The diversity-constrained model was solved using the nonlinear optimization software GRG2 (Lasdon and Waren, 1986). In this example the optimal stand

Table 8
Diversity-constrained diameter distribution and target distribution following Martin (1982)

Diameter class (inches)	j	Martin[1]			Diversity constrained			d_j[3]	Value per tree[4]
		$N_j(t-1)$	φ_j	$T_j(\varphi^{\#})^2$	$N_j(t-1)$	π_j	$T_j(\pi^{\#})^2$		
6	1	·126.6	0.633	0.367	102.3	0.412	0.588	0.221	0.14
8	2	46.5	0.233	0.135	75.3	0.303	0.285	0.150	0.34
10	3	17.1	0.086	0.049	58.5	0.236	0.049	0.0	0.60
12	4	6.3	0.032	0.018	7.8	0.032	0.018	0.0	4.23
14	5	2.3	0.012	0.006	2.9	0.012	0.006	0.0	6.77
16	6	0.8	0.004	0.002	1.0	0.004	0.002	0.0	9.45
18	7	0.3	0.002	0.001	0.4	0.002	0.001	0.0	12.31
20	8	0.1	0.001	0.0	0.1	0.001	0.0	0.0	16.05
22	9	0.0	0.0	0.0	0.0	0.0	0.0	0.0	19.46
Total trees (no.)		200.0			248.3				
BPA (ft^2)		59.5			89.8				
LEV ($)		95.1			210.2				

[1] Martin (1982) 5%, good site guide reoptimized according to his specifications using GRG2.
[2] Note that $\varphi = \varphi^{\#}$ and $\pi = \pi^{\#}$ in this example.
[3] $d_j = T_j(\pi^{\#}) - T_j(\varphi^{\#})$.
[4] Tree values are from Martin's (1982) 5%, good site guide; values for the 24- and 26-inch class are $ 23.17 and $ 27.28, respectively.

diameter distribution is desired to be at least as diverse as the target stand diameter distribution of Martin's (1982) good-site, 5%-alternative-rate-of-return stocking guide. Table 8 presents both Martin's guide and the diversity-constrained solution to (4.5). It is of interest to note that Martin's guide was originally constrained to follow a q (negative exponential) distribution and therefore $\varphi = \varphi^{\#}$. The solution vector has the same property; that is, $\pi = \pi^{\#}$ (although, π does not follow a q distribution). Additionally, comparison of $\pi^{\#}$ and $\varphi^{\#}$ shows that π is more *even* than φ. Finally, notice that the d_j are all positive, confirming that the optimal diversity-constrained distribution is intrinsically more diverse than Martin's guide. This can be seen graphically by comparing the T_j profiles in Figure 15.

4.2.2. *Concluding remarks*

Gove et al. (1994) discuss several points to consider when solving the diversity-constrained model. Two points to consider are worth mentioning here. First, in the example the model formulation kept the number of 'species' constant between the two communities. This need not be the case, however, and if this relationship is not maintained, either the community with more species will be more diverse or the two communities will not be intrinsically comparable (Section 2.3). This means that the optimal community must always be allowed to have at least as many species as the antecedent community or a feasible solution will not be found.

Finally, it is not enough simply to use the common diversity indices to constrain a model like (4.5) in the sense of equation set (4.6) because these indices may order communities differently. Indeed, even adding constraints for two different indices

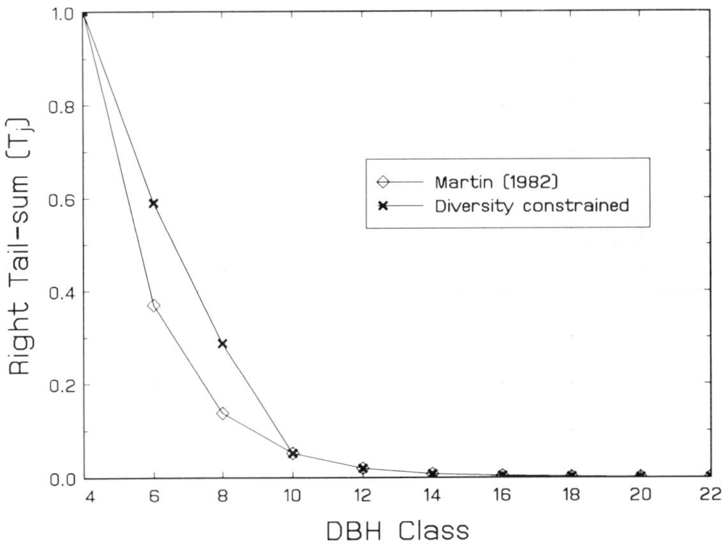

Fig. 15. Right tail-sum (T_j) diversity profiles illustrating the successful use of constraint set (4.6) on the optimal solution to model (4.5): the diversity-constrained distribution is intrinsically more diverse than Martin (1982).

will still not ensure intrinsic diversity ordering. Section 2.2 explores these concepts in more detail.

5. Predicting diversity

Numerous problems may arise precluding complete censusing of a population for the assessment of plant or total species diversity, the most common being practicality of the venture, especially on large tracts of land. Therefore, the investigator or manager is left with having to sample the population of interest and must face the biases inherent in diversity estimation when this choice is taken. Another potential problem the manager may face when asked for an assessment of diversity is the lack of adequate taxonomic skills necessary to fulfill the request. This latter concern has motivated the work discussed in this section. Here, two studies (Swindel et al., 1990, 1991) are summarized which present a modeling methodology which facilitates the prediction of total plant species diversity by grouping species into similar life forms. The method requires only that the person conducting the inventory be able to identify five plant life-forms, rather than as many as 80^+ individual plant species.

5.0.1. Predictor variables and principal components analysis

As a first step in obtaining simple descriptors of stand characteristics, the data of Appendix A were summarized separately for each watershed to obtain measured

Table 9
Total percent cover by life-form and year for both watersheds

Year	Forbs	Grasses	Grasslikes	Shrubs	Trees
Watershed 1					
1977	7.05	35.66	2.96	83.83	67.72
1980	6.86	27.66	4.35	19.14	1.93
1981	8.02	39.25	2.01	23.05	3.66
1982	9.67	41.18	1.35	34.77	6.64
1984	8.16	41.91	1.58	60.23	15.31
1986	3.63	31.57	0.41	77.39	21.83
Average ($\bar{x}_{(1)}$)	7.23	36.21	2.11	49.74	19.52
Watershed 2					
1977	3.37	30.12	3.33	79.85	72.41
1980	9.20	19.74	10.96	3.58	0.62
1981	21.43	23.21	5.95	9.39	3.13
1982	23.51	33.88	5.37	17.57	9.79
1984	25.22	39.08	3.46	31.31	18.75
1986	12.82	39.08	1.79	31.38	29.05
Average ($\bar{x}_{(2)}$)	15.93	30.85	5.14	28.85	22.29

abundances of five plant life-forms: (i) forbs; (ii) grasses; (iii) 'grasslikes' (which include sedges and rushes); (iv) 'shrubs' = shrubs + vines + ferns; and (v) trees (Table 9). The life forms in (iii) and (iv) were pooled subjectively based on graphical analysis (Swindel et al., 1991). These ($p = 5$) life forms become the variables for the subsequent principal components analysis. A $p \times p$ covariance matrix was computed for each watershed (Table 10) and these two covariance matrices were pooled

Table 10
Covariance matrices for WS 1 and WS 2 and pooled covariance matrix

	Forbs	Grasses	Grasslikes	Shrubs	Trees
S_1					
Forbs	4.1143767	7.9591900	0.5423800	−28.4420700	−11.6711100
Grasses	7.9591900	32.2733900	−3.7667400	3.7677700	−4.6118100
Grasslikes	0.5423800	−3.7667400	1.8997200	−15.6172000	2.2495000
Shrubs	−28.4420700	3.7677700	−15.6172000	781.9015100	563.4434900
Trees	−11.6711100	−4.6118100	2.2495000	563.4434900	614.6284300
S_2					
Forbs	77.3457900	23.1495900	−3.0850400	−129.6080600	−147.5416500
Grasses	23.1495900	65.3962567	−22.0454667	79.8166867	67.8348167
Grasslikes	−3.0850400	−22.0454667	10.3804667	−54.0693267	−50.3860067
Shrubs	−129.6080600	79.8166867	−54.0693267	751.5910667	725.1937867
Trees	−147.5416500	67.8348167	−50.3860067	725.1937867	712.6376167
S_{pooled}					
Forbs	40.7300833	15.5543900	−1.2713300	−79.0250650	−79.6063800
Grasses	15.5543900	48.8348233	−12.9061033	41.7922283	31.6115033
Grasslikes	−1.2713300	−12.9061033	6.1400933	−34.8432633	−24.0682533
Shrubs	−79.0250650	41.7922283	−34.8432633	766.7462883	644.3186383
Trees	−79.6063800	31.6115033	−24.0682533	644.3186383	663.6330233

using

$$\mathbf{S}_{\text{pooled}} = \frac{(n_1 - 1)\mathbf{S}_1 + (n_2 - 1)\mathbf{S}_2}{n_1 + n_2 - 2}, \tag{5.1}$$

where $\mathbf{S}_1$ and $\mathbf{S}_2$ are the covariance matrices for WS 1 and WS 2, respectively; and $n_1 = n_2 = 6$ measurement years for the two watersheds. (Note that (5.1) reduces to $\frac{1}{2}(\mathbf{S}_1 + \mathbf{S}_2)$ here because of the equal sample sizes.) The covariance matrices were calculated separately for each watershed and then pooled (rather than combining all of the data beforehand), in an attempt to preserve the differences in variation between the two watersheds due to postharvest treatments (refer to Section 3.1.2).

Swindel et al. (1990) performed principal components analysis on each watershed separately and found that the first three principal components accounted for $> 99.8\%$ of the variation on life-form abundances on each individual watershed. Table 11 presents the results of a principal components analysis for the combined watersheds using $\mathbf{S}_{\text{pooled}}$. In the combined analysis the first three principal components accounted for 98.7% of the variation in life-form abundances. The analysis presented in Table 11 is slightly different from the results of Swindel et al. (1991), who also did a combined analysis, in that a different pooling method has been used, and the principal components were not subsequently rotated here as they were by those authors. The first principal component ($\hat{\boldsymbol{e}}_1$) in Table 11 accounts for woody plant variability (i.e., shrubs and trees). The second two principal components contrast various combinations of overstory and understory life forms. The last two principal components account for so little of the variability that they are not considered in the subsequent analysis.

5.0.2. Linear regression analysis

Swindel et al. (1991) used linear regression models to predict plant species diversity on the two watersheds. The independent variables used were the principal component scores and an indicator variable for watershed. Three different regressions were run, one each for species count, Shannon's and Simpson's diversity indices as the dependent variable. For simplicity in illustration, the details to a similar analysis for prediction of species count only are presented – other diversity indices can be analyzed with the same methodology (Swindel et al., 1991).

The sample principal component scores were calculated for the first three principal components in Table 12 using the centered observations as follows

$$\hat{z}_{ij(w)} = \hat{\boldsymbol{e}}_i'(\boldsymbol{x}_{j(w)} - \bar{\boldsymbol{x}}_{(w)}). \tag{5.2}$$

Here $\hat{z}_{ij(w)}$ is the sample principal component score for the i-th principal component and the j-th measurement year on watershed w; where $i = 1, \ldots, 3$, $j = 1, \ldots, 6$, and $w = 1, 2$. In addition $\boldsymbol{x}_{j(w)}$ is the total percent cover for all life-forms in the j-th year of measurement on watershed w and $\bar{\boldsymbol{x}}_{(w)}$ is the mean life-form vector for watershed w. Table 12 presents the sample principle component scores $\hat{z}_{ij(w)}$ as calculated from the information in Tables 9 and 11 for both watersheds. To illustrate, the sample

Table 11
Principal components or eigenvectors ($\hat{e}_i$) of variation in life-form abundances, and correlations between principal component i and life-form k ($\hat{r}_{\hat{y}_i \cdot x_k}$)

Life-form	$\hat{e}_1(\hat{r}_{\hat{y}_1 \cdot x_k})$	$\hat{e}_2(\hat{r}_{\hat{y}_2 \cdot x_k})$	$\hat{e}_3(\hat{r}_{\hat{y}_3 \cdot x_k})$	$\hat{e}_4(\hat{r}_{\hat{y}_4 \cdot x_k})$	$\hat{e}_5(\hat{r}_{\hat{y}_5 \cdot x_k})$
Forbs	−0.0832 (−0.48)	−0.3148 (−0.43)	0.4412 (0.52)	0.8360 (0.56)	−0.0196 (0.00)
Grasses	0.0385 (0.20)	−0.4435 (−0.55)	0.6873 (0.74)	−0.5201 (−0.32)	0.2429 (0.04)
Grasslikes	−0.0308 (−0.46)	0.1635 (0.57)	−0.1238 (−0.38)	0.1466 (0.25)	0.9672 (0.51)
Shrubs	0.7313 (0.98)	−0.5614 (−0.18)	−0.3776 (−0.10)	0.0621 (0.01)	0.0604 (0.00)
Trees	0.6752 (0.97)	0.6020 (0.20)	0.4185 (0.12)	0.0721 (0.01)	−0.0376 (0.00)
$\hat{\lambda}_i$ [1]	1374.2447	74.6217	57.4337	18.1002	1.6840
$\hat{\gamma}_i$ [2]	90.05	4.89	3.76	1.19	0.11
$\hat{\delta}$ [3]	90.05	94.94	98.70	99.89	100.00

[1] $\hat{\lambda}_i$ = variance (eigenvalue) of the i-th principal component.
[2] $\hat{\gamma}_i$ = percent of total variance associated with the i-th principal component.
[3] $\hat{\delta}$ = cumulative percent of total variance.

Table 12
Principal component scores by watershed

Year	PC_1	PC_2	PC_3
	$\hat{z}_{ij(1)}$		
1977	57.4480183	10.3180675	6.7405152
1980	−34.6140397	10.8610558	−2.1214787
1981	−30.1649627	3.8201716	5.8930182
1982	−19.6246484	−2.4481163	4.8511479
1984	4.9947449	−11.3317419	−1.3262916
1986	21.9608875	−11.2194367	−14.0369109
	$\hat{z}_{ij(2)}$		
1977	72.2089103	5.5204234	−4.1005957
1980	−33.1570473	9.1324455	−10.8532166
1981	−27.9427868	1.1746148	−3.5957492
1982	−17.2085877	−4.8892784	4.4255380
1984	−0.9942336	−10.3648630	7.5521000
1986	7.0937451	−0.5733424	6.5719236

principal component score for principal component 2 in 1977 on WS 1 is calculated as

$$\hat{z}_{21(1)} = \hat{e}_2'(\boldsymbol{x}_{1(1)} - \bar{\boldsymbol{x}}_{(1)})$$

$$= \begin{pmatrix} -0.3148 \\ -0.4435 \\ 0.1635 \\ -0.5614 \\ 0.6020 \end{pmatrix}' \left[\begin{pmatrix} 7.0 \\ 35.66 \\ 2.96 \\ 83.83 \\ 67.72 \end{pmatrix} - \begin{pmatrix} 7.23 \\ 36.21 \\ 2.11 \\ 49.74 \\ 19.52 \end{pmatrix} \right]$$

$$= 10.318.$$

The full model was fitted using the scores from all three principal components (Table 12) and an indicator variable for watershed as independent variables:

$$\Delta_{\mathrm{SC}} = \beta_0 + \beta_1 \mathrm{WS} + \beta_2 \hat{z}_{1j(w)} + \beta_3 \hat{z}_{2j(w)} + \beta_4 \hat{z}_{3j(w)} + \varepsilon, \tag{5.3}$$

where $\beta_0, \ldots, \beta_4$ are regression coefficients to be estimated from the data, WS = 0 if WS 1 or 1 if WS 2, and Δ_{SC} is the species count index of diversity. After fitting model (5.3) it was determined that $\hat{z}_{2j(w)}$ could be dropped from the model, with the result that the final fitted model became

$$\hat{\Delta}_{\mathrm{SC}} = 76.83333 + 17.83333\ \mathrm{WS} - 0.11669\hat{z}_{1j(w)} + 0.74200\hat{z}_{3j(w)}, \tag{5.4}$$

with $R^2 = 0.87$. A plot of the observed and predicted values in Figure 16 shows that

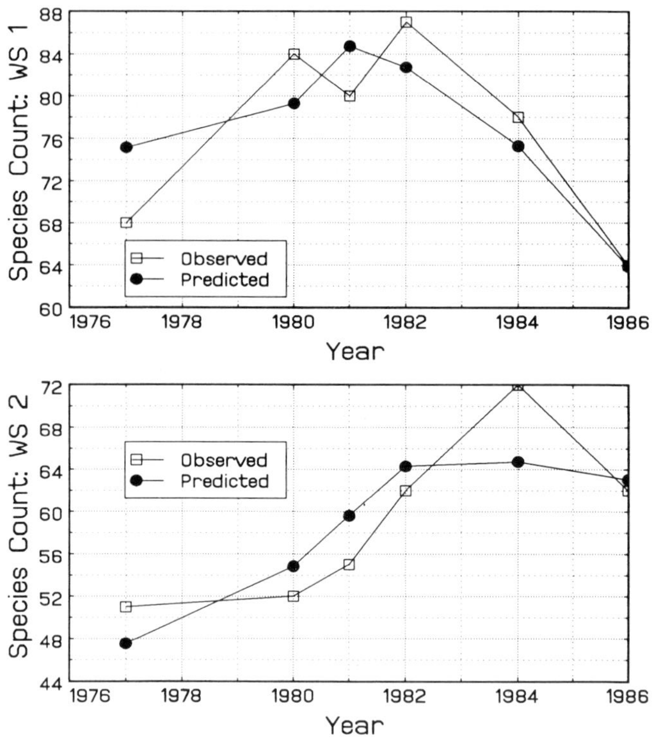

Fig. 16. Observed and predicted values from model (5.4) of the species count index on WS 1 (top) and WS 2 (bottom).

the equation does a good job of predicting species count on the calibration dataset. Unfortunately, more current data are not available on these watersheds to judge future predictions.

The work described here has great potential for practical applications because the approach allows an enormous reduction in the dimensionality of the data required to assess plant species diversity. Direct measurement of plant species diversity requires measurement of the abundances of s categories of plants, where s is the total number of plant species encountered in the community. However, (5.3) requires assessment of the abundances of just five categories of plants – individual species need not be identified. Measurement of the crown cover of plants in five categories is quick and easy and assessments can be done nondestructively; estimation of species count requires even less effort. Therefore, assessments can be made on permanent transects, and temporal trends in repeated surveys are not confounded with the considerable spatial variation typical in forests.

5.0.3. Concluding remarks

The results of fitted model (5.4) are consistent with intuition about plant community structure and its effect on plant species diversity in forested ecosystems. For example, the fitted model showed that diversity was inversely related to tree and shrub abundance ($\hat{z}_{1j(w)}$). Intuitively, it would seem that forest influences that diminish the overstory, including shrubs, and increase the herbaceous understory (e.g., tree harvesting, prescribed burning, patchy tree mortality due to pest epidemics) would tend to increase plant species diversity. Conversely, influences that have the opposite effect on stand structure (e.g., dense tree spacing, maturing plantations, weed control, over-grazing) would tend to diminish plant species diversity.

The model illustrated reduces the taxonomic skills required to categorize field data properly, and dramatically reduces bookkeeping required for data collection, storage, retrieval, and analysis. For sites very different from those described here, or for stands treated very differently, or in different ecosystems, empirical models might be developed using a methodology similar to the one described here.

Appendix A.1.

Percent crown cover for WS 1, Bradford County, Florida[1]

Life form and species	1977–78	1980	1981	1982	1984	1986
Trees						
Gordonia lasianthus	0.42			0.04	0.08	
Ilex cassine	0.42	0.04	0.23	0.04	0.15	1.08
Ilex myrtifolia	0.81	0.04	0.08	0.08	0.69	0.62
Magnolia virginiana	0.08					0.12
Nyssa sylvatica	3.15	1.54	1.92	1.62	0.77	
Persea borbonia	1.38	0.04	0.27	0.31	0.27	1.04
Pinus elliottii	57.15	0.27	0.58	4.12	12.08	17.46
Pinus palustris	2.00					
Quercus laurifolia	2.27		0.46	0.27	1.15	1.08
Quercus nigra	0.04		0.12	0.12	0.12	0.31
Quercus virginiana				0.04		0.12
Shrubs						
Aronia arbutifolia	0.19	0.15	0.88	0.46	0.31	0.85
Asimina pygmaea	0.50	0.08	0.04	0.12		0.04
Befaria racemosa	0.04					
Gaylussacia dumosa	1.73	0.38	0.73	0.81	1.88	3.35
Gaylussacia frondosa	3.77	1.12	1.54	2.31	3.46	5.35
Hypericum cistifolium			0.15	0.12	0.31	
Hypericum fasciculatum	0.40	0.15	0.73	1.12	0.38	0.12
Hypericum myrtifolium		0.15	0.31	0.15	0.12	0.01
Hypericum spp.		0.15				
Ilex coriacea	9.50	1.12	1.27	2.23	5.04	6.81
Ilex glabra	29.58	7.65	7.54	14.35	25.27	32.62
Kalmia hirsuta	0.08		0.04	0.12	0.15	0.16
Lyonia ferruginea	0.58	0.15	0.31	0.35	0.38	1.19
Lyonia lucida	4.46	0.42	0.69	1.85	5.19	7.50
Myrica cerifera	4.38	0.46	0.58	1.12	3.23	3.85
Quercus minima	0.23	0.12	0.35	0.35	0.08	0.35
Quercus pumila	0.88	0.50	0.35	0.27	0.50	0.19
Rhododendron spp.	0.04					
Rubus spp.	0.38	2.88	2.23	2.85	3.73	1.20

(continued on next page)

Appendix A.1 continued

Life form and species	1977–78	1980	1981	1982	1984	1986
Serenoa repens	15.38	1.15	1.19	1.73	1.73	3.12
Vaccinium fuscatum	4.81	0.35	0.81	0.81	2.12	1.77
Vaccinium myrsinites	4.88	1.38	2.08	2.38	3.35	2.47
Vines						
Gelsemium sempervirens					0.27	0.35
Smilax spp.	0.15	0.12	0.15	0.19	0.38	0.44
Toxicodendron radicans						0.04
Vitis rotundifolia	1.23		0.15	0.15	0.46	0.92
Grasses						
Amphicarpum muhlenbergianum		0.27	0.77	0.42	0.19	
Andropogon capillipes	2.69	5.15	13.62	19.92	28.65	18.19
Andropogon spp.	4.12	2.69	4.81	4.15	4.15	1.85
Aristida lanosa	5.31	0.31	1.50	1.73	2.42	7.08
Aristada spiciformis		0.08	0.31	0.08	0.12	
Aristada stricta	1.62	0.42	0.46	0.42	0.38	0.23
Aristada virgata	1.04	0.23	0.19	0.04	0.04	0.08
Axonopus spp.					0.04	
Ctenium spp.	0.15					
Panicum anceps		0.08	0.35	0.42	0.27	0.04
Panicum hemitomon				0.04		
Panicum spp.[2]	3.73	15.69	15.08	11.69	1.50	0.64
Paspalum spp.	0.85	1.08	0.62	0.15	0.69	0.23
Sorghastrum secundum	0.27	0.04				
Sporobolus curtissii	15.88	1.62	1.54	2.12	3.46	3.23
Grasslikes						
Carex spp.		0.19	0.08	0.12	0.15	0.08
Cyperus spp.		0.12				
Eleocharis spp.			0.19	0.19		
Fimbristylis puberula		0.08				
Juncus spp.		0.15	0.08		0.08	
Rhynchospora spp.	2.96	3.12	1.58	0.92	1.35	0.33
Scleria baldwinii		0.23	0.08	0.12		
Scleria reticularis		0.38				
Scleria spp.		0.08				
Forbs						
Ageratina aromatica		0.08				
Asclepias pedicellata				0.04		
Aster dumosa		0.42	0.23	0.27	0.08	
Aster reticulatus	0.19		0.04		0.04	
Aster squarrosus		0.08	0.08			
Bartonia paniculata		0.04				
Calopogon pallidus					0.04	
Cassia fasciculata		0.04	0.04	0.88	0.15	
Centella asiatica	0.42	0.92	0.92	1.00	0.62	0.01
Centrosema virginianum	0.04	0.08	0.04	0.04	0.04	
Cirsium spp.				0.04		
Crotalaria rotundifolia				0.12	0.04	
Diodia spp.		0.04		0.04	0.04	
Elephantopus spp.	0.50	0.15	0.31	0.50	0.46	0.15
Erechtites hieracifolia		0.04				
Erigeron vernus	0.04	0.08	0.04			
Erigeron spp.				0.04		

Appendix A.1 continued

Life form and species	1977–78	1980	1981	1982	1984	1986
Eriocaulon spp.	0.88		0.15			
Eryngium baldwinii		0.08	0.08	0.04		
Eryngium yuccifolium					0.08	
Eupatorium capillifolium		0.08		0.04		
Eupatorium compositifolium		0.04	0.08	0.27	0.27	0.38
Eupatorium pilosum						0.04
Eupatorium recurvans	0.62	0.46	0.92	0.73	1.31	0.89
Euphorbia corollata	0.04					
Euthamia minor	0.12	0.31		0.23	0.42	0.04
Galactia spp.	0.08	0.08	0.19	0.15	0.19	0.12
Galium hispidulum					0.04	
Gratiola hispida		0.04		0.08		
Gratiola pilosa		0.08	0.38	0.04		
Heterotheca graminifolia	0.27	0.23	0.27	0.19	0.12	
Houstonia procumbens				0.11		
Hypericum suffruticosum		0.04	0.35	0.19	0.04	
Hypoxis juncea		0.04				
Lachnanthes caroliniana		0.12	0.12	0.08	0.04	0.01
Lachnocaulon spp.						0.12
Lechea legettii		0.04	0.04	0.04		
Liatris spp.		0.08		0.04		
Lilium spp.	0.04					
Ludwigia virgata	0.04	0.31	0.19	0.27	0.31	0.13
Oldenlandia uniflora		0.69	0.19	0.42	0.04	0.01
Physostegia purpurea		0.12			0.04	
Platanthera spp.				0.04		
Pluchea spp.		0.04	0.04	0.04		
Polygala lutea		0.08	0.12		0.12	
Polygala setacea	0.04	0.04	0.04		0.12	0.01
Polypremum procumbens		0.12	0.38	0.65		
Proserpinaca pectinata			0.04			
Pterocaulon virgatum	0.31	0.38	0.42	0.35	0.81	0.47
Rhexia alifanus				0.12	0.04	0.01
Rhexia virginica		0.04	0.04			
Rhexia spp.	0.42	0.65	0.73	1.31	1.08	0.67
Sabatia spp.	0.04					
Solidago fistulosa				0.08	0.19	0.12
Syngonanthus flavidulus	0.42	0.35	0.96	0.42	0.46	0.04
Tephrosia hispidula	0.15	0.12	0.08	0.15	0.12	0.04
Tragia urens	0.58		0.04	0.04		0.01
Trilisa odoratissima	0.04		0.04			
Trilisa paniculata			0.12	0.12	0.19	0.12
Viola septemloba	0.08			0.08	0.04	
Xyris spp.	1.69	0.23	0.31	0.38	0.58	0.24
Ferns						
Lycopodium spp.	0.04					
Osmunda cinnamomea	0.04	0.12	0.04	0.12	0.08	0.19
Pteridium aquilinum	0.77	0.46	0.81	0.62	1.58	3.85
Woodwardia virginica	0.15	0.08	0.08	0.19	0.23	0.65
Totals	197.22	59.94	75.99	93.61	127.19	134.83

[1] Nomenclature follows USDA Soil Conservation Service (1982) and Little (1979).

[2] *Panicum* spp. includes *Dichanthelium* spp.

Appendix A.2.

Percent crown cover for WS 2, Bradford County, Florida[1]

Life form and species	1977–78	1980	1981	1982	1984	1986
Trees						
Ilex cassine	0.33					
Ilex myrtifolia	2.96	0.04		0.25	0.54	1.59
Nyssa sylvatica	2.88	0.08	0.17	0.21	0.33	0.46
Persea borbonia	0.29	0.04			0.13	
Pinus elliottii	58.67	0.46	2.96	9.33	17.75	27.00
Pinus palustris	5.08					
Quercus nigra	0.12					
Taxodium distichum	2.08					
Shrubs						
Aronia arbutifolia	0.17	0.04		0.04	0.04	0.13
Asimina pygmaea						0.04
Gaylussacia dumosa	0.75	0.04	0.04	0.04	0.13	0.21
Gaylussacia frondosa	0.88	0.04	0.08	0.04	0.13	0.33
Hypericum cistifolium		0.17	0.92	0.08	0.58	0.17
Hypericum fasciculatum	0.62	0.92	5.12	8.25	7.38	1.13
Hypericum myrtifolium		0.25	0.58	0.33	0.13	0.08
Hypericum spp.		0.25	0.08			
Ilex coriacea	2.25				0.08	0.17
Ilex glabra	42.17	0.71	1.42	6.00	14.96	20.05
Lyonia ferruginea	0.71		0.04	0.17	0.29	0.17
Lyonia lucida	1.42					0.04
Myrica cerifera	18.71		0.08		0.13	0.13
Quercus minima		0.04			0.04	0.08
Rubus spp.	0.33	0.54	0.54	1.29	2.96	1.46
Serenoa repens	3.46			0.04		0.13
Vaccinium fuscatum	4.71				0.04	0.21
Vaccinium myrsinites	1.00	0.29	0.33	0.58	0.67	0.75
Vines						
Gelsemium sempervirens						0.04
Smilax spp.	0.17	0.04	0.04	0.13	0.17	0.05
Toxicodendron radicans	0.58					0.04
Vitis rotundifolia	0.62		0.04	0.42	3.21	5.84
Grasses						
Amphicarpum muhlenbergianum			1.50	3.75	3.33	0.50
Andropogon capillipes	4.75	0.46	1.71	5.13	16.71	23.29
Andropogon spp.	4.08	0.58	1.29	1.92	6.04	6.50
Aristida lanosa	8.29		0.08	0.25	0.75	0.71
Aristida spiciformis			0.17	0.33	0.46	0.17
Aristida stricta	1.46					
Aristida virgata	0.29	0.04		0.08	0.33	0.06
Axonopus spp.					0.42	0.21
Erianthus giganteus	0.25					
Panicum anceps			0.04	0.04	0.08	0.21
Panicum hemitomon			0.29	0.38		
Panicum spp.[2]	1.12	18.50	17.92	21.25	10.21	6.59
Paspalum spp.	0.29	0.12	0.21	0.21	0.25	0.42
Sporobolus curtissii	9.42	0.04		0.54	0.50	0.42
Sporobolus floridanus	0.17					

Appendix A.2 continued

Life form and species	1977–78	1980	1981	1982	1984	1986
Grasslikes						
Carex spp.	0.75			0.08	0.17	0.17
Cyperus spp.		0.08				
Eleocharis spp.			0.29	0.42		
Fimbristylis autumnalis		0.25	0.12			
Fimbristylis puberula				0.04		
Fuirena breviseta		0.17	0.04	0.08	0.04	
Juncus spp.		0.21	0.08	0.04	0.13	0.09
Rhynchospora spp.	2.54	9.25	5.17	4.17	3.08	1.53
Scleria baldwinii		0.04	0.25	0.21		
Scleria ciliata					0.04	
Scleria reticularis		0.96		0.33		
Scleria spp.	0.04					
Forbs						
Aster dumosus			0.04	0.04		
Bigelowia nudata	0.08					
Cassia fasciculata				0.04	0.33	0.04
Centella asiatica	0.38	2.67	10.12	13.37	12.79	3.30
Centrosema virginianum					0.04	
Crotalaria rotundifolia					0.08	
Diodia spp.		0.12	0.12	0.25	0.25	0.33
Drosera spp.		0.04			0.17	0.04
Erigeron vernus					0.13	
Eriocaulon spp.	0.17		0.04			0.04
Eupatorium capillifolium		0.46	0.96	0.54	0.17	
Eupatorium compositifolium		0.08	0.08	0.17	0.38	
Eupatorium pilosum					0.13	
Eupatorium recurvans	0.17	0.33	0.54	0.33	0.83	1.21
Euthamia minor	0.08		0.08	0.46	0.42	0.17
Gratiola brevifolia						0.01
Gratiola hispida						0.01
Gratiola pilosa		0.29	0.71	0.08		
Gratiola ramosa					0.04	
Hypericum suffruticosum		0.17	0.83	0.08	0.13	0.01
Lachnanthes caroliniana	0.17	0.25	0.54	0.50	0.50	0.55
Lachnocaulon spp.						0.51
Lechea leggettii			0.08	0.08	0.08	
Lechea minor					0.04	
Lobelia floridana					0.04	
Lobelia paludosa		0.04		0.21	0.17	
Ludwigia lanceolata				0.04		
Ludwigia virgata		0.17	0.12	0.21	0.33	0.26
Oldenlandia uniflora		1.29	0.04	0.33	0.17	0.05
Physostegia purpurea	0.12					
Pluchea spp.			0.21	0.08	0.25	0.21
Polygala cymosa	0.04					
Polygala lutea		0.04	0.08		0.21	0.04
Polypremum procumbens		0.46	0.38	1.29		
Proserpinaca pectinata		0.04	0.08	0.04		0.01
Pterocaulon virgatum	0.50	0.04	0.08		0.04	

(continued on next page)

$$I(t) = 7.07933\left(\frac{\text{BPA}}{\text{TPA}}\right)^{-1.40072},$$

$$U_i(t) = 0.00330 N_i(t-1)^{0.88218} S Y_{1i}^{0.48383} \times \exp(-0.00286\,\text{BPA}), \quad i = 1, \ldots, s, \tag{C.1}$$

$$M_i(t) = 0.04109 N_i(t-1), \quad i = 1, \ldots, s,$$

where

$I(t)$ = Ingrowth: the number of trees per acre growing into the 6-inch diameter class during the growth period.

$U_i(t)$ = Upgrowth: the number of trees per acre growing from diameter class i to $i+1$ during the growth period.

$M_i(t)$ = Mortality: the number of trees dying in diameter class i during the growth period.

$N_i(t-1)$ = The number of trees in diameter class i at the *beginning* of the growth period.

$N_i(t)$ = The number of trees in diameter class i at the *end* of the growth period. These are calculated as follows

$$N_i(t) = N_i(t-1) + I(t) - U_i(t) - M_i(t), \quad \text{for } i = 1,$$

$$N_i(t) = N_i(t-1) + U_{i-1}(t) - U_i(t) - M_i(t), \quad \text{for } i = 2, \ldots, s \text{ and}$$

$$N_{s+1}(t) = U_s(t), \quad \text{for the last diameter class.}$$

s = The number of diameter classes; $s = 9$ here.

Y_{1i} = Two-inch diameter class midpoint diameters at breast height (DBH = 4.5 ft), beginning at 6 inchcs (i.e., 6, 8, ..., 22).

Y_{2i} = $\pi Y_{1i}^2/(2 \times 12)^2 = 0.005454\,Y_{1i}^2$ = The tree basal (cross-sectional) areas in square feet corresponding to the Y_{1i} (tree cross-sections generally are assumed circular for basal area calculations).

S = Site index: a measure of site quality; $S = 55$ here.

BPA = $\sum_{i=1}^{s} Y_{2i} N_i(t-1)$ = Basal area per acre in square feet.

TPA = $\sum_{i=1}^{s} N_i(t-1)$ = Number of trees per acre.

References

Adams, D. M. (1976). A note on the interdependence of stand structure and best stocking in a selection forest. *Forest Sci.* **22**, 180–184.

Adams, D. M. and A. R. Ek (1974). Optimizing the management of uneven-aged forest stands. *Canad. J. Forest. Res.* **4**, 274–286.

Arnold, B. C. (1987). *Majorization and the Lorenz Order: A Brief Introduction.* Springer, New York.

Boyce, S. G. (1985). Forestry decisions. USDA Forest Service General Technical Report SE-35.

Boyce, S. G. and N. D. Cost (1978). Forest diversity – New concepts and applications. USDA Forest Service Research Paper SE-194.

Conde, L. F., B. F. Swindel and J. E. Smith (1983a). Plant species cover, frequency, and biomass: Early responses to clearcutting, chopping, and bedding in *Pinus elliottii* flatwoods. *Forester Ecol. Management* **6**, 307–317.
Conde, L. F., B. F. Swindel and J. E. Smith (1983b). Plant species cover, frequency, and biomass: Early responses to clearcutting, burning, windrowing, discing, and bedding in *Pinus elliottii* flatwoods. *Forest Ecol. Management* **6**, 319–331.
Ek, A. R. (1974). Nonlinear models for stand table projection in northern hardwood stands. *Canad. J. Forest Res.* **4**, 23–27.
Federal Register (1982). *National Forest System Land and Resource Management Planning.* **47**(190), 43026–43052.
Gove, J. H. and S. E. Fairweather (1992). Optimizing the management of uneven-aged forest stands: A stochastic approach. *Forest Sci.* **38**(3), 623–640.
Gove, J. H., C. W. Martin, G. P. Patil, D. S. Solomon and J. W. Hornbeck (1992). Plant species diversity on even-aged harvests at the Hubbard Brook Experimental Forest: 10-year results. *Canad. J. Forest Res.* **22** (11), 1800–1806.
Gove, J. H., G. P. Patil and C. Taillie (1994). A mathematical programming model for maintaining structural diversity in uneven-aged forest stands with implications to other formulations. *Ecol. Model.*
Gove, J. H., D. S. Solomon, S. E. Fairweather and G. P. Patil (1991). Maximizing the diameter class diversity of uneven-aged northern hardwood stands. In: M. A. Buford, comp., *Proc. 1991 Sympos. on Systems Analysis in Forest Resources.* USDA Forest Service General Technical Report SE-74, 39–45.
Grassle, J. F., G. P. Patil, W. K. Smith and C. Taillie, eds. (1979). *Ecological Diversity in Theory and Practice.* International Co-operative Publishing House, Fairland, MD.
Gregoire, T. G. (1984). The jackknife: An introduction with applications in foresty data analysis. *Canad. J. Forest Res.* **14**, 493–497.
Hunter, M. L., Jr. (1990). *Wildlife, Forests, and Forestry: Principles of Managing Forests for Biological Diversity.* Prentice-Hall, Englewood Cliffs, NJ.
Hurlbert, S. H. (1971). The nonconcept of species diversity: A critique and alternative parameters. *Ecology* **52**(4), 577–586.
Lasdon, L. S. and A. D. Waren (1986). GRG2 user's guide. Department of General Business, School of Business Administration, University of Texas, Austin, TX.
Little, E. L., Jr. (1979). *Checklist of United States Trees (Native and Naturalized)*, Agricultural Handbook, Vol. 541. USDA Forest Service, Washington, DC.
MacArthur, R. H. and J. W. MacArthur (1961). On bird species diversity. *Ecology* **42**(3), 594–598.
Magurran, A. E. (1988). *Ecological Diversity and its Measurement.* Princeton Univ. Press, Princeton, NJ.
Martin, C. W. and J. W. Hornbeck (1989). Revegetation after strip cutting and block clearcutting in northern hardwoods: A 10-year history. USDA Forest Service Research Paper NE-625.
Martin, G. L. (1982). Investment-efficient stocking guides for all-aged northern hardwood forests. Agriculture and Life Sciences Research Report R3129. University of Wisconsin, Madison, WI.
McMinn, J. W. (1991). Biological diversity research: An analysis. USDA Forest Service General Technical Report SE-71.
Morris, L. A., W. L. Pritchett and B. F. Swindel (1983). Displacement of nutrients into windrows during site preparation of a flatwood forest. *Soil Sci. Soc. Amer. J.* **47**(3), 591–594.
Niese, J. N. and T. F. Strong (1992). Economic and biodiversity trade-offs in managed northern hardwoods. *Canad. J. Forest Res.* **22**(11), 1807–1813.
Office of Technology Assessment (1987). Technologies to maintain biological diversity. OTA-F-330 Congress of the United States, Washington, DC.
Palmer, M. W. (1991). Estimating species richness: The second-order jackknife reconsidered. *Ecology* **72**(4), 1512–1513.
Patil, G. P. and C. Taillie (1979a). An overview of diversity. In: J. F. Grassle, G. P. Patil, W. K. Smith and C. Taillie, eds., *Ecological Diversity in Theory and Practice.* International Co-operative Publishing House, Fairland, MD, 3–27.
Patil, G. P. and C. Taillie (1979b). A study of diversity profiles and orderings for a bird community in the vicinity of Colstrip, Montana. In: G. P. Patil and M. L. Rosenzweig, eds., *Contemporary Quantitative*

Ecology and Related Econometrics. International Co-operative Publishing House, Fairland, MD, 23–48.

Patil, G. P. and C. Taillie (1982). Diversity as a concept and its measurement. *J. Amer. Statist. Assoc.* **77**, 548–567.

Pielou, E. C. (1974). *Population and Community Ecology.* Gordon and Breach, New York.

Pielou, E. C. (1975). *Ecological Diversity.* Wiley, New York.

Pielou, E. C. (1977). *Mathematical Ecology.* Wiley, New York.

Quenouille, M. H. (1956). Notes on bias in estimation. *Biometrika* **43**, 353–360.

Solomon, D. L. (1979). A comparative approach to species diversity. In: J. F. Grassle, G. P. Patil, W. K. Smith and C. Taillie, eds., *Ecological Diversity in Theory and Practice.* International Co-operative Publishing House, Fairland, MD, 29–35.

Swindel, B. F., R. C. Abt and J. E. Smith (1990). Principal components of variation in community structure of managed forests. *J. Environ. Management* **31**, 361–366.

Swindel, B. F., L. F. Conde and J. E. Smith (1984). Species diversity: Concept, measurement, and response to clearcutting and site-preparation. *Forest Ecol. Management* **8**, 11–22.

Swindel, B. F, and L. R. Grosenbaugh (1988). Species diversity in yound Douglas-fir plantations compared to old growth. *Forest Ecol. Management* **23**, 227–231.

Swindel, B. F., R. Lloyd, G. P. Patil, C. Taillie and L. Conde (1987). U.S. National Forest Management Act, forest ecosystems diversity, and diversity profiles. *Coenoses* **2**(1), 1–8.

Swindel, B. F., J. E. Smith and R. C. Abt (1991). Methodology for predicting species diversity in managed forests. *Forest Ecol. Management* **40**, 75–85.

Taillie, C. (1979). Species equitability: A comparative approach. In: J. F. Grassle, G. P. Patil, W. K. Smith and C. Taillie, eds., *Ecological Diversity in Theory and Practice.* International Co-operative Publishing House, Fairland, MD, 51–62.

USDA Soil Conservation Service (1982). *National List of Scientific Plant Names*, Vol. 1. List of Plant Names. SCS-TP-159, Washington, DC.

Zahl, S. (1977). Jackknifing an index of diversity. *Ecology* **58**, 907–913.

G. P. Patil and C. R. Rao, eds., *Handbook of Statistics, Vol. 12*

13

Ornithological Statistics

Philip M. North

Abstract: A personalised view of ornithological statistics is presented and the subject area is reviewed and put into historical context. The review is not exhaustive, but covers as much material as is possible in just one chapter. A section on bird counts is followed by one on bird movements, within which a recent project involving bird observatories data is described in some detail. Finally there is coverage of analysis for ring recovery data, capture–recapture data, and radio-tracking information. The historical development of this last area is charted carefully, since such great advances have been made since the 1960s and 1970s, and because it has been by far the most actively developing area of ornithological statistics.

1. Introduction

In this chapter I will attempt to give an overview of how statistics is applied to the particular area of ornithology, and to trace the historical development of the subject over the past few decades. The view presented here will inevitably be a somewhat personal one. My own involvement in this area stems from 1975, when I came to the University of Kent at Canterbury. Indeed, the work in this area at the University of Kent, which has gone on since that time, serves as a useful background to illustrate the various areas of development in the subject. My view will tend to be dominated by developments in Britain and, to a lesser extent, Europe, but no one involved in the field of ornithological statistics could fail to be well aware of the extensive developments in America during the past two decades in methodology for handling capture–recapture and ring recovery (from dead birds) data.

The recent series of EURING (European Union for Bird Ringing) meetings (Wageningen, The Netherlands, 1986; Sempach, Switzerland, 1989; Montpellier, France, 1992) have proved to be extremely successful in charting the development of methods relating to ring recovery data, capture–recapture data and radio-tracking information. Involvement of Americans at each of the meetings has ensured that they have adopted a worldwide, rather than just European view of matters. For useful reviews of the subject area see the respective sets of Proceedings (North, 1987a, 1990a; Lebreton and North, 1992).

The other main area in which ornithologists have had recourse to the use of statistics is in the context of bird counts and, specifically, censuses. A very useful and (at the time) comprehensive review of this area is given in the Proceedings of the Symposium on Estimating Numbers of Terrestrial Birds, Asilomar, California, United States, 1980 (Ralph and Scott, 1981). There has also been a regular series of Proceedings of conferences on Bird Census and Atlas work: see, for example, Taylor, Fuller and Lack (1985).

When one talks to people for the first time about ornithological statistics, they are often surprised at how ornithology and statistics come together. But there should be no surprise in that. Ornithologists collect vast amounts of data and the interpretation of such data leads us, by definition, into the realms of statistics. That is not to say that the full range of statistical principles has always been adopted. Emphasis has been on analysis, rather than design – at least until recently, when one has increasingly seen properly designed and executed studies. Earlier on, data were often collected without any real purpose. There was no sense of knowing at the data collection stage what the questions of interest were, nor of the form of the statistical analysis that would be carried out to answer those questions. Happily, things are rather different now. That is not to say, however, that there are not still large sets of data around waiting to be analysed – whatever that may mean.

What has helped enormously in getting ornithological statistics on to a sounder methodological basis is the contact between ornithologists and statisticians. As with any other application area of statistics, it is essential to establish a dialogue between the statistician and the application subject expert. This should be established right at the outset and should be regular and ongoing. Wherever modelling techniques are involved (that is, in much of the work now) such dialogue is an essential part of the iterative process whereby the most effective models are developed.

2. Counts of birds

2.1. *Background*

How many breeding Blackbirds are there in the United Kingdom? To the layman it may seem that it should be relatively straightforward to answer such a question. However, a little thought reveals that this is actually a very difficult question to answer. But why would anyone want to know the answer to the question anyway? There is much interest in conservation issues nowadays. In Europe there has been growing support for 'green' politics, for example. Another question of interest might then be, what is the effect of pesticide use on certain species of birds? (Birds of prey received particular attention in this context in the UK.) This might be reformulated as, how many Peregrine Falcons are there in the UK in 1992, relative to the number in 1960? This is the way that we are likely to want to assess the effect of habitat change on birds – i.e., it is the *relative* numbers that are of interest, not the absolute numbers. This type of question has been of considerable interest in the UK, for example, because of the very marked changes in habitat brought about by modern farming methods.

2.2. The Common Birds Census in the United Kingdom

Considerations like the above led to the setting up of the Common Birds Census, a breeding bird census (strictly speaking, a sample survey) carried out annually in the UK, to monitor population *fluctuations* in common breeding bird species. It therefore tells us something about population dynamics, rather than population levels per se. This study is administered by the British Trust for Ornithology (BTO), an ornithological organisation in the UK, carrying out scientific research. The census (or survey) is carried out on farmland, woodland and in other habitats. It is one of many studies carried out by the BTO. Similar studies are carried out in other countries by the respective organisations there.

2.2.1. Design features

The Common Birds Census is carried out on a collection (the sample, in the survey) of plots of land – the census plots – which are visited regularly during the breeding season, so that relevant information can be recorded. If a statistician were given the problem of constructing an index for the number of breeding birds of a particular species in Britain, he or she would immediately think in terms of a sample survey, and be concerned with the design of that. The sample would be constructed from an identifiable sampling frame. But, in reality, in schemes such as this one, it is not possible to adhere at all rigidly to design principles, even though it is clear that is what *should* be done. The Common Birds Census, like many BTO schemes, and schemes in other countries, is carried out by voluntary observers. This means that there are limits to the extent to which design considerations can be imposed upon the observers. Properly designed experiments or surveys really need to be restricted to smaller scale studies, for which tighter organisation is possible. In practice we have to settle for what is feasible. This is one of the underlying features of much of ornithological statistics, resulting in the fact that many of the data are collected in a somewhat ad hoc way. Thus, in the case of the Common Birds Census, it is really a matter of accepting study plots into the scheme wherever they are offered. When study plots drop out of the scheme (this happens in real life, even if it would not be in the ideal sampling scheme!) every attempt is made to replace them with other plots which will not change the overall balance of the sample. So, if all that is required is a population *index* to measure *change*, we perhaps do not have to worry any further about any inadequacy in the design. But, of course, we do – and we would not expect it to be otherwise. The next concern is whether the ad hoc sample is representative of the whole of Britain. It almost certainly is not. The locations of the study plots are likely to be governed more by where in the country there are concentrations of people, rather than of birds. Now, if changes in population level are similar across the whole country, even that will not matter. But differential weather conditions are likely to dictate that this will not be the cases, and without a fully representative sample, that element is not measurable anyway.

The foregoing discussion is not meant to suggest that national schemes such as the Common Birds Census are without value. Far from it, they have made a significant contribution to the knowledge about bird population dynamics and responses to environmental factors, both with regard to habitat, and to weather, for example. It is simply that there has to be a coming to terms with what is practical in reality, and a recognition, from a statistical point of view, that it has not been possible to address design considerations as much as we would have liked. (There is further discussion later of aspects of experimental design, in the context of ring recovery/capture–recapture/radio-tracking studies.) Of course, it is true in any practical application of statistics that it is not feasible (and, indeed, there is no point in trying) to impose a mathematically attractive design simply for the sake of it. I have dwelt on this aspect somewhat because this potential problem does seem to be particularly prevalent in the ornithological context, compared with others.

2.2.2. *Statistical applications illustrated by the Common Birds Census*

The Common Birds Census is a worthwhile case study to consider further here, since it leads us into interesting applications of statistics.

2.2.2.1. Cluster analysis. Firstly, we are led into an application of cluster analysis. The Common Birds Census (in common with other such schemes carried out in Europe) is based on a mapping technique. When observers visit their study plots they record all the birds seen and/or heard on a map of the plot. This is done on a number of occasions during the breeding season. At the end of the series of visits, species maps are compiled, collecting together on separate maps for each species, all the records for those species. An example for the Robin is shown in Figure 1.

The problem now is to determine, for each species, how many territories were being held on the study plot. To address this problem, groupings (clusters) of records are sought, taking account not only of the relative positions of the records in the plot, but also of relevant behavioural information (e.g., a male singing from a songpost, a bird carrying nest material, etc.). In practice, the records are interpreted by ornithological experts who estimate not only how many territories there are, but where their boundaries are. This is actually more informative than is required for the purposes of the study, and the assessment is done by hand.

On both counts it seems sensible to seek an objective, automatic method which primarily addresses the question of interest, namely, how many territories are there? In an attempt to do this, North (1977, 1978, 1979a, 1980a) presented an adaptation of single link cluster analysis and demonstrated its use with Common Birds Census data. Although falling a little way short of being fully automatic there is good prospect that with further development and refinement the method could be very useful as an automatic means of territory assessment.

As we have seen, it can be argued that the nonautomatic method uses too much information for what is needed. A much simpler approach, which can also be useful as the basis of an index, is the point sampling method. In this approach, the observer

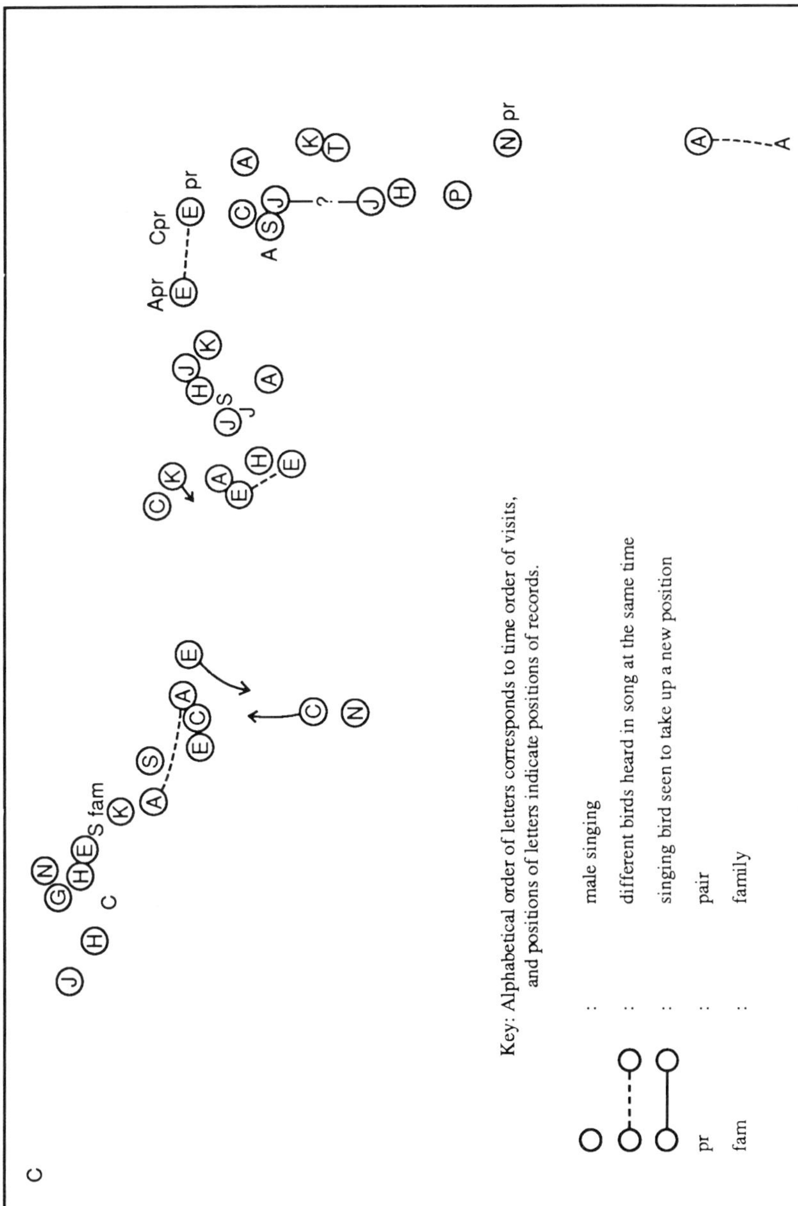

Fig. 1. Records of Robins on part of a Common Birds Census plot.

stands still and records all the birds that he or she sees or hears in a given time period. See, for example, Braae and Nøhr (1985). See also Buckland (1987) and, for a more general and extensive review of distance sampling methods, Buckland, Anderson, Burnham and Laake (1992).

2.2.2.2. A simple model for numbers of territories in successive years. From any given study plot, in any given year, the outcome from the mapping method is a set of estimated territory numbers for the breeding territorial species on that plot. Over a period of years a sequence of territory numbers is obtained for each species. Such sequences lend themselves to further stsatistical analysis. For example, North (1983) fitted simple birth–death–maturation models to the data. These models were of the following form:

$$x_{n+2} = Ax_{n+1} + Bx_n$$

for territory-holding birds of a single species, on one plot in successive years, where x_n is the mean number of birds in year n. Now, suppose the corresponding mean number of nonterritory holding birds is y_n. Note that birds have a well-defined breeding season, and deaths do not occur at the same rate throughout the year (Lack, 1966, Chapters 4 and 7; Batten, 1978). As a simplication of the real situation we therefore assume that each year from one breeding season to the next consists of three separate periods of events for the birds, viz. (i) a birth period, (ii) a death period and (iii) a maturation period (as in North and Morgan, 1979, Appendix 2). During this last period nonterritory-holding birds (henceforth referred to as 'juveniles') 'mature' to become and remain territory-holders (henceforth referred to as 'adults').

In the simplest model we suppose that, on average, in period (i) each adult gives rise to λ juveniles, in period (ii) each adult gives rise to 1-μ adults, and in period (iii) each juvenile gives rise to v adults and 1-v juveniles, where $\lambda \geqslant 0, 0 \leqslant \mu \leqslant 1, 0 \leqslant v \leqslant 1$. Because there is heavy mortality among birds in their very early life, quite different from the mortality among older birds, we assume that λ is a recruitment rate, which refers to birds which are born and survive to the next breeding season. We assume that the parameter values are constant over time, a further approximation to reality – weather conditions, for example, may affect the parameter values (e.g., North and Morgan, 1979). For the present model, in one year the number pair $\{x_n, y_n\}$ undergoes the following transitions:

$$\{x_n, y_n\} \to \{x_n, y_n + \lambda x_n\} \to \{(1-\mu)x_n, y_n + \lambda x_n\}$$
$$\to \{(1-\mu)x_n + v(y_n + \lambda x_n), (1-v)(y_n + \lambda x_n)\},$$

giving

$$x_{n+1} = (1 - \mu + \lambda v)x_n + vy_n,$$
$$y_{n+1} = (1-v)\, y_n + \lambda(1-v)x_n.$$

These equations are similar in form to the equations of Cooke and Leon (1976,

Section 1.2), and lead to

$$x_{n+2} = (2 - \mu - v + \lambda v)x_{n+1} - (1 - \mu)(1 - v)x_n.$$

This equation is similar to the deterministic model, for adult female birds, of Capildeo and Haldane (1954) who, however, assume a fixed number of years to maturity and derive a difference equation which is not necessarily of second order. For the simple model, defined by the equations for x_{n+1} and y_{n+1}, it is assumed that no juveniles die after the recruitment stage. If v is close to unity, as we would expect for most species considered here, this may not be unreasonable. However, when the esstimated values of A and B from the first equation are set equal to $(2 - \mu - v + \lambda v)$ and $-(1 - \mu)(1 - v)$ respectively, it is found that where regions exist within which λ, μ, v may be within range together, the possible values of v are very small. (Note that if we set $((2 - \mu - v + \lambda v) - (1 - \mu)(1 - v) = 1$, as the empirical evidence suggests, then either $v = 0$ or $\lambda = \mu$.) This prompts us to consider an alternative model in which deaths of juveniles older than one year are possible, and the transitions now become

$$\{x_n, y_n\} \to \{x_n, y_n + \lambda x_n\} \to \{(1 - \mu)x_n, (1 - \mu)y_n + \lambda x_n\}$$
$$\to \{(1 - \mu)x_n + v[(1 - \mu)y_n + \lambda x_n)], (1 - v)[(1 - \mu)y_n + \lambda x_n]\},$$

giving

$$x_{n+1} = (1 - \mu)x_n + v[(1 - \mu)y_n + \lambda x_n],$$
$$y_{n+1} = (1 - \mu)(1 - v)y_n + \lambda(1 - v)x_n,$$

leading to

$$x_{n+2} = [(1 - \mu)(2 - v) + \lambda v]x_{n+1} - (1 - \mu)^2(1 - v)x_n.$$

This is an example of an attempt to build simple models to give some insight into the biological mechanisms underlying the data. Nearly all that is done in the name of ornithological statistics (indeed, statistics generally) has modelling as its basis. Of course, models are a simplification of reality (if not, they would not be models, they would just represent a reflection of the data, in all their details). Sometimes (in fact often, in a biological context) the models are too simple to describe the data adequately, and/or the data are too sparse to support any more sophisticated models. This is probably true of the above application. A fuller account of statistical analyses of sequences such as those above is given by Boddy (1985), who explored the relevance of density-dependent considerations and also viewed the data in the time series context.

2.2.2.3. An index of interyear change. Yet another approach to the analysis of such sequences is given by Mountford (1982, 1985). His work is very much in the context of the construction and use of index values to monitor population fluctuations. In

looking at the whole community would be too simple to be at all realistic. The application this time concerns a wader and wildfowl community in an estuarine environment. The practical interest here centred on how it was necessary to monitor the numbers of waders and wildfowl in the study area, in order to be able to detect population changes. Sampling design is important again here. The physical event that was in mind at the time as leading to such changes was the building of a dam across the estuary. We see here the important application of ornithological statistics in the context of environmental impact assessment.

In this study, reported by Davenport, Jeffers and North (1990), we are back to the sort of design considerations discussed earlier in this chapter. The situation now is that there is the willingness and ability on the part of voluntary observers to undertake a certain amount of regular counting of birds in the estuary. (A related question here is, with what precision can that counting be undertaken? Counting moving flocks of thousands of waders is not an easy task!) The design problem becomes how to make the most effective use of that effort, and to devise a counting scheme that most effectively addresses the particular requirements of the study – so not necessarily counting everything, simply because 'it would be interesting to have the information', and 'it must be possible to make some use of the data subsequently'. (Such sentiments have all too often been behind much ornithological data collection in the past.) It is quite possible in such a situation that a few important indicator species will convey most of the information that there is to be had.

2.3.1. An example from the Liverpool Bay area in the UK

The study reported by Davenport, Jeffers and North (1990) concerns the analysis of regular wader and wildfowl counts in the estuaries of the Liverpool Bay area in the UK. Of course, even though we are interested in the wader and wildfowl *communities* in the estuaries, we would still begin an exploratory analysis of the data by examining single species profiles over time. We then have diagrams like Figure 2 to look at.

Another type of diagram that it is useful to inspect is a cusum (cumulative sum) chart. Such a chart has been much used in the past in quality control applications in manufacturing industry, but its use does carry over effectively to other applications. The cusum chart corresponding to Figure 2 is shown in Figure 3. Interest here focuses on the turning point, as giving an indication that some change of interest was taking place at about that time. Now, the cusum line necessarily returns to the starting level at the end, by the definition of the cusum (in Figure 3, cumulative sum deviations from the mean number over the period are plotted), so there is some subjectivity in assessing the degree of importance of the observed turning point. However, with practice, this can be a useful way of exploring the data, and can show up features not so obviously apparent from the standard time profile.

Based on the standard time profiles, one way of beginning to address the multivariate nature of the data, which is potentially useful for an exploratory analysis, is to plot many time profiles (for different species) on the same diagram – see Figure 4 for an example from the Dee estuary, involving five species.

The profiles are all mean-adjusted and standardised, so that the comparisons

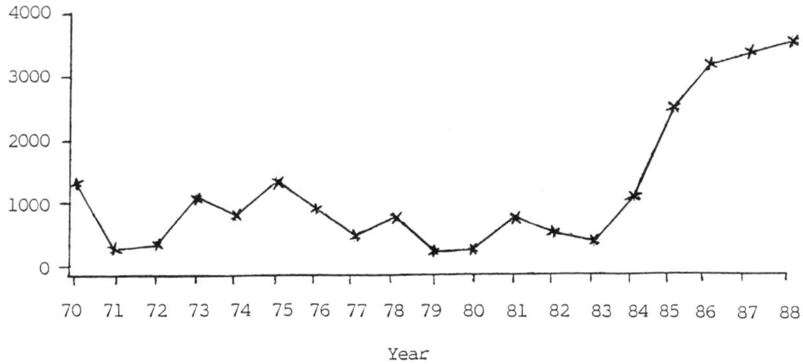

Fig. 2. Numbers of Dunlins recorded on the Alt estuary.

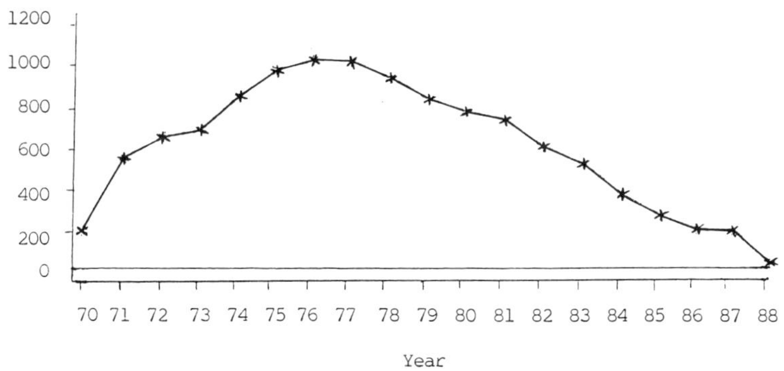

Fig. 3. Cusum chart for numbers of Curlews recorded on the Dee estuary.

between them are meaningful. At worst, such diagrams simply look a mess. At best, they can give useful information concerning overall trend and individual deviations from that trend. Figure 4 reveals that the fluctuations over time of the numbers of the five species concerned are broadly similar (this would have been anticipated, since the grouping of the five species together was done in consultation with ornithologists, to include birds of similar feeding habits), and that there is a downward trend in the numbers recorded. However, the peaks in the numbers of Sanderling in 1982 and 1985 are rather out of phase with the behaviour of the other species. This would merit further investigation, to try to determine why.

The most obvious multivariate technique to apply to the data for exploratory purposes is principal component analysis. An example for the Alt estuary is presented in Table 2, based on the correlation matrix in Table 1.

We see that an encouraging amount of the total variation is explained by the first few principal components, which appear to have quite simple interpretations. Plots

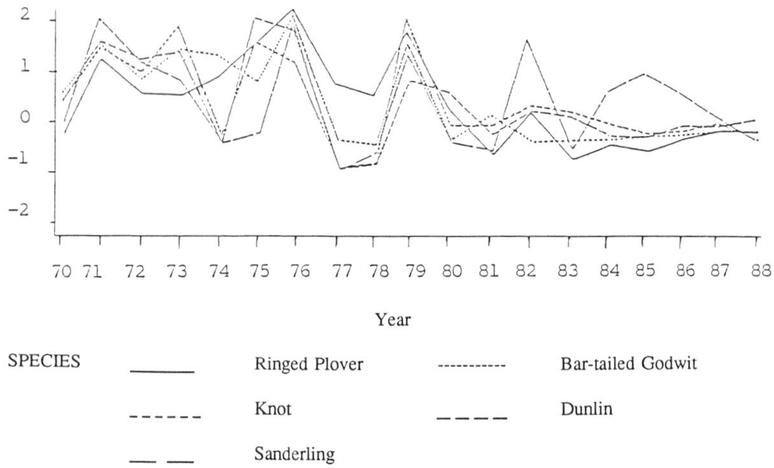

Fig. 4. Numbers recorded on the Dee estuary (mean adjusted and standardised).

of pairs of the first few principal components against each other, with minimum spanning trees superimposed to check for distortion, can then be useful in determining interesting groupings of species (or, from another viewpoint, years). If there is any need to formalise this part of the process, a method of cluster analysis can be employed.

In the context of the particular study being reported upon here, the multivariate analysis had a direct bearing on the recommendations regarding frequency of counting and detection of change (i.e., on the recommended sampling *design*). The biggest reduction in monitoring time could be achieved by considering groups of species (as identified by the multivariate analysis) rather than individual species. Note that in the majority of cases considered, increasing the frequency of counts beyond the 3 winter counts currently carried out would have little effect on the estimate of the standard errors of the numbers of birds. This emphasises the point that simply collecting more information for the sake of it is not necessarily effective. (See Davenport, Jeffers and North, 1990, for full details.) However, it does seem that increasing the frequency of counts can result in useful reductions in the time necessary for monitoring, for some species.

2.3.2. An example from The Netherlands

The work described in the preceding sections has constituted the necessary preparatory work before developing models for the British data. Such modelling work can proceed from what has been done so far, and can build on the foundations already laid. Extensive wader count data are also available from The Netherlands, and the work described in this section refers to modelling of Dutch data that has been quite recently carried out.

Table 1
Correlation matrix for species recorded on the Alt estuary

Species	1	2	3	4	5	6	9	12	16	19	20	21
1		−0.27	0.15	0.74	−0.20	0.20	−0.28	0.64	0.77	0.55	0.79	−0.15
2			0.20	−0.05	−0.02	−0.27	−0.23	0.23	−0.06	0.06	−0.28	0.27
3				0.28	0.34	−0.26	−0.62	0.37	0.02	0.04	0.07	−0.58
4					−0.31	0.14	−0.42	0.81	0.80	0.91	0.90	−0.08
5						0.26	−0.16	−0.33	−0.16	−0.44	−0.37	−0.44
6							0.34	−0.19	0.15	0.00	0.13	−0.04
9								−0.57	−0.39	−0.40	−0.37	0.37
12									0.79	0.77	0.69	0.08
16										0.79	0.81	0.08
19											0.84	0.11
20												−0.14

Table 2
Principal components for the Alt estuary (based on the correlation matrix of Table 1)

Species	Component number											
	1	2	3	4	5	6	7	8	9	10	11	12
1	0.85	0.10	0.44	−0.08	−0.27	1.00	−0.78	−0.12	0.68	−0.32	0.20	0.16
2	−0.04	−0.27	−1.00	0.96	0.69	0.05	−1.00	0.26	0.20	0.15	−0.06	−0.02
3	0.27	−1.00	−0.04	−0.04	0.80	0.54	0.92	−0.07	0.10	0.12	−0.52	0.42
4	1.00	0.03	0.09	0.09	0.70	−0.13	0.54	0.13	0.46	−0.07	−0.04	−1.00
5	−0.35	−0.67	0.62	0.83	−1.00	−0.12	0.39	0.49	0.40	0.13	0.50	−0.02
6	0.02	0.37	0.94	1.00	0.86	−0.10	−0.02	−0.64	−0.34	−0.04	−0.06	0.14
9	−0.56	0.78	0.26	0.00	0.79	0.56	0.33	1.00	−0.12	0.12	0.08	0.11
12	0.93	−0.12	−0.41	0.16	−0.09	0.42	0.33	−0.04	−1.00	0.03	0.89	−0.04
16	0.93	0.16	0.12	0.39	−1.00	0.06	−0.12	0.42	−0.57	−0.02	−1.00	−0.03
19	0.94	0.21	−0.20	0.08	0.30	−0.76	0.32	0.35	0.31	−0.57	0.20	0.65
20	0.96	0.17	0.29	−0.28	0.12	−0.32	−0.28	0.03	0.25	1.00	0.14	0.28
21	−0.1	0.8	−0.7	0.5	−0.7	0.3	0.9	−0.4	0.5	0.2	−0.1	0.1
Percentage of total variation explained	43.9	19.4	15.4	7.8	4.1	3.5	2.3	1.8	0.9	0.5	0.3	0.1
Cumulative percentage	43.9	63.3	78.7	86.5	90.6	94.1	96.4	98.2	99.1	99.6	99.9	100.0

Van Latesteijn and Lambeck (1986) describe the analysis of Oystercatcher data at the Grevelingen estuary to assess the impact of its closure in 1971 on the local wintering population of that species. To do this they use the methods of time series analysis, and, in particular, they adopt the Box–Jenkins approach to modelling the data. In order to use this approach one ideally needs at least 50 observations in the time sequence, the data must be collected at regular intervals and there should be no gaps in the data sequence. Many standard statistical computer packages contain a time series analysis module which allows the easy fitting of time series models (Box and Jenkins, 1976) to data – though the user needs to know what he or she is doing to fit appropriate models within the family to the data. PROPHET is a specialist time series analysis package developed latterly at the University of Kent, and includes both standard and nonstandard features.

2.3.2.1. The data. The floods of 1953 which also affected England affected The Netherlands very severely. Because of this it was decided to close off several estuaries in the Delta area of the south-western Netherlands. The Roggenplaat is an intertidal area in the western part of the Oosterschelde, the central estuary in the Delta region. In 1964 the Dutch Delta Institute for Hydrobiological Research began a highly standardised monthly census of Roggenplaat waders to determine whether the engineering projects in the adjacent estuaries would affect the numerical density of waders in the still undisturbed Oosterschelde area. The Grevelingen estuary, one of two to be closed (in 1971) since 1964, was by far the most important feeding area for waders – the midwinter maximum numbers there were of the order of 50 000 birds. The Oystercatcher is the dominant wader at the Roggenplaat. A plot of the data is shown in Figure 5 and, as one would expect, they display a very marked seasonal effect.

2.3.2.2. Box–Jenkins models for the data. Seasonal models are included in the Box–Jenkins family and are straightforward to fit to data within standard computer packages – once the data have been suitably differenced (possibly including season differencing) and/or transformed, and it has been decided which of the many models available to fit. Van Latesteijn and Lambeck (1986) worked with mean-corrected (by subtracting general monthly means from the data) and seasonally differenced data

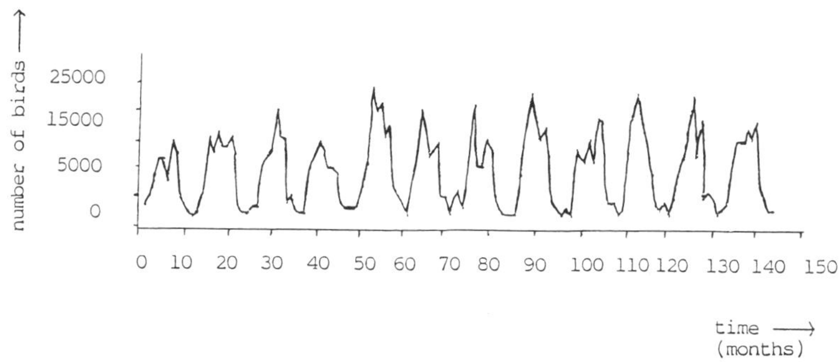

Fig. 5. The time series of the Oystercatcher counts, revealing a pronounced seasonal effect.

(differencing carried out at lag 12, to capture corresponding months in the annual cycle). They fitted a number of different models to the data and eventually chose to work with a mixed, first-order autoregressive, first-order moving average model (i.e., an ARMA (1,1) model, in standard notation) fitted to the mean-corrected data. However, this may not be the only model that would be suitable for the data.

2.3.2.3. Intervention analysis. A technique known as intervention analysis enables one to investigate the effects on a time series of some external event – in this case the closure of the Grevelingen estuary. A binary variable is added to the model. This variable takes the value 0 initially, then takes the value 1 at some time t_0 and thereafter, thus representing the abrupt effect of the intervention. Van Latesteijn and Lambeck (1986) describe how various sub-models were fitted to the Oystercatcher data, by varying the value of t_0. The optimal model was obtained by having t_0 correspond to an intervention in May 1971. This fitted well with the facts of the situation, since the last gap in the Grevelingen estuary dam was closed on May the 3-rd, 1971. The model was then able to predict that the direct impact of this action on the Roggenplaat area was that the Oystercatcher population would increase there by 8.4%.

3. Bird movements

3.1. Background

Much of the earlier interest in ornithological data centred on bird movements. Birdwatchers in Northern Europe know how exciting spring and autumn can be when birds are on passage and all sorts of unusual species might turn up almost anywhere. There are recognised spots in Europe which birdwatchers frequent at certain times of year to witness spectacular movements of birds of prey. Parts of the island of Majorca in spring are full, it seems, with Englishmen carrying binoculars and telescopes, all with one purpose in mind – watching birds and their movements. Bird rings (or bands, in American terminology) were first used to gather information about bird movements. (The emphasis now, however, is much more on the use of rings in bird *survival* studies – more of this later.) Indeed, much of what is now known about bird migration routes is only known because of the use of rings, and because of the reporting of ring recoveries. Complete summaries of ring recoveries are published annually for national ringing schemes, and some of the individual recoveries can be spectacular (though not likely to be representative of the normal behaviour for the species). Much of the interpretation of the ring recovery information is purely descriptive, with no use of formal statistical procedures. Such interpretation can still, however, be extremely informative: see, e.g., Mead (1974).

3.2. Dispersal

The recent paper by North (1988) reviews bird dispersal statistics. The title of the review paper reveals how little has been done in this area – not least of all, because it is a methodologically difficult area to work in. One conclusion of the paper was that modelling work could help to throw light on questions of biological interest. Furthermore, reflecting the sentiments already expressed earlier in this chapter, it was suggeted that further data collection should come from well-designed studies with properly defined aims. It is indicative that at the recent EURING conference in Montpellier, France (referred to earlier), it was only possible to attract one paper on bird movements, despite attempts to set aside a whole session for papers on dispersal.

3.3. Models for bird movements

As we have seen, very little modelling work has been undertaken, addressing movement information from bird ring recovery data. North (1980b) extended the work of Lloyd (1974) in connection with the analysis of distances moved by Razorbills away from the breeding colony. Lloyd (1974) and North (1980b) point out the difficulties associated with the use of ring recovery data as the basis for movement studies – especially for seabirds, such as the Razorbills that they were considering. These difficulties include the following:

(i) the probability of recovery of a ring from a dead bird is dependent upon the place of recovery;

(ii) the probability of recovery of a ring from a dead bird is dependent upon the cause of death;

(iii) birds may not be recovered where they die;

(iv) we do not know whether a bird recovered dead somewhere was still moving and intending to move further, or whether it had already stopped moving, at the place of recovery, and died there and remained there.

North (1980b) presented an analysis of the distances travelled by Razorbills, the analysis constituting a summary at the distance level rather than an in-depth attempt to model the detailed mechanics of the birds' movement patterns. An approach more akin to this was adopted by Kendall (1974) in his attempt to describe the movement feats of Manx Shearwaters by means of a variation on the theme of Brownian motion. Exploratory analysis of the Razorbill recoveries showed how important it is to take cause of death into account. If not, a quite distorted picture of the birds' geographical distribution emerges. North (1985a) presented some possible modelling approaches to describe the Razorbills' movements at a more detailed level. Random walk models, and stepping stone models (see Renshaw, 1972, 1974, 1977) were considered.

3.4. Models for migrant birds passing through bird observatories

The preceding sections have concerned attempts to model birds' actual movement behaviour. A rather different approach is to consider numbers of birds passing through some fixed point on their passage, and to attempt to interpret (model) those numbers. This is exactly the situation that pertains at coastal bird observatories in Britain. Here is another example of data being collected in a somewhat ad hoc way, and of having been so collected for many years, without any formal statistical analysis of those data having ever been carried out (or envisaged). At the British observatories, data are collected by a number of individuals, including the wardens, who typically go out on to the recording area around the observatory at, or soon after, dawn. The numbers to be recorded as the log entries for the day are then agreed at evening meetings of those observers. These numbers are not, then, simply counts. Such data had been collected over many years (about 60), with no particular purpose, and Darby (1986) was essentially the first person to attempt statistical analysis of such data in Britain. Here is an example of another area of data collection that would be well served by adhering to some formal sample design principles.

Interest in carrying out an analysis of the British bird observatories' data stemmed from the desire to use migration data to form a population index that could be used alongside, or instead of (for nonterritorial species), the breeding bird population index introduced earlier in this chapter. A stimulus for this work was provided by similar work in Canada carried out and reported by Hussell (1981). He used spring counts of migrants at the Long Point Bird Observatory, Ontario and he applied multiple regression analysis to try to explain the variability in counts by time and weather factors, and other variables. Some success was achieved with this approach, albeit at the cost of ease of interpretation, high powers of some of the variables needing to be included in the complex regression relationships obtained. The Canadian work showed that variability attributable to year provided a relatively reliable index of annual migration volume. The same success was not achieved, however, when a similar approach was applied to the British data.

The following account of modelling attempts applied to the British data is based on the work of Darby (1986). When carrying out his work at the University of Kent, he worked not only with the current author, but also with Professor Byron Morgan, the whole project being carried out in conjunction with the British Trust for Ornithology. The results of the work, which are summarised in what follows, were reported on behalf of the team by the current author at the International Biometric Conference in Namur, Belgium in 1988 (Darby, Morgan and North, 1988).

3.4.1. Poisson-binomial models

In priciple it is possible to use the total numbers of birds passing through observatories in a particular migration season as the basis for a migration index. However, this approach is rather crude. It does not take account of, for example:

- birds which remain on the observatory recording area for some time, rather than passing straight through;
- weather conditions.

The Poisson-binomial models we shall consider here take into account the first of these factors.

For the purpose of the first model we assume that:

(1) birds arrive during the early hours of the day (i.e., in the time following midnight);

(2) birds remain until a few hours after dawn, by which time they should have been recorded;

(3) then, independently, birds depart, or stay until the next day.

Of course, these assumptions constitute to some extent a simplification of reality, but that is a feature of modelling. The hope is that they do not constitute a serious distortion of reality. Although the model is simple, it does take some account of the different tendencies of birds to stay or pass through.
The model has the form

$$X_n = Y_n + I_n$$

in which
X_n is the count on day n;
Y_n is the number of birds that have stayed from day $n-1$;
I_n is the influx of birds that have stayed prior to the start of day n.
We assume that all the Y_n birds were recorded on day $n-1$, so $Y_n \leqslant X_{n-1}$.
In order to make further progress with the model, we make some distributional assumptions. These are that:
(1) Y_n has a binominal distribution, with index X_{n-1} and parameter p;
(2) I_n has a Poisson distribution, with parameter λ.
For the purposes of the model it is assumed that immigration occurs at a constant rate over time throughout the migration season. This assuption may be reasonable for at least the middle part of the migration season.

The conditional distribution of X_n, given X_{n-1}, can be expressed in the following way:

P($X_n = x_n$, given that $X_{n-1} = x_{n-1}$) is the sum over values of r from 0, to the minimum of x_n and x_{n-1}, of the following product of probabilities:

P(r birds are recorded that have stayed since the previous day)
× P($x_n - r$ newly arrived birds are recorded).

The first part of the product is given by the binomial distribution. The second part of the product is given by the Poisson distribution. The details are as follows.

3.4.1.1. Details of the Poisson-binomial model. The conditional distribution is expressed by the following:

$$\mathrm{P}(X_n = x_n | X_{n-1} = x_{n-1})$$

$$= \sum_{r=0}^{\min(x_n, x_{n-1})} \mathrm{P}(r \text{ birds are recorded that have stayed since the previous day}) \times \mathrm{P}(x_n - r \text{ newly arrived birds are recorded})$$

$$= \sum_{r=0}^{\min(x_n, x_{n-1})} \mathrm{P}(Y_n = r | X_{n-1} = x_{n-1})\, \mathrm{P}(I_n = x_n - r)$$

$$= \sum_{r=0}^{\min(x_n, x_{n-1})} \binom{X_{n-1}}{r} p^r (1-p)^{x_{n-1}-r} \frac{e^{-\lambda} \lambda^{x_n - r}}{(x_n - r)!}.$$

The likelihood function, given a series of counts over n successive days, can be expressed by

$$P(x_2, x_3, \ldots, x_n \mid x_1) = P(x_n \mid x_{n-1}) P(x_{n-1} \mid x_{n-2}) \cdots P(x_2 \mid x_1)$$
$$= \prod_{i=2}^{n} P(x_i \mid x_{i-1}).$$

As we have seen already, the resulting distribution has previously been used by Upton and Lampitt (1981), also in an ornithological context, but with Common Birds Census data.

There are two unknown parameters in the model – λ and p – and their estimates (e.g., by maximum likelihood) need to be found by some iterative method such as the Nelder–Mead simplex method, when fitting the model to data.

Even this simple model appears to give a reasonable description of the data in the examples considered.

For an indication of the results of fitting the model to some example data sets, see Table 3.

3.4.1.2. Notes on the results. (1) Ornithologist have suggested that birds might be more likely to stay around the recording area in autumn than in spring. (The biological argument here is that in spring birds are particularly eager to reach their breeding sites.) But this is not borne out by the results, i.e., p does not appear to be higher in autumn than in spring. This illustrates the need for caution concerning preconceived ideas about the results.

(2) λ varies widely over the years.

(3) As is to be expected, there is high, negative estimated correlation between $\hat{p}$ and $\hat{\lambda}$ throughout.

3.4.1.3. A migration index. In relation to the construction of a migration index the aim of fitting the Poisson-binomial model to the data has been to try to prevent those birds that have stayed in the recording are from contributing more than once to the index. A simple index is then given by $n\lambda$, where n is the number of days in the migration season. However, this does not seem to offer a great improvement over the simple total number of birds.

3.4.2. Extension of the Poisson-binomial model

In developing the Poisson-binomial model to describe the migration counts, it was recognised that it was a simplification of reality and possibly too much so. The following are extensions which aim to make the new model more realistic.

(1) The binomial parameter p, rather than being constant as in the simplest model, is allowed to take one of two values each day. This reflects the different propensity that individual birds may have to stay, each day. This may be, for example, because of different weather conditions, since birds are likely to await suitable conditions before moving on. We can assume that p follows a two-state Markov chain, with

Table 3
Some results from fitting the Poisson-binomial model to data from Dungeness

Species	Year	Spring						Autumn					
		$\hat{p}$	$\hat{\lambda}$	$s(\hat{p})$	$s(\hat{\lambda})$	$c(\hat{p}\hat{\lambda})$	n	$\hat{p}$	$\hat{\lambda}$	$s(\hat{p})$	$s(\hat{\lambda})$	$c(\hat{p}\hat{\lambda})$	n
Sedge Warbler	1968	0.73	0.84	0.07	0.19	−0.38	36	0.48	3.87	0.03	0.29	−0.47	75
	1969	0.81	0.28	0.08	0.09	−0.17	44	0.26	3.03	0.04	0.23	−0.35	76
	1970	0.25	0.34	0.16	0.12	−0.27	30	0.13	5.12	0.03	0.30	−0.48	83
	1971	0.17	1.46	0.07	0.20	−0.23	42	0.22	2.09	0.04	0.17	−0.41	97
	1972	0.16	1.04	0.06	0.14	−0.28	61	0.22	0.73	0.08	0.11	−0.34	76
Willow Warbler	1968	0.30	4.06	0.04	0.34	−0.30	44	0.66	5.54	0.02	0.29	−0.45	107
	1969	0.11	6.40	0.04	0.44	−0.54	53	0.42	6.61	0.02	0.34	−0.42	85
	1970	0.27	7.46	0.03	0.43	−0.37	55	0.34	8.87	0.02	0.43	−0.45	73
	1971	0.00	2.87	0.02	0.14	−0.99	70	0.45	8.43	0.02	0.41	−0.41	73
	1972	0.28	5.61	0.02	0.28	−0.32	89	0.45	4.37	0.03	0.29	−0.37	73
Whitethroat	1968	0.46	8.60	0.03	0.56	−0.42	42	0.67	5.33	0.02	0.33	−0.50	88
	1969	0.33	0.27	0.16	0.09	−0.23	37	0.48	0.55	0.07	0.10	−0.31	77
	1970	0.35	1.37	0.06	0.18	−0.31	54	0.52	1.41	0.04	0.15	−0.30	86
	1971	0.05	1.69	0.05	0.18	−0.40	63	0.41	2.48	0.03	0.18	−0.36	108
	1972	0.37	0.74	0.12	0.15	−0.12	37	0.46	1.30	0.05	0.15	−0.41	89
Redstart	1968	0.55	0.72	0.09	0.17	−0.41	38	0.46	1.82	0.04	0.16	−0.41	100
	1969	0.48	0.03	0.37	0.03	−0.05	32	0.41	0.70	0.06	0.10	−0.15	84
	1970	0.00	3.93	0.00	0.35	−0.99	33	0.22	0.49	0.09	0.09	−0.26	68
	1971	0.15	1.25	0.12	0.25	−0.46	29	0.25	0.27	0.11	0.07	−0.18	59
	1972	0.32	0.76	0.08	0.13	−0.30	55	0.57	0.59	0.05	0.08	−0.21	108

transition probabilities to be estimated too. When this model is fitted to the migrant data it is usually found that the two estimated values of p are close to zero and one. This seems to be intuitively reasonable.

(2) We would not expect the same number of birds to arrive at a coastal bird observatory, throughout the migration season. This suggests that we allow p to vary during the migration season, in some appropriate way. One possibility is to let λ have a polynomial form. What is computationally more attractive, and what also seems to be biologically realistic is as follows:

λ increases linearly from zero up until time t_1, at the beginning of the migration season, then remains constant until time t_2, after which λ decreases linearly back to zero.

See Figure 6 for an example of fitting this model to some of the data available.

3.4.3. Models for the data which incorporate weather information

The incorporation of weather information may help to explain the variation in counts better and to provide a yet more realistic model for the data.

3.4.3.1. A model accounting for 'bird-days' and 'non-bird-days'. The idea here is to use local weather data to distinguish between days on which there are high counts and days on which there are low counts. This is done on the basis of a discriminant analysis. (Note that it is not obvious that local weather data are the right ones to use. It might be more appropriate to try to use weather data relating to the areas in which the birds started their journeys. This illustrates the need to think very carefully about the appropriate data to use in connection with modelling work.)

In what follows the data from just the middle 80% (a figure chosen somewhat arbitrarily) of the migration periods are used, in an attempt to concentrate on the period when the migration is in a 'steady-state'.

For the model (the simplest weather-related one to be considered here) each daily

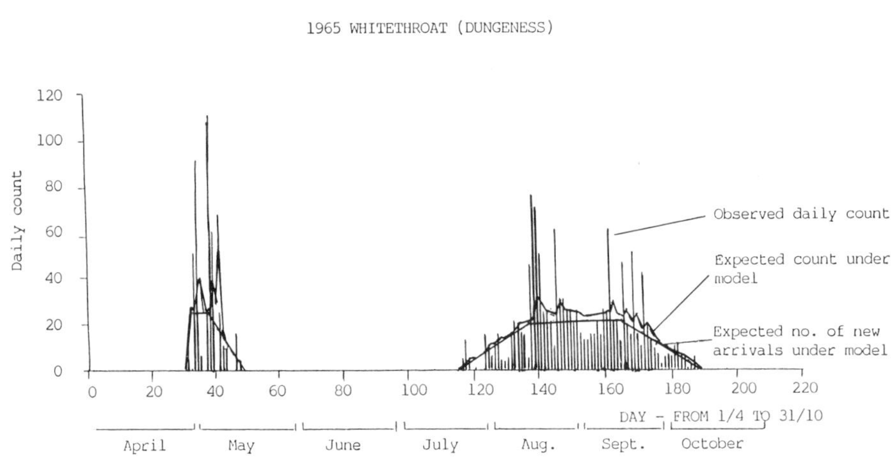

Fig. 6. Example of fitting the model with varying parameter λ.

count is considered to be an observation from one of two possible distributions of counts. These are taken to be Poisson distributions.

For each day a discriminant score is calculated from the weather information (which is made up of many different variables). One parameter of the model is then a fixed threshold value of the discriminant function. The way that this comes into operation is that if the score is less than, or greater than or equal to, the threshold value on a particular day, then that day is designed as a bird-day or a non-bird-day, respectively. This, in turn, determines which Poisson distribution of counts operates for that day.

The model can therefore be represented as follows:

$$P(X_n = x) = \begin{cases} \dfrac{e^{-\lambda_B}\lambda_B^x}{x!}, & W_n < W_0, \\ \dfrac{e^{-\lambda_N}\lambda_N^x}{x!}, & W_n \geqslant W_0, \end{cases}$$

in which

X_n = count on day n;
λ_B = mean count on bird-days;
λ_N = mean count on non-bird-days;
W_n = observed discriminant score on day n;
W_0 = threshold discriminant score.

For the simplest model we assume that the daily counts are independent of each other.

An informal assessment indicates that the model fits the data quite well. λ_B is estimated more precisely than λ_N, implying that if the model is correct, we are likely to obtain a more accurate picture of migration volume on bird-days than on non-bird-days. The ideas underlying this model lead us naturally to the next one that we consider.

3.4.3.2. Extension of the Poisson-binomial model to include weather information. We now combine the ideas of the Poisson-binomial model and the simple weather-driven model. The resulting model can be denoted as follows:

$$X_n = Y_n + \begin{cases} I_n & \text{if } W_n < W_0, \\ I'_n & \text{if } W_n \geqslant W_0, \end{cases}$$

in which

X_n = count on day n;
Y_n = has a binomial distribution with index X_{n-1} and parameter p;
I_n = has a Poisson distribution with parameter λ_B;
I'_n = has a Poisson distribution with parameter λ_N;
W_n = discriminant score for day n;
W_0 = threshold discriminant score.

The model has four parameters, namely λ_B, λ_N, p, W_0.

Table 4
Some results from fitting the 4-parameter weather-driven model to data from Dungeness

Species	Year	$\hat{p}$	$\hat{\lambda}_N$	$\hat{\lambda}_B$	$\hat{W}_0$
Sedge Warbler	1968	0.35	4	8	−3
	1969	0.14	3	3	−4
	1970	0.16	3	8	0
	1971	0.00	4	5	−3
	1972	0.03	2	6	−3
Willow Warbler	1968	0.11	5	6	−3
	1969	0.00	7	11	0
	1970	0.22	0	8	2
	1971	0.01	2	4	0
	1972	0.03	2	6	−3
Whitethroat	1968	0.18	6	8	−3
	1969	0.39	4	8	−2
	1970	0.38	1	5	−1
	1971	0.00	4	8	−3
	1972	0.28	3	5	0
Redstart	1968	0.42	2	14	−1
	1969	0.00	1	4	−3
	1970	0.00	5	6	2
	1971	0.07	1	2	0
	1972	0.27	1	9	−4

Note: Estimated asymptotic standard errors for $\hat{p}$ are of the order of 0.05, and those for the $\hat{\lambda}$ range from 0 to 1.

For results of fitting the model to data, see Table 4. As expected, the fit of the model to the data is somewhat better than that of the previous model.

3.4.3.3. Further generalisations of the weather-driven models. The parameter of the Poisson distribution determining the daily influx is now allowed to be a continuous function of the discriminant score, rather than taking on one of just two possible values. In creating a more general model we also reduce the number of parameters by one. (This may be compared with other modelling situations where the incorporation of a functional relationship to incorporate extraneous information can reduce the number of model parameters.) The Poisson parameter is taken to be a linear function of the discriminant score. The model may then be denoted by

$$X_n = Y_n + I_n$$

as in the simplest model, and in which Y_n again has a binomial distribution with index X_{n-1} and parameter p, but in which now

I_n has a Poisson distribution with parameter $\alpha + \beta W_n$.

The three parameters of this model are therefore p, α and β.

Table 5
Some results from fitting the 3-parameter weather-driven model to data from Dungeness

Species	Year	Spring			Autumn		
		$\hat{p}$	$\hat{\alpha}$	$\hat{\beta}$	$\hat{p}$	$\hat{\alpha}$	$\hat{\beta}$
Sedge Warbler	1968	0.34	3.9	0.0	0.41	6.7	0.0
	1969	0.06	2.9	0.0	0.27	4.6	0.9
	1970	0.21	7.9	0.0	0.15	7.6	0.0
	1971	0.14	3.4	0.6			
	1972	0.14	1.7	0.0			
Willow Warbler	1968	0.08	7.3	0.0	0.54	17.1	0.5
	1969	0.21	7.1	0.2	0.30	15.3	1.0
	1970	0.32	11.5	4.4	0.19	17.0	0.0
	1971	0.00	4.1	0.4			
	1972	0.25	13.5	0.0			
Whitethroat	1968	0.16	16.9	1.3	0.52	12.6	0.0
	1969	0.51	2.9	0.0	0.27	1.5	0.2
	1970	0.37	2.2	0.5	0.46	3.7	0.0
	1971	0.00	4.0	1.6			
	1972	0.30	3.0	0.0			
Redstart	1968	0.46	1.7	0.0	0.34	3.1	0.5
	1969	0.00	0.8	0.0	0.29	2.2	1.1
	1970	0.09	6.2	3.4	0.30	1.7	0.0
	1971	0.07	1.5	0.0			
	1972	0.24	1.0	0.3			

Note: Estimated standard errors of $\hat{p}$ are of the order < 0.1 and of $\hat{\alpha}$, $\hat{\beta}$ are approximately 1.

The results from fitting this model to the data (see Table 5) show that β is often estimated to be zero, or close to it. This might indicate (depending on the scale of measurement of W_n) that the daily counts do not, after all, depend on the weather (if the form of the model is correct) or that the form of the model being postulated is incorrect for the data. The earlier weather-driven model is possibly the better one to pursue, and is also probably the more biologically reasonable one. Note here the lack of usefulness in developing mathematically aesthetic models which do not have a sound biological foundation!

3.4.4. Conclusion

In conclusion it must be said that the models described here do not seem to offer a good prospect of providing useful migration indices for monitoring breeding population fluctuations (although one of the problems here is that the reference point is another set of indices, the Common Birds Census indices, whose properties themselves are not known in detail). However, they do serve to illustrate how models can be built from ornithological considerations, to give good descriptions of data, and then useful insights into those data.

4. The use of marked birds in the study of population dynamics: ring recoveries, live recaptures and radio tracking

4.1. Background

We have seen already how ringing data on birds have been used to provide information on movements of birds. More recently, the emphasis has shifted towards using such information to study avian population dynamics, particularly survival. (Not only does a ring recovery tell you something about a bird's movement, it also tells you something about its time of death – and age at death, if age at ringing is known.) Then the emphasis has shifted away from just ring recovery studies to ones which include at least as a part, if not as a whole, a capture–recapture study. Then, even more recently, as the technology has improved, radio-tracking studies involving birds have become more widespread.

The development of the methodology in these areas has been going on apace since the 1970s – indeed it can truly be said to have been explosive. In the earlier phases of the development, ornithological applications could be seen to be following the methodological leads from other application areas (particularly in capture–recapture and radio-tracking work, where nearly all the early applications were to animals other than birds). The subject has now developed to the extent where the methodological development is being led by the ornithological applications (this has essentially always been true of ring recovery work – but note applications to fish (tags) also – but is especially true of capture–recapture work now). For many people working in the area, ornithological statistics is now synonymous with the topic covered by this section.

The progress of the subject has been charted through a series of meetings, which have now established a regular, ongoing and important forum for ornithologists to meet with statisticians. The first of these meetings, entitled Statistics in Ornithology (though, in truth, it only covered some aspects of Statistics in Ornithology), was held under the auspices of the Mathematical Ecology Group (MEG) of the British Region of the Biometric Society and of the British Ecological Society in May 1982, in London, England (see Morgan and North, 1985). This meeting, and the published proceedings, which also contained some papers additional to the ones presented, covered ornithological statistics generally, but even so contained a substantial proportion of papers on survival, using recovery (mainly) and capture–recapture data. The next meeting had a more focused theme, being entitled 'Ringing recovery analytical methods'. It was organised through EURING (the European Union for Bird Ringing) and MEG, again, and held in Wageningen, The Netherlands in March 1986: see North (1987a). Although this meeting had a European background, even in this early one in the series there was participation by American statisticians. This additional dimension of interaction (American/European, as well as ornithological/statistical) has been extremely important in the series, since it reflects the importance of the developments in the subject on both sides of the Atlantic. The next meeting, again jointly organised through EURING and MEG, broadened its theme somewhat, under the title 'The statistical investigation of avian population dynamics using data from ringing recoveries and live recaptures of marked birds', and was held in Sempach, Switzerland in April 1989: see North (1990a). Most recently, the next meeting in the series, under a wider theme again – The use of marked individuals in the study of bird population dynamics: Models, methods and software (EURING 92, for short) – was held in Montpellier, France in April 1992: see Lebreton and North (1992). The American input to the meetings (indeed, the last meeting was truly international, even outside Europe) has continued, to the extent that in 1994 it is planned to take the EURING series to America, the meeting to be held in the Patuxent Wildlife Research Center, Laurel, Maryland, USA.

After the London meeting, the series has gone through the progression of concentrating almost exclusively on ring recovery data (Wageningen), through to including also capture–recapture data, and just touching on radio-tracking data (Sempach), and finally to a balance between ring recovery data, capture–recapture data and radio-tracking information (Montpellier). This has been a true reflection of the development of the subject. The meetings have provided state-of-the-art definitions of the status of the subject at each of the time points in its progression. The proceedings of the meetings therefore themselves provide fairly comprehensive reviews of the subject, and lead to extensive bibliographies. Within the proceedings the reviews by North and Morgan (1985), North (1987b) and North (1990b) provide a background to the development of the subject. As the subject has developed, there has been, as part of that development, a computer revolution, in the sense of a proliferation of computer software written specifically for the purpose, and evidence of increasingly widespread use of that software. A useful review here is provided by Lebreton, Bacou and Banco (1992).

4.2. Ring recovery data

4.2.1. History

The pioneering work in the estimation of bird survival rates from ring recoveries was carried out by Lack (1946, 1951) and followed up by Haldane (1955). Lack's approach required complete data sets, in the sense that all rings that are going to be recovered from cohorts of ringed birds have already been recovered – effectively this means that none of the ringed birds can still be alive. The Haldane approach did not have this requirement – i.e., it could handle incomplete data, in the sense that some rings from the studied birds remain to be recovered in the future, because some of the birds are still alive – and (necessarily) uses an iterative approach to the computation of the maximum likelihood estimate of the constant annual survival rate. At that early stage in the development of the methodology, the models were very simple, and were based on far-reaching assumptions – namely that the annual adult survival rate was constant and that the ring recovery rate was constant throughout. Conversely, in the development of capture–recapture methodology, even the early, seminal papers by Jolly (1965) and Seber (1965), leading to the Jolly–Seber model which has been fundamental to much that followed, allowed for different types of variation in the parameters. At that time, though, with few exceptions, the applications of capture–recapture methods were not ornithological (most ring recoveries were from dead birds and for many species the recovery rates were – and still are – extremely low).

Further development of methodology for handling ring recovery data was dormant for about 15 years after Haldane's work, most ornithologists seemingly content to use the available methods quite unquestioningly, and statisticians not yet having shown interest in becoming involved. Cormack's work, given in the statistical appendix to the paper by Fordham and Cormack (1970), gave an indication of what was to come. His model allowed for age-specific survival rates, which were estimated using a full maximum likelihood approach, incorporating information on the number of birds ringed, as well as those recovered. The 1970s heralded a period of considerable activity. Seber (1971) developed a computationally attractive approach to the age-specific survival analysis, though the problems outlined by North and Cormack (1981) effectively rendered the method impractical to use. Seber (1970) also considered a model in which survival was calendar year-specific. The American development then went ahead apace and quantum leaps forward were achieved by the models presented by Brownie and Robson (1976) and by the models and software presented by Brownie et al. (1978), updated in Brownie et al. (1985). These later models allowed for age-specific and year-specific effects in both the survival and recovery rates, allowing great flexibility in building models that match up with biological complexity – but requiring large data sets to be able to take advantage of that facility.

4.2.2. Form of the data

What form do the data to be modelled take? Ringing (banding) studies involve catching, ringing and releasing a sample of birds from some population at regular intervals. (In bird ringing studies, intervals are usually one calendar year.) Suppose that there are k ringing occasions, and let $N_1, \ldots, N_k$ be the numbers ringed and released back into the population. (These are the cohorts of ringed birds.) Equal time intervals between ringings are assumed. (Recently, some attention has been paid to the importance of this assumption.) Then the ring recovery data can be represented by R_{ij}, where R_{ij} is the number of ring recoveries in recovery year (hunting season) j from birds originally ringed in year i.

Such data can be displayed in matrix form as follows (see Brownie et al., 1985, p. 3):

Year ringed	Number ringed	Recoveries by recovery year (hunting season)				
		1	2	3	$\cdots$	l
1	N_1	R_{11}	R_{12}	R_{13}	$\cdots$	R_{1l}
2	N_2		R_{22}	R_{23}	$\cdots$	R_{2l}
3	N_3			R_{33}	$\cdots$	R_{3l}
$\vdots$	$\vdots$					$\vdots$
k	N_k					R_{kl}

Here, k is the number of years of ringing, and l is the number of years of recovery ($l \geqslant k$). Often, $k = l$, leading to a triangular array.

If both adults and juveniles are ringed, there will be one such array for each age group. Most European data refer to birds ringed as nestlings, and in many of those data sets, the numbers ringed, N_i, are not known. Note also that the recovery matrix is often presented in a reorganised way, according to recoveries by year of life. Suppose now, then, that for birds ringed as nestlings, n_{ij} is the number of birds recovered in their j-th year of life which were ringed in year i. Then the array looks as follows for the triangular array case in which number of years of recovery equals the number of years of ringing (see Brownie et al., 1985, p. 11):

Year ringed	Number ringed	Recoveries by year of life						
		1	2	3	$\cdots$	$k-2$	$k-1$	k
1	N_1	n_{11}	n_{12}	n_{13}	$\cdots$	$n_{1,k-2}$	$n_{1,k-1}$	n_{1k}
2	N_2	n_{21}	n_{22}	n_{23}	$\cdots$	$n_{2,k-2}$	$n_{2,k-1}$	
3	N_3	n_{31}	n_{32}	n_{33}	$\cdots$	$n_{3,k-2}$		
$\vdots$	$\vdots$	$\vdots$						
k	N_k	n_{k1}						

We will see a particular example of this a little later, for the Grey Heron.

4.2.3. Model structures

The general model structures (for R_{ij} arrays) in the Brownie et al. (1985) approach can then be described as follows, where S_i is the survival rate in recovery year i and f_i is the recovery rate in year i. (The terms in the body of the array are the cell occupancy probabilities.)

Year ringed	Number ringed	Model structure, by year of recovery						
		1	2	3		k		l
1	N_1	f_1	S_1f_2	$S_1S_2f_3$	$\cdots$	$S_1S_2\cdots S_{k-1}f_k$	$\cdots$	$S_1S_2\cdots S_{l-1}f_l$
2	N_2		f_2	S_2f_3		$S_2\cdots S_{k-1}f_k$		$S_2\cdots S_{l-1}f_l$
3	N_3			f_3		$S_3\cdots S_{k-1}f_k$		$S_3\cdots S_{l-1}f_l$
$\vdots$	$\vdots$					$\vdots$		$\vdots$
k	N_k					f_k		$S_k\cdots S_{l-1}f_l$

Different models in the family then come about from making different assumptions about the S_i.

In other modelling work – e.g., Morgan and Freeman (1989) – the emphasis is on age-specific survival rates (as in the Grey Heron study mentioned earlier). For some time there was a European/American division over the application of the methodology. American data tended to consist of recoveries from hunted species – wildfowl – and to constitute large data sets. British and European data, more generally, were mainly from nonhunted species, and consisted almost entirely of ring recoveries from dead birds. Data sets were typically small. Another feature of European data sets was that the number of birds ringed in the first place was often not known. The American models therefore found little early usage in Europe. This was compounded by the fact that the European data tended always to consist of recoveries of rings from birds marked as nestlings, whereas the American models catered for birds ringed at different ages – and , indeed, the authors even questioned whether any useful analysis could be carried out on data from ringed nestlings alone.

4.2.4. Incorporation of extraneous information

The work by North and Morgan (1979) (see also North, 1979b; Mead, North and Watmough, 1979) on Grey Heron survival was important especially in that it showed, for the first time, how extraneous information (in this case, a measure of winter weather severity) could be incorporated into the modelling procedure to provide a biologically plausible, yet more parsimonious, description of the data.

The model has just four parameters, however many years of ringing are included. The recovery and reporting rate is assumed to be constant (this can actually be relaxed to being cohort-specific only, which does have some biological plausibility as an assumption), though when cell occupancy probabilities in the recovery matrix are conditioned on the total number of recoveries from each cohort of ringed birds, this parameter effectively drops out of the model (and is not one of the four parameters mentioned). The four parameters are α, β (constants $\geqslant 0$), S_2, the second year survival probability, and S, the constant annual survival probability from the third year of life onwards. The constants α, β define a functional form for the first year survival probability in year i, S_{1i}, in terms of an average winter temperature in year i, t_i, as follows:

$$S_{1i} = (1 + e^{\beta - \alpha t_i})^{-1}$$

(which, by virtue of this logistic relationship, is constrained to take values only between 0 and 1, as required of a probability). A number of different measures of winter severity were investigated in the study.

Interestingly, Cavé (1983) used winter rainfall data from tropical West Africa to provide a functional form for survival probabilities for a related species, the Purple Heron, which migrates from Northern Europe to that area to spend the (European) winter there.

4.2.5. Recent developments

Further development of the ring recovery methods became beset by some difficulties from then on. Statisticians began to express concern about the validity of the assumptions underlying the models used to analyse ring recovery data. (This sort of questioning, and further development, in an evolutionary cycle of modelling, is, of course, a very healthy means of progress.) In particular, the assumption of a constant recovery probability has been questioned – see Burnham and Anderson (1979) and Anderson et al. (1981). Lakhani and Newton (1983) and Lakhani (1985) warned users of age-specific survival models for recoveries from ringed nestlings about the problems underlying the use of a mathematically convenient, apparently biologically reasonable and seemingly innocuous assumption about the equality of the last two annual survival rates in the study. As a result, ornithologists were left with the prospect of having collected sets of data for many species but which could not be analysed by any satisfactory means. This situation persisted for a number of years. The work of Morgan and Freeman (1989) and Freeman and Morgan (1990) has recently provided a way out of the impasse. (A Bayesian approach provides another possibility – work on this is ongoing. For an ornithological application of the Bayesian technique in another context, see Underhill, 1990.) See Freeman and North (1990) for an application of the model and the principles, which at the same time provides an update to the Grey Heron work described earlier. See Table 6 for a listing of the data, for illustration. The approach seems at first sight to be paradoxical, by tackling a problem of nonidentifiability of parameters through the introduction of a model with even more parameters. The first year survival probability is assumed to be year-specific. This has the effect of releasing the matrix of cell occupancy probabilities for the ring recovery data from collapsing, row upon row (since with just age-specific survival probabilities and constant recovery rate, the cell occupancy probabilities for recoveries in the first, second, third ... years of life are the same for each cohort of ringed birds), the cell occupancy probabilities becoming distinctly different for each cohort – and the model being given even greater biological plausibility.

The strategy for fitting these models to data has undergone a change of emphasis, with the potentially greater complexity of the models being fitted. Earlier approaches (see, e.g., North and Morgan, 1979) adopted a step-up approach, building from the simplest possible models and aiming at the greatest parsimony, whereas the recommended approach now (see, e.g., Freeman and Morgan, 1992) is to start with the most complex model possible, within the family of models being considered, and to step down to the most parsimonious model which still gives a good description of the data, checks on the stepping down being carried out at each stage by means of formal likelihood ratio testing. This strategy, of starting with a model including parameters for all the possible sources of biological variation in the data, and then simplifying the model to achieve the greatest possible parsimony, is, as we shall see, precisely the same as is adopted in the capture–recapture modelling – see, for example, the advice for using the package SURGE.

4.2.6. A case study

The updated Grey Heron study reported in Freeman and North (1990) uses this approach, but ends up with a rather simple model for the data. An interesting feature here is that exploratory data analysis (which should always be carried out, in order to check the various assumptions being made in the modelling) revealed the need to model recoveries from birds ringed in Scotland separately from those ringed in the

Table 6
Recoveries of British and Irish Herons ringed as nestlings 1955–85

Year	R_i	M_{ij}																	
55	348	33	5	0	0	3	4	0	0	1									
56	141	17	5	5	2	1	1	1											
57	265	31	12	2	3	3	2	0	0	0	0	0	0	0	0	1	1		
58	80	12	3	2	1	1	0	1											
59	335	34	22	7	7	1	2	1	2										
60	98	5	6	4	0	2	0	0	0	0	1								
61	170	23	7	2	1	1													
62	115	7																	
63	45	3	2	1	0	1	0	0	0	0	0	1							
64	115	7	2	0	0	0	0	0	0	1	0	0	1						
65	137	7	1	1	0	0	1	0	0	0	3	1	0	0	0	0	0	1	
66	131	1	9	3	1	1	0	2	0	0	0	0	1	0	0	0	0	0	1
67	371	32	8	5	4	1	2	1	0	2	0	1	1	0	0	0	0	1	
68	379	35	11	2	0	4	0	2	1	1									
69	503	54	19	6	0	0	1	2	0	3	0	0	0	0	0	0	1		
70	476	39	17	8	6	3	3	2	3	1	2								
71	446	31	17	4	6	1	2												
72	282	24	14	4	0	1	1	0	0	2									
73	236	31	5	4	2	0	0	1	0	1									
74	282	21	6	4	3	1	3	1											
75	450	29	7	3	0	1	0	0	1	1	0	1							
76	589	50	7	4	1	1	0	1	1										
77	464	32	13	4	2	1	1												
78	237	20	4	0	2	2													
79	434	32	13	12	0	3	4												
80	648	35	13	7	0	2	1												
81	740	43	5	6	1														
82	1014	41	12	3	4														
83	777	31	7	7															
84	1110	63	21																
85	814	49																	

R_i: Number of nestlings ringed in year *i*.
M_{ij}: Number of recoveries in their *j*-th year of life of birds in year *i*.

rest of Britain and Ireland. The fact that we end up with simple models to describe the data does not mean that the biology is very simple – it certainly cannot be – and that they therefore reflect what is really going on, but rather that the data are so sparse as not to be able to support any more sophisticated models. This is a common feature of working with such data.

In the case of the Scottish birds (1975–85), the model reduces to one with just three parameters, namely S_1, S and λ (first year survival probability, constant annual survival probability after the first year of life and constant recovery and reporting probability, respectively), although there is a suggestion that the recovery and reporting probabilities should be year-dependent. A similar model is also found to fit the recovery data from the birds from the rest of Britain and Ireland, over the same time period. However, the work of North and Morgan (1979) suggests that the process is not really as simple as that.

4.2.7. *Increased statistical sophistication*

With the maturing of ornithological statistics, greater statistical sophistication has been brought to bear on the data analyses – the full range of statistical tools is being made available to the ornithologist. For example, asymptotic results are no longer accepted unthinkingly. Confidence intervals can be derived from profile likelihoods – see, e.g., Morgan and Freeman (1989). We can expect to see more use of these in the future. When asymptotic results are used (e.g., the distribution of the maximum likelihood estimate of a model parameter), the test statistics in likelihood ratio tests can be checked using computer simulations – much more accessible to users now. Indeed, such simulations can be used to generate test statistics, via Monte Carlo testing. It is not always necessary to have recourse to Monte Carlo techniques for power, bias and variance calculations, since the computations can often be carried out theoretically – see Burnham (1990), for example. Indeed, it is a measure of the advance of the subject that both ornithologists and statistician now realise that they need to worry about such concepts. Current computing power also makes resampling methods such as boot-strapping accessible, and we are beginning to see ornithological applications.

4.3. *Capture–recapture data*

4.3.1. *History*

So far, we have charted the development of methodology for handling ring recovery data. Computer software such as BROWNIE and ESTIMATE is available for using the flexible range of models proposed by Brownie et al. (1985). Current state-of-the-art modelling for recovery data from birds ringed as nestlings is embodied in the paper by Morgan and Freeman (1989) and papers leading from it – e.g., Freeman and Morgan (1990), Freeman and North (1990). Application of the principles relies heavily, in practice, on the use of the computer package SURVIV (White, 1983, 1992).

We now turn to development of capture–recapture methods, and applications in ornithology. As indicated earlier, there were only limited early applications in ornithology – e.g., Boyd (1956) applied capture–recapture methods to data on Pink-Footed Geese and Orians (1958) – with a statistical appendix by P. H. Leslie – did so in a study of Manx Shearwaters (see also, North, 1981, for further modelling of multiple recapture data for this species). Long (1975) applied a capture–recapture approach to Reed Warbler data. But the 1980s and now the 1990s have seen explosive advances in the development of the methodology and in applications to ornithology. Lack of readily available computer software was probably a deterrent to usage in the early stages of development. That has now changed: see, for example, Table 1 in North (1990b) and the recent review by Lebreton et al. (1993).

Application of the range of models available is now not only desirable, but practical. A package such as SURGE (Lebreton and Clobert, 1986 – see also, Pradel et al., 1990, for a case study) now has an extensive worldwide user base.

As we have seen, the methodology for capture–recapture studies was initially developed quite separately from that for avian survival studies from ring recovery data. But a bird ring recovery study is essentially only a special form of a capture–recapture study, in which the first and only recapture is of a dead ringed bird. It is a natural extension to consider the case where multiple recaptures of live birds ('controls') are also recorded, and hence to the case where standard methods for multiple recapture data are available. The data to be analysed take the following form:

Suppose we think of the N animals in a population as being numbered $1, 2, \ldots$ up to N (where N is unknown, and to be estimated – though note the shift of emphasis in recent capture–recapture work away from estimation of population size to, for example, estimation of survival). Suppose a capture–recapture study consists of t capture occasions. Then a matrix X_{ij}, $i = 1, 2, \ldots, N$, $j = 1, 2, \ldots, t$ can be set up to describe the complete capture history of the animals. This leads to an array of the following form:

$$X_{ij} = \begin{bmatrix} X_{11} & X_{12} & X_{13} & \cdots & X_{1t} \\ X_{21} & X_{22} & X_{23} & \cdots & X_{2t} \\ X_{31} & X_{32} & X_{33} & \cdots & X_{3t} \\ \vdots & \vdots & \vdots & \vdots & \vdots \\ X_{M_{t+1}1} & X_{M_{t+1}2} & X_{M_{t+1}3} & \cdots & X_{M_{t+1}t} \\ \vdots & \vdots & \vdots & \vdots & \vdots \\ X_{N1} & X_{N2} & X_{N3} & \cdots & X_{Nt} \end{bmatrix}.$$

In this array all the X_{ij} are such that

$$X_{ij} = \begin{cases} 1 & \text{if the } i\text{-th animal is caught on the } j\text{-th occasion,} \\ 0 & \text{otherwise.} \end{cases}$$

Of course, some of the animals are, in general, never caught and they lead to rows of zeroes in the representative array. Thus, in the above, the first M_{t+1} rows correspond to those animals that are caught at lease once in the study, and therefore contain one or more ones. The remaining rows consist entirely of zero elements.

We find avian applications referred to by Clobert et al. (1985), Cormack (1985), Parkin and White-Robinson (1985) and Rothery (1985). Cormack's work is interesting (see also Cormack, 1981 and Cormack, 1992) in the sense that he has approached the modelling of capture–recapture data via log-linear models fitted by the general purpose statistical package GLIM. Cormack (1993) makes a case for preferring to use a general purpose package rather than a specifically purpose-designed package (such as SURGE). In similar vein, see Burnham (1989) for the use of the statistical computer package SAS.

4.3.2. Unequal catchability

Many capture–recapture models for estimating survival and abundance have been based on the assumption of equal catchability (Seber, 1982; see also Seber, 1986, an essential review paper for anyone interested in this area). But, just as the assumption of a constant recovery rate in ring recovery studies has been called into serious question, so has the assumption of equal catchability in recapture studies. Early on, Cormack (1966) provided a test for the assumption and models allowing for unequal capture probabilities were developed by Pollock (1975, 1981): see also North (1981) and Rothery (1985), who also discusses the use of computer simulation to allow for small sample sizes. Sometimes individuals in a population may be totally uncatchable, possibly because of trap shyness or being beyond the reach of the trap, for example. Aebischer (1986) described how one can obtain estimates of the uncatchable portions of target populations by adopting a double-sampling technique.

The text by Burnham et al. (1987) provides a comprehensive review, extending well beyond the fisheries applications that the title suggests. A further review is provided by Pollock et al. (1990).

4.3.3. Parsimonious models

The Jolly–Seber model (Jolly, 1965; Seber, 1965) mentioned earlier, is applicable to populations open to both birth or immigration and death or migration, and allows for time-specific survival and capture rates. It therefore has many parameters to be estimated. There has been interest in more parsimonious developments of the model (thus allowing greater precision in parameter estimates): see Jolly (1982), Jolly and Dickson (1980), Crosbie and Manly (1985). Brownie et al. (1986) extended Jolly's (1982) models to allow for age-dependent survival and capture rates.

4.3.4. Strategies for modelling

Clobert and Lebreton (1985) described the analysis of data from resightings of marked birds. This paper was important because, like the North and Morgan (1979) paper for Grey Heron ring recovery data, it described the modelling of survival as a function of winter temperature. See Brownie (1987) for further reference to this work. The Clobert and Lebreton approach is based on the model proposed by Cormack (1964) for use in estimating time-specific survival and capture rates from resighting data. (This has led Lebreton and co-authors to referring to the Cormack–Jolly–Seber model as the cornerstone of much of the capture–recapture work.)

Papers by Sandland and Kirkwood (1981) and Clobert et al. (1985) describe reduced parameter versions of Cormack's (1964) model. The Cormack (1964) and Jolly (1982) approaches are combined by Brownie et al. (1986).

Clobert and Lebreton (1987) suggest a strategy for handling capture–recapture data, as follows:

(1) fit a survival/recapture model to data on marked individuals, with the greatest flexibility possible;

(2) obtain, if needed (and if possible), estimates of population size, as described by Brownie (1987) – though note that there are difficulties with the variances.

They show when this strategy reduces to that of Jolly–Seber and when to that of Cormack. Clobert et al. (1987) provide flexibility in modelling survival and recapture, by combining the role of age (Pollock, 1981) and the role of external variables (Clobert and Lebreton, 1985), together with further possibilities, into a Cormack–Jolly–Seber framework. Modelling within this framework can be carried out using the computer package SURGE (Lebreton and Clobert, 1986).

The strategy is brought up-to-date in the important and extensive review, including ornithological applications, by Lebreton, Burnham, Clobert and Anderson (1992). The authors recommend the following procedure, with obvious similarities to the preferred strategy for handling ring recovery data:

(1) Start with a global model compatible with the biology of the species studied and with the design of the study, and assess its fit.

(2) Select a more parsimonious model using Akaike's information criterion (AIC) to limit the number of formal tests.

(3) Test for the most important biological questions by comparing this model with neighbouring ones using likelihood ratio tests.

(4) Obtain maximum likelihood estimates of model parameters together with estimates of precision.

Note the importance of the AIC value in this sequence, another example of introduction of statistical tools to the ornithological (and more generally biological) arena. The authors used the computer package RELEASE for data summarisation and goodness-of-fit testing, and the package SURGE for iterative model-fitting and the computation of likelihood ratio tests, in their work.

4.3.5. Combination of methods

We have seen how methodology has been developed separately for handling ring recovery data and for live recaptures of birds (and other animals). Until the early 1980s these approaches were thought of as being quite separate, and carried out as such. But there has been an increasing realisation of the need to combine information wherever possible, to make the most efficient use of available data relevant to, for example, survival. It is not surprising that there should be such a realisation at the same time as the use of meta analysis in medical applications, for example, has been so much in vogue. There has therefore been a coming together of the two approaches in ornithology, and they are no longer thought of as separate. They are increasingly used alongside each other in the same studies and thought is now given to how existing software can handle, or be developed to handle, both types of data. Recent collaborative work between CNRS Montpellier and the University of Kent, for example (Lebreton and Morgan, pers. comm.) has demonstrated how SURGE can handle ring recovery data.

As examples of the combined approach, Buckland (1982) described analyses of data concerning Fulmars and Galahs and Buckland and Baillie (1987) considered the analysis of live recaptures from a mist-netting programme with the incorporation of ring recovery information from dead birds. Interestingly, they suggested that for most passerines the recovery data are unlikely to yield very much useful additional information, unless they are taken from a substantially larger ringing programme than are the live recapture data. This is because of the low recovery rate of rings from many species of birds. Despite this warning, we can anticipate increasingly widespread use of combined studies (live recaptures and dead recoveries).

4.4. *Radio-tracking data*

Another type of data that is becoming increasingly useful in ornithological studies is radio-tracking data. Some years ago this would have been unthinkable, except for large species of birds, because the technology had not developed to the point where very small transmitters could be produced. They now can be, opening up very extensive possibilities for the use of radio-tracking techniques for birds, even of small species. The paper by Kenward (1993) gives some indication of the possibilities. One concern which needs to be addressed is whether the attachment of a radio transmitter to a bird affects the very parameter it is designed to be giving information on – for example survival. If so, such studies would clearly be flawed. The great attraction of radio-tracking data is the possibility of having a continuous records of a bird's movements.

The availability of yet another type of data calls for further methodological development, in order to keep abreast of the trends. For example, Pollock, Winterstein and Conroy (1989) show how survival analysis techniques already used in medical and engineering studies can be used to analyse data from animal radio-tagging studies. Indeed, the body of methods developed for the study of survival in human populations (e.g., Cox and Oakes, 1984) should, in principle, be applicable to radio-tracking data on birds (and animals). This line of approach is taken by Skalski, Hoffman and Smith (1993).

More generally, extensive and long-term studies of the population of a single species of bird may allow Leslie matrix type models to be developed. Modern computing power enables extensive simulations of stochastic models of the population's development to be run, if necessary. An example for the Screech Owl (where the interest was in the effects of a rodenticide which was being considered for use in orchards on the owl population) is described by North (1985b) and North, Boddy and Forrester (1988).

4.5. *Experimental design*

We now find ourselves with three types of data for studying avian survival. Some studies will incorporate all three types of data. There is even more need now for proper statistical principles of experimental design to be adopted, through the involvement of qualified statisticians at an early stage in ornithological studies. (There has been an encouraging growth of such collaboration in the last decade or so.) North (1988) called for such an approach in the context of dispersal studies. Consideration needs to be given to such issues as sample size, and we should expect to see power calculations in the design of studies and the interpretation of data increasingly in the future – see, for example, the discussion of the EXPECT option in RELEASE (Burnham et al., 1987, pp. 214–217). There is clearly much scope for the use of designed experiments, including controlled, randomised experiments. See, for example, Burnham et al. (1987) and references therein, such as Stromborg, Grue, Nichols, Hepp, Hines and Bourne (1988). See also Haramis, Nichols, Pollock and Hines (1986). All of this has been developed already to a sophisticated degree in, for example, the conduct of clinical trials in a pharmaceutical context. Avian studies also would benefit from the discipline of carefully setting out aims and objectives, considering the appropriate mix of studies (dead recoveries, live recaptures and radio-tracking) to be used – which need to be related to available resources and costs – assessing the appropriate sample sizes, and setting all of this out in a formal study protocol.

5. Conclusion

As indicated in the Introduction, this has been a somewhat personalised review of ornithological statistics seen primarily, in the earlier sections at least, from a British/European perspective. The material presented here can only possibly be a small part of what could, in principle, be presented in total, concerning ornithological statistics. The whole subject is now so broad that a whole book would need to be devoted to this alone, rather than the one chapter here.

The section on the use of marked birds (rings, radio transmitters) is broader in its background, and truly reflects the trans-Atlantic development work that has gone on. This part of the topic can be thought of as reflecting ornithological statistics as a whole, for it is this area which is the most active in the subject's development. It looks likely to continue that way for some time to come, too. Not least of all, this is because many statisticians have become involved in this particular area, and are contributing prominently to its further development.

The involvement of the statisticians should ensure that sound statistical principles are increasingly adopted, especially aspects of design, including choice of sample size and consideration of statistical power. Indeed, we will hopefully increasingly see a situation in the ornithological context similar to that pertaining in clinical trials work within and for the pharmaceutical industry. Here, all the specifications for the study are set out, in advance of the study itself, in a written protocol. Justification for the sample size chosen is given, and the power properties of the study to be carried out are given. This needs to happen in the ornithological arena too, in order that ornithological statistics continues to mature into a properly conducted scientific discipline.

One thing is for sure. Ornithology has for some time now been a rich source of fascinating problems for statisticians to become involved in, through the great volume of data that there has been, awaiting quantitative interpretation. There is going to be no shortage of such fascinating problems for many years to come, as ornithology has come to the forefront as a statistical application area in continuing methodological research and development. But the good thing is, for all the development of the science and what that has meant for those of us involved in it, simple birdwatching itself remains an incomparable joy for us enthusiasts. It seems that a necessary criterion for entry to the worldwide club of ornithological statisticians is a passion for birdwatching itself. It is surely this that has provided the motivation for us, and the satisfaction in what has been achieved. Professional scientific training and enthusiasm for a hobby have proved to be good partners in the development of this particular subject.

Appendix

Scientific names of birds mentioned in the text

Manx Shearwater	*Puffinus puffinus*
Grey Heron	*Ardea cinerea*
Purple Heron	*Ardea purpurea*
Pink-Footed Goose	*Anser brachyrhynchus*
Peregrine Falcon	*Falco peregrinus*
Oystercatcher	*Haematopus ostralegus*
Ringed Plover	*Charadrius hiaticula*
Dunlin	*Calidris alpina*
Knot	*Calidris canutus*
Sanderling	*Calidris alba*
Curlew	*Numenius arquata*
Bar-Tailed Godwit	*Limosa lapponica*
Razorbill	*Alca torda*
Reed Warbler	*Acrocephalus scirpaceus*
Sedge Warbler	*Acrocephalus schoenobaenus*
Whitethroat	*Sylvia communis*
Willow Warbler	*Phylloscopus trochilus*
Redstart	*Phoenicurus phoenicurus*
Robin	*Erithacus rubecula*
Blackbird	*Turdus merula*
Great Tit	*Parus major*

References

Aebischer, N. J. (1986). Estimating the proportion of uncatchable animals in a population by double-sampling. *Biometrics* **42**, 973–979.

Anderson, D. R., A. P. Wywialowski and K. P. Burnham (1981). Tests of the assumptions underlying life table methods for estimating parameters for cohort data. *Ecology* **62**, 1121–1124.

Batten, L. A. (1978). The seasonal distribution of recoveries and causes of Blackbird mortality. *Bird Study* **25**, 23–32.

Boddy, A. W. (1985). Statistics and the Common Bird Census. Unpublished M. Sc. Dissertation, University of Kent at Canterbury, UK.

Box, G. E. P. and G. M. Jenkins (1976). *Time Series Analysis. Forecasting and Control.* Holden Day, San Francisco, CA.

Boyd, H. (1956). Statistics of the British population of the Pink Footed Goose. *J. Animal Ecol.* **25**, 253–273.

Braae, L. and H. Nøhr (1985). An examination of the reliability of the Danish point count method. In: K. Taylor, R. J. Fuller and P. C. Lack, eds., *Bird Census and Atlas Studies. Proc. 8th Internat. Conf. on Bird Census and Atlas work.* British Trust for Ornithology, Tring.

Brownie, C. (1987). Recent models for mark-recapture and mark-resighting data. *Biometrics* **43**, 1017–1019.

Brownie, C., D. R. Anderson, K. P. Burnham and D. S. Robson (1978). *Statistical Inference from Band Recovery Data – A Handbook*, Publication 131. U.S. Fish and Wildlife Service Resource, Washington, DC.

Brownie, C., D. R. Anderson, K. P. Burnham and D. S. Robson (1985). *Statistical Inference from*

Band Recovery Data – A Handbook, Publication 156. 2nd ed., U.S. Fish and Wildlife Service Resource Washington, DC.

Brownie, C., J. E. Hines and J. D. Nichols (1986). Constant-parameter capture-recapture models. *Biometrics* **42**, 561–574.

Brownie, C. and D. S. Robson (1976). Models allowing for age-dependent survival rates for band-return data. *Biometrics* **32**, 305–323.

Buckland, S. T. (1982). A mark – recapture survival analysis. *J. Animal Ecol.* **51**, 833–847.

Buckland, S. T. (1987). On the variable circular plot method of estimating animal density. *Biometrics* **43**, 363–84.

Buckland, S. T., D. R. Anderson, K. P. Burnham and J. L. Laake (1992). *Distance Sampling: Estimating Abundance of Biological Populations*. Chapman and Hall, London.

Buckland, S. T. and S. R. Baillie (1987). Estimating survival rates from organised mist-netting programmes. In: P. M. North, ed., Ringing Recovery Analytical Methods, *Acta Ornithol.* **23**(1), 89–100.

Burnham, K. P. (1989). Numerical survival rate estimation for capture–recapture models using SAS PROC NLIN. In: L. McDonald, B. Manly, J. Lockwood and J. Logan, eds., *Estimation and Analysis of Insect Populations*. Springer, New York, 416–435.

Burnham, K. P. (1990). Survival analysis of recovery data from birds ringed as young: Efficiency of analyses when numbers ringed are not known. In: P. M. North, ed., The Statistical Investigation of Avian Population Dynamics Using Data from Ringing Recoveries and Live Recaptures of Marked Birds, *The Ring* **13**(1–2), 115–132.

Burnham, K. P. and D. R. Anderson (1979). The composite dynamic method as evidence for age-specific waterfowl mortality. *J. Wildlife Management* **43**, 336–366.

Burnham, K. P., D. R. Anderson, G. C. White, C. Brownie and K. H. Pollock (1987). *Design and Analysis Methods for Fish Survival Experiments Based on Release–Recapture*. Monograph 5. Amer. Fisheries Soc., Bethesda, MD.

Capildeo, R. and J. B. S. Haldane (1954). The mathematics of bird population growth and decline. *J. Animal Ecol.* **23**, 215–223.

Cavé, A. J. (1983). Purple Heron survival and drought in tropical West-Africa. *Ardea* **71**, 217–224.

Clobert, J. and J.-D. Lebreton (1985). Dépendance de facteurs de milieu dans les estimations de taux de survie par capture–recapture. *Biometrics* **41**, 1031–1037.

Clobert, J. and J.-D. Lebreton (1987). Response to C. Brownie, Recent models for mark–recapture and mark-resighting data. *Biometrics* **43**, 1019–1022.

Clobert, J., J.-D. Lebreton and D. Allaine (1987). A general approach to survival rate estimation by recapture or resightings of marked birds. *Ardea* **75**, 133–142.

Clobert, J., J.-D. Lebreton, M. Clobert-Gillet and H. Coquillart (1985). The estimation of survival in bird populations by recaptures or sightings of marked individuals. In: B. J. T. Morgan and P. M. North, eds., *Statistics in Ornithology*. Springer, Berlin, 197–213.

Cooke, D. and J. A. Leon (1976). Stability of population growth determined by 2 × 2 Leslie matrix with density dependent elements. *Biometrics* **32**, 435–442.

Cormack, R. M. (1964). Estimates of survival from sighting of marked animals. *Biometrika* **51**, 429–438.

Cormack, R. M. (1966). A test for equal catchability. *Biometrika* **22**, 330–342.

Cormack, R. M. (1981). Loglinear models for capture–recapture experiments on open populations. In: R. W. Hiorns and D. Cooke, eds., *The Mathematical Theory of the Dynamics of Biological Populations*. Academic Press, London, 197–215.

Cormack, R.M. (1985). Examples of the use of GLIM to analyse capture–recapture studies. In: B. J. T. Morgan and P. M. North, eds., *Statistics in Ornithology*. Springer, Berlin, 243–273.

Cormack, R. M. (1993). The flexibility of GLIM analyses of multiple recapture or resighting data. In: J.-D. Lebreton and P. M. North, eds., *The Use of Marked Individuals in the Study of Bird Population* Birkhäuser, Basel, 39–49.

Cox, D. R. and D. Oakes (1984). *Analysis of Survival Data*. Chapman and Hall, London.

Crosbie, S. F. and B. F. J. Manly (1985). A new approach for parsimonious modelling of capture–mark–recapture experiments. *Biometrics* **41**, 385–398.

Darby, K. V. (1986). A statistical examination of British bird observatories data. Unpublished Ph.D. Thesis, University of Kent at Canterbury, UK.

Darby, K. V., B. J. T. Morgan and P. M. North (1988). Modelling numbers of migrating birds recorded at a coastal observatory, accounting for weather conditions. *Abstracts 14th Internat. Biometric Conference*, Namur, Belgium, p. 128.

Davenport, T. R., J. N. R. Jefers and P. M. North (1990). Monitoring requirements–changes to estuary birds. Unpublished Report, Applied Statistics Research Unit, University of Kent at Canterbury, U.K.

Fordham, R. A. and R. M. Cormack (1970). Mortality and population change of Dominican Gulls in Wellington, New Zealand, with a statistical appendix. *J. Animal Ecol.* **39**, 13–27.

Freeman, S. N. and B. J. T. Morgan (1990). Studies in the analysis of ring-recovery data. In: P. M. North, ed., The Statistical Investigation of Avian Population Dynamics Using Data from Ringing Recoveries and Live Recaptures of Marked Birds. *The Ring* **13**(1–2), 271–288.

Freeman, S. N. and B. J. T. Morgan (1992). A modelling strategy for recovery data for birds ringed as nestlings. *Biometrics* **48**, 217–236.

Freeman, S. N. and P. M. North (1990). Estimation of survival rates of British, Irish and French Grey Herons. In: P. M. North, ed., The Statistical Investigation of Avian Population Dynamics Using Data from Ringing Recoveries and Live Recaptures of Marked Birds, *The Ring* **13**(1–2), 139–166.

Haldane, J. B. S. (1955). The calculation of mortality rates from ringing data. In: *Proc. 11th Internat. Ornithol. Congr.*, Basel, 454–458.

Haramis, G. M., J. D. Nichols, K. H. Pollock and J. E. Hines (1986). The relationship between body mass and survival of wintering Canvasbacks. *Auk* **103**, 506–514.

Hussell, D. J. T. (1981). The use of migration counts for monitoring bird population levels. In: C. J. Ralph and J. M. Scott, eds., *Estimating the Numbers of Terrestrial Birds*, Studies in Avian Biology, Vol. 6. Cooper Ornithological Society, Lawrence, KS, USA, 92–102.

Jeffers, J. N. R. (1987). *An Introduction to Systems Analysis: With Ecological Applications.* Edward Arnold, London.

Jolly, G. M. (1965). Explicit estimates from capture–recapture data with both deaths and immigration – stochastic model. *Biometrika* **52**, 225–247.

Jolly, G. M. (1982). Mark–recapture models with parameters constant in time. *Biometrics* **38**, 301–321.

Jolly, G. M. and J. M. Dickson (1980). Mark–recapture suite of programs. In: M. M. Barritt, and D. Wishert, eds., *COMPSTAT 1980: Proc. Computational Statistics*, Vol 4. Psysica, Vienna (for the Internat. Assoc. Statist. Comput.), 570–576.

Kendall, D. G. (1974). Pole-seeking Brownian and bird navigation. *J. Roy. Statist. Soc. Ser. B* **36**, 365–417.

Kenward, R. (1993). Modelling raptor populations: To ring or to radio-tag? In: J.-D. Lebreton and P. M. North, eds., *The Use of Marked Individuals in the Study of Bird Population.* Birkhäuser, Basel, 157–166.

Lack, D. (1946). Do juvenile birds survive less well than adults? *British Birds* **39**, 258–264.

Lack, D. (1951). Population ecology in birds. In: *Proc. 10th Internat. Ornithol. Congr.*, Uppsala, 409–448.

Lack, D. (1964). A long-term study of the Great Tit (*Parus major*). *J. Animal Ecol.* **33**, 159–173.

Lack, D. (1966). *Population Studies of Birds.* Clarendon Press, Oxford.

Lakhani, K. H. (1985). Inherent difficulties in estimating age-specific bird survival rates from ring recoveries. In: B. J. T. Morgan and P. M. North, eds., *Statistics in Ornithology.* Springer, Berlin, 311–321.

Lakhani, K. H. and I. Newton (1983). Estimating age-specific bird survival rates from ring recoveries – can it be done? *J. Animal Ecol.* **52**, 83–91.

Lebreton, J.- D., K. P. Burnham, J. Clobert and D. R. Anderson (1992). *Modelling Survival and Testing Biological Hypotheses Using Marked Animals: A Unified Approach with Case Studies,* Ecological Monographs, Vol. 62 (1). Ecological Society of America 67–118.

Lebreton, J.-D., A. M. Reboulet and G. Banco (1993). An overview of software for terrestrial vertebrate population dynamics. In: J.-D. Lebreton and P. M. North, eds., *The Use of Marked Individuals in the Study of Bird Population.* Birkhäuser, Basel, 357–370.

Lebreton, J.-D. and J. Clobert (1986). *User's Manual for Program SURGE.* C.E.F.E./C.N.R.S., Montpellier, France.

Lebreton, J.-D. and P. M. North, eds. (1993). *The Use of Marked Individuals in the Study of Bird Population Dynamics.* Birkhäuser, Basel.

Lloyd, C. S. (1974). Movement and survival of British Razorbills. *Bird Study* **21**, 102–116.
Long, R. (1975). Mortality of Reed Warblers in Jersey. *Ringing Migration* **1**, 28–32.
Mead, C. J. (1974). The results of ringing auks in Britain and Ireland. *Bird Study* **21**, 45–86.
Mead, C. J., P. M. North and B. R. Watmough (1979). The mortality of British Grey Herons. *Bird Study* **26**, 13–22.
Morgan, B. J. T. and S. N. Freeman (1989). A model with first-year variation for ring-recovery data. *Biometrics* **45**, 1087–1101.
Morgan, B. J. T. and P. M. North, eds. (1985). *Statistics in Ornithology*. Springer, Berlin.
Mountford, M. D. (1982). Estimation of population fluctuations with applications to the Common Bird Census. *Appl. Statist.* **31**, 135–143.
Mountford, M. D. (1985). An index of population change with application to the Common Bird Census. In: B. J. T. Morgan and P. M. North, eds., *Statistics in Ornithology*. Springer, Berlin, 121–132.
North, P. M. (1977). A novel clustering method for estimating numbers of bird territories. *Appl. Statist.* **26**, 149–155.
North, P. M. (1978). How many bird territories are there on a farm? A statistical approach to an ornithological problem. *Math. Spectrum* **10**, 44–48.
North, P. M. (1979a). A novel clustering method for estimating numbers of bird territories: An Addendum. *Appl. Statist.* **28**, 300–301.
North, P. M. (1979b). Relating Grey Heron survival rates to winter weather conditions. *Bird Study* **26**, 23–28.
North, P. M. (1980a). Some examples of data analysis in ornithology. E. Diday et al., eds., *Data Analysis and Informatics*. North-Holland, Amsterdam, 681–699.
North, P. M. (1980b). An analysis of Razorbill movements away from the breeding colony. *Bird Study* **27**, 11–20.
North, P. M. (1981). Models for a simple capture–recapture procedure to estimate population size. *Biometrics* **37**, 661–672.
North, P. M. (1982). Comparing changes from one year to the next in bird communities. *Ornis Scand.* **13**, 200–205.
North, P. M. (1983). Modelling animal numbers of bird territories on farmland plots. *Biometric. J.* **25**, 431–441.
North, P. M. (1985a). Models for Razorbill movements. In: B. J. T. Morgan and P. M. North, eds., *Statistics in Ornithology*. Springer, Berlin, 77–84.
North, P. M. (1985b). A computer modelling study of population dynamics of the Screech Owl (*Otus asio*). *Ecol. Model.* **30**, 105–143.
North, P. M., ed. (1987a). Ringing Recovery Analytical Methods. Proc. EURING Technical Conf. and Meeting of the Mathematical Ecology Group of the Biometric Society (British Region) and British Ecological Society, *Acta Ornithol.* **23**(1).
North, P. M. (1987b). Ring recovery analytical methods: A brief review. In: P. M. North, ed., Ringing Recovery Analytical Methods, *Acta Ornithologica* **23**(1), 7–12.
North, P. M. (1988). A brief review of the (lack of) statistics of bird dispersal. *Acta Ornithol.* **24**, 63–74.
North, P. M., ed. (1990a). The Statistical Investigation of Avian Population Dynamics Using Data from Ringing Recoveries and Live Recaptures of Marked Birds. Proc. EURING Technical Conf. and Meeting of the Mathematical Ecology Group of the Biometric Society (British Region) and British Ecological Society, *The Ring* **13**(1–2).
North, P. M. (1990b). Analysis of avian ring recovery and live recapture data: Where have we come from Wageningen to Sempach? In: P. M. North, ed., The Statistical Investigation of Avian Population Dynamics Using Data from Ringing Recoveries and Live Recaptures of Marked Birds, *The Ring* **13**(1–2), 11–22.
North, P. M., A. W. Boddy and D. R. Forrester (1988). A computer simulation study of stochastic models to investigate the population dynamics of the Screech Owl (*Otus asio*) under increased morality. *Ecol. Model.* **40**, 233–263.
North, P. M. and R. M. Cormack (1981). On Seber's method for estimating age-specific bird survival rates from ringing recoveries. *Biometrics* **37**, 103–112.

North, P. M. and J. N. R. Jeffers (1991). *Modelling – A Basis for Management or an Illusion? Scientific Management of Temperate Communities for Conservation.* Blackwell Scientific Publications Ltd. Oxford, 523–541.
North, P. M. and B. J. T. Morgan (1979). Modelling Heron survival using weather data. *Biometrics* **35**, 667–681.
North, P. M. and B. J. T. Morgan (1985). The general context of statistics in ornithology. In: B. J. T. Morgan and P. M. North, eds., *Statistics in Ornithology*. Springer, Berlin, 1–9.
Orians, G. H. (1958). A capture–recapture analysis of a Shearwater population (with a statistical appendix by P. H. Leslie). *J. Animal Ecol.* **27**, 71–84.
Parkin, D. T. and R. White-Robinson (1985). The effect of age on survival in the Canada Goose (*Branta canadensis*) in Nottinghamshire. In: B. J. T. Morgan and P. M. North, eds., *Statistics in Ornithology*. Springer, Berlin, 323–340.
Pollock, K.H. (1975). A *K*-sample tag–recapture model allowing for unequal survival and catchability. *Biometrika* **62**, 577–583.
Pollock, K. H. (1981) Capture–recapture models allowing for age-dependent survival and capture rates. *Biometrics* **37**, 521–529.
Pollock, K. H., J. D. Nichols, C. Brownie and J. E. Hines (1990). *Statistical Inference for Capture–Recapture Experiments.* Wildlife Monographs Vol. 107.
Pollock, K. H., S. R. Winterstein and M. J. Conroy (1989). Estimation and analysis of survival distributions for radio-tagged animals. *Biometrics* **45**, 99–109.
Pradel, R., J. Clobert and J.-D. Lebreton (1990). Recent developments for the analysis of capture–recapture multiple data sets. An example concerning two Blue Tit populations. In: P. M. North, ed., The Statistical Investigation of Avian Population Dynamics Using Data from Ringing Recoveries and Recaptures of Marked Birds, *The Ring* **13**(1–2), 193–204.
Ralph, C. J. and J. M. Scott, eds. (1981). *Estimating Numbers of Terrestrial Birds*, Studies in Avian Biology, Vol. 6. Cooper Ornithological Society, Lawrence, KS, USA.
Renshaw, E. (1972). Birth, death and migration processes. *Biometrika* **59**, 49–60.
Renshaw, E. (1974). Stepping stone models for population growth. *J. Appl. Probab.* **11,** 16–31.
Renshaw, E. (1977). Velocities of propagation for stepping stone models of population growth. *J. Appl. Probab.* **14**, 591–597.
Rothery, P. (1985). Estimation of age-specific survival of Hen Harriers (*Circus c. cyaneus*) in Orkney. In: B. J. T. Morgan and P. M. North, eds., *Statistics in Ornithology*. Springer, Berlin, 341–354.
Sandland, R. L. and G. P. Kirkwood (1981). Estimation of survival in marked populations with possibly dependent sighting probabilities. *Biometrika* **68**, 531–541.
Seber, G A. F. (1965). A note on the multiple-recapture census. *Biometrika* **52**, 249–259.
Seber, G. A. F. (1970). Estimating time-specific survival and reporting rates for adult birds from band returns. *Biometrika* **57**, 313–318.
Seber, G. A. F. (1971). Estimating age-specific survival rates from bird-band returns when the reporting rate is constant. *Biometrika* **58**, 491–497.
Seber, G. A. F. (1982). *The Estimation of Animal Abundance and Related Parameters.* 2nd ed., Griffin, London.
Seber, G. A. F. (1986). A review of estimating animal abundance. *Biometrics* **42**, 267–292.
Skalski, J. R., A. Hoffman and S. G. Smith (1993). Testing the significance of Individual and cohort level covariates in animal survival studies. In: J.-D. Lebreton and P. M. North, eds., *The Use of Marked Individuals in the Study of Bird Population.* Birkhäuser, Basel, 9–28.
Stromborg, K. L., C. E. Grue, J. D. Nichols, G. R. Hepp, J. E. Hines and H. C. Bourne (1988). Postfledging survival of European Starlings exposed as nestlings to an organophosphorus insecticide. *Ecology* **69**, 590–601.
Taylor, K., R. J. Fuller and P. C. Lack, eds. (1985). *Bird Census and Atlas Studies. Proc. 8th Internat. Conf. on Bird Census and Atlas Work.* BTO, Tring.
Underhill, L. G. (1990). Bayesian estimation of the size of closed populations. In: P. M. North, ed., The Statistical Investigation of Avian Population Dynamics Using Data from Ringing Recoveries and Live Recaptures of Marked Birds. *The Ring* **13**(1–2), 235–254.

Upton, G. J. G. and G. A. Lampitt (1981). A model for interyear change in the size of bird populations. *Biometrics* **37**, 113–127.

Van Latesteijn, H. C. and R. H. D. Lambeck (1986). The analysis of monitoring data with the aid of time-series analysis. *Environ. Monitor. Assess.* **7**, 287–297.

White, G. C. (1983). Numerical estimation of survival rates from band recovery and biotelemetry data. *J. Wildlife Management* **47**, 716–728.

White, G. C. (1992). PC SURVIV User's Manual, Version 1.4 Department of Fishery and Wildlife Biology, Colorado State University, Fort Collins, CO.

G. P. Patil and C. R. Rao, eds., *Handbook of Statistics, Vol. 12*

14

Statistical Methods in Developmental Toxicology

Paul J. Catalano and Louise M. Ryan

Abstract: This chapter describes some of the statistical issues surrounding analysis of data from controlled developmental toxicity experiments in animals. After discussion of relevant background information, a typical experimental design is reviewed with discussion of the main statistical considerations involved in analyzing data from these experiments. Discussion follows to popular approaches for dose-response modeling, including both likelihood based and quasi-likelihood models. Attention is given to the general problem of handling multiple outcomes, a common occurrence in developmental toxicology. The issue of conducting risk assessment is then taken up by first reviewing the current method of defining allowable doses based on determining a no observed adverse effect level (NOAEL), or the dose preceding the lowest dose that differs significantly from controls. This is contrasted with more quantitative approaches such as determining a benchmark dose. Some recent advances in the area of risk assessment for multiple outcomes are also reviewed.

1. Background

Human exposure to chemicals, pollutants, and other foreign substances has become an important public health problem of our time. Of particular interest is the impact of such compounds on the unborn. While risk assessment and control has its roots in the evaluation of potential carcinogens, regulatory agencies such as the US Environmental Protection Agency (EPA) and the US Food and Drug Administration (FDA) have become increasingly aware of the importance in studying and regulating non-cancer risks. Chief among these is the assessment of developmental impacts, because exposure in utero has been shown to be a sensitive pathway to serious health problems (US EPA, 1991; Holmes, 1992). The human tragedies resulting from exposure to substances such as DES and thalidomide in the 1950s stand as clear examples of the potential harm of foreign substances to the developing human. In addition to medications, voluntary exposures such as cigarette smoke, drugs, alcohol as well as many workplace and home exposures (for example, lead and mercury) are important areas for concern.

Developmental toxicology refers to the assessment and evaluation of foreign compounds and their impacts during human development. While epidemiological studies have been useful in identifying specific developmental toxicants (Holmes, 1992; Kline et al., 1989), human data are typically too sparse to investigate most substances effectively. Furthermore, human exposure is often unreliably measured and many studies involve several recorded and unrecorded confounding factors, making unequivocal assessment difficult. Hence controlled experiments in laboratory animals contribute valuable information about potential toxic effects during development. Since such studies involve considerable time and expense, as well as hundreds and in some cases thousands of animals, it is essential that appropriate, efficient statistical models are used to assess risk.

2. Experimental design

This section reviews typical design protocols for developmental toxicity experiments and outlines the endpoints measured.

2.1. Protocols

Several experimental protocols are used in reproductive and developmental studies. Three test designs (segments I, II, and III) were established by the US FDA in 1966 and correspond to assessing specific types of effect (Food and Drug Administration, 1966). After briefly describing the other protocols, this chapter will focus mainly on segment II experiments.

The segment I, or fertility and reproduction study, is designed to assess male and female fertility and general reproductive processes. This protocol is typically conducted in one species (usually the rat) and involves exposing males for 60 days and females for 14 days prior to mating. After mating, females are exposed until midpregnancy (Kimmel and Price, 1990). The segment III design, or peri- and postnatal study, focuses on effects later in gestation and involves exposing pregnant animals (usually rats) from day 15 of gestation through lactation (Kimmel and Price, 1990). Other designs include multigenerational studies where animals are exposed throughout two or three generations, and Dominant Lethal (genotoxicity) assays where developmental effects on the gametes are investigated by exposing the male animals prior to mating.

The protocol for the segment II design focuses on the period during major organogenesis and structural development. These experiments have been often referred to as 'teratology' studies, since historically, the primary goal was to study malformation. Two species of rodents (mice and rats) or rabbits are usually chosen as the experimental animals because they breed easily and have a short gestation. Administration of exposure is by the intended clinical or environmental route(s) to humans. Timed-pregnant animals (dams) are exposed during the critical period of major organogenesis (days 6–15 for mice and rats; 6–19 for rabbits) and sacrificed just prior to normal delivery at which time the uterus is removed and the contents are thoroughly examined.

Dose levels for the segment II design consist of a control group and three or four different does groups exposed to the test substance. EPA standards suggest that the lowest dose produce no observable maternal or developmental toxicity, and that dose levels are increased gradually to a maximum dose designed to produce some toxicity but not more than 10 percent maternal deaths (US EPA, 1991). At present, doses are chosen at equally spaced intervals on a linear or log-spaced scale, measured in milligrams per kilogram body weight (mg/kg bw). Ordinarily between 20 and 30 pregnant dams are randomized to each dose group and control and typical control littersizes (number of live-born offspring) range from 8 in the rabbit to 12 and 14 in mice and rats, respectively.

While the dam is the unit of randomization and maternal toxicity is an important factor in these studies, interest centers primarily on the outcomes of the offspring. That test animals are multiparous with several sites of implantations (initial conceptus) and littersizes are species-dependent are important features of developmental toxicity experiments. Average control littersizes are around 12 for mice, 14 for rats, and 8 for rabbits. Clustering of offspring within litters accounts for the well-known 'litter effect', or tendency for littermates to respond more similarly than offspring from different litters and much statistical research effort has been devoted to methods for appropriately accounting for heterogeneity among litters, especially with regard to quantal outcomes. Failure to account for clustering in the data analysis can lead to serious underestimation of the variances of dose effect parameters and, hence, inflated test statistics. Several methods that have been proposed to account for such intralitter correlation will be reviewed in Section 3.

2.2. *Maternal and fetal outcomes*

A variety of maternal and fetal outcomes are recorded in segment II developmental toxicity experiments. Collected maternal data include initial (pre-exposure) maternal weight and maternal weight changes at several time points during gestation, weight at sacrifice and weight of the uterus and contents. Maternal mortality and maternal liver weight are also available as well as observations relative to maternal toxicity (e.g., 'animal stopped eating'). Although there is evidence of a relationship between measurable maternal and developmental toxicity (Skalko and Johnson, 1987), analysis and risk assessment tend to focus on fetal endpoints. Many studies summarize maternal data but rarely incorporate such information into characterizing the dose-response or in conducting risk assessment.

Figure 1a shows a dissected mouse with her uterus displayed. The picture shows clearly why the placement of offspring within the rodent or rabbit uterus has often been likened to 'peas in a pod'. Figure 1b shows a closer view of a section of the uterus from which a fetus has been extracted. This figure shows the tiny placenta attaching the fetus to the mother. The small dark spot next to the fetus corresponds to a resorption, or implanted embryo that did not survive gestation.

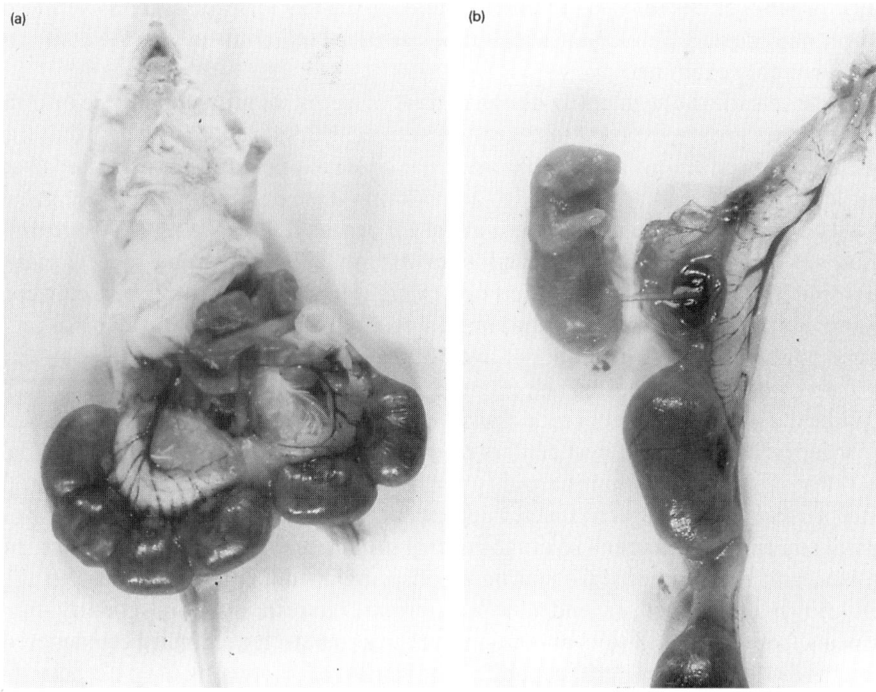

Fig. 1. Dissected mouse: (a) structure of uterus, (b) detail showing fetus, placenta, and resorption site.

Detailed examination of the uterine contents produces several discrete and continuous fetus-level outcomes. In addition to the number of implantations, litter-specific counts of resorptions (early fetal deaths) and fetal deaths (deaths occurring later in gestation) are available. Viable fetuses (animals that would be born live if the dams were followed to term) are extensively examined for many different malformation. To simplify statistical analysis, it is common to distinguish three malformation categories: external, visceral, and skeletal. These categories are based not so much on biological considerations, but from the perspective of data collection techniques. External malformations are those visible by naked eye to the teratologist. Such malformations might include missing limbs or cleft palate. Skeletal malformations will include those affecting the animal's skeletal structure and are detected through specialized staining techniques. Visceral malformations are those affecting internal organs such as the heart, brain, or lung, and are detectable after dissection. Clearly, a particular malformation can be classified in more than one way. Depending on individual laboratory procedures, techniques for skeletal examination may preclude visceral examination, and animals may need to be randomly assigned to one type of examination or the other. In addition to these discrete outcomes, fetal weight is also recorded for each viable fetus.

Since deleterious events can occur at several points in development, an interesting aspect of segment II studies lies in the staging or hierarchy of possible adverse fetal outcomes. A toxic insult early in gestation may result in a resorbed fetus. Adverse events observable later in gestation include fetal death. Fetuses that survive the entire gestation period but are still affected may be born with a variety of malformations or may suffer from low birth weight. The occurrence of all these outcomes can be conceptualized as a series of time-dependent conditional risks. Early in development, each implantation is susceptible to a risk of resorption. If the implant survives the risk of resorption, the developing fetus is then at risk of fetal death. Finally, conditional on surviving to birth, a live fetus is at risk for malformations and low birth weight. This structure is illustrated graphically in Figure 2. Since toxic agents can affect one or many of these risks, it is essential that a risk assessment procedure accounts for all of the adverse events. In fact, the EPA places concern on adverse outcomes occurring in any of four general areas of abnormal development: death, malformation, growth alteration, and functional deficit (Kimmel and Price, 1990).

Table 1 shows summary data from a study of diethylene glycol dimethyl ether (DYME) in mice (Price et al., 1987). This compound exhibits strong dose related

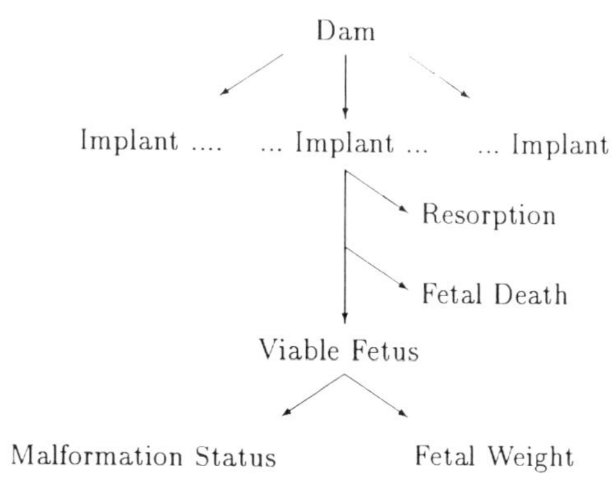

Fig. 2. Fetal outcomes in developmental toxicity studies.

Table 1
Summary data from DYME developmental toxicity study in mice

Dose (g/kg)	Dams	Implants	Non-Live*		Littersize		Malformed†		Weight (g)	
			No.	%	Mean	S.D.	No.	%	Mean	S.D.‡
0	21	296	14	4.7	13.43	2.42	1	0.4	1.000	0.109
0.063	20	242	17	7.0	11.25	3.43	0	0.0	0.967	0.116
0.125	24	312	22	7.1	12.08	2.36	7	2.4	0.910	0.113
0.250	23	299	38	12.7	11.35	2.17	59	22.6	0.793	0.097
0.500	23	285	144	50.5	6.13	3.06	132	93.6	0.562	0.083

* Indicates number of dead or resorbed animals.
† Indicates number of animals exhibiting 1 or more malformations.
‡ Ignoring clustering.

effects in each of the outcomes. Fetal death/resorption (column 5) is strongly dose related with rates increasing from 4.7% in controls to 50.5% at the maximum dose. Littersize (column 6) denotes the average number of live animals at each dose. The dose related increase in death and resorption accounts for the decrease in average littersize with dose. Malformation rates range from 0.4% in the control group to 93.6% at the highest dose level, and effects on fetal weight also appear substantial with a monotonic decrease in average weight from 1.00 g in the control group to 0.562 g at 0.5 g/kg. These data will be used in later sections to illustrate various modeling approaches.

Appropriate statistical methods to incorporate multiple outcomes must allow for the possible correlations among fetal endpoints. Occurrence of resorption and fetal death also directly reduces the number of viable fetuses (littersize) and it is believed (Brent and Jensh, 1967) that littersize itself may influence outcomes among live fetuses. For example, animals from large litters competing for essential maternal resources may be prone to reduced weight or malformation. It has also been shown (Ryan et al., 1991) that there is a strong empirical correlation between malformation and fetal weight among live fetuses. Thus, ultimately, statistical methods to evaluate suspect developmental toxicants must account for the multiplicity of adverse outcomes and also recognize the structure inherent in them. Before discussing recent work in the analysis of multiple outcomes, methods for accounting for litter effects in the case of a simple binary outcome will be first presented.

3. Accounting for litter effects: Single binary outcome

Methods for correlated outcomes in the context of developmental toxicity have received much attention within the last fifteen years or so, especially with regard to models for binary data. This is partly because unlike situations involving continuous data where the multivariate normal distribution is an obvious choice, distributions for correlated discrete outcomes do not arise naturally as extensions to common univariate distributions. This section discusses several approaches for handling the intra-litter correlation when assessing does effects involving binary endpoints such as resorption, fetal death or malformation. Although some authors have discussed the problem of estimating response rates (Haseman and Kupper, 1979), we focus on the issues of dose group comparisons and dose-response modeling. Methods based on standard maximum likelihood techniques will be discussed, as well as more recent developments involving quasi-likelihood models where specification of moment equations, rather than a full likelihood, is used to define the model and estimate parameters. Rank-based and resampling methods have also been proposed for the developmental toxicity setting. Some of these methods are reviewed in Piegorsch and Haseman (1991).

Let us first establish some notation. Let m_{ij} and n_{ij}, $i = 0, \ldots, I$; $j = 1, \ldots, N_i$ denote, respectively, the number of implantations and live fetuses for the j-th litter in dose group i receiving dose d_i. Note that the control group is denoted by a 0 subscript so that $d_0 = 0$, and N_i indicates the number dams randomized to dose group i. The number of dead or resorbed fetuses for the j-th litter in dose group i is therefore

$$m_{ij} - n_{ij}.$$

For any dichotomous outcome, let the binary variate y_{ijk} take the value 1 if the k-th fetus ($k = 1, \ldots, m_{ij}$ or n_{ij}) in litter i, j exhibits the outcome of interest and 0 otherwise. Thus if y denotes malformation, the malformation count for the j-th litter at dose i is

$$s_{ij} = \sum_{k=1}^{n_{ij}} y_{ijk}$$

with binomial proportion estimator

$$\hat{p}_{ij} = \frac{s_{ij}}{n_{ij}}.$$

The relevant denominator is n_{ij} since malformations occur only among live fetuses. When y denotes resorption or fetal death, the appropriate denominator is the number of implantations m_{ij}.

3.1. *Likelihood-based methods*

Several authors (see for example, Williams, 1975; Haseman and Kupper, 1979; Paul, 1982; Rosner, 1982; Ochi and Prentice, 1984; Rosner, 1984; Prentice, 1986; Paul, 1987) discuss likelihood generalizations of standard distributions to account for clustering. One of the first proposals along these lines was the use of the beta-binomial distribution by Williams (1975). He considered modifying the usual binomial distribution

$$\Pr(S_{ij} = s \mid n_{ij}, p_{ij}) = \binom{n_{ij}}{s} p_{ij}^{s}(1 - p_{ij})^{n_{ij} - s} \tag{1}$$

by allowing the litter probabilities p_{ij} to vary within a dose group according to a Beta (α_i, γ_i) distribution. The unconditional distribution of S is then

$$\Pr(S_{ij} = s \mid n_{ij}) = \binom{n_{ij}}{s} \frac{B(\alpha_i + s, n_{ij} + \gamma_i - s)}{B(\alpha_i, \gamma_i)} \tag{2}$$

where $B(\alpha_i, \gamma_i) = \Gamma(\alpha_i)\Gamma(\gamma_i)/\Gamma(\alpha_i + \gamma_i)$. This distribution has unconditional mean

$$E(S_{ij}) = n_{ij}\left(\frac{\alpha_i}{\alpha_i + \gamma_i}\right) = n_{ij}\mu_i$$

and variance

$$\operatorname{var}(S_{ij}) = \frac{n_{ij}\alpha_i\gamma_i(\alpha_i + \gamma_i + n_{ij})}{(\alpha_i + \gamma_i)^2(\alpha_i + \gamma_i + 1)}$$

$$= n_{ij}\mu_i(1 - \mu_i)\left\{1 + (n_{ij} - 1)\frac{\phi_i}{1 + \phi_i}\right\},$$

where

$$\phi_i = \frac{1}{\alpha_i + \gamma_i}.$$

Note that for $k \neq k'$,

$$\rho_i = \operatorname{corr}(Y_{ijk}, Y_{ijk'}) = \phi_i/(1 + \phi_i)$$

is the intralitter correlation induced by introducing the beta distribution on the p_{ij}.

Estimation of the parameters can be accomplished via maximum likelihood. For the purposes of relating outcome to dose level, it is useful to reparametrize the beta distribution in terms of its mean and variance functions (Williams, 1975). The mean parameters μ_i can be described (linked) to dose (or log dose) in terms of a regression equation. General discussion of the form of the regression equation is discussed in Section 4.

The correlation parameters ρ_i define separate intralitter correlations for each of the $I + 1$ dose groups. In cases where a common intralitter correlation can be assumed across dose groups some efficiency can be gained by estimating only a single correlation parameter ρ. Caution should be exercised, however, for as discussed by Kupper et al. (1986), assuming a common correlation when litter effects are dose-dependent can lead to bias in estimation of mean parameters.

While the beta-binomial offers a flexible extension of the binomial distribution, it has several disadvantages. First, the model is not easy to fit, and specialized software is needed. Second, the required parametric assumptions are strong and the beta-binomial incorporates only positive (extra-binomial) correlation although Prentice (1986) reparametrizes this distribution and shows that negative correlation is possible. Other generalizations of the binomial that also allow for negative intralitter correlation include the so-called correlated binomial model (Kupper and Haseman, 1978),

$$\Pr(S_{ij} = s \mid n_{ij}) = \binom{n_{ij}}{s} p_i^s (1 - p_i^{n_{ij} - s})\left\{1 + \frac{\delta_i}{2p_i^2(1 - p_i)^2}[(s - n_{ij}p_i)^2 + s(2p_i - 1) - n_{ij}p_i^2]\right\}, \tag{3}$$

where δ_i characterizes the intralitter covariance, and an extension proposed by Altham (1978)

$$\Pr(S_{ij} = s_{ij} \mid n_{ij}) = \frac{\binom{n_{ij}}{s_{ij}} p_i^{s_{ij}} (1 - p_i^{n_{ij}-s_{ij}}) \gamma_i^{s_{ij}(n_{ij}-s_{ij})}}{\sum_j \binom{n_{ij}}{s_{ij}} p_i^{s_{ij}} (1 - p_i^{n_{ij}-s_{ij}}) \gamma_i^{s_{ij}(n_{ij}-s_{ij})}},$$

where γ_i determines the degree of positive ($0 < \gamma_i < 1$) or negative ($\gamma_i > 1$) correlation.

Paul (1987) presents a model that incorporates separate parameters to describe both the intralitter correlation and the heterogeneity of outcome rates (p_i) between litters within a dose group. He allows for a second level of correlation by assuming that litter specific response rates from a correlated binomial model (3) follow a beta distribution using the same logic as in constructing the beta-binomial distribution from the uncorrelated binomial. Paul terms this three parameter distribution a beta-correlated binomial and views it as a generalization of the binomial, the beta-binomial, and the correlated binomial distributions, and recommends that the model be used for separate assessment of intralitter and interlitter sources of variability.

Another approach to accounting for the litter effect is to assume that variability in litter response rates within a dose group results from heterogeneity in factors such as genetic makeup, metabolism, and environmental influences. In an interesting paper on dose-response modeling of developmental outcomes, Rai and Van Ryzin (1985) suggested incorporating litter specific covariates to characterize this heterogeneity. In particular, they suggested a model that allowed the response probability to depend on litter size (n_{ij}) by characterizing the dose-response with a 'one-hit' model (developed originally in the context of carcinogenicity testing),

$$\lambda(d_i) = 1 - \exp[-(\beta_0 + \beta_1 d_i)]$$

and incorporating a proportional hazard-type term involving litter size into the probability of response,

$$\mu(d_i) = \lambda(d_i) \exp[-n_{ij}(\tau_0 + \tau_1 d_i)].$$

Rai and Van Ryzin argue that addition of litter size (other litter specific factors could also be included) may be adequate to account for the litter effect. However, several authors (Williams, 1987; Carr and Portier, 1991) have criticized this approach, claiming that litter size and other recorded litter specific variables are unable to completely account for all of the heterogeneity encountered in practice. Carr and Portier observe that the moderate association between developmental outcomes and litter size is insufficient to explain the extent of extra-binomial variation encountered in developmental toxicity data.

3.2. Generalized estimating equations

In recent years, considerable attention has been given to the area of quasi-likelihood modeling. As discussed by Wedderburn (1974), quasi-likelihood models generalize the ideas of weighted least squares and method of moments by setting weighted sums of the observed data equal to their expected values under the assumed model for the mean and, in the case of normally distributed outcomes, this approach is equivalent to likelihood-based inference. McCullagh and Nelder (1989) provide general discussion of this topic as well as references to much of the foundational work. Basically, quasi-likelihood models require no strong distributional assumptions other than specification of the mean $\mu(\boldsymbol{\beta})$ as a function of regression parameters $\boldsymbol{\beta}$ and variance function $V(\mu)$. In the case of uncorrelated observations, mean parameters $\boldsymbol{\beta}$ are estimated by solving a set of 'score-like' equations of the form

$$\sum_{i=1}^{I} \sum_{j=1}^{N_i} \frac{\partial \mu(d_i)}{\partial \boldsymbol{\beta}} \frac{(s_{ij} - n_{ij}\mu(d_i))}{\mathrm{var}(s_{ij})} = 0 \tag{4}$$

which can be written in vector notation as

$$\mathbf{D}^{\mathrm{T}}\mathbf{V}^{-1}(\mathbf{S} - \boldsymbol{\mu}) = 0,$$

where $\mathbf{D} = \partial \mu / \partial \boldsymbol{\beta}$, and $\mathbf{V}$ is a diagonal matrix of variances. The asymptotic covariance matrix of $\hat{\boldsymbol{\beta}}$, the solution to (4), is

$$\mathrm{var}(\hat{\boldsymbol{\beta}}) = (\mathbf{D}^{\mathrm{T}}\mathbf{V}^{-1}\mathbf{D})^{-1} \tag{5}$$

and has a form very similar to the covariance of regression parameters from weighted linear least squares.

Williams (1982) presents one of the earliest applications of quasi-likelihood methods to developmental toxicity data. Using the same moment structures as for the beta-binomial model (2), Williams noted that consistent estimators of the parameters characterizing the mean response rate $\mu(d)$ can be found by solving the following equations:

$$\sum_{i=1}^{I} \sum_{j=1}^{N_i} \frac{\partial \mu}{\partial \boldsymbol{\beta}} \frac{(s_{ij} - n_{ij}\mu(d_i))}{n_{ij}\mu(d_i)(1 - \mu(d_i))[1 + \phi(n_{ij} - 1)]} = 0.$$

Williams shows how this model can easily be fitted in standard statistical packages, such as GLIM, with a simple iterative algorithm. The scale parameter ϕ is estimated at each iteration using a weighted residual sum of squares moment-type estimator. McCullagh and Nelder (1989) propose an even simpler approach to fitting models to correlated binary data, treating the term $[1 + \phi(n_{ij} - 1)] = \rho$ as constant. The advantage of this latter approach is the parameter estimates themselves are exactly the same as from a standard binomial analysis, ignoring the correlation. One simply has to adjust the estimated variances of the parameters by the factor ρ. In practice, many people tend to prefer the Williams approach over the simpler McCullagh and Nelder (1989) model, since the assumption of a constant intra-litter correlation seems more realistic than assuming ρ is constant, especially in the common situation where litter size is correlated with dose. However, many of the same criticisms of the beta-binomial approach still apply. In practice, it is quite unlikely that intra-litter correlations will be constant (Kupper et al., 1986), so that more sophisticated techniques than those proposed by Williams (1982) and McCullagh and Nelder (1989) are needed.

Liang and Zeger (1986) and Zeger and Liang (1986) described a general method for the analysis of correlated data. This approach, most often referred to as generalized estimating equations (GEE), has gained widespread popularity and includes the simpler models of Williams (1982) and McCullagh and Nelder (1989) as special cases. Like the simpler models, the GEE approach requires specification of only the mean and variance functions of the data. An important addition in the method proposed by Liang and Zeger is the inclusion of an empirical, robust variance estimator that provides correct inferences regarding model parameters and their variances, even in cases where the assumed variance structure is misspecified. There is still incentive to correctly model the variance, however, since it can be shown to improve statistical efficiency. While some quasi-likelihood methods (such as the Williams approach) model the litter counts s_{ij} application of the GEE method with the empirical variance adjustment to development toxicity data necessitates working with the fetus-specific binary outcomes y_{ijk}. Letting $\boldsymbol{y}_{ij} = (y_{ij1}, \ldots, y_{ijn_{ij}})^{\mathrm{T}}$ denote the vector of outcomes for the j-th litter in treatment group i, the generalized estimating equations can be written in vector form as:

$$\sum_{i=1}^{I} \sum_{j=1}^{N_i} \mathbf{D}_{ij}^{\mathrm{T}} \mathbf{V}_{ij}^{-1} (\boldsymbol{y}_{ij} - \boldsymbol{\mu}_i) = \mathbf{0}, \tag{6}$$

where, similar to (5), $\boldsymbol{\mu}_i$ is the mean vector for the outcome $\boldsymbol{y}_{ij}$ (in this case, $\boldsymbol{\mu}_i$ is simply a vector of length n_{ij} with each element equal to $\mu(d_i)$), $\mathbf{D}_{ij}$ represents the $n_{ij} \times p$ design matrix $\partial\boldsymbol{\mu}/\partial\boldsymbol{\beta}$, where $\boldsymbol{\beta}$ represents the p-vector of unknown parameters characterizing $\mu(d_i)$, and $\mathbf{V}_{ij}$ is the $n_{ij} \times n_{ij}$ assumed covariance matrix of $\boldsymbol{y}_{ij}$. Liang and Zeger (1986) suggest writing $\mathbf{V}_{ij}$ in the form

$$\mathbf{V}_{ij} = \mathbf{A}_{ij}^{1/2} \mathbf{R}(\boldsymbol{\alpha}) \mathbf{A}_{ij}^{1/2}, \tag{7}$$

where $\mathbf{A}_{ij}$ is the diagonal matrix

$$\mathbf{A}_{ij} = \mathrm{diag}\,[\mathrm{var}(y_{ijk})] = \mathrm{diag}[\mu(d_i)(1 - \mu(d_i))],$$

and $\mathbf{R}(\boldsymbol{\alpha})$ is a suitable correlation matrix for the vector of fetal outcomes $\boldsymbol{y}_{ij}$ indexed by a separate set of parameters $\boldsymbol{\alpha}$. Like the scale parameter ϕ in the Williams model, the $\boldsymbol{\alpha}$ parameters are estimated iteratively using moment-type estimators based on the residuals at each iteration. While the form of $\mathbf{R}(\boldsymbol{\alpha})$ can be arbitrarily complex, an

Weibull model

$$\mathrm{P}(d_i) = 1 - \exp[-(\beta_0 + \beta_1 d_i^{\beta_2})] \tag{16}$$

although (12), (13) and (14) are easily adapted to this family by adding a power parameter. Another variation of (11) is to consider models of the form

$$\mathrm{P}(d) = \delta + (1 - \delta)F(\beta_0 + \beta_1 \log d), \tag{17}$$

where δ represents the background (control) response rate (Gart et al., 1986).

These models and generalizations of them have been used to model developmental toxicity data with varying degrees of success. For example, Chen and Kodell (1989) use (16) with a beta-binomial distribution to model the probability that a fetus is affected (resorbed, dead, or malformed). Ryan (1992) considers (17), using a probit model (13) including a power parameter on dose. Yet while all of these models are mathematically reasonable, it should be noted that none of them offer any explanation of biological mechanisms or route of developmental impact.

Figure 3 shows the results from fitting several dose-response models to the prenatal death (non-live) data from the DYME study. In the figure, points illustrate litter-specific incidence and dose-specific death rates are denoted by crosses. An independence working correlation was used in each case. The three fitted models include the linear logistic model (12), the Weibull model (16), and the probit model with multiplicative background (13). Note that both the Weibull and the probit models give virtually identical dose-response curves. This finding illustrates the general phenomenon that most choices in the families (11) and (15) provide good fits to observed data.

4.2. Multivariate approaches

Several multivariate approaches for characterizing risk have been investigated in recent years. Methods for dose-response modeling of several adverse events are important for a variety of reasons. First, incorporating multiple outcomes into a single model helps to control the type I error rate which can become inflated when conducting several tests for dose effects across separate univariate models. While methods for multiple comparisons such as Bonferroni adjustment can be used, they tend to be conservative, especially where outcomes are correlated, a common occurrence in developmental toxicity. Second, construction of dose-response models for multiple outcomes provides more flexibility in defining the dose response and allows investigation of relationships among adverse outcomes. Third, for the purposes of risk assessment, multivariate models allow a natural framework for calculation of safe dose levels. Expressions for overall risk of any adverse event can be quantified and used for low dose extrapolation. Methods based on joint modeling of multivariate outcomes more appropriately take into account risks from several sources and provide more conservative estimates of risk than the common univariate approach of regulating on the most sensitive endpoint (Ryan, 1992).

Lefkopoulou et al. (1989) consider GEE methodology for modeling litter propor-

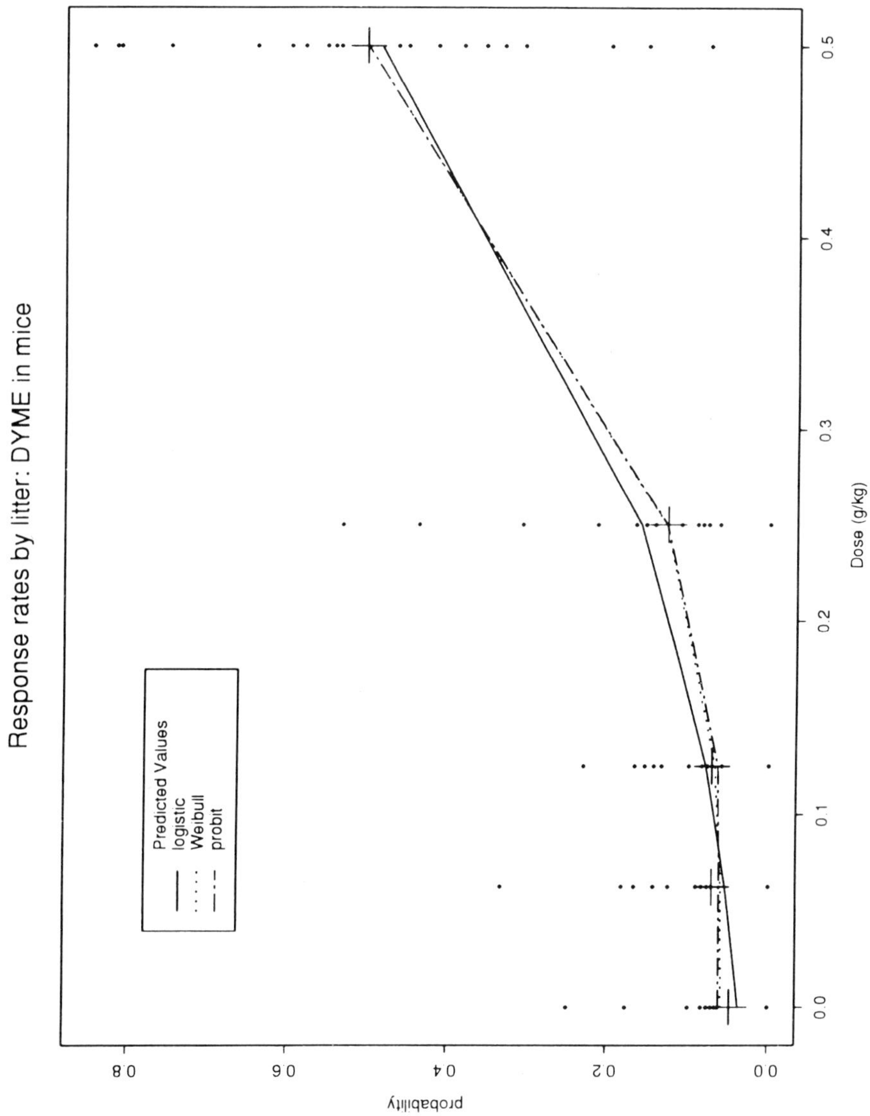

Fig. 3. Illustration of logistic, Weibull, and probit fits to the DYME data.

tions among several malformation categories. They propose a linear logistic model for the mean of L different malformation responses

$$\text{logit } p_{il} = \alpha_l + \mathbf{X}_i \boldsymbol{\beta}_l, \tag{18}$$

where i denotes litter, and $l = 1, \ldots, L$ denotes malformation outcome (external, visceral, or skeletal, for example). The terms α_l in (18) correspond to individual control rates for each malformation type and $\boldsymbol{\beta}_l$ contains regression parameters characterizing the effect of dose and other litter-specific covariates for outcome l. For outcomes that show similar response to dose, Lefkopoulou et al. suggest replacing $\boldsymbol{\beta}_l$ in (18) with a common dose parameter $\boldsymbol{\beta}$. To account for within- and between-fetus variability, they use the GEE correlation matrix $\mathbf{R}$ from (7) and a scale parameter ϕ to characterize the litter effect.

Chen et al. (1991) propose a likelihood-based model for fetal death/resorption and malformation based on a Dirichlet-trinomial distribution. They assume that conditional on the response probabilities p (for death/resorption) and $q/(1-p)$ (for malformation) the data follow a trinomial distribution. To account for litter heterogeneity, they model the litter-specific response probabilities p_{ij} and q_{ij} as being drawn from a Dirichlet density

$$f(p_{ij}, q_{ij}) = \frac{\Gamma(\alpha_i + \beta_i + \gamma_i)}{\Gamma(\alpha_i)\Gamma(\beta_i)\Gamma(\gamma_i)}\, p_{ij}^{\alpha_i - 1} q_{ij}^{\beta_i - 1} (1 - p_{ij} - q_{ij})^{\gamma_i - 1}$$

with $\alpha_i > 0$, $\beta_i > 0$, and $\gamma_i > 0$. The resulting marginal distribution for the litter counts is termed Dirichlet-trinomial. This formulation is analogous to the construction of the beta-binomial distribution (2) except that the response is a vector of three components (resorbed/dead, malformed, normal) rather than two (abnormal, normal) in the beta-binomial. In fact, the marginal distributions of death/resorption, and malformation (conditional on the number of live fetuses) from this model each have beta-binomial distributions. For this reason, Chen et al. refer to this as a 'double beta-binomial' model.

Ryan (1992) extends the ideas of Chesson (1986) and Chen et al. (1991) in constructing a model for fetal death and malformation based on a continuation ratio formulation of the multinomial distribution, and shows how this approach can be used to estimate safe dose levels in the case of multiple adverse outcomes. She considers independent background models (17) using an extreme value distribution (14) separately for risks of fetal death, $P_1(d_i)$ and malformation, $P_2(d_i)$. The model is then used to quantify overall risk by considering the probability of being abnormal

$$\Pr(\text{dead or malformed}) = 1 - [1 - P_1(d_i)][1 - P_2(d_i)]. \tag{19}$$

Ryan constructs a quasi-likelihood for this model by considering a standard multinomial likelihood except that the variance is multiplied by a scale parameter to account for overdispersion induced by the litter effect. Parameter estimates are therefore identical to a likelihood analysis, except that parameter variances are inflated by the estimated scale parameter. By using (19) as a basis for extrapolation, this formulation allows determination of doses at levels of low overall risk while allowing flexibility of dose response separately for each outcome.

Ryan (1992) argues that multiple outcomes should be carefully accounted for in quantitative risk assessment. In a simplified setting with only fetal death and malformation, Ryan proposed that risk assessment be performed on the basis of the probability of being either dead *or* malformed (19). As an illustration, Ryan applied this method to four developmental toxicity studies in mice. These studies were conducted by the Research Triangle Institute under contract to the National Toxicology Program (NTP), and investigated the chemicals di(2-ethyhexyl) phthalate (DEHP) (Tyl et al., 1988), ethylene glycol (EG) (Price et al., 1985), triethylene glycol dimethyl ether (TGDM) (George et al., 1987) and dietheylene glycol dimethyl ether (DYME) (Price et al., 1985). Table 2 shows the results of estimating the ED_{05} for each chemical using three different extrapolation procedures based on (a) a binary model with prenatal death as the outcome and implantations as the denominator; (b) a binary model with malformations as the outcome and live births as the denominator; (c) based on using the quantity above in (19), which will be referred to as the 'trinomial' model.

In each case, a probit model that incorporates a power parameter on dose,

$$P(d_i) = \Phi(\beta_0 + \beta_1 d_i^{\beta_2}),$$

was fit to the data. Several interesting features emerge from the table. First, the estimated ED_{05} based on the trinomial model are always lower than those based on the binomial models on prenatal death or malformation alone. Furthermore, if one of the individual binary model estimates is much lower than the other (as, for example, in EG and to some extent TGDM) then the estimate based on the trinomial model is close to the minimum of the two. However, in cases where the two individual binary predictions are close to each other (for example, DEHP), the estimate based on the trinomial model tends to be lower than the minimum derived from the two binary models.

In practice, benchmark doses are estimated by 95% lower confidence intervals on the ED_{05}. Ryan (1992) showed that estimates based on the quantity (19) should be

Table 2
ED_{05} (g/kg/day) values under three extrapolation models

Study	Estimated ED_{05}		
	Death	Malformation	Trinomial
EG	1.723	0.308	0.302
DEHP	0.079	0.070	0.054
TGDM	0.577	0.868	0.506
DYME	0.216	0.157	0.138

Table 3
Estimated ED_{05} and lower 95% confidence limits for binary and trinomial models

Study	Binary outcome approach		Trinomial approach	
	Estimated ED_{05}	Lower 95% CI	Estimated ED_{05}	Lower 95% CI
EG	0.213	(0.036)	0.302	(0.114)
DEHP	0.062	(0.037)	0.054	(0.034)
TGDM	0.423	(0.027)	0.506	(0.0)
DYME	0.120	(0.088)	0.138	(0.109)

more precise. For the four chemicals studied by NTP, the lower 95% confidence intervals based on the trinomial approach were compared to those from a standard analysis of the binary outcome 'dead or malformed' versus normal. Table 3 illustrates the results.

Catalano and Ryan (1992) consider models for continuous as well as discrete endpoints. They derive dose-response models for fetal weight (Y_1) and malformation (Y_2) by assuming a continuous bivariate distribution for weight and a latent malformation variable. They show that dichotomizing the latter quantity to obtain the observable binary malformation indicator leads to a pair of regression models, a linear regression on weight and a probit model for malformation, that can be fit using a two-step approach. The first step involves fitting a model for weight

$$Y_{1ijk} = \alpha_0 + \alpha_1 d_i^{\alpha_2} + \varepsilon_{1ijk} \tag{20}$$

and calculating individual $e_{ijk} = Y_{1ijk} - \hat{Y}_{1ijk}$ and average $\bar{e}_{ij}$ weight residuals. The second step involves a probit model for malformation that conditions on the regression for fetal weight

$$\Pr(Y_{2ijk} = 1) = \Phi(\beta_0 + \beta_1 d_i^{\beta_2} + \beta_3 \bar{e}_{ij} + \beta_4 e_{ijk}). \tag{21}$$

The fetal weight residual terms in the malformation mean result from assuming a common correlation structure between fetal weight and malformation for outcomes occurring within a fetus (corresponding to β_4) and between fetuses from the same litter (corresponding to β_3). Using a GEE approach, estimation of parameters in (20) and (21) is accomplished by solving separate estimating equations for each model, obviating the complication of computations involving dissimilar outcomes. Catalano and Ryan develop a joint test for the dose parameters α_1 and β_1, and also show how tests regarding the weight residual parameters β_3 and β_4 lead to inferences regarding the extent of between and within-fetus correlation.

To illustrate the method, Table 4 shows output from fitting (20) and (21) to the DYME data. Notice first that, as one would expect, the dose coefficient for weight is negative, while the dose coefficient for malformation is positive. Both dose effects are very significant, and a joint test for dose effect (H_0:$\alpha_1 = \beta_1 = 0$) yields a strongly significant χ_2^2 value of 172.1. The estimated power parameters $\hat{\alpha}_2$ and $\hat{\beta}_2$ are close to 1

Table 4
Output from fitting the Catalano and Ryan model to the DYME data

Coefficient	Est.	S.E.	Z
Weight model			
Intercept ($\hat{\alpha}_0$)	1.007	0.0176	57.20
Dose ($\hat{\alpha}_1$)	−0.992	0.0841	−11.80
Dose power ($\hat{\alpha}_2$)	1.129	0.1304	8.66
Common weight correlation	0.601		
Malformation model			
Intercept ($\hat{\beta}_0$)	−3.357	0.6505	−5.16
Dose ($\hat{\beta}_1$)	10.442	1.272	8.21
Dose power ($\hat{\beta}_2$)	1.038	0.2631	3.94
Average weight residual ($\hat{\beta}_3$)	−4.926	1.530	−3.22
Weight residual ($\hat{\beta}_4$)	−3.260	1.152	−2.83
Common malformation correlation	0.019		

indicating that there is little deviation from the linear model for weight and the linear model on the probit scale for malformation. Note also that the coefficient for the average weight residual is larger in magnitude (−4.9) than the coefficient for the individual weight residual (−3.3) indicating in these data that the between-fetus correlation for weight and malformation is slightly larger than the within-fetus weight and malformation correlation. Note also the relatively large degree of correlation among weight (0.601) compared to the estimated malformation correlaion (0.019). This phenomenon is common when modeling outcomes of different type; continuous measures tend to exhibit higher correlation than binary outcomes.

In Catalano et al. (1993), the results of Ryan (1992) and Catalano and Ryan (1992) have been extended to incorporate outcomes involving death, malformation, and birth weight. They construct a model for risk assessment based on the probability Pr(d) that a fetus is resorbed, dead, malformed, or of low birth weight.

5. Risk assessment

This section describes methods currently used for risk assessment in developmental toxicity and also considers more quantitative alternatives based on dose-response modeling.

5.1. The NOAEL approach

While quantitative risk assessment for carcinogenic endpoints has been well studied and is generally well established (Gart et al., 1986), statistical methods for assessing risk with regard to non-cancer endpoints have not been explored as thoroughly (Kimmel, 1990; Rodricks et al., 1986). The current approach to risk assessment used by both the EPA and the FDA for developmental toxicity testing centers on determination of a no observed adverse effect level (NOAEL). The NOAEL is defined as the experimental dose level immediately below the lowest dose that produces a statistically or biologically significant increase in adverse effect over controls. An 'acceptably safe' daily dose level for humans is then derived by dividing the NOAEL by a safety factor of usually 100 or 1000 to account for sensitive subgroups of the population and extrapolation from animal data to human risk. The EPA refers to this safe daily concentration as the reference dose (RfD) while the FDA uses the term allowable daily intake (ADI). In the event that the lowest experimental dose shows significant changes from control, it is termed a LOAEL (lowest observed adverse effect level) and an extra factor of 10 is used in the safety factor to determine the RfD or ADI.

The use of the NOAEL-safety factor approach to determining allowable doses is a point of controversy among risk assessors and regulators because it has a number of serious statistical drawbacks (Crump, 1984; Gaylor, 1983, 1989; Kaplan et al., 1987; Kimmel, 1990; Kimmel and Gaylor, 1988; Leisenring and Ryan, 1992). For instance, because the NOAEL is by definition one of the expermental doses, reference dose estimates can vary by orders of magnitude under repeated experimentation yet this statistical variation is ignored in reporting RfD values. Estimation of the NOAEL is anti-conservative with regard to sample size. Since the NOAEL is based on comparison to control levels, large studies have higher power to detect small changes and therefore produce low NOAELs whereas more highly variable, smaller studies tend towards higher, less conservative, NOAELs. This unfortunate feature has led regulatory agencies to require minimum sample sizes (20 to 30 dams per dose group) in order to ensure interpretable findings. Also, NOAEL estimates ignore the inherent dose response relationship of the data. Consider Figure 4, adapted from

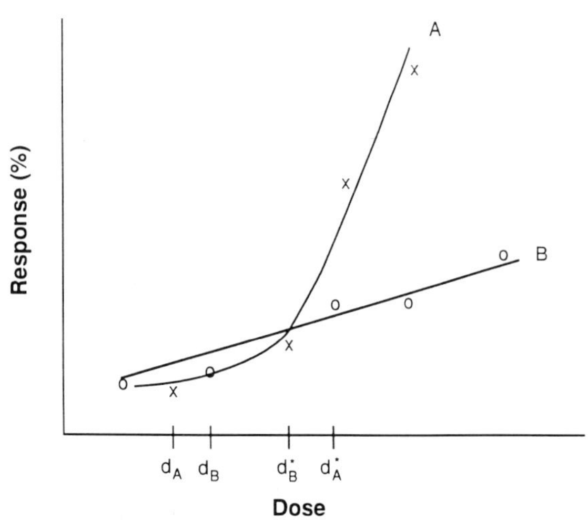

Fig. 4. Two hypothetical dose-response curves.

Table 5
NOAEL values from two hypothetical experiments

Outcome	NOAEL (g/kg/day)	
	Compound *A*	Compound *B*
Resorption	0.2	0.9
Death	0.3	0.9
Malformation	0.3	0.9
Weight	0.3	0.1

Crump (1984), which describes two hypothetical experiments: *A*, with an initially flat but increasingly steep dose response, and *B*, with a shallow but linear dose response. Because the curve is initially so flat in experiment *A*, one would expect that the *true* no effect level would be higher for *A* than for *B*. However, because the responses at d_A^* and d_B^* are the first to differ significantly from controls, the NOAEL for experiment B, d_B, is larger than that of experiment A, d_A. Finally, the most important disadvantage to this approach is that neither the NOAEL nor the RfD provide any estimate or upper bound of corresponding risk. Leisenring and Ryan (1992) derived the distributional properties of the NOAEL and found that associated risk levels are highly variable and often unacceptably high. Such uncertainty in the NOAEL makes it nearly impossible to compare compounds or conduct risk management.

A more general problem with the NOAEL approach is that it is typically based on analysis of individual outcomes rather than as part of a risk assessment procedure that considers the entire development process. Often the NOAEL for a compound is determined by the most sensitive endpoint, i.e., the outcome producing the lowest NOAEL. While it may appear conservative, this technique de-emphasizes subtle changes occurring across several endpoints. For example, consider the summary results presented in Table 5 for two hypothetical compounds with separate analyses for each endpoint. Current standard would set the NOAELs for compounds *A* and *B* at 0.2 g/kg/day and 0.1 g/kg/day, respectively. Yet clearly compound *A* shows broader effects across the range of outcomes measured and may in fact pose a greater developmental threat than compound *B*, which will be more conservatively regulated. Thus agents that cause a broad range of developmental problems may be set at unacceptably high 'safe' doses.

5.2. *Quantitative risk assessment*

The idea of using dose-response models to determine safe dose levels began in carcinogenicity testing where quantitative models have been used to find a virtually safe dose (VSD) or dose corresponding to a small risk, say 10^{-6}, above background. Recently, several authors have discussed the use of dose-response models to characterize risk at low doses as an alternative to the NOAEL for non-cancer endpoints. For example, Rai and Van Ryzin (1985) adopted the VSD in risk assessment for developmental outcomes. Crump (1984) advocates fitting a reasonably flexible dose-response model to the data and then using the estimated model to find a lower confidence limit on the dose corresponding to a specified level of excess risk. Crump refers to this lower bound to as a benchmark dose (BD). As with the NOAEL, Crump advises that the benchmark dose be divided by a safety factor of 100 or 1000 to account for inter- and intra-species differences.

Calculation of the BD, requires specification of a dose-response function and, in the case of non-zero control rates, an excess risk function. For binary data, assume that an appropriate model leads to a risk function $P(d)$, while for continuous data the notation $\mu(d)$ denotes the mean response function. The excess risk function, $r(d)$ defines the actual extrapolation level (e.g., 10% or 1%) and is therefore a very important quantity from a regulatory standpoint. A reasonable model for the excess risk is

$$r(d) = P(d) - P(0), \tag{22}$$

referred to as additional risk. This function places equal weight at all levels of background risk. Another choice for excess risk is

$$r(d) = \frac{P(d) - P(0)}{1 - P(0)}. \tag{23}$$

As described by Crump (1984), function (23) places greater weight upon outcomes with larger background levels since the same absolute change in the risk function corresponds to greater changes in excess risk as P(0) increases. However, for small background rates (22) and (23) are very similar. For continuous outcomes the analogous extrapolation quantity is the percent increase above background

$$r(d) = \frac{\mu(d) - \mu(0)}{\mu(0)},$$

called the extra response (Crump, 1984). To standardize this quantity, one could replace the denominator with the standard deviation in the control group.

In order to calculate the BD for an extra risk of q ($q = 0.05$, say), solve the extra risk function for dose d^* and find its lower confidence limit d_B. To illustrate this calculation for a binary endpoint, consider (22) as the excess risk function, and let the risk function be defined by

$$P(d) = h(\beta_0 + \beta_1 d^{\beta_2}),$$

where h is some specified link function (logit or probit, for example). Then the estimated dose corresponding to an additive excess risk of q is

$$\begin{aligned} \hat{d}^* &= P^{-1}[P(0) + q] \\ &= \left[\frac{h^{-1}(h(\hat{\beta}_0) + q) - \hat{\beta}_0}{\hat{\beta}_1}\right]^{\hat{\beta}_2^{-1}}. \end{aligned} \tag{24}$$

For example, in the case where P(d) is Weibull (16), (24) is

$$\hat{d}^* = \left[\frac{-\log(e^{-\hat{\beta}_0} + q) - \hat{\beta}_0}{\hat{\beta}_1}\right]^{\hat{\beta}_2^{-1}}.$$

There are several approaches to calculating the BD for an excess risk level q. One involves using the delta method and exploiting the asymptotic normality of the estimated parameters from the model. By letting $\boldsymbol{\beta}$ denote the parameter vector of the dose-response function, the asymptotic variance of $\hat{d}^*$ can be estimated by

$$\operatorname{var}(\hat{d}^*) = \left(\frac{\partial d^*}{\partial \boldsymbol{\beta}}\right)^{\mathrm{T}} \operatorname{var}(\hat{\boldsymbol{\beta}}) \left(\frac{\partial d^*}{\partial \boldsymbol{\beta}}\right) \tag{25}$$

evaluated at the estimated parameter values. Often h^{-1} is difficult to obtain, in which case the variance can be obtained using the excess risk function directly so that

$$\operatorname{var}(\hat{d}^*) = \left(\frac{\partial r}{\partial d}\right)^{-2} \left(\frac{\partial r}{\partial \boldsymbol{\beta}}\right)^{\mathrm{T}} \operatorname{var}(\hat{\boldsymbol{\beta}}) \left(\frac{\partial r}{\partial \boldsymbol{\beta}}\right). \tag{26}$$

Using either (25) or (26), the BD is then

$$\hat{d}_B = \hat{d}^* - 1.645\sqrt{\operatorname{var}(\hat{d}^*)}.$$

As an illustration, applying (26) for the Weibull model (16) yields

$$(\beta_1 \beta_2 d^{\beta_2 - 1})^{-2} \begin{bmatrix} 1 - \exp(\beta_1 d^{\beta_2}) \\ d^{\beta_2} \\ \beta_1 d^{\beta_2} \log d \end{bmatrix}^{\mathrm{T}} \operatorname{var}(\hat{\boldsymbol{\beta}}) \begin{bmatrix} 1 - \exp(\beta_1 d^{\beta_2}) \\ d^{\beta_2} \\ \beta_1 d^{\beta_2} \log d \end{bmatrix}.$$

Several authors (Crump, 1984; Crump and Howe, 1983; Krewski and Van Ryzin, 1981) suggest that there are drawbacks to using the delta method for calculating confidence bounds especially in low dose extrapolation settings. A more preferable approach involves finding confidence bounds using the asymptotic distribution of the likelihood ratio statistic

$$2(L_{\max} - L_1) \tag{27}$$

under the null hypothesis, where $L_{\max}$ is the unconstrained maximum value of the log-likelihood function and L_1 is value of the log-likelihood subject to some constraint (Cox and Hinkley, 1974). To continue the example, the likelihood involves parameters β_0, β_1, and β_2 and therefore (27) will follow a χ_3^2 distribution. The benchmark dose (95% lower confidence limit) for excess risk q using the likelihood ratio method is obtained by finding values of β_0, β_1, and β_2 that satisfy

$$2(L_{\max} - L_1) = (1.645)^2 \tag{28}$$

and minimize (24). This method is unavailable, however, in quasi-likelihood settings since there is no analogue to the likelihood ratio statistic.

Kimmel and Gaylor (1988) describe the advantages of using dose-response models over the NOAEL for low dose risk extrapolation. They propose a related but slightly different technique for calculating lower limits on dose for small levels excess risk by determining the dose corresponding to an upper confidence limit on the excess risk function rather than a lower confidence limit on dose producing a given excess risk. They define the effective dose at q (ED_q) to be the dose causing q excess response, and the lower limit on this dose (LED_q) to be the dose corresponding to the upper confidence limit of excess risk at q. These quantities are represented graphically in Figure 5 for $q = 0.1$. While qualitatively similar to the benchmark dose, the LED is based on the asymptotic distribution of the excess dose-response function rather than the asymptotic distribution of the estimated dose. Thus, to obtain the lower confidence limit on dose, they set the upper confidence limit on the excess risk function equal to q and solve for the LED. That is, solve

$$\hat{r}(d) + 1.645\,\mathrm{se}(\hat{r}(d)) = q$$

for dose. Again, the delta method can be used to obtain the standard error of the excess risk function and for most dose-response functions the solution must be obtained numerically. As previously discussed, a more appropriate method may be to use the likelihood ratio statistic to determine the upper confidence limit on excess risk. That is, a 95% upper confidence curve for excess risk can be obtained by maximizing (22) subject to (28) at each dose.

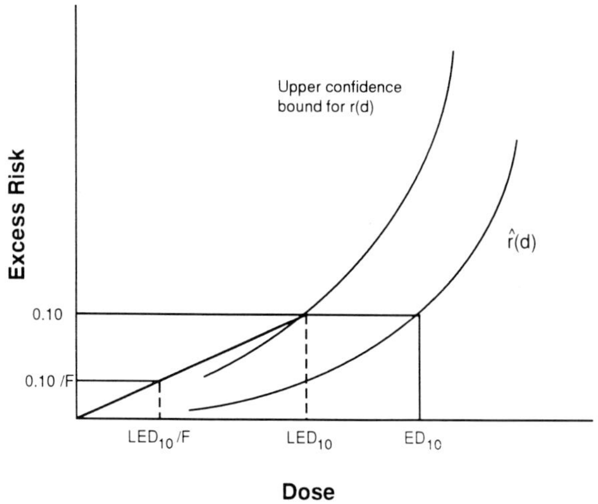

Fig. 5. Illustration of ED_{10} and LED_{10}.

Gaylor (1989) illustrates this procedure by fitting exponential polynomial models

$$P(d) = 1 - \exp[-(\beta_0 + \beta_1 d + \beta_2 d^2 + \cdots + \beta_k d^k)]$$

to proportions of affected fetuses for ten sets of data. By comparing methods, Gaylor found the NOAEL to be 2 to 6 times larger than the LED_{01}.

6. Discussion

This chapter has presented an overview of some of the important statistical issues that arise in the analysis of developmental toxicology studies. Due to space limitations, the chapter has focused on the broad topics of testing for and modeling dose response, and related issues of quantitative risk assessment. Many other issues remain open for further research however.

Design is likely to be an important topic for further study. At present, the choice of numbers and levels of the chosen exposure groups depends largely on conventions developed during the years when NOAELs and safety factors were the tool of choice for risk assessment. As quantitative methods based on dose-response models become more popular, the question of how to choose dose levels will need to be addressed. Many people who have applied dose-response models to toxicological data have noticed that parameter values can be highly correlated when only a few dose groups are represented. Dose groups need to be carefully spaced in order to provide useful and reliable information about the shape of dose-response curves.

Dose-response models of the future may rely more heavily on input from toxicologists about actual delivered doses, based on carefully constructed physiological and pharmacological models. Ongoing research into biologically based models that explore the mechanisms of toxicity may also be useful in suggesting biomarkers that can be used to improve the precision and reliability of risk assessment. Research is also ongoing in the area of in vitro assays.

Questions of model adequacy and goodness of fit are important ones. At present, appropriate methods are largely unavailable for the types of correlated outcomes encountered in developmental toxicology.

Finally, methods to incorporate historical control information may be useful. This may be particularly important, for example, when defining benchmark doses for continuous outcomes such as fetal weight.

Acknowledgements

The authors wish to thank Dr. David Lovell for the photographs in Figure 1, Dr. Bernard Schwetz at the National Toxicology Program for access to the computerized data base used in the examples, and Daniel Scharfstein at the Harvard School of Public Health for computational assistance.

References

Altham, P. M. E. (1978). Two generalizations of the binomial distribution. *Appl. Statist.* **27**, 162–167.

Brent, R. L. and R. P. Jensh (1967). Intra-uterine growth retardation. *Adv. in Teratol.* **2**, 139–227.

Carr, G. J. and C. J. Portier (1991). An evaluation of the Raj and Van Ryzin dose-response model in teratology. *Risk Anal.* **11**, 111–120.

Catalano, P. J. and L. M. Ryan (1992). Bivariate latent variable models for clustered discrete and continuous outcomes. *J. Amer. Statist. Assoc.* **87**, 651–658.

Catalano, P. J., D. O. Scharfstein, L. M. Ryan, C. A. Kimmel and G. L. Kimmel (1993). A statistical model for fetal death, fetal weight, and malformation in developmental toxicity studies. *Teratology* **47**, 281–290.

Chen, J. J. and R. L. Kodell (1989). Quantitative risk assessment for teratologic effects. *J. Amer. Statist. Assoc.* **84**, 966–971.

Chen, J. J., R. L. Kodell, R. B. Howe and D. W. Gaylor (1991). Analysis of trinomial responses from reproductive and developmental toxicity experiments. *Biometrics* **47**, 1049–1058.

Chesson, J. (1986). Developmental and reproductive toxicity data analysis. Draft final report, Battelle Columbus Laboratories – Washington Operations. Task 2-6, Office of Pesticides and Toxic Substances, US Environmental Protection Agency, Washington, DC, Contract 68-02-4246.

Cox, D. R. and D. V. Hinkley (1974). *Theoretical Statistics.* Chapman and Hall, London.

Crump, K. S. (1984). A new method for determining allowable daily intakes. *Fund. Appl. Toxicol.* **4**, 854–871.

Crump, K. S. and R. B. Howe (1985). Review of methods for calculating confidence limits in low dose extrapolation. In: D. B. Clayson, D. R. Krewski, and I. C. Munroe, eds., *Toxicological Risk Assessment.* CRC Press, Boca Raton, FL.

Food and Drug Administration (1986). Guidelines for reproduction and studies for safety evaluation of drugs for human use. Bureau of Drugs. Rockville, MD.

Gart, J. J., D. Krewski, P. N. Lee, R. E. Tarone and J. Wahrendorf (1986). *Statistical Methods in Cancer Research,* Vol. III: The Design and Analysis of Long-Term Animal Experiments. Oxford Univ. Press, Oxford, UK.

Gaylor, D. W. (1983). The use of Safety factors in controlling risk. *J. Toxicol. Environ. Health.* **11**, 329–336.

Gaylor, D. W. (1989). Quantitative risk analysis for quantal reproductive and developmenttal effects. *Environ. Health Perspect.* **79**, 243–246.

George, J. D., C. J. Price, C. A. Kimmel and M. C. Marr (1987). The developmental toxicity of triethylene glycol dimethyl ether in mice. *Fund. Appl. Toxicol.* **9**, 173–181.

Haseman, J. K. and L. L. Kupper (1979). Analysis of dichotomous response data from certain toxicological experiments. *Biometrics* **35**, 281–293.

Holmes, L. B. (1992). Fetal environmental toxins. *Pediatrics Rev.* **13**(10).

Kaplan, N., D. Hoel, C. Portier and M. Hogan (1987). An evaluation of the safety factor approach in risk assessment. Banbury Report 26: Developmental Toxicology: Mechanisms and Risk. Cold Spring Harbor Laboratory.

Kimmel, C. A. (1990). Quantitative approaches to human risk assessment for noncancer health effects. *Neurotoxicology* **11**, 201–210.

Kimmel, C. A. and D. W. Gaylor (1988). Issues in qualitative and quantitative risk analysis for developmental toxicology. *Risk Anal.* **8**, 15–20.

Kimmel, C. A. and C. J. Price (1990). Developmental toxicity studies. In: *Handbook of In Vivo Toxicity Testing.* Academic Press, New York, 271–300.

Kline, J., Z. Stein and M. Susser (1989). *From Conception to Birth: Epidemiology of prenatal Development.* Oxford Univ. Press, New York.

Krewski, D. and J. Van Ryzin (1981). Dose-response models for quantal response toxicity data. In: J. Sxorgo, D. Dawson, J. N. K. Rao and E. Saleh, eds., North-Holland, New York.

Kupper, L. L. and J. K. Haseman (1978). The use of a correlated binomial model for the analysis of certain toxicological experiments. *Biometrics* **34**, 69–76.

Kupper, L. L., C. Portier, M. D. Hogan and E. Yamamoto (1986). The impact of litter effects in dose-response modeling in teratology. *Biometrics* **42**, 85–98.

Lefkopoulou, M., D. Moore and L. Ryan (1989). The analysis of multiple correlated binary outcomes: Application to rodent teratology experiments. *J. Amer. Statist. Assoc.* **84**, 810–815.

Leisenring, W. and L. M. Ryan (1992). Statistical properties of the NOAEL. *Regul. Toxicol. Pharmacol.* **15**, 161–171.

Liang, K.-Y and S. L. Zeger (1986). Longitudinal data analysis using generalized linear models. *Biometrika* **73**, 13–22.

McCullagh, P. and J. A. Nelder (1989). *Generalized Linear Models.* Chapman and Hall, London.

Ochi, Y. and R. L. Prentice (1984). Likelihood inference in a correlated probit regression model. *Biometrika* **71**, 531–543.

Paul, S. R. (1982). Analysis of affected foetuses in teratological experiments. *Biometrics* **38**, 361–370.

Paul, S. R. (1987). On the beta-correlated binomial (bcb) distribution – a three parameter generalization of the binomial distribution. *Comm. Statist. Theory Methods* **16**, 1473–1478.

Piegorsch, W. W. and J. K. Haseman (1991). Statistical methods for analyzing developmental toxicity data. *Teratog. Carcinog. Mutag.* **11**, 115–133.

Prentice, R. L. (1986). Binary regression using an extended beta-binomial distribution, with discussion of correlation induced by covariate measurements errors. *J. Amer. Statist. Assoc.* **81**, 321–327.

Price C. J., C. A. Kimmel, J. D. George and M. C. Marr (1987). The developmental toxicity of diethylene glycol dimethyl ether in mice. *Fund. Appl. Toxicol.* **8**, 115–126.

Price C. J., C. A. Kimmel, R. W. Tyl and M. C. Marr (1985). The developmental toxicity of ethylene glycol in rats and mice. *Toxicol. Appl. Pharmacol.* **81**, 113–127.

Rai, K. and J. Van Ryzin (1985). A dose-response model for teratological experiments involving quantal responses. *Biometrics* **41**, 1–9.

Rodricks, J. V., R. G. Tardiff, S. M. Brett, R. M. Putzrath and C. St. Hilaire (1986). Elements of toxicology and chemical risk assessment. Environ Corporation. Washington, DC.

Rosner, B. (1982). Statistical methods in ophthalmology: An adjustment for the intraclass correlation between eyes. *Biometrics* **38**, 105–114.

Rosner, B. (1984). Multivariate methods in ophthalmology with applications to other paired data situations. *Biometrics* **40**, 1025–1035.

Ryan, L. M. (1992). Quantitative risk assessment for developmental toxicity. *Biometrics* **48**, 163–174.

Ryan, L. M., P. J. Catalano, C. A. Kimmel and G. L. Kimmel (1991). Relationship between fetal weight and malformation in developmental toxicity studies. *Teratology* **44**, 215–223.

Skalko, R. G. and E. M. Johnson, Mods. (1987). Concensus workshop on the evaluation of maternal and developmental toxicity Work Group I report: End points of maternal and developmental toxicity. *Teratog. Carcinog. Mutag.* **7**, 307–310.

Tyl, R. W., M. C. Price, M. C. Marr and C. A. Kimmel (1988). Developmental toxicity evaluation of dietary di(2-ethylhexyl)pathalate in Fisher 344 rats and CD-1 mice. *Fund. Appl. Toxicol.* **10**, 395–412.

US Environmental Protection Agency (1991). Guidelines for developmental toxicity and risk assessment. *Federal Register* **56**, 63798–63826.

Wedderburn, R. W. M. (1974) Quasi-likelihood functions, generalized linear models, and the Garss-Newton method. *Biometrika* **61**, 439–447.

Williams, D. A. (1975). The analysis of binary responses from toxicology experiments involving reproduction and teratogenicity. *Biometrics* **31**, 949–952.

Williams, D. A. (1982). Extra-binomial variation in logistic linear models. *Appl. Statist.* **31**, 144–148.

Williams, D. A. (1987). Dose-response models for teratological experiments. *Biometrics* **43**, 1013–1016.

Zeger, S. L. and K.-Y. Liang (1986). Longitudinal data analysis for discrete and continuous outcomes. *Biometrics* **42**, 121–130.

G. P. Patil and C. R. Rao, eds., *Handbook of Statistics, Vol. 12*

15

Environmental Biometry: Assessing Impacts of Environmental Stimuli Via Animal and Microbial Laboratory Studies

Walter W. Piegorsch

Abstract: Environmental biometry is the study and application of statistical methods in environmental studies on biological systems. In the areas of toxicology and environmental health, the damaging effects of environmental chemicals or other stimuli are often studied in animal and microbial systems. Data from such experiments are analyzed via various statistical approaches, each depending on the nature of the endpoint and aspects of the particular assay under study. Among major endpoints of interest, those of carcinogenicity, mutagenicity, and teratogenicity often garner greatest attention, and analyses for these three endpoints are discussed in this chapter. Statistical issues of interest include dose-response estimation and testing, including settings where variance heterogeneity and overdispersion are present. Similarities among the various endpoints and their biometric analyses are illustrated with data on toxicological response to 1,3-butadiene, a gaseous toxin found in cigarette smoke and automobile exhaust.

1. Introduction

Issues in environmental biometry span a large range of topics, each involved with the role statistical methods play in environmental studies of biological and physico-chemical systems. These methods include environmental monitoring (Gilbert, 1987), statistics for spatial sampling and modeling (Cressie, 1991), compartmental and biomathematical modeling (Matis and Wehrly, 1979; Jacquez, 1985; Castillo-Chavez et al., 1988) and statistics in environmental health (Hoel, 1979; Blot et al., 1985; Hills and Alexander, 1989). Other chapters in this volume touch upon one or more of these issues. Herein, attention is directed at biometrical methods useful in analyzing laboratory studies of environmental chemicals or other stimuli in animal and microbial systems. These systems are often employed to identify and study the toxicological and pharmacological effects of environmental stimuli.

A number of different forms of toxic response may be considered in environmental biometric studies, including induction of cancer (carcinogenesis), birth defects (teratogenesis) and other reproductive disease, or mutations in somatic and germ cells (mutagenesis). Other, more selective endpoints are also possible (Hayes, 1989; Mehlman, 1989; Luster et al., 1992).

The use of non-human systems when studying the effects of environmental chemicals is common. Animal systems employed to study toxic response to such chemicals generally involve rodents or other small mammals. Processes studied in whole animals include carcinogenicity after both short and long term exposures (Haseman, 1984), teratogenicity (Piegorsch and Haseman, 1991) and heritable mutagenicity (Lockhart et al., 1992). Somatic cell mutagenicity can be studied as well (Margolin, 1985), both in vivo (Murphy et al., 1992) and via in vitro cell cultures (Margolin et al., 1986; Murphy et al., 1988).

Mutagenic and other genotoxic effects can also be studied in microbial assays; typically these involve bacteria (Collings et al., 1981; Margolin et al., 1989) or yeast (Piegorsch et al., 1989) cultures. Such microbial assays combine ease of use and lower costs (Zeiger et al., 1985) with shorter time scales for study (since, e.g., bacteria reproduce rapidly, providing information on mutagenic responses over a period of days or weeks, rather than months or years for whole animals). They can be employed as initial screens for potential toxic effects of environmental chemicals, although it has been noted that microbial short-term tests can only augment, and not replace, whole animals assays for monitoring and assessing toxicity of environmental chemicals (Tennant et al., 1987; DeMarini et al., 1989). Both animal and microbial bioassays are therefore used in laboratory studies of environmentally-induced damage.

Statistical issues associated with these bioassays are many and varied. They include concerns with study design (Hoel and Jennrich, 1979; Mackey and MacGregor, 1979; Crump, 1982) and optimal animal allocations (Piegorsch and Bailer, 1989; Murphy et al., 1992), data analysis of treatments vs. control differences (Gad and Weil, 1989), dose-response analysis (Hasselblad, 1981; Krewski and Van Ryzin, 1981), incorporation of historical controls (Dempster et al., 1983; Margolin and Risko, 1984), correction for multiple comparisons (Meng and Dempster, 1987; Heyse and Rom, 1988; Piegorsch, 1991; Hothorn, 1993), and quantitative risk assessment (Hoel et al., 1983; Hoel, 1985: Chen and Kodell, 1989). In depth discussions of all these areas is beyond the scope of this chapter, although some of these issues are discussed in greater detail below, and in other chapters in this volume. The goal herein is to describe and illustrate some of the biometric methods necessary for analysis of laboratory experiments with environmental stimuli, emphasizing their practical application. Attention will be directed at analysis of potential dose response of a chemical agent or other stimulus for the three major processes of carcinogenicity, teratogenicity and mutagenicity. Important issues within this context will be seen to change, depending upon which toxicological process is being discussed. For example, in carcinogenicity testing, dose-response analysis may require adjustment for intercurrent mortality, since the competing risk of death to the experimental units may diminish the power of standard trend tests for dose response. Similar concerns concerning competing toxicity occur in mutagenicity testing, but the effect there is to infuse non-monotone curvature into the dose response. Different adjustments for trend testing are needed for these two separate cases. In teratogenicity testing, major concerns over the potential for overdispersion in discrete outcomes (i.e., variability greater than that predicted by the standard statistical models) are noted, and trend tests that account for or are independent of such overdispersion are presented. (The overdispersion issue is also prevalent in mutagenicity testing.) Statistical methods for identifying and dealing with these various problems are discussed in the sections below.

2. Biometrical methods for rodent carcinogenicity experiments

Major efforts in toxicological screening of environmental chemicals are directed at carcinogenicity studies in rodents (Gart et al., 1986; Robins et al., 1989). The goal is to identify those environmental stimuli capable of inducing cancers in mammalian systems; rodent species such as rats and mice are commonly employed in the laboratory.

The basic design of a rodent carcinogenicity study involves T dose levels (usually, $T = 3$ or 4), including a zero-dose control, of a test chemical, administered to say, $n = 50$ animals per dose group. To measure cancer incidence, the number of animals with the tumor of interest Y_i is recorded at each of T dose levels ($i = 1,\ldots,T$). Thus the response is quantal: Y_i/n. Of interest is whether these proportions exhibit an increasing trend with respect to dose.

2.1. *Trend test for dose response*

Assuming Y_i has a binomial sampling model with sample size parameter n (or n_i, if the number of animals per dose varies) and probability parameter p_i ($i = 1,\ldots, T$), statistical analysis for trend over dose may be performed via the well-known Cochran–Armitage trend test (Margolin, 1988) for proportions,

$$Z_{\mathrm{CA}} = \frac{\sum_{i=1}^{T}(x_i - \bar{x})Y_i}{\sqrt{\bar{p}\bar{q}\sum_{i=1}^{T} n_i(x_i - \bar{x})^2}},$$

where x_i is the value of the i-th dose, or some measure associated with the i-th dose (such as log-dose),

$$\bar{x} = \frac{\sum_{i=1}^{T} n_i x_i}{\sum_{i=1}^{T} n_i},$$

$$\bar{p} = \frac{\sum_{i=1}^{T} Y_i}{\sum_{i=1}^{T} n_i},$$

and $\bar{q} = 1 - \bar{p}$. The statistic is referred to a standard normal distribution, where large positive values of Z_{CA} are indicative of significant dose response. In the case of just $T = 2$ groups (treatment vs. control), the trend test collapses to a simple large-sample comparison between two proportions (Margolin et al., 1983).

2.2. *Adjusting for survival differences*

A major caveat associated with use of Z_{CA} is that the test does not adjust for differences in mortality rates across dose groups. In practice, if the carcinogenicity study is being performed over a short time frame – such as 3–6 months – intercurrent mortality may not occur and adjustment for such is not critical. If the study involves a longer, chronic exposure period – such as 15–24 months – intercurrent mortality may act to diminish the power of the trend test (Bailer and Portier, 1988). In such cases, survival-adjusted tests are computed to assess increasing tumor incidence.

Different biometric methods exist for testing increasing tumor incidence in long-term exposure studies. One may begin by first testing for survival differences among the dose groups. Cox's (1972) life table test is apropriate here, although Tarone's (1975) extension for linear trend (in survival) may also be used. If survival difference are identified, tests incorporating adjustments for these differences are advocated (Peto et al., 1980). These vary according to the differing level of fatal effect the tumor has on the test animal.

When it can be supposed that the tumor instantaneously leads to death of the animal (relative to the time scale over which measurments are taken), life table tests (Cox, 1972; Tarone, 1975) are applied to the data by simply substituting 'time to observation of the tumor' for 'death' in the analysis. This provides a comparison of the time-specific tumor incidences between control and treated groups.

When it is not reasonable to attribute animal mortality to the tumor under study, alternative survival-adjusted tests are employed. These methods hinge primarily on the converse assumption that tumors observed in animals dying before the end of the study are strictly nonlethal, or 'incidental'. Incidental tumor tests based on statistical regression models, such as a logistic score test (Dinse and Lagakos, 1983), are generally recommended in this setting. Indeed, logistic regression analysis of incidental tumors possesses the desirable property that it can incorporate tests for trend via a slope coefficient associated with the dose covariate. It is, in fact, more powerful than other models in its ability to discern treatment vs. control differences in tumor incidence (Dinse, 1985). The method adjusts for intercurrent mortality by employing the individual death times as covariables in the linear predictor. It can be extended to include any other important explanatory factors as covariables (Dinse and Haseman, 1986).

In some cases (e.g., rare tumors), the use of historical control data can add valuable information when assessing dose response for tumor incidence. Such information can be misused, however, and care must be taken when including historical data in the statistical analysis. For example, nomenclature conventions and diagnostic criteria must be carefully calibrated so as to allow meaningful comparisons among similar neoplastic effects (Haseman et al., 1984). Also, outside sources of variability, such as laboratory differences or changes in tumorigenic susceptibility of a strain or species over time, must be isolated and controlled (Haseman et al., 1989). Left unrecognized, such factors can lead to extra-binomial variability in the Y_i, reducing the sensitivity of the statistical analysis (Tarone, 1982; Dempster et al., 1983).

For some tumor types, it may be difficult to categorize the tumor as either 'instantaneously' lethal or incidental to animal mortality. The use of either of these extreme assumptions for tumor-induced lethality allows for easily-implemented, survival-adjusted tests of dose response, as described above. For tumors that may have only partial effects on animal mortality, however, survival-adjusted statistical methods are more difficult to construct and to apply. They traditionally have required additional assumptions about the context(s) of observation in which the tumor was recorded (e.g., whether an observed tumor in a particular animal contributed to that animal's death) and/or additional, interim sacrifices in order to draw statistical inferences on the chemical's carcinogenicity (McKnight and Crowley, 1984; Dinse, 1988). More recent methods have attempted to avoid making specific assumptions about the tumor's lethality, or avoid requiring multiple sacrifice times in order to produce survival-adjusted inferences (Portier and Bailer, 1989). These newer methodologies are still being evaluated in practical, experimental settings, and may not yet be ready for general use. Nonetheless, they represent promising areas of research for statisticians and biometricians interested in quantitative methods for cancer research.

2.3. *Example: 1,3-butadiene*

As an illustration of these methods, consider the following data on murine response to the environmental chemical 1,3-butadiene. This chemical is an intermediate in industrial polymer production, and is also present in a variety of airborne compounds, including cigarette smoke, gasoline vapor and automobile exhaust (National Toxicology program, 1984). Of interest is its carcinogenic potential to mammals after inhalation exposure, particularly in targets such as the lung and respiratory tract. In a study conducted by the US National Toxicology Program (NTP), female mice were exposed to 1,3-butadiene via inhalation over a 61 week period. Three dose groups were used: a zero dose control ($x_1 = 0$) and two test groups exposed to $x_2 = 625$ ppm and $x_3 = 1250$ ppm of the chemical. Data on observed mortality and tumor incidence for these groups appear in Table 1.

The quantal response data from Table 1 yield the following observed rates for lung tumors: at $x_1 = 0$ ppm, $\frac{3}{49}$ (6%); at $x_2 = 625$ ppm, $\frac{12}{48}$ (25%); and at $x_3 = 1250$ ppm, $\frac{23}{49}$ (47%). The Cochran–Armitage trend test for these data gives $Z_{CA} = 4.604$, with one-sided P-value less than 0.001. This is very strong evidence in favor of an increasing carcinogenic potential with increasing dose for this chemical.

Analysis of the mortality patterns in these data via Cox's test (1972) shows significant ($P < 0.001$) differences in survival between the control group and either exposed group. Thus, adjustment for intercurrent mortality should be considered in

Table 1
Mortality and lung tumor incidence from inhalation study of 1,3-butadiene in female mice[1]

Dose = 0 ppm		Dose = 625 ppm		Dose = 1250 ppm	
Survival time	Tumor indicator	Survival time	Tumor indicator	Survival time	Tumor indicator
2	0	2	0	20	0
6	0	2	0	22	0
30	0	29	0	40	0
61	0	29	0	40	0
61	0	32	0	43	0
61	0	33	0	47	0
61	0	37	0	50	1
61	0	38	0	53	0
61	0	39	0	53	0
61	0	40	0	55	0
61	0	41	0	55	0
61	0	46	0	56	0
61	0	48	0	56	0
61	0	48	0	58	1
61	1	52	0	58	1
61	0	52	0	58	1
61	0	53	0	59	0
61	0	53	0	59	1
61	0	54	0	60	1
61	0	55	0	61	0
61	0	56	0	61	0
61	0	56	0	61	0
61	0	56	1	61	0
61	0	56	0	61	1
61	0	56	0	61	0
61	0	56	0	61	0
61	0	56	0	61	1
61	0	57	1	61	1
61	0	58	0	61	0
61	0	58	1	61	1
61	0	59	0	61	1
61	0	60	0	61	1
61	0	60	1	61	1
61	0	61	0	61	0
61	0	61	0	61	1
61	1	61	1	61	0
61	0	61	1	61	1
61	0	61	0	61	0
61	0	61	0	61	1
61	0	61	1	61	1
61	1	61	1	61	0
61	0	61	0	61	1
61	0	61	0	61	1
61	0	61	1	61	1
61	0	61	1	61	1
61	0	61	0	61	0
61	0	61	1	61	1
61	0	61	1	61	0
61	0			61	1

[1] Data from NTP (1984). Survival times in weeks. Tumor indicator is 1 if animal exhibited tumor, 0 otherwise.

the dose-response analysis. In mice of the strain used in the NTP experiment, lung tumors are not considered to be instantaneously lethal, so that a survival-adjusted incidental tumor analysis is more appropriate than the alternative lethal tumor/life table analysis. Employing a logistic regression (Dinse and Lagakos, 1983) of tumor incidence on dose and survival yields a slope coefficient for dose of $b = 0.222$, with asymptotic standard error 0.080. A test for trend compares the ratio of these quantities to a standard normal reference distribution. The statistic is 4.829, which, again, is highly significant ($P < 0.001$). Alternatively, a likelihood ratio test for any significant effect due to dose yields a statistic of 28.83. Referenced to a χ^2 distribution with 1 degree of freedom, this statistic also indicates a significant dose effect ($P < 0.001$). In passing, one should note that a life table analysis of these data would also be strongly significant ($P < 0.001$), so that any of the trend analyses discussed herein detect a significant dose effect for these data.

3. Biometrical methods for reproductive toxicity experiments

Animal studies in reproductive and developmental toxicology involve evaluation of the teratogenic or heritable mutagenic potential of some environmental stimulus. The associated assay is conducted to ascertain whether the stimulus will affect development of the embryo or of the parental gametes. As with carcinogenicity testing, large-scale reproductive toxicity testing of environmental chemicals if often performed in rodents. If the experimental protocol directs chemical exposure to occur at some point after fertilization, the reproductive damage involves teratogenesis (Manson and Kang, 1989) (or possibly zygote mutagenesis; cf. Generoso et al., 1991). Alternatively, if the protocol directs exposure for the male or female parent prior to mating, the damage generally involves heritable mutagenesis (Haseman and Soares, 1976; Lockhart et al., 1992).

Reproductive toxicity due to chemical exposure is assessed via examination of the offspring from laboratory matings between male and female rodents. After successful conception, the pregnant female is housed and fed, then either sacrificed for uterine examination after a few weeks, or allowed to take her litter to term, depending on the specific reproductive endpoint under study. Each female's litter is examined conceptus-by-conceptus to arrive at some quantitative summary measure of the chemical's reproductive toxicity. A commonly employed measure is the per-litter ratio

$$\frac{\text{number of damaged conceptuses}}{\text{total conceptuses examined}},$$

where 'damage' is defined in the context of the experiment. (It can be some form of malformation such as a cleft pallet, or a resorption, or other form of fetal death, etc.) When the chemical exposure occurs at different doses, statistical analysis for a positive dose response is of interest. Although attention herein is directed at discrete outcomes for teratogenic processes, it is important to note that continuous response variables (such as per-conceptus weights) may also be recorded and analyzed. Many of the issues salient to discrete outcomes are, in fact, important in consideration of continuous outcomes; these are discussed in other sources (Healy, 1972; Dempster et al., 1984).

3.1. Litter effects and extra-binomial variability

Recognizing that the experimental unit is the parent that was exposed to the chemical is important in quantitative evaluation of reproductive toxicity data (Haseman and Hogan, 1975). For a typical teratogenesis experiment, this unit is the female mouse (or, alternatively, her entire litter), not the individual conceptus. The individual conceptuses represent correlated observations, and should not be analyzed without accommodation for this 'litter effect' if the endpoint of interest is binary, so that the per-litter observation involves the number Y_{ij} of conceptuses exhibiting developmental damage in the j-th litter at the i-th dose ($i = 1, \ldots, T$; $j = 1, \ldots, J_i$), then a binomial sampling assumption for this count is invalid. (One may still assume, however, that some binary response probability, p_i is of interest at each dose level i.) Generally, the litter effect acts to induce extra-binomial variability in the Y_{ij}, requiring a more complex sampling model, or some form of non-parametric or semi-parametric analysis. Ways to account for litter effects during data analyses are discussed in Sections 3.3–3.4.

In any analysis of per-litter data, the actual litter sizes N_{ij} are themselves random variables. Unfortunately, few analyses are able to account for the stochastic nature of the N_{ij}, and do not take this into account. In such cases, the analyses' inferences are understood to be conditional on the observed set of N_{ij}. Some attempts have been made to model the variability in N_{ij} and incorporate this into the overall analysis of per-litter data. For example, one could assume an independent Poisson sampling distribution on the N_{ij} (Rai and Van Ryzin, 1985), and model parental and other toxic effects on litter size via some non-linear model. This approach has met with limited success (Faustman et al., 1989), although there is some question as to its ability to capture correctly the actual sampling variability in the N_{ij} (Williams, 1987). (It is commonly observed, e.g., that litter sizes exhibit *under*dispersion relative to the Poisson model; see McCaughran and Arnold, 1976, or Lockhart et al., 1992.) Further, modeling the toxic effects of environmental chemicals on the N_{ij} appears much more difficult than first thought (Carr and Portier, 1991); additional research into this area is necessary before these models for N_{ij} can be recommended for use.

3.2. Testing for extra-binomial variability

To test for extra-binomial variability in per-litter data, a number of test statistics are available. For example, at each dose ($i = 1, \ldots, T$) one can employ the binomial variance test (Cochran, 1954)

$$X_{Vi}^2 = \sum_{j=1}^{J_i} \frac{(Y_{ij} - \bar{p}_i N_{ij})^2}{N_{ij}\bar{p}_i(1 - \bar{p}_i)}$$

or a locally most powerful $C(\alpha)$-test (Tarone, 1979)

$$X_{Ci}^2 = \frac{\left\{ \dfrac{\sum_{j=1}^{J_i} (Y_{ij} - \bar{p}_i N_{ij})^2}{\bar{p}_i(1-\bar{p}_i)} - \sum_{j=1}^{J_i} N_{ij} \right\}^2}{2\sum_{j=1}^{J_i} N_{ij}(N_{ij}-1)},$$

where, for both statistics, $\bar{p}_i$ is a consistent estimator of p_i (as $\min\{J_i\} \to \infty$), such as

$$\bar{p}_i = \frac{\sum_{j=1}^{J_i} Y_{ij}}{\sum_{j=1}^{J_i} N_{ij}}.$$

The X_{Vi}^2 statistic is referred to a χ^2 distribution with $J_i - 1$ degrees of freedom, while the X_{Ci}^2 statistic is referred to a χ^2 distribution with 1 degree of freedom. For a summary measure of departure from simple binomial variability, the test statistics can be summed over independent doses; e.g., $X_V^2 = \sum_{i=1}^{T} X_{Vi}^2$ is distributed as χ^2 with $\sum_{i=1}^{T} J_i - T$ degrees of freedom.

Both test procedures possess good statistical power to detect departures from binomial variability (Margolin, personal communication), with the $C(\alpha)$-test being asymptotically optimal against certain forms of extra-binomial variability. In particular, the $C(\alpha)$-test is optimal against the so-called *beta-binomial* distribution (Tarone, 1979), a standard alternative to the binomial in toxicological experiments where hierarchical sampling induces extra-binomial variation (Williams, 1975; Haseman and Kupper, 1979). See Section 3.4, below.

3.3. *Trend tests for dose response*

For testing dose response in the presence of extra-binomial variability, Piegorsch and Haseman (1991) recommend straightforward rank-based or permutation-based dose-response analyses, especially in those settings where the sampling distribution of the Y_{ij} (and, for that matter, the N_{ij}) is difficult to identify (Lockhart et al., 1992). Such analyses include the Jonckheere–Terpstra trend test (Lehmann, 1975, Section 5.5B) on the ranks of the observed proportions, or a permutation test for trend (Mantel, 1963) also applied to the proportions Y_{ij}/N_{ij}. For the Jonckheere–Terpstra analysis, one assumes only that the observed proportions Y_{ij}/N_{ij} have median values m_i ($i = 1, \ldots, T$), i.e. $\Pr[Y_{ij}/N_{ij} > m_i] = \frac{1}{2}$, where typically m_i will be a function of p_i. Thus we ignore the fact that Y/N is a quotient of two separate random variables and concentrate directly on the single random variable, the proportion Y/N itself. As such, the conditionality on the N_{ij} is not required for this analysis. There is, however, the additional requirement that the variances (or, more generally, the *shapes* of the distributions) of the random proportions be equivalent. This homogeneity assumption may not hold in all cases; it is most often invalidated when large disparities exist among the N_{ij}. (When large disparities occur, the applicability of the Jonckheere–Terpstra test is brought into question.)

To test the (null) hypothesis H_0: $m_1 = m_2 = \cdots = m_T$ against the specific alternative hypothesis $H_1: m_1 \leqslant \cdots \leqslant m_T$ (with strict inequality somewhere among the m_i), the Jonckheere–Terpstra test applies successive Mann–Whitney comparisons among the ranks of the observed proportions Y_{ij}/N_{ij}. Corrections are recommended if there are tied observations (Lehmann, 1975, pp. 235–236), especially at a single rank value (Lin and Haseman, 1976).

If concern exists in replacing the actual observed proportions with their rank values, thereby incurring potential loss in information (Mantel, 1979), a permutation test for trend can be employed instead (Agresti et al., 1990). In large samples, asymptotic normality of various permutation test statistics often holds (Sen, 1985). For instance, consider the trend statistic $T_P = \sum_{i=1}^{T} \sum_{j=1}^{J_i} x_i\, Y_{ij}/N_{ij}$ (Mantel, 1963), with mean

$$\mathrm{E}[T_P] = \frac{\sum_{i=1}^{T} J_i x_i}{\sum_{i=1}^{T} J_i} \left(\sum_{i=1}^{T} \sum_{j=1}^{J_i} \frac{Y_{ij}}{N_{ij}} \right)$$

and variance

$$\mathrm{var}(T_P) = \left\{ \sum_{i=1}^{T} J_i x_i^2 - \frac{(\sum_{i=1}^{T} J_i x_i)^2}{\sum_{i=1}^{T} J_i} \right\} \frac{\left\{ \sum_{i=1}^{T} \sum_{j=1}^{J_i} \frac{Y_{ij}^2}{N_{ij}^2} - \frac{(\sum_{i=1}^{T} \sum_{j=1}^{J_i} \frac{Y_{ij}}{N_{ij}})^2}{\sum_{i=1}^{T} J_i} \right\}}{(\sum_{i=1}^{T} J_i) - 1}.$$

A large-sample statistic for testing trend is simply $Z_P = (T_P - \mathrm{E}[T_P])/\sqrt{\mathrm{var}(T_P)}$. The one-sided test rejects when Z_P is greater than an upper-α quantile from a standard normal reference distribution.

3.4. Regression analyses

When additional covariates (besides dose) are recorded along with the observed proportions, it may be of interest to incorporate these variables into the analysis via some form of regression methodology. As noted above, a generalization of the standard binomial distribution is the beta-binomial model:

$$\Pr[Y_{ij} = y \,|\, \mu_i, \varphi_i, N_{ij}] = \binom{N_{ij}}{y} \frac{\Gamma(y + \mu_i/\varphi_i)\Gamma(N_{ij} - y + \{1 - \mu_i\}/\varphi_i)\Gamma(\varphi_i^{-1})}{\Gamma(\mu_i/\varphi_i)\Gamma(\{1 - \mu_i\}/\varphi_i)\Gamma(\varphi_i^{-1} + N_{ij})},$$

$(i = 1, \ldots, T;\ j = 1, \ldots, J_i)$, where $\Gamma(\cdot)$ is the gamma function. The mean of Y_{ij} is $\mathrm{E}[Y_{ij}|\mu_i, \varphi_i, N_{ij}] = N_{ij}\mu_i$, and the variance of Y_{ij} is $\mathrm{var}[Y_{ij}|\mu_i, \varphi_i, N_{ij}] = N_{ij}\mu_i \{1 - \mu_i\}\{1 + (N_{ij} - 1)\varphi_i/[1 + \varphi_i]\}$. The dispersion parameter φ_i measures departure from binomial sampling: at $\varphi_i = 0$, binomial variability is retained. This model can be reparameterized to yield an intra-litter correlation: $\rho_i = \varphi_i/[1 + \varphi_i]$ (Kupper et al., 1986). Regression under this sampling distribution can involve a logistic model on dose or log-dose, say

$$\mu_i = \frac{1}{1 + \exp\{-\beta_0 - \beta_1 x_i\}}$$

(Kupper et al., 1986). Corresponding significance tests for dose response are performed by comparing the maximum likelihood estimate of β_1 divided by its estimated standard error against the upper-α point of a standard normal distribution. Significant positive departure from $\beta_1 = 0$ suggests increasing dose response.

In those cases where this beta-binomial model is known or suspected to be inappropriate, useful alternatives involve semi-parametric regression models. Instead of specific distributional characterizations, only limited assumptions on the mean and variance are required. These are then employed to construct so-called *quasi-likelihood* estimating equations for the mean parameters (McCullagh and Nelder, 1983, Section 8).

An important feature of the quasi-likelihood approach is its ability to incorporate overdispersion into the required moment assumptions. To wit, suppose the variance of Y_{ij} takes the form

$$\operatorname{var}(Y_{ij} \mid p_i, \rho, N_{ij}) = N_{ij}\, p_i(1 - p_i)\{1 + \rho(N_{ij} - 1)\},$$

where ρ is some dispersion parameter (Moore, 1986). At $\rho = 0$ we recover the usual binomial mean and variance, while $\rho \neq 0$ corresponds in our setting to an intra-litter correlation effect. This form is similar to the variance function of a beta-binomial random variable, except that no specific family of distributions (beta-binomial or otherwise) is assumed for the Y_{ij}. (The correlation parameter ρ is assumed constant across doses here, although this constraint can be lifted with little additional computational complexity.) Thus, this approach is, in a sense, distribution-free.

Moore (1986) suggests a generalization of this variance model to allow for other forms of overdispersion. If a logistic model, $p_i = 1/\{1 + \exp(-\beta_0 - \beta_1 x_i)\}$, is postulated, then this construction corresponds to a model proposed by Williams (1982) for overdispersed binary data. The resulting calculations to find maximum quasi-likelihood (MQL) estimates and large-sample standard errors for the three parameters β_0, β_1, and ρ require computer implementation. (These calculations are easily extended to estimate regression parameters for any number of covariates.) As above, a ratio of the MQL estimate of β_1 to its estimated MQL standard error is compared to the upper-α point of a standard normal distribution. Significant positive departure indicates an increasing trend under the logistic model. (An added benefit of the MQL model is that inference on the dispersion parameter ρ is available directly. One can employ the MQL estimate of ρ and its estimated MQL standard error to test $\rho = 0$, or form a confidence interval about ρ; significant departure suggests overdispersion.)

A further generalization of the ML estimation scheme involves so-called generalized estimating equations (GEEs), where the equations used to calculate the regression estimators are modified to take different types of variance structures into account. The method was motivated originally by the problem of estimating regression effects with data recorded longitudinally on the same experimental subject (say, over time), and it attempts to take the longitudinal correlation within subjects into account (Zeger et al., 1988). Applied to developmental and reproductive toxicity data, one views the multiple conceptuses per female as a longitudinal vector of observations, and applies the GEEs to achieve inference on any covariates of interest. This includes testing for dose response (Lefkopoulou et al., 1989). Again, instead of specific assumptions on distributions, only limited assumptions on the mean and variance are required.

For the simple case of one predictor variable x_i representing an increasing measure of dose or log-dose, the test statistic is

$$Z_{\text{GEE}} = \frac{\sum_{i=1}^{T}(x_i - \bar{x})\sum_{j=1}^{J_i} Y_{ij}}{\sqrt{\sum_{i=1}^{T}(x_i - \bar{x})^2\sum_{j=1}^{J_i}(Y_{ij} - N_{ij}\bar{p})^2}},$$

where

$$\bar{p} = \frac{\sum_{i=1}^{T}\sum_{j=1}^{J_i} Y_{ij}}{\sum_{i=1}^{T}\sum_{j=1}^{J_i} N_{ij}}.$$

Z_{GEE} is compared to the upper-α point of a standard normal distribution. Significant positive departure indicates an increasing trend with dose.

Since the GEE methodology operates at the per-conceptus level by viewing the conceptuses as longitudinal data, it can include per-conceptus covariates into the regression analysis (Prentice, 1988). When adjusting for or testing such covariates is of interest, the GEE approach can be recommended.

To compare these various methodologies – and a few other; see, e.g., Boos (1993) – Lockhart et al. (1992) performed a series of Monte Carlo simulations examining the small sample size and power of each of the above-noted tests of dose response. Their results suggested that when no extra-binomial variability was present in the proportion data, any of the procedures discussed above would perform adequately. They favored the trend test based on Z_{CA} from Section 2, since it is known to be locally optimal against any form of monotone increasing dose response (Tarone and Gart, 1980). When extra-binomial variability was present in the data, few differences among those test statistics designed to incorporate or independent of the overdispersion were evidenced in the Lockhart et al. simulations (Methods based on the simple binomial sampling model performed much worse than these others in the extra-binomial case, as was expected.) Since it is not always clear which form of overdispersed sampling distribution – such as beta-binomial – is valid or appropriate for reproductive toxicity data, especially for studies of heritable mutagenicity, Lockhart et al. favored distribution-free approaches such as the permutation test in Section 3.3. If, however, additional covariates are recorded (e.g., time, strain, etc.), they also favored use of the MQL or GEE regression methods (Williams, 1982; Prentice, 1988) for assessing significance of any covariates of interest, including dose.

3.5. Example: 1,3-butadience (continued)

Continuing with the example of the airborne toxin 1,3-butadiene from Section 2.3, consider the following data on that chemical's teratogenic potential. The data in Table 2 represent a study of the lethality to rat fetuses after their mother's exposure to

Table 2
Developmental toxicity data after female rats' inhalation exposure to 1,3-butadiene

Dose = 0 ppm		Dose = 200 ppm		Dose = 1000 ppm		Dose = 8000 ppm	
Number dead	Total implants	Number dead	Total implants	Number dead	Total implants	Number dead	Total implants
0	11	0	13	0	13	0	12
1	9	0	14	1	14	1	13
0	13	0	15	1	12	5	13
0	15	0	3	0	13	0	17
0	15	2	14	0	13	1	14
0	11	0	12	0	15	0	15
0	12	1	9	0	10	1	11
0	14	4	14	2	13	0	14
2	14	0	10	0	13	1	12
1	14	0	12	1	15	1	13
1	17	0	17	0	15	9	16
1	14	0	14	0	11	0	16
0	14	1	10	0	17	0	14
0	15	0	16	1	17	2	16
0	13	5	16	0	10	0	15
0	12	1	13	2	17	1	13
0	12	0	9	2	15	0	8
0	16	1	10	0	13	0	14
0	14	0	15	3	14	0	15
3	15	0	16	0	16	0	16
3	17	2	15	1	15	1	13
1	13	0	15	2	16	0	14
0	14			0	17	0	13
1	16						
0	14						
1	16						
0	9						
1	10						
0	2						
0	9						
0	14						
0	14						
1	14						
0	4						
0	15						
0	16						

Litters examined:
$J_1 = 36$ $J_2 = 22$ $J_3 = 23$ $J_4 = 23$
Pooled proportions:
$\bar{p}_1 = 0.036$ $\bar{p}_2 = 0.060$ $\bar{p}_3 = 0.049$ $\bar{p}_4 = 0.073$

Table 3
Tests for extra-binomial variability for developmental toxicity data

Dose d_i	Litters J_i	Binomial variance statistic $X^2_{V_i}$ (df $= J_i - 1$)	$C(\alpha)$ statistic X^2_{Ci} (df $= 1$)
0	36	42.26 (P = 0.186)	1.81 (P = 0.179)
200	22	46.27 (P = 0.001)	22.22 (P = 2.4×10^{-6})
1000	23	26.83 (P = 0.218)	0.397 (P = 0.529)
8000	23	90.95 (P < 10^{-8})	135.41 (P < 10^{-8})
Sum	104	206.31 (P < 10^{-8})	159.84 (P < 10^{-8})

the chemical via inhalation during the first two-weeks of gestation. The recorded variables are: Y_{ij} equal to the number of fetuses found dead at the two-week sacrifice in the j-th litter of the i-th dose group, and N_{ij} equal to the number of implants per litter ($i = 1, \ldots, 4$; $J = 1, \ldots, J_i$) as given in Table 2. Summary statistics for each dose are given at the bottom of the table.

Applying the tests for extra-binomial variability described above to these data yields the results given in Table 3. As seen therein, significant extra-binomial variability is indicated by either the binomial variance test or the $C(\alpha)$ procedure. As such, some adjustment for this overdispersion must be considered in any tests of dose response for these data. (Given the large spacings between the original doses, one should also perform any trend calculations on a log-dose scale; cf., Margolin et al., 1986.) Applying a distribution-free approach, the permutation test for trend of these data yields a test statistic of $Z_P = 1.41$, with associated one-sided P-value of 0.079. Alternatively, a MQL fit using a logistic regression with dose as a single explanatory variable yields an estimated slope coefficient of $b_{MQL} = 0.288$, with estimated large-sample standard error 0.265. The ratio of the two quantities is 1.39; referred to a standard normal distribution, the associated one-sided P-value is 0.082. If one felt that a beta-binomial sampling model were appropriate for these data, then the corresponding logistic regression analysis yields an estimated slope parameter of $b_{\beta b} = 0.250$, with estimated large-sample standard error 0.265. Here, the ratio of the two quantities is 0.946, with associated P-value 0.172. Thus under any of these analyses, the possibility of an increasing response with dose is suggested, although the trend is not clearly significant.

4. Biometrical methods for genotoxicity experiments

Endpoints in genetic toxicology are perhaps the most varied of any toxicological studies, since both the mechanisms and identification of genetic damage are so numerous. Indeed, there exist well over 100 different genotoxicological assays, employing all sorts of animal and microbial systems (Hollestein et al., 1979; Waters et al., 1987). Statistical themes associated with these assays include proper identification and selection of the sampling distribution (Margolin et al., 1989; Lockhart et al., 1992), construction of appropriate biomathematical models to characterize dose response and other features of genetic damage (Margolin et al., 1981; Ager and Haynes, 1987; Haynes, 1989), sample size determinations (Bishop and Kodell, 1980; Margolin et al., 1983; Murphy et al., 1992), and the selection of statistical approaches associated with these models and methods (Margolin, 1985). As above, a complete overview of these issues is beyond the scope of this chapter, although a useful introduction is given in the compilation by Kirkland (1989). Herein, attention will be directed at some of the basic issues of dose-response analysis often encountered in environmental mutagenesis experiments, with the goal of illustrating features that differ from and that are shared with the experimental paradigms for environmental carcinogenicity and teratogenicity testing discussed above.

4.1. Microbial systems in environmental mutagenesis

Perhaps the most well-known of all modern genotoxicity assays is the so-called *Salmonella* mutagenicity assay (Ames et al., 1975), employing the bacteria *Salmonella typhimurium* to identify damage to DNA after exposure to environmental stimuli. The assay is based on development of auxotrophic strains of the bacterium that are unable to synthesize histidine, an amino acid required for growth. This production deficiency can be reversed into a production capability via point mutations at selected sites on the bacterial genome. DNA damage is indicated by mutation of the bacteria from the auxotrophic state to the prototrophic, self-sustaining state. Thus, the mutated cells can *grow* in a micro-environment (such as a Petri plate) containing only minimal amounts of histidine; greater mutant yield at higher exposures to the environmental chemical suggests that mutagenesis increases with increasing dose. (A rather ironic concept, but, experimentally, very useful.) Observational accuracy of the assay is enhanced by use of a selective medium for the growth environment, so that only prototrophic mutant colonies may grow after exposure to the environmental toxin.

The basic observations in such a microbial assay are the mutated colony counts Y_{ij} for the j-th replicate plate at the i-th dose ($i = 1, \ldots, T; j = 1, \ldots, R_i$). Commonly, the number of replicate plates is taken to be constant, e.g., $R_i = 3 \, \forall i$. (Table 4 presents an example of data from a *Salmonella* assay, again using the chemical 1,3-butadiene from Section 2.3) A natural candidate for the sampling distribution of the Y_{ij} is the Poisson distribution for count variables. Indeed, many processes that drive the

Table 4
Salmonella assay results for 1,3-butadiene in strain TA1535

Dose (ppm)	Mutants per plate Y_{ij}			Average plate count $\bar{Y}_i$
0.000	20	31	27	26.0
0.002	100	92	89	93.7
0.007	147	123	178	149.3
0.014	216	170	181	189.0
0.020	176	154	183	171.0
0.030	154	153	149	152.0

Poisson sampling assumption are active with the colony count data seen here. These include: (i) the microbes' responses are independent; (ii) each locally organized group of microbes experiences the same environment (ignoring controlled variables such as dose), (iii) the plated number of microbes is large, say 10^8, and (iv) the probability that a plated microbe gives rise to a visible mutant colony is small, say between 10^{-5} and 10^{-9} (Margolin et al., 1981). A fifth assumption crucial to adoption of the Poisson distribution, however, is that both the environment and the number of microbes plated should remain relatively stable across replicate plates (within a dose group). If this is not the case, extra-Poisson variability results, and one is forced to model or otherwise account for the consequent overdispersion in the Y_{ij}.

4.2. *Testing for extra-Poisson variability*

Extra-Poisson variability in a set of ostensibly homogeneous observations is assessed via a variance to mean comparison similar to the binomial variance test discussed above. The test employs Fisher's dipersion statistic (Margolin et al., 1981, equation 2):

$$X_{\mathrm{D}i}^2 = \sum \frac{(Y_{ij} - \bar{Y}_i)^2}{\bar{Y}_i},$$

where the per-dose means $\bar{Y}_i$ and the sum are taken over $j = 1, \ldots, R_i$ $(i = 1, \ldots, T)$. This test is $C(\alpha)$-optimal against the so-called negative binomial distribution (Neyman and Scott, 1966), a standard alternative to the Poisson model that is discussed in greater detail below. Large values of $X_{\mathrm{D}i}^2$ relative to a χ^2 distribution with $R_i - 1$ degrees of freedom imply significant extra-Poisson variability. Extensions of this statistic to multiple test groups are available: simply aggregate the individual $X_{\mathrm{D}i}^2$ statistics into a sum of per-dose contributions in similar fashion to that used for overdispersed binomial proportions (e.g., X_{V}^2), above (Margolin, 1985):

$$X_{\mathrm{D}}^2 = \sum_{i=1}^{T} X_{\mathrm{D}i}^2.$$

This aggregate statistic is distributed asymptotically as χ^2 with $\sum_{i=1}^{T} R_i - T$ degrees of freedom.

A common application of these tests is to large, controlled databases where similar control trials and replicate plates are available. Margolin et al. (1981) reported on such data, and were able to demonstrate that the Poisson model is generally inadequate to describe statistical variability when analyzing *Salmonella* mutagenicity experiments. Further research has confirmed and supported this extra-Poisson variability for both *Salmonella* mutagenicity data (Margolin et al., 1989), and for other genotoxicological endpoints such as frequencies of sister chromatid exchanges in mammalian cells (Carrano and Moore, 1982; Margolin et al., 1986) and numbers of implants in rodent heritable mutagenesis assays (Lockhart et al., 1992).

The need to test for potential extra-Poisson variability is clear when analyzing count data from environmental mutagenicity experiments: standard trend tests for Poisson-distributed counts (Tarone, 1982) become too sensitive in the presence of overdispersion, increasing type I error rates above the nominally-specified α level. Thus, the data analyst should be aware of the nature and level of any overdispersion in the data before selecting a test for dose response.

4.3. *Trend tests for dose response*

In those environmental mutagenicity experiments where no extra-Poisson variability is evidenced, a standard form of Cochran–Armitage trend test for dose response may be employed. Under Poisson sampling, the test is known to be optimal against any monotone (e.g., non-increasing) form of dose response (Tarone, 1982; Margolin et al., 1986). Given the observed counts Y_{ij} ($j = 1, \ldots, R_i$; $i = 1, \ldots, T$), and denoting x_i as some increasing score (dose, log-dose, etc.) associated with treatment level i, the trend statistic takes the form

$$Z^* = \frac{\sum_{i=1}^{T} x_i [(\sum_{j=1}^{R_i} Y_{ij}) - R_i \bar{Y}]}{\sqrt{\bar{Y} S_x^2}}$$

in which

$$S_x^2 = \sum_{i=1}^{T} R_i (x_i - \bar{x})^2,$$

$$\bar{x} = \frac{\sum_{i=1}^{T} R_i x_i}{\sum_{i=1}^{T} R_i}$$

and

$$\bar{Y} = \frac{\sum_{i=1}^{T} \sum_{j=1}^{R_i} Y_{ij}}{\sum_{i=1}^{T} R_i}.$$

Significant dose-response is suggested when Z^* is larger than an appropriate normal distribution quantile.

If the response is overdispersed and the Poisson model is no longer valid, a form of Cochran–Armitage trend test may still be constructed, assuming the sampling distribution can nonetheless be specified. In particular, a common extension of the Poisson sampling model that accounts for overdispersion is the negative binomial distribution, which has been applied successfully to overdispersed microbial mutagenicity data such as is seen with the *Salmonella* assay (Margolin et al., 1989). The negative binomial has a probability function of the form

$$\Pr[Y_{ij} = y \mid \mu_i, \delta] = \frac{\Gamma(y + \delta^{-1})}{y! \, \Gamma(\delta^{-1})} \left(\frac{\delta \mu_i}{1 + \delta \mu_i} \right)^y (1 + \delta \mu_i)^{-1/\delta},$$

where $E[Y_{ij}|\mu_i, \delta] = \mu_i$, and the dispersion parameter δ measures departure from the Poisson model. In particular, $\mathrm{var}[Y_{ij}|\mu_i, \delta] = \mu_i + \delta\mu_i^2$. The Poisson distribution is the limiting case as $\delta \to 0$.

Under negative binomial sampling, a trend test possessing similar optimal attributes as the Z^*-test may be readily constructed (Margolin, 1985): simply replace $\sqrt{\bar{Y}S_x^2}$ in Z^* with an updated estimator based on the negative binomial variance,

$$Z^*_{\mathrm{NB}} = \frac{\sum_{i=1}^{T} x_i[(\sum_{j=1}^{R_i} Y_{ij}) - R_i\bar{Y}]}{\sqrt{\bar{Y}(1+\hat{\delta}\bar{Y})S_x^2}}.$$

The dispersion estimator $\hat{\delta}$ may be constructed via the method of moments, maximum likelihood, or maximum quasi-likelihood (Piegorsch, 1990). The simplest of these is the method of moments: equate the observed variance at each dose with the expression for $\mathrm{var}[Y_{ij}|\mu_i, \delta]$, using $\bar{Y}_i$ to estimate μ_i. Averaging over all T dose groups yields

$$\hat{\delta}_{\mathrm{MM}} = \frac{1}{T}\sum_{i=1}^{T}\frac{s_i^2 - \bar{Y}_i}{\bar{Y}_i^2},$$

where s_i^2 is the per-dose sample variance,

$$s_i^2 = \frac{1}{R_i - 1}\sum_{j=1}^{R_i}(Y_{ij} - \bar{Y}_i)^2.$$

Typically, the dispersion parameter δ is assumed non-negative. Thus, a concern often raised with the method of moments estimator is that it can suggest negative values for δ. There are settings, however, where it is valid to interpret $\delta < 0$ as evidence of underdispersion (Piegorsch, 1990), and such cases can occur in environmental mutagenesis (Lockhart et al., 1992). Unfortunately, this is not often the case for *Salmonella* data, and the fact that $\hat{\delta}_{\mathrm{MM}}$ can be negative may invalidate use of this estimator for such data.

A common alternative to the method of moments is maximum likelihood (ML) estimation. ML estimates of δ are calculated from the negative binomial log-likelihood

$$l(\mu_i, \delta) \propto -N \log \Gamma\left(\frac{1}{\delta}\right) + \sum_{i=1}^{T}\sum_{j=1}^{R_i}\left\{\log \Gamma\left(Y_{ij} + \frac{1}{\delta}\right) + Y_{ij}\log(\mu_i\delta) - \left(Y_{ij} + \frac{1}{\delta}\right)\log(1 + \mu_i\delta)\right\}.$$

The ML estimates for the μ_i are simply the per-dose means $\bar{Y}_i\ (i = 1, \ldots, T)$. Substituting these estimates into $\partial l/\partial\delta = 0$ yields a non-linear equation in δ:

$$\frac{\partial l}{\partial \delta} \propto N\,\Psi\left(\frac{1}{\delta}\right) + \sum_{i=1}^{T}\sum_{j=1}^{R_i}\left\{\log(1 + \delta\bar{Y}_i) - \Psi\left(Y_{ij} + \frac{1}{\delta}\right)\right\} = 0,$$

where $\Psi(\cdot)$ is the digamma function (Bernardo, 1976). The solution of this equation gives $\hat{\delta}_{ML}$. This estimator for δ may be used in Z^*_{NB}. As above, significant dose response is suggested when Z^*_{NB} is larger than an appropriate standard normal distribution quantile.

4.4. *Trend test under non-monotone dose response*

An important, additional feature exemplified by the data in Table 4 is a curious downturn in the response at higher doses. This form of non-monotone dose response is common in the *Salmonella* assay, and is often observed in other environmental mutagenesis assays as well. These include microbial assays for chromosomal damage in yeast (*Saccharomyces cerevisiae*) and mold (*Aspergillus nidulans*) (Piegorsch et al., 1989; Piegorsch, 1992), and animal tissue assays for cytogenic damage, such as cell transformation assays with hamster embryo cells, or in vitro chromosome aberration assays with human lymphocytes (Richardson et al., 1989).

In the *Salmonella* assay, the downturn phenomenon is driven by a number of possible mechanisms, the most common of which is a consequence of the experimental scheme employed to identify the mutational events. Since mutagenesis is identified by growth of the prototrophs on a selective medium, any other mechanism that hinders growth will compete with mutagenesis. Thus, e.g., high exposures of the environmental stimulus can lead to cell death or perhaps chemical/threshold induced increases in DNA-repair. These factors can conspire to *reduce* the yield of mutated cells by killing or neutralizing the microbes before mutations can be observed. The result is a downturn at the higher doses, i.e., an 'umbrella response' (Mack and Wolfe, 1981). As might be expected, simple tests for monotone trend (such as Z^* or Z^*_{NB} above) can exhibit large losses in power to detect any increase in dose response in the presence of a downturn (Collings et al., 1981). Alternative methods should be considered in this case.

Non-linear biomathematical and empirical models for this downturn phenomenon are available, which, when coupled with the negative binomial sampling model on the Y_{ij}, involve fairly advanced non-linear statistical estimation (Margolin et al., 1981; Breslow, 1984). An alternative to fitting complex parametric models of this sort involves the rather intuitive concept of identifying *statistically* the point of maximal departure from background mutation levels, and then performing a simple trend test up to that point. A number of different methodologies have been suggested for such as approach (Bernstein et al., 1982; Simpson and Margolin, 1986), most of which analyze the ranks of the observations via non-parametric techniques (Piegorsch, 1992). This is done in order to avoid the additional problem of specifying or incorporating a form of overdispersion in the sampling distribution.

A non-parametric approach developed by Simpson and Margolin (1986) employs such a testing strategy to recursively estimate the point of maximal response (the 'umbrella point'), and then test for a significant increase in dose response up to that point. One begins by calculating the two-sample Mann–Whitney statistics (Lehmann, 1975, Section 1.2), W_i, comparing the i-th dose group with – collectively – all preceding dose groups, starting with $i = T$ and working backwards. That is, W_T tests whether the response information at $i = T$ derivates significantly above the pooled response information at $i = 1, \ldots, T-1$. If so, an estimate h of the umbrella index at maximal response is $h = T$. If not, one drops the information at $i = T$, and repeats the process at $i = T-1$. In this way, the test employs the values of $W_i (i = T, T-1, \ldots, 2)$ to estimate the umbrella index via recursive pretesting. In effect, it selects this estimate as the largest dose index h such that W_h is larger than a specified critical value c_h; i.e.,

$$h = \max \{ i \in \{2, \ldots, T\}: W_i > c_i \},$$

where $c_2 = 0$ and $c_i = c_i(q)$ $(i = 3, \ldots, T)$ is the q-quantile of the distribution of W_i under the null hypothesis of no dose response $(H_0: \mu_1 = \cdots = \mu_T)$. The tuning parameter $q \in (0, 1)$ must be prespecified, and is usually set at or near $q = 0.5$ (Simpson and Margolin, 1990; Piegorsch, 1992).

Once the umbrella index h is estimated, the recursive procedure calculates the conditional Jonckheere–Terpstra trend statistic $U_h = W_1 + \cdots + W_h$ (Lehmann, 1975, Section 5.5B), and rejects H_0 in favor of an increasing (up to $i = h$) dose response when U_h is larger than the $(1-\pi)$-quantile of the distribution of U_i under H_0 $(i = 2, \ldots, T)$, for $0 < \pi < 1$. For fixed q and prespecified significance level α, a conservative choice for π is given as $\pi = \alpha(1-q)/(1-q^{T-1})$ (Simpson and Margolin, 1986).

4.5. *Example: 1,3-butadiene (continued)*

Continuing with the example of the airborne toxin 1,3-butadiene from Section 2.3, consider the following data on that chemical's mutagenic potential. The data in Table 4 represent *Salmonella* mutagenic response to the chemical after gaseous exposure. As discussed above, an increase in the number of observed, prototrophic colonies suggests a mutagenic effect, even in the presence of a downturn at higher doses.

To begin, consider a test of whether the data exhibit any overdispersion, relative to the basic Poisson sampling model. Employing an aggregated dispersion statistic X_D^2 to the data in Table 4 gives $X_D^2 = 22.134$ on 12 degrees of freedom (P = 0.004). Strong departure from Poisson variability is evidenced. If the negative binomial sampling model were considered as an adequate replacement to the Poisson for these data, then a trend test for increasing dose response based on Z_{NB}^* could be employed. The trend statistic is $Z_{NB}^* = 1.285$, with one sided P = 0.099. Thus some suggestion of an increasing trend is given.

For these data, however, a downturn in the dose response is evident. Although not severe (see Figure 1), this downturn could affect the trend test's ability to identify properly a positive dose response. Application of the recursive test for umbrella response is therefore appropriate: setting the tuning parameter to $q = 0.5$, the umbrella index is estimated as $h = 6$, with corresponding rank-based trend statistic given by $U_6 = 105.5$. Rejection occurs at significance levels as low as $\alpha = 0.005$, providing strong evidence of a positive dose response for these mutagenicity data.

In testing for increasing dose response under umbrella alternatives, the tuning

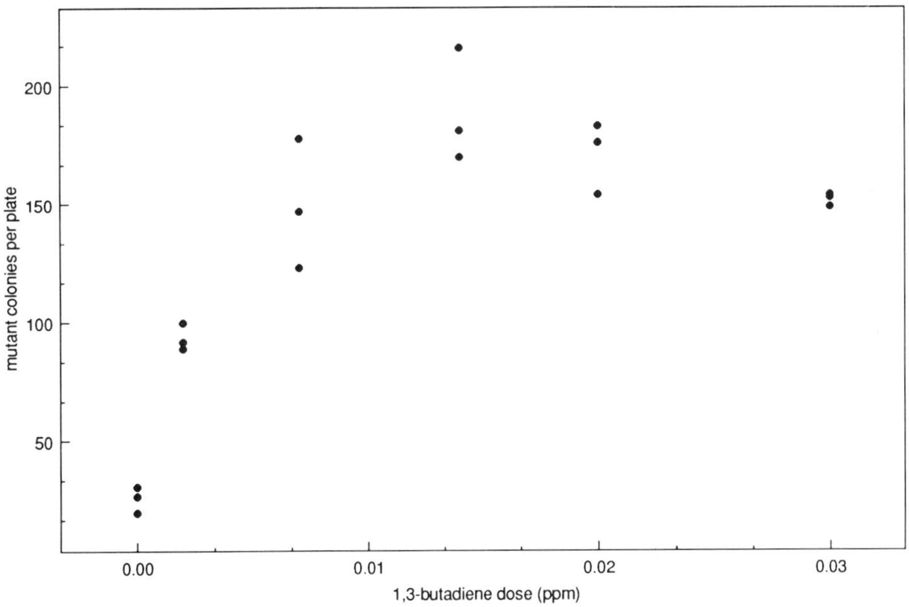

Fig. 1. *Salmonella* mutagenic response to 1,3-butadiene.

parameter can be used to increase sensitivity to earlier upturns by raising its value to, say, $q = 0.6$ or $q = 0.7$. If such sensitivity were specified a priori, the associated analysis for the 1,3-butadiene data would show an even more significant result: at either $q = 0.6$ or $q = 0.7$, the umbrella index is estimated as $h = 5$, with corresponding rank-based trend statistic given by $U_5 = 81.0$ ($P < 0.001$).

Acknowledgements

Thanks are due to Drs. Joseph K. Haseman and David M. Umbach for their careful readings of the manuscript and to Dr. Barry H. Margolin for many helpful comments.

References

Ager, D. D. and R. H. Haynes (1987). Mathematical description of the interactions between cellular inactivating agents. *Radiat. Res.* **110**, 129–141.

Agresti, A., C. R. Mehta and N. R. Patel (1990). Exact inference for contingency tables with ordered categories. *J. Amer. Statist. Assoc.* **85**, 6453–6458.

Ames, B. N., J. McCann and E. Yamasaki (1975). Methods for detecting carcinogens and mutagens with the *Salmonella*/mammalian microsome mutagenicity test. *Mutat. Res.* **31**, 347–364.

Bailer, A. J. and C. J. Portier (1988). Effects of treatment-induced mortality and tumor-induced mortality on tests for carcinogenicity in small samples. *Biometrics* **44**, 417–431.

Bernardo, J. M. (1976). Algorithm AS103. The digamma function. *Appl. Statist.* **25**, 315–317.

Bernstein, L., J. Kaldor, J. McCann and M. C. Pike (1982). An empirical approach to the statistical analysis of mutagenesis data from the Salmonella test. *Mutat. Res.* **97**, 267–281.

Bishop, J. B. and R. L. Kodell (1980). The heritable translocation assay: Its relationship to assessment of genetic risk for future generations. *Teratog. Mutag. Carcinog.* **1**, 305–322.

Blot, W. J., T. Hirayama and D. G. Hoel (1985). *Statistical Methods in Cancer Epidemiology.* Radiation Effects Research Foundation, Hiroshima.

Boos, D. D. (1992). Analysis of dose-response data in the presence of extra-binomial variation. *Appl. Statist.* **42**, 173–184.

Breslow, N. (1984). Extra-Poisson variability in log-linear models. *Appl. Statist.* **33**, 38–44.

Carr, G. J. and C. J. Portier (1991). An evaluation of the Rai and van Ryzin dose-response model in teratology. *Risk Anal.* **11**, 111–120.

Carrano, A. V. and D. H. Moore (1982). The rationale and methodology for quantifying sister chromatid exchanges in humans. In: J. A. Heddle, ed., *Mutagenicity: New Horizons in Genetic Toxicology*, Academic Press, New York, 267–304.

Castillo-Chavez, C., S. A. Levin and F. Gould (1988). Physiological and behavioral adaptation to varying environments: A mathematical model. *Evolution* **42**, 986–994.

Chen, J. J. and R. L. Kodell (1989). Quantitative risk assessment for teratological effects. *J. Amer. Statist. Assoc.* **84**, 966–971.

Cochran, W. G. (1954). Some methods for strengthening the common χ^2 tests. *Biometrics* **10**, 417–451.

Collings, B. J., B. H. Margolin and G. W. Oehlert (1981). Analyses for binomial data, with applications to the fluctuations test for mutagenicity. *Biometrics* **37**, 775–794.

Cox, D. R. (1972). Regression models and life tables. *J. Roy. Statist. Soc. Ser. B* **34**, 187–220.

Cressie, N. A. C. (1991). *Statistics for Spatial Data.* Wiley, New York.

Crump, K. (1982). Designs for discriminating between binary dose response models with application to animal carcinogenicity experiments. *Comm. Statist. Theory Methods* **11**, 375–394.

DeMarini, D. M., J. Lewtas and H. E. Brockman (1989). Utility of short-term tests for genetic toxicity. In: M. A. Mehlman, ed., *Benchmarks: Alternative Methods in Toxicology.* Princeton Scientific Publishing, Princeton, NJ, 205–216.

Dempster, A. P., C. M. Patel, M. R. Selwyn and A. J. Roth (1984). Statistical and computational aspects of mixed model analysis. *Appl. Statist.* **33**, 203–214.

Dempster, A. P., M. D. Selwyn and B. J. Weeks (1983). Combining historical and randomized controls for assessing trends in proportions. *J. Amer. Statist. Assoc.* **78,** 221–227.

Dinse, G. E. (1985). Testing for a trend in tumor prevalence rates: I. Nonlethal tumors. *Biometrics* **41,** 751–779.

Dinse, G. E. (1988). A prevalence analysis that adjusts for survival and tumour lethality. *Appl. Statist.* **37**, 435–445.

Dinse, G. E. and J. K. Haseman (1986). Logistic regression analysis of incidental-tumor data from animal carcinogenicity experiments. *Fund. Appl. Toxicol.* **6**, 44–52.

Dinse, G. E. and S. W. Lagakos (1983). Regression analysis of tumour prevalence data. *Appl. Statist.* **32**, 236–248.

Faustman, E. M., D. G. Wellington, G. P. Smith and C. A. Kimmel (1989). Characterization of a development toxicity dose-response model. *Environ. Health Perspect.* **79**, 229–241.

Gad, S. C. and C. S. Weil (1989). Statistics for toxicologists. In: A. W. Hayes, ed., *Principles and Methods of Toxicology.* 2nd ed., Raven Press, New York, 435–484.

Gart, J. J., D. Krewski, P. N. Lee, R. E. Tarone and J. Wahrendorf (1986). *Statistical Methods in Cancer Research,* Vol. III. The Design and Analysis of Long-Term Animal Experiments. IARC Scientific Publications, Lyon, France.

Generoso, W. M., A. G. Shourbaji, W. W. Piegorsch and J. B. Bishop (1991). Developmental responses of zygotes exposed to similar mutagens. *Mutat. Res.* **250**, 439–446.

Gilbert, R. O. (1987). *Statistical methods for Environmental Pollution Monitoring.* Van Nostrand Reinhold, New York.

Haseman, J. K. (1984). Statistical issues in the design, analysis and interpretation of animal carcinogenicity studies. *Environ. Health Perspect.* **58**, 385–392.

Haseman, J. K. and M. D. Hogan (1975). Selection of the experimental unit in teratology studies. *Teratology* **12**, 165–172.

Haseman, J. K., J. E. Huff and G. A. Boorman (1984). Use of historical control data in carcinogenicity studies in rodents. *Toxicol. Pathol.* **12**, 126–135.

Haseman, J. K., J. E. Huff, G. N. Rao and S. L. Eustis (1989). Sources of variability in rodent carcinogenicity studies. *Fund. Appl. Toxicol.* **12**, 793–804.

Haseman, J. K. and L. L. Kupper (1979). Analysis of dichotomous response response data from certain toxicological experiments. *Biometrics* **35**, 281–293.

Haseman, J. K. and E. R. Soares (1976). The distribution of fetal death in control mice and its implications on statistical tests for dominant lethal effects. *Mutat. Res.* **41**, 277–288.

Hasselblad, V. (1981). Modeling dose response relatioships for health effects data in: D. L. Thomsen Jr, ed., *Environmetrics 81: Selected papers.* SIAM, Philadephia, PA, 179–194.

Hayes, A. W. (1989). *Principles and Methods of Toxicology.* 2nd ed., Raven Press, New York.

Haynes, R. H. (1989). Mutugenesis and mathematics: The allure of numbers. *Environ. Molec. Mutag.* **14**, 200–205.

Healy, M. J. R. (1972). Animal litters as experimental units. *Appl. Statist.* **21**, 155–159.

Heyse, J. F. and D. Rom (1988). Adjusting for multiplicity of statistical tests in the analysis of carcinogenicity studies *Biometrical J.* **30**, 883–896.

Hills, M. and F. Alexander (1989). Statistical methods used in assessing the risk of disease near a source of possible environmental pollution: A review,. *J. Roy. Statist. Soc. Ser. A* **152**, 353–363.

Hoel, D. G. (1979). Statistical approaches to toxicological data. Environ. Health Perspect. **32**, 267–272.

Hoel, D. G. (1985). Mathematical dose-response models and their application to risk estimation. In: V. B. Vouk, G. C. Butler, D. G. Hoel and D. B. Peakall, eds., *Methods for Estimating Risk of Chemical Injury: Human and Non-Human Biota and Ecosystems.* Wiley, New York, 347–359.

Hoel, D. G., N. L. Kaplan and M. W. Anderson (1983). Implication of nonlinear kinetics on risk estimation in carcinogenesis. *Science* **219**, 1032–1037.

Hoel, P. G. and R. I. Jennrich (1979). Optimal designs for dose response experiments in cancer research. *Biometrika* **66**, 307–316.

Hollstein, M., J. McCann, F. Angelosanto and W. Nichols (1979). Short-term tests for carcinogens and mutagens *Mutat. Res.* **65**, 133–226.

Hothorn, L. (1993). Multiple comparisons in long-term toxicological studies. *Environ. Health Perspect.* **101**(7), in press.

Jacquez, J. A. (1985). *Compartmental Analysis in Biology and Medicine.* 2nd ed., Univ. of Michigan Press, Ann Arbor, MI.

Kirkland, D. J. (1989). *Statistical Evaluatiion of Mutagenicity Test Data.* Cambridge Univ. Press, Cambride.

Krewski, D. and J. Van Ryzin (1981). Dose response models for quantal response toxicity data. In: (M. Csörgö, D. A. Dawson, J. N. K. Rao and A. K. M. E. Saleh, eds., *Statistics and Related Topics.* North-Holland, Amsterdam, 201–231.

Kupper, L. L., C. J., Portier, M. D. Hogan and E. Yamamoto (1986). The impact of litter effects on dose-responses modeling in teratology. *Biometrics* **42**, 85–98.

Lefkopoulou, M., D. Moore and L. Ryan (1989). The analysis of multiple correlated binary outcomes: Applicatioins to rodent teratology experiments. *J. Amer. Statist. Assoc.* **84**, 810–815.

Lehmann, E. L. (1975). *Nonparametrics: Statistical Methods Based on Ranks.* Holden-Day, San Francisco, CA.

Lin, F. O. and J. K. Haseman (1976). A modified Jonckheere test against ordered alternatives when ties are present at a single extreme value. *Biom. Z.* **18**, 623–631.

Lockhart, A.-M., W. W. Piegorsch and J. B. Bishop (1992). Assessing overdispersion and dose response in the male dominant lethal assay. *Mutat. Res.* **272**, 35–58.

Luster, M. I., C. J. Portier, D. G. Pait, K. L. White, C. Gennings, A. E. Munson, et al. (1992). Risk assessment in immunotoxicology. 1. Sensitivity and predictability of immune tests. *Fund. Appl. Toxicol.* **18**, 200–210.

Mack, G. A and D. A. Wolfe (1981). K-sample rank tests for umbrella alternatives. *J. Amer. Statist. Assoc.* **76**, 175–181.

Mackey, B. E. and J. T. MacGregor (1979). The micronucleus test: Statistical design and analysis. *Mutat. Res.* **64**, 195–204.
Manson, J. M. and Y. J. Kang (1980). Test methods for assessing female reproductive and development toxicology. In: A. W. Hayes, ed., *Principles and Methods of Toxicology*. 2nd ed., Raven Press, New York, 311–360.
Mantel, N. (1963). Chi-square tests with one degree of freedom, extensions of the Mantel–Haenzel procedures. *J. Amer. Statist. Assoc.* **58**, 690–700.
Mantel, N. (1979). Ridit analysis and related ranking procedures – Use at your own risk. *Amer. J. Epidemiol.* **109**, 25–29.
Margolin, B. H. (1985). Statistical studies in genetic toxicology: A perspective from the U. S. National Toxicology Program. *Environ. Health Perspect.* **63**, 187–194.
Margolin, B. H. (1988). Test for trend in proportions. In: S. Kotz, N. L. Johnson and C. B. Read, eds., *Encyclopedia of Statistical Sciences Vol. 9*. Wiley, New York, 334–336.
Margolin, B. H., B. J. Collings and J. J. Mason (1983). Statistical analysis and sample-size determinations for mutagenicity experiments with binomial response. *Environ. Mutag.* **5**, 705–716.
Margolin, B. H., N. Kaplan, and E. Zeiger (1981). Statistical analysis of the Ames *salmonella*/microsome test. *Proc. Nat. Acad. Sci. USA* **76**, 3779–3783.
Margolin, B. H., B. S. Kim and K. J. Risko (1989). The Ames *salmonella*/microsome mutagenicity assay: Issues of inference and validation. *J. Amer. Statist. Assoc.* **84**, 651–661.
Margolin, B. H., M. A. Resnick, J. Y. Rimpo, P. Archer, S. M. Galloway, A. D. Bloom and E. Zeiger (1986). Statistical analyses for in vitro cytogenetic assays using Chinese hamster ovary cells. *Environ. Mutag.* **8,** 183–204.
Margolin, B. H. and K. J. Risko (1984). The use of historical control data in laboratory studies. *Proc. Internat. Biometrics Conf.* **12**, 21–30.
Matis, J. H. and T. E. Wehrly (1979). Stochastic models of compartmental systems. *Biometrics* **35,** 199–220.
McCaughran, D. A. and D. W. Arnold (1976). Statistical models for numbers of implantation sites and embryonic deaths in mice. *Toxicol. Appl. Pharmacol.* **38,** 325–333.
McCullagh, P. and J. A. Nelder (1983). *Generalized Linear Models*. Chapman and Hall, London.
McKnight, B. and J. Crowley (1984). Tests for differences in tumour incidence based on animal carcinogenesis experiments. *J. Amer. Statist. Assoc.* **79,** 639–648.
Mehlman, M. A. (1989). *Benchmarks: Alternative Methods in Toxicology*. Princeton Scientific Publishing, Princeton, NJ.
Meng, C. Y. K. and A. P. Dempster (1987). A Bayesian approach to the multiplicity problem for significance testing with binomial data. *Biometrics* **43**, 301–311.
Moore, D. F. (1986). Asymptotic properties of moment estimators for overdispersed counts and proportions. *Biometrika* **73**, 583–588.
Murphy, S. A., W. J. Caspary and B. H. Margolin (1988). A statistical analysis for the mouse lymphoma cell forward mutation assay. *Mutat. Res.* **203**, 145–154.
Murphy, S. A., R. R. Tice, M. G. Smith and B. H. Margolin (1992). Contributions to the design and analysis of in vivo SCE experiments. *Mutat. Res.* **271**, 39–48.
National Toxicology Program (1984). *Toxicology and carcinogenesis studies of 1,3-butadiene in* $B6C3F_1$ *mice*. Technical Report, number 288, U. S. Department of Health and Human Services, Public Health Service, Washington, DC.
Neyman, J. and E. L. Scott (1966). On the use of C(α) optimal tests of composite hypotheses. *Bull. Inst. Internat. Statist.* **41**, 477–497.
Peto, R., M. Pike, N. Day, R. Gray, P. Lee, S. Parish, et al. (1980). Guidelines for simple, sensitive significance tests for carcinogenic effects in long-term animal experiments. *IARC Monogr. Suppl.* **2**, 311–426.
Piegrosch, W. W. (1990). Maximum likelihood estimation for the negative binomial dispersion parameter. *Biometrics* **46**, 863–867.
Piegorsch, W. W. (1991). Multiple comparisons for analyzing dichotomous response data. *Biometrics* **47**, 45–54.

Piegorsch, W. W. (1992). Non-parametric methods to asses non-monotone dose response: Applications to genetic toxicology. In: I. A. Salama and P. K. Sen, eds., *Order Statistics and Nonparametrics: Theory and Applications*. North-Holland, Amsterdam, 419–430.

Piegorsch, W. W. and A. J. Bailer (1989). Optimal allocations for estimating area under curves for studies employing destructive sampling. *J. Pharma. Biopharma.* **17**, 493–507.

Piegorsch, W. W. and J. K. Haseman (1991). Statistical methods for analyzing developmental toxicity data. *Teratog. Carcinog. Mutag.* **11**, 115–133.

Piegorsch, W. W., F. K., Zimmermann, S. Fogel, S. G. Whittaker and M. A. Resnick (1989). Quantitative approaches for assessing chromosome loss in *Saccharomyces cerevisiae:* General methods for analyzing downturns in dose response. *Mutat. Res.* **224**, 11–29.

Portier, C. J. and A. J. Bailer (1989). Testing for increased carcinogenicity using a survival-adjusted quantal response test. *Fund. Appl. Toxicol.* **12,** 731–737.

Prentice, R. L. (1988). Correlated binary regression with covariates specific to each binary observation. *Biometrics* **44**, 1033–1048.

Rai, K. and J. Van Ryzin (1985). A dose-response model for teratological experiments involving quantal response. *Biometrics* **41**, 1–9.

Richardson, C., D. A. Williams, G. Amphlett, B. Phillips, J. A. Allen and D. O. Chanter (1989). Analysis of data from in vitro cytogenetic assays. In: D. J. Kirkland, ed., *Statistical Evaluation of Mutagenicity Test Data*. Cambridge Univ. Press, Cambridge, 141–154.

Robins, J. F., W. W. Piegorsch and R. L. Schueler (1989). Methods of testing for carcinogenicity. In: A. W. Hayes, ed., *Principles and Methods of Toxicology*. 2nd ed., Raven Press, New York, 251–274.

Sen, P. K. (1985). Permutational central limit theorems. In: S. Kotz, N. L. Jonhson and C. B. Read, eds., *Encyclopedia of Statistical Sciences*, Vol. 6. Wiley, New York, 683–687.

Simpson, D. D. and B. H. Margolin (1986). Recursive nonparametric testing for dose-response relationships subject to downturns at high doses. *Biometrika* **73**, 589–596.

Simpson, D. G. and B. H. Margolin (1990). Nonparametric testing for dose-response curves subject to downturns: Asymptotic power considerations. *Ann. Statist.* **18**, 373–390.

Tarone, R. E. (1975). Test for trend in life table analysis. *Biometrika* **62,** 679–682.

Tarone, R. E. (1979). Testing the goodness of fit of the binomial distribution. *Biometrika* **66**, 585–590.

Tarone, R. E. (1982). The use of historical control information in testing for a trend in Poisson means. *Biometrics* **38**, 457–462.

Tarone, R. E. and J. J. Gart (1980). On the robustness of combined tests for trends in proportions *J. Amer. Statist. Assoc.* **75**, 110–116.

Tennant, R. W., B. H. Margolin, M. D. Shelby, E. Zeiger, J. K. Haseman, J. Spalding, W. Caspary, et al. (1987). Prediction of chemical carcinogenicity in rodents from in vitro genetic toxicity assays. *Science* **236**, 933–941.

Waters, M. D., H. F. Stack, A. L. Brady, P. H. M. Lohman, L. Haroun and H. Vainio (1987). Activity profiles for genetic and related tests, Appendix I. In: *IARC Monographs on the Evaluation of Carcinogenic Risks to Humans. Genetic and Related Effects: An Updating of Selecting IARC Monographs from Volumes 1 to 42*. International Agency for Research on Cancer, Lyon, France, 687–696.

Williams, D. A. (1975). The analysis of binary responses from toxicological experiments involving reproduction and teratogenicity. *Biometrics* **31**, 949–952.

Williams, D. A. (1982). Extra-binomial variation in logistic linear models. *Appl. Statist.* **31**, 144–148.

Williams, D. A. (1987). Dose-response models for teratological experiments. *Biometrics* **43**, 1013–1016.

Zeger, S. L., K.-Y. Liang and P. S. Albert (1988). Models for longitudinal data: A generalized estimating equation approach. *Biometrics* **44**, 1049–1060.

Zeiger, E., K. J. Risko and B. H. Margolin (1985). Strategies to reduce the cost of mutagenicity screening with the *Salmonella* assay. *Environ. Mutag.* **7**, 901–911.

G. P. Patil and C. R. Rao, eds., *Handbook of Statistics, Vol. 12*

16

Stochasticity in Deterministic Models

J. J. M. Bedaux and S. A. L. M. Kooijman

Abstract: Three fundamentally different ways to incorporate stochastic elements in deterministic models are discussed: additive random errors (the standard approach), random parameters, and randomization of the input. The incorporation is illustrated with a one-compartment model for the kinetics of a compound in an organism. Maximum likelihood estimation has been worked out for random parameters and applied to an empirical data set. The second-order properties of the output process were derived for three different randomizations of inputs. It is concluded that the standard approach of additive random errors is the most simple, flexible and elegant one, but it usually makes little scientific sense. Both other approaches better relate to actual sources of stochasticity, but rapidly lead to rather complicated statistical analyses.

1. Introduction

In the environmental sciences mathematical models are indispensable. Nowadays much effort is done to build realistic models to describe complex systems in relatively simple formulas. Generally, the resulting formulas are differential equations based on ideas about the mechanisms underlying the observed phenomena and leading to completely deterministic models. When these models are confronted with experimental data to test their adequacy or to estimate parameters, there will always be some discrepancy between model and data. Usually little attention is given to the origin of this discrepancy. It is just called measurement error or noise, suggesting that, in case of a true model, the discrepancy could be diminished or even removed by better experimentation. This vision is seldom correct. In most cases there is some inherent stochasticity one can never get rid of. Stochastic data ask for stochastic models, so the deterministic model must be changed or extended to a stochastic model. Several ways to incorporate stochastic elements in deterministic models will be discussed.

To be more specific, let us look at the following example. Suppose we have a mathematical model describing some biological process, e.g., the dynamics of a pollutant in an animal. The model can be formulated as a differential equation

$$Y'(t) = F(Y(t), t, \theta). \tag{1}$$

In this equation θ represents a vector of parameters with biological interpretations (e.g., elimination time or ingestion rate). If we want to estimate parameters from data (t_1, Y_1), $(t_2, Y_2), \ldots, (t_n, Y_n)$, we have to make the model stochastic. The usual regression approach is to add a stochastic variable to the model function Y, based on the idea of 'additive noise'. See Figure 1a. The deviations between observations Y_i

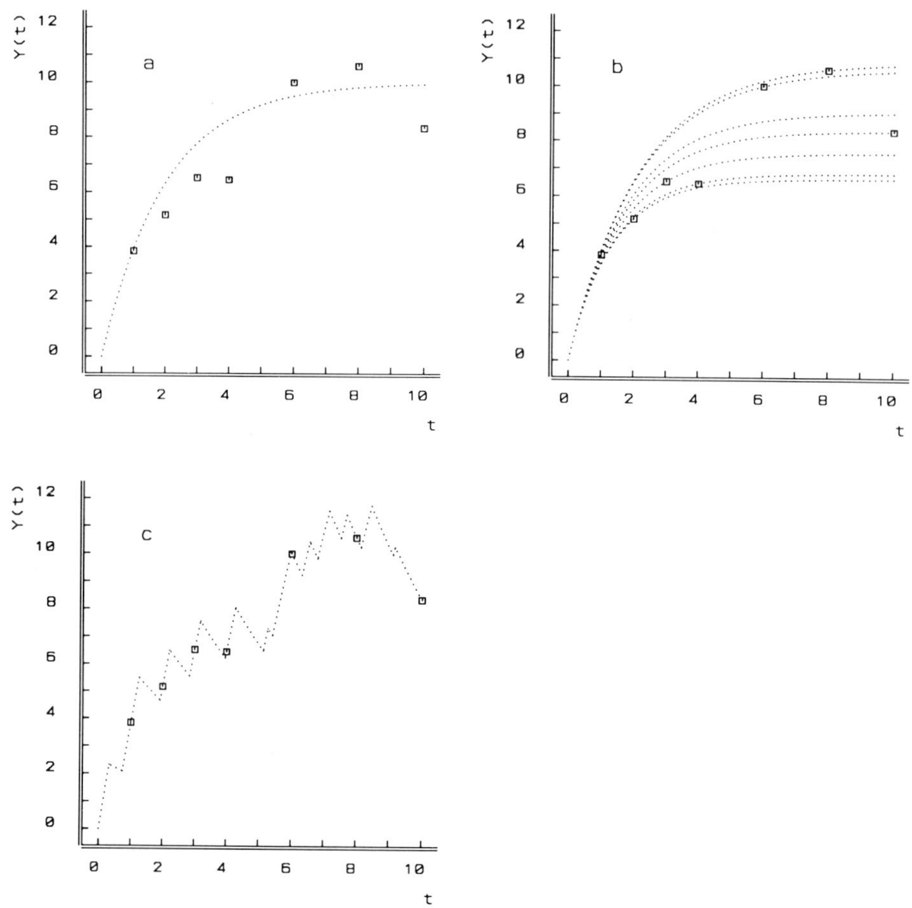

Fig. 1. Confrontation of one data set with the model $Y'(t) = -rY(t) + m$ in three different ways. (a) Regression model. Stochastic deviations between model function and data. (b) Stochastic parameters. Each data point has its own model function. (c) Randomized input. The parameter m follows an alternating Poisson process.

and model function $Y(t_i)$ are considered as stochastic variables $\underline{\varepsilon}_i$,

$$\underline{Y}_i = Y(t_i) + \underline{\varepsilon}_i \quad \text{with } \underline{\varepsilon}_i \sim \mathrm{N}(0, \sigma_i^2) \tag{2}$$

or

$$\underline{Y}_i \sim \mathrm{N}(Y(t_i), \sigma_i^2). \tag{3}$$

A lot of methods are available to estimate parameters. We will shortly describe some standard methods in the next section.

The approaches mentioned above are based on a set of fixed parameter values, i.e., they all use one model function for the whole data set. In other words, every data point is supposed to obey the same model function. This is not always realistic. For instance, in case of one observation per animal this is a very questionable assumption. Animals are not equal, due to genetic or site-specific differences for instance. A more realistic approach is then to give each animal its own parameter values. A random sample of animals results in a random sample of parameter values and so in a random sample of model functions. In other words, the parameters are considered to be stochastic. See Figure 1b. This is what Zeger et al. (1988) call a subject-specific approach, as opposed to a population-averaged approach.

The idea of stochastic parameters in pharmacokinetic compartment models is quite old (Jacquez, 1985). McArthur (1988) reviews several parameter estimation methods, which can all be described more or less as (extended, or weighted) least squares procedures. We have worked out a maximum likelihood procedure in a one-compartment model and investigated the statistical properties of the method.

A third possibility to implement stochasticity is to consider the animal as a deterministic system with a stochastically changing input. Especially if data are from one individual this can be realistic. Three different stochastic processes are studied as input to a first order linear system: discrete random increments input, dependent increments input and alternating Poisson input.

In practice we expect that combinations of these approaches are realistic. Our first aim is to compare the different approaches in a practical way to judge their merits.

2. Regression methods

One of the oldest methods of estimation is the so called 'fitting by eye'. After a suitable transformation a straight line is drawn through the data points. This quick and dirty graphical method is seldom used nowadays, except as an easy tool to get starting values for more sophisticated methods.

Least squares estimation is used very often, in various degrees of complexity: ordinary, weighted, generalized or extended least squares (see Van Houwelingen, 1988). In case of homoscedastic and normally distributed noise, ordinary least squares estimation is just a special case of maximum likelihood estimation, as can be easily seen from the log-likelihood function

$$l = -\frac{n}{2}\ln(2\pi\sigma^2) - \frac{1}{2\sigma^2}\sum_{i=1}^{n}(Y(t_i) - Y_i)^2. \tag{4}$$

Maximization of l with respect to the parameters of the model function $Y(t)$ means minimization of the sum of squares $\sum_{i=1}^{n}(Y(t_i) - Y_i)^2$.

In many cases the variance of the noise increases with the model function. We can deal with this problem by assuming for instance a constant coefficient of variation, that is the variance of $\underline{\varepsilon}_i$ is proportional with $Y^2(t_i)$, or Poisson variation, where the variance of $\underline{\varepsilon}_i$ is proportional with $Y(t_i)$. The least squares approach is to minimize a weighted sum of squared deviations with weights inversely proportional to the variances:

$$\sum_{i=1}^{n} \frac{(Y_i - Y(t_i))^2}{Y(t_i)^2} \quad \text{or} \quad \sum_{i=1}^{n} \frac{(Y_i - Y(t_i))^2}{Y(t_i)}. \tag{5}$$

This minimization is not equivalent with standard weighted least squares estimation because the weights depend on parameter values of the model function and so change in every iteration. An easier but not equivalent way of estimation is to take weights proportional to Y_i^{-2} or Y_i^{-1}. See Jobson and Fuller (1980). The log-likelihood function for the first possibility is

$$l = -\frac{n}{2}\ln(2\pi s^2) - \sum_{i=1}^{n} \ln Y(t_i) - \frac{1}{2s^2}\sum_{i=1}^{n} (1 - Y_i/Y(t_i))^2, \tag{6}$$

where s^2 is the proportionality constant of var $\underline{\varepsilon}_i$ and $Y^2(t_i)$. Maximizing this expression is equivalent with minimizing

$$\sum_{i=1}^{n} \left(\frac{(Y_i - Y(t_i))^2}{Y(t_i)^2} + \ln Y(t_i)^2 \right). \tag{7}$$

This last procedure is called extended least squares.

The normal distribution is mostly used to describe the 'error'. Of course other distributions can be used as well, e.g., the Poisson distribution (in case of count data) or the lognormal distribution (multiplicative noise). If the distribution does not have a location parameter (like Poisson and lognormal) we cannot speak of additive noise. In that case we do not describe the deviations between data and model as stochastic variables like in (2) but the data themselves are stochastic variables with the model function as expected value, like in (3). Maximum-likelihood estimation can be used to estimate parameters.

3. Stochastic parameters

In this section we will work out the maximum likelihood estimation procedure for a one-compartment model with a stochastic rate parameter and constant input.

Consider the following experiment. A sample of insects is fed with Cadmium-contaminated food for some time. After this period the animals will release Cadmium. From then on every day an insect is randomly chosen and its Cd-concentration is measured. The Cd-concentration $Q(t)$ is supposed to follow a simple first-order kinetics, i.e., it obeys the differential equation

$$Q'(t) = -kQ(t) + a. \tag{8}$$

In this particular example the Cd-input a equals zero after pollution has stopped. The Cd-concentration is then given by

$$Q(t) = Q_0 e^{-kt}. \tag{9}$$

The exponential parameter k is related with physiological properties of the animal: the ability to release Cadmium. The parameter value is of course not the same for different animals. Observing different animals therefore means observing different model functions. The same reasoning holds for Q_0. At the end of the accumulation period the Cd-concentrations will not be the same for all animals. Animals with low k-values will probably have high Cd-concentrations, as can be seen from the accumulation curve

$$Q(t) = \frac{a}{k} - \left(\frac{a}{k} - Q_0\right) e^{-kt}. \tag{10}$$

If the accumulation period is quite long, the equilibrium value will be reached, so the initial value of the elimination function will be a/k. So every data Q_i has its own parameter values k_i and Q_{0i}. At first it may seem a bad ratio between number of data and number of parameters. But, just as in the case of equation (2), where the n different values of ε_i are tied together by assuming them to originate from a single stochastic variable, we now assume k_i and Q_{0i} to be realizations of stochastic variables $\underline{\xi}$ and $\underline{\eta}$. The model for the data is then given by

$$\underline{Q}_i = \underline{\eta}_i e^{-\underline{\xi}_i t_i}. \tag{11}$$

To estimate parameters, we must construct the pdf of $\underline{Q}_i$. Therefore we will have to choose a joint pdf for $\underline{\xi}$ and $\underline{\eta}$. It is difficult to give general rules. Every special case must be considered separately. We shall discuss only one case. As we have seen earlier we may assume that $\underline{\eta}_i = a/\underline{\xi}_i$, where a is the input during the accumulation period. If we assume this input to be constant with respect to animals and time, $\underline{\xi}$ and $\underline{\eta}$ are maximally dependent. So we can write

$$\underline{Q}_i = \frac{a}{\underline{\xi}_i} e^{-\underline{\xi}_i t_i} \tag{12}$$

or more generally

$$\underline{Q}_i = h(\underline{\xi}_i; t_i, a). \tag{13}$$

If $\underline{\xi}$ has pdf f then $\underline{Q}_i$ has pdf

$$g(Q)=f(h^{-1}(Q))\cdot\left|\frac{dh^{-1}(Q)}{dQ}\right| \tag{14}$$

if $h^{-1}(Q)$ exists, where $h^{-1}(Q)$ is the inverse function of $h(\xi)$. The existence of $h^{-1}(Q)$ can easily be checked. The log-likelihood is then given by

$$l=\sum_i \ln g(Q_i)=\sum_i \ln f(h^{-1}(Q_i))+\sum_i \ln\left|\frac{dh^{-1}(Q)}{dQ}\right|_{Q=Q_i}. \tag{15}$$

We will work this out for a special choice of f, viz. the lognormal density. The reason to take the lognormal distribution is mainly because of its being positive. If $\underline{\xi}$ is lognormally distributed, then $\ln(\underline{\xi})\sim N(\mu,\sigma^2)$ for some values μ and σ^2. The pdf g of $\underline{Q}_i$ is then given by

$$g(Q)=\frac{1}{h^{-1}(Q)\sqrt{2\pi\sigma^2}}\exp\left(-\frac{\{\ln(h^{-1}(Q))-\mu\}^2}{2\sigma^2}\right)\left|\frac{dh^{-1}(Q)}{dQ}\right|. \tag{16}$$

The last part of this expression can be calculated explicitly,

$$\frac{dh^{-1}(Q)}{dQ}=\left[\frac{dh(x)}{dx}\right]^{-1}_{x=h^{-1}(Q)}=\left[-\frac{a}{x^2}e^{-xt}-\frac{at}{x}e^{-xt}\right]^{-1}. \tag{17}$$

Substitution of (17) in (16) results in

$$l=-\frac{n}{2}\ln\sigma^2-\frac{\sum(\ln x_i-\mu)^2}{2\sigma^2}-n\ln a+\sum x_it_i-\sum\ln\left(\frac{1}{x_i}+t_i\right), \tag{18}$$

where $x_i=h^{-1}(Q_i)$. Differentiating with respect to μ and σ^2 leads to the following expressions:

$$\hat{\mu}=\frac{\sum\ln x_i}{n}, \tag{19}$$

$$\hat{\sigma}^2=\frac{\sum(\ln x_i-\hat{\mu})^2}{n}. \tag{20}$$

So $\hat{\mu}$ and $\hat{\sigma}^2$ can be expressed in terms of the x_i and so in terms of $\hat{a}$. The partial derivative of l with respect to a leads to

$$\frac{\partial l}{\partial a}=-\frac{1}{\sigma^2}\sum(\ln x_i-\mu)\frac{1}{x_i}\frac{\partial x_i}{\partial a}-\frac{n}{a}+\sum\left(\frac{t_i^2x_i^2+t_ix_i+1}{x_i+t_ix_i^2}\right)\frac{\partial x_i}{\partial a}. \tag{21}$$

From (12) it can be easily derived that

$$\frac{\partial x_i}{\partial a} = \frac{x_i}{a(1 + t_i x_i)}. \tag{22}$$

Substitution of (22) into (21) leads to a third equation

$$\frac{1}{\hat{\sigma}^2} \sum \frac{\ln x_i - \hat{\mu}}{1 + t_i x_i} + \sum \frac{t_i x_i}{(1 + t_i x_i)^2} = 0, \tag{23}$$

where $\hat{\mu}$ and $\hat{\sigma}^2$ are given by (19) and (20). Equation (23) determines the three estimators $\hat{\mu}$, $\hat{\sigma}^2$ and $\hat{a}$. Notice that the parameter a is absent in the three equations. However, it is implicitly present in the x_i.

Of course the three equations cannot be solved analytically. Numerical methods have to be used. With a given data set $(t_1, Q_1), (t_2, Q_2), \ldots, (t_n, Q_n)$ and a starting value for a we have to solve the n equations (12) for x_i with, for instance, a Newton–Raphson procedure. Then the results can be substituted in (19), (20) and (23). The last equation asks again for a Newton–Raphson approach. The total procedure, then, is a nested Newton–Raphson one.

To calculate the asymptotic covariance matrix we have to develop the second-order derivatives of the log-likelihood. Straightforward calculation leads to the following six expressions

$$\begin{gathered}
\frac{\partial^2 l}{\partial \mu^2} = -\frac{n}{\sigma^2}, \qquad \frac{\partial^2 l}{\partial \mu \partial \sigma^2} = -\frac{\sum (\ln x_i - \mu)}{\sigma^4}, \\
\frac{\partial^2 l}{\partial (\sigma^2)^2} = \frac{n}{2\sigma^4} - \frac{\sum (\ln x_i - \mu)^2}{\sigma^6}, \qquad \frac{\partial^2 l}{\partial a \partial \mu} = \frac{1}{a\sigma^2} \sum \frac{1}{1 + t_i x_i}, \\
\frac{\partial^2 l}{\partial a \partial \sigma^2} = \frac{1}{a\sigma^4} \sum \frac{(\ln x_i - \mu)}{1 + t_i x_i}, \\
\frac{\partial^2 l}{\partial a^2} = \frac{1}{a^2 \sigma^2} \sum \frac{(\ln x_i - \mu)(1 + 2t_i x_i) - 1}{(1 + t_i x_i)^2} + \frac{1}{a^2} \sum \frac{t_i x_i}{(1 + t_i x_i)^2} \\
- \frac{1}{a^2} \sum \frac{t_i x_i (1 - t_i x_i)}{(1 + t_i x_i)^4}.
\end{gathered} \tag{24}$$

Expectations of the first three expressions are $-n/\sigma^2$, 0 and $-n/2\sigma^4$ respectively. Expectations of the last three expressions cannot be evaluated easily.

3.1. Simulation results

We first explored the performance of the estimation method with simulated data. So we generated for a range n, of sample sizes, 1000 samples $(t_1, Q_1), (t_2, Q_2), \ldots, (t_n, Q_n)$

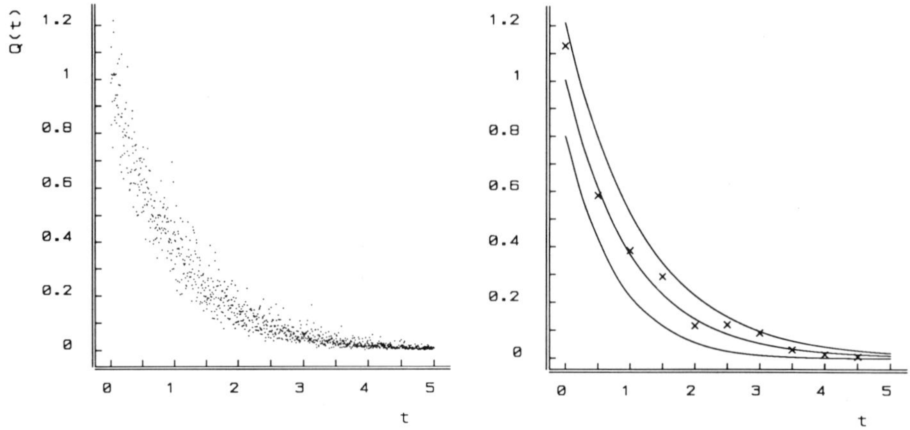

Fig. 2. Left: A large sample of simulated data with parameter values $a=1, \mu=0$ and $\sigma^2=0.01$. Right: The middle curve represents the expected concentration as a function of time. The upper and lower curve represent the expected values plus and minus $2 \times$ standard deviation. All three curves are calculated by simulation. The crosses represent a simulated sample.

according to (12) with chosen values for a, μ and σ^2 and an equidistance series of t-values. Figure 2 shows some generated data.

Maximum likelihood estimation was applied on each sample. The results are given in Table 1. To study the distribution of the estimators the asymptotic covariance matrix was evaluated. Expected values of the second-order derivatives as given by (24) were estimated by simulation. See Table 2.

Inspection of Table 1 and Table 2 leads to the following remarks. The estimator of a shows a small positive bias shrinking with increasing n. For μ there is no apparent bias. For σ^2 there is a negative bias, much larger than can be explained by the n in the denominator of (20) in stead of $n-1$. Empirical and theoretical (asymptotic)

Table 1
Results of ML estimation based on simulated data. Mean values, standard deviations and correlation coefficients of $\hat{a}$, $\hat{\mu}$ and $\hat{\sigma}^2$ are shown. The real values of the parameters were $a=1$, $\mu=0$ and $\sigma^2=0.01$

n	5	10	20	50	100	300	1000
mean($\hat{a}$)	1.0209	1.0090	1.0085	1.0023	1.0011	0.9993	1.0004
sd($\hat{a}$)	0.1991	0.1535	0.1142	0.0741	0.0527	0.0312	0.0171
mean($\hat{\mu}$)	−0.0008	−0.0029	0.0006	0.0002	−0.0006	−0.0008	−0.0001
sd($\hat{\mu}$)	0.1206	0.0887	0.0622	0.0404	0.0281	0.0168	0.0090
mean($\hat{\sigma}^2$)	0.0058	0.0081	0.0091	0.0097	0.0098	0.0099	0.0100
sd($\hat{\sigma}^2$)	0.0049	0.0042	0.0030	0.0020	0.0014	0.0008	0.0005
corr($\hat{a}$, $\hat{\mu}$)	0.9244	0.9271	0.9319	0.9341	0.9363	0.9420	0.9376
corr($\hat{a}$, $\hat{\sigma}^2$)	−0.0565	−0.1195	−0.1457	−0.1528	−0.1820	−0.2049	−0.1712
corr($\hat{\mu}$, $\hat{\sigma}^2$)	−0.0377	−0.1097	−0.1682	−0.1351	−0.1796	−0.1920	−0.1561

Table 2
Asymptotic standard deviations and correlations of ML estimators for $a = 1$, $\mu = 0$ and $\sigma^2 = 0.01$, based on 10 000 samples of size n

n	5	10	20	50	100	300	1000
sd($\hat{a}$)	0.1883	0.1485	0.1122	0.0742	0.0533	0.0310	0.0171
sd($\hat{\mu}$)	0.1180	0.0861	0.0624	0.0403	0.0287	0.0166	0.0091
sd($\hat{\sigma}^2$)	0.0064	0.0045	0.0032	0.0020	0.0014	0.0008	0.0005
corr($\hat{a}$, $\hat{\mu}$)	0.9254	0.9300	0.9337	0.9364	0.9374	0.9377	0.9381
corr($\hat{a}$, $\hat{\sigma}^2$)	−0.1111	−0.1312	−0.1446	−0.1576	−0.1590	−0.1615	−0.1622
corr($\hat{\mu}$, $\hat{\sigma}^2$)	−0.1032	−0.1208	−0.1338	−0.1476	−0.1494	−0.1511	−0.1523

standard deviations show great resemblance for values of n larger than 10. The same holds true for the correlation coefficient of a and μ. The other two correlation coefficients, involving σ^2 perform much worse.

3.2. *Application to experimental data*

We applied the method to an ecotoxicological data set (Janssen et al., 1991) which is presented in Figure 3a. The data represent Cadmium concentrations (pmol/mg) measured in an accumulation/elimination experiment. The experimental animal was the springtail *Orchesella cincta.* In the original article simple compartment models were used to describe the data. Parameter estimates were obtained by the ordinary least squares method.

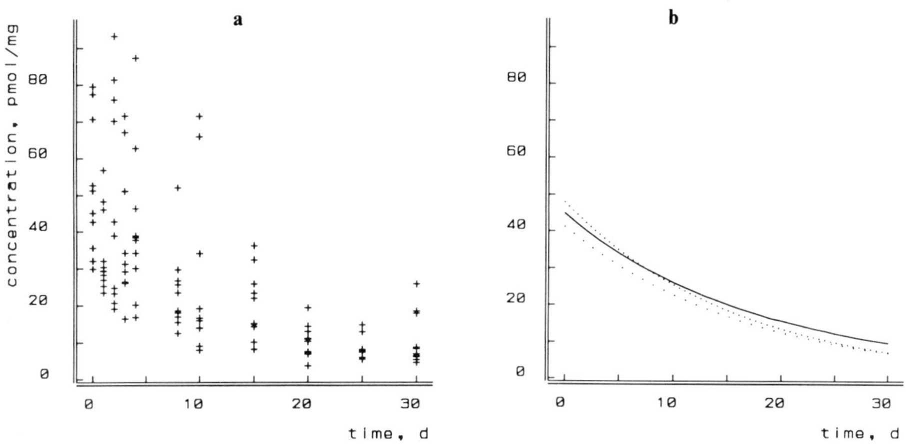

Fig. 3. Confrontation of Cd-elimination data with the model $Y'(t) = -rY(t) + m$ in three different ways. (a) Data of the elimination period. (b) Solid curve: Mean model function with stochastic parameters. Dotted curve: Model function obtained by ordinary least squares. Sparsely dotted curve: Model function obtained by linear regression after logarithmic transformation.

Estimation with ordinary least squares using the model function $Q(t) = (a/k)\,e^{-kt}$ results in $\hat{a} = 3.052 \pm 0.6053$, $\hat{k} = 0.06369 \pm 0.009902$ and $\hat{\sigma}^2 = 258.52 \pm 34.86$. Ordinary least squares presupposes homogeneity of variances, which is very doubtful by inspection of Figure 3a. A Bartlett test shows significant differences ($p = 0.000$). An alternative approach is linear regression after logarithmic transformation. Homogeneity is not rejected ($p = 0.169$). Parameter estimates are then $\hat{a} = 2.4717 \pm 0.3451$, $\hat{k} = 0.05993 \pm 0.00479$ and $\hat{\sigma}^2 = 0.2481 \pm 0.0335$.

In the stochastic parameter approach we found the following parameter estimates (point estimate $\pm$ asymptotic standard error): $\hat{a} = 2.524 \pm 0.4240$, $\hat{\mu} = -2.8179 \pm 0.1188$ and $\hat{\sigma}^2 = 0.1183 \pm 0.01729$, which means that the expectation of $\underline{\xi}$, the stochastic elimination-rate parameter, equals 0.06337 d^{-1} and the standard deviation equals 0.02246 d^{-1}. The stochastic variable $\underline{\eta}$, the initial concentration, has expectation 44.84 pmol/mg and standard deviation 15.89 pmol/mg.

The mean curve, that is $\mathscr{E}\underline{Q}(t)$ for the parameter estimates, was calculated by simulation. It is represented in Figure 3b, together with fitted model functions of the other approaches.

Comparison of the methods is difficult, because of the difference in interpretations of the parameters. Especially the σ^2 parameters have a completely different interpretation. As far as comparison makes sense parameter values are more or less in the same range.

Residuals were calculated and analyzed to get information about the validity of the statistical assumptions. In the OLS-case the residuals are the differences between model function and data. They are assumed to originate from one normal distribution. This was tested with a Kolmogorov–Smirnov test. The test showed a significant deviation from normality ($p = 0.000$). In the logarithmic regression case the residuals are the differences between the logarithmic model function and the logarithmic data. They are assumed to be normally distributed. The KS-test rejected normality ($p = 0.012$). In the stochastic parameter approach the 'residuals' are the numbers x_i from (23), which are assumed to be lognormally distributed. The logarithms of these residuals also gave significance with the KS-test ($p = 0.012$). All residuals are shown in Figure 4.

4. Randomizing the input

In this section we will discuss a third possibility of introducing stochastic elements in a deterministic model. The model of the previous section, a one-compartment model like (8) will again be used. However, stochasticity is added in a completely different way. We focus on the other parameter, the input instead of the exponential parameter. Parameters are not realizations of one stochastic variable, but of a stochastic process. This is realistic if we observe one individual in a changing environment. As a biological example consider so-called biomonitoring experiments. Animals, for instance mussels, are exposed to polluted water. Then the concentration of the pollutant is measured in the mussel. This provides information about the dynamics and the bio-availability of the pollutant. Suppose we observe a

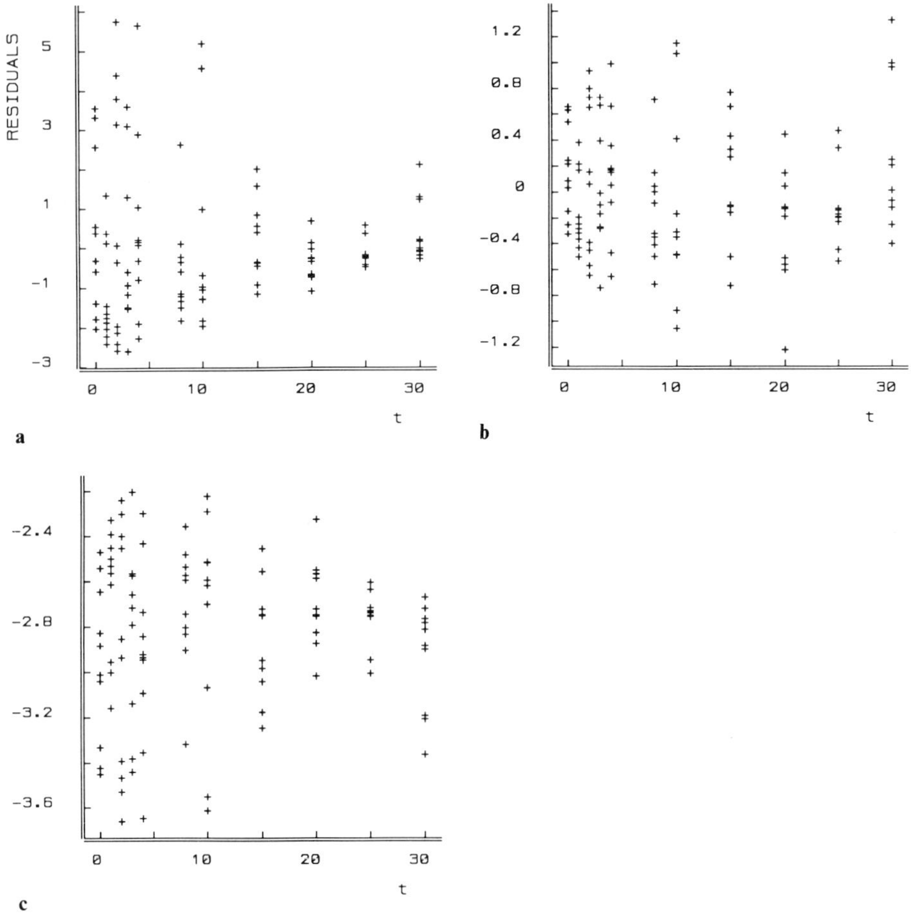

Fig. 4. Residuals as defined in the text. (a) Ordinary least squares. (b) Logarithmic regression. (c) Stochastic parameters (logarithmic).

process described by the differential equation

$$Q'(t) = -kQ(t) + a(t) \quad \text{with } Q(0) = Q_0, \tag{25}$$

where the input a changes stochastically in time. The changes in the input are more or less followed by the output $Q(t)$, especially if k is large compared with the rate of change of a. Thus output and input are both realizations of stochastic processes.

Equation (25) can easily be solved. The solution is

$$Q(t) = Q_0 \mathrm{e}^{-kt} + \int_0^t \mathrm{e}^{-k(t-s)} a(s)\,\mathrm{d}s. \tag{26}$$

Now suppose that $\underline{a}(t)$ is a stationary stochastic process. Then $\underline{Q}(t)$ will at last also be stationary, because on the long run the first term of (26) will disappear. We will only study the stationary system, so we will further neglect Q_0 and write

$$\underline{Q}(t)=\int_{-\infty}^{t} e^{-k(t-s)}\underline{a}(s)\,ds=\int_{0}^{\infty} e^{-k\tau}\underline{a}(t-\tau)\,d\tau. \tag{27}$$

By the linearity of the system some statistical properties of $\underline{Q}(t)$ can be derived directly from those of $\underline{a}(t)$. For instance expectations, variances, auto- and cross-covariance functions can be derived by interchanging of integrations (Kwakernaak and Sivan, 1972; Jacquez, 1985). Thus we get for $\mathcal{E}\underline{Q}$, the expectation of $\underline{Q}(t)$,

$$\mathcal{E}\underline{Q}=\int_{0}^{\infty} e^{-k\tau}\,\mathcal{E}\underline{a}(t-\tau)\,d\tau=\mathcal{E}\underline{a}\int_{0}^{\infty} e^{-k\tau}\,d\tau=\frac{1}{k}\mathcal{E}\underline{a} \tag{28}$$

for $\mathrm{acf}_{\underline{Q}}(h)$, the autocovariance function of $\underline{Q}$,

$$\begin{aligned}\mathrm{acf}_{\underline{Q}}(h)&=\mathrm{cov}(\underline{Q}(t+h),\underline{Q}(t))\\&=\int_{0}^{\infty}\int_{0}^{\infty}\mathrm{acf}_{\underline{a}}(h-\tau_1+\tau_2)e^{-k\tau_1}e^{-k\tau_2}\,d\tau_1\,d\tau_2,\end{aligned} \tag{29}$$

and for $\mathrm{ccf}_{\underline{Q},\underline{a}}(h)$, the crosscovariance function of $\underline{Q}$ and $\underline{a}$,

$$\mathrm{ccf}_{\underline{Q},\underline{a}}(h)=\mathrm{cov}(\underline{Q}(t+h),\underline{a}(t))=\int_{0}^{\infty}\mathrm{acf}_{\underline{a}}(h-\tau)e^{-k\tau}\,d\tau. \tag{30}$$

Once the crosscovariance function has been derived, the autocovariance function can also be obtained via the relation

$$\mathrm{acf}_{\underline{Q}}(h)=\frac{\mathrm{ccf}_{\underline{Q},\underline{a}}(h)+\mathrm{ccf}_{\underline{Q},\underline{a}}(-h)}{2k} \tag{31}$$

which can be easily derived from

$$\frac{d}{dt}\mathcal{E}(\underline{Q}(t+h),\underline{Q}(t))=0. \tag{32}$$

Application of (31) instead of (29) means one integration less. An interesting consequence of (31) is that

$$\mathrm{var}\,\underline{Q}(t)=\frac{\mathrm{ccf}_{\underline{Q},\underline{a}}(0)}{k}=\frac{\mathrm{cov}\,(\underline{Q}(t),\underline{a}(t))}{k}. \tag{33}$$

This implies that the correlation coefficient between input and output equals the ratio of their variation coefficients, so

$$\operatorname{cor}(\underline{Q}(t), \underline{a}(t)) = \frac{\operatorname{cv} \underline{Q}(t)}{\operatorname{cv} \underline{a}(t))}. \tag{34}$$

If the correlation between input and output is low, the scatter of the input is relatively high. Few samples from the output then tell as much as a large amount of samples from the input, as far as statistics is concerned.

In the following subsections we will discuss three simple stochastic processes as inputs to the deterministic system (25). Firstly a memoryless input, changing independently at equidistant time points. This will be called 'discrete random increment input.' Then we build in a kind of memory in the input by accumulating the random increments. This will be called 'dependent increments input.' Thirdly an alternating Poisson input. For all three inputs the statistical properties of the output will be derived and discussed.

4.1. Discrete random increments

The mathematically most simple type of stochastic input is when the continuous function $a(t)$ is approximated by a step function which changes only at discrete time points t_i, $i = 1, 2, \ldots$, that are a fixed time interval d apart. That is, $a(t)$ is constant over a time interval $(t_i, t_{i+1}]$, at value a_i. Now we assume that the values a_i represent trials taken independently from some probability density function with expectation $\mu_{\underline{a}}$ and variance $\sigma_{\underline{a}}^2$. This is reasonable if we think of a mussel in a river, where the well-mixed water surrounding the mussel is completely replaced in an interval of length d.

Since the concentrations in the water are constant within each time interval, the mean and variance of $\underline{a}(t)$ equal $\mu_{\underline{a}}$ and $\sigma_{\underline{a}}^2$. The mean concentration in the mussel $\mathscr{E}\underline{Q}$ equals $\mu_{\underline{a}}/k$. The autocovariance function of $\underline{a}(t)$ equals

$$\operatorname{acf}_{\underline{a}}(h) = \begin{cases} 0 & \text{for } |h| > d, \\ \sigma_{\underline{a}}^2(1 - |h|/d) & \text{for } |h| \leqslant d. \end{cases} \tag{35}$$

It decreases thus linearly to zero at $|h| = d$.

The crosscovariance function turns out to be, following (30),

$$\operatorname{ccf}_{\underline{Q},\underline{a}}(h) = \begin{cases} 0 & \text{for } h < -d, \\ \dfrac{\sigma_{\underline{a}}^2}{k}\left(1 + \dfrac{h}{d} + \dfrac{r\mathrm{e}^{-kh} - 1}{kd}\right) & \text{for } -d \leqslant h < 0, \\ \dfrac{\sigma_{\underline{a}}^2}{k}\left(1 - \dfrac{h}{d} + \dfrac{1 - (2 - r)\mathrm{e}^{-kh}}{kd}\right) & \text{for } 0 \leqslant h < d, \\ \dfrac{\sigma_{\underline{a}}^2}{k}\dfrac{(1 - r)^2}{rkd}\mathrm{e}^{-kh} & \text{for } h \geqslant d. \end{cases} \tag{36}$$

The autocovariance function of $\underline{Q}(t)$ is then easily derived by (29). The result is

$$\operatorname{acf}_{\underline{Q}}(h)=\begin{cases}\dfrac{\sigma_{\underline{a}}^2}{k^2}\left(1-\dfrac{|h|}{d}+\dfrac{r\mathrm{e}^{k|h|}-(2-r)\mathrm{e}^{-k|h|}}{2kd}\right) & \text{for } |h|<d,\\[2ex] \sigma_{\underline{a}}^2\dfrac{(1-r)^2}{2rkd}\mathrm{e}^{-k|h|} & \text{for } |h|\geqslant d.\end{cases} \tag{37}$$

For $h=0$ one gets the variance of Q,

$$\operatorname{var} Q=\frac{\sigma_{\underline{a}}^2}{k^2}\left(1-\frac{1-r}{kd}\right). \tag{38}$$

Correlation functions instead of covariance functions can be easily derived now. The difference is only a scaling factor.

4.2. *Dependent increments input*

Usually the input $a(t)$ of a system shows more 'memory'. A simple way to achieve this is to take the mathematical form of the output Q in the previous section as input. That is, we assume in fact a two-compartment model

$$\begin{aligned} a'(t)&=-k_1a(t)+b(t),\\ Q'(t)&=-k_2Q(t)+a(t), \end{aligned} \tag{39}$$

where $b(t)$ is constant over an interval $(t_i, t_i+d]$ at value b_i independently drawn from a stochast.

In biomonitoring context this can be realistic if we consider a mussel in a lake with a constant volume and an inflow and outflow of (polluted) water. If the concentration of the pollutant in the inflowing river is $b(t)$ and in the lake $a(t)$ then (39) applies.

In steady state we have, analogously with (28),

$$\mathscr{E}\underline{a}=\mathscr{E}\underline{b}/k_1 \quad \text{and} \quad \operatorname{var}\underline{a}=\frac{\sigma_{\underline{b}}^2}{k_1^2}\left(1-\frac{1-r_1}{k_1d}\right). \tag{40}$$

The autocovariance function of a is given by

$$\operatorname{acf}_{\underline{a}}(h)=\begin{cases}\dfrac{\sigma_{\underline{b}}^2}{k_1^2}\left(1-\dfrac{|h|}{d}+\dfrac{r_1\mathrm{e}^{k_1|h|}-(2-r_1)\mathrm{e}^{-k_1|h|}}{2k_1\mathrm{d}}\right) & \text{for } |h|<d,\\[2ex] \dfrac{\sigma_{\underline{b}}^2(1-r_1)^2}{k_1^2 2r_1k_1d}\mathrm{e}^{-k_1|h|} & \text{for } |h|\geqslant d.\end{cases} \tag{41}$$

The mean value of the output again equals $\mathscr{E}\underline{Q}=\mathscr{E}\underline{a}/k_2=\mathscr{E}\underline{b}/(k_1k_2)$. The evaluation of (30) is straightforward and results in

$$\frac{k_1^2k_2}{\sigma_{\underline{b}}^2}\operatorname{ccf}_{\underline{Q},\underline{a}}(h)=\begin{cases}\dfrac{k_2(1-r_1)^2\,\mathrm{e}^{k_1h}}{2r_1(k_2+k_1)k_1d} & \text{for } h<-d,\\[2ex] 1+\dfrac{h}{d}-\dfrac{1}{k_2d}-\dfrac{k_2(2-r_1)\mathrm{e}^{k_1h}}{2(k_2+k_1)k_1d}+\dfrac{k_2r_1\mathrm{e}^{-k_1h}}{2(k_2-k_1)k_1d} & \\[2ex] \quad+\dfrac{k_1^2r_2\mathrm{e}^{-k_2h}}{(k_1^2-k_2^2)k_2d} & \text{for } -d\leqslant h<0,\\[2ex] 1-\dfrac{h}{d}+\dfrac{1}{k_2d}-\dfrac{k_2(2-r_1)\mathrm{e}^{k_1h}}{2(k_2-k_1)k_1d}+\dfrac{k_2r_1\mathrm{e}^{-k_1h}}{2(k_2+k_1)k_1d} & \\[2ex] \quad-\dfrac{k_1^2(2-r_2)\mathrm{e}^{-k_2h}}{(k_1^2-k_2^2)k_2d} & \text{for } 0\leqslant h<d,\\[2ex] \dfrac{k_2(1-r_1)^2\mathrm{e}^{-k_1h}}{2r_1(k_2-k_1)k_1d}+\dfrac{k_1^2(1-r_2)^2\mathrm{e}^{-k_2h}}{r_2(k_1^2-k_2^2)k_2d} & \text{for } h\geqslant d\end{cases} \tag{42}$$

and by (31) and (42) we get for the autocovariance function of $Q(t)$,

$$\frac{k_1^2k_2^2}{\sigma_{\underline{b}}^2}\operatorname{ccf}_{\underline{Q}}(h)=\begin{cases}1-\dfrac{|h|}{d}+\dfrac{k_2^2r_1\mathrm{e}^{k_1|h|}-k_2^2(2-r_1)\mathrm{e}^{-k_1|h|}}{2(k_2^2-k_1^2)k_1d} & \\[2ex] \quad+\dfrac{k_1^2r_2\mathrm{e}^{k_2|h|}-k_1^2(2-r_2)\mathrm{e}^{-k_2|h|}}{2(k_1^2-k_2^2)k_2d} & \text{for } |h|\leqslant d, \quad (43)\\[2ex] \dfrac{k_2^2(1-r_1)^2\mathrm{e}^{-k_1|h|}}{2r_1(k_2^2-k_1^2)k_1d}+\dfrac{k_1^2(1-r_2)^2\mathrm{e}^{-k_2|h|}}{2r_2(k_1^2-k_2^2)k_2d} & \text{for } |h|>d. \quad (44)\end{cases}$$

These expressions illustrate that even the most simple stochastic process results in considerable complexity. The crosscovariance function is not symmetrical in the delay h, as illustrated in Figure 5. It has a peak at

$$h=(k_2-k_1)^{-1}\ln\left(\frac{1-r_2}{1-r_1}\right)^2\frac{r_1}{r_2}\frac{2k_1^2}{k_2(k_1+k_2)}$$

if it happens to be larger than d.

4.3. *Alternating Poisson input*

A somewhat arbitrary element in the stochastic processes of the previous sections is the fixed time interval in which the input is taken to be constant. Now we will discuss

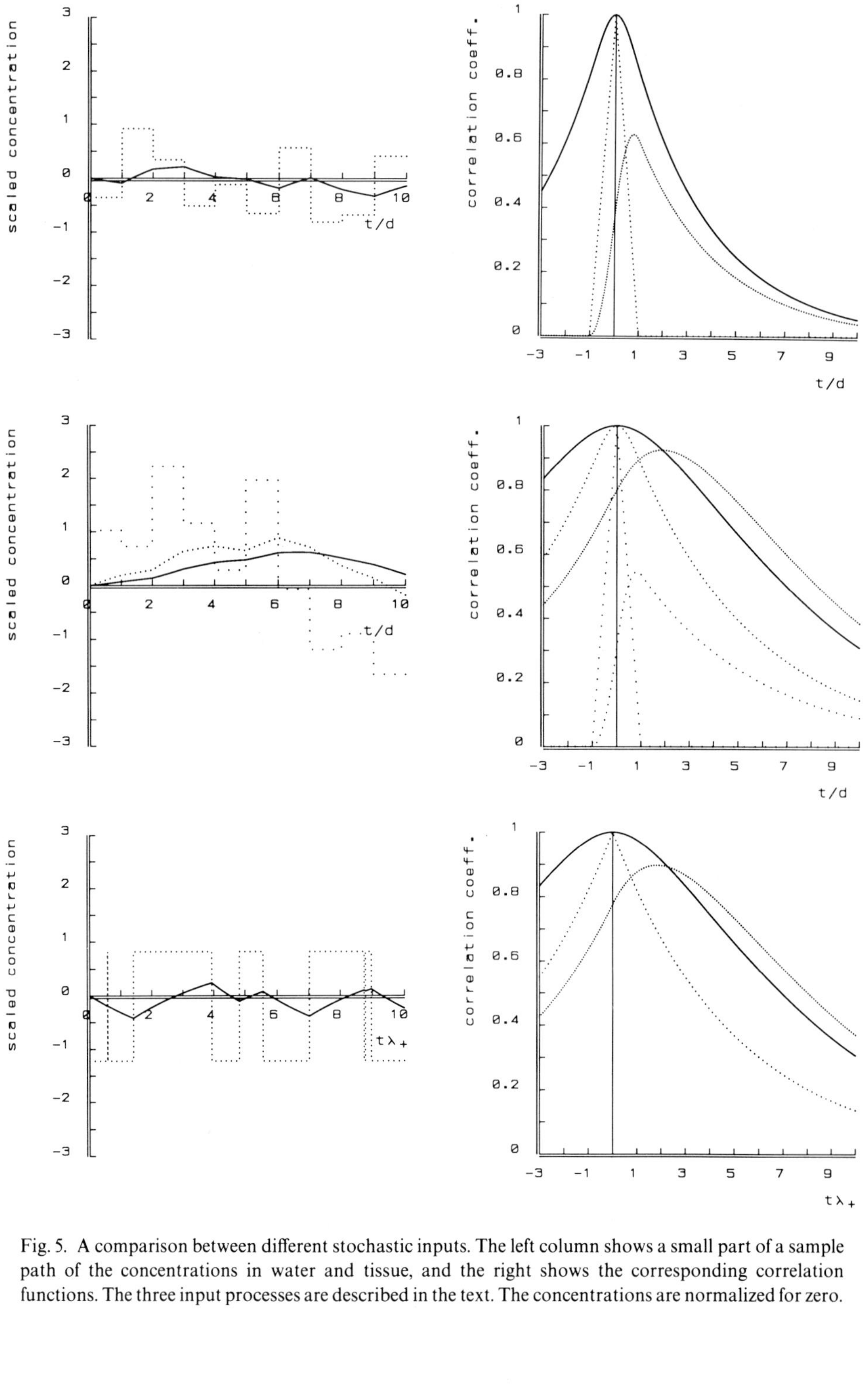

Fig. 5. A comparison between different stochastic inputs. The left column shows a small part of a sample path of the concentrations in water and tissue, and the right shows the corresponding correlation functions. The three input processes are described in the text. The concentrations are normalized for zero.

the situation where the time intervals are stochastic: the alternating Poisson process. Firstly we will give a biological justification.

Suppose that the environment can be partitioned into two types of patches. In one type, the concentration of the compound is a_0, in the other it is a_1. Suppose that the animal stays a time $\underline{t}_0$ in a patch with concentration a_0, then a time $\underline{t}_1$ in a patch with concentration a_1, then again a time $\underline{t}_0$ in a patch with concentration a_0, etc. Suppose next that the residence times follow an exponential distribution, with parameters λ_0 and λ_1 respectively. So Prob $\{\underline{t}_i > t\} = \mathrm{e}^{-\lambda_i t}$, $i = 0, 1$. Think of it as an animal walking randomly over a patchy soil, or an animal that is only exposed to the compound when feeding, or that a mussel is exposed to a flow of badly mixed batches of water. The rates λ relate to the size of the patches, relative to the travelling rate of the animal. More in particular, λ_i^{-1} is the mean residence time in a patch with concentration a_i. Such a process is called an alternating Poisson process or a random telegraph process (if $a_0 = 0$).

The mean of the alternating Poisson process is $\mathscr{E}a(t) = (a_0\lambda_1 + a_1\lambda_0)/\lambda_+$, where $\lambda_+ \equiv \lambda_0 + \lambda_1$. The variance is var $\underline{a}(t) = (a_1 - a_0)^2 \lambda_0 \lambda_1 \lambda_+^{-2}$ and the autocovariance function

$$\mathrm{acf}_{\underline{a}}(h) = \mathrm{var}\,\underline{a}\mathrm{e}^{-\lambda_+|h|}. \tag{45}$$

The mean concentration in the tissue is again $\mathscr{E}\underline{Q} = \mathscr{E}\underline{a}/k$.

The autocorrelation function of the concentrations in the tissue and the crosscorrelation function with those in the environment are

$$\frac{1}{\mathrm{var}\,\underline{a}}\mathrm{ccf}_{\underline{Q},\underline{a}}(h) = \begin{cases} \dfrac{\mathrm{e}^{\lambda_+ h}}{k + \lambda_+} & \text{for } h < 0, \\[2ex] \dfrac{(k+\lambda_+)\mathrm{e}^{-\lambda_+ h} - 2\lambda_+\mathrm{e}^{-kh}}{(k+\lambda_+)(k-\lambda_+)} & \text{for } h \geqslant 0 \end{cases} \tag{46}$$

and by (31) and (46) we get for the autocovariance function of $Q(t)$

$$\frac{1}{\mathrm{var}\,\underline{a}}\mathrm{ccf}_{\underline{Q}}(h) = \frac{k\mathrm{e}^{-\lambda_+ h} - \lambda_+\mathrm{e}^{-kh}}{k(k+\lambda_+)(k-\lambda_+)}. \tag{47}$$

The variance of $Q(t)$ is given by var $\underline{Q}$ = var $\underline{a}/k(k + \lambda_+)$, leading to a simple formula for the autocorrelation of $Q(t)$,

$$\mathrm{cor}_{\underline{Q}}(h) = \frac{k\mathrm{e}^{-\lambda_+|h|} - \lambda_+\mathrm{e}^{-k|h|}}{k - \lambda_+}. \tag{48}$$

In the present case the stochastic behaviour of the input is completely given. So the output is also completely determined. Besides mean values and variances the probability density function itself can be derived. It turns out to be a beta distribution given by

$$f_{\underline{Q}}(Q) = \frac{\Gamma(\lambda_+/k)}{\Gamma(\lambda_0/k)\Gamma(\lambda_1/k)} \frac{\left(Q - \frac{a_0}{k}\right)^{\lambda_0/k-1} \left(\frac{a_1}{k} - Q\right)^{\lambda_1/k-1}}{\left(\frac{a_1 - a_0}{k}\right)^{\lambda_+/k-1}} \quad \text{for } \frac{a_0}{k} \leqslant Q \leqslant \frac{a_1}{k}. \tag{49}$$

See Appendix for the derivation.

For special choices for $\mathscr{E}\underline{d}_i$ and var $\underline{d}_i$ of the dependent increments process of the last subsection can the mean and variance of the concentration in the environment be made identical. The first useful observation is that the autocovariance functions of both processes, (43) and (47) are close to each other for $k_1 = \lambda_+$.

5. Summary and conclusions

We briefly studied alternatives to incorporate stochasticity in deterministic models. The elegance of regression methods makes them very attractive to apply, but they do not always make scientific sense. Randomization of parameters is only feasible in particular cases and in particular parameters, at least if maximum likelihood estimation is applied. Randomization of the input only leads to explicit results in steady states. Both approaches require computer simulation methods to implement statistical inference.

In many experiments there are several sources of stochasticity simultaneously: measurement-errors as well as differences between individuals and changing parameters. One can easily extend a deterministic model to cover all sources simultaneously, for instance in the one-compartment model by combining (2), (12) and (26) to get

$$\underline{Q}_i = \underline{Q}_{0_i} \mathrm{e}^{-\underline{\xi}_i t_i} + \int_0^{t_i} \mathrm{e}^{-\underline{\xi}_i (t_i - s)} \underline{a}(s)\, \mathrm{d}s. \tag{50}$$

Statistical inference, however, appears to be a hard job, if not impossible. A lot of work still has to be done to make these approaches generally applicable.

Appendix

We will start to discuss some properties of beta distributions, that appear in the context of alternating Poisson processes. The beta distribution $B(a, b)$ is characterized by the following pdf

$$f_{\underline{y}}(y) = \frac{\Gamma(a+b)}{\Gamma(a)\Gamma(b)} y^{a-1}(1-y)^{b-1}, \quad \text{where } 0 < y < 1. \tag{51}$$

LEMMA 1. *Let $\underline{y}$ and $\underline{z}$ be independent stochastic variables, $\underline{y} \sim B(a+1, b)$ and $\underline{z} = \underline{u}^{1/a}$, where $\underline{u}$ is the standard uniform distribution. Then $\underline{y}\,\underline{z} \sim B(a, b+1)$.*

PROOF. The joint density $f_{\underline{y},\underline{z}}$ of y and z is given by

$$f_{\underline{y},\underline{z}}(y, z) = \frac{\Gamma(a+b+1)}{\Gamma(a+1)\Gamma(b)} y^a(1-y)^{b-1} a z^{a-1}, \quad \text{where } 0 < y, z < 1. \tag{52}$$

Let $\underline{v} = \underline{y}\,\underline{z}$. Then the joint density of $\underline{y}$ and $\underline{v}$ is given by

$$\phi_{\underline{y},\underline{v}}(y, v) = \frac{a\Gamma(a+b+1)}{\Gamma(a+1)\Gamma(b)} v^{a-1}(1-y)^{b-1}, \quad \text{where } 0 < v < 1, v < y < 1. \tag{53}$$

The marginal density of $\underline{v}$ is obtained by integrating (53) over y. It is easy to see that the result is $B(a, b+1)$.

LEMMA 2. *Let $\underline{y}$ and $\underline{z}$ be independent stochastic variables, $\underline{y} \sim B(a, b+1)$ and $\underline{z} = \underline{u}^{1/b}$, where $\underline{u}$ is the standard uniform distribution. Then $1 - (1 - \underline{y})\underline{z} \sim B(a+1, b)$.*

PROOF. This follows immediately from Lemma 1 and the quasi-symmetry of the beta distribution: The distribution of $B(b, a)$ is the same as that of $\{1 - B(a, b)\}$.

LEMMA 3. *$B(a, b)$ is a mixture of $B(a+1, b)$ and $B(a, b+1)$ with weights $a/(a+b)$ and $b/(a+b)$, respectively.*

PROOF. Straightforward.

Consider the following simple one-compartment model

$$Q'(t) = k(c - Q(t)) \tag{54}$$

with solution

$$Q(t) = Q_0 e^{-kt} + k \int_{t_0}^{t} c(s) e^{-ks} ds. \tag{55}$$

We will now investigate the output of an alternating Poisson input c. For the sake of notational simplicity the input c randomly jumps from 1 to 0 and back again. The residence times in the two states are independently exponentially distributed random variables with parameters λ_0 and λ_1. The mean input rate is thus given by $\bar{c} = (1/\lambda_1)/(1/\lambda_0 + 1/\lambda_1) = \lambda_0/(\lambda_0 + \lambda_1)$.

The solution Q jumps alternately between the two classes of functions

$$Q(t) = C e^{-kt} \quad \text{for } c = 0 \tag{56}$$

and

$$Q(t) = 1 - (1 - C)\,e^{-kt} \quad \text{for } c = 1 \tag{57}$$

as shown in Figure 1c.

We will at first concentrate on the turning points $Q_1, Q_2, \ldots$. They obey the following scheme:

$$Q_{2i} = Q_{2i-1}\,e^{-k\tau_{2i}}, \tag{58}$$

$$Q_{2i+1} = 1 - (1 - Q_{2i})\,e^{-k\tau_{2i+1}}, \tag{59}$$

where τ_{2i} and τ_{2i+1} are exponentially distributed random variables with parameters λ_0 and λ_1, respectively. It can be easily seen that $e^{-k\tau_{2i}}$ can be written as x^{k/λ_0}, where x is a standard uniformly distributed variable, and $e^{-k\tau_{2i+1}}$ can be written as x^{k/λ_1}. The variables Q_{2i-1}, τ_{2i} and τ_{2i+1} are independent. Suppose $Q_{2i-1} \sim B(1 + \lambda_0/k, \lambda_1/k)$. Then Lemmas 1 and 2 say that $Q_{2i} \sim B(\lambda_0/k, 1 + \lambda_1/k)$ and $Q_{2i+1} \sim B(1 + \lambda_0/k, \lambda_1/k)$. So Q_{2i-1} and Q_{2i+1} follow the same beta distribution, which is therefore the stationary distribution of the upper points. In the same way one can see that the lower points follow a beta distribution $B(\lambda_0/k, 1 + \lambda_1/k)$.

Now for the distribution of $\underline{Q}$ in an arbitrarily chosen point t_∞, long after the start of the process. At time t_∞ the output is increasing or decreasing with probability

$$\frac{\mathscr{E}\underline{\tau}_{2i+1}}{\mathscr{E}\underline{\tau}_{2i+1} + \mathscr{E}\underline{\tau}_{2i}} = \frac{\lambda_0}{\lambda_+} \quad \text{or} \quad \frac{\lambda_1}{\lambda_+}$$

respectively. The time t passed since the last upper or lower point is exponentially distributed with parameter λ_0 or λ_1, respectively, because of length-biased sampling (Cox and Lewis, 1966). So, given that Q is increasing, $\underline{Q}_\infty \sim B(1 + \lambda_0/k, \lambda_1/k)$ and, given that Q is decreasing, $\underline{Q}_\infty \sim B(\lambda_0/k, 1 + \lambda_1/k)$. The unconditional distribution of $\underline{Q}_\infty$ is a mixture of these two beta distributions with weights λ_0/λ_+ and λ_1/λ_+, respectively. Following Lemma 3 we can conclude that $\underline{Q}_\infty \sim B(\lambda_0/k, \lambda_1/k)$. In the general case, where $\underline{c}$ jumps between c_0 and c_1 it can be easily seen that $(\underline{Q}_\infty - c_0)/(c_1 - c_0) \sim B(\lambda_0/k, \lambda_1/k)$ so the pdf of $\underline{Q}_\infty$ is given by

$$f_{\underline{Q}_\infty}(Q) = \frac{\Gamma(\lambda_+/k)}{\Gamma(\lambda_0/k)\Gamma(\lambda_1/k)}(Q - c_0)^{\lambda_0/k - 1}(c_1 - Q)^{\lambda_1/k - 1}, \quad \text{where } c_0 < Q < c_1. \tag{60}$$

References

Cox, D. R. and P. A. W. Lewis (1966). *The Statistical Analysis of Series of Events.* Methuen, London.

Jacquez, J. A. (1985). *Compartmental Analysis in Biology and Medicine.* Elsevier, Amsterdam.

Janssen, M. P. M., A. Bruins, T. H. De Vries and N. M. Van Straalen (1991). Comparison of cadmium kinetics in four soil arthropods species. *Arch. Environ. Contamin. Toxicol.* **20**, 305–312.

Jobson, J. D. and W. A. Fuller (1980). Least squares estimation when the covariance matrix and parameter vector are functionally related. *J. Amer. Statist. Assoc.* **75**, 176–181.
Kwakernaak, H. and R. Sivan (1972). *Linear Optimal Control Systems.* Wiley, New York.
McArthur, R. D. (1988). Parameter estimation in a two-compartment population pharmacokinetic model with destructive sampling. *Math. Biosci.* **91**, 157–173.
Van Houwelingen, J. C. (1988). Use and abuse of variance models in regression. *Biometrics* **44**, 1073–1081.
Zeger, L. Z., K. Y. Liang and P. S. Albert (1988). Models for longitudinal data: A generalized estimating equation approach. *Biometrics* **44**, 1049–1060.

G. P. Patil and C. R. Rao, eds., *Handbook of Statistics, Vol. 12*

17

Compartmental Models of Ecological and Environmental Systems

J. H. Matis and T. E. Wehrly

Abstract: This chapter is an expository progress report on the development of stochastic compartmental models as applied to ecological and environmental systems. The standard deterministic and stochastic multicompartment models are reviewed, and their utility for describing ecological and environmental systems is demonstrated with new examples. An approach using phase-type distributions is outlined for generalized compartmental models with arbitrary retention times. The approach is illustrated with shrimp migration data. New developments of compartmental models with births are described and applied to data on the spread of Africanized honey bees.

Introduction

Compartmental models have been widely used to model ecological and environmental systems over the past forty years. Ecosystem modeling, simulation, and analysis began in the late fifties and early sixties, following early pioneering work on tracer kinetics, with the use of compartmental models to mimic storages and flows of radioisotopes, stable materials, and energy in food chains and food webs. Ambitious efforts followed in the late sixties to mid-seventies to represent ecosystems with large scale computer simulation models based on the concept of interconnected deterministic compartments. Stochasticity was, with few exceptions, absent and statistical inference in such models was largely ignored. The collection of papers in Matis et al. (1979) provide a perspective on compartmental analysis of ecosystem models through the late seventies.

In the eighties, computer simulation models of compartmental systems became increasingly sophisticated and realistic. They provide useful insight, but are often unwieldy as simple management tools. On the other hand, small scale and low order classical compartmental models became user-friendly and increasingly relevant, particularly with the incorporation of a stochastic framework. This chapter represents a progress report on the development of stochastic compartmental models as applied to ecological and environmental systems through the eighties and early nineties.

The chapter is expository and heuristic. The first four sections review the one compartment and multicompartment models viewed from both a deterministic and a stochastic foundation. The application of these models is illustrated with new examples from the authors' research, which supplement previous examples in Matis and Wehrly (1979). The purpose of these first four sections is to illustrate the utility of the basic stochastic approach. Most of Section 4 and almost all of Section 5 through 8 present new developments through the eighties and early nineties. Sections 5 and 6 discuss compartmental models with arbitrary retention times. Section 7 extends the model to include reproduction and Section 8 sketches a few other recent generalizations. Obviously in a field where the range of models is so broad and where the breadth of possible applications seems astronomical, it is not possible to present an expository yet comprehensive review of all recent results. Nor is that the intent, rather the overall objective of this paper is to review some new developments of practical interest, and to serve as a catalyst for further discussion of the role of stochastic compartmental analysis of ecological and environmental systems.

1. The one-compartment deterministic model

Consider first the one-compartment deterministic model. It requires the following notation:

NOTATION. Let

(1) $X(t)$ denote the amount (or mass) of substance of interest in the compartment at time t, and

(2) $\dot{X}(t)$ denote the derivative of $X(t)$.

ASSUMPTION. Assume that the rate of elimination, or turnover rate k is proportional to the amount of substance in the compartment.

The assumption leads directly to the following model:

MODEL.

$$\dot{X}(t) = -kX(t). \tag{1.1}$$

This differential equation has the familiar solution:

SOLUTION.

$$X(t) = X(0)\exp(-kt), \tag{1.2}$$

where $X(0)$ is the intial amount. $X(0)$ is often a known parameter, e.g., it might represent the amount of 'labeled' substance (such as radioactive tracer) or biomass of individuals (such as marked animals) introduced into the system in an experiment.

VARIATIONS. Variations are many, but we note only two common ones.

(i) The compartment might have an 'uptake', or 'input', of substance over time. Letting U denote uptake, which for present simplicity is assumed to be time invariant, the model is

$$\dot{X}(t) = U - kX(t) \tag{1.3}$$

with solution

$$X(t) = \frac{U}{k}(1 - \exp(-kt)). \tag{1.4}$$

(ii) Often the measured response is not the absolute amount of substance, but rather some variable proportional to the amount $X(t)$. In a common application, V denotes the 'volume' of a compartment, and one measures the concentration, defined as $C(t) = X(t)/V$. In this case, model (1.2) becomes

$$C(t) = C(0)\exp(-kt). \tag{1.5}$$

Letting $k_u = U/V$ denote a proportional uptake rate, model (1.4) becomes

$$C(t) = \frac{k_u}{k}(1 - \exp(-kt)). \tag{1.6}$$

ILLUSTRATION (*Mercury accumulation in fish*). The literature contains numerous examples of the use of these simple one-compartment models (see, e.g., Jacquez, 1985; Godfrey, 1983). As an illustration, Matis et al. (1991) use this simple model for the bioaccumulation of mercury in fish. Data on mercury uptake were recorded separately for each of 21 mosquito fish on each of $n = 6$ days. Model (1.6) fits the data well for most of the fish. As a particular example, fish 5 had estimated mercury (Hg) concentrations of 0, 0.287, 0.450, 0.553, 0.653 and 0.760 (μg/g dry wt) on days 0 to 4 and on day 6 respectively. Any standard nonlinear least squares program may be used to fit the model. We used KINETICA (Allen and Matis, 1992) and obtained parameter estimates $k_u = 0.322$ μg/g dry wt/day and $k = 0.387$ day^{-1}, with a mean squared error (MSE) of 2.1×10^{-4}. The equilibrium concentration of Hg in the fish is estimated from (1.6) to be $C(\infty) = 0.832$ μg/g dry wt.

2. The standard one-compartment stochastic model

The present stochastic model is the so-called 'particle' model where the substance of interest is viewed as a set of particles, e.g., mercury molecules. Obviously the number of particles in this application is very large, however in others dealing with animals population dynamics the number may be small. The notation is as follows:

NOTATION. Let

(1) $P(t)$ denote the probability that a particular particle introduced at time 0 is still in the compartment at time t, with $\dot{P}(t)$ as its time derivative, and

(2) $X(t)$ denote the number of particles introduced at $t = 0$ which are still present at time t.

ASSUMPTIONS. Assume

(1) the conditional probability that any random particle present at time t leaves by $t + \Delta t$, where Δt is small, is

$$k\Delta t, \tag{2.1}$$

(2) all $X(0)$ particles are independent.

Note that the constant k in (2.1) is a probability intensity coefficient which defines a probability of particle transfer rather than a proportional elimination rate as in Section 1.

MODEL. A differential equation model called the Kolmogorov equation, may be derived for $P(t)$ using basic probability arguments. The equation for this model is

$$\dot{P}(t) = -kP(t), \tag{2.2}$$

which is a stochastic analog to the deterministic model (1.2). The solution to the model with initial condition $P(0) = 1$ is the following:

SOLUTION.

$$P(t) = \exp(-kt). \tag{2.3}$$

$P(t)$ may be viewed as the common 'survival' probability of the particles. Because all of the particles are independent by assumption, the number of particles which survive to time t follows a binomial distribution. This fact leads to the following corollary:

COROLLARY. The distribution and mean of the particle count $X(t)$ are

(1) $$X(t) \sim \text{binomial}\,[n = X(0), p = P(t)], \tag{2.4}$$

(2) $$\mathrm{E}[X(t)] = X(0)\,P(t) = X(0)\exp(-kt). \tag{2.5}$$

Of course, the fundamental difference between the stochastic model for $X(t)$ in (2.4) and the deterministic model in (1.2) arises from the assumed chance mechanism in (2.1) which generates so-called 'process' uncertainty, or process error. The form of the process error could be derived but it is not of present practical interest. The observed data from a system assumed to follow this stochastic model may be fitted to the mean value function in (2.5), which is identical to the model in (1.2).

This stochastic model may also be defined on the basis of its retention time (also called transit time or sojourn time) distributions. In some ways, this conceptualization of the inherent chance mechanism is more satisfactory because it relies on a continuous time probability distribution, rather than on a conditional flow probability in discretized time units as in (2.1). The basic notions associated with a continuous probability distribution are needed first.

ADDITIONAL NOTATION. Let

(3) R be the retention time of a random particle in the system,

(4) $F(a) = \text{Prob}[R \leqslant a]$ be the distribution function of R, i.e., the probability that the particle will leave prior to attaining 'age' a in the compartment,

(5) $f(a) = \mathrm{d}F(a)/\mathrm{d}a$ be the density function,

(6) $S(a) = \text{Prob}[R > a] = 1 - F(a)$ be the survivorship function, i.e., the probability that the particle 'survives' to age a in the compartment, and

(7) $\lambda(a) = f(a)/S(a)$ be the 'hazard' rate, defined as in (2.1), such that the conditional probability of a particle leaving by $a + \Delta a$ given that it survives to age a is $\lambda(a)\Delta a$.

Suppose one assumes the exponential distribution for R. This gives:

ALTERNATIVE ASSUMPTION. Let $R \sim$ exponential distribution (k), which implies

(1) $$F(a) = 1 - \exp(-ka), \tag{2.6}$$

(2) $$f(a) = k \exp(-ka), \tag{2.7}$$

(3) $$S(a) = \exp(-ka), \tag{2.8}$$

(4) $$\lambda(a) = k. \tag{2.9}$$

It is clear that the survivorship function $S(a)$ in (2.8) is equivalent to the previous function $P(t)$ solved in (2.3). Also the constant hazard rate in (2.9) is equivalent to the assumption in (2.1). The particles of this model are said to 'lack a memory', with the practical consequence that the transfer mechanism of the compartment must not discriminate on the basis of the accrued age of a particle in the system. For convenience, a compartment with this assumption will subsequently be called an exponential compartment.

ADDITIONAL RESULTS CONCERNING MOMENTS. The alternative focus of defining the model based on its retention time may be exploited to determine various statistical moments of R, such as its mean and variance. This leads to the following well-known results for the assumed exponential retention time R.

COROLLARY. *The mean and variance of R are*

(1) $$E[R] = \int a f(a)\, \mathrm{d}a = k^{-1}, \tag{2.10}$$

(2) $$V[R] = \int a^2 f(a)\, \mathrm{d}a - (E[R])^2 = k^{-2}. \tag{2.11}$$

The mean retention time in (2.10), which later will also be defined as the mean residence time or MRT, is widely used in practice. Though at present just a simple function of k, its utility is largely due to the fact that the concept of a MRT is universal to all compartmental models, and hence it is an important variable for comparing particle dynamics from different models. The variance in (2.11) is inherent in the stochastic formulation of the model.

VARIATION. The model describing particle elimination only and based on assumption (2.1) is known as a pure death process. Often the assumption of Poisson entries is added, whereby specifically in any small interval Δt one has

$$\text{Prob}\{\text{new particle entering the compartment}\} = U\Delta t. \tag{2.12}$$

The model is known as an immigration-death process. If for convenience $X(0) = 0$, it is easily shown that the expected number of particles and the expected concentration are then

$$E[X(t)] = \frac{U}{k}(1 - \exp(-kt)), \tag{2.13}$$

$$E[C(t)] = \frac{k_u}{k}(1 - \exp(-kt)), \tag{2.14}$$

which are analogs of (1.4) and (1.6).

ILLUSTRATION (*Mercury accumulation in fish – continued*). In Matis et al. (1991) the data on mercury accumulation are reinterpreted based on this stochastic formulation. As a particular example, the mercury molecules for fish 5, which was previously discussed, had an estimation mean retention time of $(0.387)^{-1} = 2.58$ days and an estimated variance of 6.68. This group of fish had a 20-fold difference in size, yet one interesting finding was that the estimated mean retention time was not correlated with the size of the fish.

3. The multicompartment deterministic model

Consider now the standard multicompartment model. The subsequent generalization holds for any n-compartment model. However, for simplicity, the concepts will often be illustrated only for the two-compartment case. The following expanded set of notation is now useful:

NOTATION. Let

(1) $X_i(t)$ be the amount of substance in compartment i at time t,

(2) $\mathbf{X}(t) = [X_1(t), \ldots, X_n(t)]'$ be the column-vector of amounts at time t,

(3) k_{ij}, for $i = 0, 1, \ldots, n; j = 1, \ldots, n; i \neq j$; denote the fractional flow rate to i from j, where 0 represents the system exterior,

(4) $k_{jj} = -\sum_{i \neq j} k_{ij}$, denote the total outflow rate from j,

(5) $\mathbf{K} = (k_{ij})$ be the $n \times n$ matrix of k_{ij} coefficients,

(6) $\lambda_1, \ldots, \lambda_n$ be the eigenvalues of $\mathbf{K}$, with $\mathbf{\Lambda}$ as the diagonal matrix of λ_i, and

(7) $T_1, \ldots, T_n$ be the corresponding eigenvectors of $\mathbf{K}$ with $n \times n$ eigenvector matrix $\mathbf{T} = (T_1, \ldots, T_n)$.

One could now assume a linear compartment model, namely one where each derivative, $\dot{X}_i(t)$, is a linear function of the $X_i(t)$. For example, in the most general two-compartment model one assumes the following:

$$\dot{X}_1(t) = -(k_{01} + k_{21})X_1(t) + k_{12}X_2(t), \quad \text{and}$$

$$\dot{X}_2(t) = k_{21}X_1(t) - (k_{02} + k_{12})X_2(t).$$

Such linear compartment models may be expressed in matrix form as follows:

MODEL.

$$\dot{\mathbf{X}}(t) = \mathbf{K}\mathbf{X}(t). \tag{3.1}$$

Clearly, model (3.1) is the matrix generalization of model (1.2). The above vector of differential equations has the following solution:

SOLUTION.

$$\mathbf{X}(t) = \exp(\mathbf{K}t)\,\mathbf{X}(0) \tag{3.2}$$

with the matrix exponential defined as $\exp(\mathbf{K}t) = \mathbf{I} + \sum_{i=1}^{\infty} \mathbf{K}^i t^i / i!$.

Two corollaries are helpful in the practical application of solution (3.2). In most environmental applications, one may assume that the system is 'open' and at least 'weakly connected' so that no compartment is a sink (i.e., 'closed'). These are necessary for the following corollary:

COROLLARY 1. *Assuming the above regularity conditions and distinct λ_i, one can show*

(1) *the λ_i have negative real parts, and*

(2) $$\mathbf{X}(t) = \mathbf{T}\exp(\mathbf{\Lambda}t)\mathbf{T}^{-1}\mathbf{X}(0), \tag{3.3}$$

where $\exp(\mathbf{\Lambda}t)$ *is a diagonal matrix with elements* $\exp(\lambda_i t)$.

All two-compartment and most three-compartment models have real eigenvalues. This motivates the following corollary:

COROLLARY 2. *Assuming that the eigenvalues in Corollary 1 are distinct and real, formula (3.3) implies that*

$$X_i(t) = \sum_{j=1}^{n} A_{ij}\exp(\lambda_j t), \quad \text{for } i = 1, \ldots, n. \tag{3.4}$$

Corollary 2 defines the 'sums of exponentials' models which generalize solution (1.2) and which are direct consequences of the assumed linear (or homogeneous) compartmental model. The A_{ij} and λ_j are often called the 'macro-parameters' and they are usually involved functions of the k_{ij} 'micro-parameters'. The equivalent solutions based on the k_{ij} parameters are given explicitly for the common two- and three-compartment models in many books, including Godfrey (1983) and Jacquez (1985).

Often only a single set of observations is available, either from a single compartment directly or from the pooled elimination records. In this simple case, solution (3.4) may be used as a regression model either for total tracer, $X_i(t)$, or for concentration, $C_i(t) = X_i(t)/V_i$. Multiple data sets from various sites could also be fitted simultaneously, as illustrated in Grant et al. (1991). In this case the regression models in (3.4) would be specified in terms of the microparameters k_{ij} and V_i.

Usually environmental systems are highly structured and one postulates a given number of compartments directly from the mechanistic description of the kinetics. However, a common alternative practice when the system is less structured is to determine the assumed number of compartments in the underlying system from the number of exponential terms in (3.4) required to adequately describe experimental data.

VARIATION. The model in (3.1) could be expanded to include inputs. Consider the general linear systems model given by Kalman et al. (1969) as

$$\begin{aligned} \dot{\mathbf{X}}(t) &= \mathbf{K}(t)\mathbf{X}(t) + \mathbf{B}(t)\mathbf{U}(t), \\ \mathbf{Y}(t) &= \mathbf{C}(t)\mathbf{X}(t), \end{aligned} \tag{3.5}$$

where $\mathbf{U}(t)$, $\mathbf{Y}(t)$ and $\mathbf{X}(t)$ are vectors of inputs to, outputs from, and amounts in the compartments, respectively, and $\mathbf{K}(t)$, $\mathbf{B}(t)$ and $\mathbf{C}(t)$ are conformable matrices which define the model. For present convenience, we assume that $\mathbf{K}(t)$ and $\mathbf{U}(t)$ are time-invariant, and that inputs to and outputs from the system occur directly with individual compartments. These assumptions simplify the model to

$$\dot{\mathbf{X}}(t) = \mathbf{K}\mathbf{X}(t) + \mathbf{U}, \tag{3.6}$$

where $\mathbf{U}$ is an n-vector of inputs. The solution is

$$\mathbf{X}(t) = \exp(\mathbf{K}t)\mathbf{X}(0) + \int \exp[\mathbf{K}(t-s)\mathbf{U}\,\mathrm{d}s \tag{3.7}$$

from whence it follows, as before, that the solutions are 'sums of exponentials' for distinct λ_i. The equilibrium values of the system, denoted $\mathbf{X}^*$, are easily obtained from (3.6) or (3.7) as

$$\mathbf{X}^* = -\mathbf{K}^{-1}\mathbf{U}. \tag{3.8}$$

Equation (3.8) may be interpreted as specifying the transformations of individual inputs in $\mathbf{U}$ to equilibrium amounts, or 'storages', in $\mathbf{X}^*$ for the various compartments. Such relationships are extended to yield a general ecological input–output analysis, called a linear environ analysis (see, e.g., Barber et al., 1979 and Patten, 1992).

ILLUSTRATIONS. (i) *Mercury accumulation in fish – continued.* An alternative model for mercury accumulation, proposed also by Spacie and Hamelink (1985), conceptualizes two compartments for describing mercury dynamics in fish. An initial compartment receives Hg from and eliminates Hg to the environment. A second compartment consists of tissue with more permanent Hg binding, i.e., it is a 'storage' compartment. This model is assumed to have three parameters, with coefficient matrices

$$\mathbf{U} = \begin{bmatrix} k_u \\ 0 \end{bmatrix}, \quad \mathbf{K} = \begin{bmatrix} -(k_{01}+k_{12}) & k_{12} \\ k_{12} & -k_{12} \end{bmatrix}.$$

Note both transfer rates among compartments are assumed to be k_{12} for parameter parsimony. The model was fitted to data using the KINETICA software package (Allen and Matis, 1992).

As a particular example, this model was fitted for the fish 5 data. The parameter estimates are $k_u = 0.408$, $k_{01} = 0.875$ and $k_{12} = 0.432$ with $\text{MSE} = 0.433 \times 10^{-4}$. The equilibrium concentration is estimated to be $C(\infty) = 0.933$ μg Hg/g dry wt. This two-compartment model reduced the MSE substantially for 7 of the 21 fish.

(ii) *Hydrology in Okefenokee Swamp.* A very different environmental application of compartmental modeling is its use for describing dynamics in large-scale ecosystems, for example in the hydrology model of Patten and Matis (1982). The

Okefenokee Swamp watershed area was partitioned into four large-scale water storage compartments, namely the upland surface, the upland groundwater, the swamp surface, and the swamp subsurface compartments. Previous studies have quantified the water flow through this watershed, and the above study approximated these empirical dynamics with the following **K** matrix (in units of yr^{-1}):

$$\mathbf{K} = \begin{bmatrix} -48.5623 & 0.4444 & 0 & 0 \\ 37.5165 & -3.1739 & 0 & 0 \\ 9.5934 & 0.0147 & -2.8020 & 0.0004 \\ 0 & 0.2795 & 0.0030 & -0.0716 \end{bmatrix}.$$

The objective of the model was to provide a management tool for studying the impact of water inputs to and outputs from the various compartments. For example, a constant unit input (of $10^9\ \text{m}^3/\text{yr}$) of rainfall into the upland surface (compartment 1) would result in the following storage dynamics (in units of $10^9\ \text{m}^3$ of water):

$$X_1(t) = 0.023 - 0.020\exp(-48.9t) - 0.003\exp(-2.80t),$$

$$X_2(t) = 0.273 + 0.017\exp(-48.9t) - 0.290\exp(-2.80t),$$

$$X_3(t) = 0.081 + 0.004\exp(-48.9t) - 4.194\exp(-2.80t) + 4.109\exp(-2.81t) - 0.0002\exp(-0.07t),$$

$$X_4(t) = 1.069 - 0.0001\exp(-48.9t) + 0.004\exp(-2.80t) + 0.025\exp(-2.81t) - 1.098\exp(-0.07t).$$

These curves are given in Figure 1 for the above equations. The equilibrium amounts resulting from this unit input are clearly $X_1^* = 0.023$, $X_2^* = 0.273$, $X_3^* = 0.081$ and $X_4^* = 1.069$. These equilibrium amounts could also be obtained directly as the first column of $-\mathbf{K}^{-1}$, as apparent in (3.8). Similarly, one could obtain the overall impact of other unit ($10^9\ \text{m}^3/\text{yr}$) inputs to and outputs from an ecosystem model of this type as part of a linear environ analysis. Such considerations may be useful for managing ecological systems, as for example, river flow from the Okefenokee Swamp watershed.

(iii) *Migration of marine shrimp.* The estimation of migration and mortality rates of fisheries populations is a challenging problem. Grant et al. (1991) study a marked population of white shrimp which were released from two areas of an estuary (Moses Lake) in Galveston Bay off the Gulf of Mexico. Shrimp can migrate between the two areas (compartments) but could leave only from the second compartment, which leads to the following **K** matrix:

$$\mathbf{K} = \begin{bmatrix} -k_{21} & k_{12} \\ k_{21} & -(k_{12} + k_{02}) \end{bmatrix}.$$

Tagged shrimp were recovered from each area over 25 day elapsed time periods,

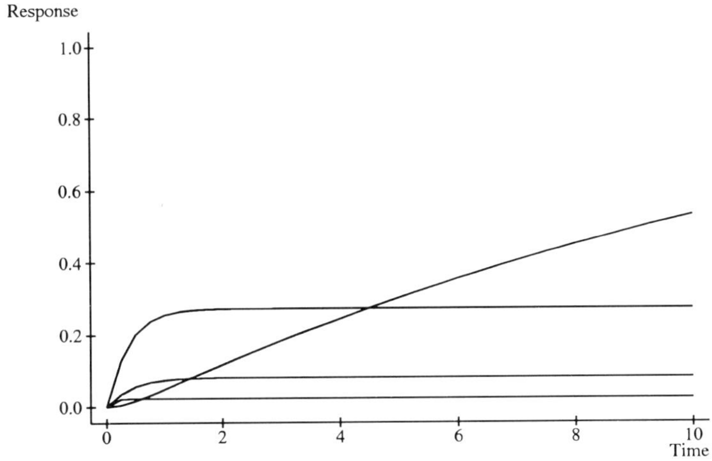

Fig. 1. Dynamic effect of constant unit input into compartment 1 on storages in four compartments of Okefenokee Swamp watershed.

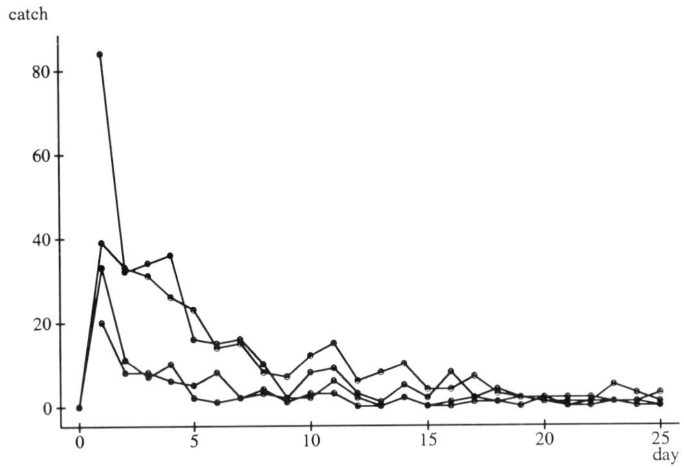

Fig. 2. Observed catch per unit effort in Moses Lake versus elapsed time for shrimp released in area 1, $Y_{11}(t)$ and $Y_{21}(t)$, and in area 2, $Y_{12}(t)$ and $y_{22}(t)$.

leading to the four data sets illustrated in Figure 2. It is possible to estimate the three rate parameters from any of the individual data sets, however, all four sets were fitted simultaneously to yield a single model incorporating the common spatial structure. This was implemented by utilizing a four-compartment structure, with each of the release areas having two comparments with the above **K** submatrix. The resulting fitted curves are given in Figure 3. The five parameter estimates with their estimated standard errors in parentheses are $C_{11}(0) = 123.8$ (6.35), $C_{22}(0) = 53.8$ (6.29), $k_{21} = 0.838$ (0.131), $k_{12} = 0.702$ (0.179) and $k_{02} = 0.532$ (0.040). The mean square error is 17.5.

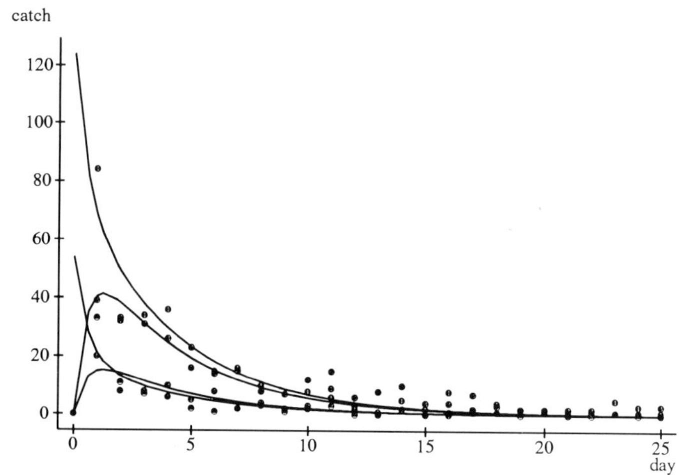

Fig. 3. Fitted abundance curve using a standard multicompartment model for four shrimp data sets from Moses Lake.

4. The standard multicompartment stochastic model

Consider a stochastic formulation of the n-compartment model. The following definitions are necessary:

NOTATION. Let

(1) $P_{ij}(t)$ be the probability that a particle starting in j at time 0 will be in i at time t,

(2) $\mathbf{P}(t) = [P_{ij}(t)]$ be an $n \times n$ matrix of probabilities,

(3) $X_{ij}(t)$ be the number of the $X_{jj}(0)$ particles starting in j at time 0 which are in i at time t, and

(4) $\mathbf{X}_j(t) = [X_{1j}(t), \ldots, X_{nj}(t)]'$ be the random vector denoting the numbers of the $X_{jj}(0)$ particles in the various compartments at time t.

For subsequent convenience, the present formulation tracks separately the distribution of particles for each compartment of origin. The assumptions defining the chance (or probability) mechanism for this model are:

ASSUMPTIONS. Assume

(1) the conditional probability that, for small Δt, a random particle in j at time t will be in i at time $t + \Delta t$ is

$$k_{ij}\Delta t, \tag{4.1}$$

(2) all particles are independent.

These assumptions define a particular stochastic model called a Markov process. As in Section 2, the k_{ij} constants may be regarded as rate coefficients which define the instantaneous transfer probabilities. A set of differential equations generalizing (2.2) may be constructed for any n, but for simplicity will be illustrated only for $n = 2$. The complete set of these differential equations, i.e., the Kolmogorov equations, for the presently defined $n = 2$ compartment model is (see, e.g., Chiang, 1980)

$$\dot{P}_{1j}(t) = -(k_{01} + k_{21})\, P_{1j}(t) + k_{12} P_{2j}(t),$$
$$\dot{P}_{2j}(t) = k_{21} P_{2j}(t) - (k_{02} + k_{12}) P_{2j}(t)$$

for $j = 1, 2$. Using the matrix definition of $\mathbf{K}$ given in Section 3, these equations give the following model which holds for any n:

MODEL.

$$\dot{\mathbf{P}}(t) = \mathbf{K}\mathbf{P}(t). \tag{4.2}$$

This matrix model for probabilities has similar form to that of (3.1). The initial condition is $\mathbf{P}(0) = \mathbf{I}$, which leads to the following solution.

SOLUTION.

$$\mathbf{P}(t) = \exp(\mathbf{K}t). \tag{4.3}$$

A number of corollaries follow which generalize the previous results. One concerns the probabilities, the other the counts.

COROLLARY 1. *Assume that the system is open and weakly connected. Then*
(1) *if the λ are distinct,* $\mathbf{P}(t) = \mathbf{T}\exp(\Lambda t)\mathbf{T}^{-1}$, *and*
(2) *if the λ are distinct and real,*

$$P_{ij}(t) = \sum_l A_{ijl} \exp(\lambda_l t) \quad \textit{for } i, j = 1, \ldots, n. \tag{4.4}$$

COROLLARY 2. *The distribution and means for the random vector* $\mathbf{X}_j(t)$ *are*
(1) $\mathbf{X}_j(t) \sim \text{multinomial}\,[n = X_{jj}(0), P_{1j}(t), \ldots, P_{nj}(t)]$,
(2) $\mathrm{E}[X_{ij}(t)] = X_{jj}(0) P_{ij}(t)$. (4.5)

Clearly, these corollaries generalize the one-compartment results in (2.3) to (2.5). The mean value function, defined in (4.5) and derived from stochastic assumptions, is identical to (2.4) which was derived from deterministic assumptions. In practical applications, one could fit experimental data to identical sum of exponentials functions in (3.4) or (4.5), the difference being only in the interpretation of results. However, it will be shown presently that the stochastic model, in addition to its theoretical appeal of a broader conceptual foundation, also leads to additional methodology of interest in describing system dynamics.

Many new random variables may be defined for a stochastic multicompartment model to describe the dynamics of particle transfer among the various compartments. Some of the more useful variables are as follows:

ADDITIONAL NOTATION. Let

(5) R_{ij} denote the retention time during a single visit in j of a particle whose next transfer will be to i,

(6) R_j denote the retention time during a single visit of a random particle in j prior to its next transfer out of j (either to another compartment or to the exterior),

(7) N_{ij} be the number of visits that a particle starting in j will make to i prior to its exit from the system,

(8) S_{ij} be the total residence time that a particle originating in j will accumulate in i during all of its N_{ij} visits, i.e., $S_{ij} = \sum_1^{Nij} R_{ij}$,

(9) $S_j = \sum_i S_{ij}$ be the residence time in the system of a particle originating in j.

Consider now the reformulation of the assumption in (4.1):

ALTERNATIVE ASSUMPTION. Let $R_{ij} \sim$ exponential distribution (k_{ij}), for all i and j.

As before, the assumption implies that each hazard rate is age-independent, more specifically each rate is the constant k_{ij}. This gives rise to the following corollary:

COROLLARY 3. *The probability that a particle in j will transfer to i on its next departure is*

$$p_{ij} = k_{ij} / \sum_{l \neq j} k_{lj} = \frac{-k_{ij}}{k_{jj}}. \tag{4.6}$$

Note that this probability concerning the destination of particle transfer is also independent of particle age.

ADDITIONAL RESULTS CONCERNING MOMENTS. The statistical moments of the random variables R_i, S_{ij}, and N_{ij} are also very useful. The present development will consider only the expected values, i.e., first moments. Expressions for the variances of these random variables are given in Matis, Wehrly and Metzler (1983). The following definitions are required:

FURTHER NOTATION. Let

(10) $\mathbf{E}(R)$ be a diagonal matrix of expected retention times, with elements $\mathrm{E}[R_i]$.

(11) $\mathbf{E}(S)$ be an $n \times n$ matrix of mean residence times, with elements $\mathrm{E}[S_{ij}]$,

(12) $\boldsymbol{E}(S)$ be an n-vector of expected residence times, with elements $\mathrm{E}[S_i]$,

(13) $\mathbf{E}(N)$ be an $n \times n$ matrix of expected number of visitations, with elements $\mathrm{E}[N_{ij}]$.

The following is an immediate generalization of (1.10):

THEOREM 1.

$$\mathbf{E}(R) = -\mathbf{K}_{\mathrm{D}}^{-1}, \tag{4.7}$$

where $\mathbf{K}_{\mathrm{D}}$ *is the matrix of diagonal elements of* $\mathbf{K}$.

This theorem gives the means for the R_j transit times. The negative inverse in (4.7) follows from the fact that the diagonal elements of $\mathbf{K}$ are negative by construction. The mean for any individual conditional transit time R_{ij} may be found directly from (2.10), i.e., $E[R_{ij}] = k_{ij}^{-1}$.

The following results are given and illustrated in Matis, Wehrly and Metzler (1983):

THEOREM 2. *The mean residence times* (MRT) *are*

(1) $\mathbf{E}(S) = -\mathbf{K}^{-1}$, *and* (4.8)

(2) $E(S) = \mathbf{E}(S)\mathbf{1}$, *where* $\mathbf{1}$ *is a vector of ones.* (4.9)

Clearly (4.8) is also a matrix analog to (2.10).

The following result is given in Matis and Wehrly (1990):

THEOREM 3. Let $\mathbf{K}^{\circ}$ *be a normalized* $\mathbf{K}$ *matrix such that* $k_{jj}^{\circ} = 0$ *and* $k_{ij}^{\circ} = -k_{ij}/k_{jj}$. *Then the matrix of expected numbers of visits is*

$$\mathbf{E}(N) = (\mathbf{I} - \mathbf{K}^{\circ})^{-1}. \tag{4.10}$$

The off-diagonal elements defined as k_{ij}° in the theorem are the age-independent transfer probabilities, i.e., the p_{ij} in (4.6). This theorem is useful in describing the extent of cycling of particles in the system.

APPLICATION OF MOMENT FORMULAS. The formulas (4.8) to (4.10) involve only simple matrix inversion, multiplication and subtraction, and therefore the formulas are very easy to apply. These statistical moments for stochastic compartmental models have been very useful in practical applications where the compartmental structure is assumed known. Their advantages are many, including the following:

(1) They are very easy to calculate from the k_{ij} estimates.

(2) Unlike half lives, they are easy to interpret for any compartment model and they are additive.

(3) The formulas are valid for all linear compartmental models regardless of the paths of elimation.

(4) The formulas can give the mean residence time of the tracer at specific target compartments specified by the model.

(5) The residence time moments are often sensitive response variables to use in the statistical analysis of designed experiments, as illustrated in Matis, Wehrly and Metzler (1983).

For these reasons, the authors expect that the use of these statistical moments from stochastic compartment models will increase substantially over the years.

VARIATION. The stochastic multicompartment model may also be extended to include unit inputs, or immigration, as in (2.12). The underlying stochastic model would then be the well-known Markov migration process with immigration and emigration (see, e.g., Chiang, 1980). As noted previously, one qualitative difference in the solution for this stochastic model and the corresponding deterministic model in Section 3 is that the solutions for the stochastic model are expected counts of particles, such as in (4.5). However, for purposes of parameter estimation one would fit the same regression model to data under either formulation. The details of an environ analysis for a stochastic model are given in Matis and Patten (1979). One chief difference in statistical methodology between the stochastic and deterministic models is that the former provides a natural framework for an analysis of residence time moments.

ILLUSTRATION. (i) *Mercury accumulation in fish – continued.* Equations (4.8) and (4.9) enable one to define further the dynamics of the two-compartment storage model. For example, under this model a mercury molecule entering fish 5 is expected to spend 2.29 days in the fish prior to its depuration.

(ii) *Hydrology of Okefenokee Swamp – continued.* The first column of the negative inverse matrix in (4.8) gives the mean residence times of a unit of water in the upland surface area, i.e., $E[S_{i1}]$ for $i = 1, \ldots, 4$; as 0.023, 0.273, 0.081 and 1.069 yr., respectively. Obviously the numerical values are identical to the equilibrium storage values associated with a unit input which were previously obtained in Section 3. Additional useful description of system dynamics is available from (4.7) and (4.10), including such items as expected water cycling between the compartments. The use of cycling is illustrated in the following example.

(iii) *Migration of marine shrimp – continued.* The additional insight into system dynamics may also be illustrated using the shrimp migration model. Substituting the estimated rate coefficients into (4.7) to (4.10) one finds the following:

$$E[R] = \begin{bmatrix} 1.208 & 0 \\ 0 & 0.810 \end{bmatrix}, \qquad E[S] = \begin{bmatrix} 2.801 & 1.593 \\ 1.878 & 1.878 \end{bmatrix},$$

$$E[N] = \begin{bmatrix} 2.32 & 1.32 \\ 2.32 & 2.32 \end{bmatrix}. \tag{4.11}$$

In summary, based on all four data sets fitted simultaneously, a shrimp introduced into area 1 is expected to visit compartments 1 and 2 an average of 2.32 times apiece, with mean single durations of 1.208 and 0.810 respectively, for a combined estimated $(2.80 + 1.88) = 4.68$ days in Moses Lake prior to entering Galveston Bay. Under the same model, a shrimp introduced into compartment 2 is predicted to visit the compartments an average of 1.32 and 2.32 times, respectively, with mean durations of 1.208 and 0.810, as above, for a combined estimate of $(1.59 + 1.88) = 3.47$ days in the system.

5. One-compartment model based on nonexponential retention time distributions

The standard stochastic one-compartment model assumes that the retention time of a random particle in the system follows the exponential distribution in (2.6) and hence has a time-invariant hazard rate, as given in (2.1) and (2.9). Consider now a one-compartment model with a nonexponential retention time, hence with an age-varying hazard rate as well. The model could be represented simply by specifying some assumed nonexponential retention time distribution. Alternatively, one could use the hazard rate formulation to specify this generalized model by generalizing (2.2) as follows:

MODEL.

$$\dot{P}(a) = -k(a)P(a). \tag{5.1}$$

Compartmental analysis based on this general one-compartment model is in many ways equivalent to statistical survival analysis, which has been widely used in biomedical and engineering applications to model the retention times, or 'survival times', of units in a population. However, the nature of the data ordinarily observed in survival studies is different from that observed in compartmental modeling. In survival analysis, for example, one might record the actual, censored survival times of a sample of cancer patients. In compartmental analysis, the individual exit times are usually either not conveniently available or are too numerous to record individually; instead one counts (or estimates) the number of 'surviving' particles at fixed times. Nevertheless, although the data and hence the estimation techniques would differ, the models in survival analysis may be used as a foundation for a generalized compartmental analysis based on nonexponential retention times for single compartments. The utility of the compartmental approach is apparent as the model is expanded subsequently to a multicompartment system to incorporate mechanistic structure.

The causal reasons for nonexponential retention times, and hence age-varying hazard rates, may be numerous. Two general generic reasons for such retention times in environmental and ecological modeling are 'noninstant mixing' and 'compartment heterogeneity'.

The Weibull, gamma and Erlang distributions, three of the more common nonexponential retention time distributions, are discussed in Johnson and Kotz (1970). One could fit these distributions, or their survivorship curves, directly to data and estimate the appropriate parameters which define their hazard rate functions. For present simplicity, consider first the Erlang distribution.

ASSUMPTION. Let $R \sim \text{Erlang}\,(n, \lambda)$; $n = 1, 2, \ldots; \lambda > 0$.

The model has the following density, survivorship function, hazard rate, mean and variance:

(1) $$f(a) = \lambda^n a^{n-1} \exp(-\lambda a)/(n-1)!, \tag{5.2}$$

(2) $$S(a) = \exp(-\lambda a) \sum_{i=0}^{n-1} (\lambda a)^i/i!, \tag{5.3}$$

(3) $$k(a) = \left[\frac{\lambda^n a^{n-1}}{(n-1)!}\right] \Big/ \sum_{i=0}^{n-1} \frac{(\lambda a)^i}{i!}, \tag{5.4}$$

(4) $$\mathrm{E}[R] = \frac{n}{\lambda}, \tag{5.5}$$

(5) $$\mathrm{V}[R] = \frac{n}{\lambda^2}. \tag{5.6}$$

Two qualitative features are apparent in the hazard rate function in (5.4). Firstly, for $n > 1$, the rate function at age 0 is $k(0) = 0$, after which the rate increases. This provides an initial dampening of the passage probability of newly introduced particles. Secondly, the rate asymptotes to λ as the age a increases. This implies that the age discrimination within the compartment diminishes, either relatively rapidly or slowly depending on n, as the retention time increases. Both of the qualitative features are characteristic of data from nonhomogeneous compartments and/or compartments with noninstantaneous initial mixing which are common in environmental and ecological modeling.

The density and survival functions in (5.2) and (5.3) are easy to fit to data directly. This simplicity of application, together with the qualitative characteristics of its hazard rate function and its tractable moments in (5.5) and (5.6), has led to the widespread usage of the Erlang model.

The Erlang distribution has the useful mathematical property that the sum of n independent exponential (λ) random variables has the Erlang (n, λ) distribution. This result enables one to generate the desired Erlang distributed retention times by passing though a sequence of identical exponential compartments, each with parameter λ. In effect, one may define a set of n_i exponential 'pseudo-compartments' for each Erlang distributed random variable in the model. These pseudo-compartments are not intended to have any mechanistic interpretation, rather they exist merely as a mathematical artifice to solve the model. The utility of this approach is that the Erlang compartment model may be solved using the previous theory of Section 4. The following notation is useful:

NOTATION. Let

(1) $P^*_{uv}(t)$ denote the probability that a particle starting in pseudo-compartment v at time 0 will be in pseudo-compartment u at time t, for $u, v = 1, 2, \ldots, n$.

(2) $\mathbf{P}^*(t) = [P^*_{uv}(t)]$ be the matrix of occupancy probabilities for the pseudo-compartments.

(3) $\mathbf{K}^*$ denote the structured matrix of rate coefficients among the pseudo-compartments.

It follows from (4.2) that the model for the expanded (linear) pseudo-compartment representation of the Erlang one-compartment model may be expressed as follows:

MODEL.

$$\dot{\mathbf{P}}^*(t) = \mathbf{K}^*\mathbf{P}^*(t). \tag{5.7}$$

It also follows from previous theory in (4.3) that the solution is:

SOLUTION.

$$\mathbf{P}^*(t) = \exp(\mathbf{K}^*t). \tag{5.8}$$

In applying the above solution, one must consider the fact that the coefficient matrix $\mathbf{K}^*$ has a very structured form which leads to equal eigenvalues. Consequently the $P^*_{uv}(t)$ will not be the sums-of-exponentials solutions given in (4.4) but instead will involve powers of time.

After solving for the $P^*_{uv}(t)$ solutions, the solution for the $P(t)$ occupancy probability in (5.1) may be expressed as the sum

$$P(t) = \sum_{v=1}^{n} P^*_{v1}(t). \tag{5.9}$$

This solution is equivalent to the survival function given in (5.3). Therefore the use of pseudo-compartments represents an alternative methology for fitting the Erlang model to data and hence estimating the parameters n and λ.

The 'pseudo-compartment' approach may be generalized through the concept of phase-type (or PH) distributions developed by Neuts (1981). A PH distribution is defined as the distribution of time until absorption in a finite-state Markov process with n transient states and one absorbing state. Mathematically, the survival function $S(t)$ of a PH distribution could be modeled as

$$S(t) = (\beta_1, \ldots, \beta_n)\exp(\mathbf{A}t)(1, \ldots, 1)',$$

where the β_j and the a_{ij} elements in coefficient matrix $\mathbf{A}$ are parameters. Conceptually, a PH distribution may be represented as the solution to a stochastic compartmental model. Clearly the exponential distribution in (2.3) is a simple PH distribution, and the Erlang distribution is also a PH distribution because it may be regarded as a survival time in a sequence of identical compartments.

Consider now the following general assumption:

ASSUMPTION. Let the retention time R have some arbitrary nondegenerate distribution.

The specification of a nondegenerate distribution rules out R being merely a fixed time delay. Neuts (1981) has shown that *any* nondegenerate distribution of a positive variable may be represented as a PH distribution. Therefore, the previous assumption may be restated, with no loss of generality, as:

ASSUMPTION. Let the retention time R follow some phase-type distribution.

This mathematical transformation enables one to represent any generalized one-compartment models as an expanded stochastic linear system described in Section 4.

One problem using PH distributions is that PH representation of a given distribution may not be unique. Also, the selection of some particular PH distribution to describe an arbitrary, completely specified distribution remains a challenging problem. Johnson and Taaffe (1989) prove that the first k moments of any nondegenerate distribution can be matched by a mixture of Erlang distributions of sufficiently high common order. They also illustrate a method for matching the first three moments of a distribution. Many other approaches are also possible and under current investigation, e.g., Faddy (1989) discusses the use of a particular class of PH distributions for population modeling.

VARIATION. The incorporation of Poisson immigration is immediate using results of Section 4.

ILLUSTRATION (*Mercury accumulation in fish – continued*). Clearly, numerous parametric and nonparametric models could be fitted to the mercury accumulation data. One mechanistic concept which differs from previous models is the assumption that *all* newly introduced mercury is initially delayed for metabolic processing, and hence has an 'age-dependent' hazard rate with shape characteristics given by the Erlang distribution in (5.4). This would differ qualitatively from the model of no initial delay in Section 2, and the model of the conceptual storage of *part* of the new material in Section 4. This proposed one-compartment Erlang model did not improve the fit of data for fish 5, but did reduce the MSE substantially for 6 of the 21 fish. This fish in this subset tended to be relatively small in size.

6. Multicompartment models based on arbitrary retention time distributions

Consider now the multicompartment generalization of the survival models in Section 5. The overall objective for such generalization may be described using a colorful analogy stated by Cohen (1980). In reviewing some previous work in the preceding five sections, Cohen suggested that some basic issues were neglected. He writes: 'First, how does one decide what are the compartments? To carve a chicken nicely, one must cut at the joints. How does one find the joints of an ecosystem? Second, is there any evidence to support, or any procedure to test, the memoryless or ahistorical assumption built into every compartmental model?' The first sections of this chapter divided the 'chicken' on the basis of simplifying mathematical assumptions which might not coincide with the natural joints. Section 5 analyzed the 'chicken' as a whole. The purpose of this section is to generalize the mathematical tools such that the 'chicken' could be divided at any arbitrary points, including at the desired natural joints. In order to achieve this objective, the assumption of a memoryless process is abandoned. The new approach addresses both of Cohen's concerns.

From a modeling perspective, the principal objective of survival analysis in Section 5 is to find nonexponential families of survival distributions which are mathematically tractable and correlate well with observed data. On the other hand, the principal objective of the standard multicompartment models in Section 4 is very different; they aim not only to describe the observed data but also to provide a rough, mechanistic description of the process underlying the data. Of course, one chief limitation of the previous multicompartment models is their assumption of exponential retention time distributions. The generalized multicompartment models in this section are the combination of the previous approaches. A mechanistic system of compartments is envisaged for the particles transferring *between* the compartments. However, the processes determining the retention time of the particles *within* each compartment are viewed as being more complex than the previous simple conceptualizations leading to constant hazard rates. The present multicompartment generalization does not attempt to describe the mechanics of the processes within each compartment, but rather to describe their net observed effect through age-varying hazard rates and nonexponential retention time distributions. In summary, the present approach attempts to characterize fully the mechanistic flow pattern between compartments (e.g., using chicken joints) and to use empirical models with the smallest possible number of parameters to describe the within-compartment processes.

A general stochastic framework for such modeling is the semi-Markov process which was originally proposed in compartmental context by Purdue (1975), was developed in an ecosystem context by Purdue (1979) and Marcus (1979), and has more recently been investigated by Mehata and Selvam (1986). The semi-Markov model assumptions are as follows:

ASSUMPTIONS. Let

(1) R_j retention time have the arbitrary distribution F_j for $j = 1, \ldots, n$; and

(2) α_{ij} be an age-invariant transfer probability from j to i.

The α_{ij} transfer probabilities correspond to the p_{ij} given in (4.1). Therefore the mechanism determining the sequential location of particles in the compartmental structure is the same as that given in Section 4; only the retention time distributions within the compartments are different. In principle, by using the PH distribution concept, compartmental submodels may be used to estimate the underlying retention time distributions within each generalized compartment. This results in an expanded compartmental model which may be used for fitting the data.

This new approach to modeling using generalized compartments may be summarized as follows:

MODELING APPROACH. (1) Represent the underlying mechanistic model (e.g., ecological or environmental structure) with a set of generalized compartments (e.g., spatial or trophodynamic components) and their interrelationships (i.e., flow diagram).

(2) If the form of the $F_j(t)$ distributions are known, express them using PH distributions. If the forms are not known, estimate them using PH distributions. In either case, specify the retention time distribution using submodels of pseudo-compartments.

(3) Estimate the parameters and solve for the occupancy probabilities $\mathbf{P}^*(t)$ of the pseudo-compartments in the expanded linear system, i.e., use (4.3) to find

$$\mathbf{P}^*(t) = \exp(\mathbf{K}^* t). \tag{6.1}$$

(4) Transform to obtain the occupancy probabilities $\mathbf{P}(t)$ of generalized compartments, i.e., use linear combinations to find

$$\mathbf{P}(t) = \mathbf{A}\mathbf{P}^*(t)\mathbf{B}, \tag{6.2}$$

where $\mathbf{A}$ and $\mathbf{B}$ are matrices of zeroes and ones.

The approach is illustrated below for the shrimp migration problem discussed in Matis et al. (1992a) where the retention time distributions were restricted to the Erlang distribution. In some applications, the retention time distribution has a 'long tail' which is not adequately approximated using an Erlang distributions. For example, Matis and Wehrly (1990) use a sequence of two Erlang distribution to generate the long-tailed distribution describing the retention time of calcium in human bone. Research is in progress to develop restricted families of parameter-parsimonious PH distributions to approximate certain long-tailed retention time distributions such as the inverse Gaussian. Research is also in progress to incorporate such approximations into KINETICA to provide user-friendly computer software to implement this general compartmental modeling strategy.

ILLUSTRATION. *Migration of marine shrimp – continued.* The shrimp data in Grant et al. (1991) which were fitted using a standard multicompartment stochastic model were reanalyzed in Matis et al. (1992a) using the present generalized approach. The model was developed as follows:

Step 1. The two spatial subdivisions of the lake defined the compartments as before, with possible transfers between the compartments but departures only from the second.

Step 2. The generic 'transfer' of shrimp between compartments involves physical movement, or migration, with an associated travel time. Hence a newly-arrived shrimp, might have a different transfer probability than a shrimp which has already been in the compartment for some time for many reasons. One is that a newly arrived shrimp might have an initial, local exploratory behavior before proceeding with migratory activity. Another is that the newly arrived shrimp might face a longer travel distance. Clearly, any such relationship between the elapsed time, or 'age', of the shrimp in a compartment and its conditional transfer probability would invalidate the assumption of exponential retention times in Section 4.

Consequently, it was assumed that shrimp transfers might have an initial delay which would gradually diminish with increasing time spent in the compartment. The Erlang hazard rate in (5.4) has such qualitative properties and was utilized for this generalized model. The shape parameters, n_1 and n_2 respectively, for the Erlang distribution of each compartment were determined empirically by fitting models with combinations of small integer values of n_1 and n_2, and then selecting the model with the smallest MSE. The model with $n_1 = 5$ and $n_2 = 1$, for which MSE = 13.77, was selected. The apparent initial delay in compartment 1 only could be interpreted as reflecting the larger size, and slower flow in parts, of this compartment.

Step 3. The expanded linear system has 6 pseudo-compartments for each site of origin, for a combined 12 compartments for the simultaneous fitting of the four data sets. The $\mathbf{K}^*$ matrix has pairs of complex eigenvalues, hence the $P_{ij}^*(t)$ solutions have damped oscillations. The estimates from KINETICA for the initial concentrations are $C_{11}(0) = 1114.1$ and $C_{22}(0) = 50.6$, for the scale parameters are $\lambda_1 = 4.19$ and $\lambda_2 = 1.28$, and for the conditional transfer probabilities are $\alpha_{02} = 0.61$ and $\alpha_{12} = 0.39$.

Step 4. The estimated occupancy probabilities are obtained by combining solutions for the pseudo-compartments, as indicated in (6.2). The solutions are given in Matis et al. (1992a). Residence time and related moments of the pseudo-compartments could also be combined in a similar manner. By way of comparison, under this model a shrimp introduced into area 1 is expected to visit compartments 1 and 2 an average of 2.56 times apiece, with mean single durations of 1.19 and 0.78 respectively, for a combined expected residence time of $(3.05 + 1.99) = 5.04$ days. The comparable figures for a shrimp released in area 2 are averages of 1.56 and 2.56 visits, respectively, for a total expected time spent of $(1.86 + 1.99) = 3.85$ days.

In summary, this generalized compartmental model has a more realistic conceptual basis for shrimp migration. It also provides a better fit for the data, though obviously there is no claim that it is either the best fitting nor a unique PH model.

7. Compartmental models with births

There is a rich literature on modeling population dynamics, yet as Levin (1986) observes 'it has only been in the last few years that theoretical studies have begun to take account of the fundamental importance of geographic distribution.' An early classic paper in geographic distribution is Skellam (1951) which proposes a diffusion model for describing population dynamics over a homogeneous area. Renshaw (1991) reviews the stepping-stone model for describing population dynamics over a heterogeneous environment. The stepping-stone model may be regarded as a

Differential equations for higher order cumulants are also given, which may be solved recursively.

Wehrly et al. (1993) propose constructing approximate bivariate distributions of $(X_1(t), X_2(t))$ by combining negative binomial distributions. Let the notation $\mathrm{NB}(k, p)$ denote the negative binomial distribution with probability mass function

$$p(x) = \binom{x+k-1}{x} p^k q^x, \quad \text{for } x = 0, 1, 2, \ldots, 0 < p < 1, \; q = 1 - p, \; k > 0.$$

Let Y_1, Y_2 and Y_3 be independent negative binomial random variables which may be combined to model the bivariate random variables (X_1, X_2). As a specific example, let $Y_1 \sim \mathrm{NB}(k_1, p_1)$, $Y_2 \sim \mathrm{NB}(k_2, p_2)$ and $Y_3 \sim \mathrm{NB}(k_3, p_2)$ and consider the partial sums $X_1 = Y_1 + Y_2$ and $X_2 = Y_2 + Y_3$. The cumulants for this particular model of (X_1, X_2) are

$$\kappa_{10} = \frac{k_1 q_1}{p_1} + \frac{k_2 q_2}{p_2}, \qquad \kappa_{01} = \frac{(k_2 + k_3) q_2}{p_2}, \tag{7.5}$$

$$\kappa_{20} = \frac{k_1 q_1}{p_1^2} + \frac{k_2 q_2}{p_2^2}, \qquad \kappa_{11} = \frac{k_2 q_2}{p_2^2}, \qquad \kappa_{02} = \frac{(k_2 + k_3) q_2}{p_2^2}.$$

The values of the first and second order cumulants of the models in (7.1) or (7.4) are obtained by solving the differential equations and then equated to the cumulants in (7.5). One could then solve for the five parameter values, k_1, k_2, k_3, p_1 and p_2, which define the bivariate approximation. The higher order cumulants may be used to assess the fit of an approximation, and hence provide a basis for selecting the best approximation from a class of bivariate distributions.

APPLICATION. *Spread of Africanized honey bees – AHB.* Otis (1980) presents data on AHB populations which he observed on established, unmanaged colonies in French Guiana from August 1976 to July 1977. The data, also reported in Otis (1991), are regarded as the best available data on the dynamics of AHB population growth, and arguably as the best data on the demography of any social insect. Based on his data, Otis develops a detailed stochastic simulation model for AHB population size from August to July. Let $X(t)$ denote size at time t. Otis' model starts with $X(0) = 1$ (on August 1) and uses Monte Carlo methods to simulate the birth, death and migration experience of the original colony and its successor colonies. The simulation model obviously contains some necessary assumptions concerning the dynamics of the system. However the model incorporates all of the qualitative properties observed in French Guiana, and it is free of assumptions motivated primarily by mathematical tractability.

Wehrly et al. (1993) use a expanded compartmental model represented in (7.4) to approximate the bivariate distribution of population sizes. The model was adapted to incorporate as much ecological realism as possible in the following ways. The observed seasonality during the 10-month interval was incorporated by subdividing the interval into three periods, and using piece-wise constant rates. The effect of multiple births was approximated by increasing the birth rate to yield the desired cumulative effect at the end of each period. However, the time intervals between events were assumed to be exponentially distributed, as is common practice, even though the assumption is inconsistent with the observed data.

The compartment midpoints were assumed to be 100 km apart. The model was parameterized using Otis' observed estimates, and the cumulant functions are obtained. As an example of the solutions, the mean value functions for the first period are

$$\begin{aligned}\kappa_{10}(t) &= -0.295 + 0.706\exp(0.1215t) + 0.589\exp(0.2805t),\\ \kappa_{01}(t) &= 0.117 - 0.706\exp(0.1215t) + 0.589\exp(0.2805t).\end{aligned} \tag{7.6}$$

Similar, though longer, sums-of-exponentials solutions are available for all cumulants.

The solutions at the end of the first period may be substituted as initial values into the equations for the subsequent period. Figure 5 illustrates the solution for the mean values over the 10-month period. The mean value function of the first compartment closely matches Otis' population results at a single site.

A wavefront may be defined to be at the compartmental midpoint when the mean value is 1.0. For the present example, the wavefront is at compartment 2, and hence has traveled 100 km, at $t = 4.55$ mo. The resulting velocity of 22 km/mo is consistent with estimates of AHB spread in South America.

The approximate multivariate distribution was obtained by solving equations (7.5). The resulting distribution at $t = 10$ is given in Figure 6. One can show that Figure 6 represents an adequate approximation for the underlying bivariate distribution of population sizes from (7.4) by comparing the third and fourth order cumulants of the approximation to the exact ones derived from (7.4).

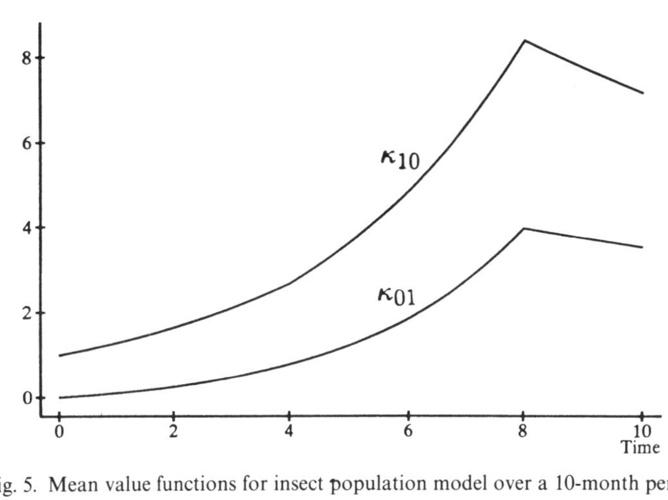

Fig. 5. Mean value functions for insect population model over a 10-month period.

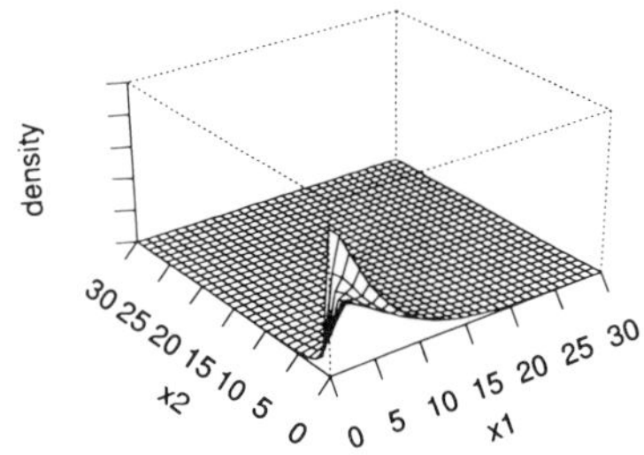

Fig. 6. Three-dimensional plot of approximate bivariate distribution for population sizes, X_1 and X_2, at the end of 10 months.

The strength of this new Markov model is that it describes the spread of an insect population in time and space, using only a few key population parameters relating to reproduction, migration and death. Hence the model is easy to use to predict the effects of various possible management programs. The authors feel that it will become a useful complement to some large-scale computer simulation models in current use for predicting insect spread.

8. Other compartmental models

Most of the applications previously described relate to ecological systems in which compartments represent various spatial entities. The standard multicompartment models are widely used for this purpose, but the models have stringent assumptions. As previously noted, Section 6 presented an approach for relaxing the assumption of constant, age-invariant hazard rates. Recent research has also focused on relaxing two other key assumptions, namely that of independent particles and that of identical particles. Small-scale models have been developed to address each of these areas, and convenient tools for their widespread application are under current investigation.

Consider first the assumption of independent particles in (4.1). It has long been recognized that when the initial numbers of particles $X_i(0)$ are very large, the coefficients of variation of the total numbers of independent particles in individual compartments will be small. Among several models constructed to give larger coefficients of variation are the urn models of Bernard (1977) and Purdue (1981), and the clustering models of Matis and Wehrly (1981). Ball and Donnelly (1988) relate variability in compartmental models to interparticle correlations. Ensor (1989) generalizes the clustering model to a two-compartment system.

Consider also the assumption of identical particles, in which each particle has the same rate coefficients in Section 4 or identical hazard rates in Section 6. Matis and Wehrly (1979) develop a one-compartment model with heterogeneous particles, and illustrate the resulting mixture model to describe vanadium depuration in marine organisms. Matis et al. (1989) extend the results to include one- and multicompartment models for heterogeneous particles with Erlang retention time distributions and gamma mixing distributions. The effect of heterogeneity is illustrated with passage data from ruminants. Other applications concerning comparative digestive physiology and habitat-related nutritional requirements of ruminant and herbivore species, including wildlife, have followed (see, e.g., Huston et al., 1986; Langer, 1984; Spalinger and Robbins, 1992). The model is extended to other mixture distributions by Ebaseh-Onofa and Matis (1992).

9. Comments

Cohen's (1980) review of some previous applications of compartmental models has already been noted. In his review, Cohen identifies three issues which he felt were largely neglected; namely (1) deciding what are the compartments in any particular application, (2) justifying the assumption of exponential compartments, and (3) testing the predictions of compartmental models. The research through the eighties and early nineties reported in this paper addresses many of these concerns. Of course, the exact definition of what constitutes a compartment for any given application is still largely an art requiring expert judgment of subject-matter specialists. However, there is now a much larger methodological toolbox from which one may choose a compartmental representation of an ecological or environmental system. Perhaps the most striking addition is methodology which enables one to utilize compartments with any desired retention time distribution. All of the models presented are amenable for statistical analysis, indeed most may be fitted to observed data using such user-friendly software as KINETICA. Hence a broad spectrum of statistical inference, including testing model predictions, is possible, as illustrated in many of the examples and references. New models and applications are continually being developed, and the authors expect an increasing use of compartmental models for describing ecological and environmental systems.

References

Allen, D. M. and J. H. Matis (1992). KINETICA, a program for kinetic modeling in biological sciences. Technical Report #317, Department of Statistics, University of Kentucky, Lexington, KY.

Ball, F. and P. Donnelly (1988). A unified approach to variability in compartmental models. *Biometrics* **44**, 685–694.

Barber, M. C., B. C. Patten and J. T. Finn (1979). Review and evaluation of input–output flow analysis for ecological applications. In: J. H. Matis, B. C. Patten and G. C. White, eds., *Compartmental Analysis of Ecosystem Models*. International Cooperative Publishing House, Fairland, MD.

Bernard, S. R. (1977). An urn model study of variability within a compartment. *Bull. Math. Biol.* **39**, 463–470.

Chiang, C. L. (1980). *An Introduction to Stochastic Processes and Their Applications*. Krieger, Huntington, NY.

Cohen, J. E. (1980). Review of compartmental analysis of ecosystem models. *Quart. Rev. Biol.* **55**, 453–454.

Ebaseh-Onofa, B. O. and J. H. Matis (1992). Compartmental models with Erlang distributed residence times and random rate coefficients. *Bull. Math. Biol.* **54**, 929–938.

Ensor, J. E. Jr. (1989). A Two-compartment, Irreversible Flow Model with Clustering. Ph.D. Dissertation. Texas A&M Univ. College Station, TX.

Faddy, M. J. (1989). Compartmental models in the analysis of populations. In: L. McDonald, B. Manly, J. Lockwood and J. Logan, eds., *Estimation and Analysis of Insect Populations.* Springer, Berlin.

Godfrey, K. (1983). *Compartmental Models and Their Applications.* Academic Press, New York.

Grant, W. E., J. H. Matis and T. H. Miller (1991). A stochastic compartmental model for migration of marine shrimp. *Ecol. Model.* **54**, 1–15.

Huston, J. E., B. S. Rector, W. C. Ellis and M. L. Allen (1986). Dynamics of digestion in cattle, sheep, goats and deer. *J. Animal Sci.* **62**, 208–220.

Jacquez, J. A. (1985). *Compartmental Analysis in Biology and Medicine.* 2nd ed., Univ. of Michigan Press, Ann Arbor, MI.

Johnson, M. A. and M. R. Taaffe (1989). Matching moments to phase distributions: Mixtures of Erlang distributions of common order. *Comm. Statist. Stochast. Models* **5**, 711–743.

Johnson, N. L. and S. Kotz (1970). *Continuous Univariate Distributions.* Wiley, New York.

Kalman, R. E., P. L. Falb and M. A. Arbib (1969). *Topics in Mathematical System Theory.* McGraw-Hill, New York.

Langer, P. (1984). Anatomical and nutritional adaptations in wild herbivores. In: F. M. C. Gilchrist and R. I. Mackie, eds., *Herbivore Nutrition in the Subtropics and Tropics.* Science Press, Craighall, South Africa.

Levin, S. A. (1986). Population models and community structure in heterogeneous environments. In: T. A. Hallam and S. A. Levin, eds., *Mathematical Ecology.* Springer, Berlin.

Marcus, A. H. (1979). Semi-Markov compartmental models in ecology and environmental health. In: J. H. Matis, B. C. Patten and G. C. White, eds. *Compartmental Analysis of Ecosystem Models.* International Cooperative Publishing House, Fairland, MD.

Matis, J. H. and B. C. Patten (1979). Environ analysis of linear compartmental systems: The static, time invariant case. *Proc. 42 Session Internat. Statist. Inst.* **47**, 527–564.

Matis J. H. and T. E. Wehrly (1979). Stochastic models of compartmental systems. *Biometrics* **35**, 199–220.

Matis, J. H. and T. E. Wehrly (1981). Compartmental models with multiple sources of stochastic variability: The one-compartment models with clustering. *Bull. Math. Biol.* **43**, 651–664.

Matis, J. H. and T. E. Wehrly (1990). Generalized stochastic compartmental models with Erlang transit times. *J. Pharmacokin. Biopharm.* **18**, 589–607.

Matis, J. H., B. C. Patten and G. C. White, eds. (1979). *Compartmental Analysis of Ecosystem Models.* International Cooperative Publishing House. Fairland, MD.

Matis, J. H., T. E. Wehrly and C. M. Metzler (1983). On some stochastic formulations and related statistical moments of pharmacokinetic models. *J. Pharmacokin. Biopharm.* **11**, 77–92.

Matis, J. H., T. E. Wehrly and W. C. Ellis (1989). Some generalized stochastic compartment models for digesta flow. *Biometrics* **45**, 703–720.

Matis, J. H., T. H. Miller and D. W. Allen (1991). Stochastic models of bioaccumulation. In: M. C. Newman and A. W. McIntosh, *Metal Ecotoxicology: Concepts and Applications.* Lewis Publishers, Ann Arbor, MI.

Matis, J. H., W. E. Grant and T. H. Miller (1992a). A semi-Markov process model for migration of marine shrimp. *Ecol. Model.* **60**, 167–184.

Matis, J. H., T. E. Wehrly, D. M. Allen and G. W. Otis (1992b). On using stochastic compartmental models for describing insect dispersal: 1. The case of univariate distributions for Markov process models. In: J. Eisenfeld, M. Whitten and D. S. Levine, eds., *Biomedical Modeling and Simulation.* Elsevier, New York.

Mehata, K. M. and D. D. Selvam (1986). A class of general stochastic compartmental systems. *Bull. Math. Biol.* **48**, 509–524.

Neuts, M. F. (1981). *Matrix-Geometric Solutions in Stochastic Models: An Algorithmic Approach.* Johns Hopkins Univ. Press, Baltimore, MD.

Otis, G. W. (1980). The Swarming Biology and Population Dynamics of the Africanized Honey Bee. Ph. D. Dissertation. University of Kansas, Manhattan, KS.

Otis, G. W. (1991). Population biology of the Africanized honey bee. In: M. Spivak, D. J. C. Fletcher and M. D. Breed, eds., *The "African" Honey Bee.* Westview, Boulder, CO, 213–234.

Patten, B. C. (1992). Energy, emergy and environs. *Ecol. Model.* **62**, 29–69.

Patten, B. C. and J. H. Matis (1982). The water environs of Okefenokee Swamp: An application of static linear environ analysis. *Ecol. Model.* **16**, 1–50.

Purdue, P. (1975). Stochastic theory of compartments: An open, two-compartment reversible system with independent Poisson arrivals. *Bull. Math. Biol.* **37**, 269–275.

Purdue, P. (1979). Stochastic compartmental models: A review of the mathematical theory with ecological applications. In: J. H. Matis, B. C. Patten and G. C. White, eds., *Compartmental Analysis of Ecosystem Models*. International Cooperative Publishing House, Fairland, MD.

Purdue, P. (1981). Variability in a single compartment system: A note on S. R. Bernard's model. *Bull. Math. Biol.* **43**, 111–116.

Renshaw, E. (1986). A survey of stepping-stone models in population dynamics. *Adv. Appl. Prob.* **18**, 581–627.

Renshaw, E. (1991). *Modeling Biological Populations in Space and Time*. Cambridge University Press, New York.

Skellam, J. G. (1951). Random dispersal in theoretical populations. *Biometrika* **38**, 196–218.

Spacie, A. and J. L. Hamelink (1985). Bioaccumulation. In: G. M. Rand and S. R. Petrocelli, eds., *Fundamentals of Aquatic Toxicology*. Hemisphere, Washington, DC.

Spalinger, D. E. and C. T. Robbins (1992). The dynamics of particle flow in the rumen of mule deer and elk. *Physiol. Zool.* **65**, 379–402.

Stuart, A. and J. K. Ord (1987). *Kendall's Advanced Theory of Statistics*, Vol. 1. Distribution Theory. 5th ed., Oxford Press, New York.

Wehrly, T. E., J. H. Matis and G. W. Otis (1993). Approximating multivariate distributions in stochastic models of insect population dynamics. In: G. P. Patil, C. R. Rao and N. P. Ross, eds., *Multivariate Environmental Statistics*, Elsevier, Amsterdam.

G. P. Patil and C. R. Rao, eds., *Handbook of Statistics, Vol. 12*

Environmental Remote Sensing and Geographic Information Systems-Based Modeling

Wayne L. Myers

Abstract: With increasing scientific knowledge of complex environmental processes, it becomes increasingly inappropriate to base decision making in environmental management on estimation by empirical areal aggregation of data collected at sampling points. Environmental processes typically operate in terms of multiple factors with strong spatial and temporal inter-dependencies that make specification of landscape structure essential for inference. Geographic information systems (GIS) constitute a rapidly improving computer technology for compiling and analyzing the needed landscape information. Digitally processed remote sensing provides repetitive and areally comprehensive spatial information on environmental variables for augmenting GIS databases to serve the needs of monitoring. Spatial/temporal environmental process models mold the GIS data into knowledge based estimates that support decision making for environmental management in full landscape context. These complementary technologies are introduced in the context of hydrologic process modeling, and discussed relative to need for integrated future development.

1. Coverage

The coverage of this chapter is substantially transdisciplinary, and the technologies of the several disciplines are charged with various shortcomings. Therefore, it is perhaps in order to engage in a little contextual stage-setting for such broadband commentary. The Office for Remote Sensing of Earth Resources (ORSER) in the Environmental Resources Research Institute (ERRI) at Penn State University constitutes centerstage in this regard. ORSER has been immersed in remote sensing technology since NASA's ERTS/Landsat program provided the impetus for remote sensing to emerge from the shadowy domain of defense as a major civilian technology in the earth resource arena. ORSER developed one of the first computer software systems for analysis of remotely sensed digital image data. While pursuing a continuing quest for environmental applications of remote sensing, ORSER has also continued its work on prototype computer software development. Ideas arising from such prototype work have frequently found homes in commercial or public domain software packages as the technology has become operationalized.

ORSER was among the early sources of suggestions that remote sensing was not applicationally complete in itself, but demanded a companion technology encompassing capabilities that were eventually subsumed under the banner of geographic information systems (GIS). ORSER routinely develops new spatial information handling capability not yet available in commercial software according to the needs of its research projects. Such capabilities may be integrated with commercial software through toolkits, or may operate in a bridging mode between various software systems having differing domains of strength.

As part of an intercollege research institute, ORSER has the advantage of being fundamentally interdisciplinary in its research work. ORSER provides a setting where spatialists from different academic disciplines can join forces in the interest of addressing environmental problems that defy the rather arbitrary envelopes of science and technology in academia. The most recent vanguard of spatialists to join the corps of the ORSER cooperative consists of environmental modelers, particularly those concerned with hydrologic processes. Such processes are fundamentally spatial, and spatial information must reflect a certain level of landscape realism in order for the models to be parameterized appropriately.

In addition to being part of this interdisciplinary dynamic as a codirector of ORSER, the author recently experienced two years in India as technology advisor for USAID on forestry and natural resource sustainability. This brought a direct encounter with immediate need to mobilize knowledge and technology (hereafter referred to jointly as tecknowledgy) in the interest of both humanity and environment. The relative incapacity of contemporary technological separatism to support such mobilization became a source of enormous frustration. If the current time and investment required to deploy complex technologies in combination should continue, the pace of environmental degradation on a global scale presents prospects of compounding disaster. Glimmers of hope appear to emanate more from trainable tecknowledgy as knowledge-based computer systems than from restructuring education as training of technologists. Computer intelligence likewise presents the only prospect of transcending the inherent limits of human individual capability for learning. Artificial intelligence has thus become a component of ORSER's tecknowledgy arsenal.

Contemporary tecknowledgy shortfall is well expressed in some favorite phraseology of Prof. G. P. Patil, coeditor of this volume. He calls attention to a recurrent cycle of 'no information → new information → non-information' relative to environmental issues. Calls for more data do seem to be an incessant refrain surrounding environmental issues. It is illogical, however, that any information should become non-information if it is correctly set in spatial–temporal context. Any information about prior status of environment should remain relevent to the future since the earth system has intrinsic memory in the sense of its process dynamics. Unfortunately, the memory aspect is often dominant – as witness the persistence of pollutants and the time trajectory of ecological succession and recolonization. The not-so-simple fact is that humanity is terribly inept at extracting meaning (knowledge) from spatial information. When information is distilled from data, the spatial–temporal weave that forms a potential fabric for knowledge is too often ignored or destroyed.

There are ample grounds to argue that instead of having data deficits, we are virtually drowning in unused or underused data. The remotely sensed data archives of NASA, NOAA, and counterparts abroad contain truly astounding amounts of information in terms of bit content. The advent of NASA's latest endeavor (EOS = earth observation system) will further boost the rate of spatial data flow by several orders of magnitude. The question is how much of the latent information content we will succeed in extracting from these virtual volcanos of data; and even more so, how much of what is extracted will acutally be mobilized in the interest of human ecosystems.

While the scope of this chapter may seem inordinately broad, it is only by melding these several technologies to produce an amalgam of knowledge that we can hope to salvage the environmental remains and reduce the onslaughts on Gaia. Any fault-finding herein is not meant to depreciate the achievements of science and technology, but to suggest possibilities for realizing greater return on investment. The intent is both to inform and inspire. Statistics has a much greater potential role relative to environment and ecology than it now plays, but filling that role is going to require a good deal of substantially Bayesian behavior.

2. Systematics of geographic information

Spatial data is sufficiently central to the business of the world that abilities to manage and map spatial data have considerable commercial value. In the course of evolution for spatial data handling software, this commercial aspect asserted itself early-on. In the author's opinion, such commercialism has proven second only to the black hole of national security classification in retarding both development of this technology and its integration with related technologies. Despite much clamor to the contrary, contemporary versions of what are held to be geographic information systems (GIS) have as yet progressed little beyond the database management stage. In establishing perspective on which to build later in this chapter, it becomes a first order of operations to characterize commonality and differences among systems at the database level. Constructive views of databases then emerge as higher level issues, leading to the meaning of models and models of meaning.

2.1. Placing things in perspective

Environmental processes operate in or around landscapes which involve high information content in terms of bits. The very first concern in mounting massive computer capture of landscape data is to arrive at elementary units of structure which constitute atomic/molecular equivalents for landscape compounds. It is on the basis of such elementary units or primitives that landscape information can be dismantled for transfer to computers. There is a classic dichotomy at this level of primitives. One branch of the dichotomy uses so-called 'rasters', and the other deals in 'vector' elements.

On the raster side of the GIS mainstream, there occurs a regular tessellation (partition) of a reference surface into cells or pixels. The cells are usually, but not always, conceived as forming a square or possibly rectangular grid spanning the reference surface. There exists a small cult which holds that there must bee some fundamental hexagonal essence in planar space by which honeycombs constitute canonical prototypes for tessellations. In terms of cell centerpoints, however, a hexagonal tessellation can also be viewed as simply a staggered rectangular raster that skips a beat. The raster regime follows the precedents of electronic image engineering. A notable aspect of a raster is that it focuses on fragmentation of (surficial) space itself, rather on space occupying features. The space fragments serve as receptacles for recording the occurrence and character of spatial features or phenomena. Even if a feature is absent from a fragment, at least a bit of information is required to indicate the absence. The raster fragments have regularity only over a particular reference space. Under space transformations, the raster fragments can change shape and may also become irregular. Another notable concerning rasters is that topology and normally also position remain implicit.

In most important respects, the vector approach is the converse of raster. Instead of focusing on the space in which things occur, vector elements are concerned with things that occur in space. In contrast to the highly implicit character of rasters, most aspects of vector elements are explicit, including position. The term 'vector' arises from the fact that boundaries of areal features and traces of linear features are represented as directed line segments in terms of a reference coordinate system. Some aspects of topology may, however, be left implicit. Adjacency is one of the things that is sometimes left wholly or partially implicit. The vector primitives are isolated points and 'chains' of line segments representing (usually partial) linears or boundaries.

Although tessera are tiny tiles in the dictionary sense, GIS cadres have drawn a distinction between tessellation and tiling. In the jargon of GIS, tiling concerns zonation of a spatial universe in the interest of controlling size of data files. Thus a tiling may be superimposed on a tessellation to accommodate constraints of computer memory media and file capacity. These same constraints are faced with both rasters and vectors, so the tiling issue is more concerned with places for information than with mode of placement.

The bicameral character of spatial placement in GIS demands some mention of relative advantage. With respect to economy and precision of spatial representation, there is vector advantage. Rasters lack intracellular precision for location. Whether this becomes important depends on both size of cell and application. There is intrinsic inflexibility in a raster, however, since all types of features become subject to the same resolution limit. Except for very high frequency spatial variation, vectors also tend to consume less computer memory media. The primary appeal of a raster lies in simplicity of concept. This conceptual simplicity carries over into simplicity for certain types of data processing operations.

2.2. Attributes of GIS

When the placement of features has been determined either by raster or vector, the next concern is to attach attribute data (the 'variates' of statistics) that characterize the features categorically and/or quantitatively. Such attachment may be either direct or indirect in both raster and vector modes. Indirection is relatively routine for the vector context. Indirection has been used quite selectively in the raster context; there being no apparent reason for the selectivity other than reluctance to complicate the conceptual simplicity of raster databases.

Although this need not be so, indirection for attributes has been associated with multiplicity of attributes for individual spatial units. In the vector context, the explicit nature of the spatial unit usually implies that the unit will have several associated attributes that arise from the unit taxonomy. For example, taxonomy and segregation of soil units implies a degree of intraunit uniformity relative to texture, fertility, permeability, load bearing capacity, etc. A single attribute indicating soil type can serve as an index to a table of other attributes associated with the soil type. Different spatial units can thus share a common table of soil property attributes, thereby reducing redundancy of information storage. In the case of soil, indirection makes equally good sense for rasters as vectors. Each raster cell can have a soil type code, which serves as an index to a single lookup table of soil properties by soil type.

Some of the issues surrounding indirection are more subtle than the foregoing paragraph would indicate. One such issue is need to delink spatial occurrence from attribute indexing. For example, several polygons of the same soil type may occur in a given region. If soil type is the only attribute attached directly to a soil polygon, then spatial analysis (search) is the only way to differentiate the several occurrences of the same soil type. Since spatial analysis is more complex and time consuming than indexing, such a database schema carries inherent disadvantages for many conceivable applications.

The foregoing disadvantage is negated by second order indirection, whereby the thing attached directly to a spatial unit is only a unit identification number. This provides entry to an indexing table which gives soil type by polygon identification number. Soil type thus becomes a 'relation' for linkage of attribute tables in a relational database. Assuming row-wise entry to the indexing table, there will be as many rows in the indexing table as there are soil polygons. Attaching both polygon ID and soil type directly to the spatial unit thus involves the same amount of data storage as does the indexing table. The indexing table, however, has the advantage of invertibility. Only the indexing table need be accessed to construct an inverted index specifying all polygons of a particular type. With direct attachment it becomes necessary to access every primary polygon entry for constructing an inverted index. Except for very small databases, the time required for the two different access scenarios becomes substantial.

A somewhat different issue of attribution in relation to space is very important. This is an issue that rasters presently avoid in a rather unsophisticated sort of way since it concerns explicit constructs. It is the issue of joint attribution for spatial units when one attribute does not necessarily imply another attribute. For the soil example above, the nature of the taxonomy was such that several attributes were jointly determined. Although not quite so simple, the issue can be described in terms of relatively stable spatial hierarchies. One such hierarchy is that of farms and fields. Fields always 'belong to' a particular farm, and since the field is the unit of management it is also normally the unit of environmental consequence. There are alternatives in constructing a database schema for fields and farms. One alternative is for the field to be the sole spatial unit, with a particular farm being a shared attribute among several fields. Working with farms apart from fields thus involves a scenario of indexing all fields with the same farm attribute and doing the required spatial analysis to remove intrafarm field boundaries. A second schema involves entirely separate database components for fields and farms, while doing spatial intersection analysis for joint field/farm questions. A third schema is to have separate field and farm spatial entities with a complete suite of attributes for each that make the relations between fields and farms explicit so that only indexing is required to answer any question regarding fields and farms.

Contemporary geographic information systems generally force a choice of schema in dealing with issues of the type raised in the previous paragraph. Such issues introduce a larger question of need of accommodate multi-schematic structure in spatial information systems. This is an aspect that points out inadequacies of existing spatial software systems, and suggests new directions in system design.

2.3. Constructive concerns in GIS

Having considered the manner in which landscape information is dismantled for transfer to computerized geographic information systems, the next matter is one of reassembling those information elements as computer constructs that speak to environmental problems and enable environmental analysis. The preceding discussion of issues in attribution provides a good beginning since schema and constructs are overlapping concepts.

In terms of contemporary constructive capability, the raster context remains more simplistic than the vector. Many of the existing raster systems have virtually no constructive capability, with the only feature organization being 'layers' that are more or less synonymous with themes or variables. A layer consists of a complete set of cellwise values for a variable . If the variable codetermines other variables (as for soil type), there may be lookup tables that embody several virtual layers. Layers may be 'compressed' where 'runs' of like-values occur along the row (or column) by using a run-flag/run-value/run-length encoding. For these relatively unsophisticated systems, even something as simple as locating a particular soil type will require a full search of a raster layer.

Raster systems can, however, be more substantially constructive. The most popular modality for doing so is through quadtrees. Quadtrees are a method of indexing through recursive symmetric decomposition. A reference region is successively quartered to levels where the pieces represent either occurrence or non-occurrence of a construct. Quadtrees serve to specify placement of a construct, and other modes of indexing can also serve this purpose. A system that admits constructs as quadtrees may or may not accommodate the building of compound constructs without making new quadtrees. This is the question of higher-order constructs as addressed subsequently in this section.

It is possible for a vector system to remain at the point and line segment level in its database while doing virtually all of its work analytically. The constant overburden of analysis would, however, make the system appear extremely ponderous. Therefore, most vector systems handle constructs in databases to some level. Like raster systems, vector systems can segregate information thematically as 'layers' or 'coverages'. These are essentially digital equivalents of map overlays on transparent material. Although there is no compelling reason to the contrary, most vector systems restrict their database constructs to formation within the layer.

The simplest of intralayer vector constructs is a chain of line segments. From there the progression moves to polygons and networks. A polygon is usually constructed as an indexed set of chains. It is, of course, possible to form a polygon as a single chain wherein the beginning and the end are coincident. Where polygons share common borders, however, the single chain approach involves redundancy of a sort that is most often deemed unacceptable. Networks can be constructed entirely in terms of chains, but a redundancy of convenience often casts networks as dual constructs involving both points and chains. Junctions for chains constitute the point component of the network. In network jargon, the junction points are called 'nodes'. The set of nodal points thus provide a sort of framework on which to hook the chains. A chain connecting two nodes is called an 'arc'. The adjacency aspect of topology is often made explicit by recording the polygons on either side of arcs as a kind of attribute.

Networks are quite flexible constructs for representing landscape information. They can portray transportation routes, drainage channels, zonations, and even surfaces. As a surface portrayal, the polygonal elements of a network become facets of the surface. Capability for constructing layered networks is often well received by environmental managers and scientists accustomed to thinking in terms of static map documents. When asked about information needs, their desires tend to run toward system queries that elicit a map as a response. Cartographic capability thus becomes a criterion of system suitability. Vector-based systems consistently outperform raster-based systems in the cartographic domain. Cartographic products from vector-based systems have the look of conventional maps, whereas spatial displays from raster-based systems are more in the nature of images.

When viewed against the familiar backdrop of map documents, the contemporary query-oriented GIS projects a good image. Customizing maps to issues of interest is only a first order benefit. Combinatorial maps coupling different themes can be created with relative ease, to a level that has been termed map algebra. Automated cartometry is also in the benefit package, with area and length determinations becoming routine. The 'analysis' capability often extends to creation of corridors, putting buffers around features, and even some tactical work such as finding low impact routes. By pre-GIS standards, this must indeed be a mature technology.

The adequacy of current GIS depends on the frame of reference. When the frame of reference is conventional maps, then the constructs are more than adequate. When the frame of reference is landscapes, then the constructs remain very primitive. Try asking a GIS about channel conformation in stream reaches or to find cove sites in complex terrain. Even such relatively simple concepts as hills, ravines, and toeslopes are likely to make GIS technicians reluctant to offer service. The solution will probably be to make a map with the system, have a human interpret the map, and then digitize the interpretations so that they become available via system query. Environmental processes hinge on rather complex constructs. When the goal is to model environmental process dynamics, the standing of the present GIS diminishes considerably.

Even in a relatively simple query context, current GIS can seem only weakly constructive if one takes the cup-half-empty view of the critic rather than the cup-half-full view of the advocate. The cause for concern is the tyranny of the database schema. The database schema must make appropriate provision for a particular query to be admissible. Furthermore, the query must be cast in a form that the system can interpret. Permissible casting of queries is also largely a function of database schema. Consequently, information does not become equally accessible to individuals from different disciplinary backgrounds. Utility of a system thus tends to depend in large part on those who stack its informational provisions on the shelves of the system. GISs do offer some recourse for schematic deficiency in the form of analytical capability. Analytical access to a system, however, presently remains very much the province of technicians trained in the idiosyncrasies of the particular system. In this sense, contemporary GIS remain very much database oriented while purporting to be analysis oriented.

So as to emphasize the positive rather than the negative, focus here now shifts from present deficiencies to needs and potentials for further development. The primary need is for adaptiveness in GISs whereby a system 'learns' by interaction with users. While a user's first encounter with a system may well entail need for a technical escort, familiarity should breed content rather than contempt. This demands much greater capability for imposing structure on the system dynamically in the course of interaction with users. In any given instance, it will remain necessary to make demands on the system be consonant with the structural state of the system. The outcome of such an encounter should, however, become a new system construct that is usable in subsequent encounters. Such an emergent construct requires a symbolic means of identification, whether this be a user-supplied term or icon. The system should also accommodate symbolic synonyms so that different users can cast queries in the terms with which they are most comfortable. In no case should it be necessary that a user descend to more primitive levels once ascendance has been accomplished in a prior instance.

A brief excursion into architecture of some existing GIS should serve both to indicate how constructs might be developed dynamically, and also why statisticians should not expect to conduct extensive environmental analysis in the context of conventional statistical software systems. Without addressing the question of whether popularity is indicative of virtue, it can be noted that a very popular GIS and a moderately popular GIS were both developed around existing relational database software packages. Limited spatial constructs such as polygons become entries in the database shell with attributes being handled as relations. In both cases, a companion software system was developed to handle the spatial placement information. The question thus arises as to why it was viewed as inappropriate to incorporate the placement information directly in the relational database shell. The answer lies largely in placement information occurring as lists of variable length, particularly for the vector context. Such information does not fit commodiously into the record-oriented tableaus of relational database shells. For similar reasons, conventional statistical software systems are inconvenient at best and more often impractical as platforms for handling large volumes of highly structured environmental spatial information that branches extensively into lists. In addition to being substantially listless, statistical software tends to have little in the way of relational capability and is geared toward relatively small volumes of data by GIS standards.

Concerning the need for higher order landscape constructs to be developed dynamically, neither the current locational nor relational sides of the GIS house appear to provide appropriate accommodation. Rather than being simple lists, these will likely be recursively structured lists of lists. It is software such a LISP (*l*ist *p*rocessing) which serves the artificial intelligence community that provides the required capability. Therefore, prospects arise for a future third software sector in GIS to address this need.

3. Insights from imagery

Remote sensing takes place when data are acquired from a source without establishing physical contact between the source and the sensor that acquires the data. In the absence of physical contact, there must be a signal that serves as a carrier between the source and the sensor. The signal is captured by the sensor in some manner. The captured signal may be stored, relayed, or subjected to immediate processing for information extraction. The nature of source and signal are major determinants of system scenarios. Source–signal–sensor interactions lie primarily in the domain of physics, and remain relevant to signal processing. Environmental sources are multitude and usable signals are multiple. Sensor systems may produce images as analog outputs, or may produce data in digital form that can be used to drive image generators (image data). There are also many types of sensors for which the output does not lend itself to imaging. Among these many possibilities, attention will focus here on systems involving electromagnetic signals and image or image data output. These are among the most frequently used and readily available sources of environmental information.

3.1. Reasons for rasters

The origins of image-based remote sensing lie in photography, and image analysis remained in the photographic domain until the advent of computers and computer-readable magnetic recording media. During this period, extraction of information from images was the sole province of human interpreters. Human interpreters use a combination of geometric and tonal clues along with scene context to pursue a logical process involving convergence of evidence to a tentative or definite conclusion. The pace and precision of human interpretation depends on both the skill and knowledge of the individual interpreter. The skill aspects come in facility with interpretive equipment, sensitivity to image clues, and logical agility. The knowledge aspect comes in the interpreter's prior experience with similar scene contexts. Due to the variability of these human elements, some are led to characterize human image interpretation as substantially subjective. The subjectivity can be argued, but the process can certainly be systematized in varying degree through interpretive keys, libraries of documented image samples, and so forth. More difficult to contest is the painstaking nature of human image handling.

Incentives for electronic imaging came independently of interpretation issues. Obvious incentives include image transmission in the manner of television and obviating the need to retrieve films from difficult places like orbiting statellites. However, the combination of electronic imaging and computer processing also offers the possibility of employing computers for information extraction from image data. Rasters are so fundamental to electronic imaging that they become almost axiomatic in the handling of image data. Likewise, the course of evolution in electronic imaging more or less dictated parallel pathways for exploration of computer-based image interpretation. Early incentive to address high-resolution images by computers was restricted largely to theoretical possibilities of vision for not yet existent robots in the then esoteric context of artificial intelligence (AI).

Early image data for environmental remote sensing arose from signal integration over raster elements (pixels) representing ground resolution of approximately 80 m (Landsat MSS = multi-spectral scanner). At such coarse grain, most of the geometry for familiar landscape features like trees and buildings is lost. Potential for recognition thus lay mostly in the spectral domain. Absence of geometric detail became a subject of continual concern to the earth-resource/environmental community. It therefore became a high priority for improvement of sensing systems. The second generation Landsat (TM = thematic mapper) stepped up resolution to the 30 m level. The French SPOT system further improved this to the 10 m level. While individual trees still do not resolve at the 10 m level, other familiar landscape features like roads, rivers and larger buildings are fairly readily discernible. Grouping of trees in forest canopies and variability in herbaceous plant cover likewise emerge.

Any particular temporal instance of image data from say Landsat or SPOT can be made to fit a multivariate mold. These systems have detectors working in several spectral bands, which become the variables. Although things do not always align so nicely initially, the pixels (raster cells) provide the observational framework in terms of data cases. The instantaneous fields of view for the pixels may represent centered samples of the raster equivalent ground scene cells, or may be partially overlapping from a cell to neighboring cells. This aspect may also differ between bands. There are thus problematic aspects associated with incorporating any scene sampling behavior of the sensors directly in the multivariate concept of such rasters. Questions of sampling and independence can be neatly avoided by conceiving the complete raster lattice for a scene as being a multivariate population of individual raster cells wherein coupled row and column numbers index the individuals comprising the population. Reasons for similarities and differences among individuals of the population are several, including area encompassed by fields of view. Among the several influences, the spatial coherence that landscape structure lends to the scene will induce a tendency for greater similarity between raster neighbors than for randomly chosen pairs.

The physics of the environment likewise almost ensures some degree of redundancy among the spectral-band variables that constitute layers of the raster lattice. This becomes even more so when image analysts pursue their propensity for introducing 'artificial' bands by formulating additional variates as band ratios, differences, etc. Landsat MSS provides 4 or 5 bands, depending on the satellite in the series. Landsat TM provides 7 bands. SPOT provides 4 bands. It is noteworthy that native pixel size is not necessarily the same for all bands of a given sensor. Interband differences of pixel size are handled by 'resampling' to obtain compatible rasters. Such resampling becomes another source of redundancy in the image data, whether done by interpolation or simple pixel replication. Sequences of images from repeating orbits of satellite borne sensors can also be registered and incorporated in a composite raster as additional raster layer variables. Temporally induced redundancy thus enters the image data picture as well. Considering the various sources and nature of redundancy in image data, the redundancy is as much a part of the image information as is the variability.

3.2. *Patterns of recognition*

For both human and computerized extraction of scene information, it is important that image data express the spectral properties of scene elements with fidelity. Electronic imaging systems are prone to 'artifacts' unrelated to the environmental features of interest. Modern sensors involve arrays of duplicate detectors that should produce identical outputs given identical inputs. Each of the detectors is designed and aligned to sample a particular scene-relative portion of the incoming signal. Duplicate detectors may be 'out of tune' in the calibration sense, or have slight inconsistencies of alignment. The platform that carries the sensing system may undergo minor attitude variations in roll, pitch and yaw. There is always some degree of electronic 'chatter' in the system that has no particular spatial or temporal pattern. There are occasional electronic 'spikes' that render the output of particular detectors unusable for one or two cycles of the system timer. There are likewise atmospheric sources of 'distraction' such as haze.

The foregoing image data artifacts constitute 'noise' to the information extraction process that is best removed insofar as possible through a series of 'preprocessing' operations that acts as 'filters' to delete unwanted artifacts selectively while allowing the desired information to pass through with as little disturbance as possible. Since 'corrections' of this nature tend to be specific to the sensing system rather than the

Unfortunately, errors in image analysis are not quite so simple as just right or wrong. Any given thematic question posed for response by image analysis usually involves a set of spatially distributed states (categories or conditions) as an answer. The spatial distribution aspect makes it possible for the answer to be exactly correct for every category on a scene-wise basis (percent of total), and yet for a map of the answer to show the wrong state for every pixel in the scene. Such an answer is likely to be useless or worse than useless since landscape structure requires that interventions be location-specific. Doing something in the wrong place is likely to worsen a problem, and half truths become a way to argue convincingly that problems are not yet well enough defined to justify action. Different ones among the several categories may be more important than others, giving rise to need for category-specific criteria of allowable error. Furthermore, it is usually of little help to receive an answer to a different question than the one which was posed. This latter becomes a very real issue with unsupervised analysis which does not admit prespecification of output categories.

The specificity of conventional supervised/unsupervised image analysis is limited by the same empiricism which lends them their generality. The same parameters of pattern recognition that regularize the computer-based approach also demand expertise in their calibration. Experience, knowledge, and external sources of landscape information become essential to refinement, validation and credibility. Errors become not only the primary indicators of failure, but also provide the criteria for success. Errors are chargeable to humans, but computers can be either instruments of redemption or deceit. Computer workstations make it possible to generate spatial displays from image data rapidly and naively. The tyranny of such displays is that they become the stuff of media campaigns for persuasion of gullible publics enthralled with the elusive aura of objectivity emanating from high-tech computer systems. Enhancement implies calling attention to some aspects of image data while suppressing (or concealing) other aspects. Unfortunately, raising the profile of errors as information on process performance runs counter to self-interest of technologists since it penetrates the haze of technological infallibility. Likewise, explication involves revealing assumptions and simplifications that affect generalization.

3.4. Knowledge of systems

With respect to environment and earth resources, the status of computer-based remote sensing is similar to that of GIS – considerable progress has been achieved, but much needed capability remains to be developed for operational use. Much of the shortfall concerns inability to make logical use of knowledge and information that is external to the image data. Many attempts to incorporate such ‘ancillary’ landscape information have been largely without a logical foundation, since they amounted to nothing more than dumping the data into the empirical image mill in hopes that things would somehow sort themselves out. Systematic attempts to capture the landscape logic of the human interpreter in computers have been notably absent until relatively recently. Such absence is undoubtedly due in large measure to lack of apparent precedents coupled with situation specificity of solutions. There are striking parallels with preference for contorting an application context to fit a classical statistical mold rather than refitting statistical equipment to facilitate handling of the application context.

Gradual descent of artificial intelligence from rarefied theoretical atmospheres is beginning to change the picture. Given prior study of a particular landscape as recorded in a GIS, many outcomes of empirical statistical classifiers become highly improbable. Further preference for some outcomes over others arises from the physics of interaction between electromagnetic radiation and environmental materials. Radiation physics provides the pilings and landscape ecology the girders on which to hang logical chains of conditions that hold classifiers away from many confusing influences in a particular landscape setting. One drawback is that both superstructure and suspension must be engineered to the requirements of individual landscapes. Monitoring a given region by remote sensing thus becomes an undertaking analogous to the exhaustive study that underlies a geologic mapping project for a region. Further complicating such an effort is the predominance of affinities over absolutes. Landscape processes are stochastic in differing degrees, so that exceptions to rules are ever present. It would seem that such situations constitute open invitations to Bayesian participation.

The structural compatibility between image rasters and raster GIS lend new relevance to the latter as a transfer medium for injecting landscape logic into digital image analysis. Among the many possibilities in this regard are 'logic layers' that indicate where applicable conditions pertain. Traditional raster GIS will probably provide only a loose fit to the new context in critical respects, however, so some raster innovation will be in order. For example, sequential scans of entire logical rasters would be akin to crossing an ocean in a rowboat. Highly structured indexed access to rasters will therefore become essential.

The foregoing suggestions call for computerized committees of image interpreter expert systems. Expert system approaches enter even more directly into the prospects for further factoring humans out of image analysis scenarios. It does make a difference who runs the workstation for computer-based image analysis. The nature of the difference becomes evident under the admission that image analysis is an iterative endeavor rather than a monotonic progression. An iteration can begin with selection of some training sets for categories of interest, extracting signatures, setting classification thresholds, and running a classifier. Given a trial classification, training sets and check sets are used to construct an error matrix. The error matrix reflects errors of omission and commission (calling one thing another). Errors of omission can be reduced by making classification thresholds less stringent, which tends to increase errors of commission. Conversely, errors of commission can be reduced by making classification thresholds more stringent, which tends to increase errors of omission. In many instances, both types of errors can be reduced simultaneously by using multiple signatures for a given thematic category. The operator/analyst assesses the potential for improvement, makes judgmentally appropriate adjustments in signatures and thresholds, then develops a revised classification. This, in turn, is subjected to error analysis; and the cycle repeats. Under an expert analyst, the improvement in error structure diminishes rapidly in a few cycles. Under a novice analyst, the iterative process is only weakly convergent and pursuit of improvement is usally abandoned at a point well short of what the expert achieves in much less time.

Since the process is parametric under criteria of error structure and available training/validation information, it becomes possible to conduct knowledge engineering on the behavior of the expert giving rise to a process rule base and heuristic stopping criteria. However, the absence of such computer intelligence in contemporary image analysis software is indicative of the work remaining to be done in this arena. Simply stated, the task for the future is to emulate the human facility for logical use of ancillary information in the course of image analysis. A major deterrent to progress in this regard lies in the structure of available expert system shells. It has already been noted that standard statistical software is largely unsuitable for image analysis because of orientation toward relative small volumes of data. Standard expert system software is unsuitable for similar reasons. Such limitations impact thematic feature interpreters much more strongly than robotic process control for iterative image analysis.

3.5. Mechanized monitoring

Coefficients of agreement like Cohen's kappa are computed from error matrices and used to test for significance of differences between two independent interpretations of an image (Rosenfield and Fitzpatrick-Lins, 1986), with the objective being to detect interpreter bias. As computer intelligent image analysis advances to the point of supporting commands like

INTERPRET imagename FOR themeset

without further decisions by humans, prospects arise for interesting variations on old ideas to achieve automated environmental monitoring. If all human decisions are made prior to interpretation and the same rule base is used, then replicate interpretations of image data will be completely invariant. Consider then, the implications of completely consistent interpretations of (independently acquired) replicate images for an area. Any differences in the interpretations must then be due entirely to differences in the images. 'Error' matrices for the interpretations relative to a GIS database for the area thus support a statistical test of difference between imaging conditions. Given the ability to remove imaging errors by preprocessing, what remains is a test of change in the area relative to themeset between instances of imaging. Repetitive coverage by satellite-borne sensors thus provides for automatic monitoring of landscapes relative to thematic interests, whereby the computer raises an alarm when significant differences are found but otherwise just continues its watch.

The foregoing scenario for automatic monitoring uses GIS baseline information to enable a statistical test of temporal change. The test itself is area-wide, but a difference image for interpretations leads to location-specific investigation of changes detected. The artificial intelligence work needed to attain that level of capability is relatively modest, and achieving it should only be a matter of focused project funding. A more futuristic alternative use for the GIS information is to drive the computer intelligent interpretation through a series of complex comparative hypotheses involving both Bayesian strategies and heuristics. For this approach, the GIS information (in the map sense) constitutes a complex null hypothesis for the area. A newly acquired image constitutes evidence to be considered for possible rejection of the null hypothesis in a location-specific sense rather than an area-wide sense.

Under the assumption of no change in thematic distribution, an analysis of variance (between and within categories) can be performed on the image data with the GIS serving to allocate each pixel to a category. If there has been a change in thematic distribution of categories, then the areas of change should become outliers relative to the within-category aspect of the analysis of variance. If the change is a shift from one thematic category to another, a changed area should be consistent with the within-category statistics for a different category. Hence, the scenario thus far consists of image analysis of variance under the GIS view, search for outliers indicating change, and comparative analysis of outliers with other categories for most likely nature of change.

The foregoing scenario thus yields a suite of contingent change hypotheses. Further evidence regarding the tenability of such hypotheses is to be found in the GIS by expectation of pairwise adjacency for particular categories. Each hypothesized change implies a restructuring of such pairwise occurrences. The image data thus become subject to a second-level analysis of evidence for consistency of hypothesized pairwise structure. Hypotheses found to be consistent at this level can be cast as a map and reported to human clients for further verification by other means such as ground check. Likewise, a map can be produced for those areas of suspected change having no consistent explanation for such change. It will then require human investigation to determine what has transpired in these areas.

These suggestions for new approaches to environmental monitoring constitute a major advance over the current generic approach to computerized image analysis. The nature of the advance is that landscape knowledge becomes cumulative. Whereas contemporary computer-based scenarios make very limited use of existing landscape knowledge, the projected scenarios make extensive use of such knowledge. There are also additional benefits. One such benefit is that subjective human influences on environmental investigations become subject to systematic computerized investigation. Another benefit is that human involvement can be focused on the subjective aspects of environmental investigations for which computers are less well suited and perhaps even well excluded.

4. Models and roles

With GIS and remote sensing having been discussed at some length, attention now turns to models and their roles in environmental analysis along with the roles of GIS and remote sensing relative to environmental models. An immediate problem arises, however, from the multifarious usage of the word model. The word carries a universal sense of abstraction, but the universality of abstraction leaves the issue unresolved. Virtually everyone abstracts to some degree, and the word model seems to carry a connotation of formality that most prefer to the word abstraction itself. In most instances of scientific usage, there is also some sense of template-like specificity whereby substitutions in the slots of the template permit a class context to be addressed. This is an important aspect for purposes of environmental analysis, and its absence would imply a lack of utility for environmental investigation and management. Therefore, only constructs of a conceptual template nature will be considered here as falling in the domain of environmental models.

Myers (1991) provided a review of remote sensing and GIS for modeling hydrologic response of watersheds and river basins. Distinctions among models proposed in that review are applicable here, and the hydrologic context is also suitable for illustrating many of the points to be made. Hydrologic processes impinge on most aspects of environmental analysis and management, including issues of environmental contaminant transport.

4.1. Meaning of models

The first concern, then, is to achieve some further typing of models. Typing is not done here for typing sake in the manner of the systematist, but rather to make operationally different modeling situations more easily recognizable. An initial dichotomy of models will be termed formulation vs. emulation. The great majority of statistical and many of the current GIS models fall under the heading of formulation, with regression models being prototypic in this regard. Formulations provide a structure of assumption and association, often expressing some things in terms of other things. In contrast, emulations focus on the chemical/physical and temporal dynamics of real-world phenomena. The intent of emulators is to simulate processes occurring in space and time. Emulations vary in detail and fidelity, but their concern with the 'how' invites further hypothesizing and testing in a way that formulations do not. It should be understood that not all so-called simulations qualify as emulations. Many simulations involve little more than sampling assumed distributions of second-order entities under rules of association.

In the context of environmental analysis and management, a formulation has the pragmatic role of a descriptive. The formulation applies to a particular locality and purpose. Extension of the formulation beyond its context of derivation constitutes extrapolation outside the range of data. In contrast, an emulation is an explanatory representation of a phenomenon in terms of parameters. An emulation should be extendable to a new location provided that the same phenomenon is operative and the parameters are adjusted (calibrated) appropriately to the circumstances. This is the crux of claims that emulation models represent science, whereas formulations represent empiricism. Modelers who deal in emulations often welcome GIS involvement as a source of data for calibration, but resist attempts to represent GIS as embodying the science of modeling.

Emulations are destined to metamorphose from hallmarks of scientific cults to primary tools of environmental analysis and management. If such a potential rise to prominence is not seen as being completely out of the question, then it becomes advantageous to have a caste structure for emulators which prevents those of lower order from masquerading as aristocracy. A similar caste structure for GIS would do much to promote truth in advertising among commercial vendors. In the case of emulators, however, one will be dealing with questions of relative credibility among competing models. Some very important questions of environmental policy can hinge on which model wins the competition for credibility.

Emulators are like mirrors of phenomena which reflect somewhat blurry images. The greater the mechanistic detail of an emulator relative to its parent phenomenon, the deeper the scientific understanding embedded in the model and the better its pedigree. However, pedigree also depends on winning performance in competitions. No emulator should be allowed to enter the ring of environmental management until its purported capabilities have been thoroughly validated. With these considerations in mind, highly mechanistic models of environmental process dynamics can be designated as alpha-emulators comprising the caste of highest order. In the hydrologic context, these are the finite-element and finite-difference models that work with large systems of partial differential equations to mimic the physics of fluid dynamics and solute transport under the influence of absorption and desorption. Mercer and Faust (1986) present an overview of such models. Unfortunately, aristocracy is expensive even in the world of models. Problems of relatively modest scope can make even large computers labor long and hard when approached through these types of models. Boundary conditions also become an issue.

In proceeding to designate a beta level of emulators, one encounters the spatial concept of parameter lumping. Novotny and Chester (1989) provide a discussion of lumped-parameter vs. distributed-parameter models. Parameter lumping involves delineating areal units having some degree of process integrity, and then deriving areally averaged parameters for each unit that determine the aggregate behavior of the unit relative to the process of interest. Emulators that involve more or less localized parameter lumping over areal segments with sequential interaction between segments can be considered as beta-emulator models. The AGNPS models of agricultural nonpoint source pollution (Young, Onstad, Bosch and Anderson, 1989) is illustrative of beta-level emulators in the context of hydrology and pollution. AGNPS is an event-based computer model that projects peak flows along with loading of both sediment and nutrients. AGNPS uses a square grid of cells for discretization of a watershed, with provision for subdivision of cells having greater localized variability.

Gamma-emulators populate a twilight zone between emulation and formulation. Discretization often, but not necessarily, takes the form of a simple raster. Parameter lumping takes place on a cell-wise basis. Interaction between cells is lacking, except for possible use of vicinity information in calculating (cell) unit parameters. Such models frequently take the form of a weighted linear composite among factors. Gamma-emulator models are rather far removed from the details of process dynamics, and thus tend to constitute more of a process index than a process projection. The gamma-sector of the emulator world is both populous and popular, providing a host of inexpensive and expedient 'environmental indexes' which play much the same indicator role in environmental diagnosis as does body temperature for physicians. Such expedients are seductive, and should be approached with caution. While usefully indicative, they are not particularly informative. As an illustration, the universal soil loss equation (USLE) has become quite popular in the GIS community for addressing matters of soil erosion. USLE treats soil erosion as a six-factor multiplicative effect involving rainfall, soil properties, topography, vegetation, and land management practices (Wischmeier and Smith, 1965). USLE has been much used as a method of ranking sites relative to erosion potential. As a quantitative predictor of actual erosion, however, it is at best questionable.

Proponents of modeling as a scientific lifestyle tend to view the gamma level with some disdain. On the other hand, the much-touted analytical capability of contemporary GIS does not yet provide much in the way of direct support for modeling beyond the gamma level. GIS-based environmental modeling work has thus been conducted largely at the gamma level. The view in this writing is one of environmental work being sufficiently varied to accommodate roles for all types modeling. There are, however, niches to which particular types of modeling are better suited than others.

4.2. Process of modeling

Computer progamming for alpha- and beta-level emulators has traditionally involved customized database designs giving the model rapid access which minimizes disruption in the flow of computationally intensive model activity. Efficient usage of computer resources is likely to remain a major concern for such models, which implies that alpha- and beta-modelers will show a continuing reluctance to relinquish control of their database designs. Opposing this reluctance, however, is the obvious applicability of GIS databases to problems of discretizing and calibrating the models. Discretization concerns organization of space. Organizing space for a finite-element model involves locating the nodes of a network. Organizing space for a finite-difference model involves configuring the cells of a rectangular grid. Nodes of a finite-element network need not be regularly distributed, and rectangular cells of a finite-difference grid can vary in both size and shape. Kernodle and Philip (1988) list six ways that a vector-based GIS has proven helpful in the development and support of a finite-difference alpha-emulator of a regional groundwater system for planning purposes. These involve various aspects of discretization, calibration, and cartographic support for application. Automatic compilation of information for cells and ability to revise the structure of the finite-difference grid are emphasized as advantages.

To be negotiated, then, is the protocol for bringing spatial data from a given GIS to bear on the needs of a particular alpha/beta emulator. The first efforts to move in this integrative direction take on the character of interdisciplinary applied research. Since hydrology was not a primary driver of early evolution in GIS, special tools may be needed in the GIS kit to install connectors for the hydrologic context. The more elaborate GIS systems include a programmers 'toolkit' which constitutes something of a blacksmith's shop. Toolkits are a means by which computer programmers gain access to proprietary software without being able to see the source code. Use of such toolkits is a matter of modular computer programming, and not just everyday GIS technician work. Typically the initial blacksmith work is done rather hurriedly under pressure to demonstrate to feasibility of addressing the new context, and requires involvement of the person most knowledgeable about the intricacies of the GIS whose time is already overdrawn. Initial capability to interface GIS with the model thus takes on the character of prototype for which usage is less than efficient and straightforward. Research work of this nature has been conducted at ORSER to link the AGNPS model with the ERDAS GIS.

Technology interface prototypes make good subjects for journal articles and presentations at professional meetings, but tend to become major sources of annoyance and frustration for routine use. The expert's experts whose ad hoc involvement provided the prototype capability are unavailable for routine assistance. Awkwardness and inefficiencies in the prototype waste computer resources, waste human time, and raise the general level of anxiety in the GIS facility. The modeler returns to the thrones of science, expecting that the brief descent for prototype work will be repaid in terms of service. Claims of operational capability come under suspicion, and there arises dissatisfaction from all quarters. Such a climate breeds an abundance of opinion on how things could work better, and the stage is set for some serious software engineering.

If the particular type of modeling work holds prospects for a substantial new clientele of GIS, there may be correspondingly large investments in the software engineering effort that lead to a new GIS subsystem as a product. Johnson (1989) describes the MAPHYD project which illustrates this level of investment, and resulted in a GIS-based computer environment dedicated to the needs of hydrologic modeling. More commonly, however, the new activity represents a modest augmentation of clientele rather than a major facility refocus. The typical response in such cases is to revamp, streamline and repackage the prototype for greater ease of use and reduced dependence on expertise. GIS macro-languages support this kind of R&D work. AI-based 'scripting' systems represent a new-age approach to generic handling of such situations that is not limited to the GIS setting or to any particular type of technology.

The GIS world is better equipped to deal with gamma-level spatial modeling and formulation work in a much less convulsive manner. Raster-based systems are suitable, or at least readily adaptable, for situations that require mathematical or logical compositing across thematic layers on a pixel-by-pixel basis. Many, if not most, raster systems also accommodate a limited amount of vicinity evaluation activity. These two kinds of capabilities go a long way toward satisfying the needs of gamma-emulation and formulation tasks. The work of Evans and Myers (1990) with the DRASTIC model (Aller et al., 1985) illustrates this kind of approach to the problem of potential for groundwater pollution. Petersen et al. (1991) have used a similar but more elaborate strategy for ranking the major watersheds of Pennsylvania relative to need for remediation of agricultural nonpoint source pollution. The latter effort involved joint use of both raster-based and vector-based GIS systems, which provides more capability than either type of system used alone.

4.3. Objects and orientation

The current incapability of GIS for building complex constructs reflecting landscape structure has been noted earlier in this exposition. This incapability looms large for prospective utility of GIS in environmental modeling. The deficiency is one of a fairly fundamental nature, such that linear incremental augmentations of structuring capability become more cosmetic than curative. The whole orientation of thinking about GIS and models is at issue, with data orientation being the current norm. Object orientation constitutes a recent turnabout for sofware engineering and stands as the antithesis of data orientation. The contrast of approach is important, and deserves some attention here.

Consider the problem of developing a beta-level rainfall runoff model for complex terrain, approaching the problem from a data-oriented point of view. In so doing, one might well consider a raster cell as the empirical data unit. Given this choice, the task becomes one of defining an appropriate set of variables (each cell has each variable) and tracking the value of each variable in each cell over the course of a precipitation event as a series of time slices. For any given time slice, there is some stochastic distribution of rainfall intensity (rate of delivery from above), with a given intensity having associated with it a frequency distribution of droplet size and consequent force of impact due to gravity. Surface cover (vegetation, litter, etc.) absorbs some of the energy and redistributes the delivery spatially as stemflow, throughfall, and so on. There is also horizontal delivery and discharge to adjacent raster elements which depends on slope steepness and direction along with the status of local reservoirs (particle absorption, soil pore space, surface ponding) and rate of downward percolation into the subsurface soil. The differences between sheet, rill, and channelized overland flow must also be considered. The whole process is modeled on an element-wise basis by sampling the delivery distribution function, polling those adjacent cells that deliver, noting their contributions, pulsing some downflow to substrate, posting some outflow to receiving cells, and adjusting the contents of internal cell reservoirs accordingly. The rates at which cells receive and discharge must also be adjusted dynamically accordingly to conditions in the cell and in its neighbors (feedback).

Although by no means a simple undertaking, this is certainly a feasible modeling task. There is also no reason why the resulting program should not be well-structured in terms of program modularity. As a modeling exercise, this scenario may be reasonably satisfying. Questions start to arise, however, when one tries to validate the model against historical data. Questions become compounded when one further proposes to use the model as a means of experimenting with implications for the environment. A class of questions for validation concerns the nature of precipitation events. Some precipitation events do have a large-area nature that can be reasonably approximated as relatively simple stochastic processes. Many more precipitation events, however, involve local storm tracks with spatial/temporal distributions of intensity that are complex indeed for casting as element-level stochastic processes. What sort of effort is involved in posing questions relative to effects of shifting land-uses and land-cover patterns? What are the opportunities for streamlining the computations by parameter lumping while preserving the fidelity of model results for the particular physiographic setting? How easily can one impose and study the likely effect of prospective engineering interventions relative to drainage systems and channels? What is involved in using the model to study implications for wetland flora and fauna of changing climatic patterns? How can a GIS be of service in answering such questions? The point is that elemental approaches do not invite such inquiry.

Consider, then, the difference in view associated with object orientation. Precipitation events become spatially distributed objects that move in space and time under the influence of regional weather patterns as modified by local physiographic features such as mountains, ridges, and water bodies. These sorts of physiographic features become objects that modify storms emanating from regional weather patterns. The nature of the interaction is specific to the combination of storm and physiographic feature. A given landscape can be partitioned into hydrologic runoff units (objects) relative to the effect of landform on hydrologic behavior. Forests, fields, wetlands, etc. become ecological objects which modify the hydrologic behavior of landform units. Channels become drainage objects that receive and transport input from runoff objects. The system as a whole is now modeled as a set of interacting objects, each of which has particular roles in the system. Changes can be made at the object level to represent questions about environmental influences, as long the integrity of object roles and interactions between objects is preserved.

The shift toward object orientation in modeling holds promise of near quantum advance in capability for environmental analysis. If GIS is to be a suitable companion technology or perhaps even platform for object-oriented modeling, then there will have to be major paradigm shifts toward object views of and access to GIS information. The current object concepts in software engineering are helpful, but not sufficient. Versions of programming languages like C and PASCAL now encompass objects with inheritance and polymorphism. Not so fully developed are compound object constructs. Recursively defined compound objects are still more easily accommodated in the languages of artificial intelligence like LISP and PROLOG. Continued evolution of software objects could obviate many limitations of present computing capacity as parallel and massively parallel computers eventually become routinely available. Objects will need to be well enough entitized that they become capable of influencing the way processors are allocated in the computer(s). Myers (1993) has outlined a metanetwork approach to GIS which embodies the levels of object orientation under consideration here.

4.4. CHEMistry and ALCHEMy

The heading of this section may well test the tolerance of the reader for acronyms, but it does so in the interest of calling attention to the gulf of modeling that separates current efforts from the needs of environmental analysis and management. Most serious modeling work still focuses on a single process or directly dependent processes, as in the case of modeling contaminant transport by atmospheric and/or hydrologic systems. Such models as do exist tend to be better studied than tested, and preparation for emulating something in a real landscape borders on the arcane. Nevertheless, contaminant transport and hydrologic models are beginning to influence regulatory policy. Global circulation models (in company with expanding ozone holes) are likewise becoming primary drivers for massive public investment in global change research.

Since real landscapes are multiprocess environments, attempts at model validation soon run afoul of extra-process noise that rivals or even dwarfs the signal being sought. Since most landscape settings do not admit experimental design to eliminate confounding influences, the apparent recourse lies in calling upon other models to account for distractions in the fashion of covariance analysis. This changes the nature of the validation from a simple hypothesis to a joint hypothesis scenario in which the credibility of one model depends on that of other models operating simultaneously. At least for the near future, it also remains prohibitively expensive to operate a multitude of models at the alpha level for routine environmental analysis and management. Therefore, need arises for arranging gamma/beta/alfa-hierarchies of models for a process whereby parameter variation experiments with the models permit the use of more mechanistic models in calibrating less mechanistic models. CHEM (composite/hierarchical environmental models) is used here as an acronym for such prospective composite modeling scenarios, and their utilization thus becomes CHEMistry.

Under present technological procedures, mobilizing a single model for environmental analysis becomes something of a project requiring considerable expertise. By comparison, the prospect of routinely fielding multiple models becomes almost logistically intractable. It thus becomes imperative to have much of the model configuration work transpire automatically under self-guiding computer intelligence. The need is for what might be described as committees of expert systems wherein each expert system configures a model, and then the expert systems hold a conference to compare configurations and reach consensus on mutually acceptable adjustments to the initial configurations. Such computer intelligent model control systems would constitute adaptively linked composite/hierarchical environmental models (ALCHEM), and their work can be considered ALCHEMy. The CHEMistry and ALCHEMy acronyms are fairly indicative of the future complexity facing environmental science, and these acronyms are certainly no more inane than the way computer scientists refer to object oriented programming as OOP or naming an advanced computer language as SmallTalk.

5. Frame of reference

This chapter covers extensive ground, some of which has been so well worked as to be almost infertile and other that has barely been broken. The intent has been to show that several spatial information technologies and scientific disciplines must be combined and jointly extended to meet the need posed by present and future problems of environmental analysis and management. Artificial intelligence, advanced computer technology, and innovative software engineering will be essential bonding agents in the glue that holds the projected technoplex together. A paradoxical aspect of the situation is that compound complexity must appear simple in order to be operable. If and when success of the order needed has been achieved, no single human expert will command enough knowledge to claim complete authority over the multisystem. In the interim, there are no shortages of either individual or team challenges for any of the factions involved. The urgency of the need should discourage complacency, and celebration of individual achievement should await global stabilization. In case it has not already become obvious, a central message has been a plea that empiricism not lay waste to collectively held knowledge.

The voluminous nature of both formal and gray literatures underlying these topics would have doomed any attempt at comprehensive citation to inadequacy, and the variegated nature of the intended audience would have insured differing critiques of omission from every sector. So as not to interrupt the flow of argument, therefore, examples have been drawn from the ORSER experience where possible and from other sources where particularly appropriate. Complete absence of citations for a given context reflects the author's view that the respective issue has not previously been very well stated. References of a general nature are included below for GIS, remote sensing, knowledge-based systems, and landscape ecology to assist those seeking an entree to these sectors of the literature.

References

Aller, L., T. Bennett, J. Lehr and R. Petty (1985). *DRASTIC: A Standardized System for Evaluating Ground Water Pollution Potential Using Hydrogeologic Settings*. National Water Well Assoc., Worthington, OH.

Antenucci, J., K. Brown, P. Croswell, M. Kevany and H. Archer (1991). *Geographic Information Systems: A Guide to the Technology*. Van Nostrand Reinhold, New York.

ASPRS (1987). *Geographic Information Systems for Resource Management: A Compendium*. Amer. Soc. for Photogram. and Remote Sensing, Falls Church, VA.

ASPRS (1989). *Fundamentals of GIS: A Compendium*. Amer. Soc. for Photogram. and Remote Sensing, Falls Church, VA.

Barrett, E. and L. Curtis (1992). *Indroduction to Environmental Remote Sensing*. 3rd ed., Chapman and Hall, New York.

Burrough, P. (1986). *Principles of Geographic Information Systems for Land Resources Assessment*. Oxford Uni. Press, New York.

Evans, B. and W. Myers (1990). A GIS-based approach to evaluate regional groundwater pollution potential with DRASTIC. *J. Soil Water Cons.* **45**(2), 242–245.

Johnson, L. (1989). MAPHYD – A digital map-based hydrologic modeling system. *Photogr. Engrg. Remote Sens.* **55**(6), 911–917.

Kernodle, J. and R. Philip (1988). Using a geographic information system to develop a ground-water flow model. In: J. Mclean and A. Johnson, eds., *Regional Aquifer Systems of the United States – Aquifers of the Western Mountain Area*. AWRA Monograph Series, Vol. 14. Amer. Water Resources Assoc., Bethesda, MD, 191–202.

Mercer, J. and C. Faust (1986). *Ground-Water Modeling*. 2nd ed., National Water Well Assoc., Worthington, OH.

Myers, W. (1991). Use of remote sensing and geographic information systems for modeling hydrologic response of watersheds and river basins. In: M. J. Gromiec and A. J. McDonnell, eds., *Water Resources Management and Water Pollution Control – Perspectives from Poland and the United States.* Proc. Conf., Warsaw, 23–25 Sept. 1991. Polish Institute of Meteorology and Water Management, Warsaw, Poland, and Environmental Resources Research Institute of The Pennsylvania State University, University Park, PA, 155–171.

Myers, W. (1993). A meta-network approach to higher-order spatial constructs and analysis of scale effects. In: G. P. Patil and C. R. Rao, eds., *Multivariate Environmental Statistics,* Vol. 6. North-Holland/Elsevier, New York.

Myers, W., B. Evans, G. Baumer and D. Heivly (1989). Synergism between human interpretation and digital pattern recognition in prepation of thematic maps. *ISPRS J. Photogr. Remote Sens.* **44,** 85–96.

Novotny, V. and G. Chester (1989). Delivery of sediment and pollutants from nonpoint sources: A water quality perspective. *J. Soil Water Cons.* **44**(6), 568–576.

Petersen, G., J. Hamlett, G. Baumer, D. Miller, R. Day and J. Russo (1991). Evaluation of agricultural nonpoint pollution potential in Pennsylvania using a geographic information system. Report ER-9105, Environmental Resources Research Institute, The Pennsylvania State University, University Park, PA.

Plant, R. and N. Stone (1991). *Knowledge-Based Systems in Agriculture.* McGraw-Hill, New York.

Rhind, D. and N. Mounsey, eds. (1990). *Understanding GIS.* Taylor and Francis, London.

Rosenfield, G. H. and K. Fitzpatrick-Lins (1986). A coefficient of agreement as a measure of thematic classification accuracy. *Photogr. Engrg. Remote Sens.* **52** (2), 223–227.

Starr, J. and J. Estes (1990). *Geographic Information Systems: An Introduction.* Prentice-Hall, Englewood Cliffs, NJ.

Tomlin, C. D. (1990). *Geographic Information Systems and Cartographic Modeling.* Prentice-Hall, Englewood Cliffs, NJ.

Wischmeier, W. and D. Smith (1965). *Predicting Rainfall-Erosion Losses from Cropland East of the Rocky Mountains,* Handbook 282. US. Department of Agriculture, Washington, DC.

Young, R., C. Onstad, D. Bosch and W. Anderson (1989). AGNPS: A nonpoint-source pollution model for evaluating agricultural watersheds. *J. Soil Water Cons.* **44**(2), 168–173.

G. P. Patil and C. R. Rao, eds., *Handbook of Statistics, Vol. 12*

19

Regression Analysis of Spatially Correlated Data: The Kanawha County Health Study

Christl A. Donnelly, James H. Ware and Nan M. Laird

Abstract: We describe a linear regression model for a spatially correlated dependent variable when a covariate is also spatially correlated and measured with error. Maximum likelihood estimates and likelihood-based confidence intervals for the regression parameters of the linear model are obtained and the method is extended to logistic regression analysis of grouped binomial data by methods analogous to empirical logistic regression. When the independent variable is spatially correlated, the measurement error variance and the other parameters of the regression model can be estimated without additional assumptions or data. The methods are used to characterize the association between outdoor concentrations of volatile organic compounds and respiratory health of school-children attending 73 elementary schools in Kanawha County, West Virginia. Results are compared to those from two-stage estimation procedures in which the dependent variable is regressed on the expectation of the true covariate conditional on the observed covariate values.

1. Introduction

The aim of this chapter is to present an approach to maximum likelihood estimation of and testing for the effect of an environmental covariate x on the response variable y in the presence of measurement error. Methods will be presented for both linear and empirical logistic regression models. The importance of measurement error methodology in the analysis of environmental health studies is great since measurement error is omnipresent in environmental exposure data. Results are compared to those from two-stage estimation procedures in which the dependent variable is regressed on the expectation of the true covariate conditional on the observed covariate values.

The classical structural measurement error problem consists of a sample of n independent and identically distributed pairs (z_j, y_j). The observed covariate z_j is the true covariate value x_j (a random variable with variance σ_x^2), plus mean-zero measurement error with variance σ_δ^2; the response y_j has mean $\beta_0 + \beta_1 x_j$ and variance σ_e^2 conditional on x_j. The naive approach of the linear regression of y on z yields an attenuated slope estimate with expected value $(\sigma_x^2/(\sigma_x^2 + \sigma_\delta^2))\beta_1$ (Fuller, 1987). An unbiased estimate of the slope can be obtained if σ_x^2, σ_δ^2, or $\sigma_x^2/(\sigma_x^2 + \sigma_\delta^2)$, the reliability coefficient, is assumed to be known. This additional information is typically obtained from a separate validation or calibration study. Without additional information, the test of H_0: $\beta_1 = 0$, although valid, has reduced power.

In Section 4, we present a linear regression estimation procedure, based on Gaussian assumptions, for the case where the data pairs (z_j, y_j) are *not* independently distributed. The correlation structure in the data is utilized to make the problem identifiable. This approach is, in effect, a form of internal validation. This interpretation is clear if, for example, the correlation in the data arose because the data consisted of two pairs of measurements on each of $n/2$ independent sampling units. The interpretation is also valid for a dataset in which the data are obtained from n correlated units, as in our example.

Due to the complexity of likelihood and quasi-likelihood methods in non-linear measurement error problems (e.g., Carroll et al., 1984; Armstrong, 1985), a number of approximation methods have been proposed. The validity of many of these approximations depends on the magnitude of the measurement error being 'small' (Stefanski and Carroll, 1985; Whittemore and Keller, 1988). Pierce et al. (1992) and Whittemore (1989) proposed the regression of the outcome variable on James–Stein estimates (James and Stein, 1961; Stein, 1955; Stein, 1981) of the true covariate values, in both the linear and non-linear regression settings. Similarly, Rosner, Willett and Spiegelman (1989) proposed the use of covariate expected values (conditional on parameter estimates of linear regression of z on x) in ordinary logistic regression. These expected value substitution approaches are applicable to small magnitude measurement error; they require either knowledge of the measurement error variance or repeated measures of z_j and ignore additional uncertainty introduced by the estimation of parameters needed to obtain the covariate expected values. Using validation study data, Rosner, Willett and Spiegelman (1989) alternatively proposed the use of a Taylor series expansion about $(\boldsymbol{x} - \bar{x}\mathbf{1}) = \mathbf{0}$ to estimate the logistic regression parameters.

In Section 5, we propose the use of empirical logistic regression to apply the methods developed for linear regression to grouped binary data with correlated covariate values. This methodology is appropriate for all magnitudes of measurement error and does not require validation data.

Section 2 introduces an environmental health study in which health outcomes were modeled as a function of air pollution concentrations. Laboratory studies of the air pollution monitors suggested the presence of considerable measurement error. Section 3 presents a measurement error model linking the true and observed pollution concentration. Section 6 presents and compares the results of the analysis of the example dataset both ignoring and modeling covariate measurement error.

2. The Kanawha County study

Kanawha County, West Virginia, which includes Charleston, is one of the country's leading centers of chemical manufacturing. Many plants are located close to residential neighborhoods. Longstanding concerns of county residents about the health effects of exposure to chemical emissions were heightened by the industrial disaster in Bhopal, India in 1984 and a subsequent chemical leak in Institute, West Virginia.

In the fall of 1986, investigators from the Harvard School of Public Health began a series of studies of chemical air pollution and health in Kanawha County. One of these studies, conducted in the fall of 1988, combined an epidemiologic survey of all third through fifth grade children attending Kanawha County schools with measurement of 15 volatile organic compounds (VOCs) at each of the 75 elementary schools.

The epidemiologic study was designed to investigate the association between measures of chemical air pollution exposure and symptom rates at the 75 schools. The investigators recognized that both the symptom rates and the exposure measurements might exhibit spatial correlation and that the exposure measurements might also be subject to measurement error. The desire to develop regression methods which accounted for the spatial correlation and the measurement error motivated this research.

2.1. Respiratory symptom rates

Respiratory symptom rates were assessed from health questionnaires administered to the parents of all third, fourth and fifth graders attending public elementary schools in Kanawha County. Elementary school children were chosen as the target population to facilitate a school-based study design and to avoid occupational exposure to the chemicals under study. The chronic symptoms included in the questionnaire were chronic cough, chronic phlegm, bronchitis, persistent wheeze, and attacks of shortness of breath. A composite indicator variable for lower respiratory symptoms was recorded as positive if any of the five chronic symptoms was reported. Among 74 schools, the return rate for the questionnaire was 97%; no questionnaire data was available for the remaining school.

2.2. Measurement of volatile organic compounds

Concentrations of 15 VOCs were measured by placing a passive organic vapor monitor (OVM) badge at each school for eight weeks and then subjecting the badges to analysis by ion chromatography. Concentration data was available for 73 of the 74 schools with questionnaire data. The 15 concentrations were then used to produce two summary measures of VOC exposure. The concentration of petroleum-related compounds was defined as the sum of concentrations of five VOCs typically associated with petroleum burning or off-gassing. The concentration of process-related VOCs was defined as the sum of the concentrations of 10 compounds believed to be uniquely associated with chemical manufacturing. The measured process-related compounds included 1,1,1-trichloroethane, carbon tetrachloride, 1-butanol, chloroform, perchloroethylene, methyl isobutyl ketone, 1, 2-dichloroethane, styrene, mesityl oxide, and 2-ethoxyethyl acetate. The analyses reported here consider only the sum of the concentrations of process-related VOCs.

Topographically, Kanawha County is transected by the Kanawha River for about 82 km from the southeast to the northwest. The river valley is 0.8 to 2.5 km wide and about 120 m deep. Most of the plants are located on the valley floor and the valley is subject to prolonged inversions. Thus, the investigators assumed that VOC concentrations would be influenced by location of the school on the valley floor or outside the valley and by proximity to the plants. Before beginning the data analysis, schools were divided into four proximity-to-industry regions:

Region 1: In valley, near.
Region 2: In valley, far.
Region 3: Out of valley, near.
Region 4: Out of valley, far.

The potential effect of proximity-to-industry group on VOC concentration suggested the need to allow dependence of the exposure measurement on other covariates, as described in the next section.

3. Specification of measurement error model

The goal of the study was to characterize the relationship between the symptom rates and the true air pollution concentrations. Due to the large amount of measurement error associated with OVM dosimeters, it is necessary to specify a model linking the true and observed pollution concentrations.

The data were collected from n locations. These data can be thought of as a sample of random variables which might exhibit both 'large-scale-variation' (Cressie, 1991), variation explained by non-random covariates (in our case, region), and 'small-scale-variation' (Cressie, 1991) which has mean zero. Each true and observed pollution concentration can be represented as a sum of the 'large-scale-variation' and 'small-scale-variation'. The true concentration for the j-th location can be expressed as:

$$x_j = \boldsymbol{w}_j\boldsymbol{\pi} + v_j$$

for $j = 1$ to n where $\boldsymbol{w}_j$ is a $1 \times p$ vector specifying region and $\boldsymbol{\pi}$ is a $p \times 1$ parameter vector. We define $\boldsymbol{r}_v$ to be the correlation matrix of the vector $\boldsymbol{v} = (v_1, \ldots, v_n)'$, so that $\mathrm{E}(\boldsymbol{v}) = \boldsymbol{0}$ and $\mathrm{var}(\boldsymbol{v}) = \sigma_v^2\boldsymbol{r}_v$, where the correlation between v_j and $v_{j'}$ is a function of the distance between schools j and j'. The mean of $\boldsymbol{x} = (x_1, \ldots, x_n)'$ is $\boldsymbol{w\pi}$ and $\mathrm{var}(\boldsymbol{x}) = \sigma_v^2\boldsymbol{r}_v$ where $\boldsymbol{w} = (\boldsymbol{w}_1', \ldots, \boldsymbol{w}_n')'$.

The observed pollution concentration for the j-th location can be expressed as the sum of the true concentration and independent (or measurement) error:

$$z_j = \boldsymbol{w}_j\boldsymbol{\pi} + v_j + \delta_j$$

for $j = 1$ to n. We define $\boldsymbol{\delta} = (\delta_1, \ldots, \delta_n)'$, with $\mathrm{E}(\boldsymbol{\delta}) = \mathbf{0}$ and $\mathrm{var}(\boldsymbol{\delta}) = \sigma_\delta^2 \boldsymbol{I}$, and define the vector $\boldsymbol{z} = (z_1, \ldots, z_n)'$. The mean of $\boldsymbol{z}$ is $\boldsymbol{w\pi}$ and $\mathrm{var}(\boldsymbol{z}) = \sigma_v^2 \boldsymbol{r}_v + \sigma_\delta^2 \boldsymbol{I}$.

Assuming that the correlation between v_j and $v_{j'}$ depends solely upon a measure of distance (denoted $h_{jj'}$) between the locations of measurements, one type of function which specifies the correlation matrix is the semi-variogram (denoted $\gamma(h_{jj'})$), where

$$\begin{aligned}\gamma(h_{jj'}) &= \tfrac{1}{2}\mathrm{E}[(r_j - r_{j'})^2] \\ &= \sigma_v^2(1 - [\boldsymbol{r}_v]_{jj'}) + \sigma_\delta^2,\end{aligned}$$

$j < j'$ and $r_j = z_j - (\boldsymbol{w}_j \boldsymbol{\pi})$.

Isaaks and Srivastava (1989) present a number of semi-variogram models. Two of the most commonly used are the exponential and Gaussian models:

$$\begin{aligned}&\text{Exponential} \quad \gamma(h_{jj'}) = \sigma_\delta^2 + \sigma_v^2(1 - \rho_v^{h_{jj'}}),\\ &\text{Gaussian} \quad \gamma(h_{jj'}) = \sigma_\delta^2 + \sigma_v^2(1 - \rho_v^{h_{jj'}^2}),\end{aligned}$$

where $0 < \rho_v < 1$ and $j < j'$. Note that the exponential semi-variogram is the continuous analogue of the first order auto-regressive correlation model. The off-diagonal elements of $\boldsymbol{r}_v$ are of the form $[\boldsymbol{r}_v]_{jj'} = \rho_v^{h_{jj'}}$ assuming an exponential semi-variogram model and of the form $[\boldsymbol{r}_v]_{jj'} = \rho_v^{h_{jj'}^2}$ assuming a Gaussian semi-variogram model.

The classical semi-variogram estimator is

$$\hat{\gamma}(h_{jj'}) = \tfrac{1}{2}(\hat{r}_j - \hat{r}_{j'})^2,$$

where $j < j'$ and $\hat{r}_j = z_j - (\boldsymbol{w}_j \hat{\boldsymbol{\pi}})$. Given an assumed model, Cressie (1985) suggests a weighted least squares method to estimate the semi-variogram parameters. The method depends, however, upon a small number of h values, which can be achieved by grouping. Maximum likelihood estimation is feasible for any distribution of h values.

4. Linear regression with covariate measurement error

Define the dependent variable y such that

$$y_j = \beta_0 + \beta_1 x_j + \eta_j + e_j$$

for $j = 1$ to n. We define $\boldsymbol{r}_\eta$ to be the correlation matrix of the vector $\boldsymbol{\eta} = (\eta_1, \ldots, \eta_n)'$, so that $\mathrm{E}(\boldsymbol{\eta}) = \mathbf{0}$ and $\mathrm{var}(\boldsymbol{\eta}) = \sigma_\eta^2 \boldsymbol{r}_\eta$ where the correlation between η_j and $\eta_{j'}$ is a function of the distance between schools j and j'. We define the vector $\boldsymbol{e} = (e_1, \ldots, e_n)'$, so that $\mathrm{E}(\boldsymbol{e}) = \mathbf{0}$ and $\mathrm{var}(\boldsymbol{e}) = \sigma_e^2 \boldsymbol{I}$. This model cannot be fit using standard regression methods since the true pollution concentrations $\boldsymbol{x}$ are not observed. One possible approach to this problem is two-stage estimation in which the true values of the concentrations (x) are predicted from observed concentrations

(z) and the predicted values are then used as known covariate values. A second approach is simultaneous parameter estimation based on all observed data.

4.1. Two-stage estimation

Pierce et al. (1992) and Whittemore (1989) suggested the 'best approach' in this setting is to replace each x_j value with the estimated value conditional on the j-th observed covariate value z_j and use this estimated quantity as the covariate for analysis. Since both the true and observed covariate values are correlated in our setting, we use $\hat{x}_j$, the estimated expected value conditional on all observed covariate values ($\mathrm{E}(x_j|z)$).

Assuming $\sigma_v^2 = 0$, the best linear unbiased predictor (BLUP) of x_j is

$$\mathrm{E}(x_j \mid \boldsymbol{z}) = \boldsymbol{w}_j \hat{\boldsymbol{\pi}},$$

where $\hat{\boldsymbol{\pi}} = (\boldsymbol{w}'\boldsymbol{w})^{-1}(\boldsymbol{w}'\boldsymbol{z})$. If $\sigma_v^2 > 0$, the BLUP of x_j is of the form

$$\mathrm{E}(x_j|\boldsymbol{z}) = \boldsymbol{w}_j\hat{\boldsymbol{\pi}} + \sigma_v^2[\boldsymbol{r}_v]_{j\cdot}(\sigma_v^2\boldsymbol{r}_v + \sigma_\delta^2\boldsymbol{I})^{-1}(\boldsymbol{z} - \boldsymbol{w}\hat{\boldsymbol{\pi}}), \tag{1}$$

where

$$\hat{\boldsymbol{\pi}} = (\boldsymbol{w}'(\sigma_v^2\boldsymbol{r}_v + \sigma_\delta^2\boldsymbol{I})^{-1}\boldsymbol{w})^{-1}\boldsymbol{w}'(\sigma_v^2\boldsymbol{r}_v + \sigma_\delta^2\boldsymbol{I})^{-1}\boldsymbol{z}$$

and $[\boldsymbol{r}_v]_{j\cdot}$ is the j-th row of the matrix $\boldsymbol{r}_v$. This predictor is known as the universal kriging predictor in the geostatistical literature (Cressie, 1986; Isaaks and Srivastava, 1989; Cressie, 1990). Kriging prediction has been used in a number of environmental exposure studies to characterize the spatial distribution of the exposure variable (e.g., Eynon and Switzer, 1983; Venkatram, 1988; Wartenberg, Uchrin and Coogan, 1991).

It remains only to specify estimates of the variance components. Assuming $\boldsymbol{v}$ and $\boldsymbol{\delta}$ are normally distributed with means $\mathbf{0}$ and variances $\sigma_v^2\boldsymbol{r}_v$ and $\sigma_\delta^2\boldsymbol{I}$, respectively, $\boldsymbol{z}$ is normally distributed with mean $\boldsymbol{w\pi}$ and variance $\sigma_v^2\boldsymbol{r}_v + \sigma_\delta^2\boldsymbol{I}$. The log likelihood of the data $\boldsymbol{z}$, denoted l_z, is

$$l_z = -\tfrac{1}{2}\log|\boldsymbol{V}_z| - \tfrac{1}{2}(\boldsymbol{z} - \boldsymbol{w\pi})'\boldsymbol{V}_z^{-1}(\boldsymbol{z} - \boldsymbol{w\pi}),$$

where $\boldsymbol{V}_z = \sigma_v^2\boldsymbol{r}_v + \sigma_\delta^2\boldsymbol{I}$.

Mardia and Marshall (1984) present a maximum likelihood method for the simultaneous estimation of both mean parameters $\boldsymbol{\pi}$ and variance components. Let $\boldsymbol{\theta}$ denote a vector of the variance components. Define the vector $\boldsymbol{L}_\theta$ such that the q-th element of $\boldsymbol{L}_\theta$ is the derivative of $\ln(L)$ with respect to the q-th element of $\boldsymbol{\theta}$,

$$(\boldsymbol{L}_\theta)_q = -\tfrac{1}{2}\mathrm{tr}\left(\boldsymbol{V}_z^{-1}\frac{\partial \boldsymbol{V}_z}{\partial \boldsymbol{\theta}_q}\right) + \tfrac{1}{2}(\boldsymbol{z} - \boldsymbol{w\pi})'\boldsymbol{V}_z^{-1}\frac{\partial \boldsymbol{V}_z}{\partial \boldsymbol{\theta}_q}\boldsymbol{V}_z^{-1}(\boldsymbol{z} - \boldsymbol{w\pi}).$$

Define the matrix $\boldsymbol{B}_\theta$ to be the information matrix of $\ln(L)$ with respect to $\boldsymbol{\theta}$. The qs-th element of $\boldsymbol{B}_\theta$ is

$$(\boldsymbol{B}_\theta)_{qs} = \tfrac{1}{2}\operatorname{tr}\left(\boldsymbol{V}_z^{-1} \frac{\partial \boldsymbol{V}_z}{\partial \boldsymbol{\theta}_q} \boldsymbol{V}_z^{-1} \frac{\partial \boldsymbol{V}_z}{\partial \boldsymbol{\theta}_s} \right).$$

The $\boldsymbol{B}_\theta$ matrix and $\boldsymbol{L}_\theta$ vector are used to update the estimate of $\boldsymbol{\theta}$. A starting value $\boldsymbol{\theta}_0$ is chosen. The current estimate of $\boldsymbol{\theta}$ is used to calculate

$$\hat{\boldsymbol{\pi}}_M = (\boldsymbol{w}'[\boldsymbol{V}_z(\boldsymbol{\theta}_M)]^{-1}\boldsymbol{w})^{-1}\boldsymbol{w}'[\boldsymbol{V}_z(\boldsymbol{\theta}_M)]^{-1}\boldsymbol{z}.$$

The estimate of $\boldsymbol{\theta}$ is updated as $\boldsymbol{\theta}_{M+1} = \boldsymbol{\theta}_M + \boldsymbol{B}_\theta^{-1}\boldsymbol{L}_\theta$. This estimate of $\boldsymbol{\theta}$ is used to update the estimate of $\boldsymbol{\pi}$ iteratively until convergence of the estimates is obtained.

4.2. Simultaneous maximum likelihood estimation

A different approach is to jointly estimate the parameters of both the measurement error model and the regression model. Assuming $\boldsymbol{v}$, $\boldsymbol{\delta}$, $\boldsymbol{\eta}$ and $\boldsymbol{e}$ are normally distributed with means $\boldsymbol{0}$ and variances $\sigma_v^2\boldsymbol{r}_v$, $\sigma_\delta^2\boldsymbol{I}$, $\sigma_\eta^2\boldsymbol{r}_\eta$ and $\sigma_e^2\boldsymbol{I}$, respectively, $\boldsymbol{z}$ and $\boldsymbol{y}$ are normally distributed with mean and variance

$$\begin{pmatrix} \boldsymbol{z} \\ \boldsymbol{y} \end{pmatrix} \sim \mathrm{MVN}\left[\begin{pmatrix} \boldsymbol{w}\boldsymbol{\pi} \\ \boldsymbol{1}\beta_0 + \boldsymbol{w}\boldsymbol{\pi}\beta_1 \end{pmatrix}, \begin{pmatrix} \sigma_v^2\boldsymbol{r}_v + \sigma_\delta^2\boldsymbol{I} & \beta_1\sigma_v^2\boldsymbol{r}_v \\ \beta_1\sigma_v^2\boldsymbol{r}_v & \beta_1^2\sigma_v^2\boldsymbol{r}_v + \sigma_\eta^2\boldsymbol{r}_\eta + \sigma_e^2\boldsymbol{I} \end{pmatrix} \right].$$

The log likelihood of the data $(\boldsymbol{z}, \boldsymbol{y})$, denoted l, is

$$l = -\tfrac{1}{2}\log|\boldsymbol{V}| - \tfrac{1}{2}(\boldsymbol{Y} - \boldsymbol{W}\boldsymbol{\Pi})'\boldsymbol{V}^{-1}(\boldsymbol{Y} - \boldsymbol{W}\boldsymbol{\Pi})$$

where

$$\boldsymbol{Y} = \begin{pmatrix} \boldsymbol{z} \\ \boldsymbol{y} \end{pmatrix},$$

$$\boldsymbol{W} = \begin{pmatrix} \boldsymbol{0} & \boldsymbol{w} \\ \boldsymbol{1} & \beta_1\boldsymbol{w} \end{pmatrix},$$

$$\boldsymbol{\Pi} = \begin{pmatrix} \beta_0 \\ \boldsymbol{\pi} \end{pmatrix}$$

and

$$\boldsymbol{V} = \begin{pmatrix} \sigma_v^2\boldsymbol{r}_v + \sigma_\delta^2\boldsymbol{I} & \beta_1\sigma_v^2\boldsymbol{r}_v \\ \beta_1\sigma_v^2\boldsymbol{r}_v & \beta_1^2\sigma_v^2\boldsymbol{r}_v + \sigma_\eta^2\boldsymbol{r}_\eta + \sigma_e^2\boldsymbol{I} \end{pmatrix}.$$

Computation of the data likelihood does not require the perhaps computationally prohibitive inversion of a $2n \times 2n$ matrix. The inverse of a symmetric partitioned matrix, presented by Searle (1982), is

$$\begin{pmatrix} a & b \\ b & c \end{pmatrix}^{-1} = \begin{pmatrix} a^{-1} + a^{-1}b(c - ba^{-1}b)^{-1}ba^{-1} & -a^{-1}b(c - ba^{-1}b)^{-1} \\ -(c - ba^{-1}b)^{-1}ba^{-1} & (c - ba^{-1}b)^{-1} \end{pmatrix}.$$

The Mardia and Marshall (1984) estimation procedure is carried out for a given β_1 value in order to estimate $\boldsymbol{\pi}$, β_0 and the variance components. The profile likelihood with regard to β_1 is obtained via direct search. The simultaneous ML estimates are the value of β_1 which maximizes the profile likelihood and the corresponding conditional ML estimates of the remaining parameters.

Let $\boldsymbol{\theta}$ denote a vector of the variance components (excluding β_1). Define the vector $\boldsymbol{L}_\theta$ such that the i-th element of $\boldsymbol{L}_\theta$ is the derivative of $\ln(L)$ with respect to the q-th element of $\boldsymbol{\theta}$,

$$(\boldsymbol{L}_\theta)_q = -\tfrac{1}{2}\mathrm{tr}\left(\boldsymbol{V}^{-1}\frac{\partial \boldsymbol{V}}{\partial \boldsymbol{\theta}_q}\right) + \tfrac{1}{2}(\boldsymbol{Y} - \boldsymbol{W\Pi})'\boldsymbol{V}^{-1}\frac{\partial \boldsymbol{V}}{\partial \boldsymbol{\theta}_q}\boldsymbol{V}^{-1}(\boldsymbol{Y} - \boldsymbol{W\Pi}).$$

Define the matrix $\boldsymbol{B}_\theta$ to be the information matrix of $\ln(L)$ with respect to $\boldsymbol{\theta}$. The qs-th element of $\boldsymbol{B}_\theta$ is

$$(\boldsymbol{B}_\theta)_{qs} = \tfrac{1}{2}\mathrm{tr}\left(\boldsymbol{V}^{-1}\frac{\partial \boldsymbol{V}}{\partial \boldsymbol{\theta}_q}\boldsymbol{V}^{-1}\frac{\partial \boldsymbol{V}}{\partial \boldsymbol{\theta}_s}\right).$$

The $\boldsymbol{B}_\theta$ matrix and $\boldsymbol{L}_\theta$ vector are used to update the estimate of $\boldsymbol{\theta}$. A starting value $\boldsymbol{\theta}_0$ is chosen. The current estimate of $\boldsymbol{\theta}$ is used to calculate

$$\hat{\boldsymbol{\Pi}}_M = (\boldsymbol{W}'[\boldsymbol{V}(\boldsymbol{\theta}_M)]^{-1}\boldsymbol{W})^{-1}\boldsymbol{W}'[\boldsymbol{V}(\boldsymbol{\theta}_M)]^{-1}\boldsymbol{Y}.$$

The estimate of $\boldsymbol{\theta}$ is updated as $\boldsymbol{\theta}_{M+1} = \boldsymbol{\theta}_M + \boldsymbol{B}_\theta^{-1}\boldsymbol{L}_\theta$. This estimate of $\boldsymbol{\theta}$ is used to update the estimate of $\boldsymbol{\Pi}$ iteratively until convergence of the estimates is obtained.

A likelihood ratio test may be used to test the hypothesis $H_0{:}\beta_1 = 0$. Let Λ denote the likehood ratio statistic, such that

$$\log(\Lambda) = \max(l_{\beta_1=0}) - \max(l).$$

Under the null hypothesis, the asymptotic distribution of $-2\log(\Lambda)$ is chi-square with 1 degree of freedom. This property may be used to create likelihood-based $(1-\alpha)\%$ confidence intervals (CIs) for β_1 of the form $(\beta_{1,\alpha/2}, \beta_{1,(1-\alpha/2)})$ where

$$\max(l_{\beta_{1,\alpha/2}}) = \max(l_{\beta_{1,(1-\alpha/2)}}) = \max(l) - \tfrac{1}{2}\chi^2_{(1-\alpha),\,1\,\mathrm{df}}.$$

The likelihood estimation procedures assume that the v, δ, η and e random variables are normally distributed. Testing these assumptions requires estimates of these variables. Conditional on the parameter estimates, the BLUP of small scale variation in z is of the form:

$$\hat{\psi}_j = (\hat{v}_j + \hat{\delta}_j) = z_j - \boldsymbol{w}_j\hat{\boldsymbol{\pi}},$$

and the BLUP of small scale variation in y is of the form

$$\hat{\omega}_j = (\hat{\eta}_j + \hat{e}_j) = y_j - (\hat{\beta}_0 + \hat{\beta}_1(\boldsymbol{w}_j\hat{\boldsymbol{\pi}} + \hat{v}_j)),$$

where

$$\hat{v}_j = (\hat{\sigma}_v^2[\hat{\boldsymbol{r}}_v]_{j\cdot} \quad \hat{\beta}_1\hat{\sigma}_v^2[\hat{\boldsymbol{r}}_v]_{j\cdot})\begin{pmatrix} \hat{\sigma}_v^2\hat{\boldsymbol{r}}_v + \hat{\sigma}_\delta^2\boldsymbol{I} & \hat{\beta}_1\hat{\sigma}_v^2\hat{\boldsymbol{r}}_v \\ \hat{\beta}_1\sigma_v^2\boldsymbol{r}_v & \hat{\beta}_1^2\hat{\sigma}_v^2\hat{\boldsymbol{r}}_v + \hat{\sigma}_\eta^2\hat{\boldsymbol{r}}_\eta + \hat{\sigma}_e^2\boldsymbol{I} \end{pmatrix}^{-1}$$
$$\times \begin{pmatrix} \boldsymbol{z} - \boldsymbol{w}\hat{\boldsymbol{\pi}} \\ \boldsymbol{y} - (\hat{\beta}_0\boldsymbol{1} + \hat{\beta}_1\boldsymbol{w}\hat{\boldsymbol{\pi}}) \end{pmatrix} \tag{2}$$

and $[\hat{\boldsymbol{r}}_v]_{j\cdot}$ is the j-th row of $\hat{\boldsymbol{r}}_v$. The Cholesky decompositions of the matrices $(\hat{\sigma}_v^2\hat{\boldsymbol{r}}_v + \hat{\sigma}_\delta^2\boldsymbol{I})^{-1}$ and $(\hat{\sigma}_\eta^2\hat{\boldsymbol{r}}_\eta + \hat{\sigma}_e^2\boldsymbol{I})^{-1}$ may be used to orthonormalize the variables ψ and ω, respectively. The orthonormalized variables, denoted Ψ and Ω, may then be used to test the null hypothesis that the data are distributed normally.

One group of tests of this hypothesis involves functions of the second, third and fourth moments about the mean (D'Agostino, 1970; D'Agostino and Pearson, 1973; Cox and Hinkley, 1974). The k-th moment of s is estimated such that

$$m_k = \frac{\sum_{j=1}^{n}(s_j - \bar{s}_.)^k}{n},$$

where $\bar{s}_. = (1/n)\sum_{j=1}^{n} s_j$. A measure of skewness is $\sqrt{b_1}$ where $\sqrt{b_1} = m_3/m_2^{2/3}$ and a measure of kurtosis is b_2 where $b_2 = m_4/m_2^2$. Define H such that

$$H = \frac{nb_1}{6} + \frac{n(b_2 - 3)^2}{24}.$$

Under the null hypothesis, the test statistic H is asymptotically chi-square distributed (with 2 degrees of freedom).

5. Application to empirical logistic regression

In the case of grouped binomial data, empirical logistic regression allows the application of linear regression methodology to discrete data. Assume that the symptom rate for the j-th group (p_j) is a function of a group-level covariate (x_j), such that

$$\log\left(\frac{p_j}{1 - p_j}\right) = \beta_0 + \beta_1 x_j + \eta_j$$

for $j=1$ to n. The observed binomial data can be summarized by the number of responses at the j-th site N_j and the number of positive responses at the j-th site M_j such that for the j-th group

$$y_j = \log\left(\frac{M_j/N_j}{1 - M_j/N_j}\right) = \beta_0 + \beta_1 x_j + \eta_j + e_j,$$

where e_j is uncorrelated variability. The variance of e_j is typically assumed to be of the form $1/N_j p_j(1-p_j)$. Since the values of p_j are not known and depend upon the random variables v_j and η_j, we approximate the variance of $\boldsymbol{e}$ by letting it be of the form σ_e^2 diag $(1/N_j)$ and estimating σ_e^2.

Assuming $\boldsymbol{v}$, $\boldsymbol{\delta}$, $\boldsymbol{\eta}$ and $\boldsymbol{e}$ are normally distributed with means $\mathbf{0}$ and variances $\sigma_v^2 \boldsymbol{r}_v$, $\sigma_\delta^2 \boldsymbol{I}$, $\sigma_\eta^2 \boldsymbol{r}_\eta$ and $\sigma_e^2 \text{diag}(1/N_j)$, respectively, then $\boldsymbol{z}$ and $\boldsymbol{y}$ are normally distributed with mean and variance

$$\begin{pmatrix} \boldsymbol{z} \\ \boldsymbol{y} \end{pmatrix} \sim \text{MVN}\left[\begin{pmatrix} \boldsymbol{w\pi} \\ \mathbf{1}\beta_0 + \boldsymbol{w\pi}\beta_1 \end{pmatrix}, \begin{pmatrix} \sigma_v^2 \boldsymbol{r}_v + \sigma_\delta^2 \boldsymbol{I} & \beta_1 \sigma_v^2 \boldsymbol{r}_v \\ \beta_1 \sigma_v^2 \boldsymbol{r}_v & \beta_1^2 \sigma_v^2 \boldsymbol{r}_v + \sigma_\eta^2 \boldsymbol{r}_\eta + \sigma_e^2 \text{diag}(1/N_j) \end{pmatrix}\right]. \tag{3}$$

6. Results

The goal of this analysis is to investigate the association between the rate of lower respiratory symptoms (LRS) and the sum of the concentrations of the 10 process-related VOCs (SUM) at each school. Initially, SUM is assumed to be measured without error. An empirical logistic regression is performed assuming a model of the form

$$y_j = \beta_0 + \beta_1 x_j + e_j,$$

where y_j is the logit transformed rate of LRS and x_j is the observed SUM at school j. Assuming the variance of $\boldsymbol{e}$ is equal to $\sigma_e^2 \text{diag}(1/N_j)$, the weighted least squares parameter estimates are $\beta_0 = -1.2026$ and $\beta_1 = 0.0312$ (p-value $= 0.064$, 95% CI $= (-0.0013, 0.0636)$). (See model NE in Table 1.) These estimates correspond to an estimated OR of 1.064 for a 2 μg/m³ change in SUM. (See model NE in Table 2.) The interquartile range of SUM is approximately equal to 2 μg/m³; all odds ratios (ORs) will be standardized to this magnitude of concentration change.

The mean value of SUM at each school is modeled as a function of the proximity-to-industry region. Thus, the design matrix for the j-th school is of the form

$$\boldsymbol{w}_j = [I_{1j} \quad I_{2j} \quad I_{3j} \quad I_{4j}],$$

where I_{ij} is an indicator variable for the j-th school being located in the i-th region. We perform a two-stage analysis assuming the true value of SUM is solely a function

Table 1
Parameter estimates assuming no concentration measurement error (model NE) and performing two-stage (models TS1 and TS2) and simultaneous (models S1, S2, S3, and S4) estimation procedures in the presence of concentration measurement error

Model	β_1	β_0	σ_e^2	p-value
NE	0.0312	−1.2026	6.6090	0.064

Model	*Stage 2*			*Stage 1*				
	β_1	β_0	σ_e^2	π	ρ_v	σ_v^2	σ_δ^2	p-value
TS1	0.0727	−1.3918	6.6691	5.6778 4.8461 3.7652 3.9141			2.4407	0.095
TS2	0.0733	−1.3950	6.6344	5.6919 4.7977 3.6552 3.8768	0.6505	0.2515	2.2062	0.075

Model	β_1	β_0	σ_e^2	π	ρ_v	σ_v^2	σ_δ^2	log likelihood	*p-value*
S_1	0	−1.0590	6.7488	5.6778 4.8461 3.7652 3.9141			2.4407	−9.4867	
S2	0.0865	−1.4558	6.3882	5.4989 4.9917 3.6428 3.9510			2.4583	−7.7442	0.062
S3	0	−1.0590	6.7488	5.6919 4.7977 3.6552 3.8768	0.6505	0.2515	2.2062	−9.3047	
S4	0.0903	−1.4756	5.8914	5.7576 5.0211 3.7116 3.9354	0.0615	0.6301	1.9102	−6.8646	0.027

Table 2
Odds ratio estimates and likelihood-based confidence intervals for a 2 μg/m^3 increase in pollution concentration assuming no concentration measurement error (model NE) and performing two-stage (models TS1 and TS2) and simultaneous (models S2 and S4) estimation procedures in the presence of concentration measurement error

Model	OR estimate for Δ2 μg/m^3	95% CI
NE	1.064	(0.997, 1.136)
TS1	1.157	(0.977, 1.369)
TS2	1.158	(0.988, 1.358)
S2	1.189	(0.991, 1.659)
S4	1.198	(1.026, 1.617)

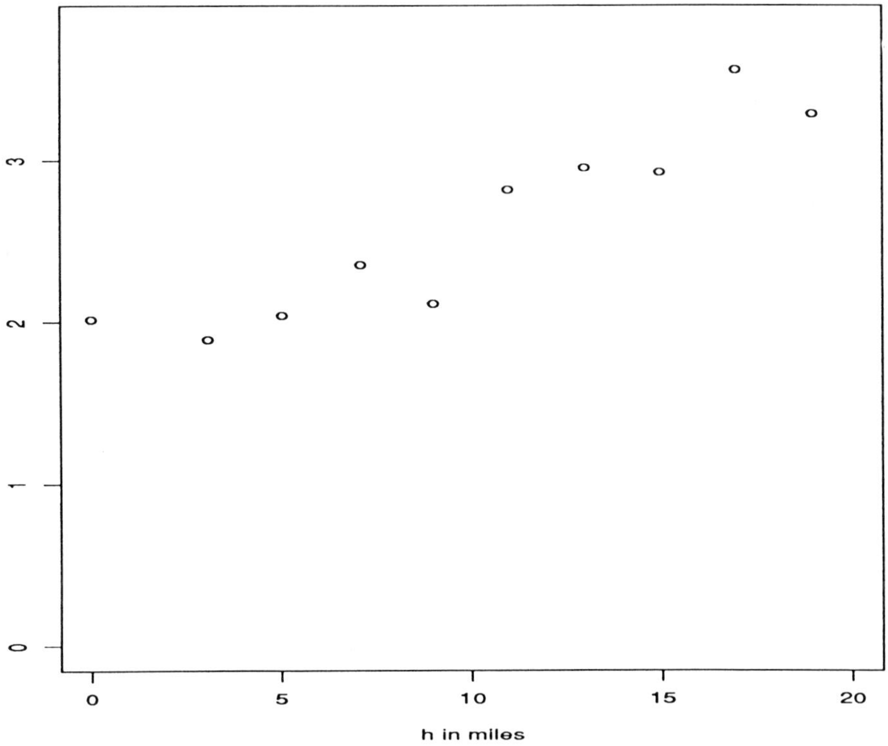

Fig. 1. Grouped empirical semi-variogram for z plotted as a function of the mean distance between each pair of schools.

of region (i.e., $\sigma_v^2 = 0$) such that

$$z_j = \boldsymbol{w}_j\boldsymbol{\pi} + \delta_j$$

for $j = 1$ to n where z_j is the observed SUM at school j, $E(\boldsymbol{\delta}) = \mathbf{0}$ and $\text{var}(\boldsymbol{\delta}) = \sigma_\delta^2 \boldsymbol{I}$. The true value of SUM for each school (x) is estimated as $\hat{x}_j = \boldsymbol{w}_j\hat{\boldsymbol{\pi}}$ where $\hat{\boldsymbol{\pi}}$ is estimated by multiple regression of z on $\boldsymbol{w}$. (See model TS1 in Table 1 for parameter estimates.) An empirical logistic regression is performed assuming x_j is equal to the estimate $\hat{x}_j$ for all schools. The second-stage parameter estimates are $\beta_0 = -1.3918$ and $\beta_1 = 0.0727$ (p-value $= 0.095$, 95% CI $= (-0.0114, 0.1569)$). Note that this CI is created ignoring the variability in the estimate of $\boldsymbol{\pi}$. These estimates correspond to an estimated OR of 1.157 for a 2 μg/m³ change in SUM. (See model TS1 in Table 2).

The mean empirical semi-variogram of the residuals ($\hat{r}_j = z_j - \boldsymbol{w}_j\hat{\boldsymbol{\pi}}$) (where $h_{jj'}$ values are grouped in two-mile intervals) is plotted in Figure 1. The increasing trend in the semi-variogram values indicates the presence of correlated variability about the mean in the true values (x). The correlation between v_j and $v_{j'}$ is assumed to be of the form $\rho_v^{h_{jj'}}$ corresponding to the exponential semi-variogram model. The maximum likelihood estimates of $\boldsymbol{\pi}$, ρ_v, σ_v^2 and σ_δ^2 are presented in Table 1 as model TS2. The true value of SUM for each school (x) is estimated as $\hat{x}_j$ using equation (1). An empirical logistic regression is performed assuming x_j is equal to the estimate $\hat{x}_j$ for all schools. The second-stage parameter estimates are $\beta_0 = -1.3954$ and $\beta_1 = 0.0734$ (p-value $= 0.075$, 95% CI $= (-0.0062, 0.1530)$). Note that the CI is created ignoring the variability in the estimates of $\boldsymbol{\pi}$, σ_v^2, and ρ_v. These estimates correspond to an estimated OR of 1.158 for a 2 μg/m³ change in SUM. (See model TS2 in Table 2.)

The simultaneous maximum likelihood estimation of parameters is undertaken assuming the observed covariates and dependent variables have a multivariate normal distribution presented in equation (3). Initially both σ_v^2 and σ_η^2 are assumed to equal zero. Parameter estimates are presented in Table 1 as models S1 (for β_1 constrained to equal zero) and S2. The maximum likelihood estimates are $\beta_0 = -1.4558$ and $\beta_1 = 0.0865$ (p-value $= 0.062$), and the likelihood-based 95% CI for β_1 of $(-0.0045, 0.2532)$. These estimates correspond to an estimated OR of 1.189 for a 2 μg/m³ change in SUM. (See model S2 in Table 2.)

In a second set of analyses, σ_η^2 is assumed to equal zero, and σ_v^2 and ρ_v are left unconstrained with $[\mathbf{r}]_{jj'} = \rho_v^{h_{jj'}}$ where $h_{jj'}$ is the distance in miles between schools j and j'. Parameter estimates are presented as models S3 (for β_1 constrained to equal zero) and S4. The maximum likelihood estimates are $\beta_0 = -1.4756$ and $\beta_1 = 0.0903$ (p-value $= 0.027$), and the likelihood-based 95% CI for β_1 is $(0.0128, 0.2404)$. These estimates correspond to an estimated OR of 1.198 for a 2 μg/m³ change in SUM. (See model S4 in Table 2.) There is no evidence of correlated variability in the empirical logit values about the mean function of SUM.

Assuming model S4, the H values for $\hat{\Psi}$ and $\hat{\Omega}$ are 131.59 and 2.82 (p-values < 0.001 and 0.244), respectively. (Since σ_η^2 is constrained to equal zero, $\hat{\Omega}_j$ is simply equal to $\hat{e}_j/\sqrt{\hat{\sigma}_e^2/N_{j\cdot}}$.) A univariate analysis of $\hat{\Psi}_j$ reveals that the school with the highest $\hat{\Psi}$ value has an estimated δ value 4.7 standard deviations from zero.

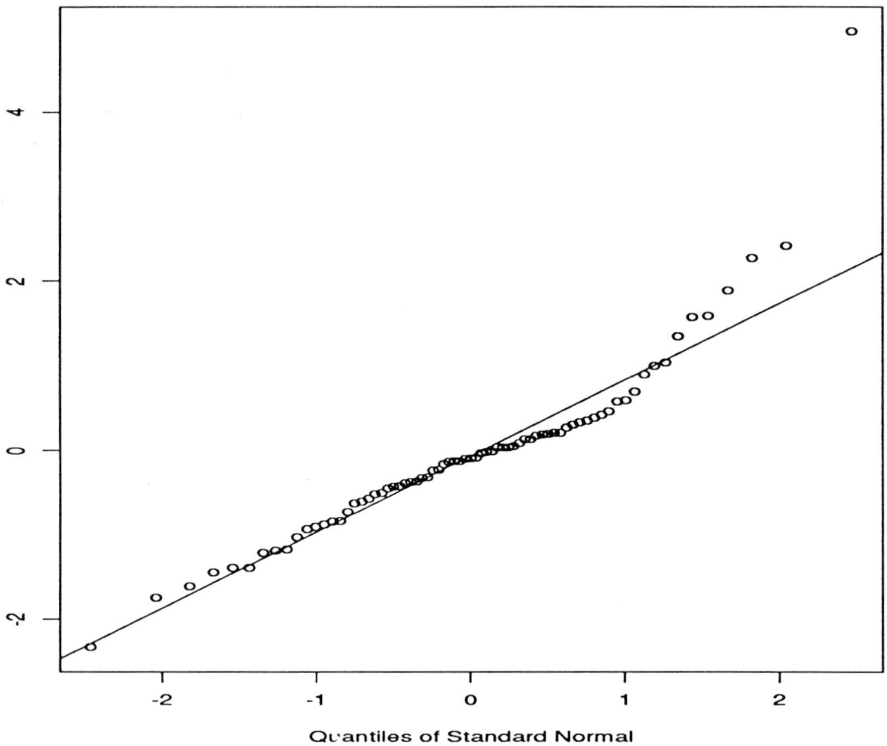

Fig. 2. Empirical quantile–quantile plot for $\hat{\Psi}$, the orthonormalized sum of $\hat{v}$ and $\hat{\delta}$.

Removing this school from the calculations yields H values for $\hat{\Psi}$ and $\hat{\Omega}$ of 3.96 and 3.03 (p-values 0.138 and 0.220), but yields similar parameter estimates. The quantile–quantile plots (Figures 2 and 3) confirm the conclusions that $\hat{\Omega}$ is approximately normally distributed and $\hat{\Psi}$ is approximately normally distributed with the exception of one unusually high value.

As anticipated, based on laboratory tests, the measurement error in the observed SUM values is considerable. The expected sum of squares for the observed covariate values z is of the form

$$\mathrm{E}[(z-\bar{z}\,\mathbf{1})'(z-\bar{z}\,\mathbf{1})]=\left(\boldsymbol{\pi}'\boldsymbol{w}'\boldsymbol{w}\boldsymbol{\pi}-\frac{1}{n}(\mathbf{1}'\boldsymbol{w}\boldsymbol{\pi})^2\right)+\sigma_v^2\left(n-\frac{1}{n}\mathbf{1}'\boldsymbol{r}_v\mathbf{1}\right)+\sigma_\delta^2(n-1),$$

where $\bar{z}_{.}=(1/n)\sum_{j=1}^{n}z_j$. The parameter estimates can be used to estimate the sum of squares due to the fixed effects

$$s_\pi=\left(\hat{\boldsymbol{\pi}}'\boldsymbol{w}'\boldsymbol{w}\hat{\boldsymbol{\pi}}-\frac{1}{n}(\mathbf{1}'\boldsymbol{w}\hat{\boldsymbol{\pi}})^2\right),$$

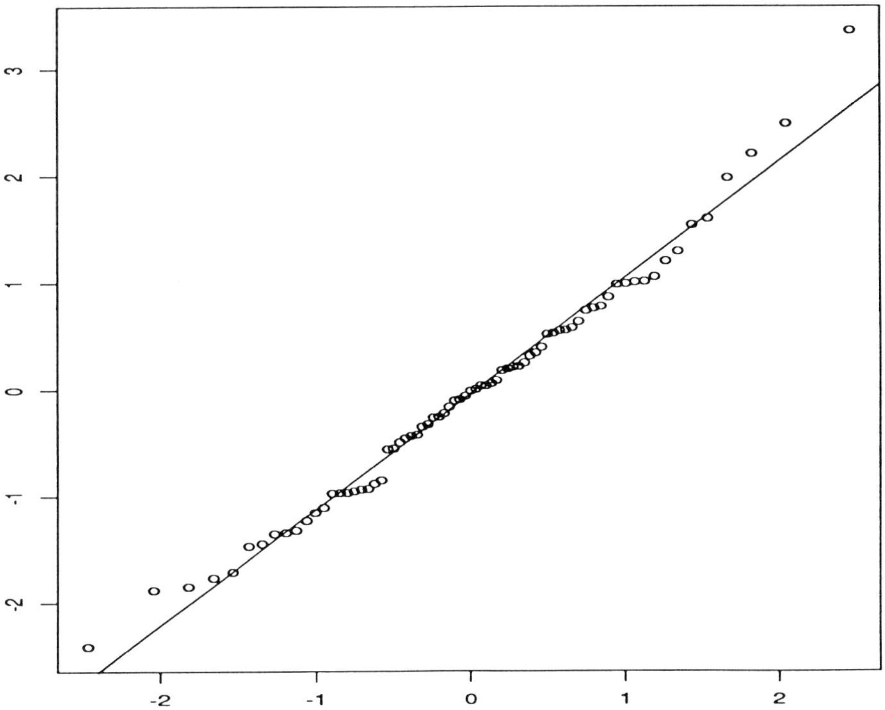

Fig. 3. Empirical quantile–quantile plot for $\hat{\Omega}$ normalized $\hat{e}$.

the sum of squares due to correlated variability in the true covariate values,

$$s_v = \hat{\sigma}_v^2\left(n - \frac{1}{n}\mathbf{1}'\hat{\boldsymbol{r}}_v\mathbf{1}\right),$$

and the sum of squares due to covariate measurement error,

$$s_\delta = \hat{\sigma}_\delta^2(n-1)$$

and the total sum of squares $s_{\text{total}} = s_\pi + s_v + s_\delta$. The observed sum of squares for z is 215.861. The estimated sums of squares (s_π, s_v and s_δ) and the fraction of the observed variability attributed to variability in the true covariate values, $(s_\pi + s_v)/s_{\text{total}}$, are presented in Table 3 for the two-stage (models TS1 and TS2) and simultaneous (models S2 and S4) estimation procedures. Note that the model with the highest fraction $(s_\pi + s_v)/s_{\text{total}}$ is the only model to yield a significant result at the 0.05 level of significance. Although the percent of variability in z due to error is quite large for all of the measurement error models, the percent deviations of the observed z values from the estimated x values, $\hat{x}_j = \boldsymbol{w}_j\hat{\boldsymbol{\pi}} + \hat{v}_j$ (equation (2)), are comparable to those

stations at the same T points in time. Let $Z_{it} = Z(x_i, t)$ denote the observation taken at location x_i at time t, $i = 1, 2, \ldots, N$, $t = 1, 2, \ldots, T$. We consider models of the form

$$Z(x, t) = \mu(x, t) + E_\tau(x, t) + E_\varepsilon(x, t),$$

where $\mu(x, t)$ represents a mean field which may include temporal and spatial trends, $E_\tau(x, t)$ is a mean zero space–time process which is L_2-continuous in space, and $E_\varepsilon(x, t)$ represents measurement error and small scale spatial variability. The process E_τ is assumed to be temporally stationary with a spatial covariance function that depends smoothly on geographic coordinates. In some applications (e.g., Haslett and Raftery, 1989; Oehlert, 1992) it is possible to further decompose this field into a spatially pervasive long-memory temporal process, and a component depicting short-term and limited range space–time features.

We assume that the distribution of the measurement error process E_ε is independent of location x and time t, as well as being independent of the process E_τ. In terms of covariance, the presence of the measurement error process induces a discontinuity in the spatial covariance, in that

$$\operatorname{var} Z(x, t) - \lim_{y \to x} \operatorname{cov}(Z(x, t), Z(y, t)) = \operatorname{var} E_\varepsilon(x, t).$$

This nonzero quantity on the diagonal, as it were, is called the *nugget effect* in geostatistics (e.g., Journel and Huijbregts, 1978).

There are two types of approaches to the specification and analysis of spatial covariance structure for space–time monitoring data generated by models such as this. One approach, which may be called the geostatistical approach, concentrates on a single snapshot or time-slice of the field at one time t. The other, call it the meteorological approach, assumes that there are several time-slices, either considered as independent realizations of the field, or at least as temporally stationary realizations. With the exception of Sections 2 and 3, we will largely pursue the meteorological approach in this chapter. Haslett (1989, Section 2) gave an extensive discussion of this distinction.

In this chapter we will largely ignore the (nontrivial) problem of modeling the mean field μ. While this may often be the focus of the research, the particular aspect of interest to us is the estimation of the spatial covariance. Among recent work on estimating mean acidic deposition fields are Le and Petkau (1988), Loader and Switzer (1992) and Oehlert (1992). We shall henceforth assume (without changing our notation) that the estimated mean has been removed from the data. In addition we assume that the space–time covariance function is *separable* in the sense that it factors into a product of a spatial covariance and a temporal covariance. This is a fairly strong assumption, but in our experience the temporal dependence of the spatial covariance is often mainly a seasonal dependence, and can be dealt with, if necessary, by analyzing the seasons separately (see Section 10.1 for an alternative treatment).

In the analysis of most space–time processes underlying environmental studies, there is little reason to expect spatial covariance structures to be homogeneous over the spatial scales of interest. Certainly, the heuristic arguments underlying assumptions of stationarity in time series do not readily apply to spatial processes. For example, in studies of the dispersion of atmospheric pollutants, landscape or topography affects weather patterns, and thereby the transportation of pollutants. Empirical spatial covariances may be computed using time series of residuals from some form of estimated location (and possibly time) dependent process means. These covariances may reflect paths of pollutant transportation (for data averaged over appropriate temporal scales), and should therefore be heterogeneous insofar as the landscape varies from place to place.

Prior to the 1980s, heterogeneity was most often addressed (when recognized) by partitioning a spatial field into relatively homogeneous regions. Anisotropy in the covariance structure was recognized in both the geostatistical literature (Journel and Huijbregts, 1978) and the meteorological literature (Thiébaux, 1977, 1991) and treated as a slowly varying function of the azimuth (cf., Section 2 below).

In what follows we describe several different approaches to estimating the spatial covariance function. The simplest assumption, namely that the covariance structure is homogeneous and isotropic, is the basis of much of the geostatistical work. We outline, very briefly, some of the methods applicable to this model in Section 2. We also indicate how one may extend this to an anisotropic, but still homogeneous, model. In Sections 3 and 4 we present two approaches to the estimation of heterogeneous covariances by methods of smoothing. The first is a purely geostatistical approach, while the second follows the meteorological path. Section 5 outlines an empirical Bayes approach to incorporating prior assumptions as to the structure of the covariance, while in Section 6 we present a general method of decomposition of the covariance in terms of certain orthogonal functions. A newer approach, using a deformation of the geographic domain into a space in which the methods of Section 2 apply, is presented in Section 7. This method combines flexibility with ease of interpretation. It is generally difficult to display covariance fields, and in Section 8 we present a method that we find useful for this purpose. The use of covariance estimates in estimation and network redesign is the topic of Section 9, while extensions to genuine space–time covariance models and to multivariate processes are outlined in Section 10.

2. Homogeneous approaches

In the theory of temporal stochastic processes, it has been quite useful to study *stationary* processes, which, in the strict sense of the word, are invariant under temporal shifts, i.e., whose distribution does not depend on the time origin. If in addition such processes have covariance functions that decrease with increasing separation, statistically valuable tools like the central limit theorem will apply (see, e.g., Brillinger, 1981, for a theoretical discussion). In the case of spatial processes, the property of invariance under shifts is usually called *homogeneity*. Matters are especially simple if the distribution of the process is also invariant under rotations, in which case it is called *isotropic*.

Define the *dispersion* between two sites located at x and y, respectively, by

$$\delta(x, y) = \mathrm{var}(Z(x,t) - Z(y,t)) = \mathrm{var}Z(x,t) + \mathrm{var}Z(y,t) - 2\mathrm{cov}(Z(x,t), Z(y,t)). \tag{2.1}$$

Knowing the variance field, one can compute the covariance from the dispersion and vice versa. If the spatial field is homogeneous, δ is a function only of the vector $x - y$, rather than the absolute locations x and y. If the field is also isotropic, δ is a function only of the length of this vector. Hence we can write

$$\delta(x, y) = h(|x - y|).$$

The function h is often called the *variogram* (terminology originating in the field of geostatistics). In the atmospheric science literature this is called the *structure function*. The choice of variogram h must be made so as to ensure that the corresponding covariance function is valid, i.e., nonnegative definite. Some popular choices of variograms are (cf. Cressie, 1991, Chapter 2.3.1):

Linear $\quad h(s) = \sigma_0^2 + \sigma^2 s,$

Exponential $\quad h(s) = \sigma_0^2 + \sigma^2(1 - \exp(-\alpha s)),$

Gaussian $\quad h(s) = \sigma_0^2 + \sigma^2(1 - \exp(-\alpha s^2)),$

Spherical $\quad h(s) = \begin{cases} \sigma_0^2 + \sigma^2(1.5\alpha s - 0.5(\alpha s)^3), & 0 < s \leqslant \alpha, \\ \sigma_0^2 + \sigma^2, & s > \alpha. \end{cases}$

Here σ_0^2 is the nugget effect. For isotropic homogeneous spatial random fields, contours of constant spatial dispersion, relative to a fixed reference point, are concentric circles.

The assumptions of homogeneity and isotropy may be natural in some geological applications, such as mining. More importantly, they may be necessary in order to estimate the variogram. If there is only one realization of the process, rather than the repeated observations typical for monitoring networks, the variogram has to be estimated empirically using average squared distance between values of process at similar distances, i.e., averages over all pairs of points that are within a bandwidth parameter of being the same distance apart. As in many other statistical applications, the bandwidth parameter determines the tradeoff between bias and variability. A specific variogram model can then be fitted to this empirical variogram using (weighted) least squares or other methods (Cressie, 1991, Sections 2.4 and 2.6). For example, Cressie et al. (1990) fitted a spherical variogram to yearly hydrogen deposition from 19 stations in the UAPSP acid precipitaion monitoring network in 1982, and separately for 1983 data. The two fits were rather different, and neither fit the corresponding empirical variogram very well. Using more data from the same network, and a temporal perspective, Guttorp et al. (1992) deduced substantial deviations from isotropy and homogeneity (cf. Section 9.1).

In environmental applications isotropy and homogeneity are usually reasonable assumptions only over small areas. Orographic effects, synoptic wind patterns, and the irregular mesoscale structure of precipitation fields all contribute to anisotropy in regional patterns of acidic precipitation. A first step towards a more realistic model is to drop the isotropy assumption, while maintaining the spatial homogeneity. A simple way of doing this is to divide the directions up into zones, and estimate the variogram separately in each directional zone (so-called zonal anisotropy). Many homogeneous variogram models have elliptical anisotropy (also called geometric anisotropy), i.e., the contours of constant dispersion relative to a reference point are concentric ellipses. We can write

$$h(u) = \gamma((u'Bu)^{1/2}),$$

where u is the vector $x - y$ and γ an appropriately chosen monotone function (here we do not consider hole effects or periodic variograms which are not monotone). In this case a linear transformation of the station coordinates, namely

$$x \to f(x) = Ax,$$

where $A'A = B$, results in a problem with isotropic covariance structure. Thus, most variogram models commonly used in geostatistical practice assuming homogeneity can be expressed as

$$\delta(x, y) = h(x - y) = \gamma(|f(x) - f(y)|). \tag{2.2}$$

In other words, a linear transformation of the original map yields a map on which dispersion corresponds to distance.

Let d_{ij} be the empirical dispersion between the observed values of the field at monitoring locations x_i and x_j. In the case of monitoring data this can be estimated using the sample covariance between the monitoring stations and equation (2.1). It may then be reasonable to separate out the two estimation problems: the linear map A and the variogram γ. For a given class of variogram functions, depending on a low-dimensional parameter θ, one would minimize over A and θ, either simultaneously or separately in an alternating iterative fashion, the quantity (Cressie, 1985)

$$\sum_{i,j}\left(\frac{d_{ij} - \gamma(|A(x_i - x_j)|;\theta)}{\gamma(|A(x_i - x_j)|;\theta)}\right)^2. \tag{2.3}$$

Under the assumption of homogeneity there is a Fourier approach to covariance estimation. For simplicity we consider only one time-slice t with observations $Z(x_i, t)$, $i = 1, \ldots, N$. Write

$$\sigma^2\rho(u) = \operatorname{cov}(Z(x,t), Z(x+u,t)),$$

where $\sigma^2 = \mathrm{var}Z(x,t)$, and $\rho(u)$ is the spatial correlation function for separation vector u. The *normalized wave-number spectrum* $\phi(w)$ is the Fourier transform of the correlation function, i.e.,

$$\phi(w) = \int \rho(u) \exp(\mathrm{i}(u \cdot w))\, \mathrm{d}w.$$

As in the time series case there is a spectral representation of the process Z (e.g., Bartlett, 1966, Section 6.50), namely

$$Z(x, t) = \int \exp(\mathrm{i}(x \cdot w))\, \mathrm{d}\zeta(w),$$

where $\zeta(w)$ is a complex-valued process orthogonal in both components of w with

$$\mathrm{E}|\zeta(w)|^2 = \sigma^2 \phi(w).$$

This suggests basing an estimate of ϕ (and, by back-transformation, ρ) on the discrete Fourier transform of the data. Brillinger (1985) expands on this. If the observations are on a regular grid (as would be the case, for example, for some remote sensing applications) it is possible to use FFT methods. However, for landbased irregularly spaced monitoring data the corresponding estimation problem is difficult. The methods of Brillinger (1972) apply.

3. Moving windows

Although most environmental processes are heterogeneous in both mean and covariance structure, the process may often be considered approximately homogeneous over subregions. With this in mind Haas introduced a technique called 'moving-window kriging' in which the spatial covariance structure (represented in terms of a variogram model) is estimated in a circular window about each target point at which a spatial (kriging) estimate is to be computed (Haas, 1990a,b). This approach, which provides an alternative to partitioning the overall region of interest into a small number of non-overlapping subregions, appears to be the only method suggested to deal with heterogeneity in the context of a single realization of a spatial process.

Haas designed his procedure for estimation of annual wet acid deposition using data from the NADP/NTN monitoring network. At each point of a relatively fine grid on which spatial estimates are desired he selects a window containing at least 35 sites. He then enlarges the radius of the window, as necessary, until there is at least one pair of sites represented in each of the distance bins over which a variogram is to be estimated (generally with at least two pairs in the shortest spatial lag bin of 50 km), and so that the nonlinear regression estimation of an isotropic, spherical variogram model converges. See Haas (1990b) for further discussion of the rationale underlying these heuristics. Haas uses only isotropic variogram models, although his general approach could accommodate anisotropic models as presented here in Section 2.

Haas' approach is intuitively attractive, albeit computationally demanding. (In Haas, 1990b, he computed variogram parameters by nonlinear least squares in windows centered on each of 2000 grid points – and solved 2000 kriging systems for the point estimates!) Perhaps the major source of concern in this procedure is the ad hoc approach to choice of window sizes and the difficulty in assessing the uncertainty in the estimated parameters of the spatially varying variogram models. Cross-validatory choice of window size does not appear feasible. Nonetheless, for point estimation with single realizations the general approach is quite sensible. This is a purely spatial (or geostatistical) approach, rather than a meteorological approach.

4. A kernel smoothing method

Oehlert (1992) introduced a kernel method for estimating spatial covariance, based on the idea that spatial covariance (or dispersion) should vary relatively smoothly. Define a kernel κ for two pairs of locations (x_1, x_2) and (x_3, x_4) by

$$\kappa((x_1, x_2), (x_3, x_4)) = \begin{cases} \exp(-(d(x_1, x_3)^2 + d(x_2, x_4)^2/\lambda_\kappa^2) \\ \quad \text{if } x_1 \neq x_2 \text{ and } x_3 \neq x_4, \text{ or} \\ \quad x_1 = x_2 \text{ and } x_3 = x_4, \\ 0 \quad \text{otherwise,} \end{cases}$$

where $d(x, y)$ is the Euclidean distance between x and y, and λ_k is a smoothing parameter. Given appropriately detrended and deseasonalized residuals and a corresponding sample spatial covariance matrix $S = [S_{ij}]$, based on monitoring N stations at locations $x_1, \ldots, x_N$, the smoothed covariance matrix $\hat{S}$ is given by

$$\hat{S}_{ij} = \frac{\sum_{k,l} S_{kl} \kappa((x_i, x_j), (x_k, x_l))}{\sum_{k,l} \kappa((x_i, x_j), (x_k, x_l))}.$$

Notice that this method smooths variances and covariances separately, but using the same smoothing parameter. A small value of λ_k essentially returns the raw matrix S. For the purposes in Oehlert's paper, namely estimation of spatial means and trend slopes, the choice of λ_k was not critical. He proposed to choose it so that a plot of the rescaled variances (observed divided by smoothed) were linear on a chi-square plot, with degrees of freedom chosen appropriately for Gaussian theory. It is not clear whether or not this procedure would always yield a valid (i.e., positive definite) covariance structure. When applied to annual sulfate data from several acid rain monitoring networks in the northeastern United States and southeastern Canada (APIOS-C, MAPS3S, NADP/NTN and UAPSP), the resulting covariance estimate showed distinct, but weak, covariance structure, with strongest covariance along the

NE/SW axis, consistent with the prevailing southwesterly winds, and weakest along a NNW/SSE axis.

5. An empirical Bayes approach

Loader and Switzer (1992) suggested an empirical Bayes type of shrinkage approach to smoothing the sample spatial covariance matrix S. Specifically, assuming an underlying Gaussian distribution for monitoring data, they assigned the inverted Wishart conjugate prior distribution for the unknown true $N \times N$ covariance matrix Σ with density

$$f_{\Sigma}(\Sigma \mid \tilde{C}, m) = \frac{|\tilde{C}|^{m/2} |\Sigma|^{(m+p+1)/2}}{2^{mp/2} \Gamma_p(m/2)} \exp(-\tfrac{1}{2}\mathrm{tr}(\Sigma^{-1}\tilde{C})) \tag{5.1}$$

with prior mean $C = \tilde{C}/(m - p - 1)$ and $\Gamma_p(t) = \pi^{p(p-1)/4} \prod_{j=1}^{p} \Gamma(t - \frac{1}{2}(j-1))$. The conditional distribution of S given Σ is then a Wishart distribution and the posterior mean of Σ has the form

$$\mathrm{E}(\Sigma \mid S) = \lambda S + (1 - \lambda) C, \tag{5.2}$$

where $\lambda = n/(n + m - p - 1)$.

In practice, the prior mean C is specified by fitting a parametric model to the observed S_{ij}. In analyzing monthly average sulfate concentrations from the UAPSP network over the two-year period 1982–83 they fitted a simple exponential model, $C(x, y) = \sigma^2 \exp(-\alpha s)$, where $s = |x - y|$. The parameter m of the prior distribution, which specifies the shrinkage parameter λ, was estimated under an empirical Bayes paradigm from the marginal density of S after integrating with respect to the prior.

Extension of this approach to covariances involving unmonitored sites is somewhat problematic. For a single unmonitored location x, they followed a suggestion of Switzer (1989) in estimating the vector of covariances by $\sigma(x) = (\sigma(x, x_1), \ldots, \sigma(x, x_N))$,

$$\hat{\sigma}(x) = \hat{\Sigma} C^{-1} c(x),$$

where $\hat{\Sigma}$ is the $N \times N$ posterior mean given in (5.2) and C and $c(x)$ are the $N \times N$ matrix of covariances and the N-vector of covariances, respectively, from the specified a priori parametric model. Note that when taking $x = x_i$, one of the actual monitoring sites, this simply returns the i-th column of $\hat{\Sigma}$, so the extension is faithful to the empirical Bayes estimates for the observed locations.

This approach does not lend itself easily to providing covariance estimates for an arbitrary number of unmonitored locations. Loader and Switzer suggested a sequential approach which, fortunately, is invariant to the order of the sequence of new sites, but for which the specification of the diagonal elements (variances) for the new sites can cause problems with positive-definiteness. This framework provides no general mechanism for estimation of covariances involving an arbitrary pair of unmonitored sites.

A disturbing aspect of this extension to unmonitored sites is that it does not affect spatial point estimates. Assuming known mean and covariance fields, $\mu(x, t)$ and $\sigma(x, y)$, respectively, the Gaussian-based maximum likelihood estimate of the random field $Z(x, t)$ at an unmonitored site x is

$$\hat{Z}(x, t) = \mu(x, t) + \sigma(x)' \Sigma^{-1}(Z(\cdot, t) - \mu(\cdot, t)).$$

Substituting the estimators given above we find that

$$\hat{\sigma}(x)' \hat{\Sigma}^{-1} = c(x)' C^{-1},$$

i.e., the estimated value at location x depends only on the (prior) parametric covariance estimate, not on the empirical Bayes estimate $\hat{\Sigma}$. The variance of the estimated value, however, does depend on $\hat{\Sigma}$.

For the application to 19 UAPSP acid deposition monitoring sites in the eastern US, plots of interpolated covariances displayed essentially no local anisotropy. This stands in contrast to the nature of the spatial covariance structure estimated using similar data in Sections 4 and 9.1. Switzer, in a recent technical report (Monestiez and Switzer, 1991), written subsequent to the article by Loader and Switzer (despite the publication dates), used a multi-dimensional scaling modeling approach similar to that introduced by Sampson and Guttorp (1992) and discussed in Section 7.

6. Empirical orthogonal functions

The first, and seemingly the most general, published analytic approach to heterogeneity in space–time process is provided by an analysis of a space–time random field in terms of a Karhunen–Loève decomposition (see, e.g., Adomian, 1983, Section 2.10) or *empirical orthogonal functions* (EOFs, see, e.g., Preisendorfer, 1988). While the focus of most of the applications of this approach has been on the interpretation of these EOFs (or spatial principal components), Obled and Creutin (1986; see also Creutin and Obled, 1982) demonstrated that an EOF analysis could be used for spatial estimation and that it readily accommodated anisotropic and even heterogeneous covariance or correlation structure. We will begin here with the simpler analysis in terms of spatial EOFs or principal components as presented, for example, in Preisendorfer (1988).

The Karhunen–Loève decomposition of the (mean-centered) field $Z(x, t)$ in terms of spatial and temporal components is

$$Z(x, t) = \sum_{k=1}^{\infty} F_k(x) A_k(t), \tag{6.1}$$

where $F_k(x)$ represents the k-th spatial component (EOF) and $A_k(t)$ represents the temporally varying amplitude of the k-th spatial component. Most of the atmospheric science literature addresses this decomposition only for a finite number of monitoring sites $x_i, i = 1, 2, \ldots, N$ and observation times $t, t = 1, 2, \ldots, T$. In this case the decomposition evaluated at the sampling sites x_i involves a sum with N_T terms, $N_T = \min(N, T)$. Let Z denote the $N \times T$ matrix of observations $[Z(x_i, t)]$. With the usual scaling conventions we compute the $N \times N_T$ matrix $F = [F_{ik}] = [F_k(x_i)]$ as eigenvectors of the sample spatial covariance matrix, $S = (1/T)ZZ'$ so that

$$F'F = I.$$

These components are thus analytically orthogonal in space. The $N_T \times T$ matrix of temporal components $A \times [A_{kt}] = [A_k(t)]$ are statistically orthogonal or uncorrelated and satisfy

$$\frac{1}{T} A'A = \Lambda,$$

where Λ is the diagonal matrix of eigenvalues of S. Because these coefficients of the F_i in the expansion of $Z(x, t)$ are uncorrelated, this is a spectral representation.

The spatial covariance matrix S is then represented in terms of the EOFs by the usual spectral decomposition

$$S = F'\Lambda F = \sum_{k=1}^{N_\tau} \lambda_k F_k F_k'. \tag{6.2}$$

The sample estimates for station pairs S_{ij} can be approximated (or smoothed) by retaining a smaller number of spatial components. Note that this type of model for the observed spatial covariances, unlike the nonparametric model of Oehlert (Section 4 above) and the empirical Bayes approach of Loader and Switzer (Section 5 above) makes no explicit reference to spatial distance and takes no account of the relative spatial locations of the observations. For this reason it does not provide estimates of spatial covariance at unmonitored locations without extending or spatially interpolating the EOFs which are computed only at the monitoring sites

Extension of an EOF analysis in order to address spatial estimation appeared in Cohen and Jones (1969) and Buell (1972, 1978). Here we present the development of Creutin and Obled (1982) and Obled and Creutin (1986). In the general case the Karhunen–Loève expansion represents $Z(x, t)$ for arbitrary spatial location x and time t as the sum in equation (6.1) with the eigenfunctions $F_k(x)$ being solutions of the Fredholm integral equation

$$\int S(x, y) F_k(x)\, dx = \lambda_k F_k(y)$$

with the analytic orthogonality constraint

$$\int F_k(x)F_l(x)\,\mathrm{d}x = \delta_{kl},$$

where δ_{kl} is the Kronecker delta function.

Generalizing the finite decomposition of the sample covariance matrix S given in (6.2) above, we can represent the covariance function as

$$S(x, y) = \sum_{k=1}^{\infty} \lambda_k F_k(x) F_k(y).$$

A solution to this general problem requires a numerical approximation to the Fredholm integral equation. Given N sampling sites we can estimate at most N eigenfunctions. Obled and Creutin solved this by approximating the true field $Z(x, t)$ (and the eigenfunctions) in terms of a vector space spanned by an elementary set of generating functions $e_1(x), \ldots, e_p(x)$. The simplest approach (yielding the solution proposed by Cohen and Jones, 1969) is to take generating functions to be piecewise constant non-overlapping functions with $e_i(x)$ set equal to one on the Thiessen polygon P_i containing x_i and zero elsewhere. In this case, since the $e_i(x)$ are orthogonal, the numerical problem to be solved remains a symmetric $N \times N$ eigenproblem as in the case of the principal component solution above, but with weights on the elements of the covariance matrix reflecting the areas of the associated Thiessen polygons. Other generating functions, like the facet-like linear ones used by Obled and Creutin (1986), considerably complicate the definition of the eigenproblem to be solved. See their work for further details, as well as Preisendorfer (1988, Section 2d).

Cruetin and Obled applied this EOF approach to spatial estimation for a network of 101 rainfall monitoring stations and 81 rainfall events over a period of 20 years. They fitted the EOF model to 73 stations and reserved 28 stations for testing the model. In their analysis they argue that it is appropriate to retain 39 components in the expansion. The resulting spatial covariance model is certainly complex. It would appear to be difficult to study and interpret the structure of this spatial covariance model.

Some insight can be gained by considering the vector

$$F^p(x_i) = (\sqrt{\lambda_1}F_1(x_i), \sqrt{\lambda_2}F_2(x_i), \ldots, \sqrt{\lambda_p}F_p(x_i))$$

as coordinates for a representation of location x_i in p dimensions. Then

$$|F^p(x) - F^p(y)|^2 = \sum_{k=1}^{p} \lambda_k (F_k(x) - F_k(y))^2$$
$$= \sum_{k=1}^{p} \lambda_k F_k(x)^2 + \sum_{k=1}^{p} \lambda_k F_k(y)^2 - 2\sum_{k=1}^{p} \lambda_k F_k(x) F_k(y).$$

The right-hand side is the p-dimensional EOF model for the spatial dispersion, i.e., the dispersion can be estimated by

$$\hat{\delta}(x, y) = |F^p(x) - F^p(y)|^2. \tag{6.2}$$

Thus the spatial covariance or spatial dispersion structure can be understood in terms of distances between stations represented as points in a high-dimensional Euclidean space. We will contrast this model with equation (2.2) and a related model in Section 7 below.

7. Multidimensional scaling

Sampson and Guttorp (1992) proposed an approach to estimation of heterogeneous spatial covariance that allows estimates between any two points of the field, whether monitored or not. Recall from equation (2.2) that most homogeneous variograms are of the form

$$\delta(x, y) = \gamma(|f(x) - f(y)|),$$

where f is a linear transformation (the notation in this chapter is somewhat different from that in Sampson and Guttorp, 1992). In order to produce a large class of heterogeneous dispersions, we allow f to vary over the class of smooth invertible functions. The basic concept is that dispersion is a smooth function of geographic coordinates and is locally anisotropic. By transforming the geographic map appropriately we get a new representation where dispersion is a monotone function of distance, i.e., where the dispersion is homogeneous and isotropic. The EOF representation (6.2) is of this form, with $f \equiv F^p$ being a mapping from $\mathbb{R}^2$ to $\mathbb{R}^p$, and $\gamma(s) = s^2$. The approach of Sampson and Guttorp described in the remainder of this section is more flexible. We suspect that it will lead to far simpler (lower-dimensional) models in most cases, but extensive comparisons have not yet been carried out.

The first step in implementing this idea is to transform the geographic map of the monitoring stations – the 'G-plane' – to a map (possibly in dimension higher than 2) where dispersion increases (approximately) monotonically with distance – the 'D-image'. This can be accomplished using non-metric multi-dimensional scaling (MDS). The D-image configuration of stations with coordinates $\tilde{x}_i$ is chosen so that interpoint distances $h_{ij} = |\tilde{x}_i - \tilde{x}_j|$ minimize the stress measure

$$\min_{\psi} \frac{\sum_{i<j}(\psi(d_{ij}) - h_{ij})^2}{\sum_{i<j} h_{ij}^2}, \tag{7.1}$$

where the minimum is taken over all monotone functions ψ, so that $\psi(d_{ij})$ represents a least squares monotone regression of the h_{ij} on estimated spatial dispersions d_{ij} (Mardia, Kent and Bibby, 1979). In order to ensure the validity of the resulting dispersion estimate when the variance field is not constant, we compute the d_{ij} based on the *correlation dispersion*, defined by

$$\delta^*(x, y) = 2 - 2\text{corr}(Z(x, t), Z(y, t)),$$

which can be estimated from the empirical correlation matrix.

We next compute a smoothly invertible function f that maps the G-plane into the D-image. Generally (except for applications with extremely high covariances, such as that in Sampson and Guttorp, 1992) an interpolating function that maps geographic coordinates x_i, $i = 1, \ldots, n$ into the coordinates $\tilde{x}_i$ returned by the MDS algorithm will not be invertible. Although it is natural to initialize the MDS algorithm with the geographic coordinates, the final coordinates $\tilde{x}_i$ may be quite jumbled in comparison to the geographic configuration because of noise in the empirical d_{ij}. In effect, the MDS algorithm fits the d_{ij} too accurately, in that the mean squared error in the variogram scatter of h_{ij} versus d_{ij} is less than the sampling error in the d_{ij}.

To correct both these problems we compute the function f using thin-plate smoothing splines (Wahba, 1990), one for each D-image coordinate, and write $f_{\mathrm{d}}(x) = x^*$. In the limit, as the spline smoothing parameter gets arbitrarily large, the mapping $f_{\mathrm{d}} : \mathbb{R}^2 \to \mathbb{R}^2$ is linear and the resulting spatial dispersion model becomes homogeneous (as in Section 2). We choose the smoothing parameter to ensure that the spread in the plot of $h_{ij} = |x_i^* - x_j^*|$ against d_{ij} is similar in size to the sampling variability of the d_{ij}, as computed for the Gaussian case.

A plot of MDS distances h_{ij} versus empirical dispersions d_{ij} indicates the quality of this spatial model. To this plot we fit a general model for isotropic homogeneous dispersions (i.e., a variogram model), since by construction the measurements, when referred to the D-image rather than the G-plane, correspond to a homogeneous and isotropic random field. Here it is possible to use any valid variogram model, as well as the spectral approach of Section 2. In the spirit of nonparametric estimation we note, following Matérn (1986, Section 2.3), that every isotropic correlation function in $\mathbb{R}^n$, continuous except at the origin, can be written in the form

$$\rho(h) = \sigma_0^2 + G(0) + \int_0^\infty \Lambda_k(hw)\, \mathrm{d}G(w), \tag{7.2}$$

where

$$\Lambda_k(v) = k!(2/v)^k J_k(v),$$

J_k is the Bessel function of the first kind with $k = (n-2)/2$, and $G(x)$ is an arbitrary distribution function of a nonnegative random variable. Results of Böhning (1982) generalize to produce an algorithm for the fitting of this model. Sampson and Guttorp (1992, Appendix) applied the algorithm to the special case of a mixture of Gaussian variogram models. The Gaussian model is obtained from (7.2) by choosing

$$\mathrm{d}G(w) \propto w^{n-1} \exp\left(\frac{-w^2}{4a^2}\right)$$

(Matérn, 1986, Section 2.4). The mixture of Gaussian variograms is the general form for a variogram model that is valid in any dimension (Schoenberg, 1938, pp. 817 ff.). In Guttorp et al. (1993) a simple exponential variogram was used. A mixture of exponential variograms yields the class of completely monotone variogram functions (Matérn, 1986, Section 2.4; he also outlines how this is obtained from (7.2)). An improved algorithm for these mixture problems is given by Lesperance and Kalbfleisch (1992).

We estimate the spatial variance field $V(x)$ separately using another thin-plate smoothing spline, $f_v: \mathbb{R}^2 \to \mathbb{R}^1$. The smoothing parameter is chosen so that the difference between raw and smoothed variances at each station behaves like the standard error of the raw variances for a Gaussian situation.

The composition of the computed smoothing spline f_d, the fitted variogram γ, and the estimated variance field f_v yields our model for the spatial covariance structure. In order to estimate the dispersion between two points x and y (monitored or unmonitored), one first computes their locations in the D-image, $x^* = f_d(x)$ and $y^* = f_d(y)$, and then estimates their correlation dispersion by $\gamma(|x^* - y^*|)$. In order to compute the corresponding covariance estimate $\widehat{\text{cov}}(x, y)$ on the original (non-standardized) scale of measurement, let $\hat{V}_x = f_v(x)$. Then the estimated covariance is given by

$$\widehat{\text{cov}}(x, y) = (\hat{V}_x \hat{V}_y)^{1/2}(1 - \tfrac{1}{2}\gamma(|x^* - y^*|)).$$

The resulting covariance estimate is nonnegative definite as required by general theory. Cross-validation or resampling methods can be used to estimate the accuracy of these estimates.

Monestiez and Switzer (1991) noted that the separation of the scaling from the estimation of the variogram was unnatural, and developed a technique in which the nonmetric scaling is replaced by a metric scaling. Given a parametric class of variogram models $\gamma(h; \theta)$, an initial set of parameters θ, and an initial configuration of points x_i^*, their algorithm has two steps:

A-step. Move the N points x_i^* to minimize the least squares criterion

$$\sum_{i,j} (d_{ij} - \gamma(|x_i^* - x_j^*|; \theta))^2.$$

B-step. For the new pattern of points estimate the value of θ from the new distances using the same least squares criterion.

Return to the A-step, unless the changes in both A- and B-steps are below a convergence criterion.

Cressie's weighted least squares criterion (2.3) could readily be substituted in the Monestiez–Switzer algorithm. Monestiez and Switzer also suggested a three-stage algorithm, including the fitting of mean and variance field as a C-step, but noted that this may encounter problems of instability.

8. Display of covariance fields

Display and interpretation of the resulting mapping f is important in our analysis. We have found the method of *biorthogonal grids* introduced by Bookstein (1978) in the context of morphometrics, valuable in this regard (Sampson et al., 1991). To describe this method, it is again useful to start with the homogeneous case. Linear or affine transformations of the geographic plane, $f(x) = Ax$, appropriate for a homogeneous spatial dispersion structure, are characterized by a single pair of principal axes given by the eigenvectors of $B = A'A$. The geographic direction or axis corresponding to the largest eigenvalue is the direction in which the geographic plane is (relatively) most stretched. The degree of such stretching, i.e., the ratio of distance in the deformed image to distance in the geographic plane, is given precisely by the square root of the largest eigenvalue (the first singular value of A), which is called the first *principal strain*. Given two locations x_a and x_b such that $|x_a - x_b| = c$, the spatial dispersion is greatest when x_a and x_b fall on a line parallel to this principal axis since this maximizes $|Ax_a - Ax_b|$. Conversely, for a given geographic separation, spatial dispersion is least (spatial correlation greatest) when x_a and x_b are aligned parallel to the principal axis of smallest eigenvalue. The ratio of the two principal strains is a measure of anisotropy. The affine derivative matrix A, together with the variogram function γ, completely characterize homogeneous spatial dispersion models with elliptical anisotropy.

We describe smooth heterogeneous spatial dispersion structures as those for which the principal axes of the affine derivative matrix of the mapping f vary smoothly from place to place. For any differentiable transformation, $f = (f_1(x), f_2(x))$, we can evaluate the affine derivative matrix,

$$\begin{pmatrix} \partial f_1/\partial x_1 & \partial f_2/\partial x_1 \\ \partial f_1/\partial x_2 & \partial f_2/\partial x_2 \end{pmatrix},$$

at any point $x = (x_1, x_2)$. These derivatives are easily computed for the thin-plate spline mappings used in the approach of Section 7. The (local) principal axes computed from this matrix are differentials at 90°. It is a theorem that through almost every point of a differentiable transformation pass just two differentials which are at 90° both before and after the transformation (see Bookstein, 1978). The integral curves of these differentials, sampled finitely, form a curvilinear grid whose intersections are at 90° in both images. These are called the biorthogonal grids for the transformation because there are two such grids, one in each image (or plane), corresponding curve for curve and intersection for intersection. They provide a useful depiction of the spatial variation in the (local) principal axes and principal strains of the nonlinear transformation. For discussion of the way in which these curves reflect the meteorological process(es) driving the random field being modeled, see the applications in Section 9.1 below and in Sampson et al. (1991).

9. Some applications

There are many different applications in environmental science that require estimated spatial covariances. In this section we discuss, very briefly, two such cases, namely how to estimate unobserved values of a spatial field, and how to redesign an existing monitoring network.

9.1. Spatial estimation

One of the main uses of monitoring data is for estimating the level of the pollutant at an unobserved site, or, more generally, over an area with or without monitoring stations in it. The kriging approach in geostatistics (pioneered by Krige, 1951, and developed by Matheron, 1963b; cf. Cressie, 1991), and the parallel theory of objective analysis in meteorology (due to Gandin, 1963; cf. Thiébaux and Pedder, 1986), were developed with this as their main goal. Using least squares ideas, the optimal linear predictor of the field at a point x, based on monitored values at points $x_1, \ldots, x_N$ (and assuming a known constant mean) is

$$\hat{Z}(x) = \sum \lambda_i z_i.$$

Since we are here referring to observations corresponding to one point in time, the time argument has been dropped. The weights λ_i are computed from the covariances; in vector form

$$\boldsymbol{\lambda} = (\lambda_1, \ldots, \lambda_N)' = \Sigma^{-1}\sigma(x),$$

where Σ is the estimated covariance for the monitored sites, and $\sigma(x)$ is the estimated vector of covariances between the monitored sites and the site x. These are estimated from the entire record of monitoring data. Provided that the variance field of $Z(x)$ is known (rather than estimated) the variance of the estimated value is

$$\operatorname{var} Z(x) - \sigma(x)'\Sigma^{-1}\sigma(x).$$

It is common in geostatistical applications to assume an unknown mean which may be either constant or parametrized as a varying function of, for example, longitude and latitute. In these cases unbiasedness constraints are imposed on the linear estimator, and the resulting kriging system of linear equations for estimation of the weights λ can be expressed in terms of the dispersion function, rather than the covariance function. Indeed, the variance of the spatial process (and hence the covariance function) need not be assumed finite (Matheron, 1963a). In environmental monitoring applications, as well as in most meteorological and climatological applications, it is usually possible to estimate the mean field from historical data or from a theoretical model. We therefore limit ourselves to the case of a specified mean. For further discussion of this issue, see Haslett (1989, Section 2.1).

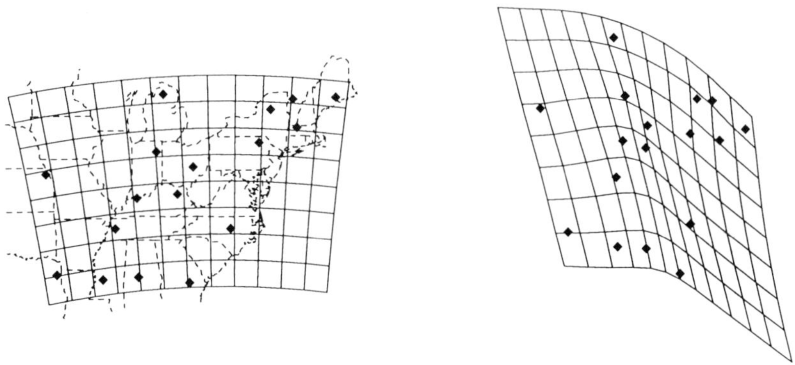

Fig. 1. Geographic map of the UAPSP wet deposition network (left) and the deformation of a regular grid into the D-image (right) for hydrogen ions using winter data, computed using the method of Section 7.

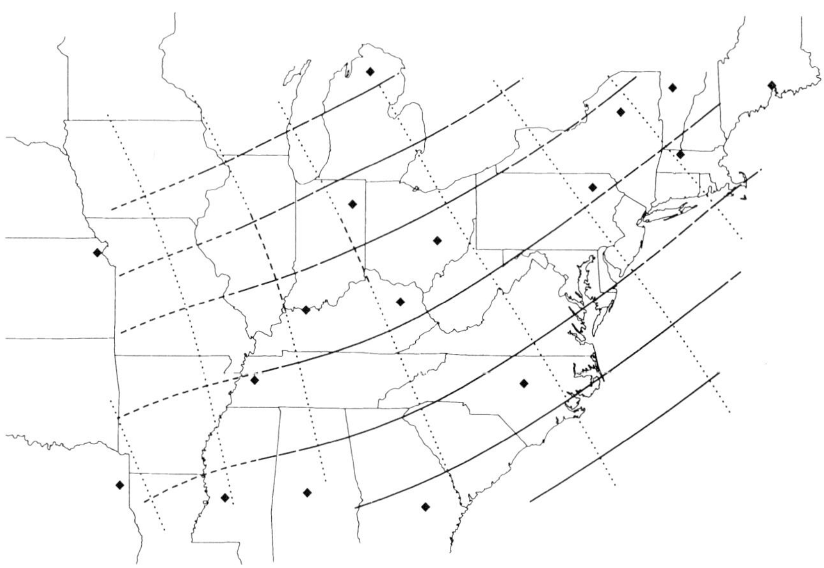

Fig. 2. Biorthogonal grid depicting the estimated spatial dispersion for hydrogen ions using winter data from the UAPSP wet deposition network. Line types encode the values of the principal strains of the local affine derivate matrix for the thin-plate spline mapping with solid lines indicating the directions of strongest correlation (smallest principal strain). Solid, long dashed, short dashed, and dotted lines are determined by splitting the range of principal strains (0.49 to 2.05) at 0.75, 1.00 and 1.55.

In order to illustrate methodology, we briefly review results in Newman et al. (1992). Here the spatial covariance was estimated based on logarithms of hydrogen deposition for the UAPSP wet deposition network, using four-weekly summaries of winter data from 1982–1987 using 17 stations.

For an analysis of the estimated correlation dispersion, Figure 1 shows how a regular geographic grid gets deformed in the D-image, according to the computed thin-plate spline mapping f_d. Figure 2 displays the biorthogonal grid corresponding to the estimated correlation dispersion. This analysis is based on residuals from a space–time mean model containing spatially smooth periodic seasonal factors and station-by-station smooth temporal trends, also smoothly spatially interpolated.

The curves running generally SW–NE in Figure 2 indicate the directions of the local principal axes of smallest principal strain, i.e., the directions of weakest dispersion or strongest correlation. Four different line types encode the value of the principal strain with solid curves representing the smallest principal strains and dotted curves representing the largest principal strains. The correlation structure is heterogeneous because these grids are in fact curved, and because the degree of spatial correlation and local anisotropy varies from place to place. Near the point where Massachusetts, Vermont and New York meet, the principal strain ratio is approximately $2.05/0.81 = 2.5$. In Virgina this ratio is $1.98/0.58 = 3.4$, in central Illinois it is $1.48/0.96 = 1.7$, while in Alabama the principal axes have rotated slightly and the strain ratio is $1.83/1.05 = 1.7$. In addition, there is generally increasing trend in the variance field from west to east, adding to the heterogeneity of the covariance field. The structure appears similar to that reported by Oehlert (1992; cf. Section 4) for annual sulfate deposition, although the heterogeneity of the correlation structure is not discussed or depicted in his paper. This deformation is consistent with the prevailing wintertime meteorology.

The effect of the deformation is indicated in Figure 3, with the raw correlation dispersions plotted against geographic distance on the left, and against D-image distance on the right. The scatter is appreciably tighter, and the improvement in terms of root mean squared error is 13%.

Figure 4 shows the estimated hydrogen field (after taking antilogarithms) for March of 1983 based on an isotropic covariance structure for the residuals (left) and the heterogeneous covariance approach from Section 7 (right). In both approaches a Gaussian variogram was judged appropriate.

While much of the structure results from the (common) estimated mean field, and thus is grossly similar, there are some differences in details. In particular, the field obtained from the isotropic approach is flatter, and has less directional structure.

A clearer image of the difference between the estimates is shown in Figure 5, where the kriging weights resulting from the two estimators are shown. These weights are relative to a location in West Virginia, marked with an $\times$ on the figures. The importance of the SE–NW axis of strong covariance is clear.

9.2. Redesign of monitoring networks

A important purpose of a monitoring network is to detect potential change in key environmental parameters. However, the designer of a long term monitoring network cannot fully foresee all of the benefits that may be derived from the network by its future users. Environmental engineers, resource developers, biologists, human health agents, etc., will need the data for a variety of purposes, some of which will not even have been identified. In addition to the detection problem mentioned above, there is a need for inference about changes in areal averages, and about the location of extreme changes. Caselton and Zidek (1984) suggested dealing with these multiple objectives by an approach which may be suboptimal in specific cases, but has overall merits for these types of networks. Recent applications to redesign of acid precipitation networks are Guttorp et al. (1992, 1993). We discuss redesign rather than design, since statistical considerations are unlikely to be given much emphasis in the early stages of designing a monitoring network.

Let Z denote a random field of measurable quantities indexed by potential site labels i. We decompose Z into the gauged sites $G = (Z_i, i \in M)$ and the ungauged sites U. One approach is to choose M in order to maximize the amount of information in G about U. Caselton and Zidek (1984) suggested Shannon's index of information transmission $I(U, G) = \mathrm{E}(\log(f(U \mid G)/f(U)))$, where $f(U \mid G)$ is the conditional density of U given G and $f(U)$ is the a priori density of U. Assuming that the random field to be monitored results in multivariate normal measurement, this information criterion is just $I(U, G) = -\frac{1}{2}\log|I - R|$, where I is the identity matrix and R the diagonal matrix with elements the squared canonical correlation coefficients between the sites (variables) in U and those in G. These correlation coeffieints must be computed from some form of model for the spatial correlation structure such as that proposed in Section 7. Guttorp et al. (1992) demonstrated this type of analysis.

A different but related approach adopts the general viewpoint that the purpose of the network is to reduce uncertainty–uncertainty about the values to be observed on the random field, and uncertainty in the parameters of the underlying models that specify the distribution of the random field. Caselton et al. (1992) proposed this approach and discussed its relation to the information criterion. The uncertainty about a random X (e.g., random field) is taken to be the entropy of its distribution,

$$H(X) = \mathrm{E}[-\log f(X)/h(X)],$$

where $f(x)$ is the density of X and $h(x)$ is a possibly improper reference density representing 'complete ignorance'. Here, entropy is calculated conditional on available data D. In Guttorp et al. (1993) this rationale is applied in the context of augmenting an existing acid deposition monitoring network with new sites. It is shown that the appropriate optimization criterion, called MEAS (for measurement), is the uncertainty about the additional sites to be gauged, or

$$\mathrm{MEAS} = \mathrm{E}[-\log(f(G \mid D)/h(G)) \mid D].$$

Using a hierarchical Bayesian approach, with the mean and spatial covariance matrix for the random field described by the normal-inverted Wishart conjugate prior distribution, it is shown that this can be evaluated in terms of the entropy of multivariate t-distributions. More specifically, the optimization criterion can be taken as

$$|\Psi_{\mathrm{add|curr}}|,$$

the determinant of the prior mean covariance matrix of the sites (variables) to be

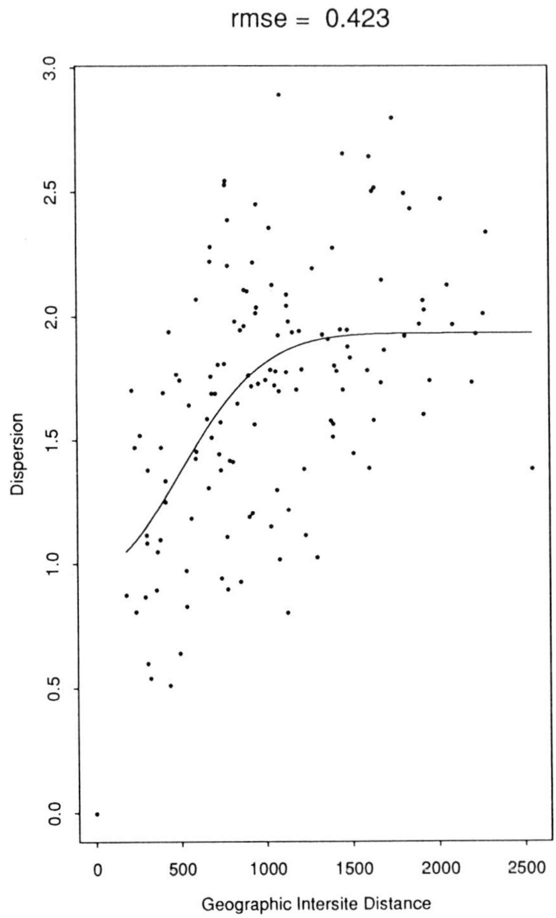

Fig. 3. Raw correlation dispersions as a function of geographic distance.

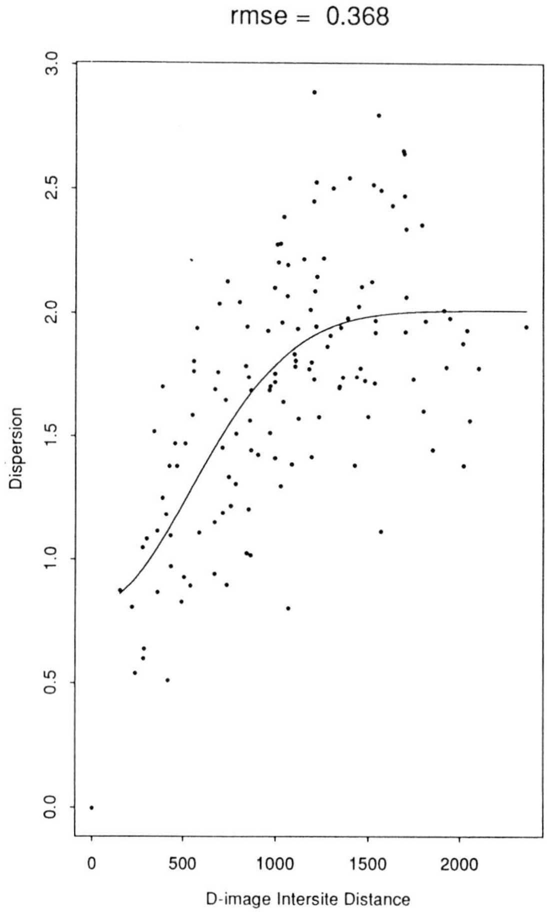

Fig. 3 (*Continued*). Raw correlation dispersions as a function of D-image distance.

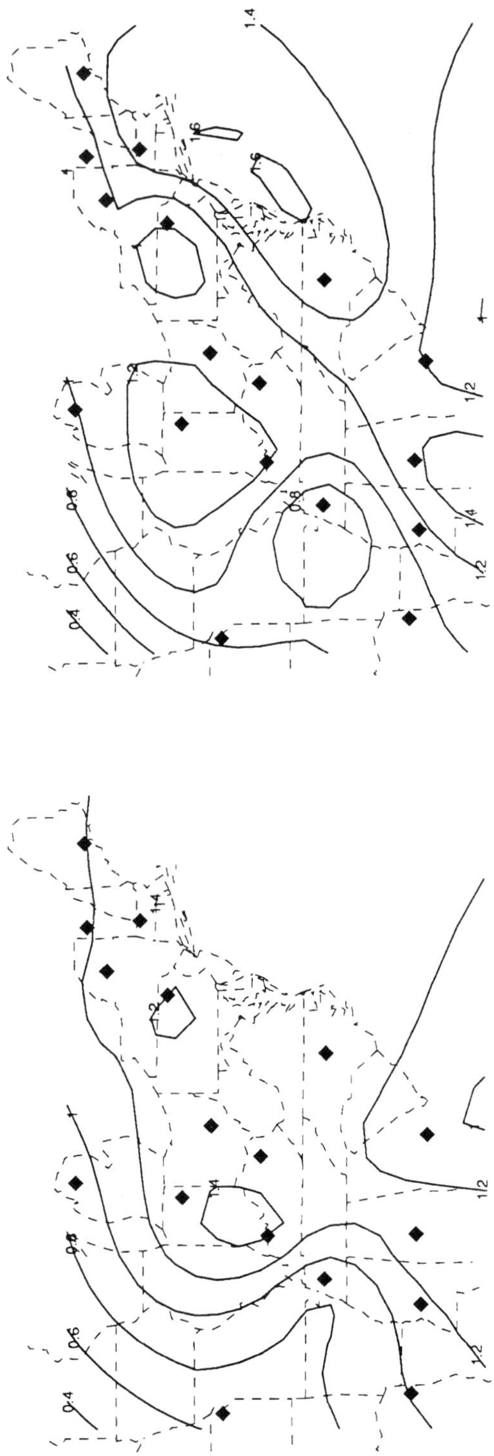

Fig. 4. Estimated hydrogen field for March of 1983. The left panel is based on an isotropic Gaussian variogram, while the right is using a heterogeneous covariance estimate.

Fig. 5. The kriging weights for the isotropic model (left) and the heterogeneous model (right). The weights are for estimating deposition in March, 1983, at the (unobserved) site marked with an ×.

*add*ed to the monitoring network conditional on the *curr*ently gauged sites. In their analysis, Guttorp et al. (1993) adopted an empirical approach in specifying the mean Ψ of the inverted-Wishart prior for the spatial covariance matrix (over all the gauged and ungauged sites) using a spatial covariance model computed as described in Section 7.

10. Temporal and multivariate extensions

In this section we comment on extensions of the analysis of heterogeneous spatial correlation structure to accommodate temporal correlation and multivariate data. Neither of these areas has been addressed in the literature to date.

10.1. Models for space–time correlation structure

The definition and interpretation of spatial covariance structure for environmental processes is closely linked to the temporal scale of the analysis. The discussion above has essentially assumed that replicate observations in time are independent, an assumption that is certainly inappropriate for many applications. For example, air quality data such as ozone are often recorded hourly and manifest short lag temporal correlation structure as well as very clear diurnal and weekly patterns. All of these components must be accommodated in any procedure that attempts to estimate ozone concentration at a point in time and a location (point or grid cell) in space. Geostatistical methods for dealing with temporal structure, such as those introduced by Seguret (1989), and Wackernagel (Wackernagel, 1988; Rouhani and Wackernagel, 1990), can now be considered in concert with current approaches to modeling heterogeneous spatial covariance structure. Temporal periodicities (such as the diurnal and weekly cycles noted above) may be considered part of a 'trend' and filtered out of an analysis (Seguret, 1989), or they may be incorporated as components in the correlogram or temporal variogram (Rouhani and Wackernagel, 1990). For those components not filtered out one may use the framework of Rouhani and Wackernagel (1990) who modeled the cross-variogram relating observations at sites i and j,

$$\gamma_{ij}(\tau) = \mathrm{E}[(Z_i(t) - Z_i(t+\tau)) \times (Z_j(t) - Z_j(t+\tau))],$$

as a sum of cross-variograms at different temporal scales

$$\gamma_{ij}(\tau) = \sum_{u=1}^{K} \gamma_{ij}^{u}(\tau) = \sum_{u=1}^{K} b_{ij}^{u} g_u(\tau).$$

In this decomposition the elementary temporal variogram functions $g_u(\tau)$ must be conditionally negative definite and the matrices of coefficients b_{ij}^{u}, for fixed u, must be positive semi-definite. Periodic behavior, if not filtered out, is represented with 'hole function' variograms. Analyses of ozone air quality might utilize one, two, or three components, representing short-term nonperiodic covariance, diurnal, and weekly cycles, respectively.

The matrices of coeffients $B_u = [b^u_{ij}]$, called the *coregionalization matrices*, represent components of the spatial covariance structure at different temporal scales. Indeed, the spatial covariance matrix S can be expressed as a mixture of covariance structures at these different temporal scales

$$S = \sum_{u=1}^{K} B_u.$$

In order to compute spatial estimates at locations x that are not monitored, it is necessary to extend this model to express the components b^u_{ij} more generally as functions of spatial location, $b^u(x_i, x_j)$, permitting extrapolation to arbitrary locations x_a and x_b. The modeling approach of Section 7 can provide this necessary extension, although it has not yet been implemented in this context.

An extension of the spectral theory in Section 2 to space–time processes which are homogeneous in space and stationary in time is straightforward. Under the assumption of separable covariance, the spectrum factors into a product of a spatial and a temporal spectrum, each of which may be analyzed separately. However, if the assumption of separability is dropped the joint space–time spectrum can be a useful parameter function. To the extent that spectral estimation is straightforward (for example, if the spectrum has a given parametric form or if observations are available on a regular grid), these estimates can be applied in the D-space. North and Nakamoto (1989) use this to compare rain estimation designs for spatially homogeneous regions.

10.2. Multivariate models

Most environmental analysis problems are multivariate. In the study of acidic deposition three primary chemical species, hydrogen, sulfate and nitrate, are usually addressed. In some situations there may be only one particular variable of primary interest (e.g., rainfall), but even in these cases there are generally one or more important spatial covariates. Multivariate problems are discussed in the geostatistical literature under the heading of co-kriging (always assuming homogeneity). General modeling for heterogeneous covariance structure has not yet been addressed. We discuss here two different aspects of this problem.

For the multivariate case we will introduce a superscript and denote observations by $Z^k(x_i, t)$ for monitoring sites $i = 1, 2, \ldots, N$, times $t = 1, 2, \ldots, I$, and variables $k = 1, 2, \ldots, K$. From these data we may compute sample covariances $S_{ij,kl}$ corresponding to variables k and l measured at sites i and j, respectively, If all of the measured variables are driven by the same environmental process, then it may be reasonable to expect that they will all have the same spatial covariance structure. Write the observation at time t as a $K \times N$ random matrix $Z_t = [Z^k_i]$. Then,

$$\operatorname{cov}(Z^k_i, Z^l_j) = S_{ij} C_{kl},$$

where S_{ij} are the elements of a spatial covariance matrix underlying all the variables and C_{kl} is the among-variables covariance matrix. In matrix notation the $NK \times NK$ covariance matrix factorizes as

$$\text{cov}(Z) = S \otimes C. \tag{10.1}$$

Mardia and Goodall (1993) adopt this approach and, assuming that this factorized covariance structure holds for some power transformation of the measured variables Z^k, carry out maximum likelihood estimation of the transformation as in Box and Cox (1964). Notice that the factorized model above specifies a symmetric cross-covariance structure,

$$\text{cov}(Z_i^k, Z_j^l) = \text{cov}(Z_j^k, Z_i^l).$$

However, such cross-covariances are not necessarily symmetric, and for many applications this assumption is inappropriate.

Assuming that the structure (10.1) is valid one may use the techniques of Sections 7 and 8 to analyze and depict the spatial covariance structure underlying the matrix S. This approach clearly provides a simpler model for the full set of covariances and, if judged an appropriate model, a useful basis for further calculations such as co-kriging.

The first generalization to consider is the usual suggestion for an EOF analysis of multivariate spatial data (see, e.g., Preisendorfer, 1988, Chapter 4). A principal components analysis is carried out on the $T \times NK$ data matrix consisting of rows given by the elements of the matrices Z_t as defined above. In this way all of the spatial covariances and cross-covariances are expressed in terms of a reduced number, say p, of dimensions and these covariances can be interpreted as in Section 6, as distances among points in p dimensions. However, now every site is represented by K points, one for each measured variable.

Along the lines of Section 7, we may instead consider spatial cross-dispersions

$$d_{ij}^{kl} = \text{var}(Z_i^k - Z_j^l),$$

allowing us to write

$$d_{ij}^{kl} = \text{var}(Z_l^k) + \text{var}(Z_j^l) - 2\text{cov}(Z_i^k, Z_j^l).$$

In the homogeneous case this has been called the pseudo cross-variogram (Myers, 1991). As before, our modeling will be most sensible in terms of standardized variables so that the right-hand side is $2 - 2\,\text{corr}(Z_i^k, Z_j^l)$. The d_{ij}^{kl} may be considered pseudo-distances on the sites indexed by i and j ('pseudo' since the corresponding $N \times N$ matrix is not generally symmetric). Consider the case of a bivariate spatial field, $K = 2$. Although the $N \times N$ matrix $d^{12} = [d_{ij}^{12}]$ is not symmetric, we have $d^{12} = d^{21\prime}$ and the $2N \times 2N$ matrix

$$\begin{bmatrix} d^{11} & d^{21\prime} \\ d^{21} & d^{22} \end{bmatrix}$$

is symmetric. This suggests applying the multidimensional scaling approach to the $2N$ entries represented in this matrix, each monitoring site represented once for each of the two variables. There are now two maps or deformations to consider, one from the original geographic coordinates to the coordinates of the sites associated with measurement of field 1, $Z^1(x, t)$, and one to the coordinates of the sites associated with measurement of field 2, $Z^2(x, t)$. Interpolated mappings (using thin-plate splines or other procedures) could be used to evaluate the spatial cross-dispersion for sites other than those at which the two spatial fields were monitored.

In this case the deformation associated with measurement of field 1 is computed to represent both the spatial correlation structure of field 1 and the spatial cross-correlation structure from field 1 to field 2. Similarly, the deformation associated with field 2 represents the spatial correlation structure of field 2 together with the spatial cross-correlation structure from field 2 to field 1. A comparison of these two mappings, expressed as a comparison between the two multidimensional scaling representations of the sites (using a Procrustes-metric) would provide a basis for judging the appropriateness of the factorized model which assumes a common mapping for both the within-field spatial correlation structures and the between field cross-correlation structures.

This approach does not require the two fields to be monitored at the same sites. Field 1 may be monitored at N_1 sites and field 2 monitored at N_2 sites and the entire analysis carried out on the basis of the resulting square matrix of order $N_1 + N_2$. It should be interesting to see what insight these procedures provide when applied to real data.

Acknowledgements

We are grateful to John Haslett for several comments that improved the presentation in this chapter. Ken Newman did most of the analysis reported in Section 9.1, and helped with some of the figures. We also thank Gary Oehlert and Colin Goodall for providing preprints of their papers.

References

Adomian, G. (1983). *Stochastic Systems*. Academic Press, New York.

Bartlett, M. S. (1966). *Stochastic Processes*. 2nd ed., Cambridge Univ. Press, Cambridge.

Böhning, D. (1982). Convergence of Simar's algorithm for finding the maximum likelihood estimate of a compound Poisson process. *Ann. Statist.* **10**, 1006–1008.

Bookstein, F. L. (1978). *The Measurement of Biological Shape Change*, Lecture Notes in Biomathematics, Vol. 24. Springer, New York.

Box, G. P. and D. R. Cox (1964). An analysis of transformations. *J. Roy. Statist. Soc. Ser. B*, **36**, 211.

Brillinger, D. R. (1972). The spectral analysis of stationary interval functions. In L. M. Le Cam, J. Neyman and E. L. Scott, eds., *Proc. 6th Berkeley Sympos. Math. Statist. Probab.* Univ. of California Press, Berkeley, CA, 483–513.

Brillinger, D. R. (1981). *Time Series: Data Analysis and Theory*. Expanded ed., Holden-Day, San Francisco, CA.

Brillinger, D. R. (1985): Fourier inference: Some methods for the analysis of array and nongaussian series data. *Water Res. Bull.* **21**, 743–756.

Buell, E. C. (1972). Integral equation representation for factor analysis. *J. Atmos. Sci.* **28**, 1502–1505.

Buell, E. C. (1978). The number of significant proper functions of two-dimensional fields. *J. Appl. Meteor.* **17**, 717–722.

Caselton, W. F. and J. V. Zidek (1984). Optimal monitoring network designs. *Statist. Probab. Lett.* **2**, 223–227.

Caselton, W. F., L. Kahn and J. V. Zidek (1992). Quality data networks that minimize entropy. In: A. Walden and P. Guttorp, eds., *Statistics in Environmental and Earth Sciences.* Edward Arnold, London, 10–38.

Cohen, A. and R. H. Jones (1969). Regression on a random field. *J. Amer. Statist. Assoc.* **64**, 1172–1182.

Cressie, N. (1985). Fitting variogram models by weighted least squares. *Math. Geol.* **17**, 565–586.

Cressie, N. (1991). *Statistics for Spatial Data.* Wiley, New York.

Cressie, N., C. A. Gotway and M. O. Grondona (1990). Spatial prediction from networks. *Chemometrics* **7**, 251–271.

Creutin, J. D. and C. Obled (1982). Objective analyses and mapping techniques for rainfall fields: An objective comparison. *Water Resources Res.* **18**, 413–431.

Gandin, L. S. (1963). *Objective Analysis of Meteorological Fields.* Israel Program for Scientific Translations, Jerusalem.

Guttorp, P., P. D. Sampson and K. Newman (1992). Nonparametric estimation of non-stationary spatial covariance structure with application to monitoring network design. In: A. Walden and P. Guttorp, eds., *Statistics in Environmental and Earth Sciences.* Edward Arnold, London, 39–51.

Guttorp, P., N. D. Le, P. D. Sampson and J. V. Zidek (1993). Using entropy in the redesign of an environmental monitoring network. In: N. K. Bose, G. P. Patil and C. R. Rao, eds., *Multivariate Environmental Statistics.* North-Holland, New York, 167–176.

Haas, T. C. (1990a). Kriging and automated variogram modeling within a moving window. *Atmosph. Environ.* **24**(A), 1759–1769.

Haas, T. C. (1990b). Lognormal and moving window methods of estimating acid deposition. *J. Amer. Statist. Assoc.* **85**, 950–963.

Haslett, J. (1989). Space time modelling in meteorology – A review. *Bull. Internat. Statist. Inst.* **51**(1), 229–246.

Haslett, J. and A. E. Raftery (1989). Space-time modelling with long-memory dependence: Assessing Ireland's wind resource (with discussion). *J. Roy. Statist. Soc. Ser. C* **38**, 1–50.

Journel, A. G. and C. J. Huijbregts (1978). *Mining Geostatistics.* Academic Press, New York.

Krige, D. G. (1951). A statistical approach to some basic mine valuation problems on the Witwatersrand. *J. Chem. Metall. Mining Soc. S. Africa* **52**, 119–139.

Le, N. D. and A. J. Petkau (1988). The variability of rainfall acidity revisited. *Canad. J. Statist.* **16**, 15–38.

Lesperance, M. L. and J. D. Kalbfleisch (1992). An algorithm for computing the nonparametric MLE of a mixing distribution. *J. Amer. Statist. Assoc.* **87**, 120–126.

Loader, C. and P. Switzer (1992). Spatial covariance estimation for monitoring data. In: A. Walden and P. Guttorp, eds., *Statistics in Environmental and Earth Sciences.* Edward Arnold, London, 52–70.

Mardia, K. V. and C. R. Goodall (1993). Factorized models in spatial modeling. In: N. K. Bose, G. P. Patil and C. R. Rao, eds., *Multivariate Environmental Statistics.* North-Holland, New York.

Mardia, K. V., J. T. Kent and J. M. Bibby (1979). *Multivariate Analysis.* Academic Press, New York.

Matérn, B. (1986). *Spatial Variation,* Lecture Notes in Statistics, Vol. 36. 2nd ed., Springer, New York.

Matheron, G. (1963a). Principles of geostatistics. *Econom. Geol.* **58**, 1246–1266.

Matheron, G. (1963b). *Traité de Geostatistique Appliquée, Tome II: Le Krigeage.* Editions Technip, Paris.

Monestiez, P. and P. Switzer (1991). Semiparametric estimation of nonstationary spatial covariance models by metric multidimensional scaling. SIMS Tech. Rep. 165, Department of Statistics, Stanford University.

Myers, D. E. (1991). Pseudo-cross variograms, positive definiteness, and cokriging. *Math. Geol.* **23**, 805–816.

Newman, K., P. Guttorp and P. D. Sampson (1993). A non-stationary spatial covariance structure for estimation of acid deposition. Technical Report, Department of Statistics, University of Washington.

North, G. R. and Nakamoto (1989). Formalism for comparing rain estimation designs. *J. Atmosph. Ocean. Tech.* **6**, 985–992.

Obled, Ch. and J. D. Creutin (1986). Some developments in the use of empirical orthogonal functions for mapping meteorological fields. *J. Appl. Meteor.* **25**, 1189–1204.

Oehlert, G. W. (1992). Regional trends in sulfate wet deposition. *J. Amer. Statist. Assoc.*, to appear.

Preisendorfer, R. W. (1988). *Principal Component Analysis in Meteorology and Oceanography*. Elsevier, Amsterdam.

Rouhani, S. and H. Wackernagel (1990). Multivariate geostatistical approach to space-time data analysis. *Water Resources Res.* **26**, 585–591.

Sampson, P. D. and P. Guttorp (1992). Nonparametric estimation of nonstationary spatial covariance structure. *J. Amer. Statist. Assoc.* **87**, 108–119.

Sampson, P. D., S. Lewis, P. Guttorp, F. L. Bookstein and C. Hurley (1991). Computation and interpretation of deformations for landmark data in morphometrics and environmetrics. In: *Proc. 23rd Sympos. on the Interface between Computing Science and Statistics*. Interface Foundation, Fairfax Station, 534–541.

Schoenberg, I. J. (1938). Metric spaces and completely monotone functions. *Ann. Math.* **39**, 811–841.

Seguret, S. A. (1989). Filtering periodic noise by using trigonometric kriging. In: M. Armstrong, ed., *Geostatistics*, Vol. 1. Kluwer Academic, Dordrecht, 481–491.

Switzer, P. (1989). Non-stationary spatial covariance estimated from monitoring data. In: M. Armstrong, ed., *Geostatistics*, Vol. 1. Kluwer Academic, Dordrecht, 127–138.

Thiébaux, H. J. (1977). Extending estimation accuracy with anisotropic interpolation. *Mon. Weather Rev.* **105**, 691–699.

Thiébaux, H. J. (1991). Statistics and the environment: The analysis of large-scale earth-oriented systems. *Environmetrics* **2**, 5–24.

Thiébaux, H. J. and M. A. Pedder (1986). *Spatial Objective Analysis with Applications in Atmospheric Science*. Academic Press, London.

Wackernagel, H. (1988). Geostatistical techniques for interpreting multivariate spatial information. In: C. F. Chung, ed., *Quantitative Analysis of Mineral and Energy Resources*. Reidel, Hingham, 393–409.

Wahba, G. (1990). *Spline Models for Observational Data*. Society for Industrial and Applied Mathematics, Philadelphia, PA.

G. P. Patil and C. R. Rao, eds., *Handbook of Statistics, Vol. 12*

21

Meta-Analysis in Environmental Statistics

Vic Hasselblad

Abstract: The synthesis of evidence from different experiments or investigations is a challenging problem for environmental science. This chapter reviews standard methods for such synthesis, including combining p-values, effect sizes, and methods for combining contingency tables. Recent developments such as random effects models and the confidence profile method are also described. For each method, published environmental examples are given.

1. Introduction

One of the most challenging problems in environmental science is the synthesis of evidence from different experiments or investigations. There are several books and review articles written on the general topic of evidence synthesis or meta-analysis. Two review articles which are most pertinent to environmental studies are those of Hedges (1987) and Laird and Mosteller (1990). This chapter will review standard methods, include some recent advances, and give environmental examples of each method.

By far the most common approach to the synthesis of evidence is the use of the global subjective judgment. One or more experts gather all the evidence they want to consider, synthesize it, and describe their conclusions, often in the form of a review article or guideline document. According to Thomas Chalmers, 'The days of the expert supposedly putting the state of the field into a review article are numbered' (Mann, 1990).

Over the past decade a set of quantitative methods has been developed to synthesize evidence in order to estimate the outcomes of different actions. This set of methods has come to be known as meta-analysis. The methods include (1) combining p-values, first described by Fisher (1932), (2) effect sizes as developed by Glass (1980) and Hedges (1981), (3) contingency table methods such as that of Mantel and Haenszel (1959) and Peto et al. (1977), (4) inverse variance weighted method as described by Hald (1952), (5) random effects models as described by DerSimonian and Laird (1986), and (6) a general method of combining evidence as described by Eddy et al. (1992).

These methods are slowly being added to environmental research, having originated in the fields of chemistry, physics, clinical medicine and psychological testing (see Hedges, 1987). This chapter will discuss the methods listed above, giving applications of each method to environmental studies.

2. Descriptive methods – Evidence tables

The first step in moving away from the global subjective judgment is the creation of an evidence table. Although this step is obvious to individuals performing meta-analyses, it has not always been the first step in a decision making process. An evidence table, as defined by Eddy (1992), is a table that describes for each study the design, sample size(s), outcomes measured, types of subjects, interventions compared, potential biases, experimental conditions, and observed outcomes. Evidence tables have been used for years. For example, Pearson (1904) presented evidence on the relation between deaths and the presence of vaccination scars in cases of Small-pox. One early example of an evidence table in the environmental field is found in the first air quality criteria document for hydrocarbons (US Department of Health, Education, and Welfare, 1970), a table on the dose-response relationships of various plants to ethylene. All current EPA criteria and assessment documents contain several evidence tables.

In some cases, it may be desirable to list all studies in an evidence table, regardless of quality. However, the evidence table is often the first step in a synthesis of studies measuring similar health endpoints. In cases where there is varying quality, inclusion criteria may be necessary. Chalmers et al. (1981) have described standards for a good randomized control trial (RCT) and have developed a scoring system to allow for a quantitative assessment of reported trials. The scoring system includes factors such as the method of blinding, the presence of a biostatistician, and loss to follow-up. Any score can then be used to include or exclude a study, or can be used as a weighting factor. However, the work of Orwin and Cordray (1985) suggests that direct ratings of study quality have low reliability.

The inclusion or exclusion of a study is a special case of weighting, where the weights are either 1 or 0. Laird and Mosteller (1990) give a brief discussion of weighting. As difficult as weighting or scoring is for RCTs, it is much more difficult for environmental studies. Not only are many different study designs used, the definitions of exposures (or treatments) are often made with significant measurement error. General guidelines for inclusion or exclusion of environmental studies may not be feasible.

Any evidence table is valid only if it considers all available information. The published literature is particularly susceptible to the claim that it is unrepresentative of all studies that may have been conducted (the publication bias problem) (Hedges, 1987). There is a strong tendency of the published literature to over represent statistically significant results. In some cases this may be less of a problem for environmental studies. The large epidemiological studies are very expensive and are usually published regardless of the results. Even animal studies can cost several hundred thousand dollars and the results are important whether they are positive or negative. However, the use of existing data bases for case-control studies does not require the same level of effort, and the location of all such relevant studies may be difficult. For additional discussion, see Hedges (1987).

The use of an evidence table can be an end unto itself. The compilation of studies may be so overwhelmingly positive or negative that further meta-analyses are not necessary. This is more likely to be the case when the object of the meta-analysis is the determination statistical significance rather than the estimation of a health effect.

3. Combining *p*-values

There are two situations where combining *p*-values may be appropriate. First, some studies do not report any effect measures but do report *p*-values. Second, the study designs or treatment levels may be so different that combining effect measures would be inappropriate. In these cases, combining *p*-values is a method for computing an overall test of significance.

Assume that there are m independent studies of an effect, each having a *p*-value corresponding to the test of the null hypothesis against a specified one-sided alternative $p_j, j = 1, 2, \ldots, m$. One method for combining *p*-values from these studies was originally given by Fisher (1932), and thus the method has been referred to as Fisher's method (Rosenthall 1984). Fisher noted that the distribution of $-2/\ln(p_j)$ is distributed as a chi-square variable with 2 degrees of freedom. Thus the sum

$$X^2 = -2 \sum_{j=1}^{m} \ln(p_i) \tag{3.1}$$

is a chi-square distribution with $2m$ degrees of freedom under the null hypothesis. The *p*-value for the combined test can be calculated from standard chi-square tables.

Several other methods of combining *p*-values have been published, and are discussed by Hedges (1987). These include the minimum p method proposed by Tippett (1931), the inverse normal method by Stouffer et al. (1949) and the logit method described by George (1977).

These methods have the advantage that the studies being combined need not have the same design or health endpoint, but there are two important drawbacks. First, the tests do not weight the studies according to their uncertainty or sample size. Second, they do not give any estimates of the magnitude of the effect. For these two reasons, methods of combining *p*-values are seldom used as meta- analytic tools. The following is an example of the use of Fisher's method for the meta-analysis of environmental studies.

Schwartz et al. (1985) summarized six studies of cognitive effects from low level lead exposure in children. The results of these studies were controversial because they suggested adverse health effects from lead exposures much lower than previously

Table 1
Epidemiological studies of cognitive effects from low level lead exposure in children as given by Schwartz et al. (1985)

Study	Sample sizes		Internal lead levels		IQ deficit	p-value	$-2\ln p$
	Control	Exposed	Control	Exposed			
McBride et al. (1982)	86	86	<9[a]	19–30[a]	1.2[c]	0.25	2.77
Yule et al. (1981)	20	21	7–10[a]	17–32[a]	7.6[d]	0.027	7.22
Smith et al. (1983)	145	155	<2.5[b]	>8[b]	2.3[d]	0.067	5.41
Yule and Lansdown (1983)	80	82	7–12[a]	13–24[a]	1.8[d]	0.13	4.08
Winneke et al. (1982)	26	26	2.4[b]	7[b]	5.0[b]	0.10	4.82
Needleman et al. (1979)	100	58	<10[b]	>20[b]	4.5[d]	0.03	7.01

[a] Blood lead levels in μg/dl.
[b] Teeth lead levels in ppm.
[c] Peabody picture vocabulary IQ test.
[d] Wechsler intelligence scale for children – revised.

thought to be of concern. Furthermore, some of the studies were statistically significant and others were not. The important consideration was the possible existence of effects at low lead levels rather than the estimation of the actual deficit in cognitive abilities. The studies are in Table 1. As calculated by Schwartz et al. (1985), the combined chi-square was 31.31 with 12 degrees of freedom, giving a one sided p-value of 0.00177. Thus the studies strongly suggest an effect of low level lead exposure on cognitive abilities. This was a key point in demonstrating the benefits of reducing lead in gasoline, and was part of EPA's regulatory impact analysis.

4. Effect sizes

In certain situations there are several studies measuring outcomes which are similar but not exactly the same. Or the studies measure the same endpoint, but under different circumstances. One solution to this problem is to create an outcome measure which does not depend on the scale of measurement. The method of creating such a measure has been termed the method of 'effect sizes'.

4.1. *Continuous outcomes*

Several authors have described the method of effect sizes for combining evidence (Cohen, 1977; Glass, 1980; Hedges, 1981; Rosenthal and Rubin, 1982; Rosenthal, 1984). Effect sizes are most commonly used for continuous outcome experiments. The effect size of an experiment d is defined as

$$d = \frac{M_t - M_c}{S}, \tag{4.1.1}$$

where M_t and M_c are the sample means of the treated and control arms respectively and S is the estimate of the standard deviation of single observation. S could be the estimate of the standard deviation in the control arm, or it could be a pooled estimate. Assuming a normal distribution for the individual observations with equal variances in each arm of the experiment, S is given by

$$S^2 = \frac{(n_t - 1)S_t^2 + (n_c - 1)S_c^2}{n_t + n_c - 2}, \tag{4.1.2}$$

where S_t^2 is the sample variance of the treated arm and S_c^2 is the sample variance of the control arm. The variance of d is

$$\text{var}(d) \equiv v = \left(1 + \frac{t^2}{2v}\right) \Big/ \left[\frac{(n_t n_c)}{(n_t + n_c)}\right], \tag{4.1.3}$$

where t is the standard student-t test statistic for testing the hypothesis of no effect of the study and $v = n_t + n_c - 2$.

Now assume that there are m studies, so that the d and v are indexed by j, $j = 1, 2, \ldots, m$. The combined effect measure is the inverse variance weighted average of the d_j:

$$\bar{d} = \sum_{j=1}^{m} w_j d_j \Big/ \sum_{j=1}^{m} w_j, \tag{4.1.4}$$

where $w_j = 1/v_j$.

The homogeneity of the d_j can be tested using the statistic

$$X^2 = \sum_{j=1}^{m} w_j (d_j - \bar{d})^2, \tag{4.1.5}$$

where $\bar{d}$ is defined as before. The statistic X^2 has an approximate chi-square distribution with $m - 1$ degrees of freedom.

4.2. *Dichotomous outcomes*

Rosenthal (1984) gives formulas for the use of effect sizes for dichotomous outcome variables. In the dichotomous case, d is given by

$$d = p_t - p_c. \tag{4.2.1}$$

The estimate of v is

$$v = \frac{n_t p_c (1 - p_c) + n_c p_t (1 - p_t)}{n_t n_c}, \tag{4.2.2}$$

where p_t and p_c are the observed proportions in the two arms of the study.

Laird and Mosteller (1990) have also described the use of effect sizes for dichotomous outcome variables. They defined d as before, but defined v as

$$v = p_c(1 - p_c). \tag{4.2.3}$$

4.3. An example

The following example shows the use of effect sizes to estimate the combined effect of studies of environmental lead exposure on nerve conduction velocity (NCV). A summary of the studies was given by Davis and Svendsgaard (1990). The studies are occupational exposure studies and as a result the actual exposures and lengths of exposure varied greatly. The summary by Davis and Svendsgaard gave several measures of NCV for each study, but median sensory NCV is perhaps the most representative of the measures. Table 2 shows the median sensory NCV effect size for each study. The result of combining the studies in Table 2 using effect sizes gives an overall value of -0.428 which has a corresponding one sided p-value less than 10^{-5}. Thus the studies show an overall effect assuming a fixed effects model. Davis and Svendsgaard (1990) give a detailed discussion of the analysis of the complete data set and also use a empirical Bayes approach (see Section 7.4) with effect sizes. Although the confidence intervals are slightly wider, the conclusion is the same.

4.4. Discussion

Effect sizes have the advantage that they can combine 'apples and oranges' if the outcome measures are reasonably similar. For example, one study of lung function

Table 2
Effect sizes for median sensory nerve conduction velocity studies of lead exposure

Study	Exposure group	Effect size	Std. error
Seppäläinen et al. (1983)	0 yr	−0.116	0.295
	1 yr	−0.916	0.441
	2 yr	−0.697	0.535
	4 yr	−0.393	0.640
Wang et al. (1985)	PbB 22 μg/dl	−0.377	0.199
	PbB 61 μg/dl	−0.333	0.199
Seppäläinen et al. (1975)		−0.706	0.220
Seppäläinen et al. (1975)		0.633	0.414
Rosén et al. (1983)		0.000	0.359
Araki et al. (1986)		−0.952	0.422
Bordo et al. (1982a, b)		−1.004	0.243
Jeyaratnam et al. (1985)		0.100	0.179
Paulev et al. (1979)		−0.398	0.339
Chen et al. (1985)		−0.548	0.280
Singer et al. (1983)		−0.766	0.275
Buchtal and Behse (1979)		−1.093	0.340
Overall effect size		−0.428	0.067

might measure forced expiratory volume (FEV) and another measure vital capacity (VC). These two measures are closely related, but the values of VC are larger by definition. The method of effect sizes incorporates the variance of the estimate of each study and is therefore an improvement over combining p-values. The disadvantage of the method is that the result is unitless, and as such cannot give us direct information about the magnitude of the effect seen.

5. Methods for contingency tables

Many studies are designed to compare the rate of a health outcome in a treated group θ_t with the rate in a control group θ_c. The results are usually summarized by a standard 2×2 table:

	Treated	Not treated
Successes	A	B
Failures	C	D

Let $N = A + B + C + D$. We will assume that there are m studies given in the form of a 2×2 tables. We will subscript the letters A, B, C, D, and N by j to indicate results from the j-th study ($j = 1, 2, \ldots, m$). The obvious estimates for θ_t and θ_c are

$$\hat{\theta}_t = \frac{A}{A + C} \quad \text{and} \quad \hat{\theta}_c = \frac{B}{B + D}. \tag{5.1}$$

Generally, the parameter of interest from a contingency table is the odds ratio Ψ defined as

$$\Psi = \frac{\theta_t(1 - \theta_c)}{\theta_c(1 - \theta_t)}. \tag{5.2}$$

Although the difference in rates $\theta_t - \theta_c$, and the relative risk, θ_t/θ_c, are often used as effect measures, the odds ratio is used for more complicated models involving dichotomous outcomes. Three methods for calculating a combined estimate of an odds ratio from 2×2 contingency tables are (1) the Mantel–Haenszel method, (2) Peto's method, and (3) the maximum likelihood method. The methods all assume a common odds ratio between studies and assume that the binomial model holds for each study.

5.1. Mantel–Haenszel method for combining odds ratios

The Mantel–Haenszel method was first described by Mantel and Haenszel (1959) and was used to calculate a combined odds ratio for stratified case-control studies. Mantel (1963) later reported that the method could be used for a wider class of problems, including prospective studies. Assuming a common odds ratio, the method can be used to combine results from different studies if those results are in the form of 2 × 2 tables. The estimates obtained using this method are usually very close to maximum likelihood estimates. The Mantel–Haenszel estimate is

$$\hat{\Psi} = \left[\sum_{j=1}^{m} \frac{A_j \times D_j}{N_j} \right] \Bigg/ \left[\sum_{j=1}^{m} \frac{B_j \times C_j}{\mathrm{N}_j} \right]. \tag{5.1.1}$$

A relatively simple estimate of the variance of $\log(\hat{\Psi})$ was given by Kleinbaum et al. (1982),

$$\operatorname{var}(\ln(\hat{\Psi})) \approx 1 \Bigg/ \sum_{j=1}^{m} 1 \Bigg/ \left(\frac{1}{A_j} + \frac{1}{B_j} + \frac{1}{C_j} + \frac{1}{D_j} \right). \tag{5.1.2}$$

Note that the formula for the estimate has a problem if any of the B_j or C_j are zero. For this reason, all values of A_j, B_j, C_j, and D_j are often increased by $\frac{1}{2}$ in equations (5.1.1) and (5.1.2).

5.2. *Peto's method for combining odds ratios*

Peto's method combines results from separate trials by adding the observed and expected cell counts. It has the advantage that it can be used even if some of the cells in the table are zero. Peto et al. (1977) give the formula for the combined odds ratio. Let v_j be defined as

$$v_j = \left[\frac{(A_j + B_j)(A_j + C_j)}{N_j} \right] \left[1 - \frac{(A_j + B_j)}{N_j} \right] \left[B_j + D_j \right] \Bigg/ \left(N_j - 1 \right). \tag{5.2.1}$$

Then

$$\hat{\Psi} = \left[\sum_{j=1}^{m} A_j - \frac{(A_j + B_j)(A_j + C_j)}{N_j} \right] \Bigg/ \left[\sum_{j=1}^{m} v_j \right]. \tag{5.2.2}$$

Note that this can be thought of as the sum of the observed values A_j minus their expected value $\mathrm{E}(A_j)$ given the null hypothesis, with the entire quantity normalized by the variance. The variance of $\log(\Psi)$ is estimated by

$$\operatorname{Var}(\ln(\Psi)) \approx 1 \Bigg/ \sum_{j=1}^{m} v_j. \tag{5.2.3}$$

Yusef et al. (1985) gave the justification for the use of the quantity $A - \mathrm{E}(A)$ as the statistic to use for the pooling of the information. An asymptotically efficient test of H_0: $\beta = 0$ versus H_a: $\beta < 0$ can be based on the slope of the log-likelihood function evaluated at $\beta = 0$ (Cox and Oates, 1984). For this particular problem, it can be shown that the slope is proportional to $A - \mathrm{E}(A)$.

5.3. Maximum likelihood estimates applied to 2 × 2 contingency tables

The maximum likelihood estimate is based on the likelihood of the m studies (with the combinatorial terms dropped)

$$L \propto \prod_{j=1}^{m} \theta_{cj}^{B_j}(1-\theta_{cj})^{D_j}\theta_{tj}^{A_j}(1-\theta_{tj})^{C_j}, \tag{5.3.1}$$

subject to

$$\Psi = \theta_{tj}(1-\theta_{cj})/\theta_{cj}(1-\theta_{tj}) \quad \text{for all } j.$$

Given this model, the likelihood is a function of $k+1$ parameters. The maximum likelihood estimate for Ψ and its asymptotic variance can be calculated using standard maximum likelihood methods (see Section 8). As a function of an increasing number of studies, the maximum likelihood estimator is not consistent because the number of parameters is increasing with the number of studies. Hauck (1984) derived a conditional maximum likelihood estimator for 2 × 2 contingency tables which is superior to the maximum likelihood estimator in both bias and precision.

5.4. An example

The US Environmental Protection Agency (1990) summarized several studies of the health effects of passive smoking. Many of these studies were described earlier by the National Research Council (1986). There were 19 case-control studies giving estimates of the odds ratio of death from long cancer in nonsmoking women as a function of the smoking status of their spouse. Of these 21 studies, seven of them used the same definition of exposure. This exposure was defined to be smoking by the spouse of at least one cigarette per day. The seven studies with a common definition of exposure are in Table 3.

An overall chi-square test for homogeneity of the seven studies gives a value of 3.42 for 6 degrees of freedom ($p = 0.755$), giving no evidence of heterogeneity. The results from combining the seven studies using the three methods just described are in Table 4. The combined estimates suggest that smoking by the spouse results in a risk of lung cancer in women 1.7 times that of women who's spouse does not smoke. If all 19 studies are included, as was done by US EPA (1990), the estimated risk is 1.42. Using either estimate, there is a large increased risk of lung cancer in women from sidestream smoke.

5.5. Discussion

For 2 × 2 contingency tables with moderate sample sizes (all cells with counts of 5 or more), the method of combining studies makes very little difference. For smaller sample sizes, the models will give slightly different answers. Hauck (1984) compared the bias and precision of estimates using maximum likelihood, conditional maximum likelihood and Mantel–Haenszel. For $m = 10$, both the Mantel–Haenszel

Table 3
Case-control studies of lung cancer deaths in nonsmoking women with husband's smoking status used as the exposure variable

Study	Location		Results		Odds ratio 95% conf. interval
			Cases	Controls	
Akiba et al.	Hiroshima, Nagasaki, Japan	Unexp.	21	82	1.577
		Exposed	73	188	(0.874, 2.629)
Correa et al.	Louisiana, US	Unexp.	8	72	2.312
		Exposed	14	61	(0.812, 5.248)
Garfinkel et al.	US	Unexp.	44	157	1.340
		Exposed	90	245	(0.868, 1.980)
Geng et al.	Tianjin, China	Unexp.	20	52	2.293
		Exposed	34	41	(1.084, 4.287)
Humble et al.	New Maxico, US	Unexp.	5	71	2.703
		Exposed	15	91	(0.819, 6.729)
Lam et al.	Hong Kong	Unexp.	84	183	1.675
		Exposed	115	152	(1.157, 2.349)
Trichopoulos et al.	Athens, Greece	Unexp.	24	109	2.228
		Exposed	38	81	(1.185, 3.829)

Table 4
Combined estimates of the odds ratio using three different methods

Method	Estimate	95% confidence limits	
		Lower	Upper
Maximum likelihood	1.666	1.360	2.042
Mantel–Haenszel	1.667	1.360	2.043
Peto's formula	1.649	1.353	2.009

estimator and the conditional maximum likelihood estimator were superior to the maximum likelihood estimator. For $m = 5$, all three methods gave very similar bias and precision. Because the maximum likelihood estimate and the conditional maximum likelihood estimate are difficult to compute, the Mantel–Haenszel estimate is a logical choice for most problems. The underlying model is slightly different for Peto's method, and there may be situations where this method is preferred.

6. General fixed effects models

A fixed parameter is one which remains the same from study to study. The opposite of this is a parameter which varies randomly from study to study, and this kind of parameter is often analyzed using a random effects model (see Section 7). There are two general methods for combining evidence on a single fixed parameter. The first of these, the variance weighted method, has been used for many years. The second method is the calculation of the maximum likelihood estimate from the joint likelihood function.

6.1. *Inverse variance weighted method*

One of the oldest methods of combining estimates of a parameter is the inverse variance weighted method. This method has been used for years, having been discussed by Birge (1932) and Cochran (1937). A simple description of the method was given by Hald (1952). Assume that there are m studies, each giving estimates $\hat{\theta}_j$ of a parameter where $j = 1, \ldots, m$. Then the minimum variance for any estimate which is a linear combination of the $\hat{\theta}_j$ is

$$\hat{\theta} = \left[\sum_{j=1}^{m} w_j \hat{\theta}_j\right] \Big/ \left[\sum_{j=1}^{m} w_j\right], \tag{6.1.1}$$

where $w_j = 1/\text{var}[\hat{\theta}_j]$. The variance of the estimate is

$$\text{var}(\hat{\theta}) = 1 \Big/ \left[\sum_{j=1}^{m} w_j\right]. \tag{6.1.2}$$

The following statistic will test the null hypothesis that all $\hat{\theta}_j$ are estimating the same parameter

$$X^2 = \sum_{j=1}^{m} w_j (\hat{\theta}_j - \hat{\theta})^2, \tag{6.1.3}$$

which is approximately distributed as a chi-square distribution with $m - 1$ degrees of freedom.

6.2. *Maximum likelihood method*

The second method is the application of the confidence profile method (CPM) to a single fixed parameter (see Section 8). When applied to a single fixed parameter, the CPM method is equivalent to the maximum likelihood estimation, or to a standard Bayesian method if a prior distribution is used.

6.3. *An example*

The studies summarized below gave evidence on the effect of nitrogen dioxide (NO_2) on respiratory disease in children. The studies were all prospective, but the model used for analysis of the data varied from study to study. All studies were adjusted so that the estimated effect was for an increase of about 30 μg/m^3 in NO_2 exposure. The

Table 5
Summary of studies of the effects of nitrogen dioxide on respiratory illness in children

Author	Model for analysis	Estimated odds ratio	95% confidence intervals
Melia et al. (1977)	Multiple logistic	1.31	(1.16, 1.48)
Melia et al. (1979)	Multiple logistic	1.24	(1.09, 1.42)
Melia et al. (1980)	Multiple logistic	1.53	(1.04, 2.24)
Melia et al. (1982)	Multiple logistic	1.11	(0.83, 1.49)
Ware et al. (1984)	Two arm experiment	1.08	(0.97, 1.19)
Neas et al. (1990)	Multiple logistic	1.47	(1.17, 1.86)
Ekwo et al. (1983)	Multiway cont. tbl.	1.10	(0.79, 1.53)
Dijkstra et al. (1990)	Multiple logistic	0.94	(0.66, 1.33)
Keller et al. (1979)	Two arm experiment	1.10	(0.79, 1.54)

effect was summarized as an odds ratio for the increase in respiratory disease in children. The nine studies giving estimates for children in the age range of 6 to 12 years are given in Table 5. The studies were reviewed and combined by Hasselblad et al. (1992).

Likelihoods for the results of nine studies are in Figure 1. Taken separately, only 4 of the 9 studies were statistically significant at the 0.05 level. The combined estimate using the variance weighted method is an odds ratio of 1.193 with 95% confidence limits of 1.123 to 1.266. The combined estimate using the maximum likelihood

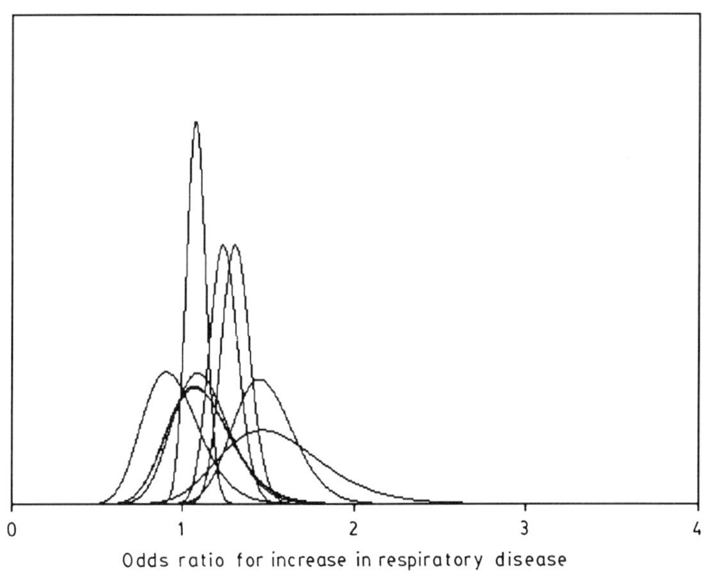

Fig. 1. Likelihood functions for studies of respiratory disease in children.

method is an odds ratio of 1.190 with 95 percent confidence limits of 1.121 to 1.261, and is also shown in Figure 1. Either estimate suggests that an increase of about 30 $\mu g/m^3$ in NO_2 exposure will result in an increase of about 20% in the reported number of cases of respiratory disease in elementary school children.

6.4. Discussion

For all but very small sample sizes, the results using the inverse variance weighted method and the maximum likelihood estimates will be extremely similar. The inverse variance weighted method may be preferred for its simplicity of computation. The use of likelihood functions has the advantage that it allows for the visualization of the estimates via marginal likelihood functions or posterior distributions.

7. Random effects models

The use of random effects models is not a recent innovation. They have been referred to as two-stage or hierarchical models. The idea of a random effects model is that the parameter or parameters of 'mother nature' do not remain constant from study to study. Instead, they vary randomly, and are in fact random variables sampled from some distribution. The problem then becomes to estimate (or derive a posterior distribution for) some function of these parameters.

7.1. DerSimonian and Laird model

DerSimonian and Laird (1986) give formulas for combining effect measures using a random effects model. For example, if the effect measure is the odds ratio, then define y_j as

$$y_j = \log(s_{tj} f_{cj} / f_{cj} s_{tj}). \tag{7.1.1}$$

The model partitions the variation of the Y_j into two sources: (1) the variation τ^2 resulting from nature choosing different parameter values for each study, (2) the sampling variation from the study itself given a particular value from nature. Let

$$\bar{y}_w = \sum_{j=1}^{m} w_j y_j \Big/ \sum_{j=1}^{m} w_j, \tag{7.1.2}$$

where w_j is defined as

$$w_j = 1 \Big/ \left[\frac{s_{tj} + f_{tj}}{s_{tj} f_{tj}} + \frac{s_{cj} + f_{cj}}{s_{cj} f_{cj}} \right],$$

and define

$$Q=\sum_{j=1}^{m} w_j(y_j-\bar{y}_w)^2. \tag{7.1.3}$$

Estimate τ^2 by

$$\hat{\tau}^2=\max\left[0,[Q-(m-1)]\bigg/\left[\sum_{j=1}^{m} w_j-\left(\sum_{j=1}^{m} w_j^2\bigg/\sum_{j=1}^{m} w_j\right)\right]\right]. \tag{7.1.4}$$

Now define u_j as

$$u_j=\frac{1}{[(1/w_j)+\hat{\tau}^2]}. \tag{7.1.5}$$

Compute

$$\hat{\Psi}=\sum_{j=1}^{m} u_j y_j\bigg/\sum_{j=1}^{m} u_j \tag{7.1.6}$$

and

$$\operatorname{var}(\hat{\Psi})=1\bigg/\sum_{j=1}^{m} u_j. \tag{7.1.7}$$

This partitioning makes no distributional assumptions. However, DerSimonian and Laird also give some maximum likelihood estimates which assume approximate normality of the y_j and the values given by mother nature. This model is described in the following section.

7.2. Normal–normal model

In the normal–normal model we assume that the observation y_j (this could be the estimate of a parameter as in the previous section) is sampled from a normal distribution with mean μ_j and variance σ_j^2, $j=1, 2, \ldots, m$. We assume that σ_j^2 is known and that μ_j is a random value from a normal distribution with mean v and variance τ^2. The likelihood is proportional to

$$L\propto\exp\left[-\sum_{j=1}^{m}\frac{(y_j-v)^2}{(\tau^2+\sigma_j^2)}+\tfrac{1}{2}\ln(\tau^2+\sigma_j^2)\right]. \tag{7.2.1}$$

One general prior for v, τ^2 is the gamma prior

$$\pi(v,\tau^2)\propto(\tau^2)^{\alpha-1}e^{-\lambda\tau^2/2}. \tag{7.2.2}$$

For $\alpha=\frac{1}{2}$ and small values of λ, this prior approaches $1/\tau$, the scale invariant prior. In practice, an informative prior with a $\alpha=\lambda=\frac{1}{2}$ works very well in that if there are a small number of studies, the prior favors a value of zero for τ^2. Otherwise, a small number of studies will result in a very diffuse posterior.

The Jeffreys' prior (Berger, 1985) is proportional to

$$\pi(v, \tau^2) \propto \left[\left(\sum_{j=1}^{n} \frac{1}{\sigma_j^2 + \tau^2}\right)\left(\sum_{j=1}^{n} \frac{1}{(\sigma_j^2 + \tau^2)^2}\right)\right]^{1/2}. \tag{7.2.3}$$

If the σ_j^2 are all equal to σ^2, then equation (7.2.3) reduces to

$$\pi(v, \tau^2) \propto \left[\frac{1}{\sigma^2 + \tau^2}\right]^{3/2},$$

or

$$\pi(v, \tau^2) = \frac{2}{\sigma}\left[\frac{1}{\sigma^2 + \tau^2}\right]^{3/2}. \tag{7.2.4}$$

The natural noninformative prior,

$$\pi(v, \tau^2) \propto 1/\tau, \tag{7.2.5}$$

leads to problems. The integrals do not converge near $\tau = 0$, and the integrals do not coverge as $\tau \to \infty$ unless there are at least four studies (likelihoods). For this reason, a gamma prior (equation (7.2.2)) is preferable, and if necessary, can be used with very small values of α and λ.

The joint posterior for v and τ^2 is proportional to the product of equations (7.2.1) and (7.2.2)

$$\pi(v, \tau^2) \propto \exp\left[-\sum_{j=1}^{m}\left[\frac{(x_j - v)^2}{\tau^2 + \sigma_j^2} + \tfrac{1}{2}\ln(\tau^2 + \sigma_j^2)\right]\tau^{2\alpha-2}e^{-\lambda\tau^2/2}\right]. \tag{7.2.6}$$

In most cases, σ_j^2 is not known. It is standard practice to replace σ_j^2 by its estimate. This should not be a problem for estimates made from reasonably large samples.

The major difficulty in computing a marginal posterior for v is that the posterior must be integrated over τ^2

$$\pi(v) = \int_0^\infty \pi(v, \tau^2)\, d\tau^2. \tag{7.2.7}$$

Thinking of $\pi(v, \tau^2)$ as a function of $\tau^2\, f(\tau^2)$, this function can be approximated by

$$\int_0^\infty f(\tau^2)\, d\tau^2 \approx \ln(r) \sum_{i=-1}^{n} cr^i f(cr^i)$$

$$\approx c\ln(r)\left[f(c) + \sum_{i=-n}^{n} cr^{-i} f(cr^{-i}) + r^i f(cr^i)\right], \tag{7.2.8}$$

where c is chosen to be near the center of the expected answer, s is an approximate standard deviation, and

$$r = e^{3[\ln(1+c^2/s^2)]^{1/2}/n}.$$

7.3. Other models

Several other random effects models have appeared in the environmental literature. One such model is the beta-binomial model which assumes a beta distribution for the parameter p of a binomial model. An example of this model was given by Williams (1975) for the analysis of responses from toxicological experiments involving reproduction and teratogenicity. A second model is the mixing of the location parameter of a lognormal distribution by another lognormal distribution. Patil et al. (1987) give an example using this model to fit recruitment data from oceanic fish stocks. Random effects models can be used with effect size measures. Hedges and Olkin (1985) give formulas very similar to those of DerSimmonian and Laird (1986).

7.4. Comparison to empirical Bayes estimation

An empirical Bayes problem is one where relationships among the parameters of the model allow the use of the data to estimate parts of the prior distribution. For problems with several observations (likelihoods), the results are usually quite similar to the random effects model solution. Several examples of empirical Bayes analyses are referenced by Morris (1983).

Efron and Morris (1975) give examples of empirical Bayes methods used to derive James–Stein estimators. These estimators will have smaller variances than standard estimators under certain assumptions. The following example, although not an environmental one, illustrates the concepts. The example is a study of the prevalence of toxoplasmosis in 36 cities of El Salvador. The prevalence rates for each study were adjusted using an indirect method of adjustment so that these 36 rates y_j had an overall mean of 0. Thus it was assumed that mother nature sampled from a normal distribution with mean 0 and variance τ^2. If τ^2 is estimated, then the James–Stein estimator for each y_j is

$$\hat{y}_j = \frac{\hat{\tau}^2 y_j}{\sigma_j^2 + \hat{\tau}^2}, \quad j = 1, 2, \ldots, 36,$$

where σ_j^2 is the known variance of y_j. Thus τ^2 is used to 'shrink' the estimate of y_j towards the overall mean. Under certain assumptions it can be shown that this estimator has a smaller variance than the original estimator.

The key estimate in this problem is the estimate for τ^2. The likelihood for τ^2 using an empirical Bayes model is shown in Figure 2 along with the random effects model posterior for τ^2 assuming two different gamma priors. The results from all three models are similar, which is to be expected since the sample size ($n = 36$) was relatively large.

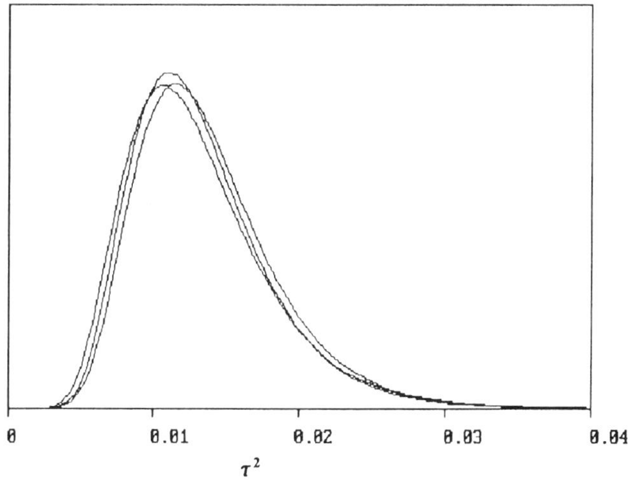

Fig. 2. Likelihood and posteriors for τ^2.

Thus the empirical Bayes and random effects model (hierarchical Bayes solution) can be quite similar. Berger (1985) concludes that '... it appears that the hierarchical Bayes approach is the superior methodology for general applications. When [m] is large, of course, there will be little difference between the two approaches, and whichever is more convenient can then be employed.'

7.5. *An example*

The same example of respiratory disease in children from exposure to NO_2 (see Section 6.3) can be used to demonstrate the use of random effects models. A test of homogeneity of the nine studies gives a chi-square of 13.55 for 8 degrees of freedom ($p = 0.094$). Thus there is some evidence for a lack of homogeneity in the nine studies, although it is not statistically significant at the 0.05 level.

Estimates of odds ratios are known to be approximately lognormal. By transforming the x-axis, we can use a normal–normal model to fit the nine studies. Estimates from the normal–normal model and the model of DerSimonian and Laird (1986) are shown in Table 6. For comparison, the maximum likelihood (ML) estimates for the fixed effects model are also given.

The estimates from either random effects model leads to the same conclusions as does the posterior from the standard maximum likelihood estimates, although the confidence limits (credible set values) are slightly wider for the random effects model. The implication is that even if each study is not measuring exactly the same health endpoint, the overall parameter which they are estimating lies to the right of one. Thus there is convincing evidence for the effect of increased NO_2 exposure on respiratory disease in children in spite of possible heterogeneity.

Raudenbush and Bryk (1985) give formulas for empirical Bayes estimates of effect sizes using the same random effects model, and Davis and Svendsgaard (1990) used

Table 6
Combined estimates of the odds ratio using three different models

Method	Estimate	95% confidence limits	
		Lower	Upper
Fixed effect (ML)	1.190	1.121	1.261
Normal–normal	1.200	1.102	1.307
DerSimonian and Laird	1.192	1.090	1.304

this model to estimate an effect size for the same studies described in Section 4.3. Allowing for the heterogeneity resulted in the one sided *p*-value being reduce to 0.0025.

7.6. Discussion

Hierarchical models are seldom used as the primary method of analysis. They are very useful when there is concern about the lack of homogeneity of the studies. In these cases it may be possible to drop the assumption of homogeneity and to make conclusions about the average effect by appropriately modeling the lack of homogeneity.

8. Multiple parameter models – The confidence profile method

The confidence profile method is a very general method for combining virtually any kind of evidence about various parameters, as long as those parameters can be described in some model. It was first described by Eddy (1989), and a more complete description was given by Eddy et al. (1982). In general, the method can be divided into four steps, but the later steps can give information requiring the modification of earlier steps, and as a results, the process may be considered iterative. The four basic steps are: (1) define the problem precisely; (2) obtain all available evidence; (3) define the model parameters and their relationships; and finally, (4) solve the model using any of several different methods described later.

The third step is the heart of the confidence profile method. Models consist of three elements: (1) basic parameters; (2) functional parameters; and (3) likelihood functions relating evidence to basic or functional parameters. These elements are described below.

8.1. Basic parameters

Basic parameters are those parameters which appear in the model which are not functions of any other parameters. Examples of basic parameters are (1) the probability of having at least one respiratory illness with a background exposure of 10 $\mu g/m^3$; (2) the mean diastolic blood pressure of US men aged 50; or (3) the expected number of respiratory illnesses in infants during the first year of life. There may or may not be any direct evidence about a basic parameter. For convenience, we will denote the basic parameters as $\theta_1, \theta_2, \ldots, \theta_k$. It is possible that some of the θ_j could be multivariate in nature. For examples, a multinomial distribution requires a multivariate parameter.

If a Bayesian analysis is to be used, then all basic parameters must have prior distributions. Noninformative priors for these parameters can be derived in a variety of ways (see Berger, 1985).

8.2. *Functional parameters*

Each functional parameter θ_j is defined as a function of the basic parameters $\theta_1, \theta_2, \ldots, \theta_k$, and previously defined functional paramaters $\theta_{k+1}, \ldots, \theta_{j-1}$

$$\theta_j = f_j(\theta_1, \theta_2, \ldots, \theta_{j-1}), \quad j = k+1, \ldots, p. \tag{8.2.1}$$

where p is the total number of parameters in the model. Functional parameters are useful in defining the likelihood functions, but the models are technically functions of k parameters. Although the function can be any mathematical expression, certain functions are very common when evaluating medical evidence. One class of common functions are effect measures. Denote θ_c as the parameter of the control group and θ_t as the parameter of the treated group. The following are three different effect measures:

$$\text{Difference} \quad \theta_{\mathrm{d}} = \theta_{\mathrm{t}} - \theta_{\mathrm{c}},$$

$$\text{Ratio} \quad \theta_{\mathrm{r}} = \frac{\theta_{\mathrm{t}}}{\theta_{\mathrm{c}}},$$

$$\text{Odds ratio} \quad \theta_{\mathrm{or}} = \frac{\theta_{\mathrm{t}}(1-\theta_{\mathrm{c}})}{\theta_{\mathrm{c}}(1-\theta_{\mathrm{t}})}.$$

A second class of functions arise from the problems of errors in measurements. Most of these are variations of the following model: Let $\theta_{\mathrm{e}1}$ denote the probability that a success is labeled a failure, and let $\theta_{\mathrm{e}2}$ denote the probability that a failure is labeled a success. The formula for the observed success rate θ' as a function of the true rate θ is given by

$$\theta' = (1 - \theta_{\mathrm{e}1})\theta + \theta_{\mathrm{e}2}(1 - \theta).$$

Other functions arise in the area of survival analysis. These functions are too numerous to describe.

8.3. *Likelihood functions*

Likelihood functions connect observed evidence to basic and functional parameters. Denote the likelihood of the i-th observation from the s-th experiments ($s = 1, 2, \ldots, S$)

as $L_s(X_i|\theta_1, \theta_2, \ldots, \theta_p)$. Remember that the likelihood functions are technically functions of the k basic parameters. For a detailed description of likelihood functions, see Barnard and Sprout (1982) and Berger and Wolpert (1984).

8.4. The model

Once the basic parameters, functional parameters, and likelihoods are defined, the model has been formulated. The log-likelihood, L', for the model (assuming independence) is

$$L' = \sum_{s=1}^{S} \sum_{i=1}^{n_s} \log L_s(X_i|\theta_1, \ldots, \theta_p). \tag{8.4.1}$$

If priors have been defined, then the k-dimensional posterior (thought of as a function of p parameters) is

$$\pi(\theta_1, \ldots, \theta_p) = \left[\prod_{s=1}^{S} \prod_{i=1}^{n_s} L_s(X_i|\theta_1, \ldots \theta_p) \right] \left[\sum_{j=1}^{k} \pi_j(\theta_j) \right]. \tag{8.4.2}$$

Although equation (8.4.1) defines the log-likelihood function, and equation (8.4.2) defines the posterior distribution, these functions are not useful for most applications. The calculation of summary statistics or marginal distributions are necessary, and these are discussed in the section on solution methods.

8.5. Influence diagrams

Influence diagrams were developed as a Bayesian computer-aided modeling tool by Howard and Matheson (1984). It is an alternative to decision trees for Bayesian decision analysis. Shachter (1986) constructed an algorithm for solving Bayesian decision analysis problems. Even when influence diagrams are not used to arrive at a solution, they provide an excellent graphical representation of the conditional relationship between the parameters and evidence.

8.6. Solution methods

As described earlier, the general solution of the model can be calculated by maximizing the log-likelihood function in equation (8.4.1) or to calculating the m-dimensional posterior in equation (8.4.2). There are four different methods for solving the model: (1) exact solution in certain special cases; (2) approximate solutions; (3) maximum likelihood methods; and (4) Monte Carlo simulation. The choice depends on the complexity of the model, the accuracy required, and the amount of computational power and time available. No distinction need be made between a Bayesian or a non-Bayesian analysis, except that a Bayesian solution requires the specification of priors.

8.6.1. Exact solution methods

There are a few situations where exact solutions for marginal likelihoods or posteriors are available. Several of these arise when we have observations from a univariate or multivariate normal distribution. Most of these results were given by DeGroot (1970) and include (1) the univariate normal with the variance unknown; (2) the multivariate normal with the covariance unknown; (3) the linear regression model; and (4) the binomial likelihood with a beta prior. Box and Tiao (1973) show that a normal prior and a normal likelihood result in a normal posterior. This is useful when normal approximations for the likelihood function are used. Other exact solutions can be calculated numerically for models with a small number of parameters.

8.6.2. Approximate solution methods

Some distributions can be closely approximated by other distributions, so that distributions of certain functional parameters can also be approximated. Beta distributions can be approximated by use of logarithmic or log-odds transformations. These transformations result in nearly normal distributions for moderate values of α and β, so that functional computations can be made. Computing the distributions of the sum or difference of two variables each independently distributed as a student-t is an extremely difficult problem, and is known as the Behrens–Fisher problem. Welch (1938) has published an approximate solution to the problem which is quite accurate for all but the most unusual cases.

8.6.3. Maximum likelihood methods

The maximum likelihood solution for the model can be found without the specification of any priors. The point estimate for the maximum likelihood solution is that value of $\theta = \theta_1, \theta_2, \ldots, \theta_m$ which maximizes equation (8.4.1). Denote this value as $\hat{\theta}$. There are two common solution methods for obtaining this estimate: the standard Newton–Raphson method (Burden and Faires, 1985) and the modified Gauss–Newton method (Berndt et al., 1974). The posterior mode and credible set limits can be calculated in the same manner (see Shachter et al., 1990), except that the prior distribution is included in the function to be maximized.

8.6.4. Monte Carlo methods

Monte Carlo sampling solution methods have the advantage that posterior distributions for any functions of the parameters can be calculated to any desired degree of accuracy. The problem is that the generation of enough random numbers to obtain this accuracy could take years. In Monte Carlo sampling, a random vector of the basic parameters is generated from which all functional parameters are calculated. Denote these values as θ^i, where the superscript i denotes the i-th simulation. From these values, the likelihood function L_i (or posterior distribution) is calculated. To estimate a particular function $g(\theta)$, compute

$$\hat{g}(\theta) = \sum_{i=1}^{n} L_i g(\theta^i) \Big/ \sum_{i=1}^{n} L_i,$$

where n is the number of random vectors generated.

The difficulty with this method is that most of the L_i will be extremely small, and the number of observations needed to get a precise estimate of $g(\theta)$ will be very large.

The obvious solution is to sample more where the likelihood is larger, a method known as importance sampling. The distribution of the importance function should mimic the posterior density. One obvious choice is to start with the approximate multivariate normal likelihood function (or posterior) for $\theta_1, \dots, \theta_k$ estimated by maximum likelihood methods. The sample is made from the k-dimensional multivariate normal. Define w_i as

$$w_i = \frac{L(\theta^i)}{I(\theta^i)},$$

where $\theta^i = \theta_1^i, \dots, \theta_k^i$ and $I(\theta^i)$ is the value of the multivariate normal density evaluated at θ^i. $\hat{g}(\theta)$ is then given by

$$\hat{g}(\theta) = \sum_{i=1}^{n} g(\theta^i) \Big/ \sum_{i=1}^{n} w_i(\theta^i).$$

The variance of the estimate can be computed from the weighted variance of the individual estimates. A detailed discussion of various Monte Carlo methods to obtain marginal density distributions was given by Gelfand and Smith (1990).

Obviously, a key to the method is the ability to quickly generate multivariate normal observations with the specified mean vector and covariance matrix. Note that if $\boldsymbol{V}$ is a positive definite matrix, then $\boldsymbol{V}$ can be written as $\boldsymbol{C}'\boldsymbol{C}$. If we start with k independent normals with mean 0 and variance 1 (denote the vector as $\boldsymbol{X}$), then the distribution of $\boldsymbol{\mu} + \boldsymbol{CX}$ will have mean vector $\boldsymbol{\mu}$ and covariance matric $\boldsymbol{C}'\boldsymbol{C} = \boldsymbol{V}$.

8.7. *An example*

The determination of 'safe' levels of ambient air toxic substances is an extremely difficult problem. EPA has the responsibility for determining such levels, and has labeled these values RfCs. The following is an application of the CPM to determine an RfC for chronic inhalation exposure of n-hexane. The general approach is to determine that concentration which will produce a specified adverse health effect. This methodology is related to the 'benchmark' concentration method of Crump (1984).

The result from each study is a posterior distribution for the dose producing the specified health effect. Thus the distributions are all on the same scale and can be compared directly. With extreme caution, the distributions can be combined. The computations are illustrated by an example detailed by Jarabek and Hasselblad (1991).

Sanagi et al. (1980) conducted an occupational study (of humans) exposed to 58 ppm n-hexane and compared them with an unexposed population. The health endpoint was nerve conduction velocity (NCV), and the specified health effect was a decrease in NCV of 2 m/se. The model for the Sanagi et al. study is

$$y_i = \alpha + \beta x_i + \varepsilon_i,$$

where y is the NCV value, x is the concentration of n-hexane, α and β are the basic parameters, and ε_i is normally distributed with mean 0 and variance σ^2. The prior distribution for α is the uniform distribution over the entire real line and the prior for β is the uniform distribution over the negative half line (n-hexane is assumed not to improve NCV). The function parameter of interest is that concentration θ_c which will produce a decrease of 2 m/s: $\theta_c = 2/\beta$. Under the assumptions, the marginal posterior distribution for θ_c will be an inverted truncated student-t distribution. This distribution, using the data of Sanagi et al., is in Figure 3.

The second study is an experiment by Dunnick et al. (1989) which showed increased nasal turbinate lesions in female mice expose to n-hexane. In this case, the EC_{10} was chosen as the biologically significant effect. The model for the data is the log-logistic model where the exposure variable x is the human equivalent concentration of n-hexane,

$$P(\text{lesion} \mid x) = \frac{1}{[1 + \exp(-\alpha - \beta \ln(x))]},$$

where α and β are the basic parameters. The prior distribution for α is the uniform distribution over the entire real line and the prior for β is the uniform distribution over the positive half line (n-hexane is assumed not to decrease lesions). The function parameter of interest is that concentration θ_c which will produce a lesion rate of 10 percent (EC_{10}),

$$\theta_c = \frac{-\alpha - \ln(9)}{\beta}.$$

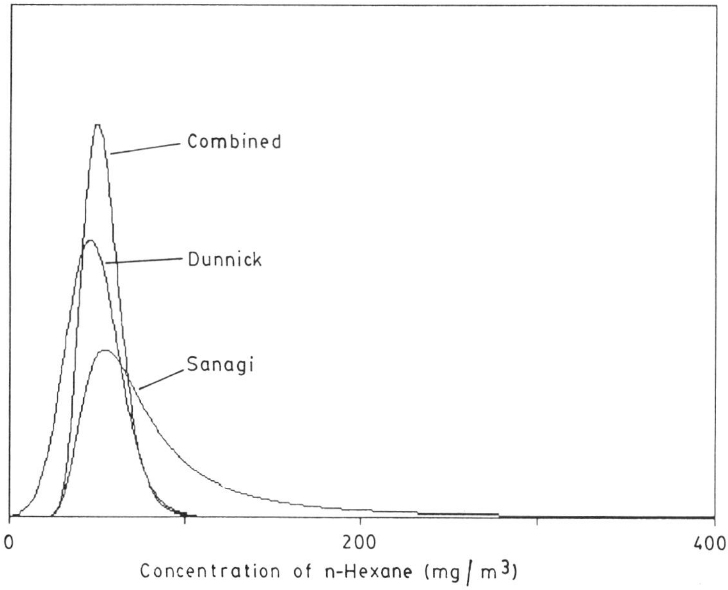

Fig. 3. Posterior for concentration producing specified effect.

An approximate solution for this problem was given by Finney (1978), but it is inaccurate for small sample sizes. The posterior distribution for θ_c based on the data of Dunnick et al. (1990) can be calculated numerically, and it is also shown in Figure 3.

The two posterior distributions are not inconsistent with each other in spite of the fact that they were done on different species and measured very different endpoints. The RfC derived from either study would be relatively similar. For mathematical interest, the posterior distribution assuming a common θ_c is also shown in Figure 3.

8.8. Discussion

The confidence profile method uses a defined model to relate multiple parameters and differing types of evidence. Any evidence which can be expressed as a likelihood function of the parameters of the model can be incorporated. Subjective priors, where necessary, may also be included. All of the assumptions of the model are explicitly given by the functional parameters and the likelihoods. The solution can be given as a joint posterior distribution or as a marginal distribution on any parameter of interest. Fixed effect models and random effect models are special cases of the method.

One advantage of CPM is that graphs of the likelihood functions or posterior distributions of the individual studies are an integral part of the analysis. According to Landwehr (1987), '... plots should be contructed and used at the very beginning of any meta-analysis, before getting into more complicated modeling, testing, and estimation'. The primary drawback of the method is that the computations can be difficult without specialized software.

References

Barnard, G. A. and A. D. Sprout (1982). Likelihood. In: *Encyclopedia of Statistical Sciences*, Vol, 3. Wiley, New York.

Berger, J. O. (1985). *Statistical Decision Theory and Bayesian Analysis*. Springer, New York.

Berger, J. O. and R. L. Wolpert (1984). *The Likelihood Principle*. Institute of Mathematical Statistics, Hayward, CA.

Berndt, E., B. Hall, R. Hall and J. Hausman (1974). Estimation and inference in nonlinear structural models. *Ann. Econom. Soc. Measure* 3, 655–665.

Birge, R. T. (1932). The calculation of errors by the method of least squares. *Phys. Rev.* **30**, 207–227.

Box, G. E. P. and G. C. Tiao (1973). *Bayesian Inference in Statistical Analysis*. Addison-Wesley, Reading, MA.

Burden, R. L. and J. D. Faires (1985). *Numerical Analysis*. Prindle, Weber and Schmidt, Boston, MA.

Charlmers, T. C., H. Smith Jr., B. Blackburn, B. Silverman, B. Schroeder, D. Reitman and A. Ambroz (1981). A method for assessing the quality of a randomized control trial. *Control. Clinic. Trials* **2**, 31–49.

Cochran, W. G. (1937). Problems arising in the analysis of a series of similar experiments. *J. Roy. Statist. Soc. Suppl.* **4**, 102–118.

Cohen, J. (1977). *Statistical Power Analysis for the Behavior Sciences*. Academic Press, New York.

Cox, D. R. and D. Oates (1984). *Analysis of Survival Data*. Chapman and Hall, London.

Crump, K. S. (1984). A new method for determining allowable daily intakes. *Fund. Appl. Toxicol.* **4**, 854–871.

Davis, J. M. and D. Svendsgaard (1990). Nerve conduction velocity and lead: A critical review and meta-analysis. In: B. L. Johnson, ed., *Advances in Neurobehavioral Toxicology: Applications in Environmental and Occupational Health.* Lewis Publications, Chelsea, MI.

DeGroot, M. H. (1970). *Optimal Statistical Decisions.* McGraw-Hill, New York.

DerSimonian, R. and N. Laird (1986). Meta analysis in clinical trials. *Control. Clinic. Trails* **7**, 177–188.

Dunnick, J. K., D. G. Graham, R. S. Yang, S. B. Haber and H. R. Brown (1990). Thirteen-week toxicity study of *n*-hexane in B6C3F1 mice after inhalation exposure. *Toxicology* **57**, 163–172.

Eddy, D. M. (1992). *Manual for Evaluating Health Practices & Designing Practice Policies.* American College of Physicians, Philadelphia, PA, 44–45.

Eddy, D. M. (1989). The confidence profile method: A bayesian method for assessing health technologies. *Oper. Res.* **37**, 210–228.

Eddy, D. M., V. Hasselblad and R. D. Shachter, (1992). *The Statistical Synthesis of Evidence: Meta-analysis by the Confidence Profile Method.* Academic Press, Boston, MA.

Efron, B. and C. Morris (1975). Data analysis using Stein's estimator and its generalizations. *J. Amer. Statist. Assoc.* **70**, 311–319.

Finney, D. J. (1978). *Statistical Methods in Biological Assay.* Griffin, London.

Fisher, R. A. (1932). *Statistical Methods for Research Workers.* 4th ed., Oliver and Boyd, London.

Gelfand, A. and A. F. M. Smith (1990). Sampling-based approaches to calculating marginal densities *J. Amer. Statist. Assoc.* **85**, 398–409.

George, E. O. (1977). Combining independent one-sided and two-sided statistical tests – Some theory and applications. Unpublished Doctoral Dissertation, University of Rochester.

Glass, G. (1980). Summarizing effect sizes. In: *New Directions for Methodology of Social and Behavioral Science: Quantitative Assessment of Research Domains.* Jossey-Bass, San Francisco, CA.

Hald, A. (1952). *Statistical Theory with Engineering Applications.* Wiley, New York.

Hasselblad, V., D. J. Kotchmar and D. M. Eddy (1992). Synthesis of environmental evidence: Nitrogen dioxide epidemiology studies. *J. Air Waste Management Assoc.* **42**, 662–671.

Hauck, W. W. (1984). A comparative study of conditional maximum likelihood estimation of a common odds ratio. *Biometrics* **40**, 1117–1123.

Hedges, L. (1981). Distribution theory for Glass's estimator of effect size and related estimators. *J. Educat. Statist.* **6**, 107–128.

Hedges, L. (1987). Statistical issues in the meta-analysis of environmental studies. In: *ASA/EPA Conf. on Interpretation of Environmental Data II. Statistical Issues in Combining Environmental Studies,* October 1–2, 1986. EPA-230-12-87-032, 30–44.

Hedges, L. V. and I. Olkin (1985). *Statistical Methods for Meta-Analysis.* Academic Press, Orlando.

Howard, R. A. and J. E. Matheson (1984). Influence diagrams. In: R. A. Howard and J. E. Matheson, ed., *Readings in the Principles and Applications of Decision Analysis.* Strategic Decision Group, Menlo Park, CA.

Jarabek, A. M. and V. Hasselbad (1991). Inhalation reference concentration methodology: Impact of dosimetric adjustments and future directions using the confidence profile method. In: *Proc. 84th Ann. Meeting of the Air & Waste Management Association,* Vancouver, Canada.

Kleinbaum, D. G., L. L. Kupper and H. Morgenstern (1982). *Epidemiologic Research.* Lifetime Learning Publications, Belmont, CA.

Laird, N. M. and F. Mosteller (1990). Some statistical methods for combining experimental results. *Internat. J. Technol. Assess. Health Care* **6**, 5–30.

Landwehr, J. M. (1987). Discussion of 'Statistical issues in the meta-analysis of Environmental Studies, In: *ASA/EPA Conf. on Interpretation of Environmental Data II. Statistical Issues in Combining Environmental Studies,* October 1–2, 1986. EPA-230-12-87-032, 47–49.

Mann, C. (1990). Meta-analysis in the breech. *Science* **249**, 476–480.

Mantel, N. (1963). Chi-square tests with one degree of freedom; Extensions of the Mantel–Haenszel procedure. *J. Amer. Statist. Assoc.* **58**, 690–700.

Mantel, N. and W. Haenszel (1959). Statistical aspects of the analysis of data from retrospective studies of disease. *J. Nat. Cancer Inst.* **22**, 719–748.

Morris, C. N. (1983). Parametric empirical bayes inference: Theory and applications. *J. Amer. Statist. Assoc.* **78**, 47–59.

National Research Council (NRC) (1986). *Environmental Tobacco Smoke: Measuring Exposures and Assessing Health Effects.* National Academy Press, Washington, DC.

Orwin, R. G. and D. S. Cordray (1985). Effects of deficient reporting on meta-analysis: A conceptual framework and reanalysis. *Psychol. Bull.* **97**, 134–147.

Patil, G. P., G. J. Babu, M. T. Bosewell, K. Chatterjee, E. Linder and C. Taillie (1987). Statistical issues in combining ecological and environmental studies with examples in marine fisheries research and management. In: *ASA/EPA Conf. on Interpretation of Environmental Data II. Statistical Issues in Combining Environmental Studies,* October 1, 1986, EPA-230-12-87-032, 30–44.

Pearson, K. (1904). Report on certain enteric fever inoculation statistics. *British Medical J.* **2**, 1243–1246.

Peto, R., M. C. Pike, P. Armitage, N. E. Breslow, D. R. Cox, S. V. Howard, N. Mantel, K. McPherson, J. Peto and P. G. Smith (1977). Design and analysis of randomized clinical trials requiring prolonged observation of each patient II. Analysis and Examples *British J. Cancer* **35,** 1–39.

Raudenbush, S. and A. S. Bryk (1985). Empirical Bayes meta-analysis. *J. Educat. Statist.* **10,** 75–98.

Rosenthal, R. (1984). *Meta-Analytic Procedures for Social Research.* Sage, Beverly Hills, CA.

Rosenthal, R. and D. B. Rubin (1982). Combining effect sizes from independent studies. *Psychol. Bull.* **92**, 500–503.

Sanagi, S., Y. Seki, K. Sugimoto and M. Hirata (1980). Peripheral nervous system functions of workers exposed to *n*-hexane at a low level. *Internat. Arch. Occup. Health* **47,** 69–70.

Schwartz, J., H. Pitcher, R. Levin, B. Ostro and A. J. Nichols (1985). Cost and benefits of reducing lead in gasoline: Final regulatory impact analysis. USEPA: EPA 230-05-85-006.

Shachter, R. (1986). Evaluating Influence Diagrams. In: A. P. Basu, ed., *Reliability and Quality Control.* Elsevier, Amsterdam, 321–344.

Shachter, R. D., D. M. Eddy and V. Hasselblad (1990). An influence diagram approach to medical technology assessment. In: R. M. Oliver and J. Q. Smith eds., *Influence Diagrams and Belief Nets.* Wiley, Chichester, England.

Stouffer, S. A., E. A. Suchman, L. C. DeVinney, S. A. Star and R. M. Williams Jr. (1949). *The American Soldier: Adjustment During Army Life,* Vol. I. Princeton Univ. Press, Princeton, NJ.

Tippett, L. H. C. (1931). *The Methods of Statistics.* Williams and Norgate, London.

US Department of Health, Education, and Welfare (1970). *Air Quality Criteria for Hydrocarbons.* National Air Pollution Control Administration, Washington DC.

US Environmental Protection Agency. (1990). *Health Effects of Passive Smoking: Assessment of Lung Cancer in Adults and Respiratory Disorders in Children,* EPA-600/6-90-006A. Office of Atmospheric and Indoor Air Programs, Washington, DC.

Welch, B. L. (1938). The significance of the difference between two means when the population variances are unequal. *Biometrika* **29**, 350–362.

Williams, D. A. (1975). The analysis of binary responses from toxicological experiments involving reproduction and teratogenicity. *Biometrics* **31,** 949–952.

Yusef, S., R. Peto, J. Lewis, R. Collins and P. Sleight (1985). Beta blockade during and after myocardial infarction: An overview of the randomized trials. *Progr. Cardiovasc. Diseases* **27**, 335–371.

G. P. Patil and C. R. Rao, eds., *Handbook of Statistics, Vol. 12*

22

Statistical Methods in Atmospheric Science

Andrew R. Solow

Abstract: This chapter describes some statistical methods that have been applied in atmospheric science. Three areas of application are considered – acid rain, ozone depletion and global warming – and within each area two papers are reviewed. The objectives of these papers fall into three broad categories: estimating a trend in multiple time series, spatial interpolation and estimating a physical parameter.

1. Introduction

The field of atmospheric science is at least as large as the field of statistics. Buried in the literature of atmospheric science are many examples of the application of statistical methods. Most of these applications have been made by nonstatisticians, with a predictably uneven degree of success. The involvement of statisticians in atmospheric science, which has certainly not been negligible, is growing. This can only improve the standard of statistical work in this field.

This chapter has been organized in the following way. Three areas of current research have been selected. These are acid rain, ozone depletion and global warming. A brief introduction to these issues is contained in Firor (1990). Within each of these areas, two papers are reviewed. The criterion for selecting these papers is that they treat a fairly broad problem in a statistically sound way. It goes without saying that these are not the only papers within these areas satisfying this criterion.

In discussing the individual papers, the decision has been made to retain (more or less) the original notation. While this makes comparisons between the papers slightly more difficult, unifying the notation would make comparisons to the individual papers more difficult. Not surprisingly, there is considerable overlap among the reviewed papers. In the final section, some comparisons are made among the approaches.

2. Acid rain

Acid rain occurs when sulfur and nitrogen oxides, emitted in the burning of coal and other fossil fuels, is converted to sulfuric and nitric acid in the atmosphere. The resulting increase in the acidity of precipitation (rain and snow) may have deleterious effects on ecosystems such as lakes and forests, as well as on manmade objects such as buildings and motor vehicles. There have been a number of statistical analyses of various aspects of the acid rain problem (see, for example, *The Canadian Journal of Statistics* **16**(1), 1988, for a number of relevant papers and references). Much of this statistical work has been administered by the SIAM Institute for Mathematics and Society (SIMS).

The first paper reviewed in this section (Shapiro and Switzer, 1989) describes a nonparametric method for extracting a trend from spatial time series of acid deposition measurements. Although the application described below is somewhat sketchy, the method is simple and powerful and deserves greater exposure. The second paper reviewed in this section (Stein, 1986) introduces a simple model for spatial time series and describes a method for interpolation to unsampled locations. While modelling of this kind comes naturally to statisticians, the same cannot be said for atmospheric (and other) scientists, who tend to apply statistical methods without considering their validity, let alone their optimality. The general approach outlined in Stein (1986) is an excellent example of how statistical modelling should be used to develop valid methods for addressing substantive issues.

2.1. *Extracting a trend from spatial time series*

One component of statistical work in acid rain has focused on extracting trends from spatial time series of acid deposition. Let $\boldsymbol{Z}(t) = (Z_i(t), i = 1, 2, \ldots, p)'$, $t = 1, 2, \ldots, n$, be a multiple time series consisting of measurements of acid deposition at p monitoring stations. Suppose that the original time series have been centered by subtracting their means. A common approach to extracting a trend from such data is to use principal component analysis (e.g., Jolliffe, 1986). The goal of principal component analysis (PCA) is to construct a set of orthogonal linear combinations of the original time series (called principal components or PCs) in which variance decreases from the maximum possible. In many cases, the first few PCs account for a large portion of the total variance in the data and are taken to represent trend. One obvious objection to this use of PCA is that it essentially assumes that the signal-to-noise rario in the data is large, which may not be the case in monitoring environmental trends. Another objection is that it does not make use of the time ordering of the data, in the sense that the weights used to form the PCs are invariant to permutation of order.

A number of alternative approaches have been developed (e.g., Box and Tiao, 1977; Pena and Box, 1987; Shapiro and Switzer, 1989). The goal of these methods is to construct a set of orthogonal linear combinations of the original time series with smoothness or predictability decreasing from the maximum possible. A particularly simple method is minimum/maximum autocorrelation factor analysis (MAFA), proposed by Shapiro and Switzer (1989). The goal of MAFA is to construct a set of orthogonal linear combinations of the original time series (called minimum/maximum autocorrelation factors or MAFs) in which lag-one autocorrelation decreases from the maximum possible.

Let $\boldsymbol{Z}$ be the n-by-p matrix whose columns are the original time series and let $\boldsymbol{D}$ be the $(n-1)$-by-p matrix whose columns are the first differences of the original time series. The p-by-p covariance matrices of the original time series and their first differences are $\boldsymbol{S} = (\boldsymbol{Z}'Z)/n$ and $\boldsymbol{V} = (\boldsymbol{D}'\boldsymbol{D})/(n-1)$, respectively. The first MAF has the form

$$Y_1(t) = \boldsymbol{a}'\boldsymbol{Z}(t). \tag{2.1}$$

Since

$$\frac{\boldsymbol{a}'\boldsymbol{va}}{\boldsymbol{a}'\boldsymbol{sa}} = 2(1-r), \tag{2.2}$$

where r is the lag-one autocorrelation of $Y_1(t)$, the weight vector $\boldsymbol{a}$ that maximizes r is proportional to the eigenvector of the matrix $\boldsymbol{S}^{-1}\boldsymbol{V}$ corresponding to the smallest eigenvalue of $\boldsymbol{S}^{-1}\boldsymbol{V}$ (which is equal to $2(1-r)$). The constant of proportionality can be chosen, for example, to ensure that $Y_1(t)$ has unit variance and is positively correlated with t. In general, the n-by-p matrix whose columns are the MAFs is given by $\boldsymbol{Y} = \boldsymbol{ZA}$, where the columns of $\boldsymbol{A}$ are proportional to the eigenvectors of $\boldsymbol{S}^{-1}\boldsymbol{V}$ arranged in increasing order of the corresponding eigenvalue.

Because the matrix $\boldsymbol{S}^{-1}\boldsymbol{V}$ is not symmetric, it is useful to have an indirect method for finding its eigenvectors. Shapiro and Switzer (1989) proposed the following method. A key feature of MAFA is that its results are invariant to affine transformations of $\boldsymbol{Z}$. It follows that Y is given by the PCs of the first differences of the scaled (i.e., unit variance) PCs of $\boldsymbol{S}$.

The original time series can be recovered from the MAFs by the inverse operation $\boldsymbol{Z} = \boldsymbol{YA}^{-1}$. Similarly, the projection of $\boldsymbol{Z}$ onto a subset of the smoothest MAFs can be found by replacing the columns of $\boldsymbol{Y}$ corresponding to the discarded MAFs by columns of 0s. The projections of the original time series formed in this way will be smoother than the original time series. An approach to determining how many MAFs to retain is discussed in Solow (1992). Further smoothing can be accomplished by smoothing the retained MAFs prior to projection.

Shapiro and Switzer (1989) illustrated the use of MAFA as a smoothing technique using spatial time series of acid deposition measurements. The data comprise 59 monthly sulfate concentrations at 5 monitoring stations for the period February 1979 to December 1983. The locations of the stations are shown in Figure 1. The time series have been deseasonalized by subtracting monthly means and linear trends have been removed by ordinary least squares regression. The top panel in Figure 2 shows the deseasonalized, detrended time series. The bottom panel shows the MAFs. The MAFs have been smoothed using a variable span smoother. The higher order MAFs are smoothed more severely than the lower order MAFs, reflecting their less structured behavior. The use of a variable span smoother

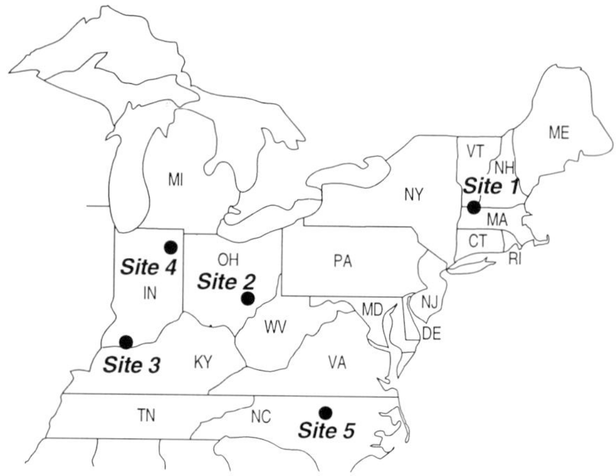

Fig. 1. Locations of 5 acid deposition monitoring sites (Shapiro and Switzer, 1989).

obviates the need to decide how many MAFs to retain, since the higher order MAFs tend to be annihilated. The smoothed MAFs are also shown in the bottom panel of Figure 2. Finally, the original time series were smoothed by projection onto the smoothed MAFs. The smoothed original series are also shown in the top panel of Figure 2.

Shapiro and Switzer (1989) went on to run a series of experiments in which different kinds of trends were added to the deseasonalized, detrended data. The results of these experiments, which will not be reviewed here, show that MAFA performs well at extracting many types of trend from multiple time series.

A word is in order about the sense in which MAFA extracts a common trend from multiple time series. In the next section, methods for estimating a global trend in stratospheric ozone from a set of monitoring stations is discussed. The underlying model in that case is that the record from each station contains the same global trend, but also local effects. In this section, the multiple time series are not assumed to contain the same trend. Instead, the trend refers to a common mode of variation.

2.2. *A simple model for spatial time series*

Like most other climate variables, acid deposition measurements are made through time at a network of monitoring stations. For a variety of reasons, interest commonly centers on using these measurements to predict values at unmonitored locations. Let $z(x, t)$ be acid deposition at location x in period t and suppose that observations $z(x_i, t_j)$, $i = 1, 2, \ldots, n$, $j = 1, 2, \ldots, k$, are available. On the basis of these nk observations, the value $z(x_0, t_\beta)$ is to be predicted, where x_0 is an unmonitored location and t_β is one of the observation times.

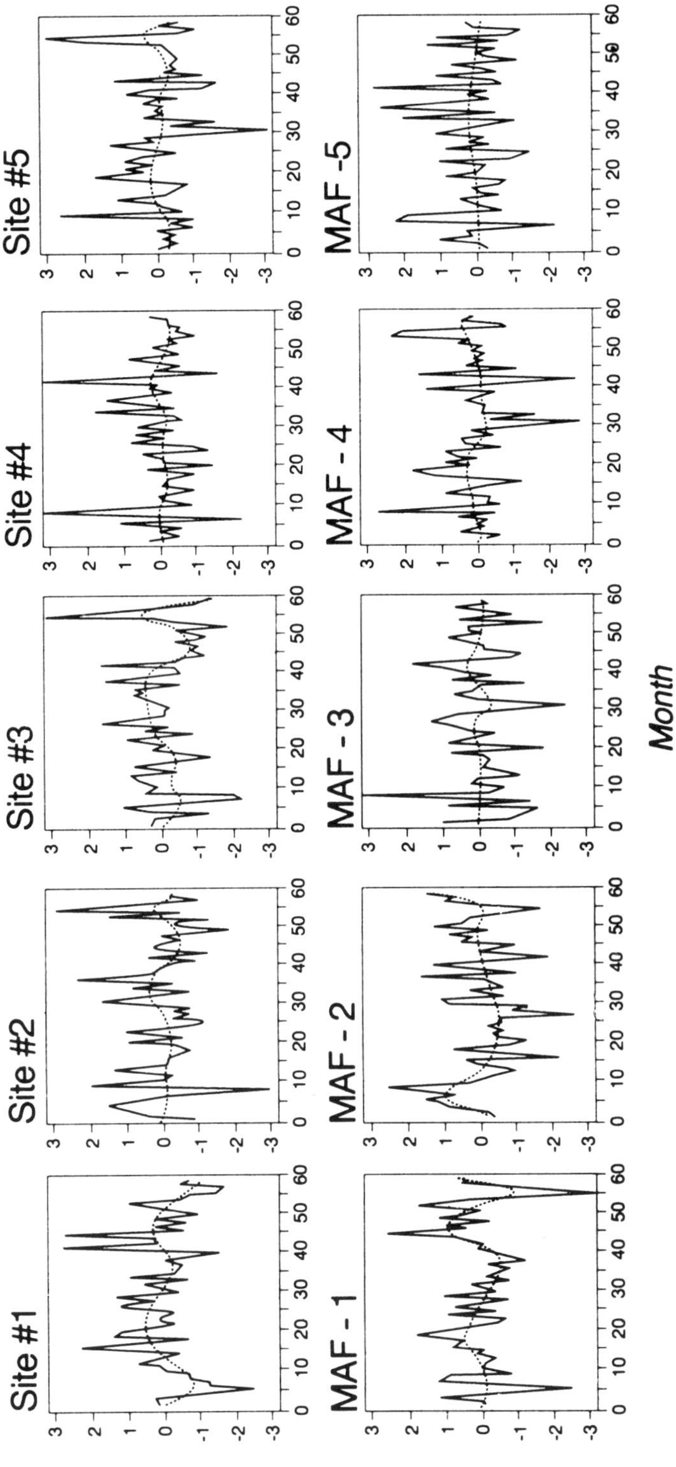

Fig.2. Deseasonalized, detrended acid deposition series (top panel) and corresponding MAF factors (bottom panel). The MAF factors have been smoothed by a variable span smoother and the original series have been smoothed by projection onto the smoothed MAF factors (Shapiro and Switzer, 1989).

To say something about the precision of a specified predictor or, more ambitiously, to choose an optimal predictor, it is necessary to have a model for the spatial and temporal variations in $z(x,t)$. Stein (1986) proposed the simple model

$$z(x, t) = m(x) + \mu(t) + e(x, t), \tag{2.3}$$

where $m(x)$ represents spatial variation independent of time, $\mu(t)$ is an unknown deterministic trend independent of location and the error field $e(x, t)$ contains the remaining spatial–temporal variations. Stein (1986) modelled the error field as a zero mean stochastic process with

$$E((e(x,t) - e(x', t))^2) = 2\gamma(|x - x'|),$$

$$E((e(x, t_i)e(x', t)_j)) = 0,$$

where $|\,|$ denotes distance and γ is called the semi-variogram. Stein (1986) actually imposed slightly different conditions to cover the possibility that the error field has infinite variance. Under these assumptions, the error field is stationary in time, homogeneous in space, and exhibits no time-lagged spatial correlations. Stein (1986) also modelled the spatial field $m(x)$ as a zero mean stochastic process, independent of $e(x, t)$, with semi-variogram $s(\,)$. Under this model, the spatial correlation between locations in the same time period is given by $g(\,) + s(\,)$, while the spatial correlation between locations in different time periods is given by $s(\,)$.

The main simplification of this model is its treatment of temporal variation. The spatial field $m(x)$ is fixed through time, while the error field $e(x, t)$ is independent through time. There is, however, no provision for intermediate types of temporal variation. This rules out spatial–temporal interactions and may not be appropriate for observations more frequent than annual.

A linear predictor of $z(x_0, t_\beta)$ has the form

$$\hat{z}(x_o, t_\beta) = \sum_{i=1}^{n} \sum_{j=1}^{k} w_{ij} z(x_i, t_j). \tag{2.4}$$

The condition for this predictor to be unbiased (over repeated realizations of both the spatial field and the error field) is

$$\sum_{i=1}^{n} w_{ij} = \delta_{j\beta}, \tag{2.5}$$

where $\delta_{j\beta} = 1$ if $j = \beta$ and 0 otherwise. The model outlined above is sufficient to express the error variance of any linear predictor of $z(x_0, t_\beta)$ in terms of $\gamma(\,)$ and $s(\,)$. If these functions are known, then the prediction weights in (2.4) can be chosen to minimize the error variance subject to (2.5) giving the best linear unbiased predictor.

Although the semi-variograms are typically unknown, they can be estimated and the estimates can be treated as correct. The effect of error in estimating the semi-variograms is ignored in what follows. In general, errors in estimating semi-variograms have a small effect on prediction, but may have a large effect on the estimation of error variance.

Let

$$z_j = (z(x_1, t_j) \quad \cdots \quad z(x_n, t_j))'$$

and

$$z = \sum_{j=1}^{k} z_j/k.$$

Under the model outlined above, the best linear unbiased predictor of $z(x_0, t_\beta)$ has the form

$$\hat{z}(x_0, t_\beta) = a'z + b'(z_\beta - z), \tag{2.6}$$

where the weight vectors a and b, which depend on the semi-variograms $\gamma(\,)$ and $s(\,)$, are given in Stein (1986). Some intuition can be gained about this predictor by considering what happens as $k \to \infty$. In that case, $\hat{z}(x_0, t_\beta)$ is essentially the sum of the prediction of the spatial field $m(x_0)$ and the error field $e(x_0, t_\beta)$. In addition to predicting $z(x_0, t_\beta)$, it is also possible to calculate the corresponding error variance.

An application of this model is given in Stein (1985). Table 1 gives 19 locations ($n = 19$) in the eastern US at which mean annual acid deposition (in μmoles $H^+ \text{cm}^{-2}$) was measured in 1982 and 1983 ($k = 2$). The measurements are also given in Table 1. Stein (1985) used the following cross-validation experiment to assess the performance of the predictor (2.6). Each of the 38 observations was predicted using the remaining 37 observations. The root mean squared prediction error was 0.95 μmoles cm^{-2}. In contrast, if each observation is predicted using only the 18 other observations from the same period (e.g., as discussed in Finkelstein (1984)), the root mean squared error is 1.45 μmoles cm^{-2}.

3. Stratospheric ozone depletion

There has been considerable concern in recent years about the depletion of stratospheric ozone (which shields the Earth's surface from ultraviolet radiation) caused by the emissions of chlorofluorocarbons (CFCs). An elementary discussion of the theory and photochemical modelling of ozone depletion is contained in Stolarski (1982). Total ozone (i.e., the total amount of ozone measured in a column extending from the Earth's surface to the top of the atmosphere) has been regularly measured at 36 locations since the late 1950s. Because the instrument used to make these measurements is the Dobson spectrophotometer, these measurements are

Table 1
Annual acid deposition (Stein, 1985)

Site	Annual acid deposition (μmole H^+/cm^2)	
	1982	1983
Turner Falls, MA	6.13	6.21
Tunkhannock, PA	4.80	5.22
Zanesville, OH	6.02	5.92
Rockport, IN	5.84	6.09
Fort Wayne, TN	5.83	4.02
Raleigh, NC	4.69	4.17
Gaylord, MI	2.48	3.39
Clearfield, KY	7.64	5.77
Alamo, TN	2.39	1.94
Winterport, ME	2.93	3.61
Uvalda, GA	2.41	2.83
Selma, AL	4.33	4.22
Clinton, MS	3.81	4.43
Marshall, TX	2.69	2.26
Lancaster, KS	2.25	2.18
Brookings, SD	0.31	0.29
Underhill, VT	5.39	6.70
Big Moose, NY	7.01	6.69
McArthur, OH	7.93	5.38

referred to as the Dobson data. Figure 3 shows some examples. There have been a number of statistical analysis aimed at extracting a global trend in total ozone from these data. In this section, two studies – Reinsel and Tiao (1987) and Bloomfield et al. (1983) – are reviewed. These studies illustrate two different approaches to estimating an overall trend from spatial time series.

3.1. *Estimating a global trend: A time domain approach*

Reinsel and Tiao (1987) estimated a global trend in total ozone by, first, estimating trends at individual stations and, second, combining the station-by-station estimates into a global estimate using a random effects (or variance components) model. Let y_t be the total ozone at a particular station in month t. The basic model adopted by Reinsel and Tiao (1987) is

$$y_t = \mu + S_t + \omega x_t + N_t, \tag{3.1}$$

where μ is the overall mean level and S_t is a seasonal component represented by a linear combination of sinusoidal terms with fundamental 12-month period. Typically, only the annual and semi-annual components were retained in the modelling.

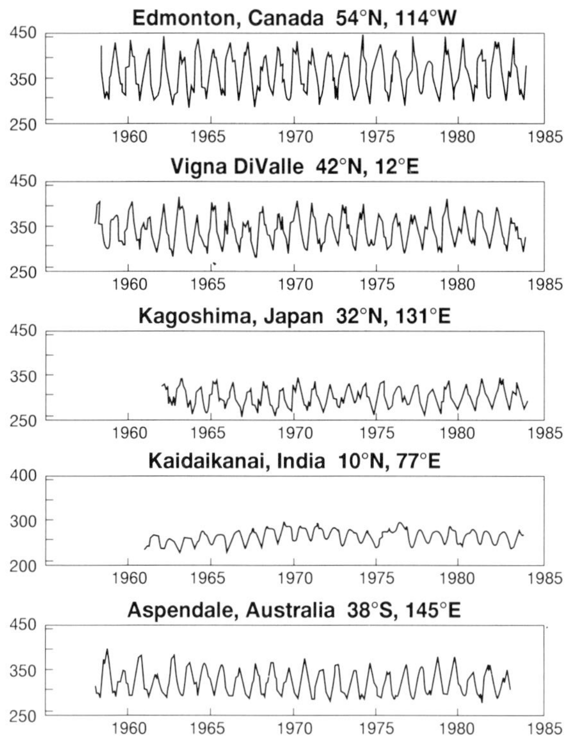

Fig. 3. Some examples of monthly Dobson data records (Reinsel and Tiao, 1987).

The variable x_t is defined as

$$x_t = \begin{cases} 0 & t < T_0, \\ (t - T_0)/12 & t \geqslant T_0, \end{cases}$$

so that ω is the annual rate of change in total ozone beginning at T_0. Since photochemical models suggest that the main effects of CFCs should have begun arround 1970, Reinsel and Tiao (1987) chose T_0 to be December 1969.

The error term N_t is assumed to follow an autoregressive process of order p (AR(p)),

$$N_t = \phi_1 N_{t-1} + \cdots + \phi_p N_{t-p} + \varepsilon_t,$$

where the variance of the zero mean innovation process ε_t is allowed to depend on the month of the year. To economize on degrees of freedom, Reinsel and Tiao (1987) assumed that var(ε_t) is equal within, but different between, groups of months.

Model (3.1) was fit to the individual station records by a combination of time series methods for identifying p (typically, $p=2$) and iteratively reweighted least squares to minimize the sum of squared residuals standardized to reflect differences in var(ε_t).

Once models were fit to each of the station records, Reinsel and Tiao (1987) estimated a global trend in total ozone in the following way. The stations were grouped into 7 geographical regions. Let ω_{ij}^* be the true percentage trend (i.e., ω/μ in terms of (3.1)) for station j within region i. These trends are assumed to follow a random effects model,

$$\omega_{ij}^* = \omega + \alpha_i + \beta_{ij}, \tag{3.2}$$

where ω is the true global trend, α_i is a random component associated with region i, with $E(\alpha_i)=0$ and $\mathrm{var}(\alpha_i)=\sigma_\alpha^2$, and β_{ij} is a random component associated with station j within region i, with $E(\beta_{ij})=0$ $\mathrm{var}(\beta_{ij})=\sigma_\beta^2$. The estimated percentage trends $\hat{\omega}_{ij}^*$ follow the model

$$\hat{\omega}_{ij}^* = \omega_{ij}^* + E_{ij}, \tag{3.3}$$

where E_{ij} is the error associated with estimating ω_{ij}^*, with $E(E_{ij})=0$ and $\mathrm{var}(E_{ij})=\sigma_{ij}^2$. Estimates of σ_{ij}^2 were based on the analyses of the individual station records.

Reinsel and Tiao (1987) used maximum likelihood estimation of variance components (Harville, 1977) to fit (3.2)–(3.3) with σ_{ij}^2 assumed to be equal to its estimate. The point estimate of ω found in this way was 0.0046% y^{-1}, with an approximate standard error of 0.0485% y^{-1}. On the basis of this analysis, Reinsel and Tiao (1987) concluded that there had been no decline in global total ozone over the period 1970–1983.

Reinsel and Tiao (1987) extended this analysis in two directions. First, they incorporated a measure of solar variability in (3.1) to account for solar influences on ozone formation. This made the point estimate of the global trend slightly negative, but still not significantly different from zero.

Second, because photochemical models suggest that ozone depletion should be greater in the upper atmosphere, Reinsel and Tiao (1987) analyzed indirect ozone measurements for the upper 5 Umkehr layers. This analysis is complicated by a relatively large number of missing observations. To estimate a model similar to (3.1) for a single Umkehr layer at an individual station, Reinsel and Tiao (1987) used a version of the following method (Reinsel and Wincek, 1987). Assuming that the error process N_t is AR(1), iteratively reweighted least squares regression of suitably weighted differences of successive (but not necessarily consecutive) ozone observations against the same weighted differences of the regressor variables (including an index of stratospheric aerosol concentration) were performed. The weighted differencing is used to eliminate serial correlation. The weights, which involve the autoregressive parameter ϕ and var(ε_t), are re-estimated in each iteration. The station results are used to estimate global trends in each layer via a random effects model similar to (3.2)–(3.3). For the period 1970–1981, the results indicate significant depletion in total ozone in layers 7, 8, and 9, but no changes in layers 5 and 6. Since the upper 3 Umkehr layers comprise only a small proportion of the atmosphere, these results (which are in general agreement with photochemical models) do not contradict the earlier results for results for total ozone.

3.2. Estimating a global trend: A frequency domain approach

Bloomfield et al. (1983) addressed the same issue as Reinsel and Tiao (1987) – namely, the estimation of a global trend in the Dobson total ozone data – using a frequency domain approach. As in Reinsel and Tiao (1987), the 36 stations were divided into 7 regions. Let $x_{t,j,k}$ be the logarithm of total ozone in month t for station k within region j. The basic model is

$$x_{t,j,k} = \mu_{t,j,k} + \alpha_t + \beta_{t,j} + \gamma_{t,j,k}, \tag{3.4}$$

where $\mu_{t,j,k}$ is the mean, α_t is a random component common to all stations, $\beta_{t,j}$ is a random component common to all stations within region j, and $\gamma_{t,j,k}$ is a station-specific random error. The random components are assumed to be stationary, with no cross-correlations among them. The regional effects $\beta_{t,j}$ are assumed to have the same spectra, as are the errors $\gamma_{t,j,k}$ within a region. The presence of regional effects allows cross-correlation to be stronger between stations within a region than between stations in different regions. The mean term in (3.4) was modelled as

$$\mu_{t,j,k} = h_t + v_{j,k}, \tag{3.5}$$

where h_t is a temporal component and $v_{j,k}$ is a spatial component independent of t. Because the original data were deseasonalized by subtracting the station-wise monthly averages, the term $v_{j,k}$ vanishes in the subsequent analysis.

Let $(Y_t, t = 1, \ldots, T)$ be a zero mean stationary time series with summable covariance function. The discrete Fourier transform of this time series is given by

$$d_y(f) = \sum_{t=1}^{T} y_t \exp(-2\pi i f t)/T,$$

where $-\frac{1}{2} > f > \frac{1}{2}$ is frequency. For large T, $\operatorname{var}(d_y(f)) \approx S_y(f)/T$, where $S_y(f)$ is the spectrum of y_t, and $d_y(f)$ and $d_y(f')$ are approximately uncorrelated if f and f' are distinct Fourier frequencies.

Under model (3.4)–(3.5), in obvious notation,

$$d_{x,j,k}(f) = d_h(f) + d_\alpha(f) + d_{\beta,j}(f) + d_{\gamma,j,k}(f). \tag{3.6}$$

Bloomfield et al. (1983) estimated the variances of some of these components (and, therefore, the corresponding spectra) from variations of the $d_{x,j,k}$ within regions

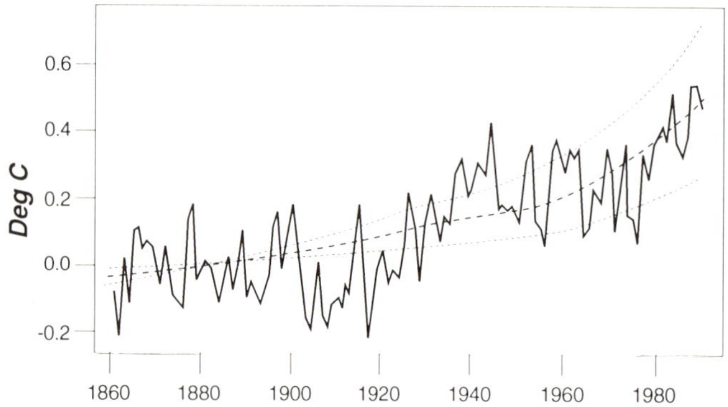

Fig. 4. Mean global surface temperature (solid) and model reconstructions for $\Delta T = 0.69°C$ (bottom), 1.39°C (middle) and 2.19°C (top) (Bloomfield, 1992).

(ARIMA(p, d, q)) for which the AIC indicates $p = 0$, $q = 1$, with a maximum likelihood estimate of $d = 0.25$.

As often happens, considerations of the number of parameters, the AIC, and the residual periodogram does not lead to an unambiguous choice of model. For estimating ΔT, Bloomfield (1992) chose to retain the ARIMA(0, d, 0) model. Using this model, the maximum likelihood estimate of ΔT is 1.39°C. The model reconstruction of historic temperature using this value of ΔT is shown in Figure 4. An approximate 0.95 confidence interval for ΔT based on the likelihood function is 0.69–2.19°C. The model reconstructions using the endpoints of this interval are also shown in Figure 4.

The approach used by Bloomfield (1992) assumes, first, that the only uncertainty is over ΔT and, second, that the secular trend in historic temperature is solely a response to known changes in greenhouse forcing. As Bloomfield (1992) noted, neither of these assumptions is correct. Uncertainties over other model parameters (and in the estimates of mean global temperature) would serve to widen the confidence interval for ΔT. Failure to account for other sources of warming and other changes in radiative forcing could lead to bias in the estimation of ΔT.

4.2. Estimating ΔT from historic temperature variability

Changes in mean global temperature are caused by smooth, secular changes in radiative forcing (e.g., due to changes in atmospheric concentrations of greenhouse gases) and also to short-term, irregular changes in radiative forcing (e.g., due to variability in solar activity, volcanic eruptions, etc.). Let the change in radiative forcing between a base year 0 and year t ($t = 1, 2, \ldots, n$) be

$$Q(t) = m_Q(t) + \varepsilon_Q(t),$$

where $m_Q(t)$ is the smooth component and $\varepsilon_Q(t)$ is the irregular component. A common model is that $\varepsilon_Q(t)$ is white noise with $\mathrm{var}(\varepsilon_Q(t)) \approx 1.0\ \mathrm{W\,m^{-2}}$ (e.g., Wigley and Raper, 1990). This model is consistent with satellite measurements of the annual global energy flux.

For a fixed value of ΔT, the modelled response of mean global temperature to this forcing is

$$M(t;\Delta T) = m_M(t;\Delta T) + \varepsilon_M(t;\Delta T),$$

where $m_M(t;\Delta T)$ can essentially be attributed to $m_Q(t)$ and $\varepsilon_M(t;\Delta T)$ can essentially be attributed to $\varepsilon_Q(t)$. This attribution ignores nonlinearities in the model. However, provided that the time scales of variability in $m_Q(t)$ and $\varepsilon_Q(t)$ are sufficiently different, this assumption (which can be assessed through model experiments) is reasonable.

Finally, let

$$T(t) = m_T(t) + \varepsilon_T(t),$$

where $m_T(t)$ is a smooth trend and $\varepsilon_T(t)$ is an irregular component.

In terms of this notation, Bloomfield (1992) assumed that $m_Q(t)$ is known and estimated ΔT by fitting $m_M(t;\Delta T)$ to the data. As noted, it is unrealistic to assume that $m_Q(t)$ is known. As alternative approach to estimating ΔT that does not make this assumption can be based on the dependence of the statistical characteristics of $\varepsilon_M(t;\Delta T)$ on ΔT. Briefly, this dependence arises from the fact that the same processes controlling the response of temperature to secular changes in radiative forcing also control its response to short-term changes in forcing. So, for example, if temperature is sensitive to secular changes in forcing, it is also sensitive to short-term changes.

Let $s(f;\Delta T)$ be the spectrum of $\varepsilon_M(t;\Delta T)$, where f is frequency. If $\varepsilon_T(t)$, $t = 1, 2, \ldots, n$, were known, then ΔT could be estimated by maximum likelihood in the frequency domain (e.g., Dzhaparidze, 1986). For a Gaussian process satisfying certain regularity conditions

$$-2\,L(\Delta T) \approx \sum_{i=1}^{n} \log s\left(\frac{i}{n};\Delta T\right) + \left(I\left(\frac{i}{n}\right)\Big/ s\left(\frac{i}{n};\Delta T\right)\right), \tag{4.2}$$

where $L(\Delta T)$ is the log-likelihood and $I(f)$ is the periodogram of $\varepsilon_T(t), t = 1, 2, \ldots, n$. Estimation of ΔT can be based on minimizing (4.2).

Solow and Patwardhan (1992) applied this approach by, first, estimating $s(f;\Delta T)$ for a range of values of ΔT by applying white noise forcing with variance $1.0\ \mathrm{W\,m^{-2}}$ to a physical model and, second, estimating $\varepsilon_T(t)$ by the residual from a nonparametric estimate of $m_T(t)$. In estimating ΔT from these residuals, $s(f;\Delta T)$ must be modified to account for the filtering effects of estimating $m_T(t)$. The estimate of ΔT found in this way is 1.4°C with approximate 0.95 confidence interval 0.9–2.3°C. These results are in remarkably with those of Bloomfield (1992). In Figure 5, the residual periodogram is plotted along with the model spectra (adjusted for the effects

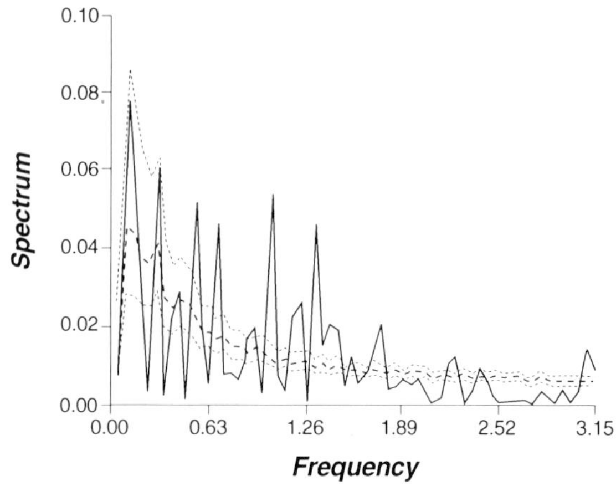

Fig. 5. Residual periodogram (solid) and modified model spectra for $\Delta T = 0.9°C$ (bottom), 1.4°C (middle) and 2.3°C (top) (Solow and Patwardhan, 1992).

of filtering) for ΔT given by the point estimate and the endpoints of the 0.95 confidence interval.

By basing estimation of ΔT on the variability of temperature around the trend, this approach avoids the problem of reconstructing secular changes in radiative forcing. On the other hand, the approach depends on the correct characterization of $\varepsilon_Q(t)$, so the burden is merely shifted.

5. Discussion

The purpose of this chapter has been to survey some statistical approaches to problems arising in atmospheric science. The problems considered here fall into three broad categories: estimation of a common trend from multiple time series, interpolation and estimation of a physical parameter. Not surprisingly, variants of these problems arise in many different contexts.

The approaches described in this chapter themselves fall into two broad categories: empirical and model-based. The approaches in Shapiro and Switzer (1989), Stein (1986), and to a large extent Reinsel and Tiao (1987) are empirical, in the sense that the data alone determine the results. This is not strictly true, of course, since the methods were developed with knowledge about the physical processes involved. Reinsel and Tiao (1987) went a step further and based their sepecification of the form of a global trend on the results of a physical model.

In contrast, the approaches of Bloomfield et al. (1983), Bloomfield (1992) and Solow and Patwardhan (1992) make direct use of physical models in parameter estimation. Solow and Patwardhan (1992) went the farthest in this respect and used the physical model to provide not only a trend model, but also a model of variability about the trend. In contrast, Bloomfield et al. (1983) and Bloomfield (1992) used a physical model for the trend alone and apply more standard time series techniques to model variability.

The random effects model used by Reinsel and Tiao (1987) and Bloomfield et al. (1983) is a useful method for combining results from different locations. A key assumption underlying this model is that the random effects have zero mean. While this is perfectly reasonable for the Dobson data – the global trend being defined as the average of trends over space – it may not always be so. For example, there have been a number of attempts to estimate the global trend in sea level by combining results from a large number of spatially distributed stations. Because only relative sea level (i.e., sea level relative the land) can be measured, and because the land itself can move up or down, this approach assumes that the local effects have zero mean. In fact, there is reason to believe that coastlines tend to sink more than they rise, so that the local effects have a positive, but unknown, mean. Failure to recognize this leads to an over-estimation of historic sea level rise.

In each of the studies reviewed here, some attempt was made to assess the robustness of the results to alternative specifications of the model. Stein (1985) considered alternative semi-variogram models. Reinsel and Tiao (1987) and Bloomfield et al. (1983) experimented with the choice of regressors and considered different error models, as did Bloomfield (1992). Solow and Patwardhan (1992) studied the effects of varying what might be called nuisance model parameters and different amounts of smoothing. Moreover, different approaches to the same problem – Reinsel and Tiao (1987) and Bloomfield et al. (1983) on ozone depletion and Bloomfield (1992) and Solow and Patwardhan (1992) on global warming – gave similar results, providing a further test of robustness.

References

Bloomfield, P. (1992). Trends in global temperature. *Climat. Change* **21**, 1–16.

Bloomfield, P., G. Oehlert, M. L. Thompson and S. Zeger (1983). A frequency domain analysis of trends in Dobson total ozone records. *J. Geophys. Res.* **88**, 8512–8522.

Box, G. E. P. and G. C. Tiao (1977). A canonical analysis of multiple time series. *Biometrika* **64**, 355–365.

Dzhapharidze, K. O. (1986). *Parameter Estimation and Hypothesis Testing in Spectral Analysis.* Springer, New York.

Finkelstein, P. L. (1984). The spatial analysis of acid precipitation data. *J. Climate Appl. Meteorol.* **23**, 52–62.

Firor, J. (1990). *The Changing Atmosphere.* Yale Univ. Press, New Haven, CT.

Harville, D. A. (1977). Maximum likelihood approaches to variance component estimation and to related problems. *J. Amer. Statist. Assoc.* **72**, 320–338.

Houghton, J. T., G. J. Jenkins and J. J. Ephraums (1990). *Climate Change: The IPCC Scientific Assessment.* Cambridge Univ. Press, Cambridge.

Jolliffe, I. T. (1986). *Principal Component Analysis.* Springer, New York.

Pena, D. and G. E. P. Box (1987). Identifying a simplifying structure in time series. *J. Amer. Statist. Assoc.* **82**, 836–843.

Reinsel, G. C. and G. C. Tiao (1987). Impact of chlorofluoromethanes on stratospheric ozone. *J. Amer. Statist. Assoc.* **82**, 20–30.

Reinsel, G. C. and M. A. Wincek (1987). Asymptotic distribution of parameter estimates for nonconsecutively observed time series. *Biometrika* **74**, 115–124.

Shapiro, D. E. and P. Switzer (1989). Extracting time trends from multiple monitoring sites. Technical Report 132, Department of Statistics, Stanford University, Stanford, CA.

Solow, A. R. (1992). Detecting change in the composition of a multi-species community. *Biometrics*, in press.

Solow, A. R. and A. Patwardhan (1992). Some model-based inference about global warming. *Environmetrics*, submitted.

Stein, M. (1985). A simple model for spatial–temporal processes with an application to estimation of acid deposition. Technical Report 82, Department of Statistics, Stanford University, Stanford, CA.

Stein, M. (1986). A simple model for spatial–temporal processes. *Water Resources Res.* **22**, 2107–2110.

Stolarski, R. S. (1982). Fluorocarbons and stratospheric ozone: A review of current knowledge. *Amer. Statist.* **36**, 303–311.

Wigley, T. M. L. and S. C. B. Raper (1990). Natural variability of the climate system and the detection of the greenhouse effect. *Nature* **344**, 324–327.

G. P. Patil and C. R. Rao, eds., *Handbook of Statistics, Vol. 12*

Statistics with Agricultural Pests and Environmental Impacts

Linda J. Young and Jerry H. Young

Abstract: A review of the problems arising from indiscriminate insecticide use is given. Common statistical distributions encountered when sampling pest species within agroecosystems, and possible biological models giving rise to these distributions are discussed. Iwao's patchiness regression and Taylor's power law are described. Sequential estimation of pest population densities is reviewed. Sequential hypothesis tests of population densities for deciding whether chemical control is needed are discussed.

1. Introduction

Modern agriculture relies heavily on pesticides such as insecticides, herbicides, fungicides and acaracides. Increasing concern for the environment has led some to call for the reduction, if not the elimination, of chemical applications in agricultural production. While recognizing the need for more judicious application, most, if not all, agriculturists, agronomists, and entomologists cannot envision modern agriculture without the use of some pesticides. This chapter focuses on the statistical methods that are useful in assessing the need for pesticides. These methods will be discussed in the context of insect control. Their applicability to other pests will then be discussed. A brief review of the impact of indiscriminate use of insecticides will be given first to motivate the need for the statistical methods.

2. Insecticide use

Agricultural pesticide use began in 1790 when teas of tobacco (*Nicotiana tabaccum* L.) containing the alkaloid nicotine were recommended for aphid control without damage to the host plant (see Metcalf, 1986). The first synthetic pesticide, Paris green or copper acetoarsenite, was applied to reduce Colorado potato beetle (*Leptinotarsa decemlineata* Say) populations on potatoes in about 1865. In 1921, the first aerial

Table 1
Crop loss as a percentage of total production*

Crop	1900–1904	1910–1935	1942–1951	1951–1960
Cotton	10	14.9	15	19
Corn	8	11.8	3.5	12
Apple	20	10.4	14	13
Tobacco	–	–	11	11
Potato	10	22	15.6	14
Cabbage	10	20	8	17
Alfalfa	10	5	9.3	15

* Taken from USDA data (Marlatt, 1904; Hyslop, 1983; USDA, 1954, 1965) by Metcalf (1986).

application of insecticide was conducted when lead arsenate dust was used for the control of the catalpa sphinx (*Ceratomia catalpae* (Boisd.)) (Metcalf et al., 1962).

From these small beginnings, the rate of insecticide application has increased dramatically, especially since 1945. Today, about 10 000 commercial insecticides are registered for market in the United States. The volume of insecticides produced in the United States has increased from 13.6 million kg in 1928 to more than 300 million kg in 1980 (Metcalf, 1980). As a result of this massive effort, some insect pests no longer significantly affect agricultural production. However, as may be seen in Table 1, the victory is far from complete.

Obviously, insecticides have not been the cure-all that was first hoped. What went wrong? The primary difficulty has been that pest species develop resistance at an essentially exponential rate until high level resistance or legal restrictions have slowed further usage. Although acquired resistance or tolerance of insect pests to insecticide action was first demonstrated in 1914 when San Jose scale (*Quadraspidiòtus perniciòsus* (Comstock)) showed resistance to lime sulfur spray (Metcalf, 1955), resistance began receiving significant scientific attention only after resistant strains of the house fly (*Músca doméstica* L.) appeared in Sweden and Denmark in 1946, the bedbug (*Cìmex lectulàrius* L.) in Hawaii in 1947, the mosquitoes *Cúlex pípiens* L. in Italy and *Aèdes sollícitans* (Walker) in Florida in 1947, and the human body louse (*Pedículus humànus humànus* L.) in Korea and Japan in 1951 (Brown and Pal, 1971). The proliferation of new insecticides and their expanding use led to a rapidly increasing number of documented cases of insect resistance to insecticides, increasing from 224 species in 1970 to 447 species in 1984 (see Metcalf, 1986). By 1980, 260 insect pests of crops, forests, and stored products had developed resistance to one or more insecticides. Because these are documented cases, the extent of the problem is probably understated.

Not only do insects develop resistance to a particular insecticide, they often develop resistance to chemically related insecticides or to a variety of classes of insecticides. The resistance genes are persistent. Even after insecticide removal, a changed background of residual inheritance in the genome remains. Consequently, the species regains its resistance as soon as the insecticide is reapplied (Brown, 1979).

Predators and parasites of pest species have, for the most part, been slower to develop resistance. Therefore, after insecticide application, the target pest species or a nontarget secondary pest species may have a resurgence in population. For example, in the early 1900s, the cotton boll weevil (*Anthónomus grándis* Boheman), the pink bollworm (*Pectinóphora gossypiélla* (Saunders)) and the cotton fleahopper (*Pseudatomoscelis seriatus* (Reuters)) were the primary pests of cotton. The tobacco budworm (*Heliothis virescens* (Fabricius)) and the tobacco bollworm (*Heliothis zèa* (Boddie)) were secondary pests and occurred only sporadically. However, the organic insecticides applied to control the boll weevil also heavily decimated the insect enemies of the bollworm and budworm, and these pests have become primary pests of cotton. A series of insecticides have provided control of the early life stages of the bollworm and budworm. However, they develop resistance to these insecticides, permitting a resurgence in the population. The cycle of extensive use of a new insecticide, followed by development of resistance, followed by another new insecticide has been repeated so often that these pests are resistant to most of the insecticides registered for use on cotton in the United States (Adkisson et al., 1982).

Microbial action has caused insecticide application to soils at planting time for control of soil insects to become less effective. As an example, the residual action against soil insects has been inadequate and treated corn fields, after repeated insecticide applications, have experienced damaged roots, extensive lodging and yield reductions (Felsot et al., 1982). At least some of these problems may be alleviated by reduced use of broad-spectrum insecticides in favor of insecticides highly sensitive to target pests without appreciably damaging natural enemy populations.

In addition to difficulties in controlling pest populations by extensive insecticide applications, indiscriminate insecticide usage has, at times, resulted in environmental catastrophes. One of the worst cases occurred in 1957 when 7 226.8 ha of Illinios farm land was sprayed with 2.2–3.4 kg/ha of dieldrin in a Japanese beetle (*Popíllia japónica* Newman) eradication program. 'Everything was treated including farms, houses, streams, and roads' (Luckmann and Decker, 1960). The effects on wildlife and domestic animals throughout the treated area were devastating. Insecticide-related deaths were recorded for 19 species of birds and 12 species of animals. Some of the species were almost annihilated; others suffered heavy losses. Scott et al. (1959) estimated that 1840 birds and 440 mammals were killed by a single application of dieldrin in the town of Sheldon. Beef animals were poisoned, and 90% of the cats and some dogs on the treated farms died. This and other examples of ecological catastrophe resulted largely from the use of organochlorines. These insecticides produce the greatest impact on the targeted species, but their uncontrollable nature also results in perturbations throughout the various trophic levels leading to humans, their domestic animals, and wildlife. Consequently, stringent Federal regulation of their use has been imposed in the United States: DDT (1973), aldrin and dieldrin (1974), heptachlor and chlordane (1976), chlordecone (1980), endrin (1981), mirex (1982) and toxaphene (1983).

When insecticides were first introduced, they were effective and inexpensive. This led to a philosophy of spraying if any pest species was observed within a cropping system. However, as described above, the weaknesses of this approach became evident. Therefore, integrated pest management (IPM) was introduced as an ecologically sound approach that would rectify the overuse and misuse of insecticides. Insecticide use may be an important part of IPM, but efforts are made to integrate insecticide applications with other biotic and abiotic factors that regulate the population dynamics of the pest species. While the concept of IPM is extremely appealing, its practice has been less than ideal. Researchers are searching for methods of quantifying the effect of pest and beneficial insect species on the various life stages of a crop. Currently, the emphasis is on precise estimation of population density and hypothesis tests of population density with specified error rates. An overview of some of these methods, their strengths and weaknesses comprise the remainder of this chapter.

3. Distributions

Precise methods of estimation and hypothesis testing assume knowledge of the underlying probability distribution. Several models of insect behavior have been proposed that would give rise to various probability distributions. For simplicity of discussion, the micro-habitat of the insect will be taken to be the plant, and this will be the sampling unit.

The first, and perhaps the most common, model of insect distribution may be illustrated by considering the distribution of two insects on two plants. Suppose each permutation of the insects on the plants is equiprobable (see Figure 1). Let X be the number of insects on a plant. One of two probability mass functions will be observed for X. For the first mass function, the probabilities of $X=0$ and $X=2$ are each 0.5 (arrangements (c) and (d) in Figure 1). For the other, the probability is 1 that $X=1$ (arrangements (a) and (b) in Figure 1). If the number of insects C and the number of plants N approach infinity while the average number of insects on a plant $\mu = C/N$ remains constant, the limiting distribution for X is Poisson with mass

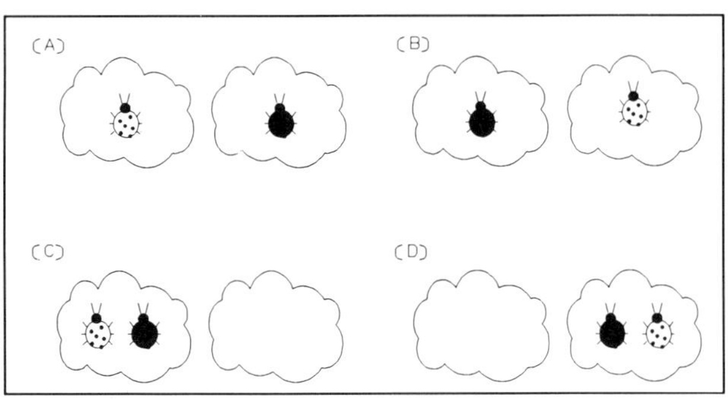

Fig. 1. The four possible permutations of two insects on two plants. The insects are shaded differently so that they may be distinguished.

function

$$P(X=x)=\frac{\mu^x}{x!}e^{-x}, \quad x=0,1,2,\ldots. \tag{1}$$

Here the mean and variance are each equal to μ. Perhaps because this derivation agrees with many intuitive concepts of randomness, the Poisson and randomness are used interchangeably in much of the biological literature. In agreement with the model, a random spatial pattern is also associated with the Poisson. However, Patil and Stiteler (1974) have noted that the Poisson does not necessarily imply spatial randomness. Suppose the number of insects on a plant is binomial with parameters n and p, but n varies from plant to plant according to the Poisson distribution with mean μ. Then the distribution of the number of insects on a randomly selected plant is Poisson with mean $p\mu$, but the spatial distribution is not random unless the variation of n is spatially random. The Poisson rarely adequately describes insect counts. This has led several authors to comment on the rarity of randomness (Taylor et al., 1978, 1979; Taylor, 1984).

The negative binomial distribution often fits insect data well and may be derived from several reasonable biological models. As an example, suppose an insect lays her eggs on a plant according to the Poisson distribution with parameter λ. However, due to varying attractiveness of the plants to the insect, λ varies according to the gamma distribution with parameters r and θ. Then the probability of observing $X=x$ insects on a randomly selected plant would be negative binomial with probability mass function

$$P(X=x)=\binom{k+x-1}{x}\left(\frac{k}{\mu+k}\right)^k\left(\frac{\mu}{\mu+k}\right)^x, \quad x=0,1,2,\ldots, \tag{2}$$

where $k=r$ and $\mu=r/\theta$. Here the mean is μ, and the variance is $\sigma^2=\mu+\mu^2/k$. The negative binomial is referred to as an aggregated, contagious, or over-dispersed distribution because the variance exceeds the mean. If the variation in λ is spatially related, then the spatial pattern of the insects would not be random. Because the aggregation of the insects occurs on a larger scale than the sampling unit, this model is an example of an apparent contagion, or a compound Poisson process (Gurland, 1959).

The negative binomial parametrization given above is due to Anscombe (1949) and is the one most commonly used in biology. In elementary probability courses, the negative binomial is often introduced as arising as the number of failures prior to the r-th success when conducting independent Bernoulli trials each with the probability p of success. Here $r=k$ and $p=k/(\mu+k)$.

Unfortunately, as with the Poisson, the fact that the negative binomial adequately describes the data does not necessarily provide insight into the underlying biological mechanisms giving rise to that distribution. The negative binomial can be established based on different biological assumptions (see Boswell and Patil, 1970). For example, suppose a moth lays her eggs in clusters on a plant. If the number of clusters

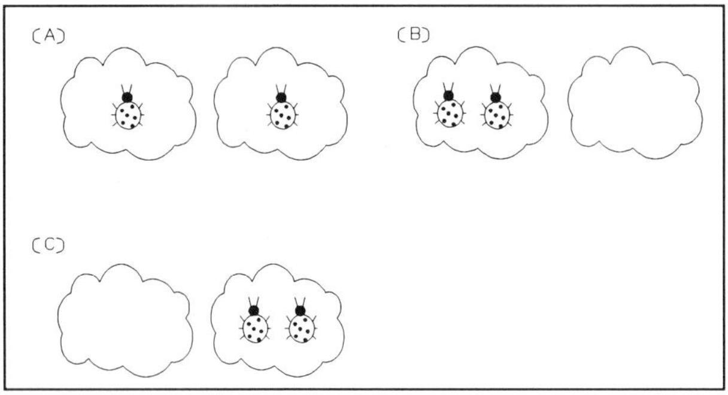

Fig. 2. The three distinguishable spatial patterns possible for two insects on two plants. Both insects have identical shading, emphasizing that they are indistinguishable.

on a plant has the Poisson distribution with mean λ while the number of eggs in a cluster has the logarithmic distribution with parameter θ, the distribution of eggs on a plant is negative binomial with

$$\mu = \frac{-\lambda\theta}{(1-\theta)\ln(1-\theta)} \quad \text{and} \quad k = \frac{-\lambda}{\ln(1-\theta)}. \tag{3}$$

Because the aggregation is occurring within the plant, a scale smaller than the sampling unit, this is an example of a true contagion, or a generalized Poisson, process and can be associated with spatial randomness.

Another model of insect distribution, based on Bose–Einstein statistics, was first developed to model the distribution of particles in a phase space. Again consider the example of two insects on two plants, but now assume that each distinguishable spatial pattern is equiprobable (see Figure 2). As in the earlier case, two probability mass functions of X, the number of insects on a plant, are possible. The first has a probability of 0.5 of observing $X = 0$ or $X = 2$ when a plant is selected at random (arrangements (b) and (c) in Figure 2). For the other, the probability of $X = 1$ is one (arrangement (a)). Whereas the first mass function can occur in two ways, as in the case of the Poisson, the second can now occur in only one way because arrangements (a) and (b) in Figure 1 result in the same spatial pattern. Therefore, the first mass function has probability two-thirds, and the second has probability one-third. If the number of insects C and number of plants N approach infinity while the average number of insects on a plant $\mu = C/N$ remains fixed, the geometric distribution is the limiting distribution for X. The geometric is a special case of the negative binomial distribution with $k = 1$. This approach is equivalent to maximizing entropy subject to the constraints that given numbers of plants and insects are in the agroecosystem. Several insect species have been found to be in the geometric distribution when the micro-habitat of the insect is the sampling unit (Young and Willson, 1987; Willson et al., 1987).

Anscombe (1950) and Evans (1953) have reviewed a number of other mathematical models that have been developed to describe the distribution of organisms within an ecosystem. As with the negative binomial distribution, the Thomas, Neyman's type A, and the Pólya–Aeppli distributions are referred to as aggregated or contagious distributions because the variance exceeds the mean.

The Thomas (1949) distribution is based on the assumptions that the number of colonies on a sampling unit has the Poisson distribution and that the number of organisms in a colony is a Poisson variable increased by 1. It is well suited to situations in which one 'parent' as well as the 'offspring' comprise a colony (Johnson and Kotz, 1969).

Neyman (1939) proposed a model to explain the distribution of larvae in a randomly chosen area in a field. Suppose the female lays eggs in clusters on a plant. If the number of clusters on a plant has the Poisson distribution with mean λ and the number of eggs in a cluster has a Poisson distribution with mean ϕ, then the number of eggs on a plant has the Neyman type A distribution with probability mass function

$$P(X=x)=\frac{\phi^x}{x!}\,e^{-\lambda}\sum_{j=0}^{\infty}\frac{j^x}{j!}(\lambda\, e^{-\phi})^j, \quad x=0,1,2,\ldots. \tag{4}$$

Here the mean is $\lambda\phi$ and the variance is $\lambda\phi(1+\phi)$. Skellam (1958) has shown that Neyman's type A is particularly applicable when the organisms occur in compact clusters, but that it is a good approximation under certain conditions when the clustering is less compact. It has been used extensively in biological applications.

The Pólya–Aeppli distribution, described by Pólya (1931), is similar to the Thomas distribution in that the number of colonies on a sampling unit is assumed to be Poisson. However, the number of organisms within a colony is assumed to be geometrically distributed. The Pólya–Aeppli distribution tends to fit well when the population simultaneously invades an area and produces offspring (Southwood, 1978).

The binomial distribution is most often used in presence–absence sampling. Suppose the probability of each fruit being damaged by insects is p, and n fruit are selected at random. The probability of exactly $X=x$ of the n fruit being damaged is

$$P(X=x)=\binom{n}{x}p^x q^{n-x}, \quad x=0,1,2,\ldots n, \tag{5}$$

where $q=1-p$. For the binomial distribution, the mean is np and the variance is npq. The binomial is referred to as a regular or under-dispersed distribution because the mean is less than the variance.

Each of the models given thus far is based on a biological mechanism that gives rise to a distribution at a given point in time. They do not describe the dynamics of a system through time. However, the model based on maximizing entropy can be extended so that it is a stochastic model of insect distribution. To make this extension, an immigration model is perhaps the best starting point. Suppose that insects are sequentially entering the agroecosystem. Further assume that the probability that the next insect to enter the system with N plants will select plant i for habitation is

$$P_i = \frac{1 + \omega x_i}{N + \omega \sum_{j=1}^{N} x_j}, \tag{6}$$

where N is the number of plants and x_i is the number of insects currently on plant i. If ω is zero, then every permutation of the insects is equiprobable, and the Poisson is the limiting distribution. If ω is one, then each distinguishable spatial pattern of the insects is equiprobable, and the geometric is the limiting distribution (Ijiri and Simon, 1975). The negative binomial with $k = 1/\omega$ is the limiting distribution for $\omega > 0$. If $-1 < \omega < 0$, the binomial is the limiting distribution. This is an application of Pólya's urn model (Feller, 1968). Movement and various limiting factors such as restricted carrying capacity, sexual preference, and predator–prey relationships can also be incorporated into the model (see Young and Willson, 1987; Willson et al. 1987; Young and Young, 1989; Young and Young, 1991).

The adequacy of a model for describing insect distribution is generally judged using chi-square goodness-of-fit tests (Gates, 1989). Hinz and Gurland (1970) developed a more powerful test for deciding between the negative binomial and Neyman type A distributions. Bhalerao et al. (1980) modified the quadratic form of the test proposed by Hinz and Gurland to obtain a more powerful test. These tests do not verify that the model accurately reflects the underlying biological mechanisms giving rise to that distribution. As pointed out by Pielou (1977), model verification of compound processes requires two independent assessments. For example, in counts of soil bacteria, Jones et al. (1948) found the number of bacterial colonies per microscopic field agreed well with the Poisson and the number of bacteria per colony were well modeled by the logarithmic distribution. As anticipated, the observed frequencies of bacteria per microscopic field were found to be in agreement with the negative binomial distribution. Such examples of model verification are rare in the literature.

From an IPM perspective, statistical models can help provide insight into the biology of the agroecosystems. As these models improve, the effects of biotic and abiotic intervention may become clearer. Extensive models, often based on regression relationships, have been created to model crop development throughout the season for some of the primary crops. However, at the present time, no model is accurate or long-term enough that it can be used without field validation within each agroecosystem in every year. This requires sampling. Depending on the goal of sampling, population density estimation or hypothesis testing may be needed.

4. Sampling and estimating population density

The population density is the mean number of insects per unit area in the sampling region. There are two types of density estimates: absolute and relative. In absolute density estimation, data are collected in such a manner that the mean number of insects per unit area of the habitat can be estimated. A plant (when the number of plants for a given area is known), a square-foot quadrat, or a linear foot of row are common sampling units within agroecosystems when absolute density estimation is desired.

Relative population density estimates are based on measurements that cannot be directly related to a unit of area within the sampling region. Traps, such as light traps or pheromone traps, and sweep nets are examples of relative density sampling units. As the population density increases, more insects are measured within the sampling unit; fewer insects are recorded as the population density is decreased. However, no estimate of the population density per unit area is available. Often attempts are made to relate relative to absolute population density estimates through regression. The success of such efforts depends on the strength of the correlation between the two measures.

The micro-habitat of the insect is an ideal sampling unit if it is a practical one. For many insects, the plant, or a portion of the plant, serves as a micro-habitat. These provide well-defined discrete sampling units with biological meaning. This is not always possible as is the case with soil insects. For these, a unit of soil must be extracted and the insect distribution depends on the size of the sampling unit.

Improved sampling methods can often be developed if the form of the distribution is known. This requires experience with the insect and cropping system of interest. In spite of the extensive variation within agroecosystems, the same insect distributions do occur consistently. The Thomas, Neyman type A and Pólya–Aeppli distribution have not been used extensively in sampling programs. This is probably due to the difficulty in estimating the parameters (see Johnson and Kotz, 1969, and Douglas, 1955). The binomial, Poisson, and negative binomial distributions are the most frequently used and will be considered here. Three measures of the precision of the estimated population density will be addressed: estimation with a specified coefficient of variation C, estimation within a proportion D of μ with $(1-\alpha)100\%$ confidence and estimation within H units with $(1-\alpha)100\%$ confidence. Fixed-sample-size procedures will be considered first, followed by the sequential methods.

Suppose the purpose of sampling is to obtain a precise estimate of population density with a specified coefficient of variation of the sample mean C; that is,

$$\mathrm{CV}(\bar{X}) = C = \frac{\sigma/\sqrt{n}}{\mu}. \tag{7}$$

The optimal sample size may be expressed as

$$n = \left(\frac{\sigma}{\mu C}\right)^2. \tag{8}$$

This expression takes varying forms depending of the relationship in mean and variance for a given population distribution as shown in Table 2 for the binomial, Poisson and negative binomial distributions. It should be noted that in the binomial parametrization given in (5), n is used as a parameter of the binomial, but n has also been used to denote sample size. Because each experimental unit is examined for the

Table 2
Optimal sample sizes when estimating the population density with a specified CV($\bar{X}$) = C(C), within $D\mu$ with confidence $(1-\alpha)100\%$ (D) and within H with confidence $(1-\alpha)100\%$ confidence (H)

	C	D	H
Binomial	$\frac{1-p}{pC^2}$	$\frac{(1-p)z_{\alpha/2}^2}{pD^2}$	$p(1-p)\left(\frac{z_{\alpha/2}}{H}\right)^2$
Poisson	$\frac{1}{\mu C^2}$	$\frac{1}{\mu}\left(\frac{z_{\alpha/2}}{D}\right)^2$	$\mu\left(\frac{z_{\alpha/2}}{H}\right)^2$
Negative binomial	$\frac{1}{C^2}\left(\frac{1}{\mu}+\frac{1}{k}\right)$	$\left(\frac{z_{\alpha/2}}{D}\right)^2\left(\frac{1}{\mu}+\frac{1}{k}\right)$	$\left(\mu+\frac{\mu^2}{k}\right)\left(\frac{z_{\alpha/2}}{H}\right)^2$
Iwao's patchiness regression	$\frac{1}{C^2}\left(\frac{\alpha+1}{\mu}+(\beta-1)\right)$	$\left(\frac{z_{\alpha/2}}{D}\right)^2\left(\frac{\alpha+1}{\mu}+(\beta-1)\right)$	$[(\alpha+1)\mu+(\beta-1)\mu^2]\left(\frac{z_{\alpha/2}}{H}\right)^2$
Taylor's power law	$\frac{a\mu^{b-2}}{C^2}$	$\frac{az_{\alpha/2}^2\mu^{b-2}}{C^2}$	$a\mu^b\left(\frac{z_{\alpha/2}}{H}\right)^2$

specified characteristic of interest, the binomial parameter n is one and will be taken to be one during the remainder of this chapter. Therefore, n in the equations will continue to represent the sample size.

A confidence interval for the population density may be based on a random sample of size n from a population with mean μ and variance σ^2. By invoking the Central Limit Theorem, a confidence interval has the form

$$\mathrm{P}\left(\bar{X}-z_{\alpha/2}\frac{\sigma}{\sqrt{n}}<\mu<\bar{X}+z_{\alpha/2}\frac{\sigma}{\sqrt{n}}\right)=1-\alpha, \tag{9}$$

where $(1-\alpha)$ is the specified confidence level and $z_{\alpha/2}$ is the $(1-\alpha/2)$ quantile of the standard normal distribution. If estimation within $D\mu$ or within H with $(1-\alpha)100\%$ confidence is desired, the optimal sample size may be shown to be

$$n=\left(\frac{z_{\alpha/2}\sigma}{D\mu}\right)^2 \tag{10}$$

or

$$n=\left(\frac{z_{\alpha/2}\sigma}{H}\right)^2 \tag{11}$$

respectively. Table 2 gives the optimal sample size in terms of the parameters for the binomial, Poisson and negative binomial distributions.

Equations (7) through (11) and those in Table 2 involve at least one unknown parameter. 'Reasonable' values may be used for these unknown parameters to determine approximate sample sizes. However, one can never be sure whether a predetermined fixed sample size will result in estimates of the population density μ with the desired level of precision.

When attempting to estimate the mean using a confidence interval no longer than a predetermined length, Stein (1945) suggested replacing μ and σ^2 in (11) with some preliminary estimates of the mean and variance, say $\bar{X}_0$ and s_0^2, based on an initial sample of size n_0. Thus,

$$n = \left(\frac{z_{\alpha/2} s_0}{H}\right)^2. \tag{12}$$

However, the choice of n_0 may be difficult. In addition, it is not assured that

$$n \geqslant \left(\frac{z_{\alpha/2} s_n}{H}\right)^2, \tag{13}$$

where s_n^2 is the sample estimate of the variance based on n observations.

Chow and Robbins (1965) suggested updating the estimates of mean and variance after each observation and sampling until equation (13) is statisfied in order to estimate the mean within H units with $(1-\alpha)100\%$ confidence. They were able to show that this method is both asymptotically consistent and efficient for any population distribution with finite variance.

Nadás (1969) suggested a similar approach when estimation of the mean within a specified proportion D of the mean with a stated level of confidence is desired. He proposed to sample until

$$n \geqslant \left(\frac{z_{\alpha/2} s_n}{D\bar{X}_n}\right)^2, \tag{14}$$

where $\bar{X}_n$ and s_n^2 are sample estimates of mean and variance, respectively, based on the n observations in the sample. Nadás showed that the above procedure is asymptotically consistent and efficient.

Rojás (1964) introduced sequential CV control sampling for soil insects. By estimating the parameters in equation (8) and updating these estimates after each observation, he was led to propose sampling until the first time

$$n \geqslant \left(\frac{s_n}{\bar{X}_n C}\right)^2. \tag{15}$$

Willson and Folks (1983) building on the work of Mukhopadhyay (1978) and Ghosh and Mukhopadhyay (1979) showed that this procedure is asymptotically consistent and efficient.

The procedures outlined in equations (13) to (15) are intuitive, nonparametric extensions of fixed-sample-size results. Although they are based on normality through the Central Limit Theorem, the methods are asymptotically consistent and efficient for any distribution with finite mean and variance. However, in each case, when working with moderate sample sizes, Monte Carlo studies have shown that for the negative binomial distribution with k-values in the range most frequently encountered in population studies, the estimate is not as precise as stated. The difference in the nominal and observed levels of precision differs depending on μ and k. For example, suppose we want to estimate the mean with a $\mathrm{CV}(\bar{X}) = 0.25$. When $(\mu, k) = (1,1)$ and $(5,5)$, simulations indicate that the actual $\mathrm{CV}(\bar{X})$ is 0.42 and 0.32, respectively (Willson, 1981). Similar results hold for the other two procedures. Also, the need to update estimates of the mean and variance after each observation requires the use of some form of computing equipment in the field. Although this is certainly feasible, it may not be as desirable as a method that does not require such computations. Consequently, the properties of the population distribution being sampled have been used to devise more precise sampling programs.

Sequential sampling plans are often based on heuristic extensions of optimal fixed-sample-size procedures. For example, for each plan in Table 2, the parameters of the distribution could be estimated after each observation, and sampling could continue until the equation, with the estimates substituted for the parameters, is satisfied. The stopping rules in Table 3 were obtained in this manner and are well-accepted by IPM researchers. Unfortunately, the statistical properties of these heuristic procedures are not always known.

Consider precise estimation of the binomial parameter p. The purpose of sampling is to estimate the binomial proportion p with specified precision. In each case given in Table 2, p is replaced with its uniformly minimum variance unbiased estimator (UMVUE) $\bar{X}_n$, the mean of the first n observations, to obtain the stopping rule. This stopping rule is rewritten in terms of the sum of the first n observations T_n and given in Table 3. For the rule to be well-defined and nontrivial, sampling is conducted until at least one zero and at least one one have been observed.

Because of its wide applicability, estimation of the binomial parameter p has been extensively studied, including work by Robbins and Siegmund (1974), Cabilio and Robbins (1975), Cabilio (1977) and Martinsek (1983). The three cases presented in Table 3 are given here because of their wide usage in entomological applications (see Karandinos, 1976, and Ruesink, 1980, for examples).

Starr and Woodroofe (1968) proved that for a sequence of constants $\{c_n\}$ and stopping rules of the form $\{N = \text{smallest integer such that } n \geqslant n_0 \text{ and } \bar{X}_n \geqslant c_n\}$, $\mathrm{E}(\bar{X}_n) \geqslant \mu$; similarly, if the stopping rule is of the form $\{N = \text{smallest integer such that } n \geqslant n_0 \text{ and } \bar{X}_n \leqslant c_n\}$, $\mathrm{E}(\bar{X}_n) \leqslant \mu$. Therefore, when the stopping rule is based on the sum exceeding or being smaller than a specified constant as in Table 3, the estimate of population density may be biased. However, simulation studies for the distributions considered here indicate that the bias is small and is not of practical importance in most cases.

Suppose now that it is known that the population distribution is Poisson. The purpose of sampling is to estimate the population density μ with a specified level of precision. The UMVUE of μ, $\bar{X}$, is used in each of the Poisson equations in Table 2 to obtain the stopping rules in Table 3, expressed in terms of the total. Vardi (1979) has considered a related problem in estimating a Poisson mean.

Table 3
Stopping rules when estimating the population density with a specified $\mathrm{CV}(\bar{X}) = C$ (*C*), within $D\mu$ with confidence $(1-\alpha)100\%$ confidence (*D*), and within *H* with confidence $(1-\alpha)100\%$ confidence (*H*)

	C	*D*	*H*
Binomial	$0 < T_n < n$ and $T_n \geqslant \dfrac{n}{nC^2+1}$	$0 < T_n < n$ and $T_n \geqslant \dfrac{nz_{\alpha/2}^2}{nD^2 + z_{\alpha/2}^2}$	$0 < T_n < n$ and either $T_n \geqslant \dfrac{n}{2} + n\sqrt{\dfrac{1}{4} - \dfrac{nH^2}{z_{\alpha/2}^2}}$ or $T_n \geqslant \dfrac{n}{2} - n\sqrt{\dfrac{1}{4} - \dfrac{nH^2}{z_{\alpha/2}^2}}$
Poisson	$n \geqslant 2$ and $T_n \geqslant \dfrac{1}{C^2}$	$n \geqslant 2$ and $T_n \geqslant \left(\dfrac{z_{\alpha/2}}{D}\right)^2$	$T_n > 0$ and $T_n \leqslant \left(\dfrac{nH}{z_{\alpha/2}}\right)^2$
Negative binomial	$n > \dfrac{1}{C^2k} - \dfrac{1}{k}$ and $T_n \geqslant \dfrac{nk}{(nk+1)C^2 - 1}$	$n > \dfrac{z_{\alpha/2}^2}{kD^2} - \dfrac{1}{k}$ and $T_n \geqslant \dfrac{nkz^2}{(nk+1)D^2 - z^2}$	$T_n > 0$ and $T_n \leqslant -\dfrac{nk}{2} + n\sqrt{\dfrac{k^2}{4} + \dfrac{nH^2}{z_{\alpha/2}^2}}$
Iwao's patchiness regression	$n > \dfrac{\beta - 1}{C^2}$ and $T_n \geqslant \dfrac{n(\alpha+1)}{nC^2 - \beta + 1}$	$n > \dfrac{(\beta-1)z_{\alpha/2}^2}{D^2}$ and $T_n \geqslant \dfrac{n(\alpha+1)z_{\alpha/2}^2}{nD^2 - (\beta-1)z_{\alpha/2}^2}$	$T_n > 0$ and $T_n \leqslant -\dfrac{n(\alpha+1)}{2(\beta-1)} + n\sqrt{\dfrac{nH^2}{(\beta-1)z_{\alpha/2}^2} + \left(\dfrac{\alpha+1}{2(\beta-1)}\right)^2}$

For the negative binomial, the knowledge of k is a determining factor in the development of a sampling program. Sometimes, a biological model of the ecosystem may imply a value for k. If the model has been verified previously, then this value of k may be used in the development of a sampling scheme. In other cases, k has been found to be stable through time and for varying population densities. This constant value of k could then be appropriately used in a sampling program. If knowledge of k, or a precise estimate of k, is not available, then k must be estimated at the time of sampling. Estimation of the mean when k is not known is considered first. Next, estimation of the mean when k is assumed known is addressed. Because knowledge of k permits more precise estimation of the mean, methods of obtaining an estimate of k for use in these sampling schemes will be considered.

Intuitive sequential sampling schemes for the negative binomial distribution could be developed by updating the estimates of μ and k after each observation and sampling until one of the equations (depending on the measure of precision being used) in Table 2 is satisfied. If the method-of-moments estimator (MME) of k is used, these sampling procedures are equivalent to the nonparametric methods. Because the maximum likelihood estimator (MLE) of k does not have a closed form, it has not been used in sequential procedures.

Consider now the case for which either k is known or a precise estimate of k is available. Kuno (1969, 1972) presented a sequential procedure for estimating the mean of a distribution with variance $(a\mu + b\mu^2)$, where a and b are known. The negative binomial arises as a special case with $a = 1$ and $b = 1/k$. The goal of sampling is to estimate μ with a specified $\mathrm{CV}(\bar{X}) = C$. Using the mean and variance estimators $\bar{X}_n$ and $(a\bar{X}_n + b\bar{X}_n^2)$, respectively, he proposed sampling until

$$n \geqslant \frac{1}{C^2}\left(\frac{a}{\bar{X}_n} + b\right). \tag{16}$$

This stopping rule may be rewritten in terms of a minimum sample size and the total needed to stop, T_n. Kuno (1972) considered the bias of the resulting estimator $\bar{X}_n$. However, this procedure results in larger sample sizes than are needed to attain the desired precision (Willson, 1981). Willson and Folks (1983) and Willson and Young (1983) recommended using the UMVUE of σ^2, V_n, in the estimation process. V_n has the form

$$V_n = \frac{nk}{nk+1}\bar{X}_n + \frac{n}{nk+1}\bar{X}_n^2. \tag{17}$$

The stopping rule based on V_n may be rewritten in terms of a minimum sample size and the total as given in Table 3. Often tables are constructed giving the total needed to stop at several levels of precison, say $C = 0.1, 0.15, 0.20$ and 0.25. Extensive simulations studies indicate this procedure results in an estimate of the mean with the desired precision for a C as moderate as 0.3.

Binns (1975) considered estimating a negative binomial mean within $D\mu$ with $(1-\alpha)100\%$ confidence when k is known. He used $(\bar{X}_n + \bar{X}_n^2/k)$ to estimate the variance. Employing a finite population correction factor on nk, he proposed sampling until

$$nk > a^2 + \tfrac{1}{2} \quad \text{and} \quad T_n \geqslant a^2 + \frac{a^4}{nk - 0.5 - a^2}, \tag{18}$$

where $a = z_{\alpha/2}^2/D$. Denoting by (t, r_t) the point of intersection of the sample path and the stopping boundary $(N - 1 < t \leqslant N)$, μ is estimated by

$$\hat{\mu} = \frac{kr_t}{tk - 0.5}. \tag{19}$$

The asymptotic distribution of the natural logarithm of the estimate was determined to be normal with mean $\ln(\mu)$ and variance $(D/z_{\alpha/2})^2$.

Willson and Folks (1983) proposed using V_n as the estimate of σ^2, obtaining the stopping rule in Table 3 for estimating the mean within $D\mu$ with $(1-\alpha)100\%$ confidence. This procedure was shown to be asymptotically consistent and efficient. Simulation studies indicate that with a D as moderate as 0.30 and 95% confidence, the desired precision is achieved. The sample sizes based on the Willson and Folks (1983) procedure are less than those of Binns (1975).

Kuno (1969, 1972) considered estimation of the mean with a specified standard error. Willson (1981) developed an asymptotically consistent and efficient procedure, based on V_n, for estimating the mean within H with a specified level of confidence. However, the level of confidence is well below the stated one for moderate D. Using $\bar{X}_n + \bar{X}_n^2/k$ as the estimate of σ^2 results in the stopping rule given in Table 3. However, the confidence level tends to be below the stated one for many values of D of practical interest. For example, if the mean is one and k is about one, D should not exceed 0.2. More study is needed to determine how large D can be in other regions of the parameter space without reducing the confidence below the stated level.

In each of the negative binomial sampling procedures that are based on a knowledge of k, using a value smaller than the true k results in an estimate of the population mean that is more precise than the stated level. If a range of possible values of k is available, a conservative approach would be to use one of the smallest, if not the smallest, value in the range as the k-value. If the value of k used in developing the sampling program is within 1 of the true parameter value, these procedures continue to produce estimates with a level of precision close to the stated one. Therefore, if a fairly reliable estimate of k is available, it would be better to use that than to turn too quickly to the nonparametric approaches presented earlier. Methods of obtaining a precise estimate of k will be considered next.

Sometimes, biological models may indicate the presence of the negative binomial distribution with a specified k value. Goodness-of-fit tests can be used to verify that the negative binomial with that value of k does, in fact, adequately describe the data. More frequently, the estimates of k from several data sets tend to be *close* to each other. From these, we may want to obtain an estimate of a common k value for the ecosystem or for a series of ecosystems and to test whether that common value of k seems to hold over all data sets. Several methods of estimating a common k have been proposed (Anscombe, 1950; Southwood, 1978). The method proposed by Bliss and Owen (1958) is the most widely used and will be discussed first. The likelihood ratio test for a common k (Bliss and Fisher, 1953) will also be discussed. Finally, a multistage method of estimating k (Willson et al., 1984) will be presented.

Suppose t data sets have been collected and a common k is to be estimated. Let n_i, s_i, and $\bar{X}_i$ denote the sample size, estimated standard deviation and estimated mean, respectively, for the i-th data set, $i = 1, 2, \ldots, t$. The estimation process begins by computing

$$x_i' = \bar{X}_i^2 - \frac{s_i^2}{n_i} \quad \text{and} \quad y_i' = s_i^2 - \bar{X}_i \tag{20}$$

for each of the t data sets. The regression line of y' on x' passes through the origin and has slope $1/\hat{k}_c$. The precision of the estimate may be increased by weighting each population inversely to the variance; that is,

$$w_i = \frac{0.5(n_i - 1)k_c^4}{\mu_i^2(\mu_i + k_c)^2\left(k_c(k_c + 1) - \frac{(2k_c - 1)}{n_i} - \frac{3}{n_i^2}\right)}. \tag{21}$$

Because the weight involves unknown parameters, including k_c, the estimation process must be iterative. An unweighted regression provides an initial estimate of k_c. Then estimates of the parameters are used to obtain the weight w_i for each population $i, i = 1, 2, \ldots, t$. Next the weighted estimate of k_c is computed as the reciprocal of the slope of the weighted regression of y' on x' constrained to pass through the origin. If the weighted estimate of k_c differs appreciably from its initial unweighted estimate, then the weights should be recomputed using the weighted estimate of k_c, and a new weighted estimate of k_c, determined. This process is repeated until the change in the estimate of k_c is negligible. The estimate of a common k may change very little during these iterations. However, the iterations are essential for a valid test of homogeneity.

The test of homogeneity is a test of the null hypothesis that there is a common k for all t populations. Bliss and Owen proposed two tests for the test of this hypothesis. The first test is based on the fact that the error sum of squares from the weighted regression through the origin is approximately χ^2 with $(t - 1)$ degrees of freedom. If the value of the test statistic, the error sum of squares associated with a linear regression through the origion, exceeds the $(1 - \alpha)$ quantile of a χ^2 distribution with $(t - 1)$ degrees of freedom, the null would be rejected, and the use of a common k would not be appropriate for this set of t populations.

The second test requires four quantities that may be obtained from the weighted linear regression constrained to pass through the origin and the weighted linear regression with an intercept. First, the sum of squares associated with the regression constrained to pass through the origin is denoted by B_0^2. For the regression that passes through the origin, CSS represents the corrected total sum of squares, and C is the sum of squares associated with the intercept. The regression sum of squares when the line is not constrained to pass through the origin is B^2. The test may be arranged as an analysis of variance:

Effect of	DF	Sum of squares	Mean square	F
Slope, $1/k_c$	1	B_0^2	B_0^2	B_0^2/s^2
Computed intercept against 0	1	$C + B^2 - B_0^2$	I_0	I_0/s^2
Error	$t - 3$	$\text{CSS} - B^2$	s^2	–

If a common value of k, k_c, is justified, the F-value in the first row should be significant and that in the second row should not be significant. A significant F in the second row indicates a progressive change in k.

The two tests proposed by Bliss and Owen do not always result in the same conclusion as to the appropriateness of a common k. The second test seems to be preferred, but a definitive comparison of these and the other techniques that will be discussed next has yet to be conducted.

Bliss and Fisher (1953) suggested using maximum likelihood to estimate a k common to several negative binomial populations. To test the hypothesis of a common k, the likelihood ratio λ is formed. Under the null hypothesis, $-2\ln(\lambda)$ has an approximate χ^2 distribution with $(t-1)$ degrees of freedom. The null hypothesis would be rejected at the α significance level if $-2\ln(\lambda)$ exceeded the $(1-\alpha)$ quantile value of a χ^2 with $(t-1)$ degrees of freedom.

Maximum likelihood estimation and the likelihood ratio test for a common k have not been used much in biological applications. The reason for this probably lies in the extensive computations involved. However, with today's computing equipment, this should no longer be a major concern. A limited Monte Carlo study (Willson, 1981) indicated that the likelihood ratio test controlled the size of the test better than the regression technique while at the same time giving more power.

A third method of estimating k is a multistage method given by Willson and Young (1983). This method estimates k in the field at the time data are taken. In this method, five observations are taken, and k is estimated using the method of moments estimator (MME). Five more observations are taken, and k is estimated based on all ten observations. This process of adding five more values to the sample and determining the MME of k is continued until the last two estimates of k differ by no more than 0.05. Then the last MME of k is taken as the final estimate. Computer simulations and field studies have shown this to be equal or superior to the method-of-moments or maximum likelihood estimators based on comparable fixed sample sizes (Willson et al., 1984).

The sequential methods discussed thus far are either nonparametric and make no assumption about the relationship in mean and variance or parametric and assume the distribution, and hence the relationship in mean and variance, is known. Another approach that is frequently cited attempts to relate the mean and variance empirically without assuming a parametric form of the model. Iwao's patchiness regression and Taylor's power law are two such approaches.

Lloyd (1967) proposed to define mean crowding as the 'mean number per individual of other individuals in the same quadrat.' He further determined that this index of mean crowding (x^*) could be approximated by

$$x^* \approx \mu + \frac{\sigma^2}{\mu} - 1. \tag{22}$$

Given a parametric model, mean crowding can be evaluated.

Based on empirical studies, Iwao (1968) proposed that mean crowding had the following relationship to mean density:

$$x^* = \alpha + \beta\mu. \tag{23}$$

If the relationship in (23) holds, then from (22), we could conclude that

$$\sigma^2 = (\alpha + 1)\mu + (\beta - 1)\mu^2. \tag{24}$$

The negative binomial, geometric, Poisson and binomial are special cases of this more general relationship in mean and variance. However, other distributions could also have this same relationship in mean and variance in at least some regions of the parameter space. Even though the relationship in mean and variance is assumed to be of a certain form, the parametric form of the distribution is not assumed to be known.

Given a series of data sets collected from an ecosystem of interest, a regression of estimated mean crowding on the sample mean provides estimates of α and β and is called Iwao's patchiness regression. The r^2 from the regression has historically been used as the measure of the adequacy of the model. An alternative method of testing the adequacy of this model is to regress the predicted values of σ^2 based on the model on the observed estimates of σ^2 from the samples. If the model is accurate, a line passing through the origin with a slope of one should fit within error.

Often most of the data sets used to establish the relationship between mean and variance produce small estimated means. The few data sets with large estimated means tend to greatly influence the model. Therefore, a study of the influence of each point on the regression of $\hat{x}^*$ on $\bar{x}$ may also be appropriate. In addition, the adequacy of the model for different ranges of the mean may be investigated by performing the regression of the model-based variance estimates on the sample-based variance estimates, as mentioned above, within each of the ranges of interest.

Kuno's (1969, 1972) procedures mentioned earlier were developed for the relationship in mean and variance expressed by Iwao's patchiness regression. Similar extensions could be made for the other measures of precision as shown in Table 3. Kuno considered the bias of the estimate, but other statistical properties have not been studied.

Based on empirical data, Taylor (1961, 1965, 1970, 1971) suggested that the relationship in mean and variance could be expressed by

$$\sigma^2 = a\mu^b \tag{25}$$

for numerous species. The relationship expressed in (25) has come to be described as Taylor's power law. The parameter b is referred to as an index of aggregation characteristic of the species under consideration (Taylor, Woiwod and Perry, 1978, 1979). The parameter a is considered to be a sampling factor. While a may change with environment or sampling unit, b is considered fixed for any species although Banerjee (1976) disputes this. The parameter b is often used as a measure of aggregation. A b of 1 is indicative of randomness, a b greater than 1 implies aggregation, and a b less than 1 indicates more regularity. However, these terms can only be appropriately applied to the numerical distribution of the organisms and not to the spatial patterns of those organisms.

Several samples are required to obtain estimates of a and b in Taylor's power law. Estimation is based on the regression of the natural logarithm of the estimated variance on the natural logarithm of the estimated mean. The r^2 from regression is used to assess the strength of the relationship, with high r^2 values being common. Graphs of this relationship are usually on the log–log scale. This approach to evaluating the adequacy of Taylor's power law in describing the relationship in mean and variance has a primary weakness. Once a and b are estimated, inference concerning the population is based on the estimated mean and variance and not on the natural logarithms of those estimates. The r^2 from regression is for the log–log scale and may be a poor indicator of the fit on the untransformed scale. Tests of the adequacy of the model should be conducted in the same manner as suggested for Iwao's patchiness regression. Based on these alternative approaches, Taylor's power law may not be as strong as is often believed. This is especially true when a high proportion of the samples have small means and variances, making a study of the influence of points on the estimated relationship extremely important.

Now suppose that the relationship in mean and variance is expressed satisfactorily by Taylor's power law. Procedures for estimating the mean with a specified precision can be developed as in Table 3 (see Ruesink, 1980, as an example). However, the statistical properties of these procedures have not been studied.

5. Hypothesis testing

Sequential estimation procedures are most often used by researchers in the study of population dynamics. In IPM, a decision as to whether chemical control is needed must be made. Fixed-sample-size or sequential hypothesis testing is used for this purpose. The sequential hypothesis testing considered here requires some prior knowledge of the population distribution. This permits most computations to be completed in advance of sampling and to be stored in handheld calculators, laptop computers, or printed on cards or sheets. Wald's (1945, 1947) sequential probability ratio test was the earliest sequential test, and it is still the one most frequently used. It will be described first. Lorden's 2-SPRT is a more recent development that has some exciting possibilities for tests of hypotheses concerning population density and will be described after the SPRT.

Wald (1947) introduced a sequential method of testing a simple null versus a simple alternative hypothesis with specified error probabilities. He called this sequential process the sequential probability ratio test (SPRT). The SPRT was used as a basis for inspecting wartime weapons and evaluating war research problems. Oakland (1950) and Morgan et al. (1951), presented early applications of the SPRT in the biological sciences. The SPRT was used increasingly in biology after Waters (1955) presented details and examples of its use in forest insect studies. By the 1960s the SPRT was widely used for a variety of pests and cultivars in insect pest management. Today, the SPRT is most often used to determine whether or not a specified region needs to be treated with pesticides (see Onsager, 1976; Pieters and Sterling, 1974; Allen et al., 1972; Rudd, 1980).

The SPRT assumes that the parametric form of the random variable's distribution is known. The binomial, Poisson, and negative binomial distributions will be considered. For the negative binomial, the parameter k is assumed known; if it is unknown, then it must be estimated before the test of hypotheses can begin. The equations associated with the normal distribution with a known variance will be given because these are used in some biological applications. Because Iwao's patchiness regression is based on a relationship in mean and variance and not a parametric distribution, this approach is not appropriate for use in Wald's SPRT. To meet their perceived need for a distribution characterized by Taylor's power law, Perry and Taylor (1985) evolved the Adès distribution; the Adès variable Y_i results from a standard gamma variate, logarithmically transformed, and then raised to the power $2/(2-r)$ and discretized. Kemp (1987) suggested an alternative would be to satisfy the requirements of Taylor's power law by constraining the negative binomial, Neyman type A, Pólya–Aeppli, or Poisson with zeroes distribution. Of these, only the negative binomial will be considered here.

Once the sampling unit and associated random variable are clearly defined, the hypotheses to be tested must be stated:

$$H_0: \theta = \theta_0,$$
$$H_1: \theta = \theta_1 \ (> \theta_0). \tag{26}$$

For the Poisson, negative binomial, and normal distributions, θ is μ. θ is p for the binomial. In most IPM programs, θ_0 is the safety level (the largest population density for which no chemical control is needed), and θ_1 is the economic threshold (the population density at which control is needed to prevent an increasing pest population from reaching the economic injury level). Sometimes, the purpose of sampling is to determine whether the population density of a predator species is high enough to protect the crop from economic damage in which case treatment occurs if the population drops below a given threshold. The probabilities of type I and type II errors are α and β, respectively.

The foundation of hypothesis testing for fixed-sample sizes will be reviewed briefly, and then extended to sequential testing. The Neyman–Pearson lemma states that, with a fixed sample size n and type I error probability α, the test that minimizes β is the likelihood ratio test. The likelihood ratio is

$$\lambda_n = \frac{\prod_{i=1}^{n} f_x(x_i; \theta_1)}{\prod_{i=1}^{n} f_x(x_i; \theta_0)}, \tag{27}$$

where f_x is the probability mass function associated with the random variable X of interest. The critical region that would lead to rejection of H_0 is given by $\lambda > C_n$. C_n is obtained using the predetermined values of α and n . Consideration of β may impact the choice of n and/or α, but once α and n are set, the type II error rate β is determined.

Wald presented an alternative approach by setting α and β and then minimizing the sample size n. Thus, the final sample size N becomes a random variable of interest. A sequence of random observations $(X_1, X_2, X_3, \ldots)$ is drawn from one of the two hypothesized distributions resulting in a sequence of likelihood ratios, λ_n, $n = 1, 2, 3, \ldots$.

A decision is made and the test terminated after the n-th observation if $\lambda_n \geqslant A$ (accept H_1) or $\lambda_n \leqslant B$ (accept H_0) where A and B are constants depending on the predetermined α and β. Exact computations for A and B are extremely difficult. Thus, the approximations

$$A \approx \frac{1-\beta}{\alpha} \quad \text{and} \quad B \approx \frac{\beta}{1-\alpha} \tag{28}$$

are used. The approximations for A and B result in error rates α' and β' that differ from the desired error rates α and β. However, it may be shown that the total of the actual error rates, α' and β', is no larger than the sum of the specified error rates, α and β; that is, $\alpha' + \beta' \leqslant \alpha + \beta$.

Sampling continues as long as $B < \lambda_n < A$. In this form, computations would be required after each observation in order to determine whether the stopping criterion is met. However, by taking the natural logarithm, and rearranging terms, the following decision procedure results: (1) terminate the test and accept the alternative hypothesis of $T_n \geqslant h_2 + Sn$; (2) terminate the test and accept the null hypothesis if $T_n \leqslant h_1 + Sn$; or (3) continue sampling if $h_1 + Sn < T_n < h_2 + Sn$.

Notice that the decision boundaries are straight lines. They are parallel because both have slope S. The intercepts are h_1 and h_2 for the lower and upper boundaries, respectively (see Figure 3). S, h_1 and h_2 depend on the distribution, the hypothesized parameters, and the specified error rates (see Table 4, see also Waters, 1955; Fowler and Lynch, 1987). In order to attain the desired error rates U, the upper limit, should always be rounded up and L, the lower limit, should always be rounded down. It is customary to program these systems into handheld calculators or laptop computers or to construct charts or graphs as seen in the partial one given below:

Sequential sampling for negative binomial distribution, $k = 3$, $\alpha = 0.1$, $\beta = 0.1$, economic threshold = 10, safety level = 7

Sample number	Lower limit	Running total	Upper limit	Sample number	Lower limit	Running total	Upper limit
1	nd*	–	8	9	19	–	31
2	nd	–	11	10	22	–	34
3	2	–	14	11	24	–	37
4	5	–	17	12	27	–	40
5	7	–	20	13	30	–	42
6	10	–	23	14	33	–	45
7	13	–	25	15	36	–	48

* nd stands for no decision.

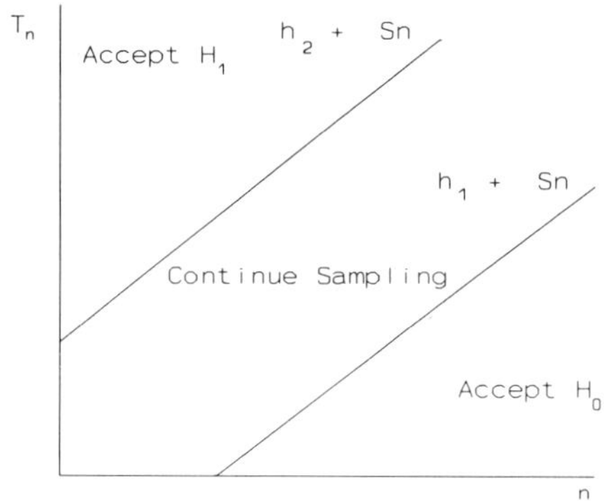

Fig. 3. Decision lines for Wald's SPRT.

Table 4
Intercepts and slopes for Wald's SPRT for the binomial (B), negative binomial (NB), normal (N) and Poisson (P) distributions

Distribution	h_1	h_2	s
B	$\dfrac{b}{\ln\left[\dfrac{p_1 q_0}{p_0 q_1}\right]}$	$\dfrac{a}{\ln\left[\dfrac{p_1 q_0}{p_0 q_1}\right]}$	$\dfrac{\ln\left[\dfrac{q_0}{q_1}\right]}{\ln\left[\dfrac{p_1 q_0}{p_0 q_1}\right]}$
NB	$\dfrac{b}{\ln\left[\dfrac{\mu_1(\mu_0+k)}{\mu_0(\mu_1+k)}\right]}$	$\dfrac{b}{\ln\left[\dfrac{\mu_1(\mu_0+k)}{\mu_0(\mu_1+k)}\right]}$	$k\dfrac{\ln\left[\dfrac{\mu_1+k}{\mu_0+k}\right]}{\ln\left[\dfrac{\mu_1(\mu_0+k}{\mu_0(\mu_1+k)}\right]}$
N	$\dfrac{b\sigma^2}{\mu_1-\mu_0}$	$\dfrac{a\sigma^2}{\mu_1-\mu_0}$	$\dfrac{\mu_1+\mu_0}{2}$
P	$\dfrac{b}{\ln\left[\dfrac{\mu_1}{\mu_0}\right]}$	$\dfrac{a}{\ln\left[\dfrac{\mu_1}{\mu_0}\right]}$	$\dfrac{\mu_1-\mu_0}{\ln\left[\dfrac{\mu_1}{\mu_0}\right]}$

Here $a = \ln((1-\beta)/\alpha)$ and $b = \ln(\beta/(1-\alpha))$.

Wald's SPRT is used extensively in disease control and in professional scouting programs, such as those for cotton, fruits, vegetables and forests. Extensive field evaluation has shown that the use of Wald's SPRT saves in the time spent sampling when compared to fixed-sample-size methods (Wald, 1945; Waters, 1955). The greatest time savings result from decisions being made quickly when the populations are either considerably above the economic threshold or considerably below the safety level. Sampling will take more time when the population mean density is near or between the economic threshold and the safety level. Thus, sequential hypothesis testing results in a greater allocation of resources when decision-making is most critical.

The safety level is often the most difficult to determine. In many scouting programs, a region is monitored throughout the growing season. Periodically, tests are conducted to determine whether treatment is needed. If the decision is made that no treatment is necessary, the scout wants to be assured that the population is not likely to exceed the economic threshold for some period of time; that is, the population density should be enough below the economic threshold that the likelihood of the population attaining that threshold before the region is sampled again is small. The largest population density that would offer this assurance is the safety level. Time and labor constraints and the relationship between pest density and damage are two of the primary factors considered when establishing this safety level. As the difference in the economic threshold and safety level increases, the distance between the decision lines decreases and the number of observations needed to reach a decision decreases.

Many consultants and other pest management personnel believe that it is more serious to make a type II error than a type I error. The primary reason for this is that the costs are usually more severe due to a lack of needed treatment than the costs of applying a treatment when it is not really needed. Further, many producers who are paying for pest management advice may believe they are getting a greater return on their money if they are told a particular action is needed instead of being told that no action is required. With the increase in environmental concerns, this approach is debatable and should be reevaluated.

Knowledge, or a precise estimate, of k is required to determine the boundaries of the SPRT for the negative binomial. The robustness of the SPRT to misspecification of k has been studied (Hubbard and Allen, 1991). Improperly specifying k does not yield error rates that are seriously different from those attained at the true value of k. If k is underestimated, the test is conservative. As k becomes larger, the effect of misspecification is less pronounced. Some evidence indicates that k may depend on the mean (Bliss and Owen, 1958). A reasonable approach would be to consider the estimates of k for means near and between the two hypothesized values for the mean. Use a common k if one can be found within that region.

The evaluation of Wald's SPRT is generally based on the operating characteristic (OC) and average sample number (ASN) functions. The OC function $L(\theta)$ is the probability that the null hypothesis will be accepted given any parameter θ in the parameter space. In constructing the test, the user specifies $L(\theta_0) = 1 - \alpha$ and $L(\theta_1) = \beta$. Wald (1947) proved that, for general θ,

$$L(\theta) \approx \frac{A^{h(\theta)} - 1}{A^{h(\theta)} - B^{h(\theta)}}, \tag{29}$$

where A and B are given in (28) and $h(\theta)$ is the solution of

$$\sum_x \left(\frac{f(x;\theta_1)}{f(x;\theta_0)} \right)^{h(\theta)} f(x;\theta) = 1 \quad \text{or} \quad \int_{-\infty}^{\infty} \left(\frac{f(x;\theta_1)}{f(x;\theta_0)} \right)^{h(\theta)} f(x;\theta) = 1 \tag{30}$$

Table 5
The operating characteristic function $L(\theta)$; average sample number function $E_\theta(N)$; and the relationship between θ and $h(\theta)$, when $h(\theta) \neq 0$, for the binomial (B), negative binomial (NB), normal (N) and Poisson (P) distributions

Distribution	θ	$L(\theta)$	$E_\theta(N)$	$\theta = f[h(\theta)]$
B	p	$\frac{A^{h(p)}-1}{A^{h(p)}-B^{h(p)}}$	$\frac{bL(p)+a[1-L(p)]}{p\ln\left[\frac{p_1(1-p_0)}{p_0(1-p_1)}\right]+\ln\left[\frac{1-p_1}{1-p_0}\right]}$	$p=\frac{1-\left[\frac{1-p_1}{1-p_0}\right]^{h(p)}}{\left[\frac{p_1}{p_0}\right]^{h(p)}-\ln\left[\frac{1-p_1}{1-p_0}\right]^{h(P)}}$
NB	μ	$\frac{A^{h(\mu)}-1}{A^{h(\mu)}-B^{h(\mu)}}$	$\frac{bL(\mu)+a[1-L(\mu)]}{k\ln\left[\frac{\mu_0+k}{\mu_1+k}\right]+\mu\ln\left[\frac{\mu_1(\mu_0+k)}{\mu_0(\mu_1+k}\right]}$	$\mu=k\frac{1-\left[\frac{\mu_0+k}{\mu_1+k}\right]^{h(\mu)}}{\left[\frac{\mu_1(\mu_0+k)}{\mu_0(\mu_1+k)}\right]^{h(\mu)}-1}$
N	μ	$\frac{A^{h(\mu)}-1}{A^{h(\mu)}-B^{h(\mu)}}$	$\frac{bL(\mu)+a[1-L(\mu)]}{\left[\frac{\mu_1+\mu_0}{\sigma^2}\right]\left[\mu-\frac{\mu_1\,\mu_0}{2}\right]}$	$\mu=\frac{\mu_1+\mu_0-h(\mu)(\mu_1-\mu_0)}{2}$
P	μ	$\frac{A^{h(\mu)}-1}{A^{h(\mu)}-B^{h(\mu)}}$	$\frac{bL(\mu)+a[1-L(\mu)]}{\mu\ln\left[\frac{\mu_1}{\mu_0}\right]+\mu_0-\mu_1}$	$\mu=\frac{(\mu_1-\mu_0)h(\mu)}{\left[\frac{\mu_1}{\mu_2}\right]^{h(\mu)}-1}$

Here $a=\ln((1-\beta)/\alpha)$ and $b=\ln(\beta/(1-\alpha))$.

depending on whether $f(x;\theta)$ is discrete or continuous, respectively. With the exception of $h(\theta_0)=1$ and $h(\theta_1)=-1$, it is difficult to solve for $h(\theta)$ given θ in (30). Therefore, Wald (1947) suggested soving for θ given h values, rather than vice versa. For $h(\theta)\neq 0$, the solution for θ is given in Table 5 (see Waters, 1955, and Fowler and Lynch, 1987). Then (29) is used to compute $L(\theta)$. This is also given in Table 5 for completeness. When $h(\theta)=0$, then θ is the solution of

$$E\left[\ln\left(\frac{f(x;\theta_1)}{f(x;\theta_0)}\right)\right]=0 \tag{31}$$

and

$$L(\theta)\approx\frac{a}{a-b}, \tag{32}$$

where a and b are $\ln(A)$ and $\ln(B)$, respectively. The θ for $h(\theta)=0$ and the corresponding $L(\theta)$ are given in Table 6. By computing $L(\theta)$ for h values ranging from -2 to 2 in increments of 0.2, the OC function can usually be adequately described.

Table 6
The operating characteristic value $L(\theta)$; average sample number value $E_\theta(N)$; and θ, when $h(\theta)=0$ for the binomial (B), negative binomial (NB), normal (N) and Poisson (P) distributions

Distribution	θ	$L(\theta)$	$E_\theta(N)$	θ when $h(\theta)=0$
B	p	$\frac{a}{a-b}$	$\frac{-ab}{\ln\left[\frac{p_1}{p_0}\right]\ln\left[\frac{1-p_0}{1-p_1}\right]}$	$p=\frac{\ln\left[\frac{1-p_0}{1-p_1}\right]}{\ln\left[\frac{p_1(1-p_0)}{p_0(1-p_1)}\right]}$
NB	μ	$\frac{a}{a-b}$	$\frac{-ab}{\left[\frac{\mu(\mu+k)}{k}\right]\left[\ln\left(\frac{\mu_1(\mu_0+k)}{\mu_0(\mu_1+k)}\right)\right]^2}$	$\mu=k\frac{\ln\left[\frac{\mu_1+k}{\mu_0+k}\right]}{\ln\left[\frac{\mu_1(\mu_0+k)}{\mu_0(\mu_1+k)}\right]}$
N	μ	$\frac{a}{a-b}$	$\frac{-ab}{\frac{(\mu_1-\mu_0)^2}{\sigma^2}}$	$\mu=\frac{\mu_1+\mu_0}{2}$
P	μ	$\frac{a}{a-b}$	$\frac{-ab}{\mu\left[\ln\left(\frac{\mu_1}{\mu_0}\right)\right]^2}$	$\mu=\frac{(\mu-\mu_0)}{\ln\left[\frac{\mu_1}{\mu_0}\right]}$

Here $a=\ln((1-\beta)/\alpha)$ and $b=\ln(\beta/(1-\alpha))$.

One of the disturbing features of the SPRT for many first-time users is that the indecision zone lies between two parallel lines. Wald (1945, 1947) did prove that the probability of making a terminating decision was one regardless of the true value of θ. However, this does not guarantee that the sample size needed to reach a decision will be within the time and labor constraints of the sampler. Therefore, the ASN function $E_\theta(N)$, the average number of observations needed to make a terminating decision for each θ in the parameter space Θ, is important in judging the feasibility of a proposed sampling scheme. Wald (1947) approximated the ASN function using

$$\mathrm{E}_\theta(N)\approx\frac{bL(\theta)+a(1-L(\theta))}{\mathrm{E}\left[\ln\left(\frac{f_x(x;\theta_1)}{f_x(x;\theta_0)}\right)\right]} \tag{33}$$

for $h(\theta)\neq 0$ and

$$\mathrm{E}_\theta(N)\approx\frac{-ab}{\mathrm{E}\left[\ln\left(\frac{f_x(x;\theta_1)}{f_x(x;\theta_0)}\right)\right]} \tag{34}$$

for $h(\theta) = 0$. The forms (33) and (34) would take for the binomial, negative binomial, Poisson and normal distribution are shown in Tables 5 and 6 (see Waters, 1955; Fowler and Lynch, 1987).

Wald's equations for the OC and ASN functions are based on the assumption that sampling will stop on the boundary. This rarely occurs. Usually, there is some *overshooting* of the boundary with the size of the overshoot being the absolute difference in the final total T_N and the boundary at that N. The approximations tend to overstate the error probabilities and understate the average sample size. The difference in the true and approximate error probabilities is not of practical significance in most cases. However, Corneliussen and Ladd (1970) found that Wald's formula under-represented the ASN at the maximum by about 20% when sampling from the binomial distribution. Similar results were obtained by Fowler and Lynch (1987) and Seebeck (1989) for the Poisson and negative binomial distributions. In addition, the distribution of the sample size tends to be skewed highly to the right.

A realistic assessment of the ASN function and possible sample size range is needed. Corneliussen and Ladd (1970) developed an algorithm for computing the exact OC and ASN functions for the binomial. Fowler and Lynch (1987) suggested Monte Carlo studies should be conducted to evaluate the performance of a test over the parameter space before it is implemented. Seebeck (1989) extended the work of Corneliussen and Ladd (1970) to include other members of the exponential family with special emphasis on the Poisson and negative binomial distributions. Because exact calculations or simulations of the sample size distribution are computationally more involved, Wald's approximations are good for an initial assessment. Once a reasonable test seems to have been developed, the exact properties of the test can be studied. In addition to the OC and ASN functions for various parameter values, the 95-th percentile of sample size $N_{0.95}$ is often given.

No other test of a simple versus a simple hypothesis with type I and type II error rates at most α and β, respectively, results in a smaller average sample size at the hypothesized values than Wald's sequential probability ratio test (Wald and Wolfowitz, 1948; Lehmann, 1959). However, because θ may not always assume one of the hypothesized values, the behavior of the ASN over the full parameter space is often of interest. The ASN function peaks at a point in the parameter space intermediate to the two hypothesized values, and this peak may represent a substantially larger average sample size than that at either hypothesized value.

Suppose that sampling is conducted according to the SPRT and that no decision has been made by the time the maximum possible sample size has been attained. The sampler could assume that the true parameter value is intermediate to the two hypothesized ones. If sampling is being conducted to determine whether a field should be treated, a value of the parameter between the economic threshold and safety level would indicate that a pest population may be building in numbers and that the field should be closely monitored. The sampler would return to the field sooner than the normal scouting schedule would dictate as a delay could result in economic damage. Alternatively, if no decision has been made after attaining the maximum feasible sample size, a rule can be established for deciding between the two hypotheses. This approach generally decreases the expected sample size at the hypothesized values of the parameter and increases the actual error probabilities.

Several different schemes have been proposed to modify or replace the SPRT with a closed sequential test, a test with a bounded zone of indecision resulting in a decision on or before a known maximum sample size. Weiss (1953) introduced the generalized SPRT. For the generalized SPRT, the predetermined constants A and B change at each stage of sampling instead of remaining constant as they do in the SPRT. Armitage (1957) proposed restricted SPRTs for testing the mean of the normal distribution. Anderson (1960), Billard (1977a), and Arghami and Billard (1982) developed closed sequential tests of hypotheses for normal means. Pollock and Golhar (1986) and Golhar and Pollock (1988) presented a recursive method for truncation of the SPRT that assures that the actual error probabilities are no larger than the specified ones. The 2-SPRT, proposed by Lorden (1976, 1980), is another closed test that will be presented here.

An alternative approach to minimizing the expected sample size at the hypothesized values as in the SPRT is to minimize the average sample size at an intermediate parameter value θ^*. The problem of finding such a test is known as the modified Kiefer–Weiss problem (see Kiefer and Weiss, 1957). Minimizing the ASN at the value of θ^* for which the ASN is a maximum provides a solution to the Kiefer–Weiss problem.

An asymptotic solution to the modified Kiefer–Weiss problem was given by Lorden (1976, 1980). He developed the 2-SPRT test which simultaneously performs two one-sided SPRTs. Again consider testing the hypotheses in (26). Let θ^* be a value intermediate to θ_0 and θ_1 for which the ASN is to be minimized. Define a third hypothesis

$$H_2 : \theta = \theta^*. \tag{35}$$

A one-sided hypothesis of H_2 against H_0 is conducted for possible rejection of H_0. Simultaneously, another one-sided SPRT of H_2 against H_1 is conducted for the possible rejection of H_1. The decision boundaries for this test are two converging lines that produce a triangular continuation region (see Figure 4). The 2-SPRT for a given θ^* will be developed first. Then based on work by Huffman (1983), the choice of θ^* that will minimize the ASN asymptotically will be addressed.

Let $X_1, X_2, X_3, \ldots$ be a random sample from a density of the Koopman–Darmois form; this is, $f(x;\theta) = \exp\{k(x) + \theta x - b(\theta)\}$, where $k(x)$ is a function of x alone and $b(\theta)$ is a function of θ alone. The Kullback–Leibler information numbers $I_i(\theta) = (\theta - \theta_i)\, b'(\theta) - [b(\theta) - b(\theta_i)]$, for $i = 0, 1$, where $b'(\theta)$ is the first derivative of $b(\theta)$. Further define

$$a_i(\theta') = \frac{\theta' - \theta_i}{I_i(\theta')} \tag{36}$$

for $i = 0, 1$. θ and $b(\theta)$ for the binomial, Poisson and negative binomial distributions are given in Table 7. It is desired to test the hypotheses in (26) with type I and type II error probabilities equal to α and β, respectively.

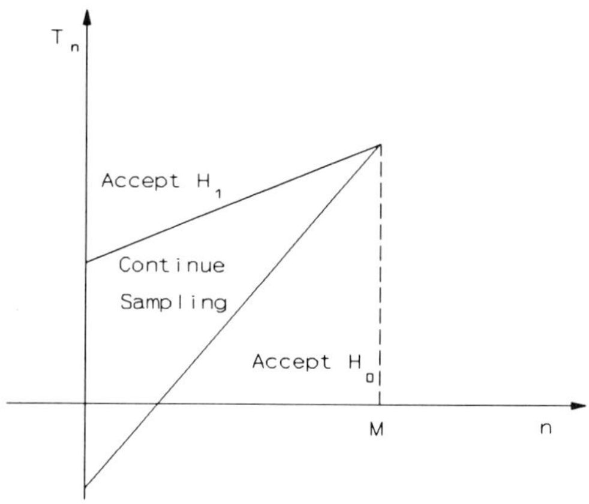

Fig. 4. Decision boundaries for the 2-SPRT.

Table 7
Quantities used for Lorden's 2-SPRT for the binomial, negative binomial and Poisson distributions

	Binomial	Poisson	Negative binomial
θ	$\ln\left(\frac{p}{q}\right)$	$\ln(\mu)$	$\ln\left(\frac{\mu}{\mu+k}\right)$
$b(\theta)$	$-\ln(q)$	μ	$k\ln(\mu+k)$
h_0	$\frac{\ln(B(\theta^*))}{\ln\left(\frac{p_1 q^*}{q_1 p^*}\right)}$	$\frac{\ln(B(\theta^*))}{\ln\left(\frac{\mu_1}{\mu^*}\right)}$	$\frac{\ln(B(\theta^*))}{\ln\left(\frac{\mu_1(\mu^*+k)}{\mu^*(\mu_1+k)}\right)}$
h_1	$\frac{\ln\left(\frac{1}{A(\theta^*)}\right)}{\ln\left(\frac{p^* q_0}{q^* p_0}\right)}$	$\frac{\ln\left(\frac{1}{A(\theta^*)}\right)}{\ln\left(\frac{\mu^*}{\mu_0}\right)}$	$\frac{\ln\left(\frac{1}{A(\theta^*)}\right)}{\ln\left(\frac{\mu^*(\mu_0+k)}{\mu_0(\mu^*+k)}\right)}$
s_0	$\frac{\ln\left(\frac{q^*}{q_1}\right)}{\ln\left(\frac{p_1 q^*}{q_1 p^*}\right)}$	$\frac{\mu_1-\mu^*}{\ln\left(\frac{\mu_1}{\mu^*}\right)}$	$\frac{k\ln\left(\frac{\mu_1+k}{\mu^*+k}\right)}{\ln\left(\frac{\mu_1(\mu^*+k)}{\mu^*(\mu_1+k)}\right)}$
s_1	$\frac{\ln\left(\frac{q_0}{q^*}\right)}{\ln\left(\frac{p^* q_0}{q^* p_0}\right)}$	$\frac{\mu^*-\mu_0}{\ln\left(\frac{\mu^*}{\mu_0}\right)}$	$\frac{k\ln\left(\frac{\mu^*+k}{\mu_0+k}\right)}{\ln\left(\frac{\mu^*(\mu_0+k)}{\mu_0(\mu^*+k)}\right)}$

Sampling will continue until (1) $T_n \geqslant h_1 n + S_1$ (accept H_1) or (2) $T_n \leqslant h_0 n + S_0$ (accept H_0). It remains to determine h_0, h_1, S_0 and S_1. The values h_1 and S_1 are determined by the one-sided test of H_2 versus H_0. Similarly, h_0 and S_0 are obtained from a one-sided test of H_1 and H_2.

Because two SPRTs are to be conducted simultaneously, two likelihood ratios will need to be formed. Let λ_{0n} and λ_{1n} denote the likelihood ratios for H_0 versus H_2 and H_1 versus H_2, respectively. The 2-SPRT takes the form: (1) terminate the test, reject H_0, and accept H_1 if $\lambda_{0n} \leqslant A$; (2) terminate the test, reject H_1, and accept H_0 if $\lambda_{1n} \leqslant B$; (3) otherwise, continue sampling.

The quantities A and B are chosen so that the desired errors rates α and β are attained when both one-sided tests are conducted simultaneously. Ignoring the excess over the boundary, Huffman (1983) recommended for practical uses that

$$A(\theta) = \frac{a_0(\theta) - a_1(\theta)}{a_0(\theta)} \alpha \quad \text{and} \quad B(\theta) = \frac{a_1(\theta) - a_0(\theta)}{a_1(\theta)} \beta. \tag{37}$$

The slopes and intercepts of the decision boundaries may now be determined. The form of the distribution permits the intercepts in the decision rule to be computed as follows:

$$h_1 = \frac{\ln\left(\dfrac{1}{A(\theta^*)}\right)}{\theta^* - \theta_0} \quad \text{and} \quad h_0 = \frac{\ln(B(\theta^*))}{\theta_1 - \theta^*}. \tag{38}$$

The slopes are calculated using

$$S_1 = \frac{b(\theta^*) - b(\theta_0)}{\theta^* - \theta_0} \quad \text{and} \quad S_0 = \frac{b(\theta_1) - b(\theta^*)}{\theta_1 - \theta^*}. \tag{39}$$

This gives a solution to the modified Keifer–Weiss problem. The maximum sample-size M for the SPRT occurs at the point of intersection of the two decision boundaries.

The maximum of the ASN function occurs at the midpoint between θ_0 and θ_1 if the distribution is symmetric about this point. However, when the distribution is not symmetric about the midpoint, the maximum of the ASN function does not occur at the midpoint between θ_0 and θ_1. In order to obtain θ^*, the value of θ such that the ASN is at a maximum, θ' is first determined such that

$$\frac{\ln(A(\theta'))^{-1}}{I_0(\theta')} = \frac{\ln(B(\theta'))^{-1}}{I_1(\theta')}. \tag{40}$$

Let n' be the common value of the two sides in (40). Denote $a_i(\theta')$ as a_i'. Find r' such that

$$\Phi(r') = \frac{a_1'}{a_1' - a_0'}, \tag{41}$$

where Φ is the distribution function for the standard normal random variable. Also, for $\theta = \theta'$,

$$\sigma' = \sqrt{\mathrm{var}_{\theta'}(X)}. \tag{42}$$

The point θ^* for which the ASN is to be minimized may be expressed as

$$\theta^* = \theta' + \frac{r'}{\sigma'\sqrt{n'}}. \tag{43}$$

Based on H_2: $\theta = \theta^*$, the error rates are adjusted using (37) with $\theta = \theta^*$. Finally, the values needed to construct the decision boundaries for Huffman's extension of the 2-SPRT are as in equations (38) and (39), with the value of θ^* found in (43) and these corrected errors rates.

Huffman's (1983) extension of Lorden's work determines the value of θ^* which minimizes the maximum expected sample size to within $o((\log \alpha^{-1})^{1/2})$ as α and β tend to zero. This provides an asymptotic solution to the Kiefer–Weiss problem. Mulekar et al. (1992) investigated the use of the 2-SPRT with the use of θ^* with negative binomial populations. Joo (1989) considered the OC and ASN functions for the negative binomial, binomial and Poisson distributions. In each case, a significant reduction in the ASN function was observed for values intermediate to the hypothesized values while controlling the type I and type II error rates.

Instead of deciding between two hypotheses, Waters (1955) developed a sequential procedure for determining whether the infestation of trees by spruce budworm larvae was light, medium, or heavy. Increasingly IPM workers are attempting to discriminate among three hypotheses instead of between two hypotheses as in the SPRT and the 2-SPRT. Consider the test of the following three hypotheses: H_0: $\theta = \theta_0$ vs. $H_1: \theta = \theta_1$ vs. $H_2: \theta = \theta_2$. Assume $\theta_0 < \theta_1 < \theta_2$. Several methods have been developed for the purpose of discriminating among the three hypotheses.

Armitage (1947) suggested simultaneously conducting two SPRTs to decide among the three hypotheses. One SPRT S_1 would be constructed to decide between H_0 and H_1. Another S_2 would discriminate between H_1 and H_2. The test has the form depicted in Figure 5. Let α_i and β_i, $i = 1, 2$, be the type I and type II error rates, respectively, associated with test S_i. Then Armitage reasoned that the probability of error given the true value of θ is θ_i is α_1, $\beta_1 + \alpha_2$, and β_2 for $i = 0$, 1 and 2, respectively. Fowler and Lynch (1987) found the error rates to differ significantly from the nominal ones. However, this remains the most commonly used simultaneous test of three hypotheses, probably due to the ease with which the sampling boundaries can be derived.

Sobel and Wald (1949) suggested performing S_1 and S_2 simultaneously and independently. Therefore, as soon as a decision can be made for one test, sampling ceases, regardless of the status of the other test. Approximations for the OC and ASN functions can be derived because of the independence of the two tests (see Wetherill and Glazebrook, 1986).

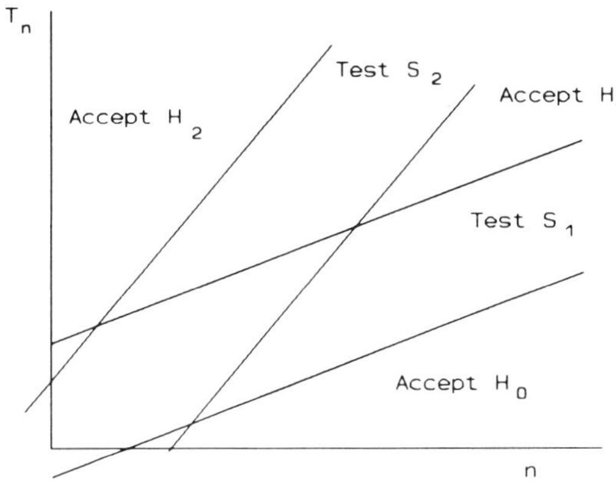

Fig. 5. Decision boundaries for Armitage's (1947) method for deciding among three hypotheses.

Armitage (1950) suggested simultaneously conducting three SPRTs. In addition to S_1 and S_2, a test S_3 of H_0 vs. H_2 is performed. Armitage presented an analogy of this test with linear discriminant function analysis. Knowledge of the properties of these boundaries is scanty (Wetherill and Glazebrook, 1986).

Billard and Vagholkar (1969) and Billard (1977b) used a computer optimization method to determine the procedure that minimizes the ASN function at a specified value of the normal mean μ and the binomial proportion p, respectively. The decision boundaries for the test are based on a geometric set of seven parameters, and the values of these are determined during the optimization process. Because determination of the boundaries require specialized software that has only been developed for the normal and binomial distributions, this test has had limited use.

Payton (1991) developed a method of adjusting the decision boundaries in Armitage's (1947) test so that the observed error rates are near the nominal ones. The procedure was developed for members of the Koopman–Darmois family of densities but has only been applied to the exponential, normal and binomial distributions.

6. Summary

Insects have developed resistance to insecticides, resulting in resurgence of populations. Destruction of predator and parasite species has also caused some secondary pests to become primary pests. This coupled with increasing environmental concerns, especially groundwater quality, has led researchers to more carefully evaluate the population density at which control should begin. To do this, precise methods of estimating population density have been developed.

The primary methods used in population density estimation are sequential procedures based on the underlying distribution or on the relationship in mean and variance. The binomial, Poisson and negative binomial distributions are used most frequently for this purpose. Iwao's patchiness regression and Taylor's power law are two methods of relating mean and variance that are often cited, and they have been used extensively in sampling programs.

The negative binomial parameter k must be known, or a precise estimate available, if the most precise methods of population density estimation are to be used. The value of k may be determined from a proposed model, but it is most likely to be obtained from historical data from an ecosystem or a set of ecosystems. Bliss and Owens' (1958) regression procedure and the maximum likelihood method suggested by Bliss and Fisher (1953) are two methods of estimating a common k. Estimation of k at the time of sampling can be based on a multistage approach discussed by Willson et al. (1984).

The decision to treat a field with insecticide is often based on a sequential hypothesis test. Wald's SPRT is most frequently the foundation for sequential hypothesis testing in entomology. Sometimes sampling must cease before a decision is made. Several sequential methods with closed boundaries have been suggested. The 2-SPRT developed by Lorden (1976, 1980) and extended by Huffman (1983) provides an asymptotic solution to the Kiefer–Weiss problem. When trying to decide among three hypotheses, Armitage's (1947) procedure is used extensively even though some statistically more appealing methods are available.

The methods most commonly used in field entomology are not always the strongest statistical procedures. Computational ease and field application are two primary factors considered in the choice of a sampling program. A less precise method will often be used if it is easier to understand and to implement. This has resulted in some ad hoc procedures being widely used without a good knowledge of the statistical properties of the methods.

The statistical methods presented in this chapter assume that observations are independently and identically distributed. For some species, the numbers of insects in a sampling unit are spatially correlated (Schotzko and O'Keefe, 1989); other species exhibit a random spatial pattern (Young and Young, 1990). When studying the gypsy moth, *Lymantria dispar* (L.), Liebhold et al. (1991) found that the range and magnitude of spatial dependence varied with year and location, with no spatial correlation being detected in some years and locations. The effect of spatial correlation on the statistical estimation and testing procedures given here has not yet been considered in integrated pest management programs. Spatial statistics (see Cressie, 1991) should provide a means of addressing this issue.

Environmental and financial concerns have also led producers to question the prophylactic use of herbicides. Recent work by Brain and Cousens (1990), Wiles et al. (1992) and Johnson et al. (1991) indicate that weed populations tend to be described well by the negative binomial distribution. Therefore, many of the methods described here would be appropriate.

References

Adkisson, P. L., G. A. Niles, J. K. Walker, L. S. Bird and H. B. Scott (1982). Controlling cotton's insect pests: A new system. *Science* **216**, 19–22.

Allen, J. C., D. Gonzalez and D. V. Gokhale (1972). Sequential sampling plans for the bollworm, *Heliothis zea*. *Environ. Entomol.* **1**, 771–780.

Anderson, T. W. (1960). A modification of the sequential probability ratio test to reduce the sample size. *Ann. Math Statist.* **31**, 165–196.

Anscombe, F. G. (1949). The statistical analysis of insect counts based on the negative binomial distribution. *Biometrics* **5**, 165–173.

Anscombe, F. J. (1950). Sampling theory of the negative binomial and logarithmic series distributions. *Biometrika* **37**, 358–382.

Arghami, N. R. and L. Billard (1982). A modification of a truncated partial sequential procedure. *Biometrika* **69**, 613–618.

Armitage, P. (1947). Some sequential tests of Student's hypotheses. *J. Roy. Statist. Soc. Suppl.* **9**, 250.

Armitage, P. (1950). Sequential analysis with more than two alternative hypotheses, and its relation to discriminant function analysis. *J. Roy. Statist. Soc. Ser. B* **12**, 137–144.

Armitage, P. (1957). Restricted sequential procedures. *Biometrika* **44**, 9–26.

Banerjee, B. (1976). Variance to mean ratio and the spatial distribution of animals. *Experientia* **32**, 993–994.

Bhalerao, N. R., J. Gurland and R. C. Tripathi (1980). A method of increasing power of a test for the negative binomial and Newman type A distributions. *J. Amer. Statist. Assoc.* **75**, 934–938.

Billard, L. (1977a). A truncated partial sequential procedure. *Biometrika* **64**, 576–572.

Billard, L. (1977b). Optimum partial sequential tests for two-sided tests of the binomial parameter. *J. Amer. Statist. Assoc.* **72**, 197–201.

Billard, L. and M. K. Vagholkar (1969). A sequential procedure for testing a null hypothesis against a two-sided alternative hypothesis. *J. Roy Statist. Soc. Ser. B* **31**, 285–294.

Binns, M. (1975). Sequential estimation of the mean of a negative binomial distribution. *Biometrika* **62**, 433–440.

Bliss, C. I. and A. R. G. Owen (1958). Negative binomial distributions with a common *k*. *Biometrika* **45**, 37–58.

Bliss, C. I. and R. A. Fisher (1953). Fitting the negative binomial distribution to biological data and note on the efficient fitting of the negative binomial. *Biometrics* **9**, 176–200.

Boswell, M. T. and G. P. Patil (1970). Chance mechanisms generating the negative binomial distribution. In: G. P. Patil, ed., *Random Counts in Scientific Work*, Vol. 1, Pennsylvania State Univ., University Park, PA, 3–22.

Brain, P. and R. Cousens (1990). The effect of weed distribution on predictions of yield loss. *J. Appl. Ecol.* **27**, 735–742.

Brown, A. W. A. (1979). Resistance as a factor in pesticide management. In: *Proc. 15th internat.Congr. Entomology*. US Department of Agriculture, Forest Service, Washington, DC, 816–884.

Brown, A. W. A. and R. Pal (1971). *Insecticide Resistance in Arthropods*. WHO Monograph Series, Vol. 38, Geneva, Switzerland.

Cabilio, P. (1977). Sequential estimation in Bernoulli trials. *Ann. Statist.* **5**, 342–356.

Cabilio, P. and H. Robbins (1975). Sequential estimation of *p* with squared relative error loss. *Proc. Nat. Acad. Sci. USA* **72**, 191–193.

Chow, Y. S. and H. Robbins (1965). On the asymptotic theory of fixed-width sequential confidence intervals for the mean. *Ann. Math. Statist.* **36**, 457–462.

Corneliussen, A. H. and D. W. Ladd (1970). On sequential tests of the binomial distribution. *Technometrics* **12**, 635–646.

Cressie, N. (1991). *Statistics for Spatial Data*. Wiley, New York.

Douglas, J. B. (1955). Fitting the Newman Type A (two parameter) contagious distribution. *Biometrics* **11**, 149–173

Evans, D. A. (1953). Experimental evidence concerning contagious distributions in ecology. *Biometrika* **40**, 186–211.

Feller, W. (1968). *An Introduction to Probability Theory and Its Applications*. Vol. 1. 3rd ed., Wiley, New York.

Felsot, A. S., J. C. Wilson, D. E. Kuhlman and K. L. Steffy (1982). Rapid dissipation of carbofuran as a limiting factor in corn rootworm control in fields with histories of continuous carbofuran use. *J. Econom. Entomol.* **75**, 1098–1103.

Fowler, G. W. and A. M. Lynch (1987). Sampling plans in insect pest management based on Wald's sequential probability ratio test. *Environ. Entoml.* **16**, 345–354.

Gates, G. E. (1989). Discrete, a computer program for fitting discrete frequency distributions. In: L. McDonald, B. Manly, J. Lockwood and J. Logan, eds., *Lecture Notes in Statistics: Estimation and Analysis of Insect Populations.* Springer, Berlin, 458–466.

Ghosh, M. and N. Mukhopadhyay (1979). Sequential point estimation of the mean when the distribution is unspecified. *Comm. Statist. Theory Methods* **8**, 637–652.

Golhar, D. Y. and S. M. Pollock (1988). On converging boundaries for the SPRT. *Sequential Anal.* **7**, 307–320.

Gurland, J. (1959). Some applications of the negative binomial and other contagious distributions. *Amer. J. Public Health* **49**, 1388–1399.

Hinz, P. and J. Gurland (1970). A test of fit for the negative binomial and other contagious distributions. *J. Amer. Statist. Assoc.* **65**, 887–903.

Hubbard, D. J. and O. B. Allen (1991). Robustness of the SPRT for a negative binomial to misspecification of the dispersion parameter. *Biometrics* **47**, 419–427.

Huffman, M. D. (1983). An efficient approximate solution to the Kiefer–Weiss problem. *Ann. Statist.* **11**, 306–316.

Hyslop, J. A. (1983). Losses occasioned by insects, mites, and ticks in the U.S. USDA Bur. Entomol. E-444.

Ijiri, Y. and H. A. Simon (1975). Some distributions associated with Bose–Einstein statistics. *Proc. Nat. Acad. Sci. USA.* **72**, 1654–1657.

Iwao, S. (1968). A new regression method for analyzing the aggregation pattern of animal populations. *Research. Popul. Ecol.* **10**, 1–20.

Johnson, G. A., D. A Mortensen, L. J. Young and A. R. Martin (1991). Preliminary results of weed distribution studies in Nebraska. In: *Proc. North Central Weed Sci. Soc. Amer.* **46**, 111–112.

Johnson, N. L. and S. Kotz (1969). *Discrete Distributions.* Wiley, New York.

Jones, P. C. T., J. W. Mollison and M. H. Quenouille (1948). A technique for the quantitative estimation of soil microorganisms. Statistical note. *J. General Microbiol.* **2**, 54–69.

Joo, L. S. (1989). A computer program to develop and evaluate a 2-SPRT for the negative binomial, binomial, and Poisson distributions. M. S. report, Oklahoma State University, Stillwater, OK.

Karandinos, M. G. (1976). Optimum sample size and comments on some published formulae. *Bull. Entomol. Soc. Amer.* **22**, 417–421.

Kemp, A. W. (1987). Families of discrete distributions satisfying Taylor's power law. *Biometrics* **43**, 693–699.

Kiefer, J. and L. Weiss (1957). Some properties of generalized sequential probability ratio tests. *Ann. Math. Statist.* **28**, 57–75.

Kuno, E. (1969). A new method of sequential sampling to obtain the population estimates with a fixed level of accuracy. *Research. Popul. Ecol.* **11**, 127–136.

Kuno, E. (1972). Some notes on population estimation by sequential sampling. *Research. Popul. Ecol.* **14**, 58–73.

Lehmann, E. L. (1959). Testing Statistical Hypotheses. Wiley, New York.

Liebhold, A. M., X. Zhang, M. E. Hohn, J. S. Elkinton, M. Ticehurst, G. L. Benzon and R. W. Campbell (1991). Geostatistical analysis of gypsy moth (Lepidoptera: Lymantriidae) egg mass populations. *Environ. Entomol.* **20**, 1407–1417.

Lloyd, M. (1967). Mean crowding. *J. Animal Ecol.* **36**, 1–30.

Lorden, G. (1976). 2-SPRT and the modified Kiefer–Weiss problem of minimizing the expected sample size. *Ann. Statist.* **4**, 281–291.

Lorden, G. (1980). Structure of sequential tests minimizing an expected sample size. *Z. Wahrsch. Veru. Geb.* **51**, 291–302.

Luckmann, W. H. and G. C. Decker (1960). A five-year report on observations in the Japanese beetle control area of Sheldon, Illinois. *J. Econom. Entomol.* **53**, 821–827.

Marlatt, C. L. (1904). Annual losses occasioned by destructive insects in the U.S. *USDA Agric. Yearbook Agric.*, 461–474.

Martinsek, A. T. (1983). Sequential estimation with squared relative error loss. *Bull. Inst. Math. Acad. Sinica* **11**, 607–623.
Metcalf, R. L. (1955). Physiological basis for insecticide resistance to insecticides. *Physiol. Rev.* **35**, 192–232.
Metcalf, R. L. (1980). Changing role of insecticides in crop protection. *Ann. Rev. Entomol.* **25**, 219–256.
Metcalf, R. L. (1986). The ecology of insecticides and the chemical control of insects. In: M. Kogan, ed., *Ecological Theory and Integrated Pest Management Practice*. Wiley, New York, 251–297.
Metcalf, C. L., W. P. Flint and R. L. Metcalf (1962). *Destructive and Useful Insects: Their Habits and Control*. McGraw-Hill, New York.
Morgan, M. E. P., P. MacLeod, E. O. Anderson and C. I. Bliss (1951). A sequential procedure for grading milk by microscopic counts. *Conn. (Storrs) Agric. Exp. Stn. Bull.* **276**, 1–35.
Mukhopadhyay, N. (1978). Sequential point estimation of the mean when the distribution is unspecified. Technical Report No. 312, University of Minnesota.
Mulekar, M. S., L. J. Young and J. H. Young (1992). Using an approximate Kiefer–Weiss solution for testing insect population densities. *Metrika* **39**, 219–226.
Nadás, A. (1969). An extension of a theorem of Chow and Robbins on sequential confidence intervals for the mean. *Ann. Math. Statist.* **40**, 667–671.
Neyman, J. (1939). On a new class of 'contagious' distributions, applicable in entomology and bacteriology. *Ann. Math. Statist.* **10**, 35–57.
Oakland, G. B. (1950). An application of sequential analysis to whitefish sampling. *Biometrics* **6**, 59–67.
Onsager, J. A. (1976). The rationale of sequential sampling, with emphasis on its use in pest management. Technical Bulletin 1526, US Department of Agriculture.
Patil, G. P. and W. M. Stiteler (1974). Concepts of aggregation and their quantification: A critical review with some new results and application. *Research. Popul. Ecol.* **15**, 238–254.
Playton, M. E. (1991). *An examination of sequential procedures for the testing of three hypotheses*. Ph.D. dissertation, Oklahoma State University, Stillwater, OK.
Perry, J. N. and L. R. Taylor (1985). Adès: New ecological families of species-specific frequency distributions that describe repeated spatial samples with a intrinsic power-law variance-mean property. *J. Animal. Ecol.* **54**, 931–953.
Pielou, E. C. (1977). *Mathematical Ecology*. Wiley, New York.
Pieters, E. P. and W. Sterling (1974). A sequential sampling plan for the cotton fleahooper, P*seudatomoscelis seriatus. Environ. Entomol.* **3**, 102–106.
Pollock, S. M. and D. Y. Golhar (1986). Efficient recursions for truncation of the SPRT. *Sequential Anal.* **5**, 253–262.
Pólya, G. (1931). Sur quelques points de la théorie des probabilités. *Ann. Inst. Henri Poincaré* **1**, 117–161.
Robbins, H. and D. Siegmund (1974). Sequential estimation of *p* in Bernoulli trials. In: E. J. Williams, ed., *Studies in Probability and Statistics: Paper in honour of Edwin J. G. Pitman*. North-Holland, Amsterdam, 103–107.
Rojás, B. A. (1964). La binomial negativa y la estimatción de itensidad de plagas en el suelo. *Fitotecnia Latinamer* **1**, 27–36.
Rudd, W. G. (1980). Sequential estimation of soybean arthropod population densities. In: M. Kogan and D. C. Herzog, eds., *Sampling Methods in Soybean Entomology*. Springer, New York, 94–104.
Ruesink, W. G. (1980). Introduction to sampling theory. In: M. Kogan and D. C. Herzog, eds., *Sampling Methods in Soybean Entomology*. Springer, New York, 61–78.
Scott, T. G., Y. L. Willis and J. A. Elis (1959). Some effects of a field application of dieldrin on wildlife. *J. Wildlife Management* **23**, 409–427.
Schotzko, D. J. and L. E. O'Keefe (1989). Geostatistical description of the spatial distribution of *Lygus hesperus* (Heteroptera: Miridae) in lentils. *J. Econom. Entomol.* **82**, 1277–1288.
Seebeck, K. (1989). A computer program to develop and evaluate a Wald's sequential probability ratio test for the parameters of three discrete distributions. M.S. Report, Oklahoma State University, Stillwater, OK.
Skellam, J. G. (1958). On the derivation and applicability of Newman's type A distribution. *Biometrika* **45**, 32–36.

Sobel, M. and A. Wald (1949). A sequential decision procedure for choosing one of three hypotheses concerning the unknown mean of a normal distribution. *Ann. Math. Statist.* **20,** 502–522.

Southwood, T. R. E. (1978). *Ecological Methods.* 2nd ed., Chapman and Hall, London.

Starr, N. and M. B. Woodroofe (1968). Remarks on a stopping time. *Proc. Nat. Acad. Sci. USA* **61**, 1215–1218.

Stein, C. (1945). A two-sample test for a linear hypothesis whose power is independent of the variance. *Ann. Math. Statist.* **16**, 243–258.

Taylor, L R. (1961). Aggregation, variance and the mean. *Nature* **189**, 732–735.

Taylor, L. R. (1965). A natural law for the spatial disposition of insects. In: *Proc. 12th Internat. Congr. Entomology*, 396–397.

Taylor, L. R. (1970). Aggregation and the transformation of counts of *Aphis fabae* Scop. on beans. *Ann. Appl. Biol.* **65**, 181–189.

Taylor, L. R. (1971). Aggregation as a species characteristic. In: G. P. Patil, E. C. Pielou and W. E. Waters, eds., *Statistical Ecology 1.* Penn State Univ. Press, Philadelphia, PA, 357–377.

Taylor, L. R. (1984). Assessing and interpreting the spatial distribution of insect populations. *Ann. Entomol.* **29**, 321–357.

Taylor, L. R., I. P. Woiwod and J. N. Perry (1978). The density-dependence of spatial behavior and the rarity of randomness. *J. Animal Ecol.* **47**, 383–406.

Taylor, L. R., I. P. Woiwod and J. N. Perry (1979). The negative binomial as a dynamic ecological model for aggregation, and the density dependence of *k*. *J. Animal Ecol.* **4**, 289–304.

Thomas, A. (1949). A generalization of Poisson's distributions. *Yokohama Math. J.* **10**, 73–96.

USDA (1954). Losses in agriculture. Report 20-1, Agric. Res. Service, Washington, DC.

USDA (1965). Losses in agriculture. Agric. Handbook 291. Agric. Res. Service, Washington, DC.

Vardi, Y. (1979). Asymptotic optimal sequential estimation: The Poisson case. *Ann. Statist.* **7**, 1040–1051.

Wald, A. (1945). Sequential tests of statistical hypotheses. *Ann. Math. Statist.* **16**, 117–186.

Wald, A. (1947). *Sequential Analysis.* Wiley, New York.

Wald, A. and J. Wolfowitz (1948). Optimum character of the sequential probability ratio test. *Ann. Math. Statist.* **19**, 326–339.

Waters, W. E. (1955). Sequential sampling in forest insect surveys. *Forest Sci.* **1**, 68–79.

Weiss, L. (1953). Testing one simple hypothesis against another. *Ann. Math. Statist.* **24**, 273–281.

Wetherill, G. B. and K. D. Glazebrook (1986). *Sequential Methods in Statistics.* 3rd ed., Chapman and Hall, New York.

Wiles, L. J., G. W. Oliver, A. C. York, H. J. Gold and G. G. Wilkerson (1992). Spatial distribution of broadleaf weeds in North Carolina soybean (*Glycine max*) fields. *Weed Sci.* **40**, 554–557.

Willson, L. J. (1981). Estimation and testing procedures for the parameters of the negative binomial distribution. Ph.D. Thesis, Oklahoma State University, Stillwater, OK.

Willson, L. J. and J. L. Folks (1983). Sequential estimation of the mean of the negative binomial distribution. *Com. Statist. Sequential Anal.* **2**, 55–70.

Willson, L. J., J. L. Folks and J. H. Young (1984). Multistage estimation compared with fixed-sample-size estimation of the negative binomial parameter *k*. *Biometerics* **40**, 109–117.

Willson, L. J. and J. H. Young (1983). Sequential estimation of insect population densities with a fixed coefficient of variation. *Environ. Entomol.* **12**, 669–672.

Willson, L J., J. H. Young and J. L. Folks (1987). A biological application of Bose–Einstein statistics. *Com. Statist. Theory Methods* **16**, 445–459.

Young, J. H. and L. J. Willson (1987). Use of Bose–Einstein statistics in population dynamics models of arthropods. *Ecol. Model.* **36**, 89–99.

Young, L. J. and J. H. Young (1989). A model of arthropod movement within agroecosystems. In: L. McDonald, B. Manly, J. Lockwood and J. Logan, eds., *Lecture Notes in Statistics: Estimation and Analysis of Insect Populations.* Springer, Berlin, 378–386.

Young, L. J. and J. H. Young (1990). A spatial view of the negative binomial parameter *k* when describing insect populations. In: *Proc. 1990 Kansas State Univ. Conf. on Applied Statistics in Agriculture*, 13–20.

Young, L. J. and J. H. Young (1991). New approaches in insect dispersal models compared to classical approaches. In: *1991 Amer. Statist. Assoc. Proc. Section on Statistics and the Environment*, 58–65.

G. P. Patil and C. R. Rao, eds., *Handbook of Statistics, Vol. 12*

A Crystal Cube for Coastal and Estuarine Degradation: Selection of Endpoints and Development of Indices for Use in Decision Making

M. T. Boswell, J. S. O'Connor and G. P. Patil

Abstract: Environmental regulators and decisionmakers would like to have a crystal ball that could predict how ecosystems would respond to stresses such as pollution or over-fishing. In this way, information on important parameters such as effects on valued species, propagation of these effects through the ecosystem and subsequent recovery after the removal of these stresses, could all be properly considered for the protection of important natural resources. In the real world, however, such information cannot be gained with certainty. Instead a conceptual cube having a series of faces, each of which represents a specific parameter that can be directly related to marine environmental degradation, has been constructed.

The emphasis in developing and testing these indices is on standardizing long-term data sets in order to develop a single variable, termed an index. This index is based on a variable that measures contamination or, ideally, contamination effects. The choice of such a variable is not easy and usually involves extensive data analysis. To be useful, the index must also be sensitive to contamination and relatively insensitive to other factors.

Choices of the demarcation points between 'no cause for concern' and 'warning' and between 'warning' and 'alarm', at least ideally, require agonizing consensus among interested parties, the decisionmaker and scientists. This chapter deals with the formulation and calibration of indices using 'statistical' reference values or benchmarks based on the natural variability inherent in a 'clean' or control region.

1. Introduction

Environmental regulators and decisionmakers would like to have a crystal ball that could predict how ecosystems would respond to stresses such as pollution or over-fishing. In this way, information on important parameters such as effects on valued species, propagation of these effects through the ecosystem, and subsequent recovery after the removal of these stresses, could all be properly considered for the protection of important natural resources. In the real world, however, such information cannot be gained with certainty.

Instead a conceptual crystal cube having a series of faces, each of which represents a specific parameter that can be directly related to marine environmental degradation, has been constructed. At present, 10 indices or faces of the cube are being tested: dietary risks from contaminants in marine foods; contaminant stress in sediments; contaminant stress in the water column; human pathogen risks; benthic species distribution and composition; fish and shellfish diseases; reproduction in fish and shellfish; mortality in eggs and larvae of fish and shellfish; reproductive success in marine birds; and oxgen depletion effects (O'Connor and Dewling, 1986).

The emphasis in developing and testing these indices is on standardizing long-term data sets in order to develop a single variable, termed an index. This index is based on a variable that measures contamination or, ideally, contamination effects. The choice of such a variable is not easy and usually involves extensive data analysis. To be useful, the index must also be sensitive to contamination and relatively insensitive to other factors. Selection of useful indices also require consultation with their principal users, decisionmakers and the public.

The use of this variable index in the crystal cube analyses is for the separation of 'concern' or 'alarm' from 'no concern' conditions. Here, concern or alarm indicates not only the violation of legislated or regulated standards, but also that the scientific community is unable to assure decisionmakers that issues of widespread public concern will not arise from specific environmental stress.

The index is calibrated so that when the number falls in the range of 0 to 1, there is 'no cause for concern'. A flag is raised as soon as the index becomes 1. The range from 1to 10 indicates 'warning', something is happening and should be investigated. The range above 10 indicates 'cause for alarm'. The index is designed to be 10 when there is scientific reason for ecological concern; the environment has been adversely affected.

The fundamental concept underlying the use of an indicator variable of environmental degradation is to compare conditions in a stressed estuary or coastal area with those of a clean region. The crystal cube with 10 faces, each representing an indicator variable of one important environmental pollution parameter, will flag cases where legal or scientific benchmarks are exceeded.

This technique will assist the environmental decisionmaker, to focus attention on those specific environmental parameters indicative of serious problems. The crystal cube is not intended to be the 'ideal' crystal ball desired by environmental manager, but it should provide a framework for evaluating and comparing different environmental measures that must be weighed not only against each other, but also against other (e.g., economic, aesthetic, etc.) considerations. Thus, the crystal cube could develop into a valuable tool to help define or delineate 'unreasonable degradation' and make environmental decisionmaking more consistent and equitable.

Choices of the demarcation points between 'no cause for concern' and 'warning' and between 'warning' and 'alarm' at least ideally, require agonizing consensus among interested parties, the decisionmaker and scientists. This chapter deals with the formulation and calibration of indices using 'statistical' reference values or bench-marks based on the natural variability inherent in 'clean' or control region.

2. Formulation of the indices

Suppose an indicator variable Y increases with pollution, then a simple formulation for the index is the ratio $I = Y/R$ of this variable to a reference value. On the other hand, if the indicator variable decreases with stress, then the index is $I = R/Y$. With this formulation the index increases, reaching a value of 1 when Y equals the reference value; however, this formulation will be modified in Section 4 to make the index reach a value of 10 when Y equals an 'alarm value'.

For the remainder of this chapter, we will assume that Y increases with environmental stress. The case with a decreasing variable is handled similarly, as is shown in the example of Section 8. There are two types of errors that can occur. An index can indicate cause for concern (or alarm) when there is no cause for concern. Conversely, the index can fail to indicate concern when there is substantial environmental stress. The statistical reference value is chosen at the upper limit of variability in an unstressed environment so that a false warning occurs, on the average, once in ten years. The index will not be useful unless there is a good chance of flagging a region with important environmental stress. Before considering an index to be useful, the power of detecting important stress which persists for several years must be reasonably large. The following table illustrates the case where for any single year the power is 70%:

Probability of detecting important environmental stress	
one time in 1 year	0.700
at least 1 time in 2 years	0.910
at least 2 times in 2 years	0.490
at least 1 time in 3 years	0.973
at least 2 times in 3 years	0.784
at least 1 time in 5 years	0.998
at least 2 times in 5 years	0.918
at least 3 times in 5 years	0.786

Of course the above probabilities are for environmental stress at the reference value R and the probabilities increase with increasing environmental stress. The testing of the indices involves estimating the parameters for regions which are known to have an important impact due to pollution and calculating the power of detecting this stress. If biologically significant values of the index variable Y can be obtained from biologists, then the power for this level of pollution can, also be calculated. If the power is not 70% or more, then the index will not be useful and modifications must be considered. One such modification is considered in Section 5.

3. Estimation of the statistical reference value

The index is Y/R where the reference value R is an estimate of the 90-th percentile of the reference (i.e., 'clean') region. When new observations are taken from an unimpacted region 10%, on the average, will be larger than the 90-th percentile. The 90-th percentile can be estimated by the 90-th percentile prediction limit R. This limit is the predicted value that will be exceeded 10% of the time, on the average, by new observations taken in the reference region. Let $X_1, X_2, \ldots, X_n$ be n observations from an unimpacted region. In order to incorporate the interannual variation as well as spatial variation, these n observations must be taken at random, one a year, for n years. Assuming these observations are independent identically distributed normal random variable, then the upper 90-th percentile prediction limit is

$$R = \bar{X}_n + S_n t_{0.10}(n-1)\left(1 + \frac{1}{n}\right)^{1/2}.$$

Note that transformations to normality can be used when data are not normal. The prediction limit estimator is fairly robust to nonnormal distribution.

It still remains to standardize the index 'alarm' value at 10. This is done in the following section.

4. Standardization of the index

The reference value R was defined above to take into account the natural variability of the measurement of environmental stress. Once R has been determined, it still remains to standardize the index so that an alarm is raised when justified. As mentioned in Section 2, decisionmakers must determine an 'alarm value' A, say. Then the index must be adjusted so that it becomes 10 when $Y = A$, while keeping the index at 1 when $Y = R$. A convenient way to do this is to raise the index to some power b, say. Then $I = (Y/R)^b$. Setting $I = 10$ when $Y = A$ gives $(A/R)^b = 10$ or $b = 1/\log_{10}(A/R)$. The index becomes $I = (Y/R)^{1/\log_{10}(A/R)}$. Note when $Y = R$ the index is 1 and when $Y = A$ the index is 10. The fraction of false alarms is still 0.10 and the power remains unchanged.

5. A more powerful index

If the natural variability is large, as is typical, a single year's observations may not be enough to determine if a region is impacted or not. It is possible to modify the index to increase the power to detect important ecological stress in the face of large natural variability. Variability can be reduced by averaging; instead of taking a single yearly value from the test region, an average $\bar{Y}_m$ of m yearly observations can be taken from the test region. This reduces the variance by a factor of m. The 90-th percentile prediction limit for the average of m observations will be smaller. Let $Y_1, Y_2, \ldots Y_m$ be m new observations taken in an unimpacted region. Here again it is important to include the year to year variation by taking observations over a number of years. The reference value is the 90-th percentile prediction limit of the distribution of the average $\bar{Y}$. Then

$$R_m = \bar{X} + S_n t_{0.10}(n-1)\left(\frac{1}{n} + \frac{1}{m}\right)^{1/2}.$$

After standardizing, the index becomes

$$I_m = \left[\frac{\bar{Y}_m}{R_m}\right]^b, \quad \text{where} \quad b = 1/\log_{10}\left[\frac{A}{R_m}\right].$$

For m greater than one, this is not the same index considered above, and it will reach one, indicating a warning situation, more often than $I = I_1$, when in a stressed environment. False warnings are still raised, on the average, only once in 10 applications; however, the power to detect important ecological stress has been increased.

6. Comparison of the power of the indices

The calculation of the power that the index exceeds one depends on the standardized difference between the means of the test region and the reference region, as well as on the number of observations. The power comes from a noncentral t distribution with noncentrality parameter

$$D = (\mu_y - \mu_x) \Bigg/ \left[\sigma\sqrt{\frac{1}{n} + \frac{1}{m}}\right]$$

and $n-1$ degrees of freedom. Values from the noncentral t distribution were evaluated by a FORTRAN77 computer program which called the IMSL routine MDTN; a table of the power for one-sided t-tests with level 0.10 is given in Appendix C. Table 6.1 compares the power with which the indices I_1, I_3 and I_6 can detect values that actually exceed their reference value. We observe that the index I_6 has power 0.70 of detecting standardized differences slightly larger than 1.0 while I_1 detects values slightly larger than 2.0.

Another, perhaps more useful, way of looking at the power of these indices is in terms of the 'barely detectable mean' which we define as the value of the true mean of the test region which has a 7% chance of being distinguished from the reference

Table 6.1
Power* for $(\mu_1 - \mu_2)/\sigma$

	0.5	1.0	2.0	3.0	4.0	5.0
I_1	0.20	0.34	0.68	0.91	0.99	1.00
I_3	0.27	0.53	0.92	1.00	1.00	1.00
I_6	0.33	0.65	0.98	1.00	1.00	1.00

* Based on a sample of size $n = 7$ from the reference region, with probability of false alarms, $\alpha = 10\%$.

Table 6.2
Barely detectable mean* for true mean = 100 in the reference region

σ	20	50	100	200	500
I_1	142	204	308	516	1140
I_2	131	178	256	412	880
I_3	127	167	234	369	772
I_4	124	161	222	344	710
I_5	123	157	214	328	670
I_6	122	154	208	317	642
I_7	121	152	204	308	620
I_8	120	150	201	302	604
I_9	120	149	198	296	591
I_{10}	119	148	196	292	580
I_{20}	117	143	186	271	528

* Based on a sample size $n = 7$ from the reference region.

mean. This will depend on the true mean μ_0 and standard deviation σ of the reference region, as well as the sample sizes. The barely detectable means for $\mu_0 = 100$ and various standard deviations σ are given in Appendix A. Table 6.2 gives a few of the barely detectable means based on a sample of size $n = 7$ from the reference region. Observe that as the number of observations in the test region (the subscript m of the index) increases, the barely detectable mean decreases.

The values in Table 6.2 are based on a mean of 100 in the reference region. However, since the calculations use the standardized differences, the table can be used to look up other values of the mean as long as the coefficient of variation σ/μ_0 is one of those tabled ($\frac{1}{5}, \frac{1}{2}, 1, 2$ or 5). For example, if $\mu_0 = 10$ and $\sigma = 20$, then look up the value for $\mu_0 = 100$ and $\sigma = 200$ and divide by 10 to get a barely detectable mean of 51.6 for the index I_1 based on 7 observations from the reference region. More extensive tables of the barely detectable mean are given in Table A.1 to A.5 of Appendix A.

To compare regions, or to compare the same region for different years, it is important not to change the index. Suppose the index I_5 has been in use, and for some test region, there are only 3 years of data available. In such a case one should not use the index I_3. Instead, one should should use the reference value for I_5, but use the average of only 3 years of data in the numerator. This will have, *on the average*, the same value as I_5. However, there will be a higher chance of false alarms and smaller power to detect significant ecological stress. The index will, with these qualifications, be comparable to I_5. Let

$$I_{m,k} = \frac{\bar{Y}_k}{\bar{X}_n + S_n t_{0.10}(n-.1)(1/n+1/m)^{1/2}}$$

be the index I_m except that the numerator is the average of k values. We observe $I_{m,m}$ is the same as I_m. Further if k is smaller than m, then the index is more variable, and if k is larger than m, then the index is less variable than I_m.

If the mean of the index variable increases and the standard deviation of the test region decreases with pollution, the effect on the index is unclear. This is because the benchmark is, more than likely, greater than the mean. But if the mean and standard deviation of the test region both increase with pollution (as is typical) then the barely detectable mean would decrease. Table 6.3 gives the barely detectable mean and probability of false alarms when the true mean is 100 and the standard deviation is 50. Three observations instead of five from the test region causes the barely detectable mean to increase from 157 to 160 and the fraction of false warnings to increase from 10% to 13%.

Table 6.3
The barely detectable mean* and (probability of false alarms) for index $I_{m,k}$ when the true mean $\mu = 100$ and standard deviation $\sigma = 50$

m	k						
	1	2	3	4	5	10	20
1	204	198	195	194	193	191	190
	(0.10)	(0.05)	(0.03)	(0.02)	(0.02)	(0.01)	(0.01)
2	185	178	175	174	173	171	170
	(0.16)	(0.10)	(0.07)	(0.06)	(0.05)	(0.03)	(0.02)
3	177	170	167	166	165	162	161
	(0.19)	(0.13)	(0.10)	(0.08)	(0.07)	(0.05)	(0.03)
4	172	165	163	161	160	158	157
	(0.22)	(0.15)	(0.12)	(0.10)	(0.09)	(0.06)	(0.04)
5	169	162	160	158	157	155	154
	(0.23)	(0.17)	(0.13)	(0.11)	(0.10)	(0.07)	(0.05)
10	163	156	153	151	150	148	147
	(0.27)	(0.21)	(0.17)	(0.15)	(0.14)	(0.10)	(0.08)
20	159	152	149	147	146	144	143
	(0.29)	(0.23)	(0.20)	(0.18)	(0.16)	(0.12)	(0.10)

* Based on a sample of size $n = 7$ from the reference region.

7. Prevalence of fin rot in winter flounder near the New York Bight

The prevalence of fin rot is an increasing function of pollution. To estimate a reference value, seven observations taken in the years 1975, 1977, 1979 and 1980–1983 in water on the continental shelf deeper than 27 m were used (O'Connor, Ziskowski and Murchelano, 1987). The arcsine transformation to normality was used throughout the calculation of this index. This reference region, by comparison to inshore waters, is clean. For this reference region $\bar{X} = 2.20$ and $S = 2.15$ giving a reference value for an index based on a single observation from test regions of

$$R_1 = \bar{X} + S_n t_{0.10}(7-1)(1 + \tfrac{1}{7})^{1/2} = 9.42.$$

The index for yearly observations is $I_1 = Y/9.42$. The index values of I_1 were calculated for areas 3, 4, 5 and 7, the New York Bight apex, Long Island New Jersey inshore areas, other inshore regions, and area 7 is Raritan Bay, respectively. These values are listed in Table 7.1

Putatively, each of these regions is polluted enough to cause fish disease. To increase the power, we look at index values $I_3 = \bar{Y}_3/R_3$, where

$$R_3 = \bar{X} + S t_{0.10}(7-1)(\tfrac{1}{3} + \tfrac{1}{7})^{1/2} = 6.86,$$

and $\bar{Y}_3$ is the average of 3 years. The index values given in the Table 7.2 are recorded using for each date the three years centered on that date. Indices I_1 and I_3 are compared for areas 3 and 7 in Figure 7.1.

The above discussion deals with the index over time for fixed regions. We also observe that these regions are known to be impacted by pollution, but the index I_3 does not show the true nature of the pollution effect. We now construct the index I_7

Table 7.1
Index values I_1

Test area	1973	1974	1975	1976	1977	1978	1979	1980	1981	1982	1983
3	2.29	1.52	0.77	0.47	1.09	0.87	–	0	0.77	0.79	0.39
4	0.85	0.58	0.37	–	–	0	–	0.47	0.58	0.45	0.42
5	–	0	–	–	–	–	0	0	0	0.55	0.49
7	1.65	0.84	0.93	0.65	0.62	0.53	–	–	–	–	–

Table 7.2
Index values I_3

Test area	1973	1974	1975	1976	1977	1978	1979	1980	1981	1982	1983
3	–	2.10	1.27	1.07	1.11	–	–	–	0.71	0.89	–
4	–	0.82	–	–	–	–	–	–	0.69	0.66	–
5	–	–	–	–	–	–	–	0	0.25	0.47	–
7	–	1.57	1.11	1.00	0.82	–	–	–	–	–	–

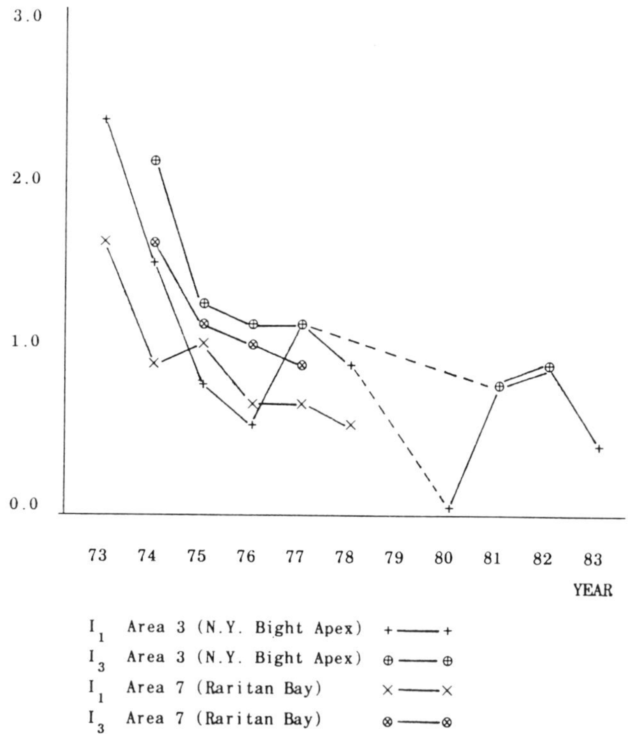

Fig. 7.1. Indices of prevalence of fin rot in fish near the New York Bight.

to use essentially all the data. There are 7 observations from the reference region. We will calculate the average $\bar{Y}_k$ over all available data. That is, for region 3, we will use $I_{7,10}$, for region 4 we will use $I_{7,8}$, for region 5 we will use $I_{7,6}$ and for region 7 we will use $I_{7,6}$. See Table 7.3. The reference value is

$$\begin{aligned} R_7 &= \bar{X}_7 + S_7 t_{0.10}(7-1)(\tfrac{1}{7}+\tfrac{1}{7})^{1/2} \\ &= 2.20 + 2.15(3.143)(\tfrac{2}{7})^{1/2} \\ &= 5.81. \end{aligned}$$

Table 7.3
Index values for I_7

	Region				7 most recent years*	
	3	4	5	7	3	4
k	10	8	6	6	7	7
$I_{7,k}$	1.63	0.84	0.31	1.58	1.01	0.66

*$I_{77} = I_7$.

Even the index I_7 does not show the true nature of pollution. Region 3 is known to be severely impacted by pollution. The power of this index is not large enough to show this effect. The problem may be, in part, due to migration of fish from polluted regions where they contract disease, to the cleaner reference region. Further, many of the fish with fin rot that are captured in the reference region show signs of recovery. The choice of a reference region further from regions known to be polluted may result in an index with more power to detect pollution when it is present. Another possible change would be to restrict the sampling to certain times when the fish that are present are thought to be resident.

If the prevalence of fin rot that is important to detect can be determined, then the power of these indices to exceed 1 for that level can be calculated using Table C.1 in Appendix C. Then we can decide how many observations are necessary to make the current formulation of the index powerful enough. Tables A.1–A.5 of Appendix A would also be useful.

8. Reproductive success of marine birds

Ospreys are marine birds nesting and fishing on the east coast. They nest in accessible areas using the same nesting sites from year to year, and they are tolerant of humans. This makes collection of data about the reproductive success of ospreys easy. Paul Spitzer has surveyed the reproductive success on the east coast for many years (see Spitzer et al., 1983). In the early 1970s there was extensive DDT pollution which gradually cleared from the environment after its use was banned. The years 1973 to 1979 are transition years when the effects of DDT were still present. The 1980s appear to be essentially free from DDT effects upon osprey reproduction (Spitzer et al., 1983).

The reproductive success of ospreys was reduced when DDT was present in the environment; the thickness of egg shells was reduced. The reproductive success is also reduced by environmental stress caused by such things as weather and food abundance. Figure 8.1 plots reproductive success of various locations on the east coast versus time. It is clear that the reproductive success varies from place to place as well as from year to year. To use reproductive success of ospreys as the basis of an index requires that all sources of natural variation be incorporated into the estimate of the reference value.

Reproductive success, intuitively, decreases with pollution. In this case the index is $I_m = R_m/\bar{Y}_m$ where

$$R_m = \bar{X} - S_n t_{0.10}(n-1)\sqrt{1/n + 1/m}$$

is estimated from a sample $X_1, X_2, \ldots, X_n$ from a reference period of time, instead of a control region. The years 1980–1984 form the reference period; the data is taken from the east coast from New York to Boston. For the control period, $\bar{X} = 1.284$ and $S = 0.2479$ giving

$$R_1 = \bar{X} + St_{0.10}(5-1)(1+\tfrac{1}{5})^{1/2} = 1.70$$

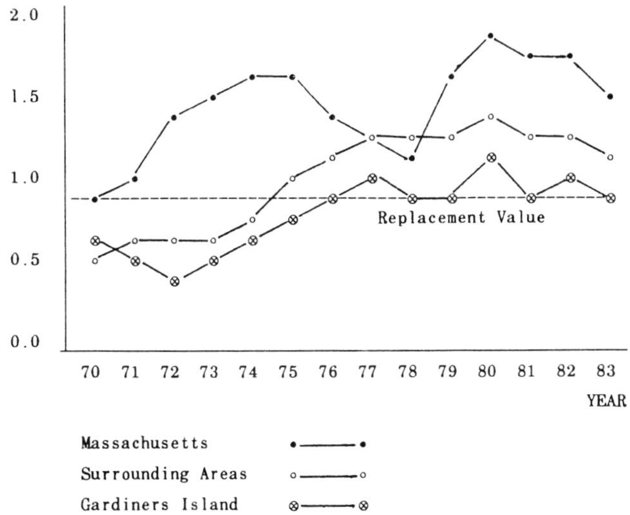

Fig. 8.1. Average osprey young per active nest (3 year running averages).

and

$$R_3 = \bar{X} + St_{0.10}(5-1)(\tfrac{1}{3}+\tfrac{1}{5})^{1/2} = 1.55.$$

In order to standardize the indices to be 10 when an alarming situation is reached, the reproduction rate that will just replace the population was estimated (Spitzer et al., 1983) and used as the 'alarm value'. The replacement rate of 0.8 young per active nest was used in this example. The index then becomes

$$I_m = (R_m/\bar{Y}_m)^{b_m} \quad \text{and} \quad (R_m/0.8)^{b_m} = 10.$$

Thus

$$b_m = 1/[\log_{10}(1.25) + \log_{10}(R_m)]$$

and

$$I_1 = (1.70/Y)^{3.05}, \qquad I_3 = (1.55/\bar{Y}_3)^{3.48}.$$

These indices use a global reference value R_m calculated using the combined data from the east coast.

The indices I_1 and I_3 are given in Tables 8.1 and 8.2 for areas: 1: the entire east coast from New York to Boston; 2: Massachusetts; 3: Gardiners Island; 4: the surrounding areas of New York.

The population of ospreys on Gardiners Island is stressed by a limited food supply (see Spitzer et al., 1983). Using a global reference value for such populations which are naturally stressed results in index values perpetually in a warning or alarm range. To see what is happening locally, a local index could be used where the reference

Table 8.1
Index I_1

Area	1969	1970	1971	1972	1973	1974	1975	1976	1977	1978	1979
1	34.98	22.78	19.68	28.02	11.20	7.20	7.20	2.02	4.61	4.22	3.57
2	26.57	4.35	11.65	4.35	0.56	1.66	1.89	0.76	18.77	3.67	1.46
3	17.92	17.92	34.98	1643.58	29.57	11.65	22.78	5.71	44.10	9.96	21.68
4	66.19	61.71	20.65	29.57	66.19	17.11	6.09	2.49	3.21	2.89	2.89

Table 8.2
Index I_3

Area	1969	1970	1971	1972	1973	1974	1975	1976	1977	1978	1979
1	–	28.27	26.16	19.51	12.90	8.09	3.96	3.40	2.85	2.74	–
2	–	10.14	5.43	1.89	1.08	0.85	0.98	1.86	2.39	3.44	–
3	–	23.37	75.93	99.46	55.01	21.72	11.07	7.17	5.63	8.79	–
4	–	51.28	38.40	39.23	36.04	17.59	5.36	2.94	2.39	2.53	–

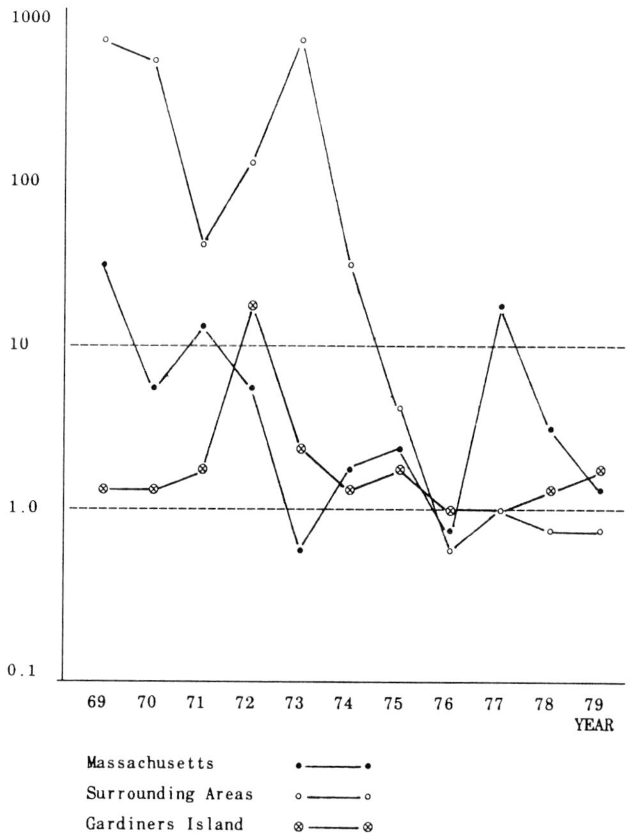

Fig. 8.2. Index I_1 of osprey reproductive success using a local reference value.

value is estimated from a similarly stressed region free from pollution effects. For ospreys a local index was calculated for each region using the years 1980 to 1984 as a reference period. The index I_1 for the three regions are shown in Figure 8.2. For a global reference value, the Gardiners Island index never approaches the value 1 as time approaches the clean period. However, the local index does approach 1. That is for Gardiners Island the local index shows the pollution effect has passed; Figure 8.2 shows that there is still environmental stress present. Occasionally, because of transient natural factors such as rain storms at critical times, large values of the index occur. Some of the peaks including 1972 for Gardiners Island can be attributed to this cause.

The above discussion deals with the index over time for fixed regions. To see the spatial extent of the region affected by pollution, as indicated by the index exceeding one, we consider the years 1969–1972 when the pollution was the greatest. The entire indexed region, has an alarming index value. These are based on the index I_4 using a global reference value.

Appendix A[1]

Table A.1
Barely detectable mean when mean $\mu = 100$ and standard deviation $\sigma = 20$ in the reference region

	n*								
	2	3	4	5	6	7	8	10	20
I_1	182	154	147	144	143	142	141	140	138
I_2	167	143	137	134	132	131	130	129	127
I_3	161	138	132	130	128	127	126	125	123
I_4	158	136	130	127	126	124	124	122	120
I_5	156	134	128	126	124	123	122	121	118
I_6	155	133	127	125	123	122	121	120	117
I_7	154	132	127	124	122	121	120	119	116
I_8	153	132	126	123	121	120	119	118	115
I_9	153	131	126	123	121	120	119	117	115
I_{10}	152	131	125	122	120	119	118	117	114
I_{20}	150	129	123	120	118	117	116	115	112

* n is the number of observations in the reference region.

[1] See Section 6 to see how to use these tables to look up values other than $\mu = 100$.

Table A.2
Barely detectable mean when mean $\mu = 100$ and standard deviation $\sigma = 50$ in the reference region

	n^* 2	3	4	5	6	7	8	10	20
I_1	306	235	219	211	207	204	202	200	195
I_2	268	207	192	185	181	178	176	174	169
I_3	253	196	181	174	170	167	165	162	157
I_4	246	189	175	168	164	161	159	156	151
I_5	241	185	171	164	160	157	155	152	146
I_6	237	183	169	161	157	154	152	149	143
I_7	235	181	167	159	155	152	150	147	141
I_8	233	179	165	158	153	150	148	145	139
I_9	231	178	164	157	152	149	147	144	137
I_{10}	230	177	163	156	151	148	146	142	136
I_{20}	225	172	158	151	146	143	140	137	129

* n is the number of observations in the reference region.

Table A.3
Barely detectable mean when mean $\mu = 100$ and standard deviation $\sigma = 100$ in the reference region

	n^* 2	3	4	5	6	7	8	10	20
I_1	511	370	337	322	314	308	304	299	290
I_2	436	314	284	270	262	256	252	247	237
I_3	407	291	262	248	240	234	230	225	215
I_4	391	279	250	236	228	222	218	212	201
I_5	381	271	242	228	220	214	210	204	192
I_6	374	266	237	223	214	208	204	198	186
I_7	369	262	233	219	210	204	200	194	181
I_8	366	258	230	216	207	201	196	190	177
I_9	363	256	228	213	204	198	194	187	174
I_{10}	360	254	226	211	202	196	191	185	172
I_{20}	349	245	216	202	192	186	181	174	158

* n is the number of observations in the reference region.

Table A.4
Barely detectable mean when mean $\mu = 100$ and standard deviation $\sigma = 200$ in the reference region

	n*								
	2	3	4	5	6	7	8	10	20
I_1	922	640	574	545	528	516	508	498	479
I_2	771	527	468	440	423	412	405	394	374
I_3	713	482	424	397	380	369	361	350	329
I_4	681	457	400	372	356	344	336	325	303
I_5	662	442	385	357	340	328	320	308	285
I_6	648	431	374	346	329	317	308	296	272
I_8	631	417	360	332	314	302	293	280	255
I_9	625	412	355	327	309	296	287	274	248
I_{10}	620	408	351	322	305	292	283	270	243
I_{20}	598	390	333	303	284	271	261	247	217

* n is the number of observations in the reference region.

Table A.5
Barely detectable mean when mean $\mu = 100$ and standard deviation $\sigma = 500$ in the reference region

	n*								
	2	3	4	5	6	7	8	10	20
I_1	2154	1450	1285	1211	1168	1140	1120	1094	1047
I_2	1778	1167	1018	949	908	880	861	834	785
I_3	1631	1054	910	841	799	772	752	724	672
I_4	1553	993	850	780	739	710	689	661	606
I_5	1504	954	811	742	699	670	649	619	562
I_6	1470	927	785	714	671	642	620	590	530
I_7	1445	907	765	694	650	620	598	567	506
I_8	1426	891	750	678	634	604	581	550	487
I_9	1412	879	737	666	621	591	568	536	471
I_{10}	1400	870	728	656	611	580	557	524	458
I_{20}	1344	824	681	607	561	528	503	467	392

* n is the number of observations in the reference region.

Appendix B[2]

Table B.1
Barely detectable mean and fraction of false alarms when mean $\mu = 100$ and standard deviation $\sigma = 20$ in the reference region

m*	k**						
	1	2	3	4	5	10	20
1	142	139	138	137	137	136	136
	(0.10)	(0.05)	(0.03)	(0.02)	(0.02)	(0.01)	(0.01)
2	134	131	130	130	129	128	128
	(0.16)	(0.10)	(0.07)	(0.06)	(0.05)	(0.03)	(0.02)
3	131	128	127	126	126	125	124
	(0.19)	(0.13)	(0.10)	(0.08)	(0.07)	(0.05)	(0.03)
4	129	126	125	124	124	123	123
	(0.22)	(0.15)	(0.12)	(0.10)	(0.09)	(0.06)	(0.04)
5	128	125	124	123	123	122	121
	(0.23)	(0.17)	(0.13)	(0.11)	(0.10)	(0.07)	(0.05)
10	125	122	121	121	120	119	119
	(0.27)	(0.21)	(0.17)	(0.15)	(0.14)	(0.10)	(0.08)
20	124	121	120	119	119	118	117
	(0.29)	(0.23)	(0.20)	(0.18)	(0.16)	(0.12)	(0.10)

*m observations from the test region.
** Index based on k observations from the test region and $n = 7$ observations from the reference region.

Table B.2
Barely detectable mean and fraction of false alarms when mean $\mu = 100$ and standard deviation $\sigma = 50$ in the reference region

m*	k**						
	1	2	3	4	5	10	20
1	204	198	195	194	193	191	190
	(0.10)	(0.05)	(0.03)	(0.02)	(0.02)	(0.01)	(0.01)
2	185	178	175	174	173	171	170
	(0.16)	(0.10)	(0.07)	(0.06)	(0.05)	(0.03)	(0.02)
3	177	170	167	166	165	162	161
	(0.19)	(0.13)	(0.10)	(0.08)	(0.07)	(0.05)	(0.03)
4	172	165	163	161	160	158	157
	(0.22)	(0.15)	(0.12)	(0.10)	(0.09)	(0.06)	(0.04)
5	169	162	160	158	157	155	154
	(0.23)	(0.17)	(0.13)	(0.11)	(0.10)	(0.07)	(0.05)
10	163	156	153	151	150	148	147
	(0.27)	(0.21)	(0.17)	(0.15)	(0.14)	(0.10)	(0.08)
20	159	152	149	147	146	144	143
	(0.29)	(0.23)	(0.20)	(0.18)	(0.16)	(0.12)	(0.10)

* m observations from the test region.
** Index based on k observations from the test region and $n = 7$ observations from the reference region.

[2] Based on index $K_{m,k}$.

Table B.3
Barely detectable mean and fraction of false alarms when mean $\mu = 100$ and standard deviation $\sigma = 100$ in the reference region

m*	k**						
	1	2	3	4	5	10	20
1	308	295	290	287	286	282	279
	(0.10)	(0.05)	(0.03)	(0.02)	(0.02)	(0.01)	(0.01)
2	270	256	251	248	246	242	239
	(0.16)	(0.10)	(0.07)	(0.06)	(0.05)	(0.03)	(0.02)
3	254	240	234	231	229	225	222
	(0.19)	(0.13)	(0.10)	(0.08)	(0.07)	(0.05)	(0.03)
4	244	231	225	222	220	216	213
	(0.22)	(0.15)	(0.12)	(0.10)	(0.09)	(0.06)	(0.04)
5	239	225	219	216	214	210	207
	(0.23)	(0.17)	(0.13)	(0.11)	(0.10)	(0.07)	(0.05)
10	225	212	206	203	201	196	193
	(0.27)	(0.21)	(0.17)	(0.15)	(0.14)	(0.10)	(0.08)
20	218	204	198	195	193	188	186
	(0.29)	(0.23)	(0.20)	(0.18)	(0.16)	(0.12)	(0.10)

* m observations from the test region.
** Index based on k observations from the test region and $n = 7$ observations from the reference region.

Table B.4
Barely detectable mean and fraction of false alarms when mean $\mu = 100$ and standard deviation $\sigma = 200$ in the reference region

m*	k**						
	1	2	3	4	5	10	20
1	516	419	480	475	471	463	459
	(0.10)	(0.05)	(0.03)	(0.02)	(0.02)	(0.01)	(0.01)
2	439	412	402	395	392	383	378
	(0.16)	(0.10)	(0.07)	(0.06)	(0.05)	(0.03)	(0.02)
3	407	380	369	363	359	350	345
	(0.19)	(0.13)	(0.10)	(0.08)	(0.07)	(0.05)	(0.03)
4	389	362	351	344	340	331	326
	(0.22)	(0.15)	(0.12)	(0.10)	(0.09)	(0.06)	(0.04)
5	377	350	339	332	328	319	314
	(0.23)	(0.17)	(0.13)	(0.11)	(0.10)	(0.07)	(0.05)
10	351	323	312	305	301	292	287
	(0.27)	(0.21)	(0.17)	(0.15)	(0.14)	(0.10)	(0.08)
20	335	308	296	290	286	276	271
	(0.29)	(0.23)	(0.20)	(0.18)	(0.16)	(0.12)	(0.10)

* m observations from the test region.
** Index based on k observations from the test region and $n = 7$ observations from the reference region.

Table B.5
Barely detectable mean and fraction of false alarms when mean $\mu = 100$ and standard deviation $\sigma = 500$ in the reference region

m*	k**						
	1	2	3	4	5	10	20
1	1140	1076	1050	1036	1027	1003	997
	(0.10)	(0.05)	(0.03)	(0.02)	(0.02)	(0.01)	(0.01)
2	947	880	853	838	829	807	795
	(0.16)	(0.10)	(0.07)	(0.06)	(0.05)	(0.03)	(0.02)
3	867	799	772	756	746	724	712
	(0.19)	(0.13)	(0.10)	(0.08)	(0.07)	(0.05)	(0.03)
4	821	754	726	710	700	678	665
	(0.22)	(0.15)	(0.12)	(0.10)	(0.09)	(0.06)	(0.04)
5	792	724	696	680	670	648	635
	(0.23)	(0.17)	(0.13)	(0.11)	(0.10)	(0.07)	(0.05)
10	726	657	629	613	603	580	567
	(0.27)	(0.21)	(0.17)	(0.15)	(0.14)	(0.10)	(0.08)
20	688	619	591	574	564	541	528
	(0.29)	(0.23)	(0.20)	(0.18)	(0.16)	(0.12)	(0.10)

* m observations from the test region.
** Index based on k observations from the test region and $n = 7$ observations from the reference region.

Appendix C

Table C.1
Power (%) for one-sided t-tests with level $\alpha = 0.10$

D	Degrees of freedom										D
	1	2	3	4	5	6	12	18	24	120	
0.0	10	10	10	10	10	10	10	10	10	10	0.0
0.1	11	12	12	12	12	12	12	12	12	12	0.1
0.2	13	13	13	14	14	14	14	14	14	14	0.2
0.3	14	15	15	16	16	16	16	16	16	16	0.3
0.4	16	17	18	18	18	18	19	19	19	19	0.4
0.5	17	19	20	20	21	21	21	21	21	22	0.5
0.6	19	21	22	23	23	23	24	24	24	25	0.6
0.7	21	24	25	26	26	26	27	28	28	28	0.7
0.8	23	26	28	29	29	30	31	31	31	31	0.8
0.9	24	29	31	32	32	33	34	34	35	35	0.9
1.0	26	31	34	35	36	36	38	38	38	39	1.0
1.1	28	34	37	38	39	40	41	42	42	43	1.1
1.2	30	37	40	42	43	43	45	46	46	47	1.2
1.3	32	40	43	45	46	47	49	50	50	51	1.3
1.4	34	43	47	49	50	51	53	53	54	55	1.4
1.5	36	46	50	52	54	54	57	57	58	58	1.5
1.6	38	49	54	56	57	58	60	61	61	62	1.6

Table C.1 continued

D	Degrees of freedom										D
	1	2	3	4	5	6	12	18	24	120	
1.7	41	52	57	59	61	62	64	65	65	66	1.7
1.8	43	55	60	63	64	65	68	68	69	70	1.8
1.9	45	58	63	66	68	69	71	72	72	73	1.9
2.0	47	61	66	69	71	72	74	75	75	76	2.0
2.1	49	64	69	72	74	75	77	78	78	79	2.1
2.2	50	66	72	75	77	78	80	81	81	82	2.2
2.3	52	69	75	78	79	80	83	83	84	84	2.3
2.4	54	72	78	80	82	83	85	86	86	87	2.4
2.5	56	74	80	83	84	85	87	88	88	89	2.5
2.6	58	76	82	85	86	87	89	90	90	90	2.6
2.7	60	78	84	87	88	89	91	91	91	92	2.7
2.8	61	80	86	89	90	91	92	93	93	93	2.8
2.9	63	82	88	90	91	92	94	94	94	95	2.9
3.0	65	84	89	92	93	93	95	95	95	96	3.0
3.1	66	86	91	93	94	94	96	96	96	96	3.1
3.2	68	87	92	94	95	95	96	97	97	97	3.2
3.3	69	89	93	95	96	96	97	97	98	98	3.3
3.4	71	90	94	96	96	97	98	98	98	98	3.4
3.5	72	91	95	96	97	97	98	98	98	99	3.5
3.6	73	92	96	97	98	98	99	99	99	99	3.6
3.7	75	93	97	98	98	98	99	99	99	99	3.7
3.8	76	94	97	98	98	99	99	99	99	99	3.8
3.9	77	95	98	98	99	99	99	99	99	100	3.9
4.0	78	96	98	99	99	99	99	100	100	100	4.0
4.1	79	96	98	99	99	99	100	100	100	100	4.1
4.2	81	97	99	99	99	100	100	100	100	100	4.2
4.3	82	97	99	99	100	100	100	100	100	100	4.3
4.4	83	98	99	100	100	100	100	100	100	100	4.4
4.5	84	98	99	100	100	100	100	100	100	100	4.5
4.6	84	98	99	100	100	100	100	100	100	100	4.6
4.7	85	98	100	100	100	100	100	100	100	100	4.7
4.8	86	99	100	100	100	100	100	100	100	100	4.8
4.9	87	99	100	100	100	100	100	100	100	100	4.9
5.0	88	99	100	100	100	100	100	100	100	100	5.0

D is the noncentrality parameter, $D = (\mu_v - \mu_x)/[\sigma\sqrt{1/n + 1/m}]$ and DF $= n - 1$.

Acknowledgements

We would like to express our thanks to Dr. Charles Taillie, for many insightful discussions. The research effort was partially supported by a NOAA research grant to the Center for Statistical Ecology and Environmental Statistics, Pennsylvania State University.

References

Boswell, M. T. and G. P. Patil (1985). Marine degradation and indices for coastal and estuarine monitoring and management. A research paper presented at the spring meetings of the American Statistical Association and the Biometric Society, ENAR, North Carolina State University, Raleigh, NC.

Boswell, M. T. and G. P. Patil (1986). Field based coastal and estuarine statistical indices of marine degradation. In: *Oceans 86 Proc.*, Vol. 3: Monitoring Strategies Symposium, IEEE, Piscataway, NJ, 929–933.

Mandel, J. (1984). Fitting straight lines when both variables are subject to error. *J. Qual. Technol.* **16**, 1–14.

Morris, P. A. (1977). Combining expert judgements. *Management Sci.* **23**, 679–693.

O'Connor, J. S. and R. T. Dewling (1986). Indices of marine degradation, their utility. *Environ. Management* **10**, 335–343.

O'Connor, J. S., J. J. Ziskowski and R. A. Murchelano (1987). Index of pollutant-induced fish and shellfish disease. NOAA Special Report, National Ocean Service, NOAA, USDC, Rockville, MD.

Patil, G. P. (1984). On constructing a crystal cube for environmental degradation. Opening Technical Remarks at the Workshop on Indices of Marine Degradation: An Overview of Managers. November 15–16, 1984, Washington, DC.

Patil, G. P. (1984). Some perspectives of statistical ecology and environmental statistics. In: S. M. Gertz and M. D. London, eds., *Statistics in Environmental Sciences*, ASTM STP 845. Amer. Soc. Testing and Materials, Philadelphia, PA, 3–22.

Patil, G. P. (1985). Fishery and forestry management: Preface. *Amer. Statist.* **39**(4), 361–362.

Patil, G. P. and C. Taillie (1985). A conceptual development of quantitative indices of marine degradation for use in coastal and estuarine monitoring and management. A research paper presented at the spring meetings of the American Statistical Association and the Biometric Society, ENAR, North Carolina State University, Raleigh, NC.

Pugh, W. L., G. P. Patil and M. T. Boswell (1986). The crystal cube for coastal and estuarine degradation. *Sea Technol.* September, 33.

Spitzer, P. R., A. F. Poole and M. Scheibel (1983). Initial population recovery of breeding ospreys in the region between New York City and Boston. In: D. M. Bird, ed., *Biology and Management of Bald Eagles and Ospreys.* Harpell Press, Ste. Anne de Bellevue, Quebec.

Verner, J., R. Pastorok, J. O'Connor, W. Severinghaus, N. Glass and B. Swindel (1985). Ecology community structure analysis in the formulation, implementation, and enforcement of law and policy. *Amer. Statist.* **39**(4) Part 2, 393–402.

G. P. Patil and C. R. Rao, eds., *Handbook of Statistics, Vol. 12*

How does Scientific Information in General and Statistical Information in Particular Input to the Environmental Regulatory Process?*

C. Richard Cothern

Abstract: The input of of statistics and science in general to the development of environmental regulations is discussed as it is followed in the United States Environmental Protection Agency (USEPA). The procedure involved in developing an environmental regulation is presented showing the specific places where scientific data and information enter. The difference between science and policy are discussed along with their relationship in the regulatory process. Specific examples of the use of statistics in the development of environmental regulations are given. The general area of quantitative risk assessment is included as an example of the broad input of science. Possible future directions in the area of scientific input to the regulatory process are discussed.

1. Introduction

Scientific data and information concerning human health and the environment, and specifically statistically derived scientific input, are necessary to understand before regulations can be developed to reduce the risks that are involved in the exposure to contaminants and other environmental insults. This information and data involves both the nature of the processes that occur and the trends that are happening. By developing an understanding of the background events, the problems that are occurring now and may occur in the future can be identified. Success in protecting public health and the environment is measured by the extent to which we understand and manage the human activities affecting them. This effort requires scientific input and planning and thus a substantial investment in monitoring, research and

* The thoughts and ideas expressed in this chapter are those of the author and are not necessarily those of the US Environmental Protection Agency.

development to provide an overall picture of what is happening and what the state of the environment is.

The development of environmental regulations is one endeavor of an agency such as the United States Environmental Protection Agency (USEPA) that can reduce pollution and associated risks. Categorically environmental regulations can involve, among other things, national environmental standards, control technology requirements and enforcement procedures.

Some would argue that 'It is axiomatic that reliable scientific data should underlie the decisions of federal regulatory agencies' (Burack, 1987). In the 1970s and since then EPA has found itself often operating at the frontiers of science. This leads to making decisions in areas where scientific uncertainty abounds.

Through regulatory activity some of the most blatant risks to the environment have been reduced. Now science is revealing some problems that we had not thought of – e.g., radon, stratospheric ozone and global climate change (warming?). We do much for environmental protection and spend about $130 billion annually. Therefore the stakes are high and we had better do it right. This makes the input of science even more important to regulatory decision making.

A key role of science at USEPA is to reduce uncertainties in environmental decision making. Science is especially necessary to characterize today's subtle and complex environmental problems that cut across all environmental media (air, water and land) and transcent national boundaries. Science is also key to determining which environmental problems pose the greatest risks to human health, ecosystems and the economy.

We often face environmental problems where the science is highly uncertain. In most cases the problems are urgent and we cannot wait for the complete answers to scientific questions before acting. We have been forced to make enormously costly decisions on uncertain scientific bases. Just a few examples of such problem areas are dioxin, electromagnetic fields, the Superfund program, the health effects of the oil fires in Kuwait, acid rain and priorities for the protection of drinking water.

This chapter contains the following topics showing the role of statistically based science in the development of environmental regulations. First, the regulatory process is examined and some specific examples of the role of statistically based science are discussed. Next, the important distinction between science and policy is examined. Third, examples of statistics to the regulatory process are given. Fourth, the input of science, in particular science used in developing risk assessments of environmental contaminants to the regulatory process, is discussed. Fifth, the importance of the scientific peer review process is reviewed. The last section involves a discussion of the future role of science and statistical inference in the development of environmental regulations.

2. The regulatory process

Congress enacts statutes, and the executive branch carries them out. Congress usually does not establish all the substantive or procedural details of the programs that it establishes in the laws that it writes. In these cases the congress delegates part of its power to write laws to the executive branch agencies giving them the power to fill in the gaps and impose specific requirements, in the form of rules.

For an agency like the USEPA, rulemaking occupies a significant portion of its personnel and other resources. Because of the enormous scope of its statutory responsibilities, EPA is constantly engaged in rulemaking. As such, rulemaking is one of the most important functions the Environmental Protection Agency performs.

There are, generally speaking, three types of rules, viz., legislative, interpretive and procedural. Legislative rules establish legal obligations. These are based on authority found in laws administered by an agency such as the Environmental Protection Agency. These rules are binding on affected parties and must be developed according to prescribed procedures. Interpretive rules are an agency's interpretation of existing legal authority. They may also be called guidelines or policy statements. Procedural rules define an agency's organization and operational features.

To give a better overall picture of the development of a regulation some of the major steps are listed here. This is only a general overview and not intended to be a complete and detailed account. The first formal step in regulatory development is a start action request which initiates the sequence of events to follow and is a formal notification to all involved that a regulatory effort has started. Discussions occur at this state among most of the parts of the agency that will be involved in the rulemaking process. Input is sought from all possible areas to insure that all facets and ramifications are considered thoroughly in the development of the regulation. A formal notification of such process is, in only some cases, published in the Federal Register as an advanced notice of proposed rulemaking (ANPRM). An ANPRM can be a short declaration or a detailed collection of the input information involved. The next most important step is the formation of a workgroup which is composed of every part of the USEPA. Involved in the workgroup are representatives from: program offices, regional offices, the Office of Research and Development, the Office of Policy, Planning and Evaluation, and any other interested parties. In a few cases, statisticians are members of the workgroup. Some members of the workgroup may be scientists. The workgroup meets regularly throughout the development of the regulation and is the main group involved in the regulatory development process.

Part of the input to the process of rulemaking is the development of background documents in the areas of economics, science, engineering, monitoring, risk analysis, risk-benefit analysis and others. The program office developing the regulation is responsible for producing these documents and for their review. The documents are produced by in-house scientists and professionals or by contractors under the guidance of in-house professionals. The documents are reviewed internally and by external experts. In addition some of the documents are reviewed by the agency's Science Advisory Board, whose role is discussed later in more detail in this chapter.

The next few steps in the development of a proposed regulation for publication in the Federal Register include a development plan and presentation to the steering committee, which is a group of senior managers from the program offices, regional offices and research laboratories. The decision package passes through internal approval stages and also needs approval from the Office of Management and Budget (OMB) before publication of the proposed regulation in the Federal Register.

In response to publication of a proposed regulation in the Federal Register there are public meetings and public comment is solicited there or in writing. The inputs from scientific review, policy review, public comments and other review comments are considered by the workgroup and a final decision package is submitted to the steering committee, internal EPA review and OMB review before the final regulation is published in the Federal Register.

Regulations are usually written by attorneys and engineers and seldom by scientists. The scientific input can surface a problem, but the development of a regulation is not necessarily a scientific process. The direct role of statistically based science enters the regulatory process, in addition to the above mentioned roles, through planning monitoring studies and surveys, interpreting and analyzing monitoring and survey data and in the development of background documentation.

Scientists are involved in the searches of the existing literature to develop criteria documents in the areas of science, risk assessment, engineering and technology and economics. Analysis of these documents can lead to a determination that more research is needed, or in some cases the regulatory demands require a scientific judgement based on these searches.

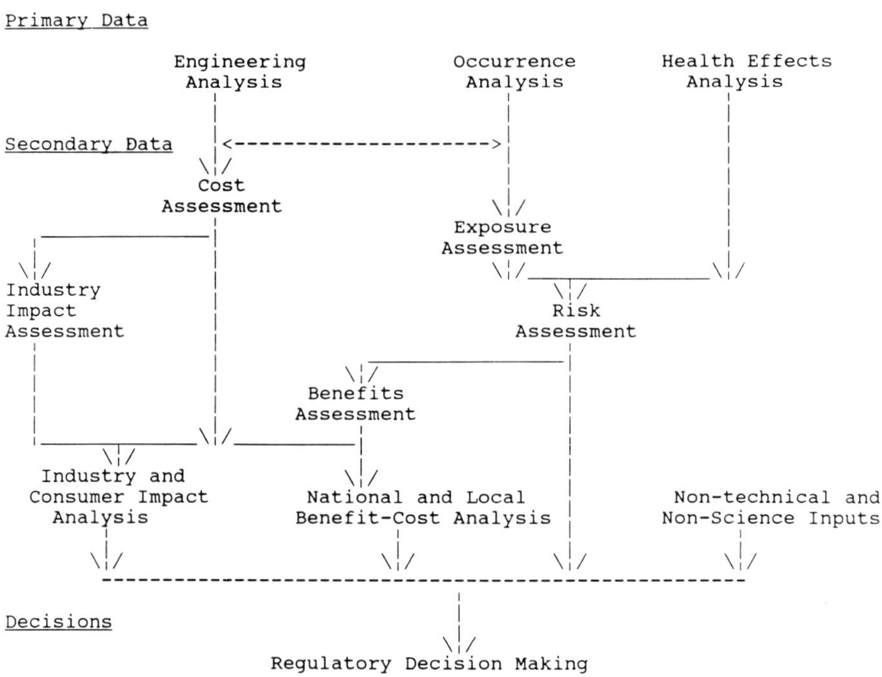

Fig. 1. Schematic diagram of the development of a regulation in the Office of Drinking Water, USEPA.

Although each program office may have a somewhat different procedure for developing a regulation, the one shown in Figure 1 from the Office of Drinking and Ground Water is typical (Schnare, 1992).

In Figure 1 the statistical and scientific inputs to the regulatory process in the development of a regulation in the area of drinking water are shown. This schematic is somewhat of a simplification used here to show how the different major inputs are involved. As can be seen the scientific inputs occur in several places. In the engineering analysis process the surveys are planned to be statistically representative of the actual requirements for removal of contaminants in drinking water. The occurrence analysis includes survey data that involves planning with statistical design. The justification of an appropriate model (delta, log normal, truncated lognormal, ...) is most important here because it determines the number of water supply systems that will be affected by the validity of probabilistic conclusions and can be considerably different, depending on the model chosen. Note that in only a few cases the statistical planning is done first – a problem in all regulation development. The impact assessment involves analyses that can be multivariate for past actions and may involve Monte Carlo calculation for future actions. In the exposure assessment, uncertainties occur due to too much missing data. In the final presentation of the data for decision making, the accuracy and precision can be presented in the form of 'uncertainty circles' on plots of cost vs. risk. These represents the range of quantitative uncertainty in the estimates. The end result of most regulatory decision making is some form of benefit–cost analysis – the quality of the information and the confidence involved in estimating parameters here is most important.

3. Science and policy

Science produces scientific facts which change in the long term, require research and involve variability, whereas policy involves regulatory reality, can change quickly and reflects the political, social and economic realities. There are many inputs to policy making which include: federal, state and local governments, political matters, economics, social concerns, scientific data and information and lobby groups such as from industry and environmental groups, the media and the public.

A rough idea of the relationship between pure scientific fact and pure policy can be seen schematically in Figure 2. On the left is pure scientific fact, a situation where the complete scientific data and information are known and with all the exactness necessary for understanding the situation. On the right is represented the situation where there is little scientific data or information and the policy will therefore have no scientific basis. An example of where there is little scientific knowledge is that of whether there is a threshold in the dose-response curve for exposure to low level radiation. There is no clear scientific evidence that there is or is not a threshold in the dose-response curve at levels of ionizing radiation to which the general public is exposed. The question, in a sense, transcends the science into the policy realm and thus Weinberg coined the word 'transcientific' to describe this situation (Weinberg, 1972). In this case a policy has been developed, in the spirit of protecting public

SCIENTIFIC INFORMATION AND ENVIRONMENTAL REGULATION

Scientific fact	Decision making	Policy
	Science and Policy Judgement	
Characteristics:		
science is known	some science known, but it is not definitive enough for clear and unambiguous decisions	science is not known
Quantitative character:		
minimum variability, high precision	more variability and wider ranges of values, quantitative estimates based on scientific facts, more anecdotal information, controversy scientists disagree	estimate exceeds credibility-e.g. much longer than human lifetime, population risk estimate is larger than population involved, risk less than one in a million-million-...
Qualitative character:		
confidence high	lower confidence	uncertainty

Fig. 2. Schematic presentation of the continuum between pure scientific fact and pure policy.

health, of assuming that there is no threshold and that, in addition, the shape of the curve is linear. These policy assumptions have a strong effect on the risk estimates made for exposure to ionizing radiation. In the area of global climate there are so many scientific complexities and unknowns that actions by governments are based primarily on policy.

Science and policy can also be characterized by their quantitative and qualitative character as shown in Figure 2. On the left is the situation where there is minimum variability and high precision and accuracy. In all these cases there will be some quantitative error or uncertainty because of the very nature of measurement of any scientific or technological quantity. On the right or policy side are quantitative examples such as when estimates of the time for a human to be affected is longer than a human lifetime, the estimate for the total population fatalities is larger than the existing level, the estimate of risk is so small as to make no sense, e.g., one millionth of a millionth of a millionth .... In these situations the scientific estimates are clearly beyond credibility and thus suspect. This is a situation where policy must be used to make the decision. In the center region of Figure 2, the scientific data is incomplete, anecdotal and sometimes contradictory. Scientists may disagree about the interpretation of the data in this area of knowledge. This circumstance is, unfortunately, the one most commonly encountered.

The qualitative character of scientific data can be high for scientific fact. However, more commonly the available data will be incomplete, not fully characteristic of the population of inference, fuzzy and thus the confidence in the estimate is lower.

Problems that characterize the center region include questions such as how to compare different health endpoints and how to balance human and ecological needs. Specific examples include: health effects of different forms of asbestos, synergisms such as between radon and cigarette smoke, relation between health effects of dioxin (2,3,7,8-tetrachlorodibenzo-*p*-dioxin) for animals and humans and how toxic perchloroethylene really is compared to other common environmental pollutants.

As can be seen in the above discussion the relationship between science and policy in regulatory decision making is a most important one. The significant perspective to remember is that statistically valid scientific data are often lacking and are only one of many inputs to the environmental regulatory process.

4. Statistical inputs to the regulatory process

Discussed in this section are a few of the many examples of the direct input of statistics into the regulatory process. This is far from a comprehensive survey and only gives some of the many situations where statistically based science is involved in the regulatory process.

There are many issues that involve input of statistical planning, thinking and concepts and due to the limited space only a few are discussed here. Others of major importance include: historical controls for bioassay studies, comparisons of multiple studies, estimates of the probability of false negatives and false positives, how to combine multiple endpoints and how to achieve consistence when there are data from multiple sources. Perhaps the most important area where statistical input is essential is in the planning and sampling of environmental data. A more thorough discussion of these and other principles can be found in Gart et al. (1979).

4.1. Examples from the enforcement of the clean air act

Three examples of the use of statistics in enforcement are discussed here: selective enforcement audit, evaluation of fuel waivers, and reformulated gasoline enforcement sampling (Nussbaum, 1992 and Nussbaum and Rachlin, 1992).

EXAMPLE 1. (*Selective enforcement audit*). This example involves the sampling of assembly line automobiles and similar vehicles for emissions. This kind of testing is done in cases where there is a prior emission problem for a particular model and make. Involved is an unannounced inspection at the assembly line or at the pier of entry for a foreign manufacturer. The sampling used is a classic and is similar to that used for military standards. If all pass or all fail the situation is clear. However, if some pass and some fail, further investigation is needed. Action is taken when there is a 95% certainty that the true rate is that 40% or more of the vehicles fail the test. This procedure has led, among other things, to manufacturers doing their own quality control sampling on the assembly line.

EXAMPLE 2. (*Evaluation of fuel waivers*). Section 211 of the Clean Air Act Amendments of 1990 is a prohibition of new fuels or new fuel additives for motor vehicles. The section contains two mutually exclusive ways around the prohibition. A waiver can be granted if a new fuel or fuel additive is 'substantially similar' to the fuel used in certifying a vehicle, or if it can be demonstrated to the USEPA that the new fuel will *not* make the vehicle exceed the emissions standards. The problem is that of demonstrating a negative. The criteria described below, in practice, involves engineering judgment, legal and policy considerations.

The following is a brief description of the statistical tests used to characterize the emission effects of a fuel or fuel additive (emphasized here is the character of the tests and additional concerns include the number of vehicles involved):

(1) *The paired difference test*: For each vehicle tested on a base fuel and on the waiver fuel additive, the difference between the waiver fuel or fuel additive emissions and the base fuel emissions is calculated. A 90% interval is constructed for the mean difference. If the resulting interval lies entirely below zero, it is indicative of no adverse effect from this waiver fuel or fuel additive. If the entire interval is above zero, it is indicative of an adverse effect from the waiver fuel or fuel additive. If the interval contains zero, there is arguably no difference between the base fuel and the waiver fuel or fuel additive with regard to emissions, provided the confidence interval is small.

(2) *The sign of difference test*: For each vehicle tested with a base fuel and the waiver fuel or fuel additive, the sign of the emission difference between the waiver fuel or fuel additive emissions and the base fuel emissions is ascertained. This test is designed to determine whether the number of vehicles demonstrating an increase (+) in emissions with the waiver fuel or fuel additive significantly (at least at the 90% confidence level) exceeded those showing a decrease (−) in emissions with the waiver fuel or fuel additive.

(3) *The deteriorated emissions test*: For each vehicle the effect the waiver fuel or fuel additive has on emissions is determined. This incremental effect, either positive or negative, is added to the 50 000-mile certification emission value of the certification vehicle which the test vehicle represented. This incremented 50 000-mile emission value is compared to emission standards to determine if it did or did not exceed the standards. Either a pass or fail is assigned accordingly. The pass/fail results are analyzed using a one-sided sign test. The test was designed such that the risk of failing would be at least 90% if 25% or more of the vehicle population that is represented in the test failed to meet federal emission standards while operated on the test fuel. For example, for a sample of eight test vehicles, all eight of these test vehicles must pass in order that the deteriorated emissions test be passed.

The first two methods of analysis are designed to determine whether the waiver fuel or fuel additive has an adverse effect on emissions as compared to the base fuel. Each characterizes a different aspect of adverse effect. The paired difference test determines the mean difference in emissions between the base fuel and the waiver fuel or fuel additive. The sign of difference test assesses the number of vehicles indicating an increase or decrease in emissions. The two tests are considered together in evaluating whether an adverse effect exists to assure that a mean difference determination is not unduly influenced by very high or very low emission results from only a few vehicles.

The deteriorated emissions test analysis indicates whether the fuel or fuel additive causes a vehicle to fail to meet emissions standards. This test examines each vehicle's performance as compared to each pollutant standard. It is useful to perform this analysis even if the first two analyses indicate the waiver fuel or fuel additive has no adverse effect. The analysis indicates whether the emissions from any particular type of vehicles or special emission control technologies are uniquely sensitive to the waiver fuel or fuel additive thus causing vehicles to fail to meet emission standards. This effect could be missed in the previous analyses which consider the emissions results as a group without distinguishing the emissions impact on subgroups.

EXAMPLE 3. (*Reformulated gasoline enforcement sampling* – from Nussbaum and Rachlin, 1992). One of the major distinctions of the Clean Air Act Amendments of 1990 from its predecessors is the recognition of the effect that fuels that are put into a vehicle have a very different effect on the automobile emissions that come out of that vehicle. While earlier versions of the act had fuel provisions, this was the first time the act called for modifications in fuel to actually reduce resultant emissions. The two major provisions are the winter oxygenated fuels program and the reformulated gasoline program.

Reformulated gasoline is a general term for a new version of gasoline that will have lower volatility and less toxic emissions. The actual formulation was not specified by the act, but left to the USEPA. The USEPA, in turn, used a relatively new technique called negotiated regulation (reg-neg) to establish the specific criteria in this case. Reg-neg brings together representatives of all affected parties to negotiate the form of regulations before they are submitted as a proposal for public comment. The 'before-hand' negotiations are designed to alert all parties of the issues involved, to elicit compromises on positions and to commit the entities not to file lawsuits that invariable delay implementation of environmental regulations. One of the agreements reached at the reg-neg included the skeleton of a sampling plan that would determine the degree of compliance with the regulation. Interestingly the results of the sample would automatically trigger more stringent standards for the regulated parties should evidence of violation be found.

The reformulated gasoline will be required beginning in 1995 in at least the nine most polluted ozone areas, with the ability of other areas to 'opt-in' to the program. It is expected that there will be many additional 'opt-in' areas. For each individual area, there are three methods to determine compliance with the standard.

First, each supplier may elect to have each gallon meet the appropriate standard for reid vapor pressure, benzene, toxic emission reductions, oxygen, NO_x emissions, heavy metal content, sulfur, T-90 and olefins.

Second, each supplier may elect to have the average emissions of reid vapor pressure, benzene, toxic emissions reduction and oxygen meet an 'average' standard. However, the reg-neg group realized that a supplier meeting a per gallon standard would probably produce at a more stringent level in order to assure that each gallon met the standard despite production variation. Thus a supplier wishing to average emissions would have a somewhat more lenient situation since some gallons could exceed the standard and the refiner still be in compliance. To avert this discrepancy, the standard to be met by those electing the averaging method ('average standard') is more stringent than the standard for per gallon compliance. For instance, for certain regions the reid vapor pressure standard for each gallon is 7.2 psi. If the averaging method is selected, an 'average' standard of 7.1 psi must be met. Further, if the averaging method is selected there is also a per gallon maximum which much not be exceeded. In the above situation, it is 7.4 psi. The average levels of the various parameters will be determined by a compliance survey conducted by the supplier.

Third, the survey requirement could be satisfied by a consortium of refiners and importers who wish to average across covered areas to plan a comprehensive survey program that should be less expensive than each party conducting a survey individually. Again the more stringent 'average' standards would pertain.

The USEPA has provided specific in conducting either of the two averaging methods of compliance. If a supplier elects to do his own survey, he must perform four surveys annually each area which is supplied with reformulated gasoline by that refiner or importer. Each supplier would be required to survey the covered area's gasoline supply, not just that supplier's fuel in the area. Since there are many refiners supplying these areas (they are major metropolitan areas), this results in a large number of surveys. Such a large number should reveal any spikes in the parameters of interest. For the comprehensive survey by the consortium, the rules differ somewhat. A total of 120 surveys are required in the first year of the program. It is expected that compliance in subsequent years will be easier after refiners and importers have developed reformulations and have some experience with the program. Accordingly the program calls for fewer required surveys in subsequent years. In fact the number of required surveys, decreasing further to 60 surveys in 1997, and down to 50 surveys for each year therafter. This corresponding reduction does not apply to those who elect the individual survey method rather than comprehensive approch. In the case of a failure to meet the 'average' standard, the USEPA borrows a page from quality control sampling and will reduce the decrease in the number of surveys that would normally take place in the year after a failure such that a higher level of inspection is possible in the failed area. There is an upward adjustment of the number of surveys required for additional areas that opt into the program.

Some further guidance reflects the USEPA's concern about the representativeness of the surveys. The surveys must be conducted to reflect various operating diversities. These include assuring samples at small and large stations; major and minor brands, downtown and suburban locations and all grades of gasoline. The number of samples within each survey must be sufficient to establish a 95% one-sided confidence interval about the sample mean to assure that the error in predicting the mean is less than the appropriate tolerance for each parameter. Thus for each survey the sample size is N where

$$N = \left(\frac{zs}{e}\right)^2,$$

where $z = 1.645$, s = sample standard deviation, e = error tolerance (specified as 0.1 psi for reid vapor pressure, 0.05% for benzene and 0.1% for oxygen). This simplistic formula assumes normality, a condition that may not be justified.

The sample standard deviation must be justified based on prior similar testing or a pilot test. Despite this stipulation on the justification of standard deviation, the actual determination of compliance will only be based on the sample mean with no allowance for any variance. For instance, if the average reid vapor pressure in the above example exceed 7.2 psi, the survey result is a failure, no matter what the variance of the data. The selection of the 'non-average' standard of 7.2 psi rather than the 'average' standard of 7.1 psi for compliance is done to avoid giving the refineries a 50% probability of failure if their true average in fact complies with the 'average' standard. While this is some relief from a type I error, the USEPA does note that if in fact the true average is right at the 'average' standard (in this case, 7.1 psi), there is still considerable risk of failing the survey. To reduce the level of such risk, either the sample size should be increased above the minimum calculated, or production targets should be somewhat better than the level of the 'average' standard.

Perhaps the most interesting stipulation in the comprehensive scheme are the results of a failure. Two distinct things happen. First, the standard gets more stringent for the following year in an area which has violated the standard in any survey. For instance in the above example, the 'average' standard moves from 7.1 psi to 7.0 psi and the per gallon maximum falls from 7.4 to 7.3 psi for the subsequent year. The second ramification is even more interesting. This tighter standard must be met by each refiner and importer serving the area, not just the refiner or importer which may have been responsible for the exceedence. Furthermore, that tighter standard applies to all of the averaged gasoline from those refiners and importers, not just the proportion of averaged gasoline going to the area of non-compliance.

4.2. Statistical meta-analysis

In a recent study at the USEPA of the scientific evidence for the health effects of lower respiratory illness in childern aged 12 and under the results of several epidemiology studies were combined in a meta-analysis. The goal was to estimate the odds ratio corresponding to an increase in the household average of 30 $\mu g/m^2$ (0.016 ppm) in NO_2 chronic exposure. This is approximately the increase seen as a result of gas stove use as compared with electric stove use in both the United States and England. Two basic models were employed. One assumes that each study estimates the same fixed, but unknown, parameter. The other model assumes that the parameter of interest is not fixed, but is itself a random variable from a distribution. The odds ratios considered are tabulated below:

Estimated odds ratio	2.5 percentile	97.5 percentile
1.31	1.16	1.48
1.24	1.09	1.42
1.53	1.04	2.24
1.11	0.83	1.49
0.63	0.36	1.10
1.08	0.97	1.19
1.47	1.17	1.86
1.14	0.86	1.50
1.10	0.79	1.53
0.94	0.66	1.33
1.10	0.79	1.54

The above odds ratios, when analyzed using meta-analysis, yield overall ratios of 1.18 (1.11 to 1.25) and 1.18 (1.08 to 1.29) using the two approaches. The basic evidence from the individual studies is somewhat mixed. All but two of the studies showed increased respiratory disease rates associated with increased exposure. A few of the individual studies were statistically significant. Combining the studies as if the end points were similar gives an estimated odds ratio of 1.20 (95% confidence limits of 1.1 and 1.3) for the effect of gas stove use on respiratory illness. When combined, the studies indicated that an increase of 30 μ/m^2 in NO_2 exposure would result in an increase of about 20% in the odds of respiratory disease, subject to the assumptions, made for the synthesis.

4.3. *Use of statistics in compliance*

In 1987 the USEPA held an entire conference dovoted to presentations concerning the design, interpretation of environmental regulations that incorporate statistically valid compliance verification procedures (USEPA, 1987a). The papers presented deal with a wide variety of environmental concerns including various novel approaches to air emissions regulations and monitoring, spatial sampling of soil, incorporation of potential health effects considerations into the design of monitoring programs and considerations in the statistical evaluation of analytical laboratory performance. One of the main focuses is sampling frequency.

4.4. *Use of trend tests and pairwise comparisons in analyzing dose-response data*

Certain practices in the application of statistical methods that are components of the weight-of-evidence for classifying carcinogens considerably increase the error of missing positive study results and thereby reduce the likelihood of identifying an actual carcinogen. While there is a considerable difference in policy action on a chemical that has a B2 classification with one classified as a C (probable vs. possible carcinogen – for more details concerning the classification system see USEPA, 1987c), their respective weights-of-evidence basically may differ only by a single 'positive' study. Therefore, the determination of whether a particular study is 'positive' is critical in regulatory policy.

In order to make unbiased decisions concerning the study results of bioassay studies, statistical tests are applied to the data. A valid statistical test has two error rates: (1) the probability of saying the effect under study exists when it does not, and (2) the probability of missing a true effect. In laboratory studies, the incidence of tumors is observed at different doses, and the presence of an effect under study is adjusted to be a relationship between the applied dose and the tumor incidences.

If 'dose-response' represents a criterion for chemical effect, the most important type to test for in identifying a relationship is one that determines whether there is a monotonic change in the intensity of the response as the applied dosage changes, i.e., are there more tumors as the dose increases. One can merely observe the study data and make a judgment as to whether the graded response is sufficient to say that a relationship exists in this data set. But, while laboratory studies are done on a relatively small number of animals their purpose is to shed light on what would be the response in the whole species. It is at this point that statistical methods are called upon to determine, with specified error levels, whether the relationship observed in the study would hold in the species. The tested (null) hypothesis is that there is actually no relationship between the chemical and the tumor response in the species. Thus, when this hypothesis is rejected by the statistical test, we say that the relationship does exist in the species population. In common parlance, this is known as a 'significant' result and the test as a 'positive' one. The test of a relationship is loosely called a 'trend' test, but is not restricted to identifying strictly linear relationships.

Typically, the error of mistakenly identifying the existence in the population of the relationship under study is set at 5%. With the level of the type I error set, the type II error (and, hence, the power) will be determined by the nature of the test, the accuracy of the assumptions underlying the test's application to these study results, the sample size, and the magnitude of the rates being tested, with the power increasing with increasing numbers of animals used in the test. In laboratory studies of carcinogenicity, the power of the trend test is the probability of identifying a relationship that actually exists in the species between a chemical and tumorogenicity. Ideally, the study is designed so that the power is at some acceptable level, such as 70%, leaving no more than a 30% chance that a true relationship will be missed, while the chance that a false relationship is inferred is only 5% – the significance level of the test.

A concern is that the power of the tests by which one determines the existence of a relaionship between chemical and tumor response is being reduced by requiring more than one test to be statistically significant. Currently, some parts of the USEPA require, in addition to a statistically significant trend test, that some 'pairwise' differences (between the response at the control and at one or more of the doses) are also found to be statistically significant, thereby regarding a statistically significant trend not to be sufficient statistically evidence to designate a study as 'positive'. All of the animals in a laboratory study are used in the trend test, but only subsets of them are used in each pairwise test. The larger number of animals that provide the basis for the test of trend increase the power of this test to identify a true relationship, whereas, 'pairwise' tests have less power and a greater likelihood of failing to identify the relationship, because each test is based on fewer animals. This increased requirement for statistically significant results decreases the overall power to identify a carcinogen, thereby increasing the chance of missing an existing effect.

A better approach would be to require a test that is capable of producing a statistically significant positive trend indication as sufficient statistical evidence of identifying a relationship between dose and effect, and to constitute a 'positive' test when there are no preexisting biological of toxicological contraindications.

4.5. A center for environmental statistics

In a broader context there is a need to collect, analyze, integrate and provide data on the overall state of the environment. The mechanisms for this would be a national center for environmental statistics similar to other federal statistical agencies. Such a center would provide for several needs not being completely fulfilled at the present. With an integrated and more complete understanding of the current statistical data the USEPA could make better policy decisions on ideas ranging from program budgets to legislative initiatives and regulations. Such a center could help analyze large quantities of data collected by the USEPA and other federal agencies to see what they say about the broader picture. We need to do a better job of analyzing data from a multimedia perspective. Most of the data collected by the USEPA are only analyzed from the perspective of only one medium (e.g., air, drinking water, soil, food, etc.). However, we know that environmental impacts of human activities often affect more than one medium. Improved analyses would assess the public in understanding environmental issues and would help the USEPA and other agencies in managing their programs. By pulling together data from various sources and making a whole picture we would expect to gain new insights. A central statistical agency for the environment would be able to provide authoritative statistical information in standard data sets and directory-type information formats. An environmental statistics center would provide a consistent baseline of data for tracking environmental conditions over time. The US is the only major developed country without a national center for environmental statistics. Such a center would enhance US influence in environmental matters in the world.

5. Risk assessment as an example of scientific input to the regulatory process

Many think that the quantitative risk assessment estimates used by the USEPA are statistically based. Part of the reason for this is the use of the jargon of statistics such as: upper 95% confidence limit, q_1^*, multistage model, and the quoting of estimates to several decimal places. However, although stastistical methods are used in the analysis, the estimates are almost never good to more than one significant figure, have confidence levels far from 95%, involve data that was not generated for use in such sophisticated models as the multistage model and usually involve extrapolations of several orders of magnitude into the unknown. A closer look at these estimates and the uncertainties involved in their generation reveals that they are not statistically based and should not be interpreted as being nearly as accurate as they seem. More detail on the nature of these estimates can be found in Cothern (1988, 1993b).

5.1. What is risk assessment?

It is perceived that risk assessment, quantitative risk assessment and comparative quantitative risk assessment are statistically based procedures for estimating and evaluating the importance of environmental contaminants in terms of their health risks and in detemining which are more important.

One tool that can help foster the evolution of an integrated and targeted national environmental policy is the concept of environmental risk (see USEPA, 1990 and Cothern et al., 1989). Each environmental problem poses some possibility of harm to human health, the ecology, the economic system or the quality of human life. That is, each problem poses some environmental risk. Risk assessment is the process by which the form, dimension and characteristics of that risk are estimated, and risk management is the process by which the risk is reduced.

The concept of environmental risk, together with its related terminology and analytical methodologies, helps people discuss disparate environmental problems with a common language. It allows many environmental problems to be measured and compared in common terms and it allows different risk reduction options to be evaluated from a common basis. Thus the concept of environmental risk can help the develop environmental policies in a consistent and systematic way.

Scientists have made some progress in developing quantitative measurements for use in comparing different risks to human health. Although current ability to access and quantify ecological risks is not as well developed, an increased capacity for comparing different kinds of risks more systematically would help determine which problems are most serious and deserving the most urgent attention. That capacity would be even more valuable as the number and seriousness of environmental problems competing for attention and resources increase.

Quantitative risk assessment of human health consequences of environmental exposures produces two numbers, viz., individual risk and population risk. Individual risk is the probability one person exposed for a length of time (usually one year or a lifetime) will develop a given malady. Population risk is the estimate of all in a given region (state, city, US) that will likely be affected in a given time period.

In 1986 the administrator of the USEPA commissioned a special task force of senior career managers and technical experts to compare the risk currently associated with major environmental problems and compare them in some quantitative way. The result was a report called 'Unfinished business: A comparative assessment of environmental problems' (USEPA, 1987b). This report was the first systematic analysis of the comparative quantitative risk of environmental problems. In conducting the project there were four important limitations. First, the universe of environmental problems was divided into 31 arbitrary pieces. Each piece represents an environmental problem area which corresponded generally to existing programs or statutes. Second, four different and independent types of risk were considered for each problem area: cancer risks, non-cancer health risks, ecological effects and welfare effects (visibility impairment, materials damage and the like). Third, the project did not consider economic or technical controllability, qualitative factors such as voluntary familiar or equitable; the benefits to society or the statutory and public mandates. Finally, the risks were identified as they exist now with current levels of control. The major results of this study were: (1) No problems rank relatively high in all four types of risk or relatively low in all four. (2) Problems that rank relatively high in three of four risk types, or at least medium in all four include: criteria air pollutants; stratospheric ozone depletion; pesticide residues on food; and other pesticide risks such as runoff and air deposition. (3) Problems that rank relatively high in cancer and non-cancer health risks but low in ecological and welfare risks include: hazardous air pollutants; indoor radon; indoor air pollution other than radon; pesticide application; exposure to consumer products; and worker exposures to chemicals. (4) Problems that rank relative high in ecological and welfare risks, but low in both health risks include: global warming; point and non-point sources of surface water pollution; and physical alteration of aquatic habitats (including estuaries and wetlands) and mining waste. (5) Areas related to ground water consistently rank medium or low.

A recent re-examination of the Unfinished business report basically confirmed the general conclusions (USEPA, 1990) in the two areas of human health and natural ecology. It was concluded that the relatively high risk problems, considering risks to natural ecology and human welfare, are: habitat alteration and destruction, species extinction (these last two were not considered in the Unfinished business study), stratospheric ozone depletion and global climate change. The relatively medium-risk problems in the ecology area are: herbicides and pesticides, toxics, nutrients, biochemical oxygen demand and turbidity in surface waters, acid deposition and airborne toxics. In the area of human health problems were identified that represented major types of human exposure known to be associated with significant impacts on human health and included: ambient air pollutants, worker exposure to chemicals in industry and agriculture, pollution indoors (including radon) and pollutants in drinking water.

From the above discussion it becomes clear that risk assessment is usually not statistically based nor does it involve statistical analyses. In fact there are a range of limitations as discussed next.

5.2. Problems and limitations in the assessment of risk

Several limitations can be listed concerning estimation of risks (Cothern and Schnare, 1986). In the area of health effects there are problems such as: no causal link exists between exposure and health endpoint; in only few cases are synergisms and antagonisms known or understood; long latent periods make epidemiology studies difficult, how to compare morbidity and mortality, is there hormesis? (chemicals essential for life) and possible existence of a threshold in the dose-response curve (repair mechanisms). In the area of exposure analysis there are problems such as: variation in uptake, distribution, excretion and metabolism, exposure from different media (air, water, food), and not knowing the specific cell that is exposed. Other limitations in making a risk estimate include: animal endpoints may not be the same as for humans; only know toxicity at very high exposure levels; and disagreement among scientists concerning unknown and partially known areas.

One of the main impediments to the development and use of quantitative risk assessment is the lack of complete information and data as input to this process. It was the goal of the symposium on 'The quantitative ranking of environmental problems according to risk: What must we yet know to accomplish this task?' to examine some of these gaps and deficiencies and determine how they might be filled (Cothern, 1993a, b). The symposium was not intended to be a complete analysis of all the missing elements, but is intended to give some idea of the breadth and depth of this problem and thus provide direction for solutions.

Support for the idea of determining the number of human lives involved and impact of environmental contamination on ecosystems in the form of quantitative risk assessment in the development of plans, budgets and research directions was given recently by the US Environmental Protection Agency's Science Advisory Board (SAB). Their report *Reducing risk* (USEPA, 1990) made several recommendations (USEPA, 1990). These recommendations are directed towards the USEPA, but are equally applicable to state, local and other environmental groups that are involved in regulatory, policy and mitigation activities. First the SAB recommended that 'EPA should target its environmental protection efforts on the basis of opportunities for the greatest risk reduction.' Here it is recognized that the country has addressed the most obvious environmental problems first (e.g., those easily detected my human senses) and now should use several criteria to determine the future direction including quantitative risk assessment (other criteria include economic, social and political concerns as well as feasibility and timing to mention some). Secondly, 'EPA should improve the data and analytical methodologies that support the assessment, comparison, and reduction of different environmental risks.' As pointed out here this is a major deficiency in the current process and has many facets and dimensions that will be discussed in this symposium/book. Third, 'EPA should reflect risk-based priorities in its strategic planning process.' These risks should include those already prevented and those that could be prevented by furure actions. Fourth, 'EPA should reflect risk-based priorities in its budget process.' The legislation determines the priorities to a large degree, however, some discretion is available and in the long run the public and hence the congress should develop a new direction that minimizes the health and ecological consequences of environmental degradation. This broader mandate is more clearly seen in the next two recommendations. Fifth, 'EPA – and the nation as a whole – should make greater use of all the tools available to reduce risk.' In order to use these tools the concept of quantitative risk needs to be more completely understood throughout our society. Sixth, 'EPA should work to improve public understanding of environmental risks and train a professional workforce to help reduce them.' An important aspect of this process is risk communication. Some problems in this area will be discussed later.

Exposure data is a good example of large areas of missing information and data. Often the existing data is anecdotal, does not include temporal or spatial variation. Another complication is that exposure to a contaminant is not exposure of the organ or body part that is susceptible since the chemical structure can be metabolized or changed after it enters the human or other living things. Only in a few cases does baseline data exist, thus it is difficult to determine whether the data is from a new source or has been there for a long time. Another complication that is seldom addressed in exposure data is its multichemical and multimedia character. An environmental contaminant can be transported in the air, water, soil, food, dust and enter humans by ingestion, inhalation and dermal routes. Data for all these variables are seldom available. Two variables that are too often missing in exposure data are the temporal and spatial character. Some contaminants have constant levels for centuries and some only exist for seconds. Seasonal variations are often important and even day-night variations can be important and are often missing. Often contaminants exist only locally and sometimes the global concentrations are important. As examples industrial pollution is often localized and stratospheric ozone depleters are global in character. In the area of ecotoxicity the loss of species, communities and systems is not clear due to lack of data. In most cases the information involves estimates by experts and no 'hard' data. Another area, habitat destruction, is difficult to measure and determine the base line data.

A number of problems occur in comparing risks due to different environmental contaminants. One of these problems is that the things that are being compared are so different in character that it seems like they cannot be compared or major contaminants that are ignored. One example of this problem in comparative risk assessment is comparing two contaminants, one of which has a cancer endpoint and the other has a non-cancer endpoint. For example compare the risk from inhaling radon that can cause lung cancer and ingesting lead which causes neurotoxicological effects such as lowering the IQ. Is early death equal to a lowered IQ? Can early death be compared in any way to a lowered IQ? Can a weighting factor be used to estimate their comparative risk? Methods exist for these intercomparisons. Another comparison that presents problems is that of human health versus ecological health. Which is more important, a human being or a bird, fish or forest? An interesting example of the interrelationship between human and ecological health is the treatment of drinking water. Some drinking water supplies treat the water and dump the sludge into the river. It is common to remove iron, copper and zinc as part of the process although they are relatively harmless to humans. However, the latter two are rather deadly to other forms of life in the river. How should the relative importance of these two areas be weighed?

One of the largest gaps in the comparative risk assessment field is the methodology to communicate the results to decision makers and the public. As seen in the reducing risk report (Appendix C, pp. 127–128) there is a disparity between the judgment of scientists concerning the comparative risk between environmental problems and the perception of the public. If the approach of basing the development of priorities in the environmental area on factors such as comparative risk, the methodology of risk communication must be developed.

There are several cross-cutting issues that are important in the development of comparative quantitative risk assessment. For example, environmental contaminants can exist in several media (air – indoor and outdoors, water, soil, food) and can be transferred from one to the other, thus potentially changing the relative risk. Other cross-cutting issues involve the artificial division among federal, state and local governments that make cooperation in addressing a single environmental contaminant difficult, differential time scales in the flow of environmental contaminants and the relative importance of an individual contaminant in the areas of human health and ecological health.

5.3. *Risk communication*

The results of comparative quantitative risk assessment must be communicated to decision makers and the public or it will have little value. However, little attention has been given to risk communication both at the formal educational level and the general public level. More effort will clearly be needed here as more information and data become available for comparative quantitative risk assessments. The focus here should be on the language used and the mathematical concepts needed such as order of magnitude, probability and uncertainty.

Perhaps an impediment (for all of us amateurs in environmental analyses) to being able to rank environmental problems quantitatively is that few think they have the training in quantitative and analytical thinking to be able to achieve this broad perspective. Few have complete knowledge about probability, statistics, uncertainty and logarithms in or out of school. Further, most do not realize that they use this knowledge in everyday life. Many would argue that it takes too much time to learn about these concepts and that they have more important things to do. It seems paradoxical that the very concepts that would help in getting perspective about life are rejected because we have too little perspective about the value of them.

What are the concepts from statistical thinking that help us all in our daily lives and are necessary to being able to sort out environmental problems (or any set of problems) in order to get some perspective about which ones we should focus on? Three basic concepts that seem important are: ranges of uncertainty, lack of absolutes and order of magnitude.

We often see headlines that some environmental risk has been overestimated or underestimated. In reality all that has probably happened is that another measurement has been made that falls within existing the range of uncertainty. Where do we get the idea that all numbers are exact? Ask any simple question – how many people were at the meeting? – or how long does it take you to get to work? – and you will quickly see that there is not exactness in daily life. So why do we expect exactness in environmental measurements? And we even expect estimates based on scientific assumptions and judgment to be exact. If an estimate is 300 plus or minus 200 and another scientist estimates the number to be 150 plus or minus 100 these numbers are in agreement – and they are not different by 100% as too many would say (you can see the headlines now!). Uncertainty is part of our daily life – why then is it so difficult for us to understand? (For more discussion see Talcott, 1992.) Surely this concept could be included in public school curricula – in almost every subject. Examples of the use of uncertainty in everyday life include: weather reports, surgeons estimates of risk and census reports. Adding this concept would be a trivial extra in terms of time and effort but a valuable addition to our understanding of the world we live in.

There is a threshold for most folk – should we worry about risks less than one in a thousand, one in a million, one in a billion. ...? Probabilistic thinking allows the development of a threshold. We could then easily realize that there is risk in everything we do and that the important thing to do is determine which are the most serious and apply our scarce resources in those areas. Thinking probabilistically would help us to conceive of the most likely situation rather that always looking for the worst case. As as example the highest daily rainfall happened in Texas in 1968 – 43 inches. Taking this as even likely would imply that in the US there could be over 15 000 inches of rain in a year. But is that likely? Of course not? How did we reach that conclusion? We used probabilistic thinking. This kind of thinking is part of our everyday life. So why not apply that kind of thinking to the analysis of environmental problems? Why do we resist this? A little education would be most helpful here.

What is needed for statistically meaningful proof of the consequences of environmental contamination? It took over ten years and millions of rats and human subjects to determine that smoking cigarettes caused lung cancer. The lifetime risk level for cigarette smokers was determined to be of the order of 10%. The risk levels for most environmental contaminants is much lower. Generally if the risk level is ten times lower it takes of the order of one hundred times more subjects exposed to produce a statistically meaningful study. Very quickly we realize that there are not enough people in the world to determine with statistical accuracy what the consequences of environmental contamination is. Asking for direct proof, bodies on the floor, of the consequences of environmental contamination is generally a transcientific question (Weinberg, 1972). It is a scientific question, and a plausible one, but not one that will likely be answered with direct evidence.

6. Scientific peer review

In the process of developing regulations to protect human health and the environment, the USEPA considers a wide variety and range of information and scientific data that sometimes appears to be conflicting. One endeavor where scientific data are important is in the assessment of the risk to human health from environmental contaminants. For no contaminant are data complete and there are usally significant gaps and uncertainties in the available data. To assess any risk based on only incomplete data, scientific judgment is needed. However, scientists can and sometimes do disagree on specifics of individual judgments that are a part of the assessment process. This disagreement can lead to controversy. Controversy is expected and is a desirable part of the intellectual process in the development of new ideas. This should not be confused with controversy such as that in non-science areas. Examples of such scientific controversies include: discussions regarding the modification of risk assessments in cases where the contaminant is a promoter, leads to a non-fatal rather than a fatal cancer, or acts by a mechanism of carcinogenesis that occurs in the test animal but not in humans.

There are two main reasons that the USEPA seeks peer review: need to improve scientific quality of risk assessments and desire to obtain a seal of approval from an independent, 'objective' group of scientists to shield and deflect criticism from adversaries.

Consider the scientific method – hypothesis, experiment planned, experiment conducted, experiment replicated, hypothesis confirmed, proved wrong or find an unexpected result. Peer review constitutes a sixth step in this methodology.

In the 1970s a variety of developments created a context for the development of what is now the Science Advisory Board: fracturing of a scientific consensus on environmental matters, explosion of public ecological awareness and concern, a new openness in scientific advising and need for independent outside scientific advise.

In the context of the relationship between science and policy it can be observed that an advisory committee such as the Science Advisory Board provides a bridge between the scientific community and policy makers in regulatory agencies.

One major role of the USEPA's Science Advisory Board (SAB) is to give informed opinion based on independent review of backgroung documents for regulatory decisions and on issues that arise in the analysis of scientific data relating to environmental health risk assessments.

The SAB reviews documents and issues, especially in cases of conflicting claims and advises the administrator of the USEPA regarding their perspective on the adequacy and reliability of the technical basis for rules and regulations.

Some of these characteristics involved in determining the importance of scientific advice include: the increasingly fragile nature of the decision making process, the increasing view that risk assessment and risk management are a continuum rather than separate entities, an increasing reliance by the USEPA on the scientific community, a longer time for decisions, and the substitution of scientific data for conservative assumptions.

In a 1983 speech by William Ruckelshaus, then administrator of the USEPA, at the National Academy of Sciences he observed that science and policy are uneasy partners. There is a tension between them that causes court fights, lawsuits and public concern. In the 1970s policy drove science. But Ruckelshaus suggested that there are two areas – risk assessment and risk management and this paradigm described a new relationship. Scientists do risk assessment and provide the policy maker or risk manager with the scientific information. Although this is not an entirely clear distinction it provides a new approach to the relationship between science and policy.

Examples of contributions from the Science Advisory Board that provided impact on policy at the USEPA are: national ambient air quality standard (NAAQS) for particulates in emphasizing the importance of particulates that are smaller than 10 microns; 1986 decision to prohibit hazardous waste incineration at sea; development of guidelines for risk assessment (cancer, chemical mixtures, development effects, mutagenicity and exposure effects); regulating airborne radionuclides, and the proactive role taken in producing the reports on future risk and relative risk.

Peer review has now become commonplace in the regulatory landscape and it helps to ensure that administrative agencies base their regulatory decisions on reliable science.

7. Future role of science in the development of environmental regulations

Two recent reports from the USEPA discuss the future role of science in the development of regulation (USEPA, 1988, 1992). These reports are titled 'Future risk' and 'Reducing risk'. The first of these reports was produced to examine strategies research planning at the USEPA regarding long-term research needs. The report states that the USEPA is more than a regulatory agency. It is also a research agency for environment along with the National Institutes for Environmental Health Sciences (NIEHS). In addition it is a technology transfer agency responsible for sharing scientific findings with industry, state and local governments.

It was observed by the Science Advisory Board (SAB) that EPA needs to change its strategy from control and clean-up to anticipation and prevention. In the first 18 years of EPA's existence the focus was on 'end-of-the-pipe' control now new approach is needed because we cannot just move pollution from one media to another, there are new problems such as: indoor air, numerous sources of ground level ozone, the overwhelming capacity of the environment to assimilate pollution, and irreversible problems such as global warming, stratospheric ozone and species reduction. From the ten recommendations contained in the 'future risk' report the ones that are important here include those relating to the need for long term research (2), focus on problem anticipation (4), a greater emphasis on ecological research (5), a more complete understanding of exposure assessments (6), more epidemiology studies (7) and a more complete determination of the inputs to our present pollution (8).

In 1991 EPA administrator, William K. Reilly established the expert panel on the role of science at EPA as an independent advisory committee to evaluate how EPA can meet the goal of using sound sciences as the foundation for agency decision making. The findings and recommendations of this panel were published in (USEPA, 1992). The direction of EPA that was used as a context for this panel were the following goals: gathering and analyzing the data needed to evaluate environmental risks and trends, measure environmental results and inform the choices of institutions and individuals throughout society; promoting and supporting innovative technological solutions to environmental problems; encouraging and conducting research that improves our understanding of health and ecological risks; providing objective, reliable, and understandable information that helps build trust in the USEPA's judgment and actions and sharing research findings and innovative technologies with other nations. The panel held three public meetings and interviewed more than 30 individuals from the USEPA, other government agencies, industry, environmental groups and other organizations. They focused on five topics that are crucial to obtaining and using sound science for credible decisions: the mission and direction of USEPA science; the quality of science at the USEPA, the quality of scientists at the USEPA; how the budget process affects science at the UPESA and how the USEPA uses science in decision making. In the following the findings and recommendations in these five areas will be discussed in order.

7.1. The mission and direction of EPA science

The panel found that EPA needs its own strong science base to provide the scientific foundations for effective decision making, that regulations are based on both technical and policy considerations. Most people outside the UPEPA do not recognize that the USEPA has made strong science a priority and that the USEPA lacks a coherent agenda for science throughout the agency. The panel recommended, among other things, that the USEPA should make a deliberate effort to create and maintain the climate, culture and incentives necessary to encourage superior science, the administrator should send broad, clear and continuing signals within the USEPA, to the wider scientific community and to the public about the importance of having high-quality science to protect health and the environment and should modify its research planning process to ensure that it leads to a coherent science agenda and operational plan for science throughout the agency.

7.2. The quality of science at EPA

Issues that the panel considered were: how can the quality of science be ensured?, to whom should contract or federal scientists be accountable?, is the science the best available?, and what type of research should be done in-house? The panel found that the perception exists that science has been asjusted to fit policy, quality assurance and peer review are essential and that the USEPA needs to heed this advice, the USEPA dose not give sufficient attention to validating the models, scientific assumptions and databases it uses, the USEPA science serves primarily the USEPA program officers and regional officers and not the broad scientific community and that the use of large level-of-effort contracts in many parts of the agency does not promote or ensure the use of high-quality science. In dealing with these findings the panel, among other things, recommended that the science function within the USEPA should be enhanced and coherently organized, that the administrator should appoint a science advisor, that quality assurance and peer review should be applied to the planning and results of all scientific and technical efforts, including research, technical studies and data collection, used for guidance or decisions at the USEPA, there be a thorough review of the use of in-house contracts by the Office of Research and Development.

7.3. The quality of the scientists at the USEPA

The important issues here included current quality of the USEPA scientists, environment and incentives to attract, nurture and keep high quality scientists. The panel found that there was an inadequate infrastructure and lack of long-term support for the USEPA scientists, many of the USEPA scientists do not perform laboratory research, but rather provide technical and regulatory support, there are insufficient incentives to reward the production of high-quality science. The panel recommended that EPA immediately recruit four to six research scientists and engineers with world-class reputations, expand the use of career development paths and ensure that EPA scientists are continually exposed to and challenged by current knowledge and methodologies.

7.4. How the budget process affects science at the USEPA

The focus here was on the relation of the budget process to scientific quality, long term strategic planning and anticipatory research efforts. The panel found that the budget does not reflect the importance of cross-media approaches, the congress, OMB and others do not understand the strategic direction of the USEPA, the USEPA science programs currently are unable to carry out prolonged research of anticipate future information needs, the technical support of regulations does not receive the recognition it deserves and that scientists outside the USEPA regard the USEPA contracts as the least desirable funding mechanism among those of all federal agencies because of inconsistencies in direction and lack of funding continuity. The panel recommended, among other things, that a comprehensive, long-term scientific strategy should drive annual budget decisions, the USEPA should place greater emphasis on cross-media planning and scientific efforts, the USEPA needs to clearly define the mission of each of its laboratories and that EPA must include long-term funding and support for anticipatory scientific studies.

7.5. How EPA uses science in decision making

In this area the panel focused on the role of science in the development of regulations and policy, the identification of university and industry science, where in the decision making process science should enter and how scientific input should relate to other inputs such as political. The panel found that the USEPA has not always ensured that contrasting, reputable scientific views are well-explored, the USEPA science is perceived as adjusted to fit policy, uncertainties make it difficult to give yes/no answers to some questions, the USEPA does not scientifically evaluate the impact of its regulations and that the interpretation and use of science is uneven and haphazard across programs and issues. The panel recommended, among other things, that the climate and culture within the USEPA should emphasize that science should never be adjusted to fit policy, the USEPA should implement a widely advertised, open process whereby the scientific view of all parties are heard prior to regulatory decisions and that EPA should implement a decision making process whereby science advice is provided as early as possible.

8. Discussion and conclusions

The former director of EPA's Science Advisory Board, Terry Yosie, observed a few years ago that we have entered the post-conservative risk assessment period at EPA. He suggested that the decision making process for environmental regulations would become increasingly fragile and that we would be trying to serve even wider constituencies. Clearly the broader use of statistically based science will help this process. He also predicted an increasing reliance by EPA on the scientific community and that we would see more substitution of data for conservative policy assumptions. As discussed here, this is happening.

As a general observation, statistically based science enters the environmental regulatory development process in many different places. As with any human endeavor, this one could be improved. There could be more planning using statistical methods of monitoring surveys and better integration of the data that has been collected. A more complete use of statistical methods and statistically based science would strengthen the environmental regulatory process. Finally, a single organization could be formed to collect, analyze and publish data and information on the state of the environment.

A note of perspective is important concerning the role of statistically based science in the environmental regulatory process. First the relationship between science and policy needs to be kept clear, including where each is being used. Also important to realize is that there are several inputs to the regulatory process including: science, policy, economics, social, political, legal, engineering and feasibility. Each of these inputs has a role and it is important for the environmental regulatory decision maker to balance all these inputs and the competition between them.

Acknowledgement

The author is grateful for the critical review comments and suggestions of John Warren.

References

Burack, T. S. (1987). Of reliable science: Scientific peer review, federal regulatory agencies and the courts. *Virginia J. Natural Resources Law* **7**(1), 27–110.

Cothern, C. R. and D. W. Schnare (1986). The limitations of summary risk management data. *Drug Metabolism Rev.* **17** (1, 2), 145–169.

Cothern, C. R., M. Mehlman and W. Marcus, eds. (1989). *Risk Assessment and Risk Management of Industrial and Environmental Chemicals.* Princeton Scientific, Princeton, NJ.

Cothern, C. R. (1993a). Introduction and overview of difficulties encountered in developing comparative rankings of environmental problems. In: C. R. Cothern, ed., *Comparative Environmental Risk Assessment.* Lewis, Chelsea, MI.

Cothern, C. R., ed. (1993b). *Comparative Environmental Risk Assessment.* Lewis, Chelsea, MI.

Gart, J. J., K. C. Chu and R. E. Tarone (1979). Statistical issues in interpretation of chronic bioassay tests for carcinogenicity. *J. Nat. Cancer Inst.* **62**(4), 1–24.

Nussbaum, B. (1992). Office of Air and Radiation, USEPA. Private communication.

Nussbaum, B. and D. L. Rachlin (1992). Presentation Ann. Meeting Amer. Statist. Assoc., Office of Air Radiation, USEPA, Boston, MA.

Schnare, D. W. (1992). Office of Ground Water and Drinking Water, US Environmental Protection Agency, Private communication.

Talcott, F. W. (1992). How certain is that environmental risk estimate? *Resources For the Future* **107**, 10–15.

USEPA (1987a). *ASA/EPA Conf. on Interpretation of Environmental Data*, IV Compliance Sampling, October 5–6. US Environmental Protection Agency, Office of Policy, Planning and Evaluation, Washington, DC, EPA-230-03-047.

USEPA (1987b). Unfinished business: A comparative assessment of environmental problems. US Environmental Protection Agency, Office of Policy, Planning and Evaluation, Washington, DC.

USEPA (1987c). The risk assessment guidelines of 1986. US Environmental Protection Agency, Office of Health and Environmental Assessment, Office of Research and Development, Washington, DC, EPA-600-8-87-045.

USEPA (1988). Future risk: Research strategies for the 1990s. Science Advisory Board, US Environmental Protection Agency, SAB-EC-88-040, Washington, DC.

USEPA (1990). Reducing risk: Setting priorities and strategies for environmental protection. Science Advisory Board, US Environmental Protection Agency, SAB-EC-90-021, Washington, DC.

USEPA (1992). Safeguarding the future: Credible science, credible decisions. EPA/600/9-92/050, US Environmental Protection Agency, Washington, DC.

Weinberg, A. (1972). Science and transcience. *Minerva* **10**, 209–212.

G. P. Patil and C. R. Rao, eds., *Handbook of Statistics, Vol. 12*

26

Environmental Regulatory Statistics

Charles B. Davis

Abstract: Doing statistics in the regulatory arena brings challenges: the work is interdisciplinary, the setting is often adversarial and the traditions, regulations and guidance are often unsatisfactory from a statistical perspective. Several areas of application are described, including ground-water monitoring at hazardous and solid waste management facilities and related rules for the chemical laboratory analysis of environmental samples. Issues and methodologies of particular relevance in regulatory statistics are discussed, including simultaneous inference, dealing with censored data, and 'detection limit' definition problems. Recommendations for improvement and research needed both in statistical methodology and in regulation are given.

1. Environmental regulatory statistics

'Environmental regulatory statistics' refers to the meeting ground between statistics and the regulatory process. This interface can occur explicitly, as in the Code of Federal Regulations (CFR) statement 'Whenever hazardous constituents ... are detected .... Detected is defined as statistically significant evidence of contamination ....' (40 CFR §264.91(a)(1)) It can also occur in implicit, less-than-obvious ways, as in 'The owner or operator must determine the ground-water flow rate and direction in the uppermost aquifer at least annually.' (40 CFR §294.98(e))

Some of the more prominent areas involved in regulatory statistics include

- ground-water monitoring at hazardous waste management facilities under 40 CFR Parts 264 and 265 and at municipal solid waste facilities under 40 CFR Part 258 (see, e.g., Loftis et al., 1987; McNichols and Davis, 1988; Huber, 1990; Starks and Flatman, 1991);
- leak detection monitoring for underground storage tanks (see, e.g., Flora, 1991);
- characterization of the contamination at Superfund and other such sites, and evaluating remediation progress and the attainment of remediation standards at such sites (see, e.g., Flatman and Yfantis, 1984; Schweitzer and Black, 1985; Millard, 1987; US EPA, 1989a);

- evaluation of new characterization/remediation methodologies for inclusion among 'approved' procedures which a cautious project manager may use without fear of legal jeopardy (see, e.g., Davis et al., 1992);
- assigning responsibility for contamination and resultant remediation costs at contaminated sites (see, e.g., Gibbons, 1990a);
- proposed monitoring of air quality under the Clean Air Act of 1990 (see, e.g., Council on Environmental Quality, 1992, and 56 FR 24468, May 30, 1991);
- the establishment of maximum concentration levels (MCLs) for environmental concentrations of toxic substances (see, e.g., ENVIRON, 1986); and
- data analysis and decision algorithms for environmental laboratory chemistry, including such matters as waste characterization protocols and laboratory quality assurance/quality control (QA/QC) and calibration (see, e.g., Currie, 1968; Hubaux and Vos, 1970; US EPA, 1986b; Clayton et al., 1987; and Gibbons et al., 1991).

Several of these areas are discussed in other chapters in this volume. This chapter will concentrate on the nature of 'doing statistics' in the regulatory arena.

1.1. Goals of this chapter

This chapter is intended for applied statisticians and others with a good background in statistics. The goals of the chapter are:

- to describe the nature of regulatory statistics;
- to describe current difficulties arising from inadequate statistics embedded in regulatory thought in the recent past and present and to suggest improvements;
- to describe specific methodologies and issues, in particular regarding simultaneous inference and the treatment of censored data;
- to make recommendations for improvements in practice and regulation and for research needed in statistical methodology;
- to suggest less obvious items which are in need of statistical attention; and
- to provide sources of information for statisticians becoming involved in regulatory statistics.

The selection of problems and examples included is not exhaustive, but rather reflects the author's experience and biases. Likewise, judgements expressed in this chapter are those of the author, and do not necessarily reflect opinions or policies of the editors, the publisher, the reviewers, or any other organization.

1.2. Features of environmental regulatory statistics

There are three main aspects which influence the practice of 'regulatory statistics'.

- *Regulatory statistics is highly interdisciplinary and problems are not always stated clearly.* The first task of a statistician working on any applied problem is to ascertain just what that problem is. In the regulatory arena, there are often conflicting demands placed on statistics or statisticians: one is charged to derive procedures which satisfy various, sometimes contradictory regulatory pronouncements while protecting the client's interests. This is to be done, of course, while maintaining one's scientific integrity. A detailed discussion is given in Section 2.1 on some of these conflicts in the areas of ground-water monitoring statistics and laboratory chemistry. The reader is urged to look as deeply as possible into the substantive issues surrounding the problem at hand before doing any calculations. Doing so generally involves background reading, asking many 'dumb questions', and reviewing previous statistical work in the area quite critically. It often happens that a statistician's 'dumb question' does not have a ready answer other than 'that's how we've always done it'; this presents yet another opportunity for developing new knowledge and understanding in the field of application involved.

In particular, in doing regulatory statistics, one must understand the applicable regulations, guidance, and traditions. If one wants to do something non-standard, one must expect resistance. The regulatory response to this author's first submission of a statistical program for monitoring at a hazardous waste management facility, which used novel techniques for facility-wide false positive rate control (Davis and McNichols, 1987) which are now beginning to appear in guidance (US EPA, 1992), was along the lines of 'This is interesting work; keep it up; but of course we can't approve it at this time.' The client's reaction, given a two week deadline for a revised submission of a multi-volume proposal after months of waiting, was less printable.

A major part of statistical consulting is always educational: educating the statistician about the problem to be solved and educating the client and other recipients of the statistician's work about the appropriateness of the solutions offered. The more closely the recommended solution seems to correspond to something which is traditional or in regulatory guidance, the more easily it will be found to be acceptable. In situations where it is inappropriate to follow the guidance or tradition, the statistician will want to be aware of that guidance and tradition and of the fact that departing from it has its consequences. Sometimes it becomes important to ask not what the best solution to a problem is, but rather how undesirable are the implications of simply following the tradition; in other words, deciding whether or not to fight all of the battles which present themselves.

- *Regulatory statistical work is often done in an adversarial climate.* In adversarial situations, one should follow Einstein's advice to make everything 'as simple as possible, but not simpler.' Complicated statistical models and analyses can impress one's peers and generate publications. However, they can also be distrusted by regulators and the regulated community alike; the more non-standard the model, the greater the neeed for patient explanation. A procedure which looks like a t-test is often more easily accepted than something more complicated; nonetheless the analysis behind the recommendation of that procedure should be adequate to ensure that it solves the problem at hand.

One aspect of this is that the parties in an adversarial situation are often not of equal technical sophistication. Where possible, a statistician involved in such a situation should adopt the role of an educator, even if the 'class' is hostile.

Another aspect is that one must work closely with clients to make sure that they understand the recommendations being made, whether or not the statistician will be present during any negotiations and/or involved personally in writing proposals or other documents. Surprises arising during a negotiation or arriving in a 'notice of deficiency' are not welcome, particularly surprises which could have been avoided by adequate preparation and communication with the client.

- *Statistically unsound regulation, guidance and tradition abound.* The difficulties caused by the previous two points are intensified considerably by the great amount of statistically unsound regulation, guidance and tradition to be found. It is quite necessary to make sure that any statistical procedures which are being considered are actually appropriate. Simply because a procedure has appeared in regulation or guidance or in a journal article is no guarantee of either its appropriateness or its correctness. Statistical work in subject matter journals often receives less stringent review than that in professional statistical journals and there have been many instances of statistical work appearing in regulation or guidance with no professional review.

For example, in ground-water monitoring there is a firm tradition of comparing several 'downgradient' monitoring wells with one 'upgradient' well. The Davis and McNichols (1987) procedure mentioned previously is a partially sequential multiple-comparisons-with-control procedure which allows for the analytical evaluation of overall significance level and individual comparison power. The article actually refers to comparisons on several occasions, but the same computations could apply to comparisons at several locations. However neat the technique may be, though, its use is appropriate only so long as inter-well comparisons are appropriate. It is the author's experience that nearly all measurements in ground water of naturally occurring chemicals and of 'indicator parameters' show significant amounts of spatial variability, whether or not the facility being monitored actually contains any sources of contamination. Therefore, significant differences between downgradient and upgradient measurements are not at all a reliable indication of contamination arising from the operation of the facility, and the procedure mentioned above is not appropriate for inter-well comparisons in most such circumstances. (The problem of dealing with spatial variability is discussed further in Section 4.1.)

Hence, even though the technique is quite useful where appropriate, appeared in a respectable journal and is directly related to procedures being considered for inclusion in future guidance (US EPA, 1992), it is up to the user to make sure that it is appropriate for any particular application; chances are that it is not.

Similarly, a previous version of the statistical rules in 40 CFR Part 264 stated that an alternate statistical procedure would be acceptable so long as it 'reasonably balances the probability of falsely identifying a non-contaminating regulated unit and the probability of failing to identify a contaminating regulated unit in a manner that is comparable to that of the statistical procedure described in paragraph (h)(1)(i) of this section (40 CFR §264.97 (h)(1)(ii), 7-1-86 edition). This statement would have some merit if the procedure described in (h)(1)(i) (the CABF t-test, discussed further in Section 2.1.1) were actually capable of achieving a balance between type I and type II error rates. Many articles appeared in the ground-water monitoring literature describing ways to carry out tests which achieve comparable nominal error rates; very few actually examined in depth whether or not these procedures are reasonable decision models for the problem at hand, and very few pointed out the extremely high type I error rates of the CABF t-test as implemented in that regulation.

These problems abound. Some of the more interesting past problems and some of the more pressing current problems are outlined in Sections 2 and 3 which follow. These would be just 'war stories' if they were not still major sources of difficulty. Perhaps these will be resolved in the near future; then one should wonder about yet unsolved or undetected problems.

2. Doing regulatory statistics

Several of the areas in which 'environmental regulatory statistics' becomes involved are discussed in other chapters of this volume. It will be instructive to examine one area in some detail in this chapter. The area chosen is ground-water monitoring at RCRA facilities. This area is chosen because it involves the most formalized set of statistical rules with the most widespread impact and because it demonstrates many of the aspects of doing regulatory statistics outlined above.

2.1. Example: Ground-water monitoring at regulated facilities

Under the Resource Conservation and Recovery Act of 1976 (RCRA), the US Environmental Protection Agency (EPA) has developed regulations for landfills and other facilities which are used to manage or dispose of hazardous wastes and of municipal solid wastes. Among these regulations are provisions for the monitoring of ground water in the vicinity of such facilities. These regulations were one of the topics discussed by Millard (1987) in his *American Statistician* article entitled 'Environmental monitoring statistics and the law: Room for improvement'.

2.1.1. Regulatory history and traditions

Statistical rules for such monitoring at hazardous waste management facilities were promulgated in the Federal Register in 1980 (45 FR 33232, May 19, 1980) and 1982 (47 FR 32350, July 26, 1982), and became part of the *Code of Federal Regulations*, cited as 40 CFR §265.90-94 (for interim status facilities) and 40 CFR §264.90-101 (for permitted facilities) respectively.

A brief history of the statistical rules follows. This history may seem curious and even a bit bizarre to statisticians who have not been involved in this area; nonetheless, it is the setting in which statistics is done in this application. These are federal rules; the regulations and guidance in force in various states were formed at various times in this history, using various sources and precedents.

Initially, all facilities were in interim (Part 265) status; therefore, the Part 265 statistical rules appeared first. These rules for detecting contamination are simple: use pH, specific conductance, total organic carbon (TOC), and total organic halides (TOX) as 'indicator parameters'; take quarterly background samples for one year from at least one upgradient well, obtaining at least four replicate measurements for each sample on each occasion; pool the resulting measurements (at least 16) to obtain the background data pool; and compute the mean and standard deviation. Upgradient means hydraulically upgradient; the water level in upgradient wells will be higher than the level in downgradient wells in the same aquifer, so that water is assumed to flow from the upgradient to the downgradient direction.

Subsequently, at least semi-annually, obtain samples from each well, upgradient and downgradient, and compare the measured values with the background mean and standard deviation through a Student's *t*-test at the 0.01 level of significance. If an upgradient well shows a significant increase (or difference, for pH), submit this information. If a downgradient well shows a significant increase (or difference, for pH), resample the wells showing the difference, split the samples in two and 'obtain analyses of all additional samples to determine whether the significant difference was a result of laboratory error.' If the resamples confirm the significant change, report that the facility may be affecting ground-water quality and immediately develop and submit an assessment plan. If the assessment finds no hazardous constituents in the ground water, return to the detection monitoring scenario described above. (Summarized from 40 CFR §265.90-94, 7-1-91 edition.)

The statistically aware reader will note several difficulties immediately:

- several variance components are usually present, including spatial, temporal, sampling and laboratory or measurement variability; the required pooling of background data would be valid only if spatial, temporal and sampling variability were negligible, which is decidedly not the case with these measurements at most facilities;
- in particular, replicate measurements made on the same sample are almost always nearly the same for pH and specific conductance and are very highly correlated through having a very small proportion of between-replicate variance for the other indicator parameters, so that the effective number of degrees of freedom involved will be more nearly three than fifteen;
- if only one upgradient well is used (the minimum required; therefore, the most common situation), the background data pool does not incorporate any component of spatial variability; whereas
- if more than one upgradient well is used, the naive data pooling described above results in a variance estimate which estimates the wrong quantity, which is nearly always too small (see Section 4.1 below);
- no provision is made for normality testing, the use of non-parametric tests, or data transformation; and
- no adjustment is made to critical values to allow for the effects of the multiple comparisons between downgradient and upgradient wells resulting from the several downgradient wells and the four analytes, or for the effect of the resample allowed.

Furthermore, the four indicator parameters selected often do not serve well as early warning indicators of incipient contamination of the ground water by leachate from the facility. The differences in measurement means for these parameters between leachate and ground water are often relatively small by comparison with background variability, so the power of statistical tests for shifts in the ground-water means for these parameters is rather less than it would be with a better choice of indicator parameters. Such a better choice could be made using site-specific information.

Not surprisingly, many facilities found 'confirmed statistically significant' changes with this program, which were later determined to be unrelated to the presence of any hazardous constituents in the ground water. The result was and continues to be a considerable financial and legal burden on the regulated community and a surely unanticipated flood of work for the regulatory agencies.

In 1986, the RCRA ground-water technical enforcement guidance document (US EPA, 1986a) (the TEGD) appeared, directed mainly toward Part 265 hazardous waste facilities. Chapter 5 of the TEGD deals with statistical analyses. After a very brief section on data reporting, the title of Section 5.2 is curious and revealing:

> Introductory topics: Available *t*-tests, definition of terms, components of variability, validity of the *t*-test assumptions, false positives versus false negatives, and the transition to permitting.

In particular, §5.2.1 discusses 'available *t*-tests'. Recall that the Part 264 regulation had appeared in 1982, including the CABF *t*-test mentioned in the opening of this chapter since this particular form of *t*-test had appeared in a related regulation, many facilities had been using it. The TEGD states:

> One alternative *t*-test, which has been recommended for use, is referred to as the averaged replicate (AR) *t*-test. The AR *t*-test is a preferred test for owner/operators to apply to their interim status detection monitoring data because it helps to reduce statistically caused false positives. Although special situations demanding alternative *t*-test procedures may arise, this document generally recommends the use of the AR *t*-test for maintaining compliance with the statistical analysis requirements of 40 CFR §265, Subpart F.

The AR *t*-test simply means that the single data value used in statistical analyses should be the average of replicate measurements, a conclusion which comes quite naturally from a standard variance components analysis.

After the discussion in §5.2.3 of 'components of variability', emphasizing spatial variability, the TEGD simply recommends installing additional upgradient wells and pooling their AR measurements in the naive manner described previously in order to incorporate spatial variability into the background data pool. It is important to allow for this spatial variability; after all, 'statistically significant' means 'extreme by comparison with natural (null hypothesis) variability' and the spatial component of variance is a large proportion of the total variance for many common indicator parameters, including pH, specific conductance, and TOC which are required in the Part 265 rules. However, simply including more upgradient wells, without taking into account the variance components structure involved, is inadequate. A detailed discussion follows in Section 4.1 below.

Moreover, recognizing the multiple comparisons nature of a monitoring program, the TEGD then recommends using critical values 'that have been adjusted to control the overall false positive probability for the management unit' over all monitoring wells and all four indicator parameters. The technique is the Bonferroni t-test (see, e.g., Miller, 1981), which simply replaces the significance level for each individual test, α with $\alpha \div$ (number of tests performed). The regulatory statement required a 0.01 significance level, so $\alpha = 0.01$ was used. Tables were provided; however, the table critical values were considerably in error, being off by as much as a factor of two or more. For example, for a minimal monitoring network using one upgradient well and three downgradient wells, the AR t-tests would use variance estimators with three degrees of freedom and would have $4 \times 4 = 16$ tests to spread $\alpha = 0.01$ over, so each individual comparison should use a critical value of $t_{3,0.01/16} = 11.984$; however, the TEGD table value is only 6.297. A sample power analysis is given in the TEGD; however, this power analysis fails to take the adjusted critical value into account and consequently overstates the power considerably.

The analysis further fails to take into account the resample and retest permitted. If multiple wells and indicator parameters make it more likely to 'fail a test' somewhere at the facility, the resample and retest makes it harder since the well has to 'fail' twice consecutively, not just once. It was found that, in spite of the failure to allow for spatial variability appropriately, these tests with their Bonferroni critical values and resamples could be rather insensitive to contamination. Hence, both false positives and false negatives were problems with the Part 265 rules.

Unfortunately, these Part 265 rules are still in force for interim status facilities. More unfortunately, they are still used as model for monitoring programs for municipal solid waste disposal facilities in at least one state.

The statistical rules for permitted status (Part 264) monitoring appeared in 1982. These rules offered a bit more flexibility. Although four background quarters were still mandated, one could argue that 'background' did not necessarily mean upgradient. During background monitoring, one still needed four samples on each occasion and at least one from each background well. This requirement was often fulfilled by using 'aliquots'. If the sample coefficient of variation (CV = relative standard deviation = $s/\bar{x}$) of the background value (sic) was less than 1.00, the statistical test should be the Cochran's approximation to the Behrens Fisher (CABF) t-test, using a significance level of 0.05, comparing measurements on each of four portions (often called aliquots) with the background data. If the test 'failed', a resample and retest using the same number of portions was to be done. If the retest failed, one concluded that a 'statistically significant change has occurred'.

An alternate statistical procedure could be suggested. It would be acceptable so long as it 'reasonably balances the probability of falsely identifying a non-contaminating regulated unit and the probability of failing to identify a contaminating regulated unit in a manner that is comparable to that of the statistical procedure described in paragraph (h)(1)(i) of this section.'

If the sample coefficient of variation was not less than 1.00, the regional administrator would specify a statistical test which was appropriate for the distribution of the background data, and which would 'provide a reasonable balance between the probability of falsely identifying a non-contaminating regulated unit and the probability of failing to identify a contaminating regulated unit.'

The regional administrator would specify appropriate monitoring parameters for the facility taking into account the coefficients of variation of the proposed monitoring parameters or constituents among other factors. Monitoring would take place at least semi-annually after the background period. Upon finding a 'statistically significant increase' (confirmed by resampling as described previously), the owner/operator must immediately sample all wells and determine the concentrations of all of the hazardous constituents listed in Appendix VII (now Appendix IX) of Part 264, at a cost currently around 2000–2500 dollars per well and submit an application for permit modification including an identification of all of the Appendix VII constituents found. It is not clear in the regulation that a failure to find any hazardous constituents would allow a simple return to the detection monitoring program.

Should hazardous constituents be found, the facility would enter into compliance monitoring. A new list of monitoring parameters would be specified and similar statistical rules would apply, except that concentration limits for the Appendix VII compounds could be specified. The Appendix VII analysis must be done at least annually on samples from all monitoring wells at the facility. If the extent of contamination should be excessive, corrective action might be required. (This and the previous four paragraphs are a summary of rules contained in 40 CFR §264.90-101, 7-1-86 edition.)

The CABF *t*-test is somewhat curious. Perhaps those who developed this regulation realized, correctly, that a monitoring data set involving only replicate or aliquot measurements taken from a single sample obtained at one time from one monitoring well would show only analytical variability, whereas a background data set containing replicate or aliquot measurements from samples from one or more wells taken at four different times would also show variability due to temporal and sampling effects. Spatial effects would also appear if more than one background well was involved. Whether the causes for the greater variability in background data sets were understood, empirical evidence was certainly available. The quite incorrect response was that this must be a 'Behrens–Fisher' problem. The CABF *t*-test uses a method of adjusting ordinary *t*-test critical values to accommodate heteroscedasticity which is fairly obscure today. (The more common approach is, of course, the separate variance estimate *t*-test with adjusted, perhaps non-integer, degrees of freedom.) An appropriate response would have been to sort out the variance components, which would have led to the AR *t*-test of the TEGD if spatial variance components were ignored.

Again, the faulty statistics in the regulation presented a considerable financial and legal burden on the regulated community. The confirmatory resample and retest were not adequate to overcome the false positives generated by $\alpha = 0.05$ statistical tests for each indicator parameter at each well with a statistical test based on a model which omits spatial variability and all other variance components except for analytical variability (i.e., measurement 'error').

A new final rule for Part 264 monitoring appeared in 1988 (53 FR 39728, Oct. 11, 1988). Discussions in the 'Advance notice of proposed rulemaking' (51 FR 29812, Aug. 20, 1986) and the 'Proposed Rule' (52 FR 31947, Aug. 24, 1987) acknowledge the

positive rates, as suggested by Davis and McNichols (1987) (see also Gibbons, 1990b, 1991a,b and Gibbons and Baker, 1991). In addition, the addendum re-emphasizes the importance of using a small number of well-chosen constituents or indicator parameters in controlling facility-wide false positive rates. The addendum is to be revised somewhat before its release in final form, which is expected to occur sometime during 1993 (personal communication, J. Brown, US EPA Office of Solid Waste, 1992).

Further comments on the final rule and the IFGD were given by Davis and McNichols (1989). Gibbons (1989) also provides comments on the final rule.

2.1.2. Developing statistical ground-water monitoring programs

As stated previously, a statistician contemplating doing work in regulatory statistics must understand the regulatory setting and traditions in which the work is done, as his or her work will certainly be reviewed against those. Therefore, the regulations and guidance documents cited in Section 2.1.1 should be considered mandatory background material and at least some of the other references cited there should be reviewed. In addition to these materials, there are the technical considerations involving the nature of the measurements. The development of an adequate model for the types of variability found in ground-water measurements is discussed in several of those references; additional discussions are given by Keith et al. (1983), Nacht (1983) and Montgomery et al. (1987).

One feature of the measurements of many ground-water monitoring parameters is the presence of 'non-detects', or observations reported as '< DL', where 'DL' stands for a reporting limit, often called a detection limit. These observations can complicate the development of programs considerably. Both the proper treatment of such observations and the notion of 'detection limit' itself are the subjects of some confusion and controversy. A detailed duscussion of these issues follows in Section 5.3; it is important to recognize, however, that the concept of 'detection limit' itself involves statistical issues. A natural reaction is to argue that any 'detection' (or at least a confirmed detection) of a non-naturally occurring toxic compound should itself be enough to trigger assessments and contemplation of remediation. Contrary to that is the experience that indicates that 'detections' can and do routinely occur as a result of laboratory contamination and as false positives due to measurement variability; see, e.g., Hurd (1987) and Gibbons (1987). It is imperative that such false positives be accounted for adequately in the statistical program in some fashion, either through explicitly studying the statistical properties of the particular measurements involved or by treating the measurement process as a 'black box' which generates sporadic 'hits' through a random (possibly Poisson) process.

Having reviewed the available data, applicable regulations and local regulatory idiosyncracies and having considered the technical sophistication of those who will be presenting and defending and of those who will be reviewing the proposed program, the statistician produces a well-written proposal, explaining the rationale for choices made, ensuring to the extent possible under the current applicable regulation that the facility-wide false positives are under control and that the program has adequate power for detecting real contamination. If it is not possible to develop such a program under the applicable regulations, then the client or others affected are advised, and alternatives, perhaps involving legal or political action, are discussed.

Most submissions will be reviewed leisurely. One common response is a request for clarifications, possibly in formal negotiation sessions. Another is a 'notice of deficiency', outlining objections to the proposal; this is typically followed by revisions and resubmissions. Statistical issues are usually not addressed in public hearings, in spite of the great impact which matters such as facility-wide significance level and power have on the efficacy of the monitoring program.

In addition to the usual technical skills and background and the understanding of the regulatory background discussed above, the statistician involved in this area can benefit from

- understanding the technical background of other professional involved (chemists, hydrologists, etc.),
- reading and thinking quite carefully and critically, especially when statistical terms are used by non-statisticians, as meanings are not always shared,
- exercising as much diplomacy as possible, especially with regard to the various pressures which can affect adversaries in this area which can become emotionally exciting, and
- developing work habits which accommodate short deadlines.

3. Deficient statistical regulation and guidance

The statistical rules and traditions described previously for use at RCRA-regulated hazardous waste facilities were adopted in new regulations for municipal solid waste landfills by the final rule for 40 CFR Part 258 announced October 9, 1991 (56 FR 51008, Oct. 9, 1991). It would seem that the regulatory difficulties in dealing with statistical issues outlined above will have a great impact, at least in the near future, while statisticians, the regulated community, and the regulatory agencies are coming to a resolution of these problems.

3.1. Further examples

Confused statistical thinking is found in other areas. Some are amusing to statisticians, such as the following provision from draft municipal solid waste facility monitoring rules from a state which will not be identified:

> If the background value of a constituent has a sample coefficient of variation equal to or greater than 1.00 there is evidence that the background data is not normal. The owner and operator shall determine if the non-normal condition is due to a laboratory or sampling error, poor well construction, seasonal or spatial variability, or actual site conditions requiring recalculation of background. The owner and operator shall propose a method to correct the non-normal condition.

Presumably, this provision allows for data transformation, which is not discussed otherwise in the draft regulation.

It is inappropriate here to catalog such items exhaustively. A few further examples are given below, which are from current or proposed regulation and guidance and which can have serious negative effects on the soundness of regulatory efforts.

3.1.1. The TCLP statistical test

In deciding whether a waste is to be considered toxic, the 'toxic characteristic leaching procedure' (TCLP) given in Appendix II to 40 CFR Part 261 is used. The procedure is fairly complicated, which gives ample opportunity for random variability. The decision algorithm is given in *Test methods for evaluating solid waste*, known as *SW-846* (US EPA, 1986b); the following summarizes pages 13ff of Chapter 9.

First, compute 'preliminary estimates of $\bar{x}$ and s^2'. Using $\Delta = \bar{x} - \mathrm{RT}$, where RT is the regulatory threshold for the particular chemical, derive a sample size (or a revised sample size for the new sample, etc.). Collect such a sample and recompute $\bar{x}$ and s^2 using the combined samples. If $\bar{x}$ exceeds the RT, stop; the waste is considered hazardous. Otherwise, compute a t-distribution-based upper 90% confidence limit for the mean. If the upper limit is less than the RT, conclude that the waste is not hazardous. If the upper limit is not less than the RT, but $\bar{x}$ is below the RT, re-evaluate the sample size needed and iterate until a definite conclusion is reached, or until 'increased sampling effort is impractical.'

If $\bar{x}$ is greater than s^2 (sic), one should use non-transformed data. Otherwise, one should consider square root (if $\bar{x}$ is about equal to s^2) or arcsine (!) (otherwise) transformations, comparing the transformed data with the transformed RT. (The example given makes it clear that the comparison of $\bar{x}$ with s^2, rather than with s, was intended and is not simply a typographical error.)

Discussions and formulas for stratified, systematic and composite sampling are also given.

There are several points of interests in this algorithm:

- this is a test of the hypothesis $H_0: \mu \geqslant \mathrm{RT}$ against $H_1: \mu < \mathrm{RT}$, the null hypothesis being that the waste is hazardous;
- the hypothesis test procedure is actually sequential, with a stopping time determined partially by the economics of the situation, so that the actual type I and type II error rates are quite difficult to determine;
- the normality test (or test of the appropriateness of t-type confidence intervals) involving the ratio $\bar{x}/s^2$ is not dimensionless, so that the result depends on the measurement scale adopted (e.g., mg/l or μg/l) rather than on the shape of the sample distribution;
- if the normality test actually intended were based on $\bar{x}/s$, the reciprocal of the sample coefficient of variation, the test would find a sample with the shape of an exponential distribution adequately normal;
- the suggestion of the use of the arcsine transformation is completely inappropriate for other than proportion data; and
- the regulation itself requires a test of the mean, not the median; performing the test on a transformed scale accomplishes the latter, which may give different

results from the former, especially with rather skewed distributions which are common in this situation.

It is difficult to imagine that this regulation had received adequate technical review. It should be mentioned that this procedure is described in the third edition of SW-846, which is approved for use as guidance, while for regulatory activities the second edition and only the analytical chemistry methods of the third edition are in force at this time (September 1992).

Incidentally, the use of the coefficient of variation for 'normality testing' has much tradition behind it in the regulatory arena; see, e.g., the IFGD (US EPA, 1989b). The most recent draft guidance (US EPA, 1992) is moving away from this toward a more standard use of probability plots and related tests.

3.1.2. Matrix spike corrections

In the description of the TCLP in 40 CFR Part 261, App. II (7-1-91 edition), is a requirement that laboratory measurements be corrected for analytical bias as determined from 'matrix spike recoveries'. A 'matrix spike' here refers to a sample prepared by adding a known amount of analyte (the 'spike') to actual environmental materials (the 'matrix'), rather than to reagent water. This requirement would mean that one or more matrix spikes would be prepared for each analytical batch (of at most 20 samples) by spiking a reserved portion of one of the samples. The recovery would be determined by subtracting the measured concentration in the unspiked portion from that in the spiked portion and dividing by the amount of analyte added. All of the measurements in that batch were to be 'corrected' by dividing by the (estimated) recovery. Further, proposed update I of SW-846 (US EPA, 1990) would require that *all* analytical determinations, not just TCLP analyses, receive such corrections, unless the recovery was between 80% and 120% or greater than 100% (sic). This latter proposal triggered a flood of criticism from industry; see, e.g., Syhre (1991), Stephens and Paessun (1991) and Environmental Laboratory Council (1991).

In a related study, Davis et al. (1991) examined appropriate models which could be used for matrix spike and/or surrogate bias corrections and studied conditions under which those corrections would actually reduce the mean squared error of estimation of analyte concentration. Their conclusions were that even with surrogates (non-naturally occurring compounds introduced into the sample for quality control purposes) in the same analysis, the correlation between surrogate and analyte recovery would have to be reasonably high for there to be any benefit to doing any adjustments and that the simple ratio adjustment contemplated in the proposal should be avoided. The matrix spike recovery determination involves a separate analysis; whether or not the matrix spike uses the same sample, the correlation between matrix spike and analyte recoveries would rarely be expected to be high enough for corrections based on matrix spike recoveries to provide any improvement. The expected consequence of using matrix spike corrections will be increased analytical error. In addition, the imposition of a new measurement protocol which deliberately alters the bias of measurements will render former and new data less comparable.

It is expected that the US EPA has dropped or will drop the matrix spike correction requirement from one or both of these (personal communication, J. Warren, US EPA).

3.1.3. *Nonparametric tests*

Regulations and guidance seem to adopt the attitude that rank tests are a panacea which can correct for a multitude of defects in the data, including non-normality (appropriately) and heteroscedasticity (not appropriately). In fact, the assumptions of equal spread, common distribution shape and independence of observations are as necessary for two- or multi-sample rank tests as for two-sample t-tests or ANOVAs. Further, the limitations on available significance levels will hamper efforts to deal adequately with facility-wide significance levels (McNicholas and Davis, 1988).

3.1.4. *Tests of normality and homoscedasticity*

The problems with the use of the coefficient of variation as a normality test are noted above. A further problem with the application of such tests along with tests of homogeneity of variance is that these are presented as rigorous hypothesis tests using strict significance levels. It is the opinion of this author that this is inappropriate. These tests, along with such matters as selecting transformations (e.g., the power λ for a Box–Cox power family transformation) and variables or effects to include in a linear model, ought to be done in such a manner as to make the significance level of the ultimate hypothesis test as robust as possible.

For example, Shumway et al. (1989) discussed methods for choosing the Box–Cox power λ in the presence of censored data and then using the transformed data to estimate the mean on the original scale. In their study, they generated data for which no transformation, square roots and logarithms would be appropriate; their data-dependent algorithm selected one of these transformations or a fourth root. λ, μ, and σ were all selected by maximum likelihood. Sample sizes were 20 or 50, and the (left) censoring proportion was 10% or 20% (the censoring point was fixed).

For producing nominal 90% confidence intervals on the original scale, using the data-dependent transformation, they report the delta method as being the best of three methods, in that the actual coverage was 89–90% when the normal or square root transformations would be appropriate and 82–87% for the lognormal data. However, they also reported that using the delta method *and always using the logarithmic transformation* had a coverage of 88–91% whether the data were normal (coefficient of variation $= \frac{1}{3}$) or lognormal. This study thus suggests that automatically using the logarithmic transformation is the robust treatment for this type of data *for the problem of doing confidence intervals or hypothesis tests concerning the mean on the original scale*. However, one should be aware that this study concerns the one-sample problem, not the two-sample problem, and it is not clear that the same conclusions would be found in the latter problem. The conclusions are also appropriate for the sample sizes used; it would be useful to repeat their study using the rather smaller sample sizes often found in regulatory statistics. This bias toward the logarithmic transformation is adopted in the most recent draft guidance for RCRA ground-water monitoring (US EPA, 1992).

3.2. Reasons

Why do these problems with statistics in regulation and in guidance exist? There are several reasons.

- Regulations and guidance documents are often written without the benefit of sufficient informed technical support and often do not receive adequate technical review.

Examples have been noted earlier in this chapter. In some cases, there are traditions which do not make good statistical sense, such as the coefficient of variation test for normality. Some problems may result from professional jealousies between disciplines, as between chemists or chemometricians and statisticians, or even between programs within the regulatory agencies. In other cases, state personnel transcribe federal regulation or guidance with omissions and 'improvements', or even with transcription errors which then become enforceable local regulation. Even at the federal level, technical review is an expense, budgets and schedules are often tight and qualified personnel are not always available.

Proposals from industry are technically weak as often as those from the regulatory agencies. It is not a great surprise that novel proposals are sometimes received with skepticism.

- Not all problems are adequately recognized at the time regulation is written.
- Not all methodological problems have technically satisfactory solutions waiting on the shelf.

Regulation is often produced under time and cost constraints. Frequently, the person producing the regulation will use what appears to be the best available method for a particular application, whether or not it is really designed for that application. See, for example, the revealing comments by Ross (1987) on the political realities involved in producing technical regulation.

In many cases, regulation produced under these conditions is not optimal but not terrible. Sometimes, though, existing methodology is really not adequate, and further research is needed. One such example, involving again the treatment of censored data, is the use of maximum likelihood estimators of means and standard deviations with small samples, which are then simply substituted into t-type formulas for hypothesis tests. Although much work has been done on the censored data *estimation* problem (see, e.g., Chapter 14 of Gilbert, 1987), optimality results for the estimation problem do not transfer to robustness results for the hypothesis testing problem (McNichols and Davis, 1988). A further discussion of this important problem is given in Section 4.3.

Another example along these lines is the matter of 'detection limit' definitions. As will be discussed in Section 5.3, 'detection limits' are defined statistically and those definitions are technical. The definitions are not created with regulatory applications in mind and the resulting mismatch has caused difficulties.

3.3. The statistician's response

What is a statistician to do in this setting?

The first thing is to repeat a recommendation given previously: look as deeply as possible at the traditions and the technical background of the problem and do not assume that published or traditional statistical techniques are appropriate. If the regulatory requirements are not technically correct, a simple simulation study can often reveal clearly the extent of the consequences. It may be that a particular difficulty is 'not worth fighting' for a particular project. This response may be short-sighted, however, if the consequences could possibly be more severe under other conditions, since regulatory and legal decisions often rely at least as much on precedent as on principle.

If the problem at hand is worth pursuing, do not hesitate to contact those persons responsible for issuing the regulations involved. The regulatory climate is fluid and it does happen that private decisions are made which can be precedent-setting and have considerable implications for statistical practice. Such decisions may have been communicated with regional and state offices, but it is a mistake to assume that local, state, or regional regulatory personnel are aware of them.

If the matter is serious enough, it may be worth writing an article or presenting a paper. *Ground Water*, *Ground Water Monitoring Review*, *Environmental Science & Technology*, *EnvironMetrics* and *Journal of Environmental Regulation* are receptive to articles about the interface between technology and regulation in the environmental area. The Annual Symposium on Waste Testing and Quality Assurance, jointly sponsored by the American Chemical Society and the US Environmental Protection Agency, usually meeting in July in or around Washington, DC, presents a forum for papers on technical areas impacting RCRA and other regulatory efforts and provides sessions unveiling the latest in regulatory thinking on related issues.

4. Particular issues and methods

There are a few matters of statistical practice which deserve special attention in this chapter, as they are of particular concern in regulatory statistics. These have been mentioned previously; greater detail is provided here.

4.1. Spatial variability

As discussed in Section 2.1, the regulatory tradition has been to give at most passing attention to the presence of naturally occurring spatial variability in the measurements of indicator parameters used in monitoring programs.

Three of the four Part 265 indicator parameters are among those having the highest proportions of spatial variability. In a recent analysis of data from an East Coast site which had not received any waste, the spatial variance component accounted for around 90% of total variance for pH and specific conductance and around 47% for TOC. At a facility in the Midwest, the spatial component for specific conductance contributed around 96%. At a larger Midwestern facility, with monitoring networks in three different strata, the spatial component for TOC contributed 27%, 34% and 20% in the three strata.

Other potential indicator parameters, which have been used for Part 264 monitoring programs, exhibit similar patterns. At the same facilities, the estimated spatial variance proportion ranged from 69% to 98% for chloride, 76% to 97% for sodium, 85% to 91% for sulfate, 68% to 80% for iron and was around 75% for manganese. These facilities were not selected to be exceptional in this regard; they were simply those whose data was most recently examined by this author. At each of these facilities, the only analytes examined which did not show such predominant spatial variance components were those whose measurements were all or nearly all non-detects. In each case, the variance components analysis was preceded by an appropriate data transformation, chosen to make standard deviations approximately homogeneous across the facility or stratum and to make the intra-well data distributions approximately normal.

In spite of this ubiquitous spatial variability, the regulatory tradition of performing 'upgradient–downgradient' comparisons for detecting possible contamination remains very strong at this time (September 1992). For example, in a draft state rule implementing the federal Part 258 regulations which extend ground-water monitoring programs to municipal solid waste facilities, we find this definition:

> 'Background' means *the* concentration or level of a substance which exists in the environment at or regionally proximate to a site that is not attributable to any release at or regionally proximate to the site. (emphasis supplied)

It is extremely simplistic to think in terms of *the* background concentration of a substance if the spatial variance component contributes 90% or 60% or even 30% of the total variance. Using just one background well for statistical comparisons is virtually guaranteed to produce unanticipated high false positive or false negative rates, depending on whether the concentration at the background well happens to be relatively low or high compared with the concentrations at the downgradient wells. At many facilities, it will produce both high false positive rates and high false negative rates at different downgradient wells.

Using multiple upgradient wells with the naive pooling suggested in the TEGD does not work; the naive variance estimator underestimates the appropriate variance. Assume the homoscedastic linear model

$$Y_{ij} = \mu + W_i + T_j + S_{ij}, \tag{4.1}$$

where Y_{ij} represents the measurement (or mean of aliquot measurements, as appropriate) from the i-th background well on the j-th background monitoring occasion. μ is the overall mean of all measurements at all wells. The random effects are the well-to-well differences represented by W_i, the parallel time-to-time component represented by T_j and the sum of sampling variability and laboratory analytical error represented by S_{ij}. Each of these has a variance, denoted by σ_W^2, σ_T^2 and σ_S^2 respectively; those variances can be estimated by a variance components analysis.

The naive pooled variance estimate suggested in guidance will use the sum of squares

$$\sum_{i=1}^{B}\sum_{j=1}^{Q}(Y_{ij}-\bar{Y}_{..})^2$$
$$=\sum_{i=1}^{B}\sum_{k=1}^{Q}(\bar{Y}_{i.}-\bar{Y}_{..})^2+\sum_{i=1}^{B}\sum_{j=1}^{Q}(\bar{Y}_{.j}-\bar{Y}_{..})+\sum_{i=1}^{B}\sum_{j=1}^{Q}(Y_{ij}-\bar{Y}_{i.}-\bar{Y}_{.j}+\bar{Y}_{..})^2$$
$$=\mathrm{SS}W+\mathrm{SS}T+\mathrm{SS}S. \tag{4.2}$$

By writing

$$\sum_{i=1}^{B}\sum_{k=1}^{Q}(\bar{Y}_{i.}-\bar{Y}_{..})^2=Q\sum_{i=1}^{B}(W_i-\bar{W}_.+\bar{S}_{i.}-\bar{S}_{..})^2, \tag{4.3}$$

we see that the expected value of the well-to-well sum of squares is

$$\mathrm{E}[\mathrm{SS}W]=(B-1)(Q\sigma_W^2+\sigma_S^2). \tag{4.4}$$

Similarly, the other two expected sums of squares are

$$\mathrm{E}[\mathrm{SS}T]=(Q-1)(B\sigma_T^2+\sigma_S^2) \quad \text{and} \tag{4.5}$$

$$\mathrm{E}[\mathrm{SS}S]=(B-1)(Q-1)(\sigma_S^2). \tag{4.6}$$

Adding these and dividing by the naive degrees of freedom $(BQ-1)$ gives the expected value of the naive variance estimator as

$$\frac{BQ-Q}{BQ-1}\sigma_W^2+\frac{BQ-B}{BQ-1}\sigma_T^2+\sigma_S^2. \tag{4.7}$$

Of course, the variance of Y_{ij} is $\sigma_W^2+\sigma_T^2+\sigma_S^2$. The naive variance estimator can underestimate this considerably; with $B=2$ and $Q=4$ (the minimal values which satisfy the regulations and have more than one background well), the coefficient of σ_W^2 is $\frac{4}{7}$; where spatial variability is the greatest proportion of the overall variance, the naive variance estimator underestimates the intended quantity by nearly half.

If it is indeed necessary to incorporate spatial variability by using multiple upgradient or background wells, a more sophisticated variance estimator is appropriate. A new measurement at a different well is presumably independent of the background measurements and has the same marginal variance; hence,

$$\mathrm{var}(Y_{\mathrm{new}})=\sigma_W^2+\sigma_T^2+\sigma_S^2.$$

The natural predictor of future measurements is $\bar{Y}_{..}$, the mean of the background measurements

$$\mathrm{var}(\bar{Y}_{..}) = \frac{1}{B}\sigma_W^2 + \frac{1}{Q}\sigma_T^2 + \frac{1}{BQ}\sigma_S^2.$$

Thus, the prediction variance is

$$\mathrm{var}(Y_{\mathrm{new}} - \bar{Y}_{..}) = \frac{B+1}{B}\sigma_W^2 + \frac{Q+1}{Q}\sigma_T^2 + \frac{BQ+1}{BQ}\sigma_S^2. \tag{4.8}$$

This can be estimated as a linear combination of mean squares from the variance components analysis,

$$\text{prediction variance estimator} = \frac{B+1}{BQ}\mathrm{MS}W + \frac{Q+1}{BQ}\mathrm{MS}T + \frac{(B-1)(Q-1)-2}{BQ}\mathrm{MS}S. \tag{4.9}$$

It is customary to approximate the distribution of such a linear combination of mean squares by a multiple of a chi-square distribution, matching the first two moments. The expected value is already equal to the desired value; the degrees of freedom are then fit to match the second moment of the approximating chi-square distribution. The degree of freedom formula requires the values of the variance components, or at least of their ratios; those are typically estimated from data. The result is that to estimate

$$\mathrm{VAR} = \sum_k a_k \mathrm{E}[\mathrm{MS}_k] \tag{4.10}$$

one uses

$$\hat{V} = \sum_k a_k \mathrm{MS}_k \tag{4.11}$$

with estimated fitted degrees of freedom

$$\hat{D} = \frac{\hat{V}^2}{\sum_k (a_k \mathrm{MS}_k)^2/d_k}, \tag{4.12}$$

where d_k is the degrees of freedom of the mean square MS_k.

One would then use a test statistic of the form

$$t_{\hat{D}} = \frac{Y_{\mathrm{new}} - \bar{Y}_{..}}{\sqrt{\hat{V}}}. \tag{4.13}$$

If $\hat{V}$ has approximately a chi-square distribution, one expects that $t_{\hat{D}}$ will then have approximately a t distribution with the fitted degrees of freedom. The approximate degrees of freedom analyses here and in the following section follow from Hocking

(1985, §8.4.2). Davis and McNichols (1991) studied the performance of such procedures in the simpler situation where σ_T^2 is assumed to be zero. They reported that the nominal significance level was maintained fairly well by such procedures so long as $B \geqslant 3$.

This sort of analysis depend on several assumptions. All of the random effects are assumed to be normally distributed, possibly after transformation, and the W_i are assumed to be an independent random sample from some population of well effects. These assumptions are difficult to verify empirically with the amounts of data typically available. Furthermore, the latter is not particularly plausible a priori. Rather, one expects the chemical properties of the soils from which the water is drawn to exhibit at least some spatial correlation. Davis and McNichols (1991) found the significance level of such tests to be rather sensitive to the assumption of a lack of spatial correlation and recommended against using these techniques whenever possible.

An alternative is to treat the well-to-well variability as a fixed effect since the well means are determined by nature and should not change between monitoring events. This approach leads to the use of intra-well comparisons or of two-way mixed models in which the interaction term is of interest. Such models are not balanced, as there are generally different numbers of background and downgradient wells, so the analysis is not always a plug-in problem for a computer package. Blocking out the spatial variability can also enhance the power of the statistical tests by making the background variance much smaller.

At the present time regulations are being implemented which extend ground-water monitoring requirements previously required for hazardous waste management facilities to municipal solid waste facilities. Most existing solid waste facilities have not collected background data which will allow for a demonstration that downgradient wells are free of contamination, and most have the traditional minimum of one upgradient well (giving zero degrees of freedom for spatial variability). The consequence will be a great many reports that municipal landfills are leaking. The leaks may or may not exist, but the ubiquity of spatial variability which cannot be controlled for statistically for lack of adequate data ensures that the reports will be there.

4.2. *Simultaneous inference*

It is the author's opinion that the regulatory community has yet to adequately come to grips with this area, as of September 1992. The problem is that the individual statistically based decisions are not isolated but are part of a larger decision problem. In RCRA ground-water monitoring at hazardous and solid waste facilities, the facility will have to enter assessment unless *all* of the individual tests for all indicator parameters at all monitoring wells accept a null hypothesis of 'no change' or 'no leakage'. In CERCLA (Comprehensive Environmental Response, Compensation and Liability Act of 1980, or Superfund) remediation activities, the remediation activity should continue until *all* upper confidence limits for mean concentrations of the various compounds at all areas of the site are below the designed regulatory threshold. In air monitoring, one is often concerned about the *largest* concentration measured during a specified time period.

In each of these situations, *all* (equivalently, the largest) of a set of measurements or test statistics should be below some critical value to avoid adverse regulatory consequences. If the critical value is exceeded, the consequences are the same whether one or several or all test statistics exceed it. In this setting, it is appropriate for that critical value to take into account the fact that it is a critical value for the maximum and thereby control the *facility-wide false positive rate.*

There are two main techniques which are useful for accomplishing this:

- spreading out the false positive rate over component decisions and
- using 'combined' or omnibus statistical tests.

4.2.1. Partitioning false positive rates

The use of the Bonferroni t-test in the TEGD (US EPA, 1986a) has been discussed in Section 2.1.1. In English, the Bonferroni inequality states that the chance of a false positive on at least one of a number of tests is no greater than the sum of the individual significance levels. This is true no matter what the dependence structure of the individual component tests and hence is conservative.

So long as the overall significance level being partitioned is small, the Bonferroni technique is not overly conservative. Nonetheless, there are somewhat sharper bounds due to Sidak and to Slepian (see Miller, 1981, pp. 254–255) for multiple t-tests which state that the independence assumption is itself conservative in various situations; see also §8.5 of Miller (1981). Indeed, so long as incidences of false positives are not negatively correlated, one expects that the independence assumption should be conservative, although there is not a theorem so broad as this. (Recall that the Bonferroni inequality becomes equality when and only when the incidences of false positives are mutually exclusive; that is, at most one of the measurements can have a false positive.)

The extent of conservatism involved was studied by Milke and Huitric (1991). From their graphs, this author infers that the power penalty of a monitoring program based on the independence assumption over one which would take the correlation between variables into account would be not large for the particular variables involved and the situation studied.

Assuming that the goal of controlling the overall significance level in one of these situations is agreed upon and is met by such partitioning, the significance level for each individual test becomes smaller as the number of quantities measured and the number of locations at which they are measured becomes larger. With decreasing individual test significance levels and correspondingly increasing critical values, the power of each individual test decreases. It is therefore desirable to limit the number of tests done by decreasing the number of quantities measured and the number of locations at which they are measured. This is acknowledged in the most recent draft guidance (US EPA, 1992), although there are often political pressures on decision makers to 'do more monitoring.' With a small number of appropriate quantities, chosen to have the greatest potential for indicating problems at the facility in question and to be not too highly correlated, with the amounts of data typically available, it will probably be difficult to do anything more effective than simply using Bonferroni's inequality to partition α over such quantities.

With measurements of the same quantity at different locations, there is often some positive correlation between locations in time. Assuming a normalizing transformation, if variances were known, Slepian's inequality (see Miller, 1981, p. 255) would indicate that the independence assumption for t-type statistics for different locations (for after–before differences) would again be conservative. One expects that this will be the case even if variances are estimated. In any event, the independence assumption individual significance levels will be close to the Bonferroni individual significance levels, so that any error made here cannot be large.

A more interesting question would be whether there might be some way of taking the correlation structure into account in order to obtain more powerful tests. Let the measurements be denoted Y_{ij}, assuming a homoscedastic linear model

$$Y_{ij} = \mu + \omega_i + T_j + S_{ij}, \tag{4.14}$$

where μ is the overall mean as usual, ω_i is the (now fixed) effect due to location differences, T_j is the (random) effect due to the common part of the time variability at all locations which causes the positive correlation and S_{ij} is the variability due to sampling, transportation, measurement, etc. The T_j are assumed i.i.d. $N(0, \sigma_T^2)$, independent of the S_{ij} which are assumed i.i.d. $N(0, \sigma_S^2)$. This model would be appropriate for data such as the fluoride measurements made during quarterly background observations at sixteen wells at a RCRA hazardous waste management facility, shown in Figure 1.

A variance components analysis on these background observations gives the following result:

Source	DF	SS	MS	F	P	E.M.S.	Variance component
Wells	15	4.967	0.331	24.4	0.000		
Up–down	1	0.732	0.732				
Other	14	4.235	0.302				
Time	7	2.426	0.347	25.6	0.000	$\sigma_S^2 + 16\sigma_T^2$	0.021
Error	105	1.424	0.014			σ_S^2	0.014
Total	127	8.817					

The cause of the pronounced parallelism between measurements at different wells is not known. Other indicator parameters do not share these patterns, and there is no evidence of leakage from the facility. One possible source is day-to-day variation in laboratory response.

The well-to-well differences are highly statistically significant ($P = 0.000$). When these differences are partitioned into the one degree of freedom constrast between upgradient and downgradient wells and the other fourteen degrees of freedom, the up–down contrast mean square is about 2.5 times that of the other, indicating that

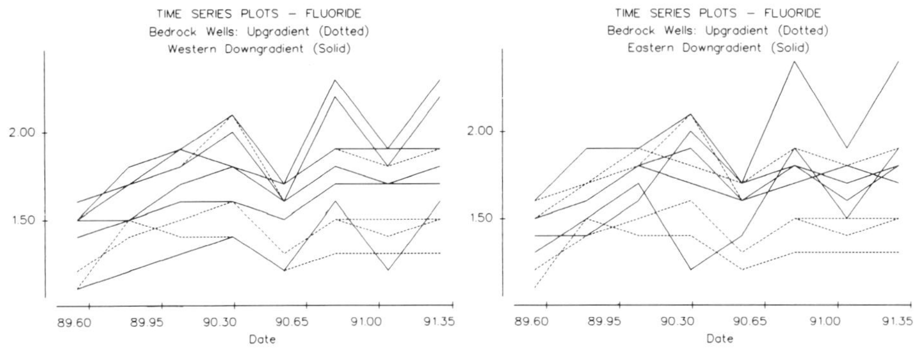

Fig. 1. Fluoride measurements at a RCRA hazardous waste facility. Date = 90.00 corresponds to 12/13/89.

perhaps the well measurements are more alike within groups than between groups. This could possibly reflect spatial autocorrelation since the upgradient wells and downgradient wells are on opposite sides of the facility.

Residual analyses show no reason to doubt the assumptions given. Note, though, that with only seven degrees of freedom it is difficult to assess the assumption that the T_j are independent random variables without trend or temporal autocorrelation.

The goal is to develop a statistical procedure which will use this background data along with new monitoring observations taken at one point in time in the future to determine whether a statistically significant increase in fluoride concentrations has occurred between the background period and the future monitoring occasion. Since there already exists highly statistically significant spatial variability, the procedure must control for this spatial variation. The simplest way to do so is to use strictly intra-well comparisons with the facility-wide significance level spread over all monitoring parameters and wells. This particular facility has been affected by the 'more monitoring is better' syndrome, with the result that there are 28 indicator parameters and a total of 83 monitoring locations, so that $24 \times 83 = 2324$ statistical tests will be performed during every monitoring period.

If a facility-wide α of 0.05 is spread over all comparisons, the conservative Bonferroni inequality would require that each individual comparison use a significance level of 0.00002151, whereas the (still almost surely conservative) independence assumption would require an individual comparison significance level of 0.00002207. If an omnibus test for all wells is used, the significance level should be 0.001786 or 0.001830 respectively per monitoring parameter.

It is appropriate and environmentally conservative in evaluating such programs to consider the power as the probability of detecting a change in the level of just one compound at just one monitoring location. Attaining satisfactory power with such a low per-comparison significance level can be a challenge. Explicitly allowing for one or more resamples and retests can aid. Power curves are shown in Figure 2 for t-type prediction interval tests (i.e., to 'pass the test' the future observation must be less than

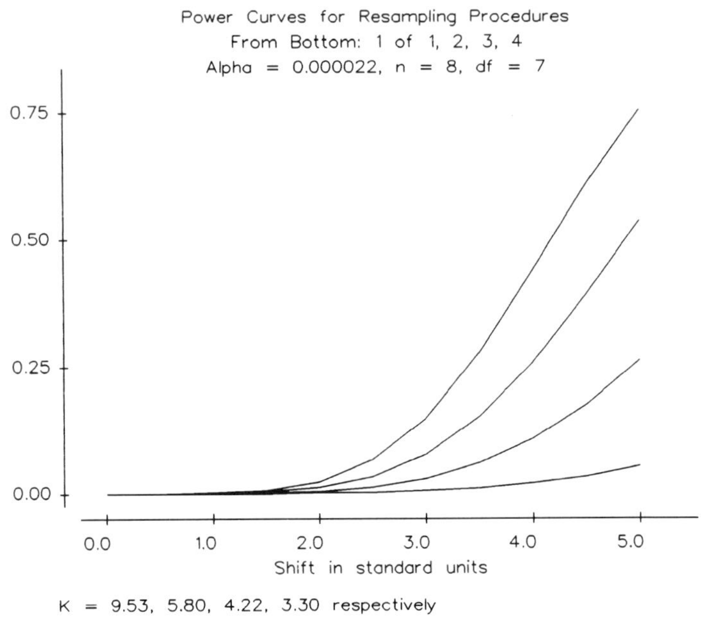

Fig. 2. Power curves for resampling strategies.

$\bar{Y}+KS$), for ordinary t-tests ($K=t_{7,0.000022}\sqrt{1+\frac{1}{8}}$ with power determined from non-central t subroutines) and for tests allowing 2, 3 or 4 chances to get one observation 'inbounds' (i.e., allowing for 1, 2 or 3 resamples) (K and power determined using the algorithms of Davis and McNichols, 1987). Even allowing for three resamples, the power is not very satisfactory here. More resamples would help, but there are limits to the number of statistically independent resamples obtainable during a semi-annual or quarterly monitoring event.

One would perhaps like to pool variance estimates across wells to increase the degrees of freedom and thereby enhance the power. However, the presence of the highly significant T_j effect will make the effective number of degrees of freedom available from such pooling rather lower than anticipated. The total intra-well variance is the sum of variance components:

$$\sigma^2_{\text{Total}}=\sigma^2_T+\sigma^2_S. \tag{4.15}$$

In this case, the naive estimator is the correct unbiased estimator, but with approximate degrees of freedom given by

$$\hat{D}=\frac{(Q-1)(F+W-1)^2}{F^2+W-1}, \tag{4.16}$$

where Q represents the number of quarters, W the number of wells and F the F-ratio for the test of the hypothesis that no time main effect in present. Observe that if $F = 1$, this gives the naive degrees of freedom $(Q-1)W$; as F increases from 1, $\hat{D}$ decreases; and as $F \to \infty$, $\hat{D} \to (Q-1)$, the degrees of freedom of the time main effect. (With $F < 1$, one will not consider this analysis applicable.) In the example under consideration, $F = 25.55$, $Q = 8$ and $W = 16$, so the approximate degrees of freedom would be $\hat{D} = 17.24$. This is a modest increase over the strictly intra-well seven degrees of freedom and falls well short of the naive 112 pooled degrees of freedom.

Caveat. Such pooling requires a credible assumption of homogeneity of variances across wells. Plots of residuals from a one-way ANOVA with main effect well against well means and of well standard deviations against well means can offer quick checks of the reasonableness of this assumption. Very often environmental data has spread increasing with level so the use of a variance stabilizing transformation is often appropriate before such pooling.

Alternately, to enhance power, one could consider blocking out the time component of variability. In the example under consideration there are five upgradient wells; let $\bar{U}_j$ be the mean of the measurements from the upgradient wells on the j-th monitoring occasion (past or future) and $D_{ij} = Y_{ij} - \bar{U}_j$ be the difference between each measurement and the upgradient mean. Under the model (4.14),

$$\bar{U}_j = \mu + \tilde{\omega} + T_j + \tilde{S}_j \tag{4.17}$$

and

$$D_{ij} = \omega_i - \tilde{\omega} + S_{ij} - \tilde{S}_j. \tag{4.18}$$

Here the tilde ($\tilde{}$) indicates the upgradient or background average of the effect in question, ω or S. The time component of variability is blocked out; the variance of D_{ij} will be $\frac{6}{5}\sigma_S^2$ for downgradient wells and $\frac{4}{5}\sigma_S^2$ for upgradient wells. If this is less than the total intra-well variance $\sigma_{\text{Total}}^2 = \sigma_T^2 + \sigma_S^2$, which is surely the case in this example, a more powerful test can be constructed by using background D_{ij} to form prediction intervals for future D_{ij}. The D_{ij} will still be positively correlated between wells by virtue of the common $\tilde{S}_j$; however, the common T_j will have been blocked out. With the values of the variance components estimated above, the inter-well correlation reduces from 0.605 (the ratio of the time variance component to the total variance) to either $\frac{1}{6}$ (for downgradient wells) or $-\frac{1}{5}$ (for upgradient wells). The conservatism of the Bonferroni critical values is thereby largely removed.

There are advantages and disadvantages to using this more sophisticated analysis. The first advantage is the increase in power. This increase in power will not show up on a power curve such as that above, where all shifts are measured in standard units, since the standard unit itself will have changed by virtue of eliminating the time variance component. If the horizontal axis of the power curve is re-expressed in real units, using the appropriate estimated standard deviations, the increase in power will be apparent. (The new draft guidance (US EPA, 1992) sets power standards in standard units; arguing otherwise may take some doing unless that draft is revised.) An additional advantage is that, by blocking out the time main effect, the assumption that it represents an independent random process becomes moot, since tests based on the D_{ij} will not be affected by trends or temporal autocorrelation in the T_j.

There are some disadvantages as well. The first is that the resulting test no longer looks quite like the tolerance or prediction intervals discussed in the IFGD or other guidance documents and could meet with some suspicion or resistance on this account. A sharp reviewer would observe that contamination which might affect upgradient and downgradient wells equally would be undetected by such a program. This is an accurate observation, but that scenario is highly unlikely. If resampling plans are used, a resample of a D_{ij} measurement involves resampling several wells rather than just one. Finally, given that the significance level is maintained, an owner/operator may prefer a less powerful test. In the above situation, the 'disadvantage' of greater power to such an owner/operator would be offset by the advantage of not relying on a lack of trends or autocorrelation in the time component of variable.

In the final analysis, the lack of simplicity of the more powerful test makes it more suspect in an adversarial situation.

One additional warning is in order. The use of resamples and retests in order to contain the facility-wide false positive rate depends on future observations at a given well being independent of each other. If the methods discussed in Section 4.1 are used to incorporate spatial variability into the prediction variance, the spatial variance component will be the same for a given well, no matter how many times one resamples. If that spatial component is relatively large, the original sample and resample measurements will be highly correlated and the use of resamples will not have the desired effect of reducing false positive rates. (To obtain an independent resample when the spatial effect is part of the variability modeled, one would need to drill a new well.) This is one more argument against the techniques presented in Section 4.1.

4.2.2. Omnibus tests

As suggested previously, it is the author's opinion that little is to be gained by using methods other than Bonferroni's inequality or the independence computation to spread a facility-wide false positive rate across quantities being measured. For one thing, almost all statistical tests for environmental regulation are one-sided tests, whereas multivariate tests are generally two-sided, although Carson and Gupta (1992) have explored versions of the Hotelling T^2 test designed to detect positive differences. Also, the difficulties of dealing with non-detects are adequately problematic with univariate tests.

However, it is possible that one could be forced to make every statistical test have a certain significance level, regardless of how many or how few there are. Although one would hope not to find oneself in such a situation, the only defense in such a case might be to attempt to create a single multivariate test for the entire facility. Such a test for the entire facility. Such a test would be definitely contrary to the intent of RCRA ground-water monitoring regulation which clearly requires that each indicator parameter be handled separately.

A particularly troublesome situation arises in this regard with the analysis of toxic compounds which do not appear in nature. Frequently extended lists of these appear in the monitoring programs mandated for RCRA facilities; for example, Michigan routinely requires two organics scans involving 26 different compounds, along with numerous other inorganic analytes and indicator parameters, and the new solid waste facility monitoring rules in 40 CFR Part 258 require routine monitoring of 47 organic compounds along with other constituents. As will be discussed in Section 5.3, 'detection limits' for these laboratory analyses are defined in statistical terms, but those definitions do not explicitly take into account the numbers of tests to be performed on any one occasion. Moreover, detection limit determinations are generally based on data which is at best a surrogate for real environmental measurement data.

Accordingly, various efforts have been made to essentially reduce the tests on all organic compounds to one statistical test with one false positive rate. Such efforts include tests of proportions (US EPA, 1989b), Poisson prediction or tolerance limits for numbers of 'hits' (Gibbons, 1987), and even Poisson limits for the sum of the measurements, adding in non-detects at half the detection limit (Gibbons, 1987 and US EPA, 1992). Indeed, the exemption of prediction and tolerance limits from the performance standard in the current 40 CFR 264 is at least partially in recognition of the work of Gibbons cited here.

The use of such omnibus tests could be viewed as an end run around the regulatory requirement that the statistical tests for all constituents be conducted separately. Whether this is the intention or not, the practical and economic difficulties involved in producing site-specific trigger values sensitive to facility-wide false positive rate considerations for laboratory measurements of such trace compounds may well make techniques such as these attractive.

Other forms of omnibus tests exist; indeed, the ordinary ANOVA is one. For example, in the one-way layout with group means $\sim \mathrm{N}(\mu_i, \sigma^2/n_i)$ being independent of each other and independent of a variance estimator S^2 with $\nu S^2/\sigma^2 \sim \chi^2(\nu)$, the usual null hypothesis of equal group means is equivalent to $\bigcap_c \sum c_i \mu_i = 0$, where the coefficient vectors $\boldsymbol{c}$ range over all *contrasts* ($\sum c_i = 0$). The individual hypothesis for each particular contrast can be tested by a t-test:

$$t_c = \frac{\sum c_i \bar{Y}_i}{S\sqrt{\sum c_i^2/n_i}}. \tag{4.19}$$

By writing this as

$$t_c = \frac{\sum \frac{c_i}{\sqrt{\pi_i}} \sqrt{\pi_i}\,(\bar{Y}_i - \bar{Y}.)}{S\sqrt{\sum c_i^2/n_i}} \tag{4.20}$$

and recognizing that the number is just the dot product between vectors whose components are respectively $c_i/\sqrt{n_i}$ and $\sqrt{n_i}(\bar{Y}-\bar{Y}.)$, and then applying the Cauchy–Schwarz inequality for sizes of dot products, we obtain

$$\max_c t_c^2 = \frac{\sum n_i(\bar{Y}-\bar{Y}.)^2}{S^2} \propto F. \tag{4.21}$$

In other words, the usual F statistic is just a multiple of the square of the biggest $|t_c|$ and the F distribution is equivalent to the distribution of the maximum of $|t_c|$ values.

There are two weaknesses in the use of the ANOVA in environmental monitoring even when a lack of inherent variability between groups makes such a procedure reasonable. One is that each t_c is a two-tailed test and most often it is appropriate to do one-tailed tests for increases (in downgradient wells, or downwind, etc.). The other is that very often when the ANOVA is used, very specific contrasts are intended, such as comparing each downgradient well with the upgradient well or wells. The ANOVA does not have its power optimized for one-sided values of specified contrasts. In fact, it can occur that the ANOVA F will be significant because of a contrast which is of little or no interest for regulatory purposes. In many cases, a more appropriate test would involve multiple comparisons with control (MCC) tests. This would be the case if (a) the assumption of no inherent spatial variability were valid and (b) one wanted to compare each downgradient well measurement with an upgradient background data set. The classic test of Dunnett (1955) carries this out; the test statistic is

$$\max_i \frac{\bar{Y}_i - \bar{Y}_0}{S\sqrt{2/n}} \tag{4.22}$$

in the one-tailed case and

$$\max_i \frac{|\bar{Y}_i - \bar{Y}_0|}{S\sqrt{2/n}} \tag{4.23}$$

in the two-tailed case. Here $\bar{Y}_0$ is the mean of the reference or control group. Dunnett's formula and tables require equal group sample size (n); extensions to unequal sample sizes are available (see Miller, 1981).

Each component test looks like a t-test here, which is helpful in an adversarial situation. Davis and McNichols (1987) (see also Chou and Owen, 1986a) extended this test to allow for resamples. Steel (1959) produced a rank version of Dunnett's test and Chou and Owen (1986b) extended this rank test to allow for resamples (see also Gibbons, 1990b; Davis and McNichols, 1993).

It should be mentioned that some regulators are uncomfortable with the idea that an initial measurement and one or more resamples could exceed a prediction limit, but that the final resample producing an 'inbounds' measurement should be enough to accept a null hypothesis of no increase. For example, California regulation [§66264.97(e)(8)(E)(1) of Title 22 of *Barclay's California Code of Regulations*] specifies that if resamples are used there should be at least two resamples and *all* resample measurements should be inbounds in order to accept a null hypothesis of no contamination. However, so long as the facility-wide significance level is maintained, these programs can be shown to provide rather lower power than the '*m* chances to get 1 measurement inbounds' programs advocated previously.

The California regulation cited in the previous paragraph also specifies how critical values are to be computed for such tests. The critical value apparently is intended to achieve approximately a 0.05 significance level over all monitoring parameters and wells, except that each individual comparison significance level cannot be lower than 0.00167365. This exception applies only for facilities with rather large monitoring programs. The nominal 0.05 significance level assumes independence of all tests; it does not take into account the fact that the resample measurements are compared with the same background data, with the consequence that the actual significance level is somewhat higher than the nomial 0.05. The excess diminishes as the background sample size increases. Modifications of the algorithms given by Davis and McNichols (1987) can be used to evaluate the significance level and power of the California 'first or all resamples inbounds' procedures. For a non-parametric version of these 'first or all resamples inbounds' procedures (see US EPA, 1992; Gibbons, 1990a; Davis and McNichols, 1993).

4.3. Non-detect data

The other major technical problem involves dealing with data sets including both actual measurements and notations '$<$ DL', DL being a stated reporting limit. These data are termed 'censored'. The censoring is said to be type I censoring when the censoring point is fixed in advance. (In type II censoring, the censoring proportion, not the censoring point, is fixed.) With environmental samples, the type I censoring is virually always one-sided, although the censoring point may vary from measurement to measurement.

There are two aspects of this problem. One is that the various types of limits are often determined using statistical computations. A discussion of the models used in those computations is given below in Section 5.3, along with some recommendations. These considerations are of considerable interest; however, they are rarely under the control of the statistician doing data analysis with environmental data.

The other aspect is in dealing with such censored data. Many articles and a book have appeared recently on this topic; see, e.g., articles by Cohen (1959), Gleit (1985), Gilliom and Helsel (1986), Helsel and Gilliom (1986), Shunway et al. (1989), Newman et al. (1989) (who provide a PC program called UNCENSOR), El-Shaarawi (1989), El-Shaarawi et al. (1989), Helsel (1990) and Atwood et al. (1990) and the book by Schneider (1986); see also Gilbert (1987).

There are three types of problems considered:

- point estimation of the mean and possibly other parameters;
- performing hypothesis tests about the mean and possibly other parameters, which is in principle equivalent to producing interval estimates; and
- comparing two or more populations.

Most of the articles which have appeared have dealt with the point estimation problem, using criteria of bias, variance and mean squared error to assess and compare estimators. Simple substitution methods tend to perform less well than more sophisticated likelihood-based methods at this task. Estimation is the appropriate problem to address for many purposes. However, in detection monitoring and many other regulatory settings, the goal is hypothesis testing, not estimation. The differences between these are important. One must recognize that the MSE optimality (or any other optimality) of an estimator gives no guarantee that a hypothesis test constructed using that estimator will meet the first requirement for a statistical hypothesis test, which is that the significance level be controlled with the finite sample sizes actually in use.

The one-sample hypothesis testing problem receives some attention from Shumway et al. (1989), who look at actual versus nominal confidence levels for confidence limits derived using their techniques. Their goal is to estimate the mean of the underlying distribution on the original scale with both point and interval estimates. The interval estimates would 'invert' to become hypothesis tests appropriate for a one-sample problem such as comparing means with a regulatory threshold. Their paper addresses the multiple problems of normalizing transformations and dealing with detection limits simultaneously.

Most other discussions of hypothesis testing involve two-sample non-parametric tests, either ordinary rank tests with ties at the lowest value '< DL', or methods adopted from survival analysis to deal with data sets with multiple detection limits. For large data sets, or for situations where the per-comparison significance level need not be small in order to control facility-wide false positive rates, these techniques will probably be adequate. However, the two-sample problem of comparisons of monitoring data with background data in a regulatory setting where very small significance levels per comparison are needed differs in important ways from the one-sample estimation or even the one-sample hypothesis testing problem, in that we are not really concerned with the true background and monitoring means, but rather are interested in whether there has been an increase in levels.

Hence, the adverse effects of the simple substitution techniques on estimating means and standard deviations of the original populations reported by many authors become almost irrelevant. One way to look at it is the following. The underlying conceptual distribution may look like (A) in Figure 3. With the data censored at a detection limit, the effective distribution is (B), where the vertical line $\uparrow$ represents a point mass at DL/2 containing the probability content of the < DL region. The plot (C) further represents the data after rounding somewhat coarsely, which is typical for these measurements.

Our conceptual model is (A); for many purposes, it is appropriate to use the best techniques available for estimating or testing the parameters of (A), even if the data really look like (C). However, for the two-sample monitoring problem, there may be no harm done in changing our consideration from (A) to (C), and simply testing whether the (C) data from the monitoring observations is significantly different from the (C) data from the background data set.

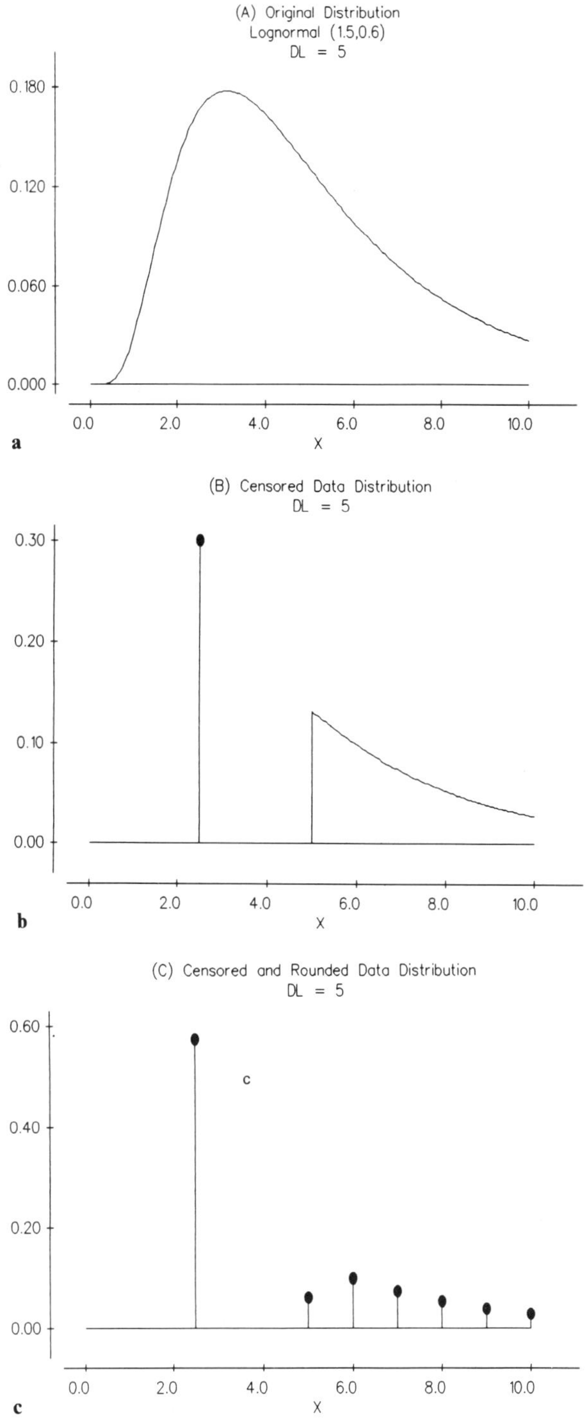

(A) Original Distribution
Lognormal (1.5,0.6)
DL = 5
0.180
0.120
0.060
0.000
0.0
2.0
4.0
6.0
8.0
10.0
X
a
(B) Censored Data Distribution
DL = 5
0.30
0.20
0.10
0.00
0.0
2.0
4.0
6.0
8.0
10.0
X
b
(C) Censored and Rounded Data Distribution
DL = 5
0.60
0.40
0.20
0.00
c
0.0
2.0
4.0
6.0
8.0
10.0
X
c

Fig. 3.

When evaluating hypothesis testing procedures, the first requirement is that the significance level of the test be controlled. (Some authors, particularly in non-statistical journals and in regulatory guidance, demonstrate a technique using a particular data set, but give no information about the significance level which would result from using that technique.) Then, given that the significance level is controlled, one wants the test to be adequately powerful. Controlling the significance level is generally not troublesome if one knows the model that generates the data; if by no other means, one can simulate the process and 'calibrate' the test statistic. With real-world non-detect data, though, the underlying (A) model is not known. Therefore one should choose procedures which would control the significance level as well as possible, over a range of (A) models which appear plausible given the information and data available. Once the significance level is under control, power comparisons allow one to choose among potential treatments.

Some study of this problem is reported by McNichols and Davis (1988). There and in unpublished follow-up studies, they have found that substituting half the detection limit seems to do as well as any simple technique. Using this simple technique, it seems possible to maintain reasonable control over significance levels, under some circumstances, for non-detect proportions up to around 80%. In particular, using resamples seems not only to aid in controlling facility-wide false positive rates, but also in aiding the robustness of the resulting tests. This result is vastly different from the conclusions of others and current guidance (US EPA, 1989b), but the contrary conclusions generally result from work on estimation rather than hypothesis testing.

Success in the estimation problem depends on believing a form for the distribution of (A), and so likelihood-based methods such as Cohen's method or probability plot extrapolation (a graphical version of best linear unbiased estimators, or BLUEs) will do better than naive substitution methods.

On the contrary, when comparing (C) data with (C) data, the distributions are not even continuous, even when the (A) distribution is normal. t-tests are quite robust against non-normality and with any substitution method, the mean of the substituted data is an unbiased estimator of the true (C) mean and the sample S^2 is an unbiased estimator of the true (C) variance regardless of the shape of the (C) distribution . As a result, substitution methods often perform more reliably than do methods which depend on a assumed likelihood for the underlying (A) distribution.

One might object that choosing a value to substitute after seeing what the results of the resulting statistical test might be, with the result that substitution methods could lend themselves to abuses in a regulatory situation. This objection would apply, however, to any situation in which one has some choice of hypothesis testing procedures available. Actually, in a regulatory situation it should be the case that the test to be performed is based on the apparent distributional properties of background data or other data available before the samples on which the ultimate decisions are to be made are obtained. In such a situation, this objection is less valid, especially if the choice of test is made on criteria of validity and robustness of the significance level of the resulting test.

'Cohen's method', being actually Cohen's (1959) algorithm for performing censored data maximum likelihood estimation of the means and standard deviations for normal distributions, was adapted for the TEGD (US EPA, 1986a) and is still recommended in the IFGD (US EPA, 1989b). The adaptation involves using these MLEs of means and standard deviations as if they were the usual estimators, and plugging them into t-tests and related normal distribution-based procedures. McNichols and Davis (1988) found that with small samples, this adaptation tended to produce much higher significance levels than expected. In further unpublished research they found that adjustments based on estimated asymptotic standard errors taking into account the non-detect proportion tended to produce even higher significance levels.

Further research is needed in this area. For example, the expected non-detect proportion will generally vary from well to well along with the mean of the underlying (A) distribution for measurements of naturally occurring trace level constituents. It is often reasonable to assume that the relative standard deviation (i.e., the coefficient of variation) is constant; this information should be of use in providing reliable statistical tests for those wells with higher non-detect proportions.

4.4. Simulation

Many of the situations which arise in dealing with statistics in a regulatory setting are not readily amenable to analytical solution. Even those that admit to a 'pretty' solution when assumptions are valid need to be checked for robustness. Simulations can provide answers and assurances needed. Again, the desire for elegant solutions for inclusion in articles needs to be suppressed often in interests of providing timely acceptable answers.

4.5. Graphical displays

The use of graphical displays for communication and for checking the assumptions behind any statistical procedure used is simply good statistical practice. Unfortunately, more of this is needed in regulatory statistics work. Well-chosen plots can often point out the advantage and/or difficulties of alternative statistical treatments.

5. Recommendations

A number of recommendations have already been mentioned or are implicit in the previous discussions. Two additionals items deserve brief mention here. In addition, a more extended discussion is presented of the matter of 'detection' and other limits, with a call for further study and research.

5.1. Laboratory quality assurance/quality control

The Environmental Protection Agency, recognizing the great need for demonstrably reliable data for its programs, has instituted an idea called 'data quality objectives' (DQOs). 'One goal' is that decision makers who will depend on data should consider the uses to which the data will be put and the options available for data quality (and trade-offs with cost) before embarking on a data collection effort. In other words, they should become aware of the implications of data quality, or the lack thereof, in the planning stages of a project. In order to carry through this idea, there must be programs for ensuring that the data that are obtained actually will meet the desired and expected DQOs.

The various laboratory methods descriptions in SW-846 (US EPA, 1986b) and elsewhere include quality control procedures, including such matters as calibration and calibration check protocols. Still, time pressures can encourage shortcuts and there are some questions about the efficacy of some QC procedures in actually ensuring the quality of measurements of real environmental samples. Statisticians who deal routinely with environmental data recognize that transcription and typographical errors, failures to account for dilutions appropriately and other gross errors routinely affect a small portion of the data.

It would be appropriate to encourage research into QA/QC: determining which procedures or programs are most effective and most effective and which could be reduced or eliminated with minimal loss of overall quality in environmental analyses; detecting, assessing and eliminating undesirable laboratory short-cuts; and documenting and publicizing appropriate uses of QA/QC results.

QA/QC research has its challenges. In particular, designing studies which can accurately assess the potential impact of different protocols on the quality of routine data is tricky, in that non-blind or even single-blind performance evaluation studies have tended to give data quality different from that obtained from double-blind studies which more nearly mimic routine analyses of environmental samples. When testing new protocols, it may well be quite difficult or even impossible to simulate double-blind routine analyses. Even when possible, conducting double-blind studies can be extremely expensive; nonetheless, effort is needed. For further discussion, see Garner et al. (1991).

5.2. Technical review of statistical regulations and proposals

The US EPA has some resources for the technical review of new or proposed regulations which have statistical components or impacts. In addition, notices of proposed rule-making for major regulatory efforts appear in the Federal Register, and sessions at the Annual Quality Assurance and Waste Testing Symposia outline new regulatory directions. These notices and sessions often attract volunteered technical review from outside of the regulatory community. However, these resources have not always been put to good use. In some situations, internal review has been sought much later than appropriate, or not sought at all. In others, volunteered technical review has had little impact or has been ignored or misunderstood. State agencies typically have fewer resources for this sort of technical review than does the US EPA. Nonetheless, both state and federal agencies could consider soliciting external reviews, even gratis. Affected industry groups should seriously consider funding technical reviews of the highest caliber available.

Proposals for statistical procedures to be used at specific facilities or in specific projects are generally reviewed first at the local level. If the techniques are not clearly in accordance with guidance, state or federal assistance can be sought, or advice can be sought from outside consultants. This author's experience is that the sophistication of the reviews obtained from local or state agencies and their consultants can vary widely. Unenlightened reviews from anonymous reviewers can be rather frustrating in adversarial situations. More progress can be made with more proficient reviewers, especially if opportunity for dialogue to clear up technical questions and misunderstandings is provided. This latter openness is to be encouraged.

5.3. Detection and other limits

The issues and methodologies discussed previously with 'purely statistical' questions. However, in this author's opinion, one of the more pressing areas where research is needed involves the definitions of 'detection limits' and other limits. This area involves problematic statistical concepts, firmly held chemometric traditions and confusion in both the regulated and regulatory communities. Progress will require open-minded collaboration of investigators from all sides of the issue.

The conceptual framework involved in the traditions is attributed to Currie (1968). In re-examining numerous and inconsistent definitions of detection limit, Currie defined three 'limits':

- L_C: 'decision limit': a signal level (response) above which an *observed* signal may be reliably recognized as being 'detected'.
- L_D: 'detection limit': the *true* signal level (concentration) which will reliably produce observed signals which lead to detection; and
- L_Q: 'determination limit' the *true* signal level (concentration) which will be expected to provide measurements of adequate precision for quantitative determination (as opposed to qualitative detection).

Note that L_C, against which an observed signal is compared, properly has units of raw instrument response, such as area under a curve or ion counts or millivolts, whereas L_D and L_Q properly have units of concentration of chemicals, such as μg/l or mg/kg. Both the observed signal and L_C can, of course, be referred back to concentration units using a calibration line or curve. Note also that the definition of L_D is meaningless in the absence of a definition of 'detection'; Currie intended that 'detected' should mean that the the observed signal is above L_C. One expects that generally L_C (in concentration units) $< L_D < L_Q$. L_Q is now more often referred to as a 'quantitation limit'.

It is most curious that the term 'detection limit' does *not* refer to the decision threshold in this trilogy. This labelling of a technical concept with an English term whose common and technical meanings do not quite correspond has generated a good deal of confusion, rather like the confusion between the common and technical meanings of the terms 'normal' and 'significant'. In practice, the value used as the decision threshold almost never corresponds to Currie's 'decision limit' L_C. Although reasons for this are open to conjecture, it appears that the conceptual errors involved in using, say, L_Q as a regulatory threshold may simply balance other errors made in carrying out Currie's program.

A translation of these concepts into statistical terms follows. Let ξ represent the true concentration of an analyte, with measured response Y. Y is a random quantity.

The distribution of Y depends on ξ; let $\mu(Y;\xi)$ and $\sigma(Y;\xi)$ denote the mean and standard deviation of the distribution of Y, given ξ. Based on calibration data, a line or curve $F(\xi)$ will be derived to estimate $\mu(Y;\xi)$. Given a response Y, the reported concentration will then be $X = F^{-1}(Y)$. X and F are random as well, and F is often based on rather small amounts of data which is in some ways not representative of routine analyses. Let $\mu(X;\xi)$ and $\sigma(X;\xi)$ denote the mean and standard deviation of the distribution of X, given ξ. Then

- L_C is conceptually a critical value for a test of the null hypothesis that $\xi = 0$, using the observed data Y;
- L_D is conceptually the ξ-value on the power curve above which Prob $(Y \geqslant L_C)$, or perhaps Prob $(X \geqslant F^{-1}(L_C))$, is adequately high; and
- L_Q is conceptually the ξ-value above which $\sigma(Y;\xi)$, or perhaps $\sigma(Y;\xi)/\mu(Y;\xi)$ (relative standard deviation), or perhaps $\sigma(X;\xi)/\mu(X;\xi)$, is adequately small.

Commonly the significance level for L_C is given as 0.01, the power for L_D is given as 0.99 and the relative standard deviation target for L_Q is given as 10%.

In his 1968 article, Currie examines the implications of various assumptions about the relationships between the standard deviations and means, including the effects of assuming $\sigma = \sigma(Y;\xi)$ constant, or assuming $\sigma(Y;0) = 0$ or assuming $\sigma^2(Y;\xi) = \mu(Y;\xi)$ (Poisson statistics). His calibration curves are all straight lines through the origin ($F(\xi) = K\xi$) with known slope. He is particularly concerned with radioactivity measurements, which often involve the difference between two measured quantities, a sample and a background, and for which the Poisson model may be attractive for theoretical reasons. With the Poisson model, of course, estimating σ is simplified.

Hubaux and Vos (1970) examine the effects on Currie's decision and detection limits of the uncertainty in the calibration line. They replace L_C and L_D with y_C and x_D, emphasizing the fact that the decision limit is on the vertical (response) axis and the detection limit is on the horizontal (concentration) axis of the plot of calibration data. Their calibration line, fit by OLS regression, does not necessarily have a zero intercept. They point out that y_C and x_D are random, being based on calibration data with random errors and that their expected values are dependent on the sample size of the calibration data set and on the set of ξ values chosen for the calibration. Their calibration sample sizes are single digit integers, as is typical: gas chromatograph/mass spectrometer (GC/MS) analyses take around one hour each and larger calibration data sets are impractical.

Glaser et al. (1981), in work performed at the US EPA's environmental monitoring and support laboratory in Cincinnati in response to the needs generated by the Clean Water Act of 1977, first cite the still existing confusion about detection limits in the literature. They then define a concept:

> The method detection limit [MDL] is defined as the minimum concentration of a substance that can be identified, measured, and reported with 99% confidence that the analyte concentration is greater than zero and is determined from replicate analyses of a sample of a given matrix containing analyte.

This appears to agree with Currie's L_D, except for the inclusion of the word 'measured'. However, when one examines the formulas, this agreement requires essentially that $L_C = 0$. In Currie's system, this ought to happen only if $\sigma(Y;0) = 0$. There is some evidence that $\sigma(Y;\xi)$ is an increasing function of ξ; for example, this would be predicted by the Poisson model. However, $\sigma(Y;0) = 0$ requires an absolute lack of variability in the electrical signals obtained in the instrument while analyzing blank samples.

Glaser et al. give some discussion of the possibility that the standard deviation varies, even in ploynomial fashion, with the true concentration. However, 'for obvious economic reasons', they ultimately settle on a constant and define

$$\mathrm{MDL} = t_{(N-1,0.01)} \times S_C, \tag{5.1}$$

where S_C is the sample standard deviation of seven replicate analyses of samples 'at a fixed concentration' C (comparable to ξ). That fixed concentration is to be from one to five times the estimated (i.e., anticipated) MDL. Samples containing that concentration could result from spiking reagent water or could be actual environmental samples. An iterative procedure for deciding what that fixed concentration is to be is also presented. In that case, the calculated MDL is used as the new spiking level and a new S_C is calculated. The process is to be iterated until successive values of S_C are not significantly different, using an F test with 6 and 6 degrees of freedom and a 5% significance level. Notice that this implies that the 'constant' standard deviation is not expected to be constant over too wide an interval.

An alternate interpretation of equation (5.1) is possible, in that if $\mathrm{E}[X] = \xi$, a lower 99% confidence limit of true concentration given a single value (and using the historical S_C) would on the average roughly equal or exceed zero if that true concentration exceeded the MDL. This would be equivalent to a hypothesis test of $H_0\colon \xi \leqslant 0$ accomplished through inverting confidence intervals. If $\sigma(Y;\xi)$ is really increasing in ξ, though, this test ought to be conservative. However, often $\mathrm{E}[X] < \xi$.

Observe that if $S_C = K\,C$, iteration on equation (5.1) has no fixed point unless the constant $K = 1/t_{(6,.99)} = 1/3.1427 = 32\%$; that is, the relative standard deviation is exactly 32% in which case *any* true concentration is a fixed point (and hence an MDL), give or take sampling variability in the S_C. This is not an unreasonable value for some kinds of semivolatile organic analyses. If $S_C = K\sqrt{C}$, as for the Poisson model allowing for rescaling or changes of units, the fixed point becomes $\mathrm{MDL} = (3.1427\ K)^2$.

Glaser's definition has been adopted; see, e.g. (US EPA, 1990). The minimum required sample size is three.

Even aside from the uncertainties about and simplifying assumptions regarding the true relationship between $\sigma(Y;\xi)$ and ξ, there are several rather statistically unsatisfactory aspects of this definition and algorithm. For one thing, the measurements whose standard deviation is used in the MDL computation are non-blind analyses, obtained by analysts who know the spiked concentration. The variability obtained with double-blind routine analyses is rather larger than with samples identified as performance evaluation or 'research' samples, according to several sources. MDLs computed this way are obtained from replicate analyses of a single sample. Using the single sample blocks out variability in ξ so long as the sample is homogeneous and using the appropriate matrix (water or soil or other substance in which the analyte is found) can account for differences in sensitivity between matrices. However, using a single sample on one occasion effectively ignores such variance sources as sample preparation, inter-analyst variability, and drifting of the calibration over time. Both of these have been found to be large in performance evaluation studies.

On the other hand, in regulatory work the MDL is often used a decision threshold. If the calibration were reliable enough to assume that $E[X] \approx \xi$, and if the MDL were expressed not on the Y scale but on the X scale through the inverse of F, using the MDL as a decision limit ought to imply that a true concentration ξ equal to the MDL would not have a 99% chance of being 'detected', but approximately a 50% chance. This is certainly not intended by Currie (1968) and probably not by Glaser et al. (1981).

The logical and conceptual errors, therefore, go in both directions. Further, in practice one wonders how accurately MDLs can be or are matched to the matrix for routine environmental samples; recall that each MDL determination takes from three to seven hours for GC/MS analyses even without iteration, which are not economically productive hours for the laboratory.

Clayton et al. (1987) first point out the fallacy involved in using the detection limit as a decision limit. They return to Currie's 1968 concept, referring explicitly to 'detection limits with specified assurance probabilities'. They then present a methodology for estimating L_D, taking into account the fact that the calibration line and error variance are unknown and must be estimated from data. Their derivation assumes that a linear calibration curve (not necessarily through the origin) is appropriate, that the error variance is constant, at least in the region of interest, and that the error distribution is normal. They state that the assumption of constant error variance can often be accomplished by transforming the response Y. A compensating transformation will then be needed for X in order for the straight-line calibration to be appropriate. There is no problem with computing the decision limit and the detection limit on a transformed scale and then referring them back to the original scale. They also discuss the influence of the design of the calibration experiment on the decision and detection limits obtained.

Gibbons et al. (1991) extend the work of Clayton et al. to include increasing variance and multiple future detection decisions. They assume that the error variance will be proportional to the concentration (roughly, the Poisson model). Using this assumption, they use either weighted least squares or a square root transformation. In either case they derive formulas on the original scale which can be used without explicit transformation.

For the extension to dealing with multiple future detection decisions, they call on β-content, $1-\gamma$ confidence tolerance intervals, using $\beta = 99\%$. In order to do this, they rely on the analogy between tolerance and prediction limits to provide a surrogate α^*, which is then used in the L_D estimate computation. However, they avoid stating explicitly what the decision limit should be, on either the Y or X scales, to achieve the desired assurance that a true concentration of at least L_D on any specific occasion should have a 99% chance of leading to a 'detection'. Again, if the models are correct otherwise and L_D is used as the decision limit, the power should be expected to be around 50% not 99% when the true concentration ξ equals L_D.

Currie's L_Q (1968), the determination or quantitation limit, has its own, somewhat separate history. Gibbons et al. (1992) write:

> US EPA defines the PQL (practical quantitation limit) as 'the lowest level achievable by good laboratories within specified limits during routine laboratory operating conditions'. This rather vague definition has been operationally defined by USEPA in three ways; first, as 5 or 10 times the method detection limit, or second, as the concentration at which 75% of the laboratories in an inter-laboratory study report concentrations $\pm$ 20% of the true value, or third, as the concentration at which 75% of the laboratories report concentrations within $\pm$ 40% of the true value. The first operational definition is arbitrary and depends completely on the validity of the corresponding method detection limit ... (citing Clayton et al.). The second and third operational definitions are somewhat better; however, the inter-laboratory studies are often done at a single concentration ... in experienced government laboratories that 'knew they were being tested with standard samples in distilled water without matrix interferences'. USEPA points out that 'the resulting PQLs are optimistic'.

PQLs had been published in 40 CFR Part 264, Appendix VII (later designated Appendix IX). In the announcement of the 1988 revision of 40 CFR Part 264, US EPA writes:

> The Appendix IX rule (52 FR 25942, July 9, 1987) listed practical quantitation limits (PQLs) that were established from 'test methods for evaluating solid waste' (SW-846). ... the PQLs listed were EPA's best estimate of the practical sensitivity of the applicable method for RCRA ground-water monitoring purposes. However, some of the PQLs may be unattainable because they are based on general estimates for the specific substance. Furthermore, due to site specific factors, these limits may not be reached. For these reasons the agency feels that the PQLs listed in Appendix IX are not appropriate for establishing a national baseline value for each constituent for determining whether a release to ground-water has occurred. ... In the event that a laboratory cannot achieve the suggested [Appendix IX] PQL, the owner or operator may submit a justification stating the reasons why these values cannot be achieved. ... the regional administrator may choose to estabilish facility specific PQLs based on the technical limitations of the contracting laboratory.

Furthermore, 40 CFR Part 264.97(i)(5), in describing performance standards for statistical tests to be used in ground-water monitoring, states that

> any practical quantitation limit (PQL) approved by the Regional Administrator ... that is used in the statistical method shall be the lowest concentration level that can be reliably achieved within specified limits of precision and accuracy during routine laboratory operating conditions that are available to the facility.

Gibbons et al. then proceed to provide PQL analyses which parallel the MDL analyses cited previously, returning to Currie's definition of L_Q as the point above which the relative standard deviation is no less than 10% or perhaps 20%. They produce both point and upper confidence interval estimates of the PQL. They rely on the square root as a variance stabilizing transformation, citing comments by Clayton et al. One curious technical wrinkle is that although decision limits and MDLs pass simply through transformations, relative standard deviations do not, and so they must consider also the propagation of variance for the non-linear transformation.

Several questions arise. The first is, what happened to the decision limits so emphasized by Currie and by Clayton et al.? The answer seems to be that for regulatory purposes, the effective decision threshold is often not Currie's L_C or even L_D (the MDL), but the PQL, a surrogate for L_Q, which is almost always much larger. This substitution should almost surely, it would seems, lower the significance level and the power substantially over what they should be with Currie's paradigm.

One counter to that argument is that using the PQL as the decision threshold saves the agencies and the facilities the embarrassment of having decided that something is present without being able to reliable quantify it. After all, if it is toxic and it is present, some sort of action is needed, the minimal action being to monitor it and see if the level increases. To do this, the current level should be determined, which cannot reliably be done if it is less than the PQL.

Another counter to that argument is that laboratory- and site-specific decision limits and MDLs are almost always evaluated from limited studies, which are not double-blind and do not incorporate all of the variance components which actually affect real environmental measurements. The extra protection against false positives gained from using PQLs rather than decision limits as defined by Currie or by Clayton et al. is illusory, if the decision limits are based on inappropriate studies which are more highly controlled than will be the case with routine analyses.

Yet another reason for not relying on Currie's L_D is that the significance level for laboratory analyses is determined on a per-analysis basis, whereas monitoring at a regulated facility typically involves hundreds if not more analyses on each occasion. The regulatory community has not yet fully come to grips with simultaneous inference ideas for ordinary kinds of measurements; adding this discussion to the other detection limit complexities, and possibly involving the chemists as well, may be too difficult. It is much simpler to retain the higher PQL as the decision limit to compensate.

It is even possible that there has been enough confusion about the various terms and definitions that those writing regulations simply used the wrong acronym.

At any rate, almost all references to limits in federal regulation have been to PQLs. In the recent draft addendum to the IFGD, the suggestion is made that when assigning ranks for nonparametric tests, lower ranks be assigned to observations less than the MDL than to those observations between the MDL and PQL (US EPA, 1992). By contrast, some states reject the use of PQLs as decision thresholds. In principle, the lower values should be used, if they could be based on the 'right' studies.

Those studies are, unfortunately, virtually impossible to carry out. It is possible to design and carry out double-blind performance evaluation studies, incorporating matrix effects and time components of variability; see, e.g., Sadowski et al. (1990) and Davis et al. (1990a,b). However, in order to actually calculate L_C and estimate L_D, one needs real trace level data. Trace measurements in the data obtained from double-blind studies will include only notations '$<$DL', where DL is a reporting limit which is usually in the vicinity of the Appendix IX PQL. The censoring at the reporting limit DL is typically done by the instrumentation rather than by sentient beings. Although it would be possible to set the instrument thresholds lower, that would result in non-routine, non-double-blind, analyses. Heisenberg's principle holds: in attempting to examine the measurement process, one alters the process.

Where does this whole area stand at this time? Further research and discussion are needed. This will be a sensitive area for several reasons. One is that industry has become accustomed to the extra protection against false positives that using PQLs as decision thresholds provides. Another is that good data for studies in this area is extremely expensive.

The research questions which seem to this author to be unanswered include the following:

- What is the actual relationship between $\sigma(Y; \xi)$ and $\mu(Y; \xi)$? Various assumptions have been used, primarily that σ is approximately constant, or that $\sigma^2 = K\mu$. Gibbons et al. (1991) use the latter, which would imply that a square root transformation ought to be approximately variance stabilizing; however, Figure 2 of their article seems to indicate that the variance is still increasing with the square-root transformed data.

In the articles and reports by Sadowski et al. and Davis et al., for most of the eleven volatile organic compounds studied the relationship was more nearly like $\sigma = K\mu$ (i.e., constant relative standard deviation) in the range of true concentrations from just above the Appendix IX PQL to several times that. Some variables showed a higher relative standard deviation around the Appendix IX PQL.

One could argue for such models on a priori grounds, even over larger concentration ranges. For one thing, with $\xi = 0$, there is still electronic noise and the variability due to random variations in the automatic integration routines in the instrumentation should be relatively more prominent than with stronger signals; this should lead to a positive σ (0). For another, concentrated samples are diluted before analysis; if the variability added by the dilution step is relatively minor, this would argue towards constant relative standard deviations.

However, actually estimating and fitting a model for $\sigma = G(\mu)$ would be quite data intensive and consequently expensive, as standard deviations are difficult to estimate reliabily. Again, there is the Heisenberg problem; to be relevant to actual environmental measurements, such studies should use double-blind measurements obtained during routine analyses.

A further complication to this research is that a constant relative standard deviation model would render the concept of L_Q and hence of PQL meaningless. Such a finding could be quite unpopular among the many accustomed to thinking in terms of PQLs.

- Until now, with the exception of the article by Gibbons et al. (1991), the detection limit question has considered false positive rates (significance levels) on a per-analysis basis. (These authors produce MDLs for a decision process which provides tolerance intervals for 99% of future observations with 99% confidence.)

The decisions to be made on the basis of these laboratory determinations depends on the regulatory situation involved. In some cases, the per-analysis significance level analysis is appropriate. In many regulatory situations, though, it would be appropriate to consider facility-wide significance levels. However, chemists, regulators, and the public may balk at the idea that a given laboratory result might signal the 'detection' of a toxic compound in ground water at one facility but not at a neighboring facility with a more extensive monitoring network. Some research might be needed here, but certainly if the concept of simultaneous inference were to be applied to decision limit questions, a good deal of educational effort would be needed.

- These studies can be very expensive. In principle, one should repeat them for each different matrix, for each laboratory. However, one expects that similar matrices analyzed on similar instruments should have similar statistical properties. How 'similar' is 'close enough' for these purposes?

6. Other roles for statistics/statisticians

It has been said that the role of applied statistics is to help decision makers deal with randomness. There are many opportunities for randomness to affect regulatory decisions. In many cases, the regulations do not explicitly involve statistics or statistical concepts. Some examples are given here.

- In ground-water monitoring at RCRA facilities, the owner/operator is often instructed to inform the agency if the gradient direction or flow 'has changed'. Gradients are determined from well water height measurements which have variance components. Therefore there are statistical implications to the phrase 'has changed'.
- Much work has been done on the statistical properties of sampling schemes, both in the context of spatial sampling to obtain data for site assessment and verification of remediation and in the context of sampling of heterogeneous materials such as soil.
- Difficult questions arise in attempting to perform a decent analysis of the uncertainties involved in monitoring well network design. These networks are typically established through a combination of professional hydrologists' expert

opinions and negotiation with regulatory agencies. There are opportunities for statistical input to these decision processes.

- As are many things in the legal/regulatory world, data quality objectives (DQOs) are often based on precedent in order to be defensible. One should think in terms of decision quality rather than of data quality. For example, when doing field screening for constituents which tend to have highly variable concentrations, it may be well rather more effective to use larger numbers of less expensive, less precise field measurements than smaller numbers of more expensive (and more legally defensible) measurements. Doing so would require two efforts, however; the technical analysis of the sampling plan would be only the first step. Convincing the decision maker that legal jeopardy would not be the major reward for using the superior, but less traditional, plan, would remain.
- In a remediation effort, the goal is often (loosely) to clean up contamination so that the resulting concentrations of the offending chemicals are not greater than 'background'. 'Background' can be an interesting concept, when dealing with compounds with interesting patterns of spatial variability. There is opportunity for statistical involvement both in characterizing 'background' and in deciding when the remediation effort has succeeded.

7. Conclusion

To summarize briefly:

- Regulatory statistics is an emerging sub-specialty of applied statistics.
- Understanding of the traditions and the legal/political/regulatory climate is nearly as important for work in this area as is technical skill.
- As with other forms of applied statistics, producing elegant papers is not so important as solving the problems at hand, and the problems at hand are business/scientific/legal problems.
- Communication is essential as is a willingness to work in adversarial situations.
- Statistical regulation and guidance are often deficient; understanding the regulatory history behind the problem at hand can be advantageous. Sometimes these deficiencies indicate a need for research in statistical methodology or in other areas.
- A few crucial issues and methodologies have been outlined in some detail, including in particular simultaneous inference questions, censored data problems and 'detection limit' definition problems. Suggestions for practice and for research have been given.
- Statisticians working in regulatory statistics can often find randomness that has not previously been thought of as the province of statistics.
- Opportunities to improve the state of the art abound.

The discussion in this chapter has largely ignored the role of geostatistics. This omission is not the result of any deliberate bias, but rather reflects this author's experience with regulatory statistics oriented toward hypothesis testing, rather than data analysis and spatial modeling. The concepts of geostatistics and spatial data analysis are indeed useful to those working in this area.

Acknowledgements

Previous versions of this chapter were reviewed by James Brown (Office of Solid Waste, US EPA, Washington, DC), John Flueck (University of Nevada, Las Vegas), George Flatman (Environmental Monitoring Systems Laboratory, Las Vegas, US EPA), Forest Garner (Lockheed Engineering & Science Company, Las Vegas, NV), Robert Gibbons (University of Illinois, Chicago, IL), Roger McNichols (University of Toledo and Environmetrics & Statistics Limited), Brian Schumacher (Environmental Monitoring Systems Laboratory, Las Vegas, US EPA) and an anonymous reviewer. The author expresses his appreciation to these people, and also to numerous colleagues and clients for many candid and fruitful discussions of the issues, problems and methodologies involved.

References

In addition to the books, articles, reports and papers listed below, there are references in the text to the *Code of Federal Regulations* (CFR) and the *Federal Register* (FR), published by the US Government Printing Office, Washington, DC and to *Barclay's California Code of Regulations.*

Atwood, C. L., L. G. Blackwood, G. A. Harris and C. A. Loehr (1990). Recommended methods for statistical analysis of data containing less-than-detectable measurements. Unpublished Technical Report, Idaho National Engineering Laboratory.

Carson, J. H. Jr. and A. K. Gupta (1992). Multivariate non-normal statistics in site characterization and evaluation. Technical Report No. 92-11, Department of Mathematics and Statistics, Bowling Green State University, Bowling Green, OH.

Chou, Y.-M. and D. B. Owen (1986a). One-sided simultaneous lower prediction intervals for l future samples from a normal distribution. *Technometrics* **28**, 247–251.

Chou, Y.-M. and D. B. Owen (1986b). One-sided distribution-free simultaneous prediction limits for p future samples. *J. Quality Technol.* **18**, 96–98.

Clayton, C. A., J. W. Hines and P. D. Elkins (1987). Detection limits with specified assurance probabilities. *Anal. Chem.* **59**, 2506–2514.

Cohen, A. C. C. (1959). Simplified estimators for the normal distribution when samples are singly censored or truncated. *Technometrics* **1**, 217–237.

Council on Environmental Quality (1992). *Environmental Quality: 22nd Annual Report of The Council of Environmental Quality.* US Government Printing Office, Washington, DC.

Currie, L. A. (1968). Limits for qualitative detection and quantitative determination: Application to radiochemistry. *Anal. Chem.* **40**, 586–593.

Davis, C. B., V. A. Ecker, S. N. Deming and S. Billets (1992). Experimental design, statistical and data quality considerations in the field evaluation of innovative environmental monitoring technologies (Abstract). Presented at: *43rd Pittsburgh Conf. Exposition on Analytical Chemistry and Applied Spectroscopy*, New Orleans, LA, March 9–13.

Davis, C. B., F. C. Garner, J. A. Berges and L. C. Butler (1991). Analysis-specific techniques for estimating precision and accuracy using surrogate recovery (Abstract). In: *Proc. 7th Ann. Waste Testing and Quality Assurance Sympos.* July 8–12, 1992, Washington, DC, I-84-85.

Davis, C. B. and R. J. McNichols (1987). One-sided intervals for at least p of m observations from a normal population on each of r future occasions. *Technometrics* **29**, 359–370.

Davis, C. B. and R. J. McNichols (1989). Comments on 'Statistical analysis for ground-water monitoring data at RCRA facilities: Interim final guidance'. Submitted to RCRA Docket F-89-HWGWS-FFFF.

Davis, C. B. and R. J. McNichols (1991). On dealing with spatial variability in a regulatory situation (Abstract). Presented at: *3rd Internat. Conf. on Environmetrics*, October 7–10, 1991, Madison, WI.

Davis, C. B. and R. J. McNichols (1993). Nonparametric simultaneous prediction limits. Submitted.

Davis, C., R. McNichols, L. Sherman and P. Sadowski (1990a). RCRA detection monitoring statistical analysis for volatile organic constituents: Part II, Experimental results and statistical techniques. In: *Proc. 6th Ann. Waste Testing and Quality Assurance Symposium*, July 16–20, 1990, Washington, DC, I-97-111.

Davis, C., R. McNichols, L. Sherman and P. Sadowski (1990b). RCRA detection monitoring statistical analysis for volatile organic constituents: Part II, Experimental results and statistical techniques; Further details. Unpublished technical report.

Dunnett, C. W. (1955). A multiple comparisons procedure for comparing several treatments with a control. *J. Amer. Statist. Assoc.* **60**, 1096–1121.

El-Shaarawi, A. H. (1989). Inferences about the mean from censored water quality data. *Water Resources Res.* **25**, 685–690.

El-Shaarawi, A. H., P. B. Kauss, M. K. Kirby and M. Walsh (1989). Inferences about the variability of means from censored data. *Environ. Monitor. Assess.* **12**, 295–304.

ENVIRON (1986). *Elements of Toxicology and Chemical Risk Assessment: A Handbook for Nonscientists, Attorneys and Decision Makers.* ENVIRON Corporation, Washington, DC.

Environmental Laboratory Council (1991). Position paper, laboratory demonstrations and supplemental comments regarding problems associated with EPA's proposal to require accuracy adjustment based on spike recovery correction. Submitted to RCRA DOCKET F-90-WTRX FFFF, March 21, 1991.

Flatman, G. T. and A. A. Yfantis (1984). Geostatistical strategy for soil sampling: the survey and the census. *Environ. Monitor. Assess.* **4**, 335–349.

Flora, J. D. (1971). Statistics and environmental regulations. *EnvironMetrics* **2**(2), 129–137.

Garner, F. C., J. M. Nocerino and J. Warren (1991). Are we getting all we can from our quality assurance? *EPA Statist.* **7**, 1–2,4.

Gibbons, R. D. (1987). Statistical models for the analysis of volatile organic compounds in waste disposal sites. *Ground Water* **25**, 572–580; see also the discussion by C. B. Davis and R. J. McNichols in *Ground Water* **26**, 649–651 (1988).

Gibbons, R. D. (1989). A comment on 'Statistical methods for evaluating ground-water monitoring from hazardous waste facilities' – Final rule. *Ground Water* **27**, 252–254.

Gibbons, R. D. (1990a). Contaminant source analysis (Abstract). Presented at: *2nd Internat. Conf. on Statistical Methods for the Environmental Sciences*, Como, Italy, September 27–30.

Gibbons, R. D. (1990b). A general statistical procedure for ground-water detection monitoring at waste disposal facilities. *Ground Water* **28**, 235–243.

Gibbons, R. D. (1991a). Some additional nonparametric prediction limits for ground-water detection monitoring at waste disposal facilities. *Ground Water* **29**, 729–736.

Gibbons, R. D. (1991b). Statistical tolerance limits for ground-water monitoring. *Ground Water* **29**, 563–570.

Gibbons, R. D. and J. Baker (1991). The properties of various statistical prediction intervals for ground-water detection monitoring. *J. Environ. Sci. Health* **A26**, 535–553.

Gibbons, R. D., N. E. Grams, F. H. Jarke and K. P. Stoub (1992). Practical quantitation limits. *Chemometrics Intell. Laboratory Systems* **12**, 225–235.

Gibbons, R. D., F. H. Jarke and K. P. Stoub (1991). Detection limits: For linear calibration curves with increasing variance and multiple future detection decisions. In: D. E. Tatsch, ed., *Waste Testing and Quality Assurance: Vol 3.* American Society for Testing and Materials, Philadelphia, PA.

Gilbert, R. O. (1987). *Statistical Methods for Environmental Pollution Monitoring.* Van Nostrand Reinhold, New York.

Gilliom, R. J. and D. R. Helsel (1986). Estimation of distributional parameters for censored trace level water quality data: I. Estimation techniques. *Water Resources Res.* **22**, 135–146.

Glaser, J. A., D. L. Foerst, G. D. McKee, S. A. Quave and W. L. Budde (1981). Trace analyses for wastewaters. *Environ. Sci. Technol.* **15**, 1426–1435.

Gleit, A. (1985). Estimation for small normal data sets with detection limits. *Environ. Sci. Technol.* **19**, 1201–1206.

Helsel, D. R. and R. J. Gilliom (1986). Estimation of distributional parameters for censored trace level water quality data: II. Verification and applications. *Water Resources Res.* **22**, 146–155.

Helsel, D. R. (1990). Less than obvious: Statistical treatment of data below the detection limit. *Environ. Sci. Technol.* **24**, 1766–1774.

Hocking, R. R. (1985). *The Analysis of Linear Models.* Brooks/Cole, Monterey, CA.

Hubaux, A. and G. Vos (1970). Decision and detection limits for linear calibration curves. *Anal. Chem.* **42**, 849–855.

Huber, W. A. (1990). Selecting a statistical methodology for groundwater monitoring at RCRA facilities. *Hazmat World*, November.

Hurd, M. (1987). Evaluating the impact of land disposal: The review of organic analytical data. Unpublished Technical Report, Environmental Testing and Certification Corp., Edison, NJ.

Keith, S. J., L. G. Wilson, H. R. Fitch and D. M. Esposito (1983). Sources of spatial temporal variability in ground water quality data and methods of control. *Ground Water Monitor. Rev.* Spring, 21–32.

Loftis, J. C., J. Harris and R. H. Montgomery (1987). Detecting changes in ground water quality at regulated facilities. *Ground Water Monitor. Rev.*, Winter, 72–76.

McNichols, R. J and C. B. Davis (1988). Statistical issues and problems in ground water detection monitoring at hazardous waste facilities. *Ground Water Monitor. Rev.* Fall, 135–160.

Milke, M. and R. Huitric (1991). Implications of co-dependence of parameters for statistical tests of water quality (Abstract). In: *Proc. 3rd Internat. Conf. on Environmetrics*, October 7–10, 1991, Madison, WI.

Millard, S. P. (1987). Environmental monitoring statistics and the law: Room for improvement. *Amer. Statist.* **41**, 249–253.

Miller, R. G. (1981). *Simultaneous Statistical Inference.* 2nd ed., Springer, New York.

Montgomery, R. J., J. Loftis and J. Harris (1987). Statistical characteristics of ground-water quality variables. *Ground Water* **25**, 176–184.

Nacht, S. (1983). Ground-water monitoring system considerations. *Ground Water Monitor. Rev.* Spring, 33–39.

Newman, M. C., P. M. Dixon, B. B. Looney and J. E. Pinder, III (1989). Estimating mean and variance for environmental samples with below detection limit observations. *Water Resources Bull.* **25**, 905–916.

Ross, N. P. (1987). Comment on 'Environmental monitoring statistics and the law: Room for improvement'. *Amer. Statist.* **41**, 254–256.

Sadowski, P., J. Hamilton, L. Sherman, C. Davis, R. McNichols, J. Lenssen and H. Burkholder (1990). RCRA detection monitoring statistical analysis for volatile organic constituents: Part I, A case study. In: *Proc. 6th Ann. Waste Testing and Quality Assurance Sympos.* July 16–20, 1990, Washington, DC, I-86-96.

Schneider, H. (1986). *Truncated and Censored Samples from Normal Populations.* Marcel Dekker, New York.

Schweitzer, G. E. and S. C. Black (1985). Monitoring statistics: An important tool for groundwater and soil studies. *Environ. Sci. Technol.* **19**, 1026–1030.

Shumway, R. H., A. S. Azari and P. Johnson (1989). Estimating mean concentrations under transformation for environmental data with detection limits. *Technometrics* **31**, 347–356.

Starks, T. H. and G. T. Flatman (1991). RCRA ground-water monitoring decision procedures viewed as quality control schemes. *Environ. Monitor. Assess.* **16**, 19–37.

Steel, R. G. D. (1959). A multiple comparison rank sum test: Treatments versus control. *Biometrics* **15**, 560–572.

Stephens, M. W. and M. A. Paessun (1991). Bias correction: Evaluation of effects on environmental samples. In: *Proc. 7th Ann. Waste Testing and Quality Assurance Sympos.*, July 8–12, 1992, Washington, DC, I-66-80.

Syhre, D. (1991). Application of bias correction. In: *Proc. 7th Ann. Waste Testing and Quality Assurance Sympos.* July 8–12, 1992, Washington, DC, I-93-115.

US EPA (1986a). *RCRA Ground-Water Monitoring Technical Enforcement Guidance Document.* Office of Solid Waste and Emergency Response, US Environmental Protection Agency, Washington, DC.

US EPA (1986b). *Test Methods for Evaluating Solid Waste.* 3rd ed. Office of Solid Waste and Emergency Response, United States Environmental Protection Agency, Washington, DC.

US EPA (1989a). *Methods for Evaluating the Attainment of Cleanup Standards.* Vol. 1: Soils and Solid Media. Office of Solid Waste and Emergency Response, US Environmental Protection Agency, Washington, DC.

US EPA (1989b). *Statistical Analysis of Ground-Water Monitoring Data at RCRA Facilities: Interim Final Guidance.* Office of Solid Waste, US Environmental Protection Agency, Washington, DC.

US EPA (1990). *Test Methods for Evaluating Solid Waste. 3rd ed., Proposed Update I.* Office of Solid Waste and Emergency Response, US Environmental Protection Agency, Washington, DC.

US EPA (1992). *Statistical Analysis of Ground-Water Monitoring Data at RCRA Facilities: Addendum to Interim Final Guidance.* Office of Solid Waste, US Environmental Protection Agency, Washington, DC. Currently available as part of: *Statistical Training Course for Ground-Water Monitoring Data Analysis,* EPA/530-R-93-003, which may be obtained through the RCRA Docket (202/260-9327).

G. P. Patil and C. R. Rao, eds., *Handbook of Statistics, Vol. 12*

27

An Overview of Statistical Issues Related to Environmental Cleanup*

Richard O. Gilbert

Abstract: Statisticians face many interesting issues and challenges in planning and implementing environmental restoration (ER) and cleanup activities at contaminated sites. Key factors are teamwork and thorough planning of the entire ER process using structured approaches such as data quality objectives (DQOs) and the observational approach. These tools in combination incorporate methods for developing optimum sampling plans, a statistical-testing framework and a strategy for developing contingency plans to handle deviations from probable conditions at a site. Other issues and challenges are: (1) appropriately quantifying and incorporating the uncertainty of risk-based and background-based standards when testing for attainment of those standards, (2) developing statistical procedures for severely censored (incomplete) data sets, nonnormally distributed data, and data correlated in space and/or time, (3) further developing innovative sampling designs such as interpenetrating sampling, adaptive cluster sampling, ranked set sampling and composite sampling, (4) appropriately soliciting and incorporating expert opinion information about models, model parameters and site history into the development of models and cleanup standards and (5) developing and merging computer intensive techniques such as geostatistics, geographical information systems, and parameter uncertainty analyses of contaminant source, transport and risk models, into integrated analysis systems.

1. Introduction

The cleanup and environmental restoration of sites and facilities contaminated with hazardous chemicals and/or radionuclides is a critical problem facing the United States. The magnitude of the problem is staggering in its scope and cost.

* Work conducted for the Basic Environmental Compliance and Monitoring Program (BECAMP), U.S. Department of Energy, Nevada Field Office, Las Vegas, Nevada.

Contaminated sites number in the thousands. Many billions of dollars will be needed to determine the types and extent of contamination, to assess risks to human health and the environment from exposures to contamination, to develop cleanup technologies, to conduct cleanup activities and to evaluate the effectiveness of these activities.

The success of the cleanup effort will require the acquisition and analysis of large amounts of data to develop information for making critical cleanup decisions such as 'Is cleanup needed at this site?'; 'Is cleanup progressing satisfactorily?'; and 'Has the cleanup standard at this site been achieved?' Efficient and appropriate statistical sampling designs and data analyses are needed to cost-effectively make these decisions with known levels of confidence using data of known quality.

This chapter presents an overview of statistical issues and challenges with regard to environmental restoration (ER) and cleanup activities. Some suggested statistical research needs are also provided. To permit a more focussed discussion, most of our attention is directed toward statistical aspects of assessing the attainment of cleanup standards for environmental media such as soil and groundwater. However, much of the discussion applies to a broader spectrum of data gathering and analysis needs.

In Section 2 we discuss data quality objectives (DQOs) and the observational approach, which can be used to plan environmental restoration and cleanup activities. Several types of cleanup standards and associated statistical testing procedures are discussed in Section 3. Null and alternative hypotheses that may be used when testing for the attainment of a cleanup standard are reviewed in Section 4. Statistical design and data analysis issues are discussed in Section 5. Section 6 summarizes some of the challenges facing statisticians. This chapter was supported by the Basic Environmental Compliance and Monitoring Program (BECAMP), U.S. Department of Energy, Nevada Field Office, Las Vegas, Nevada.

2. Planning environmental restoration and cleanup

Statisticians should be actively involved in the planning of ER and cleanup studies. As discussed in USEPA (1992, Chapter 3), the study of a contaminated site begins with preliminary planning that includes a site visit, rapid review of historical and environmental data, identifying contaminants of concern, identifying applicable regulatory and remediation limits and developing a conceptual site model. After these initial scoping activities are completed, attention turns to developing detailed work plans using DQOs and the observational approach. These methods are briefly discussed in this section.

2.1. *Data quality objectives (DQOs)*

As stated in USEPA (1992, p. 29), DQOs '... are the full set of constraints needed to design a study, including a specification of the level of uncertainty that a data user is willing to accept in the decision'. The DQO process consists of applying in sequence the following action steps to establish uncertainty constraints or requirements on decisions and an optimum design for obtaining data (USEPA, 1992, Figure 3.2):

- State the problem to be solved.
- Identify the decision to be made.
- Identify inputs to the decision.
- Narrow the boundaries of the study.
- Develop a decision rule (null and alternative hypotheses, test statistics, allowed probabilities of making false positive and false negative decisions, etc.).
- Develop uncertainty constraints to determine the number and quality of data required.
- Optimize the sampling design for obtaining data.

A statistician should be on the team implementing these steps. His/her knowledge of sampling designs and data-analysis procedures is needed to develop efficient sampling designs. As illustrated by Neptune et al. (1990), the statistician can develop sampling designs that ensure false positive and false negative decision error rates will not exceed allowable or required levels. Ryti and Neptune (1991), USEPA (1992), USEPA (1987a,b, 1989a,c, 1991) and Barth et al. (1989) provide further discussion of the DQO process.

2.2. The observational approach

The observational approach is a process for initiating remedial action at a site without full characterization of the nature and extent of contamination. The lack of knowledge about conditions at the site is taken into account by making contingency plans for handling deviations from probable conditions if they occur during remedial action.

The observational approach requires planning to identify lack-of-knowledge uncertainty and to determine both possible and probable conditions at the site. The method is intended to reduce cleanup costs by accepting greater uncertainty and allowing earlier selection of a remedial action approach based on probable conditions at the waste site. Contingency plans are prepared in the event that deviations from the probable conditions occur during remedial action. Advantages and limitations of the method are discussed by Peck (1969) for engineering applications. Application of the observational approach to ER is discussed by Smyth and Quinn (1991), Brown et al. (1989) and Myers and Gianti (1989).

3. Types of cleanup standards and related testing approaches

In this section we discuss statistical issues related to evaluating the attainment of several types of standards.

3.1. Risk-based standards

A risk-based standard is an absolute number (e.g., an average concentration) derived on the basis of a formal or informal risk assessment. Maximum concentration limits for drinking water are an example of risk-based standards. For this type of standard, samples are collected at the remediated site and the resulting measurements statistically analyzed and compared in some way with the risk-based (constant) standard value. Examples of statistical procedures for this case are one-sided confidence intervals on the mean concentration (Exner et al., 1985, and USEPA, 1989a) and a test for whether a proportion of the site exceeds the standard (USEPA, 1989a).

In reality, a risk-based standard has uncertainty. This uncertainty exists because the standard is derived using inherently uncertain conceptual and/or mathematical models that relate exposure and risk to environmental concentrations. The parameters of these models are usually assigned values on the basis of limited information about their true values and how those values change over space and time.

An important issue is whether statistical testing procedures should take into account the uncertainty of risk-based standards. That is, should the standard be considered to be a random variable rather than a constant? Typically, a risk-based standard is considered to be a constant. Statisticians may argue that this approach is not correct. However, it is likely that for government agencies to adopt the random variable approach, a convincing case will need to be made that the added complexities of treating a risk-based standard as a random variable will lead to more defendable, reliable and cost-effective cleanup decisions. This is an area where comparative power (probability of correctly rejecting the null hypothesis) and cost-benefit studies may be useful.

One approach to incorporating uncertainty in risk-based standards at specific sites is to conduct computer-simulation parameter uncertainty analyses to assess the uncertainty of environmental contaminant transport, dose and risk model predictions used in deriving risk-based standards for specific sites. The environmental transport and dose models used by Kennedy (1992) are an example of models that might be appropriate for parameter uncertainty analyses. Kennedy (1992) outlines work in progress for the U.S. Nuclear Regulatory Commission to incorporate radionuclide transport and exposure models of contamination conditions in buildings and soil in the development of decommissioning criteria (standards). If uncertainty analyses were used, a distribution (probability density function) of model predictions would be produced rather than a single prediction. The distribution would represent the uncertainty in model predictions due to lack of knowledge and/or space/time variability in model parameter values. The explicit estimation of uncertainty in model predictions and resulting standards could add credibility to the standards-setting process, assuming that the methods used to estimate the uncertainty in model parameters and in the conceptual model are credible. Uncertainty analysis concepts and methods are discussed by, e.g., IAEA (1989), Morgan and Henrion (1990), Finkel (1990), Elderkin and Kelly (1990), Liebetrau and Scott (1991) and Hoffman and Gardner (1983).

3.2. Background-based standards

A background-based standard is derived on the basis of samples collected in a background area, that is, in an area that is free of contamination from the remediated site and that is similar to the remediated site in terms of important physical, chemical and biological characteristics. The background standard is sometimes defined as a single concentration value derived by analyzing a set of background measurements. Examples of such standards are the arithmetic mean, the median, some multiple of the mean or median, an estimated percentile (e.g., the 90-th percentile) of the background distribution, or an estimated upper confidence limit for a specified percentile of the background distribution. After this type of standard is estimated, it is typically considered to be a constant.

Once the constant background standard is set, the data from the site are analyzed and compared with the standard. For example, the upper confidence limit on the mean of the site measurements (perhaps obtained assuming the data are normally or lognormally distributed), might be compared with the background standard. Alternatively, the upper tolerance limit for, say, the 90-th percentile of the distribution of site measurements might be used, again assuming normal or lognormal data. In addition, each site measurement might be compared with the background standard to determine the frequency and magnitude of measurements that exceed the standard. If either the frequency or magnitude of exceedances is too great, the standard has not been achieved.

The above testing methodology is in the statistical guidance issued by the Washington State Department of Ecology (WSDOE) (WSDOE, 1992). That philosophy appears to be predicted on using the null hypothesis that contaminant concentrations at the site exceed the cleanup level (see Section 4). WSDOE (1992, p. 23) states that most statistical tests (for comparing data with a constant value) are not appropriate for this null hypothesis. While that may be true, one can avoid the problem by changing the null hypothesis to 'the site is clean' and use two-sample statistical tests to compare the background and site areas. For example, Gilbert and Simpson (1990, 1992) use that null hypothesis and recommend that the Wilcoxon Rank Sum (WRS) test (Conover, 1980) and the Quantile test (Johnson et al., 1987) be used to test for the attainment of background standards. Alternatively, the two-sample t-test could be used if the data are normally distributed and uncorrelated and there are no data below detection limits. See Section 4 for additional discussion of null and alternative hypotheses.

In the author's view, there is a need to evaluate more thoroughly the performance of the above hypothesis and testing approaches in real life situations. Some type of comparative statistical test-performance studies performed for actual ER sites might be helpful in identifying the most useful approaches. Some type of multi-site study jointly funded by the DOE, NRC and EPA might be useful.

3.3. Background plus risk standard

Frequently, a 'background plus risk' standard is used, that is, a standard stated in terms of 'X units of concentration above background', where the value of X is determined using risk assessment methods. Examples are the EPA standard for ^{226}Ra (radium) in soil at inactive uranium processing sites (USEPA, 1983) and DOE's soil cleanup standards (Kennedy, 1992). If X is considered to be a constant, this type of standard reduces to a background-based standard by simply subtracting X from each site measurement. Gilbert et al. (1987) estimate the power of three decision rules (tests) when both X and background are specified as constants when the contaminant is ^{226}Ra concentrations in soil at uranium mill-tailing sites. Studies of the power of statistical tests for attainment of cleanup standards are needed for the case where both the background and risk components of the standard are random variables.

3.4. Interim cleanup standard

An interim cleanup standard specifies the amount or percentage reduction that is required during a specified time period in which the cleanup technology is applied. For this case, two sets of data are required; one collected from the contaminated site before cleanup begins and one at the appropriate time during or following remediation at the site. Patil and Taillie (1990) discuss likelihood ratio tests and nonparametric tests for this situation when both sets of data are modeled as gamma distributions. They use the null hypothesis of no difference between contaminate distributions at the two sampling times. O'Brien et al. (1991) extend their work by deriving a uniformly most powerful unbiased test and a large-sample approximation of this test. These tests are two-sample tests and hence fall under the same testing philosophy as in Gilbert and Simpson (1990, 1992) for the background-based case (Section 3.2).

3.5. Hot-measurement standards (HMS)

A hot-measurement risk-based or background-based standard may also be used, either by itself or in conjunction with other standards. A hot-measurement standard is a concentration value that must not be exceeded by any single measurement at any location in the site being evaluated. Exceedance of this value may trigger collecting additional samples and possible doing additional cleanup. An example of a background-based HMS is the upper confidence limit on an upper quantile of the distribution of background measurements, e.g., the upper 95-th confidence limit on the 95-th percentile of the background distribution. This type of standard is discussed by USEPA (1989b) for background groundwater concentrations.

3.6. Hot-spot standard

A hot-spot standard is similar to a hot-measurement standard except that the standard applies to a small contiguous area rather than to each measurement. Examples of this kind of standard are given in Berger (1992) and Kennedy (1992). For purposes of statistical testing, it appears that hot-spot standards have been considered to be constants rather than values with inherent uncertainty arising from either uncertain risk models or variability of background concentration measurements.

4. Form of the hypotheses

It is important to consider carefully what null hypothesis (H_0) and alternative hypothesis (H_a) should be used when conducting a statistical test to assess the attainment of a cleanup standard. The issue is important because the cost of the ER effort and the level of protection provided to the public by a statistical test procedure depends on which hypotheses are used.

In USEPA (1989a), Berger (1992) and WSDOE (1992), the hypotheses

$$\begin{array}{ll} H_0: & \text{Cleanup standard not achieved} \\ H_a: & \text{Cleanup standard achieved} \end{array} \qquad (1)$$

are used to test for the attainment of a risk-based standard for soil or solid media at a remediated Superfund site. The rationale for these hypotheses is that the site must be shown decisively to be clean (within specified type I and type II error rates) before the H_0 is rejected and the cleanup standard is declared to be attained.

In contrast to equation (1), the report USEPA (1989a) indicates that the hypotheses

$$\begin{array}{ll} H_0: & \text{Cleanup standard achieved} \\ H_a: & \text{Cleanup standard not achieved} \end{array} \qquad (2)$$

are appropriate at the initial stages of site investigation when there is no a priori reason to believe that the site is contaminated above the standard. Gilbert and Simpson (1990, 1992) use these hypotheses (equation (2)) when testing to evaluate the attainment of reference-based (background-based) standards. Their rationale is that to do otherwise, i.e., to use the hypothesis in equation (1), would require that most site measurements be less than the background-area measurements before the site could be declared as having attained the background standard. They believe that this requirement is unreasonable and may be unachievable or very expensive. It should be noted that Patil and Taillie (1990) and O'Brien et al. (1991) use the hypotheses in equation (2) to test the attainment of an interim standard (Section 3.4) that the cleanup technology is operating properly.

If the hypotheses in equation (2) are used, then the power of the statistical test should be as near to one as possible for the values of the parameter covered in the alternative hypothesis. Then an incomplete interim or final remediation effort will be detected with high probability.

The hypotheses in equation (1) might be used in preference to those in equation (2) to provide additional protection to the public that cleanup to standards has indeed been achieved. To illustrate, suppose the cleanup standard is considered to be a constant and we use the site mean to determine if the cleanup standard has been attained. Then, using the hypotheses in equation (1) would be consistent with requiring that the upper (one-sided) 95 percent (say) confidence limit on the mean must be less than the cleanup standard before accepting that the site has attained the standard. However, if the hypotheses in equation (2) are used, the site would be accepted as having met the cleanup standard even if the upper confidence limit exceeded the standard, unless the lower confidence limit also exceeded the standard. This example illustrates the importance of carefully considering which hypotheses to use.

5. Some statistical design and data analysis issues

Environmental contaminant data tend to violate the usual assumptions needed for many standard statistical techniques. For example, environmental data are often correlated and nonnormally distributed. Also, measurements at very low concentrations may not be available because they were reported by the analytical laboratory as 'not detected' or 'below the limit of detection'. These and other problems are discussed in this section.

5.1. Censored data sets

Many environmental contaminant data sets are 'censored on the left' because some of the concentration measurements are reported by the analytical laboratory as being less than the limit of detection (LOD). These values may be reported by the analytical laboratory as ND (not detected), zeros, or a numerical less-than value (e.g., < 3, where 3 is the limit of detection).

Censored data sets complicate testing for the attainment of cleanup standards. Some nonparametric tests can handle censored data sets more easily than tests that require the data to have a particular distribution. For example, the Quantile test (Gilbert and Simpson, 1990) for testing the attainment of background standards makes use of only the upper tail of the pooled (background plus site) data. Hence, the lack of actual measurements in the lower tail is not a problem as long as those measurements can be assumed to be less than the smallest measurement in the pooled sample used by the test.

More generally, a nonparametric test may require that the measurements for all samples reported as nondetects be smaller than those reported as detects, or be below a threshold such as the LOD. Neither of these assumptions may be true in practice, as illustrated by Lambert et al. (1991). There is a need to develop nonparametric tests that avoid this problem. Also, users of data that are near the LOD need to become familiar with laboratory measurement processes and the meaning of such terms as 'limit of detection', 'method detection limit', 'instrument detection limit', and the 'limit of quantitation'. Keith (1991) discusses these concepts and provides guidelines for reporting, presenting and handling data near the limit of detection.

5.2. Nonnormally distributed data

Measurements of radionuclides and hazardous chemicals in environmental media are frequently not normally (Gaussian) distributed. This may be especially true for samples collected from background areas and from contaminated sites before remediation. Nonparametric statistical tests such as the WRS test and the quantile test may be used in place of classical tests such as the t-test if the distribution form is not Gaussian or known with assurance and the data are not correlated.

Goodness-of-fit tests for normal, lognormal, or other specified distributions are often used to help decide whether to use a parametric or nonparametric test for attainment of a cleanup standard. However, the power of such tests to detect small departures from the hypothesized distribution will not be very large if the number of samples is too small. Hence, for small data sets, there appears to be no compelling reason to conduct a goodness-of-fit test. It makes more sense to use a powerful nonparametric procedure.

5.3. Sampling designs

Some classical sampling designs that may be used to determine the location where soil samples are collected in the background area and/or the site are discussed by Thompson (1992) and Gilbert (1987). These designs include simple random sampling and stratified random sampling as well as systematic, two-stage and double sampling. Gilbert and Simpson (1990, 1992) recommend sampling on an equilateral triangular grid for testing attainment of background standards. However, new sampling designs should be investigated to meet the special needs of cleanup studies. We briefly discuss three designs that may be useful.

5.3.1. Interpenetrating designs

At some sites it may be important to track the progress being made in cleanup over time, i.e., to detect and characterize trends over time in concentration levels. This may be particularly important for ground-water cleanup procedures that may require long periods of time to achieve cleanup standards. One design approach to consider for this case is an interpenetrating design wherein the sampling locations selected at different times rotate among a prespecified set of spatially overlapping (interpenetrating) locations on a triangular grid. The number of rotations would be prespecified and all locations would be sampled by the time the rotation was completed. The rotation cycle could then be repeated on a defined schedule so that the change in concentrations at each location could be assessed. This general approach is discussed by Overton et al. (1990) with regard to EPA's Environmental Monitoring and Assessment Program. Tests for attainment of cleanup standards could be repeatedly made over time. Similarly, tests for trend in concentration levels could be made once sufficient data over time became available.

5.3.2. Adaptive cluster sampling

Thompson (1990, 1991a,b, 1992) developed adaptive cluster sampling designs that entail collecting additional samples in the vicinity of a location where previously collected sample(s) indicates additional information is needed. At a remediated Superfund site, one or more of the samples collected according to a standard probability-based statistical design may have contaminant concentrations close to or exceeding a cleanup standard. In this case, additional samples may be collected in the immediate vicinity to determine whether a local hot spot has been found, and if so, its probable size and shape. However, problems of bias arise when estimating means and testing for the attainment of cleanup standards if the new data are simply combined with prior data obtained at locations selected using a standard statistical design. Thompson develops estimators of means and totals that are unbiased for the adaptive cluster sampling designs he considers. Extension of his work to environmental cleanup situations should be considered.

5.3.3. Ranked set sampling

Patil et al. (1992) provide an annotated bibliography of ranked set sampling (RSS) publications. We quote from the abstract of their paper:

> McIntyre (1952) has noted that, in many sampling situations, there is enough information available to enable sampling units to be ranked with reasonable accuracy according to the character of interest but without actually quantifying the units with respect to that character. The ranking may be done on the basis of visual inspection, prior information, results of earlier sampling episodes, rank-correlated covariates, expert-opinion/expert-systems, etc., or some combination of such methods. McIntyre proposed a sampling scheme in which this ranking capability is exploited to produce an estimator of the population mean which is unbiased and, for a given number of quantifications, has a smaller variance than the estimator employed in simple random sampling.

They then add,

> McIntyre's method, ... consists of (i) identifying a pool of sampling units from the target population, (ii) randomly partitioning the pool into disjoint subsets each having a pre-assigned size, ... (iii) ranking each subset, and (iv) quantifying exactly one suitably selected member from each ranked subset. (Ellipses are the author's.)

The application of RSS to cleanup studies should be studied. Patil et al. (1992) suggest several RSS research areas including extending RSS methods to estimation of population parameters other than the mean, hypothesis testing, multivariate populations and structured populations (trend, spatial pattern, etc.).

5.4. Correlated data

Most statistical tests assume the data are not correlated. This assumption is frequently not true for environmental data. That is, measurements at one spot in place and time will tend to be similar in value to measurements made at nearby spots. Correlated data effect the type I and type II errors rates of statistical tests. The magnitude of these effects depends on the particular correlation structure present. Also, the number of samples required to estimate the mean concentration with specified confidence and accuracy will change depending on the correlation structure. Gilbert (1987, Chapter 4) illustrates the impact of spatial and temporal correlation on the number of measurements needed to estimate a mean. Flatman et al. (1988) discuss geostatistical approaches to the design of sampling plans. Cressie (1991, Chapter 5) reviews spatial design considerations. Borgman and Quimby (1988) discuss sampling for tests of hypothesis when data are correlated in space and time.

The correlation structure may change over time as cleanup of the site progresses. Before remedical action, the emphasis is frequently on estimating the spatial pattern of contamination. At that stage, geostatistical kriging techniques that estimate the correlation structure and use it to estimate the spatial pattern are frequently used, as discussed, e.g., by Cressie (1991). However, the cleanup process will in many cases change or perhaps even eliminate the correlation structure, whereas measurements from the background area may still be correlated. Hence, we may have to deal with comparing correlated background data with uncorrelated site data.

If cleanup efforts are not completely successful, the residual contamination above standards may be present in various patterns in one or more places about the site depending on the method of cleanup. The power of various statistical methods to detect patterns likely to arise from various cleanup methods should be studied. The number of samples required might be determined using models that take correlation structures into account, perhaps incorporating risk-based models as suggested by Skiles et al. (1991), but with the additional feature of incorporating correlation among the parameters of the risk models. The performance of nonparametric indicator and probability kriging techniques (Journel, 1983, 1984, 1988; Flatman et al., 1985; USEPA, 1989a) for these and other situations should be evaluated. The use of robust statistical procedures rather than geostatistical models to handle spatial correlation is another area that should be explored.

We note that almost 20 years ago (in the mid-1970s) kriging techniques were used by the DOE to approximate patterns of plutonium contamination in soil on a real-time day-to-day basis to direct the removal of plutonium-contaminated soil from islands of the Enewetak Atoll (Barnes, 1979; Friesen, 1982). This experience in estimating and using spatial correlation structures to direct cleanup efforts should be reviewed for its applicability to present-day cleanup efforts.

6. Some challenges facing statisticians

Many challenges face statisticians who work on environmental cleanup problems; some are as follows:

- Learn to work effectively in teams composed primarily of nonstatisticians.
- Promote the use of structured planning tools such as data quality objectives and the observational approach.
- Assure that the proper environmental restoration decision analysis framework is used (hypotheses, test statistics, type I and type II decision error rates, etc.).

- Take time to develop an understanding of the measurement process to understand the meaning of measurements reported as below the limit of detection.
- Continue developing new sampling designs such as interpenetrating sampling, adaptive cluster sampling, and ranked set sampling.
- Continue developing statistical methods appropriate for data that are correlated and nonnormally distributed.
- Develop methods for analyzing data sets that are severely censored on the left.
- Assure that expert opinion is properly elicited, analyzed and incorporated into models.
- Develop methods for merging models of contaminant source and transport with geostatistical estimation approaches (suggested by Myers, 1990).
- Merge space/time statistical analysis procedures with geographical information systems to display, organize and analyze space/time data.
- Use uncertainty analyses to assess uncertainty in environmental transport, dose and risk model predictions and cleanup standards.
- Develop and merge computer intensive techniques such as geostatistics, geographical information systems and parameter uncertainty analyses of contaminant source, transport and risk models into integrated analysis systems.

The challenges are great and the opportunities for interesting research are abundant.

References

Barnes, M. G. (1979). Some statistical lessons learned in Enewetak. *TranStat, Statistics for Environmental Transuranic Studies*, PNL-SA-7655, Pacific Northwest Laboratory, Richland, WA.

Barth, D. S., B. J. Mason, T. H. Starks and K. W. Brown (1989). *Soil Sampling Quality Assurance User's Guide*, EPA 600/8-89/046. 2nd ed. Environmental Monitoring Systems Laboratory, U.S. Environmental Protection Agency, Las Vegas, NV.

Berger, J. D. (1992). *Manual for Conducting Radiological Surveys in Support of License Termination*, DRAFT, NUREG/CR-5849. U.S. Nuclear Regulatory Commission, Washington, DC.

Borgman, L. E. and W. F. Quimby (1988). Sampling for tests of hypothesis when data are correlated in space and time. In L. H. Keith, ed., *Principles of Environmental Sampling*. American Chemical Society, Washington, DC, 25–43.

Brown, S. M., D. R. Lincoln and W. A. Wallace (1989). *Application of the Observational Method to Remediation of Hazardous Waste Sites*. CHEM HILL, Bellevue, WA.

Conover, W. J. (1980). *Practical Nonparametric Statistics*. 2nd ed., Wiley, New York.

Cressie, N. A. C. (1991). *Statistics for Spatial Data*. Wiley, New York.

Elderkin, C. E. and G. N. Kelly (1990). *Proc. CEC/USDOE Workshop on Uncertainty Analysis*, PNL-SA-18372. Pacific Northwest Laboratory, Richland, WA.

Exner, J. H., W. D. Keffer, R. O. Gilbert and R. R. Kinnison (1985). A sampling strategy for remedial action at hazardous waste sites: Clean-up of soil contaminated by tetrachlorodibenzo-p-dioxin. *Hazard. Waste Hazard. Mater.* **2**, 503–521.

Finkel, A. M. (1990). *Confronting Uncertainty in Risk Management, A Guide for Decision-Makers*. Center for Risk Management, Resources for the Future, Washington, DC.

Flatman, G. T., K. T. Brown and J. W. Mullings (1985). Probabilistic spatial contouring of the plume around a lead smelter. In: *Proc 6th Nat. Conf. on Management of Uncontrolled Hazardous Waste Sites, Superfund '85*. Hazardous Material Control Research Institute, Greenbelt, MD, 442–448.

Flatman, G. T., E. J. Englund and A. A. Yfantis (1988). Geostatistical approaches to the design of sampling regimes. In: L. H. Keith, ed., *Principles of Environmental Sampling*. American Chemical Society, Washington, DC, 74–84.

Friesen, B. (1982). *Enewetak radiological support project final report*, NVO-213. U.S. Department of Energy, Field Office, Las Vegas, NV.

Gilbert, R. O. (1987). *Statistical Methods for Environmental Pollution Monitoring*. Van Nostrand Reinhold, New York.

Gilbert, R. O. and J. C. Simpson (1990). Statistical sampling and analysis issues and needs for testing attainment of background-based cleanup standards at superfund sites. In: H. Lacayo, R. J. Nadeau, G. P. Patil and L. Zaragoza, eds., *Proc. Workshop on Superfund Hazardous Waste: Statistical Issues in Characterizing a Site: Protocols, Tools, and Research Needs*. Center for Statistical Ecology and Environmental Statistics, Pennsylvania State University, University Park, PA, 1–16.

Gilbert, R. O. and J. C. Simpson (1992). *Statistical Methods for Evaluating the Attainment of Cleanup Standards*, Vol. 3: *Reference-Based Standards for Soils and Solid Media*, PNL-7409 REV 1. Pacific Northwest Laboratory, Richland, WA.

Gilbert, R. O., K. R. Baker, R. A. Nelson, R. H. Miller and M. L. Miller (1987). *Sampling errors associated with soil composites used to estimate mean Ra-226 concentrations at an UMTRA remedial action site*, PNL-6262. Pacific Northwest Laboratory, Richland, WA.

Hoffman, F. O. and R. H. Gardner (1983). Evaluation of uncertainties in radiological assessment models. In: J. E. Till and H. R. Meyer, eds., *Radiological Assessment*, NUREG/CR-3332, U.S. Nuclear Regulatory Commission, Washington DC, 11-1–11-55.

IAEA (1989). *Evaluating the reliability of predictions made using environmental transfer models*. International Atomic Energy Agency Safety Series No. 100, Vienna.

Johnson, R. A., S. Verrill and D. H. Moore II (1987). Two-sample rank tests for detecting changes that occur in a small proportion of the treated population. *Biometrics* **43**, 641–655.

Journel, A. G. (1983). Nonparametric estimation of spatial distributions. *Math. Geol.* **15**(3), 445–468.

Journel, A. G. (1984). *Indicator approach to toxic chemical sites* (EPA Project No. CR-811235-02-0-1984). Stanford University, Stanford, CA.

Journel, A. G. (1988). Nonparametric geostatistics for risk and additional sampling assessment. In: L. H. Keith, ed., *Principles of Environmental Sampling*. American Chemical Society, Washington, DC, 46–72.

Keith, L. H. (1991). *Environmental Sampling and Analysis, A Practical Guide*. Lewis Publishers, Chelsea, MI.

Kennedy, Jr., W. E. (1992). *Residual radioactivity criteria*, PNL-SA-21061. Pacific Northwest Laboratory, Richland, WA.

Lambert, D., B. Peterson and I. Terpenning (1991). Nondetects, detection limits, and the probability of detection. *J. Amer Statist. Assoc.* **86**, 266–277.

Liebetrau, A. M. and M. J. Scott (1991). Strategies for modeling the uncertain impacts of climate change. *J. Policy Model.* **13**(2), 185–204.

McIntyre, G. A. (1952). A method for unbiased selective sampling, using ranked sets. *Austral. J. Agricult. Res.* **3**, 385–390.

Morgan, M. G. and M. Henrion (1990). *Uncertainty, a Guide to Dealing with Uncertainty in Quantitative Risk and Policy Analyses*. Cambridge University Press, Cambridge.

Myers, W. L. (1990). Discussion of the paper 'Statistical sampling and analysis issues and needs for testing attainment of background-based cleanup standards at superfund sites', by R. O. Gilbert and J. C. Simpson. In: H. Lacayo, R. J. Nadeau, G. P. Patil and L. Zaragoza, eds., *Proc. Workshop on Superfund Hazardous Waste: Statistical Issues in Characterizing a Site: Protocols, Tools, and Research Needs*. Center for Statistical Ecology and Environmental Statistics, Pennsylvania State University, University Park, PA, 17–18.

Myers, R. G. and S. J. Gianti (1989). The observational approach for site remediation at federal facilities. *Superfund '89 Proc. 10th Nat. Conf.* November 27–28, Washington, DC.

Neptune, D., E. P. Brantly, M. J. Messner and D. I. Michael (1990). Quantitative decision making in superfund: A data quality objectives case study. *Hazard. Mater. Control* **3**, 18–27.

O'Brien, R., B. K. Sinha and W. P. Smith (1991). A statistical procedure to evaluate clean-up standards. *J. Chemometrics* **5**, 249–261.

Overton, W. S., D. White and D. L. Stevens Jr. (1990). Design Report for EMAP, environmental monitoring and assessment program, EPA/600/3-91/053. Environmental Research Laboratory, U.S. Environmental Protection Agency, Corvallis, OR.

Patil, G. P. and C. Taillie (1990). Evaluating the attainment of interim cleanup standards. In: H. Lacayo, R. J. Nadeau, G. P. Patil and L. Zaragoza, eds., *Proc. Workshop on Superfund Hazardous Waste: Statistical Issues in Characterizing a Site: Protocols, Tools, and Research Needs.* Center for Statistical Ecology and Environmental Statistics, Pennsylvania State University, University Park, PA, 180–210.

Patil, G. P., A. K. Sinha and C. Taillie (1992). Ranked set sampling· Annotated bibliography, First Draft. Center for Statistical Ecology and Environmental Statistics, Pennsylvania State University, University Park, PA.

Peak, C. E. (1969). Advantages and limitations of the observational method in applied soil mechanics. *Geotechnique* **19**, 171–187.

Ryti, R. T. and D. Neptune (1991). Planning issues for superfund site remediation. *Hazard. Mater. Control* **4**, 47–53.

Skiles, J. L., A. Redfearn and R. K. White (1991). Determining the number of samples required for decisions concerning remedial actions at hazardous waste sites. In: *Environmental Remediation '91*, U.S. Department of Energy, Field Office, Richland, WA, 195–198.

Smyth, J. D. and R. D. Quinn (1991). The observatiion approach in environmental restoration, PNL-SA-18817. Pacific Northwest Laboratory, Richland, WA.

Thompson, S. K. (1990). Adaptive cluster sampling, *J. Amer. Statist. Assoc.* **85**, 1050–1058.

Thompson, S. K. (1991a). Adaptive cluster sampling: Designs with primary and secondary units. *Biometrics* **47**, 1103–1115.

Thompson, S. K. (1991b). Stratified adaptive cluster sampling. *Biometrika* **78**, 389–397.

Thompson, S. K. (1992). *Sampling.* Wiley, New York.

USEPA (1983). Standard for remedial actions at inactive uranium processing sites; Final rule (40 CFR Part 192). *Federal Register* **48** (3), 590–604.

USEPA (1987a). Data quality objectives for remedial response activities, development process, EPA/540/G-87/003. U.S. Environmental Protection Agency, Washington, DC.

USEPA (1987b). Data quality objectives, example scenario, EPA/540/G-87/004. U.S. Environmental Protection Agency, Washington, DC.

USEPA (1989a). Methods for evaluating the attainment of cleanup standards, Vol. 1, Soils and Solid Media, EPA/230/02-89/042. Statistical Policy Branch, U.S. Environmental Protection Agency, Washington, DC.

USEPA (1989b). Statistical analysis of ground-water monitoring data at RCRA facilities, interim final guidance. Office of Solid Waste, U.S. Environmental Protection Agency, Washington, DC.

USEPA (1989c). Risk assessment guidance for Superfund, Vol. I: Human health evaluation manual (Part A), interim final, EPA/540/1-89/002. Office of Emergency and Remedial Response, U.S. Environmental Protection agency, Washington, DC.

USEPA (1991). Data quality objectives process for planning environmental data collection activities. Draft. Quality Assurance Management Staff, Office of Research and Development, U.S. Environmental Protection Agency, Washington, DC.

USEPA (1992). Characterizing heterogeneous wastes: Methods and recommendations, EPA/600/R-92/033, U.S. Environmental Protection Agency, Washington, DC.

WSDOE (1992). Statistical guidance for ecology site managers. Washington State Department of Ecology, Olympia, WA.

G. P. Patil and C. R. Rao, eds., *Handbook of Statistics, Vol. 12*

28

Environmental Risk Estimation and Policy Decisions

Herbert Lacayo Jr.

Abstract: Estimators of risk in environmental work are frequently chosen for intuitive, historical or mathematical reasons that are unrelated to government policy and sometimes are unfortunately even suboptimal. We give simple examples to support this claim. Specifically we show how sample size and economic loss affect the choice of an estimator. The author concludes that it is appropriate to consider more than one estimation procedure.

1. Introduction

The purpose of this chapter is to stimulate a discussion among EPA and other environmental scientists on choosing good statistical estimators. The idea is to choose those procedures which are reasonable with respect to the analytical features of the problem at hand and are 'best' with respect to the underlying policy objectives of the government. The main contention implied below is that decision makers and managers should be encouraged to examine the respective situation of any estimation problem they may have and to be allowed to choose, with consultation if necessary, *the most appropriate estimator from among several legitimate candidates.*

Examples are presented to convince the reader that in certain cases the estimator used will have a direct bearing on the risk a manager will take. Statistically there is little new in this chapter and, in fact, most of the material was created more than 40 years ago. This is rather a synthesis of theory that is slowly (perhaps too slowly) making its way into current teaching and applied work.

The chapter is written largely in statistical language and aimed at EPA and associated private sector statisticians. However, a strong attempt has been made to reach the general scientist or engineer by making the examples reasonably accessible.

2. The 'best' estimator

Consider the following hypothetical situation. A quart of soil or water is tested to see if it passes a contamination test. The test only tells us either that the sample passed the test and is therefore considered uncontaminated (clean); or the sample failed the test and is considered contaminated (dirty). It is necessary to estimate from a small sample the fraction of contaminated quarts of soil from among an extremely large number (i.e., an effectively infinite number) of such batches. This fraction is denoted by p. Formally we state: p is the probability that a quart of soil chosen at random is contaminated.

In what follows p is equivalenty considered the same as the true fraction of the land that is contaminated (i.e., dirty). For illustrative purposes suppose that:

> The estimate is based on only five samples. And the only information we receive from the sample is that either it passes (P) or it fails (F).

Appropriate notation is introduced as follows. Denote the outcome, X, of a sample quart either passing or failing by

$$X = 0 \quad \text{if the sample passes (is clean),}$$
$$X = 1 \quad \text{if the sample fails (is dirty).}$$

Under the above set-up one is asked to decide the 'best' procedure for estimating the fraction (p) of contamination from among the following three frequently used methods (where D1, D2, D3 are used to denote the method or procedure and where $d1$, $d2$, $d3$ are used to denote the estimator of p associated with each method).

(D1) Estimate p by

$$\begin{aligned} d_1 &= \sum x_i/n \\ &= k/n, \end{aligned}$$

where k is the number of failed quarts in a sample of size n.

Here p is estimated by the well-known ratio of the total number of failed quarts divided by the sample size.

(D2) Estimate p by

$$\begin{aligned} d_2 &= [\sum x_i + 1]/[n+2] \\ &= (k+1)/(n+2). \end{aligned}$$

This is the ratio of the total number of failed quarts plus 1 all divided by (sample size plus 2). Here the estimate is constructed so as to minimize risk under the assumption that all possibilities of p from 0 to 1 are equally likely. This results in a slight bias toward 0.5.

(D3) Estimate p by

$$d_3 = [\sum x_i + \tfrac{1}{2}\sqrt{n}/2]/[n + \sqrt{n}]$$
$$= (k + \tfrac{1}{2}\sqrt{n}/2)/(n + \sqrt{n}).$$

In words the estimate is given by [(Total number of failed samples) plus (the square root of the sample size) /2)] all divided by (the sample size plus the square root of the sample size). This estimate is constructed so as to minimize an individual or company or government agency's chances of sustaining a very large loss. This estimate also results in a slight bias.

Which is the 'best' estimation method?

You guessed it! All three methods are 'best' depending on the assumptions and on implied policy and management objectives. For example, if consistency of procedure, familiarity, and/or adherence to estimators with good large sample properties are highly valued then D1 could well be the 'best' procedure. D1 is known as the maximum likelihood estimate (MLE) for p.

Roughly, the MLE is based on the intuitive notion that having observed a sample, the MLE is constructed so that when it is used for p it yields the highest probability in the assumed model for the observed sample. The MLE has many desirable properties. For example, it is invariant in the sense that the 'MLE of $h(p)$' is the same as 'h (MLE of p)'. It has large sample (i.e., asymptotic) properties of efficiency, normality and consistency. This type of estimator is commonly used without questioning whether or not it best applies to the problem at hand. Quoting the well-known statistician Lucien Le Cam: 'In view of R. A. Fisher's vast influence, it is perhaps not surprising that the presumed superiority of the method (of MLE) is still for many an article of faith promoted with religious fervor.' Along this line it is illustrated below (Figures 1a–1c and 2–4) that for very small sample sizes the MLE may be inappropriate.

The procedure D2 represents a move into a broader reality. It may happen that a consensus has been reached in which it is agreed that squared error loss realistically represents economic, social or political reality. By this we mean that the expected loss to society due to implementing decision d (when the true state of nature is p) is represented by the loss function

$$W = c(p - d)^2,$$

where

d is our decision (i.e., our choice for the estimate of p),

p is the true state of nature (i.e., the true fraction of polluted land),

c is an appropriate constant so that W is in dollars.

For example, if $d = 0.4$ and $c = 400$ while in fact $p = 0.3$, then the loss is

$$W = \$400(0.3 + 0.4)^2 = \$4.$$

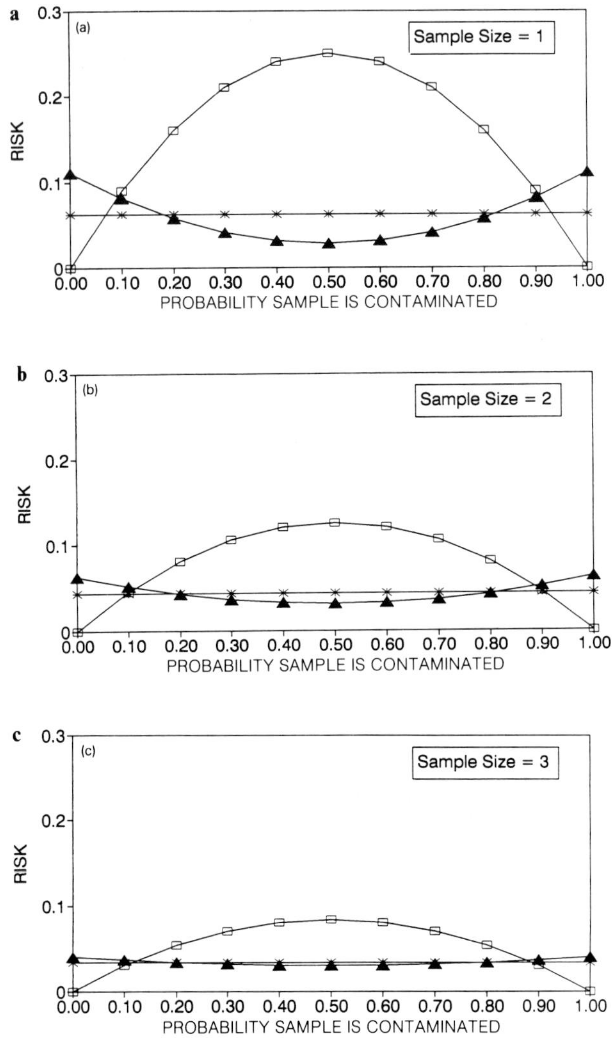

Fig. 1. Hazardous waste site. Risk under squared error loss; (a) sample size = 1, (b) sample size = 2, (c) sample size = 3. —□— MLE, —▲— Bayes, —∗— Minimax.

However, if $d = 0.9$ while in fact $p = 0.3$, then the loss is

$$W = \$400(0.3 - 0.9)^2 = \$144.$$

The idea behind the above loss function is that if the management decision or estimate d is close to the real value of p (i.e., close to the state of nature), then the loss will be small, say \$4. However, if the decision d differs greatly from p (as in $d = 0.9$ vs $p = 0.3$), then the loss is large, say, \$144.

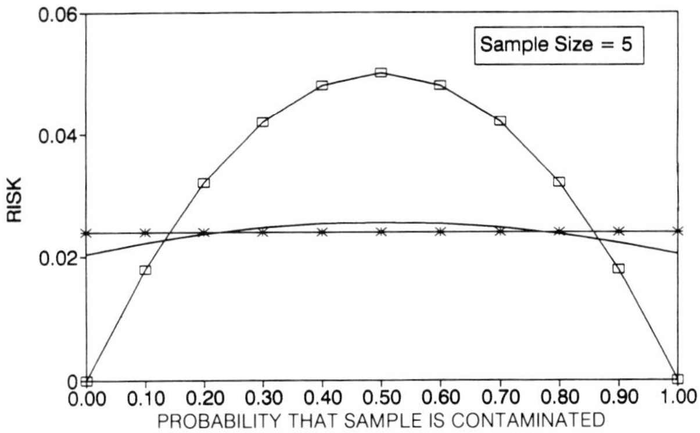

Fig. 2. Hazardous waste site. Risk under squared error loss; sample size = 5, —□— MLE, ——— Bayes, —✱— Minimax.

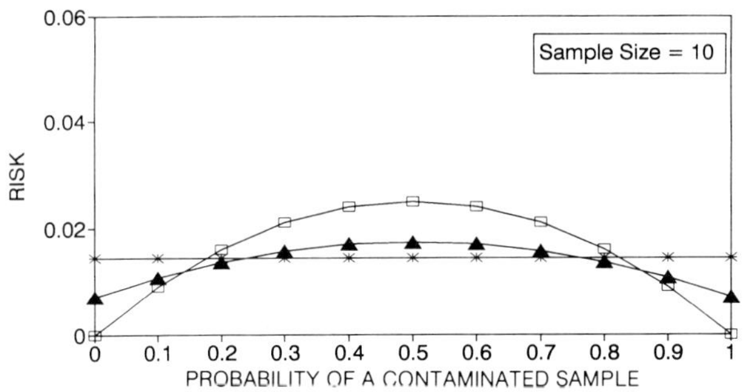

Fig. 3. Hazardous waste site. Risk under squared error loss; sample size = 10, —□— MLE, —▲— Bayes, —✱— Minimax.

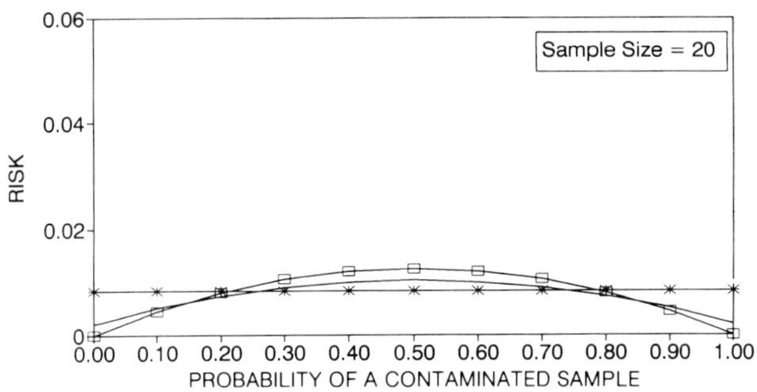

Fig. 4. Hazardous waste site. Risk under squared error loss; sample size = 20, —□— MLE, ——— Bayes, —✱— Minimax.

Another way of saying this is that the above function W works so that a small mistake in estimating p will result in only a small (but quardratically proportional) loss while a large mistake will result in a large (quadratically proportional) loss.

Under the above loss function W the procedure D2 is called a Bayes estimate under a uniform 'so-called noninformative' prior (i.e., the state of nature p is assumed a priori to be uniformly distributed from 0 to 1 since there is no reason to give preference to any state). Under the MLE (D1) procedure the state of nature p is considered a fixed parameter of the distribution. However, in a Bayesian analysis, such as D2, the state of nature p is treated as a variable with it own distribution. By using the assumed distribution of p together with data, if any, an enhanced estimate of the p parameter is sometimes possible. The term Bayes comes from the use of Bayes formula that is used to calculate that estimate.

If you are given detailed information on the economics of the situation, the procedure D2 may very well be the 'best' procedure. For example, D2 is optimal (i.e., best) in the sense that:

If you are in total ignorance of the likely value of p, the area under the plotted risk function is a minimum for the above type risk function. See Figures 1a, 1b and 1c where sample sizes of 1, 2 and 3 are used to make this point more obvious. But the same is true for higher sample sizes as can be seen in Figures 2–4.

Formally, D2 in this case minimizes the average or expected loss W under the condition that p is uniformly distributed from 0 to 1.

Observe the large difference between the estimators d_1 and d_2 of p for very small sample size n. For example, if n is 2 and k is 0, then d_1 gives the estimate for p of 0 while d_2 gives $\frac{1}{4}$. This is a substantial difference. It is just these differences that can cause large differences in loss calculations. This is illustrated in the figures given below.

Procedure D3 is similar in spirit to D2 in that it assumes the same loss function W and the noninformative uniform prior on p. It differs in that it strives to minimize your risk for the worst possible case. For example, a decision maker may want an estimator that protects against the worst-case scenario in a toxic site containment problem. Formally, an estimator of this type, such as d_3, is called a Minimax estimator. A Minimax estimator (i.e., a Minimax decision rule) is constructed to have the smallest possible maximum risk so the maximum possible risk for any other estimator (i.e., any other decision rule) is at least as big as that for the Minimax estimator.

Although such estimators are sometimes difficult to find, they have several features that should concern a decision maker. Foremost is that they can protect one against the worst risk. This is important in situations where a worst-case analysis is desired. For example, the worst risk that a planner wishes to guard against in a situation may be more important than having an overall low average risk (this situation is illustrated later in Figure 13 comparing the Bayes and Minimax estimates). The government implicitly does many such worst-case studies where the worst realization in a sequence of events is assumed to occur.

Some additional comments on the potential usefulness of Minimax estimators will be helpful. In addition to protecting against the worst risk, the estimators sometimes result in an 'equalizer rule'. That is a rule in which the risk is the same forall values (see the Minimax risk in Figure 2) of the parameter space (i.e., all allowable values of p). If this overall risk is low, then it is sometimes advantageous to use a estimator that results in constant risk. For example, a constant risk factor lowers the overall variability (and hence frequently lowers cost) in contingency planning.

In classical statistics, estimators such as the MLE are constructed to satisfy a variety of criteria. Most of these criteria are related to theoretical conveniences or to some plausible or intuitive appeal such as unbiasedness or invariance. Some are 'best' with respect to reduced or minimum error in some subclass of estimators ... ; for example, such an estimator which is unbiased and at the same time has the smallest variance among all unbiased estimators; or for example, a best linear unbiased estimate (which is the unbiased linear estimator that has the smallest variance among all linear unbiased estimators). But mostly, such estimators appealing though they may be mathematically or statistically, seldom have any direct relation to Government policy and objectives. In what follows, the risk to the decision makers of D1 through D3 will be compared on the basis of squared errors loss. Squared error loss is not necessarily the most appropriate loss function to use for site cleanup. However, it is used as a starting point in this chapter for the following reasons. First, its close relation to the variance of distribution. Second, it is very commonly used in the literature. Third, it illustrates quite well the fact that different estimators can yield importantly different risks. Last, and unfortunately not least, squared error loss is mathematically tractable.

Risk r is formally defined as a function of the decision d and the state of nature p,

$$r = \mathrm{E}\{W\},$$

where the expectation $\mathrm{E}\{\,\}$ of the loss W is taken with respect to the underlying distribution of X; in this case the binomial. $\mathrm{E}\{W\}$ is interpreted to mean the long term avearge loss of W. In the examples presented here, the binomial distribution is the standard probability distribution associated with the number k of failed samples out of a total of n independent samples where the probability of the failure of a single sample is p:

$$\mathrm{P}\left(\sum x_i = k\right) = \binom{n}{k} p^k (1-p)^{n-k}.$$

In words, the risk r of deciding d (i.e., using estimator d) when p is the state of nature, is defined to be the expectation of the loss function W where W has been defined above. For example, given a state of nature (i.e., a value of p) and several estimators (say d_1, d_2, d_3), one is then able to compare these estimators in terms of their risk r.

This comparison will show that for a given p value the estimator with smallest risk r is a function of sample size. What estimator should be used is a function not only of sample size but also of policy objectives.

3. The 'best' estimator as a function of sample size

Given the above simplified situation, the calculation formulas of the risk functions (with respect to squared error loss) of estimators d_1 through d_3 for sample size n are:

(R1) $$r_{d_1}(p) = p(1-p)/n,$$

(R2) $$r_{d_2}(p) = [1 + (n-4)\,p(1-p)]/(n+2)^2,$$

(R3) $$r_{d_3}(p) = 1/(4(n^5+1)^2).$$

These formulas are straightforward. For example, R1 is derived as follows: Denote d_1 by $Y = \sum x_i/n$. Then

$$\begin{aligned} r_{d_1}(p) &= \mathrm{E}[(p-Y)^2] \\ &= p^2 - 2p\mathrm{E}[Y] + \mathrm{E}[Y^2]. \end{aligned}$$

But $\mathrm{E}[Y\} = p$, and

$$\begin{aligned} \mathrm{E}[Y^2] &= \mathrm{var}(Y) + (\mathrm{E}[Y])^2 \\ &= p(1-p)/n + p^2. \end{aligned}$$

Hence,

$$\begin{aligned} r_{d_1}(p) &= p^2 - 2p^2 + p(1-p)/n + p^2 \\ &= p(1-p)/n. \end{aligned}$$

It is informative to compare risk functions R1 through R3 for various values of p and n. This is illustrated in Figures 2–10.

Before proceeding with a discussion of these Figures, we note that d_1, the MLE, is the only unbiased estimate of p among d_1, d_2 and d_3. However, we will see that

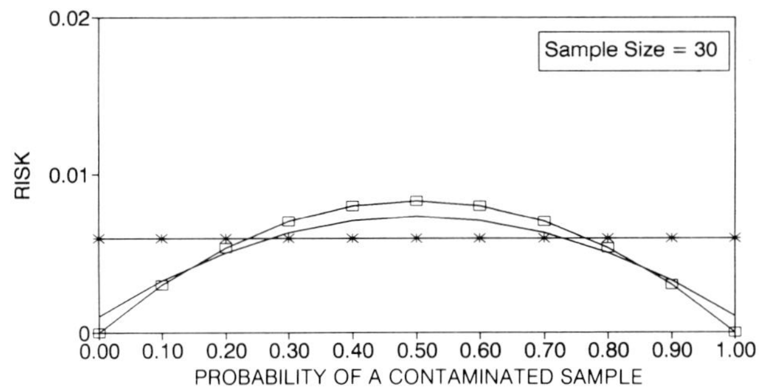

Fig. 5. Hazardous waste site. Risk under squared error loss; sample size = 30, —⊞— MLE, —— Bayes, —✱— Minimax.

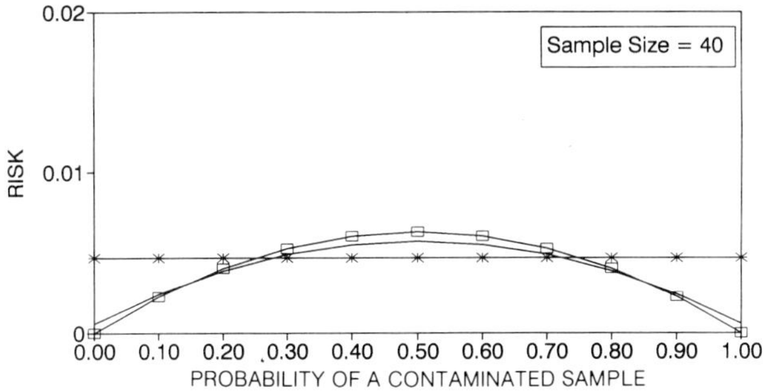

Fig. 6. Hazardous waste site. Risk under squared error loss; sample size = 40, —⊟— MLE, —— Bayes, —✱— Minimax.

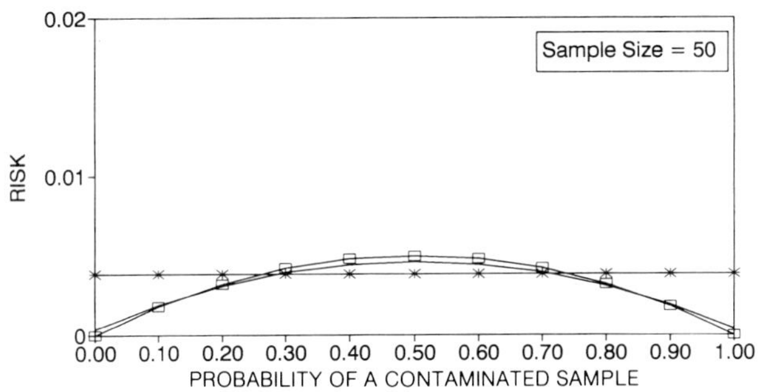

Fig. 7. Hazardous waste site. Risk under squared error loss; sample size = 50, —⊟— MLE, —— Bayes, —✱— Minimax.

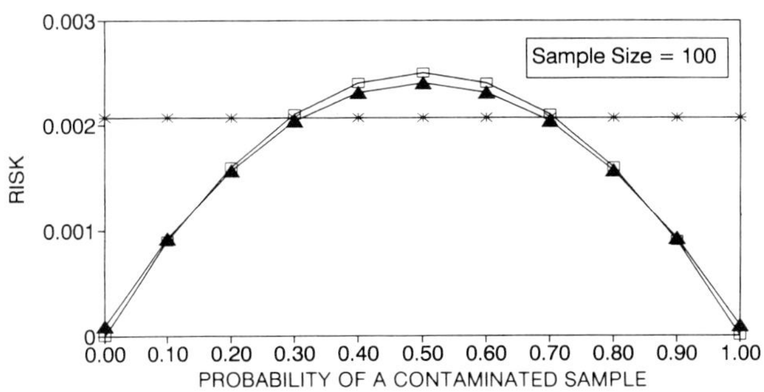

Fig. 8. Hazardous waste site. Risk under squared error loss; sample size = 100, —⊟— MLE, —▲— Bayes, —✱— Minimax.

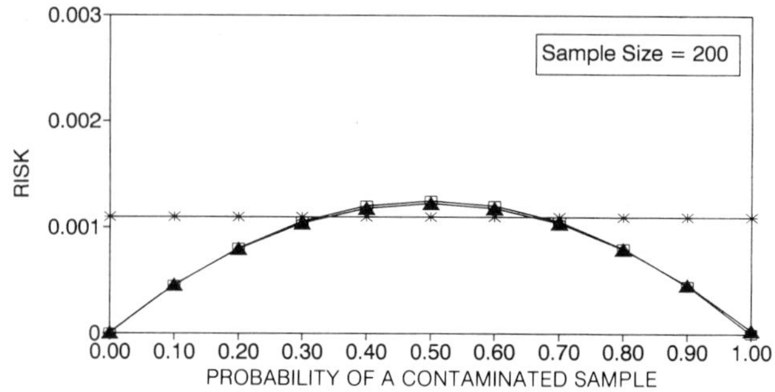

Fig. 9. Hazardous waste site. Risk under squared error loss; sample size = 200, —⊟— MLE, —▲— Bayes, —✱— Minimax.

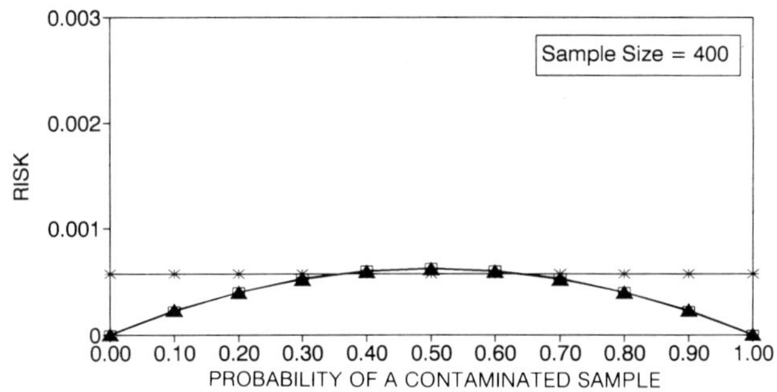

Fig. 10. Hazardous waste site. Risk under squared error loss; sample size = 400, —⊟— MLE, —▲— Bayes, —✱— Minimax.

unbiasedness by itself is not an overriding virtue in all cases. In any event, too large a bias will result in a large loss by virtue of our definition of W; and so badly biased estimators will perform poorly.

We start with Figure 2 which compares the risk of estimators d_1 to d_3 as a function of squared error loss when the sample size is $n = 5$. The interpretation depends on what is known about the situation. It seems clear that d_1, the MLE, is a poor choice for an estimator since it is outperformed by d_2, the Bayes and d_3, the Minimax estimators over the range $0.1 < p < 0.9$. Therefore, the Bayes estimator d_2 and the Minimax estimator d_3 appear to be the estimators of choice for $n = 5$ unless there are ancillary reasons to believe that p is very close to 0 or 1.

As the sample size increases from 5 to 20 (Figures 2–4) the situation changes somewhat. The Bayes estimator is still preferable to the MLE in the sense that the Bayes has smaller risk over about 70% of the range of p. The Minimax risk is lower than the MLE risk over about 60% of the range of p. In this case it happens that the area under the Bayes risk is minimum while that of the MLE is the largest of the three. The areas under the risk curves are easily calculated. For example, if $n = 10$ we have

$$\text{area(R1)} = \int p(1-p)/10\,\mathrm{d}p$$
$$= 0.0167.$$

Similarly

$$\text{area(R2)} = 0.0081 \quad \text{and} \quad \text{area(R3)} = 0.0144.$$

When the sample size increases from 30 to 50 (Figures 5–7) the choices between Minimax, Bayes and MLE become harder to make with perhaps the Minimax having the edge due to its constant risk. As the sample sizes become very large (Figures 8–10), the MLE and the Bayes estimates tend to converge and are better than the Minimax estimator except on a short region near $p = 0.5$.

At small sample sizes it may be advantageous, when using conservative policies, to use the Minimax estimation since it affords protection against very high risks. Its use may also be helpful when the Minimax estimator gives a constant risk (i.e., is an equalizer rule) and that estimate is a subpart of an overall process where keeping variability to a minimum is desirable. Without considering policy objective, Figures 8–10 indicate that for 'large' sample sizes the Minimax becomes less attractive than the Bayes and MLE; and these last two become operationally equivalent since the Bayes converges to the MLE. But if the sample size is'large' then most traditional estimators will yield the same conclusions and simple estimates suffice.

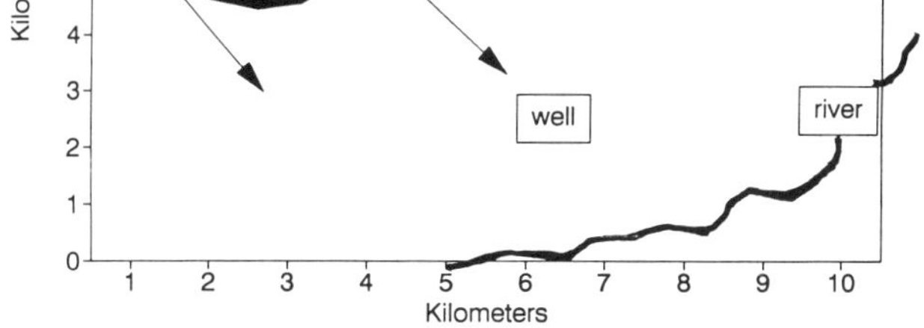

Fig. 11. Hypothetical superfund site.

4. A hazardous waste site

Assume that a hazardous waste site has undergone cleanup treatment by the responsible party and the record of decision associated with that site states that the down gradient ground water will be sampled 12 times per year. It is assumed that independent random sampling is possible in that time frame and that appropriate measures have been taken to insure that soil media on water flow rates have been taken into account.

Figure 11 is an outline of the site and will be used for illustrative purposes. The small site is located at the foot of some mountains that start a watershed for the region. Several wells have been placed down gradient of the site by hydrologists and those locations that are optimal for the detection of any possible leakage.

For illustrative purposes, a simplified analysis of the site is carried out under the following conditions.

(1) Assume that the number k of failed samples from n observed samples follows a binomial distribution. This follows naturally if the test used returns only the fact that the sample either failed or passed a certain test. For example, it either killed certain bacteria in a petri dish (failed) or it did not kill the bacteria (passed).

(2) Assume squared error loss. This is the same loss function as given above and used in Figures 2 through 10.

(3) Assume a decision must be made with respect to choosing an estimator and that further sampling is not allowed.

(4) Assume MLE, Bayes and Minimax are the only estimators that management will consider.

Then the risk situation will look like Figure 12. There are no surprises here. The Bayes and Minimax estimators on a purely risk basis appear superior to the MLE at the mid-ranges of the possible values for p while the MLE provides smaller risk for low and high values of p. For example, if $p = 0.1$ then the MLE, Bayes and Minimax risks are respectively 0.0075, 0.0088 and 0.0125, showing that MLE has the least risk and Minimax the greatest. However, at $p = 0.5$, then the risks are 0.0208, 0.0153 and

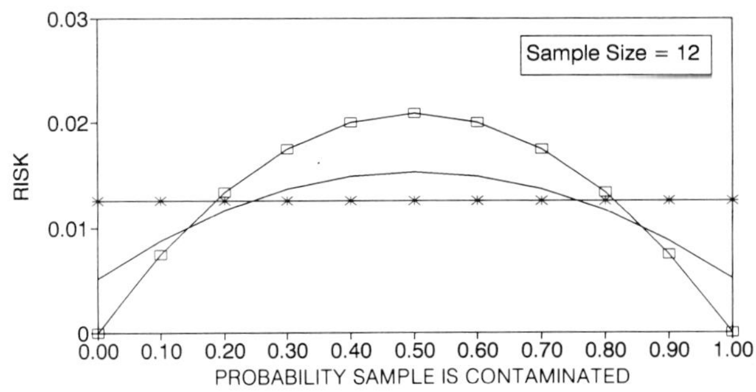

Fig. 12. Hazardous waste site. Risk under squared error loss; sample size = 12, —⊟— MLE, —— Bayes, —✱— Minimax.

0.0125 respectively, contrarily showing that the MLE has the greatest risk and the Minimax the smallest.

Can policy enter into the choice of estimators? Does it make any difference? It can and it does. The following situation, which has somewhat more realistic cost assumptions than those used to generate Figures 2 through 10 will be assumed for a risk analysis of the site.

Consider the following scenario:

(1) A cleanup standard CS has been specified. A CS in this chapter is equivalent to a specific target contamination level that cleanup remediation efforts try to meet or exceed. For example, if the CS $= 0.1$ and testing indicates (via the value of some estimator d as given by, say, d_1, d_2, d_3, or some other estimator) that the contamination p is greater than 0.1, then the CS has not been met.

(2) The squared error loss (i.e., $C(p-d)^2$) function used for illustrative purposes is unrealistic, not only due to the symmetry about p but also because it does not directly incorporate the CS. The following type of loss function will be used for an analysis of the hazardous waste site where the cleanup standard is CS $= 0.1666$.

(a) The following site cleanup costs are identified. First there is the cost associated with extra cleanup of the site if the site manager decides that remediation efforts have failed and more site remediation is deemed necessary. For this example, such cost constant in W is specified by

$$\begin{aligned} C_1 &= \text{extra cleanup costs} \\ &= 100. \end{aligned}$$

Second, there is the cost associated with adverse health or environmental effects if the site manager decides the site cleanup efforts have met the CS, but, in fact, the site is still contaminated. For this example the cost constant in W is specified by

$$\begin{aligned} C_2 &= \text{health costs} \\ &= 1000. \end{aligned}$$

(b) The underlying distribution is still binomial.

(c) The loss function W is specified by the following four possible situations where the state of nature (i.e., the real contamination level) is denoted by p and the estimate is denoted by d (where d in this analysis will come from one of the procedures D1, D2, or D3).

If p and $d \leqslant$ CS, then loss $= 0$ since the site is really clean and the site manager decides that it is.

If $p \leqslant$ CS and CS $\leqslant d$, then loss $= C_1$ since the site is clean but the site manager wrongly decides that the site is dirty and needs to be recleaned.

If CS $\leqslant p$ and d, then loss $= C_1$ since the site is dirty and the site manager rightly decides that the site is dirty and needs to be recleaned.

If $d \leqslant$ CS and CS $< p$, then loss $= C_2$ since the site is dirty but the site manager wrongly decides that the site is clean and society incurs a health or environmental cost.

The above loss function is translated into the corresponding risk function below:

$$r^d(p) = C_1 * P(d > CS) \qquad \text{for } p \leqslant CS,$$
$$r^d(p) = C_1 * P(d > CS) + C_2 * P(d < CS) \qquad \text{for } p > CS.$$

The risk function, $r^d(p)$ was calculated for the estimator d_1, d_2 and d_3, given previously. The results are graphed in Figure 13 for sample size 12 and $CS = \frac{2}{12}$.

Figure 13 (note also the situation carries forth in Figures 14 and 15 for higher sample sizes and a slightly lower CS) can be viewed in a variety of ways. Consider the following interpretation. If the site manager is confident that cleanup has been effective in the sense that she or he believes that site contamination is likely to be low, then using the MLE appears 'best' due to its relatively low risk for values of p less

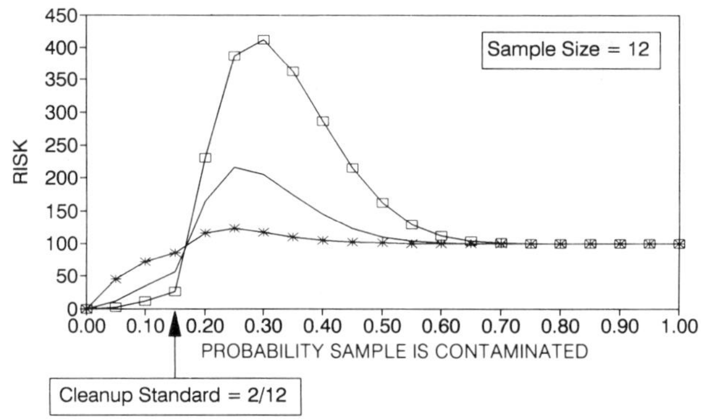

Fig. 13. Hazardous waste site. Economic health risk; sample size = 12, —□— MLE, ——— Bayes, —∗— Minimax.

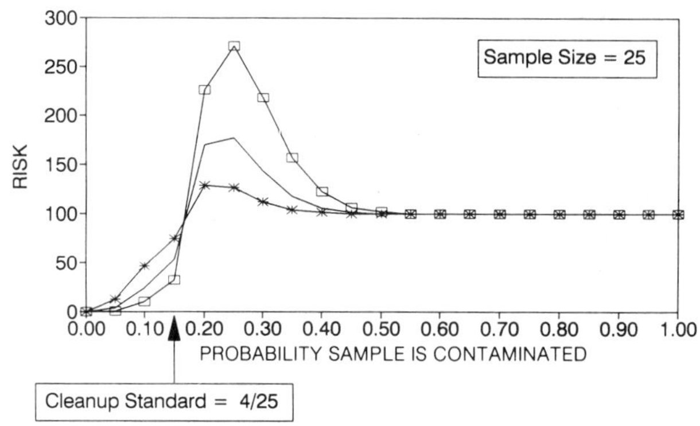

Fig. 14. Hazardous waste site. Economic health risk; sample size = 25, —□— MLE, ——— Bayes, —∗— Minimax.

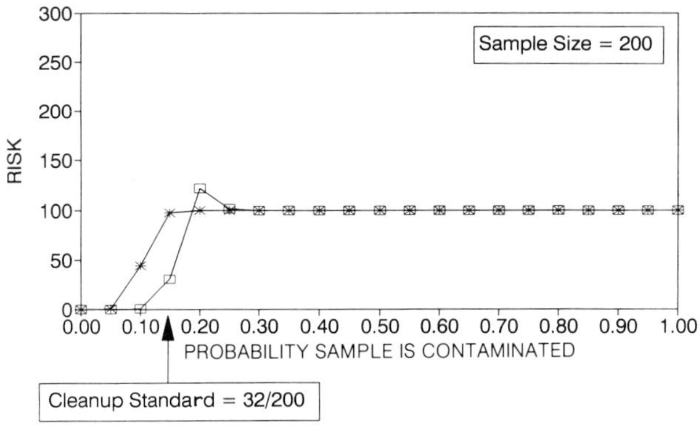

Fig. 15. Hazardous waste site. Economic and health risk; sample size = 200, —⊟— MLE, —*— Minimax.

than 0.1666. However, if policy mandates that risk above 150 should be avoided (or for that matter if it is desirable for the overall risk to be low) then the Minimax is the estimator of choice. We found this to be a surprising result since Minimax estimators are so seldom used in environmental work. It will be of great interest to see if further analysis sustains the above conclusion.

Let us end this chapter with one last thought. There is of course another risk, the greatest risk of all. This is the risk to the site manager if he or she makes a wrong decision with respect to attainment of cleanup standards.

References

Berger, J. O. (1985). *Statistical Decision Theory and Bayesian Analysis.* Springer, New York.

Kiefer, J. C. (1987). *Introduction to Statistical Inference.* Springer, New York.

Lacayo, H., R. Nadeau, G. P. Patil and L. Zaragoza, eds. (1990). *Proc. Workshop on Superfund Hazardous Waste.* Center for Statistical Ecology and Environmental Statistics of the Pennsylvania State University.

Le Cam, L. (1986). *Asymptotic Methods in Statistical Decision Theory.* Springer, New York.

U. S. Environmental Protection Agency (1989). *Methods for Evaluating the Attainment of Cleanup Standards*, Vol 1: Soils and Solid Media. Office of Policy Planning and Evaluation, Washington, DC (EPA 230/02-89-042).

U. S. Environmental Protection Agency (1992). *Methods for Evaluating the Attainment of Cleanup Standards.* Vol 1: Office of policy Planning and Evaluation, Washington, DC (EPA 230-R-92-014).

Biographical Information

Michael AKRITAS is Professor in the Department of Statistics at the Pennsylvania State University. His research interests have been in the analysis of censored data in general, and in the fields of astronomy and ecology in particular. (Chapter 7.)

Jacques BEDAUX is Assistent Professor in Biostatistics at the Department of Theoretical Biology of the Vrije Universiteit Amsterdam. He is involved in several ecotoxicological projects and in the School of Environmental Sciences. He has a long experience as statistical consultant, where most problems relate to environmental research. (Chapter 16.)

Dr. M.T. BOSWELL has been a longtime colleague and collaborator in the Penn State Center for Statistical Ecology and Environmental Statistics. (Chapter 24.)

Harold E. BURKHART is a Professor of Forest Biometrics at Virginia Polytechnic Institute and State University (VPI & SU), Blacksburg, VA, USA. He has been a faculty member at VPI & SU since 1969; he has published extensively on forest growth and yield prediction and on forest inventory and sampling. (Chapter 11.)

Paul J. CATALANO is Assistant Professor, Department of Biostatistics, Harvard School of Public Health and Dana-Farber Cancer Institute, Boston, MA 02115. Dr. Catalano's doctoral thesis focused on dose-response modeling involving discrete and continuous outcomes in developmental toxicity studies. His research interests include multivariate dose-response models for laboratory assays, methods for correlated data, and methods for risk assessment. (Chapter 14.)

C. Richard COTHERN is on the Center For Environmental Statistics Development Staff and is responsible for symposia, conferences, advisory committees and research planning. He is a past executive secretary of the US Environmental protection Agency's Science Advisory Board and teaches environmental chemistry at the George Washington University. He has edited ten books on topics including: environmental radon, radionuclides in drinking water, environmental risk assessment, environmental lead and trace substances in environmental health. He has over 70 published papers in these areas along with nuclear physics, environmental physics and surface chemistry. (Chapter 25.)

Charles B. DAVIS received a Ph.D. in Statistics from the University of New Mexico in 1976. He then joined the Department of Mathematics of The University of Toledo, where he formed and headed the M.S. program in statistical consulting. After becoming involved in environmental regulatory statistics during the 1986–87 debate over revisions to the RCRA ground-water monitoring statistical rules, he and Roger J. McNichols formed Environmetrics & Statistics Limited in 1990 to provide consulting, analytical, and research services in environmental regulatory statistics and related areas. (Chapter 26.)

Christl A. DONNELLY is currently a Temporary Lecturer at the Department of Mathematics and Statistics at the University of Edinburgh. She completed this work as a doctoral student in the Department of Biostatistics at Harvard School of Public Health. Her research interests have focussed on the analysis of correlated longitudinal and spatial data. (Chapter 19.)

Dr. R. O. GILBERT is a Staff Scientist in the Statistical Design and Analysis Group, Analytic Sciences Department, Pacific Norhtwest Laboratory, Richland, Washington. He has applied statistics to environmental pollution problems for 25 years. Dr. Gilbert is author of the book *Statistical Methods for Environmental Pollution Monitoring* published by Van Nostrand Reinhold in 1987. He is also the author or co-author of more than 25 journal papers and 60 reports. (Chapter 27.)

Dr. S. D. GORE is Associate Professor in Statistics and a Faculty Associate in the School of Environmental Sciences at the University of Poona, Pune, India. He is currently a Visiting Associate Professor in Statistics and Visiting Faculty Associate at the Penn State Center for Statistical Ecology and Environmental Statistics. (Chapter 4.)

Dr. J. H. GOVE serves as Research Forester at the USDA Forest Service, Northeastern Forest Experiment Station, P.O. Box 640, Durham, New Hampshire 03824, USA. His interest in diversity-related research centers on developing practical quantitative methods for integrating diversity measurement into forest management decisionmaking and policy analysis. (Chapter 12.)

Timothy G. GREGOIRE is an Associate Professor of Forest Biometrics at Virginia Polytechnic Institute and State University (VPI & SU), Blacksburg, VA, USA. He has been at VPI & SU since 1985, during which time he has specialized in forest sampling techniques and on error specification in forest growth and yield models. (Chapter 11.)

Dr. P. GUTTORP is Associate Professor of statistics at the University of Washington. He has been working on problems in acidic deposition since 1985. He wrote the chapters on *Spatial statistics in ecology* and *Spatial statistics in environmental sciences* in the recent National Research Council report on *Spatial Statistics and Digital Image Analysis.* His principle research interests are in spatial statistics and stochastic models. (Chapter 20.)

Vic HASSELBLAD has been collaborating for several years with David M. Eddy at Duke University in the area of meta-analysis. The results of this research have produced several papers and the book, *Meta-Analysis by the Confidence Profile Method.* Vic Hasselblad has been working with the U.S.E.P.A. on applications of the method to toxicological problems. (Chapter 21.)

J. Stuart HUNTER is Professor Emeritus, School of Engineering and Applied Science, Princeton University. He is a consultant to industry and government, the founding Editor of *Technometrics* and the 1993 President of the American Statistical Society. He has contributed to the environmental science research literature since the early 1970s and is currently a member of the editorial advisory board of *Environmetrics.* (Chapter 1.)

Bas KOOIJMAN is Full Professor and head of the Department of Theoretical Biology of the Vrije Universiteit Amsterdam and consultant of the TNO Institute for Environmental Sciences. His scientific contributions to the subject of statistics and modelling in relation with environmental problems include a large number of papers on topics such as the statistical analysis of toxicity and mutagenicity tests, of air pollution and biomonitor data, toxicokinetics and toxic effects, the extrapolation from laboratory to field situations, environmental safety factors. His department collaborates intensively with the National Institute of Public Health and Environmental Protection on risk assessment. (Chapter 16.)

Dr. H. LACAYO has been a Senior Statistician in the Office of Policy, Planning, and Evaluation of the United States Environmental Protection Agency. Among other things, he is involved with statistical issues and approaches relevant to hazardous waste site characterization, remediation, and evaluation. He recently chaired a first EPA Conference on the subject, and has been a co-editor of its proceedings volume. (Chapter 28.)

Nan M. LAIRD is Henry Pickering Walcott Professor of Biostatistics and Chair of the Department of Biostatistics at Harvard School of Public Health. Her principle research interests include the analysis of longitudinal data, algorithms for missing and incomplete data, and empirical Bayes estimation. (Chapter 19.)

Walter LIGGETT received his Ph.D. in applied mathematics from Rensselaer Polytechnic Institute. His experience with the application of statistics to environmental studies has been acquired largely in the Division of Environmental Planning of the Tennessee Valley Authority and in the Statistical Engineering Division of the National Institute of Standards and Technology, where he has been since 1979. The emphasis at TVA was on studies of the environmental impact of energy production, and the emphasis at NIST has been on support of work by chemists on the improvement of environmental measurements. Liggett has served for several years on American Statistical Association committees concerned with the environment. (Chapter 3.)

Dr. Gianfranco LOVISON is Associate Professor in the Department of Statistics, University of Padova, Italy. His main research interests are in categorical data analysis, multivariate analysis and environmental sampling and modeling. He developed his interest in composite sampling during his visit at the Center for Statistical Ecology and Environmental Statistics of the Pennsylvania State University. (Chapter 4.)

Bryan MANLY studied at the City University in London, England, and worked as an industrial statistician before taking his first academic appointment at the University of Salford in 1967. In 1969 he moved to the University of Papua New Guinea, and in 1972 to the University of Otago in New Zealand, where he still is. His research interests are in most areas of statistics as they relate to biology, and he has major publications in the areas of natural selection, stage-structured populations, computer intensive methods, and, most recently, the general principles involved in the design, execution, and analysis of research studied. (Chapter 10.)

Jim MATIS is Professor of Statistics at Texas A&M University. His research interests include the theory and application of compartmental modeling. Recent applications to ecological and environmental systems include environ analysis and African bee invasion. (Chapter 17.)

Dr. W. L. MYERS has multifaceted interests and experience in remote sensing, GIS, multivariate analysis, artificial intelligence, and computer technologies in the context of forestry, landscape ecology, and natural resource sustainability. A motivating factor in his current research is inadequacy of conventional technologies for mobilizing knowledge to deal effectively with natural resource problems that he encountered during two years as technology advisor to USAID on forestry and natural resource sustainability in India. (Chapter 18.)

Philip M. NORTH, B.Sc., M.Sc., Ph.D., C.Stat., has been a keen birdwatcher since an early age and it was the combination of that hobby with his scientific training as a statistician that brought him to the University of Kent in 1975. His doctoral thesis was on Statistics in Ornithology. After spending 2 years as a lecturer at the University of Kent at Canterbury, England he was involved in the setting up of the Applied Statistics Research Unit, of which he has been the Director (and, latterly, Managing Director) since 1984. The Unit counts environmental statistics amongst its activities and has run a successful course in Ecological Statistics repeatedly. Philip North has published extensively on various aspects of the subject of ornithological statistics and has been involved in the organisation of a series of meetings in Europe on this theme. (Chapter 13.)

Dr. J. S. O'CONNOR is at present Ocean Policy Co-ordinator in the Marine and Wetlands Protection Branch of the United States Environmental Protection Agency in New York. He was formerly a Senior Ecologist and a Biometrician in the Ocean Assessment Division of the National Oceanic and Atmospheric Administration, Washington, DC. (Chapter 24.)

W. Scott OVERTON is Professor of Statistics at Oregon State University. He has designed numerous surveys in wildlife biology and forestry, and was responsible for the sampling plans for EPA's National Surface Water Surveys. He was a principal architect of the proposed design for EPA's Environmental Monitoring and Assessment Program. (Chapter 9.)

Dr. G. P. PATIL is Distinguished Professor of Mathematical Statistics and Founder Director of the Penn State Center for Statistical Ecology and Environmental Statistics. He has received Ph.D. in Mathematics, Michigan; D. Sc. in Statistics, Indian Statistical Institute; Hon. D. Sc. in Biological Sciences, Parma; and Hon. D. Litt, Poona. He has been a founder member and sponsor of statistical ecology sections of the International Association for Ecology and the Ecological Society of America, and of the Statistics and the Environmental Section of the American Statistical Association. He is Editor-in-Chief of the *Journal of Environmental Statistics* published by Van Nostrand Reinhold and serves on the International Editorial Advisory Board of *Environmetrics*, a Journal of the International Environmetric Society. He has been a recipient of a most significant paper award of American Fisheries Society and a distinguished statistical ecologist award of the International Association for Ecology. He has been a Fellow of the Institute of Mathematical Statistics, American Statistical Association, American Association of the Advancement of Science, International Statistical Institute, and a Founder Fellow of the National Institute of Ecology, India. (Chapters 2, 4, 5, 7, 12, 24; Editor.)

Walter W. PIEGORSCH was a Mathematical Statistician in the Statistics and Biomathematics Branch, U.S. National Institute of Environmental Health Sciences, Research Triangle Park, NC. His research focuses on modeling and analysis for experiments in molecular biology and genetic toxicology, with application to environmental health studies. His current affiliation is the Department of Statistics, University of South Carolina, Columbia, SC, USA. (Chapter 15.)

Dr. C. R. RAO is Eberly Professor of Statistics and Director of the Penn State Center for Multivariate Analysis, and National Professor, India. He has received Ph. D. and Sc. D. from Cambridge University, and has been awarded numerous honorary doctorates from universities around the world. He is a Fellow of Royal Society, UK; Fellow of Indian National Science Academy; Foreign Honorary Member of American Academy of Arts and Science; and Founder Fellow of the Third World Academy of Sciences. He has been elected Fellow and President of International Statistical Institute, Biometric Society and Institute of Mathematical Statistics. He has made outstanding contributions to virtually all important topics of theoretical and applied statistics, and many results bear his name. He has been Editor of *Sankhyā* and the *Journal of Multivariate Analysis*, and serves on international advisory boards of several professional journals, including *Environmetrics* and the *Journal of Environmental Statistics*. (Editor.)

Louise M. RYAN is Associate Professor, Department of Biostatistics, Harvard School of Public Health and Dana-Farber Cancer Institute, Boston, MA 02115. Dr. Ryan has worked in the areas of carcinogenicity and developmental toxicity experiments and has published papers in both toxicological and statistical journals. Some specific areas of her research include work on the development of statistical methods for data analysis, including the use of quasi-likelihood methods for dose-response data, and application of these techniques to quantitative risk assessment. (Chapter 14.)

Thomas RUSCITTI received his B.S. and M.S. degrees in mechanical engineering, an M.S. in statistics and has worked for ten years on instrumentation and test issues. (Chapter 7.)

Dr. P. D. SAMPSON is Research Associate Professor at the University of Washington. He directs the department consulting program, and has principal research interests in multivariate analysis, morphometrics and spatial statistics. He has been working on problems in acidic deposition since 1985. He participated in the first study that resulted in unequivocal proof that emission changes can result in deposition changes. (Chapter 20.)

G. A. F. SEBER is Professor and Head of the Statistics Department, The University of Auckland, Auckland, New Zealand. (Chapter 6.)

Dr. A. K. SINHA, Professor of Statistics at Patna University, Patna, India is at present a Visiting Professor and Research Associate in the Penn State Center for Statistical Ecology and Environmental Statistics, Department of Statistics, The Pennsylvania State University. (Chapter 5.)

Dr. Eric P. SMITH is a Professor of Statistics and member of the University Center for Environmental and Hazardous Materials Studies at Virginia Polytechnic Institute and State University. His research interests include the analysis of environmental data, ecological statistics, uncertainty in risk analysis, multivariate methods and simulation modeling. He is an Associate Editor of *Environmetrics* and an editorial collaborator of the *Journal of Environmental Statistics*. (Chapter 8.)

Andrew R. SOLOW is an Associate Scientist at the Woods Hole Oceanographic Institution. He is a member of the Scientific Working Group of the Intergovernmental Panel on Climate Change (IPCC), as well as a number of advisory panels for U.S. climate change research. (Chapter 22.)

Stephen STEHMAN is Assistant Professor in the Faculty of Forestry at the College of Environmental Science and Forestry of the State University of New York (SUNY-ESF). In addition to providing statistical consulting at SUNY-ESF, he has consulted on sampling problems with EPA and the U.S. Fish and Wildlife Service. (Chapter 9.)

Dr. B. F. SWINDEL was retired Research Forester with the USDA Forest Service, Southeastern Forest Experiment Station; and subsequently, Consulting Statistician in Gainsesville, Florida, USA. Approaching the diversity issue with a background in quantitative silviculture, he was one of the first scientists in forestry to do substantive research in diversity as it relates to forest management practices. Dr. Swindel passed away unexpectedly during this project and we dedicate chapter 12 to his memory. (Chapter 12.)

Dr. C. TAILLIE serves as Senior Research Associate, Center for Statistical Ecology and Environmental Statistics, Department of Statistics, The Pennsylvania State University. He has a continuing interest in the overall issue of environmental sampling, observational economy, and statistical decision making. (Chapters 5, 12.)

S. K. THOMPSON is Associate Professor in the Department of Statistics and a faculty associate of the Center for Statistical Ecology and Environmental Statistics, The Pennsylvania State University. He has previously served on the faculties of the University of Auckland and the University of Alaska, Fairbanks. (Chapter 6.)

James H. WARE is Frederick Mosteller Professor of Public Health and Dean for Academic Affairs at the Harvard School of Public Health. He is senior statistician for the Six Cities Study of Air Pollution and Health and a co-investigator in the Environmental Epidemiology Program at HSPH. His principal methodological interests are longitudinal analysis, measurement error models, and the analysis of clustered data. (Chapter 19.)

Tom WEHRLY is Professor of Statistics at Texas A&M University. His research interests include the theory and application of compartmental modeling. Recent applications to ecological and environmental systems include environ analysis and African bee invasion. (Chapter 17.)

J. H. YOUNG is Professor Emeritus of the Department of Entomology at Oklahoma State University. He has extensive research experience in sampling agroecosystems and the application of integrated pest management systems. (Chapter 23.)

L. J. YOUNG is an Associate Professor in the Department of Biometry at the University of Nebraska-Lincoln. She consults and participates in collaborative research with scientists in UNL's Institute of Agriculture and Natural Resources. (Chapter 23.)

Subject Index

Handbook of Statistics
Contents of Previous Volumes

Volume 1. Analysis of Variance
Edited by P. R. Krishnaiah
1980 xviii + 1002 pp.

Volume 2. Classification, Pattern Recognition and Reduction of Dimensionality
Edited by P. R. Krishnaiah and L. N. Kanal
1982 xxii + 903 pp.

Volume 3. Time Series in the Frequency Domain
Edited by D. R. Brillinger and P. R. Krishnaiah
1983 xiv + 485 pp.

Volume 4. Nonparametric Methods
Edited by P. R. Krishnaiah and P. K. Sen
1984 xx + 968 pp.

Volume 5. Time Series in the Time Domain
Edited by E. J. Hannan, P. R. Krishnaiah and M. M. Rao
1985 xiv + 490 pp.

Volume 6. Sampling
Edited by P. R. Krishnaiah and C. R. Rao
1988 xvi + 594 pp.

Volume 7. Quality Control and Reliability
Edited by P. R. Krishnaiah and C. R. Rao
1988 xiv + 503 pp.

Volume 8. Statistical Methods in Biological and Medical Sciences
Edited by C. R. Rao and R. Chakraborty
1991 xvi + 554 pp.

Volume 9. Computational Statistics
Edited by C. R. Rao
1993 xix + 1045 pp.

Volume 10. Signal Processing and its Applications
Edited by N. K. Bose and C. R. Rao
1993 xvii + 992 pp.

Volume 11. Econometrics
Edited by G. S. Maddala, C. R. Rao and H. D. Vinod
1993 xx + 783 pp.